Field Theory of Condensed Matter and Ultracold Gases

Volume 1

Field Theory of Condensed Matter and Ultracold Gases

Volume 1

Nicolas Dupuis

CNRS & Sorbonne Université, France

World Scientific

NEW JERSEY • LONDON • SINGAPORE • BEIJING • SHANGHAI • HONG KONG • TAIPEI • CHENNAI • TOKYO

Published by

World Scientific Publishing Europe Ltd.
57 Shelton Street, Covent Garden, London WC2H 9HE
Head office: 5 Toh Tuck Link, Singapore 596224
USA office: 27 Warren Street, Suite 401-402, Hackensack, NJ 07601

British Library Cataloguing-in-Publication Data
A catalogue record for this book is available from the British Library.

FIELD THEORY OF CONDENSED MATTER AND ULTRACOLD GASES
Volume 1

ISBN 978-1-80061-390-4 (hardcover)
ISBN 978-1-80061-391-1 (ebook for institutions)
ISBN 978-1-80061-392-8 (ebook for individuals)

For any available supplementary material, please visit
https://www.worldscientific.com/worldscibooks/10.1142/Q0409#t=suppl

Desk Editor: Rhaimie B Wahap

Typeset by Stallion Press
Email: enquiries@stallionpress.com

À mes parents, Marie-Noël et Bernard Dupuis

Preface

Theoretical studies of condensed matter and ultracold gases are difficult primarily because of the very large number of interacting particles. Whenever the particles are correlated over distances much larger than the microscopic scales, collective effects become important at low energies. This implies that many physical systems exhibit different behaviors on different energy scales and the relevant degrees of freedom which permit a simple description of the low-energy (macroscopic) properties may be different from the microscopic ones. Quantum many-particle systems are thus often characterized by "emergent" properties — a prominent example being superfluidity or superconductivity — that involve all particles and are not related in any simple way to the microscopic physical laws.[1]

Historically the two cornerstones of condensed-matter physics have been adiabatic continuity and broken symmetries.[2] The principle of adiabatic continuity refers to the fact that some quantum many-body systems are related to simpler, non-interacting, systems with which they share the same symmetry properties and the same quantum numbers of the excitations. Elementary excitations ("quasi-particles") in the interacting system can be obtained from the excitations of the non-interacting system by adiabatically switching on the interactions. The principle of adiabatic continuity can often be justified by perturbation theory even though its domain of validity may extend beyond the region where perturbation theory converges. The best known example is Landau's Fermi-liquid theory which relates the normal phase of three-dimensional metals to the non-interacting electron gas. The adiabatic continuity principle breaks down when a phase transition is induced by varying a parameter of the system such as the temperature or an external field, leading to a singularity in the free energy that cannot be taken into account within perturbation theory. This leads to a substantial rearrangement of the ground state and the elementary excitations may be different from those of the parent non-interacting phase.

---

[1] See P. W. Anderson, More is different, *Science* **177**, 393 (1972).

[2] P. W. Anderson, *Basic Notions of Condensed Matter Physics* (Advanced Book Classics, Menlo Park, California, 1984).

The adiabatic continuity principle is the reason why perturbation theory, and its formulation in terms of Feynman diagrams, has played such an important role in the theory of quantum many-body systems. Phase transitions are usually signaled by a divergence of perturbation theory; when associated with spontaneous symmetry breaking they can be described by introducing the proper order parameter whose fluctuations, in the simplest cases, can be ignored or taken into account perturbatively. Mean-field theories have been extremely successful in understanding many phase transitions, e.g. the superfluidity of weakly interacting bosons, the superconductivity of metals or the magnetic instabilities in various systems. In addition to perturbation theory and mean-field theory one should also mention the renormalization group (RG). Besides its key role in understanding second-order phase transitions, the RG has been instrumental not only in improving plain perturbation theory but also in constructing low-energy effective descriptions of quantum many-body systems thus explaining the observed universality in condensed matter. Some strongly correlated systems have nevertheless remained out of reach of these "traditional" approaches and specific methods had to be designed, for instance, the dynamical mean-field theory (DMFT) — originally to understand the Mott metal–insulator transition in lattice fermion models — or the non-perturbative functional RG.

This book aims at introducing the physical concepts and theoretical methods necessary for the study of quantum (non-relativistic) many-body systems with a particular emphasis on field theory and functional integral methods. The focus is on the study of model Hamiltonians, where the microscopic physics and characteristic energy scales are encoded into a few effective parameters, rather than first-principle methods which start from a realistic Hamiltonian at the microscopic level and make material-specific predictions. We not only revisit some standard theories of condensed matter and ultracold gases — such as Landau's Fermi-liquid theory, the electron gas in the random-phase approximation, Stoner's theory of ferromagnetism in itinerant fermion models, Bogoliubov's theory of superfluidity in dilute Bose gases, the Bardeen–Cooper–Shrieffer (BCS) theory of superconductivity, one-dimensional quantum fluids (Luttinger liquids), quantum magnetism in the framework of the Heisenberg model — but also discuss recent theoretical progress (notably for strongly correlated systems) such as the two-particle self-consistent (TPSC) approach or the DMFT in the framework of the Hubbard model, as well as the non-perturbative functional RG and some of its applications to interacting fermion and boson systems.

The reasons why the methods of field theory are appropriate for quantum many-body systems are numerous. For example, expressing the partition function as a functional integral leads to a simple derivation of Feynman diagrammatic rules in perturbation theory. In the case of spontaneous symmetry breaking, the functional integral formalism enables an easy implementation of mean-field theories and provides us with a practical framework for the study of the fluctuations of the order parameter. But the real power of field theory comes from its generality and flexibility: It is a universal language that can be used whenever we deal with classical or quantum systems where the degrees of freedom are described by fields. Functional integrals have made it possible to build a unified field theory of collective phenomena in quite different physical systems. In all cases, the partition function can be expressed as a functional integral in d space or $d+1$ spacetime dimensions, the difference between systems being encoded in the nature of the field (scalar, vector, etc.), the symmetry of the action (e.g. relativistic (Lorentz) *vs* Galilean invariance) or the range of the interactions. Thus, quantum field theory describes not only elementary particles and their interactions but also quantum interacting systems that can be found in condensed matter or ultracold gases. This is truly remarkable given the different energy scales involved: 10^{-7} meV for an ultracold atomic gas, 1 eV for conduction electrons in a solid, 1 GeV for quantum chromodynamics (QCD), 125 GeV for the Higgs boson. This explains why similar phenomena can occur in high- and low-energy physics; e.g. the gapped amplitude mode in a superconductor is the analog of the Higgs boson in high-energy physics. The functional integral also shows that from a theoretical (practical) point of view there is no genuine difference between a classical and a quantum system since the latter can be seen as a "classical" system in $d+1$ dimensions with a temporal $(d+1)$th dimension of length $\hbar\beta = \hbar/k_BT$.[3] The analogy is often very fruitful, for instance, in the study of quantum phase transitions. Nowadays, quantum field theory plays an important role in many studies of strongly correlated condensed-matter systems: high-T_c superconductors, heavy-fermion systems, low-dimensional organic conductors, graphene, quantum Hall systems and topological materials, as well as cold atomic gases in optical lattices or in the vicinity of a Feshbach resonance.

Content of the Book

The first volume of the book is divided into two parts. The first chapter of the first part introduces the basics of many-body theory: coherent states for fermions and bosons, the functional integral in the Euclidean formalism and the perturbative calculation of the partition function and the correlation functions (Feynman diagrams). It also addresses the one-particle irreducible (1PI) and two-particle irreducible (2PI) effective actions and the Luttinger–Ward functional as well as the saddle-point approximation and loop expansion, and the quantization of the electromagnetic field. Chapters 2 and 3 deal with symmetries and the general properties of correlation functions.

This first part provides the reader with all the technical background necessary to understand the various chapters of the second part: Fermi-liquid theory, the electron gas, magnetism in lattice fermion systems, superfluidity and superconductivity, quantum magnetism and the Baym–Kadanoff formalism. These chapters address standard subjects in condensed matter and ultracold gases, usually (but not always) discussed in the language of functional integrals, as well as more recent theoretical progress such as the TPSC theory describing antiferromagnetic fluctuations in the (weak-coupling) Hubbard model, the DMFT of the Mott transition and the related self-energy-functional theory, or the BCS-BEC crossover in a dilute Fermi gas near a Feshbach resonance. All chapters in the second part are self-contained and can be read independently.

---

[3] This analogy holds in the Euclidean formalism where the (Minkowskian) real time is analytically continued to the (Euclidean) imaginary time. In this formalism, there is no real difference between quantum field theory and statistical field theory.

The second volume of the book is concerned with Wilson's RG, in particular its modern implementation known as the non-perturbative functional RG, and some of its applications to quantum phase transitions and interacting boson and fermion systems, e.g. the superfluid–Mott-insulator transition of lattice bosons, one-dimensional Bose and Fermi fluids (Luttinger liquids) and the competition between antiferromagnetism and superconductivity in two-dimensional fermion systems.

The reader is expected to be familiar with elementary quantum mechanics and statistical physics and some acquaintance with condensed-matter physics and ultracold gases may also be useful. No prior knowledge of field theory or the many-body problem is required.

Some parts, which address particular examples, more technical points or less important issues, appear in small print and are indented. They are not essential for understanding the main text and may be skipped in first reading.

Acknowledgments

The writing of this book was a unique opportunity to realize how much science is a long-term collective effort. Depending on the chapters I have relied on previous books[4] on the many-body problem, review or research articles as well as my own research. Given the broad range of subjects addressed in the book, it was not possible to provide the reader with an exhaustive list of references and I apologize to people whose work has not been properly cited.

I am indebted to many colleagues and students who have provided me with feedback on some parts of the book, as well as many others who over the years have increased my depth of understanding of the many-body problem. In particular, I would like to thank Patrick Azaria, Karol Borejsza, Claude Bourbonnais, Romain Daviet, Thibault Debelhoir, Bertrand Delamotte, Lucas Désoppi, Jean-Noël Fuchs, Philippe Lecheminant, Xavier Leyronas, Gilles Montambaux, Dominique Mouhanna, Christoph Nickel, Yuki Nojiri, Stéphane Pairault, Ludovic Pricoupenko, Adam Rançon, Félix Rose, Krishendu Sengupta, Julian Talbot, Gilles Tarjus, Sofian Teber, Matthieu Tissier, André-Marie Tremblay, Julien Vidal, Félix Werner, Christof Wetterich and Nicolás Wschebor.

Nicolas Dupuis

[4]To cite a few among the oldest ones: A. A. Abrikosov, L. P. Gor'kov, and I. E. Dzyaloshinski, *Methods of Quantum Field Theory in Statistical* (Physics Dover, Mineola, New York, 1965, 1975); P. Nozières, *Theory of Interacting Fermi Systems* (Addison Wesley, Reading, Massachusetts, 1964, 1997); A. L. Fetter and J. D. Walecka, *Quantum Theory of Many-Particle Systems* (Dover, Mineola, New York, 1971, 2003); G. D. Mahan, *Many-Particle Physics* (Kluwer Academic/Plenum Publishers, New York, New York, 1981, 2000).

Contents

Preface vii

Acknowledgments xi

Acronyms xix

List of Symbols and Notations xxi

Part 1: Functional Integrals, Symmetries, and Correlation Functions 1

1. Functional Integrals 3

1.1 Introduction 5
1.2 Second Quantization and Green Functions 28
1.3 Coherent States 42
1.4 Functional Integral in Many-Particle Systems 51
1.5 Perturbation Theory and Feynman Diagrams 59
1.6 Generating Functionals 78
1.7 Saddle-Point Approximation and Loop Expansion 103
1.8 Perturbation Theory at Zero Temperature 115
1.9 Quantization of the Electromagnetic Field 123
Appendix 1.A Review of Classical Mechanics 133
Appendix 1.B Review of Quantum Mechanics 146
Appendix 1.C Review of Statistical Physics 151
Appendix 1.D Functionals and Functional Derivatives 168
Appendix 1.E Gaussian Integrals and Wick's Theorem 170
Appendix 1.F Matsubara Frequency Sums 178
Appendix 1.G Perturbation Theory in the Operator Formalism 183
Bibliography 191

2. Symmetries 193

2.1 Symmetries, Invariances and Conservation Laws 194
2.2 Examples of Symmetry Transformations 199
2.3 Symmetries and Functional Integrals 214
Appendix 2.A Ward Identities in a Fermion System 229
Bibliography 233

3. Correlation and Response Functions 235

3.1 Linear Response Theory 236
3.2 Properties of Correlation Functions 240
3.3 Density–Density Response Function 258
3.4 Electromagnetic Response Function 261
3.5 One-Particle Green Function 283
3.6 Spontaneous Symmetry Breaking and Goldstone's Theorem 290
Bibliography 309

Part 2: Quantum Many-Body Systems (I) 311

4. Fermi-Liquid Theory 313

4.1 The Quasi-Particle Concept 314
4.2 Thermodynamics 322
4.3 Non-Equilibrium Properties 324
4.4 Microscopic Basis of Fermi-Liquid Theory 339
4.5 Fermi Liquids in Low Dimensions 362
Appendix 4.A Second-Order Self-energy 363
Bibliography 367

5. The Electron Gas 369

5.1 The Jellium Model 370
5.2 The Random-Phase Approximation 371
5.3 Physical Properties of the RPA 378
5.4 One-Particle Properties 387
5.5 Beyond the RPA 393
5.6 Wigner Crystallization 395
Bibliography 397

6. Magnetism in Lattice Fermion Systems 399

6.1 The Hubbard Model 400
6.2 Charge and Magnetic Instabilities 405
6.3 Stoner's Theory of Ferromagnetism 409
6.4 Slater's Theory of Antiferromagnetism 421

Appendix 6.A Hartree–Fock Correlation Functions 447
Appendix 6.B SU(2) Gauge Fields 450
Bibliography 453

7. Superfluidity and Superconductivity 457

7.1 Superfluidity and Long-Range Current–Current Correlations 458
7.2 Spontaneous U(1) Symmetry Breaking and Superfluidity 462
7.3 Gross–Pitaevskii Theory 483
7.4 Bogoliubov Theory 490
7.5 BCS Mean-Field Theory of Superconductivity 505
7.6 Collective Modes in Superconductors 525
7.7 BCS–BEC Crossover 539
Appendix 7.A Elements of Scattering Theory 553
Bibliography 563

8. Quantum Magnetism 569

8.1 The Heisenberg Model 570
8.2 Random-Phase Approximation 573
8.3 Spin Waves: Holstein–Primakoff Bosons 577
8.4 Spin-Coherent-State Functional Integral 581
8.5 Schwinger-Boson Mean-Field Theory 597
Appendix 8.A A Single Spin in an External Field 604
Appendix 8.B Spin-Coherent States 606
Bibliography 609

9. Baym–Kadanoff Formalism 611

9.1 Conserving Approximations 612
9.2 Mott Transition and Dynamical Mean-Field Theory 628
9.3 Self-Energy Functional Theory 637
Appendix 9.A Brinkman–Rice Theory: Gutzwiller Approximation 645
Bibliography 653

Index 657

Contents of Volume II

Part 3: The Renormalization Group

10. Renormalization Group and Critical Phenomena
11. The Non-Perturbative Functional Renormalization Group

Part 4: Quantum Many-Body Systems (II)

12. Quantum Phase Transitions
13. Interacting Bosons
14. The Bose–Hubbard Model and the Superfluid–Mott-Insulator Transition
15. Interacting Fermions, Bosonization and Renormalization Group

Acronyms

1PI one-particle irreducible
2PI two-particle irreducible

BCS Bardeen–Cooper–Shrieffer
BEC Bose–Einstein condensation
BKT Berezinskii–Kosterlitz–Thouless

C-DIA cellular dynamical impurity approximation
C-DMFT cellular dynamical mean-field theory
CPT cluster-perturbation theory

DΓA dynamical vertex approximation
DIA dynamical impurity approximation
DMFT dynamical mean-field theory

FLEX fluctuation-exchange approximation

NLσM nonlinear σ model

ODLRO off-diagonal long-range order

QCP quantum critical point
QPT quantum phase transition

RG renormalization group
RPA random-phase approximation

TPSC two-particle self-consistent

UV ultraviolet

VCA variational cluster approximation

List of Symbols and Notations

The following list is not intended to be exhaustive. It includes symbols and notations of frequent occurrence or special importance.

h Planck's constant
$\hbar = h/2\pi$ usually set to unity
k_B Boltzmann's constant, usually set to unity
$\beta = 1/k_B T$
c_l speed of light

Space coordinate and (imaginary) time

d space dimension
$\mathcal{V} = L^d$ volume
$\mathbf{r}$ d-dimensional space coordinate
t real (Minkowskian) time
τ imaginary (Euclidean) time
$x = (\mathbf{r}, t)$ or $(\mathbf{r}, \tau)$
N number of particles (or number of sites in a lattice model)
$\langle \mathbf{r}, \mathbf{r}' \rangle$ nearest-neighbor sites on a hypercubic lattice

Momenta and Matsubara frequencies

$\mathbf{k}$, $\mathbf{p}$, $\mathbf{q}$ momentum
$\frac{1}{\mathcal{V}} \sum_{\mathbf{k}} \underset{(\mathcal{V}\to\infty)}{=} \int \frac{d^d k}{(2\pi)^d} = \int_{\mathbf{k}}$
$\frac{1}{N} \sum_{\mathbf{k}} \underset{(N\to\infty)}{=} \int \frac{d^d k}{(2\pi)^d} = \int_{\mathbf{k}}$ with the sum/integral over $\mathbf{k}$ restricted to the first Brillouin zone (lattice model)
$\epsilon_{\mathbf{k}} = \frac{\mathbf{k}^2}{2m}$
$\xi_{\mathbf{k}} = \epsilon_{\mathbf{k}} - \mu$
$\omega_n = 2n\pi/\beta$ bosonic Matsubara frequency
$\omega_n = (2n+1)\pi/\beta$ fermionic Matsubara frequency
$k = (\mathbf{k}, i\omega_n)$

Fermion and boson fields (corresponding operators are denoted by $\hat{\psi}(\mathbf{r},\tau)$, etc.)
$\psi(\mathbf{r},\tau) = \frac{1}{\sqrt{\beta V}} \sum_{\mathbf{k},\omega_n} e^{i(\mathbf{k}\cdot\mathbf{r}-\omega_n\tau)}\psi(\mathbf{k}, i\omega_n)$
$\psi^*(\mathbf{r},\tau) = \frac{1}{\sqrt{\beta V}} \sum_{\mathbf{k},\omega_n} e^{-i(\mathbf{k}\cdot\mathbf{r}-\omega_n\tau)}\psi^*(\mathbf{k}, i\omega_n)$
$\psi(\mathbf{k}, i\omega_n) = \frac{1}{\sqrt{\beta V}} \int_0^\beta d\tau \int d^d r\, e^{-i(\mathbf{k}\cdot\mathbf{r}-\omega_n\tau)}\psi(\mathbf{r},\tau)$ Fourier transformed field
$\psi^*(\mathbf{k}, i\omega_n) = \frac{1}{\sqrt{\beta V}} \int_0^\beta d\tau \int d^d r\, e^{i(\mathbf{k}\cdot\mathbf{r}-\omega_n\tau)}\psi^*(\mathbf{r},\tau)$
$G(\mathbf{r},\tau) = -\langle\psi(\mathbf{r},\tau)\psi^*(0,0)\rangle = \frac{1}{\beta V} \sum_{\mathbf{k},\omega_n} e^{i(\mathbf{k}\cdot\mathbf{r}-\omega_n\tau)}G(k)$ one-particle Green function
$G(k) = -\langle\psi(k)\psi^*(k)\rangle = \int_0^\beta d\tau \int d^d r\, e^{-i(\mathbf{k}\cdot\mathbf{r}-\omega_n\tau)}G(\mathbf{r},\tau)$

Miscellaneous
$\Theta(x)$ Heaviside step function
$n_F(x) = \frac{1}{e^{\beta x}+1}$ Fermi-Dirac distribution function
$n_B(x) = \frac{1}{e^{\beta x}-1}$ Bose-Einstein distribution function
$[\hat{A},\hat{B}]_- = [\hat{A},\hat{B}] = \hat{A}\hat{B} - \hat{B}\hat{A}$ commutator of operators $\hat{A}$ and $\hat{B}$
$[\hat{A},\hat{B}]_+ = \hat{A}\hat{B} + \hat{B}\hat{A}$ anticommutator
$|\mathrm{vac}\rangle$ vacuum state (or quantum vacuum state)
$\hat{\mathbf{x}}$, $\hat{\mathbf{y}}$, $\hat{\mathbf{z}}$ unit vectors in the x, y, and z directions
$\sigma_x = \begin{pmatrix} 0 & 1 \\ 1 & 0 \end{pmatrix}$, $\sigma_y = \begin{pmatrix} 0 & -i \\ i & 0 \end{pmatrix}$, $\sigma_z = \begin{pmatrix} 1 & 0 \\ 0 & -1 \end{pmatrix}$ Pauli matrices

Part 1

Functional Integrals, Symmetries, and Correlation Functions

Chapter 1

Functional Integrals

Contents

1.1 Introduction . . . 5

1.1.1 Path integral in quantum mechanics . . . 6

1.1.2 Functional integral in statistical physics . . . 21

1.2 Second Quantization and Green Functions . . . 28

1.2.1 Fock space . . . 28

1.2.2 Operators in second-quantized form . . . 35

1.2.3 Green functions . . . 38

1.3 Coherent States . . . 42

1.3.1 Boson coherent states . . . 43

1.3.2 Grassmann variables . . . 46

1.3.3 Fermion coherent states . . . 48

1.4 Functional Integral in Many-Particle Systems . . . 51

1.4.1 Coherent-state functional integral . . . 51

1.4.2 Non-interacting particles . . . 54

1.5 Perturbation Theory and Feynman Diagrams . . . 59

1.5.1 Wick's theorem . . . 59

1.5.2 Partition function and thermodynamic potential . . . 61

1.5.3 Green functions . . . 67

1.5.4 (Anti)symmetrized vertices . . . 72

1.6 Generating Functionals . . . 78

1.6.1 Connected Green functions . . . 78

1.6.2 Effective action and 1PI vertices . . . 83

1.6.3 Luttinger–Ward functional and 2PI vertices . . . 94

1.7 Saddle-Point Approximation and Loop Expansion . . . 103

1.7.1 Steepest-descent and stationary-phase approximations . . . 103

1.7.2 Saddle-point approximation and loop expansion in field theory . . . 105

1.8 Perturbation Theory at Zero Temperature . . . 115

1.8.1 Fermions . . . 116

1.8.2 Bosons . . . 120
1.8.3 The zero-temperature limit . . . 122
1.9 Quantization of the Electromagnetic Field . . . 123
1.9.1 Lagrangian and Hamiltonian formalism . . . 123
1.9.2 Canonical quantization in the Coulomb gauge . . . 127
1.9.3 Functional integral in the Coulomb gauge . . . 130
Appendix 1.A Review of Classical Mechanics . . . 133
1.A.1 Lagrangian formalism . . . 134
1.A.2 Hamiltonian formalism . . . 137
1.A.3 Classical field theory . . . 139
Appendix 1.B Review of Quantum Mechanics . . . 146
1.B.1 The postulates of quantum mechanics . . . 146
1.B.2 Quantum fields . . . 150
Appendix 1.C Review of Statistical Physics . . . 151
1.C.1 Macrostates — Density operators . . . 151
1.C.2 Statistical entropy — Boltzmann–Gibbs distribution . . . 153
1.C.3 Statistical ensembles . . . 156
1.C.4 The classical limit . . . 164
1.C.5 Non-interacting quantum gases . . . 166
Appendix 1.D Functionals and Functional Derivatives . . . 168
1.D.1 Grassmann variables . . . 170
Appendix 1.E Gaussian Integrals and Wick's Theorem . . . 170
1.E.1 Real variables . . . 170
1.E.2 Complex variables . . . 172
1.E.3 Grassmannian integrals . . . 173
1.E.4 Gaussian functional integrals . . . 176
Appendix 1.F Matsubara Frequency Sums . . . 178
Appendix 1.G Perturbation Theory in the Operator Formalism . . . 183
1.G.1 Zero-temperature formalism . . . 183
1.G.2 Finite-temperature formalism . . . 187
Bibliography . . . 191

This chapter aims at introducing the general framework for the study of macroscopic quantum systems. The formalism is based on the functional integral, an extension of Feynman's path integral to systems with an infinite number of degrees of freedom. It enables to derive standard results, such as perturbation expansions, in an economical way and set up non-perturbative approaches. It also provides a unified view of many concepts and theoretical methods that can be found in various fields of physics: condensed-matter physics, quantum optics, high-energy physics, etc.

We first derive the functional integral representation of the partition function of a system of interacting quantum particles (bosons or fermions) (Secs. 1.2–1.4). This representation is based on second-quantized fields and coherent states. We discuss the perturbative calculation of the partition function and the correlation functions (Green functions), and its representation in terms of Feynman diagrams (Secs. 1.5 and 1.6). These diagrams provide an intuitive picture of perturbation theory in terms of elementary interaction processes taking place in the system. They are also a very efficient tool to partially resum the perturbation series to infinite order. We show how the perturbation expansion may be organized about a broken-symmetry state when a symmetry is spontaneously broken (Sec. 1.7). We also discuss the quantization of the electromagnetic field in the functional integral formalism (Sec. 1.9).[1]

Although Feynman diagrams play an essential role in many-body physics, both as a pictorial representation of perturbation expansions and an efficient computational tool, they also suffer serious drawbacks. On the one hand, they often lead to uncontrolled approximations. On the other hand, diagrammatic resummations are usually restricted to systems with weak interactions. In many cases, however, one is interested in strongly correlated systems where the interactions are strong and perturbative expansions likely to fail. We shall see how the generating functionals of one-particle and two-particle-irreducible vertices introduced in Sec. 1.6 can be used to set up non-perturbative approaches. The effective action (Sec. 1.6.2) is at the basis of the formulation of the non-perturbative functional renormalization group (RG) approach (Chapter 11) while the Luttinger–Ward functional (Sec. 1.6.3) is the starting point of the dynamical mean-field theory (DMFT) and various cluster expansions (Chapter 9).

This chapter is somewhat formal but constitutes the backbone of the book. The prerequisite knowledge of classical, quantum and statistical mechanics is summarized in Appendices 1.A–1.C. Other appendices provide technical details on functional calculus, Gaussian integrals and Matsubara sums as well as a discussion of perturbation theory in the operator formalism.

1.1 Introduction

Elementary quantum mechanics is usually formulated in the framework of canonical quantization where observables of the classical theory become quantum mechanical operators acting on state vectors in a Hilbert space (Appendix 1.B). In this introductory section, we discuss an alternative formulation of quantum mechanics based on the Feynman path integral. As in classical mechanics, the concept of trajectories (or paths) plays a central role and the Feynman path integral yields a very intuitive picture of quantum mechanics. For a single quantum particle or a system with few degrees of freedom, there is no real advantage to use the path integral formalism, except for pedagogical purposes, the usual approach based on the Schrödinger equation being usually more efficient. However, for systems with a large number of degrees of freedom and in particular in statistical physics, the path integral formalism — or, more precisely, the functional integral formalism — is often more appropriate.

[1]The coherent-state functional integral for quantum spin systems is discussed in Sec. 8.4.

We start the discussion with the study of a quantum particle moving in a static potential (Sec. 1.1.1). Then we consider the functional integral formalism for the quantum harmonic string (Sec. 1.1.2). This example already exhibits many key aspects of the functional integral formalism of quantum many-particle systems that will be introduced in the following sections and used throughout the book.

1.1.1 *Path integral in quantum mechanics*

To introduce the concept of path integral in quantum mechanics, we consider a particle of mass m in a one-dimensional space. Its classical dynamics is governed by the Lagrangian

$$L(q,\dot{q}) = \frac{1}{2}m\dot{q}^2 - V(q), \tag{1.1}$$

assuming that the particle moves in a time-independent potential $V(q)$. q denotes the position of the particle and $\dot{q} = \partial_t q$. The trajectory of the particle is obtained from the Euler–Lagrange equation (Appendix 1.A)

$$\frac{d}{dt}\frac{\partial L(q,\dot{q})}{\partial \dot{q}} = \frac{\partial L(q,\dot{q})}{\partial q} \quad \text{i.e.} \quad m\ddot{q} = -\frac{\partial V(q)}{\partial q}. \tag{1.2}$$

Alternatively, one can consider the Hamiltonian $H(p,q)$ defined by

$$\begin{aligned} p &= \frac{\partial L(q,\dot{q})}{\partial \dot{q}} = m\dot{q}, \\ H(p,q) &= p\dot{q} - L(q,\dot{q}) = \frac{p^2}{2m} + V(q), \end{aligned} \tag{1.3}$$

where p is the conjugate momentum of the variable q. The classical dynamics of the particle then follows from Hamilton's equations of motion,

$$\begin{aligned} \dot{p} &= -\frac{\partial H(p,q)}{\partial q} = -\frac{\partial V(q)}{\partial q}, \\ \dot{q} &= \frac{\partial H(p,q)}{\partial p} = \frac{p}{m}. \end{aligned} \tag{1.4}$$

1.1.1.1 *Canonical quantization*

Canonical quantization is achieved by promoting the canonical variables p and q of the classical theory to operators $\hat{p}$ and $\hat{q}$ satisfying the commutation relations

$$[\hat{q},\hat{p}] = i\hbar. \tag{1.5}$$

A classical variable $A(p,q)$ becomes an operator $\hat{A} \equiv A(\hat{p},\hat{q})$. In particular, the quantum Hamiltonian reads

$$\hat{H} = \frac{\hat{p}^2}{2m} + V(\hat{q}). \tag{1.6}$$

A physical state of the system is described by a state vector $|\psi(t)\rangle$ of a Hilbert space $\mathcal{H}$. The time evolution of $|\psi(t)\rangle$ is given by the Schrödinger equation

$$i\hbar\partial_t|\psi(t)\rangle = \hat{H}|\psi(t)\rangle, \tag{1.7}$$

the solution of which can be written in terms of the evolution operator $\hat{U}$,

$$|\psi(t)\rangle = \hat{U}(t)|\psi(t=0)\rangle, \qquad i\hbar\partial_t\hat{U}(t) = \hat{H}\hat{U}(t). \tag{1.8}$$

The Hamiltonian being time independent, one simply has

$$\hat{U}(t) = e^{-\frac{i}{\hbar}\hat{H}t}. \tag{1.9}$$

The probability amplitude to find the particle at position q_f at time t_f is given by

$$\psi(q_f,t_f) = \langle q_f|\psi(t_f)\rangle = \langle q_f|\hat{U}(t_f-t_i)|\psi(t_i)\rangle = \int dq_i\, U(q_f,q_i;t_f-t_i)\psi(q_i,t_i). \tag{1.10}$$

$U(q_f,q_i;t_f-t_i) = \langle q_f|\hat{U}(t_f-t_i)|q_i\rangle$ is referred to as the propagator, since it expresses the probability amplitude for the particle to propagate from point q_i to point q_f in a time $t_f - t_i$. If we know a basis $\{|n\rangle, \epsilon_n\}$ of eigenstates of the Hamiltonian $\hat{H}$, we can write the propagator as

$$\begin{aligned} U(q_f,q_i;t_f-t_i) &= \langle q_f|e^{-\frac{i}{\hbar}\hat{H}(t_f-t_i)}|q_i\rangle = \sum_n \langle q_f|n\rangle e^{-\frac{i}{\hbar}\epsilon_n(t_f-t_i)}\langle n|q_i\rangle \\ &= \sum_n e^{-\frac{i}{\hbar}\epsilon_n(t_f-t_i)}\varphi_n(q_f)\varphi_n^*(q_i) \end{aligned} \tag{1.11}$$

(using the closure relation $\sum_n |n\rangle\langle n| = 1$), where $\varphi_n(q) = \langle q|n\rangle$ denotes the wavefunction in the coordinate representation. $U(q_f,q_i;t_f-t_i)$ provides information about wavefunctions and energy levels of the Hamiltonian.

1.1.1.2 *Path integral*

The propagator $U(q_f,q_i;t)$ can also be represented as a "path integral". We first consider an infinitesimal time ϵ. To leading order in ϵ, one has[2]

$$U(q_f,q_i;\epsilon) = \langle q_f|e^{-i\hat{H}\epsilon}|q_i\rangle \simeq \langle q_f|e^{-i\epsilon\frac{\hat{p}^2}{2m}}e^{-i\epsilon V(\hat{q})}|q_i\rangle \tag{1.12}$$

(we now set $\hbar = 1$). The propagator is then calculated by inserting the closure relation $\sum_p |p\rangle\langle p| = 1$,[3]

$$\begin{aligned} U(q_f,q_i;\epsilon) &= \sum_p \langle q_f|e^{-i\epsilon\frac{\hat{p}^2}{2m}}|p\rangle\langle p|e^{-i\epsilon V(\hat{q})}|q_i\rangle \\ &= \int \frac{dp}{2\pi}\exp\left\{-i\epsilon\left[\frac{p^2}{2m}+V(q_i)\right]+ip(q_f-q_i)\right\} \\ &= \left(\frac{m}{2\pi i\epsilon}\right)^{1/2}\exp\left\{i\epsilon\left[\frac{m}{2}\frac{(q_f-q_i)^2}{\epsilon^2}-V(q_i)\right]\right\}. \end{aligned} \tag{1.13}$$

---

[2]This follows from $e^{\epsilon\hat{A}+\epsilon\hat{B}} = 1+\epsilon(\hat{A}+\hat{B})+\mathcal{O}(\epsilon^2) = e^{\epsilon\hat{A}}e^{\epsilon\hat{B}}e^{\mathcal{O}(\epsilon^2)}$ where the $\mathcal{O}(\epsilon^2)$ term is given by the commutator $-\frac{\epsilon^2}{2}[\hat{A},\hat{B}]$ (Baker–Hausdorff formula).

[3]We use the normalization conventions $\langle q|p\rangle = L^{-1/2}e^{ipq}$, $\int dq|q\rangle\langle q| = 1$ and $\sum_p |p\rangle\langle p| = 1$. $q \in [0,L]$ is a continuous position variable and $p = n\frac{2\pi}{L}$ ($n \in \mathbb{Z}$) a discrete momentum variable corresponding to periodic boundary conditions ($e^{ipL} = 1$). In the thermodynamic limit $L \to \infty$, one has $\frac{1}{L}\sum_p \to \int_{-\infty}^{\infty}\frac{dp}{2\pi}$ (Appendix 1.B).

In order for the integral over p to converge, we assume ϵ to contain a small negative imaginary part. The argument in the exponential is nothing but i times the infinitesimal action $S(q_f, q_i; \epsilon)$ corresponding to the straight line trajectory (at constant velocity) between q_i and q_f in the infinitesimal time ϵ,

$$U(q_f, q_i; \epsilon) = \left(\frac{m}{2\pi i\epsilon}\right)^{1/2} e^{iS(q_f, q_i; \epsilon) + \mathcal{O}(\epsilon^2)} \tag{1.14}$$

(writing explicitly that the error in the exponential is of order ϵ^2).

To calculate the propagator $U(q_f, q_i; t_f - t_i)$ for an arbitrary time $t_f - t_i$, we split $t_f - t_i$ into N equal steps of size $\epsilon = (t_f - t_i)/N$; we shall eventually take the limit $N \to \infty$ ($\epsilon \to 0$). Inserting the closure relation $\int dq|q\rangle\langle q| = 1$ at each intermediate time, we obtain

$$\begin{aligned} U(q_f, q_i; t_f - t_i) &= \langle q_f|e^{-i\hat{H}\epsilon} \cdots e^{-i\hat{H}\epsilon}|q_i\rangle \\ &= \int \prod_{k=1}^{N-1} dq_k \langle q_f|e^{-i\hat{H}\epsilon}|q_{N-1}\rangle\langle q_{N-1}|e^{-i\hat{H}\epsilon}|q_{N-2}\rangle \cdots \langle q_1|e^{-i\hat{H}\epsilon}|q_i\rangle \\ &= \int \prod_{k=1}^{N-1} dq_k \prod_{k=1}^{N} U(q_k, q_{k-1}; \epsilon), \end{aligned} \tag{1.15}$$

where $q_0 = q_i$ and $q_N = q_f$. The expression (1.14) introduces an error of order ϵ^2 at each time step, i.e. a total error of order ϵ. In the limit $N \to \infty$, one therefore finds[4]

$$U(q_f, q_i; t) = \lim_{N\to\infty} \left(\frac{mN}{2\pi it}\right)^{N/2} \int \prod_{k=1}^{N-1} dq_k \, e^{iS[q]}, \tag{1.16}$$

where

$$S[q] = \sum_{k=1}^{N} S(q_k, q_{k-1}; \epsilon) = \epsilon \sum_{k=1}^{N} \left[\frac{m}{2}\frac{(q_k - q_{k-1})^2}{\epsilon^2} - V(q_{k-1})\right] \tag{1.17}$$

is the action associated with the discrete trajectory $(q_0, q_1, \ldots, q_N)$ with $q_0 = q_i$ and $q_N = q_f$ (Fig. 1.1). In the limit $N \to \infty$, we may write symbolically

$$\begin{aligned} \epsilon \sum_{k=1}^{N} \frac{m}{2}\frac{(q_k - q_{k-1})^2}{\epsilon^2} &\to \int_{t_i}^{t_f} dt \frac{m}{2}\dot{q}^2, \\ \epsilon \sum_{k=1}^{N} V(q_{k-1}) &\to \int_{t_i}^{t_f} dt \, V(q) \end{aligned} \tag{1.18}$$

[4]The extension of (1.16), (1.17) to three dimensions is straightforward. The variable q becomes a three-dimensional vector $\mathbf{q}$ and the factor $(mN/2\pi it)^{N/2}$ must be replaced by $(mN/2\pi it)^{3N/2}$.

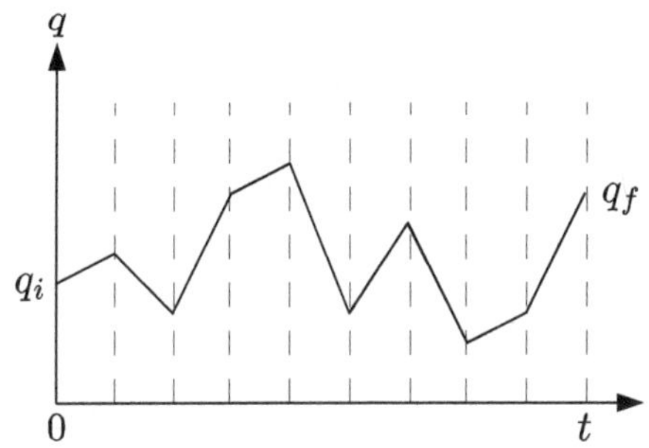

Fig. 1.1 A typical discrete trajectory contributing to the action $S[q]$.

and denote by $q(t)$ the "trajectory" $(q_0, \ldots, q_N)$ with $q(t_i) = q_i$ and $q(t_f) = q_f$. This notation does not imply continuity or differentiability of the trajectory $(q_0, q_1, \ldots, q_N)$ in the limit $N \to \infty$. If we define the integration measure as

$$\mathcal{D}[q] = \lim_{N \to \infty} \left(\frac{mN}{2\pi i \hbar t} \right)^{N/2} \prod_{k=1}^{N-1} dq_k, \tag{1.19}$$

then the propagator reads

$$U(q_f, q_i; t_f - t_i) = \int_{q(t_i)=q_i}^{q(t_f)=q_f} \mathcal{D}[q] \, e^{\frac{i}{\hbar} S[q]}, \tag{1.20}$$

where

$$S[q] = \int_{t_i}^{t_f} dt \, L(q, \dot{q}) \tag{1.21}$$

is the classical action corresponding to the trajectory $q(t)$. Note that the integration measure $\mathcal{D}[q]$ contains a factor that diverges in the limit $N \to \infty$. In (1.20), we have momentarily restored Planck's constant $\hbar$ since it plays an essential role in the discussion of the classical limit (see the following). Thus, the matrix element of the evolution operator $\hat{U}(t_f - t_i)$ can be represented as a sum over all possible paths going from $q(t_i) = q_i$ to $q(t_f) = q_f$ and weighted by the classical action $S[q]$ times $i/\hbar$. The Feynman path integral (1.20) provides a nice illustration of the superposition principle of quantum mechanics: When a process can take place in more than one way, its probability amplitude is given by the sum of the amplitudes for each way. Although no explicit reference to quantum mechanical operators is made, the Feynman path integral is an exact representation of the evolution operator $\hat{U}(t) = e^{-i\hat{H}t}$ and may be used as the starting point for the formulation of quantum mechanics [2].

The kinetic term selects sufficiently regular trajectories such that $[q(t+\epsilon) - q(t)]^2/\epsilon$ remains finite in the limit $\epsilon \to 0$. When this condition is not fulfilled, the kinetic energy is large and the exponential factor $e^{\frac{i}{\hbar} S[q]}$ strongly oscillates when considering nearby trajectories and therefore averages to zero. Thus, typical paths are continuous but not necessarily differentiable, and reminiscent of Brownian motion.[5] Since there is no reason for

[5]For a detailed discussion of the relation between path integrals and Brownian motion, see Ref. [3].

$(q_{k+1}-q_k)/\epsilon$ to remain finite in the limit $\epsilon \to 0$, the interpretation of $\dot{q}$ may appear delicate. We show in the following that for the free particle and the harmonic oscillator, the path integral can be computed from the sum over discrete trajectories (taking the limit $N \to \infty$ at the end of the calculation) or using "blindly" the continuous-time formulation.[6]

1.1.1.3 *Classical limit*

The important trajectories are those for which the action varies weakly when the path $q(t)$ is slightly deformed, i.e. those which are near the classical trajectory $q_c(t)$ whose action is stationary,

$$\left.\frac{\delta S[q]}{\delta q(t)}\right|_{q=q_c} = 0. \tag{1.22}$$

Non-stationary trajectories imply large oscillations of the action and therefore average to zero (the Feynman paths interfere destructively). More precisely, the propagator $U(q_f, q_i; t_f - t_i)$ is dominated by the trajectories $q(t)$ whose action $S[q]$ differs from the classical action $S_c = S[q_c]$ by a term of order $\hbar$: $|S - S_c| \lesssim \hbar$. When $|S_c| \gg \hbar$, these trajectories are very close to the classical trajectory. In the opposite limit, the condition $|S - S_c| \lesssim \hbar$ is fulfilled by trajectories very different from the classical one. Formally, the classical limit corresponds to the limit $\hbar \to 0$.

To obtain the propagator in the limit $\hbar \to 0$, we write $q(t) = q_c(t) + r(t)$ (assuming there is only one classical trajectory) and expand the action to second order in $r(t)$,

$$\begin{aligned} &U(q_f, q_i; t_f - t_i) \\ &\quad \simeq e^{\frac{i}{\hbar}S[q_c]} \int_{r(t_i)=0}^{r(t_f)=0} \mathcal{D}[r] \exp\left\{ \frac{i}{2\hbar} \int_{t_i}^{t_f} dt\, dt' \left.\frac{\delta^2 S[q]}{\delta q(t)\delta q(t')}\right|_{q=q_c} r(t) r(t') \right\}. \end{aligned} \tag{1.23}$$

The integral is Gaussian and can be done exactly (Appendix 1.E),

$$U(q_f, q_i; t_f - t_i) \simeq e^{\frac{i}{\hbar}S[q_c]} \det\left(\frac{1}{2\pi i\hbar} \left.\frac{\delta^2 S[q]}{\delta q(t)\delta q(t')}\right|_{q=q_c} \right)^{-1/2}. \tag{1.24}$$

The procedure we have followed to obtain (1.24) is known as the stationary-phase approximation. It gives the leading term of an expansion in powers of $\hbar$ that can be systematically carried out to higher orders (see Sec. 1.7).

[6]We shall see that except for Gaussian actions, we rarely compute path or functional integrals, but rather the ratio I/I_0 between the (path or functional) integral of interest and a reference (path or functional) integral whose calculation is trivial. For instance, in perturbation theory, the ratio I/I_0 is given by Feynman diagrams whose calculation follows well-established rules and has nothing to do with the actual computation of the functional integral.

1.1.1.4 *Time-ordered product of operators*

It is also possible to consider matrix elements including the operator $\hat{q}(t) = e^{i\hat{H}t}\hat{q}e^{-i\hat{H}t}$ (in the Heisenberg representation),[7]

$$\langle q_f, t_f|\hat{q}(t)|q_i, t_i\rangle = \langle q_f|e^{-i\hat{H}(t_f-t)}\hat{q}e^{-i\hat{H}(t-t_i)}|q_i\rangle. \tag{1.25}$$

By splitting the time intervals $t_f - t$ and $t - t_i$ into infinitesimal time steps, one easily obtains the following path integral representation

$$\langle q_f, t_f|\hat{q}(t)|q_i, t_i\rangle = \int_{q(t_i)=q_i}^{q(t_f)=q_f} \mathcal{D}[q]\, q(t)e^{iS[q]}. \tag{1.26}$$

Products of operators at different times are also represented by a path integral. For example,[8]

$$\langle q_f, t_f|T\hat{q}(t)\hat{q}(t')|q_i, t_i\rangle = \int_{q(t_i)=q_i}^{q(t_f)=q_f} \mathcal{D}[q]\, q(t)q(t')e^{iS[q]}, \tag{1.27}$$

where T is the time-ordering operator defined by

$$T\hat{q}(t)\hat{q}(t') = \Theta(t-t')\hat{q}(t)\hat{q}(t') + \Theta(t'-t)\hat{q}(t')\hat{q}(t). \tag{1.28}$$

Path integrals naturally represent expectation values of time-ordered products of operators.

1.1.1.5 *Hamiltonian form of the path integral*

Equations (1.20) and (1.21) define the Lagrangian form of the path integral, where the action is expressed in terms of the Lagrangian and the functional integral involves a sum over the paths $q(t)$. Alternatively, one can keep explicitly the variable p appearing in (1.13) and write the propagator as

$$\begin{aligned} U(q_f, q_i; t_f - t_i) &= \lim_{N\to\infty} \int \prod_{k=1}^{N-1} dq_k \int \prod_{k=1}^{N} \frac{dp_k}{2\pi} e^{\sum_{k=1}^N \left[ip_k(q_k-q_{k-1}) - i\epsilon\frac{p_k^2}{2m} - i\epsilon V(q_{k-1})\right]} \\ &\equiv \int_{q(t_i)=q_i}^{q(t_f)=q_f} \mathcal{D}[p,q]\, e^{iS[p,q]}, \end{aligned} \tag{1.29}$$

where

$$\begin{aligned} \mathcal{D}[p,q] &= \lim_{N\to\infty} \prod_{k=1}^{N-1} dq_k \prod_{k=1}^{N} \frac{dp_k}{2\pi}, \\ S[p,q] &= \lim_{N\to\infty} \sum_{k=1}^{N} \left[p_k(q_k - q_{k-1}) - \epsilon\frac{p_k^2}{2m} - \epsilon V(q_{k-1})\right] \\ &\equiv \int_{t_i}^{t_f} dt\, [p\dot{q} - H(p,q)]. \end{aligned} \tag{1.30}$$

---

[7]More generally, one could consider the matrix element $\langle q_f, t_f|\hat{A}(t)|q_i, t_i\rangle$ with $\hat{A}$ any operator.
[8]Equation (1.27) is easily proven by considering the two cases $t > t'$ and $t < t'$.

Equations (1.29) and (1.30) give the Hamiltonian form of the path integral, where the action $S[p,q]$ is a function of the position q as well as the momentum p of the particle. Note that q and p are independent variables in the functional integral (1.29) and only q is fixed at the initial (t_i) and final (t_f) times. When the particle travels from q_{k-1} to q_k, its momentum is given by p_k (which is independent of the velocity $(q_k - q_{k-1})/\epsilon$) so that the path is not a classical trajectory. The interpretation of the propagator as a sum over classical paths is therefore lost in the Hamiltonian form of the path integral. Nevertheless, for $\hbar \to 0$, the path integral is dominated by the trajectories satisfying

$$\frac{\delta S}{\delta p(t)} = \frac{\delta S}{\delta q(t)} = 0, \tag{1.31}$$

and one recovers Hamilton's equations of motion.

1.1.1.6 *Euclidean path integral*

Suppose we want to calculate the partition function

$$Z = \operatorname{Tr} e^{-\beta \hat{H}} = \int dq \langle q | e^{-\beta \hat{H}} | q \rangle \tag{1.32}$$

of the quantum particle considered above. Z can be obtained from the matrix elements of the evolution operator $e^{-i\hat{H}t}$ evaluated at the imaginary time $t = -i\beta$ (or $t = -i\beta\hbar$ if we restore $\hbar$). Real-time dynamics and quantum statistical mechanics are thus related by the transformation $t \to -i\tau$ known as a Wick rotation, i.e. a rotation of angle $-\pi/2$ in the complex time plane. The concept of imaginary time plays an essential role in quantum statistical physics. The propagator in imaginary time, $U(q_f, q_i; -i\tau) = \langle q_f | e^{-\hat{H}\tau} | q_i \rangle$ is given by (1.11) with the replacement $t \to -i\tau$. It can be represented as a path integral following the same steps as before[9]:

$$\begin{aligned} U(q_f, q_i; -i\tau) &= \lim_{N\to\infty} \left(\frac{mN}{2\pi\tau} \right)^{N/2} \int \prod_{k=1}^{N-1} dq_k \, e^{-\epsilon \sum_{k=1}^{N} \left[\frac{m}{2} \frac{(q_k - q_{k-1})^2}{\epsilon^2} + V(q_{k-1}) \right]} \\ &= \int_{q(0)=q_i}^{q(\tau)=q_f} \mathcal{D}[q] \, e^{-S_E[q]}. \end{aligned} \tag{1.33}$$

The imaginary-time path integral is defined by a sum over trajectories from $q(0) = q_i$ to $q(\tau) = q_f$ weighted by the action

$$S_E[q] = \int_0^\tau d\tau' \left[\frac{m}{2} \dot{q}^2 + V(q) \right]. \tag{1.34}$$

[9]Contrary to the real-time propagator, there is no need here to add a small imaginary part to the infinitesimal time ϵ (see the remark after (1.13)). From a mathematical point of view, the Euclidean formalism is preferable to the real-time one.

$S_E[q]$ is referred to as the Euclidean action since (imaginary) time and space enter the action in a symmetric way.[10,11,12] The kinetic and potential terms have the same sign in the Euclidean Lagrangian, since the kinetic term changes sign under the Wick rotation $\tau = it$. As in the real-time formalism, the kinetic term selects trajectories such that $[q(\tau + \epsilon) - q(\tau)]^2/\epsilon$ remains finite in the limit $\epsilon \to 0$ — paths that do not fulfill this condition are exponentially suppressed — so that the paths contributing to the propagator are continuous (but not necessarily differentiable). The real-time action (1.21) can be recovered from S_E performing a Wick rotation,

$$S_E[q] \xrightarrow[(\tau=it)]{\text{Wick rotation}} i \int_0^t dt' \left[-\frac{m}{2}\dot{q}^2 + V(q) \right] = -iS[q]. \tag{1.35}$$

Using the preceding results, we can express the partition function of the quantum particle,

$$Z = \int dq\, U(q, q, -i\beta) = \int_{q(\beta)=q(0)} \mathcal{D}[q]\, e^{-S_E[q]}, \tag{1.36}$$

as a path integral in imaginary time over all periodic trajectories of period β. Similarly, we can write the expectation value of the imaginary-time operator $\hat{q}(\tau) = e^{\tau\hat{H}}\hat{q}e^{-\tau\hat{H}}$ as

$$\langle \hat{q}(\tau) \rangle = \frac{1}{Z}\mathrm{Tr}\big[e^{-\beta\hat{H}}\hat{q}(\tau)\big] = \frac{1}{Z}\int_{q(\beta)=q(0)} \mathcal{D}[q]\, q(\tau) e^{-S_E[q]}. \tag{1.37}$$

For the time-ordered product $\langle T_\tau \hat{q}(\tau)\hat{q}(\tau')\rangle$, we obtain

$$\langle T_\tau \hat{q}(\tau)\hat{q}(\tau')\rangle = \frac{1}{Z}\int_{q(\beta)=q(0)} \mathcal{D}[q]\, q(\tau)q(\tau') e^{-S_E[q]}. \tag{1.38}$$

The imaginary-time-ordering operator T_τ is defined analogously to its real-time counterpart [Eq. (1.28)].

If $\beta \to 0$ (i.e. $T \to \infty$) or $\hbar \to 0$, the propagator $U(q, q; -i\beta\hbar)$ can be calculated with a single time step $N = 1$ (we restore $\hbar$ in this paragraph). This leads to

$$\begin{aligned} Z_{\text{cl}} &= \frac{1}{\hbar}\sqrt{\frac{m}{2\pi\beta}} \int dq\, e^{-\beta V(q)} \\ &\equiv \int_{-\infty}^{\infty} \frac{dp}{2\pi\hbar} \int_{-\infty}^{\infty} dq\, \exp\left\{ -\beta\left[\frac{p^2}{2m} + V(q) \right] \right\}, \end{aligned} \tag{1.39}$$

which is nothing but the classical partition function. At finite temperature $T < \infty$, the classical approximation (1.39) is valid if the potential $V(q)$ is slowly varying on a length scale of the order of the de Broglie wavelength $\xi_{\text{th}} \sim \hbar/\sqrt{mT}$. In Sec. 1.1.2, we shall further

[10] The Minkowski metric $ds^2 = dt^2 - dx^2$ becomes $-d\tau^2 - dx^2$ in the Euclidean formalism.

[11] The imaginary time is sometimes referred to as the Euclidean time.

[12] The Euclidean action will often be the starting point for the description of quantum many-body systems; in quantum mechanics, it is used for the study of tunneling.

discuss the connection between high-temperature limit and classical behavior of a quantum system.

Equation (1.39) shows a very important property of classical systems: Thermodynamics and dynamics decouple. The particle's position and momentum being independent variables, one can integrate out the momentum (which yields an additive contribution to the free energy) and write the partition function in terms of the position variable only. By contrast, in quantum systems, coordinate and momentum variables are non-commuting operators so that statics and dynamics are not independent.[13] This is the reason why the partition function can be related to the evolution operator in imaginary time [Eq. (1.36)].

Example: free particle and harmonic oscillator. Let us first consider the propagator of a free particle,

$$U(q_f, q_i; t) = \lim_{N\to\infty} \left(\frac{mN}{2\pi i t}\right)^{N/2} \int \prod_{k=1}^{N-1} dq_k \exp\left\{ i\frac{m}{2\epsilon} \sum_{k=1}^{N} (q_k - q_{k-1})^2 \right\}, \tag{1.40}$$

where $q_0 = q_i$ and $q_N = q_f$. One can successively integrate out q_1, q_2, etc. The integral over q_1 gives

$$\begin{aligned} U(q_f, q_i; t) &= \frac{1}{\sqrt{2}} \lim_{N\to\infty} \left(\frac{mN}{2\pi i t}\right)^{(N-1)/2} \\ &\times \int \prod_{k=2}^{N-1} dq_k \exp\left\{ i\frac{m}{4\epsilon}(q_2 - q_i)^2 + i\frac{m}{2\epsilon} \sum_{k=3}^{N} (q_k - q_{k-1})^2 \right\}. \end{aligned} \tag{1.41}$$

Integrating then over q_2 yields

$$\begin{aligned} U(q_f, q_i; t) &= \frac{1}{\sqrt{3}} \lim_{N\to\infty} \left(\frac{mN}{2\pi i t}\right)^{(N-2)/2} \\ &\times \int \prod_{k=3}^{N-1} dq_k \exp\left\{ i\frac{m}{6\epsilon}(q_3 - q_i)^2 + i\frac{m}{2\epsilon} \sum_{k=4}^{N} (q_k - q_{k-1})^2 \right\}, \end{aligned} \tag{1.42}$$

and we easily deduce the final result

$$U(q_f, q_i; t) = \left(\frac{m}{2\pi i t}\right)^{1/2} \exp\left(i\frac{m(q_f - q_i)^2}{2t} \right). \tag{1.43}$$

We recognize in the exponential i times the classical action: The classical path gives the exact propagator for a free particle. The result (1.43) can also be directly obtained from the definition (1.11) of the propagator using the eigenstates $\varphi_p(q) = L^{-1/2} e^{ipq}$ with $\epsilon_p = p^2/2m$.

[13]Heuristically, the existence of $\hbar$ implies that any energy scale E that enters thermodynamics necessarily determines a time scale $\hbar/E$.

Let us now consider the harmonic oscillator action

$$S[q] = \frac{m}{2}\int_{t_i}^{t_f} dt\,(\dot{q}^2 - \omega^2 q^2). \tag{1.44}$$

Although it is possible to do the calculation of the evolution operator in discrete time (see, e.g. Ref. [5]), we take the continuum-time limit. (The discrete time calculation is carried out in the following for the partition function.) To compute the propagator, we write $q(t) = q_c(t)+r(t)$. The classical trajectory q_c satisfies the equation of motion $\ddot{q} + \omega^2 q = 0$ and reads

$$q_c(t) = A\cos(\omega t) + B\sin(\omega t) \tag{1.45}$$

with the boundary conditions

$$q_i = A\cos(\omega t_i) + B\sin(\omega t_i) \quad \text{and} \quad q_f = A\cos(\omega t_f) + B\sin(\omega t_f), \tag{1.46}$$

while the classical action is given by

$$S[q_c] = \frac{m\omega}{2\sin(\omega(t_f - t_i))}\left[(q_f^2 + q_i^2)\cos\big(\omega(t_f - t_i)\big) - 2q_i q_f\right]. \tag{1.47}$$

Since

$$S[q] = S[q_c] + \frac{m}{2}\int_{t_i}^{t_f} dt\,\left(\dot{r}^2 - \omega^2 r^2\right) \tag{1.48}$$

with $r(t_i) = r(t_f) = 0$, one obtains

$$U(q_f, q_i; t_f - t_i) = e^{iS[q_c]}\int \mathcal{D}[r]\exp\left\{i\frac{m}{2}\int_{t_i}^{t_f} dt\,\left(\dot{r}^2 - \omega^2 r^2\right)\right\}. \tag{1.49}$$

Taking advantage of the boundary conditions $r(t_i) = r(t_f) = 0$, one can expand $r(t)$ in a Fourier series (assuming that this is possible; see the discussion at the end of Sec. 1.1.1.2)

$$r(t) = \sum_{l=1}^{\infty} a_l \sin\left(\frac{l\pi(t - t_i)}{T}\right) \tag{1.50}$$

with $T = t_f - t_i$. The action then reads

$$S[r] = \frac{mT}{4}\sum_{l=1}^{\infty} a_l^2\left(\frac{l^2\pi^2}{T^2} - \omega^2\right) \tag{1.51}$$

and the integration over $r(t)$ becomes an integration over the coefficients a_l,

$$\mathcal{D}[r] \to J(T)\prod_{l=1}^{\infty}\int_{-\infty}^{\infty} da_l, \tag{1.52}$$

with a Jacobian $J(T)$ that depends on T but not on ω. We then find

$$\begin{aligned}\int \mathcal{D}[r]\, e^{iS[r]} &\propto \prod_{l=1}^{\infty} \int_{-\infty}^{\infty} da_l \, \exp\left[i\frac{mT}{4}\left(\frac{l^2\pi^2}{T^2} - \omega^2\right) a_l^2\right] \\ &\propto \prod_{l=1}^{\infty} \left(\frac{l^2\pi^2}{T^2} - \omega^2\right)^{-1/2} \\ &\propto \left(\frac{\omega T}{\sin(\omega T)}\right)^{1/2}, \end{aligned} \tag{1.53}$$

where at each step the proportionality constant is independent of ω. We have used the identity

$$\prod_{l=1}^{\infty} \left(1 - \frac{z^2}{l^2\pi^2}\right) = \frac{\sin z}{z}. \tag{1.54}$$

We then have

$$U(q_f, q_i; T) = \mathcal{N}(T) \left(\frac{\omega T}{\sin(\omega T)}\right)^{1/2} e^{iS[q_c]}. \tag{1.55}$$

For $\omega \to 0$, we must reproduce the free particle result (1.43), which fixes the constant $\mathcal{N}(T) = (m/2i\pi T)^{1/2}$, so that

$$\begin{aligned} U(q_f, q_i; t_f - t_i) &= \left(\frac{m\omega}{2\pi i \sin(\omega(t_f - t_i))}\right)^{1/2} \\ &\times \exp\left\{ i \frac{m\omega}{2\sin(\omega(t_f - t_i))} \left[(q_f^2 + q_i^2)\cos\big(\omega(t_f - t_i)\big) - 2q_i q_f\right]\right\}. \end{aligned} \tag{1.56}$$

By setting $t_f - t_i \to -i\beta$, one obtains the imaginary-time propagator

$$\begin{aligned} U(q_f, q_i, -i\beta) &= \left(\frac{m\omega}{2\pi \sinh(\beta\omega)}\right)^{1/2} \\ &\times \exp\left\{ -\frac{m\omega}{2\sinh(\beta\omega)} \left[(q_f^2 + q_i^2)\cosh(\beta\omega) - 2q_i q_f\right]\right\} \end{aligned} \tag{1.57}$$

and the partition function

$$Z = \int_{-\infty}^{\infty} dq\, U(q, q; -i\beta) = \frac{e^{-\beta\omega/2}}{1 - e^{-\beta\omega}} \tag{1.58}$$

of the harmonic oscillator.

Alternatively, one can calculate the partition function directly in the Euclidean formalism with discrete times by writing

$$q_k = \frac{1}{\sqrt{N}} \sum_{l=0}^{N-1} a_l \, e^{-i\omega_l \tau_k}, \tag{1.59}$$

where $\tau_k = k\epsilon = k\beta/N$, $\omega_l = l2\pi/\beta$, $a_l^* = a_{N-l}$ and $a_0 = a_0^*$ (q_k is real). The action reads

$$S_E[a] = \frac{m}{\epsilon}\sum_{l=0}^{N-1}\Big\{[1-\cos(\omega_l\epsilon)] + \frac{1}{2}\epsilon^2\omega^2\Big\}a_l^* a_l, \tag{1.60}$$

and the partition function is given by a product of Gaussian integrals (Appendix 1.E),

$$\begin{aligned} Z &= \lim_{N\to\infty}\left(\frac{m}{2\pi\epsilon}\right)^{N/2}\int\prod_{l=0}^{N-1} da_l^*\, da_l\, e^{-S_E[a]} \\ &= \lim_{N\to\infty}\left(\frac{1}{2\epsilon}\right)^{N/2}\prod_{l=0}^{N-1}\left\{\frac{1}{\epsilon}[1-\cos(\omega_l\epsilon)] + \frac{1}{2}\epsilon\omega^2\right\}^{-1/2} \end{aligned} \tag{1.61}$$

(the exponent 1/2 is due to the constraints $a_l^* = a_{N-l}$ and $a_0^* = a_0$). Writing $\cosh\theta = 1 + \epsilon^2\omega^2/2$ and using the identity

$$\prod_{l=0}^{N-1}[\cosh\theta - \cos(\omega_l\epsilon)] = 2^{1-N}[\cosh(N\theta) - 1], \tag{1.62}$$

one reproduces the result (1.58).

Let us finally try to calculate the partition function in the Euclidean formalism with the continuum-time limit. Assuming that a Fourier series decomposition is justified,

$$q(\tau) = \frac{1}{\sqrt{\beta}}\sum_{l=-\infty}^{\infty} a_l e^{-i\omega_l\tau} \tag{1.63}$$

and $a_l^* = a_{-l}$ ($q(\tau)$ is real), the Euclidean action becomes

$$S_E[a] = \frac{m}{2}\sum_{l=-\infty}^{\infty}\left(\omega_l^2 + \omega^2\right)|a_l|^2, \tag{1.64}$$

which leads to the partition function

$$Z = \int \mathcal{D}[q]\, e^{-S_E[q]} \propto \prod_{l=-\infty}^{\infty}\frac{1}{\sqrt{\omega_l^2+\omega^2}}, \tag{1.65}$$

using the result for Gaussian integration (Appendix 1.E). The product in the rhs vanishes. To get a finite result for the partition function, one must take into account the infinite factor included in the integration measure $\mathcal{D}[q]$ (that was dropped in (1.65)). This infinite factor being ω independent, the derivative of $\ln Z$ with respect to ω is however well defined,

$$\frac{\partial \ln Z}{\partial\omega} = -\sum_{l=-\infty}^{\infty}\frac{\omega}{\omega_l^2+\omega^2} = -\frac{\beta}{2\tanh(\beta\omega/2)} \tag{1.66}$$

(the method to compute the sum over ω_l in (1.66) is explained in Appendix 1.F), so that

$$Z = \frac{\mathcal{N}}{\sinh(\beta\omega/2)}, \tag{1.67}$$

where the constant $\mathcal{N}$ is independent of ω. Since $\mathcal{N}$ is dimensionless, it cannot be a function of β alone and is therefore a pure number. It can be determined by requiring $\lim_{\beta\to\infty} Z = e^{-\beta\epsilon_0}$ with $\epsilon_0 = \omega/2$ the energy of the ground state. This yields $\mathcal{N} = 1/2$ and in turn the result (1.58).

1.1.1.7 *Path integral with vector potential*

To quantize a classical Hamiltonian $H(p, q)$, it is not always sufficient to replace the classical variables p and q by the corresponding quantum operators. When this correspondence rule yields products of the non-commuting operators $\hat{p}$ and $\hat{q}$, the order of the operators must be determined by additional conditions, such as the hermiticity of the Hamiltonian. These difficulties also show up in the path integral formulation: The path integral must be carefully defined in order to respect the proper ordering of the operators.

As an example, we consider a free particle moving in a three-dimensional space in the presence of a magnetic field $\mathbf{B} = \boldsymbol{\nabla} \times \mathbf{A}$. Its dynamics is governed by the Lagrangian

$$L(\mathbf{q}, \dot{\mathbf{q}}) = \frac{1}{2} m\dot{\mathbf{q}}^2 + e\dot{\mathbf{q}} \cdot \mathbf{A}(\mathbf{q}) \tag{1.68}$$

(e is the charge of the particle). The conjugate momentum is defined by $\mathbf{p} = \partial L/\partial\dot{\mathbf{q}} = m\dot{\mathbf{q}} + e\mathbf{A}(\mathbf{q})$ and the classical Hamiltonian reads

$$H(\mathbf{p}, \mathbf{q}) = \frac{[\mathbf{p} - e\mathbf{A}(\mathbf{q})]^2}{2m}. \tag{1.69}$$

In order to determine the quantum Hamiltonian, one has to choose the order of the operators in the product $\hat{\mathbf{p}} \cdot \mathbf{A}(\hat{\mathbf{q}})$. By requiring the Hamiltonian to be Hermitian, one obtains

$$\hat{H} = \frac{1}{2m}\left[\hat{\mathbf{p}}^2 - e\hat{\mathbf{p}} \cdot \mathbf{A}(\hat{\mathbf{q}}) - e\mathbf{A}(\hat{\mathbf{q}}) \cdot \hat{\mathbf{p}} + e^2\mathbf{A}(\hat{\mathbf{q}})^2\right] = \frac{[\hat{\mathbf{p}} - e\mathbf{A}(\hat{\mathbf{q}})]^2}{2m}. \tag{1.70}$$

This expression can also be deduced from gauge invariance, which implies that the Hamiltonian can only be a function of the gauge-invariant combination $\hat{\mathbf{p}} - e\mathbf{A}(\hat{\mathbf{q}})$. Any other quantification choice would add to the Hamiltonian a term proportional to the commutator $e[\hat{\mathbf{p}}, \mathbf{A}(\hat{\mathbf{q}})] = -ie\boldsymbol{\nabla} \cdot \mathbf{A}(\hat{\mathbf{q}})$ which would violate both hermiticity and gauge invariance.

Because of the additional term $e\dot{\mathbf{q}} \cdot \mathbf{A}(\mathbf{q})$ in the Lagrangian (1.68), we expect the action $S(\mathbf{q}_k, \mathbf{q}_{k-1}; \epsilon)$ along the infinitesimal trajectory $(\mathbf{q}_k, \mathbf{q}_{k-1})$ to write

$$S(\mathbf{q}_k, \mathbf{q}_{k-1}; \epsilon) = \frac{m}{2}\frac{(\mathbf{q}_k - \mathbf{q}_{k-1})^2}{\epsilon} + e(\mathbf{q}_k - \mathbf{q}_{k-1}) \cdot \mathbf{A}(\mathbf{q}), \tag{1.71}$$

but it is not clear whether $\mathbf{A}(\mathbf{q})$ should be evaluated at $\mathbf{q}_k$, $\mathbf{q}_{k-1}$ or somewhere in between. Since typical paths satisfy $|\mathbf{q}_k - \mathbf{q}_{k-1}| \sim \sqrt{\epsilon}$, different discretizations of $\dot{\mathbf{q}} \cdot \mathbf{A}(\mathbf{q})$ lead to changes in the action of order ϵ which cannot be ignored. The proper choice can be obtained by noting that the hermiticity of the Hamiltonian implies $U(\mathbf{q}_k, \mathbf{q}_{k-1}; \epsilon)^* = U(\mathbf{q}_{k-1}, \mathbf{q}_k; -\epsilon)$ and therefore $S(\mathbf{q}_k, \mathbf{q}_{k-1}; \epsilon) = -S(\mathbf{q}_{k-1}, \mathbf{q}_k; -\epsilon)$. The only possibility is then to evaluate the vector potential at the midpoint $(\mathbf{q}_k + \mathbf{q}_{k-1})/2$. The choice $[\mathbf{A}(\mathbf{q}_k) + \mathbf{A}(\mathbf{q}_{k-1})]/2$ (or any other combination symmetric in $\mathbf{q}_k, \mathbf{q}_{k-1}$) is also possible, since two symmetric combinations differ by a term $\mathcal{O}[(\mathbf{q}_k - \mathbf{q}_{k-1})^2] = \mathcal{O}(\epsilon)$ and the corresponding actions by a negligible term $\mathcal{O}(\epsilon^{3/2})$.

The "midpoint" rule that follows from the hermiticity of the Hamiltonian is the only choice compatible with gauge invariance. In the gauge transformation $\mathbf{A} \to \mathbf{A} + \boldsymbol{\nabla}\Lambda$, the action along the path $(\mathbf{q}_i, t_i) \to (\mathbf{q}_f, t_f)$ changes by a term

$$e\int_{t_i}^{t_f} dt\, \dot{\mathbf{q}} \cdot \boldsymbol{\nabla}\Lambda(\mathbf{q}) = e\int_{\mathbf{q}_i}^{\mathbf{q}_f} d\mathbf{q} \cdot \boldsymbol{\nabla}\Lambda(\mathbf{q}) = e[\Lambda(\mathbf{q}_f) - \Lambda(\mathbf{q}_i)]. \tag{1.72}$$

The corresponding change in the propagator,

$$U(\mathbf{q}_f, \mathbf{q}_i; t_f - t_i) \to e^{ie\Lambda(\mathbf{q}_f)} U(\mathbf{q}_f, \mathbf{q}_i; t_f - t_i) e^{-ie\Lambda(\mathbf{q}_i)}, \tag{1.73}$$

reflects the change in the phase of the wavefunctions

$$\varphi(\mathbf{q}) \to e^{ie\Lambda(\mathbf{q})} \varphi(\mathbf{q}) \tag{1.74}$$

[see Eq. (1.11)]. Up to a trivial phase factor, the propagator is independent of the function Λ (gauge invariance). In the path integration formulation, the change in the action is given by

$$e\sum_{k=1}^{N} (\mathbf{q}_k - \mathbf{q}_{k-1}) \cdot \boldsymbol{\nabla}\Lambda(\mathbf{u}_k), \tag{1.75}$$

if we choose to evaluate the vector potential in (1.71) at the point $\mathbf{u}_k$. To see whether (1.72) and (1.75) agree, we assume Λ to be a well-behaved function and rewrite (1.72) as

$$\begin{aligned} \Lambda(\mathbf{q}_f) - \Lambda(\mathbf{q}_i) &= \sum_{k=1}^{N} [\Lambda(\mathbf{q}_k) - \Lambda(\mathbf{q}_{k-1})] \\ &= \sum_{k=1}^{N} \Big\{ (\mathbf{q}_k - \mathbf{q}_{k-1}) \cdot \boldsymbol{\nabla}\Lambda(\mathbf{u}_k) \\ &\quad + \frac{1}{2}(1 - 2\theta)[(\mathbf{q}_k - \mathbf{q}_{k-1}) \cdot \boldsymbol{\nabla}]^2 \Lambda(\mathbf{u}_k) + \mathcal{O}[(\mathbf{q}_k - \mathbf{q}_{k-1})^3] \Big\}, \end{aligned} \tag{1.76}$$

where $\mathbf{u}_k = \mathbf{q}_{k-1} + \theta(\mathbf{q}_k - \mathbf{q}_{k-1})$ (with $0 \le \theta \le 1$). Since $|\mathbf{q}_k - \mathbf{q}_{k-1}|^2$ is of order ϵ for typical paths, the second term in the rhs of the last equation does not vanish in the limit $\epsilon \to 0$ unless $\theta = 1/2$, i.e. $\mathbf{u}_k = (\mathbf{q}_k + \mathbf{q}_{k-1})/2$ (midpoint rule). With this choice, Eqs. (1.72) and (1.75) agree.

We conclude this section by showing that the midpoint rule can also be obtained from an explicit calculation of the propagator $U(\mathbf{q}_k, \mathbf{q}_{k-1}; \epsilon)$. We rewrite the latter as

$$U(\mathbf{q}_k, \mathbf{q}_{k-1}; \epsilon) = \frac{1}{(2i\pi)^{3/2}} \int d^3u\, e^{\frac{i}{2}\mathbf{u}^2} \langle \mathbf{q}_k | e^{-i\bar{\epsilon}\mathbf{u}\cdot[\hat{\mathbf{p}} - eA(\hat{\mathbf{q}})]} | \mathbf{q}_{k-1} \rangle, \tag{1.77}$$

where $\bar{\epsilon} = \sqrt{\epsilon/m}$. Contrary to what was done previously, it is not possible to factorize the exponential. This would introduce an error of order $\bar{\epsilon}^2 = \epsilon/m$, whereas we need the error to be smaller than ϵ to construct the path integral over a finite time interval $t_f - t_i$. This difficulty can be circumvented by using the relation

$$e^{\bar{\epsilon}(\hat{A}+\hat{B})} = e^{\frac{\bar{\epsilon}}{2}\hat{B}} e^{\bar{\epsilon}\hat{A}} e^{\frac{\bar{\epsilon}}{2}\hat{B}} + \mathcal{O}(\bar{\epsilon}^3), \tag{1.78}$$

which can be checked by a direct expansion of the exponentials. We then deduce

$$\langle \mathbf{q}_k | e^{-i\bar{\epsilon}\mathbf{u}\cdot[\hat{\mathbf{p}}-eA(\hat{\mathbf{q}})]} | \mathbf{q}_{k-1} \rangle = e^{\frac{i}{2}e\bar{\epsilon}\mathbf{u}\cdot\mathbf{A}(\mathbf{q}_k)} \langle \mathbf{q}_k | e^{-i\bar{\epsilon}\mathbf{u}\cdot\hat{\mathbf{p}}+\mathcal{O}(\epsilon^{3/2})} | \mathbf{q}_{k-1} \rangle e^{\frac{i}{2}e\bar{\epsilon}\mathbf{u}\cdot\mathbf{A}(\mathbf{q}_{k-1})}, \tag{1.79}$$

where the $\mathcal{O}(\epsilon^{3/2})$ error can now be ignored. The matrix element in (1.79) is calculated by inserting the closure relation $\sum_{\mathbf{p}} |\mathbf{p}\rangle\langle\mathbf{p}| = 1$,

$$\begin{aligned} \langle \mathbf{q}_k | e^{-i\bar{\epsilon}\mathbf{u}\cdot\hat{\mathbf{p}}} | \mathbf{q}_{k-1} \rangle &= \sum_{\mathbf{p},\mathbf{p}'} \langle \mathbf{q}_k | \mathbf{p} \rangle \langle \mathbf{p} | e^{-i\bar{\epsilon}\mathbf{u}\cdot\hat{\mathbf{p}}} | \mathbf{p}' \rangle \langle \mathbf{p}' | \mathbf{q}_{k-1} \rangle \\ &= \int \frac{d^3p}{(2\pi)^3} e^{i\mathbf{p}\cdot(\mathbf{q}_k-\mathbf{q}_{k-1}-\bar{\epsilon}\mathbf{u})} = \frac{1}{\bar{\epsilon}^3}\delta\left(\frac{\mathbf{q}_k-\mathbf{q}_{k-1}}{\bar{\epsilon}} - \mathbf{u}\right), \end{aligned} \tag{1.80}$$

which gives

$$\begin{aligned} U(\mathbf{q}_k, \mathbf{q}_{k-1}; \epsilon) = \left(\frac{m}{2i\pi\epsilon}\right)^{3/2} \exp\biggl\{ i\epsilon \biggl[\frac{m}{2} \frac{(\mathbf{q}_k - \mathbf{q}_{k-1})^2}{\epsilon^2} \\ + \frac{e}{2} \frac{\mathbf{q}_k - \mathbf{q}_{k-1}}{\epsilon} \cdot (\mathbf{A}(\mathbf{q}_k) + \mathbf{A}(\mathbf{q}_{k-1})) \biggr] \biggr\}. \end{aligned} \tag{1.81}$$

We recognize in the exponential i times the action (1.71) with the vector potential $\mathbf{A}(\mathbf{q})$ approximated by $[\mathbf{A}(\mathbf{q}_k) + \mathbf{A}(\mathbf{q}_{k-1})]/2$ which, as discussed above, is equivalent to the mid-point rule.[14]

1.1.1.8 *Many-particle system*

With little more effort, we can now write a path integral for the partition function — or the propagator — of a system with N particles interacting via a two-body potential $v(q_i - q_j)$ (we consider a one-dimensional system for simplicity). The Hilbert space is defined as the subspace of the tensor product $\mathcal{H} \otimes \cdots \otimes \mathcal{H}$ ($\mathcal{H}$ denotes the one-particle Hilbert space) containing all N-particle states properly symmetrized or antisymmetrized according to the quantum statistics of the particles. The partition function can be written as (see Sec. 1.2)

$$Z = \frac{1}{N!} \sum_{P \in S_N} \epsilon_P \int dq_1 \cdots dq_N \, (q_1 \cdots q_N | e^{-\beta\hat{H}} | q_{P(1)} \cdots q_{P(N)}), \tag{1.82}$$

where $|q_1 \cdots q_N) = |q_1\rangle \otimes \cdots \otimes |q_N\rangle$. The sum in (1.82) is over all permutations P of $\{1, \ldots, N\}$. ϵ_P equals one for bosons and the signature of the permutation P for fermions.

[14] One can also directly verify that $U(\mathbf{q}_k, \mathbf{q}_{k-1}; \epsilon)$ yields the time evolution of the wavefunction as obtained from the Hamiltonian (1.70) [3]. This provides a definitive proof of the validity of the path integral.

Proceeding as in the one-particle case and inserting the closure relation

$$\int dq_1 \cdots dq_N \, |q_1 \cdots q_N)(q_1 \cdots q_N| = 1 \tag{1.83}$$

at each time step, we obtain the path integral

$$Z = \frac{1}{N!} \sum_{P \in S_N} \epsilon_P \int_{q_i(\beta)=q_{P(i)}(0)} \mathcal{D}[q] \, e^{-S_E[q]} \tag{1.84}$$

with the Euclidean action

$$S_E[q] = \int_0^\beta d\tau \left[\frac{m}{2} \sum_{i=1}^N \dot{q}_i^2 + \sum_{\substack{i,j=1 \\ (i<j)}}^N v(q_i - q_j) \right]. \tag{1.85}$$

This expression bears an obvious similarity with that obtained for a single particle. It is however not well suited for the study of a many-particle system. In Sec. 1.4, we shall introduce a path integral or, more precisely, a functional integral representation of the partition function that turns out to be much more convenient. This functional integral is based on the coherent states introduced in Sec. 1.3 and involves an integration over all configurations of a field weighted by an appropriate action. This turns out to be an essential aspect of (quantum) statistical physics and field theory: Systems with an infinite number of degrees of freedom are naturally described by fields, rather than the set of coordinates of all particles. Before embarking on the study of quantum many-body systems, we consider in the following section an elementary example of field theory.

1.1.2 *Functional integral in statistical physics*

In this section, we consider the functional integral approach to the quantum harmonic string. This simple example already exhibits many essential aspects of the field theory of (more complicated, i.e. anharmonic) quantum systems.[15]

1.1.2.1 *The classical harmonic string*

We consider a one-dimensional harmonic string of length L and mass per unit length ρ. We denote by $\phi(x,t)$ the displacement of the infinitesimal mass ρdx located between x and $x + dx$ at equilibrium. The dynamics of the chain is governed by the Lagrangian[16]

$$\begin{aligned} L[\phi] &= \int_0^L dx \, \mathcal{L}(\partial_x \phi, \dot{\phi}), \\ \mathcal{L}(\partial_x \phi, \dot{\phi}) &= \frac{1}{2}\rho\dot{\phi}^2 - \frac{1}{2}\kappa(\partial_x \phi)^2, \end{aligned} \tag{1.86}$$

[15] In a particle language, a "harmonic" system (i.e. a system whose Lagrangian is quadratic in its variables) corresponds to non-interacting particles, whereas an "anharmonic" system includes interactions between particles. These particles do not necessarily correspond to the bare particles but can refer to elementary excitations (a concept that will be explained in the forthcoming chapters).

[16] In high-energy physics, the analog of the Lagrangian (1.86) in four spacetime dimensions is known as the Klein–Gordon Lagrangian.

where $\mathcal{L}$ is referred to as the Lagrangian density. The first term in the rhs of (1.86) corresponds to the kinetic energy while the second one gives the energy cost associated with the deformation of the string. The harmonic string can be seen as the low-energy or long-wavelength limit of a one-dimensional "crystal" consisting of masses $m = \rho a$ separated by a distance a (at equilibrium) and connected by springs with stiffness κ/a.[17]

The Euler–Lagrange equation

$$\frac{d}{dt}\frac{\partial\mathcal{L}(\partial_x\phi,\dot{\phi})}{\partial\dot{\phi}} + \frac{d}{dx}\frac{\partial\mathcal{L}(\partial_x\phi,\dot{\phi})}{\partial(\partial_x\phi)} = 0 \tag{1.87}$$

gives the wave equation

$$\rho\ddot{\phi} - \kappa\partial_x^2\phi = 0. \tag{1.88}$$

The solutions are plane waves $\phi(x,t) \propto e^{i(kx-c|k|t)}$ + c.c. propagating with the velocity $c = \sqrt{\kappa/\rho}$, where $k = p\frac{2\pi}{L}$ (p integer) for periodic boundary conditions $\phi(x+L,t) = \phi(x,t)$.

1.1.2.2 *The quantum harmonic string*

The conjugate momentum of $\phi(x,t)$ is

$$\Pi(x,t) = \frac{\partial\mathcal{L}(\partial_x\phi,\dot{\phi})}{\partial\dot{\phi}(x,t)} = \rho\dot{\phi}(x,t) \tag{1.89}$$

and the Hamiltonian is defined by

$$H = \int_0^L dx\left[\Pi\dot{\phi} - \mathcal{L}(\partial_x\phi,\dot{\phi})\right] = \int_0^L dx\left[\frac{\Pi^2}{2\rho} + \frac{1}{2}\kappa(\partial_x\phi)^2\right]. \tag{1.90}$$

To quantize the string, we promote ϕ and Π to operators satisfying the commutation relations

$$[\hat{\phi}(x),\hat{\Pi}(x')] = i\delta(x-x'). \tag{1.91}$$

It is convenient to introduce the Fourier-transformed operators

$$\begin{aligned}\hat{\phi}(k) &= \frac{1}{\sqrt{L}}\int_0^L dx\, e^{-ikx}\hat{\phi}(x) = \hat{\phi}^\dagger(-k),\\ \hat{\Pi}(k) &= \frac{1}{\sqrt{L}}\int_0^L dx\, e^{-ikx}\hat{\Pi}(x) = \hat{\Pi}^\dagger(-k),\end{aligned} \tag{1.92}$$

satisfying the commutation relations

$$[\hat{\phi}(k),\hat{\phi}(k')] = [\hat{\Pi}(k),\hat{\Pi}(k')] = 0 \quad \text{and} \quad [\hat{\phi}(k),\hat{\Pi}^\dagger(k')] = i\delta_{k,k'}. \tag{1.93}$$

[17]The Lagrangian of the one-dimensional crystal reads $L(q_i,\dot{q}_i) = \frac{1}{2}\sum_{i=1}^n[m\dot{q}_i^2 - k_s(q_{i+1}-q_i)^2]$, where q_i denotes the displacement of the ith atom with respect to its equilibrium position. The normal modes (phonons) of this system propagate with frequencies $\omega_k = \sqrt{2k_s/m}(1-\cos ka)^{1/2}$. In the long wavelength limit $|k|a \ll 1$, one recovers the spectrum $\omega = c|k|$ of the harmonic continuous string.

This allows us to write the Hamiltonian as a sum of harmonic oscillators,

$$\hat{H} = \sum_k \left[\frac{1}{2\rho} \hat{\Pi}^\dagger(k)\hat{\Pi}(k) + \frac{1}{2}\rho\omega_k^2 \hat{\phi}^\dagger(k)\hat{\phi}(k) \right], \tag{1.94}$$

where $\omega_k = c|k|$ and the sum is over all vectors $k = p2\pi/L$ ($p \in \mathbb{Z}$) satisfying the boundary conditions $e^{ikL} = 1$. $\hat{H}$ is diagonalized by introducing the ladder operators

$$\begin{aligned} \hat{a}(k) &= \sqrt{\frac{\rho\omega_k}{2}} \left[\hat{\phi}(k) + \frac{i}{\rho\omega_k} \hat{\Pi}(k) \right], \\ \hat{a}^\dagger(k) &= \sqrt{\frac{\rho\omega_k}{2}} \left[\hat{\phi}^\dagger(k) - \frac{i}{\rho\omega_k} \hat{\Pi}^\dagger(k) \right]. \end{aligned} \tag{1.95}$$

Using the commutation relations

$$[\hat{a}(k), \hat{a}(k')] = [\hat{a}^\dagger(k), \hat{a}^\dagger(k')] = 0, \qquad [\hat{a}(k), \hat{a}^\dagger(k')] = \delta_{k,k'}, \tag{1.96}$$

one easily finds

$$\hat{H} = \sum_k \omega_k \left(\hat{a}_k^\dagger \hat{a}_k + \frac{1}{2} \right). \tag{1.97}$$

The eigenstates of the harmonic oscillator indexed by k are defined by

$$\begin{aligned} |n_k\rangle &= \frac{1}{\sqrt{n_k!}} (\hat{a}_k^\dagger)^{n_k} |0\rangle, \\ \omega_k \left(\hat{a}_k^\dagger \hat{a}_k + \frac{1}{2} \right) |n_k\rangle &= \omega_k \left(n_k + \frac{1}{2} \right) |n_k\rangle, \end{aligned} \tag{1.98}$$

where $|0\rangle = |n_k = 0\rangle$ is the normalized vacuum state: $\hat{a}_k|0\rangle = 0$ and $\langle 0|0\rangle = 1$. The ladder operators $\hat{a}_k$ and $\hat{a}_k^\dagger$ satisfy

$$\begin{aligned} \hat{a}_k|n_k\rangle &= \sqrt{n_k}|n_k - 1\rangle, \\ \hat{a}_k^\dagger|n_k\rangle &= \sqrt{n_k + 1}|n_k + 1\rangle, \end{aligned} \tag{1.99}$$

and can be seen as annihilation and creation operators of an elementary excitation (phonon) with momentum k. $n_k = \langle n_k|\hat{a}_k^\dagger \hat{a}_k|n_k\rangle$ denotes the number of phonons in the state $|n_k\rangle$. The eigenstates of the Hamiltonian (1.97) are obtained from the tensor product of the states $|n_k\rangle$,

$$\begin{aligned} |n_{k_1} \cdots n_{k_i} \cdots\rangle &= |n_{k_1}\rangle \otimes \cdots \otimes |n_{k_i}\rangle \otimes \cdots = \prod_i \frac{(\hat{a}_{k_i}^\dagger)^{n_{k_i}}}{\sqrt{n_{k_i}!}} |\text{vac}\rangle, \\ \hat{H}|n_{k_1} \cdots n_{k_i} \cdots\rangle &= \left[\sum_j \left(n_{k_j} + \frac{1}{2} \right) \omega_{k_j} \right] |n_{k_1} \cdots n_{k_i} \cdots\rangle, \end{aligned} \tag{1.100}$$

where $|\text{vac}\rangle$ is the vacuum, i.e. the normalized state defined by $\hat{a}_k|\text{vac}\rangle = 0$ (for any k). The eigenstate $|n_{k_1} \cdots n_{k_i} \cdots\rangle$ has a total number of phonons $n = \sum_i n_{k_i}$. The Hilbert space can be written as the direct sum $\mathcal{H} = \mathcal{H}_0 \oplus \mathcal{H}_1 \oplus \cdots \oplus \mathcal{H}_n \oplus \cdots$ where $\mathcal{H}_n$ is the Hilbert space with n phonons. This type of Hilbert space, generically referred to as a Fock space, is at the basis of the second-quantization formalism described in Sec. 1.2.

1.1.2.3 *Functional integral formulation*

A functional integral formulation can be obtained following the same procedure as in the case of the path integral discussed in Sec. 1.1.1. $\hat{\phi}(x)$ and $\hat{\Pi}(x)$ play the role of the position and momentum operator $\hat{q}$ and $\hat{p}$ of the single particle (the only difference being that they are labeled by x). One can introduce states $|\phi\rangle$ and $|\Pi\rangle$ satisfying $\hat{\phi}(x)|\phi\rangle = \phi(x)|\phi\rangle$ and $\hat{\Pi}(x)|\Pi\rangle = \Pi(x)|\Pi\rangle$. Since both sets of states $\{|\phi\rangle\}$ and $\{|\Pi\rangle\}$ form a complete basis of the Hilbert space, we have the closure relations

$$\begin{aligned} \mathcal{N} \lim_{a\to 0} \int \prod_{l=0}^{L/a} d\phi(la)\, |\phi\rangle\langle\phi| &= 1, \\ \mathcal{N}' \lim_{a\to 0} \int \prod_{l=0}^{L/a} d\Pi(la)\, |\Pi\rangle\langle\Pi| &= 1. \end{aligned} \tag{1.101}$$

Here, we have discretized the chain; the continuous variable x becomes a discrete variable $x = la$. The harmonic chain is recovered in the limit $a \to 0$. This step is necessary to properly define the closure relations (1.101) as well as the functional integration measure [Eq. (1.110)]. $\mathcal{N}$ and $\mathcal{N}'$ are unimportant normalization constants that contribute a factor to the partition function but do not affect the expectation values. They will be dropped in the following.

To write the partition function as a functional integral, we use the first closure relation (1.101) to obtain

$$Z = \sum_n \langle n|e^{-\beta\hat{H}}|n\rangle = \int d\phi \sum_n \langle n|e^{-\beta\hat{H}}|\phi\rangle\langle\phi|n\rangle = \int d\phi \langle\phi|e^{-\beta\hat{H}}|\phi\rangle, \tag{1.102}$$

where $d\phi \equiv \prod_{l=0}^{L/a} d\phi(la)$ and $\{|n\rangle\}$ denotes a complete basis of the Hamiltonian $\hat{H}$. We now proceed as in Sec. 1.1.1 and split the imaginary time β into N infinitesimal steps $\epsilon = \beta/N$,[18]

$$Z = \int \prod_{k=1}^{N} d\phi_k \prod_{k=1}^{N} \langle\phi_k|e^{-\epsilon\hat{H}}|\phi_{k-1}\rangle, \tag{1.103}$$

with $\phi_0 = \phi_N$. For $\epsilon \to 0$, we can approximate

$$\langle\phi_k|e^{-\epsilon\hat{H}}|\phi_{k-1}\rangle = \langle\phi_k|\exp\Big\{-\frac{\epsilon}{2\rho}\int dx\, \hat{\Pi}^2\Big\}\exp\Big\{-\frac{\epsilon\kappa}{2}\int dx\, (\partial_x\hat{\phi})^2\Big\}|\phi_{k-1}\rangle. \tag{1.104}$$

Inserting the second closure relation (1.101), we obtain

$$\begin{aligned} &\langle\phi_k|e^{-\epsilon\hat{H}}|\phi_{k-1}\rangle \\ &\quad = \int d\Pi_k \exp\Big\{-\frac{\epsilon}{2\rho}\int dx\, \Pi_k^2 - \frac{\epsilon\kappa}{2}\int dx\, (\partial_x\phi_{k-1})^2\Big\}\langle\phi_k|\Pi_k\rangle\langle\Pi_k|\phi_{k-1}\rangle \\ &\quad = \int d\Pi_k \exp\Big\{-\frac{\epsilon}{2\rho}\int dx\, \Pi_k^2 - \frac{\epsilon\kappa}{2}\int dx\, (\partial_x\phi_{k-1})^2 + i\int dx\, \Pi_k(\phi_k - \phi_{k-1})\Big\}, \end{aligned} \tag{1.105}$$

[18]The discrete time variable k should not be confused with the momentum variable k used in (1.92) or (1.113).

where we have used

$$\langle \phi_k | \Pi_k \rangle = \exp\left\{ i \int dx\, \Pi_k \phi_k \right\} \tag{1.106}$$

(up to an irrelevant multiplicative constant). Inserting (1.105) into (1.103) and taking the continuum-time limit, the partition function

$$Z = \int_{\phi(x,\beta)=\phi(x,0)} \mathcal{D}[\Pi, \phi] \exp\left\{ -\int_0^\beta d\tau \int_0^L dx \left[\frac{\Pi^2}{2\rho} + \frac{\kappa}{2}(\partial_x \phi)^2 - i\Pi\dot{\phi} \right] \right\} \tag{1.107}$$

is written as a functional integral in imaginary time over the real fields $\phi(x,\tau)$ and $\Pi(x,\tau)$.[19] Performing the Gaussian functional integral over Π,[20] one obtains

$$Z = \int_{\phi(x,\beta)=\phi(x,0)} \mathcal{D}[\phi]\, e^{-S_E[\phi]} \tag{1.108}$$

with the Euclidean action

$$S_E[\phi] = \frac{1}{2} \int_0^\beta d\tau \int_0^L dx \left[\rho\dot{\phi}^2 + \kappa(\partial_x \phi)^2 \right]. \tag{1.109}$$

The integration measure in (1.108) is defined as

$$\mathcal{D}[\phi] = \lim_{N\to\infty} \lim_{a\to 0} \prod_{k=1}^{N} \prod_{l=0}^{L/a} d\phi(la, k\beta/N) \tag{1.110}$$

(up to a multiplicative factor). Performing a Wick rotation ($\tau = it$), we see that the real-time action is given by the classical action $S[\phi] = \int dt \int_0^L dx\, \mathcal{L}(\partial_x \phi, \dot{\phi})$.

Since the field $\phi(x,\tau)$ is periodic in time, it can be expanded as a Fourier series,

$$\begin{aligned} \phi(x,\tau) &= \frac{1}{\sqrt{\beta}} \sum_{\omega_n} e^{-i\omega_n \tau} \phi(x, i\omega_n), \\ \phi(x, i\omega_n) &= \frac{1}{\sqrt{\beta}} \int_0^\beta d\tau\, e^{i\omega_n \tau} \phi(x,\tau), \end{aligned} \tag{1.111}$$

where the discrete frequency

$$\omega_n = \frac{2\pi}{\beta} n \qquad (n \in \mathbb{Z}) \tag{1.112}$$

is called a Matsubara (imaginary) frequency. The Fourier transformed field $\phi(k, i\omega_n)$ diagonalizes the action,

$$S_E[\phi] = \frac{\rho}{2} \sum_{k,\omega_n} \phi(-k, -i\omega_n) \left(\omega_n^2 + \omega_k^2 \right) \phi(k, i\omega_n), \tag{1.113}$$

so that the partition function (1.108) is given by a product of Gaussian integrals. Note that $\phi(-k, -i\omega_n) = \phi^*(k, i\omega_n)$ since $\phi(x,\tau)$ is real.

[19] Equation (1.107) is similar to (1.29).

[20] Gaussian functional integration is discussed in Sec. 1.E.

The functional integral approach to the quantum harmonic string already exhibits many essential aspects of the field theory of (more complicated) quantum systems:

- Equation (1.108) is an exact representation of the partition function. Since the action is quadratic in the field, the functional integral can be done exactly[21]:

$$Z = \prod_k \frac{e^{-\frac{\beta}{2}\omega_k}}{1 - e^{-\beta\omega_k}}, \tag{1.114}$$

  which is the standard result for a set of decoupled harmonic oscillators. More generally, all results that can be obtained from the Hamiltonian (1.97) can also be derived from the functional integral (1.108). When the action is not quadratic, e.g. when it contains a term

$$S_{\text{int}}[\phi] = \int_0^\beta d\tau \int_0^L dx\, V(\phi) \tag{1.115}$$

  with $V(\phi)$ a potential of order 4 or higher, the functional integral cannot be done exactly and one has to resort to approximate strategies, such as a perturbative expansion (Sec. 1.5) or a saddle-point approximation (Sec. 1.7).
- Requiring the action to be stationary, $\delta S_E/\delta\phi(x,\tau) = 0$, one obtains the classical equation of motion (1.88) in imaginary time: $\rho\partial_\tau^2\phi + \kappa\partial_x^2\phi = 0$.
- The continuous string model cannot be valid at arbitrarily small length scales. For example, if it corresponds to the low-energy description of a one-dimensional crystal (as discussed at the beginning of this section), it is valid only for momenta k much smaller than $1/a$ where a is the lattice spacing. The dispersion ω_k is therefore bounded by a maximum value $\omega_0 = \omega_{k_{\max}} = \omega_{k\sim 1/a}$. At high temperatures, when $\omega_0 \ll T$, the partition function is dominated by the zero Matsubara frequency $\omega_{n=0} = 0$. The action

$$S_E[\phi] \simeq \frac{\rho}{2}\sum_k \phi(-k, -i\omega_{n=0})\omega_k^2\phi(k, i\omega_{n=0}) = \frac{\kappa}{2T}\int_0^L dx(\partial_x\phi)^2 \tag{1.116}$$

  (with $\phi(x) \equiv \sqrt{T}\phi(x, i\omega_{n=0})$) then yields the classical partition function

$$Z_{\text{cl}} = \int \mathcal{D}[\phi] \exp\left\{-\frac{\kappa}{2T}\int_0^L dx(\partial_x\phi)^2\right\}. \tag{1.117}$$

  A quantum system at sufficiently high temperature behaves classically. When $T \sim \omega_0$, the system enters the quantum regime and non-zero Matsubara frequencies cannot be ignored. More precisely, the harmonic oscillator with frequency ω_k behaves classically if $\omega_k \ll 2\pi T$. In that case, the non-zero Matsubara frequencies can be ignored when computing the partition function: $Z \simeq T/\omega_k$ (this result is easily obtained from (1.66)–(1.67) by retaining only the term $\omega_l = 0$), which gives the mean energy

[21]See the calculation of the partition function of the harmonic oscillator in Sec. 1.1.1 [in particular, Eqs. (1.64) and (1.65)].

$E = T$ in agreement with the equipartition theorem. We recover the well-known result that a harmonic oscillator behaves classically when the mean occupation factor $n_B(\omega_k)$ is much larger than unity. Thus, the quantum-mechanical behavior of the system shows up in the imaginary-time dependence of the field.

- In the limit $T \to 0$ ($\beta \to \infty$), the imaginary time τ varies in the interval $[0, \infty[$ and the discrete Matsubara frequency ω_n becomes a continuous variable ω. By rescaling the time variable, we can rewrite the action as

$$S_E[\phi] = \frac{\kappa}{2c} \int d^2r [\boldsymbol{\nabla}\phi(\mathbf{r})]^2 \tag{1.118}$$

(we assume both $\beta \to \infty$ and $L \to \infty$), where $\mathbf{r} = (x, c\tau)$. The action of the quantum one-dimensional string reduces to the action of a classical two-dimensional string. More generally, zero-temperature d-dimensional quantum systems behave as $(d+1)$-dimensional classical systems (d space dimensions and one (imaginary) time dimension); however, the action may not exhibit the ("relativistic") SO$(d+1)$ spacetime symmetry as in (1.118).

- The fact that the excitation energies ω_k can be directly read off from the partition function (1.108), (1.113) illustrates the connection between thermodynamics and dynamics in a quantum system. The relevant tool to study the dynamics of the system is the (imaginary-time) propagator of the field ϕ — also called the (imaginary-time) Green function — defined as the correlation function

$$\begin{aligned} G(k,\tau) &= \langle \phi(k,\tau)\phi(-k,0)\rangle \\ &= \frac{1}{Z}\int \mathcal{D}[\phi]\, \phi(k,\tau)\phi(-k,0)\, e^{-S_E[\phi]}. \end{aligned} \tag{1.119}$$

It can also be written as the time-ordered correlation function

$$\begin{aligned} G(k,\tau) &= \langle T_\tau \hat\phi(k,\tau)\hat\phi(-k,0)\rangle \\ &= \Theta(\tau)\langle \hat\phi(k,\tau)\hat\phi(-k,0)\rangle + \Theta(-\tau)\langle \hat\phi(-k,0)\hat\phi(k,\tau)\rangle, \end{aligned} \tag{1.120}$$

where T_τ is the time-ordering operator defined in Sec. 1.1.1 and $\langle \cdots \rangle = Z^{-1}\mathrm{Tr}(e^{-\beta \hat H} \cdots)$. The imaginary-time operator $\hat\phi(k,\tau)$ is defined by

$$\hat\phi(k,\tau) = e^{\tau \hat H}\hat\phi(k)e^{-\tau \hat H}. \tag{1.121}$$

The action (1.113) being quadratic, $G(k, i\omega_n)$ can be calculated exactly,[20]

$$G(k, i\omega_n) = \langle \phi(k, i\omega_n)\phi(-k, -i\omega_n)\rangle = \frac{1}{\rho}\frac{1}{\omega_n^2 + \omega_k^2}. \tag{1.122}$$

To obtain the dynamics in real time, one performs an analytic continuation $i\omega_n \to \omega + i\eta$ to real frequencies ω. The infinitesimal $\eta \to 0^+$ is crucial for the analytic continuation to be well defined in the general case (Chapter 3). The resulting correlation function

$$G^R(k,\omega) = G(k,\omega+i\eta) = -\frac{1}{\rho}\frac{1}{(\omega+i\eta)^2-\omega_k^2} \tag{1.123}$$

is called the retarded Green function (Sec. 1.2.3). Its poles $\omega = \pm\omega_k - i\eta$ directly give the excitation energies of the quantum string. The excitation spectrum also shows up in the spectral function

$$A(k,\omega) = \Im[G^R(k,\omega)] = \frac{\pi}{2\rho\omega_k}[\delta(\omega-\omega_k)-\delta(\omega+\omega_k)]. \tag{1.124}$$

This is a very general result: Green functions — or correlation functions — contain information about the excitation spectrum of the system and are therefore the main quantities of interest besides the free energy $-\frac{1}{\beta}\ln Z$. We shall see in Chapter 3 that correlation functions also determine the linear response of the system to an external field.

1.2 Second Quantization and Green Functions

In quantum statistical physics, one is interested in systems with a large number of particles. The state vectors describing the possible physical states must be symmetric or antisymmetric under the exchange of two particles, depending on the statistics of the particles (bosons or fermions). Furthermore, the number of particles fluctuates when the system can exchange particles with a reservoir — as, for instance, in the grand canonical ensemble. Second quantization provides a natural formalism to describe many-particle systems (the main results are summarized in Table 1.1). It yields the concept of quantum field and in turn the possibility to express the partition function as a functional integral over suitably defined coherent states (Secs. 1.3.1, 1.3.3 and 1.4).

1.2.1 *Fock space*

The Fock space is defined as the direct sum

$$\mathcal{H} = \mathcal{H}_0 \oplus \mathcal{H}_1 \oplus \cdots = \mathop{\oplus}_{n=0}^{n=\infty} \mathcal{H}_n, \tag{1.125}$$

where $\mathcal{H}_n$ is the Hilbert space corresponding to a system of n particles. $\mathcal{H}_0$ contains only the vacuum state $|\text{vac}\rangle$ and $\mathcal{H}_1$ is the Hilbert used when studying the quantum mechanics of a single particle.

Let us first focus on the Hilbert space $\mathcal{H}_n$. $\mathcal{H}_n$ contains all n-particle states properly symmetrized (bosons) or antisymmetrized (fermions). It is a subspace of the direct product $\bar{\mathcal{H}}_n = \mathcal{H}_1 \otimes \mathcal{H}_1 \otimes \cdots \otimes \mathcal{H}_1$. If we know an orthonormal basis $\{|\alpha\rangle\}$ of $\mathcal{H}_1$, then we can obtain an orthonormal basis of $\bar{\mathcal{H}}_n$ from the tensor product

$$|\alpha_1\cdots\alpha_n) = |\alpha_1\rangle \otimes |\alpha_2\rangle \otimes \cdots \otimes |\alpha_n\rangle. \tag{1.126}$$

Table 1.1 Main results from second quantification.

Many-body states	$\vert\alpha_1\cdots\alpha_n) = \vert\alpha_1\rangle\otimes\cdots\otimes\vert\alpha_n\rangle$ $\vert\alpha_1\cdots\alpha_n\} = \frac{1}{\sqrt{n!}}\sum_P \varsigma^P\vert\alpha_{P(1)}\cdots\alpha_{P(n)})$ $\vert\alpha_1\cdots\alpha_n\rangle = \frac{1}{(\prod_{i=1}^p n_{\alpha_i}!)^{1/2}}\vert\alpha_1\cdots\alpha_n\}$
Closure relation	$\sum_{n=0}^{\infty}\frac{1}{n!}\sum_{\alpha_1\cdots\alpha_n}\vert\alpha_1\cdots\alpha_n\}\{\alpha_1\cdots\alpha_n\vert = 1$
Creation and annihilation operators	$\hat\psi^\dagger_\alpha\vert\alpha_1\cdots\alpha_n\} = \vert\alpha\alpha_1\cdots\alpha_n\}$ $\hat\psi_\alpha\vert\alpha_1\cdots\alpha_n\} = \sum_{i=1}^{n}\varsigma^{i-1}\delta_{\alpha,\alpha_i}\vert\alpha_1\cdots\hat\alpha_i\cdots\alpha_n\}$ $\hat\psi^\dagger_\alpha\vert\alpha_1\cdots\alpha_n\rangle = \sqrt{n_\alpha+1}\vert\alpha\alpha_1\cdots\alpha_n\rangle$ $\hat\psi_\alpha\vert\alpha_1\cdots\alpha_n\rangle = \frac{1}{\sqrt{n_\alpha}}\sum_{i=1}^{n}\varsigma^{i-1}\delta_{\alpha,\alpha_i}\vert\alpha_1\cdots\hat\alpha_i\cdots\alpha_n\rangle$
(Anti)commutation relations	$[\hat\psi_\alpha,\hat\psi_{\alpha'}]_{-\varsigma} = [\hat\psi^\dagger_\alpha,\hat\psi^\dagger_{\alpha'}]_{-\varsigma} = 0$ $[\hat\psi_\alpha,\hat\psi^\dagger_{\alpha'}]_{-\varsigma} = \delta_{\alpha,\alpha'}$
Field operators	$\hat\psi_\sigma(\mathbf{r}) = \sum_\alpha\langle\mathbf{r},\sigma\vert\alpha\rangle\hat\psi_\alpha$ $\hat\psi^\dagger_\sigma(\mathbf{r}) = \sum_\alpha\langle\alpha\vert\mathbf{r},\sigma\rangle\hat\psi^\dagger_\alpha$ $[\hat\psi_\sigma(\mathbf{r}),\hat\psi_{\sigma'}(\mathbf{r}')]_{-\varsigma} = [\hat\psi^\dagger_\sigma(\mathbf{r}),\hat\psi^\dagger_{\sigma'}(\mathbf{r}')]_{-\varsigma} = 0$ $[\hat\psi_\sigma(\mathbf{r}),\hat\psi^\dagger_{\sigma'}(\mathbf{r}')]_{-\varsigma} = \delta_{\sigma,\sigma'}\delta(\mathbf{r}-\mathbf{r}')$
One-body operator	$\hat V^{(1)} = \sum_{\alpha,\alpha'}\langle\alpha\vert\hat V\vert\alpha'\rangle\hat\psi^\dagger_\alpha\hat\psi_{\alpha'}$ $= \int d^drd^dr'\sum_{\sigma,\sigma'}\langle\mathbf{r},\sigma\vert\hat V\vert\mathbf{r}',\sigma'\rangle\hat\psi^\dagger_\sigma(\mathbf{r})\hat\psi_{\sigma'}(\mathbf{r}')$
Two-body operator	$\hat V^{(2)} = \frac{1}{2}\sum_{\alpha_1\cdots\alpha_2'}(\alpha_1\alpha_2\vert\hat V\vert\alpha_1'\alpha_2')\hat\psi^\dagger_{\alpha_1}\hat\psi^\dagger_{\alpha_2}\hat\psi_{\alpha_2'}\hat\psi_{\alpha_1'}$

Note that we use the notation $|\cdots)$ (with a curly bracket) to denote this n-particle state. The overlap of two vectors is simply

$$(\alpha_1'\cdots\alpha_n'|\alpha_1\cdots\alpha_n) = \langle\alpha_1'|\alpha_1\rangle\cdots\langle\alpha_n'|\alpha_n\rangle = \delta_{\alpha_1',\alpha_1}\cdots\delta_{\alpha_n',\alpha_n} \tag{1.127}$$

and the completeness of the basis gives the closure relation

$$\sum_{\alpha_1\cdots\alpha_n}|\alpha_1\cdots\alpha_n)(\alpha_1\cdots\alpha_n| = 1, \tag{1.128}$$

where "1" in the rhs means the unit operator in $\bar{\mathcal{H}}_n$. A wavefunction in coordinate space is obtained by multiplying the ket $|\alpha_1 \cdots \alpha_n)$ by the bra $(\mathbf{r}_1 \cdots \mathbf{r}_n| = \langle \mathbf{r}_1| \otimes \cdots \otimes \langle \mathbf{r}_n|$,

$$(\mathbf{r}_1 \cdots \mathbf{r}_n|\alpha_1 \cdots \alpha_n) = \varphi_{\alpha_1}(\mathbf{r}_1) \cdots \varphi_{\alpha_n}(\mathbf{r}_n), \tag{1.129}$$

where $\varphi_\alpha(\mathbf{r}) = \langle \mathbf{r}|\alpha\rangle$ is the coordinate-representation wavefunction of a particle in the quantum state $|\alpha\rangle$.

To obtain a basis of $\mathcal{H}_n$, we must (anti)symmetrize the tensor product (1.126),

$$\begin{aligned} |\alpha_1 \cdots \alpha_n\} &= \sqrt{n!}\mathcal{P}_\zeta|\alpha_1 \cdots \alpha_n) \\ &= \frac{1}{\sqrt{n!}} \sum_{P \in S_n} \zeta^P |\alpha_{P(1)} \cdots \alpha_{P(n)}). \end{aligned} \tag{1.130}$$

$\mathcal{P}_+$ ($\mathcal{P}_-$) is a symmetrization (antisymmetrization) operator and $\zeta = +1$ for bosons and -1 for fermions. The sum in (1.130) is over all permutations P of $\{1 \cdots n\}$. ζ^P equals one for bosons and the signature of the permutation P for fermions. One can easily verify that $\mathcal{P}_\zeta$ is a projector, i.e. $\mathcal{P}_\zeta^2 = \mathcal{P}_\zeta$. For fermions, the state $|\alpha_1 \cdots \alpha_n\}$ vanishes if two of the α_i's are identical (Pauli principle). The states $|\alpha_1 \cdots \alpha_n\}$ form a basis of $\mathcal{H}_n$. The closure relation (1.128) in $\bar{\mathcal{H}}_n$ becomes a closure relation in $\mathcal{H}_n$,

$$\begin{aligned} 1 &= \sum_{\alpha_1 \cdots \alpha_n} \mathcal{P}_\zeta |\alpha_1 \cdots \alpha_n)(\alpha_1 \cdots \alpha_n| \mathcal{P}_\zeta \\ &= \frac{1}{n!} \sum_{\alpha_1 \cdots \alpha_n} |\alpha_1 \cdots \alpha_n\}\{\alpha_1 \cdots \alpha_n|, \end{aligned} \tag{1.131}$$

and the overlap between two states is given by

$$\begin{aligned} \{\alpha'_1 \cdots \alpha'_n|\alpha_1 \cdots \alpha_n\} &= n!(\alpha'_1 \cdots \alpha'_n|\mathcal{P}_\zeta^2|\alpha_1 \cdots \alpha_n) \\ &= \sum_P \zeta^P (\alpha'_1 \cdots \alpha'_n|\alpha_{P(1)} \cdots \alpha_{P(n)}). \end{aligned} \tag{1.132}$$

The permutations P contributing to the sum must satisfy $\alpha'_1 = \alpha_{P(1)}, \ldots, \alpha'_n = \alpha_{P(n)}$. Since for fermions, all α_i's must be different, there is only one such permutation and the overlap reduces to

$$\{\alpha'_1 \cdots \alpha'_n|\alpha_1 \cdots \alpha_n\} = (-1)^P \qquad \text{(fermions)}. \tag{1.133}$$

In the case of bosons, several particles can occupy the same one-body state, and any permutation which interchanges particles in the same state contributes to the sum in (1.132). The number of these permutations is $n_{\alpha_1}! \cdots n_{\alpha_p}!$ if we denote by n_{α_i} the number of bosons in the one-body state $|\alpha_i\rangle$ (with $\alpha_1, \ldots, \alpha_p$ all different). We therefore find

$$\{\alpha'_1 \cdots \alpha'_n|\alpha_1 \cdots \alpha_n\} = n_{\alpha_1}! \cdots n_{\alpha_p}! \qquad \text{(bosons)}. \tag{1.134}$$

A normalized basis of $\mathcal{H}_n$ is obtained by normalizing the states $|\alpha_1 \cdots \alpha_n\}$,

$$\begin{aligned} |\alpha_1 \cdots \alpha_n\rangle &= \frac{1}{(\prod_{i=1}^p n_{\alpha_i}!)^{1/2}} |\alpha_1 \cdots \alpha_n\} \\ &= \frac{1}{(n! \prod_{i=1}^p n_{\alpha_i}!)^{1/2}} \sum_P \zeta^P |\alpha_{P(1)} \cdots \alpha_{P(n)}). \end{aligned} \tag{1.135}$$

The coordinate-representation wavefunction is given by

$$(\mathbf{r}_1 \cdots \mathbf{r}_n | \alpha_1 \cdots \alpha_n) = \frac{1}{(n! \prod_{i=1}^p n_{\alpha_i}!)^{1/2}} \sum_P \zeta^P \varphi_{\alpha_{P(1)}}(\mathbf{r}_1) \cdots \varphi_{\alpha_{P(n)}}(\mathbf{r}_n)$$

$$= \begin{cases} \dfrac{1}{\sqrt{n!}} \det(\varphi_{\alpha_i}(\mathbf{r}_j)) & \text{(fermions)}, \\ \dfrac{1}{(n! \prod_{i=1}^p n_{\alpha_i}!)^{1/2}} \operatorname{per}(\varphi_{\alpha_i}(\mathbf{r}_j)) & \text{(bosons)}. \end{cases} \tag{1.136}$$

We thus obtain a basis of permanents for bosons and determinants for fermions. The latter are known as Slater determinants.

It is also possible to define an (anti)symmetrized many-body state in the coordinate representation,

$$|\mathbf{r}_1\sigma_1, \ldots, \mathbf{r}_n\sigma_n\} = \frac{1}{\sqrt{n!}} \sum_{P \in S_n} \zeta^P |\mathbf{r}_{P(1)}\sigma_{P(1)}, \ldots, \mathbf{r}_{P(n)}\sigma_{P(n)})$$

$$= \frac{1}{V^{n/2}} \sum_{\mathbf{k}_1 \cdots \mathbf{k}_n} e^{i(\mathbf{k}_1 \cdot \mathbf{r}_1 + \cdots + \mathbf{k}_n \cdot \mathbf{r}_n)} |\mathbf{k}_1\sigma_1, \ldots, \mathbf{k}_n\sigma_n\}, \tag{1.137}$$

although the states $|\mathbf{r}\sigma\rangle$ are not normalized. σ denotes the spin index (as well as other discrete indices if necessary) and $\mathbf{k}$ the momentum.

1.2.1.1 *Creation and annihilation operators*

To complete the description of many-particle systems, we need to introduce operators which couple $\mathcal{H}_n$ to $\mathcal{H}_{n\pm1}$, i.e. which add or remove a particle from the system. The creation operator $\hat{\psi}^\dagger_\alpha$ is defined by

$$\begin{aligned} \hat{\psi}^\dagger_\alpha |\alpha_1 \cdots \alpha_n\} &= |\alpha\alpha_1 \cdots \alpha_n\}, \\ \hat{\psi}^\dagger_\alpha |\alpha_1 \cdots \alpha_n\rangle &= \sqrt{n_\alpha + 1} |\alpha\alpha_1 \cdots \alpha_n\rangle. \end{aligned} \tag{1.138}$$

$\hat{\psi}^\dagger_\alpha$ adds a particle in the one-body state α and (anti)symmetrizes the resulting many-body state. When acting on the vacuum state, it gives the one-body state $|\alpha\rangle$: $\hat{\psi}^\dagger_\alpha|\text{vac}\rangle = |\alpha\rangle$. More generally,

$$\begin{aligned} |\alpha_1 \cdots \alpha_n\} &= \hat{\psi}^\dagger_{\alpha_1} \cdots \hat{\psi}^\dagger_{\alpha_n} |\text{vac}\rangle, \\ |\alpha_1 \cdots \alpha_n\rangle &= \frac{1}{(\prod_{i=1}^p n_{\alpha_i}!)^{1/2}} \hat{\psi}^\dagger_{\alpha_1} \cdots \hat{\psi}^\dagger_{\alpha_n} |\text{vac}\rangle. \end{aligned} \tag{1.139}$$

The annihilation operator $\hat{\psi}_\alpha$ is defined as the adjoint of the creation operator $\hat{\psi}^\dagger_\alpha$. From (1.138), we find

$$\begin{aligned} \{\alpha_1 \cdots \alpha_n | \hat{\psi}_\alpha | \alpha'_1 \cdots \alpha'_m\} &= \{\alpha'_1 \cdots \alpha'_m | \hat{\psi}^\dagger_\alpha | \alpha_1 \cdots \alpha_n\}^* \\ &= \{\alpha\alpha_1 \cdots \alpha_n | \alpha'_1 \cdots \alpha'_m\}. \end{aligned} \tag{1.140}$$

The result is finite only if $n = m - 1$ so that $\hat{\psi}_\alpha$ removes a particle from the system. This means in particular that $\hat{\psi}_\alpha|\text{vac}\rangle = 0$. Using the closure relation in the Fock space $\mathcal{H}$,

$$\sum_{n=0}^{\infty} \sum_{\alpha_1\cdots\alpha_n} \frac{1}{n!}|\alpha_1\cdots\alpha_n\}\{\alpha_1\cdots\alpha_n| = 1, \tag{1.141}$$

we obtain

$$\begin{aligned}\hat{\psi}_\alpha|\alpha'_1\cdots\alpha'_m\} &= \sum_{n=0}^{\infty} \sum_{\alpha_1\cdots\alpha_n} \frac{1}{n!}\{\alpha_1\cdots\alpha_n|\hat{\psi}_\alpha|\alpha'_1\cdots\alpha'_m\}|\alpha_1\cdots\alpha_n\} \\ &= \frac{1}{(m-1)!}\sum_{\alpha_1\cdots\alpha_{m-1}} \{\alpha\alpha_1\cdots\alpha_{m-1}|\alpha'_1\cdots\alpha'_m\}|\alpha_1\cdots\alpha_{m-1}\} \\ &= \frac{1}{(m-1)!}\sum_{P\in S_m} \zeta^P\langle\alpha|\alpha'_{P(1)}\rangle|\alpha'_{P(2)}\cdots\alpha'_{P(m)}\},\end{aligned} \tag{1.142}$$

where we have used (1.127) and (1.132). The permutation $(P(2),\dots,P(m)) \to (1,\dots,P(1)-1,P(1)+1,\dots,m)$ has a signature $\zeta^{P+P(1)-1}$,[22] so that we finally obtain[23]

$$\begin{aligned}\hat{\psi}_\alpha|\alpha'_1\cdots\alpha'_m\} &= \frac{1}{(m-1)!}\sum_{P\in S_m} \zeta^{P(1)-1}\delta_{\alpha,\alpha'_{P(1)}}|\alpha'_1\cdots\alpha'_{P(1)-1},\alpha'_{P(1)+1}\cdots\alpha'_m\} \\ &= \sum_{i=1}^{m}\zeta^{i-1}\delta_{\alpha,\alpha'_i}|\alpha'_1\cdots\hat{\alpha}'_i\cdots\alpha'_m\},\end{aligned} \tag{1.143}$$

where $\hat{\alpha}'_i$ indicates that α'_i is omitted. The analogous result for the normalized states reads

$$\hat{\psi}_\alpha|\alpha'_1\cdots\alpha'_m\rangle = \frac{1}{\sqrt{n_\alpha}}\sum_{i=1}^{m}\zeta^{i-1}\delta_{\alpha,\alpha'_i}|\alpha'_1\cdots\hat{\alpha}'_i\cdots\alpha'_m\rangle. \tag{1.144}$$

For bosons, it is sometimes convenient to introduce the states

$$|n_{\alpha_1}\cdots n_{\alpha_p}\rangle = \frac{1}{(\prod_{i=1}^p n_{\alpha_i}!)^{1/2}}\left(\hat{\psi}^\dagger_{\alpha_1}\right)^{n_{\alpha_1}}\cdots\left(\hat{\psi}^\dagger_{\alpha_p}\right)^{n_{\alpha_p}}|\text{vac}\rangle, \tag{1.145}$$

where n_{α_i} denotes the occupation number of the one-body state α_i (with $\alpha_1\cdots\alpha_p$ all different). These states form an orthonormal basis of the Fock space $\mathcal{H}$ and satisfy the closure relation

$$\sum_{n_{\alpha_1}=-\infty}^{\infty}\sum_{n_{\alpha_2}=-\infty}^{\infty}\cdots|n_{\alpha_1}\cdots n_{\alpha_i}\cdots\rangle\langle n_{\alpha_1}\cdots n_{\alpha_i}\cdots| = 1 \tag{1.146}$$

[22]The permutation $(P(1),P(2),\dots,P(m)) \to (P(1),2,\dots,P(1)-1,P(1)+1,\dots,m)$ can be written as the permutation P^{-1} $[(P(1),P(2),\dots,P(m)) \to (1,2,\dots,m)$, signature $\zeta^P]$ followed by $P(1)-1$ transpositions $[(1,2,\dots,m) \to (P(1),2,\dots,P(1)-1,P(1)+1,\dots,m)$, signature $\zeta^{P(1)-1}]$.

[23]One can easily verify that (1.143) reproduces (1.140).

as well as

$$\begin{aligned}\hat{\psi}^\dagger_\alpha|n_{\alpha_1}\cdots n_{\alpha_p}\rangle &= \sqrt{n_\alpha+1}|n_{\alpha_1}\cdots(n_\alpha+1)\cdots n_{\alpha_p}\rangle,\\ \hat{\psi}_\alpha|n_{\alpha_1}\cdots n_{\alpha_p}\rangle &= \textstyle\sum_{i=1}^p \delta_{\alpha,\alpha_i}\sqrt{n_\alpha}|n_{\alpha_1}\cdots(n_\alpha-1)\cdots n_{\alpha_p}\rangle,\end{aligned}\qquad \text{(bosons)}. \tag{1.147}$$

They are therefore eigenstates of the operator $\hat{n}_\alpha = \hat{\psi}^\dagger_\alpha\hat{\psi}_\alpha$ which measures the number of particles in the one-body state $|\alpha\rangle$,

$$\hat{n}_\alpha|n_{\alpha_1}\cdots n_{\alpha_p}\rangle = \sum_{i=1}^p \delta_{\alpha,\alpha_i} n_\alpha |n_{\alpha_1}\cdots n_{\alpha_p}\rangle. \tag{1.148}$$

1.2.1.2 *(Anti)commutation relations*

The symmetry properties of the states $|\alpha_1\cdots\alpha_n\}$ under the exchange of two particles have important consequences for the operators $\hat{\psi}_\alpha, \hat{\psi}^\dagger_\alpha$. We have

$$\begin{aligned}\hat{\psi}^\dagger_\alpha\hat{\psi}^\dagger_{\alpha'}|\alpha_1\cdots\alpha_n\} &= |\alpha\alpha'\alpha_1\cdots\alpha_n\}\\ &= \zeta|\alpha'\alpha\alpha_1\cdots\alpha_n\}\\ &= \zeta\hat{\psi}^\dagger_{\alpha'}\hat{\psi}^\dagger_\alpha|\alpha_1\cdots\alpha_n\}.\end{aligned} \tag{1.149}$$

Since (1.149) holds for any state $|\alpha_1\cdots\alpha_n\}$, we deduce

$$[\hat{\psi}^\dagger_\alpha, \hat{\psi}^\dagger_{\alpha'}]_{-\zeta} = [\hat{\psi}_\alpha, \hat{\psi}_{\alpha'}]_{-\zeta} = 0 \tag{1.150}$$

(the second equation is deduced from the first by taking the adjoint). $[\cdot,\cdot]_-$ and $[\cdot,\cdot]_+$ denote the commutator and the anticommutator, respectively. By comparing

$$\begin{aligned}\hat{\psi}_\alpha\hat{\psi}^\dagger_{\alpha'}|\alpha_1\cdots\alpha_n\} &= \hat{\psi}_\alpha|\alpha'\alpha_1\cdots\alpha_n\}\\ &= \delta_{\alpha,\alpha'}|\alpha_1\cdots\alpha_n\} + \sum_{i=1}^n \zeta^i\delta_{\alpha,\alpha_i}|\alpha'\alpha_1\cdots\hat{\alpha}_i\cdots\alpha_n\}\end{aligned} \tag{1.151}$$

and

$$\begin{aligned}\hat{\psi}^\dagger_{\alpha'}\hat{\psi}_\alpha|\alpha_1\cdots\alpha_n\} &= \hat{\psi}^\dagger_{\alpha'}\sum_{i=1}^n \zeta^{i-1}\delta_{\alpha,\alpha_i}|\alpha_1\cdots\hat{\alpha}_i\cdots\alpha_n\}\\ &= \sum_{i=1}^n \zeta^{i-1}\delta_{\alpha,\alpha_i}|\alpha'\alpha_1\cdots\hat{\alpha}_i\cdots\alpha_n\},\end{aligned} \tag{1.152}$$

we also obtain

$$[\hat{\psi}_\alpha, \hat{\psi}^\dagger_{\alpha'}]_{-\zeta} = \delta_{\alpha,\alpha'}. \tag{1.153}$$

Equations (1.150) and (1.153) are fundamental relations satisfied by the creation and annihilation operators $\hat{\psi}^\dagger_\alpha, \hat{\psi}_\alpha$. Recall the crucial assumption that $\{|\alpha\rangle\}$ forms an orthonormal basis of the single-particle Hilbert space $\mathcal{H}_1$.

1.2.1.3 *Change of basis*

Creation and annihilation operators in another basis $\{|\tilde{\alpha}\rangle\}$ can be straightforwardly obtained from the operators $\hat{\psi}^\dagger_\alpha, \hat{\psi}_\alpha$. Using $|\tilde{\alpha}\rangle = \sum_\alpha \langle\alpha|\tilde{\alpha}\rangle|\alpha\rangle$, we deduce

$$\begin{aligned}\hat{\psi}^\dagger_{\tilde{\alpha}}|\alpha_1\cdots\alpha_n\} &= |\tilde{\alpha}\alpha_1\cdots\alpha_n\} \\ &= \sum_\alpha \langle\alpha|\tilde{\alpha}\rangle|\alpha\alpha_1\cdots\alpha_n\} \\ &= \sum_\alpha \langle\alpha|\tilde{\alpha}\rangle\hat{\psi}^\dagger_\alpha|\alpha_1\cdots\alpha_n\},\end{aligned} \tag{1.154}$$

and therefore

$$\begin{aligned}\hat{\psi}^\dagger_{\tilde{\alpha}} &= \sum_\alpha \langle\alpha|\tilde{\alpha}\rangle\hat{\psi}^\dagger_\alpha, \\ \hat{\psi}_{\tilde{\alpha}} &= \sum_\alpha \langle\tilde{\alpha}|\alpha\rangle\hat{\psi}_\alpha.\end{aligned} \tag{1.155}$$

The transformed operators satisfy the (anti)commutation relations

$$\begin{aligned}&[\hat{\psi}^\dagger_{\tilde{\alpha}}, \hat{\psi}^\dagger_{\tilde{\alpha}'}]_{-\zeta} = [\hat{\psi}_{\tilde{\alpha}}, \hat{\psi}_{\tilde{\alpha}'}]_{-\zeta} = 0, \\ &[\hat{\psi}_{\tilde{\alpha}}, \hat{\psi}^\dagger_{\tilde{\alpha}'}]_{-\zeta} = \langle\tilde{\alpha}|\tilde{\alpha}'\rangle.\end{aligned} \tag{1.156}$$

If the new basis is orthonormal, the (anti)commutation relations are preserved; the transformation $\{\hat{\psi}^\dagger_\alpha, \hat{\psi}_\alpha\} \to \{\hat{\psi}^\dagger_{\tilde{\alpha}}, \hat{\psi}_{\tilde{\alpha}}\}$ is unitary.

1.2.1.4 *Field operators*

The "field operators" are defined by

$$\begin{aligned}\hat{\psi}_\sigma(\mathbf{r}) &= \sum_\alpha \langle\mathbf{r},\sigma|\alpha\rangle\hat{\psi}_\alpha = \frac{1}{\sqrt{V}}\sum_\mathbf{k} e^{i\mathbf{k}\cdot\mathbf{r}}\hat{\psi}_\sigma(\mathbf{k}), \\ \hat{\psi}^\dagger_\sigma(\mathbf{r}) &= \sum_\alpha \langle\alpha|\mathbf{r},\sigma\rangle\hat{\psi}^\dagger_\alpha = \frac{1}{\sqrt{V}}\sum_\mathbf{k} e^{-i\mathbf{k}\cdot\mathbf{r}}\hat{\psi}^\dagger_\sigma(\mathbf{k}),\end{aligned} \tag{1.157}$$

where the sum extends over all states α of the orthonormal basis. The last relations are obtained by choosing the momentum-representation basis $\{|\mathbf{k},\sigma\rangle\}$. The field operators satisfy the (anti)commutation relations

$$\begin{aligned}&[\hat{\psi}^\dagger_\sigma(\mathbf{r}), \hat{\psi}^\dagger_{\sigma'}(\mathbf{r}')]_{-\zeta} = [\hat{\psi}_\sigma(\mathbf{r}), \hat{\psi}_{\sigma'}(\mathbf{r}')]_{-\zeta} = 0, \\ &[\hat{\psi}_\sigma(\mathbf{r}), \hat{\psi}^\dagger_{\sigma'}(\mathbf{r}')]_{-\zeta} = \delta_{\sigma,\sigma'}\delta(\mathbf{r}-\mathbf{r}').\end{aligned} \tag{1.158}$$

From Eqs. (1.138) and (1.143), we deduce that

$$\begin{aligned}&\hat{\psi}^\dagger_\sigma(\mathbf{r})|\mathbf{r}_1\sigma_1,\ldots,\mathbf{r}_n\sigma_n\} = |\mathbf{r}\sigma,\mathbf{r}_1\sigma_1,\ldots,\mathbf{r}_n\sigma_n\}, \\ &\hat{\psi}_\sigma(\mathbf{r})|\mathbf{r}_1\sigma_1,\ldots,\mathbf{r}_n\sigma_n\} = \sum_{i=1}^n \zeta^{i-1}\delta(\mathbf{r}-\mathbf{r}_i)|\mathbf{r}_1\sigma_1,\ldots,\widehat{\mathbf{r}_i\sigma_i},\ldots,\mathbf{r}_n\sigma_n\},\end{aligned} \tag{1.159}$$

where $\widehat{\mathbf{r}_i\sigma_i}$ means that $\mathbf{r}_i\sigma_i$ is omitted. Thus, the field operator $\hat{\psi}^\dagger_\sigma(\mathbf{r})$ adds a spin-σ particle at point $\mathbf{r}$ and (anti)symmetrizes the resulting many-body state.

1.2.2 *Operators in second-quantized form*

Any operator acting within the Fock space can be written in terms of the creation and annihilation operators. In the following, we discuss in detail one- and two-body operators before generalizing to n-body operators.

An operator $\hat{V}^{(1)}$ is a one-body operator if it acts on each particle separately,[24]

$$\begin{aligned}\hat{V}^{(1)}|\alpha_1\cdots\alpha_n) &= \sum_{i=1}^{n}\hat{V}_i|\alpha_1\cdots\alpha_n),\\ \hat{V}^{(1)}|\alpha_1\cdots\alpha_n\} &= \frac{1}{\sqrt{n!}}\sum_P \zeta^P\sum_{i=1}^{n}\hat{V}_i|\alpha_{P(1)}\cdots\alpha_{P(n)}),\end{aligned} \tag{1.160}$$

where $\hat{V}_i$ operates only on the ith particle. $\hat{V}_1|\alpha_1\cdots\alpha_n) = (\hat{V}|\alpha_1\rangle)\otimes|\alpha_2\rangle\otimes\cdots\otimes|\alpha_n\rangle$, etc. Let us first choose a basis where the operator $\hat{V}$ is diagonal,

$$\hat{V}|\alpha\rangle = \langle\alpha|\hat{V}|\alpha\rangle|\alpha\rangle \equiv V_\alpha|\alpha\rangle. \tag{1.161}$$

We then have

$$\begin{aligned}\hat{V}^{(1)}|\alpha_1\cdots\alpha_n\} &= \frac{1}{\sqrt{n!}}\sum_P \zeta^P\sum_{i=1}^{n}V_{\alpha_{P(i)}}|\alpha_{P(1)}\cdots\alpha_{P(n)})\\ &= \sum_{i=1}^{n}V_{\alpha_i}|\alpha_1\cdots\alpha_n\}\\ &= \sum_\alpha V_\alpha\hat{n}_\alpha|\alpha_1\cdots\alpha_n\},\end{aligned} \tag{1.162}$$

where the sum extends over the complete set of one-body states α and $\hat{n}_\alpha = \hat{\psi}^\dagger_\alpha\hat{\psi}_\alpha$. We deduce

$$\hat{V}^{(1)} = \sum_\alpha\langle\alpha|\hat{V}|\alpha\rangle\hat{n}_\alpha. \tag{1.163}$$

Using (1.155), we obtain the general expression for one-body operators,

$$\hat{V}^{(1)} = \sum_{\alpha,\alpha'}\langle\alpha|\hat{V}|\alpha'\rangle\hat{\psi}^\dagger_\alpha\hat{\psi}_{\alpha'}, \tag{1.164}$$

valid in any complete basis. In particular,

$$\hat{V}^{(1)} = \int d^drd^dr'\sum_{\sigma,\sigma'}\langle\mathbf{r},\sigma|\hat{V}|\mathbf{r}',\sigma'\rangle\hat{\psi}^\dagger_\sigma(\mathbf{r})\hat{\psi}_{\sigma'}(\mathbf{r}') \tag{1.165}$$

---

[24]Recall that in first quantization, one would write $\hat{V}^{(1)} = \sum_{i=1}^{n} V(\hat{\mathbf{r}}_i, \hat{\mathbf{p}}_i, \hat{\sigma}_i)$ where the sum extends over the n particles of the system.

in terms of the field operators. Note that the expressions (1.164) and (1.165) make no reference to the total number of particles actually present in the system (to be compared with the first-quantized form of a one-body operator acting on an n-particle state[24]).

Examples. The density $\hat{n}(\mathbf{r})$ of particles at point $\mathbf{r}$ reads $\sum_{i=1}^n \delta(\mathbf{r}-\hat{\mathbf{r}}_i)$ in first quantization and

$$\begin{aligned}\hat{n}(\mathbf{r}) &= \int d^d r_1 d^d r_2 \sum_{\sigma,\sigma'} \langle \mathbf{r}_1,\sigma|\delta(\mathbf{r}-\hat{\mathbf{r}})|\mathbf{r}_2,\sigma'\rangle \hat{\psi}^\dagger_\sigma(\mathbf{r}_1)\hat{\psi}_{\sigma'}(\mathbf{r}_2) \\ &= \sum_\sigma \hat{\psi}^\dagger_\sigma(\mathbf{r})\hat{\psi}_\sigma(\mathbf{r}) \end{aligned} \tag{1.166}$$

in second-quantized form. The number of particles operator is given by

$$\hat{N} = \int d^d r\, \hat{n}(\mathbf{r}) = \sum_\alpha \hat{\psi}^\dagger_\alpha \hat{\psi}_\alpha \tag{1.167}$$

for any complete orthonormal basis of one-body states $|\alpha\rangle$. For a one-body potential $\hat{V}^{(1)} = \sum_{i=1}^n V(\hat{\mathbf{r}}_i)$, one finds

$$\begin{aligned}\hat{V}^{(1)} &= \int d^d r_1 d^d r_2 \sum_{\sigma,\sigma'} \langle \mathbf{r}_1,\sigma|V(\hat{\mathbf{r}})|\mathbf{r}_2,\sigma'\rangle \hat{\psi}^\dagger_\sigma(\mathbf{r}_1)\hat{\psi}_{\sigma'}(\mathbf{r}'_2) \\ &= \int d^d r\, V(\mathbf{r}) \sum_\sigma \hat{\psi}^\dagger_\sigma(\mathbf{r})\hat{\psi}_\sigma(\mathbf{r}) = \int d^d r\, V(\mathbf{r})\hat{n}(\mathbf{r}). \end{aligned} \tag{1.168}$$

The kinetic energy operator $\hat{T} = \sum_{i=1}^n \hat{\mathbf{p}}_i^2/2m$ is most easily calculated in the plane-wave basis,

$$\hat{T} = \sum_{\substack{\mathbf{k},\mathbf{k}' \\ \sigma,\sigma'}} \langle \mathbf{k},\sigma|\frac{\hat{\mathbf{p}}^2}{2m}|\mathbf{k}',\sigma'\rangle \hat{\psi}^\dagger_\sigma(\mathbf{k})\hat{\psi}_{\sigma'}(\mathbf{k}') = \sum_{\mathbf{k},\sigma} \frac{\mathbf{k}^2}{2m} \hat{\psi}^\dagger_\sigma(\mathbf{k})\hat{\psi}_\sigma(\mathbf{k}). \tag{1.169}$$

In direct space, this gives

$$\hat{T} = -\int d^d r \sum_\sigma \hat{\psi}^\dagger_\sigma(\mathbf{r}) \frac{\boldsymbol{\nabla}^2}{2m} \hat{\psi}_\sigma(\mathbf{r}) = \frac{1}{2m} \int d^d r \sum_\sigma \boldsymbol{\nabla}\hat{\psi}^\dagger_\sigma(\mathbf{r}) \cdot \boldsymbol{\nabla}\hat{\psi}_\sigma(\mathbf{r}). \tag{1.170}$$

The last line is obtained from an integration by part assuming that the fields vanish at infinity (or satisfy periodic boundary conditions). For the current operator $\hat{\mathbf{j}}(\mathbf{r}) = \frac{1}{2m}\sum_i [\hat{\mathbf{p}}_i \delta(\mathbf{r}-\hat{\mathbf{r}}_i) + \delta(\mathbf{r}-\hat{\mathbf{r}}_i)\hat{\mathbf{p}}_i]$, one obtains

$$\begin{aligned}\hat{\mathbf{j}}(\mathbf{r}) &= \frac{1}{2m} \int d^d r_1 d^d r_2 \sum_\sigma \langle \mathbf{r}_1|\hat{\mathbf{p}}\delta(\mathbf{r}-\hat{\mathbf{r}}) + \delta(\mathbf{r}-\hat{\mathbf{r}})\hat{\mathbf{p}}|\mathbf{r}_2\rangle \hat{\psi}^\dagger_\sigma(\mathbf{r}_1)\hat{\psi}_\sigma(\mathbf{r}_2) \\ &= -\frac{i}{2m} \sum_\sigma \left[\hat{\psi}^\dagger_\sigma(\mathbf{r})\boldsymbol{\nabla}\hat{\psi}_\sigma(\mathbf{r}) - \text{h.c.}\right]. \end{aligned} \tag{1.171}$$

The most general form of the grand canonical Hamiltonian of non-interacting particles is $\hat{H}_0 = \sum_\alpha \xi_\alpha \hat{\psi}^\dagger_\alpha \hat{\psi}_\alpha$ in the diagonal basis $\{|\alpha\rangle\}$, where $\xi_\alpha = \epsilon_\alpha - \mu$ denotes the energy of the one-body state $|\alpha\rangle$ with respect to the chemical potential μ.

An operator $\hat{V}^{(2)}$ is a two-body operator if it acts only on pairs of particles,

$$\begin{aligned} \hat{V}^{(2)}|\alpha_1\cdots\alpha_n) &= \sum_{\substack{i,j=1\\(i<j)}}^{n} \hat{V}_{ij}|\alpha_1\cdots\alpha_n), \\ \hat{V}^{(2)}|\alpha_1\cdots\alpha_n\} &= \frac{1}{\sqrt{n!}}\sum_P \zeta^P \sum_{\substack{i,j=1\\(i<j)}}^{n} \hat{V}_{ij}|\alpha_{P(1)}\cdots\alpha_{P(n)}), \end{aligned} \tag{1.172}$$

where $\hat{V}_{ij}$ acts only on particles i and j. In the basis $\{|\alpha)\}$ where $\hat{V}$ is diagonal,

$$\hat{V}|\alpha\beta) = (\alpha\beta|\hat{V}|\alpha\beta)|\alpha\beta) \equiv V_{\alpha\beta}|\alpha\beta), \tag{1.173}$$

one has

$$\begin{aligned} \hat{V}^{(2)}|\alpha_1\cdots\alpha_n\} &= \frac{1}{\sqrt{n!}}\sum_P \zeta^P \sum_{\substack{i,i=1\\(i<j)}}^{n} V_{\alpha_{P(i)}\alpha_{P(j)}}|\alpha_{P(1)}\cdots\alpha_{P(n)}) \\ &= \sum_{\substack{i,j=1\\(i<j)}}^{n} V_{\alpha_i\alpha_j}|\alpha_1\cdots\alpha_n\}, \end{aligned} \tag{1.174}$$

which leads to

$$\begin{aligned} \hat{V}^{(2)}|\alpha_1\cdots\alpha_n\} &= \frac{1}{2}\sum_{i,j=1}^{n}\left(V_{\alpha_i\alpha_j} - \delta_{i,j}V_{\alpha_i\alpha_i}\right)|\alpha_1\cdots\alpha_n\} \\ &= \frac{1}{2}\sum_{\alpha,\beta} V_{\alpha\beta}\left(\hat{n}_\alpha\hat{n}_\beta - \delta_{\alpha,\beta}\hat{n}_\alpha\right)|\alpha_1\cdots\alpha_n\}, \end{aligned} \tag{1.175}$$

where the sum extends over all states α, β of the complete basis of the one-body states. Using the (anti)commutation relations (1.150), (1.153), one finds $\hat{n}_\alpha\hat{n}_\beta - \delta_{\alpha,\beta}\hat{n}_\alpha = \hat{\psi}^\dagger_\alpha\hat{\psi}^\dagger_\beta\hat{\psi}_\beta\hat{\psi}_\alpha$, which leads to

$$\hat{V}^{(2)} = \frac{1}{2}\sum_{\alpha,\beta}(\alpha\beta|\hat{V}|\alpha\beta)\hat{\psi}^\dagger_\alpha\hat{\psi}^\dagger_\beta\hat{\psi}_\beta\hat{\psi}_\alpha. \tag{1.176}$$

Transforming to an arbitrary basis, we finally find the general expression of a two-body operator,

$$\begin{aligned} \hat{V}^{(2)} &= \frac{1}{2}\sum_{\alpha_1\cdots\alpha_2'}(\alpha_1\alpha_2|\hat{V}|\alpha_1'\alpha_2')\hat{\psi}^\dagger_{\alpha_1}\hat{\psi}^\dagger_{\alpha_2}\hat{\psi}_{\alpha_2'}\hat{\psi}_{\alpha_1'} \\ &= \frac{1}{2}\int d^d r_1\cdots d^d r_2' \sum_{\sigma_1\cdots\sigma_2'}(\mathbf{r}_1\sigma_1,\mathbf{r}_2\sigma_2|\hat{V}|\mathbf{r}_1'\sigma_1',\mathbf{r}_2'\sigma_2')\hat{\psi}^\dagger_{\sigma_1}(\mathbf{r}_1)\hat{\psi}^\dagger_{\sigma_2}(\mathbf{r}_2)\hat{\psi}_{\sigma_2'}(\mathbf{r}_2')\hat{\psi}_{\sigma_1'}(\mathbf{r}_1'). \end{aligned} \tag{1.177}$$

Note the respecting order of α_1' and α_2' in the matrix element and in the product of annihilation operators. It is sometimes convenient to use a variant of (1.177) where the matrix element is (anti)symmetrized,

$$\hat{V}^{(2)} = \frac{1}{4} \sum_{\alpha_1 \cdots \alpha_2'} \{\alpha_1\alpha_2|\hat{V}|\alpha_1'\alpha_2'\} \hat{\psi}^\dagger_{\alpha_1}\hat{\psi}^\dagger_{\alpha_2}\hat{\psi}_{\alpha_2'}\hat{\psi}_{\alpha_1'} \tag{1.178}$$

with

$$\{\alpha_1\alpha_2|\hat{V}|\alpha_1'\alpha_2'\} = (\alpha_1\alpha_2|\hat{V}|\alpha_1'\alpha_2') + \zeta(\alpha_1\alpha_2|\hat{V}|\alpha_2'\alpha_1'). \tag{1.179}$$

Example. For a two-body potential $\hat{V}^{(2)} = \sum_{i<j} v(\hat{\mathbf{r}}_i - \hat{\mathbf{r}}_j)$ (in first-quantized form), one finds

$$\begin{aligned}
\hat{V}^{(2)} &= \frac{1}{2} \int d^d r_1 \cdots d^d r_2' \sum_{\sigma_1 \cdots \sigma_2'} (\mathbf{r}_1\sigma_1, \mathbf{r}_2\sigma_2|v(\hat{\mathbf{r}} - \hat{\mathbf{r}}')|\mathbf{r}_1'\sigma_1', \mathbf{r}_2'\sigma_2') \\
&\quad \times \hat{\psi}^\dagger_{\sigma_1}(\mathbf{r}_1)\hat{\psi}^\dagger_{\sigma_2}(\mathbf{r}_2)\hat{\psi}_{\sigma_2'}(\mathbf{r}_2')\hat{\psi}_{\sigma_1'}(\mathbf{r}_1') \\
&= \frac{1}{2} \int d^d r_1 d^d r_2 \sum_{\sigma_1,\sigma_2} v(\mathbf{r}_1 - \mathbf{r}_2)\hat{\psi}^\dagger_{\sigma_1}(\mathbf{r}_1)\hat{\psi}^\dagger_{\sigma_2}(\mathbf{r}_2)\hat{\psi}_{\sigma_2}(\mathbf{r}_2)\hat{\psi}_{\sigma_1}(\mathbf{r}_1).
\end{aligned} \tag{1.180}$$

Here, we have used the fact that $v(\hat{\mathbf{r}} - \hat{\mathbf{r}}')$ is diagonal in the coordinate representation: $v(\hat{\mathbf{r}} - \hat{\mathbf{r}}')|\mathbf{r}_1\sigma_1, \mathbf{r}_2\sigma_2) = v(\mathbf{r}_1 - \mathbf{r}_2)|\mathbf{r}_1\sigma_1, \mathbf{r}_2\sigma_2)$. In Fourier space, Eq. (1.180) becomes

$$\hat{V}^{(2)} = \frac{1}{2\mathcal{V}} \sum_{\substack{\mathbf{k},\mathbf{k}',\mathbf{q} \\ \sigma,\sigma'}} v(\mathbf{q})\hat{\psi}^\dagger_\sigma(\mathbf{k}+\mathbf{q})\hat{\psi}^\dagger_{\sigma'}(\mathbf{k}'-\mathbf{q})\hat{\psi}_{\sigma'}(\mathbf{k}')\hat{\psi}_\sigma(\mathbf{k}), \tag{1.181}$$

where $v(\mathbf{q})$ is the Fourier transform of $v(\mathbf{r})$.

To conclude this section, we give without proof the expression of an n-body operator in second-quantized form,

$$\begin{aligned}
\hat{V}^{(n)} &= \frac{1}{n!} \sum_{\{\alpha_i,\alpha_i'\}} (\alpha_1 \cdots \alpha_n|\hat{V}|\alpha_1' \cdots \alpha_n')\hat{\psi}^\dagger_{\alpha_1} \cdots \hat{\psi}^\dagger_{\alpha_n}\hat{\psi}_{\alpha_n'} \cdots \hat{\psi}_{\alpha_1'} \\
&= \frac{1}{(n!)^2} \sum_{\{\alpha_i,\alpha_i'\}} \{\alpha_1 \cdots \alpha_n|\hat{V}|\alpha_1' \cdots \alpha_n'\}\hat{\psi}^\dagger_{\alpha_1} \cdots \hat{\psi}^\dagger_{\alpha_n}\hat{\psi}_{\alpha_n'} \cdots \hat{\psi}_{\alpha_1'}.
\end{aligned} \tag{1.182}$$

1.2.3 *Green functions*

1.2.3.1 *Retarded Green function*

When studying the quantum mechanics of a single particle in Sec. 1.1.1, we introduced the propagator

$$G^R(\mathbf{r},\mathbf{r}';t) = -i\Theta(t)\langle\mathbf{r}|e^{-i\hat{H}t}|\mathbf{r}'\rangle = -i\Theta(t)\sum_n e^{-i\epsilon_n t}\varphi_n(\mathbf{r})\varphi_n^*(\mathbf{r}') \tag{1.183}$$

describing the propagation of a particle from $(\mathbf{r}', 0)$ to $(\mathbf{r}, t)$. We ignore for the time being the spin of the particle. To make causality explicit, we have added the factor $\Theta(t)$: The particle can propagate only forward in time. The sum in (1.183) extends over a complete set $\{\varphi_n(\mathbf{r}), \epsilon_n\}$ of one-body eigenstates. G^R is called the retarded propagator or retarded Green function. Its Fourier transform is defined by

$$G^R(\mathbf{r}, \mathbf{r}'; \omega) = \int_{-\infty}^{\infty} dt\, e^{i(\omega+i\eta)t} G^R(\mathbf{r}, \mathbf{r}'; t) = \sum_n \frac{\varphi_n(\mathbf{r})\varphi_n^*(\mathbf{r}')}{\omega + i\eta - \epsilon_n}, \tag{1.184}$$

where the infinitesimal $\eta \to 0^+$ is introduced to ensure the convergence of the integral.[25] Thus, the energies ϵ_n of the eigenstates $|n\rangle$ show up as poles of the retarded Green function while the residues give information about the wavefunctions $\varphi_n(\mathbf{r}) = \langle \mathbf{r}|n\rangle$. Since $|\mathbf{r}\rangle = \hat{\psi}^\dagger(\mathbf{r})|\text{vac}\rangle$, G^R can be rewritten as

$$G^R(\mathbf{r}, \mathbf{r}'; t) = -i\Theta(t)\langle \text{vac}|\hat{\psi}(\mathbf{r})e^{-i\hat{H}t}\hat{\psi}^\dagger(\mathbf{r}')|\text{vac}\rangle. \tag{1.185}$$

In a many-body system, it is tempting to generalize (1.185) as

$$\begin{aligned} G^R(\mathbf{r}, \mathbf{r}'; t) &= -i\Theta(t)\langle 0|\hat{\psi}(\mathbf{r})e^{-i\hat{H}t}\hat{\psi}^\dagger(\mathbf{r}')|0\rangle \\ &= -i\Theta(t)\langle 0|\hat{\psi}(\mathbf{r}, t)\hat{\psi}^\dagger(\mathbf{r}', 0)|0\rangle, \end{aligned} \tag{1.186}$$

where we take $\epsilon_0 = 0$ for the energy of the many-body ground state $|0\rangle$.

$$\begin{aligned} \hat{\psi}(\mathbf{r}, t) &= e^{i\hat{H}t}\hat{\psi}(\mathbf{r})e^{-i\hat{H}t}, \\ \hat{\psi}^\dagger(\mathbf{r}, t) &= e^{i\hat{H}t}\hat{\psi}^\dagger(\mathbf{r})e^{-i\hat{H}t} \end{aligned} \tag{1.187}$$

are the field operators in the Heisenberg picture. In the grand canonical formalism, the term $-\mu\hat{N}$ is included in the Hamiltonian. Contrary to (1.183), the definition (1.186) does not satisfy $G^R(\mathbf{r}, t^+; \mathbf{r}', t) = -i\delta(\mathbf{r} - \mathbf{r}')$ — an expected result for a "propagator" since within an infinitesimal time, the particle has no time to propagate. The latter relation is however satisfied if we choose

$$G^R(\mathbf{r}, \mathbf{r}'; t) = -i\Theta(t)\langle 0|\left[\hat{\psi}(\mathbf{r}, t), \hat{\psi}^\dagger(\mathbf{r}', 0)\right]_{-\varsigma}|0\rangle. \tag{1.188}$$

This expression reflects the fact that in a many-body system, we can not only add but also remove a particle. At finite temperature, the ground state expectation value must be replaced by a statistical average and we can define the retarded Green function as

$$\begin{aligned} G^R(\alpha, \alpha'; t) &= -i\Theta(t)\langle [\hat{\psi}_\alpha(t), \hat{\psi}^\dagger_{\alpha'}]_{-\varsigma}\rangle \\ &= -i\Theta(t)\frac{1}{Z}\text{Tr}\left\{e^{-\beta\hat{H}}\left[\hat{\psi}_\alpha(t), \hat{\psi}^\dagger_{\alpha'}\right]_{-\varsigma}\right\} \end{aligned} \tag{1.189}$$

for any complete set $\{|\alpha\rangle\}$ of one-body states. α includes a spin index as well as other internal indices if necessary.

---

[25] The physical meaning of η is further discussed in Chapter 3.

For a non-interacting system with grand canonical Hamiltonian $\hat{H}_0 = \sum_\alpha \xi_\alpha \hat{\psi}^\dagger_\alpha \hat{\psi}_\alpha$ ($\xi_\alpha = \epsilon_\alpha - \mu$) in the diagonal basis, the field operators satisfy the equations of motion

$$\begin{aligned} \partial_t \hat{\psi}_\alpha(t) &= i\big[\hat{H}_0, \hat{\psi}_\alpha(t)\big] = -i\xi_\alpha \hat{\psi}_\alpha(t), \\ \partial_t \hat{\psi}^\dagger_\alpha(t) &= i\big[\hat{H}_0, \hat{\psi}^\dagger_\alpha(t)\big] = i\xi_\alpha \hat{\psi}^\dagger_\alpha(t) \end{aligned} \tag{1.190}$$

and are therefore given by

$$\begin{aligned} \hat{\psi}_\alpha(t) &= e^{-i\xi_\alpha t}\hat{\psi}_\alpha, \\ \hat{\psi}^\dagger_\alpha(t) &= e^{i\xi_\alpha t}\hat{\psi}^\dagger_\alpha. \end{aligned} \tag{1.191}$$

From (1.189), we then deduce

$$G_0^R(\alpha, t) = -i\Theta(t)e^{-i\xi_\alpha t}\langle[\hat{\psi}_\alpha, \hat{\psi}^\dagger_\alpha]_{-\zeta}\rangle = -i\Theta(t)e^{-i\xi_\alpha t} \tag{1.192}$$

and

$$G_0^R(\alpha, \omega) = \int_{-\infty}^{\infty} dt\, e^{i(\omega+i\eta)t} G^R_{0,\alpha}(t) = \frac{1}{\omega + i\eta - \xi_\alpha}. \tag{1.193}$$

As in the single-particle case [Eq. (1.184)], the poles of the retarded Green function give the single-particle excitation energies, i.e. the energy of the many-body states obtained by adding (or removing) a particle to the equilibrium state. In Chapter 3, we shall study in detail the properties of the retarded Green function in interacting systems and show that its poles are indeed related to the single-particle excitation energies.

1.2.3.2 *Matsubara Green function*

The retarded Green function is not the one that is calculated in the first place. Instead, one usually focuses on the Matsubara Green function

$$\begin{aligned} G(\alpha, \alpha'; \tau) &= -\langle T_\tau \hat{\psi}_\alpha(\tau)\hat{\psi}^\dagger_{\alpha'}\rangle \\ &= -\Theta(\tau)\langle \hat{\psi}_\alpha(\tau)\hat{\psi}^\dagger_{\alpha'}\rangle - \zeta\Theta(-\tau)\langle \hat{\psi}^\dagger_{\alpha'}\hat{\psi}_\alpha(\tau)\rangle, \end{aligned} \tag{1.194}$$

where the time-ordering operator T_τ is defined by the second line. For fermions, the interchange of the two operators when $\tau < 0$ implies a sign change. The imaginary-time operators in (1.194) are defined by

$$\begin{aligned} \hat{\psi}_\alpha(\tau) &= e^{\tau\hat{H}}\hat{\psi}_\alpha e^{-\tau\hat{H}}, \\ \hat{\psi}^\dagger_\alpha(\tau) &= e^{\tau\hat{H}}\hat{\psi}^\dagger_\alpha e^{-\tau\hat{H}}. \end{aligned} \tag{1.195}$$

For $\tau = it$, they reduce to the real-time Heisenberg operators (1.187). Note that $\hat{\psi}^\dagger_\alpha(\tau)$ is not the adjoint $[\hat{\psi}_\alpha(\tau)]^\dagger$ of $\hat{\psi}_\alpha(\tau)$. For $\tau > 0$, $G(\alpha, \alpha'; \tau)$ describes the creation of a particle and its propagation in the system, while for $\tau < 0$ it describes the annihilation of a particle and the propagation of the corresponding "hole". There is no need to specify here how the equal-time ($\tau = 0$) Green function is defined. As we will see, $G(\alpha, \alpha'; 0)$ will always be interpreted as either $G(\alpha, \alpha'; 0^+)$ or $G(\alpha, \alpha'; 0^-)$.

We shall see later that we need to know $G(\alpha,\alpha';\tau)$ only for $\tau\in[-\beta,\beta]$. For $-\beta<\tau<0$, one obtains

$$\begin{aligned} G(\alpha,\alpha';\tau) &= -\zeta\langle\hat{\psi}^\dagger_{\alpha'}\hat{\psi}_\alpha(\tau)\rangle \\ &= -\zeta e^{\beta\Omega}\mathrm{Tr}\big[e^{-\beta\hat{H}}\hat{\psi}^\dagger_{\alpha'}e^{\tau\hat{H}}\hat{\psi}_\alpha e^{-\tau\hat{H}}\big] \\ &= -\zeta e^{\beta\Omega}\mathrm{Tr}\big[e^{\tau\hat{H}}\hat{\psi}_\alpha e^{-(\tau+\beta)\hat{H}}\hat{\psi}^\dagger_{\alpha'}\big] \\ &= -\zeta e^{\beta\Omega}\mathrm{Tr}\big[e^{-\beta\hat{H}}\hat{\psi}_\alpha(\tau+\beta)\hat{\psi}^\dagger_{\alpha'}\big] \\ &= \zeta G(\alpha,\alpha';\tau+\beta) \end{aligned} \tag{1.196}$$

(with $\Omega=-\frac{1}{\beta}\ln Z$ the thermodynamic potential) using the cyclic invariance of the trace. A similar reasoning gives

$$G(\alpha,\alpha';\tau)=\zeta G(\alpha,\alpha';\tau-\beta) \tag{1.197}$$

for $0<\tau<\beta$. Since $G(\alpha,\alpha';\tau)$ is (anti)periodic for $\tau\in[-\beta,\beta]$ with period β, it can be expanded in a Fourier series,

$$\begin{aligned} G(\alpha,\alpha';\tau) &= \frac{1}{\beta}\sum_{\omega_n} e^{-i\omega_n\tau}G(\alpha,\alpha';i\omega_n), \\ G(\alpha,\alpha';i\omega_n) &= \int_0^\beta d\tau\, e^{i\omega_n\tau}G(\alpha,\alpha';\tau), \end{aligned} \tag{1.198}$$

where $\sum_{\omega_n}=\sum_{n=-\infty}^{\infty}$ and the discrete frequencies

$$\omega_n=\begin{cases} \dfrac{2\pi}{\beta}n & \text{(bosons)}, \\[2ex] \dfrac{2\pi}{\beta}\left(n+\frac{1}{2}\right) & \text{(fermions)} \end{cases} \tag{1.199}$$

are called Matsubara frequencies.

The Matsubara Green function can be easily calculated for a non-interacting system with Hamiltonian $\hat{H}_0=\sum_\alpha\xi_\alpha\hat{\psi}^\dagger_\alpha\hat{\psi}_\alpha$. Inserting $\hat{\psi}_\alpha(\tau)=e^{-\xi_\alpha\tau}\hat{\psi}_\alpha$ and $\hat{\psi}^\dagger_\alpha(\tau)=e^{-\xi_\alpha\tau}\hat{\psi}^\dagger_\alpha$ in (1.194), we obtain

$$G_0(\alpha,\tau)=-e^{-\xi_\alpha\tau}\big[\Theta(\tau)(1+\zeta n_\alpha)+\Theta(-\tau)\zeta n_\alpha\big], \tag{1.200}$$

where the occupation number $n_\alpha=\langle\hat{\psi}^\dagger_\alpha\hat{\psi}_\alpha\rangle$ is given by the Fermi–Dirac function $n_F(\xi_\alpha)$ for fermions and the Bose–Einstein function $n_B(\xi_\alpha)$ for bosons. In frequency space

$$G_0(\alpha,i\omega_n)=\frac{1}{i\omega_n-\xi_\alpha}. \tag{1.201}$$

Comparing (1.193) and (1.201), we see that the retarded Green function can be obtained from the Matsubara one by an analytic continuation,

$$G_0^R(\alpha,\omega)=G_0(\alpha,i\omega_n\to\omega+i\eta). \tag{1.202}$$

This result turns out to be valid also in interacting systems. The retarded Green function, which contains physical information about the system, can be deduced from the Matsubara

Green function which is the natural quantity to consider in (finite-temperature) interacting systems. We postpone a detailed study of the properties of Green functions to Chapter 3 and in the following sections of this chapter discuss how we can define and compute Green functions in the functional integral formalism.

1.2.3.3 *Many-particle Green function*

The Green function we have discussed so far is called the one-particle Green function (or propagator) since it describes the propagation of a single particle. One can also define higher-order Green functions

$$\begin{aligned} G^{(2n)}(\alpha_1\tau_1, \dots, \alpha_n\tau_n; \alpha'_n\tau'_n, \dots, \alpha'_1\tau'_1) \\ = (-1)^n \langle T_\tau \hat\psi_{\alpha_1}(\tau_1)\cdots\hat\psi_{\alpha_n}(\tau_n)\hat\psi^\dagger_{\alpha'_n}(\tau'_n)\cdots\hat\psi^\dagger_{\alpha'_1}(\tau'_1)\rangle. \end{aligned} \tag{1.203}$$

The superscript $(2n)$ corresponds to the number of operators rather than the number n of particles propagating in the system. We shall often denote the one-particle Green function $G^{(2)}$ merely by G. As in (1.194), the operator T_τ chronologically orders the operators from left to right with the latest time to the left,

$$T_\tau \hat A_{\alpha_1}(\tau_1)\cdots\hat A_{\alpha_n}(\tau_n) = \zeta^P \hat A_{\alpha_{P(1)}}(\tau_{P(1)})\cdots\hat A_{\alpha_{P(n)}}(\tau_{P(n)}) \tag{1.204}$$

$(\tau_{P(1)} > \tau_{P(2)} > \cdots > \tau_{P(n)})$, where ζ^P equals one for bosons and the signature of the permutation P for fermions.

1.2.3.4 *Real-time Green function*

Finally, we can define Green functions in real time,

$$\begin{aligned} G^{(2n)}(\alpha_1 t_1, \dots, \alpha_n t_n; \alpha'_n t'_n, \dots, \alpha'_1 t'_1) \\ = (-i)^n \langle T \hat\psi_{\alpha_1}(t_1)\cdots\hat\psi_{\alpha_n}(t_n)\hat\psi^\dagger_{\alpha'_n}(t'_n)\cdots\hat\psi^\dagger_{\alpha'_1}(t'_1)\rangle, \end{aligned} \tag{1.205}$$

where T is the time-ordering operator (not to be confused with the temperature) defined like its imaginary-time counterpart T_τ, and $\hat\psi_\alpha(t), \hat\psi^\dagger_\alpha(t)$ the Heisenberg picture operators (1.187). The relation between the various Green functions introduced in this section will be studied in Secs. 3.2.7 and 3.5.

1.3 Coherent States

The quantum mechanical behavior of a single particle is usually described in terms of the position and momentum operators $\hat{\mathbf{q}}$ and $\hat{\mathbf{p}}$. When constructing a path integral representation of the propagator or the partition function, it therefore appears natural to introduce the closure relations $\int d^dq|\mathbf{q}\rangle\langle\mathbf{q}| = 1$ and/or $\sum_{\mathbf{p}} |\mathbf{p}\rangle\langle\mathbf{p}| = 1$ at each time step (Sec. 1.1.1). Since second quantization is the natural formalism for studying many-particle systems, we should try to construct the path integral using a closure relation based on the eigenstates of the creation or annihilation operators.

Now, it is clear that an arbitrary state

$$|\psi\rangle = \sum_{\alpha_1\cdots\alpha_n} C_{\alpha_1\cdots\alpha_n}|\alpha_1\cdots\alpha_n\rangle \tag{1.206}$$

cannot be an eigenstate of the creation operator $\hat{\psi}^\dagger_\alpha$. $|\psi\rangle$ has necessarily a component with a minimum number n_0 of particles. Applying $\hat{\psi}^\dagger_\alpha$ to this component will either increase n_0 by 1 or yield zero if the single-particle state α is already occupied and the particles are fermions. In all cases, the minimum numbers of particles in $|\psi\rangle$ and $\hat{\psi}^\dagger_\alpha|\psi\rangle$ differ. On the other hand, there is *a priori* nothing that prevents $|\psi\rangle$ to be an eigenstate of $\hat{\psi}_\alpha$. Suppose that $|\psi\rangle$ is an eigenstate of the annihilation operators,

$$\hat{\psi}_\alpha|\psi\rangle = \psi_\alpha|\psi\rangle \quad \forall\alpha. \tag{1.207}$$

Then the (anti)commutation relations

$$[\hat{\psi}_\alpha, \hat{\psi}_\beta]_{-\zeta} = 0 \tag{1.208}$$

imply

$$\begin{aligned}\hat{\psi}_\alpha\hat{\psi}_\beta|\psi\rangle &= \hat{\psi}_\alpha\psi_\beta|\psi\rangle = \psi_\alpha\psi_\beta|\psi\rangle \\ &= \zeta\hat{\psi}_\beta\hat{\psi}_\alpha|\psi\rangle = \zeta\hat{\psi}_\beta\psi_\alpha|\psi\rangle = \zeta\psi_\beta\psi_\alpha|\psi\rangle,\end{aligned} \tag{1.209}$$

i.e.

$$[\psi_\alpha, \psi_\beta]_{-\zeta} = 0. \tag{1.210}$$

Thus, for fermions, the eigenvalues cannot be ordinary numbers. We shall need to introduce anticommuting variables called Grassmann numbers.[26] Postponing the case of fermions to Secs. 1.3.2 and 1.3.3, we now restrict the discussion to the case of bosons.[27]

1.3.1 *Boson coherent states*

It is straightforward to verify that the so-called boson coherent states

$$\begin{aligned}|\psi\rangle &= \exp\Big\{\sum_\alpha \psi_\alpha\hat{\psi}^\dagger_\alpha\Big\}|\text{vac}\rangle \\ &= \sum_{n_{\alpha_1},n_{\alpha_2},\dots=0}^{\infty} \frac{(\psi_{\alpha_1}^{n_{\alpha_1}}\cdots\psi_{\alpha_p}^{n_{\alpha_p}}\cdots)}{(n_{\alpha_1}!\cdots n_{\alpha_p}!\cdots)^{1/2}}|n_{\alpha_1}\cdots n_{\alpha_p}\cdots\rangle\end{aligned} \tag{1.211}$$

(ψ_α is a c-number) are eigenstates of the annihilation operators,[28]

$$\hat{\psi}_\alpha|\psi\rangle = \psi_\alpha|\psi\rangle. \tag{1.212}$$

---

[26]Note that when ψ_α is an anticommuting variable, (1.209) holds regardless of whether ψ_α commutes or anticommutes with the operator $\hat{\psi}_\beta$.

[27]The properties of coherent states are summarized in Table 1.2 on page 50.

[28]Expand the exponential in (1.211) and use $[\hat{\psi}_\alpha, (\hat{\psi}^\dagger_\beta)^n]_- = \delta_{\alpha,\beta}n(\hat{\psi}^\dagger_\alpha)^{n-1}$.

The action of the creation operator $\hat{\psi}^\dagger_\alpha$ on the coherent states is given by

$$\hat{\psi}^\dagger_\alpha|\psi\rangle = \frac{\partial}{\partial\psi_\alpha}|\psi\rangle. \tag{1.213}$$

By taking the Hermitian conjugate in (1.212) and (1.213), we deduce

$$\begin{aligned} \langle\psi|\hat{\psi}^\dagger_\alpha &= \langle\psi|\psi^*_\alpha, \\ \langle\psi|\hat{\psi}_\alpha &= \frac{\partial}{\partial\psi^*_\alpha}\langle\psi|, \end{aligned} \tag{1.214}$$

where

$$\langle\psi| = \langle\text{vac}|\exp\left\{\sum_\alpha \psi^*_\alpha\hat{\psi}_\alpha\right\} \tag{1.215}$$

is the bra corresponding to the ket $|\psi\rangle$ and ψ^*_α the complex conjugate of ψ_α.

The overlap between two states is given by

$$\langle\psi|\psi'\rangle = \exp\left\{\sum_\alpha \psi^*_\alpha\psi'_\alpha\right\}. \tag{1.216}$$

Another crucial property of the coherent states is their overcompleteness in the Fock space,[29] which results in the closure relation

$$\int\prod_\alpha \frac{d\psi^*_\alpha d\psi_\alpha}{2i\pi} e^{-\sum_\alpha|\psi_\alpha|^2}|\psi\rangle\langle\psi| = 1, \tag{1.217}$$

where

$$\frac{d\psi^*_\alpha d\psi_\alpha}{2i\pi} = \frac{1}{\pi}d\Re(\psi_\alpha)d\Im(\psi_\alpha). \tag{1.218}$$

For a single degree of freedom, writing $\psi = \rho e^{i\theta}$, one finds

$$\begin{aligned} \int\frac{d\psi^* d\psi}{2i\pi}e^{-|\psi|^2}|\psi\rangle\langle\psi| &= \frac{1}{\pi}\int_0^\infty \rho d\rho\int_0^{2\pi} d\theta\, e^{-\rho^2}\sum_{n,m=0}^\infty \frac{\rho^{n+m}e^{i(n-m)\theta}}{(n!m!)^{1/2}}|n\rangle\langle m| \\ &= 2\int_0^\infty \rho d\rho\, e^{-\rho^2}\sum_{n=0}^\infty\frac{\rho^{2n}}{n!}|n\rangle\langle n| \\ &= \sum_{n=0}^\infty |n\rangle\langle n| = 1. \end{aligned} \tag{1.219}$$

This result is generalized to many degrees of freedom by writing the coherent state $|\psi\rangle$ in terms of the states $|n_{\alpha_1}\cdots\rangle$ [Eq. (1.211)] and using the closure relation (1.146) in Fock space.

[29]The basis is said to be overcomplete because the coherent states are not pairwise orthogonal.

Let us now consider an operator $\hat{A} \equiv A(\hat{\psi}^\dagger_\alpha, \hat{\psi}_\alpha)$ where all creation operators stand to the left of the annihilation operators. Such an operator is said to be normal ordered. Its matrix element between coherent states takes the simple form

$$\langle\psi|A(\hat{\psi}^\dagger_\alpha, \hat{\psi}_\alpha)|\psi'\rangle = e^{\sum_\alpha \psi^*_\alpha \psi'_\alpha} A(\psi^*_\alpha, \psi'_\alpha). \tag{1.220}$$

To conclude this section, we note that the occupation numbers n_α in the coherent state $|\psi\rangle$ are Poisson-distributed with mean values $|\psi_\alpha|^2$,

$$|\langle n_{\alpha_1} \cdots n_{\alpha_p} \cdots |\psi\rangle|^2 = \prod_\alpha \frac{|\psi_\alpha|^{2n_\alpha}}{n_\alpha!}. \tag{1.221}$$

The total number of particles has the average value

$$N = \langle \hat{N} \rangle = \frac{\langle\psi|\hat{N}|\psi\rangle}{\langle\psi|\psi\rangle} = \sum_\alpha |\psi_\alpha|^2 \tag{1.222}$$

(recall that $\hat{N} = \sum_\alpha \hat{\psi}^\dagger_\alpha \hat{\psi}_\alpha$) and variance

$$\sigma^2 = \left\langle \left(\hat{N} - \langle\hat{N}\rangle\right)^2 \right\rangle = \langle \hat{N} \rangle. \tag{1.223}$$

In the thermodynamic limit where $N \to \infty$, $\sigma/N = 1/\sqrt{N} \to 0$ and the particle number distribution becomes sharply peaked around N, reflecting the fact that a product of Poisson distributions approaches a normal distribution.

The physical meaning of the boson coherent states can be understood from the study of a one-dimensional harmonic oscillator,

$$\hat{H} = \frac{\hat{p}^2}{2m} + \frac{1}{2} m\omega^2 \hat{x}^2, \tag{1.224}$$

with $[\hat{x}, \hat{p}] = i$. The ladder operators

$$\hat{a} = \sqrt{\frac{m\omega}{2}} \left(\hat{x} + i \frac{\hat{p}}{m\omega} \right), \qquad \hat{a}^\dagger = \sqrt{\frac{m\omega}{2}} \left(\hat{x} - i \frac{\hat{p}}{m\omega} \right) \tag{1.225}$$

satisfy canonical bosonic commutation relations, $[\hat{a}, \hat{a}^\dagger] = 1$, and enable us to rewrite the Hamiltonian as

$$\hat{H} = \omega \left(\hat{a}^\dagger \hat{a} + \frac{1}{2} \right). \tag{1.226}$$

The states $|n\rangle = \frac{(\hat{a}^\dagger)^n}{\sqrt{n!}} |\text{vac}\rangle$ are eigenstates with eigenvalues $\left(n + \frac{1}{2}\right)\omega$.

Let us now consider the normalized coherent state

$$|z\rangle = e^{-\frac{|z|^2}{2} + z\hat{a}^\dagger} |\text{vac}\rangle. \tag{1.227}$$

A simple calculation yields

$$\begin{aligned} \langle z|\hat{x}|z\rangle &= \sqrt{\frac{2}{m\omega}} \Re(z), \qquad \langle z|\hat{x}^2|z\rangle = \frac{1}{2m\omega} [4\Re(z)^2 + 1], \\ \langle z|\hat{p}|z\rangle &= \sqrt{2m\omega} \Im(z), \qquad \langle z|\hat{p}^2|z\rangle = \frac{m\omega}{2} [4\Im(z)^2 + 1] \end{aligned} \tag{1.228}$$

and therefore

$$\Delta x^2 = \frac{1}{2m\omega}, \qquad \Delta p^2 = \frac{m\omega}{2}. \tag{1.229}$$

Thus, the coherent state $|z\rangle$ yields the minimum position and momentum fluctuations compatible with the uncertainty relation $\Delta x \Delta p \geq \hbar/2$ (we have restored Planck's constant).

In order to further study the relation between coherent states and the classical limit of the harmonic oscillator, we consider the dynamics of the eigenstates $|n\rangle$ in the Schrödinger picture,

$$e^{-i\hat{H}t}|n\rangle = e^{-i(n+\frac{1}{2})\omega t}|n\rangle. \tag{1.230}$$

From (1.230), we deduce the time evolution of the coherent states,

$$e^{-i\hat{H}t}|z\rangle = e^{-\frac{|z|^2}{2}} \sum_{n=0}^{\infty} \frac{z^n}{\sqrt{n!}} e^{-i\hat{H}t}|n\rangle = e^{-\frac{i}{2}\omega t}|e^{-i\omega t}z\rangle. \tag{1.231}$$

Apart from the unimportant phase factor $e^{-\frac{i}{2}\omega t}$, the dynamics of the coherent state $|z\rangle$ is simply obtained by replacing z by $z(t) = ze^{-i\omega t}$. One then finds that the expectation values

$$\begin{aligned} \langle \hat{x} \rangle (t) &= \langle z(t)|\hat{x}|z(t)\rangle = \sqrt{\frac{2}{m\omega}} z(0) \cos(\omega t), \\ \langle \hat{p} \rangle (t) &= \langle z(t)|\hat{p}|z(t)\rangle = \sqrt{2m\omega} z(0) \sin(\omega t) \end{aligned} \tag{1.232}$$

(we assume $z(0) = z$ to be real) satisfy the classical equations of motion $\dot{x} = \partial H/\partial p = p/m$ and $\dot{p} = -\partial H/\partial x = -m\omega^2 x$. The amplitude $|z(0)|$ of the oscillations is related to the expectation value of $\hat{N} = \hat{a}^\dagger \hat{a}$,

$$\langle \hat{N} \rangle = \langle z(t)|\hat{a}^\dagger \hat{a}|z(t)\rangle = |z(t)|^2 = |z(0)|^2. \tag{1.233}$$

The time independence of $\langle \hat{N} \rangle$ follows from $[\hat{H}, \hat{N}] = 0$ (see Appendix 1.B and Chapter 2). When $\langle \hat{N} \rangle = |z(0)|^2 \gg 1$, $|\langle \hat{x} \rangle (t)| \gg \Delta x$ and $|\langle \hat{p} \rangle (t)| \gg \Delta p$ for most times t (with Δx and Δp given by (1.229)). A coherent state with large occupation number $\langle \hat{N} \rangle = |z(0)|^2$ therefore corresponds to a quasi-classical behavior of the harmonic oscillator. Boson coherent states are sometimes referred to as quasi-classical states.[30]

1.3.2 *Grassmann variables*

A Grassmann algebra is an algebra of anticommuting variables called Grassmann variables (or numbers).[31] It is defined by a set $\{\psi_1, \dots, \psi_n\}$ of n generators whose product is anticommutative,

$$[\psi_i, \psi_j]_+ = \psi_i \psi_j + \psi_j \psi_i = 0. \tag{1.234}$$

[30] Coherent states were introduced in quantum optics by Glauber [29] to describe the electromagnetic field; see the discussion on page 129 in Sec. 1.9.

[31] Recall that an algebra is a vector space endowed with a multiplication rule.

This implies that the square ψ_i^2 of any generator vanishes. An element of the Grassmann algebra is a polynomial of first order in the generators,

$$f(\psi_1,\ldots,\psi_n) = c_0 + \sum_{k=1}^{n} \sum_{i_1\cdots i_k=1}^{n} c_{i_1\cdots i_k}\psi_{i_1}\cdots\psi_{i_k}, \tag{1.235}$$

where the coefficients c_0 and $c_{i_1\cdots i_k}$ are complex and we assume that the indices $i_1 < i_2 < \cdots < i_k$ are ordered. For example, the elements of a Grassmann algebra with only one or two generators are

$$\begin{aligned} f(\psi) &= c_0 + c_1\psi, \\ f(\psi_1,\psi_2) &= c_0 + c_1\psi_1 + c_2\psi_2 + c_{12}\psi_1\psi_2. \end{aligned} \tag{1.236}$$

Differentiation with respect to Grassmann numbers is defined by

$$\frac{\partial}{\partial\psi_i}\psi_j = \delta_{i,j}. \tag{1.237}$$

As with ordinary numbers, this is a linear operation. The anticommutation relations (1.234) imply

$$\left[\partial_{\psi_i},\partial_{\psi_j}\right]_+ = 0. \tag{1.238}$$

The operator ∂_{ψ_i} is therefore nilpotent, $\partial_{\psi_i}^2 = 0$, reflecting the fact that the function $f(\psi_1,\ldots,\psi_n)$ is at most of order 1 in each variable. Equation (1.237) defines a left differentiation,

$$\partial_{\psi_i}\psi_j\psi_i = -\partial_{\psi_i}\psi_i\psi_j = -\psi_j \qquad (i \neq j). \tag{1.239}$$

For ∂_{ψ_i} to act on ψ_i, one must first shift ψ_i to the left of the monomial. The chain rule for differentiation then reads

$$\frac{\partial}{\partial\psi_i}f(g) = \frac{\partial g}{\partial\psi_i}\frac{\partial f}{\partial g}. \tag{1.240}$$

Contrary to ordinary variables, the order of the terms in the rhs matters.

For Grassmann numbers, integration and differentiation are identical,

$$\int d\psi_i\, f(\psi_1,\ldots,\psi_n) = \frac{\partial}{\partial\psi_i}f(\psi_1,\ldots,\psi_n). \tag{1.241}$$

This property ensures that two fundamental properties of ordinary integrals over functions vanishing at infinity are satisfied. The integral of an exact differential form is zero,

$$\int d\psi_i \frac{\partial}{\partial\psi_i}f(\psi_1,\ldots,\psi_n) = 0, \tag{1.242}$$

and the integral over ψ_i of $f(\psi_1,\ldots,\psi_n)$ does not depend on ψ_i so that its derivative vanishes,

$$\frac{\partial}{\partial\psi_i}\int d\psi_i\, f(\psi_1,\ldots,\psi_n) = 0. \tag{1.243}$$

Both properties (1.242) and (1.243) follow from the definition (1.241) and the nilpotence of the differential operator ∂_{ψ_i}. For a Grassmann algebra with a single generator, one has

$$\begin{aligned} f(\psi) &= c_0 + c_1\psi, \\ \frac{\partial f}{\partial \psi} &= \int d\psi\, f = c_1. \end{aligned} \tag{1.244}$$

Other properties of Grassmann integration are discussed in Appendix 1.E.

It will often be convenient to consider Grassmann algebras with $2n$ generators $\{\psi_1, \dots, \psi_n, \psi_1^*, \dots, \psi_n^*\}$. One can then define a (complex) conjugation for Grassmann variables,

$$\left.\begin{aligned} (\psi_i)^* &= \psi_i^*, \\ (\psi_i^*)^* &= \psi_i, \\ (c\psi_i)^* &= c^*\psi_i^*, \\ (\psi_i\psi_j)^* &= \psi_j^*\psi_i^*, \end{aligned}\right. \qquad (c \in \mathbb{C}). \tag{1.245}$$

The general element of a Grassmann algebra with two generators $\{\psi, \psi^*\}$ reads

$$f(\psi, \psi^*) = c_0 + c_1\psi^* + c_2\psi + c_{12}\psi^*\psi. \tag{1.246}$$

Using the differentiation and integration rules, one easily finds

$$\begin{aligned} \frac{\partial f}{\partial \psi} &= \int d\psi\, f = c_2 - c_{12}\psi^*, \\ \frac{\partial f}{\partial \psi^*} &= \int d\psi^*\, f = c_1 + c_{12}\psi, \\ \frac{\partial^2 f}{\partial \psi^* \partial \psi} &= \int d\psi^* d\psi\, f = -c_{12}. \end{aligned} \tag{1.247}$$

1.3.3 *Fermion coherent states*

Grassmann variables enable us to define coherent states similar to the ones introduced for bosons. Given a set $\{\hat{\psi}_\alpha^\dagger, \hat{\psi}_\alpha\}$ of creation and annihilation operators, we define a Grassmann algebra by associating a generator ψ_α to $\hat{\psi}_\alpha$ and a generator ψ_α^* to $\hat{\psi}_\alpha^\dagger$ for any α. We also require

$$\begin{aligned} &[\psi_\alpha^{(*)}, \hat{\psi}_\beta^{(\dagger)}]_+ = 0, \\ &(\psi_\alpha \hat{\psi}_\beta)^\dagger = \hat{\psi}_\beta^\dagger \psi_\alpha^*, \quad (\psi_\alpha \hat{\psi}_\beta^\dagger)^\dagger = \hat{\psi}_\beta \psi_\alpha^*, \quad \text{etc.} \end{aligned} \tag{1.248}$$

A fermion coherent state is defined by the ket

$$|\psi\rangle = \exp\Big\{-\sum_\alpha \psi_\alpha \hat{\psi}_\alpha^\dagger\Big\}|\text{vac}\rangle = \prod_\alpha (1 - \psi_\alpha \hat{\psi}_\alpha^\dagger)|\text{vac}\rangle \tag{1.249}$$

and the corresponding bra

$$\langle\psi| = \langle\text{vac}|\exp\left\{\sum_\alpha \psi^*_\alpha\hat{\psi}_\alpha\right\} = \langle\text{vac}|\prod_\alpha(1+\psi^*_\alpha\hat{\psi}_\alpha). \tag{1.250}$$

It is straightforward to show that the properties (1.212)–(1.214) obtained previously for the boson coherent states can be transposed to the fermionic case,[32]

$$\begin{aligned}\hat{\psi}_\alpha|\psi\rangle &= \psi_\alpha|\psi\rangle,\\ \hat{\psi}^\dagger_\alpha|\psi\rangle &= -\frac{\partial}{\partial\psi_\alpha}|\psi\rangle,\\ \langle\psi|\hat{\psi}^\dagger_\alpha &= \langle\psi|\psi^*_\alpha,\\ \langle\psi|\hat{\psi}_\alpha &= \frac{\partial}{\partial\psi^*_\alpha}\langle\psi|.\end{aligned} \tag{1.251}$$

The overlap between two states is given by

$$\langle\psi|\psi'\rangle = \exp\left\{\sum_\alpha \psi^*_\alpha\psi'_\alpha\right\} \tag{1.252}$$

and for a normal-ordered operator $\hat{A} = A(\hat{\psi}^\dagger_\alpha, \hat{\psi}_\alpha)$,

$$\langle\psi|A(\hat{\psi}^\dagger_\alpha, \hat{\psi}_\alpha)|\psi'\rangle = e^{\sum_\alpha \psi^*_\alpha\psi'_\alpha} A(\psi^*_\alpha, \psi'_\alpha). \tag{1.253}$$

The overcompleteness of the coherent state basis reflects into the closure relation

$$\int\prod_\alpha d\psi^*_\alpha d\psi_\alpha\, e^{-\sum_\alpha \psi^*_\alpha\psi_\alpha}|\psi\rangle\langle\psi| = 1. \tag{1.254}$$

This relation can be proven by considering the matrix element of the lhs of (1.254) between two arbitrary states of the Fock space. From

$$\begin{aligned}\langle\alpha_1\cdots\alpha_n|\psi\rangle &= \langle\text{vac}|\hat{\psi}_{\alpha_n}\cdots\hat{\psi}_{\alpha_1}|\psi\rangle = \psi_{\alpha_n}\cdots\psi_{\alpha_1},\\ \langle\psi|\beta_1\cdots\beta_m\rangle &= \langle\psi|\hat{\psi}^\dagger_{\beta_1}\cdots\hat{\psi}^\dagger_{\beta_m}|\text{vac}\rangle = \psi^*_{\beta_1}\cdots\psi^*_{\beta_m}\end{aligned} \tag{1.255}$$

and $\langle\text{vac}|\psi\rangle = 1$, we deduce

$$\begin{aligned}&\int\prod_\alpha d\psi^*_\alpha d\psi_\alpha\, e^{-\sum_\alpha \psi^*_\alpha\psi_\alpha}\langle\alpha_1\cdots\alpha_n|\psi\rangle\langle\psi|\beta_1\cdots\beta_m\rangle\\ &= \int\prod_\alpha d\psi^*_\alpha d\psi_\alpha \prod_\alpha(1-\psi^*_\alpha\psi_\alpha)\psi_{\alpha_n}\cdots\psi_{\alpha_1}\psi^*_{\beta_1}\cdots\psi^*_{\beta_m}.\end{aligned} \tag{1.256}$$

[32]For instance, the first equation in (1.251) is obtained from $\hat{\psi}_\alpha|\psi\rangle = \hat{\psi}_\alpha\prod_\beta(1-\psi_\beta\hat{\psi}^\dagger_\beta)|\text{vac}\rangle = \prod_{\beta\neq\alpha}(1-\psi_\beta\hat{\psi}^\dagger_\beta)\hat{\psi}_\alpha(1-\psi_\alpha\hat{\psi}^\dagger_\alpha)|\text{vac}\rangle = \prod_{\beta\neq\alpha}(1-\psi_\beta\hat{\psi}^\dagger_\beta)\psi_\alpha|\text{vac}\rangle = \prod_{\beta\neq\alpha}(1-\psi_\beta\hat{\psi}^\dagger_\beta)\psi_\alpha(1-\psi_\alpha\hat{\psi}^\dagger_\alpha)|\text{vac}\rangle = \psi_\alpha|\psi\rangle$.

Four types of integrals appear in (1.256),

$$\int d\psi_\alpha^* d\psi_\alpha (1-\psi_\alpha^*\psi_\alpha) \begin{pmatrix} \psi_\alpha\psi_\alpha^* \\ \psi_\alpha \\ \psi_\alpha^* \\ 1 \end{pmatrix} = \begin{pmatrix} 1 \\ 0 \\ 0 \\ 1 \end{pmatrix}. \tag{1.257}$$

Thus, each α must belong to both $\{\alpha_i\}$ and $\{\beta_i\}$ or be absent from both sets for (1.256) to be non-zero. In this case, $n=m$ and $\beta_i = \alpha_{P(i)}$ with $P \in S_n$ a permutation, and Eq. (1.256) reduces to

$$\begin{aligned} &\int \prod_\alpha d\psi_\alpha^* d\psi_\alpha \prod_\alpha (1-\psi_\alpha^*\psi_\alpha)\psi_{\alpha_n}\cdots\psi_{\alpha_1}\psi_{\alpha_{P(1)}}^*\cdots\psi_{\alpha_{P(m)}}^* \\ &= \int \prod_\alpha d\psi_\alpha^* d\psi_\alpha \prod_\alpha (1-\psi_\alpha^*\psi_\alpha)(-1)^P\psi_{\alpha_n}\cdots\psi_{\alpha_1}\psi_{\alpha_1}^*\cdots\psi_{\alpha_m}^* = (-1)^P. \end{aligned} \tag{1.258}$$

This result should be compared with

$$\langle \alpha_1\cdots\alpha_n|\beta_1\cdots\beta_m\rangle = \delta_{n,m}\sum_{P\in S_n}(-1)^P \delta_{\alpha_i,\beta_{P(i)}}, \tag{1.259}$$

which proves the closure relation (1.254).

To conclude this section, we note that fermion coherent states are not physical states of the system; they do not belong to the Fock space. For instance, the expectation value of the particle number operator $\hat N = \sum_\alpha \hat\psi_\alpha^\dagger\hat\psi_\alpha$ in the coherent state $|\psi\rangle$,

$$\frac{\langle\psi|\hat N|\psi\rangle}{\langle\psi|\psi\rangle} = \sum_\alpha \psi_\alpha^*\psi_\alpha, \tag{1.260}$$

is not a real number.

Table 1.2 Main properties of coherent states. ψ_α denotes a complex variable for bosons ($\zeta=1$) and a Grassmann number for fermions ($\zeta=-1$). $\mathcal{N}=2i\pi$ for bosons and $\mathcal{N}=1$ for fermions.

Definition	$\|\psi\rangle = e^{\zeta\sum_\alpha \psi_\alpha\hat\psi_\alpha^\dagger}\|\mathrm{vac}\rangle$ $\langle\psi\| = \langle\mathrm{vac}\|e^{\sum_\alpha \psi_\alpha^*\hat\psi_\alpha}$
	$\hat\psi_\alpha\|\psi\rangle = \psi_\alpha\|\psi\rangle, \qquad \langle\psi\|\hat\psi_\alpha^\dagger = \langle\psi\|\psi_\alpha^*$
Overlap	$\langle\psi\|\psi'\rangle = e^{\sum_\alpha \psi_\alpha^*\psi'_\alpha}$
Closure relation	$\int\prod_\alpha \frac{d\psi_\alpha^* d\psi_\alpha}{\mathcal{N}} e^{-\sum_\alpha \|\psi_\alpha\|^2}\|\psi\rangle\langle\psi\| = 1$
Trace	$\mathrm{Tr}(\hat A) = \int\prod_\alpha \frac{d\psi_\alpha^* d\psi_\alpha}{\mathcal{N}} e^{-\sum_\alpha \|\psi_\alpha\|^2}\langle\zeta\psi\|\hat A\|\psi\rangle$
Normal-ordered operator	$\langle\psi\|A(\hat\psi_\alpha^\dagger,\hat\psi_\alpha)\|\psi'\rangle = e^{\sum_\alpha \psi_\alpha^*\psi'_\alpha}A(\psi_\alpha^*,\psi'_\alpha)$

1.4 Functional Integral in Many-Particle Systems

We are now in a position to derive a functional integral representation of the partition function of a many-particle system following the general procedure outlined in Sec. 1.1.

1.4.1 *Coherent-state functional integral*

1.4.1.1 *Partition function*

Let us first express the partition function Z in the coherent-state basis (the following derivation actually holds for the trace of any operator). Using the closure relation (1.217) or (1.254), we can write

$$Z = \sum_n \langle n|e^{-\beta\hat{H}}|n\rangle = \sum_n \int d(\psi^*,\psi)\, e^{-\sum_\alpha \psi^*_\alpha \psi_\alpha} \langle n|\psi\rangle\langle\psi|e^{-\beta\hat{H}}|n\rangle, \tag{1.261}$$

where $\hat{H}$ is the grand canonical Hamiltonian of the system (including the term $-\mu\hat{N}$), and $\{|n\rangle\}$ a basis in the Fock space. $|\psi\rangle$ is a coherent state, and ψ_α a c-number for bosons and a Grassmann variable for fermions. We use the notation

$$d(\psi^*,\psi) = \begin{cases} \prod_\alpha \dfrac{d\psi^*_\alpha d\psi_\alpha}{2i\pi} & \text{(bosons)}, \\ \prod_\alpha d\psi^*_\alpha d\psi_\alpha & \text{(fermions)}. \end{cases} \tag{1.262}$$

For bosons, we can interchange the two terms $\langle n|\psi\rangle$ and $\langle\psi|e^{-\beta\hat{H}}|n\rangle$ in (1.261) and then use $\sum_n |n\rangle\langle n| = 1$. For fermions, this interchange must be realized with care. Let us write $e^{-\beta\hat{H}}|n\rangle = \sum_m a_m|m\rangle$. Since $\hat{H}$ conserves the particle number, the kets $|n\rangle$ and $|m\rangle$ must have the same number of particles. Writing

$$\begin{aligned} |n\rangle &= \hat{\psi}^\dagger_{\alpha_1}\cdots\hat{\psi}^\dagger_{\alpha_p}|\text{vac}\rangle, \\ |m\rangle &= \hat{\psi}^\dagger_{\gamma_1}\cdots\hat{\psi}^\dagger_{\gamma_p}|\text{vac}\rangle, \end{aligned} \tag{1.263}$$

we obtain

$$\begin{aligned} \langle n|\psi\rangle\langle\psi|m\rangle &= \psi_{\alpha_p}\cdots\psi_{\alpha_1}\psi^*_{\gamma_1}\cdots\psi^*_{\gamma_p} \\ &= \zeta^{p^2}\psi^*_{\gamma_1}\cdots\psi^*_{\gamma_p}\psi_{\alpha_p}\cdots\psi_{\alpha_1} \\ &= \langle\zeta\psi|m\rangle\langle n|\psi\rangle \end{aligned} \tag{1.264}$$

(note that $\zeta^{p^2} = \zeta^p$). We therefore deduce

$$\langle n|\psi\rangle\langle\psi|e^{-\beta\hat{H}}|n\rangle = \sum_m a_m\langle\zeta\psi|m\rangle\langle n|\psi\rangle = \langle\zeta\psi|e^{-\beta\hat{H}}|n\rangle\langle n|\psi\rangle \tag{1.265}$$

and the partition function can be written as

$$Z = \int d(\psi^*,\psi)\, e^{-\sum_\alpha \psi^*_\alpha\psi_\alpha}\langle\zeta\psi|e^{-\beta\hat{H}}|\psi\rangle. \tag{1.266}$$

We are now in a position to write the partition function as a coherent-state functional integral. Dividing the imaginary time β into $N = \beta/\epsilon$ steps and inserting $N-1$ times the closure relation (1.217) or (1.254), we obtain

$$Z = \int \prod_{k=1}^{N} d(\psi_k^*, \psi_k)\, e^{-\sum_{k=1}^{N}\sum_\alpha \psi_{k,\alpha}^*\psi_{k,\alpha}} \prod_{k=1}^{N} \langle \psi_k | e^{-\epsilon \hat{H}} | \psi_{k-1} \rangle, \tag{1.267}$$

where we have relabeled $\zeta\psi^{(*)} \equiv \psi_N^{(*)}$ in Eq. (1.266) and introduced the variables (not to be integrated over) $\psi_0^{(*)} = \zeta\psi_N^{(*)} \equiv \psi^{(*)}$. If the Hamiltonian $\hat{H} = H(\hat{\psi}_\alpha^\dagger, \hat{\psi}_\alpha)$ is normal ordered,

$$\langle \psi_k | e^{-\epsilon\hat{H}} | \psi_{k-1} \rangle = \langle \psi_k | \psi_{k-1} \rangle e^{-\epsilon H(\psi_k^*, \psi_{k-1})} + \mathcal{O}(\epsilon^2), \tag{1.268}$$

one finds

$$Z = \int \prod_{k=1}^{N} d(\psi_k^*, \psi_k) \exp\left\{ -\epsilon \sum_{k=1}^{N} \left[\sum_\alpha \psi_{k,\alpha}^* \frac{(\psi_{k,\alpha} - \psi_{k-1,\alpha})}{\epsilon} + H(\psi_{k,\alpha}^*, \psi_{k-1,\alpha}) \right] \right\} \tag{1.269}$$

for $\epsilon \to 0$. As for the Feynman path integral (Sec. 1.1), in the limit $N \to \infty$ it is convenient to represent the set $\{\psi_{0,\alpha}, \ldots, \psi_{N,\alpha}\}$ by a continuous trajectory $\psi_\alpha(\tau)$ with τ an imaginary time varying between 0 and β and the boundary conditions

$$\psi_\alpha(\beta) = \zeta\psi_\alpha(0), \qquad \psi_\alpha^*(\beta) = \zeta\psi_\alpha^*(0). \tag{1.270}$$

We can then (symbolically) write

$$\begin{aligned} \epsilon \sum_{k=1}^{N} &\to \int_0^\beta d\tau, \\ \psi_{k,\alpha}^* \frac{(\psi_{k,\alpha} - \psi_{k-1,\alpha})}{\epsilon} &\to \psi_\alpha^*(\tau)\partial_\tau \psi_\alpha(\tau), \\ H(\psi_{k,\alpha}^*, \psi_{k-1,\alpha}) &\to H\big(\psi_\alpha^*(\tau^+), \psi_\alpha(\tau)\big), \end{aligned} \tag{1.271}$$

so that the partition takes the final form

$$Z = \int_{\substack{\psi(\beta)=\zeta\psi(0) \\ \psi^*(\beta)=\zeta\psi^*(0)}} \mathcal{D}[\psi^*, \psi]\, e^{-S[\psi^*,\psi]} \tag{1.272}$$

with the Euclidean action[33]

$$S[\psi^*, \psi] = \int_0^\beta d\tau \left[\sum_\alpha \psi_\alpha^*(\tau)\partial_\tau\psi_\alpha(\tau) + H\big(\psi_\alpha^*(\tau^+), \psi_\alpha(\tau)\big) \right]. \tag{1.273}$$

The notation τ^+ reminds us that the creation field is always evaluated infinitesimally later than the annihilation field [see (1.271)]. For fermions, the notation $\partial_\tau\psi_\alpha(\tau)$ is purely symbolic since there is no sense in which $\psi_{k,\alpha} - \psi_{k-1,\alpha}$ is small. The symbol $\partial_\tau\psi_\alpha(\tau)$ must be

[33]For simplicity, we denote the Euclidean action merely by S (and not S_E).

understood as $\lim_{\epsilon\to 0}\frac{1}{\epsilon}(\psi_{k,\alpha}-\psi_{k-1,\alpha})$. The measure in the functional integral (1.272) is defined as

$$\mathcal{D}[\psi^*,\psi]=\lim_{N\to\infty}\prod_{k=0}^{N}d(\psi_k^*,\psi_k) \tag{1.274}$$

and must be supplemented with the boundary conditions $\psi(\beta)=\zeta\psi(0)$ and $\psi^*(\beta)=\zeta\psi^*(0)$.

For a system with a non-interacting one-body Hamiltonian $\hat{h}_0$ and a two-body interaction $\hat{v}$, the action reads

$$\begin{aligned} S[\psi^*,\psi]=&\int_0^\beta d\tau\Big\{\sum_{\alpha,\alpha'}\psi_\alpha^*(\tau)\left[\delta_{\alpha,\alpha'}\partial_\tau+\langle\alpha|\hat{h}_0|\alpha'\rangle\right]\psi_{\alpha'}(\tau)\\ &+\frac{1}{2}\sum_{\alpha_1,\alpha_2,\alpha_1',\alpha_2'}(\alpha_1\alpha_2|\hat{v}|\alpha_1'\alpha_2')\psi_{\alpha_1}^*(\tau^+)\psi_{\alpha_2}^*(\tau^+)\psi_{\alpha_2'}(\tau)\psi_{\alpha_1'}(\tau)\Big\}. \end{aligned} \tag{1.275}$$

1.4.1.2 *Green functions*

Let us now consider the Green function $G(\alpha\tau,\alpha'\tau')=-\langle T_\tau\hat{\psi}_\alpha(\tau)\hat{\psi}^\dagger_{\alpha'}(\tau')\rangle$ introduced in Sec. 1.2.3. For $\tau>\tau'$, it reads

$$G(\alpha\tau,\alpha'\tau')=-\frac{1}{Z}\mathrm{Tr}\left[e^{-(\beta-\tau)\hat{H}}\hat{\psi}_\alpha e^{-(\tau-\tau')\hat{H}}\hat{\psi}^\dagger_{\alpha'}e^{-\tau'\hat{H}}\right]. \tag{1.276}$$

We now split the time intervals $\beta-\tau$, $\tau-\tau'$ and τ' in infinitesimal steps $\epsilon=\beta/N$. If we assume that τ corresponds to the discrete time $\tau_{k_1}=k_1\epsilon$ and τ' to $\tau_{k_2}=k_2\epsilon$, we obtain

$$\begin{aligned} G(\alpha\tau,\alpha'\tau')=&-\frac{1}{Z}\int\prod_{k=1}^{N}d(\psi_k^*,\psi_k)\,e^{-\sum_{k=1}^N\sum_\alpha\psi_{k,\alpha}^*\psi_{k,\alpha}}\langle\psi_N|e^{-\epsilon\hat{H}}|\psi_{N-1}\rangle\cdots\\ &\cdots\langle\psi_{k_1+1}|e^{-\epsilon\hat{H}}\hat{\psi}_\alpha|\psi_{k_1}\rangle\cdots\langle\psi_{k_2}|\hat{\psi}^\dagger_{\alpha'}e^{-\epsilon\hat{H}}|\psi_{k_2-1}\rangle\cdots\langle\psi_1|e^{-\epsilon\hat{H}}|\psi_0\rangle, \end{aligned} \tag{1.277}$$

with the boundary conditions $\psi_N=\zeta\psi_0$ and $\psi_N^*=\zeta\psi_0^*$. We deduce

$$\begin{aligned} G(\alpha\tau,\alpha'\tau')&=-\frac{1}{Z}\int\prod_{k=1}^{N}d(\psi_k^*,\psi_k)\,\psi_{k_1,\alpha}\psi_{k_2,\alpha'}^*e^{-S(\{\psi_k^*,\psi_k\})}\\ &\equiv-\frac{1}{Z}\int\mathcal{D}[\psi^*,\psi]\,\psi_\alpha(\tau)\psi_{\alpha'}^*(\tau')e^{-S[\psi^*,\psi]}, \end{aligned} \tag{1.278}$$

where $S(\{\psi_k^*,\psi_k\})$ is the action defined in (1.269) and $S[\psi^*,\psi]$ its continuum-time limit (1.273). For $\tau<\tau'$, we can follow a similar reasoning. Starting from

$$G(\alpha\tau,\alpha'\tau')=-\frac{\zeta}{Z}\mathrm{Tr}\left[e^{-(\beta-\tau')\hat{H}}\hat{\psi}^\dagger_{\alpha'}e^{-(\tau'-\tau)\hat{H}}\hat{\psi}_\alpha e^{-\tau\hat{H}}\right], \tag{1.279}$$

one obtains

$$G(\alpha\tau,\alpha'\tau') = -\frac{\zeta}{Z}\int \mathcal{D}[\psi^*,\psi]\,\psi^*_{\alpha'}(\tau')\psi_\alpha(\tau)e^{-S[\psi^*,\psi]}$$
$$= -\frac{1}{Z}\int \mathcal{D}[\psi^*,\psi]\,\psi_\alpha(\tau)\psi^*_{\alpha'}(\tau')e^{-S[\psi^*,\psi]}. \qquad (1.280)$$

Thus, the Green function can be written as a functional integral. This result generalizes to the n-particle Green function (1.203) and any statistical average of a time-ordered product of operators,

$$\langle T_\tau A_1(\hat\psi^\dagger_\alpha(\tau_1),\hat\psi_\alpha(\tau_1))\cdots A_n(\hat\psi^\dagger_\alpha(\tau_n),\hat\psi_\alpha(\tau_n))\rangle$$
$$= \frac{1}{Z}\mathrm{Tr}\left[e^{-\beta\hat H}T_\tau A_1(\hat\psi^\dagger_\alpha(\tau_1),\hat\psi_\alpha(\tau_1))\cdots A_n(\hat\psi^\dagger_\alpha(\tau_n),\hat\psi_\alpha(\tau_n))\right]$$
$$= \frac{1}{Z}\int \mathcal{D}[\psi^*,\psi]\,A_1(\psi^*_\alpha(\tau_1),\psi_\alpha(\tau_1))\cdots A_n(\psi^*_\alpha(\tau_n),\psi_\alpha(\tau_n))\,e^{-S[\psi^*,\psi]}. \qquad (1.281)$$

1.4.2 *Non-interacting particles*

For non-interacting particles with grand canonical Hamiltonian $\hat H_0 = \sum_\alpha \xi_\alpha \hat\psi^\dagger_\alpha\hat\psi_\alpha$, the action reads

$$S_0 = \sum_\alpha\sum_{k=1}^{N}\left[\psi^*_{k,\alpha}(\psi_{k,\alpha}-\psi_{k-1,\alpha}) + \epsilon\xi_\alpha\psi^*_{k,\alpha}\psi_{k-1,\alpha}\right]. \qquad (1.282)$$

In order to diagonalize S_0, we expand the fields in Fourier series,

$$\psi_{k,\alpha} = \frac{1}{\sqrt N}\sum_{n=0}^{N-1} a_{n,\alpha}e^{-i\omega_n\tau_k}, \qquad \psi^*_{k,\alpha} = \frac{1}{\sqrt N}\sum_{n=0}^{N-1} a^*_{n,\alpha}e^{i\omega_n\tau_k}, \qquad (1.283)$$

where $\tau_k = k\epsilon$ and the Matsubara frequencies ω_n are defined in (1.199). The a_n's are complex variables for bosons and Grassmann numbers for fermions. Since the transformation $\psi^{(*)} \to a^{(*)}$ has unit Jacobian,[34] one obtains

$$Z_0 = \lim_{N\to\infty}\int\prod_{n=0}^{N-1} d(a^*_n,a_n)\exp\left\{-\sum_\alpha\sum_{n=0}^{N-1}|a_{n,\alpha}|^2\left[1+(\epsilon\xi_\alpha-1)e^{i\omega_n\epsilon}\right]\right\}$$
$$= \lim_{N\to\infty}\prod_\alpha\prod_{n=0}^{N-1}\left[1+(\epsilon\xi_\alpha-1)e^{i\omega_n\epsilon}\right]^{-\zeta} \qquad (1.284)$$

using the results of Appendix 1.E. Writing $1-\epsilon\xi_\alpha = e^{-\theta_\alpha}$ and using the identity

$$\prod_{n=0}^{N-1}\left(1-e^{i\omega_n\epsilon-\theta}\right) = 1-\zeta e^{-N\theta}, \qquad (1.285)$$

[34]See Appendix 1.E.3 for change of variables in a Grassmannian integral.

we finally deduce

$$Z_0 = \lim_{N\to\infty} \prod_\alpha \left(1 - \zeta e^{-N\theta_\alpha}\right)^{-\zeta} = \prod_\alpha \left(1 - \zeta e^{-\beta\xi_\alpha}\right)^{-\zeta}, \tag{1.286}$$

which is the familiar result for non-interacting particles.

The Fourier transform of the one-particle Green function is obtained from

$$\begin{aligned}
-\langle a_{n,\alpha} a^*_{n,\alpha}\rangle &= -\lim_{N\to\infty} \frac{1}{Z} \int \prod_{n=0}^{N-1} d(a^*_n, a_n)\, a_{n,\alpha} a^*_{n,\alpha}\, e^{-S_0} \\
&= \lim_{N\to\infty} \left[(1-\epsilon\xi_\alpha)e^{i\omega_n\epsilon} - 1\right]^{-1} \\
&= \frac{1}{\epsilon} \frac{1}{i\omega_n - \xi_\alpha}.
\end{aligned} \tag{1.287}$$

Up to a factor $1/\epsilon$, we recover the Green function $G_0(\alpha, i\omega_n) = (i\omega_n - \xi_\alpha)^{-1}$ of the non-interacting system [Eq. (1.201)]. Equation (1.287) yields the Green function

$$\begin{aligned}
G_0(\alpha, \tau_{k_1} - \tau_{k_2}) &= -\lim_{N\to\infty} \frac{1}{N} \sum_{n=0}^{N-1} \langle a_{n,\alpha} a^*_{n,\alpha}\rangle e^{-i\omega_n(\tau_{k_1} - \tau_{k_2})} \\
&= -\lim_{N\to\infty} \frac{1}{N} \sum_{n=0}^{N-1} \sum_{p=0}^{\infty} a^p e^{i\omega_n\epsilon(p-k_1+k_2)} \\
&= -\frac{1}{2} \lim_{N\to\infty} \sum_{p=0}^{\infty} \sum_{m=-\infty}^{\infty} a^p \delta_{p-k_1+k_2, mN} [1 + \zeta + (1-\zeta)(-1)^m]
\end{aligned} \tag{1.288}$$

with $a = 1 - \epsilon\xi_\alpha$ and $\omega_n\epsilon = 2\pi n/N$ (bosons) or $2\pi(n+1/2)/N$ (fermions). We have used the fact that the sum over n vanishes unless $p - k_1 + k_2 = mN$ with m integer. Furthermore, the constraints $p \geq 0$ and $0 \leq k \leq N-1$ imply $m \geq 1$ if $\tau_{k_1} - \tau_{k_2} < 0$ and $m \geq 0$ if $\tau_{k_1} - \tau_{k_2} \geq 0$, so that we finally obtain

$$G_0(\alpha, \tau_{k_1} - \tau_{k_2}) = \begin{cases} -\lim\limits_{N\to\infty} \dfrac{\zeta a^{N+k_1-k_2}}{1 - \zeta a^N} = -\zeta n_\alpha e^{-\xi_\alpha(\tau_{k_1} - \tau_{k_2})} & \text{if } \tau_{k_1} < \tau_{k_2}, \\[2ex] -\lim\limits_{N\to\infty} \dfrac{a^{k_1-k_2}}{1 - \zeta a^N} = -(1 + \zeta n_\alpha) e^{-\xi_\alpha(\tau_{k1} - \tau_{k2})} & \text{if } \tau_{k_1} \geq \tau_{k_2}, \end{cases} \tag{1.289}$$

where

$$n_\alpha = \frac{1}{e^{\beta\xi_\alpha} - \zeta} \tag{1.290}$$

if the occupation number $n_B(\xi_\alpha)$ for bosons and $n_F(\xi_\alpha)$ for fermions. One may rewrite (1.289) as

$$G_0(\alpha, \tau) = -e^{-\xi_\alpha\tau} \left[\Theta(\tau)(1 + \zeta n_\alpha) + \Theta(-\tau)\zeta n_\alpha\right], \tag{1.291}$$

where $\tau = \tau_{k_1} - \tau_{k_2} \in [-\beta, \beta]$, in agreement with (1.200).

It is also possible to calculate the partition function Z_0 directly from (1.282) without introducing the Fourier-transformed variables $a_{n,\alpha}^{(*)}$. To this end, we write the action as

$$S_0 = \sum_{\alpha} \sum_{k,k'=1}^{N} \psi_{k,\alpha}^{*} S_{k,k'}^{(\alpha)} \psi_{k',\alpha}, \tag{1.292}$$

where $S^{(\alpha)}$ is the $N \times N$ matrix

$$S^{(\alpha)} = \begin{pmatrix} 1 & 0 & \cdots & \cdots & 0 & -\zeta a \\ -a & 1 & 0 & & & 0 \\ 0 & -a & \ddots & \ddots & & \vdots \\ \vdots & \ddots & \ddots & \ddots & \ddots & \vdots \\ \vdots & & \ddots & -a & 1 & 0 \\ 0 & \cdots & \cdots & 0 & -a & 1 \end{pmatrix} \tag{1.293}$$

with $a = 1 - \frac{\beta}{N}\xi_\alpha$. Expanding with respect to the first line, we obtain

$$\lim_{N\to\infty} \det S^{(\alpha)} = \lim_{N\to\infty} \left[1 + (-\zeta a)(-1)^{N-1}(-a)^{N-1}\right] = 1 - \zeta e^{-\beta\xi_\alpha}, \tag{1.294}$$

and finally

$$Z_0 = \lim_{N\to\infty} \prod_{\alpha} \left[\det S^{(\alpha)}\right]^{-\zeta} = \prod_{\alpha} \left(1 - \zeta e^{-\beta\xi_\alpha}\right)^{-\zeta}. \tag{1.295}$$

Similarly, we can calculate the one-particle Green function

$$\begin{aligned} & G_0(\alpha, \tau_{k_1} - \tau_{k_2}) \\ &= -\lim_{N\to\infty} \frac{1}{Z_0} \int \prod_{k=1}^{N} d(\psi_k^*, \psi_k)\, \psi_{k_1,\alpha} \psi_{k_2,\alpha}^{*} \exp\Big\{-\sum_{k,k'=1}^{N} \sum_{\alpha} \psi_{k,\alpha}^{*} S_{k,k'}^{(\alpha)} \psi_{k',\alpha}\Big\} \\ &= -\lim_{N\to\infty} \left(S^{(\alpha)-1}\right)_{k_1,k_2}, \end{aligned} \tag{1.296}$$

where $\tau_k = k\beta/N$. The inverse matrix $S^{(\alpha)-1}$ being given by

$$S^{(\alpha)-1} = \frac{1}{1-\zeta a^N} \begin{pmatrix} 1 & \zeta a^{N-1} & \zeta a^{N-2} & \cdots & \cdots & \zeta a \\ a & 1 & \zeta a^{N-1} & \ddots & & \vdots \\ a^2 & a & \ddots & \ddots & \ddots & \vdots \\ \vdots & \ddots & \ddots & \ddots & \zeta a^{N-1} & \zeta a^{N-2} \\ \vdots & & \ddots & a & 1 & \zeta a^{N-1} \\ a^{N-1} & \cdots & \cdots & a^2 & a & 1 \end{pmatrix}, \tag{1.297}$$

we obtain

$$G_0(\alpha,\tau_{k_1}-\tau_{k_2}) = -\lim_{N\to\infty}\begin{cases}\dfrac{\zeta a^{N+k_1-k_2}}{1-\zeta a^N} & \text{if } \tau_{k_1}<\tau_{k_2},\\[2mm] \dfrac{a^{k_1-k_2}}{1-\zeta a^N} & \text{if } \tau_{k_1}\geq\tau_{k_2},\end{cases} \tag{1.298}$$

i.e.

$$G_0(\alpha,\tau_{k_1}-\tau_{k_2}) = \begin{cases}-\zeta n_\alpha e^{-\xi_\alpha(\tau_{k_1}-\tau_{k_2})} & \text{if } \tau_{k_1}<\tau_{k_2},\\ -(1+\zeta n_\alpha)e^{-\xi_\alpha(\tau_{k1}-\tau_{k_2})} & \text{if } \tau_{k_1}\geq\tau_{k_2}.\end{cases} \tag{1.299}$$

1.4.2.1 *Continuum-time limit*

In the continuum-time limit, the variables $a_{n,\alpha}^{(*)}$ become Fourier-transformed fields $\psi_\alpha^{(*)}(i\omega_n)$ defined by

$$\begin{aligned}\psi_\alpha(\tau) &= \frac{1}{\sqrt{\beta}}\sum_{\omega_n} e^{-i\omega_n\tau}\psi_\alpha(i\omega_n), \qquad \psi_\alpha(i\omega_n) = \frac{1}{\sqrt{\beta}}\int_0^\beta d\tau\, e^{i\omega_n\tau}\psi_\alpha(\tau),\\ \psi^*_\alpha(\tau) &= \frac{1}{\sqrt{\beta}}\sum_{\omega_n} e^{i\omega_n\tau}\psi^*_\alpha(i\omega_n), \qquad \psi^*_\alpha(i\omega_n) = \frac{1}{\sqrt{\beta}}\int_0^\beta d\tau\, e^{-i\omega_n\tau}\psi^*_\alpha(\tau),\end{aligned} \tag{1.300}$$

and the action reads

$$\begin{aligned}S_0[\psi^*,\psi] &= \int_0^\beta d\tau \sum_\alpha \psi^*_\alpha(\tau)\,(\partial_\tau+\xi_\alpha)\,\psi_\alpha(\tau)\\ &= \sum_{\alpha,\omega_n}\psi^*_\alpha(i\omega_n)(-i\omega_n+\xi_\alpha)\psi_\alpha(i\omega_n),\end{aligned} \tag{1.301}$$

while the functional integral measure is given by

$$\mathcal{D}[\psi^*,\psi] = \prod_{\alpha,\omega_n} d\big(\psi^*_\alpha(i\omega_n),\psi_\alpha(i\omega_n)\big). \tag{1.302}$$

If we try to compute directly the partition function from the continuum-time limit, we obtain

$$\begin{aligned}Z_0 &= \prod_{\alpha,\omega_n}(-i\omega_n+\xi_\alpha)^{-\zeta},\\ \Omega_0 &= -\frac{1}{\beta}\ln Z_0 = \frac{\zeta}{\beta}\sum_{\alpha,\omega_n}\ln(-i\omega_n+\xi_\alpha).\end{aligned} \tag{1.303}$$

The Matsubara sum in (1.303) is divergent; the continuum-time limit of the partition function is therefore ill-defined. To understand the origin of this difficulty, consider the mean particle number derived from the preceding equation,

$$\langle\hat{N}\rangle = -\frac{\partial\Omega_0}{\partial\mu} = -\frac{\zeta}{\beta}\sum_{\alpha,\omega_n}\frac{1}{i\omega_n-\xi_\alpha}. \tag{1.304}$$

This expression should be compared with the result directly obtained from the action (1.282),

$$\langle \hat{N} \rangle = \frac{\epsilon}{\beta} \sum_\alpha \sum_{k=1}^N \langle \psi^*_{k,\alpha} \psi_{k-1,\alpha} \rangle = \sum_\alpha \langle \psi^*_\alpha(\tau_k) \psi_\alpha(\tau_{k-1} \rangle, \tag{1.305}$$

where the last result holds for any τ_k and follows from time-translation invariance. Thus, in the continuum-time limit, the correct definition of the average particle number should be $\langle \hat{N} \rangle = \sum_\alpha \langle \psi^*_\alpha(\tau^+) \psi_\alpha(\tau) \rangle$ with $\tau^+ = \tau + \eta$ and $\eta \to 0^+$. This gives

$$\langle \hat{N} \rangle = -\zeta \sum_\alpha G_0(\alpha, -\eta) = \frac{\zeta}{\beta} \sum_{\alpha,\omega_n} G_0(\alpha, i\omega_n)\, e^{i\omega_n \eta}, \tag{1.306}$$

where the Green function

$$\begin{aligned} G_0(\alpha, i\omega_n) &= -\langle \psi_\alpha(i\omega_n) \psi^*_\alpha(i\omega_n) \rangle \\ &= -\frac{1}{Z_0} \int \mathcal{D}[\psi^*, \psi]\, \psi_\alpha(i\omega_n) \psi^*_\alpha(i\omega_n)\, e^{-S_0[\psi^*,\psi]} \\ &= \frac{1}{i\omega_n - \xi_\alpha} \end{aligned} \tag{1.307}$$

is obtained using standard results of Gaussian integration (Appendix 1.E). Performing the Matsubara sum in (1.306),[35] we obtain $\langle \hat{N} \rangle = \sum_\alpha n_\alpha$ with n_α defined in (1.290). In order for this result to be consistent with $\langle \hat{N} \rangle = -\partial \Omega_0 / \partial \mu$, we must reexpress the thermodynamic potential as

$$\Omega_0 = \frac{\zeta}{\beta} \sum_{\alpha,\omega_n} \ln(-i\omega_n + \xi_\alpha) e^{i\omega_n \eta}. \tag{1.308}$$

The Matsubara sum is now converging and leads to the usual expression of the thermodynamic potential of non-interacting bosons or fermions. The calculation is detailed in Appendix 1.F.

The difficulties encountered in the continuum-time limit of the coherent-state functional integral are very similar to those that appear in the path integral formulation of a quantum particle in a magnetic field (Sec. 1.1.1). In the formal continuum limit, the ordering of the operators in the Hamiltonian is lost, which results in ambiguities or ill-defined quantities. In practice, however, these difficulties seldom occur and can be fixed by introducing the convergence factor $e^{i\omega_n \eta}$.[36]

---

[35] See Appendix 1.F.

[36] There are essentially two instances where the convergence factor $e^{i\omega_n \eta}$ is necessary: when calculating the thermodynamic potential and when computing boson/fermion loops in perturbation theory (Sec. 1.5).

1.5 Perturbation Theory and Feynman Diagrams[37]

Interacting systems are described by non-Gaussian actions so that the functional integral cannot in general be calculated exactly. When the interactions between particles are "weak" (in a sense that is to be explained), they can be perturbatively taken into account with respect to the kinetic energy. Perturbation theory in many-body systems is based on Wick's theorem (Sec. 1.5.1) and formulated in terms of Feynman diagrams (Secs. 1.5.2–1.5.4).

1.5.1 *Wick's theorem*

We consider the partition function of a non-interacting system,

$$\begin{aligned} Z_0[J^*,J] &= \int \mathcal{D}[\psi^*,\psi] \exp\Big\{-S_0[\psi^*,\psi] + \int_0^\beta d\tau \sum_\alpha [J^*_\alpha(\tau)\psi_\alpha(\tau) + \text{c.c.}]\Big\}, \\ S_0[\psi^*,\psi] &= -\int_0^\beta d\tau d\tau' \sum_{\alpha\alpha'} \psi^*_\alpha(\tau) G_0^{-1}(\alpha\tau,\alpha'\tau')\psi_{\alpha'}(\tau'). \end{aligned} \tag{1.309}$$

J and J^* are external "sources" that couple linearly to the fields ψ^* and ψ. They are c-numbers for bosons and Grassmann variables for fermions. $Z_0[J^*,J]$ can be calculated exactly using the results (1.E.38) for Gaussian integrals,

$$Z_0[J^*,J] = Z_0 \exp\Big\{-\int_0^\beta d\tau d\tau' \sum_{\alpha,\alpha'} J^*_\alpha(\tau) G_0(\alpha\tau,\alpha'\tau') J_{\alpha'}(\tau')\Big\}, \tag{1.310}$$

where $Z_0 = Z_0[0,0]$ is the non-interacting partition function for vanishing sources.

Let us now consider the expectation value $\langle\psi_{\alpha_1}(\tau_1)\cdots\psi_{\alpha_n}(\tau_n)\psi^*_{\alpha'_n}(\tau'_n)\cdots\psi^*_{\alpha'_1}(\tau'_1)\rangle_0$ where

$$\langle\cdots\rangle_0 = \frac{1}{Z_0}\int \mathcal{D}[\psi^*,\psi]\,(\cdots)e^{-S_0[\psi^*,\psi]}. \tag{1.311}$$

This average can be expressed as the functional derivative

$$\frac{1}{Z_0}\frac{\zeta^n \delta^{2n} Z_0[J^*,J]}{\delta J^*_{\alpha_1}(\tau_1)\cdots\delta J^*_{\alpha_n}(\tau_n)\delta J_{\alpha'_n}(\tau'_n)\cdots\delta J_{\alpha'_1}(\tau'_1)}\Bigg|_{J^*=J=0}. \tag{1.312}$$

This leads to [see (1.E.44) in Appendix 1.E]

$$\begin{aligned} &\langle\psi_{\alpha_1}(\tau_1)\cdots\psi_{\alpha_n}(\tau_n)\psi^*_{\alpha'_n}(\tau'_n)\cdots\psi^*_{\alpha'_1}(\tau'_1)\rangle_0 \\ &\quad = (-1)^n \sum_{P\in S_n} \zeta^P G_0(\alpha_n\tau_n,\alpha'_{P(n)}\tau'_{P(n)})\cdots G_0(\alpha_1\tau_1,\alpha'_{P(1)}\tau'_{P(1)}), \end{aligned} \tag{1.313}$$

where the sum is over all permutations P of $\{1\cdots n\}$. The average $\langle\psi_\alpha(\tau)\psi^*_{\alpha'}(\tau')\rangle_0$ is often called a contraction and denoted by

$$-G_0(\alpha\tau,\alpha'\tau') = \langle\psi_\alpha(\tau)\psi^*_{\alpha'}(\tau')\rangle_0 = \overbracket{\psi_\alpha(\tau)\psi^*_{\alpha'}}(\tau'). \tag{1.314}$$

[37]In this section, we derive the rules of perturbation theory using the functional integral formalism. The alternative derivation, based on the operator formalism, is briefly reviewed in Appendix 1.G.

Wick's theorem states that the expectation value of a product of fields with the Gaussian action S_0 is given by the sum of all complete contractions [Eq. (1.313)],

$$\langle \psi_{\alpha_1}(\tau_1)\cdots\psi_{\alpha_n}(\tau_n)\psi^*_{\alpha'_n}(\tau'_n)\cdots\psi^*_{\alpha'_1}(\tau'_1)\rangle_0 = \sum \text{all complete contractions}. \tag{1.315}$$

A complete contraction is a configuration in which each ψ is contracted with a ψ^* and the overall sign is specified by ζ^P where P is the permutation such that ψ_{α_i} is contracted with $\psi^*_{\alpha_{P(i)}}$. One can generalize (1.315) by allowing for contractions $\overbracket{\psi\psi}$ and $\overbracket{\psi^*\psi^*}$ since the latter vanish,

$$\begin{aligned}
\overbracket{\psi_\alpha(\tau)\psi_{\alpha'}(\tau')} &= \langle\psi_\alpha(\tau)\psi_{\alpha'}(\tau')\rangle_0 = \frac{1}{Z_0}\frac{\delta^2 Z_0[J^*,J]}{\delta J^*_\alpha(\tau)J^*_{\alpha'}(\tau')}\bigg|_{J^*=J=0} = 0,\\
\overbracket{\psi^*_\alpha(\tau)\psi^*_{\alpha'}(\tau')} &= \langle\psi^*_\alpha(\tau)\psi^*_{\alpha'}(\tau')\rangle_0 = \frac{1}{Z_0}\frac{\delta^2 Z_0[J^*,J]}{\delta J_\alpha(\tau)J_{\alpha'}(\tau')}\bigg|_{J^*=J=0} = 0.
\end{aligned} \tag{1.316}$$

The expectation value of a product of fields is non-zero only if there is the same number of fields ψ and ψ^*.

An immediate consequence of Wick's theorem is that once we know the one-particle Green function G_0, we also know the higher-order Green functions in a non-interacting system. For instance, the two-particle Green function reads[38]

$$\begin{aligned}
G_0^{(4)}(\alpha_1\tau_1,\alpha_2\tau_2;\alpha'_2\tau'_2,\alpha'_1\tau'_1) &= \langle\psi_{\alpha_1}(\tau_1)\psi_{\alpha_2}(\tau_2)\psi^*_{\alpha'_2}(\tau'_2)\psi^*_{\alpha'_1}(\tau'_1)\rangle_0\\
&= \psi_{\alpha_1}(\tau_1)\psi_{\alpha_2}(\tau_2)\psi^*_{\alpha'_2}(\tau'_2)\psi^*_{\alpha'_1}(\tau'_1) + \psi_{\alpha_1}(\tau_1)\psi_{\alpha_2}(\tau_2)\psi^*_{\alpha'_2}(\tau'_2)\psi^*_{\alpha'_1}(\tau'_1)\\
&= G_0(\alpha_1\tau_1,\alpha'_1\tau'_1)G_0(\alpha_2\tau_2,\alpha'_2\tau'_2) + \zeta G_0(\alpha_1\tau_1,\alpha'_2\tau'_2)G_0(\alpha_2\tau_2,\alpha'_1\tau'_1).
\end{aligned} \tag{1.317}$$

The two particles created in the states α_1 and α_2 at times τ_1 and τ_2 propagate independently. The second term in the rhs of (1.317) comes from the indiscernability of the particles with the factor ζ reflecting the bosonic or fermionic statistics. It is convenient to represent the propagator G_0 by a directed line,

$$G_0(\alpha\tau,\alpha'\tau') =$$ 

which starts at (α',τ') (creation of a particle in the one-body state α' at time τ') and ends at (α,τ) (annihilation of a particle in the one-body state α at time τ). The two-particle Green function can be represented by the following diagram:

$$G_0^{(4)}(\alpha_1\tau_1,\alpha_2\tau_2;\alpha'_2\tau'_2,\alpha'_1\tau'_1) =$$ 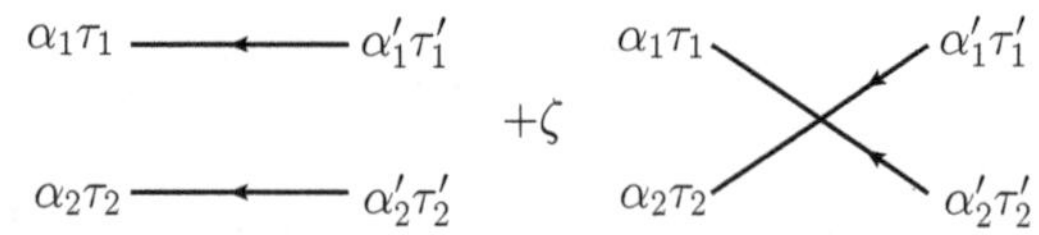

Such a diagram is called a Feynman diagram (or graph). As we shall see in the following sections, Feynman diagrams play an essential role in the perturbation theory of many-body systems.

[38] Recall that the superscript (4) denotes the number of fields appearing in the definition of $G^{(4)}$. It also corresponds to the number of external "legs" in the Feynman diagrams representing $G^{(4)}$.

1.5.2 Partition function and thermodynamic potential

We consider a many-body system with a two-body interaction,

$$Z = \int \mathcal{D}[\psi^*, \psi]\, e^{-S_0[\psi^*,\psi]-S_{\text{int}}[\psi^*,\psi]},$$

$$S_{\text{int}}[\psi^*, \psi] = \frac{1}{2}\int_0^\beta d\tau \sum_{\alpha_1,\alpha_2,\alpha_1',\alpha_2'} (\alpha_1\alpha_2|\hat{v}|\alpha_1'\alpha_2')\psi^*_{\alpha_1}(\tau)\psi^*_{\alpha_2}(\tau)\psi_{\alpha_2'}(\tau)\psi_{\alpha_1'}(\tau). \tag{1.318}$$

The perturbation expansion of the partition function Z is obtained by expanding $e^{-S_{\text{int}}}$ in powers of the two-body interaction v,

$$\frac{Z}{Z_0} = \left\langle e^{-S_{\text{int}}}\right\rangle_0 = \sum_{n=0}^\infty \frac{(-1)^n}{n!}\langle S^n_{\text{int}}\rangle_0 \tag{1.319}$$

where

$$\frac{(-1)^n}{n!}\langle S^n_{\text{int}}\rangle_0 = \frac{(-1)^n}{2^n n!}\sum_{\alpha_1,\beta_1,\gamma_1,\delta_1}\cdots\sum_{\alpha_n,\beta_n,\gamma_n,\delta_n}(\alpha_1\beta_1|\hat{v}|\gamma_1\delta_1)\cdots(\alpha_n\beta_n|\hat{v}|\gamma_n\delta_n)$$
$$\times\int_0^\beta d\tau_1\cdots d\tau_n\langle\psi^*_{\alpha_1}(\tau_1)\psi^*_{\beta_1}(\tau_1)\psi_{\delta_1}(\tau_1)\psi_{\gamma_1}(\tau_1)\cdots$$
$$\cdots\psi^*_{\alpha_n}(\tau_n)\psi^*_{\beta_n}(\tau_n)\psi_{\delta_n}(\tau_n)\psi_{\gamma_n}(\tau_n)\rangle_0. \tag{1.320}$$

The averages $\langle\cdots\rangle_0$ in (1.320) can be evaluated using Wick's theorem. For the $\mathcal{O}(v)$ term, one obtains

$$-\frac{1}{2}\sum_{\alpha_1\cdots\alpha_2'}(\alpha_1\alpha_2|\hat{v}|\alpha_1'\alpha_2')\int_0^\beta d\tau\langle\psi^*_{\alpha_1}(\tau)\psi^*_{\alpha_2}(\tau)\psi_{\alpha_2'}(\tau)\psi_{\alpha_1'}(\tau)\rangle_0$$
$$= -\frac{1}{2}\sum_{\alpha_1\cdots\alpha_2'}(\alpha_1\alpha_2|\hat{v}|\alpha_1'\alpha_2')\int_0^\beta d\tau[G_0(\alpha_1'\tau,\alpha_1\tau^+)G_0(\alpha_2'\tau,\alpha_2\tau^+)$$
$$+\zeta G_0(\alpha_1'\tau,\alpha_2\tau^+)G_0(\alpha_2'\tau,\alpha_1\tau^+)], \tag{1.321}$$

where the equal-time propagator $G_0(\alpha\tau,\alpha'\tau)$ is interpreted as $G_0(\alpha\tau,\alpha'\tau^+)$ in agreement with the proper time ordering in the interaction term [see the discussion after (1.273)]. Equation (1.321) can be represented by the two Feynman diagrams

$$-\frac{1}{2}\sum_{\alpha_1\cdots\alpha_2'}(\alpha_1\alpha_2|\hat{v}|\alpha_1'\alpha_2')\int_0^\beta d\tau\, G_0(\alpha_1'\tau,\alpha_1\tau^+)G_0(\alpha_2'\tau,\alpha_2\tau^+) =$$

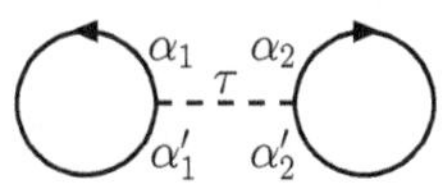

$$-\frac{\zeta}{2}\sum_{\alpha_1\cdots\alpha_2'}(\alpha_1\alpha_2|\hat{v}|\alpha_1'\alpha_2')\int_0^\beta d\tau\, G_0(\alpha_1'\tau,\alpha_2\tau^+)G_0(\alpha_2'\tau,\alpha_1\tau^+) =$$

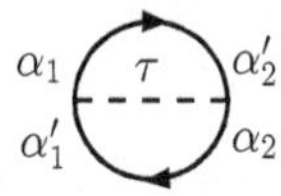

where each propagator is represented as a directed line (see preceding section) and each interaction as a vertex with two incoming and two outgoing lines corresponding to $\psi_{\alpha'_1}\psi_{\alpha'_2}$ and $\psi^*_{\alpha_1}\psi^*_{\alpha_2}$, respectively,

$(\alpha_1\alpha_2|\hat{v}|\alpha'_1\alpha'_2) =$ 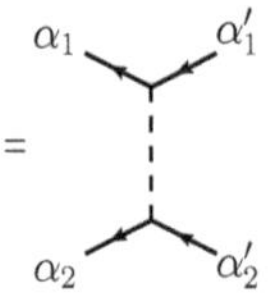

Note that unless otherwise specified, we do not write explicitly the sign and prefactor when drawing a Feynman diagram.

All terms appearing in the perturbation expansion of the partition function can be represented by Feynman diagrams. These diagrams are often referred to as vacuum fluctuation graphs since in quantum field theory they correspond to the "vacuum to vacuum transition" amplitude $\langle \text{vac}|e^{-i\hat{H}t}|\text{vac}\rangle$ (Sec. 1.8). The $\mathcal{O}(v^2)$ term

$$\frac{1}{2!2^2}\sum_{\alpha_1\cdots\delta_2}(\alpha_1\beta_1|\hat{v}|\gamma_1\delta_1)(\alpha_2\beta_2|\hat{v}|\gamma_2\delta_2)$$
$$\times\int_0^\beta d\tau_1 d\tau_2\langle\psi^*_{\alpha_1}(\tau_1)\cdots\psi_{\gamma_1}(\tau_1)\psi^*_{\alpha_2}(\tau_2)\cdots\psi_{\gamma_2}(\tau_2)\rangle_0 \tag{1.322}$$

yields the diagrams shown in Fig. 1.2. What is interesting here is that the $4! = 24$ contractions arising from (1.322) are represented by only 8 diagrams. The number of diagrams even reduces to 5 for the thermodynamic potential (see the following). Rather than starting from (1.320) and considering all possible contractions, it is therefore easier to draw all diagrams to a given order in v and compute each of these diagrams. Since we already know that to each line is associated a propagator G_0 and to each vertex a matrix element $(\alpha_1\alpha_2|\hat{v}|\alpha'_1\alpha'_2)$, our remaining task is to determine the combinatorial factor and sign of a diagram.

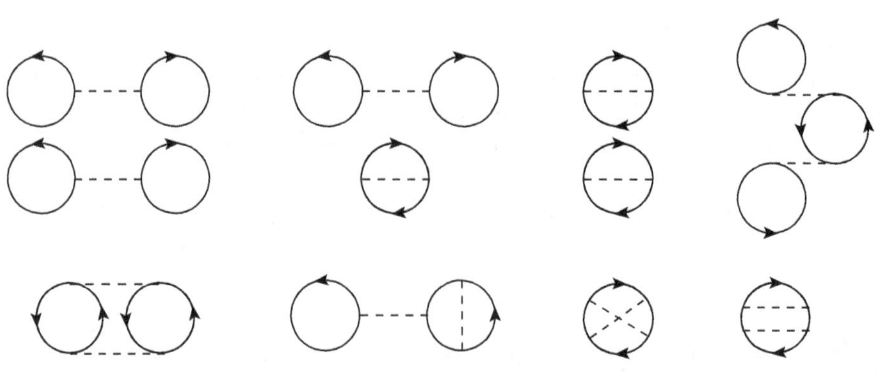

Fig. 1.2 Diagrams contributing to Z/Z_0 to second order. The first three diagrams are made of two disconnected pieces.

Let us start with the first of the first-order diagram shown after (1.321). Schematically, this diagram is obtained from

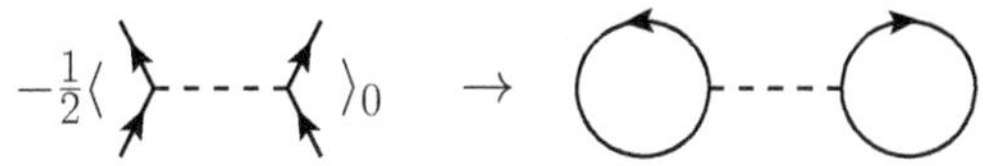

There is only one contraction that leads to the diagram so that the overall factor is 1/2. Consider now the following second-order diagram:

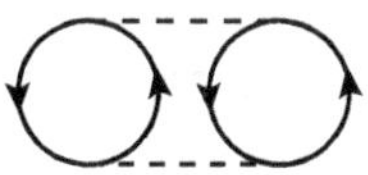

It is obtained from

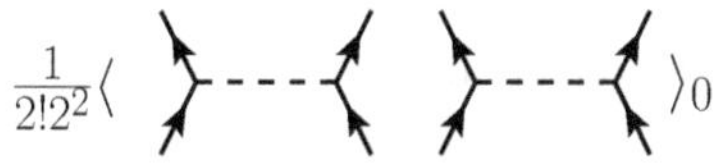

where the factor $1/2!2^2$ comes from the expansion of $e^{-S_{\text{int}}}$ to second order. There are two contractions consistent with the topology of the diagram. To see this, take one of the outgoing lines of one of the two vertices. It must be contracted with one of the incoming line of the other vertex (hence the 2 contractions). Once this is done, there is no choice in the remaining contractions to obtain the diagram so that the combinatorial factor is[39]

$$\frac{1}{2!2^2} \times 2 = \frac{1}{4}. \tag{1.323}$$

To obtain the sign of the diagram, it suffices to consider a particular contraction leading to the diagram, e.g.

$$\begin{aligned}
&\frac{1}{2!2^2} \sum_{\alpha_1 \cdots \delta_2} (\alpha_1\beta_1|\hat{v}|\gamma_1\delta_1)(\alpha_2\beta_2|\hat{v}|\gamma_2\delta_2) \int_0^\beta d\tau_1 d\tau_2 \\
&\quad \times \psi^*_{\alpha_1}(\tau_1)\psi^*_{\beta_1}(\tau_1)\psi_{\delta_1}(\tau_1)\psi_{\gamma_1}(\tau_1)\psi^*_{\alpha_2}(\tau_2)\psi^*_{\beta_2}(\tau_2)\psi_{\delta_2}(\tau_2)\psi_{\gamma_2}(\tau_2) \\
&= \frac{1}{2!2^2} \sum_{\alpha_1 \cdots \delta_2} (\alpha_1\beta_1|\hat{v}|\gamma_1\delta_1)(\alpha_2\beta_2|\hat{v}|\gamma_2\delta_2) \int_0^\beta d\tau_1 d\tau_2 \\
&\quad \times \psi_{\gamma_2}(\tau_2)\psi^*_{\alpha_1}(\tau_1)\psi_{\delta_2}(\tau_2)\psi^*_{\beta_1}(\tau_1)\psi_{\delta_1}(\tau_1)\psi^*_{\beta_2}(\tau_2)\psi_{\gamma_1}(\tau_1)\psi^*_{\alpha_2}(\tau_2),
\end{aligned} \tag{1.324}$$

where each contraction $\psi\psi^*$ can be replaced by $-G_0$ using (1.314). The sign of the diagram is therefore positive.

[39] The combinatorial factor can also be interpreted as $1/S$, where S is the symmetry factor of the graph. S is defined as the number of deformations (obtained by permuting the time labels or exchanging the legs $(\alpha_1\alpha'_1) \leftrightarrow (\alpha_2\alpha'_2)$ in a vertex $(\alpha_1\alpha_2|\hat{v}|\alpha'_2\alpha'_1)$) that leaves the diagram unchanged. See Ref. [9] for a detailed discussion.

It turns out that the sign of a diagram is determined by the number of closed loops of one-particle propagators. Consider a general diagram with a closed loop coming from the contraction

$$(\cdots)\psi^*\psi^*\psi\psi(\cdots)\psi^*\psi^*\psi\psi(\cdots)\psi^*\psi^*\psi\psi(\cdots), \tag{1.325}$$

where $(\cdots)$ denotes vertices not involved in the closed loop. Each $(\cdots)$ contains an even number of fields and therefore commutes with the fields explicitly written in (1.325). We therefore have to consider

$$\psi^*\psi^*\psi\psi\psi^*\psi^*\psi\psi\psi^*\psi^*\psi\psi. \tag{1.326}$$

The fields not included in the cycle may be shifted to the left or to the right without changing the sign so that the cycle of contractions has the form

$$\psi^*\psi\psi^*\psi\psi^*\psi = \zeta\psi\psi^*\psi\psi^*\psi\psi^* = \zeta(-1)^3 G_0 G_0 G_0. \tag{1.327}$$

Thus, the contribution of a cycle gives a sign $\zeta(-1)^{n_p}$, where n_p is the number of propagators in the cycle. Since the total number of propagators in a diagram is always even, a diagram with n interaction lines and n_L closed loops will have the overall sign $(-1)^n\zeta^{n_L}$ where the factor $(-1)^n$ simply comes from the expansion of $e^{-S_{\text{int}}}$ to nth order.

The diagrammatic rules for calculating the nth-order contribution to Z/Z_0 are therefore the following:

1. Draw all distinct diagrams with n vertices connected by directed lines.
2. To each directed line, associate $G_0(\alpha\tau,\alpha'\tau')$. An equal-time propagator, with both ends connected to the same vertex, must be interpreted as $G_0(\alpha\tau,\alpha'\tau^+)$.
3. To each vertex, associate $(\alpha_1\alpha_2|\hat{v}|\alpha_1'\alpha_2')$.
4. Sum over all indices α and integrate all times $\tau \in [0,\beta]$.
5. Multiply the result by the combinatorial factor times $(-1)^n\zeta^{n_L}$, where n_L is the number of closed loops.

It is possible to write the diagrams directly in terms of the propagators $G_0(\alpha,\alpha';i\omega_n)$. In frequency space, the interaction action (1.318) becomes

$$S_{\text{int}} = \frac{1}{2\beta}\sum_{\substack{\alpha_1,\alpha_2,\alpha_1',\alpha_2'\\ \omega_n,\omega_{n'},\omega_\nu}} (\alpha_1\alpha_2|\hat{v}|\alpha_1'\alpha_2')\psi^*_{\alpha_1}(i\omega_n+i\omega_\nu)\psi^*_{\alpha_2}(i\omega_{n'}-i\omega_\nu)\psi_{\alpha_2'}(i\omega_{n'})\psi_{\alpha_1'}(i\omega_n). \tag{1.328}$$

The sum of the frequencies associated with the propagators entering the vertex is conserved in the interaction process (frequency conservation). The preceding diagrammatic rules must therefore be modified as

2′. To each directed line, associate $G_0(\alpha,\alpha';i\omega_n)$ while satisfying frequency conservation at each vertex. For a propagator with both ends connected to the same vertex (loop), multiply the propagator by $e^{i\omega_n\eta}$.

4′. Sum over all indices α and Matsubara frequencies ω_n.
6′. Multiply the result by $\frac{1}{\beta^n}$, where n is the number of vertices.

In translation-invariant systems, it is often convenient to use the momentum basis $\{|\mathbf{k},\sigma\rangle\}$ where σ denotes the spin of the particle (as well as other internal indices if any). For particles interacting *via* a two-body interaction $v(\mathbf{r}-\mathbf{r}')$, one has

$$S_{\rm int} = \frac{1}{2\beta\mathcal{V}} \sum_{\substack{k,k',q \\ \sigma,\sigma'}} v(\mathbf{q})\psi^*_\sigma(k+q)\psi^*_{\sigma'}(k'-q)\psi_{\sigma'}(k')\psi_\sigma(k), \tag{1.329}$$

where $v(\mathbf{q})$ is the Fourier transform of $v(\mathbf{r})$ (see Sec. 1.2.2) and $k=(\mathbf{k},i\omega_n)$, $k'=(\mathbf{k}',i\omega_{n'})$, $q=(\mathbf{q},i\omega_\nu)$; ω_n and $\omega_{n'}$ are bosonic or fermionic Matsubara frequencies in accordance with the statistic of the particles but ω_ν is always a bosonic frequency. The interaction vertex has the diagrammatic representation

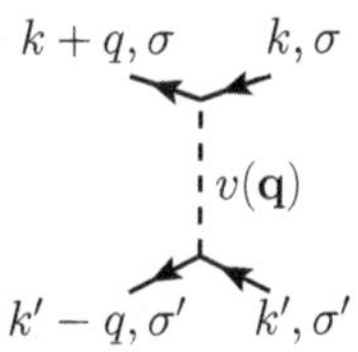

The modifications of the previous diagrammatic rules are straightforward:

2″. To each directed line, associate $G_{0,\sigma}(\mathbf{k},i\omega_n)$ while satisfying momentum, frequency and spin conservation at each vertex. For a propagator with both ends connected to the same vertex (loop), multiply the propagator by $e^{i\omega_n\eta}$.
3″. To each vertex, associate $v(\mathbf{q})$, where $\mathbf{q}$ is the transferred momentum in the interaction process.
4″. Sum over all momenta and Matsubara frequencies.
6″. Multiply the result by $\frac{1}{(\beta\mathcal{V})^n}$, where n is the number of vertices.

With these rules, the first-order correction to the partition function Z/Z_0 reads

$$\begin{aligned}
&-\frac{1}{2\beta\mathcal{V}} v(\mathbf{q}=0) \sum_{k,k',\sigma,\sigma'} G_{0,\sigma}(k)G_{0,\sigma'}(k')e^{i(\omega_n+\omega'_n)\eta} \\
&-\frac{\zeta}{2\beta\mathcal{V}} \sum_{k,k',\sigma} v(\mathbf{k}-\mathbf{k}')G_{0,\sigma}(k)G_{0,\sigma}(k')e^{i(\omega_n+\omega'_n)\eta}
\end{aligned} \tag{1.330}$$

and agrees with the previous result (1.321).

In practice, computing Feynman diagrams beyond the first or second order becomes rapidly cumbersome. However, as will become clear in the following chapters, the diagrammatic representation of the perturbation expansion often brings a physical interpretation which in turn suggests a resummation (to infinite order) of a subclass of diagrams.

One of the most familiar examples is the random-phase approximation (RPA) which amounts to summing the bubble (or ring) diagrams[40]

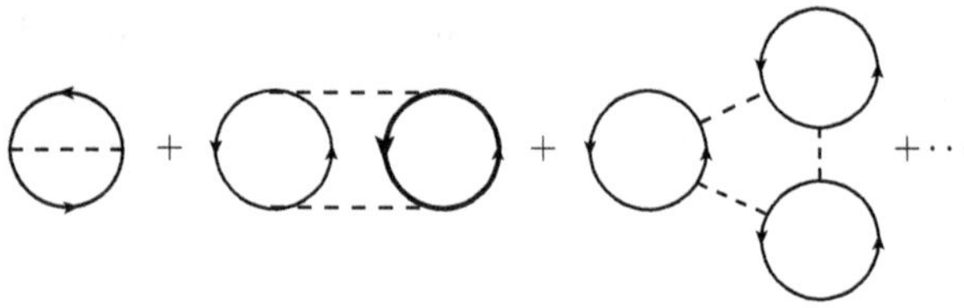

Using the diagrammatic rules, we obtain

$$\begin{aligned}\left(\frac{Z}{Z_0}\right)_{\text{bubble graphs}} &= -\frac{\zeta}{2}\mathrm{Tr}(vG_0G_0) + \frac{1}{4}\mathrm{Tr}(vG_0G_0)^2 - \frac{\zeta}{6}\mathrm{Tr}(vG_0G_0)^3 + \cdots \\ &= -\frac{1}{2}\mathrm{Tr}\ln(1+\zeta vG_0G_0), \end{aligned} \tag{1.331}$$

where Tr stands for summation over spacetime and internal indices. For a translation-invariant system, this gives

$$\left(\frac{Z}{Z_0}\right)_{\text{bubble graphs}} = -\frac{1}{2}\sum_q \ln\left[1-\zeta v(\mathbf{q})\Pi_0(q)\right], \tag{1.332}$$

where

$$\Pi_0(q) = -\frac{1}{\beta\mathcal{V}}\sum_{k,\sigma} G_{0,\sigma}(k)G_{0,\sigma}(k+q). \tag{1.333}$$

Since $\Pi_0(\mathbf{q},\tau) \propto \sum_{\mathbf{k}} G_{0,\sigma}(\mathbf{k},\tau)G_{0,\sigma}(\mathbf{k}+\mathbf{q},-\tau)$, it describes the propagation of a particle and a hole (see Sec. 1.2.3) and is therefore called a particle–hole pair propagator.

1.5.2.1 *Linked cluster theorem and thermodynamic potential*

To obtain the thermodynamics, we need to compute $\ln Z$ rather than Z. The linked cluster theorem states that $\ln(Z/Z_0)$ is given by the sum of the connected graphs contributing to Z/Z_0. A graph is said to be connected if it consists of only one piece.

To derive this theorem, we use

$$\ln\left(\frac{Z}{Z_0}\right) = \lim_{R\to 0}\frac{d}{dR}\exp\left[R\ln\left(\frac{Z}{Z_0}\right)\right] = \lim_{R\to 0}\frac{d}{dR}\left(\frac{Z}{Z_0}\right)^R. \tag{1.334}$$

To compute $(Z/Z_0)^R$, we consider R replicas of the system and introduce a set of fields $\{\psi^*_{r,\alpha},\psi_{r,\alpha}\}$ for the rth copy ($r\in[1,R]$),

$$\left(\frac{Z}{Z_0}\right)^R = \frac{1}{Z_0^R}\int\prod_{r=1}^{R}\mathcal{D}[\psi^*_r,\psi_r]\exp\left\{-\sum_{r=1}^{R}S[\psi^*_r,\psi_r]\right\}. \tag{1.335}$$

[40]Depending on the context, the RPA sums up bubble or ladder diagrams.

The rules for calculating $(Z/Z_0)^R$ are the same as before except that each propagator $G_{0,r}(\alpha\tau, \alpha'\tau')$ now carries a replica index r. Since all fields attached to a given vertex share the same replica index, all propagators joining at this vertex must also share the same replica index. It follows that each connected part of a diagram carries a single index r, which yields a factor R when summed from 1 to R. A graph with n_c connected parts is therefore proportional to R^{n_c} and the graphs proportional to R are fully connected. From (1.319) and (1.334), we then deduce

$$\ln\left(\frac{Z}{Z_0}\right) = \sum_{n=0}^{\infty} \frac{(-1)^n}{n!} \langle S_{\text{int}}^n \rangle_{0,c} = \sum \text{connected graphs} \tag{1.336}$$

(the subscript c stands for connected diagrams) and

$$\Omega = \Omega_0 - \frac{1}{\beta} \sum \text{connected graphs}, \tag{1.337}$$

where $\Omega_0 = -\frac{1}{\beta} \ln Z_0$ is the thermodynamic potential of the non-interacting system.

1.5.3 *Green functions*

Green functions can be calculated perturbatively in a similar manner and the various terms in the expansion can be represented by Feynman diagrams. Let us start with the one-particle Green function

$$G(\alpha_1\tau_1, \alpha_2\tau_2) = -\langle \psi_{\alpha_1}(\tau_1)\psi^*_{\alpha_2}(\tau_2)\rangle = -\frac{Z_0}{Z}\langle \psi_{\alpha_1}(\tau_1)\psi^*_{\alpha_2}(\tau_2) e^{-S_{\text{int}}}\rangle_0. \tag{1.338}$$

To compute G to a given order in the interaction, one must expand both Z and the average $\langle\cdots\rangle_0$ in (1.338). To first order, we use the expression (1.321) for Z/Z_0 and

$$-\langle \psi_{\alpha_1}(\tau_1)\psi^*_{\alpha_2}(\tau_2) e^{-S_{\text{int}}}\rangle_0 = G_0(\alpha_1\tau_1, \alpha_2\tau_2) + \langle \psi_{\alpha_1}(\tau_1)\psi^*_{\alpha_2}(\tau_2) S_{\text{int}}\rangle_0 \tag{1.339}$$

with

$$\begin{aligned}\langle \psi_{\alpha_1}(\tau_1)\psi^*_{\alpha_2}(\tau_2) S_{\text{int}}\rangle_0 = \frac{1}{2} &\sum_{\gamma_1\cdots\gamma_2'} (\gamma_1\gamma_2|\hat{v}|\gamma_1'\gamma_2') \\ &\times \int_0^\beta d\tau \langle \psi_{\alpha_1}(\tau_1)\psi^*_{\alpha_2}(\tau_2)\psi^*_{\gamma_1}(\tau)\psi^*_{\gamma_2}(\tau)\psi_{\gamma_2'}(\tau)\psi_{\gamma_1'}(\tau)\rangle_0. \end{aligned} \tag{1.340}$$

Using Wick's theorem, we obtain for the last term

$$\begin{aligned} -\sum_{\gamma_1\cdots\gamma_2'} (\gamma_1\gamma_2|\hat{v}|\gamma_1'\gamma_2') \int_0^\beta d\tau \Big\{ &\frac{1}{2} G_0(\alpha_1\tau_1, \alpha_2\tau_2)\big[G_0(\gamma_1'\tau, \gamma_1\tau) G_0(\gamma_2'\tau, \gamma_2\tau) \\ &+ \zeta G_0(\gamma_1'\tau, \gamma_2\tau) G_0(\gamma_2'\tau, \gamma_1\tau)\big] + G_0(\alpha_1\tau_1, \gamma_1\tau)\big[\zeta G_0(\gamma_1'\tau, \alpha_2\tau_2) G_0(\gamma_2'\tau, \gamma_2\tau) \\ &+ G_0(\gamma_2'\tau, \alpha_2\tau_2) G_0(\gamma_1'\tau, \gamma_2\tau)\big]\Big\}, \end{aligned} \tag{1.341}$$

which can be represented by the following four diagrams:

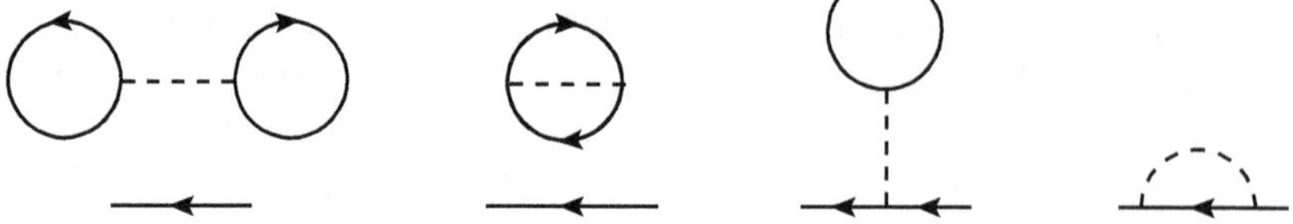

The diagrams for the one-particle Green function have two external lines (or legs) which carry the indices $(\alpha_1\tau_1)$ (outgoing line) and $(\alpha_2\tau_2)$ (incoming line). Only the indices carried by the internal lines are summed over. Combining (1.341) with the $\mathcal{O}(v)$ correction to Z/Z_0, we obtain

$$\begin{aligned} G(\alpha_1\tau_1,\alpha_2\tau_2) &= G_0(\alpha_1\tau_1,\alpha_2\tau_2) - \sum_{\gamma_1\cdots\gamma_2'} (\gamma_1\gamma_2|\hat{v}|\gamma_1'\gamma_2') \int_0^\beta d\tau\, G_0(\alpha_1\tau_1,\gamma_1\tau) \\ &\quad \times \left[\zeta G_0(\gamma_1'\tau,\alpha_2\tau_2)G_0(\gamma_2'\tau,\gamma_2\tau) + G_0(\gamma_2'\tau,\alpha_2\tau_2)G_0(\gamma_1'\tau,\gamma_2\tau)\right] \end{aligned} \quad (1.342)$$

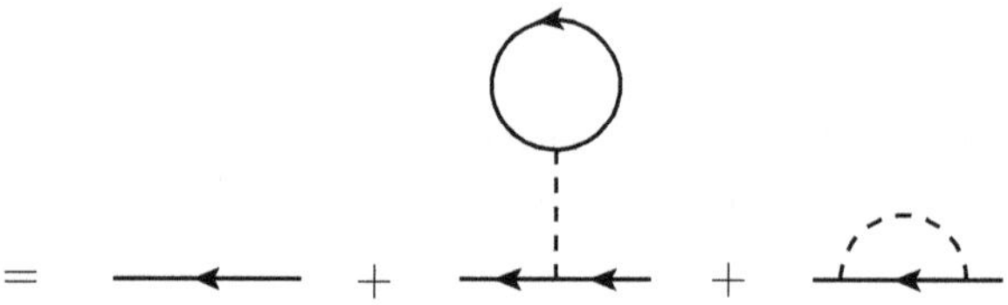

The $\mathcal{O}(v)$ correction to G_0 is given by the connected diagrams that contribute to $-\langle\psi_{\alpha_1}(\tau_1)\psi^*_{\alpha_2}(\tau_2)e^{-S_{\rm int}}\rangle_0$; the vacuum fluctuation graphs factorize out in the numerator and denominator of (1.338).

This is a general result: The Green function is given by the sum of all connected graphs. Let us prove this result for the n-particle Green function

$$G^{(2n)}(\alpha_1\tau_1,\ldots,\alpha_n\tau_n;\alpha_n'\tau_n',\ldots,\alpha_1'\tau_1') = (-1)^n\langle\psi_{\alpha_1}(\tau_1)\cdots\psi^*_{\alpha_1'}(\tau_1')\rangle. \quad (1.343)$$

We introduce R replicas of the system and consider the correlation function

$$\begin{aligned} &G_R^{(2n)}(\alpha_1\tau_1,\ldots,\alpha_n\tau_n;\alpha_n'\tau_n',\ldots,\alpha_1'\tau_1') \\ &\quad = (-1)^n \int \prod_{r=1}^R \mathcal{D}[\psi_r^*,\psi_r]\,\psi_{1,\alpha_1}(\tau_1)\cdots\psi^*_{1,\alpha_1'}(\tau_1')\,e^{-\sum_{r=1}^R S[\psi_r^*,\psi_r]} \\ &\quad = (-1)^n Z^{R-1}\int \mathcal{D}[\psi_1^*,\psi_1]\,\psi_{1,\alpha_1}(\tau_1)\cdots\psi^*_{1,\alpha_1'}(\tau_1')\,e^{-S[\psi_1^*,\psi_1]}, \end{aligned} \quad (1.344)$$

where the last line is obtained by integrating out the fields ψ_r^*,ψ_r with $r\geq 2$. Thus, $G^{(2n)}$ can be formally obtained from $G_R^{(2n)}$ in the limit $R=0$. The perturbation expansion for $G_R^{(2n)}$ is obtained as before by expanding $e^{-S_{\rm int}}$. The Feynman diagrams involve propagators carrying a replica index $r\in[1,R]$. All propagators attached to the same vertex share the same value of the replica index. Those corresponding to the external lines $\alpha_1\cdots\alpha_1'$ have

$r = 1$. Since the index r in the interaction vertex is summed over, a diagram with p disconnected parts, i.e. p parts that are not connected to the external legs, is proportional to R^p and vanishes in the limit $R = 0$ when $p \geq 1$. For the connected diagrams ($p = 0$), the integration of the fields $\psi_r^{(*)}$ with $r \in [2, R]$ in (1.344) gives a factor Z_0^{R-1} so that we finally obtain

$$G^{(2n)}(\alpha_1\tau_1, \ldots, \alpha_n\tau_n; \alpha'_n\tau'_n, \ldots, \alpha'_1\tau'_1) = \frac{(-1)^n}{Z_0} \int \mathcal{D}[\psi^*, \psi]\, \psi_{\alpha_1}(\tau_1) \cdots \psi^*_{\alpha'_1}(\tau'_1)\, e^{-S[\psi^*,\psi]}\Big|_{\text{connected diagrams}} \tag{1.345}$$

where only the connected diagrams, i.e. with all parts connected to the external lines, must be retained.

The combinatorial factor of a diagram can be computed, as for the partition function, by a direct evaluation of the number of contractions leading to that diagram. Let us consider, for instance, the following second-order diagram contributing to the one-particle Green function,

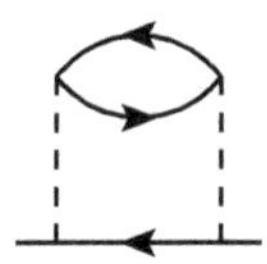

which is obtained from

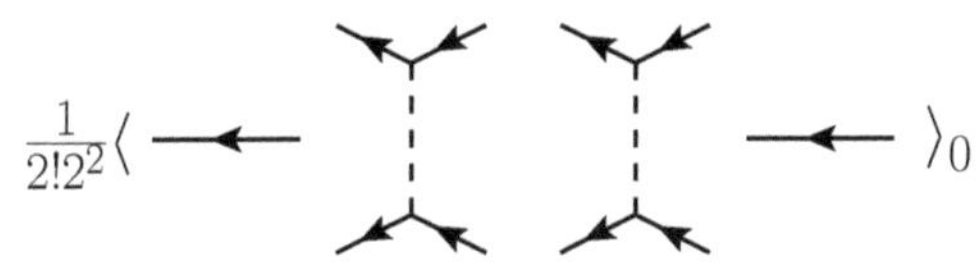

The combinatorial factor reads

$$\frac{1}{2!2^2} \times \underbrace{2!}_{\substack{\text{interchange}\\ \text{of vertices}}} \times \underbrace{2}_{\substack{\text{connection of}\\ \text{outgoing line}\\ \text{to vertex}}} \times \underbrace{2}_{\substack{\text{connection of}\\ \text{incoming line}\\ \text{to vertex}}} = 1. \tag{1.346}$$

It turns out that the combinatorial factor is always unity for the n-particle Green function. One can easily convince oneself by working out a few examples. A proof can be found in Ref. [9].

The sign of a diagram can be obtained from (1.345) and a particular contraction leading to that diagram. For the one-particle Green function, the sign of a diagram with l vertices is simply given by $(-1)^l \zeta^{n_L}$, where n_L is the number of closed loops. (The proof is similar to that given for the partition function.) In the case of fermions, a diagram for the n-particle Green function takes the additional factor ζ^P where ζ^P is the signature of the permutation such that the incoming line $(\alpha'_i\tau'_i)$ joins the outgoing line $(\alpha_{P(i)}\tau_{P(i)})$.[41]

[41] For a simple example, consider the non-interacting two-particle Green function [Eq. (1.317)].

The diagrammatic rules for calculating the lth-order contribution to the n-particle Green function $G^{(2n)}(\alpha_1\tau_1, \ldots, \alpha_n\tau_n; \alpha'_n\tau'_n, \ldots, \alpha'_1\tau'_1)$ are the following:

1. Draw all distinct connected diagrams with n incoming lines $(\alpha'_n\tau'_n \cdots \alpha'_1\tau'_1)$, n outgoing lines $(\alpha_1\tau_1 \cdots \alpha_n\tau_n)$ and l interaction vertices.
2. To each directed line, associate $G_0(\alpha\tau, \alpha'\tau'^+)$.
3. To each vertex, associate $(\alpha\alpha'|\hat{v}|\gamma\gamma')$.
4. Sum over all internal indices α and integrate all internal times $\tau \in [0, \beta]$. The indices $(\alpha'_i\tau'_i)$ and $(\alpha_i\tau_i)$ of incoming and outgoing lines must be held fixed.
5. Multiply the result by the factor $(-1)^l\zeta^{n_L+P}$, where n_L is the number of closed loops and P the permutation such that each incoming line $(\alpha'_i\tau'_i)$ terminates at $(\alpha_{P(i)}\tau_{P(i)})$.

As an example, the second-order diagrams for the one-particle Green function are shown in Fig. 1.3 and the first-order diagrams for the two-particle Green function in Fig. 1.4.

Let us illustrate the various ways to calculate a diagram with the example of a system with the following action:

$$
\begin{aligned}
S_0 &= -\int_0^\beta d\tau d\tau' \int d^d r d^d r' \sum_\sigma \psi_\sigma^*(\mathbf{r},\tau) G_{0,\sigma}^{-1}(\mathbf{r},\tau;\mathbf{r}',\tau')\psi_\sigma(\mathbf{r}',\tau'), \\
S_{\text{int}} &= \frac{1}{2}\int_0^\beta d\tau \int d^d r d^d r' \sum_{\sigma,\sigma'} v(\mathbf{r}-\mathbf{r}')\psi_\sigma^*(\mathbf{r},\tau)\psi_{\sigma'}^*(\mathbf{r}',\tau)\psi_{\sigma'}(\mathbf{r}',\tau)\psi_\sigma(\mathbf{r},\tau).
\end{aligned}
\tag{1.347}
$$

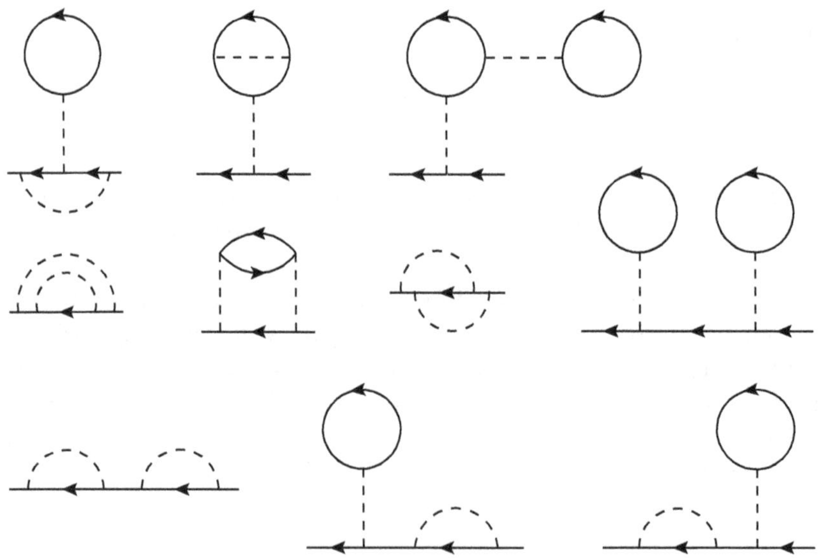

Fig. 1.3 Second-order diagrams for the one-particle Green function $G^{(2)} = G$.

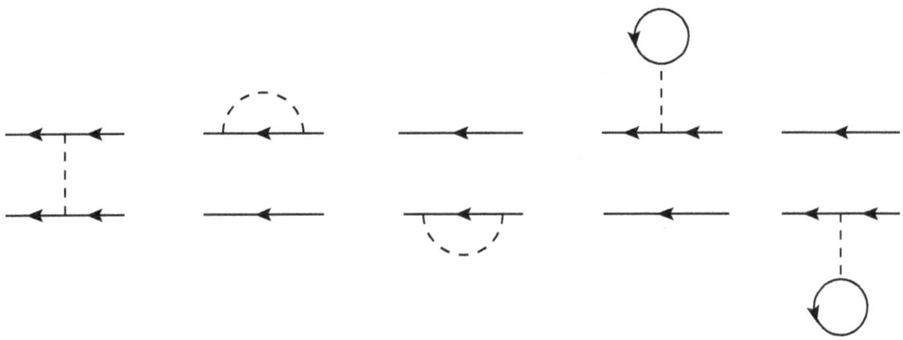

Fig. 1.4 First-order diagrams for the two-particle Green function $G^{(4)}$. Diagrams obtained by exchanging the two incoming or outgoing lines are not shown.

The first-order correction to $G_{0,\sigma}(\mathbf{r},\tau;\mathbf{r}',\tau')$ is obtained from the diagrams

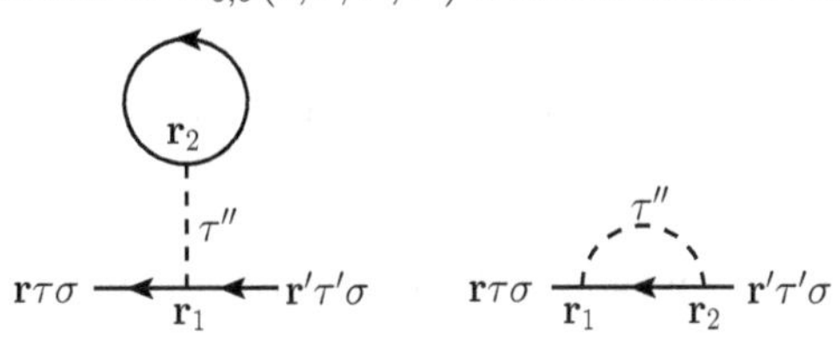

and is given by

$$-\zeta\int_0^\beta d\tau''\int d^dr_1d^dr_2\sum_{\sigma'}v(\mathbf{r}_1-\mathbf{r}_2)G_{0,\sigma}(\mathbf{r},\tau;\mathbf{r}_1,\tau'')G_{0,\sigma}(\mathbf{r}_1,\tau'';\mathbf{r}',\tau')$$
$$\times\, G_{0,\sigma'}(\mathbf{r}_2,\tau'';\mathbf{r}_2,\tau''^+)$$
$$-\int_0^\beta d\tau''\int d^dr_1d^dr_2\, v(\mathbf{r}_1-\mathbf{r}_2)G_{0,\sigma}(\mathbf{r},\tau;\mathbf{r}_1,\tau'')$$
$$\times\, G_{0,\sigma}(\mathbf{r}_1,\tau'';\mathbf{r}_2,\tau''^+)G_{0,\sigma}(\mathbf{r}_2,\tau'';\mathbf{r}',\tau'). \quad (1.348)$$

When the system is translation invariant, it is convenient to start from

$$S_0=-\sum_{k,\sigma}\psi^*_\sigma(k)G^{-1}_{0,\sigma}(k)\psi_\sigma(k),$$
$$S_{\rm int}=\frac{1}{2\beta\mathcal{V}}\sum_{\substack{k,k',q\\ \sigma,\sigma'}}v(\mathbf{q})\psi^*_\sigma(k+q)\psi^*_{\sigma'}(k'-q)\psi_{\sigma'}(k')\psi_\sigma(k). \quad (1.349)$$

The first-order correction to $G_{0,\sigma}(k)$ is represented by the two diagrams

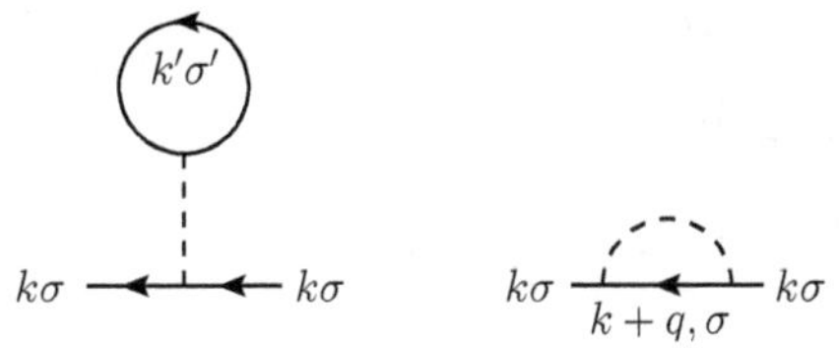

and reads

$$-\frac{\zeta}{\beta\mathcal{V}}\sum_{k',\sigma'} v(0)G_{0,\sigma}(k)^2 G_{0,\sigma'}(k')e^{i\omega'_n\eta}$$
$$-\frac{1}{\beta\mathcal{V}}\sum_{q} v(\mathbf{q})G_{0,\sigma}(k)^2 G_{0,\sigma}(k+q)e^{i\omega_{n+\nu}\eta}. \tag{1.350}$$

This result can also be obtained from the Fourier transform of (1.348).

1.5.4 *(Anti)symmetrized vertices*

Perturbation theory can be formulated in terms of the (anti)symmetrized vertex introduced in Sec. 1.2 [Eq. (1.179)]. One then starts from the action

$$S_{\rm int}[\psi^*,\psi] = \frac{1}{4}\int_0^\beta d\tau \sum_{\alpha_1,\alpha_2,\alpha'_1,\alpha'_2} \{\alpha_1\alpha_2|\hat{v}|\alpha'_1\alpha'_2\}\psi^*_{\alpha_1}(\tau)\psi^*_{\alpha_2}(\tau)\psi_{\alpha'_2}(\tau)\psi_{\alpha'_1}(\tau) \tag{1.351}$$

with a vertex which is (anti)symmetric under the exchange of the two incoming or outgoing particles,

$$\{\alpha_1\alpha_2|\hat{v}|\alpha'_1\alpha'_2\} = \zeta\{\alpha_1\alpha_2|\hat{v}|\alpha'_2\alpha'_1\} = \zeta\{\alpha_2\alpha_1|\hat{v}|\alpha'_1\alpha'_2\}. \tag{1.352}$$

Since this vertex no longer distinguishes between direct and exchange scatterings of the two incoming particles, it should be graphically represented by a dot,

$\{\alpha_1\alpha_2|\hat{v}|\alpha'_1\alpha'_2\} =$ 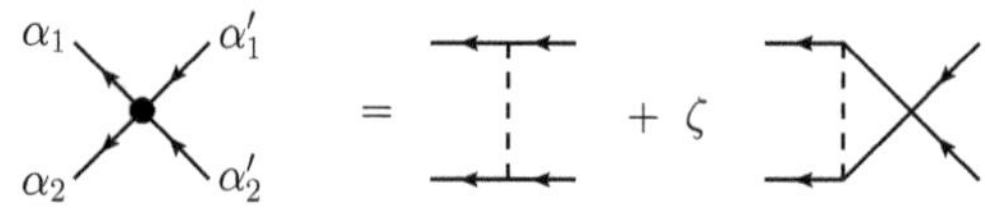

(we write explicitly that the second diagram comes with a sign ζ).

As previously, perturbation theory is obtained by expanding $e^{-S_{\rm int}}$. Using Wick's theorem, we can again express the results by diagrams made of (anti)symmetrized vertices connected by propagators. Because we do not longer distinguish between the two incoming and outgoing lines of a vertex, the number of diagrams is considerably reduced. For instance, the $\mathcal{O}(v)$ correction to Z/Z_0,

$$-\frac{1}{2}\sum_{\alpha_1,\alpha_2,\alpha'_1,\alpha'_2}\{\alpha_1\alpha_2|\hat{v}|\alpha'_1\alpha'_2\}\int_0^\beta d\tau\, G_0(\alpha'_1\tau,\alpha_1\tau^+)G_0(\alpha'_2\tau,\alpha_2\tau^+), \tag{1.353}$$

is represented by a single diagram:

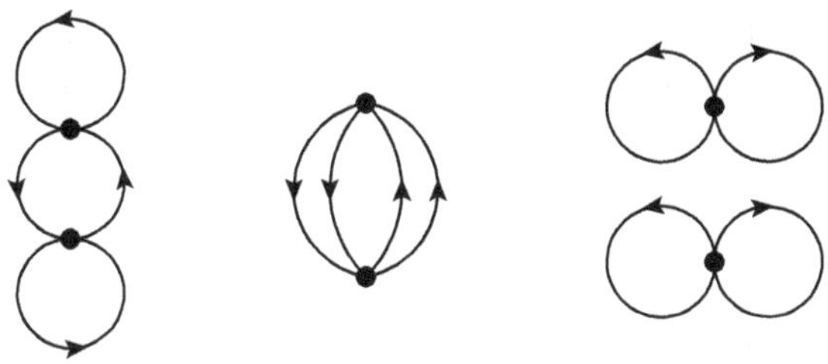

Fig. 1.5 Second-order diagrams contributing to the partition function Z/Z_0.

To second order, there are only three different diagrams (Fig. 1.5) instead of the eight to be considered when working with non-(anti)symmetrized vertices (Fig. 1.2).

The combinatorial factor is determined as before. (Note that the action $S_{\rm int}$ is now defined with a factor $\frac{1}{4}$.) For instance, the second of the three diagrams shown in Fig. 1.5 is obtained from

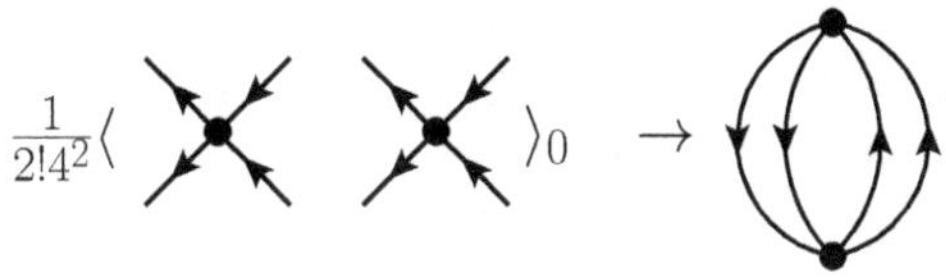

and has a combinatorial factor

$$\frac{1}{2!4^2} \times 4 = \frac{1}{8}. \tag{1.354}$$

The factor 4 in the lhs comes from the four ways to connect the two vertices in a way consistent with the topology of the diagram.

As for the sign of the diagram, it can be determined as before from a particular contraction. One may also use the fact that the diagram should have the same sign as the one obtained *via* the replacement $\{\alpha_1\alpha_2|\hat v|\alpha_1'\alpha_2'\} \rightarrow (\alpha_1\alpha_2|\hat v|\alpha_1'\alpha_2')$. For instance, if we write the preceding diagram as

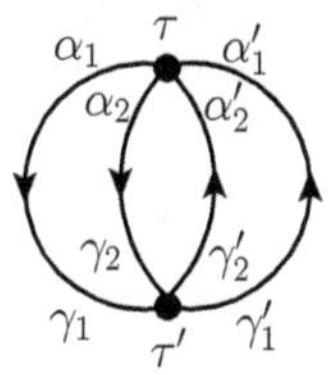

$$= \text{sign} \times \frac{1}{8} \sum_{\alpha_1\cdots\gamma_2'} \{\alpha_1\alpha_2|\hat v|\alpha_1'\alpha_2'\}\{\gamma_1\gamma_2|\hat v|\gamma_2'\gamma_1'\} \int_0^\beta d\tau d\tau' \times G_0(\gamma_1\tau', \alpha_1\tau)G_0(\gamma_2\tau', \alpha_2\tau)G_0(\alpha_2'\tau, \gamma_2'\tau')G_0(\alpha_1'\tau, \gamma_1'\tau'),$$

the replacement $\{\alpha_1\alpha_2|\hat v|\alpha_1'\alpha_2'\} \rightarrow (\alpha_1\alpha_2|\hat v|\alpha_1'\alpha_2')$ gives

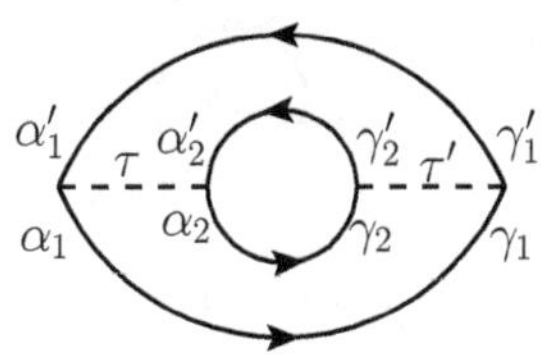

and the sign is therefore $(-1)^2\zeta^2$ since there are two vertices and two closed loops. Note that the sign of the diagram can be unambiguously defined only once we have its expression in terms of the vertices. For instance, we could write $\{\alpha_2\alpha_1|\hat{v}|\alpha_1'\alpha_2'\}$ instead $\{\alpha_1\alpha_2|\hat{v}|\alpha_1'\alpha_2'\}$ in the preceding equation and the diagram would then have a negative sign.

The diagrammatic rules for calculating a diagram are therefore the following:

1. To each directed line, associate $G_0(\alpha\tau, \alpha'\tau'^+)$.
2. To each vertex, associate $\{\alpha_1\alpha_2|\hat{v}|\alpha_1'\alpha_2'\}$.
3. Sum over all one-particle indices α and integrate all times $\tau \in [0, \beta]$.
4. Multiply the diagram by the combinatorial factor and by $(-1)^l\zeta^{n_L}$, where l is the number of vertices and n_L the number of closed loops in the diagram obtained by replacing $\{\alpha_1\alpha_2|\hat{v}|\alpha_1'\alpha_2'\}$ by $(\alpha_1\alpha_2|\hat{v}|\alpha_1'\alpha_2')$ in each vertex.

When considering the perturbative expansion of the n-particle Green function $G^{(2n)}(\alpha_1\tau_1, \ldots, \alpha_1'\tau_1')$, the last two rules must be replaced by

3′. Sum over all indices α and integrate all times τ over the interval $[0, \beta]$, while holding fixed the indices $(\alpha_i\tau_i)$ and $(\alpha_i'\tau_i')$ of the external lines.

4′. Multiply by the combinatorial factor and by $(-1)^l\zeta^{n_L}\zeta^P$, where l is the number of vertices, n_L the number of closed loops in the diagram obtained by replacing all vertices $\{\alpha_1\alpha_2|\hat{v}|\alpha_1'\alpha_2'\}$ by $(\alpha_1\alpha_2|\hat{v}|\alpha_1'\alpha_2')$ and P the permutation of the incoming and outgoing lines.

For example, the first-order contribution to the one-particle Green function $G_0(\alpha\tau, \alpha'\tau')$ is given by the diagram

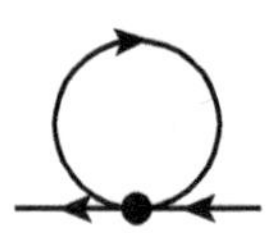

$$= -\zeta \int_0^\beta d\tau'' \sum_{\alpha_1\cdots\alpha_4} \{\alpha_1\alpha_2|\hat{v}|\alpha_4\alpha_3\} G_0(\alpha\tau, \alpha_1\tau'') G_0(\alpha_3\tau'', \alpha_2\tau'') G_0(\alpha_4\tau'', \alpha'\tau'). \quad (1.355)$$

The second-order diagrams are shown in Fig. 1.6.

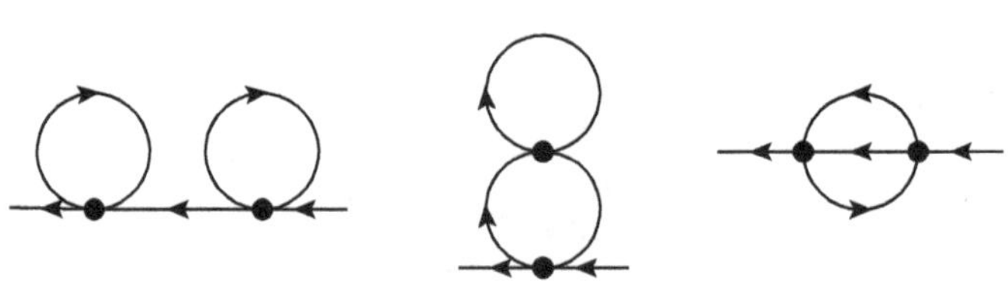

Fig. 1.6 Second-order diagrams with (anti)symmetrized vertices contributing to the one-particle Green function.

1.5.4.1 *Fully (anti)symmetrized vertices*

We define a new field

$$\psi_{\tilde{\alpha}}(\tau) = \begin{cases} \psi_\alpha(\tau) & \text{if } c = -, \\ \psi_\alpha^*(\tau) & \text{if } c = +, \end{cases} \tag{1.356}$$

where $\tilde{\alpha} = (\alpha, c)$ with $c = \pm$ a "charge" index. $\psi_{\alpha+}$ creates a particle whereas $\psi_{\alpha-}$ creates a hole (i.e. removes a particle from the system). The action can be written as

$$\begin{aligned} S_0[\psi] &= \frac{1}{2}\int_0^\beta d\tau d\tau' \sum_{\tilde{\alpha},\tilde{\alpha}'} \psi_{\tilde{\alpha}}(\tau) G_0^{-1}(\tilde{\alpha}\tau, \tilde{\alpha}'\tau')\psi_{\tilde{\alpha}'}(\tau'), \\ S_{\text{int}}[\psi] &= \frac{1}{4!}\int_0^\beta d\tau \sum_{\tilde{\alpha}_1\cdots\tilde{\alpha}_4} v_{\tilde{\alpha}_1\tilde{\alpha}_2\tilde{\alpha}_3\tilde{\alpha}_4} \psi_{\tilde{\alpha}_1}(\tau)\psi_{\tilde{\alpha}_2}(\tau)\psi_{\tilde{\alpha}_3}(\tau)\psi_{\tilde{\alpha}_4}(\tau), \end{aligned} \tag{1.357}$$

where the (anti)symmetric propagator G_0 is defined by

$$G_0(\tilde{\alpha}\tau, \tilde{\alpha}'\tau') = \zeta G_0(\tilde{\alpha}'\tau', \tilde{\alpha}\tau) = \begin{cases} -G_0(\alpha\tau, \alpha'\tau') & \text{if } c = -c' = -, \\ 0 & \text{if } c = c'. \end{cases} \tag{1.358}$$

Note that in (1.357) and (1.358), we use opposite sign conventions for $G_0(\tilde{\alpha}\tau, \tilde{\alpha}'\tau')$ and $G_0(\alpha\tau, \alpha'\tau')$. This choice turns out to be more convenient when dealing with the field $\psi_{\tilde{\alpha}}$ (in particular for the formal manipulations in Sec. 1.6). The vertex $v_{\tilde{\alpha}_1\tilde{\alpha}_2\tilde{\alpha}_3\tilde{\alpha}_4}$ is (anti)symmetric under the exchange of any two of its arguments,

$$v_{\tilde{\alpha}_1\tilde{\alpha}_2\tilde{\alpha}_3\tilde{\alpha}_4} = \zeta v_{\tilde{\alpha}_2\tilde{\alpha}_1\tilde{\alpha}_3\tilde{\alpha}_4} = \zeta v_{\tilde{\alpha}_1\tilde{\alpha}_2\tilde{\alpha}_4\tilde{\alpha}_3} = \zeta v_{\tilde{\alpha}_3\tilde{\alpha}_2\tilde{\alpha}_1\tilde{\alpha}_4}, \tag{1.359}$$

and satisfies

$$v_{\tilde{\alpha}_1\tilde{\alpha}_2\tilde{\alpha}_3\tilde{\alpha}_4} = \begin{cases} \{\tilde{\alpha}_1\tilde{\alpha}_2|\hat{v}|\tilde{\alpha}_4\tilde{\alpha}_3\} & \text{if } c_1 = c_2 = -c_3 = -c_4 = +, \\ 0 & \text{if } c_1 + c_2 + c_3 + c_4 \neq 0. \end{cases} \tag{1.360}$$

Note the position of $\tilde{\alpha}_3$ and $\tilde{\alpha}_4$ which follows from the definition (1.357) of $v_{\tilde{\alpha}_1\tilde{\alpha}_2\tilde{\alpha}_3\tilde{\alpha}_4}$.

The Wick theorem now reads

$$\langle \psi_{\tilde{\alpha}_1}(\tau_1)\cdots\psi_{\tilde{\alpha}_n}(\tau_n)\rangle = \sum \text{all complete contractions} \tag{1.361}$$

and is a consequence of the results for Gaussian integration over complex or Grassmann variables discussed in Sec. 1.E.4. The various terms appearing in the perturbation expansion of the partition function can be represented by Feynman diagrams. Since we do not distinguish between ψ_α and ψ_α^* (the difference is encoded in the charge index c of $\psi_{\tilde{\alpha}}$), the propagator should be represented by a non-directed line

$$G_0(\tilde{\alpha}\tau, \tilde{\alpha}'\tau') =$$

and the vertex by a simple dot,

$$v_{\tilde{\alpha}_1\tilde{\alpha}_2\tilde{\alpha}_3\tilde{\alpha}_4} =$$ 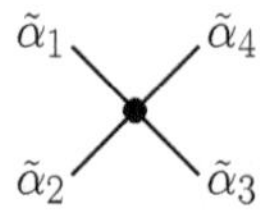

The first-order correction to the partition function reads

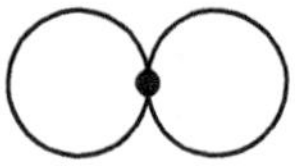
$$= -\frac{1}{8}\int_0^\beta d\tau \sum_{\tilde{\alpha}_1\cdots\tilde{\alpha}_4} v_{\tilde{\alpha}_1\tilde{\alpha}_2\tilde{\alpha}_3\tilde{\alpha}_4} G_0(\tilde{\alpha}_1\tau, \tilde{\alpha}_4\tau) G_0(\tilde{\alpha}_2\tau, \tilde{\alpha}_3\tau), \quad (1.362)$$

where the combinatorial factor comes from the 1/4! in front of the interaction and the three ways to pair the lines of the vertex. The sign is obtained from a particular contraction, e.g.

$$-\frac{1}{4!}\int_0^\beta d\tau \sum_{\tilde{\alpha}_1\cdots\tilde{\alpha}_4} v_{\tilde{\alpha}_1\tilde{\alpha}_2\tilde{\alpha}_3\tilde{\alpha}_4} \psi_{\tilde{\alpha}_1}(\tau)\psi_{\tilde{\alpha}_2}(\tau)\psi_{\tilde{\alpha}_3}(\tau)\psi_{\tilde{\alpha}_4}(\tau). \quad (1.363)$$

Equation (1.362) gives

$$\begin{aligned} &-\frac{1}{8}\int_0^\beta d\tau \sum_{\substack{\alpha_1\cdots\alpha_4 \\ c,c'=\pm}} v_{\alpha_1 c,\alpha_2 c',\alpha_3 \bar{c}',\alpha_4 \bar{c}} G_0(\alpha_1, c, \tau; \alpha_4, \bar{c}, \tau) G_0(\alpha_2, c', \tau; \alpha_3, \bar{c}', \tau) \\ &= -\frac{1}{2}\int_0^\beta d\tau \sum_{\alpha_1\cdots\alpha_4} v_{\alpha_1+,\alpha_2+,\alpha_3-,\alpha_4-} G_0(\alpha_4, -, \tau; \alpha_1, +, \tau^+) G_0(\alpha_3, -, \tau; \alpha_2, +, \tau^+) \end{aligned} \quad (1.364)$$

(with $\bar{c} = -c$) and therefore reproduces (1.353). In (1.364), we have interpreted the equal time propagator $G_0(\alpha\tau, \alpha'\tau)$ as $G_0(\alpha\tau, \alpha'\tau^+)$ for reasons previously discussed. Note that $G_0(\tilde{\alpha}\tau, \tilde{\alpha}'\tau)$ corresponds to $G_0(\tilde{\alpha}\tau, \tilde{\alpha}'\tau^+)$ if $c' = -c = +$ and to $G_0(\tilde{\alpha}\tau^+, \tilde{\alpha}'\tau)$ if $c = -c' = +$.

Similarly, for the first-order correction to the Green function $G(\tilde{\alpha}\tau, \tilde{\alpha}'\tau')$, one obtains

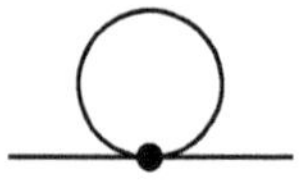

$$= -\frac{1}{2}\int_0^\beta d\tau'' \sum_{\tilde{\alpha}_1\cdots\tilde{\alpha}_4} v_{\tilde{\alpha}_1\tilde{\alpha}_2\tilde{\alpha}_3\tilde{\alpha}_4} G(\tilde{\alpha}\tau, \tilde{\alpha}_1\tau'') G(\tilde{\alpha}_2\tau'', \tilde{\alpha}_3\tau'') G(\tilde{\alpha}_4\tau'', \tilde{\alpha}'\tau'), \quad (1.365)$$

which can be shown to be equivalent to (1.355) when $\tilde{\alpha} = (\alpha, -)$ and $\tilde{\alpha}' = (\alpha', +)$.

One must be careful about the time ordering of the fields when using this formalism. To illustrate this point, let us consider non-interacting particles and express the action

$$S[\Psi^\dagger, \Psi] = \frac{1}{2}\sum_{\omega_n} \Psi^\dagger(i\omega_n)\mathcal{G}^{-1}(i\omega_n)\Psi(i\omega_n) \quad (1.366)$$

in terms of the two-component field

$$\Psi(\tau) = \begin{pmatrix} \psi(\tau) \\ \psi^*(\tau) \end{pmatrix} = \begin{pmatrix} \psi(\tau) \\ \psi^*(\tau) \end{pmatrix}, \qquad \Psi(i\omega_n) = \begin{pmatrix} \psi(i\omega_n) \\ \psi^*(-i\omega_n) \end{pmatrix}. \quad (1.367)$$

For simplicity, we consider a single one-body state with energy $\epsilon = \xi + \mu$. In frequency space, the propagator $\mathcal{G}$ is a 2×2 matrix defined by

$$\mathcal{G}^{-1}(i\omega_n) = \begin{pmatrix} -i\omega_n + \xi & 0 \\ 0 & \zeta(i\omega_n + \xi) \end{pmatrix}. \tag{1.368}$$

The functional integral gives

$$Z_0 = \int \mathcal{D}[\psi^*, \psi]\, e^{-S[\Psi^\dagger, \Psi]} = \prod_{\omega_n} [\mathcal{G}_{11}(i\omega_n)\mathcal{G}_{22}(i\omega_n)]^{\zeta/2} \tag{1.369}$$

and

$$\Omega_0 = -\frac{\zeta}{2\beta} \sum_{\omega_n} \ln[\mathcal{G}_{11}(i\omega_n)\mathcal{G}_{22}(i\omega_n)]. \tag{1.370}$$

If now one introduces the convergence factor $e^{i\omega_n \eta}$, one obtains (see Appendix 1.F)[42]

$$\frac{\zeta}{2\beta} \sum_{\omega_n} \ln(\omega_n^2 + \xi^2) e^{i\omega_n \eta} = \frac{\zeta}{\beta} \ln\left(1 - \zeta e^{-\beta\xi}\right) + \zeta \frac{\xi}{2}, \tag{1.371}$$

which differs from the exact result by $\zeta\xi/2$.

The reason for this apparent difficulty is that if we permute creation and annihilation fields, which is what we have done when writing

$$\begin{aligned} \psi^*(\tau) G_0(\tau - \tau')\psi(\tau') &= \zeta \psi(\tau') G_0(-\tau' + \tau)\psi^*(\tau) \\ &\equiv \Psi_2^\dagger(\tau')\mathcal{G}_{22}(\tau' - \tau)\Psi_2(\tau), \end{aligned} \tag{1.372}$$

one must then interpret the equal time propagator $\mathcal{G}_{22}(0)$ as $\mathcal{G}_{22}(0^+)$ rather than $\mathcal{G}_{22}(0^-)$. Accordingly, we must use the convergence factor $e^{-i\omega_n \eta}$ for the (22) component of the propagator. Thus, the correct expression of the thermodynamic potential is

$$\begin{aligned} \Omega_0 &= -\frac{\zeta}{2\beta} \sum_{\omega_n} \left\{ \ln[\mathcal{G}_{11}(i\omega_n)] e^{i\omega_n \eta} + \ln[\mathcal{G}_{22}(i\omega_n)] e^{-i\omega_n \eta} \right\} \\ &= -\frac{\zeta}{\beta} \sum_{\omega_n} \ln[\mathcal{G}_{11}(i\omega_n)] e^{i\omega_n \eta}, \end{aligned} \tag{1.373}$$

in agreement with (1.308). The mean particle number $N = -\partial\Omega_0/\partial\mu$ is then given by

$$\begin{aligned} N &= \frac{\zeta}{2\beta} \sum_{\omega_n} \left[\mathcal{G}_{11}(i\omega_n) e^{i\omega_n \eta} + \zeta \mathcal{G}_{22}(i\omega_n) e^{-i\omega_n \eta} \right] \\ &= -\frac{\zeta}{\beta} \sum_{\omega_n} \frac{e^{i\omega_n \eta}}{i\omega_n - \xi}, \end{aligned} \tag{1.374}$$

which coincides with (1.306).

---

[42] For fermions, the Matsubara frequencies appear in pairs $(\omega_n, \omega_{-n-1} = -\omega_n)$ so that $\det[\mathcal{G}^{-1}(i\omega_n)\mathcal{G}^{-1}(-i\omega_n)] = (\omega_n^2 + \xi^2)^2$ and $\det \mathcal{G}^{-1} = \prod_{\omega_n} (\omega_n^2 + \xi^2)$.

In formal manipulations of the functional integral as well as in some practical calculations, it is sometimes convenient to use the field $\psi_{\tilde{\alpha}}$ rather than ψ_α and ψ^*_α. This is particularly true when one has to deal with superfluid or superconducting systems where the (global) U(1) symmetry is broken.[43] In these systems, the one-particle Green function $G(\tilde{\alpha}\tau, \tilde{\alpha}'\tau')$ has both "normal" ($G(\alpha, c, \tau; \alpha', -c, \tau')$) and "anomalous" ($G(\alpha, c, \tau; \alpha', c, \tau')$) components and can be represented as a 2×2 matrix

$$G(\tilde{\alpha}\tau, \tilde{\alpha}'\tau') = \begin{pmatrix} G(\alpha, +, \tau; \alpha', +, \tau') & G(\alpha, +, \tau; \alpha', -, \tau') \\ G(\alpha, -, \tau; \alpha', +, \tau') & G(\alpha, -, \tau; \alpha', -, \tau') \end{pmatrix} \tag{1.375}$$

Moreover, in bosonic superfluid systems, the Green functions $G^{(2n+1)}$ with an odd number of legs do not vanish, and $G^{(1)}(\tilde{\alpha}, \tau) = \langle \psi_{\tilde{\alpha}}(\tau) \rangle$ is the order parameter of the superfluid phase (see Chapter 7).[44] Superfluid or superconducting systems cannot be described within a perturbation expansion starting from the non-interacting limit. As for any system where a symmetry is spontaneously broken, it is necessary to reorganize the perturbation theory about a broken-symmetry state. The perturbation expansion in bosonic superfluid systems is explained in detail in Sec. 1.7. Superconductors (i.e. superfluid fermion systems) are discussed in Chapter 7.

1.6 Generating Functionals

In this section, we define connected Green functions, one-particle irreducible (1PI) and two-particle irreducible (2PI) vertices. 1PI and 2PI vertices allow one to reformulate and simplify the perturbation expansion discussed in Sec. 1.5. Furthermore, we shall see in the forthcoming chapters that the corresponding generating functionals enable one to set up non-perturbative approaches to strongly correlated quantum systems (Chapters 9 and 10).

1.6.1 *Connected Green functions*

In this section, we define the connected Green functions and their generating functional. We first consider normal systems where global U(1) invariance is not broken[43] before discussing the general case where we allow the system to be superfluid.

[43]The (global) U(1) invariance, i.e. the invariance of the action in the transformation $\psi_\sigma(\mathbf{r}, \tau) \to \psi_\sigma(\mathbf{r}, \tau) e^{i\Lambda}$ and $\psi^*_\sigma(\mathbf{r}, \tau) \to \psi^*_\sigma(\mathbf{r}, \tau) e^{-i\Lambda}$, is related to the conservation of the number of particles (Sec. 2.2.7). Broken U(1) symmetry in superfluid and superconducting systems is discussed in Chapter 7, while the concept of spontaneous symmetry breaking is introduced in Chapters 2 and 3.

[44]In fermion systems, $\langle \psi_{\tilde{\alpha}}(\tau) \rangle$ always vanishes. The spontaneously broken U(1) symmetry in a superfluid fermion system manifests itself by non-zero expectation values $\langle \psi_\alpha(\tau) \psi_{\alpha'}(\tau') \rangle$ and $\langle \psi^*_\alpha(\tau) \psi^*_{\alpha'}(\tau') \rangle$ (Chapter 7).

1.6.1.1 *Normal systems*

We consider the partition function in the presence of external sources,

$$Z[J^*, J] = \int \mathcal{D}[\psi^*, \psi] \exp\Big\{-S[\psi^*, \psi] + \int_0^\beta d\tau \sum_\alpha [J^*_\alpha(\tau)\psi_\alpha(\tau) + \text{c.c.}]\Big\}, \tag{1.376}$$

where J^*, J are c-numbers for bosons and anticommuting variables for fermions. The Green functions

$$\begin{aligned} &G^{(2n)}(\alpha_1\tau_1, \dots, \alpha_n\tau_n; \alpha'_n\tau'_n, \dots, \alpha'_1\tau'_1) \\ &\quad = (-1)^n \langle \psi_{\alpha_1}(\tau_1)\cdots\psi_{\alpha_n}(\tau_n)\psi^*_{\alpha'_n}(\tau'_n)\cdots\psi^*_{\alpha'_1}(\tau'_1)\rangle \\ &\quad = (-\zeta)^n \frac{\delta^{2n}\mathcal{G}[J^*, J]}{\delta J^*_{\alpha_1}(\tau_1)\cdots\delta J^*_{\alpha_n}(\tau_n)\delta J_{\alpha'_n}(\tau'_n)\cdots\delta J_{\alpha'_1}(\tau'_1)}\bigg|_{J^*=J=0} \end{aligned} \tag{1.377}$$

can be obtained from the generating functional

$$\mathcal{G}[J^*, J] = \frac{Z[J^*, J]}{Z[0,0]}, \tag{1.378}$$

where $Z[0,0] = Z$ is the partition function in the absence of external sources. In a normal system, the only non-vanishing Green functions are of the type (1.377) with n incoming particles and n outgoing particles.

In Sec. 1.5, we showed that the Green function $G^{(2n)}$ can be obtained from the sum of all Feynman diagrams that are connected to the external legs. However, these diagrams need not be fully connected.[45] As the disconnected diagrams can be expressed in terms of lower-order Green functions ($G^{(2n-2)}$, $G^{(2n-4)}$, etc.), it is useful to deal with the "connected" Green functions $G_c^{(2n)}$ defined as the sum of all fully connected diagrams. The corresponding generating functional is $W[J^*, J] = \ln Z[J^*, J]$. This can be shown using the replica method as in the derivation of the linked cluster theorem (Sec. 1.5.2). The generating functional $\mathcal{G}[J^*, J]^R$ of the Green functions of R replicas of the system can be written as a functional integral over R distinct fields. The connected diagrams are proportional to R (Sec. 1.5.2), while a diagram with n disconnected pieces is of order R^n. The connected diagrams can therefore be obtained from the functional

$$\lim_{R\to 0} \frac{\partial}{\partial_R} \mathcal{G}[J^*, J]^R = \ln \mathcal{G}[J^*, J] = \ln Z[J^*, J] - \ln Z[0,0], \tag{1.379}$$

or more simply $W[J^*, J] = \ln Z[J^*, J]$ since $\ln Z[0,0]$ is independent of the external sources.

A direct calculation gives

$$G_c(\alpha_1\tau_1, \alpha'_1\tau'_1) = -\zeta \frac{\delta^2 W[J^*, J]}{\delta J^*_{\alpha_1}(\tau_1)\delta J_{\alpha'_1}(\tau'_1)}\bigg|_{J^*=J=0} = G(\alpha_1\tau_1, \alpha'_1\tau'_1) \tag{1.380}$$

[45] For instance, in a non-interacting system, none of the diagrams contributing to $G^{(4)}$ are fully connected [see Eq. (1.317) and the corresponding figure].

and

$$\begin{aligned} G_c^{(4)}(\alpha_1\tau_1, \alpha_2\tau_2; \alpha_2'\tau_2', \alpha_1'\tau_1') &= \frac{\delta^4 W[J^*, J]}{\delta J_{\alpha_1}^*(\tau_1)\delta J_{\alpha_2}^*(\tau_2)\delta J_{\alpha_2'}(\tau_2')\delta J_{\alpha_1'}(\tau_1')}\bigg|_{J^*=J=0} \\ &= G^{(4)}(\alpha_1\tau_1, \alpha_2\tau_2; \alpha_2'\tau_2', \alpha_1'\tau_1') \\ &\quad - G(\alpha_1\tau_1, \alpha_1'\tau_1')G(\alpha_2\tau_2, \alpha_2'\tau_2') \\ &\quad - \zeta G(\alpha_1\tau_1, \alpha_2'\tau_2')G(\alpha_2\tau_2, \alpha_1'\tau_1') \end{aligned} \tag{1.381}$$

in a normal system where $\langle\psi_\alpha(\tau)\rangle = \langle\psi_\alpha^*(\tau)\rangle = 0$. The last equation can be graphically represented as

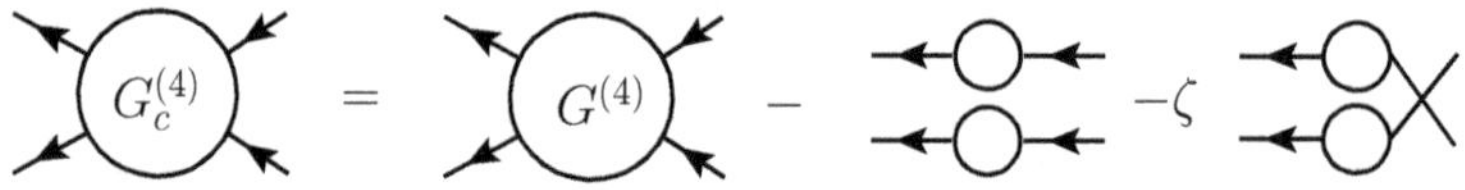

(writing explicitly the sign of the diagrams), where the empty circles stand for the one-particle Green function $G \equiv G^{(2)}$. Clearly, $G_c^{(4)}$ is given by the sum of all fully connected diagrams.

The connected Green functions are similar to the cumulants of a probability distribution $P(x)$ with moment-generating function

$$Z(j) = \int_{-\infty}^{\infty} dx\, P(x)e^{jx} = \sum_{n=0}^{\infty} \frac{j^n}{n!}\langle x^n\rangle. \tag{1.382}$$

The nth-order cumulant $\langle x^n\rangle_c$ is defined by

$$\langle x^n\rangle_c = \frac{\partial^n}{\partial j^n}W(j)\Big|_{j=0}, \qquad W(j) = \ln Z(j) = \sum_{n=1}^{\infty} \frac{j^n}{n!}\langle x^n\rangle_c. \tag{1.383}$$

One easily finds

$$\begin{aligned} \langle x\rangle_c &= \langle x\rangle, \\ \langle x^2\rangle_c &= \langle x^2\rangle - \langle x\rangle^2 = \langle(x - \langle x\rangle)^2\rangle, \\ \langle x^3\rangle_c &= \langle x^3\rangle - 3\langle x^2\rangle\langle x\rangle + 2\langle x\rangle^3 = \langle(x - \langle x\rangle)^3\rangle, \\ \langle x^4\rangle_c &= \langle(x - \langle x\rangle)^4\rangle - 3\langle(x - \langle x\rangle)^2\rangle^2. \end{aligned} \tag{1.384}$$

For a Gaussian probability distribution $P(x) = (a/2\pi)^{1/2}e^{-(x-x_0)^2/2a}$,

$$W(j) = \ln\left\{\sqrt{\frac{a}{2\pi}}\int_{-\infty}^{\infty} dx\, e^{-\frac{1}{2a}(x-x_0)^2+jx}\right\} = \frac{1}{2}aj^2 + jx_0, \tag{1.385}$$

so that $\langle x\rangle_c = x_0$, $\langle x^2\rangle_c = a$ and $\langle x^n\rangle_c = 0$ for $n \geq 3$. By inverting (1.384), one can express all averages $\langle x^n\rangle$ in terms of the average value x_0 and the second-order cumulant $\langle x^2\rangle_c$ (which is nothing but Wick's theorem). The analogous result for a many-particle system is that when the action $S[\psi^*, \psi]$ is Gaussian (non-interacting particles), the only non-vanishing connected Green function is $G_c^{(2)}$, and all higher-order Green functions $G^{(n\geq 4)}$ are entirely determined by $G_c^{(2)}$.

1.6.1.2 *General case*

As pointed out in Sec. 1.5.4, in systems where the U(1) symmetry is broken, the Green functions take a more general form where the number of outgoing particles may differ from that of incoming particles. The total number of fields can even be odd in bosonic (superfluid) systems. It is then convenient to use the field $\psi_{\tilde\alpha}(\tau)$ (with $\tilde\alpha = (\alpha, c)$ and $c = \pm$ a charge index) introduced in Sec. 1.5.4 and define the Green functions as[46]

$$G^{(n)}(\tilde\alpha_1\tau_1, \dots, \tilde\alpha_n\tau_n) = \langle \psi_{\tilde\alpha_1}(\tau_1)\cdots\psi_{\tilde\alpha_n}(\tau_n)\rangle = \left.\frac{\delta^n \mathcal{G}[J]}{\delta J_{\tilde\alpha_1}(\tau_1)\cdots\delta J_{\tilde\alpha_n}(\tau_n)}\right|_{J=0}. \tag{1.386}$$

This definition will be used from now on whenever we work with the field $\psi_{\tilde\alpha}$. The generating functional is given by

$$\begin{aligned}\mathcal{G}[J] &= \frac{1}{Z[0]}\int \mathcal{D}[\psi]\exp\left\{-S[\psi] + \int_0^\beta d\tau \sum_{\tilde\alpha} J_{\tilde\alpha}(\tau)\psi_{\tilde\alpha}(\tau)\right\} \\ &= 1 + \sum_{n=1}^\infty \frac{1}{n!}\int_0^\beta d\tau_1\cdots d\tau_n \sum_{\tilde\alpha_1\cdots\tilde\alpha_n} G^{(n)}(\tilde\alpha_1\tau_1, \dots, \tilde\alpha_n\tau_n) J_{\tilde\alpha_n}(\tau_n)\cdots J_{\tilde\alpha_1}(\tau_1).\end{aligned} \tag{1.387}$$

Similarly, the connected Green functions and their generating functional $W[J] = \ln Z[J]$ are defined by

$$W[J] = \ln Z[0] + \sum_{n=1}^\infty \frac{1}{n!}\int_0^\beta d\tau_1\cdots d\tau_n \sum_{\tilde\alpha_1\cdots\tilde\alpha_n} G_c^{(n)}(\tilde\alpha_1\tau_1, \dots, \tilde\alpha_n\tau_n) J_{\tilde\alpha_n}(\tau_n)\cdots J_{\tilde\alpha_1}(\tau_1) \tag{1.388}$$

and

$$G_c^{(n)}(\tilde\alpha_1\tau_1, \dots, \tilde\alpha_n\tau_n) = \left.\frac{\delta^n W[J]}{\delta J_{\tilde\alpha_1}(\tau_1)\cdots\delta J_{\tilde\alpha_n}(\tau_n)}\right|_{J=0}. \tag{1.389}$$

For the lowest-order Green functions, one finds

$$\begin{aligned} G_c^{(1)}(\tilde\alpha\tau) &= G^{(1)}(\tilde\alpha\tau) = \langle\psi_{\tilde\alpha}(\tau)\rangle, \\ G_c^{(2)}(\tilde\alpha_1\tau_1, \tilde\alpha_2\tau_2) &= G^{(2)}(\tilde\alpha_1\tau_1, \tilde\alpha_2\tau_2) - \langle\psi_{\tilde\alpha_1}(\tau_1)\rangle\langle\psi_{\tilde\alpha_2}(\tau_2)\rangle \end{aligned} \tag{1.390}$$

and

$$\begin{aligned} G_c^{(3)}(\tilde\alpha_1\tau_1, \tilde\alpha_2\tau_2, \tilde\alpha_3\tau_3) = {} & G^{(3)}(\tilde\alpha_1\tau_1, \tilde\alpha_2\tau_2, \tilde\alpha_3\tau_3) - \langle\psi_{\tilde\alpha_1}(\tau_1)\rangle G_c^{(2)}(\tilde\alpha_2\tau_2, \tilde\alpha_3\tau_3) \\ & - \zeta\langle\psi_{\tilde\alpha_2}(\tau_2)\rangle G_c^{(2)}(\tilde\alpha_1\tau_1, \tilde\alpha_3\tau_3) - \langle\psi_{\tilde\alpha_3}(\tau_3)\rangle G_c^{(2)}(\tilde\alpha_1\tau_1, \tilde\alpha_2\tau_2) \\ & - \langle\psi_{\tilde\alpha_1}(\tau_1)\rangle\langle\psi_{\tilde\alpha_2}(\tau_2)\rangle\langle\psi_{\tilde\alpha_3}(\tau_3)\rangle. \end{aligned} \tag{1.391}$$

For a fermion system, all Green functions with an odd number of legs vanish.

[46] The sign convention in (1.386) agrees with that of (1.358).

1.6.1.3 *Equations of motion: Schwinger–Dyson equations*

The Green functions $G^{(n)}$ are not independent but are related by "equations of motion".[47] Let us consider the generating functional $\mathcal{G}[J]$ [Eq. (1.387)]. The change of variables $\psi_{\tilde\alpha}(\tau) \to \psi_{\tilde\alpha}(\tau) + \epsilon f_{\tilde\alpha}(\tau)$, with $\epsilon \to 0$ and $f_{\tilde\alpha}(\tau)$ a complex or Grassmannian field, has a unit Jacobian and gives

$$\mathcal{G}[J] = \frac{1}{Z[0]} \int \mathcal{D}[\psi]\, e^{-S[\psi+\epsilon f] + \int_0^\beta d\tau \sum_{\tilde\alpha} J_{\tilde\alpha}(\tau)[\psi_{\tilde\alpha}(\tau)+\epsilon f_{\tilde\alpha}(\tau)]}. \tag{1.392}$$

Expanding to first order in ϵ, one finds

$$\frac{1}{Z[0]} \int \mathcal{D}[\psi]\, e^{-S[\psi] + \int_0^\beta d\tau \sum_{\tilde\alpha} J_{\tilde\alpha}(\tau)\psi_{\tilde\alpha}(\tau)} \int_0^\beta d\tau \sum_{\tilde\alpha} f_{\tilde\alpha}(\tau) \left[\zeta \frac{\delta S}{\delta \psi_{\tilde\alpha}(\tau)} - J_{\tilde\alpha}(\tau) \right] = 0 \tag{1.393}$$

(for Grassmann variables, we use the chain rule (1.240) for derivation) and, since the function $f_{\tilde\alpha}(\tau)$ is arbitrary,

$$\frac{1}{Z[0]} \int \mathcal{D}[\psi]\, e^{-S[\psi] + \int_0^\beta d\tau \sum_{\tilde\alpha} J_{\tilde\alpha}(\tau)\psi_{\tilde\alpha}(\tau)} \left[\zeta \frac{\delta S}{\delta \psi_{\tilde\alpha}(\tau)} - J_{\tilde\alpha}(\tau) \right] = 0. \tag{1.394}$$

If we write the action as in (1.357), Eq. (1.394) gives

$$\begin{aligned} \frac{1}{Z[0]} \int \mathcal{D}[\psi]\, e^{-S[\psi] + \int_0^\beta d\tau \sum_{\tilde\alpha} J_{\tilde\alpha}(\tau)\psi_{\tilde\alpha}(\tau)} \Bigg[\int_0^\beta d\tau_1' \sum_{\tilde\alpha_1'} G_0^{-1}(\tilde\alpha_1\tau_1, \tilde\alpha_1'\tau_1')\psi_{\tilde\alpha_1'}(\tau_1') \\ + \frac{1}{3!} \sum_{\tilde\alpha_2'\tilde\alpha_3'\tilde\alpha_4'} v_{\tilde\alpha_1\tilde\alpha_2'\tilde\alpha_3'\tilde\alpha_4'} \psi_{\tilde\alpha_2'}(\tau_1)\psi_{\tilde\alpha_3'}(\tau_1)\psi_{\tilde\alpha_4'}(\tau_1) - \zeta J_{\tilde\alpha_1}(\tau_1) \Bigg] = 0. \end{aligned} \tag{1.395}$$

The equation of motion for the Green function $G^{(n)}$ is obtained by taking the $(n-1)$th-order functional derivative with respect to J and setting $J = 0$ in the end,

$$\begin{aligned} &\int_0^\beta d\tau_1' \sum_{\tilde\alpha_1'} G_0^{-1}(\tilde\alpha_1\tau_1, \tilde\alpha_1'\tau_1') G^{(n)}(\tilde\alpha_n\tau_n, \ldots, \tilde\alpha_2\tau_2, \tilde\alpha_1'\tau_1') \\ &\qquad + \frac{1}{3!} \sum_{\tilde\alpha_2', \tilde\alpha_3', \tilde\alpha_4'} v_{\tilde\alpha_1\tilde\alpha_2'\tilde\alpha_3'\tilde\alpha_4'} G^{(n+2)}(\tilde\alpha_n\tau_n, \ldots, \tilde\alpha_2\tau_2, \tilde\alpha_2'\tau_1, \tilde\alpha_3'\tau_1, \tilde\alpha_4'\tau_1) \\ &\qquad = \zeta \sum_{i=2}^n \delta_{\tilde\alpha_i, \tilde\alpha_1} \delta(\tau_i - \tau_1) G^{(n-2)}(\tilde\alpha_n\tau_n, \ldots, \tilde\alpha_{i+1}\tau_{i+1}, \tilde\alpha_{i-1}\tau_{i-1}, \ldots, \tilde\alpha_2\tau_2). \end{aligned} \tag{1.396}$$

The equation of motion for $G^{(2)} = G$,

$$\begin{aligned} &\int_0^\beta d\tau_1' \sum_{\tilde\alpha_1'} G_0^{-1}(\tilde\alpha_1\tau_1, \tilde\alpha_1'\tau_1') G(\tilde\alpha_1'\tau_1', \tilde\alpha_2\tau_2) \\ &+ \frac{1}{3!} \sum_{\tilde\alpha_2', \tilde\alpha_3', \tilde\alpha_4'} v_{\tilde\alpha_1\tilde\alpha_2'\tilde\alpha_3'\tilde\alpha_4'} G^{(4)}(\tilde\alpha_2'\tau_1, \tilde\alpha_3'\tau_1, \tilde\alpha_4'\tau_1, \tilde\alpha_2\tau_2) = \delta_{\tilde\alpha_1, \tilde\alpha_2} \delta(\tau_1 - \tau_2) \end{aligned} \tag{1.397}$$

[47] In the operator formalism, the equations of motion are obtained by calculating $\partial_{\tau_1} G^{(n)}(\tilde\alpha_1\tau_1 \cdots \tilde\alpha_n\tau_n) = \partial_{\tau_1} \langle T_\tau \hat\psi_{\tilde\alpha_1}(\tau_1) \cdots \hat\psi_{\tilde\alpha_n}(\tau_n) \rangle$ using $\partial_\tau \hat\psi_{\tilde\alpha}(\tau) = [\hat H, \hat\psi_{\tilde\alpha}(\tau)]$.

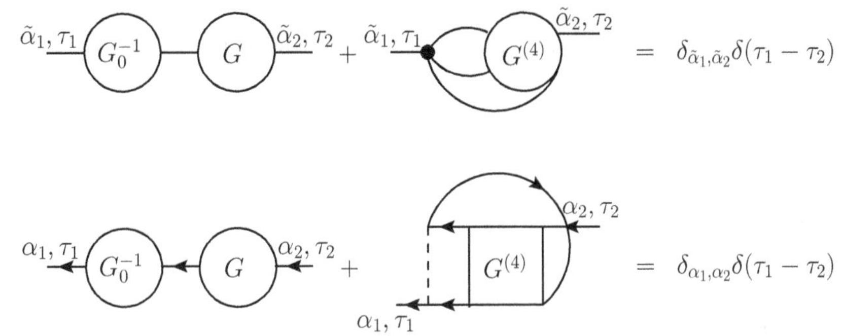

Fig. 1.7 Diagrammatic representation of the equation of motion for $G = G^{(2)}$ in the general case [Eq. (1.397)] and for a normal system [Eq. (1.398)].

is represented diagrammatically in Fig. 1.7. For a normal system, one finds (with $\tilde{\alpha}_1 = (\alpha_1, +)$ and $\tilde{\alpha}_2 = (\alpha_2, +)$)

$$\int_0^\beta d\tau_1' \sum_{\alpha_1'} G_0^{-1}(\alpha_1\tau_1, \alpha_1'\tau_1') G(\alpha_1'\tau_1', \alpha_2\tau_2) + \sum_{\alpha_2', \alpha_3', \alpha_4'} (\alpha_1\alpha_2'|\hat{v}|\alpha_4'\alpha_3') G^{(4)}(\alpha_3'\tau_1, \alpha_4'\tau_1, \alpha_2'\tau_1, \alpha_2\tau_2) = \delta_{\alpha_1,\alpha_2}\delta(\tau_1 - \tau_2) \tag{1.398}$$

(see Fig. 1.7), where (α_3', τ_1) and (α_4', τ_1) must be understood as (α_3', τ_1^-) and (α_4', τ_1^-) (with $\tau_1^- = \tau_1 - 0^+$) because of the time ordering associated with the matrix element $(\alpha_1\alpha_2'|\hat{v}|\alpha_4'\alpha_3')$.

1.6.2 *Effective action and 1PI vertices*

In this section, we introduce the effective action and the 1PI vertices considering again first normal systems before discussing the general case.[48] The effective action is further discussed in Sec. 1.7 and plays a fundamental role in the non-perturbative functional RG (Chapter 10).

1.6.2.1 *Normal systems*

In presence of external sources, the fields $\psi_\alpha, \psi_\alpha^*$ acquire non-zero expectation values,[49]

$$\begin{aligned} \phi_\alpha[\tau; J^*, J] &= \langle \psi_\alpha(\tau) \rangle = \frac{\delta W[J^*, J]}{\delta J_\alpha^*(\tau)}, \\ \phi_\alpha^*[\tau; J, J^*] &= \langle \psi_\alpha^*(\tau) \rangle = \zeta \frac{\delta W[J^*, J]}{\delta J_\alpha(\tau)}, \end{aligned} \tag{1.399}$$

[48]In this section, the effective action is defined by the Legendre transform of $\ln Z[J^*, J]$. The same terminology is used for an action $S[\psi^*, \psi]$ providing an effective description of the system, e.g. at low energies or for a subset of degrees of freedom.

[49]The notation emphasizes that $\phi_\alpha^{(*)}[\tau; J^*, J]$ is a functional of the external sources.

where $\langle \cdots \rangle$ denotes an expectation value in the presence of the external sources. ϕ_α and ϕ^*_α are c-numbers for bosonic fields and Grassmann variables for fermionic fields.

The (1PI) effective action $\Gamma[\phi^*, \phi]$ is defined as the Legendre transform of $\ln Z[J^*, J]$,

$$\Gamma[\phi^*, \phi] = -\ln Z[J^*, J] + \int_0^\beta d\tau \sum_\alpha [J^*_\alpha(\tau)\phi_\alpha(\tau) + \text{c.c.}]. \tag{1.400}$$

In (1.400), $J_\alpha(\tau)$ and $J^*_\alpha(\tau)$ must be considered as functionals of $\phi_\alpha(\tau)$ and $\phi^*_\alpha(\tau)$ obtained by inverting (1.399). $\Gamma[\phi^*, \phi]$ satisfies[50]

$$\begin{aligned}\frac{\delta\Gamma[\phi^*, \phi]}{\delta\phi_\alpha(\tau)} &= -\int_0^\beta d\tau' \sum_{\alpha'} \left(\frac{\delta J_{\alpha'}(\tau')}{\delta\phi_\alpha(\tau)} \frac{\delta W[J^*, J]}{\delta J_{\alpha'}(\tau')} + \frac{\delta J^*_{\alpha'}(\tau')}{\delta\phi_\alpha(\tau)} \frac{\delta W[J^*, J]}{\delta J^*_{\alpha'}(\tau')} \right) \\ &\quad + \int_0^\beta d\tau' \sum_{\alpha'} \left(\frac{\delta J^*_{\alpha'}(\tau')}{\delta\phi_\alpha(\tau)} \phi_{\alpha'}(\tau') + \zeta \frac{\delta J_{\alpha'}(\tau')}{\delta\phi_\alpha(\tau)} \phi^*_{\alpha'}(\tau') \right) + \zeta J^*_\alpha(\tau) \\ &= \zeta J^*_\alpha(\tau)\end{aligned} \tag{1.401}$$

and the companion equation

$$\frac{\delta\Gamma[\phi^*, \phi]}{\delta\phi^*_\alpha(\tau)} = J_\alpha(\tau). \tag{1.402}$$

The effective action is similar to the Gibbs free energy introduced in statistical mechanics. Let us consider, for instance, a classical spin system with partition function

$$Z(\mathbf{h}) = e^{-\beta F(\mathbf{h})} = \int \mathcal{D}[\mathbf{S}]\, e^{-\beta \int d^d r [\mathcal{H}(\mathbf{S}(\mathbf{r})) - \mathbf{h}\cdot\mathbf{S}(\mathbf{r})]}, \tag{1.403}$$

where $\mathcal{H}(\mathbf{S}(\mathbf{r}))$ is the Hamiltonian density and $\mathbf{h}$ an external field. The magnetization density of the system is given by

$$\mathbf{m} = \frac{1}{\mathcal{V}} \int d^d r \langle \mathbf{S}(\mathbf{r}) \rangle = \frac{1}{\beta\mathcal{V}} \frac{\partial}{\partial\mathbf{h}} \ln Z(\mathbf{h}) = -\frac{1}{\mathcal{V}} \frac{\partial F(\mathbf{h})}{\partial\mathbf{h}}. \tag{1.404}$$

The Gibbs free energy is defined as the Legendre transform of the Helmholtz free energy $F(\mathbf{h})$,

$$G(\mathbf{m}) = F(\mathbf{h}) + \mathcal{V}\mathbf{m}\cdot\mathbf{h}, \tag{1.405}$$

and satisfies

$$\frac{\partial G(\mathbf{m})}{\partial\mathbf{m}} = \mathcal{V}\mathbf{h}. \tag{1.406}$$

$F(\mathbf{h})$ and $G(\mathbf{m})$ contain the same physical information. In the absence of an external field, the magnetization of the system at thermodynamic equilibrium is obtained by requiring the Gibbs free energy to be stationary. Note that $G(\mathbf{m})$ (if calculated exactly) includes all effects of thermal fluctuations.

More generally, one can consider a spatially varying field $\mathbf{h}(\mathbf{r})$ giving rise to a nonuniform magnetization density $\mathbf{m}(\mathbf{r}) = \langle \mathbf{S}(\mathbf{r}) \rangle$. $F[\mathbf{h}]$ and $G[\mathbf{m}]$ are then functionals of $\mathbf{h}(\mathbf{r})$ and $\mathbf{m}(\mathbf{r})$, respectively. $F[\mathbf{h}]$ is the generating functional of the connected correlation functions of the spin density field $\mathbf{S}(\mathbf{r})$ whereas $G[\mathbf{m}]$ is the generating functional of the 1PI vertices (see the following).

[50] In the case of Grassmann variables, we use the chain rule (1.240) for derivation.

1.6.2.2 Self-energy and Dyson's equation

The 1PI vertices[51] are defined as the functional derivatives of the effective action,

$$\Gamma^{(2n)}(\alpha_1\tau_1,\dots,\alpha_n\tau_n;\alpha'_n\tau'_n,\dots,\alpha'_1\tau'_1)$$
$$= \left.\frac{\delta^{2n}\Gamma[\phi^*,\phi]}{\delta\phi_{\alpha'_1}(\tau'_1)\cdots\delta\phi_{\alpha'_n}(\tau'_n)\delta\phi^*_{\alpha_n}(\tau_n)\cdots\delta\phi^*_{\alpha_1}(\tau_1)}\right|_{\phi^*=\phi=0}. \tag{1.407}$$

Inverting (1.407), one obtains

$$\Gamma[\phi^*,\phi] = \Gamma[0,0] + \sum_{n=1}^{\infty}\frac{1}{(n!)^2}\int_0^\beta d\tau_1\cdots d\tau'_1 \sum_{\alpha_1\cdots\alpha'_n}\Gamma^{(2n)}(\alpha_1\tau_1,\dots,\alpha_n\tau_n;\alpha'_n\tau'_n,\dots,\alpha'_1\tau'_1)$$
$$\times\, \phi^*_{\alpha_1}(\tau_1)\cdots\phi^*_{\alpha_n}(\tau_n)\phi_{\alpha'_n}(\tau'_n)\cdots\phi_{\alpha'_1}(\tau'_1), \tag{1.408}$$

where $\Gamma[0,0] = -W[0,0] = \beta\Omega$.

Taking the functional derivatives of the first of Eqs. (1.399) with respect to $\phi_{\alpha'}(\tau')$, one obtains

$$\begin{aligned}\delta_{1,2} &= \frac{\delta\phi_1}{\delta\phi_2}\\ &= \sum_3\left(\frac{\delta J_3}{\delta\phi_2}\frac{\delta^2 W[J^*,J]}{\delta J_3\delta J_1^*} + \frac{\delta J_3^*}{\delta\phi_2}\frac{\delta^2 W[J^*,J]}{\delta J_3^*\delta J_1^*}\right)\\ &= \sum_3\left(\frac{\delta^2\Gamma[\phi^*,\phi]}{\delta\phi_2\delta\phi_3^*}\frac{\delta^2 W[J^*,J]}{\delta J_3\delta J_1^*} + \zeta\frac{\delta^2\Gamma[\phi^*,\phi]}{\delta\phi_2\delta\phi_3}\frac{\delta^2 W[J^*,J]}{\delta J_3^*\delta J_1^*}\right),\end{aligned} \tag{1.409}$$

where we use the short-hand notation $i \equiv (\alpha_i,\tau_i)$, $\delta_{i,j} = \delta_{\alpha_i,\alpha_j}\delta(\tau_i-\tau_j)$ and $\sum_i = \int_0^\beta d\tau_i \sum_{\alpha_i}$. By considering the functional derivatives of (1.399) with respect to both $\phi_{\alpha'}(\tau')$ and $\phi^*_{\alpha'}(\tau')$, one obtains four equations that can be cast in the form

$$\sum_3\begin{pmatrix}\dfrac{\delta^2\Gamma}{\delta\phi_2\delta\phi_3^*} & \zeta\dfrac{\delta^2\Gamma}{\delta\phi_2\delta\phi_3}\\ \zeta\dfrac{\delta^2\Gamma}{\delta\phi_2^*\delta\phi_3^*} & \dfrac{\delta^2\Gamma}{\delta\phi_2^*\delta\phi_3}\end{pmatrix}\begin{pmatrix}\dfrac{\delta^2 W}{\delta J_3\delta J_1^*} & \dfrac{\delta^2 W}{\delta J_3\delta J_1}\\ \dfrac{\delta^2 W}{\delta J_3^*\delta J_1^*} & \dfrac{\delta^2 W}{\delta J_3^*\delta J_1}\end{pmatrix} = \delta_{1,2}\begin{pmatrix}1 & 0\\ 0 & 1\end{pmatrix}. \tag{1.410}$$

In a normal system, the matrices in (1.410) become diagonal for $J^* = J = 0$ and one obtains

$$-\int_0^\beta d\tau_3 \sum_{\alpha_3} G_c^{(2)}(\alpha_1\tau_1,\alpha_3\tau_3)\Gamma^{(2)}(\alpha_3\tau_3,\alpha_2\tau_2) = \delta_{\alpha_1,\alpha_2}\delta(\tau_1-\tau_2), \tag{1.411}$$

i.e.

$$\Gamma^{(2)} = -G_c^{(2)-1} = -G^{(2)-1} \tag{1.412}$$

in a matrix sense (recall that $G_c^{(2)} = G^{(2)}$ when $\langle\psi_\alpha(\tau)\rangle = \langle\psi^*_\alpha(\tau)\rangle = 0$).

[51] The reason for this terminology is explained in the following.

It is conventional to express $\Gamma^{(2)}$ in terms of the self-energy Σ defined by $\Gamma^{(2)} = -G_0^{-1} + \Sigma$ or $G^{-1} = G_0^{-1} - \Sigma$ (with $G \equiv G^{(2)}$). This equation can be rewritten as

$$G = G_0 + G_0 \Sigma G \tag{1.413}$$

or, more explicitly,

$$\begin{aligned} G(\alpha\tau, \alpha'\tau') &= G_0(\alpha\tau, \alpha'\tau') \\ &\quad + \int_0^\beta d\tau_1 d\tau_2 \sum_{\alpha_1, \alpha_2} G_0(\alpha\tau, \alpha_1\tau_1) \Sigma(\alpha_1\tau_1, \alpha_2\tau_2) G(\alpha_2\tau_2, \alpha'\tau'). \end{aligned} \tag{1.414}$$

Equation (1.413) is known as Dyson's equation. For a translation-invariant system, we obtain the simple expression

$$G(\mathbf{k}, i\omega_n) = \frac{1}{G_0^{-1}(\mathbf{k}, i\omega_n) - \Sigma(\mathbf{k}, i\omega_n)} = \frac{1}{i\omega_n - \xi_{\mathbf{k}} - \Sigma(\mathbf{k}, i\omega_n)} \tag{1.415}$$

(neglecting for simplicity internal degrees of freedom such as the spin of the particle), where $\Sigma(\mathbf{k}, i\omega_n)$ is the Fourier transform of $\Sigma(\mathbf{r}, \tau; \mathbf{r}', \tau') = \Sigma(\mathbf{r} - \mathbf{r}', \tau - \tau')$. The Dyson equation can be graphically represented as

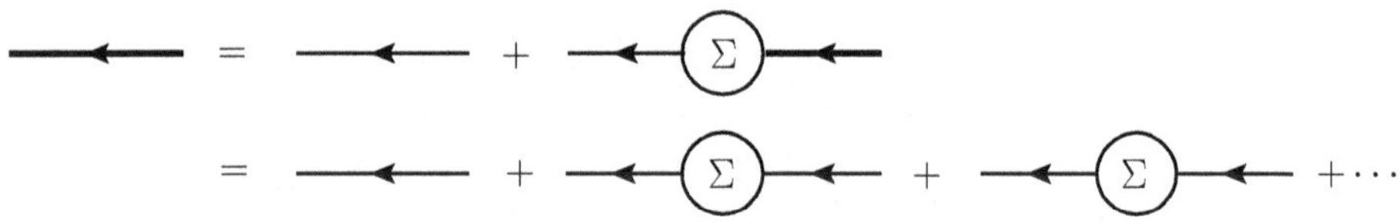

where the thick line stands for G and the thin one for G_0.

Let us introduce two definitions. (i) A diagram is n-particle irreducible (nPI) if it cannot be separated into two or more disconnected pieces by cutting n internal lines. (ii) An amputated diagram has no propagator attached to the external legs ($\alpha_i \tau_i$); the latter connect directly to an interaction vertex. With these definitions, it is easy to see that the self-energy is given by the sum of 1PI amputated diagrams. One easily verifies that all diagrams for the Green function G are indeed obtained from the Dyson equation. For instance, the self-energy diagram

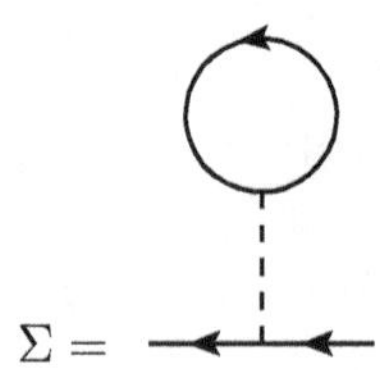

generates an infinite number of diagrams for the Green function,

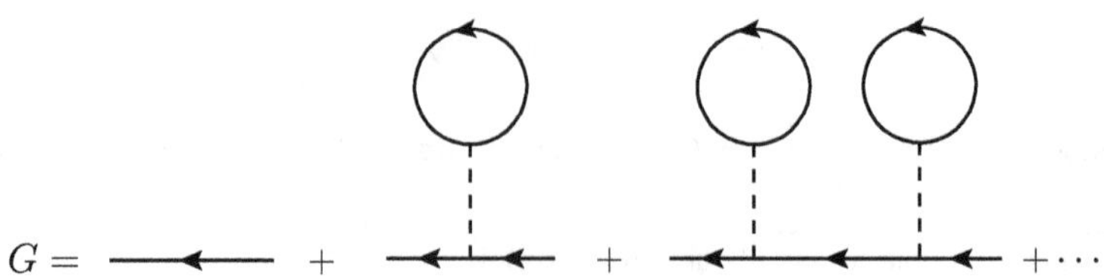

It should be understood that the external legs in the self-energy diagrams are not part of the self-energy even though they are often drawn for clarity.

The diagrammatic rules for the self-energy follow directly from those corresponding to the one-particle Green function. To calculate the nth-order contribution to the self-energy:

1. Draw all distinct 1PI amputated connected (two-leg) diagrams with n interaction vertices.
2. To each directed line, associate $G_0(\alpha\tau, \alpha', \tau'^+)$.
3. To each vertex, associate $(\gamma_1\gamma_2|\hat{v}|\gamma_1'\gamma_2')$.
4. Sum over all internal indices α and integrate all internal times $\tau \in [0, \beta]$.
5. Multiply the result by the factor $(-1)^n\zeta^{n_L}$, where n_L is the number of closed loops. (The combinatorial factor is unity since we use non-(anti)symmetrized vertices.)

The two first-order diagrams contributing to the self-energy $\Sigma(\alpha_1\tau_1, \alpha_1'\tau_1')$ are

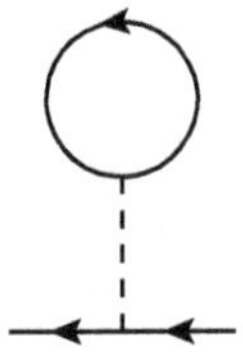

$$= -\zeta\delta(\tau_1 - \tau_1'^+) \sum_{\alpha_2,\alpha_2'} (\alpha_1\alpha_2|\hat{v}|\alpha_1'\alpha_2')G_0(\alpha_2'\tau_1, \alpha_2\tau_1^+)$$

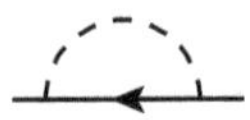

$$= -\delta(\tau_1 - \tau_1'^+) \sum_{\alpha_2,\alpha_2'} (\alpha_1\alpha_2|\hat{v}|\alpha_2'\alpha_1')G_0(\alpha_2'\tau_1, \alpha_2\tau_1^+)$$

The first-order self-energy correction can also be represented by the diagram

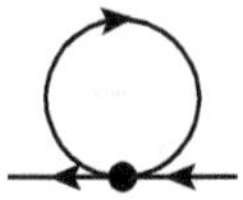

The second-order diagrams are given by those of Fig. 1.3 (with amputated legs) except the last four that are not 1PI. They are also given by the last two diagrams (with amputated legs) of Fig. 1.6.

In perturbation theory, one always computes the self-energy rather than directly the one-particle Green function. Any finite order calculation of Σ leads (*via* the Dyson equation) to the summation of an infinite number of diagrams for the Green function. Moreover, a perturbative calculation of G to finite order is likely to violate basic properties of the Green function such as causality (see Sec. 3.5).

1.6.2.3 *Higher-order vertices*

By taking functional derivatives of (1.410) with respect to J^* and J, one can relate the higher-order connected Green functions $G_c^{(n)}$ ($n \geq 4$) to the 1PI vertices. After a little algebra, one finds

$$G_c^{(4)}(\alpha_1\tau_1, \alpha_2\tau_2; \alpha_2'\tau_2', \alpha_1'\tau_1') = -\int_0^\beta du_1 du_2 du_2' du_1' \sum_{\gamma_1,\gamma_2,\gamma_2',\gamma_1'} G(\alpha_1\tau_1, \gamma_1 u_1)$$
$$\times G(\alpha_2\tau_2, \gamma_2 u_2)\Gamma^{(4)}(\gamma_1 u_1, \gamma_2 u_2; \gamma_2' u_2', \gamma_1' u_1')G(\gamma_2' u_2', \alpha_2'\tau_2')G(\gamma_1' u_1', \alpha_1'\tau_1'). \quad (1.416)$$

This expression can be graphically represented as

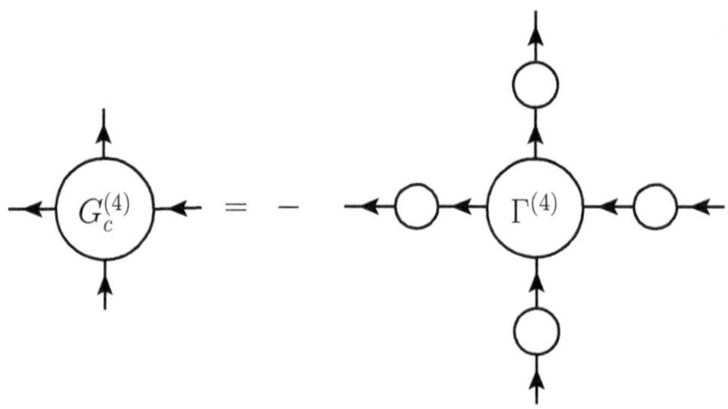

(writing explicitly the sign of the diagrams), where the empty circle stands for the one-particle Green function G, and identifies $\Gamma^{(4)}(\alpha_1\tau_1, \alpha_2\tau_2; \alpha_2'\tau_2', \alpha_1'\tau_1')$ as the four-leg (or two-particle) 1PI vertex. $\Gamma^{(4)}$ can be seen as the effective interaction vertex between two particles. To lowest order, it is given by the bare vertex $\{\alpha_1\alpha_2|\hat{v}|\alpha_1'\alpha_2'\}$, and its perturbation expansion can be represented by diagrams,

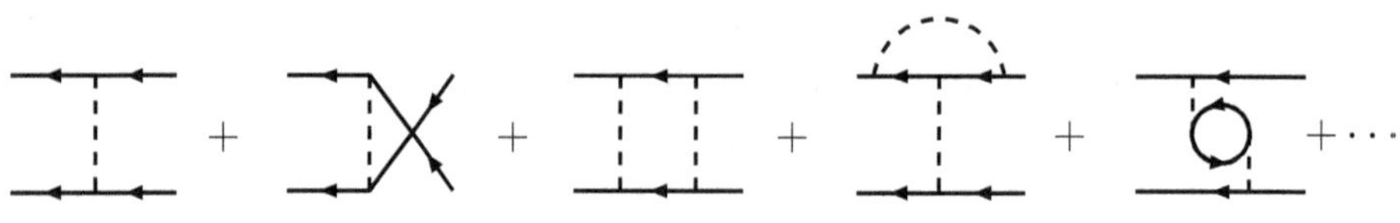

or, equivalently,

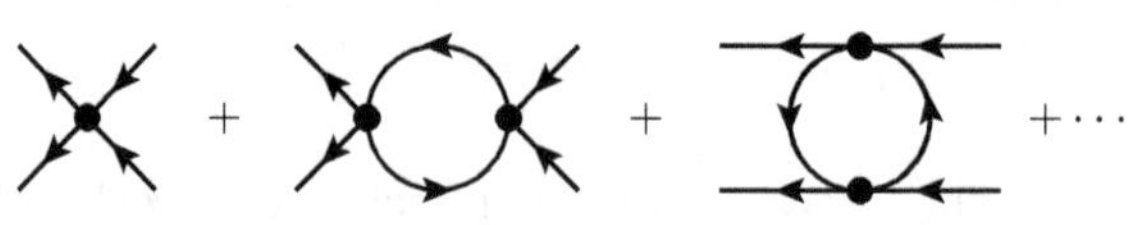

where the lines may contain any self-energy corrections. The diagrammatic rules follow from those corresponding to the two-particle Green function $G_c^{(4)}$. Since $\Gamma^{(4)}$ shares the same symmetry properties as $G_c^{(4)}$ [Eq. (1.416)], it is (anti)symmetric under the exchange of the two incoming or outgoing particles,

$$\Gamma^{(4)}(\alpha_1\tau_1, \alpha_2\tau_2; \alpha_2'\tau_2', \alpha_1'\tau_1') = \zeta\Gamma^{(4)}(\alpha_2\tau_2, \alpha_1\tau_1; \alpha_2'\tau_2', \alpha_1'\tau_1')$$
$$= \zeta\Gamma^{(4)}(\alpha_1\tau_1, \alpha_2\tau_2; \alpha_1'\tau_1', \alpha_2'\tau_2'). \quad (1.417)$$

One can also write down a perturbation expansion for the (non-(anti)symmetrized) vertex, equal to $(\alpha_1\alpha_2|\hat{v}|\alpha_1'\alpha_2')$ to lowest order,

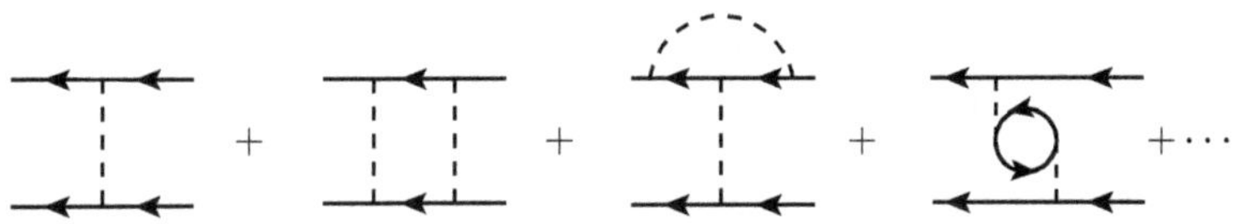

1.6.2.4 *General case*

It is possible to define fully (anti)symmetric 1PI vertices starting from the partition function

$$Z[J] = \int \mathcal{D}[\psi] \exp\Big\{-S[\psi] + \int_0^\beta d\tau \sum_{\tilde{\alpha}} J_{\tilde{\alpha}}(\tau)\psi_{\tilde{\alpha}}(\tau)\Big\} \tag{1.418}$$

and the effective action $\Gamma[\phi]$ defined as the Legendre transform

$$\begin{aligned} \phi_{\tilde{\alpha}}[\tau; J] &= \langle \psi_{\tilde{\alpha}}(\tau)\rangle = \frac{\delta W[J]}{\delta J_{\tilde{\alpha}}(\tau)}, \\ \Gamma[\phi] &= -W[J] + \int_0^\beta d\tau \sum_{\tilde{\alpha}} J_{\tilde{\alpha}}(\tau)\phi_{\tilde{\alpha}}(\tau), \end{aligned} \tag{1.419}$$

where $W[J] = \ln Z[J]$. $\Gamma[\phi]$ satisfies

$$\frac{\delta \Gamma[\phi]}{\delta \phi_{\tilde{\alpha}}(\tau)} = \zeta J_{\tilde{\alpha}}(\tau). \tag{1.420}$$

In the absence of external sources ($J = 0$), the effective action is stationary. If the stationary value $\bar{\phi}$ is non-zero, the U(1) symmetry is spontaneously broken.[52]

The 1PI vertices in the state $\bar{\phi}$ are defined by[53]

$$\Gamma^{(n)}(\tilde{\alpha}_1\tau_1, \dots, \tilde{\alpha}_n\tau_n) = \frac{\delta^n \Gamma[\phi]}{\delta\phi_{\tilde{\alpha}_n}(\tau_n)\cdots\delta\phi_{\tilde{\alpha}_1}(\tau_1)}\bigg|_{\phi=\bar{\phi}} \tag{1.421}$$

or the inverse relation

$$\begin{aligned} \Gamma[\phi] = \Gamma[\bar{\phi}] + \sum_{n=1}^{\infty} \frac{1}{n!} \int_0^\beta d\tau_1 \cdots d\tau_n \sum_{\tilde{\alpha}_1\cdots\tilde{\alpha}_n} \Gamma^{(n)}(\tilde{\alpha}_1\tau_1, \dots, \tilde{\alpha}_n\tau_n) \\ \times [\phi_{\tilde{\alpha}_1}(\tau_1) - \bar{\phi}_{\tilde{\alpha}_1}(\tau_1)] \cdots [\phi_{\tilde{\alpha}_n}(\tau_n) - \bar{\phi}_{\tilde{\alpha}_n}(\tau_n)]. \end{aligned} \tag{1.422}$$

They are (anti)symmetric under the exchange of two particles,

$$\begin{aligned} &\Gamma^{(n)}(\tilde{\alpha}_1\tau_1, \dots, \tilde{\alpha}_i\tau_i, \dots, \tilde{\alpha}_j\tau_j, \dots, \tilde{\alpha}_n\tau_n) \\ &\quad = \zeta\Gamma^{(n)}(\tilde{\alpha}_1\tau_1, \dots, \tilde{\alpha}_j\tau_j, \dots, \tilde{\alpha}_i\tau_i, \dots, \tilde{\alpha}_n\tau_n). \end{aligned} \tag{1.423}$$

---

[52]See footnote 44 on page 78.

[53]The 1PI vertices can also be defined with respect to the state $\phi = 0$, i.e. by expanding the effective action $\Gamma[\phi]$ about $\phi = 0$. In fermion systems, this is the only possibility since $\bar{\phi} = 0$ in the physical state.

From the first of Eq. (1.419), one easily deduces[54]

$$\frac{\delta^2\Gamma[\phi]}{\delta\phi\delta\phi} = \zeta\left(\frac{\delta^2 W[J]}{\delta J\delta J}\right)^{-1} \tag{1.424}$$

and, for $J = 0$ and $\phi = \bar{\phi}$,

$$\Gamma^{(2)} = G_c^{(2)\,-1}. \tag{1.425}$$

The self-energy is defined as $\Gamma^{(2)} = G_0^{-1} + \Sigma$,[55] i.e.

$$G_c^{-1} = G_0^{-1} + \Sigma, \tag{1.426}$$

and has a 2×2 matrix structure with respect to the charge index c,

$$\Sigma(\alpha\tau, \alpha'\tau') = \begin{pmatrix} \Sigma(\alpha,+,\tau;\alpha',+,\tau') & \Sigma(\alpha,+,\tau;\alpha',-,\tau') \\ \Sigma(\alpha,-,\tau;\alpha',+,\tau') & \Sigma(\alpha,-,\tau;\alpha',-,\tau') \end{pmatrix} \tag{1.427}$$

$=$ 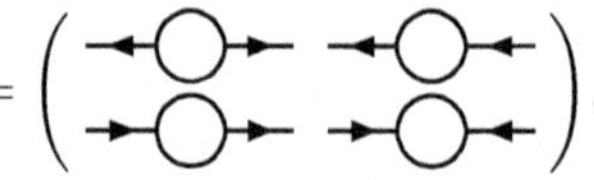

Equation (1.426) is a generalization of the Dyson equation to systems with spontaneous U(1) symmetry breaking. It is known as the Dyson–Beliaev equation. The perturbative expansion of $\Sigma(\tilde{\alpha}\tau, \tilde{\alpha}'\tau')$ in boson systems will be discussed in Sec. 1.7.

From (1.424) and using the relations[56]

$$\begin{aligned} \frac{\delta}{\delta J_{\tilde{\alpha}_{n+1}}(\tau_{n+1})} G_c^{(n)}[\tilde{\alpha}_1\tau_1, \ldots, \tilde{\alpha}_n\tau_n; J] &= \zeta^n G_c^{(n+1)}[\tilde{\alpha}_1\tau_1, \ldots, \tilde{\alpha}_{n+1}\tau_{n+1}; J], \\ \frac{\delta}{\delta J_{\tilde{\alpha}_{n+1}}(\tau_{n+1})} \Gamma^{(n)}[\tilde{\alpha}_1\tau_1, \ldots, \tilde{\alpha}_n\tau_n; J] &= \int_0^\beta d\tau \sum_{\tilde{\alpha}} G_c^{(2)}[\tilde{\alpha}_{n+1}\tau_{n+1}, \tilde{\alpha}\tau; J] \\ &\quad \times \Gamma^{(n+1)}[\tilde{\alpha}_1\tau_1, \ldots, \tilde{\alpha}_n\tau_n, \tilde{\alpha}\tau; J], \end{aligned} \tag{1.428}$$

one can express higher-order Green functions $G_c^{(n)}$ in terms of $\Gamma^{(m\leq n)}$ and $G_c^{(m<n)}$. For instance, one finds

$$\begin{aligned} G_c^{(3)}(\tilde{\alpha}_1\tau_1, \tilde{\alpha}_2\tau_2, \tilde{\alpha}_3\tau_3) = & -\zeta \int_0^\beta du_1 du_2 du_3 \sum_{\tilde{\gamma}_1,\tilde{\gamma}_2,\tilde{\gamma}_3} G_c(\tilde{\alpha}_1\tau_1, \tilde{\gamma}_1 u_1) \\ & \times G_c(\tilde{\alpha}_2\tau_2, \tilde{\gamma}_2 u_2) G_c(\tilde{\alpha}_3\tau_3, \tilde{\gamma}_3 u_3) \Gamma^{(3)}(\tilde{\gamma}_1 u_1, \tilde{\gamma}_2 u_2, \tilde{\gamma}_3 u_3). \end{aligned} \tag{1.429}$$

[54]The sign difference between (1.412) and (1.425) comes from the sign convention for the Green function $G^{(2)}(\tilde{\alpha}\tau, \tilde{\alpha}'\tau') = +\langle\psi_{\tilde{\alpha}}(\tau)\psi_{\tilde{\alpha}'}(\tau')\rangle$ (see Sec. 1.6.1).

[55]Given the definition of the Green function $G^{(2)}$,[54] the self-energy Σ in (1.426) is defined with the same sign convention as the one introduced for normal systems.

[56]The notation in (1.428) emphasizes that the Green functions $G_c^{(n)}[J]$ and the vertices $\Gamma^{(n)}[J]$, computed for arbitrary external sources, are functional of J.

This relation can be graphically represented as

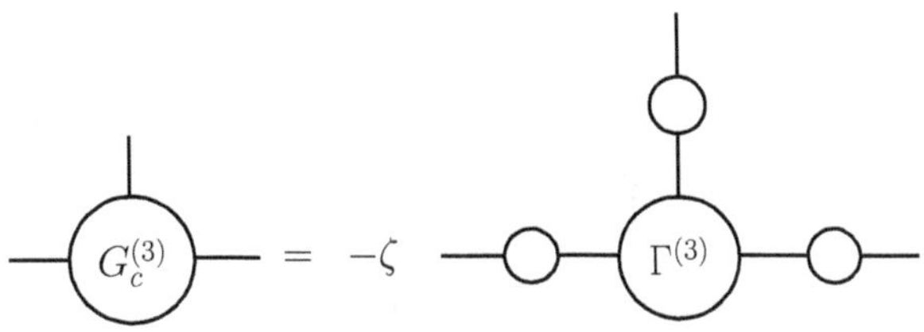

(the empty circles stand for the connected one-particle Green function) and identifies $\Gamma^{(3)}$ as the three-leg 1PI vertex. To the next order, we obtain the equation

$$
\begin{aligned}
&G_c^{(4)}(\tilde{\alpha}_1\tau_1, \tilde{\alpha}_2\tau_2, \tilde{\alpha}_3\tau_3, \tilde{\alpha}_4\tau_4) \\
&= -\int_0^\beta du_1 du_2 du_3 du_4 \sum_{\tilde{\gamma}_1 \cdots \tilde{\gamma}_4} G_c(\tilde{\alpha}_1\tau_1, \tilde{\gamma}_1 u_1) G_c(\tilde{\alpha}_2\tau_2, \tilde{\gamma}_2 u_2) \\
&\qquad \times G_c(\tilde{\alpha}_3\tau_3, \tilde{\gamma}_3 u_3) G_c(\tilde{\alpha}_4\tau_4, \tilde{\gamma}_4 u_4) \Gamma^{(4)}(\tilde{\gamma}_1 u_1, \tilde{\gamma}_2 u_2, \tilde{\gamma}_3 u_3, \tilde{\gamma}_4 u_4) \\
&\quad - \int_0^\beta du_1 du_2 du_3 \sum_{\tilde{\gamma}_1, \tilde{\gamma}_2, \tilde{\gamma}_3} \Gamma^{(3)}(\tilde{\gamma}_1 u_1, \tilde{\gamma}_2 u_2, \tilde{\gamma}_3 u_3) \\
&\qquad \times \Big[G_c^{(3)}(\tilde{\alpha}_1\tau_1, \tilde{\gamma}_1 u_1, \tilde{\alpha}_4\tau_4) G_c(\tilde{\alpha}_2\tau_2, \tilde{\gamma}_2 u_2) G_c(\tilde{\alpha}_3\tau_3, \tilde{\gamma}_3 u_3) \\
&\qquad\quad + G_c(\tilde{\alpha}_1\tau_1, \tilde{\gamma}_1 u_1) G_c^{(3)}(\tilde{\alpha}_2\tau_2, \tilde{\gamma}_2 u_2, \tilde{\alpha}_4\tau_4) G_c(\tilde{\alpha}_3\tau_3, \tilde{\gamma}_3 u_3) \\
&\qquad\quad + G_c(\tilde{\alpha}_1\tau_1, \tilde{\gamma}_1 u_1) G_c(\tilde{\alpha}_2\tau_2, \tilde{\gamma}_2 u_2) G_c^{(3)}(\tilde{\alpha}_3\tau_3, \tilde{\gamma}_3 u_3, \tilde{\alpha}_4\tau_4) \Big]. \qquad (1.430)
\end{aligned}
$$

Combining (1.429) and (1.430), we finally obtain the following diagrammatic representation of $G_c^{(4)}$:

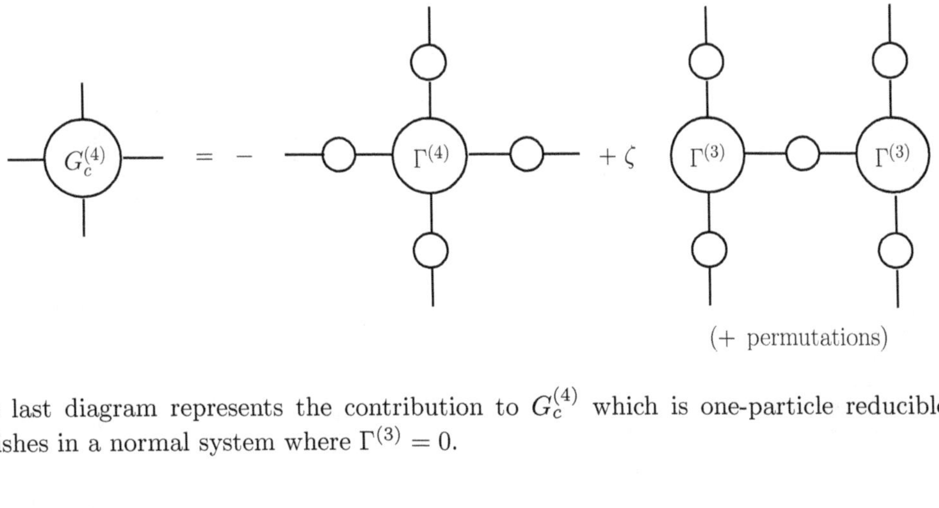

(+ permutations)

The last diagram represents the contribution to $G_c^{(4)}$ which is one-particle reducible. It vanishes in a normal system where $\Gamma^{(3)} = 0$.

1.6.2.5 *Equations of motion*

The equations of motion of the Green functions imply relations between the 1PI vertices. Equation (1.395) can be rewritten as[57]

$$\sum_{1'} G_0^{-1}(1,1')\phi_{1'} + \frac{1}{3!}\sum_{2',3',4'} v_{12'3'4'}\mathcal{G}^{(3)}(2',3',4') - \frac{\delta\Gamma}{\delta\phi_1} = 0, \tag{1.431}$$

where we use the shorthand notation $i \equiv (\tilde{\alpha}_i, \tau_i)$ and $\sum_i \equiv \int_0^\beta d\tau_i \sum_{\tilde{\alpha}_{i'}}$ and

$$\mathcal{G}^{(3)}(2',3',4') = \frac{\delta^3 \mathcal{G}[J]}{\delta J_{2'}\delta J_{3'}\delta J_{4'}}. \tag{1.432}$$

$\mathcal{G}[J] = Z[J]/Z[0]$ is the generating functional of the Green functions [Eq. (1.392)]. Using

$$\begin{aligned}\mathcal{G}^{(3)}(2',3',4') &= W^{(3)}(2',3',4') + \phi_{2'}W^{(2)}(3',4') + \zeta\phi_{3'}W^{(2)}(2',4') \\ &\quad + \phi_{4'}W^{(2)}(2',3') + \phi_{2'}\phi_{3'}\phi_{4'}\end{aligned} \tag{1.433}$$

[see Eq. (1.391)] and

$$W^{(3)}(2',3',4') = -\zeta \sum_{2'',3'',4''} W^{(2)}(2',2'')W^{(2)}(3',3'')W^{(2)}(4',4'')\Gamma^{(3)}(2'',3'',4''), \tag{1.434}$$

one finds

$$\begin{aligned}&\sum_{1'} G_0^{-1}(1,1')\phi_{1'} - \frac{\delta\Gamma}{\delta\phi_1} \\ &+\frac{1}{3!}\sum_{2',3',4'} v_{1,2',3',4'}\Big[-\zeta\sum_{2'',3'',4''} W^{(2)}(2',2'')W^{(2)}(3',3'')W^{(2)}(4',4'')\Gamma^{(3)}(2'',3'',4'') \\ &+\phi_{2'}W^{(2)}(3',4') + \zeta\phi_{3'}W^{(2)}(2',4') + \phi_{4'}W^{(2)}(2',3') + \phi_{2'}\phi_{3'}\phi_{4'}\Big] = 0.\end{aligned} \tag{1.435}$$

If we consider this equation in the physical state, where $\delta\Gamma/\delta\phi = \zeta J = 0$ and $W^{(2)} = G_c$, we obtain an equation for the order parameter $\bar{\phi}_{\tilde{\alpha}}(\tau) = \langle\psi_{\tilde{\alpha}}(\tau)\rangle$ which is diagrammatically represented in Fig. 1.8.

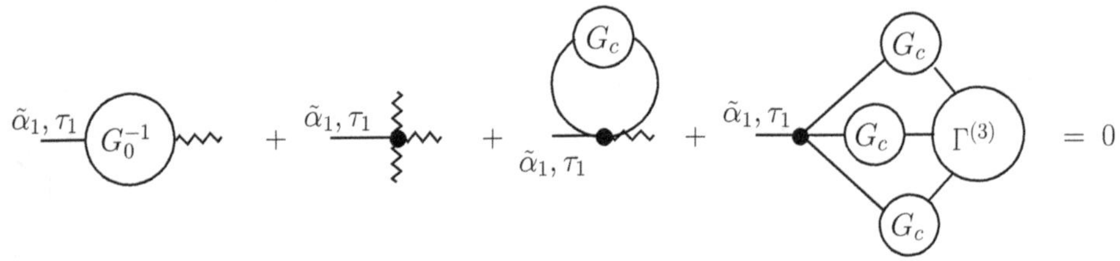

Fig. 1.8 Diagrammatic representation of Eq. (1.435) with $\delta\Gamma/\delta\phi = \zeta J = 0$. The zigzag line stands for the order parameter $\bar{\phi}_{\tilde{\alpha}}(\tau)$.

[57] To alleviate the notations, we do not indicate explicitly that $\mathcal{G}^{(3)}[2',3',4';J]$ in (1.431), $W^{(3)}[2',3',4';J]$ in (1.433), etc., are actually functionals of J.

Equation (1.435) is also the starting point for deriving the equations of motion satisfied by the 1PI vertices. If we take the functional derivative with respect to ϕ_2, we obtain

$$\begin{aligned}\Sigma(1,2) = &-\frac{\zeta}{3!}\sum_{2'\cdots4''} v_{1,2',3',4'}G_c(2',2'')G_c(3',3'')G_c(4',4'')\Gamma^{(4)}(2'',3'',4'',2)\\ &+\frac{\zeta}{2}\sum_{2'\cdots4''} v_{1,2',3',4'}\big(G_c\Gamma^{(3)}(2)G_c\big)(2',2'')G_c(3',3'')G_c(4',4'')\Gamma^{(3)}(2'',3'',4'')\\ &-\frac{\zeta}{2}\sum_{2',3',4'} v_{1,2',3',4'}\phi_{2'}\big(G_c\Gamma^{(3)}(2)G_c\big)(3',4')\\ &+\frac{1}{2}\sum_{3',4'} v_{1,2,3',4'}[G_c(3',4')+\phi_{3'}\phi_{4'}]. \end{aligned} \tag{1.436}$$

For a normal system ($\phi = 0$ when $J = 0$), the self-energy reads

$$\begin{aligned}\Sigma(1,2) = &\sum_{3',4'}\{1,3'|\hat{v}|4',2\}G(4',3')\\ &-\frac{1}{2}\sum_{2'\cdots4''}\{1,2'|\hat{v}|4',3'\}G(2'',2')G(3',3'')G(4',4'')\Gamma^{(4)}(3'',4'';2'',2) \end{aligned} \tag{1.437}$$

and is shown diagrammatically in Fig. 1.9. Equation (1.437) can be used to derive a relation between the self-energy and the expectation value of the interacting Hamiltonian. Multiplying $\Sigma(1,3)$ by $G(3,2)$ and summing over 3 gives[58]

$$\sum_3 \Sigma(1,3)G(3,2) = \sum_{2',3',4'}(1,2'|\hat{v}|4',3')G^{(4)}(3',4';2',2), \tag{1.438}$$

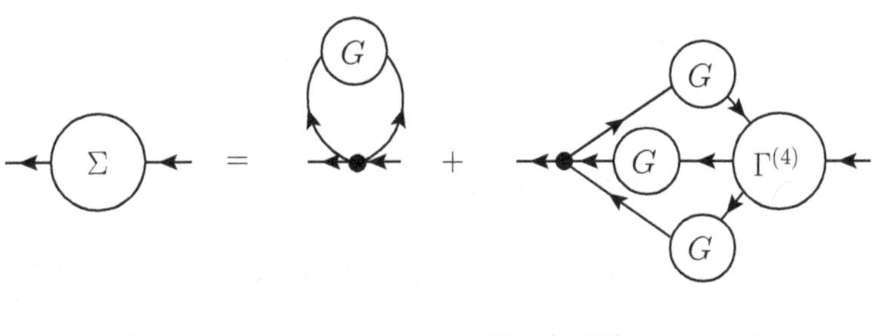

Fig. 1.9 Diagrammatic representation of Eq. (1.437) in a normal system.

[58] Equation (1.438) is derived by using the relation between $G^{(4)}$ and $\Gamma^{(4)}$.

where $G^{(4)}$ is the two-particle Green function. Taking $2 = 1^+$ (i.e. $\alpha_2 = \alpha_1$ and $\tau_2 = \tau_1^+ = \tau_1 + 0^+$) and summing over 1, we obtain[59]

$$\begin{aligned}
&\int_0^\beta d\tau_1 d\tau_3 \sum_{\alpha_1,\alpha_3} \Sigma(\alpha_1\tau_1, \alpha_3\tau_3) G(\alpha_3\tau_3, \alpha_1\tau_1^+) \\
&= \int_0^\beta d\tau_1 \sum_{\alpha_1,\alpha_2',\alpha_3',\alpha_4'} (\alpha_1\alpha_2'|\hat{v}|\alpha_4'\alpha_3') G^{(4)}(\alpha_3'\tau_1^-, \alpha_4'\tau_1^-; \alpha_2'\tau_1, \alpha_1\tau_1^+),
\end{aligned} \tag{1.439}$$

i.e.

$$\frac{1}{\beta} \sum_{\alpha,\alpha',\omega_n} \Sigma(\alpha,\alpha'; i\omega_n) G(\alpha',\alpha; i\omega_n) e^{i\omega_n 0^+} = -2\langle \hat{H}_{\rm int} \rangle. \tag{1.440}$$

Equations (1.438) and (1.440) relate the single-particle properties, *via* the self-energy, to a two-particle quantity. Note that the propagator in (1.437)–(1.440) is defined by $G(1,2) = \langle \psi_{\alpha_1}(\tau_1)\psi^*_{\alpha_2}(\tau_2)\rangle$ (with a + sign).

1.6.3 *Luttinger–Ward functional and 2PI vertices*

In this section, we introduce the Luttinger–Ward functional and the 2PI vertices. The Luttinger–Ward functional is used to set up "conserving" approximations, i.e. approximations that satisfy conservation laws, as well as non-perturbative approaches (Chapter 9).

1.6.3.1 *Normal systems*

We consider the partition function in the presence of a bilinear source which couples to a particle–hole pair operator,

$$Z[J] = \int \mathcal{D}[\psi^*,\psi] \exp\Big\{ -S[\psi^*,\psi] + \int_0^\beta d\tau d\tau' \sum_{\alpha,\alpha'} J_{\alpha\alpha'}(\tau,\tau')\psi^*_\alpha(\tau)\psi_{\alpha'}(\tau') \Big\}. \tag{1.441}$$

By taking the functional derivative with respect to the source, we obtain the one-particle Green function

$$G[\alpha\tau, \alpha'\tau'; J] = -\langle \psi_\alpha(\tau)\psi^*_{\alpha'}(\tau')\rangle = -\zeta \frac{\delta \ln Z[J]}{\delta J_{\alpha'\alpha}(\tau',\tau)}. \tag{1.442}$$

The Legendre transform of $-\ln Z[J]$, the 2PI effective action, is a functional of the propagator G,

$$\Gamma[G] = -\ln Z[J] - \zeta \int_0^\beta d\tau d\tau' \sum_{\alpha,\alpha'} J_{\alpha\alpha'}(\tau,\tau') G(\alpha'\tau', \alpha\tau) \tag{1.443}$$

(with $J \equiv J[G]$ a functional of G obtained from (1.442), assuming that $G[J]$ is invertible), which satisfies

$$\frac{\delta \Gamma[G]}{\delta G(\alpha\tau, \alpha'\tau')} = -\zeta J_{\alpha'\alpha}(\tau',\tau). \tag{1.444}$$

---

[59] We use $(1,2'|\hat{v}|4',3') = \delta(\tau_1 - \tau_2')\delta(\tau_3' - \tau_4')\delta(\tau_1^- - \tau_3')(\alpha_1\alpha_2'|\hat{v}|\alpha_4'\alpha_3')$.

The actual Green function $\bar{G}$, obtained for vanishing external sources, corresponds to the stationary value of the functional $\Gamma[G]$.

For a non-interacting system, $\Gamma[G]$ can be calculated exactly since the action is Gaussian. The partition function reads[60]

$$Z[J] = \det\left(-G_0^{-1} - J\right)^{-\zeta} = e^{-\zeta \mathrm{Tr}\ln(-G_0^{-1}-J)}, \tag{1.445}$$

and (1.442) then gives $G[J] = (G_0^{-1} + J)^{-1}$. We deduce

$$\Gamma[G] = -\zeta \mathrm{Tr}\ln(-G) + \zeta \mathrm{Tr}(G_0^{-1}G - 1). \tag{1.446}$$

For an interacting system, we write the functional Γ as

$$\Gamma[G] = -\zeta \mathrm{Tr}\ln(-G) + \zeta \mathrm{Tr}(G_0^{-1}G - 1) + \Phi[G], \tag{1.447}$$

where $\Phi[G]$ is referred to as the Luttinger–Ward functional. $\Gamma[G]$ and $\Phi[G]$ are well defined only if $G[J]$ is invertible, i.e. if there is a one-to-one correspondence between J and G. This is not necessarily the case and the Luttinger–Ward functional can be multivalued: See the discussion in Sec. 9.1.4. In the following, $\Phi[G]$ should be understood as the physical branch of the Luttinger–Ward functional.

From Eq. (1.444), we deduce

$$-\zeta G^{-1}(\alpha'\tau', \alpha\tau) + \zeta G_0^{-1}(\alpha'\tau', \alpha\tau) + \frac{\delta\Phi[G]}{\delta G(\alpha\tau, \alpha'\tau')} = -\zeta J_{\alpha'\alpha}(\tau', \tau), \tag{1.448}$$

which is nothing but the Dyson equation $G^{-1} = G_0^{-1} - \Sigma$ [Eq. (1.413)]. For $J = 0$, the self-energy is given by the first-order functional derivative of the Luttinger–Ward functional,

$$\Sigma(\alpha\tau, \alpha'\tau') = -\zeta \left.\frac{\delta\Phi[G]}{\delta G(\alpha'\tau', \alpha\tau)}\right|_{G=\bar{G}}, \tag{1.449}$$

where $\bar{G} = (G_0^{-1} - \Sigma)^{-1}$ is the solution of the equation $\delta\Gamma/\delta G = 0$. Equation (1.449) can be used to find a diagrammatic interpretation of $\Phi[G]$. The perturbation expansion of Σ with respect to G_0 (Sec. 1.6.2) can be rewritten as an expansion with respect to the full propagator G (represented here by a thick solid line),[61]

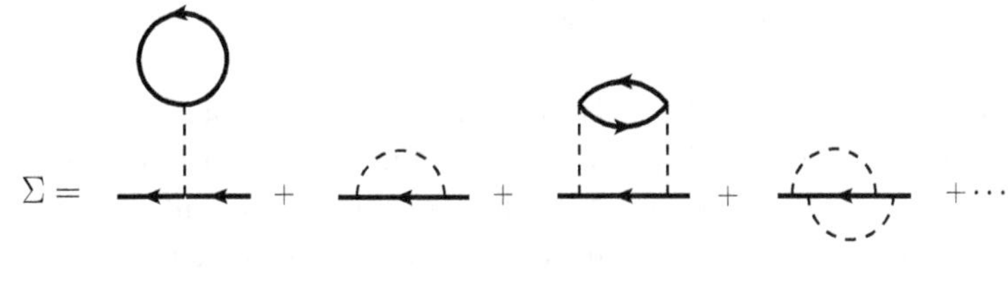

One can verify that this diagrammatic resummation makes sense by expanding G to recover the Feynman diagrams in terms of G_0 with proper factors and signs. Σ being defined as the functional derivative of the Luttinger–Ward functional $\Phi[G]$ [Eq. (1.449)], we deduce that

---

[60] We use Eq. (1.E.41).

[61] We assume that the particles interact *via* a two-body interaction $\hat{v}$.

$\Phi[G]$ is given by the sum of the 2PI Feynman diagrams defined as the graphs that cannot be separated into two disconnected pieces by cutting one or two lines[62]:

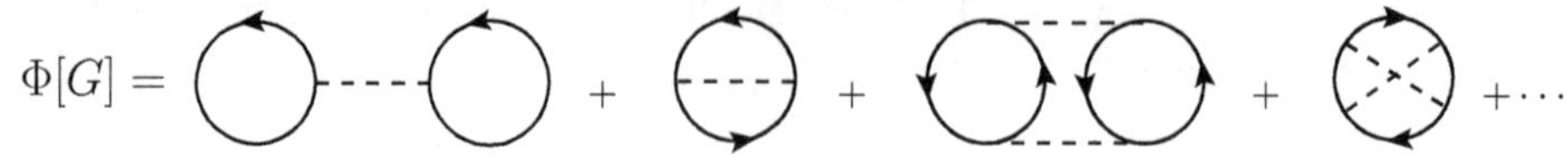

Since the functional derivation with respect to G amounts to removing a propagator G from the diagrams contributing to the Luttinger–Ward functional, the diagrammatic definition of $\Phi[G]$ reproduces the expansion of Σ in terms of G. Note that a two-particle reducible diagram for $\Phi[G]$ would give a contribution to the self-energy which is not 1PI. The diagrammatic rules for $\Phi[G]$ are the same as those for $-\ln Z$. One can check that they are consistent with the diagrammatic rules for the self-energy. For instance, to first-order $\Phi[G]$ is given by

$$\Phi[G] = \frac{1}{2}\sum_{\alpha_1\cdots\alpha_2'} (\alpha_1\alpha_2|\hat{v}|\alpha_1'\alpha_2')\int_0^\beta d\tau\, G(\alpha_1'\tau,\alpha_1\tau^+)G(\alpha_2'\tau,\alpha_2\tau^+)$$
$$+\frac{\zeta}{2}\sum_{\alpha_1\cdots\alpha_2'} (\alpha_1\alpha_2|\hat{v}|\alpha_1'\alpha_2')\int_0^\beta d\tau\, G(\alpha_1'\tau,\alpha_2\tau^+)G(\alpha_2'\tau,\alpha_1\tau^+), \tag{1.450}$$

and the first-order functional derivative $\Phi^{(1)}$ reproduces the lowest-order contribution to the self-energy with G_0 replaced by G (see page 87).

The 2PI vertices $\Phi^{(n)}$ are defined by

$$\Phi^{(n)}(\alpha_1\tau_1,\alpha_1'\tau_1';\cdots;\alpha_n\tau_n,\alpha_n'\tau_n') = \left.\frac{(-\zeta)^n\delta^n\Phi[G]}{\delta G(\alpha_1\tau_1,\alpha_1'\tau_1')\cdots\delta G(\alpha_n\tau_n,\alpha_n'\tau_n')}\right|_{G=\bar{G}}. \tag{1.451}$$

$\Phi^{(n)}$ is a vertex with n external "bosonic" legs $(\alpha_i\tau_i,\alpha_i'\tau_i')$. For example, to second-order $\Phi[G]$ is defined by the four diagrams of the preceding figure, so that $\Phi^{(1)}$ is given by the (amputated) diagrams

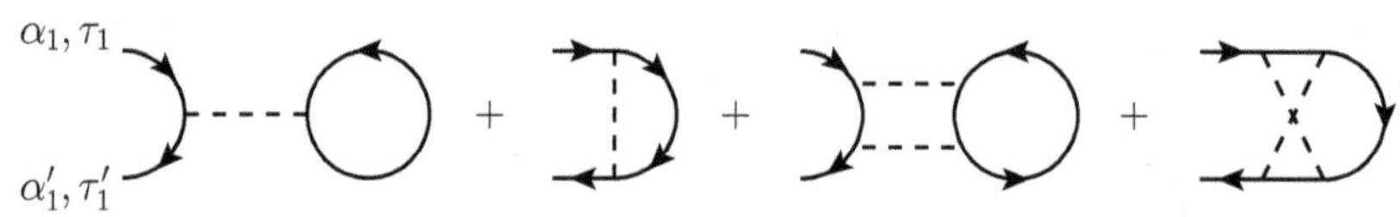

where the external legs $(\alpha_i\tau_i)$ and $(\alpha_i'\tau_i')$ are drawn nearby so that they can be seen as a single bosonic (particle–hole) leg $(\alpha_i\tau_i,\alpha_i'\tau_i')$. The diagrams contributing to $\Phi^{(2)}$ to $\mathcal{O}(v^2)$ are shown in Fig. 1.10. It is clear that $\Phi^{(1)}$, $\Phi^{(2)}$ and more generally all vertices obtained from $\Phi[G]$ are 2PI: They cannot be separated into two disconnected pieces by cutting only two internal lines if one considers each external bosonic leg as a single piece. Since the leg $(\alpha_i\tau_i,\alpha_i'\tau_i')$ corresponds to a particle–hole pair, the vertices $\Phi^{(n)}$ are said to be irreducible in the particle–hole channel. This, of course, follows from the choice of $J_{\alpha\alpha'}(\tau,\tau')$ as an

[62]The 2PI diagrams are also called skeleton diagrams or bold-line diagrams as they do not contain self-energy insertions.

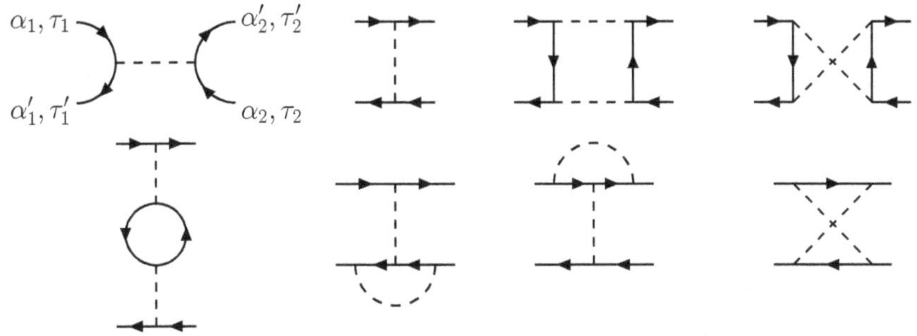

Fig. 1.10 Diagrams contributing to the 2PI vertex $\Phi^{(2)}$ to second order in the interaction.

external source that couples to a particle–hole pair. We shall see in the following section that it is also possible to define irreducible vertices in the particle–particle channel.

In the 2PI formalism, the Green functions are naturally defined by

$$W^{(n)}(\alpha_1\tau_1, \alpha_1'\tau_1'; \cdots ; \alpha_n\tau_n, \alpha_n'\tau_n') = \left. \frac{(-\zeta)^n \delta^n \ln Z[J]}{\delta J_{\alpha_1'\alpha_1}(\tau_1', \tau_1) \cdots \delta J_{\alpha_n'\alpha_n}(\tau_n', \tau_n)} \right|_{J=0}. \tag{1.452}$$

One easily verifies that

$$\begin{aligned} W^{(1)}(\alpha_1\tau_1, \alpha_1'\tau_1') &= G^{(2)}(\alpha_1\tau_1, \alpha_1'\tau_1') \equiv \bar{G}(\alpha_1\tau_1, \alpha_1'\tau_1'), \\ W^{(2)}(\alpha_1\tau_1, \alpha_1'\tau_1'; \alpha_2\tau_2, \alpha_2'\tau_2') &= G^{(4)}(\alpha_1\tau_1, \alpha_2\tau_2; \alpha_2'\tau_2', \alpha_1'\tau_1') \\ &\quad - \bar{G}(\alpha_1\tau_1, \alpha_1'\tau_1')\bar{G}(\alpha_2\tau_2, \alpha_2'\tau_2'). \end{aligned} \tag{1.453}$$

$W^{(2)}$ is a "connected" Green function with respect to the external bosonic legs $(\alpha_1\tau_1, \alpha_1'\tau_1')$ and $(\alpha_2\tau_2, \alpha_2'\tau_2')$. It plays a role similar to that of the connected Green function $G_c^{(4)}$ introduced in the 1PI formalism (Sec. 1.6.2).[63] By taking a functional derivative with respect to J in (1.444), one obtains the relation

$$\frac{\delta^2 W[J]}{\delta J \delta J} = \left(\frac{\delta \Gamma[G]}{\delta G \delta G} \right)^{-1}, \tag{1.454}$$

where the inverse should be understood in a bosonic matrix sense, the multiplication being defined by

$$(AB)_{\gamma_1\gamma_2} = \sum_{\gamma_3} A_{\gamma_1\gamma_3} B_{\gamma_3\gamma_2}, \tag{1.455}$$

with the short-hand notation

$$\gamma = (\alpha\tau, \alpha'\tau'), \qquad \sum_\gamma = \int_0^\beta d\tau d\tau' \sum_{\alpha,\alpha'}. \tag{1.456}$$

[63]Note that for a non-interacting system, one has $W^{(2)}(\alpha_1\tau_1, \alpha_1'\tau_1'; \alpha_2\tau_2, \alpha_2'\tau_2') = \zeta G_0(\alpha_1\tau_1, \alpha_2'\tau_2') \times G_0(\alpha_2\tau_2, \alpha_1'\tau_1')$ while $G_c^{(4)}(\alpha_1\tau_1, \alpha_1'\tau_1'; \alpha_2\tau_2, \alpha_2'\tau_2') = 0$.

Using

$$\left.\frac{\delta^2\Gamma[G]}{\delta G_{\gamma_1}\delta G_{\gamma_2}}\right|_{G=\bar{G}} = \Pi^{-1}_{\gamma_1\gamma_2} + \Phi^{(2)}_{\gamma_1\gamma_2} \tag{1.457}$$

[which follows from (1.447)], where

$$\Pi_{\gamma_1\gamma_2} = \zeta\bar{G}(\alpha_1\tau_1, \alpha'_2\tau'_2)\bar{G}(\alpha_2\tau_2, \alpha'_1\tau'_1) \tag{1.458}$$

is a particle–hole pair propagator, we can rewrite (1.454) as

$$W^{(2)} = \left(\Pi^{-1} + \Phi^{(2)}\right)^{-1} = \Pi - \Pi\Phi^{(2)}W^{(2)}. \tag{1.459}$$

Equation (1.459) is known as the Bethe–Salpeter equation and relates the two-particle Green function $W^{(2)}$ to the 2PI vertex $\Phi^{(2)}$:

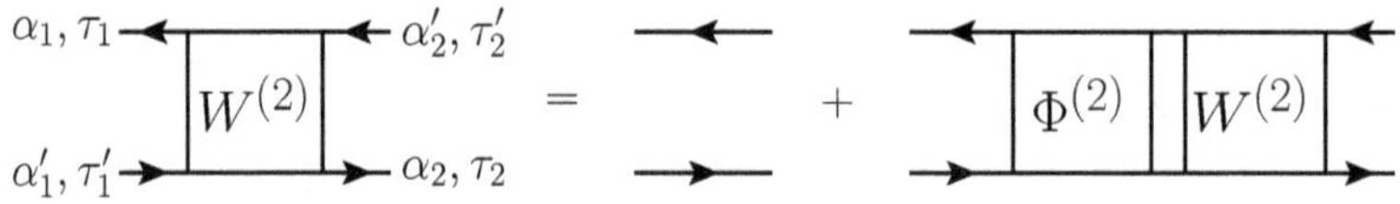

It is similar to the Dyson equation (1.413) with $\Phi^{(2)}$ and Π playing the role of the self-energy Σ and the non-interacting propagator G_0, respectively.

Using the relation between $W^{(2)}$ and $G^{(4)}_c$,

$$W^{(2)}_{\gamma_1\gamma_2} = G^{(4)}_c(\alpha_1\tau_1, \alpha_2\tau_2; \alpha'_2\tau'_2, \alpha'_1\tau'_1) + \zeta\bar{G}(\alpha_1\tau_1, \alpha'_2\tau'_2)\bar{G}(\alpha_2\tau_2, \alpha'_1\tau'_1) \tag{1.460}$$

[which follows from (1.453)] and the relation (1.416) between $G^{(4)}_c$ and the 1PI vertex $\Gamma^{(4)}$, we also obtain a relation between $\Gamma^{(4)}$ and $\Phi^{(2)}$,

$$\Gamma^{(4)}(\alpha_1\tau_1, \alpha_2\tau_2; \alpha'_2\tau'_2, \alpha'_1\tau'_1) = \left[\Phi^{(2)}\left(1 + \Pi\Phi^{(2)}\right)^{-1}\right](\alpha'_1\tau'_1, \alpha_1\tau_1; \alpha'_2\tau'_2, \alpha_2\tau_2), \tag{1.461}$$

which takes the diagrammatic form

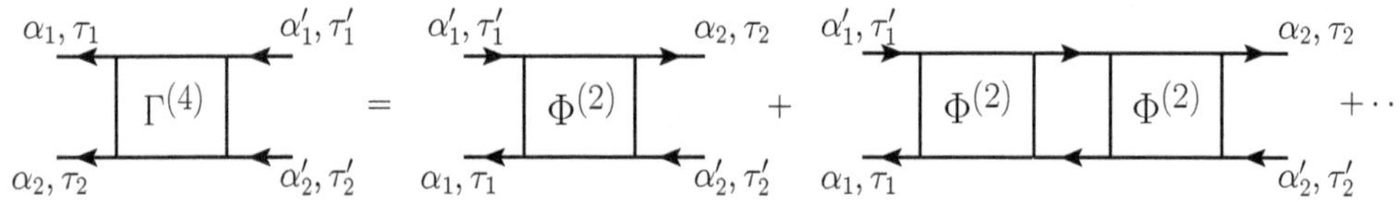

1.6.3.2 *General case*

We can generalize the preceding analysis by considering the action

$$S[\psi] - \frac{1}{2}\int_0^\beta d\tau d\tau' \sum_{\tilde{\alpha},\tilde{\alpha}'} J_{\tilde{\alpha}\tilde{\alpha}'}(\tau, \tau')\psi_{\tilde{\alpha}}(\tau)\psi_{\tilde{\alpha}'}(\tau'), \tag{1.462}$$

where $J_{\tilde{\alpha}\tilde{\alpha}'}(\tau, \tau') = \zeta J_{\tilde{\alpha}'\tilde{\alpha}}(\tau', \tau)$ is an (anti)symmetric source. With our sign convention for the Green function (Sec. 1.6.1),

$$G[\tilde{\alpha}\tau, \tilde{\alpha}'\tau'; J] = \langle\psi_{\tilde{\alpha}}(\tau)\psi_{\tilde{\alpha}'}(\tau')\rangle = \frac{\delta \ln Z[J]}{\delta J_{\tilde{\alpha}\tilde{\alpha}'}(\tau, \tau')}. \tag{1.463}$$

The 2PI effective action is now defined as

$$\Gamma[G] = -\ln Z[J] + \frac{1}{2}\int_0^\beta d\tau d\tau' \sum_{\tilde{\alpha},\tilde{\alpha}'} J_{\tilde{\alpha},\tilde{\alpha}'}(\tau,\tau')G(\tilde{\alpha}\tau,\tilde{\alpha}'\tau')$$
$$= -\frac{\zeta}{2}\mathrm{Tr}\ln(G) + \frac{\zeta}{2}\mathrm{Tr}(G_0^{-1}G - 1) + \Phi[G] \tag{1.464}$$

and satisfies

$$\frac{\delta\Gamma[G]}{\delta G_\gamma} = -G_\gamma^{-1} + G_{0,\gamma}^{-1} + \frac{\delta\Phi[G]}{\delta G_\gamma} = J_\gamma \tag{1.465}$$

with the short-hand notation $\gamma = (\tilde{\alpha}\tau, \tilde{\alpha}'\tau')$. Equation (1.465) implies that for $J = 0$, the self-energy is defined by

$$\Sigma_\gamma = \left.\frac{\delta\Phi[G]}{\delta G_\gamma}\right|_{G=\bar{G}}, \tag{1.466}$$

where $\bar{G} = (G_0^{-1} + \Sigma)^{-1}$ is solution of $\delta\Gamma/\delta G = 0$, and the Luttinger–Ward functional given by the sum of 2PI diagrams (Fig. 1.11).

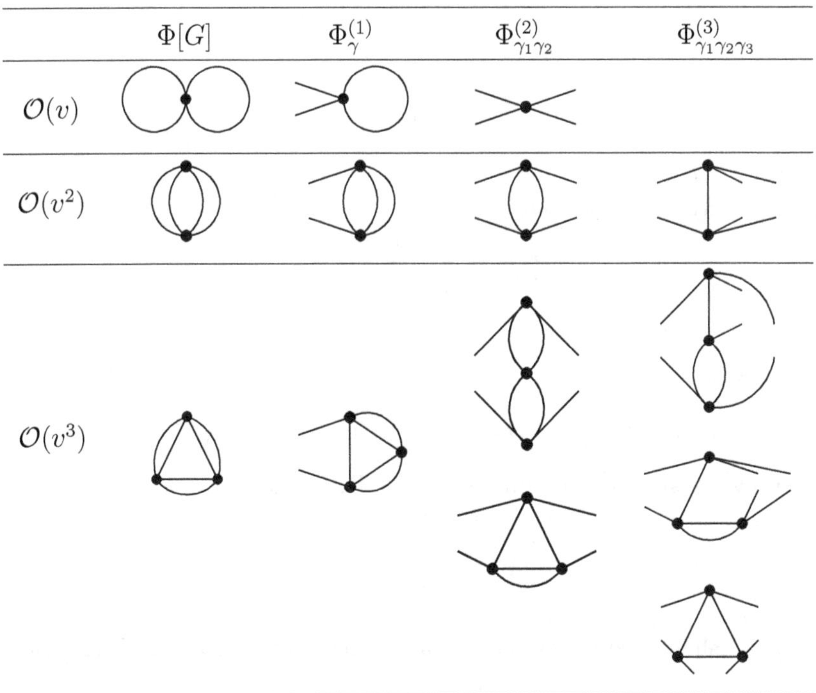

Fig. 1.11 Diagrammatic representation of the Luttinger–Ward functional $\Phi[G]$ and the 2PI vertices $\Phi^{(1)}_\gamma$, $\Phi^{(2)}_{\gamma_1\gamma_2}$ and $\Phi^{(3)}_{\gamma_1\gamma_2\gamma_3}$ to third order in the interaction.

Because of the (anti)symmetry of the source J_γ and the Green function G_γ in the exchange $(\tilde\alpha\tau) \leftrightarrow (\tilde\alpha'\tau')$ (with $\gamma = (\tilde\alpha\tau, \tilde\alpha'\tau')$), we define a "bosonic" unit matrix by

$$I_{\gamma_1\gamma_2} = \delta_{\tilde\alpha_1,\tilde\alpha_2}\delta_{\tilde\alpha'_1,\tilde\alpha'_2}\delta(\tau_1-\tau_2)\delta(\tau'_1-\tau'_2) + \zeta\delta_{\tilde\alpha_1,\tilde\alpha'_2}\delta_{\tilde\alpha'_1,\tilde\alpha_2}\delta(\tau_1-\tau'_2)\delta(\tau'_1-\tau_2), \quad (1.467)$$

so that $\delta J_{\gamma_1}/\delta J_{\gamma_2} = I_{\gamma_1\gamma_2}$. The variation of a functional, e.g. $W[J]$, involves an additional factor $1/2$,

$$\delta W[J] = \frac{1}{2}\sum_\gamma \frac{\delta W[J]}{\delta J_\gamma}\delta J_\gamma \quad (1.468)$$

and the chain rule for derivation reads

$$\frac{\delta W[J]}{\delta G_\gamma} = \frac{1}{2}\sum_{\gamma'} \frac{\delta W[J]}{\delta J_{\gamma'}}\frac{\delta J_{\gamma'}}{\delta G_\gamma}. \quad (1.469)$$

The 2PI vertices are defined by

$$\Phi^{(n)}_{\gamma_1\cdots\gamma_n} = \frac{\delta^n\Phi[G]}{\delta G_{\gamma_1}\cdots G_{\gamma_n}}\bigg|_{G=\bar G}. \quad (1.470)$$

$\Phi^{(n)}$ is a vertex with n external bosonic legs (Fig. 1.11) and is 2PI in the sense that it cannot be separated into two disconnected pieces by cutting only two internal lines. An external leg $\gamma = (\tilde\alpha\tau, \tilde\alpha'\tau')$ corresponds to a particle–hole pair if $c = -c'$ and to a particle–particle pair if $c = c'$. Since G_γ is (anti)symmetric in the exchange $(\tilde\alpha\tau) \leftrightarrow (\tilde\alpha'\tau')$, $\Phi^{(n)}$ satisfies

$$\begin{aligned} \Phi^{(n)}_{\gamma_1\cdots\gamma_i\cdots\gamma_j\cdots\gamma_n} &= \Phi^{(n)}_{\gamma_1\cdots\gamma_j\cdots\gamma_i\cdots\gamma_n}, \\ \Phi^{(n)}_{\gamma_1\cdots\gamma_i\cdots\gamma_n} &= \zeta\Phi^{(n)}_{\gamma_1\cdots\bar\gamma_i\cdots\gamma_n}, \end{aligned} \quad (1.471)$$

where $\bar\gamma_i = (\tilde\alpha'_i\tau'_i, \tilde\alpha_i\tau_i)$.

The Green functions are defined by

$$W^{(n)}_{\gamma_1\cdots\gamma_n} = \frac{\delta^n \ln Z[J]}{\delta J_{\gamma_1}\cdots\delta J_{\gamma_n}}\bigg|_{J=0}. \quad (1.472)$$

To lowest-order, they read

$$\begin{aligned} W^{(1)}_\gamma &= \bar G_\gamma, \\ W^{(2)}_{\gamma_1\gamma_2} &= G^{(4)}(\tilde\alpha_1\tau_1, \tilde\alpha'_1\tau'_1, \tilde\alpha_2\tau_2, \tilde\alpha'_2\tau'_2) - \bar G(\tilde\alpha_1\tau_1, \tilde\alpha'_1\tau'_1)\bar G(\tilde\alpha_2\tau_2, \tilde\alpha'_2\tau'_2). \end{aligned} \quad (1.473)$$

$W^{(2)}_{\gamma_1\gamma_2}$ is given by the sum of diagrams that are fully connected to the external bosonic legs γ_1 and γ_2. To relate $W^{(2)}$ to $\Phi^{(2)}$, we start from

$$\frac{\delta W}{\delta J\delta J} = \left(\frac{\delta^2\Gamma[G]}{\delta G\delta G}\right)^{-1}, \quad (1.474)$$

where the inverse should be understood in a bosonic matrix sense with the multiplication defined by

$$(AB)_{\gamma_1\gamma_2} = \frac{1}{2}\sum_{\gamma_3} A_{\gamma_1\gamma_3}B_{\gamma_3\gamma_2} \quad (1.475)$$

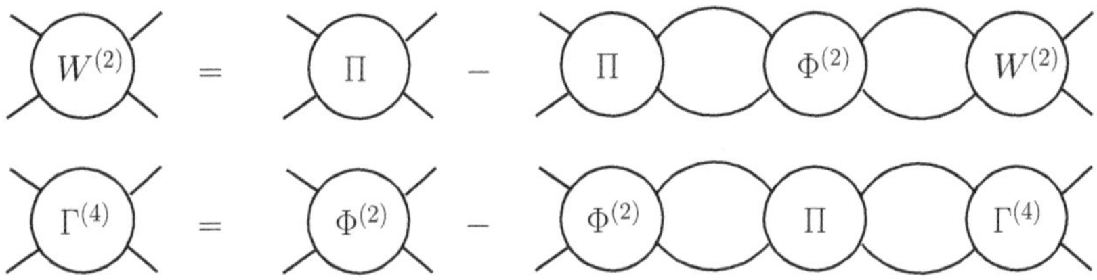

Fig. 1.12 Diagrammatic representation of the Bethe–Salpeter equations relating $W^{(2)}$ to $\Phi^{(2)}$ and the 1PI vertex $\Gamma^{(4)}$ to $\Phi^{(2)}$ [Eqs. (1.477), (1.478)].

and the identity matrix by (1.467). These definitions are consistent with the (anti)symmetry properties of the Green functions and the vertices under the exchange $(\tilde{\alpha}_i\tau_i, \tilde{\alpha}'_i\tau'_i) \leftrightarrow (\tilde{\alpha}'_i\tau'_i, \tilde{\alpha}_i\tau_i)$ of the two particles in the bosonic external leg γ_i. Defining the particle–particle pair propagator

$$\Pi_{\gamma_1\gamma_2} = \bar{G}(\tilde{\alpha}_1\tau_1, \tilde{\alpha}'_2\tau'_2)\bar{G}(\tilde{\alpha}'_1\tau'_1, \tilde{\alpha}_2\tau_2) + \zeta\bar{G}(\tilde{\alpha}_1\tau_1, \tilde{\alpha}_2\tau_2)\bar{G}(\tilde{\alpha}'_1\tau'_1, \tilde{\alpha}'_2\tau'_2), \tag{1.476}$$

we obtain the Bethe–Salpeter equation relating $W^{(2)}$ to $\Phi^{(2)}$ (Fig. 1.12),

$$W^{(2)} = \left(\Pi^{-1} + \Phi^{(2)}\right)^{-1} = \Pi - \Pi\Phi^{(2)}W^{(2)}. \tag{1.477}$$

Similarly, we can derive a relation between $\Phi^{(2)}$ and the 1PI vertex $\Gamma^{(4)}$,

$$\Gamma^{(4)} = \left(\Phi^{(2)-1} + \Pi\right)^{-1} = \Phi^{(2)} - \Phi^{(2)}\Pi\Gamma^{(4)}, \tag{1.478}$$

with the definition $\Gamma^{(4)}_{\gamma_1\gamma_2} = \Gamma^{(4)}(\tilde{\alpha}_1\tau_1, \tilde{\alpha}'_1\tau'_1, \tilde{\alpha}_2\tau_2, \tilde{\alpha}'_2\tau'_2)$ (Fig. 1.12). In a normal system, $c_1 + c'_1 + c_2 + c'_2 = 0$, and the Bethe–Salpeter equation can be projected onto the particle–hole ($c_1 + c'_1 = c_2 + c'_2 = 0$) and particle–particle ($c_1 = c'_1 = -c_2 = -c'_2$) channels. The corresponding vertices, $\Phi^{(2)}_{\rm ph}$ and $\Phi^{(2)}_{\rm pp}$, are 2PI in the particle–hole and particle–particle channels, respectively.

When dealing with bosons, it appears natural to work with a functional $\Phi[G_c, \phi]$ of both the connected Green function G_c and the classical field $\phi = \langle\psi\rangle$. Let us consider the partition function with both linear and bilinear external sources,

$$\begin{aligned} Z[J,h] = \int \mathcal{D}[\psi] \exp\Big\{ -S[\psi] + \frac{1}{2}\int_0^\beta d\tau d\tau' \sum_{\tilde{\alpha},\tilde{\alpha}'} \psi_{\tilde{\alpha}}(\tau) J_{\tilde{\alpha}\tilde{\alpha}'}(\tau,\tau')\psi_{\tilde{\alpha}'}(\tau') \\ + \int_0^\beta d\tau \sum_{\tilde{\alpha}} h_{\tilde{\alpha}}(\tau)\psi_{\tilde{\alpha}}(\tau) \Big\}, \end{aligned} \tag{1.479}$$

where $J_{\tilde{\alpha}\tilde{\alpha}'}(\tau,\tau') = J_{\tilde{\alpha}'\tilde{\alpha}}(\tau',\tau)$ is symmetric. We have

$$\begin{aligned} \phi_{\tilde{\alpha}}[\tau; J, h] &= \langle\psi_{\tilde{\alpha}}(\tau)\rangle = \frac{\delta \ln Z[J,h]}{\delta h_{\tilde{\alpha}}(\tau)}, \\ G_c[\tilde{\alpha}\tau, \tilde{\alpha}'\tau'; J, h] &= \langle\psi_{\tilde{\alpha}}(\tau)\psi_{\tilde{\alpha}'}(\tau')\rangle - \langle\psi_{\tilde{\alpha}}(\tau)\rangle\langle\psi_{\tilde{\alpha}'}(\tau')\rangle \\ &= \frac{\delta \ln Z[J,h]}{\delta J_{\tilde{\alpha}\tilde{\alpha}'}(\tau,\tau')} - \phi_{\tilde{\alpha}}(\tau)\phi_{\tilde{\alpha}'}(\tau'). \end{aligned} \tag{1.480}$$

The effective action is defined as a double Legendre transform,

$$\begin{aligned}\Gamma[G_c,\phi] = &-\ln Z[J,h] + \int_0^\beta d\tau \sum_{\tilde\alpha} h_{\tilde\alpha}(\tau)\phi_{\tilde\alpha}(\tau)\\ &+\frac{1}{2}\int_0^\beta d\tau d\tau' \sum_{\tilde\alpha,\tilde\alpha'} J_{\tilde\alpha\tilde\alpha'}(\tau,\tau')\left[G_c(\tilde\alpha\tau,\tilde\alpha'\tau')+\phi_{\tilde\alpha}(\tau)\phi_{\tilde\alpha'}(\tau')\right],\end{aligned} \tag{1.481}$$

and satisfies

$$\begin{aligned}\frac{\delta\Gamma[G_c,\phi]}{\delta\phi_{\tilde\alpha}(\tau)} &= h_{\tilde\alpha}(\tau) + \int_0^\beta d\tau' \sum_{\tilde\alpha'} J_{\tilde\alpha\tilde\alpha'}(\tau,\tau')\phi_{\tilde\alpha'}(\tau'),\\ \frac{\delta\Gamma[G_c,\phi]}{\delta G_c(\tilde\alpha\tau,\tilde\alpha'\tau')} &= J_{\tilde\alpha\tilde\alpha'}(\tau,\tau').\end{aligned} \tag{1.482}$$

It can be written as

$$\begin{aligned}\Gamma[G_c,\phi] = &-\frac{1}{2}\mathrm{Tr}\ln(G_c) + \frac{1}{2}\int_0^\beta d\tau d\tau' \phi_{\tilde\alpha}(\tau)G_0^{-1}(\tilde\alpha\tau,\tilde\alpha'\tau')\phi_{\tilde\alpha'}(\tau')\\ &+\frac{1}{2}\mathrm{Tr}(G_0^{-1}G_c - 1) + \Phi[G_c,\phi],\end{aligned} \tag{1.483}$$

where $\Phi[G_c,\phi]$ is the Luttinger–Ward functional. The equations of state (1.482) give

$$\begin{aligned}&\int_0^\beta d\tau' \sum_{\tilde\alpha'} G_0^{-1}(\tilde\alpha\tau,\tilde\alpha'\tau')\phi_{\tilde\alpha'}(\tau') + \frac{\delta\Phi[G_c,\phi]}{\delta\phi_{\tilde\alpha}(\tau)}\\ &= h_{\tilde\alpha}(\tau) + \int_0^\beta d\tau' \sum_{\tilde\alpha'} J_{\tilde\alpha\tilde\alpha'}(\tau,\tau')\phi_{\tilde\alpha'}(\tau')\end{aligned} \tag{1.484}$$

and

$$-G_c^{-1}(\tilde\alpha\tau,\tilde\alpha'\tau') + G_0^{-1}(\tilde\alpha\tau,\tilde\alpha'\tau') + \frac{\delta\Phi[G_c,\phi]}{\delta G_c(\tilde\alpha\tau,\tilde\alpha'\tau')} = J_{\tilde\alpha\tilde\alpha'}(\tau,\tau'). \tag{1.485}$$

The first equation is an equation for the order parameter $\phi_{\tilde\alpha}(\tau)$. The second one is Dyson's equation $G_c^{-1} = G_0^{-1} + \Sigma$ with the self-energy

$$\Sigma(\tilde\alpha\tau,\tilde\alpha'\tau') = \frac{\delta\Phi[G_c,\phi]}{\delta G_c(\tilde\alpha\tau,\tilde\alpha'\tau')} \tag{1.486}$$

for vanishing external sources ($J = h = 0$).

The Luttinger–Ward functional $\Phi[G_c,\phi]$ is given by the sum of the 2PI diagrams,

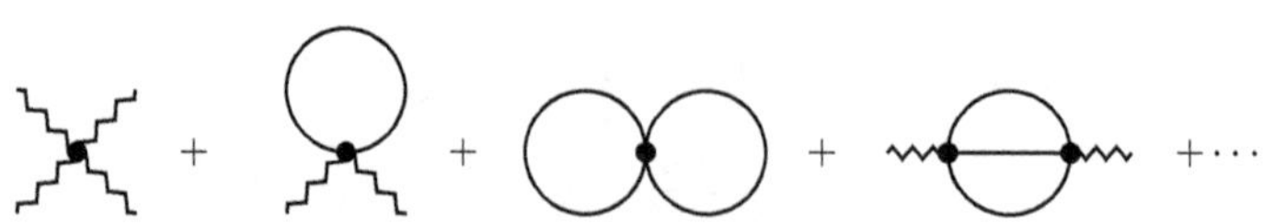

and the corresponding self-energy has the following diagrammatic expansion:

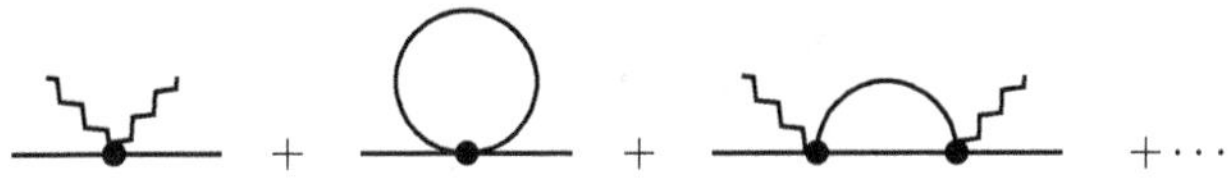

The zigzag line in these diagrams stands for the order parameter $\phi_{\tilde{\alpha}}(\tau)$. The perturbation expansion (in the 1PI formalism) about a broken symmetry state ($\phi_{\tilde{\alpha}}(\tau) \neq 0$) in boson systems is discussed in detail in Sec. 1.7.2.

1.7 Saddle-Point Approximation and Loop Expansion

In Sec. 1.5.4, we have pointed out that the perturbation theory of interacting bosons about the non-interacting limit breaks down when the field takes a finite expectation value $\langle \psi_\alpha \rangle$ which breaks the U(1) symmetry. In this section, we show how the stationary-phase approximation allows us to reorganize the perturbation expansion as a "loop expansion" about a broken-symmetry state. We start by introducing the principle of the method on the simple example of a one-dimensional integral.

1.7.1 *Steepest-descent and stationary-phase approximations*

We consider the integral

$$I(l) = \int_{-\infty}^{\infty} dt\, e^{-lf(t)}, \tag{1.487}$$

where $f(t)$ is a real function with an absolute minimum at t_0 and look for an expansion in $1/l$. For large l, the integral is dominated by the vicinity of t_0. Expanding $f(t)$ about t_0 and using $f_0' \equiv f'(t_0) = 0$, we obtain

$$\begin{aligned} I(l) &= e^{-lf_0} \int_{-\infty}^{\infty} dt\, \exp\biggl\{ -\frac{l}{2} f_0''(t-t_0)^2 - l \sum_{n=3}^{\infty} \frac{f_0^{(n)}}{n!}(t-t_0)^n \biggr\} \\ &= e^{-lf_0} \sqrt{\frac{2\pi}{f_0'' l}} \int_{-\infty}^{\infty} \frac{d\tau}{\sqrt{2\pi}} \exp\biggl\{ -\frac{\tau^2}{2} - \sum_{n=3}^{\infty} \frac{\tau^n}{n!} \frac{f_0^{(n)}}{(f_0'')^{n/2} l^{n/2-1}} \biggr\}, \end{aligned} \tag{1.488}$$

where $\tau = \sqrt{f_0'' l}(t-t_0)$. The integral can be calculated in powers of $1/l$ using Wick's theorem with the bare propagator

$$\sqcap\!\!\!\!\!\!\tau\tau = \int_{-\infty}^{\infty} \frac{d\tau}{\sqrt{2\pi}} \tau^2 e^{-\tau^2/2} = 1 \tag{1.489}$$

and the vertices

$$V_n = \frac{f_0^{(n)}}{(f_0'')^{n/2} l^{n/2-1}}. \tag{1.490}$$

By the linked cluster theorem (Sec. 1.5.2)

$$I(l) = e^{-lf_0}\sqrt{\frac{2\pi}{f_0''l}}\exp\Big\{\sum \text{connected diagrams}\Big\}. \tag{1.491}$$

Since $V_n = \mathcal{O}(l^{-n/2+1})$, the $\mathcal{O}(1/l)$ contribution is given by the three diagrams

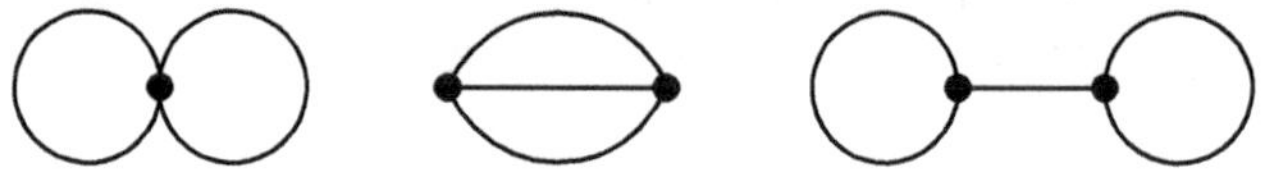

with respective combinatorial factors $-1/8$, $1/12$ and $1/8$. We deduce

$$\ln I(l) = -lf_0 + \frac{1}{2}\ln\left(\frac{2\pi}{f_0''l}\right) - \frac{1}{8l}\frac{f_0^{(4)}}{f_0''^2} + \frac{5}{24l}\frac{\left(f_0^{(3)}\right)^2}{f_0''^3} + \mathcal{O}\left(l^{-2}\right). \tag{1.492}$$

A similar analysis can be made when the function $f(t)$ is purely imaginary. In this case, for large l the integral is dominated by the vicinity of the point t_0 where the phase of the integrand is stationary ($f'(t_0) = 0$). The expansion (1.491) is then known as the stationary-phase method.

More generally, let us consider the integral

$$I(l) = \int_{\mathcal{C}} dz\, e^{-lf(z)} \tag{1.493}$$

along a path $\mathcal{C}$ in the complex plane. The complex function $f(z)$ is analytic along $\mathcal{C}$ and vanishes at the end points. Since we can deform the integration contour in the domain of analyticity of $f(z)$, we deform $\mathcal{C}$ into a new contour $\mathcal{C}'$ which satisfies the three conditions:

1. The end points of $\mathcal{C}'$ are the same as those of $\mathcal{C}$.
2. $\Re[f(z)]$ has a minimum at the point z_0 of $\mathcal{C}'$.
3. $\Im[f(z)]$ is constant along $\mathcal{C}'$ near z_0.

The last two conditions imply that z_0 is a stationary point of $f(z)$ ($f'(z_0) = 0$) and therefore a saddle point of $\Re[f(z)]$.[64] For large l, we expect the vicinity of z_0 to give the dominant contribution to $I(l)$ (condition (2)). The condition (3) ensures that $\Im[f(z)]$ does not give rise to rapid oscillations in the integrand as $l \to \infty$ which would suppress the contribution of the vicinity of z_0. To calculate $I(l)$, we then write $z = z_0 + \rho e^{i\theta}$ and expand $f(z)$ about z_0,

$$f(z) = f(z_0) + \frac{1}{2}\rho^2|f_0''|e^{i(2\theta+\theta_0)} + \sum_{n=3}^{\infty}\frac{\rho^n e^{in\theta}}{n!}f_0^{(n)}, \tag{1.494}$$

[64] Recall that an analytic complex function $f(z)$ satisfies the Cauchy–Riemann conditions $\partial_x\Re[f(z)] = \partial_y\Im[f(z)]$ and $\partial_x\Im[f(z)] = -\partial_y\Re[f(z)]$ (obtained by calculating $f'(z) = \lim_{dz\to 0}[f(z+dz) - f(z)]/dz$ with $dz = dx$ and $dz = idy$). The Cauchy–Riemann conditions imply $\boldsymbol{\nabla}^2\Re[f(z)] = \boldsymbol{\nabla}^2\Im[f(z)] = 0$, which prevents $\Re[f(z)]$ and $\Im[f(z)]$ to have an absolute minimum or maximum.

where $f_0'' \equiv f''(z_0) = |f_0''|e^{i\theta_0}$. Condition (3) implies $2\theta + \theta_0 = 0$ or π, and the existence of a minimum of $\Re[f(z)]$ at z_0 selects the solution $2\theta + \theta_0 = 0$.[65] It is easy to see that $\mathcal{C}'$ is then the path of steepest descents of $e^{-l\Re[f(z)]}$. The integral over ρ can be calculated as in the previous case using the linked cluster theorem.

1.7.2 *Saddle-point approximation and loop expansion in field theory*

We consider the partition function

$$Z[J^*, J] = \int \mathcal{D}[\psi^*, \psi]\, e^{-l\left(S[\psi^*,\psi] - \int_0^\beta d\tau \int d^dr [J^*(x)\psi(x) + \text{c.c.}]\right)} \tag{1.495}$$

($x = (\mathbf{r}, \tau)$) describing a system of (spin-zero) bosons (ψ is a complex field). The factor l is introduced to organize the perturbation expansion and will be set to unity in the end. We assume that the bosons interact *via* a contact interaction,

$$S[\psi^*, \psi] = \int_0^\beta d\tau \int d^dr \left\{ \psi^*(x) \left(\partial_\tau - \mu - \frac{\boldsymbol{\nabla}^2}{2m} \right) \psi(x) + \frac{g}{2}|\psi(x)|^4 \right\} \tag{1.496}$$

($g > 0$).

1.7.2.1 *Partition function and thermodynamic potential*

Within a saddle-point approximation, the action reads

$$S_c = \beta \mathcal{V} \left(-\mu|\psi_c|^2 + \frac{g}{2}|\psi_c|^4 \right) \tag{1.497}$$

where the "classical" field ψ_c (assumed to be time and space independent) is determined from the condition

$$\frac{\partial S_c}{\partial \psi_c} = \frac{\partial S_c}{\partial \psi_c^*} = 0 \tag{1.498}$$

in the absence of external sources ($J = J^* = 0$). When $\mu > 0$, the solution which minimizes the action is $|\psi_c| = \sqrt{\mu/g}$. As will be discussed in Chapter 7, this solution corresponds to a superfluid system and $|\psi_c|^2$ determines the density of the Bose–Einstein condensate. Expanding about the saddle-point solution, $\psi = \psi_c + \chi/\sqrt{l}$, we obtain

$$lS[\chi^*, \chi] = lS_c + S^{(2)} + \mathcal{O}(l^{-1/2}), \tag{1.499}$$

where the Gaussian part of the action

$$S^{(2)} = -\frac{1}{2} \int_0^\beta d\tau d\tau' \int d^dr d^dr' \, (\chi^*(x), \chi(x))\, \mathcal{G}_c^{-1}(x, x') \begin{pmatrix} \chi(x') \\ \chi^*(x') \end{pmatrix} \tag{1.500}$$

[65] The fact that the path of steepest descents of $e^{-l\Re[f(z)]}$ keeps $\Im[f(z)]$ constant near z_0 (stationary phase) is a consequence of the Cauchy–Riemann conditions.[64]

is defined by the inverse "classical" propagator

$$\mathcal{G}_c^{-1}(x,x') = -\delta(\tau-\tau')\delta(\mathbf{r}-\mathbf{r}') \times \begin{pmatrix} \partial_{\tau'} - \mu - \frac{\nabla^2_{\mathbf{r}'}}{2m} + 2g|\psi_c|^2 & g\psi_c^2 \\ g\psi_c^{*2} & -\partial_{\tau'} - \mu - \frac{\nabla^2_{\mathbf{r}'}}{2m} + 2g|\psi_c|^2 \end{pmatrix}. \tag{1.501}$$

In the superfluid phase ($|\psi_c| > 0$), the bosonic (connected) propagator $\mathcal{G}$ becomes a 2×2 matrix,[66]

$$\mathcal{G}(x,x') = -\begin{pmatrix} \langle\psi(x)\psi^*(x')\rangle_c & \langle\psi(x)\psi(x')\rangle_c \\ \langle\psi^*(x)\psi^*(x')\rangle_c & \langle\psi^*(x)\psi(x')\rangle_c, \end{pmatrix}, \tag{1.502}$$

where $\langle\psi(x)\psi^*(x')\rangle_c = \langle\psi(x)\psi^*(x')\rangle - \langle\psi(x)\rangle\langle\psi^*(x')\rangle$, etc. It satisfies the Beliaev–Dyson equation $\mathcal{G}^{-1}(k) = \mathcal{G}_0^{-1}(k) - \Sigma(k)$, where

$$\mathcal{G}_0(k) = \begin{pmatrix} G_0(k) & 0 \\ 0 & G_0(-k) \end{pmatrix} \tag{1.503}$$

is the non-interacting propagator and $\Sigma(k)$ the self-energy. More explicitly, the Beliaev–Dyson equations read

$$\begin{aligned} \mathcal{G}_{11}(k) &= G_0(k) + G_0(k)\Sigma_{11}(k)\mathcal{G}_{11}(k) + G_0(k)\Sigma_{12}(k)\mathcal{G}_{21}(k), \\ \mathcal{G}_{12}(k) &= G_0(k)\Sigma_{11}(k)\mathcal{G}_{12}(k) + G_0(k)\Sigma_{12}(k)\mathcal{G}_{22}(k), \end{aligned} \tag{1.504}$$

with $\mathcal{G}_{11}(k) = \mathcal{G}_{22}(-k) = -\langle\psi(k)\psi^*(k)\rangle$ and $\mathcal{G}_{12}(k) = \mathcal{G}_{21}^*(k) = -\langle\psi(k)\psi(-k)\rangle$ (Fig. 1.13). The classical propagator (1.501) corresponds to the self-energy

$$\begin{aligned} \Sigma_{11}(k) &= \Sigma_{22}(k) = 2g|\psi_c|^2, \\ \Sigma_{12}(k) &= \Sigma_{21}^*(k) = g\psi_c^2 \end{aligned} \tag{1.505}$$

(Fig. 1.14).

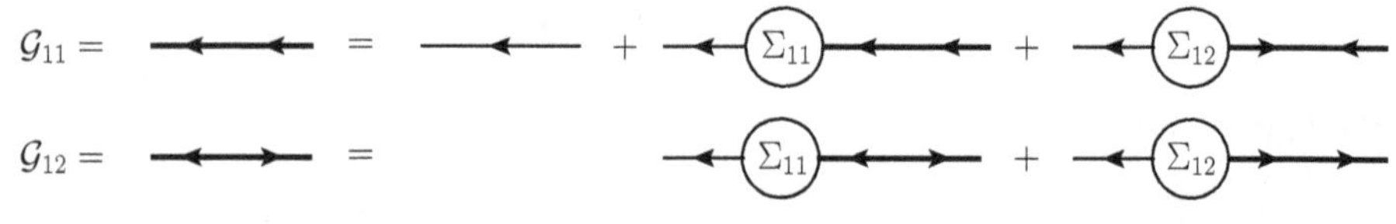

Fig. 1.13 Diagrammatic representation of the Beliaev–Dyson equations (1.504). The thin line stands for G_0 and the thick one for $\mathcal{G}$.

[66] In this section, $\mathcal{G}$ denotes the connected propagator and the subscript "c" of $\mathcal{G}_c$ stands for "classical". Note the sign convention in the definition of $\mathcal{G}$ [Eq. (1.500)].

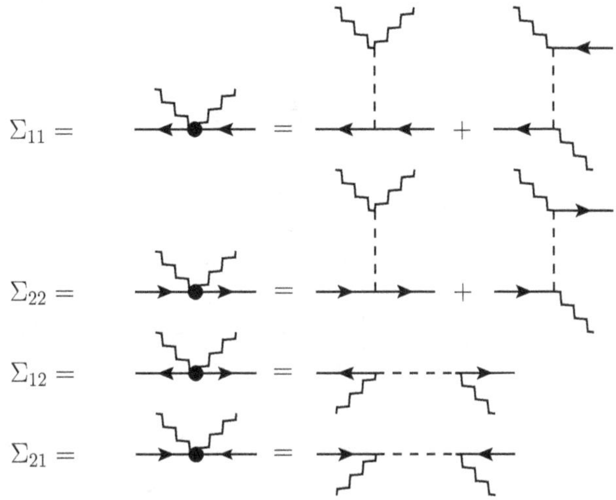

Fig. 1.14 Diagrammatic representation of the self-energies (1.505). The zigzag line stands for ψ_c or ψ_c^*. One sometimes introduces a directed zigzag line to distinguish between ψ_c (incoming line) and ψ_c^* (outgoing line).

Since the action (1.499) is Gaussian (ignoring the $\mathcal{O}(l^{-1/2})$ terms), we can carry out the functional integral (see Appendix 1.E.4),[67]

$$Z = e^{-lS_c} \left(\det \mathcal{G}_c^{-1}\right)^{-1/2}, \tag{1.506}$$

and obtain the thermodynamic potential[68]

$$\Omega = \frac{l}{\beta} S_c - \frac{1}{2\beta} \mathrm{Tr} \ln \mathcal{G}_c + \mathcal{O}(l^{-1}). \tag{1.507}$$

Let us now consider the corrections to the Gaussian action (1.499),

$$lS[\chi^*, \chi] = lS_c + S^{(2)}[\chi^*, \chi] + \frac{1}{\sqrt{l}} S^{(3)}[\chi^*, \chi] + \frac{1}{l} S^{(4)}[\chi^*, \chi], \tag{1.508}$$

with

$$\begin{aligned} S^{(3)} &= g \int_0^\beta d\tau \int d^d r \left[\psi_c \chi^*(x) |\chi(x)|^2 + \text{c.c.}\right], \\ S^{(4)} &= \frac{g}{2} \int_0^\beta d\tau \int d^d r \, |\chi(x)|^4. \end{aligned} \tag{1.509}$$

These corrections give rise to three- and four-leg vertices which can be represented by the following diagrams:

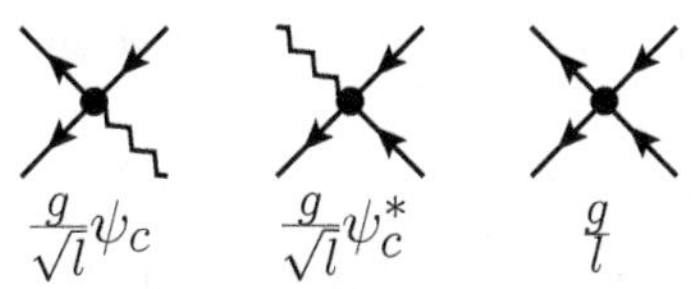

[67] Because of the 2×2 matrix structure of $\mathcal{G}_c$, $\det(-\mathcal{G}_c) = \det(\mathcal{G}_c)$.
[68] We use $\ln \det \mathcal{G} = \mathrm{Tr} \ln \mathcal{G}$ (see Appendix 1.E.4).

The partition function can be calculated treating $S^{(3)}$ and $S^{(4)}$ in perturbation and using the linked cluster theorem,

$$\begin{aligned} Z &= e^{-lS_c} \int \mathcal{D}[\chi^*, \chi]\, e^{-S^{(2)} - \frac{1}{\sqrt{l}} S^{(3)} - \frac{1}{l} S^{(4)}} \\ &= e^{-lS_c} \left(\det \mathcal{G}_c^{-1}\right)^{-1/2} \exp\left\{\sum \text{connected diagrams}\right\}. \end{aligned} \tag{1.510}$$

To order $1/l$, one finds that the partition function is given by the following diagrams:

$$\ln Z = -lS_c + \frac{1}{2}\operatorname{Tr}\ln \mathcal{G}_c + \frac{1}{l}\Big\{$$  $$\Big\},$$

where the lines in the diagrams must be drawn in all possible ways compatible with $S^{(3)}$, $S^{(4)}$, and the classical propagator $\mathcal{G}_c$. For example, the first diagram means

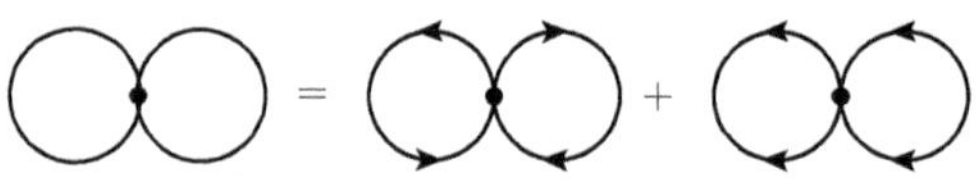

where the lines stand for the classical propagator $\mathcal{G}_c$.

The expansion in $1/l$ turns out to be an expansion in the number of loops. Consider a diagram with $n_V = n_V^{(3)} + n_V^{(4)}$ vertices, where $n_V^{(3)}$ ($n_V^{(4)}$) is the number of three-leg (four-leg) vertices. The propagator $\mathcal{G}_c$ being independent of l, the diagram is of order $l^{-n_V^{(4)} - n_V^{(3)}/2} = l^{-n_V + n_V^{(3)}/2}$. The number n_I of internal lines is given by

$$2n_I = 4n_V^{(4)} + 3n_V^{(3)} = 4n_V - n_V^{(3)} \tag{1.511}$$

(obtained by counting the total number of lines attached to the n_V vertices). Since each vertex satisfies momentum conservation, a diagram with n_V vertices has $n_V - 1$ momentum conservation constraints, so that the number of independent momentum loops is

$$n_M = n_I - (n_V - 1) = n_V - \frac{n_V^{(3)}}{2} + 1. \tag{1.512}$$

The order of the diagram, $l^{-n_V + n_V^{(3)}/2} = l^{-(n_M - 1)}$, is therefore given by the number of loops. In the absence of a condensate ($\psi_c = 0$), $n_V^{(3)} = 0$ and $\mathcal{G}_c = \mathcal{G}_0$, a diagram with n_M loops is of order $g^{n_V} = g^{n_M - 1}$. The loop expansion then corresponds to the perturbation expansion about the non-interacting limit discussed in Sec. 1.5. On the other hand, when $|\psi_c| > 0$, the classical propagator $\mathcal{G}_c$ contains an infinite resummation of interaction terms and a diagram with n_M loops contains from $n_M - 1$ to $2(n_M - 1)$ vertices. To any order, the loop expansion then corresponds to an infinite resummation of perturbation theory.

1.7.2.2 *Effective action and 1PI vertices*

We now consider the effective action defined in Sec. 1.6.2,

$$\Gamma[\phi^*, \phi] = -\frac{1}{l}\ln Z[J^*, J] + \int_0^\beta d\tau \int d^d r [J^*(x)\phi(x) + \text{c.c.}], \tag{1.513}$$

where

$$\phi(x) = \frac{1}{l}\frac{\delta \ln Z[J^*, J]}{\delta J^*(x)}, \qquad \phi^*(x) = \frac{1}{l}\frac{\delta \ln Z[J^*, J]}{\delta J(x)} \tag{1.514}$$

(with $Z[J^*, J]$ given by (1.495)). For arbitrary external sources, the saddle point equations read

$$\left.\frac{\delta S[\psi^*, \psi]}{\delta \psi(x)}\right|_{\psi^{(*)}=\psi_c^{(*)}} = J^*(x), \qquad \left.\frac{\delta S[\psi^*, \psi]}{\delta \psi^*(x)}\right|_{\psi^{(*)}=\psi_c^{(*)}} = J(x). \tag{1.515}$$

We expand about the classical state, $\psi(x) = \psi_c(x) + \frac{1}{\sqrt{l}}\chi(x)$, using

$$\begin{aligned} lS[\psi^*, \psi] &- l\int_0^\beta d\tau \int d^d r [J^*(x)\psi(x) + \text{c.c.}] \\ &= lS_c - l\int_0^\beta d\tau \int d^d r [J^*(x)\psi_c(x) + \text{c.c.}] + S^{(2)}[\chi^*, \chi] + \mathcal{O}(l^{-1/2}), \end{aligned} \tag{1.516}$$

where $S_c = S[\psi_c^*, \psi_c]$ and the Gaussian part $S^{(2)}$ of the action is defined by (1.500) and (1.501) with a time- and space-dependent classical field $\psi_c(x)$. From

$$\ln Z[J^*, J] = -lS_c + l\int_0^\beta d\tau \int d^d r [J^*(x)\psi_c(x) + \text{c.c.}] + \frac{1}{2}\text{Tr}\ln \mathcal{G}_c + \mathcal{O}(l^{-1}), \tag{1.517}$$

and (1.514), (1.515), we deduce

$$\begin{aligned} \phi(x) &= -\frac{\delta S_c}{\delta J^*(x)} + \frac{\delta}{\delta J^*(x)}\int_0^\beta d\tau' \int d^d r' [J^*(x')\psi_c(x') + \text{c.c.}] + \mathcal{O}(l^{-1}) \\ &= \psi_c(x) + \mathcal{O}(l^{-1}) \end{aligned} \tag{1.518}$$

and $\phi^*(x) = \psi_c^*(x) + \mathcal{O}(l^{-1})$. This yields

$$\Gamma[\phi^*, \phi] = S_c + \int_0^\beta d\tau \int d^d r \{J^*(x)[\phi(x) - \psi_c(x)] + \text{c.c.}\} - \frac{1}{2l}\text{Tr}\ln \mathcal{G}_c + \mathcal{O}(l^{-2}), \tag{1.519}$$

where the $\mathcal{O}(l^{-2})$ contribution comes from the two-loop and higher-order diagrams. Noting that

$$\begin{aligned} S_c &+ \int_0^\beta d\tau \int d^d r \{J^*(x)[\phi(x) - \psi_c(x)] + \text{c.c.}\} \\ &= S[\psi_c^*, \psi_c] + \int_0^\beta d\tau \int d^d r \left\{ \left.\frac{\delta S}{\delta \psi(x)}\right|_{\psi^{(*)}=\psi_c^{(*)}} [\phi(x) - \psi_c(x)] + \text{c.c.} \right\} \\ &= S[\phi^*, \phi] + \mathcal{O}(l^{-2}), \end{aligned} \tag{1.520}$$

we finally deduce

$$\Gamma[\phi^*, \phi] = S[\phi^*, \phi] - \frac{1}{2l}\mathrm{Tr}\ln(\mathcal{G}_c[\phi^*, \phi]) + \mathcal{O}(l^{-2}), \tag{1.521}$$

where $\mathcal{G}_c[\phi^*, \phi]$ is obtained from (1.501) by replacing $\psi_c(x)$ by $\phi(x)$, i.e.

$$\mathcal{G}_c^{-1}[x, x'; \phi^*, \phi] = -\begin{pmatrix} \dfrac{\delta^2 S[\phi^*, \phi]}{\delta\phi^*(x)\delta\phi(x')} & \dfrac{\delta^2 S[\phi^*, \phi]}{\delta\phi^*(x)\delta\phi^*(x')} \\ \dfrac{\delta^2 S[\phi^*, \phi]}{\delta\phi(x)\delta\phi(x')} & \dfrac{\delta^2 S[\phi^*, \phi]}{\delta\phi(x)\delta\phi^*(x')} \end{pmatrix}. \tag{1.522}$$

In (1.520) and (1.521), we have used the fact that $\phi(x) - \psi_c(x)$ is of order $1/l$.

To order $1/l$ and for vanishing external sources, the equations of state (1.401), (1.402) read

$$\begin{aligned} 0 &= \frac{\delta\Gamma}{\delta\phi^*(x)} = \frac{\delta S}{\delta\phi^*(x)} - \frac{1}{2l}\mathrm{Tr}\left(\mathcal{G}_c\frac{\delta\Sigma}{\delta\phi^*(x)}\right), \\ 0 &= \frac{\delta\Gamma}{\delta\phi(x)} = \frac{\delta S}{\delta\phi(x)} - \frac{1}{2l}\mathrm{Tr}\left(\mathcal{G}_c\frac{\delta\Sigma}{\delta\phi(x)}\right), \end{aligned} \tag{1.523}$$

where we have used $\mathcal{G}_c^{-1} = \mathcal{G}_0^{-1} - \Sigma$, with

$$\Sigma[x, x'; \phi^*, \phi] = g\delta(\mathbf{r} - \mathbf{r}')\delta(\tau - \tau')\begin{pmatrix} 2|\phi(x)|^2 & \phi(x)^2 \\ \phi^*(x)^2 & 2|\phi(x)|^2 \end{pmatrix}. \tag{1.524}$$

We therefore obtain

$$\begin{aligned} 0 = &-\int_0^\beta d\tau' \int d^d r'\, G_0^{-1}(x, x')\phi(x') + g\phi^*(x)\phi(x)^2 \\ &- \frac{g}{l}\{(\mathcal{G}_{c,11}[x, x; \phi^*, \phi] + \mathcal{G}_{c,22}[x, x; \phi^*, \phi])\phi(x) + \mathcal{G}_{c,12}[x, x; \phi^*, \phi]\phi^*(x)\} \end{aligned} \tag{1.525}$$

and a similar equation with $\phi(x) \leftrightarrow \phi^*(x)$. Equation (1.525) can be diagrammatically represented as

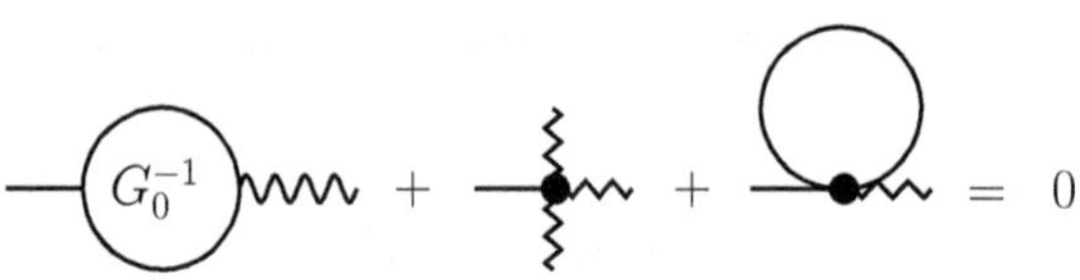

and corresponds to a one-loop approximation to the exact result (1.435) (Fig. 1.8). Here, the zigzag line stands for $\phi(x)$ or $\phi^*(x)$ and the solid one (in the loop) for the classical propagator $\mathcal{G}_c \equiv \mathcal{G}_c[\phi^*, \phi]$. The one-loop term in (1.525) gives the $\mathcal{O}(1/l)$ correction to the classical (saddle-point) result $|\phi(x)| = \sqrt{\mu/g}$. From (1.521), one can also obtain the two-point vertices

$$\begin{aligned}\Gamma^{(2)}_{11}(x,x') &= \frac{\delta^2\Gamma}{\delta\phi^*(x)\delta\phi(x')},\\ \Gamma^{(2)}_{12}(x,x') &= \frac{\delta^2\Gamma}{\delta\phi^*(x)\delta\phi^*(x')},\\ \Gamma^{(2)}_{22}(x,x') &= \frac{\delta^2\Gamma}{\delta\phi(x)\delta\phi^*(x')}\end{aligned} \tag{1.526}$$

to order $1/l$. For instance, one finds

$$\begin{aligned}\Gamma^{(2)}_{11}(x,x') = &-\mathcal{G}^{-1}_{c,11}(x,x')\\ &-\frac{1}{2l}\mathrm{Tr}\left(\mathcal{G}_c\frac{\delta^2\Sigma}{\delta\phi^*(x)\delta\phi(x')}\right)-\frac{1}{2l}\mathrm{Tr}\left(\mathcal{G}_c\frac{\delta\Sigma}{\delta\phi^*(x)}\mathcal{G}_c\frac{\delta\Sigma}{\delta\phi(x')}\right)\end{aligned} \tag{1.527}$$

with Σ the classical self-energy (1.524). The $\mathcal{O}(1/l)$ correction to $-\mathcal{G}^{-1}_{c,ij}$ gives the one-loop self-energy. It can be diagrammatically represented as

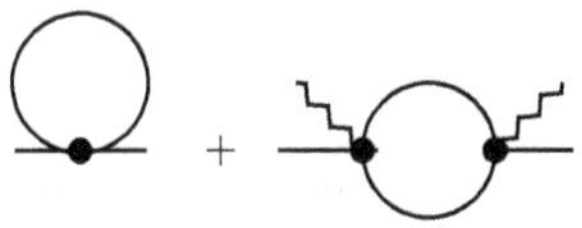

Higher-order vertices are obtained by taking the functional derivative of $\Gamma^{(2)}$ with respect to $\phi(x)$ or $\phi^*(x)$. The derivative of the (bare) three-leg vertex $g\phi^{(*)}(x)$ gives the (bare) four-leg vertex g. On the other hand, when acting on the classical propagator $\mathcal{G}_c = (\mathcal{G}_0^{-1}-\Sigma)^{-1}$, the functional derivative pulls out an external line attached to a three-leg vertex. Thus, we obtain diagrams of the type

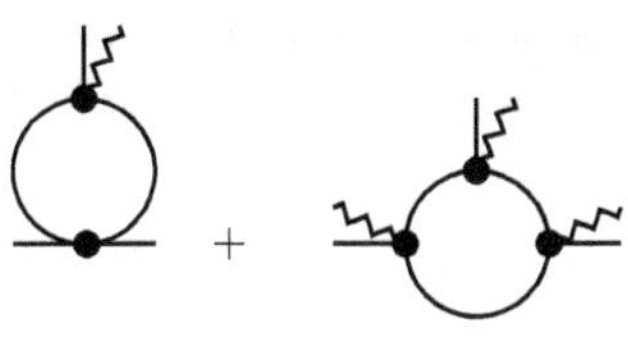

(diagrams obtained by exchanging external lines are not shown) for the one-loop correction to the three-leg vertex $\Gamma^{(3)}$, and

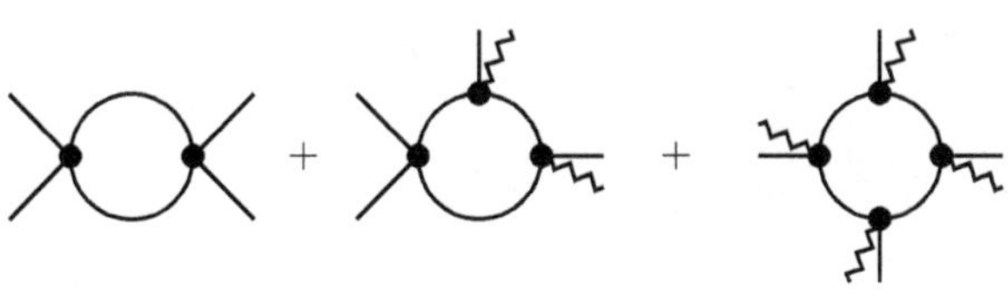

for the four-leg vertex.

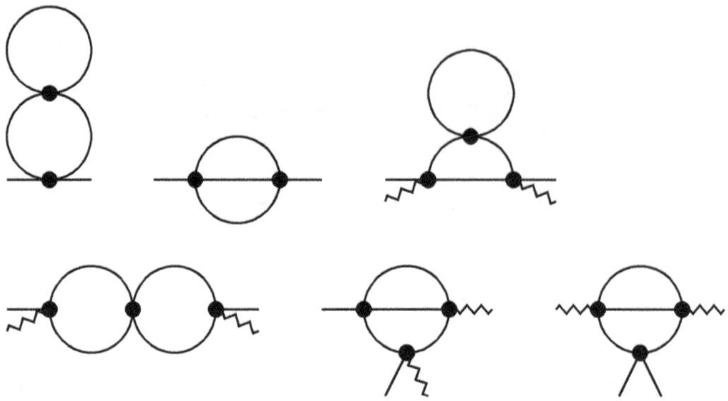

Fig. 1.15 Two-loop corrections to the self-energy.

A similar analysis can be carried out at order $1/l^2$. One finds that the effective action

$$\Gamma[\phi^*,\phi] = S[\phi^*,\phi] - \frac{1}{2l}\mathrm{Tr}\ln\mathcal{G}_c[\phi^*,\phi] + \frac{1}{l^2}\Big\{ \quad + \quad \Big\} \tag{1.528}$$

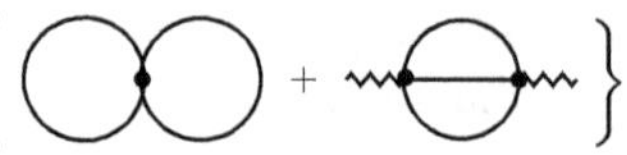

to this order is given by the 1PI two-loop diagrams (see the calculation in the following). The two-loop order vertices are obtained from the functional derivatives of (1.528). For instance, the two-loop correction to the self-energy is given by the diagrams of Fig. 1.15. Note that the absence of one-particle reducible diagrams in the effective action is a necessary condition for the self-energy to be 1PI. The one- and two-loop results easily generalize to all orders in perturbation theory: The effective action $\Gamma[\phi^*,\phi]$ may be expressed as the sum of all 1PI diagrams obtained from the classical propagator $\mathcal{G}_c[\phi^*,\phi]$ and the vertices $S^{(n)}[\phi^*,\phi]$ derived from the action $S[\phi^*,\phi]$ [5].

The effective action to two-loop order. Let us compute the effective action to order $1/l^2$. For simplicity, we consider a real field. The partition function reads

$$Z[J] = \int \mathcal{D}[\psi]\exp\Big\{-lS[\psi] + l\int dx\, J(x)\psi(x)\Big\} \tag{1.529}$$

with the action

$$S[\psi] = -\frac{1}{2}\int dx dx'\, \psi(x)G_0^{-1}(x,x')\psi(x') + S_{\rm int}[\psi]. \tag{1.530}$$

The saddle-point approximation gives

$$\frac{\delta S[\psi]}{\delta\psi(x)}\bigg|_{\psi=\psi_c} = J(x). \tag{1.531}$$

Expanding about the classical field ψ_c, $\psi = \psi_c + \chi/\sqrt{l}$, we rewrite the action as

$$\begin{aligned}
&lS[\psi] - l\int dx\, J(x)\psi(x) \\
&= lS[\psi_c] - l\int dx\, J(x)\psi_c(x) - \frac{1}{2}\int dx_1 dx_2\, \chi(x_1)G_c^{-1}(x_1,x_2)\chi(x_2) \\
&\quad + \frac{1}{3!\sqrt{l}}\int dx_1 dx_2 dx_3\, \chi(x_1)\chi(x_2)\chi(x_3)S_c^{(3)}(x_1,x_2,x_3) \\
&\quad + \frac{1}{4!l}\int dx_1 dx_2 dx_3 dx_4\, \chi(x_1)\chi(x_2)\chi(x_3)\chi(x_4)S_c^{(4)}(x_1,x_2,x_3,x_4) + \mathcal{O}(l^{-3/2}),
\end{aligned} \tag{1.532}$$

where

$$S_c^{(n)}(x_1,\dots,x_n) = \left.\frac{\delta^n S[\psi]}{\delta\psi(x_1)\cdots\delta\psi(x_n)}\right|_{\psi=\psi_c} \tag{1.533}$$

and $G_c^{-1}(x_1,x_2) = -S^{(2)}(x_1,x_2)$ is the inverse classical propagator. $Z[J]$ is calculated using the linked cluster theorem,

$$Z[J] = e^{-lS_c + l\int dx\, J(x)\psi_c(x)}\left[\det(-G_c^{-1})\right]^{-1/2}\exp\Big\{\sum \text{connected graphs}\Big\}. \tag{1.534}$$

The $\mathcal{O}(1/l)$ contribution is given by the three two-loop diagrams shown after (1.491), i.e.

$$\begin{aligned}
\ln Z[J] = & -lS_c + l\int dx\, J(x)\psi_c(x) - \frac{1}{2}\mathrm{Tr}\ln(-G_c^{-1}) \\
& - \frac{1}{8l}\int dx_1 dx_2 dx_3 dx_4\, G_c(x_1,x_2)G_c(x_3,x_4)S_c^{(4)}(x_1,x_2,x_3,x_4) \\
& - \frac{1}{l}\int dx_1 dx_2 dx_3 dy_1 dy_2 dy_3\, S_c^{(3)}(x_1,x_2,x_3)S_c^{(3)}(y_1,y_2,y_3) \\
& \qquad \times \left[\frac{1}{12}G_c(x_1,y_1)G_c(x_2,y_2)G_c(x_3,y_3)\right. \\
& \qquad \left. + \frac{1}{8}G_c(x_1,x_2)G_c(y_1,y_2)G_c(x_3,y_3)\right] + \mathcal{O}(l^{-2}).
\end{aligned} \tag{1.535}$$

Equation (1.535) gives the thermodynamic potential to two-loop order.

We can now compute the effective action $\Gamma[\phi]$. We have

$$\begin{aligned}
\phi(x) &= \frac{1}{l}\frac{\delta \ln Z[J]}{\delta J(x)} \\
&= \psi_c(x) - \frac{\delta S_c}{\delta J(x)} + \int dx'\, J(x')\frac{\delta\psi_c(x')}{\delta J(x)} - \frac{1}{2l}\frac{\delta}{\delta J(x)}\mathrm{Tr}\ln(-G_c^{-1}) + \mathcal{O}(l^{-2}) \\
&= \psi_c(x) - \frac{1}{2l}\frac{\delta}{\delta J(x)}\mathrm{Tr}\ln(-G_c^{-1}) + \mathcal{O}(l^{-2}).
\end{aligned} \tag{1.536}$$

Using

$$\frac{\delta}{\delta J(x)} \operatorname{Tr} \ln(-G_c^{-1}) = \frac{\delta}{\delta J(x)} \operatorname{Tr} \ln S_c^{(2)} = \int dy \frac{\delta \psi_c(y)}{\delta J(x)} \operatorname{Tr} \left[\frac{\delta S_c^{(2)}}{\delta \psi_c(y)} S_c^{(2)-1} \right], \quad (1.537)$$

with $\delta J(x)/\delta \psi_c(y) = -G_c^{-1}(x,y)$, we obtain

$$\begin{aligned}
\phi(x) &= \psi_c(x) - \frac{1}{2l} \int dy dy_1 dy_2 \, G_c(x,y) S_c^{(3)}(y, y_1, y_2) G_c(y_2, y_1) + \mathcal{O}(l^{-2}). \\
&= \psi_c(x) - \frac{1}{2l} \int dy dy_1 dy_2 \, G_c(x, y; \phi) S_c^{(3)}(y, y_1, y_2; \phi) G_c(y_2, y_1; \phi) + \mathcal{O}(l^{-2}),
\end{aligned} \quad (1.538)$$

where the last line is obtained by replacing G_c by $G_c(\phi) \equiv G_c[\phi]$ (and similarly $S_c^{(3)} \to S_c^{(3)}(\phi) \equiv S_c^{(3)}[\phi]$), which does not change the result to $\mathcal{O}(1/l)$. To obtain $\Gamma[\phi]$, we also need $J(x)$ to $\mathcal{O}(1/l)$,

$$\begin{aligned}
J(x) &= \frac{\delta S}{\delta \psi(x)}\bigg|_{\psi=\phi} + \int dy \, \frac{\delta^2 S}{\delta \psi(x) \delta \psi(y)}\bigg|_{\psi=\phi} [\psi_c(y) - \phi(y)] + \mathcal{O}(l^{-2}) \\
&= \frac{\delta S[\phi]}{\delta \phi(x)} - \frac{1}{2l} \int dy \, G_c^{-1}(x, y; \phi) \int dy_1 dy_2 dy_3 \, G_c(y, y_3; \phi) \\
&\qquad \times S_c^{(3)}(y_1, y_2, y_3; \phi) G_c(y_2, y_1; \phi) + \mathcal{O}(l^{-2}) \\
&= \frac{\delta S[\phi]}{\delta \phi(x)} - \frac{1}{2l} \int dy_1 dy_2 \, S_c^{(3)}(x, y_1, y_2; \phi) G_c(y_2, y_1; \phi) + \mathcal{O}(l^{-2}).
\end{aligned} \quad (1.539)$$

We deduce

$$\begin{aligned}
\Gamma[\phi] &= -\frac{1}{l} \ln Z[J] + \int dx \, J(x) \phi(x) \\
&= S_c + \frac{1}{2l} \operatorname{Tr} \ln\big(-G_c^{-1}(\phi)\big) + \int dx J(x) [\phi(x) - \psi_c(x)] \\
&\quad - (\text{two-loop diagrams}) + \mathcal{O}(l^{-3}),
\end{aligned} \quad (1.540)$$

where the contribution of the three two-loop diagrams is given in (1.535) (up to a factor $1/l$). Using (1.538), we then find

$$\begin{aligned}
\Gamma[\phi] = S[\phi] &+ \frac{1}{2l} \operatorname{Tr} \ln\big(-G_c^{-1}(\phi)\big) - (\text{two-loop diagrams}) \\
&- \frac{1}{8l^2} \int dx_1 dx_2 dx_3 dy_1 dy_2 dy_3 \, G_c(x_3, y_3; \phi) G_c(x_1, x_2; \phi) G_c(y_1, y_2; \phi) \\
&\times S_c^{(3)}(x_1, x_2, x_3; \phi) S_c^{(3)}(y_1, y_2, y_3; \phi) + \mathcal{O}(l^{-3}).
\end{aligned} \quad (1.541)$$

The last term cancels the two-loop diagram which is one-particle reducible and we finally obtain

$$\begin{aligned}\Gamma[\phi] &= S[\phi] + \frac{1}{2l}\operatorname{Tr}\ln(-G_c^{-1}(\phi)) \\ &+ \frac{1}{8l^2}\int dx_1dx_2dx_3dx_4\, S_c^{(4)}(x_1,x_2,x_3,x_4;\phi)G_c(x_1,x_2;\phi)G_c(x_3,x_4;\phi) \\ &+ \frac{1}{12l^2}\int dx_1dx_2dx_3dy_1dy_2dy_3\, S_c^{(3)}(x_1,x_2,x_3;\phi)S_c^{(3)}(y_1,y_2,y_3;\phi) \\ &\qquad \times G_c(x_1,y_1;\phi)G_c(x_2,y_2;\phi)G_c(x_3,y_3;\phi) + \mathcal{O}(l^{-3}), \end{aligned} \tag{1.542}$$

in agreement with (1.528).

1.8 Perturbation Theory at Zero Temperature

In this section, we briefly discuss perturbation theory at zero temperature for systems having a fixed number of particles.[69] The advantage of this approach is that one works directly in real time and no analytic continuation is necessary, contrary to the finite-temperature formalism in the grand canonical ensemble described in the previous sections. On the other hand, we shall see that the zero-temperature approach suffers from some drawbacks and the finite-temperature formalism is often preferable in practice.

We write the Hamiltonian of the N-particle system as $\hat H = \hat H_0 + \hat H_{\rm int}$, where $\hat H_0$ is the non-interacting Hamiltonian whose ground state $|\phi_0\rangle$ is known. If we let $|\phi_0\rangle$ evolve during a time T, we obtain the state

$$e^{-i\hat HT}|\phi_0\rangle = e^{-i\hat HT}\sum_n |n\rangle\langle n|\phi_0\rangle = \sum_n e^{-iE_nT}|n\rangle\langle n|\phi_0\rangle, \tag{1.543}$$

where $\{|n\rangle, E_n\}$ denotes an orthonormal basis of $\hat H$. Sending T to infinity with a (small) negative imaginary part, we project out the ground state $|0\rangle$ of $\hat H$,

$$\lim_{T\to\infty(1-i\eta)} e^{-i\hat HT}|\phi_0\rangle = e^{-iE_0T}|0\rangle\langle 0|\phi_0\rangle \tag{1.544}$$

($\eta = 0^+$), provided that $|0\rangle$ is non-degenerate and has a finite overlap with the non-interacting ground state $|\phi_0\rangle$. We deduce that the ground state energy of the interacting system can be obtained from

$$E_0 = \lim_{T\to\infty(1-i\eta)} \frac{i}{T}\ln\langle\phi_0|e^{-i\hat HT}|\phi_0\rangle. \tag{1.545}$$

Let us now consider the one-particle Green function[70]

$$G(\alpha t, \alpha' t') = -i\langle 0|T\hat\psi_\alpha(t)\hat\psi^\dagger_{\alpha'}(t')|0\rangle \tag{1.546}$$

[69]A thorough discussion of perturbation theory at zero temperature can be found in Chapter 3 of Ref. [9].
[70]The time-ordering operator T in (1.546) should not be confused with the time T appearing, e.g. in (1.545).

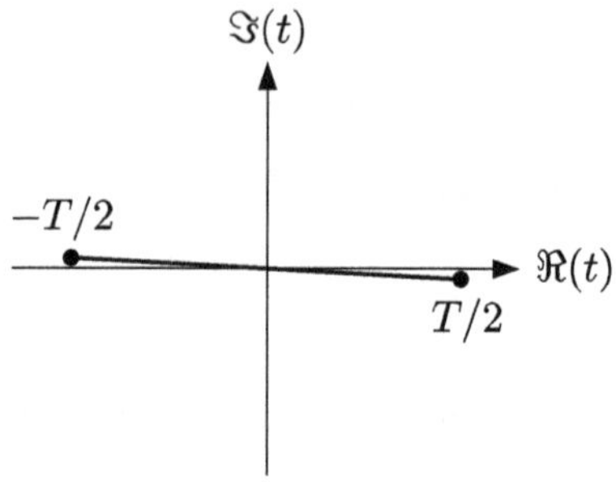

Fig. 1.16 Contour in the complex t plane used in the definition of the functional integral at zero temperature.

defined in Sec. 1.2.3 [Eq. (1.205)]. $\hat{\psi}_\alpha(t)$ and $\hat{\psi}^\dagger_\alpha(t)$ are operators in the Heisenberg picture and the index α labels the one-body states. Using

$$\begin{aligned} \lim_{T\to\infty(1-i\eta)} e^{-i\hat{H}\frac{T}{2}}|\phi_0\rangle &= e^{-iE_0\frac{T}{2}}|0\rangle\langle 0|\phi_0\rangle, \\ \lim_{T\to\infty(1-i\eta)} \langle\phi_0|e^{-i\hat{H}\frac{T}{2}} &= e^{-iE_0\frac{T}{2}}\langle\phi_0|0\rangle\langle 0|, \end{aligned} \tag{1.547}$$

we obtain

$$iG(\alpha t, \alpha' t') = \lim_{T\to\infty(1-i\eta)} \frac{\langle\phi_0|e^{-i\hat{H}\frac{T}{2}}\left(T\hat{\psi}_\alpha(t)\hat{\psi}^\dagger_{\alpha'}(t')\right)e^{-i\hat{H}\frac{T}{2}}|\phi_0\rangle}{\langle\phi_0|e^{-i\hat{H}T}|\phi_0\rangle}. \tag{1.548}$$

We show in the following that both $\langle\phi_0|e^{-i\hat{H}T}|\phi_0\rangle$ and the Green function (1.548) can be expressed as a coherent-state functional integral. We shall consider a contour in the complex t plane which is slightly tilted with respect to the real axis and extends from $-T/2$ and $T/2$ (Fig. 1.16). A time $t = \tilde{t}(1-i\eta)$ on this contour can be parameterized by a real variable $\tilde{t}$. In practice, we will distinguish between t and $\tilde{t}$ only when necessary.

1.8.1 *Fermions*

In the case of fermions, the non-interacting ground state is a Slater determinant,

$$|\phi_0\rangle = \prod_{\alpha=1}^{N} \hat{\psi}^\dagger_\alpha|\text{vac}\rangle, \tag{1.549}$$

with the N lowest one-particle states occupied. Here, we assume the one-body states $|\alpha\rangle$ to be labeled in order of increasing eigenvalues ϵ_α. We require $|\phi_0\rangle$ to be non-degenerate, which implies that there is a finite energy gap $\epsilon_{N+1} - \epsilon_N$ between the last occupied state and the first unoccupied state. We shift the energies ϵ_α such that

$$\begin{cases} \epsilon_\alpha < 0 \text{ for } \alpha \le N, \\ \epsilon_\alpha > 0 \text{ for } \alpha > N, \end{cases} \tag{1.550}$$

and define the operators

$$\hat{\chi}^\dagger_\alpha = \begin{cases} \hat{\psi}^\dagger_\alpha & (\alpha > N), \\ \hat{\psi}_\alpha & (\alpha \leq N), \end{cases} \qquad \hat{\chi}_\alpha = \begin{cases} \hat{\psi}_\alpha & (\alpha > N), \\ \hat{\psi}^\dagger_\alpha & (\alpha \leq N). \end{cases} \tag{1.551}$$

$\hat{\chi}^\dagger_\alpha$ creates an elementary excitation with respect to $|\phi_0\rangle$, i.e. a particle for $\alpha > N$ and a hole for $\alpha \leq N$. Since $\hat{\chi}_\alpha|\phi_0\rangle = 0$ for any α, $|\phi_0\rangle$ is the vacuum state with respect to the excitations of the operator $\hat{\chi}$. The Green function (1.546) is expressed as

$$G(\alpha t, \alpha' t') = \begin{cases} -i\langle 0|T\hat{\chi}_\alpha(t)\hat{\chi}^\dagger_{\alpha'}(t')|0\rangle & \text{if } \alpha > N, \\ i\langle 0|T\hat{\chi}_{\alpha'}(t')\hat{\chi}^\dagger_\alpha(t)|0\rangle & \text{if } \alpha \leq N. \end{cases} \tag{1.552}$$

1.8.1.1 *Coherent-state functional integral*

We can now construct coherent states by associating with $\hat{\chi}_\alpha$ and $\hat{\chi}^\dagger_\alpha$ a pair of complex conjugate Grassmann variables ψ_α and ψ^*_α (Sec. 1.3.3). Since $|\phi_0\rangle$ is the vacuum state with respect to the field ψ, it can be seen formally as the coherent state $|\psi\rangle$ with $\psi_\alpha = 0$. Dividing the time T into $M = \beta/\epsilon$ steps and inserting $M-1$ times the closure relation (1.217) or (1.254), one obtains

$$\begin{aligned} \langle\phi_0|e^{-i\hat{H}T}|\phi_0\rangle &= \int \prod_{k=1}^{M-1} d(\psi^*_k, \psi_k)\, e^{-\sum_{k=1}^{M-1}\sum_\alpha |\psi_{k,\alpha}|^2} \prod_{k=1}^{M} \langle\psi_k|e^{-i\epsilon\hat{H}}|\psi_{k-1}\rangle \\ &= \int \prod_{k=1}^{M-1} d(\psi^*_k, \psi_k)\, e^{-\sum_{k=1}^{M}\left[\sum_\alpha \psi^*_{k,\alpha}(\psi_{k,\alpha}-\psi_{k-1,\alpha})+i\epsilon H(\psi^*_{k,\alpha},\psi_{k,-1\alpha})\right]}, \end{aligned} \tag{1.553}$$

with $\psi_{0,\alpha} = \psi^*_{M,\alpha} = 0$. The integration measure $d(\psi^*_\alpha, \psi_\alpha)$ is defined by (1.262) and we assume the Hamiltonian $\hat{H} = H(\hat{\chi}^\dagger_\alpha, \hat{\chi}_\alpha)$ to be normal ordered. In the continuum-time limit ($M \to \infty$), we may write

$$\langle\phi_0|e^{-i\hat{H}T}|\phi_0\rangle = \int \mathcal{D}[\psi^*, \psi]\, e^{iS[\psi^*,\psi]} \tag{1.554}$$

with the action

$$S[\psi^*, \psi] = \int_{-T/2}^{T/2} dt \left[\sum_\alpha \psi^*_\alpha(t) i\partial_t \psi_\alpha(t) - H\big(\psi^*_\alpha(t), \psi_\alpha(t)\big)\right], \tag{1.555}$$

and the boundary conditions $\psi_\alpha(-T/2) = \psi^*_\alpha(T/2) = 0$. Except for the boundary conditions at $\pm T/2$, the real-time action (1.555) can be deduced from the Euclidean action (1.273) by a Wick rotation $\tau \to it$ and $-S_E \to iS$.

Similarly, we find

$$\langle\phi_0|e^{-i\hat{H}\frac{T}{2}}(T\hat{\chi}_\alpha(t)\hat{\chi}^\dagger_{\alpha'}(t'))e^{-i\hat{H}\frac{T}{2}}|\phi_0\rangle = \int \prod_{k=1}^{M-1} d(\psi^*_k, \psi_k)\, \psi_{k_1,\alpha}\psi^*_{k_2,\alpha'}\, e^{iS(\{\psi^*_k,\psi_k\})}, \tag{1.556}$$

where the action $S(\{\psi_k^*, \psi_k\})$ is defined by (1.553). $t_{k_1} = -T/2 + k_1\epsilon$ and $t_{k_2} = -T/2 + k_2\epsilon$ are the discrete times corresponding to t and t', respectively. Equation (1.556) is similar to (1.278) and can be derived in the same way. In the continuum-time limit, we therefore obtain

$$\langle 0|T\hat{\chi}_\alpha(t)\hat{\chi}^\dagger_{\alpha'}(t')|0\rangle = \lim_{T\to\infty(1-i\eta)} \frac{\int \mathcal{D}[\psi^*,\psi]\,\psi_\alpha(t)\psi^*_{\alpha'}(t')\,e^{iS[\psi^*,\psi]}}{\int \mathcal{D}[\psi^*,\psi]\,e^{iS[\psi^*,\psi]}} \tag{1.557}$$

with the boundary conditions $\psi_\alpha(-T/2) = \psi^*_\alpha(T/2) = 0$. The Green function $G(\alpha t, \alpha' t')$ can be deduced from (1.552) and (1.557).

1.8.1.2 *Non-interacting fermions*

Assuming that the non-interacting Hamiltonian is diagonal in the one-body state basis $\{|\alpha\rangle\}$,

$$\hat{H}_0 = \sum_{\alpha=1}^{N} \epsilon_\alpha \hat{\chi}_\alpha \hat{\chi}^\dagger_\alpha + \sum_{\alpha>N} \epsilon_\alpha \hat{\chi}^\dagger_\alpha \hat{\chi}_\alpha = \sum_{\alpha=1}^{N} \epsilon_\alpha (1 - \hat{\chi}^\dagger_\alpha \hat{\chi}_\alpha) + \sum_{\alpha>N} \epsilon_\alpha \hat{\chi}^\dagger_\alpha \hat{\chi}_\alpha, \tag{1.558}$$

which gives the action

$$\begin{aligned} iS_0 &= -iT\sum_{\alpha=1}^{N}\epsilon_\alpha - \sum_{k=1}^{M-1}\sum_{\alpha}\Big[\psi^*_{k,\alpha}(\psi_{k,\alpha}-\psi_{k-1,\alpha}) + i\frac{T}{M}|\epsilon_\alpha|\psi^*_{k,\alpha}\psi_{k-1,\alpha}\Big] \\ &\equiv -iT\sum_{\alpha=1}^{N}\epsilon_\alpha - \sum_{\alpha}\sum_{k,k'=1}^{M-1}\psi^*_{k,\alpha}S^{(\alpha)}_{k,k'}\psi_{k',\alpha}, \end{aligned} \tag{1.559}$$

where $S^{(\alpha)}$ is the $(M-1)\times(M-1)$ matrix

$$S^{(\alpha)} = \begin{pmatrix} 1 & 0 & \cdots & \cdots & 0 & 0 \\ -a & 1 & 0 & & & 0 \\ 0 & -a & 1 & \ddots & & \vdots \\ \vdots & \ddots & \ddots & 1 & \ddots & \vdots \\ \vdots & & \ddots & -a & 1 & 0 \\ 0 & \cdots & \cdots & 0 & -a & 1 \end{pmatrix} \tag{1.560}$$

with $a = 1 - i\frac{T}{M}|\epsilon_\alpha|$. Since $\det S^{(\alpha)} = 1$, we find

$$\langle\phi_0|e^{-i\hat{H}_0 T}|\phi_0\rangle = e^{-iT\sum_{\alpha=1}^{N}\epsilon_\alpha}\prod_\alpha \det S^{(\alpha)} = e^{-iT\sum_{\alpha=1}^{N}\epsilon_\alpha} \tag{1.561}$$

so that the non-interacting ground state energy is $E_0^{(0)} = \sum_{\alpha=1}^{N} \epsilon_\alpha$ [Eq. (1.545)]. From the inverse matrix

$$S^{(\alpha)-1} = \begin{pmatrix} 1 & 0 & 0 & \cdots & \cdots & 0 \\ a & 1 & 0 & \ddots & & \vdots \\ a^2 & a & 1 & 0 & \ddots & \vdots \\ \vdots & \ddots & a & 1 & 0 & 0 \\ \vdots & & \ddots & a & 1 & 0 \\ a^{M-2} & \cdots & \cdots & a^2 & a & 1 \end{pmatrix}, \tag{1.562}$$

we deduce

$$\langle \phi_0 | T \hat{\chi}_\alpha(t) \hat{\chi}^\dagger_{\alpha'}(t') | \phi_0 \rangle = \lim_{\substack{M\to\infty \\ T\to\infty(1-i\eta)}} S^{(\alpha)-1}_{k,k'} = \lim_{\substack{M\to\infty \\ T\to\infty(1-i\eta)}} \begin{cases} 0 & \text{if } k < k', \\ a^{k-k'} & \text{if } k \geq k', \end{cases} \tag{1.563}$$

where $t_k = -T/2 + kT/M$ and $t_{k'} = -T/2 + k'T/M$ are the discrete times corresponding to t and t', respectively. Using (1.552) and

$$\lim_{M\to\infty} a^{k-k'} = \lim_{M\to\infty} \left(1 - i\frac{T}{M}|\epsilon_\alpha|\right)^{(t_k - t_{k'})\frac{M}{T}} = e^{-i|\epsilon_\alpha|(t-t')}, \tag{1.564}$$

we finally obtain

$$iG_0(\alpha, t) = e^{-i\epsilon_\alpha t} \left[(1 - n_\alpha)\Theta(t) - n_\alpha \Theta(-t)\right], \tag{1.565}$$

where

$$n_\alpha = \begin{cases} 1 \text{ if } \alpha \leq N, \\ 0 \text{ if } \alpha > N \end{cases} \tag{1.566}$$

is the occupation of the one-body state $|\alpha\rangle$ in the non-interacting ground state $|\phi_0\rangle$.

So far, we have ignored the fact that the time $t = \tilde{t}(1 - i\eta)$ ($\tilde{t}$ real) contains a small imaginary part (Fig. 1.16). A more correct expression of the Green function would then be

$$iG_0(\alpha, \tilde{t}) = e^{-i\epsilon_\alpha \tilde{t}(1-i\eta)} \left[(1 - n_\alpha)\Theta(\tilde{t}) - n_\alpha \Theta(-\tilde{t})\right]. \tag{1.567}$$

In Fourier space, this yields

$$\begin{aligned} G_0(\alpha, \omega) &= \int_{-\infty}^{\infty} d\tilde{t}\, e^{i\omega\tilde{t}} G_0(\alpha, \tilde{t}) \\ &= \frac{1 - n_\alpha}{\omega - \epsilon_\alpha(1 - i\eta)} e^{i\tilde{t}[\omega - \epsilon_\alpha(1-i\eta)]}\Big|_0^\infty - \frac{n_\alpha}{\omega - \epsilon_\alpha(1 - i\eta)} e^{i\tilde{t}[\omega - \epsilon_\alpha(1-i\eta)]}\Big|_{-\infty}^0. \end{aligned} \tag{1.568}$$

Since $\epsilon_\alpha < 0$ for $n_\alpha = 1$ and $\epsilon_\alpha > 0$ for $n_\alpha = 0$, the infinitesimal $\eta = 0^+$ ensures the convergence of the time integral,

$$G_0(\alpha, \omega) = \frac{n_\alpha}{\omega - \epsilon_\alpha - i\eta} + \frac{1 - n_\alpha}{\omega - \epsilon_\alpha + i\eta}. \tag{1.569}$$

1.8.1.3 *Perturbation theory*

Perturbation theory is derived by expanding e^{iS} in powers of the interacting action

$$S_{\rm int}[\psi^*,\psi] = -\int_{-T/2}^{T/2} dt\, H_{\rm int}\big(\psi^*_\alpha(t),\psi_\alpha(t)\big). \tag{1.570}$$

For example, to obtain the ground state energy using (1.545), one uses

$$\begin{aligned}\langle\phi_0|e^{-i\hat{H}T}|\phi_0\rangle &= \int \mathcal{D}[\psi^*,\psi]\, e^{iS_0[\psi^*,\psi]+iS_{\rm int}[\psi^*,\psi]} \\ &= \int \mathcal{D}[\psi^*,\psi]\, e^{iS_0[\psi^*,\psi]} \sum_{n=0}^{\infty} \frac{i^n}{n!}(S_{\rm int}[\psi^*,\psi])^n. \end{aligned}\tag{1.571}$$

The average value

$$\langle (S_{\rm int}[\psi^*,\psi])^n\rangle_0 = \frac{\int \mathcal{D}[\psi^*,\psi]\,(S_{\rm int}[\psi^*,\psi])^n\, e^{iS_0[\psi^*,\psi]}}{\int \mathcal{D}[\psi^*,\psi]\, e^{iS_0[\psi^*,\psi]}} \tag{1.572}$$

can be evaluated using Wick's theorem and represented by Feynman diagrams. The diagrammatic rules follow straightforwardly from those of the finite-temperature formalism. The linked cluster theorem states that the ground state energy

$$\begin{aligned} E_0 &= \lim_{T\to\infty(1-i\eta)} \frac{i}{T}\ln\langle\phi_0|e^{-i\hat{H}T}|\phi_0\rangle \\ &= E_0^{(0)} + \lim_{T\to\infty(1-i\eta)} \frac{i}{T}\sum \text{connected graphs} \end{aligned}\tag{1.573}$$

is given by the sum of the connected diagrams (vacuum fluctuation diagrams). More generally, most results derived in Secs. 1.4–1.6 can be transposed to the zero-temperature formalism.

1.8.2 ***Bosons***

The ground state of a non-interacting system of N bosons is a Bose–Einstein condensate,

$$|\phi_0(N)\rangle = \frac{1}{\sqrt{N!}}\left[\hat{\psi}^\dagger(\mathbf{k}=0)\right]^N|\text{vac}\rangle, \tag{1.574}$$

where all the particles occupy the $\mathbf{k}=0$ one-body state. For simplicity, we consider spin-zero bosons and assume the system to be translation invariant. Since

$$\begin{aligned} \hat{\psi}(\mathbf{k}=0)|\phi_0(N)\rangle &= \sqrt{N}|\phi_0(N-1)\rangle, \\ \hat{\psi}^\dagger(\mathbf{k}=0)|\phi_0(N)\rangle &= \sqrt{N+1}|\phi_0(N+1)\rangle, \end{aligned}\tag{1.575}$$

neither $\hat{\psi}(\mathbf{k}=0)$ nor $\hat{\psi}^\dagger(\mathbf{k}=0)$ annihilates the ground state $|\phi_0(N)\rangle$, and it is not possible to separate the operators between creation and annihilation parts as in the fermion case [Eq. (1.551)].

Let us consider the rescaled $\mathbf{k} = 0$ operators

$$\hat{\xi}_0 = \frac{1}{\sqrt{\mathcal{V}}}\hat{\psi}(\mathbf{k}=0) = \frac{1}{\mathcal{V}}\int d^d r\,\hat{\psi}(\mathbf{r}),$$
$$\hat{\xi}_0^\dagger = \frac{1}{\sqrt{\mathcal{V}}}\hat{\psi}^\dagger(\mathbf{k}=0) = \frac{1}{\mathcal{V}}\int d^d r\,\hat{\psi}^\dagger(\mathbf{r}). \tag{1.576}$$

Their commutator

$$[\hat{\xi}_0, \hat{\xi}_0^\dagger] = \frac{1}{\mathcal{V}} \tag{1.577}$$

vanishes in the thermodynamic limit, while they yield a finite factor when acting on $|\phi_0(N)\rangle$,

$$\hat{\xi}_0|\phi_0(N)\rangle = \left(\frac{N}{\mathcal{V}}\right)^{1/2}|\phi_0(N-1)\rangle \simeq \left(\frac{N}{\mathcal{V}}\right)^{1/2}|\phi_0(N)\rangle,$$
$$\hat{\xi}_0^\dagger|\phi_0(N)\rangle = \left(\frac{N+1}{\mathcal{V}}\right)^{1/2}|\phi_0(N+1)\rangle \simeq \left(\frac{N}{\mathcal{V}}\right)^{1/2}|\phi_0(N)\rangle. \tag{1.578}$$

They can therefore be replaced by c-numbers as long as we consider only states where a finite fraction of the particles occupies the $\mathbf{k} = 0$ one-body state. This leads us to rewrite the boson operator as

$$\hat{\psi}(\mathbf{r}) = \xi_0 + \frac{1}{\sqrt{\mathcal{V}}}\sum_{\mathbf{k}\neq 0} e^{i\mathbf{k}\cdot\mathbf{r}}\hat{\psi}(\mathbf{k}) \equiv \xi_0 + \hat{\chi}(\mathbf{r}),$$
$$\hat{\psi}^\dagger(\mathbf{r}) = \xi_0^* + \frac{1}{\sqrt{\mathcal{V}}}\sum_{\mathbf{k}\neq 0} e^{-i\mathbf{k}\cdot\mathbf{r}}\hat{\psi}^\dagger(\mathbf{k}) \equiv \xi_0^* + \hat{\chi}^\dagger(\mathbf{r}). \tag{1.579}$$

Since $\hat{\chi}(\mathbf{r})$ annihilates the non-interacting ground state $|\phi_0(N)\rangle$, we can proceed as in the fermion case and write $\langle\phi_0(N)|e^{-i\hat{H}T}|\phi_0(N)\rangle$ as a functional integral over a complex bosonic field with the action

$$S[\psi^*, \psi] = \int_{-T/2}^{T/2} dt\left[\int d^d r\,\psi^*(\mathbf{r},t)i\partial_t\psi(\mathbf{r},t) - H\big(\psi^*(\mathbf{r},t), \psi(\mathbf{r},t)\big)\right], \tag{1.580}$$

where $\psi(\mathbf{r}, -T/2) = \psi^*(\mathbf{r}, T/2) = 0$. Again, we see that the real-time action (1.580) can be deduced from the Euclidean action (1.273) by a Wick rotation $\tau = it$. As in Sec. 1.7.2, the interaction vertices coming from the interacting Hamiltonian are of four different types depending on the number of condensate lines attached to them.

Considering $\hat{\xi}_0$ and $\hat{\xi}_0^\dagger$ as c-numbers breaks the (global) U(1) invariance, i.e. the invariance of the Hamiltonian under the transformation $\hat{\psi}(\mathbf{r}) \to e^{i\Lambda}\hat{\psi}(\mathbf{r})$ and $\hat{\psi}^\dagger(\mathbf{r}) \to e^{-i\Lambda}\hat{\psi}^\dagger(\mathbf{r})$. As a result, the total number of particles is not conserved, $[\hat{N}, \hat{H}] \neq 0$, and the canonical ensemble cannot be used. Instead, one must work in the grand canonical ensemble where the Hamiltonian is replaced by $\hat{H} - \mu\hat{N}$, with the chemical potential μ chosen such that $\langle 0|\hat{N}|0\rangle = N$. In the interacting ground state, not all the particles are in the $\mathbf{k} = 0$ one-body state, and the average value

$$\langle 0|\hat{\xi}_0^\dagger\hat{\xi}_0|0\rangle = \frac{N_0}{\mathcal{V}} = n_0 \tag{1.581}$$

is less than the total density $N/\mathcal{V}$. As long as the condensate density n_0 remains finite, the replacement of $\hat{\xi}_0$ and $\hat{\xi}_0^\dagger$ by c-numbers ξ_0 and ξ_0^* (with $|\xi_0| = \sqrt{n_0}$) is permissible. To determine the condensate density n_0, we require that

$$\frac{\partial}{\partial n_0} \langle 0|\hat{H} - \mu\hat{N}|0\rangle \Big|_\mu = 0, \tag{1.582}$$

which is an implicit equation for $n_0(\mu)$.

1.8.3 *The zero-temperature limit*

We have now two different ways to compute perturbatively the ground state energy or a correlation function. One can either use the $T = 0$ perturbation theory or consider the $T \to 0$ limit of the finite-temperature formalism discussed in Secs. 1.4–1.6. While one would naively expect these two approaches to give always the same result, this is in fact not the case. In the zero-temperature formalism, the interacting ground state $|0\rangle$ is obtained from (1.545) provided that $|0\rangle$ and $|\phi_0\rangle$ are not orthogonal. In some cases, $e^{-i\hat{H}T}|\phi_0\rangle$ is trapped in a given symmetry space selected by $|\phi_0\rangle$ and produces the lowest energy eigenstate in that space. In contrast, the finite-temperature theory obtains the state of the system from a statistical average (over a certain ensemble) which has access to all possible symmetries. In principle, it is always possible to use the zero-temperature theory, provided that one chooses a Hamiltonian $\hat{H}_0$ with a ground state $|\phi_0\rangle$ which has the right symmetries. In practice however, it is often preferable to use the (Euclidean) finite-temperature even when one is interested only in the $T = 0$ limit.

From a technical point of view, there are two differences between the zero- and finite-temperature formalisms:

1. In the finite-temperature formalism, the chemical potential μ must be adjusted so that $N = -\partial\Omega/\partial\mu$. In general, μ differs from the chemical potential μ_0 of the non-interacting system defined by $N = -\partial\Omega_0/\partial\mu_0$. Thus, the correction

$$E_0 - E_0^{(0)} = \lim_{T\to 0} [\Omega - \Omega_0 + (\mu - \mu_0)N] \tag{1.583}$$

to the ground state energy is determined not only by $\Omega - \Omega_0$ (i.e. the vacuum fluctuation diagrams) but also by the variation $\mu - \mu_0$ of the chemical potential. In the zero-temperature theory, $E_0 - E_0^{(0)}$ is directly given by the perturbation expansion of $\langle\phi_0(N)|e^{-i\hat{H}T}|\phi_0(N)\rangle$, i.e. by the vacuum fluctuation diagrams [Eq. (1.573)].
2. The finite-temperature perturbation theory yields "anomalous" diagrams which include parts such as

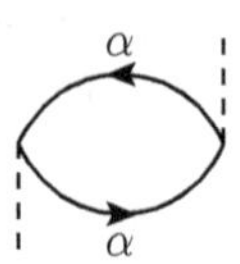

$\propto n_\alpha(1 - n_\alpha)$

where a particle and a hole propagate in the same one-body state α. Since the one-body state α is either empty ($n_\alpha = 0$) or occupied ($n_\alpha = 1$) in the zero-temperature non-interacting ground state, the anomalous diagrams vanish at $T = 0$. On the contrary, $\lim_{T\to 0} n_\alpha(1-n_\alpha)$ does not necessarily vanish in the finite-temperature formalism.[71]

For the zero- and finite-temperature perturbation theories to agree, the differences (1) and (2) must cancel.

1.9 Quantization of the Electromagnetic Field

In this section, we discuss the functional integral describing charged particles coupled to the electromagnetic field. We start by briefly reviewing the Lagrangian and Hamiltonian formalism of classical electromagnetism. The difficulties due to the redundancy of degrees of freedom are resolved by using the Coulomb gauge. We then discuss the canonical quantization in this gauge and express the partition function of the coupled system matter-field as a functional integral.

1.9.1 *Lagrangian and Hamiltonian formalism*

The standard Lagrangian of classical (three-dimensional) electrodynamics reads

$$\begin{aligned} L = L'_m + \frac{\epsilon_0}{2}\int d^3r \left\{ [\boldsymbol{\nabla}\phi(\mathbf{r},t) + \partial_t \mathbf{A}(\mathbf{r},t)]^2 - c_l^2\, [\boldsymbol{\nabla}\times \mathbf{A}(\mathbf{r},t)]^2 \right\} \\ - \int d^3r\, [\phi(\mathbf{r},t)\rho(\mathbf{r},t) - \mathbf{J}(\mathbf{r},t)\cdot \mathbf{A}(\mathbf{r},t)] \end{aligned} \tag{1.584}$$

(c_l denotes the velocity of light), where $L'_m = (m/2)\sum_i \dot{\mathbf{r}}_i^2$ is the Lagrangian of the matter in the absence of electromagnetic interactions with $\mathbf{r}_i$ and $\dot{\mathbf{r}}_i$ the position and velocity of the ith particle. The electric and magnetic fields can be expressed as a function of the electromagnetic potentials,

$$\begin{aligned} \mathbf{E}(\mathbf{r},t) &= -\boldsymbol{\nabla}\phi(\mathbf{r},t) - \partial_t \mathbf{A}(\mathbf{r},t), \\ \mathbf{B}(\mathbf{r},t) &= \boldsymbol{\nabla}\times \mathbf{A}(\mathbf{r},t). \end{aligned} \tag{1.585}$$

The field–matter coupling involves the charge and current densities

$$\begin{aligned} \rho(\mathbf{r},t) &= e\sum_i \delta(\mathbf{r}-\mathbf{r}_i(t)), \\ \mathbf{J}(\mathbf{r},t) &= e\sum_i \dot{\mathbf{r}}_i \delta(\mathbf{r}-\mathbf{r}_i(t)) \end{aligned} \tag{1.586}$$

(e denotes the charge of the particles). The Euler–Lagrange equations deduced from (1.584) reproduce Maxwell equations (Appendix 1.A).

It is tempting to quantize the theory in the usual way, i.e. by introducing the momenta $\mathbf{p}_i$ of the particles and the canonical conjugates $(\Pi_0, \boldsymbol{\Pi})$ to the electromagnetic potential $(\phi, \mathbf{A})$, and assuming the usual commutation rules between conjugate variables. However,

[71] For example, $\beta n_F(\xi_\alpha)[1-n_F(\xi_\alpha)] = -\frac{\partial n_F(\xi_\alpha)}{\partial \xi_\alpha} \to \delta(\xi_\alpha)$ for $T\to 0$.

since $\dot{\phi} = \partial_t \phi$ does not appear in the Lagrangian, $\Pi_0 = 0$. One must therefore first eliminate the scalar potential by expressing it as a function of the other dynamical variables to obtain a reduced Lagrangian which is a function of the vector potential $\mathbf{A}$ and its time derivative $\dot{\mathbf{A}}$, as well as the positions and velocities of the particles. Before quantizing, one should also get rid of the arbitrariness of $\mathbf{A}$ which follows form gauge invariance. A gauge transformation

$$\begin{aligned} \mathbf{A}'(\mathbf{r},t) &= \mathbf{A}(\mathbf{r},t) + \boldsymbol{\nabla}\Lambda(\mathbf{r},t), \\ \phi'(\mathbf{r},t) &= \phi(\mathbf{r},t) - \partial_t\Lambda(\mathbf{r},t) \end{aligned} \tag{1.587}$$

does not change the electromagnetic field and therefore leaves the Maxwell equations invariant. By choosing a particular gauge, we fix the longitudinal component $\mathbf{A}_\parallel$ of the vector potential, which is otherwise arbitrary. Once this is done, we are left with four degrees of freedom of the field at each position in space, the two components of the transverse vector potential $\mathbf{A}_\perp$ and their time derivatives. Since the longitudinal components $\mathbf{E}_\parallel$ and $\mathbf{B}_\parallel$ of the field are fixed by the Maxwell equations $\boldsymbol{\nabla}\cdot\mathbf{E} = \rho/\epsilon_0$ and $\boldsymbol{\nabla}\cdot\mathbf{B} = 0$, they are not true dynamical variables and we indeed expect any description of the electromagnetic field to contain four independent variables.

We can eliminate the scalar potential using $\boldsymbol{\nabla}\cdot\mathbf{E} = \rho/\epsilon_0$ (which follows from the Euler–Lagrange equation for ϕ), i.e. $\boldsymbol{\nabla}^2\phi + \boldsymbol{\nabla}\cdot\dot{\mathbf{A}}_\parallel = -\rho/\epsilon_0$ or

$$\phi(\mathbf{k},t) = \frac{1}{\mathbf{k}^2}\Big[i\mathbf{k}\cdot\dot{\mathbf{A}}_\parallel(\mathbf{k},t) + \frac{1}{\epsilon_0}\rho(\mathbf{k},t)\Big]. \tag{1.588}$$

The electric field is then given by

$$\begin{aligned} \mathbf{E}_\parallel(\mathbf{k},t) &= -i\mathbf{k}\frac{\rho(\mathbf{k},t)}{\epsilon_0\mathbf{k}^2}, \\ \mathbf{E}_\perp(\mathbf{k},t) &= -\dot{\mathbf{A}}_\perp(\mathbf{k},t). \end{aligned} \tag{1.589}$$

In the following, we shall work in the Coulomb gauge where $\mathbf{A}_\parallel$ vanishes,

$$\boldsymbol{\nabla}\cdot\mathbf{A}(\mathbf{r},t) = 0 \quad \text{(Coulomb gauge)}. \tag{1.590}$$

Using (1.588)–(1.590), we obtain the Lagrangian

$$\begin{aligned} L = L'_m + L_c + \int d^3r \Big\{ &\frac{\epsilon_0}{2}\Big[\big(\partial_t\mathbf{A}_\perp(\mathbf{r},t)\big)^2 - c_l^2\big(\boldsymbol{\nabla}\times\mathbf{A}_\perp(\mathbf{r},t)\big)^2\Big] \\ &+ \mathbf{J}_\perp(\mathbf{r},t)\cdot\mathbf{A}_\perp(\mathbf{r},t)\Big\}, \end{aligned} \tag{1.591}$$

where[72]

$$L_c = -\frac{1}{2}\sum_{\mathbf{k}} \frac{\rho(-\mathbf{k},t)\rho(\mathbf{k},t)}{\epsilon_0\mathbf{k}^2} = -\frac{1}{2}\int d^3r d^3r' \frac{\rho(\mathbf{r},t)\rho(\mathbf{r}',t)}{4\pi\epsilon_0|\mathbf{r}-\mathbf{r}'|} \tag{1.592}$$

is (minus) the Coulomb energy of the charge distribution $\rho(\mathbf{r},t)$.

[72]The Fourier-transformed variables are defined by $\rho(\mathbf{r},t) = \frac{1}{\sqrt{V}}\sum_{\mathbf{k}} e^{i\mathbf{k}\cdot\mathbf{r}}\rho(\mathbf{k},t)$, etc.

Let us discuss in more detail the role of the longitudinal component of the vector potential. Equation (1.588) enables to eliminate ϕ and rewrite the Lagrangian as

$$L = L'_m + L_c + \frac{\epsilon_0}{2}\sum_{\mathbf{k}}\left[\dot{\mathbf{A}}_\perp(-\mathbf{k},t)\cdot\dot{\mathbf{A}}_\perp(\mathbf{k},t) - c_l^2\mathbf{k}^2\mathbf{A}_\perp(-\mathbf{k},t)\cdot\mathbf{A}_\perp(\mathbf{k},t)\right]$$
$$+\sum_{\mathbf{k}}\Big[\mathbf{J}_\perp(-\mathbf{k},t)\cdot\mathbf{A}_\perp(\mathbf{k},t) + \mathbf{J}_\parallel(-\mathbf{k},t)\cdot\mathbf{A}_\parallel(\mathbf{k},t) - i\rho(-\mathbf{k},t)\frac{\mathbf{k}}{\mathbf{k}^2}\cdot\dot{\mathbf{A}}_\parallel(\mathbf{k},t)\Big]. \tag{1.593}$$

The Lagrange equation for $\mathbf{A}_\parallel$,

$$\frac{d}{dt}\frac{\partial L}{\partial\dot{\mathbf{A}}_\parallel(\mathbf{k},t)} = \frac{\partial L}{\partial\mathbf{A}_\parallel(\mathbf{k},t)}, \tag{1.594}$$

gives

$$i\frac{\mathbf{k}}{\mathbf{k}^2}\dot{\rho}(\mathbf{k},t) = \mathbf{J}_\parallel(\mathbf{k},t), \quad \text{i.e.} \quad \partial_t\rho(\mathbf{r},t) + \boldsymbol{\nabla}\cdot\mathbf{J}_\parallel(\mathbf{r},t) = 0. \tag{1.595}$$

This is a continuity equation, not an equation of motion for $\mathbf{A}_\parallel$. Using (1.595), we find

$$\sum_{\mathbf{k}}\Big[\mathbf{J}_\parallel(-\mathbf{k},t)\cdot\mathbf{A}_\parallel(\mathbf{k},t) - i\rho(-\mathbf{k},t)\frac{\mathbf{k}}{\mathbf{k}^2}\cdot\dot{\mathbf{A}}_\parallel(\mathbf{k},t)\Big]$$
$$= \sum_{\mathbf{k}}\frac{1}{\mathbf{k}^2}\left[\mathbf{k}\cdot\mathbf{J}_\parallel(-\mathbf{k},t)\mathbf{k}\cdot\mathbf{A}_\parallel(\mathbf{k},t) - i\rho(-\mathbf{k},t)\mathbf{k}\cdot\dot{\mathbf{A}}_\parallel(\mathbf{k},t)\right]$$
$$= \sum_{\mathbf{k}}\frac{1}{\mathbf{k}^2}\left[-\dot{\rho}(-\mathbf{k},t)i\mathbf{k}\cdot\mathbf{A}_\parallel(\mathbf{k},t) - \rho(-\mathbf{k},t)i\mathbf{k}\cdot\dot{\mathbf{A}}_\parallel(\mathbf{k},t)\right]$$
$$= -i\frac{d}{dt}\sum_{\mathbf{k}}\frac{1}{\mathbf{k}^2}\rho(-\mathbf{k},t)\mathbf{k}\cdot\mathbf{A}_\parallel(\mathbf{k},t), \tag{1.596}$$

and the Lagrangian becomes

$$L = L'_m + L_c + \frac{\epsilon_0}{2}\sum_{\mathbf{k}}\left[\dot{\mathbf{A}}_\perp(-\mathbf{k},t)\cdot\dot{\mathbf{A}}_\perp(\mathbf{k},t) - c_l^2\mathbf{A}_\perp(-\mathbf{k},t)\cdot\mathbf{A}_\perp(\mathbf{k},t)\right]$$
$$+\sum_{\mathbf{k}}\mathbf{J}_\perp(-\mathbf{k},t)\cdot\mathbf{A}_\perp(\mathbf{k},t) - i\frac{d}{dt}\sum_{\mathbf{k}}\frac{1}{\mathbf{k}^2}\rho(-\mathbf{k},t)\mathbf{k}\cdot\mathbf{A}_\parallel(\mathbf{k},t). \tag{1.597}$$

The last term being a total derivative, it does not affect the equations of motion and can therefore be removed from the Lagrangian which then becomes independent of $\mathbf{A}_\parallel$. Thus, the longitudinal component of the vector potential does not affect the dynamics of the system and is not a true dynamical variable. Of course, the possibility to arbitrarily choose $\mathbf{A}_\parallel$ is nothing but a consequence of gauge invariance.

The Hamiltonian is obtained by introducing the momenta of the particles,

$$\mathbf{p}_i = \frac{\partial L}{\partial\dot{\mathbf{r}}_i} = m\dot{\mathbf{r}}_i + e\mathbf{A}_\perp(\mathbf{r}_i,t), \tag{1.598}$$

and the conjugate momentum of $\mathbf{A}_\perp$,

$$\mathbf{\Pi}_\perp(\mathbf{r},t) = \frac{\delta L}{\delta \dot{\mathbf{A}}_\perp(\mathbf{r},t)}. \tag{1.599}$$

Since $\mathbf{A}_\perp(\mathbf{r},t)$ is transverse, we cannot vary its components independently. We must therefore obtain $\mathbf{\Pi}_\perp$ in reciprocal space, using

$$\begin{aligned}\frac{\epsilon_0}{2}\int d^3r\, \dot{\mathbf{A}}_\perp(\mathbf{r},t)^2 &= \frac{\epsilon_0}{2}\sum_{\mathbf{k}} \dot{\mathbf{A}}_\perp(-\mathbf{k},t)\cdot \dot{\mathbf{A}}_\perp(\mathbf{k},t) \\ &= \frac{\epsilon_0}{2}\sum_{\substack{\mathbf{k}\\ \epsilon=\epsilon_1,\epsilon_2}} \dot{A}_{\perp\epsilon}(-\mathbf{k},t)\dot{A}_{\perp\epsilon}(\mathbf{k},t),\end{aligned} \tag{1.600}$$

where $A_{\perp\epsilon_1}$ and $A_{\perp\epsilon_2}$ denote the two independent components of $\mathbf{A}_\perp$,

$$\mathbf{A}_\perp(\mathbf{k},t) = A_{\perp\epsilon_1}(\mathbf{k},t)\boldsymbol{\epsilon}_1 + A_{\perp\epsilon_2}(\mathbf{k},t)\boldsymbol{\epsilon}_2, \tag{1.601}$$

with $\boldsymbol{\epsilon}_1, \boldsymbol{\epsilon}_2, \mathbf{k}$ mutually orthogonal and $\boldsymbol{\epsilon}_1^2 = \boldsymbol{\epsilon}_2^2 = 1$. The conjugate momentum is then given by

$$\Pi_{\perp\epsilon}(\mathbf{k},t) = \frac{\partial L}{\partial \dot{A}_{\perp\epsilon}(-\mathbf{k},t)} = \epsilon_0 \dot{A}_{\perp\epsilon}(\mathbf{k},t) \tag{1.602}$$

[see Eq. (1.A.46) in Appendix 1.A], so that

$$\mathbf{\Pi}_\perp(\mathbf{r},t) = \epsilon_0 \dot{\mathbf{A}}_\perp(\mathbf{r},t). \tag{1.603}$$

$\mathbf{\Pi}_\perp$ is transverse: $\boldsymbol{\nabla}\cdot\mathbf{\Pi}_\perp(\mathbf{r},t) = 0$. In the Coulomb gauge, the Hamiltonian is therefore defined by

$$\begin{aligned}H &= \sum_i \mathbf{p}_i(t)\cdot\dot{\mathbf{r}}_i(t) + \int d^3r\, \mathbf{\Pi}_\perp(\mathbf{r},t)\cdot\dot{\mathbf{A}}_\perp(\mathbf{r},t) - L, \\ &= \sum_i \frac{[\mathbf{p}_i(t) - e\mathbf{A}_\perp(\mathbf{r}_i,t)]^2}{2m} + H_c + \frac{1}{2\epsilon_0}\int d^3r \left[\mathbf{\Pi}_\perp(\mathbf{r},t)^2 + \epsilon_0^2 c_l^2 \left(\boldsymbol{\nabla}\times\mathbf{A}_\perp(\mathbf{r},t)\right)^2\right],\end{aligned} \tag{1.604}$$

where $H_c = -L_c$ is the Coulomb energy. The electromagnetic field is now given by

$$\begin{aligned}&\mathbf{E}_\parallel(\mathbf{k},t) = -i\mathbf{k}\frac{\rho(\mathbf{k},t)}{\epsilon_0\mathbf{k}^2}, && \mathbf{E}_\parallel(\mathbf{r},t) = \int d^3r' \frac{\rho(\mathbf{r}',t)(\mathbf{r}-\mathbf{r}')}{4\pi\epsilon_0|\mathbf{r}-\mathbf{r}'|^3}, \\ &\mathbf{E}_\perp(\mathbf{k},t) = -\frac{1}{\epsilon_0}\mathbf{\Pi}_\perp(\mathbf{k},t), \quad \text{i.e.} && \mathbf{E}_\perp(\mathbf{r},t) = -\frac{1}{\epsilon_0}\mathbf{\Pi}_\perp(\mathbf{r},t), \\ &\mathbf{B}(\mathbf{k},t) = i\mathbf{k}\times\mathbf{A}_\perp(\mathbf{k},t), && \mathbf{B}(\mathbf{r},t) = \boldsymbol{\nabla}\times\mathbf{A}_\perp(\mathbf{r},t).\end{aligned} \tag{1.605}$$

The Hamiltonian of the particles (i.e. without the Hamiltonian $\hat{H}_{\mathrm{em}}$ of the free field given by the last line of (1.604)) can be written as

$$H = H'_m + H_c - \int d^3r\, \mathbf{j}(\mathbf{r},t)\cdot\mathbf{A}_\perp(\mathbf{r},t) + \frac{e}{2m}\int d^3r\, \rho(\mathbf{r},t)\mathbf{A}_\perp^2(\mathbf{r},t), \tag{1.606}$$

where

$$\mathbf{j}(\mathbf{r},t) = \frac{e}{m}\sum_i \mathbf{p}_i \delta(\mathbf{r}-\mathbf{r}_i) \tag{1.607}$$

is the paramagnetic part of the current

$$\mathbf{J}(\mathbf{r},t) = e\sum_i \dot{\mathbf{r}}_i \delta(\mathbf{r}-\mathbf{r}_i) = \frac{e}{m}\sum_i [\mathbf{p}_i - e\mathbf{A}_\perp(\mathbf{r}_i,t)]\delta(\mathbf{r}-\mathbf{r}_i) \tag{1.608}$$

and $H'_m = \sum_i \mathbf{p}_i^2/2m$ is the Hamiltonian of the matter in the absence of electromagnetic interactions.

1.9.2 *Canonical quantization in the Coulomb gauge*

We are now in a position to quantize the theory by promoting the variables $(\mathbf{r}_i, \mathbf{p}_i)$ and $(\mathbf{A}_\perp, \mathbf{\Pi}_\perp)$ to operators. The position and momentum operators $\hat{\mathbf{r}}_i$ and $\hat{\mathbf{p}}_i$ satisfy the usual commutations relations (Appendix 1.B). As for the electromagnetic field, there are two independent dynamical variables $A_{\perp\epsilon_1}$ and $A_{\perp\epsilon_2}$, and their conjugate momenta $\Pi_{\perp\epsilon_1}$ and $\Pi_{\perp\epsilon_2}$, for each wavevector $\mathbf{k}$. The reality of $\mathbf{A}_\perp(\mathbf{r},t)$ and $\mathbf{\Pi}_\perp(\mathbf{r},t)$ implies $A^*_{\perp\epsilon}(\mathbf{k},t) = A_{\perp\epsilon}(-\mathbf{k},t)$ and $\Pi^*_{\perp\epsilon}(\mathbf{k},t) = \Pi_{\perp\epsilon}(-\mathbf{k},t)$. The corresponding operators satisfy

$$\begin{aligned} \left[\hat{A}_{\perp\epsilon}(\mathbf{k}), \hat{A}_{\perp\epsilon'}(\mathbf{k}')\right] &= \left[\hat{\Pi}_{\perp\epsilon}(\mathbf{k}), \hat{\Pi}_{\perp\epsilon'}(\mathbf{k}')\right] = 0, \\ \left[\hat{A}_{\perp\epsilon}(\mathbf{k}), \hat{\Pi}^\dagger_{\perp\epsilon'}(\mathbf{k}')\right] &= i\delta_{\epsilon,\epsilon'}\delta_{\mathbf{k},\mathbf{k}'}, \end{aligned} \tag{1.609}$$

where $\hat{\Pi}^\dagger_{\perp\epsilon}(\mathbf{k}) = \hat{\Pi}_{\perp\epsilon}(-\mathbf{k})$ and $\hat{A}^\dagger_{\perp\epsilon}(\mathbf{k}) = \hat{A}_{\perp\epsilon}(-\mathbf{k})$. Note that we consider here the Schrödinger picture where the operators do not depend on time (Appendix 1.B). The quantum Hamiltonian $\hat{H}$ is obtained from (1.604) by replacing the classical variables by the corresponding operators. Using

$$\begin{aligned} \hat{A}_{\perp\mu}(\mathbf{k}) &= \sum_\epsilon \hat{A}_{\perp\epsilon}(\mathbf{k})\epsilon_\mu, \\ \hat{\Pi}_{\perp\mu}(\mathbf{k}) &= \sum_\epsilon \hat{\Pi}_{\perp\epsilon}(\mathbf{k})\epsilon_\mu \end{aligned} \tag{1.610}$$

$(\mu = x, y, z)$, we obtain[73]

$$\left[\hat{A}_{\perp\mu}(\mathbf{k}), \hat{\Pi}^\dagger_{\perp\mu'}(\mathbf{k}')\right] = i\delta_{\mathbf{k},\mathbf{k}'}\sum_\epsilon \epsilon_\mu\epsilon_{\mu'} = i\delta_{\mathbf{k},\mathbf{k}'}\left(\delta_{\mu,\mu'} - k_\mu k_{\mu'}/\mathbf{k}^2\right). \tag{1.611}$$

In direct space, the commutation relations (1.609) become

$$\begin{aligned} \left[\hat{A}_{\perp\mu}(\mathbf{r}), \hat{A}_{\perp\mu'}(\mathbf{r}')\right] &= \left[\hat{\Pi}_{\perp\mu}(\mathbf{r}), \hat{\Pi}_{\perp\mu'}(\mathbf{r}')\right] = 0, \\ \left[\hat{A}_{\perp\mu}(\mathbf{r}), \hat{\Pi}_{\perp\mu'}(\mathbf{r}')\right] &= i\delta^\perp_{\mu,\mu'}(\mathbf{r}-\mathbf{r}'). \end{aligned} \tag{1.612}$$

[73]Since $(\mathbf{k}, \epsilon_1, \epsilon_2)$ form an orthonormal basis in reciprocal space, $\epsilon_{1\mu}\epsilon_{1\mu'} + \epsilon_{2\mu}\epsilon_{2\mu'} + k_\mu k_{\mu'}/\mathbf{k}^2 = \delta_{\mu,\mu'}$.

The "transverse delta function"

$$\delta^{\perp}_{\mu,\mu'}(\mathbf{r}) = \int \frac{d^3k}{(2\pi)^3} e^{i\mathbf{k}\cdot\mathbf{r}} \left(\delta_{\mu,\mu'} - \frac{k_\mu k_{\mu'}}{\mathbf{k}^2} \right) = \delta_{\mu,\mu'}\delta(\mathbf{r}) - \frac{\partial}{\partial r_\mu}\frac{\partial}{\partial r_{\mu'}}\frac{1}{4\pi|\mathbf{r}|} \tag{1.613}$$

appearing in (1.612) (in place of $\delta_{\mu,\mu'}\delta(\mathbf{r}-\mathbf{r}')$) reflects the fact that the three components of $\hat{\mathbf{A}}_\perp(\mathbf{r})$ and $\hat{\mathbf{\Pi}}_\perp(\mathbf{r})$ are not independent but constrained by the transversality conditions $\boldsymbol{\nabla}\cdot\hat{\mathbf{A}}_\perp(\mathbf{r}) = \boldsymbol{\nabla}\cdot\hat{\mathbf{\Pi}}_\perp(\mathbf{r}) = 0$.

1.9.2.1 *Second quantization and photons*

Let us consider the Hamiltonian of the free field (no coupling to the matter)

$$\hat{H}_{\rm em} = \frac{1}{2\epsilon_0}\sum_{\mathbf{k},\epsilon}\left[\hat{\Pi}^\dagger_{\perp\epsilon}(\mathbf{k})\hat{\Pi}_{\perp\epsilon}(\mathbf{k}) + \epsilon_0^2 c_l^2 \mathbf{k}^2 \hat{A}^\dagger_{\perp\epsilon}(\mathbf{k})\hat{A}_{\perp\epsilon}(\mathbf{k})\right]. \tag{1.614}$$

Given the commutation relations (1.609) between $\hat{A}_{\perp\epsilon}$ and $\hat{\Pi}_{\perp\epsilon}$, this Hamiltonian describes a sum of independent harmonic oscillators, one for each mode $(\mathbf{k},\epsilon)$. It can be diagonalized by introducing the ladder operators

$$\begin{aligned} \hat{a}_\epsilon(\mathbf{k}) &= \sqrt{\frac{\epsilon_0\omega_\mathbf{k}}{2}}\left[\hat{A}_{\perp\epsilon}(\mathbf{k}) + \frac{i}{\epsilon_0\omega_\mathbf{k}}\hat{\Pi}_{\perp\epsilon}(\mathbf{k})\right], \\ \hat{a}^\dagger_\epsilon(\mathbf{k}) &= \sqrt{\frac{\epsilon_0\omega_\mathbf{k}}{2}}\left[\hat{A}^\dagger_{\perp\epsilon}(\mathbf{k}) - \frac{i}{\epsilon_0\omega_\mathbf{k}}\hat{\Pi}^\dagger_{\perp\epsilon}(\mathbf{k})\right], \end{aligned} \tag{1.615}$$

with $\omega_\mathbf{k} = c_l|\mathbf{k}|$. $\hat{a}$ and $\hat{a}^\dagger$ satisfy the commutation relations

$$\begin{aligned} \left[\hat{a}_\epsilon(\mathbf{k}), \hat{a}_{\epsilon'}(\mathbf{k}')\right] &= \left[\hat{a}^\dagger_\epsilon(\mathbf{k}), \hat{a}^\dagger_{\epsilon'}(\mathbf{k}')\right] = 0, \\ \left[\hat{a}_\epsilon(\mathbf{k}), \hat{a}^\dagger_{\epsilon'}(\mathbf{k}')\right] &= \delta_{\epsilon,\epsilon'}\delta_{\mathbf{k},\mathbf{k}'}, \end{aligned} \tag{1.616}$$

and $\hat{H}_{\rm em}$ takes the form

$$\hat{H}_{\rm em} = \sum_{\mathbf{k},\epsilon}\omega_\mathbf{k}\left(\hat{a}^\dagger_\epsilon(\mathbf{k})\hat{a}_\epsilon(\mathbf{k}) + \frac{1}{2}\right). \tag{1.617}$$

The Hamiltonian of the free field is similar to the one obtained for the quantum harmonic string in Sec. 1.1.2. The elementary excitations of the quantized field (photons) are created and annihilated by the operators $\hat{a}^\dagger$ and $\hat{a}$. The eigenstates of $\hat{H}_{\rm em}$ can be written as

$$\begin{aligned} |n_1\cdots n_i\cdots\rangle &= \prod_i \frac{\left(\hat{a}^\dagger_i\right)^{n_i}}{\sqrt{n_i!}}|\text{vac}\rangle, \\ \hat{H}_{\rm em}|n_1\cdots n_i\cdots\rangle &= \left[\sum_j \omega_j\left(n_j + \frac{1}{2}\right)\right]|n_1\cdots n_i\cdots\rangle, \end{aligned} \tag{1.618}$$

where n_i denotes the number of photons in the mode $i \equiv (\mathbf{k}_i, \epsilon_i)$ and $|\text{vac}\rangle$ the (normalized) vacuum state: $\hat{a}_i|\text{vac}\rangle = 0$. From (1.605) and (1.615), we can express the transverse field in terms of the photon operators,

$$\hat{\mathbf{E}}_\perp(\mathbf{r}) = \frac{i}{\sqrt{\mathcal{V}}}\sum_{\mathbf{k},\epsilon}\sqrt{\frac{\omega_\mathbf{k}}{2\epsilon_0}}\left[\hat{a}_\epsilon(\mathbf{k})e^{i\mathbf{k}\cdot\mathbf{r}} - \hat{a}^\dagger_\epsilon(\mathbf{k})e^{-i\mathbf{k}\cdot\mathbf{r}}\right]\boldsymbol{\epsilon},$$
$$\hat{\mathbf{B}}(\mathbf{r}) = \frac{i}{\sqrt{\mathcal{V}}}\sum_{\mathbf{k},\epsilon}\frac{1}{\sqrt{2\epsilon_0\omega_\mathbf{k}}}\left[\hat{a}_\epsilon(\mathbf{k})e^{i\mathbf{k}\cdot\mathbf{r}} - \hat{a}^\dagger_\epsilon(\mathbf{k})e^{-i\mathbf{k}\cdot\mathbf{r}}\right]\mathbf{k}\times\boldsymbol{\epsilon}. \tag{1.619}$$

Coherent states. As with any bosonic operator, we can define the (normalized) coherent state

$$|z\rangle = \exp\left\{-\sum_i\left(\frac{|z_i|^2}{2} - z_i\hat{a}^\dagger_i\right)\right\}|\mathrm{vac}\rangle, \tag{1.620}$$

where z_i is a c-number. One easily finds[74]

$$\langle z|\hat{A}_i|z\rangle = \sqrt{\frac{2}{\epsilon_0\omega_i}}\Re(z_i), \qquad \langle z|\hat{A}_i^2|z\rangle = \frac{1}{2\epsilon_0\omega_i}[4\Re(z_i)^2+1],$$
$$\langle z|\hat{\Pi}_i|z\rangle = \sqrt{2\epsilon_0\omega_i}\Im(z_i), \qquad \langle z|\hat{\Pi}_i^2|z\rangle = \frac{\epsilon_0\omega_i}{2}[4\Im(z_i)^2+1], \tag{1.621}$$

and therefore

$$\Delta A_i^2 = \frac{1}{2\epsilon_0\omega_i}, \qquad \Delta\Pi_i^2 = \frac{\epsilon_0\omega_i}{2}. \tag{1.622}$$

The coherent state minimizes the uncertainty relation $\Delta A_i\Delta\Pi_i \geq \hbar/2$ (we have restored Planck's constant) and is therefore the quantum state which is closest to the classical electromagnetic field. Since the coherent state evolves in time as

$$e^{-i\hat{H}t}|z\rangle = e^{-\frac{i}{2}\sum_i\omega_i t}|z(t)\rangle \tag{1.623}$$

[see Eq. (1.231)] with $z_i(t) = z_i e^{-i\omega_i t}$, the expectation value of the transverse field reads

$$\langle z(t)|\hat{\mathbf{E}}_\perp(\mathbf{r})|z(t)\rangle = \frac{i}{\sqrt{\mathcal{V}}}\sum_{\mathbf{k},\epsilon}\sqrt{\frac{\omega_\mathbf{k}}{2\epsilon_0}}\left[z_\epsilon(\mathbf{k})e^{i(\mathbf{k}\cdot\mathbf{r}-\omega_\mathbf{k}t)} - \mathrm{c.c.}\right]\boldsymbol{\epsilon},$$
$$\langle z(t)|\hat{\mathbf{B}}(\mathbf{r})|z(t)\rangle = \frac{i}{\sqrt{\mathcal{V}}}\sum_{\mathbf{k},\epsilon}\frac{1}{\sqrt{2\epsilon_0\omega_\mathbf{k}}}\left[z_\epsilon(\mathbf{k})e^{i(\mathbf{k}\cdot\mathbf{r}-\omega_\mathbf{k}t)} - \mathrm{c.c.}\right]\mathbf{k}\times\boldsymbol{\epsilon}. \tag{1.624}$$

One easily verifies that these mean values are solutions of the Maxwell equations. By virtue of (1.622), the fluctuations $\Delta\mathbf{E}_\perp(\mathbf{r})^2$ and $\Delta\mathbf{B}(\mathbf{r})^2$ are the smallest ones compatible with the Heisenberg uncertainty relations. When the average number of photons $n_i = \langle\hat{a}^\dagger_i\hat{a}_i\rangle = |z_i|^2$ is large, $|\langle\hat{A}_i\rangle(t)| \gg \Delta A_i$ and $|\langle\Pi_i\rangle(t)| \gg \Delta\Pi_i$ for most times t and the electromagnetic field in the mode $i = (\mathbf{k},\epsilon)$ behaves classically.

---

[74] These relations can be deduced from (1.228) and (1.229) by the correspondence $\hat{p}\to\hat{\Pi}_i$, $\hat{x}\to\hat{A}_i$, $m\to\epsilon_0$ and $\omega\to\omega_i$.

1.9.3 *Functional integral in the Coulomb gauge*

We are now in a position to write the partition function $Z = \mathrm{Tr}\, e^{-\beta\hat{H}}$ as a functional integral. In the free field case, the Hamiltonian (1.614) reduces to a sum of independent harmonic oscillators. Following the general method introduced in Sec. 1.1.1, and in particular the derivation of the functional integral of the quantum harmonic string in Sec. 1.1, one finds

$$Z_{\mathrm{em}} = \int \mathcal{D}[\mathbf{A}_\perp]\, e^{-S_{\mathrm{em}}[\mathbf{A}_\perp]}, \tag{1.625}$$

with the boundary condition $\mathbf{A}_\perp(\mathbf{r},\beta) = \mathbf{A}_\perp(\mathbf{r},0)$ and the Euclidean action

$$\begin{aligned} S_{\mathrm{em}}[\mathbf{A}_\perp] &= \int_0^\beta d\tau \sum_{\mathbf{k},\epsilon} \frac{\epsilon_0}{2}\left[|\partial_\tau A_{\perp\epsilon}(\mathbf{k},\tau)|^2 + c_l^2\mathbf{k}^2|A_{\perp\epsilon}(\mathbf{k},\tau)|^2\right] \\ &= \int_0^\beta d\tau \int d^3r\, \frac{\epsilon_0}{2}\left[\left(\partial_\tau \mathbf{A}_\perp(x)\right)^2 + c_l^2\left(\boldsymbol{\nabla}\times\mathbf{A}_\perp(x)\right)^2\right]. \end{aligned} \tag{1.626}$$

The measure in the functional integral is defined by

$$\mathcal{D}[\mathbf{A}_\perp] = \prod_{\mathbf{k},\epsilon,\tau}{}' dA^*_{\perp\epsilon}(\mathbf{k},\tau)dA_{\perp\epsilon}(\mathbf{k},\tau) = \mathcal{D}[\mathbf{A}]\prod_{\mathbf{r},\tau}\delta(\boldsymbol{\nabla}\cdot\mathbf{A}(\mathbf{r},\tau)) \tag{1.627}$$

(up to an irrelevant multiplicative constant), where the prime means that the product over $\mathbf{k}$ is restricted to half the reciprocal space in order to take into account the constraint $A^*_{\perp\epsilon}(\mathbf{k}) = A_{\perp\epsilon}(-\mathbf{k})$. The discretization of space (necessary to properly define the integration measure) introduces an irrelevant multiplicative constant in the continuum limit.

When the coupling to the matter is introduced, the Hamiltonian is given by

$$\hat{H} = \hat{H}'_m + \hat{H}_c - \int d^3r\, \hat{\mathbf{j}}(\mathbf{r},t)\cdot\hat{\mathbf{A}}_\perp(\mathbf{r},t) + \frac{e}{2m}\int d^3r\, \hat{\rho}(\mathbf{r},t)\hat{\mathbf{A}}^2_\perp(\mathbf{r},t) + \hat{H}_{\mathrm{em}} \tag{1.628}$$

[see Eq. (1.606)], where

$$\hat{H}_c = \frac{1}{2}\int d^3r d^3r' \frac{\hat{\rho}(\mathbf{r})\hat{\rho}(\mathbf{r}')}{4\pi\epsilon_0|\mathbf{r}-\mathbf{r}'|}. \tag{1.629}$$

The degrees of freedom of the particles can be described by field operators (Sec. 1.2.2). The charge and current density operators are then defined by

$$\begin{aligned} \hat{\rho}(\mathbf{r}) &= e\sum_\sigma \hat{\psi}^\dagger_\sigma(\mathbf{r})\hat{\psi}_\sigma(\mathbf{r}), \\ \hat{\mathbf{J}}(\mathbf{r}) &= -\frac{ie}{2m}\sum_\sigma\left[\hat{\psi}^\dagger_\sigma(\mathbf{r})\boldsymbol{\nabla}\hat{\psi}_\sigma(\mathbf{r}) - \mathrm{h.c.}\right] - \frac{e}{m}\hat{\rho}(\mathbf{r})\hat{\mathbf{A}}_\perp(\mathbf{r}) \\ &= \hat{\mathbf{j}}(\mathbf{r}) - \frac{e}{m}\hat{\rho}(\mathbf{r})\hat{\mathbf{A}}_\perp(\mathbf{r}), \end{aligned} \tag{1.630}$$

for spin-σ particles (Sec. 1.2.2). $\hat{\mathbf{j}}$ is the paramagnetic part of the current and $\hat{H}'_m = \hat{H}_m - \hat{H}_c$ the Hamiltonian of the particles in the absence of electromagnetic interactions. Proceeding as in Sec. 1.4, we derive from (1.628) the coherent-state functional integral

$$Z = \int \mathcal{D}[\psi^*, \psi, \mathbf{A}_\perp]\, e^{-S[\psi^*,\psi,\mathbf{A}_\perp]} \tag{1.631}$$

with the action

$$\begin{aligned} S[\psi^*, \psi, \mathbf{A}_\perp] &= S_{\text{em}}[\mathbf{A}_\perp] + S_m[\psi^*, \psi] \\ &\quad + \int_0^\beta d\tau \int d^3r \left[\frac{e}{2m}\rho(x)\mathbf{A}_\perp^2(x) - \mathbf{j}(x)\cdot\mathbf{A}_\perp(x)\right] \end{aligned} \tag{1.632}$$

and the boundary conditions $\mathbf{A}_\perp(\mathbf{r},\beta) = \mathbf{A}_\perp(\mathbf{r},0)$, $\psi_\sigma^{(*)}(\mathbf{r},\beta) = \zeta\psi_\sigma^{(*)}(\mathbf{r},0)$.

It is now possible to reintroduce the scalar potential by rewriting the Coulomb interaction as a functional integral over an auxiliary real bosonic field $\phi(x)$,[75]

$$\begin{aligned} &\exp\left\{-\frac{1}{2}\sum_k \frac{\rho(-k)\rho(k)}{\epsilon_0\mathbf{k}^2}\right\} \\ &\quad = \int \mathcal{D}[\phi]\exp\left\{-\frac{1}{2}\sum_k \left[\epsilon_0\mathbf{k}^2\phi(-k)\phi(k) + 2i\phi(-k)\rho(k)\right]\right\} \\ &\quad = \int \mathcal{D}[\phi]\exp\left\{-\frac{1}{2}\int_0^\beta d\tau \int d^3r\left[\epsilon_0\big(\boldsymbol{\nabla}\phi(x)\big)^2 + 2i\phi(x)\rho(x)\right]\right\} \end{aligned} \tag{1.633}$$

(again we neglect an irrelevant multiplicative term). Equation (1.633) is obtained using the basic results of Gaussian integration derived in Appendix 1.E. The action becomes

$$\begin{aligned} S[\psi^*, \psi, \phi, \mathbf{A}_\perp] &= S_{\text{em}}[\phi, \mathbf{A}_\perp] + S'_m[\psi^*, \psi] \\ &\quad + \int_0^\beta d\tau \int d^3r \left[\frac{e}{2m}\rho(x)\mathbf{A}_\perp^2(x) + i\phi(x)\rho(x) - \mathbf{j}(x)\cdot\mathbf{A}_\perp(x)\right], \end{aligned} \tag{1.634}$$

with the action of the electromagnetic potentials given by

$$S_{\text{em}}[\phi, \mathbf{A}_\perp] = \int_0^\beta d\tau \int d^3r\, \frac{\epsilon_0}{2}\left[\big(\boldsymbol{\nabla}\phi(x)\big)^2 + \big(\partial_\tau \mathbf{A}_\perp(x)\big)^2 + c_l^2\big(\boldsymbol{\nabla}\times\mathbf{A}_\perp(x)\big)^2\right]. \tag{1.635}$$

$S'_m = S_m - S_c$ denotes the action of the particles with no electromagnetic interactions. According to (1.634), the field $i\phi(x)$ plays the role of the scalar potential, since it couples to the particle density. The factor i cannot be eliminated by a redefinition of the ϕ field, $i\phi \to \phi$, since this would change the sign of $(\boldsymbol{\nabla}\phi)^2$ in S_{em} and makes the functional integral ill-defined. Thus, in the Euclidean formalism, the mean values of the electric and magnetic fields are

$$\begin{aligned} \langle \mathbf{E}(x)\rangle &= -\boldsymbol{\nabla}\langle i\phi(x)\rangle - i\partial_\tau\langle \mathbf{A}_\perp(x)\rangle, \\ \langle \mathbf{B}(x)\rangle &= \boldsymbol{\nabla}\times\langle \mathbf{A}_\perp(x)\rangle, \end{aligned} \tag{1.636}$$

[75]The transformation (1.633) is known as a Hubbard–Stratonovich transformation. It will be encountered many times in the following chapters (see, e.g. Sec. 5.2.3.1 or 6.3.3.1).

or, in Fourier space,

$$\begin{aligned}\langle \mathbf{E}(k)\rangle &= \mathbf{k}\langle \phi(k)\rangle - \omega_n \langle \mathbf{A}_\perp(k)\rangle,\\ \langle \mathbf{B}(k)\rangle &= i\mathbf{k}\times\langle \mathbf{A}_\perp(k)\rangle.\end{aligned} \tag{1.637}$$

According to (1.605), the mean value of the longitudinal electric field is given by

$$\langle \mathbf{E}_\parallel(k)\rangle = -i\mathbf{k}\frac{\langle\rho(k)\rangle}{\epsilon_0\mathbf{k}^2}. \tag{1.638}$$

The mean particle density can be calculated from

$$\langle\rho(k)\rangle = \left.\frac{\partial \ln Z[h]}{\partial h(-k)}\right|_{h=0}, \tag{1.639}$$

where $Z[h]$ is the partition function obtained by adding to the action the source term $-\sum_k h(-k)\rho(k)$. After the Hubbard–Stratonovich transformation (1.633), the action reads

$$\begin{aligned} S &= \frac{1}{2}\sum_k \left[\epsilon_0\mathbf{k}^2\phi(-k)\phi(k) + 2i\phi(-k)\rho(k) - 2h(-k)\rho(k)\right] + \cdots \\ &= \frac{1}{2}\sum_k \left[\epsilon_0\mathbf{k}^2\big(\phi(-k)-ih(-k)\big)\big(\phi(k)-ih(k)\big) + 2i\phi(-k)\rho(k)\right]\cdots \end{aligned} \tag{1.640}$$

(we do not write the part of the action which is independent of h). The second line of (1.640) is obtained by shifting the field $\phi\to\phi - ih$ in the functional integral. From (1.639) and (1.640), we then deduce

$$\langle\rho(k)\rangle = \epsilon_0\mathbf{k}^2\langle i\phi(k)\rangle, \qquad \langle \mathbf{E}_\parallel(k)\rangle = \mathbf{k}\langle\phi(k)\rangle \tag{1.641}$$

in agreement with (1.637).

More generally, one can define the propagators of the electromagnetic potentials,

$$\begin{aligned} D_{00}(k) &= \langle\phi(k)\phi(-k)\rangle,\\ D_{0\mu}(k) &= \langle\phi(k)A_\mu(-k)\rangle,\\ D_{\mu\nu}(k) &= \langle A_\mu(k)A_\nu(-k)\rangle = \frac{k_\mu k_\nu}{\mathbf{k}^2}D_\parallel(k) + \left(\delta_{\mu,\nu} - \frac{k_\mu k_\nu}{\mathbf{k}^2}\right)D_\perp(k)\end{aligned} \tag{1.642}$$

($\mu = x,y,z$), where $D_\parallel(k)$ and $D_\perp(k)$ are functions of $(|\mathbf{k}|, i\omega_n)$. In (1.642), we have taken advantage of the homogeneity and isotropy of the system to write the propagator $D_{\mu\nu}$ in terms of its longitudinal and transverse parts. In the absence of coupling to the particles, one finds

$$\begin{aligned} D_{00}^{(0)}(k) &= \langle\phi(k)\phi(-k)\rangle = \frac{1}{\epsilon_0\mathbf{k}^2},\\ D_\perp^{(0)}(k) &= \frac{1}{\epsilon_0(\omega_n^2 + c_l^2\mathbf{k}^2)}.\end{aligned} \tag{1.643}$$

In the Coulomb gauge, the propagator of the vector potential is transverse: $D_{\parallel}^{(0)}(k) = D_{0\mu}^{(0)}(k) = 0$. $D_{00}^{(0)}(k)$ is the Fourier transform of the Coulomb interaction, whereas $D_{\perp}^{(0)}(k)$ is the propagator of the (transverse) electromagnetic field in free space. We shall see in Sec. 3.4 that the propagators D_{00} and $D_{\perp}$ contain important information about the collective excitations of the particles, besides describing the propagation of the electromagnetic field in the matter.

In the Coulomb gauge, the electric field is split into a Coulomb field and a transverse field. Although the Coulomb interaction is instantaneous [Eq. (1.629)], the interaction between two charged particles takes place with a retardation due to the propagation of the field with velocity c_l. This retardation results from an exact compensation of the instantaneous contributions of the Coulomb and transverse fields [18]. In non-relativistic systems, the separation between an instantaneous Coulomb field and a transverse field is very natural since the magnetic interactions mediated by the vector potential are negligible when the characteristic velocity of the particles is small with respect to the velocity of light. To see this, let us integrate out the vector potential in (1.632). If we approximate $\rho(x)$ in the diamagnetic term $\rho(x)\mathbf{A}_{\perp}^2(x)$ by its mean value en, we are left with a simple Gaussian functional integral (Appendix 1.E.4) which gives a contribution[76]

$$-\frac{1}{2\epsilon_0}\sum_k \frac{\mathbf{j}_{\perp}(-k)\cdot\mathbf{j}_{\perp}(k)}{\omega_n^2 + c_l^2\mathbf{k}^2 + \omega_p^2} \tag{1.644}$$

to the action of the particles. $\mathbf{j}_{\perp}$ is the transverse part of the paramagnetic current and $\omega_p = (ne^2/\epsilon_0 m)^{1/2}$ is called the plasma frequency for reasons given in Chapter 3. For particles with characteristic velocity v, $j_{\perp} \sim v\rho$ so that the magnetic interaction (1.644) is of order v^2/c_l^2 with respect to the Coulomb interaction $\rho(k)\rho(-k)/\epsilon_0\mathbf{k}^2$. If $v \ll c_l$, the magnetic interaction can be omitted, which amounts to neglecting the retardation effects in the electromagnetic interactions between particles. When studying systems of charged particles, we will therefore in general consider the action $S_m = S'_m + S_c$ including the Coulomb interaction but not the coupling to the vector potential. Only when considering the response to an external transverse field (i.e. induced by a current distribution external to the system) or the propagation of the electromagnetic field in the matter should we include the transverse vector potential in the action (see, e.g. Sec. 3.4.2).

Appendix 1.A Review of Classical Mechanics[77]

In the following sections, we review the Lagrangian and Hamiltonian formalisms for both a finite number and a continuum of degrees of freedom. In particular, we discuss symmetries of their consequences.

[76] The functional integral must be done in Fourier space where the transverse vector potential is defined by its two independent components $A_{\perp\epsilon_1}(\mathbf{k},\tau)$ and $A_{\perp\epsilon_2}(\mathbf{k},\tau)$.

[77] A thorough discussion of classical mechanics can be found in Refs. [22, 23].

1.A.1 *Lagrangian formalism*

Let us consider a system with n degrees of freedom described by n generalized coordinates $q_1 \cdots q_n$. The latter can be represented by a point in an n-dimensional configuration space. The trajectory of the system is determined by the time dependence of the coordinates. The coordinates q_i and the velocities $\dot{q}_i = \partial_t q_i$ form the dynamical variables of the system. The accelerations $\ddot{q}_i$ can be expressed at any time as a function of these variables. The resulting equations of motion are then second-order differential equations in time.

In the Lagrangian approach, the dynamics of the system is determined by the Lagrangian $L(q_i, \dot{q}_i, t)$, a function of the dynamical variables which can also explicitly depend on time. The principle of least action stipulates that the trajectory of the system between times t_1 and t_2, with initial and final coordinates assumed to be known, corresponds to an extremum of the action

$$S = \int_{t_1}^{t_2} dt\, L(q_i, \dot{q}_i, t). \tag{1.A.1}$$

For an arbitrary variation δq_i of the coordinate q_i, the variation of the action is

$$\begin{aligned} \delta S &= \int_{t_1}^{t_2} dt \sum_i \left[\frac{\partial L}{\partial q_i} \delta q_i + \frac{\partial L}{\partial \dot{q}_i} \delta \dot{q}_i \right] \\ &= \int_{t_1}^{t_2} dt \sum_i \left[\frac{\partial L}{\partial q_i} \delta q_i - \frac{d}{dt} \left(\frac{\partial L}{\partial \dot{q}_i} \right) \delta q_i \right] + \sum_i \frac{\partial L}{\partial \dot{q}_i} \delta q_i \Big|_{t_1}^{t_2}, \end{aligned} \tag{1.A.2}$$

where we have integrated by parts. The last term in (1.A.2) vanishes since $\delta q_i(t_1) = \delta q_i(t_2) = 0$ (the initial and final coordinates $q_i(t_1)$ and $q_i(t_2)$ are fixed), and we deduce the Euler–Lagrange equations

$$\frac{d}{dt} \frac{\partial L}{\partial \dot{q}_i} - \frac{\partial L}{\partial q_i} = 0. \tag{1.A.3}$$

The resolution of these n second-order differential equations requires the knowledge of $2n$ parameters: $q_i(t_1)$ and $q_i(t_2)$.

The Lagrangian is not unique. In particular, adding to the Lagrangian a total time derivative,

$$L(q_i, \dot{q}_i, t) \to L(q_i, \dot{q}_i, t) + \frac{d}{dt} F(q_i, t) \tag{1.A.4}$$

($F(q_i, t)$ is an arbitrary function) modifies the action only by the surface term

$$F(q_i(t_2), t_2) - F(q_i(t_1), t_1) \tag{1.A.5}$$

and therefore does not affect the equations of motion ($q_i(t_1)$ and $q_i(t_2)$ being fixed). For the transformed Lagrangian to be a function of the dynamical variables q_i and $\dot{q}_i$, the function $F(q_i, t)$ should not depend on the velocities.

1.A.1.1 Symmetries and invariance

Suppose we describe the system by two sets of coordinates, $\{q_i\}$ and $\{q_i'\}$, corresponding to Lagrangians $L(q_i, \dot{q}_i, t)$ and $L'(q_i', \dot{q}_i', t)$, respectively. The transformation $q_i \to q_i'$ is called a symmetry transformation.[78] Since both sets $\{q_i\}$ and $\{q_i'\}$ refer to the same system, we must have $L(q_i, \dot{q}_i, t) = L'(q_i', \dot{q}_i', t)$ (up to a total time derivative). The equations of motion for the new set of coordinates read

$$\frac{d}{dt}\frac{\partial L'}{\partial \dot{q}_i'} = \frac{\partial L'}{\partial q_i'} \tag{1.A.6}$$

and follow from the principle of least action applied to the action $S'[q'] = \int dt\, L'(q_i', \dot{q}_i', t)$. In general, L and L' are different functions.[79] If, however, the action $S'[q']$ and $S[q']$ coincide — i.e. if $L'(q_i', \dot{q}_i', t)$ and $L(q_i', \dot{q}_i', t)$ differ only by a total derivative — then the equation of motion of q_i' [Eq. (1.A.6)] is identical to that of q_i [Eq. (1.A.3)]; the system is invariant in the symmetry transformation $q_i \to q_i'$.

Let us now consider an infinitesimal (continuous) transformation $q_i \to q_i + \epsilon f_i$ ($\epsilon \to 0$) which leaves the system invariant. To order ϵ, the variation of the Lagrangian can be written as

$$\delta L = \epsilon \sum_i \left(\frac{\partial L}{\partial q_i} f_i + \frac{\partial L}{\partial \dot{q}_i} \dot{f}_i \right) = \epsilon \frac{d}{dt} F(q_i, t), \tag{1.A.7}$$

with $F(q_i, t)$ some function, i.e.

$$\sum_i \left[\frac{\partial L}{\partial q_i} f_i + \frac{d}{dt}\left(\frac{\partial L}{\partial \dot{q}_i} f_i \right) - \frac{d}{dt}\left(\frac{\partial L}{\partial \dot{q}_i} \right) f_i \right] = \frac{d}{dt} F(q_i, t). \tag{1.A.8}$$

If q_i satisfies the equation of motion (1.A.3), Eq. (1.A.8) implies that

$$\frac{d}{dt}\left(\sum_i \frac{\partial L}{\partial \dot{q}_i} f_i - F(q_i, t) \right) = 0. \tag{1.A.9}$$

Thus, to any transformation which leaves the action invariant, there corresponds a constant of motion (Noether's theorem).

Let us illustrate Noether's theorem with a few examples. We consider a three-dimensional system.

For a system which is translation invariant $L(\mathbf{q}_i + \boldsymbol{\epsilon}, \dot{\mathbf{q}}_i, t) = L(\mathbf{q}_i, \dot{\mathbf{q}}_i, t)$, we can apply (1.A.9) with $F = 0$ and $f_{i\mu} = \delta_{\mu,x}$ if $\boldsymbol{\epsilon}$ is parallel to the x axis ($f_{i\mu} = \delta_{\mu,y}$ if $\boldsymbol{\epsilon}$ is

[78] This is the so-called passive point of view where two observers, $\mathcal{O}$ (coordinates $\{q_i\}$) and $\mathcal{O}'$ (coordinates $\{q_i'\}$), look at the same system. Alternatively, one can adopt the active point of view where a symmetry transformation is defined with a single observer $\mathcal{O}$ and two systems $\mathcal{S}$ and $\mathcal{S}'$.

[79] It is crucial here to distinguish between L and L'. In general, however, we shall follow the usual convention to denote a given dynamical variable (such as the Lagrangian or the Hamiltonian) by the same symbol in all coordinate systems.

parallel to the y axis, etc.). One then finds that the total momentum of the system, defined as

$$\mathbf{P} = \sum_i \mathbf{p}_i = \sum_i \frac{\partial L}{\partial \dot{\mathbf{q}}_i}, \tag{1.A.10}$$

is conserved. $\mathbf{p}_i$ denotes the momentum of the ith particle (see Sec. 1.A.2).

The Lagrangian of a rotation-invariant system is unchanged by the transformation $\mathbf{q}_i \to \mathbf{q}_i + \theta \mathbf{n} \times \mathbf{q}_i$, i.e. $q_{i\mu} \to q_{i\mu} + \theta \sum_{\nu,\delta} \epsilon_{\mu\nu\delta} n_\nu q_{i\delta}$ (rotation of angle θ about the axis $\mathbf{n}$); $\epsilon_{\mu\nu\delta}$ denotes the antisymmetric tensor. One then finds that the quantity

$$\sum_{i,\mu,\nu,\delta} \frac{\partial L}{\partial \dot{q}_{i\mu}} \epsilon_{\mu\nu\delta} n_\nu q_{i\delta} = \sum_{i,\mu,\nu,\delta} n_\nu \epsilon_{\nu\delta\mu} q_{i\delta} \frac{\partial L}{\partial \dot{q}_{i\mu}} = \sum_i \mathbf{n} \cdot \left(\mathbf{q}_i \times \frac{\partial L}{\partial \dot{\mathbf{q}}_i} \right) \tag{1.A.11}$$

is conserved. Since (1.A.11) holds for any unit vector $\mathbf{n}$, we conclude that the total angular momentum

$$\mathbf{J} = \sum_i \mathbf{q}_i \times \frac{\partial L}{\partial \dot{\mathbf{q}}_i} = \sum_i \mathbf{q}_i \times \mathbf{p}_i \tag{1.A.12}$$

is a constant of motion.

If the Lagrangian $L(\mathbf{q}_i, \dot{\mathbf{q}}_i)$ does not depend explicitly on time,

$$L\big(\mathbf{q}_i(t+\epsilon), \dot{\mathbf{q}}_i(t+\epsilon)\big) = L(\mathbf{q}_i + \epsilon\dot{\mathbf{q}}_i, \dot{\mathbf{q}}_i + \epsilon\ddot{\mathbf{q}}_i) = L + \epsilon\frac{dL}{dt} \tag{1.A.13}$$

to first order in ϵ. From (1.A.9) with $\mathbf{f}_i = \dot{\mathbf{q}}_i$ and $F = L$, we deduce that the Hamiltonian of the system

$$H = \sum_i \frac{\partial L}{\partial \dot{\mathbf{q}}_i} \cdot \dot{\mathbf{q}}_i - L = \sum_i \mathbf{p}_i \cdot \dot{\mathbf{q}}_i - L \tag{1.A.14}$$

(which is usually identified with the total energy) is conserved.

Let us finally consider a Galilean transformation $\mathbf{q}_i \to \mathbf{q}_i + \mathbf{v}t$ ($\mathbf{f}_i = \mathbf{v}t$) for a Lagrangian of the form

$$L = \sum_i \frac{1}{2} m \dot{\mathbf{q}}_i^2 - V, \tag{1.A.15}$$

where V depends only on the differences $\mathbf{q}_i - \mathbf{q}_j$. To $\mathcal{O}(\mathbf{v})$, the Lagrangian changes by

$$\delta L = \sum_i \left(\frac{\partial L}{\partial \mathbf{q}_i} \cdot \mathbf{v}t + \frac{\partial L}{\partial \dot{\mathbf{q}}_i} \cdot \mathbf{v} \right) = \sum_i m\dot{\mathbf{q}}_i \cdot \mathbf{v} = \mathbf{v} \cdot \frac{d}{dt} M\mathbf{R}, \tag{1.A.16}$$

where $\mathbf{R}$ is the position of the center of mass and M the total mass. We conclude that

$$\sum_i \frac{\partial L}{\partial \dot{\mathbf{q}}_i} t - M\mathbf{R} = \mathbf{P}t - M\mathbf{R} \tag{1.A.17}$$

is conserved. Since the total momentum $\mathbf{P}$ is conserved for a translation-invariant system, we deduce that the center of mass moves with a constant velocity,

$$M\frac{d\mathbf{R}}{dt} = \mathbf{P}. \tag{1.A.18}$$

1.A.2 *Hamiltonian formalism*

In the Hamiltonian approach, the state of the system is specified by n generalized coordinates q_i and n conjugate momenta p_i defined by

$$p_i = \frac{\partial L}{\partial \dot{q}_i}. \tag{1.A.19}$$

Coordinates and momenta can be represented by a point in a $2n$-dimensional phase space. The Hamiltonian $H(p_i, q_i, t)$, which is a function of coordinates and momenta, is defined as the Legendre transform of the Lagrangian,

$$H(p_i, q_i, t) = \sum_i p_i \dot{q}_i - L(q_i, \dot{q}_i, t). \tag{1.A.20}$$

Using the Euler–Lagrange equations (1.A.3), one obtains Hamilton's equations of motion,

$$\begin{aligned} \dot{q}_i &= \frac{\partial H}{\partial p_i}, \\ \dot{p}_i &= -\frac{\partial H}{\partial q_i}. \end{aligned} \tag{1.A.21}$$

The dynamics of the system is described by $2n$ first-order differential equations.

The Poisson bracket of two functions $F(p_i, q_i, t)$ and $G(p_i, q_i, t)$ is defined by

$$[F, G] = \sum_i \left(\frac{\partial F}{\partial q_i} \frac{\partial G}{\partial p_i} - \frac{\partial G}{\partial q_i} \frac{\partial F}{\partial p_i} \right). \tag{1.A.22}$$

Hamilton's equations of motion can be rewritten as

$$\begin{aligned} \dot{q}_i &= [q_i, H], \\ \dot{p}_i &= [p_i, H]. \end{aligned} \tag{1.A.23}$$

The time evolution of an arbitrary function $F(p_i, q_i, t)$,

$$\begin{aligned} \frac{dF}{dt} &= \frac{\partial F}{\partial t} + \sum_i \left(\frac{\partial F}{\partial p_i} \dot{p}_i + \frac{\partial F}{\partial q_i} \dot{q}_i \right) \\ &= \frac{\partial F}{\partial t} + \sum_i \left(-\frac{\partial F}{\partial p_i} \frac{\partial H}{\partial q_i} + \frac{\partial F}{\partial q_i} \frac{\partial H}{\partial p_i} \right) \\ &= \frac{\partial F}{\partial t} + [F, H], \end{aligned} \tag{1.A.24}$$

can also be expressed by means of its Poisson bracket with the Hamiltonian.

It is possible to describe the system in terms of new coordinates and momenta,

$$\begin{aligned} Q_i &\equiv Q_i(p_j, q_j, t), \\ P_i &\equiv P_i(p_j, q_j, t). \end{aligned} \tag{1.A.25}$$

The transformation $(q_i, p_i) \to (Q_i, P_i)$ is canonical if P_i and Q_i are canonically conjugate, i.e. if they satisfy

$$[Q_i, Q_j] = [P_i, P_j] = 0 \quad \text{and} \quad [Q_i, P_j] = \delta_{i,j}. \tag{1.A.26}$$

In this case, the Poisson bracket (1.A.22) of two functions F and G keeps the same form

$$[F, G] = \sum_i \left(\frac{\partial F}{\partial Q_i} \frac{\partial G}{\partial P_i} - \frac{\partial G}{\partial Q_i} \frac{\partial F}{\partial P_i} \right). \tag{1.A.27}$$

When the transformation (1.A.25) does not depend explicitly on time, Hamilton's equations of motion (1.A.23) are invariant, i.e. $\dot{Q}_i = [Q_i, H]$ and $\dot{P}_i = [P_i, H]$, where the Poisson bracket is calculated with the new coordinates Q_i and P_i.

1.A.2.1 *Symmetries and invariance*

For any function $G(p_i, q_i, t)$, we can define the infinitesimal (canonical) continuous transformation ($\epsilon \to 0$)

$$\begin{aligned} q_i &\to q_i - \epsilon[G, q_i] = q_i + \epsilon \frac{\partial G}{\partial p_i}, \\ p_i &\to p_i - \epsilon[G, p_i] = p_i - \epsilon \frac{\partial G}{\partial q_i}. \end{aligned} \tag{1.A.28}$$

G is called the generator of the transformation.[80] For a given transformation, the generator is unique, apart from an arbitrary additive function of t, independent of p_i and q_i. If the transformation (1.A.28) does not involve the time explicitly, G can be chosen time independent.

The system is invariant in the transformation if the equations of motion are preserved. If G is time independent, this implies that the Hamiltonian is unchanged (see the remark after (1.A.27)), i.e.

$$\delta H = \epsilon \sum_i \left(-\frac{\partial H}{\partial p_i} \frac{\partial G}{\partial q_i} + \frac{\partial H}{\partial q_i} \frac{\partial G}{\partial p_i} \right) = \epsilon[H, G] = 0. \tag{1.A.29}$$

to order ϵ. Since $\partial G/\partial t = 0$, Eqs. (1.A.24) and (1.A.29) imply

$$\frac{dG}{dt} = 0. \tag{1.A.30}$$

The generator of the transformation is a constant of motion. When G has an explicit dependence on t, the Hamiltonian cannot be invariant in the transformation (1.A.28), since the energy of a system does depend on the choice of the (moving) reference frame. Nevertheless,

[80]There is a close analogy between symmetries in the (classical) Hamiltonian formalism and symmetries in quantum mechanics. In particular, to the canonical transformation generated by $G(p_i, q_i, t)$ corresponds, in quantum mechanics, a unitary transformation generated by $G(\hat{p}_i, \hat{q}_i, t)$ (Sec. 2.1).

the condition for the equations of motion to be unchanged is still given by (1.A.30). For the first of Eq. (1.A.21) to be unchanged, we need[81]

$$\delta\dot{q}_i = \delta\frac{\partial H}{\partial p_i} \quad \text{i.e.} \quad \frac{d}{dt}\frac{\partial G}{\partial p_i} = -\left[G, \frac{\partial H}{\partial p_i}\right]. \tag{1.A.31}$$

Using (1.A.24) for the time evolution of $\partial G/\partial p_i$ and G, one finds that (1.A.31) is equivalent to

$$\frac{\partial}{\partial p_i}\frac{dG}{dt} = 0. \tag{1.A.32}$$

Similarly, from the invariance of the second of Eq. (1.A.21), one deduces

$$\frac{\partial}{\partial q_i}\frac{dG}{dt} = 0. \tag{1.A.33}$$

It follows that dG/dt must be a function of t alone. But since we can always add to G any function of t without affecting the transformation it generates, we can choose it such that $dG/dt = 0$.

It is straightforward to verify that the constants of motion obtained from the Lagrangian formalism (Sec. 1.A.1) are the generators of the corresponding symmetry transformations. The total momentum $\mathbf{P} = \sum_i \mathbf{p}_i$ satisfies

$$\begin{aligned} \mathbf{q}_i - [\mathbf{P}\cdot\delta\mathbf{a}, \mathbf{q}_i] &= \mathbf{q}_i + \delta\mathbf{a}, \\ \mathbf{p}_i - [\mathbf{P}\cdot\delta\mathbf{a}, \mathbf{p}_i] &= \mathbf{p}_i, \end{aligned} \tag{1.A.34}$$

and is therefore the generator of translations. Similarly, one finds that the total angular momentum $\mathbf{J} = \sum_i \mathbf{q}_i \times \mathbf{p}_i$ is the generator of rotations,

$$\begin{aligned} \mathbf{q}_i - [\mathbf{J}\cdot\delta\mathbf{a}, \mathbf{q}_i] &= \mathbf{q}_i + \delta\mathbf{a}\times\mathbf{q}_i, \\ \mathbf{p}_i - [\mathbf{J}\cdot\delta\mathbf{a}, \mathbf{p}_i] &= \mathbf{p}_i + \delta\mathbf{a}\times\mathbf{p}_i \end{aligned} \tag{1.A.35}$$

($\delta\mathbf{a} = \delta\theta\mathbf{n}$ for a rotation of angle $\delta\theta$ about the $\mathbf{n}$ axis), and the Hamiltonian the generator of time translations [Eq. (1.A.23)]. The generator of a Galilean transformation is given by $\mathbf{P}t - M\mathbf{R}$,

$$\begin{aligned} \mathbf{q}_i - [(\mathbf{P}t - M\mathbf{R})\cdot\delta\mathbf{v}, \mathbf{q}_i] &= \mathbf{q}_i + \delta\mathbf{v}t, \\ \mathbf{p}_i - [(\mathbf{P}t - M\mathbf{R})\cdot\delta\mathbf{v}, \mathbf{p}_i] &= \mathbf{p}_i + m\delta\mathbf{v}. \end{aligned} \tag{1.A.36}$$

1.A.3 *Classical field theory*

The Lagrangian and Hamiltonian formalisms also apply to systems with a continuum of degrees of freedom described by a field $\psi_j(\mathbf{r}, t)$ which depends on a continuous variable $\mathbf{r}$

[81]We use the fact that any function $F(p_i, q_i, t)$ changes by $\delta F = \sum_i \left(\frac{\partial F}{\partial q_i}\delta q_i + \frac{\partial F}{\partial p_i}\delta p_i\right) = -\epsilon[G, F]$ in the transformation (1.A.28).

(which is generally the space coordinate) and a discrete index j. $\psi_j(\mathbf{r},t)$ is usually assumed to be real.[82] The Lagrangian is a functional of the dynamical variables $\psi_j(\mathbf{r},t)$ and $\dot{\psi}_j(\mathbf{r},t) = \partial_t\psi_j(\mathbf{r},t)$,[83] which can be written as

$$L[\psi(t),\dot{\psi}(t)] = \int d^d r\, \mathcal{L}(\psi_j, \dot{\psi}_j, \boldsymbol{\nabla}\psi_j), \tag{1.A.37}$$

where $\mathcal{L}$ is called the Lagrangian density. For simplicity, we assume that $\mathcal{L}$ has no explicit dependence on $\mathbf{r}$ and t. The notation $L[\psi(t),\dot{\psi}(t)]$ emphasizes that the Lagrangian is a functional of $\psi_j(\mathbf{r},t)$ and $\dot{\psi}_j(\mathbf{r},t)$ at a given time t. It should be noticed that $\boldsymbol{\nabla}\psi_j(\mathbf{r},t)$ is not an independent dynamical variable but a mere function of the field $\psi_j(\mathbf{r},t)$ at different positions $\mathbf{r}$. Although the Lagrangian (1.A.37) describes a large number of physical systems, the Lagrangian density $\mathcal{L}$ may depend on spatial derivatives to all orders.

To obtain the Euler–Lagrange equations of motion, we require the action[84]

$$S[\psi] = \int_{t_1}^{t_2} dt\, L[\psi(t),\dot{\psi}(t)] = \int_{t_1}^{t_2} dt \int d^d r\, \mathcal{L}(\psi_j, \dot{\psi}_j, \boldsymbol{\nabla}\psi_j) \tag{1.A.38}$$

to be stationary under infinitesimal variations $\delta\psi_j(\mathbf{r},t)$ with the condition $\delta\psi_j(\mathbf{r},t_1) = \delta\psi_j(\mathbf{r},t_2) = 0$. Applying the rules of functional derivatives, one finds[85]

$$\begin{aligned} 0 = \frac{\delta S}{\delta\psi_j(\mathbf{r},t)} &= \frac{\delta L}{\delta\psi_j(\mathbf{r},t)} - \frac{d}{dt}\frac{\delta L}{\delta\dot{\psi}_j(\mathbf{r},t)} \\ &= \frac{\partial\mathcal{L}}{\partial\psi_j(\mathbf{r},t)} - \boldsymbol{\nabla}\cdot\frac{\partial\mathcal{L}}{\partial(\boldsymbol{\nabla}\psi_j(\mathbf{r},t))} - \frac{d}{dt}\frac{\partial\mathcal{L}}{\partial\dot{\psi}_j(\mathbf{r},t)} \end{aligned} \tag{1.A.39}$$

under the assumption that the field vanishes at infinity ($|\mathbf{r}| \to \infty$) or satisfies periodic boundary conditions. Without changing the equations of motion, we can add to the Lagrangian density a total time or space derivative,[86]

$$\mathcal{L} \to \mathcal{L} + \frac{d}{dt}\mathcal{F}_0(\psi_j(\mathbf{r},t)) + \boldsymbol{\nabla}\cdot\boldsymbol{\mathcal{F}}(\psi_j(\mathbf{r},t)). \tag{1.A.40}$$

The space integral of $\boldsymbol{\nabla}\cdot\boldsymbol{\mathcal{F}}$ gives a surface integral at infinity which vanishes. The transformed Lagrangian then differs from the original one by a total time derivative which does not affect the equations of motion.

[82] If the theory is naturally expressed in terms of a complex field ψ and its complex conjugate ψ^*, we write $\psi = \psi_1 + i\psi_2$ and $\psi^* = \psi_1 - i\psi_2$, where ψ_1 and ψ_2 are two real fields.

[83] Functionals are defined and discussed in Appendix 1.D.

[84] The knowledge of $\psi_j(\mathbf{r},t)$ over the time interval $[t_1,t_2]$ fixes the value of $\dot{\psi}_j(\mathbf{r},t)$ over the same time interval. The action $S[\psi]$ is therefore a functional of $\psi_j(\mathbf{r},t)$, whereas the Lagrangian $L[\psi(t),\dot{\psi}(t)]$ at time t is a functional of both $\psi_j(\mathbf{r},t)$ and $\dot{\psi}_j(\mathbf{r},t)$.

[85] Equation (1.A.39) is a direct consequence of Eq. (1.D.5) in Appendix 1.D.

[86] In the general case, the functions $\mathcal{F}_0$ and $\boldsymbol{\mathcal{F}}$ can depend explicitly on $\mathbf{r}$ and t.

The conjugate momentum and the Hamiltonian are defined by

$$
\begin{aligned}
\Pi_j(\mathbf{r},t) &= \frac{\delta L}{\delta\dot{\psi}_j(\mathbf{r},t)} = \frac{\partial\mathcal{L}}{\partial\dot{\psi}_j(\mathbf{r},t)},\\
H[\Pi(t),\psi(t)] &= \int d^d r \sum_j \Pi_j(\mathbf{r},t)\dot{\psi}_j(\mathbf{r},t) - L[\psi(t),\dot{\psi}(t)] \\
&= \int d^d r\, \mathcal{H}(\Pi_j,\psi_j,\nabla\psi_j),
\end{aligned}
\tag{1.A.41}
$$

where $\mathcal{H}$ is the Hamiltonian density. From (1.A.39) and (1.A.41), we obtain Hamilton's equations of motion

$$
\begin{aligned}
\dot{\psi}_j(\mathbf{r},t) &= \frac{\delta H}{\delta\Pi_j(\mathbf{r},t)} = \frac{\partial\mathcal{H}}{\partial\Pi_j(\mathbf{r},t)},\\
\dot{\Pi}_j(\mathbf{r},t) &= -\frac{\delta H}{\delta\psi_j(\mathbf{r},t)} = -\frac{\partial\mathcal{H}}{\partial\psi_j(\mathbf{r},t)} + \nabla\cdot\frac{\partial\mathcal{H}}{\partial(\nabla\psi_j(\mathbf{r},t))}.
\end{aligned}
\tag{1.A.42}
$$

It is often convenient to work with the Fourier-transformed variables

$$
\begin{aligned}
\psi_j(\mathbf{k},t) &= \frac{1}{\sqrt{\mathcal{V}}}\int d^d r\, e^{-i\mathbf{k}\cdot\mathbf{r}}\psi_j(\mathbf{r},t),\\
\Pi_j(\mathbf{k},t) &= \frac{1}{\sqrt{\mathcal{V}}}\int d^d r\, e^{-i\mathbf{k}\cdot\mathbf{r}}\Pi_j(\mathbf{r},t),
\end{aligned}
\tag{1.A.43}
$$

where $\psi_j(-\mathbf{k},t) = \psi_j^*(\mathbf{k},t)$ and $\Pi_j(-\mathbf{k},t) = \Pi_j^*(\mathbf{k},t)$ for real fields $\psi_j(\mathbf{r},t)$ and $\Pi_j(\mathbf{r},t)$. For periodic boundary conditions, $k_x = \frac{2\pi}{L_x}n_x$ ($n_x \in \mathbb{Z}$) with L_x the system size in the x direction, and similar relations for k_y, etc. One then has[87]

$$
\begin{aligned}
\Pi_j(\mathbf{k},t) &= \frac{1}{\sqrt{\mathcal{V}}}\int d^d r\, e^{-i\mathbf{k}\cdot\mathbf{r}}\frac{\delta L}{\delta\dot{\psi}_j(\mathbf{r},t)}\\
&= \frac{1}{\sqrt{\mathcal{V}}}\int d^d r\, e^{-i\mathbf{k}\cdot\mathbf{r}}\sum_{\mathbf{k}'}\frac{\partial L}{\partial\dot{\psi}_j(\mathbf{k}',t)}\frac{\delta\dot{\psi}_j(\mathbf{k}',t)}{\delta\dot{\psi}_j(\mathbf{r},t)}.
\end{aligned}
\tag{1.A.44}
$$

Since

$$
\frac{\delta\dot{\psi}_j(\mathbf{k}',t)}{\delta\dot{\psi}_j(\mathbf{r},t)} = \frac{e^{-i\mathbf{k}'\cdot\mathbf{r}}}{\sqrt{\mathcal{V}}},
\tag{1.A.45}
$$

[87]Note that as long as the system volume $\mathcal{V}$ is kept finite, $\mathbf{k}$ is not a continuous variable and $\partial L/\partial\dot{\psi}_j(\mathbf{k},t)$ is a usual derivative (not a functional derivative).

we finally obtain

$$\begin{aligned}\Pi_j(\mathbf{k},t) &= \frac{1}{\mathcal{V}}\sum_{\mathbf{k}'}\int d^d r\, e^{-i(\mathbf{k}+\mathbf{k}')\cdot\mathbf{r}}\frac{\partial L}{\partial\dot{\psi}_j(\mathbf{k}',t)} \\ &= \sum_{\mathbf{k}'}\delta_{\mathbf{k}+\mathbf{k}',0}\frac{\partial L}{\partial\dot{\psi}_j(\mathbf{k}',t)} \\ &= \frac{\partial L}{\partial\dot{\psi}_j(-\mathbf{k},t)}. \end{aligned} \tag{1.A.46}$$

1.A.3.1 *Symmetries and Noether's theorem*

Suppose we describe the system with two fields, $\psi_j(x)$ and $\psi'_j(x)$, corresponding to Lagrangian densities $\mathcal{L}(\psi_j(x)), \partial_\mu\psi_j(x))$ and $\mathcal{L}'(\psi'_j(x), \partial_\mu\psi'_j(x))$, respectively.[88] Here, $\mu = 0, x, \ldots$, $\partial_0 = \partial_t$ and $\psi_j(x) = \psi_j(\mathbf{r},t)$. The transformation $\psi_j(x) \to \psi'_j(x)$ can arise from a genuine transformation of the field (such as a rotation for a N-component vector field $\boldsymbol{\psi} = (\psi_1, \ldots, \psi_N))$ or a mere transformation of the spacetime coordinates (such as a space translation $\psi_j(\mathbf{r},t) \to \psi'_j(\mathbf{r}',t) = \psi_j(\mathbf{r},t)$ with $\mathbf{r}' = \mathbf{r} + \mathbf{a}$). Since both fields refer to the same system, we must have $\mathcal{L}(\psi_j, \partial_\mu\psi_j) = \mathcal{L}'(\psi'_j, \partial_\mu\psi'_j)$ (up to a total time or space derivative), i.e. $S'[\psi'] = S[\psi]$. If $\mathcal{L}'(\psi'_j, \partial_\mu\psi'_j)$ and $\mathcal{L}(\psi'_j, \partial_\mu\psi'_j)$ differ only by a total time or space derivative, i.e. $S'[\psi'] = S[\psi']$, then the equation of motion of $\psi'_j(x)$ is identical to that of $\psi_j(x)$ and the system is invariant in the transformation $\psi_j(x) \to \psi'_j(x)$.

To derive Noether's theorem, we express the fact that the change

$$\delta S = \int dt \int d^d r\, \{\mathcal{L}(\psi'_j(x), \partial_\mu\psi'_j(x)) - \mathcal{L}(\psi_j(x), \partial_\mu\psi_j(x))\} = 0 \tag{1.A.47}$$

in the action $S[\psi]$ vanishes if the system is invariant in the transformation $\psi_j(x) \to \psi'_j(x)$ (this follows from the two conditions $S'[\psi'] = S[\psi]$ and $S'[\psi'] = S[\psi']$). Let us consider an infinitesimal (continuous) transformation,

$$\psi_j(x) \to \psi_j(x) + i\epsilon F_j(x) \qquad (\epsilon \to 0), \tag{1.A.48}$$

which satisfies (1.A.47). The Lagrangian density $\mathcal{L}$ is invariant up to a total derivative

$$\mathcal{L}(\psi_j, \partial_\mu\psi_j) \to \mathcal{L}(\psi_j, \partial_\mu\psi_j) + \epsilon\sum_\mu \partial_\mu\mathcal{F}_\mu(\psi_j), \tag{1.A.49}$$

where $\mu = 0, x...$ Equations (1.A.48) and (1.A.49) imply

$$\sum_j\left[\frac{\partial\mathcal{L}}{\partial\psi_j(x)}iF_j(x) + \sum_\mu\frac{\partial\mathcal{L}}{\partial(\partial_\mu\psi_j(x))}i\partial_\mu F_j(x)\right] = \sum_\mu\partial_\mu\mathcal{F}_\mu(x) \tag{1.A.50}$$

[88]See footnote 78 on page 135.

$(\mathcal{F}_\mu(x) \equiv \mathcal{F}_\mu(\psi_j(x)))$, i.e.

$$\sum_j \left[\frac{\partial \mathcal{L}}{\partial \psi_j(x)} iF_j(x) + \sum_\mu \partial_\mu \left(\frac{\partial \mathcal{L}}{\partial(\partial_\mu \psi_j(x))} iF_j(x) \right) - \sum_\mu \partial_\mu \left(\frac{\partial \mathcal{L}}{\partial(\partial_\mu \psi_j(x))} \right) iF_j(x) \right] = \sum_\mu \partial_\mu \mathcal{F}_\mu(x). \tag{1.A.51}$$

If we assume that the field $\psi_j(x)$ satisfies the classical equations of motion (1.A.39), Eq. (1.A.51) becomes

$$\sum_{j,\mu} \partial_\mu \left(\frac{\partial \mathcal{L}}{\partial(\partial_\mu \psi_j(x))} iF_j(x) \right) = \sum_\mu \partial_\mu \mathcal{F}_\mu(x). \tag{1.A.52}$$

We deduce the existence of a conserved (divergenceless) current $(j_0(x), \mathbf{j}(x))$,

$$\begin{aligned} &\sum_\mu \partial_\mu j_\mu(x) = 0, \\ &j_\mu(x) = -i \sum_j \frac{\partial \mathcal{L}}{\partial(\partial_\mu \psi_j(x))} F_j(x) + \mathcal{F}_\mu(x). \end{aligned} \tag{1.A.53}$$

The continuity equation implies the existence of a conserved (time-independent) charge,

$$Q = \int d^d r \, j_0(x), \qquad \frac{dQ}{dt} = 0. \tag{1.A.54}$$

For a Lagrangian density $\mathcal{L}(\psi_j, \nabla\psi_j, \dot{\psi}_j)$ which does not depend explicitly on time, an infinitesimal time translation $\psi_j(x) \to \psi_j(x) - \epsilon\dot{\psi}_j(x)$ gives $F_j(x) = i\dot{\psi}_j(x)$ and $\mathcal{F}_\mu = -\delta_{\mu,0}\mathcal{L}$. The density

$$j_0(x) = \sum_j \Pi_j(x)\dot{\psi}_j(x) - \mathcal{L} \equiv \mathcal{H} \tag{1.A.55}$$

is the Hamiltonian density and the conserved charge

$$\int d^d r \, j_0(x) = H \tag{1.A.56}$$

is the Hamiltonian. Similarly, for a space translation $\psi_j(x) \to \psi_j(x) - \epsilon\partial_\mu\psi_j(x)$ along the μ axis, $F_j(x) = i\partial_\mu\psi_j(x)$ and $\mathcal{F}_\nu = -\delta_{\nu,\mu}\mathcal{L}$. This gives

$$\mathbf{j}_0(x) = \sum_j \Pi_j(x) \nabla\psi_j(x) \tag{1.A.57}$$

(we use the fact that the axis μ is arbitrary) and the conserved charge

$$\mathbf{P} = \int d^d r \, \mathbf{j}_0(x) = \int d^d r \sum_j \Pi_j(x) \nabla\psi_j(x) \tag{1.A.58}$$

is naturally interpreted as the momentum carried by the field.[89] Finally, for a space rotation about an arbitrary axis $\mathbf{n}$, $\psi_j(x) \to \psi_j(x) - \epsilon(\mathbf{n}\times\mathbf{r})\cdot\boldsymbol{\nabla}\psi_j(x)$ (we consider a three-dimensional system), we obtain

$$j_0(x) = \sum_j \Pi_j(x)(\mathbf{n}\times\mathbf{r})\cdot\boldsymbol{\nabla}\psi_j(x) = \sum_j \Pi_j(x)\mathbf{n}\cdot(\mathbf{r}\times\boldsymbol{\nabla})\psi_j(x), \tag{1.A.59}$$

which leads to the conservation of the angular momentum

$$\mathbf{J} = \int d^3r \sum_j \Pi_j(x)(\mathbf{r}\times\boldsymbol{\nabla})\psi_j(x). \tag{1.A.60}$$

The infinitesimal continuous transformation (1.A.48) can often be expressed with the generator G of the symmetry,

$$\psi'_j(x) = \psi_j(x) - i\epsilon G\psi(x). \tag{1.A.61}$$

For the examples discussed above, one has

$$\begin{aligned} G &\equiv P_\mu = -i\partial_\mu, \quad \text{(space translation in the } \mu \text{ direction)}, \\ G &\equiv L_{\mu\nu} = -i(x_\mu\partial_\nu - x_\nu\partial_\mu) \quad \text{(rotation in the plane } (\mu,\nu)). \end{aligned} \tag{1.A.62}$$

These examples refer to a mere transformation of the spacetime variable $x = (\mathbf{r}, t)$. But the field can also be transformed. For instance, in a space rotation (considering for simplicity a three-dimensional system), for a three-component field $\boldsymbol{\varphi}(x) = (\varphi_1(x), \varphi_2(x), \varphi_3(x))$ transforming as a vector, the total generator is

$$G = L_{\mu\nu} + S_{\mu\nu}, \tag{1.A.63}$$

where $S_{\mu\nu}$ is a skew-symmetric matrix of the Lie algebra $\mathfrak{so}(3)$. More generally, depending on the nature of the field (scalar, vector, tensor, etc.), the matrix $S_{\mu\nu}$ belongs to some representation of the group SO(3).

1.A.3.2 *Classical electromagnetism*

Classical electromagnetism in three-dimensional space follows from the Lagrangian

$$\begin{aligned} L = \frac{m}{2}\sum_i \dot{\mathbf{r}}_i^2 + \frac{\epsilon_0}{2}\int d^3r \left\{ [\boldsymbol{\nabla}\phi(\mathbf{r},t) + \partial_t\mathbf{A}(\mathbf{r},t)]^2 - c_l^2\,[\boldsymbol{\nabla}\times\mathbf{A}(\mathbf{r},t)]^2 \right\} \\ - \int d^3r\, [\phi(\mathbf{r},t)\rho(\mathbf{r},t) - \mathbf{J}(\mathbf{r},t)\cdot\mathbf{A}(\mathbf{r},t)] \end{aligned} \tag{1.A.64}$$

(c_l denotes the velocity of light). L is a function of the positions $\mathbf{r}_i$ and velocities $\dot{\mathbf{r}}_i$ of the particles and a functional of the electromagnetic potentials $\phi(\mathbf{r},t)$ and $\mathbf{A}(\mathbf{r},t)$. The

[89] Equations (1.A.58) and (1.A.60) actually correspond to the opposite of the usual definitions of momentum and angular momentum. In the relativistic formalism, one finds that the Minkowski metric $\eta_{\mu\nu} = \text{diag}(1,-1,-1,-1)$ reintroduces the usual sign in the definition of $\mathbf{P}$ and $\mathbf{J}$. For instance, for a spacetime translation $x^\mu \to x^\mu + \epsilon^\mu$, Noether's theorem gives the conservation law $\sum_\mu \partial_\mu T^\mu{}_\nu = 0$ where $T^\mu{}_\nu = \frac{\partial}{\partial(\partial_\mu\psi)}\partial_\nu\psi - \delta^\mu{}_\nu\mathcal{L}$ is the stress–energy tensor, also called the energy–momentum tensor. The conserved densities are $T^{00} = T^0{}_0 = \mathcal{H}$ and $T^{0i} = -T^0{}_i = -\Pi(x)\partial_i\psi(x)$ $(i = x, y, z)$.

electric and magnetic fields can be expressed as $\mathbf{E} = -\boldsymbol{\nabla}\phi - \partial_t \mathbf{A}$ and $\mathbf{B} = \boldsymbol{\nabla} \times \mathbf{A}$. The matter–field coupling involves the charge and current densities, $\rho(\mathbf{r}, t) = e \sum_i \delta(\mathbf{r} - \mathbf{r}_i)$ and $\mathbf{J}(\mathbf{r}, t) = e \sum_i \dot{\mathbf{r}}_i \delta(\mathbf{r} - \mathbf{r}_i)$, where e is the charge of the particles. Using[90]

$$\begin{aligned} \frac{\partial L}{\partial \dot{\mathbf{r}}_i} &= m\dot{\mathbf{r}}_i + e\mathbf{A}(\mathbf{r}_i, t), \\ \frac{\partial L}{\partial \mathbf{r}_i} &= -e\boldsymbol{\nabla}\phi(\mathbf{r}_i, t) + e(\dot{\mathbf{r}}_i \cdot \boldsymbol{\nabla})\mathbf{A}(\mathbf{r}_i, t) + e\dot{\mathbf{r}}_i \times \big(\boldsymbol{\nabla} \times \mathbf{A}(\mathbf{r}_i, t)\big), \end{aligned} \tag{1.A.65}$$

we find that the Euler–Lagrange equation

$$\frac{d}{dt}\frac{\partial L}{\partial \dot{\mathbf{r}}_i} = \frac{\partial L}{\partial \mathbf{r}_i} \tag{1.A.66}$$

reproduces the Newton–Lorentz equation

$$m\ddot{\mathbf{r}}_i = e\mathbf{E}(\mathbf{r}_i, t) + e\dot{\mathbf{r}}_i \times \mathbf{B}(\mathbf{r}_i, t). \tag{1.A.67}$$

Similarly, from

$$\begin{aligned} \frac{\delta L}{\delta \dot{\phi}(\mathbf{r}, t)} &= 0, \\ \frac{\delta L}{\delta \phi(\mathbf{r}, t)} &= -\epsilon_0[\boldsymbol{\nabla}^2\phi(\mathbf{r}, t) + \boldsymbol{\nabla} \cdot \dot{\mathbf{A}}(\mathbf{r}, t)] - \rho(\mathbf{r}, t) \end{aligned} \tag{1.A.68}$$

and

$$\begin{aligned} \frac{\delta L}{\delta \dot{\mathbf{A}}(\mathbf{r}, t)} &= \epsilon_0[\boldsymbol{\nabla}\phi(\mathbf{r}, t) + \dot{\mathbf{A}}(\mathbf{r}, t)], \\ \frac{\delta L}{\delta \mathbf{A}(\mathbf{r}, t)} &= -\epsilon_0 c_l^2 \boldsymbol{\nabla} \times \big(\boldsymbol{\nabla} \times \mathbf{A}(\mathbf{r}, t)\big) + \mathbf{J}(\mathbf{r}, t) \end{aligned} \tag{1.A.69}$$

as well as the Euler–Lagrange equations

$$\begin{aligned} \frac{d}{dt}\frac{\delta L}{\delta \dot{\phi}(\mathbf{r}, t)} &= \frac{\delta L}{\delta \phi(\mathbf{r}, t)}, \\ \frac{d}{dt}\frac{\delta L}{\delta \dot{\mathbf{A}}(\mathbf{r}, t)} &= \frac{\delta L}{\delta \mathbf{A}(\mathbf{r}, t)}, \end{aligned} \tag{1.A.70}$$

we deduce the Maxwell equations

$$\begin{aligned} \boldsymbol{\nabla} \cdot \mathbf{E}(\mathbf{r}, t) &= \frac{\rho(\mathbf{r}, t)}{\epsilon_0}, \\ \boldsymbol{\nabla} \times \mathbf{B}(\mathbf{r}, t) &= \mu_0 \mathbf{J}(\mathbf{r}, t) + \frac{1}{c_l^2}\frac{\partial \mathbf{E}(\mathbf{r}, t)}{\partial t}. \end{aligned} \tag{1.A.71}$$

[90] We have used the vector identity $\boldsymbol{\nabla}(\mathbf{A} \cdot \mathbf{B}) = (\mathbf{A} \cdot \boldsymbol{\nabla})\mathbf{B} + (\mathbf{B} \cdot \boldsymbol{\nabla})\mathbf{A} + \mathbf{A} \times (\boldsymbol{\nabla} \times \mathbf{B}) + \mathbf{B} \times (\boldsymbol{\nabla} \times \mathbf{A})$ to obtain the second of Eq. (1.A.65).

1.B Review of Quantum Mechanics[91]

In this section, we review the elementary aspects of quantum mechanics and introduce the concept of quantum fields which is discussed at length in Sec. 1.2.

1.B.1 *The postulates of quantum mechanics*

1.B.1.1 *State space*

To any physical system, we associate a Hilbert space $\mathcal{H}$ called the state space. A Hilbert space is a vector space with an inner product (or scalar product).[92] In Dirac's notation, a vector Ψ is represented by a ket $|\Psi\rangle$. Physical states are represented by rays in the Hilbert space. A ray is a set of normalized vectors which differ only by a phase factor. Thus, $|\Psi\rangle$ and $|\Psi'\rangle$ belong to the same ray if $|\Psi'\rangle = e^{i\alpha}|\Psi\rangle$ (α real). To each ket $|\Psi\rangle$, one can associate a bra $\langle\Psi|$ which is an element of the dual space $\mathcal{H}^*$ and such that the action of $\langle\Psi|$ on $|\Phi\rangle$ yields the scalar product $\langle\Psi|\Phi\rangle$ between $|\Psi\rangle$ and $|\Phi\rangle$.[93]

We can define coordinates in the state space by choosing a complete orthonormal basis, i.e. a set $\{|n\rangle\}$ of vectors satisfying

$$\langle n|m\rangle = \delta_{n,m} \tag{1.B.1}$$

and the closure relation

$$\sum_n |n\rangle\langle n| = 1, \tag{1.B.2}$$

where the symbol "1" in the rhs denotes the identity operator acting on $\mathcal{H}$.

1.B.1.2 *Observables*

With any measurable quantity A, we associate a Hermitian operator $\hat{A}$ whose normalized eigenvectors $|n\rangle$ form an orthonormal basis of the Hilbert space $\mathcal{H}$.[94] Such an operator is called an observable. In a given basis $\{|n\rangle\}$, $\hat{A}$ is defined by its matrix elements $\langle n|\hat{A}|m\rangle$.

The only possible result of the measurement of the physical quantity A is one of the eigenvalues a_n of the operator $\hat{A}$ with probability $\mathcal{P}(a_n) = \langle\Psi|\hat{P}_{a_n}|\Psi\rangle$ where $\hat{P}_a$ is the projector onto the subspace associated with the eigenvalue a (we assume the state vector to be normalized, $\langle\Psi|\Psi\rangle = 1$).[95] The expectation value of the observable $\hat{A}$ is therefore

$$\langle\hat{A}\rangle = \sum_n \mathcal{P}(a_n)a_n = \sum_n \langle\Psi|\hat{P}_{a_n}|\Psi\rangle a_n = \langle\Psi|\hat{A}|\Psi\rangle \tag{1.B.3}$$

---

[91] See, for instance, Refs. [24–27].

[92] For a vector space to be a Hilbert space, it must also be complete: If a sequence of vectors is Cauchy, then it converges to some limit in the space.

[93] The dual space $\mathcal{H}^*$ is defined as the set of all linear functionals on $\mathcal{H}$.

[94] An operator $\hat{A}$ is Hermitian if it is equal to its adjoint $\hat{A}^\dagger$, where $\hat{A}^\dagger$ is defined by $\langle\Psi|\hat{A}^\dagger|\Phi\rangle = \langle\Phi|\hat{A}|\Psi\rangle^*$. To the ket $\hat{A}|\Psi\rangle$ corresponds the bra $\langle\Psi|\hat{A}^\dagger$.

[95] $\hat{P}_{a_n} = |n\rangle\langle n|$ when the eigenvalue a_n is not degenerate ($\hat{A}|n\rangle = a_n|n\rangle$).

and its statistical fluctuations (of quantum origin) are given by the variance

$$\Delta A^2 = \langle \hat{A}^2 \rangle - \langle \hat{A} \rangle^2 = \langle \Psi | \big(\hat{A} - \langle \hat{A} \rangle\big)^2 | \Psi \rangle. \tag{1.B.4}$$

If the eigenvalue a_n is measured, then the state of the system after the measurement is given by the projection of the state vector onto the subspace associated with a_n: $|\Psi'\rangle \propto \hat{P}_{a_n} |\Psi\rangle$ (collapse of the state vector).

1.B.1.3 *Coordinate and momentum representations*

For a spinless particle moving in a d-dimensional space, a convenient basis is given by the eigenstates $|\mathbf{r}\rangle$ of the position operator $\hat{\mathbf{r}}$: $\hat{\mathbf{r}}|\mathbf{r}\rangle = \mathbf{r}|\mathbf{r}\rangle$. These states are not normalizable and therefore not physical states. They satisfy

$$\langle \mathbf{r} | \mathbf{r}' \rangle = \delta(\mathbf{r} - \mathbf{r}') \quad \text{and} \quad \int d^d r \, |\mathbf{r}\rangle \langle \mathbf{r}| = 1. \tag{1.B.5}$$

Any state can be expanded in this basis as

$$|\Psi\rangle = \int d^d r \, \Psi(\mathbf{r}) |\mathbf{r}\rangle, \tag{1.B.6}$$

where $\Psi(\mathbf{r}) = \langle \mathbf{r} | \Psi \rangle$ is the wavefunction in the coordinate representation. If the state is normalized,

$$\langle \Psi | \Psi \rangle = \int d^d r \, |\Psi(\mathbf{r})|^2 = 1, \tag{1.B.7}$$

$|\Psi(\mathbf{r})|^2$ gives the probability density for the particle to be at point $\mathbf{r}$.

Another useful basis is given by the eigenstates $|\mathbf{p}\rangle$ of the momentum operator $\hat{\mathbf{p}}$: $\hat{\mathbf{p}}|\mathbf{p}\rangle = \mathbf{p}|\mathbf{p}\rangle$. They satisfy

$$|\mathbf{p}\rangle = \int d^d r \, \langle \mathbf{r} | \mathbf{p} \rangle |\mathbf{r}\rangle = \frac{1}{\sqrt{\mathcal{V}}} \int d^d r \, e^{i\mathbf{p}\cdot\mathbf{r}} |\mathbf{r}\rangle, \tag{1.B.8}$$

or the inverse relation

$$|\mathbf{r}\rangle = \frac{1}{\sqrt{\mathcal{V}}} \sum_{\mathbf{p}} e^{-i\mathbf{p}\cdot\mathbf{r}} |\mathbf{p}\rangle, \tag{1.B.9}$$

as well as the closure relation

$$\sum_{\mathbf{p}} |\mathbf{p}\rangle \langle \mathbf{p}| = 1. \tag{1.B.10}$$

The wavefunction in the $\mathbf{p}$ representation is the Fourier transform of $\Psi(\mathbf{r})$,

$$\Psi(\mathbf{p}) = \langle \mathbf{p} | \Psi \rangle = \frac{1}{\sqrt{\mathcal{V}}} \int d^d r \, e^{-i\mathbf{p}\cdot\mathbf{r}} \Psi(\mathbf{r}). \tag{1.B.11}$$

For periodic boundary conditions, $\Psi(\mathbf{r} + L\hat{\mathbf{x}}) = \Psi(\mathbf{r} + L\hat{\mathbf{y}}) = \cdots = \Psi(\mathbf{r})$,[96] $\mathbf{p}$ is a discrete variable which takes quantized values

$$p_\mu = \frac{2\pi}{L} n_\mu, \quad n_\mu \in \mathbb{Z} \quad (\mu = x, y, \ldots). \tag{1.B.12}$$

[96] $\hat{\mathbf{x}}$ denotes the unit vector in the x direction, etc.

When the volume of the system tends to infinity, $\mathbf{p}$ becomes a continuous variable and

$$\lim_{\mathcal{V}\to\infty} \frac{1}{\mathcal{V}} \sum_{\mathbf{p}} = \int \frac{d^d p}{(2\pi)^d} \equiv \int_{\mathbf{p}} . \tag{1.B.13}$$

1.B.1.4 *Canonical quantization*

In the Hamiltonian formalism of classical mechanics, the system is described by N coordinates $q_i(t)$ and N conjugate momenta $p_i(t)$. In the quantum formalism, these variables become operators $\hat{q}_i(t), \hat{p}_i(t)$ acting on the state space $\mathcal{H}$. For any physical quantity $A(p_i, q_i, t)$ of the classical theory, one then obtains an observable $\hat{A}(t) \equiv A(\hat{p}_i(t), \hat{q}_i(t), t)$ (Heisenberg's correspondence principle). When this correspondence rule yields products of the non-commuting operators $\hat{p}_i$ and $\hat{q}_i$, the order of the operators must be determined by additional conditions such as the Hermiticity of the observable $\hat{A}(t)$.

Moreover, the Poisson bracket $[A, B]$ is replaced by the commutator of the two corresponding operators,

$$[A, B] \equiv \sum_i \left(\frac{\partial A}{\partial q_i} \frac{\partial B}{\partial p_i} - \frac{\partial A}{\partial p_i} \frac{\partial B}{\partial q_i} \right) \to \frac{1}{i\hbar}(\hat{A}\hat{B} - \hat{B}\hat{A}) \equiv \frac{1}{i\hbar}[\hat{A}, \hat{B}]. \tag{1.B.14}$$

The factor i ensures that the rhs in (1.B.14) is Hermitian if both $\hat{A}$ and $\hat{B}$ are, and the factor $\hbar$ (now written explicitly) preserves the units of the Poisson bracket. Equations (1.A.26) yield the equal-time commutation relations

$$[\hat{q}_i(t), \hat{q}_j(t)] = [\hat{p}_i(t), \hat{p}_j(t)] = 0, \qquad [\hat{q}_i(t), \hat{p}_j(t)] = i\hbar\delta_{i,j}. \tag{1.B.15}$$

The time evolution of these operators is deduced from (1.A.23),

$$\begin{aligned} \frac{d}{dt}\hat{q}_i(t) &= \frac{1}{i\hbar}[\hat{q}_i(t), \hat{H}(t)], \\ \frac{d}{dt}\hat{p}_i(t) &= \frac{1}{i\hbar}[\hat{p}_i(t), \hat{H}(t)], \end{aligned} \tag{1.B.16}$$

where $\hat{H}(t) \equiv H(\hat{p}_i(t), \hat{q}_i(t), t)$ is the quantum Hamiltonian. For an observable $\hat{A}(t) \equiv A(\hat{p}_i(t), \hat{q}_i(t), t)$, Eq. (1.A.24) yields

$$\frac{d\hat{A}(t)}{dt} = \frac{\partial \hat{A}(t)}{\partial t} + \frac{1}{i\hbar}[\hat{A}(t), \hat{H}(t)], \tag{1.B.17}$$

where $\partial\hat{A}/\partial t$ comes from the explicit time dependence of the operator $\hat{A}(t)$. From (1.B.17), we deduce that the time evolution of the expectation value $\langle \hat{A}(t) \rangle = \langle \Psi | \hat{A}(t) | \Psi \rangle$ of an observable is given by

$$\frac{d\langle \hat{A}(t) \rangle}{dt} = \left\langle \frac{\partial \hat{A}(t)}{\partial t} \right\rangle + \frac{1}{i\hbar} \langle [\hat{A}(t), \hat{H}(t)] \rangle \tag{1.B.18}$$

(Ehrenfest's equation). An observable which commutes with the Hamiltonian $\hat{H}$ and does not depend on time explicitly is a constant of motion.

So far, we have discussed the Heisenberg picture of quantum mechanics where the state vector $|\Psi_H\rangle$ is time independent while the dynamics of the system is given by the time evolution of the observables $\hat{A}_H(t)$ [Eq. (1.B.17)]. Alternatively, one can choose the Schrödinger picture where the state vector $|\Psi_S(t)\rangle$ is time dependent while the operators $\hat{A}_S(t)$ are not (except for a possible explicit time dependence). Both descriptions are related by a unitary transformation,

$$\begin{aligned} \hat{A}_H(t) &= \hat{U}^\dagger(t,t_0)\hat{A}_S(t)\hat{U}(t,t_0), \\ |\Psi_H\rangle &= \hat{U}^\dagger(t,t_0)|\Psi_S(t)\rangle \end{aligned} \tag{1.B.19}$$

($\hat{U}^\dagger(t,t_0)\hat{U}(t,t_0) = \hat{U}(t,t_0)\hat{U}^\dagger(t,t_0) = 1$), which preserves the value of any matrix element

$$\langle\Psi_H|\hat{A}_H(t)|\Phi_H\rangle = \langle\Psi_H|\hat{U}^\dagger(t,t_0)\hat{A}_S(t)\hat{U}(t,t_0)|\Phi_H\rangle = \langle\Psi_S(t)|\hat{A}_S(t)|\Phi_S(t)\rangle. \tag{1.B.20}$$

t_0 is some arbitrary fixed time. If the two descriptions coincide at t_0, then $\hat{U}(t_0,t_0) = 1$ and the equation

$$|\Psi_S(t)\rangle = \hat{U}(t,t_0)|\Psi_H\rangle = \hat{U}(t,t_0)|\Psi_S(t_0)\rangle \tag{1.B.21}$$

shows that $\hat{U}(t,t_0)$ is the time evolution operator in the Schrödinger picture. The unitary operator $\hat{U}(t,t_0)$ is further defined by its time evolution equation

$$i\hbar\frac{d}{dt}\hat{U}(t,t_0) = \hat{H}_S(t)\hat{U}(t,t_0), \tag{1.B.22}$$

where $\hat{H}_S(t)$ is the Hamiltonian in the Schrödinger picture. Equation (1.B.22) implies that the state vector $|\Psi_S(t)\rangle$ obeys the Schrödinger equation

$$i\hbar\frac{d}{dt}|\Psi_S(t)\rangle = \hat{H}_S(t)|\Psi_S(t)\rangle. \tag{1.B.23}$$

From (1.B.19) and (1.B.22), one deduces

$$\frac{d}{dt}\hat{q}_{iH}(t) = \frac{1}{i\hbar}[\hat{q}_{iH}(t),\hat{H}_H(t)] + \hat{U}^\dagger(t,t_0)\frac{d\hat{q}_{iS}(t)}{dt}\hat{U}(t,t_0) \tag{1.B.24}$$

and a similar equation for $\hat{p}_{iH}(t)$. Comparing with (1.B.16), we conclude that the operators $\hat{q}_{iS}$ and $\hat{p}_{iS}$ are time independent in the Schrödinger picture. It follows that the time dependence of an observable $\hat{A}_S(t) \equiv A(\hat{p}_{iS},\hat{q}_{iS},t)$ in the Schrödinger picture can only be explicit. For an arbitrary observable, Eqs. (1.B.19) and (1.B.22) imply

$$\begin{aligned} \frac{d}{dt}\hat{A}_H(t) &= \frac{1}{i\hbar}[\hat{A}_H(t),\hat{H}_H(t)] + \hat{U}^\dagger(t,t_0)\frac{\partial\hat{A}_S(t)}{\partial t}\hat{U}(t,t_0) \\ &= \frac{1}{i\hbar}[\hat{A}_H(t),\hat{H}_H(t)] + \frac{\partial\hat{A}_H(t)}{\partial t} \end{aligned} \tag{1.B.25}$$

in agreement with the Heisenberg equation of motion (1.B.17). When the Hamiltonian $\hat{H}_S(t) \equiv \hat{H}$ in the Schrödinger picture is time independent, the time evolution operator takes the form

$$\hat{U}(t,t_0) = \hat{U}(t-t_0) = e^{-\frac{i}{\hbar}\hat{H}(t-t_0)} \tag{1.B.26}$$

and the Hamiltonian $\hat{H}_H(t) = \hat{H}$ in the Heisenberg picture is also time independent.

1.B.2 *Quantum fields*

To quantize a classical field theory defined by its Hamiltonian density $\mathcal{H}(\Pi_j, \psi_j, \nabla\psi_j)$ (Sec. 1.A.2), we promote the field $\psi_j(\mathbf{r}, t)$ and its conjugate variable $\Pi_j(\mathbf{r}, t)$ to operators acting on the state space $\mathcal{H}$ and satisfying the equal-time commutation relations

$$\begin{aligned} [\hat{\psi}_j(\mathbf{r}, t), \hat{\psi}_{j'}(\mathbf{r}', t)] &= [\hat{\Pi}_j(\mathbf{r}, t), \hat{\Pi}_{j'}(\mathbf{r}', t)] = 0, \\ [\hat{\psi}_j(\mathbf{r}, t), \hat{\Pi}_{j'}(\mathbf{r}', t)] &= i\hbar\delta_{j,j'}\delta(\mathbf{r} - \mathbf{r}'). \end{aligned} \tag{1.B.27}$$

In the Heisenberg picture, the quantum fields $\hat{\psi}_j(\mathbf{r}, t)$ and $\hat{\Pi}_j(\mathbf{r}, t)$ are time dependent while the state of the system is described by a time-independent ket $|\Psi\rangle$. Alternatively, one can use the Schrödinger picture as explained in the preceding section.

In Sec. 1.9, we use this procedure to quantize the classical electromagnetic field.[97] A similar approach can be used to quantize a system of particles.[98] Let us first consider a single particle. Its wavefunction satisfies the Schrödinger equation (1.B.23),

$$i\hbar\frac{\partial\Psi(\mathbf{r}, t)}{\partial t} = \hat{H}\Psi(\mathbf{r}, t). \tag{1.B.28}$$

We can see (1.B.28) as a wave equation for the matter, analogous to Maxwell's equations for electromagnetic waves, where $\Psi(\mathbf{r}, t)$ is considered as a classical field (the Schrödinger field) rather than the wavefunction of a single particle. Equation (1.B.28) derives from a Lagrangian. The canonical quantization procedure leads to Hermitian conjugate quantum fields $\hat{\psi}(\mathbf{r}, t)$ and $\hat{\psi}^\dagger(\mathbf{r}, t)$ whose quantized excitations correspond to the particles of the system. While the standard commutation relations

$$\begin{aligned} [\hat{\psi}(\mathbf{r}, t), \hat{\psi}(\mathbf{r}', t)] &= [\hat{\psi}^\dagger(\mathbf{r}, t), \hat{\psi}^\dagger(\mathbf{r}', t)] = 0, \\ [\hat{\psi}(\mathbf{r}, t), \hat{\psi}^\dagger(\mathbf{r}', t)] &= \delta(\mathbf{r} - \mathbf{r}') \end{aligned} \tag{1.B.29}$$

can be used for bosons, they would violate the Pauli principle for fermions. In this case, it is necessary to use anticommutators instead of commutators in (1.B.29). This quantization procedure is often (improperly) called "second quantization" as it seems to be based on a quantization of the Schrödinger equation, which is itself a quantum equation of motion. In fact, $\psi(\mathbf{r}, t)$ should be seen as a classical field (not the wavefunction of a quantum particle) which is quantized only once.

While the classical field is the inescapable starting point when quantizing relativistic particles [7], a different quantization procedure, which does not rely on a classical field, can be used for non-relativistic particles. In this approach, one constructs the Fock space of the many-particle system directly from the one-particle Hilbert space used for the quantum mechanics of a single particle, and then directly defines the quantum fields $\hat{\psi}(\mathbf{r}, t)$ and $\hat{\psi}^\dagger(\mathbf{r}, t)$ by their action on the many-body states. This is the route followed in Sec. 1.2.

[97] An elementary example, the quantization of the classical harmonic string, is also discussed in Sec. 1.1.2.
[98] For a detailed discussion, see Ref. [24] (Chapter XIII) and [18] (Chapter II and exercise $D_{\text{II}}.8$).

1.C Review of Statistical Physics[99]

In this section, we review the basis of quantum statistical physics.

1.C.1 *Macrostates — Density operators*

In quantum mechanics, the system is described by a normalized vector $|\Psi\rangle$ of the Hilbert space $\mathcal{H}$. Such a state, where the system is completely known or ideally prepared (*via* the control of a complete set of commuting observables), is called a microstate or a pure state. In statistical physics, one is interested in macroscopic systems with a very large number of degrees of freedom. These systems are never ideally prepared or completely known and are usually described by the specification of a small number of macroscopic variables. On the microscopic scale, a partially known or incompletely prepared system is described by a macrostate (or statistical mixture), i.e. by the set of possible microstates $|\Psi_\lambda\rangle$ and the corresponding probabilities q_λ. In this statistical description, a "system" denotes an element of a statistical ensemble of systems, all prepared under the same conditions. The central assumption of statistical physics is that the time average (over a characteristic time larger than microscopic time scales) of an observable for a given system coincides with the statistical average obtained from the statistical ensemble.

1.C.1.1 *Density operator*

All the information about the macrostate is contained in the density operator (or density matrix)

$$\hat{\rho} = \sum_{\lambda} |\Psi_\lambda\rangle q_\lambda \langle\Psi_\lambda|, \tag{1.C.1}$$

where the probabilities q_λ are positive and normalized: $\sum_\lambda q_\lambda = 1$. For instance, the expectation value of an observable $\hat{A}$ is given by

$$\langle\hat{A}\rangle = \sum_{\lambda} q_\lambda \langle\Psi_\lambda|\hat{A}|\Psi_\lambda\rangle = \mathrm{Tr}(\hat{\rho}\hat{A}). \tag{1.C.2}$$

The random character of the measurement has two origins, the probabilities q_λ on the one hand and the quantum nature of the microstates $|\Psi_\lambda\rangle$ on the other hand.

Two different statistical mixtures, $\{|\Psi_\lambda\rangle, q_\lambda\}$ and $\{|\Psi'_\mu\rangle, q'_\mu\}$, can lead to the same density operator and therefore the same physical reality. Thus, the only meaningful representation of macrostates is given by density operators. A microstate $|\Psi\rangle$ corresponds to the density operator $\hat{\rho} = |\Psi\rangle\langle\Psi|$. Note that all state vectors $|\Psi\rangle$ belonging to the same ray (i.e. differing only by a phase factor) yield the same density operator.

The definition (1.C.1) gives the following properties:

1. $\hat{\rho}$ is Hermitian: $\hat{\rho}^\dagger = \hat{\rho}$.
2. $\hat{\rho}$ has unit trace: $\mathrm{Tr}(\hat{\rho}) = 1$.
3. $\hat{\rho}$ is non-negative: $\langle\Psi|\hat{\rho}|\Psi\rangle \geq 0, \ \forall\, |\Psi\rangle \in \mathcal{H}$.

[99]This section closely follows Chapters 2–5 of Ref. [28].

Conversely, any operator $\hat{\rho}$ satisfying these properties can be written in the basis of its eigenstates,

$$\hat{\rho} = \sum_m |m\rangle p_m \langle m| \qquad (p_m \in \mathbb{R}). \tag{1.C.3}$$

It follows from the properties (2) and (3) that $p_m \geq 0$ and $\sum_m p_m = 1$, which allows us to regard p_m as the probability of the microstate $|m\rangle$. A necessary and sufficient condition to have a pure state is that $\hat{\rho}^2 = \hat{\rho}$.

1.C.1.2 *Time evolution of the density operator*

In the Schrödinger picture, the time evolution of the density operator is directly obtained from the Schrödinger equation satisfied by the states $|m\rangle$ [Eq. (1.C.3)],

$$i\hbar \frac{d\hat{\rho}}{dt} = [\hat{H}, \hat{\rho}] \tag{1.C.4}$$

(Liouville–von Neumann equation). The evolution of the expectation value of an observable follows from (1.C.4),

$$\frac{d}{dt}\mathrm{Tr}(\hat{\rho}\hat{A}) = \frac{1}{i\hbar}\mathrm{Tr}([\hat{H}, \hat{\rho}]\hat{A}) + \mathrm{Tr}\left(\hat{\rho}\frac{\partial \hat{A}}{\partial t}\right) = \frac{1}{i\hbar}\langle[\hat{A}, \hat{H}]\rangle + \left\langle \frac{\partial \hat{A}}{\partial t} \right\rangle, \tag{1.C.5}$$

in agreement with Ehrenfest's equation (1.B.18). In the Heisenberg picture, the states $|m\rangle$ are time independent and $d\hat{\rho}/dt = 0$.

In statistical physics, one is often interested in a subsystem of the whole system. Knowing the density operator $\hat{\rho}$ of the system, it is possible to define a reduced density operator for the subsystem. Let us assume, for instance, that the system contains two subsystems a and b. The Hilbert space $\mathcal{H} = \mathcal{H}_a \otimes \mathcal{H}_b$ is then the tensor product of the Hilbert spaces of the subsystems. Let $\{|k_a, l_b\rangle\} = \{|k_a\rangle \otimes |l_b\rangle\}$ be a basis of $\mathcal{H}$. For an observable $\hat{A}_a \otimes 1_b$ acting only on $\mathcal{H}_a$, one has

$$\begin{aligned}
\langle \hat{A}_a \rangle &= \mathrm{Tr}[\hat{\rho}(\hat{A}_a \otimes 1_b)] \\
&= \sum_{k_a, l_b, k'_a, l'_b} \langle k_a, l_b|\hat{\rho}|k'_a, l'_b\rangle \langle k'_a, l'_b|\hat{A}_a \otimes 1_b|k_a, l_b\rangle \\
&= \sum_{k_a, l_b, k'_a} \langle k_a, l_b|\hat{\rho}|k'_a, l_b\rangle \langle k'_a|\hat{A}_a|k_a\rangle \\
&= \mathrm{Tr}(\hat{\rho}_a \hat{A}_a),
\end{aligned} \tag{1.C.6}$$

where

$$\hat{\rho}_a = \mathrm{Tr}_b(\hat{\rho}) \tag{1.C.7}$$

is the density operator in the subsystem a, obtained by tracing out the degrees of freedom in the subsystem b. The definition of a density operator for a subsystem is particularly natural in the functional integral formalism where it relies on a partial functional integration of the degrees of freedom.

1.C.2 *Statistical entropy — Boltzmann–Gibbs distribution*

The knowledge of the system based on the density operator is incomplete. The missing information can be estimated by the statistical entropy (or von Neumann entropy)

$$S = -k_B \sum_{m=1}^{M} p_m \ln p_m = -k_B \mathrm{Tr}(\hat{\rho} \ln \hat{\rho}), \tag{1.C.8}$$

where the p_m's are the probabilities of the states $|m\rangle$ that diagonalize the density operator. The factor k_B is *a priori* arbitrary. By choosing the Boltzmann constant, we ensure that at equilibrium, the statistical entropy coincides with the thermodynamic entropy (see the following). From the definition (1.C.8), we deduce the following properties:

1. S is a symmetric function of $p_1, \ldots, p_M$.
2. S is non-negative and vanishes when the system is in a pure state (one probability equals one and all others vanish): $S(1, 0, \ldots, 0) = 0$.
3. For M fixed, S is maximum when all probabilities are equal to $1/M$:

$$S\left(\frac{1}{M}, \ldots, \frac{1}{M}\right) = k_B \ln M. \tag{1.C.9}$$

4. If we divide the microstates $\{|\Psi_m\rangle\}$ into two groups of states, $A = \{|\Psi_1\rangle| \cdots |\Psi_N\rangle\}$ and $B = \{|\Psi_{N+1}\rangle| \cdots |\Psi_M\rangle\}$, then the entropy satisfies the following additivity property:

$$S = S(q_A, q_B) + q_A S\left(\frac{p_1}{q_A}, \ldots, \frac{p_N}{q_A}\right) + q_B S\left(\frac{p_{N+1}}{q_B}, \ldots, \frac{p_M}{q_B}\right), \tag{1.C.10}$$

where

$$q_A = \sum_{m=1}^{N} p_m \quad \text{and} \quad q_B = \sum_{m=N+1}^{M} p_m. \tag{1.C.11}$$

The first term in (1.C.10) corresponds to the missing information due to the choice between the sets of states A and B. The second term comes from the missing information once we know that the system is in a microstate of the set A (hence the factor q_A). p_m/q_A is the conditional probability giving the probability to find the system in the microstate $|\Psi_m\rangle$ knowing that $|\Psi_m\rangle$ belongs to the set A. The last term of (1.C.10) has a similar interpretation with the role of the sets A and B interchanged.

While these properties are a consequence of the definition (1.C.8), one can proceed the other way round and show that the expression (1.C.8) of the statistical entropy follows from the above properties, the most important one being the additivity postulate (1.C.10).

1.C.2.1 *The maximum statistical entropy principle*

When some information about the system is known exactly (e.g. the energy and the volume of the system or the number of particles in a fluid), it is taken into account through the definition of the Hilbert space. If no other information is available, it is natural to use a statistical ensemble where the probabilities of the W possible microstates are equal. In macroscopic equilibrium, the macrostate is then defined by the density operator

$$\hat{\rho} = \sum_{m=1}^{W} |m\rangle \frac{1}{W} \langle m| \tag{1.C.12}$$

and the entropy is given by

$$S = k_B \ln W. \tag{1.C.13}$$

In many cases, however, the macrostate is characterized by the knowledge of the expectation values $\langle \hat{A}_i \rangle$ of some observables $\hat{A}_i$ such as the Hamiltonian $\hat{H}$ or the total number $\hat{N}$ of particles (while other properties such as the volume may be known exactly). This happens, for instance, when the system is put into contact with a large system with which it can exchange energy or particles. This large system plays the role of a reservoir which fixes the mean value of the energy $\langle \hat{H} \rangle$ or the mean total number of particles $\langle \hat{N} \rangle$ in the system we are interested in. The density operator must then satisfy the constraints

$$\langle \hat{A}_i \rangle = \mathrm{Tr}(\hat{\rho} \hat{A}_i). \tag{1.C.14}$$

The maximum statistical entropy principle states that the density operator $\hat{\rho}$ describing the system in macroscopic equilibrium is the one which maximizes the entropy while being compatible with the known information about the system. Any other choice would mean that we assume possessing information which is actually not available and consequently would lead to biased predictions.

1.C.2.2 *Boltzmann–Gibbs distribution*

We are thus led to maximize the entropy $S = -k_B \mathrm{Tr}(\hat{\rho} \ln \hat{\rho})$ with the constraints $\mathrm{Tr}(\hat{\rho}) = 1$ and (1.C.14), and the fact that $\hat{\rho}$ must be Hermitian and non-negative. This can be done by introducing Lagrange multipliers λ_0, λ_i and requiring

$$-\mathrm{Tr}(\hat{\rho} \ln \hat{\rho}) - \sum_i \lambda_i \mathrm{Tr}(\hat{\rho} \hat{A}_i) - \lambda_0 \mathrm{Tr}(\hat{\rho}) \tag{1.C.15}$$

to be stationary. When $\hat{\rho}$ changes by $\delta \hat{\rho}$, (1.C.15) changes by

$$-\mathrm{Tr}\Big[\delta \hat{\rho} \Big(\ln \hat{\rho} + 1 + \sum_i \lambda_i \hat{A}_i + \lambda_0 \Big) \Big], \tag{1.C.16}$$

so that the stationary condition gives

$$\hat{\rho} = e^{-1-\lambda_0 - \sum_i \lambda_i \hat{A}_i}. \tag{1.C.17}$$

The density operator $\hat{\rho}$ is Hermitian, since the observables $\hat{A}_i$ are, and non-negative. Since $\mathrm{Tr}(\hat{\rho}) = 1$, we can eliminate the Lagrange multiplier λ_0. Defining the partition function

$$Z = e^{1+\lambda_0} = \mathrm{Tr}\, e^{-\sum_i \lambda_i \hat{A}_i}, \tag{1.C.18}$$

we can rewrite the density operator as

$$\hat{\rho} = \frac{1}{Z} e^{-\sum_i \lambda_i \hat{A}_i}. \tag{1.C.19}$$

This expression, which corresponds to the Boltzmann–Gibbs distribution, gives the general form of the density operator in thermodynamic equilibrium when the expectation values $\langle \hat{A}_i \rangle$ are given. To complete the derivation, we must show that Eq. (1.C.19) corresponds to a maximum of the statistical entropy (and not to a minimum). We write the entropy as

$$S(\hat{\rho}) = -k_B \langle \ln \hat{\rho} \rangle = k_B \ln Z + k_B \sum_i \lambda_i \langle \hat{A}_i \rangle \tag{1.C.20}$$

and use the fact that two non-negative operators $\hat{X}$ and $\hat{Y}$ satisfy the inequality

$$\mathrm{Tr}(\hat{X} \ln \hat{Y}) - \mathrm{Tr}(\hat{X} \ln \hat{X}) \leq \mathrm{Tr}(\hat{Y}) - \mathrm{Tr}(\hat{X}). \tag{1.C.21}$$

For $\hat{X} = \hat{\rho}'$ an arbitrary density operator and $\hat{Y} = \hat{\rho}$, this gives

$$\mathrm{Tr}(\hat{\rho}' \ln \hat{\rho}) - \mathrm{Tr}(\hat{\rho}' \ln \hat{\rho}') \leq 0 \tag{1.C.22}$$

and therefore

$$\begin{aligned} S(\hat{\rho}') &\leq -k_B \mathrm{Tr}(\hat{\rho}' \ln \hat{\rho}) \\ &\leq k_B \ln Z + k_B \sum_i \lambda_i \mathrm{Tr}(\hat{\rho}' \hat{A}_i) \\ &\leq S(\hat{\rho}), \end{aligned} \tag{1.C.23}$$

where we have used $\mathrm{Tr}(\hat{\rho}'\hat{A}_i) = \mathrm{Tr}(\hat{\rho}\hat{A}_i) = \langle \hat{A}_i \rangle$ to obtain the last inequality. Thus, the Boltzmann–Gibbs distribution provides a statistical entropy which is larger than that of any density operator satisfying the same constraints.

To determine the Lagrange multipliers λ_i as a function of the expectation values $\langle \hat{A}_i \rangle$, we use

$$\langle \hat{A}_i \rangle = \frac{1}{Z} \mathrm{Tr}\left(e^{-\sum_j \lambda_j \hat{A}_j} \hat{A}_i\right) = -\frac{\partial \ln Z}{\partial \lambda_i}. \tag{1.C.24}$$

In practice, we take the Lagrange multipliers λ_i as the parameters characterizing the equilibrium state rather than the expectation values $\langle \hat{A}_i \rangle$. (We discuss in the following their physical meaning when $\hat{A}_i$ corresponds to the Hamiltonian or the total number of particles.) The correlations and the statistical fluctuations of the constants of motion $\hat{A}_i$ can also be obtained from the derivatives of the partition function $Z(\lambda_i)$,

$$\langle \hat{A}_i \hat{A}_j \rangle - \langle \hat{A}_i \rangle \langle \hat{A}_j \rangle = \frac{\partial^2 \ln Z}{\partial \lambda_i \partial \lambda_j}. \tag{1.C.25}$$

Using (1.C.20) and (1.C.24), we can write the entropy as

$$S = k_B \ln Z - k_B \sum_i \lambda_i \frac{\partial \ln Z}{\partial \lambda_i}. \tag{1.C.26}$$

If the parameters λ_i characterizing the equilibrium change by an infinitesimal amount $d\lambda_i$,

$$d \ln Z = -\sum_i \langle \hat{A}_i \rangle d\lambda_i \tag{1.C.27}$$

and, using (1.C.26),

$$dS = k_B d\ln Z - k_B \sum_i \left(d\lambda_i \frac{\partial \ln Z}{\partial \lambda_i} - \lambda_i d\langle \hat{A}_i \rangle \right) = k_B \sum_i \lambda_i d\langle \hat{A}_i \rangle. \tag{1.C.28}$$

While Z is a function of λ_i, the natural variables for the entropy are the expectation values $\langle \hat{A}_i \rangle$, and

$$\frac{\partial S}{\partial \langle \hat{A}_i \rangle} = k_B \lambda_i. \tag{1.C.29}$$

Equations (1.C.24), (1.C.26) and (1.C.29) show that $k_B \ln Z(\lambda_i)$ and $S(\langle \hat{A}_i \rangle)$ are related by a Legendre transform.

1.C.3 *Statistical ensembles*

In this section, we discuss the three most common statistical ensembles for a quantum fluid. The main results are summarized in Table 1.3.

1.C.3.1 *Microcanonical ensemble*

In the microcanonical ensemble, the number of particles N and the volume $\mathcal{V}$ of the system are known exactly, while the energy is assumed to be between E and $E + \Delta E$. ΔE is a small energy interval as compared to the accuracy to which E is known, but large with respect to the typical energy level spacing. The density operator and the entropy are then given by (1.C.12) and (1.C.13) where $W = \mathcal{D}(E)\Delta E$ is the number of possible microstates, with

$$\mathcal{D}(E) = \mathrm{Tr}(E - \hat{H}) \tag{1.C.30}$$

the density of states of the system. Note that the dependence of $\hat{\rho}$ and S on $E, \mathcal{V}, N$ is not explicit but comes from that of W. The conjugate variables of E, N and $\mathcal{V}$

$$\left.\frac{\partial S}{\partial E}\right|_{N,\mathcal{V}} = \frac{1}{T}, \qquad \left.\frac{\partial S}{\partial N}\right|_{E,\mathcal{V}} = -\frac{\mu}{T}, \qquad \left.\frac{\partial S}{\partial \mathcal{V}}\right|_{E,N} = \frac{P}{T}, \tag{1.C.31}$$

are related to the temperature T, the chemical potential μ and the pressure P. These relations yield the usual thermodynamic definition of μ and P,

$$\mu = \left.\frac{\partial E}{\partial N}\right|_{S,\mathcal{V}} = -\frac{\left.\frac{\partial S}{\partial N}\right|_{E,\mathcal{V}}}{\left.\frac{\partial S}{\partial E}\right|_{N,\mathcal{V}}}, \qquad P = -\left.\frac{\partial E}{\partial \mathcal{V}}\right|_{S,N} = \frac{\left.\frac{\partial S}{\partial \mathcal{V}}\right|_{E,N}}{\left.\frac{\partial S}{\partial E}\right|_{N,\mathcal{V}}}. \tag{1.C.32}$$

1.C.3.2 *Canonical ensemble*

In the canonical ensemble, the number of particles and the volume are known exactly but the energy $E = \langle \hat{H} \rangle$ is known only on average. The corresponding Lagrange multiplier is denoted by β, and is related to the thermodynamic absolute temperature T by $\beta = 1/k_B T$

Table 1.3 Table of thermodynamic potentials.

Ensemble	Microcanonical	Canonical	Grand canonical
Variables	$E, N, \mathcal{V}$	$T, N, \mathcal{V}$	$T, \mu, \mathcal{V}$
Partition function	W	$Z_C = \mathrm{Tr}\, e^{-\beta \hat{H}}$	$Z_G = \mathrm{Tr}\, e^{-\beta(\hat{H} - \mu \hat{N})}$
Density operator	$\hat{\rho} = \sum_m \lvert m \rangle \frac{1}{W} \langle m \rvert$	$\hat{\rho} = \frac{1}{Z_C} e^{-\beta \hat{H}}$	$\hat{\rho} = \frac{1}{Z_G} e^{-\beta(\hat{H} - \mu \hat{N})}$
Thermodynamic	Entropy	Helmholtz free energy	Grand potential
potential	$S(E, N, \mathcal{V}) = k_B \ln W$	$F(T, N, \mathcal{V}) = -\frac{1}{\beta} \ln Z_C = E - TS$	$\Omega(T, \mu, \mathcal{V}) = -\frac{1}{\beta} \ln Z_G = E - TS - \mu N$
	$dS = \frac{1}{T}(dE - \mu dN + P d\mathcal{V})$	$dF = -SdT + \mu dN - P d\mathcal{V}$	$d\Omega = -SdT - N d\mu - P d\mathcal{V}$

(see the discussion on statistical and thermodynamic entropies on page 163). The density operator is given by

$$\hat{\rho} = \frac{1}{Z_C} e^{-\beta \hat{H}}, \tag{1.C.33}$$

where

$$Z_C(\beta, N, \mathcal{V}) = \operatorname{Tr} e^{-\beta \hat{H}} \tag{1.C.34}$$

is the canonical partition function. The entropy reads

$$S(E, N, \mathcal{V}) = k_B \ln Z_C + k_B \beta E \tag{1.C.35}$$

and satisfies

$$\left.\frac{\partial S}{\partial E}\right|_{N,\mathcal{V}} = k_B \beta \tag{1.C.36}$$

[see Eqs. (1.C.20) and (1.C.29)]. In the canonical ensemble, it is traditional to use as thermodynamic potential the Helmholtz free energy

$$F(T, N, \mathcal{V}) = E - TS = -\frac{1}{\beta} \ln Z_C(\beta, N, \mathcal{V}) \tag{1.C.37}$$

defined as the Legendre transform of the energy. Its differential reads

$$dF = -SdT + \mu dN - Pd\mathcal{V}, \tag{1.C.38}$$

where

$$S = -\left.\frac{\partial F}{\partial T}\right|_{N,\mathcal{V}}, \quad \mu = \left.\frac{\partial E}{\partial N}\right|_{S,\mathcal{V}} = \left.\frac{\partial F}{\partial N}\right|_{T,\mathcal{V}}, \quad P = -\left.\frac{\partial E}{\partial \mathcal{V}}\right|_{S,N} = -\left.\frac{\partial F}{\partial \mathcal{V}}\right|_{T,N}. \tag{1.C.39}$$

1.C.3.3 *Grand canonical ensemble*

In the grand canonical ensemble, both the energy $E = \langle \hat{H} \rangle$ and the number of particles $N = \langle \hat{N} \rangle$ are known on average. When the number of particles is not fixed, one must use a Hilbert space (the Fock space) which is constructed from the direct sum of the N-particle spaces (Sec. 1.2.1). The density operator

$$\hat{\rho} = \frac{1}{Z_G} e^{-\beta \hat{H} + \alpha \hat{N}} \tag{1.C.40}$$

is expressed in terms of the two Lagrange multipliers β and $-\alpha$, the grand canonical partition function being defined by

$$Z_G(\beta, \alpha, \mathcal{V}) = \operatorname{Tr} e^{-\beta \hat{H} + \alpha \hat{N}}. \tag{1.C.41}$$

The equilibrium entropy is given by

$$S(E, N, \mathcal{V}) = k_B \ln Z_G + k_B \beta E - k_B \alpha N \tag{1.C.42}$$

and satisfies

$$\left.\frac{\partial S}{\partial E}\right|_{N,\mathcal{V}} = k_B\beta, \qquad \left.\frac{\partial S}{\partial N}\right|_{E,\mathcal{V}} = -k_B\alpha. \tag{1.C.43}$$

From these equations, we can relate the chemical potential

$$\mu = \left.\frac{\partial E}{\partial N}\right|_{S,\mathcal{V}} = -\frac{\left.\frac{\partial S}{\partial N}\right|_{E,\mathcal{V}}}{\left.\frac{\partial S}{\partial E}\right|_{N,\mathcal{V}}} = \frac{\alpha}{\beta} \tag{1.C.44}$$

to the Lagrange multipliers, so that $\alpha = \beta\mu = \mu/k_BT$. In the grand canonical ensemble, it is traditional to use as thermodynamic potential the grand potential

$$\Omega(T,\mu,\mathcal{V}) = E - TS - \mu N = -\frac{1}{\beta}\ln Z_G(\beta,\alpha,\mathcal{V}) \tag{1.C.45}$$

defined as a double Legendre transform of the energy. Its differential reads

$$d\Omega = -SdT - Nd\mu - Pd\mathcal{V}, \tag{1.C.46}$$

where

$$S = -\left.\frac{\partial\Omega}{\partial T}\right|_{\mu,\mathcal{V}}, \qquad N = -\left.\frac{\partial\Omega}{\partial\mu}\right|_{T,\mathcal{V}}, \qquad P = -\left.\frac{\partial\Omega}{\partial\mathcal{V}}\right|_{T,\mu}. \tag{1.C.47}$$

The extensivity of $\ln Z_G$ and Ω then implies

$$\Omega(T,\mu,\mathcal{V}) = -P\mathcal{V}. \tag{1.C.48}$$

Apart from the sign, the grand potential per unit volume is equal to the pressure.

1.C.3.4 *Other examples of ensembles*

One can easily extend the preceding discussion to other physical observables. Let us, for example, consider a system of N interacting particles that can freely move in space; the expectation value of its total momentum $\hat{\mathbf{P}}$ can then be non-zero. Denoting by $-\beta\mathbf{v}$ the Lagrange multiplier associated with $\hat{\mathbf{P}}$ (we shall see that $\mathbf{v}$ is the velocity of the system), the partition function reads

$$Z(T,N,\mathcal{V},\mathbf{v}) = \mathrm{Tr}\, e^{-\beta(\hat{H}-\hat{\mathbf{P}}\cdot\mathbf{v})}, \qquad \hat{H} = \sum_i \frac{\hat{\mathbf{p}}_i^2}{2m} + \hat{V}, \tag{1.C.49}$$

where we assume that the interacting part of the Hamiltonian, $\hat{V}$, depends only on the particles' positions. Equation (1.C.49) shows that the Hamiltonian in the rest frame of the fluid (i.e. the moving frame) is $\hat{H} - \hat{\mathbf{P}}\cdot\mathbf{v}$ with $\hat{\mathbf{P}}$ the total momentum in the laboratory frame.

Using

$$\hat{H} - \hat{\mathbf{P}}\cdot\mathbf{v} = \sum_i \frac{(\hat{\mathbf{p}}_i - m\mathbf{v})^2}{2m} - \frac{1}{2}Nm\mathbf{v}^2 + \hat{V}, \tag{1.C.50}$$

one finds

$$Z(T,N,\mathcal{V},\mathbf{v}) = Z_0(T,N,\mathcal{V})e^{\beta\frac{N}{2}m\mathbf{v}^2}, \tag{1.C.51}$$

since we can transform $\hat{\mathbf{p}}_i - m\mathbf{v}$ back to $\hat{\mathbf{p}}_i$ by a unitary transformation which does not change the trace. Hence

$$F(T,N,\mathcal{V},\mathbf{v}) = -\frac{1}{\beta}\ln Z(T,N,\mathcal{V},\mathbf{v}) = F_0(T,N,\mathcal{V}) - \frac{N}{2}m\mathbf{v}^2, \tag{1.C.52}$$

where $F_0(T,N,\mathcal{V}) = -\frac{1}{\beta}\ln Z_0(T,N,\mathcal{V})$ is the free energy for $\mathbf{v}=0$. Equation (1.C.24) gives

$$\langle\hat{\mathbf{P}}\rangle = -\frac{\partial}{\partial\mathbf{v}}F(T,N,\mathcal{V},\mathbf{v})\bigg|_{T,N,\mathcal{V}} = Nm\mathbf{v}, \tag{1.C.53}$$

which shows that $\mathbf{v}$ is the velocity of the system. From (1.C.52), we deduce the chemical potential

$$\mu = \frac{\partial F}{\partial N}\bigg|_{T,\mathcal{V},\mathbf{v}} = \frac{\partial F_0}{\partial N}\bigg|_{T,\mathcal{V}} - \frac{1}{2}m\mathbf{v}^2 \equiv \mu_0 - \frac{1}{2}m\mathbf{v}^2 \tag{1.C.54}$$

and the grand potential $\Omega = F - \mu N$ in the grand canonical ensemble,

$$\begin{aligned}\Omega(T,\mu,\mathcal{V},\mathbf{v}) &= F_0(T,N,\mathcal{V}) - \frac{N}{2}m\mathbf{v}^2 - \left(\mu_0 - \frac{1}{2}m\mathbf{v}^2\right)N \\ &= \Omega_0(T,\mathcal{V},\mu_0) \\ &= \Omega_0\left(T,\mathcal{V},\mu + \frac{1}{2}m\mathbf{v}^2\right).\end{aligned} \tag{1.C.55}$$

Equation (1.C.53) gives

$$\langle\hat{\mathbf{P}}\rangle = -\frac{\partial}{\partial\mathbf{v}}\Omega(T,\mu,\mathcal{V},\mathbf{v})\bigg|_{T,\mu,\mathcal{V}} = Nm\mathbf{v}. \tag{1.C.56}$$

Another example is provided by a system which can freely rotate and whose expectation value of its angular momentum $\hat{\mathbf{J}}$ can therefore be non-zero. In that case, the Lagrange multiplier associated with $\hat{\mathbf{J}}$ is $-\beta\boldsymbol{\omega}$ and the equilibrium density operator can be written in the form

$$\frac{1}{Z}e^{-\beta(\hat{H} - \hat{\mathbf{J}}\cdot\boldsymbol{\omega})}, \tag{1.C.57}$$

where $\boldsymbol{\omega}$ is the angular velocity of the system.

1.C.3.5 *Thermodynamic limit — Equivalence between ensembles*

In the thermodynamic limit, the size of the system becomes very large with respect to the typical microscopic scales and can be considered to be infinite. Extensive variables are proportional to the volume. For instance, the entropy reads

$$S(E,N,\mathcal{V}) = \mathcal{V}s\left(\frac{E}{\mathcal{V}},\frac{N}{\mathcal{V}}\right), \tag{1.C.58}$$

where s reaches a finite limit for $\mathcal{V}\to\infty$ with $E/\mathcal{V}$, $N/\mathcal{V}$ or α,β constant.

Let us consider a system in equilibrium in the canonical ensemble. The density operator $\hat\rho$ can be written in terms of the eigenstates $\{|m\rangle,\epsilon_m\}$ of the Hamiltonian,

$$\hat\rho = \sum_m |m\rangle p_m \langle m|, \tag{1.C.59}$$

where

$$p_m = \frac{e^{-\beta\epsilon_m}}{Z_C} \tag{1.C.60}$$

is the probability for the system to be in the microstate $|m\rangle$. Thus, the system has a probability $p(E)dE$ to have an energy in the interval $[E,E+dE]$,

$$p(E) = \frac{e^{-\beta E}}{Z_C}\mathcal{D}(E), \tag{1.C.61}$$

where $\mathcal{D}(E)$ is the density of states [Eq. (1.C.30)]. $\mathcal{D}(E)$ is related to the microcanonical entropy $S(E)$ *via* $S(E) = k_B\ln[\mathcal{D}(E)\Delta E]$, where ΔE is the energy uncertainty in the microcanonical ensemble. This enables us to rewrite $p(E)$ as

$$p(E) = \frac{e^{-\beta E+k_B^{-1}S(E)}}{Z_C\Delta E} \propto e^{-\beta E+k_B^{-1}S(E)}. \tag{1.C.62}$$

In the thermodynamic limit, $\mathcal{V}\to\infty$, we thus obtain

$$p(E) \propto \exp\left\{-\mathcal{V}\left[\beta\frac{E}{\mathcal{V}} - \frac{S(E)}{k_B\mathcal{V}}\right]\right\}, \tag{1.C.63}$$

with both $S/\mathcal{V}$ and $E/\mathcal{V}$ taking a finite limit (independent of $\mathcal{V}$) when $\mathcal{V}\to\infty$. $p(E)$ is then strongly peaked about the most probable value $\bar E$ and can be approximated by a Gaussian probability distribution[100]

$$p(E) \propto \exp\left\{-\frac{1}{2k_B}\frac{\partial^2 S}{\partial E^2}\bigg|_{E=\bar E}(E-\bar E)^2\right\}. \tag{1.C.64}$$

$\bar E$ is defined by

$$\frac{1}{k_B}\frac{\partial S}{\partial E}\bigg|_{E=\bar E} - \beta = 0 \tag{1.C.65}$$

and therefore corresponds to the energy of the system in the microcanonical ensemble (recall that $S(E)$ is the entropy in the microcanonical ensemble) [Eq. (1.C.31)]. The statistical

[100]See Sec. 1.7.1 for a mathematical justification of this approximation.

fluctuations of the energy in the canonical ensemble are characterized by the variance

$$\Delta E^2 = \langle E^2 \rangle - \bar{E}^2 \sim \left(\frac{\partial^2 S}{\partial E^2}\bigg|_{E=\bar{E}} \right)^{-1}. \tag{1.C.66}$$

Since $S \propto \mathcal{V}$ and $E \propto \mathcal{V}$, we deduce $\partial^2 S/\partial E^2 \propto 1/\mathcal{V}$ and $\Delta E \propto \sqrt{\mathcal{V}}$. In the thermodynamic limit, the relative fluctuations

$$\frac{\Delta E}{\bar{E}} \propto \frac{1}{\sqrt{\mathcal{V}}} \tag{1.C.67}$$

of the energy become negligible and the microcanonical and canonical descriptions are equivalent.

Let us now consider a fluid in the grand canonical ensemble,

$$\hat{\rho} = \frac{1}{Z_G} e^{-\beta(\hat{H}-\mu\hat{N})} = \frac{1}{Z_G} \sum_m |m\rangle e^{-\beta(\epsilon_m - \mu N_m)} \langle m|, \tag{1.C.68}$$

where we have used the basis $\{|m\rangle\}$ which diagonalizes the Hamiltonian $\hat{H}$ and the particle number operator $\hat{N}$. The probability to have N particles in the system is

$$p(N) = \frac{1}{Z_G} \sum_{\substack{m \\ (N_m = N)}} e^{-\beta(\epsilon_m - \mu N_m)} = e^{\beta\mu N} \frac{Z_C}{Z_G} = \frac{1}{Z_G} e^{-\beta(F-\mu N)}, \tag{1.C.69}$$

where $Z_C(\beta, N, \mathcal{V})$ is the canonical partition function and $F(T, N, \mathcal{V}) = -\frac{1}{\beta} \ln Z_C$ the Helmholtz free energy. In the thermodynamic limit, N is large and can be considered as a continuous variable. Moreover, since F/N takes a finite limit when $N \to \infty$, we can factorize out a factor N in the argument of the exponential in (1.C.69), and approximate $p(N)$ by the Gaussian distribution function[100]

$$p(N) \propto \exp\left\{ -\beta \frac{\partial^2 F}{\partial N^2}\bigg|_{N=\bar{N}} (N - \bar{N})^2 \right\}, \tag{1.C.70}$$

where the most probable value $\bar{N}$ is defined by

$$\mu - \frac{\partial F}{\partial N}\bigg|_{N=\bar{N}} = 0. \tag{1.C.71}$$

$\bar{N}$ is therefore such that the canonical chemical potential $\partial F/\partial N$ [Eq. (1.C.39)] coincides with the grand canonical potential μ. The statistical fluctuations of N are given by

$$\Delta N^2 \sim \left(\beta \frac{\partial^2 F}{\partial N^2}\bigg|_{N=\bar{N}} \right)^{-1} \propto \bar{N}, \tag{1.C.72}$$

so that the relative fluctuations

$$\frac{\Delta N}{\bar{N}} \propto \frac{1}{\sqrt{\bar{N}}} \tag{1.C.73}$$

become negligible when $\bar{N} \to \infty$. The canonical and grand canonical descriptions are equivalent in the thermodynamic limit.

Statistical entropy *vs* thermodynamic entropy. For an isolated system, the Liouville–von Neumann equation (1.C.4) ensures that the energy is conserved,

$$\frac{dE}{dt} = \frac{1}{i\hbar}\langle[\hat{H}, \hat{H}]\rangle = 0. \tag{1.C.74}$$

If the system is not isolated, its energy varies by an amount

$$dE = \mathrm{Tr}(d\hat{\rho}\hat{H}) + \mathrm{Tr}(\hat{\rho}d\hat{H}) \tag{1.C.75}$$

in an infinitesimal transformation where the Hamiltonian and the density operator vary by $d\hat{H}$ and $d\hat{\rho}$, respectively. When $\hat{\rho}$ commutes with the Hamiltonian, it can be written as

$$\hat{\rho} = \sum_m |m\rangle p_m \langle m| \tag{1.C.76}$$

in the basis $\{|m\rangle, \epsilon_m\}$ of the eigenstates of $\hat{H}$, where p_m is the probability of the microstate $|m\rangle$. The energy of the equilibrium state is then $E = \sum_m p_m \epsilon_m$ and its variation (1.C.75) reads

$$dE = \sum_m (dp_m \epsilon_m + p_m d\epsilon_m). \tag{1.C.77}$$

The quantity

$$\delta W = \sum_m p_m d\epsilon_m = \mathrm{Tr}(\hat{\rho}d\hat{H}) \tag{1.C.78}$$

corresponds to the work received by the system in the infinitesimal transformation. On the other hand, the energy variation

$$\delta Q = \sum_m dp_m \epsilon_m = \mathrm{Tr}(d\hat{\rho}\hat{H}) \tag{1.C.79}$$

due to the change in the probabilities p_m is the heat transferred to the system.

Let us consider a system in equilibrium in the canonical ensemble. In a quasi-static transformation, the system remains in equilibrium and its statistical entropy changes by[101]

$$\begin{aligned} dS &= -k_B \mathrm{Tr}(d\hat{\rho}\ln\hat{\rho}) \\ &= k_B \mathrm{Tr}[d\hat{\rho}(\ln Z + \beta\hat{H})] \\ &= k_B \beta \mathrm{Tr}(d\hat{\rho}\hat{H}) \\ &= k_B \beta \delta Q. \end{aligned} \tag{1.C.80}$$

This result must be compared with Clausius' equation $\delta Q = TdS_{\mathrm{th}}$ where S_{th} is the thermodynamic entropy at equilibrium. If we choose $k_B\beta = 1/T$ (with T the thermodynamic absolute temperature), then $dS = dS_{\mathrm{th}}$, which implies that the statistical entropy S coincides with the thermodynamic entropy up to an additive constant [28].

[101] Since the density operator satisfies $\mathrm{Tr}(\hat{\rho}) = 1$, $\mathrm{Tr}(d\hat{\rho}) = 0$.

1.C.4 *The classical limit*

In the Hamiltonian formalism, a classical microstate is represented by a point $\{q_i, p_i\}$ in phase space. A classical macrostate is characterized by the probability law for the classical microstates,

$$\rho_N(q_1, p_1, \dots, q_N, p_N) d\tau_N, \tag{1.C.81}$$

where ρ_N is the phase density (or density in phase) and $d\tau_N$ the volume element in phase space. ρ_N is real, positive and normalized. A classical observable A then becomes a random variable and its mean value is given by

$$\langle A \rangle = \int d\tau_N \, \rho_N(q_1, p_1, \dots, q_N, p_N) A(q_1, p_1, \dots, q_N, p_N). \tag{1.C.82}$$

The normalization of $d\tau_N$ does not matter, since ρ_N is normalized. However, when one derives (1.C.82) as a limiting case of quantum statistical physics, one finds

$$d\tau_N = \frac{1}{N!} \prod_{i=1}^{N} \frac{d^3 q_i d^3 p_i}{h^3} \tag{1.C.83}$$

for a three-dimensional fluid of N particles. The factor h^{-3N} makes $d\tau_N$ dimensionless, while the factor $1/N!$ is associated with the indistinguishability of the particles. This suggests that even in classical mechanics, two points in phase space which differ from one another by a permutation of the indices of the particles represent the same physical microstate.[102]

The time evolution of the quantum density operator was derived from the Schrödinger equation and the conservation of the probability q_λ of the microstate $|\psi_\lambda\rangle$. Similarly, the probability $\rho_N d\tau_N$ is conserved along the classical trajectory defined by Hamilton's equations of motion (1.A.21). Moreover, the volume of a region in phase space remains constant when one follows the region during the temporal evolution (Liouville's theorem).[103] It follows that both $d\tau_N$ and ρ_N are conserved along the classical trajectories. From

$$\rho_N(q_i + \dot{q}_i dt, p_i + \dot{p}_i dt, t + dt) = \rho_N(q_i, p_i, t) \qquad (dt \to 0), \tag{1.C.84}$$

we then deduce

$$\frac{\partial \rho_N}{\partial t} + \sum_i \left(\frac{\partial \rho_N}{\partial q_i} \dot{q}_i + \frac{\partial \rho_N}{\partial p_i} \dot{p}_i \right) = 0, \tag{1.C.85}$$

i.e.

$$\frac{\partial \rho_N}{\partial t} = [H, \rho_N] \tag{1.C.86}$$

(Liouville's equation), where [,] denotes the Poisson bracket.

---

[102] For arbitrary conjugate variables (q_i, p_i), $N!$ is replaced by a factor $\mathcal{S}$ equal to the number of configurations in phase space which describe the same situation if we take into account the indistinguishability of the particles.

[103] Liouville's theorem follows from the fact that the infinitesimal transformation $q_i \to q_i + \dot{q}_i dt$, $p_i \to p_i + \dot{p}_i dt$ has a unit Jacobian, which is a direct consequence of Hamilton's equations of motion (1.A.21).

The classical statistical entropy is defined as

$$S = -k_B \sum_N \int d\tau_N \, \rho_N \ln \rho_N, \tag{1.C.87}$$

which can be shown to be the classical limit of (1.C.8). If the number of particles in the system is well defined, the sum over N in (1.C.87) is absent. The phase density of the system in equilibrium is obtained from the maximum entropy principle. This leads to the classical Boltzmann–Gibbs distribution

$$\rho_N = \frac{1}{Z} e^{-\sum_i \lambda_i A_i}, \tag{1.C.88}$$

where the partition function is defined by

$$Z = \sum_N \int d\tau_N \, e^{-\sum_i \lambda_i A_i}. \tag{1.C.89}$$

The observables A_i in (1.C.88) and (1.C.89) correspond to the (known) constants of motion and the λ_i's are Lagrange multipliers introduced to enforce the constraints $\langle A_i \rangle = \text{const}$. For example, the classical canonical ensemble has a phase density

$$\rho_N = \frac{1}{Z_C} e^{-\beta H_N} \tag{1.C.90}$$

with a partition function

$$Z_C(\beta, N, \mathcal{V}) = \int d\tau_N \, e^{-\beta H_N}, \tag{1.C.91}$$

while the classical grand canonical ensemble is defined by

$$\begin{aligned} \rho_N &= \frac{1}{Z_G} e^{-\beta H_N + \alpha N}, \\ Z_G(\beta, \alpha, \mathcal{V}) &= \sum_N \int d\tau_N \, e^{-\beta H_N + \alpha N} = \sum_N e^{\alpha N} Z_C(\beta, N, \mathcal{V}), \end{aligned} \tag{1.C.92}$$

where $\alpha = \beta\mu = \mu/k_B T$.

Classical ideal gas. For N non-interacting classical particles in three dimensions with Hamiltonian $H = \sum_{i=1}^N \mathbf{p}_i^2/2m$, the canonical partition function reads

$$Z_C(\beta, N, \mathcal{V}) = \frac{1}{N!} \int \prod_{i=1}^N \frac{d^3 q_i d^3 p_i}{h^3} \exp\left\{-\beta \sum_{i=1}^N \frac{\mathbf{p}_i^2}{2m}\right\} = \frac{z^N}{N!}, \tag{1.C.93}$$

where

$$z = \int \frac{d^3 q d^3 p}{h^3} e^{-\beta \mathbf{p}^2/2m} = \frac{\mathcal{V}}{\lambda_T^3} \tag{1.C.94}$$

is the partition function of a single particle and $\lambda_T = (2\pi\hbar^2\beta/m)^{1/2}$ the thermal de Broglie wavelength. We then deduce the free energy $F(T, N, \mathcal{V}) = -\frac{1}{\beta} \ln Z_C(\beta, N, \mathcal{V})$ and the pressure

$$P = -\left.\frac{\partial F}{\partial \mathcal{V}}\right|_{T,N} = \frac{N}{\beta \mathcal{V}} \tag{1.C.95}$$

and recover the equation of state $P\mathcal{V} = N k_B T$ of the classical ideal gas.

The grand canonical partition function is simply

$$Z_G(\beta,\alpha,\mathcal{V}) = \sum_{N=0}^{\infty} \frac{z^N}{N!} e^{\alpha N} = \exp(ze^{\alpha}). \tag{1.C.96}$$

This gives the grand potential

$$\Omega(T,\mu,\mathcal{V}) = -\frac{\mathcal{V}}{\beta\lambda_T^3} e^{\beta\mu} \tag{1.C.97}$$

and the mean number of particles

$$N = -\left.\frac{\partial\Omega}{\partial\mu}\right|_{T,\mathcal{V}} = -\beta\Omega, \tag{1.C.98}$$

and in turn the equation of state $P\mathcal{V} = -\Omega = Nk_BT$.

1.C.5 *Non-interacting quantum gases*

In this section, we consider a system of non-interacting quantum particles (bosons or fermions) in equilibrium in the grand canonical ensemble. We denote by $\{|\alpha\rangle, \epsilon_\alpha\}$ an orthonormal basis of one-body states which diagonalizes the one-particle Hamiltonian. The many-particle Hilbert space $\mathcal{H}$ (Fock space) is obtained from the direct sum $\oplus_{N=0}^{\infty}\mathcal{H}_N$ of the N-particle Hilbert spaces $\mathcal{H}_N$. An orthonormal basis in the Fock space is formed by the states[104]

$$|n_{\alpha_1}\cdots n_{\alpha_i}\cdots\rangle \tag{1.C.99}$$

obtained by putting n_{α_i} particles in the one-body state $|\alpha_i\rangle$ and symmetrizing or antisymmetrizing the many-body states according to the quantum statistics of the particles. These states are eigenstates of the Hamiltonian and the total number of particles operator,

$$\begin{aligned} \hat{H}|n_{\alpha_1}\cdots n_{\alpha_i}\cdots\rangle &= \left(\sum_j n_{\alpha_j}\epsilon_{\alpha_j}\right)|n_{\alpha_1}\cdots n_{\alpha_i}\cdots\rangle, \\ \hat{N}|n_{\alpha_1}\cdots n_{\alpha_i}\cdots\rangle &= \left(\sum_j n_{\alpha_j}\right)|n_{\alpha_1}\cdots n_{\alpha_i}\cdots\rangle. \end{aligned} \tag{1.C.100}$$

While for bosons there is no restriction on the occupation numbers n_α, for fermions they are constrained by the Pauli principle and can take only the value 0 or 1. Using the basis $\{|n_{\alpha_1}\cdots n_{\alpha_i}\cdots\rangle\}$, we easily calculate the grand canonical partition function,

$$Z = \operatorname{Tr} e^{-\beta(\hat{N}-\mu\hat{N})} = \sum_{\{n_\alpha\}} e^{-\beta\sum_\alpha n_\alpha\xi_\alpha}, \tag{1.C.101}$$

[104]See Sec. 1.2.1 for more details.

where $\xi_\alpha = \epsilon_\alpha - \mu$. From (1.C.101), we obtain

$$Z = \sum_{n_{\alpha_1}, n_{\alpha_2}, \cdots} e^{-\beta n_{\alpha_1}\xi_{\alpha_1}} e^{-\beta n_{\alpha_2}\xi_{\alpha_2}} \cdots = \prod_\alpha \sum_{n_\alpha} e^{-\beta n_\alpha \xi_\alpha} = \prod_\alpha \left(1 - \zeta e^{-\beta\xi_\alpha}\right)^{-\zeta}, \quad (1.C.102)$$

where $\zeta = 1$ for bosons and $\zeta = -1$ for fermions. We deduce that the grand potential reads

$$\Omega = \frac{\zeta}{\beta} \sum_\alpha \ln\left(1 - \zeta e^{-\beta\xi_\alpha}\right). \quad (1.C.103)$$

The mean number of particles is given by

$$N = -\left.\frac{\partial\Omega}{\partial\mu}\right|_T = \sum_\alpha \frac{1}{e^{\beta\xi_\alpha} - \zeta} \equiv \sum_\alpha n_\zeta(\xi_\alpha), \quad (1.C.104)$$

where $n_+(\xi_\alpha) = n_B(\xi_\alpha)$ (Bose–Einstein distribution function) and $n_-(\xi_\alpha) = n_F(\xi_\alpha)$ (Fermi–Dirac distribution function). From the grand potential, one can also deduce the mean energy

$$E = \left.\frac{\partial\beta\Omega}{\partial\beta}\right|_\mu + \mu N = \sum_\alpha \epsilon_\alpha n_\zeta(\xi_\alpha) \quad (1.C.105)$$

and the entropy

$$\begin{aligned} S = -\left.\frac{\partial\Omega}{\partial T}\right|_\mu &= \frac{1}{T}\sum_\alpha \left[\xi_\alpha n_\zeta(\xi_\alpha) - \frac{\zeta}{\beta}\ln\left(1 - \zeta e^{-\beta\xi_\alpha}\right)\right] \\ &= -k_B \sum_\alpha \left\{n_\zeta(\xi_\alpha)\ln n_\zeta(\xi_\alpha) - \zeta[1 + \zeta n_\zeta(\xi_\alpha)]\ln[1 + \zeta n_\zeta(\xi_\alpha)]\right\}. \end{aligned} \quad (1.C.106)$$

Quantum ideal gas. The grand potential of non-interacting spin-1/2 fermions read

$$\Omega = -\frac{2}{\beta}\sum_{\mathbf{k}} \ln\left(1 + e^{-\beta(\epsilon_{\mathbf{k}} - \mu)}\right) = -2\frac{\mathcal{V}}{\beta}\int_0^\infty d\epsilon\, \rho(\epsilon) \ln\left(1 + e^{-\beta(\epsilon - \mu)}\right) \quad (1.C.107)$$

(a factor 2 comes from the spin degeneracy), where $\epsilon_{\mathbf{k}} = \mathbf{k}^2/2m$ and

$$\rho(\epsilon) = \int_{\mathbf{k}} \delta(\epsilon - \epsilon_{\mathbf{k}}) = \frac{(2m)^{3/2}}{4\pi^2}\sqrt{\epsilon} \quad (1.C.108)$$

is the density of states per spin and unit volume (we assume a three-dimensional system). An elementary calculation gives the pressure $P = -\Omega/\mathcal{V}$ and the density $n = N/\mathcal{V} = \partial P/\partial\mu$,

$$P = \frac{2}{\beta\lambda_T^3} f_{5/2}(z), \qquad n = \frac{2}{\lambda_T^3} f_{3/2}(z), \quad (1.C.109)$$

where $z = e^{\beta\mu}$, λ_T is the thermal de Broglie wavelength defined after (1.C.94) and

$$f_p(z) = \frac{1}{\Gamma(p)}\int_0^\infty dx \frac{x^{p-1}}{z^{-1}e^x + 1}. \quad (1.C.110)$$

In the low-temperature dilute limit, $\mu < 0$ and $T \ll |\mu|$ (or, equivalently, $n\lambda_T^3 \ll 1$), using $f_p(z) \to z$ for $z \to 0$ one recovers the equation of state $P = nT$ of the classical ideal gas.

Similarly, for non-interacting spin-zero bosons (with chemical potential $\mu < 0$), one has

$$\Omega = \frac{\mathcal{V}}{\beta} \int_0^\infty d\epsilon\, \rho(\epsilon) \ln\left(1 - e^{-\beta(\epsilon-\mu)}\right) \tag{1.C.111}$$

and one finds

$$P = \frac{1}{\beta\lambda_T^3} g_{5/2}(z), \qquad n = \frac{1}{\lambda_T^3} g_{3/2}(z), \tag{1.C.112}$$

where

$$g_p(z) = \frac{1}{\Gamma(p)} \int_0^\infty dx \frac{x^{p-1}}{z^{-1}e^x - 1}. \tag{1.C.113}$$

Again, noting that $g_p(z) \to z$ for $z \to 0$, one recovers the ideal classical gas in the low-temperature dilute limit $\mu < 0$ and $T \ll |\mu|$. On the other hand, we see that the equation for the density cannot be satisfied at low temperatures when $g_{3/2}(z \to 1^-)/\lambda_T^3$ becomes smaller than n, i.e. below the temperature

$$T_c = \frac{2\pi}{m} \left(\frac{n}{\zeta(3/2)}\right)^{2/3} \tag{1.C.114}$$

(we use $g_{3/2}(1) = \zeta(3/2)$). This signals Bose–Einstein condensation (BEC), that is the occupation of the one-body zero-momentum state by a macroscopic number $N_0 = n_0\mathcal{V}$ of particles. The equation for the density becomes (with $\mu \to 0^-$)

$$n = n_0 + \frac{1}{\mathcal{V}} \sum_{\mathbf{k}\neq 0} n_B(\epsilon_{\mathbf{k}}) = n_0 + \frac{\zeta(3/2)}{\lambda_T^3} = n_0 + n\left(\frac{T}{T_c}\right)^{3/2}, \tag{1.C.115}$$

which gives the condensate density

$$n_0 = n\left[1 - \left(\frac{T}{T_c}\right)^{3/2}\right]. \tag{1.C.116}$$

Whereas the non-interacting fermion gas provides us in some cases with a good starting point for understanding the effect of interactions (e.g. in the case of a Fermi liquid, see Chapter 4), the non-interacting boson system is somewhat pathological and perturbation theory about the non-interacting limit is ill-behaved.[105]

1.D Functionals and Functional Derivatives

A functional is a function that maps functions to numbers. Its argument is conventionally written in square brackets rather than parentheses. The derivative of a functional $F[\phi]$ is defined in such a way that if one discretizes[106] the function $\phi(\mathbf{x})$ ($\mathbf{x}$ denotes a d-dimensional

[105] For instance, the non-interacting Bose gas at zero temperature has an infinite compressibility, but any non-zero interactions make the compressibility finite (see Chapter 7).

[106] The continuous variable $\mathbf{x}$ then becomes a discrete variable defined on a d-dimensional lattice with lattice spacing a. The continuous formulation is recovered when a tends to zero. For example, for a one-dimensional variable, $x \to x_i = ia$ and $\phi(x) \to \phi(x_i) \equiv \phi_i$, with i integer.

variable), one recovers the usual rules for the derivation of ordinary functions. We then impose

$$\frac{\delta\phi(\mathbf{x})}{\delta\phi(\mathbf{y})} = \delta(\mathbf{x}-\mathbf{y}), \tag{1.D.1}$$

which is the analog of $\partial\phi_i/\partial\phi_j = \delta_{i,j}$, and the standard rules

$$\begin{aligned}
\frac{\delta}{\delta\phi(\mathbf{x})}\left(F_1[\phi]+F_2[\phi]\right) &= \frac{\delta F_1[\phi]}{\delta\phi(\mathbf{x})} + \frac{\delta F_2[\phi]}{\delta\phi(\mathbf{x})},\\
\frac{\delta}{\delta\phi(\mathbf{x})}F_1[\phi]F_2[\phi] &= \frac{\delta F_1[\phi]}{\delta\phi(\mathbf{x})}F_2[\phi] + F_1[\phi]\frac{\delta F_2[\phi]}{\delta\phi(\mathbf{x})},\\
\frac{\delta}{\delta\phi(\mathbf{x})}f(\phi(\mathbf{y})) &= f'(\phi(\mathbf{y}))\frac{\delta\phi(\mathbf{y})}{\delta\phi(\mathbf{x})} = f'(\phi(\mathbf{y}))\delta(\mathbf{x}-\mathbf{y}).
\end{aligned} \tag{1.D.2}$$

Equation (1.D.1) also implies

$$\frac{\delta}{\delta\phi(\mathbf{x})}\boldsymbol{\nabla}\phi(\mathbf{y}) = \boldsymbol{\nabla}\frac{\delta\phi(\mathbf{y})}{\delta\phi(\mathbf{x})} = \boldsymbol{\nabla}\delta(\mathbf{x}-\mathbf{y}). \tag{1.D.3}$$

For a functional $F[\phi]$ that can be expressed as an integral of a function of $\phi(\mathbf{x})$ and $\boldsymbol{\nabla}\phi(\mathbf{x})$,

$$F[\phi] = \int d^dx\, f(\phi(\mathbf{x}), \boldsymbol{\nabla}\phi(\mathbf{x})), \tag{1.D.4}$$

one then finds

$$\begin{aligned}
\frac{\delta F[\phi]}{\delta\phi(\mathbf{x})} &= \int d^dy \left(\frac{\partial f}{\partial\phi(\mathbf{y})}\frac{\delta\phi(\mathbf{y})}{\delta\phi(\mathbf{x})} + \frac{\partial f}{\partial(\boldsymbol{\nabla}\phi(\mathbf{y}))}\cdot\frac{\delta(\boldsymbol{\nabla}\phi(\mathbf{y}))}{\delta\phi(\mathbf{x})}\right)\\
&= \int d^dy \left(\frac{\partial f}{\partial\phi(\mathbf{y})}\delta(\mathbf{x}-\mathbf{y}) + \frac{\partial f}{\partial(\boldsymbol{\nabla}\phi(\mathbf{y}))}\cdot\boldsymbol{\nabla}\delta(\mathbf{x}-\mathbf{y})\right)\\
&= \frac{\partial f}{\partial\phi(\mathbf{x})} - \boldsymbol{\nabla}\cdot\frac{\partial f}{\partial(\boldsymbol{\nabla}\phi(\mathbf{x}))},
\end{aligned} \tag{1.D.5}$$

where the last result is obtained by integrating the second term by parts assuming that the field $\phi(\mathbf{x})$ vanishes at the boundaries of its domain of definition or satisfies periodic boundary conditions.

The functional $F[\phi]$ and its functional derivative can be expanded as

$$\begin{aligned}
F[\phi] &= F[0] + \sum_{n=1}^{\infty}\frac{1}{n!}\int d^dx_1\cdots d^dx_n\, F^{(n)}(\mathbf{x}_1,\dots,\mathbf{x}_n)\phi(\mathbf{x}_1)\cdots\phi(\mathbf{x}_n),\\
\frac{\delta F[\phi]}{\delta\phi(\mathbf{x})} &= F^{(1)}(x) + \sum_{n=2}^{\infty}\frac{1}{(n-1)!}\int d^dx_1\cdots d^dx_{n-1}\, F^{(n)}(\mathbf{x}_1,\dots,\mathbf{x}_{n-1},\mathbf{x})\\
&\qquad\times\phi(\mathbf{x}_1)\cdots\phi(\mathbf{x}_{n-1}),
\end{aligned} \tag{1.D.6}$$

where the function $F^{(n)}(\mathbf{x}_1,\dots,\mathbf{x}_n)$ is symmetric under the exchange $\mathbf{x}_i \leftrightarrow \mathbf{x}_j$. We deduce

$$\begin{aligned}
F[\phi+u] = F[0] &+ \sum_{n=1}^{\infty}\frac{1}{n!}\int d^dx_1\cdots d^dx_n\, F^{(n)}(\mathbf{x}_1,\dots,\mathbf{x}_n)\\
&\times\left(\phi(\mathbf{x}_1)+u(\mathbf{x}_1)\right)\cdots\left(\phi(\mathbf{x}_n)+u(\mathbf{x}_n)\right)
\end{aligned}$$

$$
\begin{aligned}
&= F[\phi] + \int dx\, F^{(1)}(x)u(x) + \sum_{n=2}^{\infty} \frac{1}{(n-1)!} \int d^d x_1 \cdots d^d x_{n-1} \\
&\qquad \times F^{(n)}(\mathbf{x}_1, \dots, \mathbf{x}_{n-1}, \mathbf{x})\phi(\mathbf{x}_1) \cdots \phi(\mathbf{x}_{n-1})u(\mathbf{x}) + \mathcal{O}(u^2) \\
&= F[\phi] + \int d^d x\, \frac{\delta F[\phi]}{\delta \phi(\mathbf{x})} u(\mathbf{x}) + \mathcal{O}(u^2).
\end{aligned}
\tag{1.D.7}
$$

More generally, one finds

$$
F[\phi + u] = F[\phi] + \sum_{n=1}^{\infty} \frac{1}{n!} \int d^d x_1 \cdots d^d x_n\, \frac{\delta^n F[\phi]}{\delta \phi(\mathbf{x}_1) \cdots \delta \phi(\mathbf{x}_n)} u(\mathbf{x}_1) \cdots u(\mathbf{x}_n). \tag{1.D.8}
$$

1.D.1 *Grassmann variables*

Similar rules can be defined for a functional $F[\psi^*, \psi]$ of a Grassmannian function $\psi(\mathbf{x})$ and its complex conjugate $\psi^*(\mathbf{x})$, provided the anticommuting properties of $\psi(\mathbf{x})$ and $\psi^*(\mathbf{x})$ are properly taken into account. For instance,

$$
\begin{aligned}
\frac{\delta}{\delta \psi(\mathbf{x})} \int d^d y\, \psi^*(\mathbf{y})\psi(\mathbf{y}) &= -\psi^*(\mathbf{x}), \\
\frac{\delta}{\delta \psi^*(\mathbf{x})} \int d^d y\, \psi^*(\mathbf{y})\psi(\mathbf{y}) &= \psi(\mathbf{x}).
\end{aligned}
\tag{1.D.9}
$$

In particular, one has

$$
\begin{aligned}
F[\psi^* + u^*, \psi + u] &= F[\psi^*, \psi] \\
&\quad + \int d^d x \left(u^*(\mathbf{x}) \frac{\delta F[\psi^*, \psi]}{\delta \psi^*(\mathbf{x})} + u(\mathbf{x}) \frac{\delta F[\psi^*, \psi]}{\delta \psi(\mathbf{x})} \right) + \mathcal{O}(u^2),
\end{aligned}
\tag{1.D.10}
$$

where the Grassmann numbers $u^*(\mathbf{x})$ and $u(\mathbf{x})$ must appear on the left of the functional derivatives in agreement with the chain rule for derivation with anticommuting variables [Eq. (1.240)].

1.E Gaussian Integrals and Wick's Theorem

1.E.1 *Real variables*

We consider the multidimensional integral

$$
\begin{aligned}
Z(J) &= \int_{-\infty}^{\infty} dx_1 \cdots dx_n\, e^{-\frac{1}{2}\sum_{i,j} x_i A_{ij} x_j + \sum_i J_i x_i} \\
&= \int_{-\infty}^{\infty} dx_1 \cdots dx_n\, e^{-\frac{1}{2} x^T A x + J^T x},
\end{aligned}
\tag{1.E.1}
$$

where x denotes the column vector $(x_1, \dots, x_n)^T$. The n-dimensional real matrix A is assumed to be symmetric and positive definite. It can be diagonalized by an orthogonal transformation: $A = O^T D O$, with O an orthogonal matrix ($O^T O = O O^T = 1$) and D

a diagonal matrix with non-negative elements ($D_{ii} > 0$). With the change of variables $y = O(x - A^{-1}J)$ (Jacobian $|\det 0| = 1$), we obtain

$$-\frac{1}{2}x^T Ax + J^T x = -\frac{1}{2}y^T Dy + \frac{1}{2}J^T A^{-1}J \tag{1.E.2}$$

and $Z(J)$ reduces to a product of Gaussian integrals,[107]

$$\begin{aligned} Z(J) &= e^{\frac{1}{2}J^T A^{-1}J} \prod_{i=1}^{n} \int_{-\infty}^{\infty} dy\, e^{-\frac{1}{2}y^2 D_{ii}} \\ &= e^{\frac{1}{2}J^T A^{-1}J} \prod_{i=1}^{n} \left(\frac{2\pi}{D_{ii}}\right)^{1/2} \\ &= e^{\frac{1}{2}J^T A^{-1}J} (2\pi)^{n/2} (\det A)^{-1/2}, \end{aligned} \tag{1.E.3}$$

using $\det D = \det A$.

1.E.1.1 *Wick's theorem*

The mean value

$$\langle x_{i_1} \cdots x_{i_k} \rangle = \frac{\int_{-\infty}^{\infty} dx_1 \cdots dx_n\, x_{i_1} \cdots x_{i_k}\, e^{-\frac{1}{2}x^T Ax}}{\int_{-\infty}^{\infty} dx_1 \cdots dx_n\, e^{-\frac{1}{2}x^T Ax}} \tag{1.E.4}$$

can be obtained from the generating function (1.E.1),

$$\begin{aligned} \langle x_{i_1} \cdots x_{i_k} \rangle &= \frac{1}{Z(0)} \frac{\partial}{\partial J_{i_1}} \cdots \frac{\partial}{\partial J_{i_k}} Z(J) \Big|_{J=0} \\ &= \frac{\partial}{\partial J_{i_1}} \cdots \frac{\partial}{\partial J_{i_k}} e^{\frac{1}{2}J^T A^{-1}J} \Big|_{J=0}. \end{aligned} \tag{1.E.5}$$

We deduce

$$\begin{aligned} \langle x_{i_1} x_{i_2} \rangle &= A^{-1}_{i_1 i_2}, \\ \langle x_{i_1} x_{i_2} x_{i_3} x_{i_4} \rangle &= A^{-1}_{i_1 i_2} A^{-1}_{i_3 i_4} + A^{-1}_{i_1 i_3} A^{-1}_{i_2 i_4} + A^{-1}_{i_1 i_4} A^{-1}_{i_2 i_3}, \end{aligned} \tag{1.E.6}$$

and more generally

$$\begin{aligned} \langle x_{i_1} \cdots x_{i_{2k}} \rangle &= \sum_{\substack{\text{all possible pairings} \\ P \text{ of } \{i_1,\dots,i_{2k}\}}} \langle x_{i_{P(1)}} x_{i_{P(2)}} \rangle \cdots \langle x_{i_{P(2k-1)}} x_{i_{P(2k)}} \rangle \\ &= \sum_{\substack{\text{all possible pairings} \\ P \text{ of } \{i_1,\dots,i_{2k}\}}} A^{-1}_{i_{P(1)} i_{P(2)}} \cdots A^{-1}_{i_{P(2k-1)} i_{P(2k)}}. \end{aligned} \tag{1.E.7}$$

This result holds for any Gaussian probability distribution $e^{-\frac{1}{2}x^T Ax}$ with zero mean $\langle x_i \rangle = 0$. In field theory, it is known as Wick's theorem and underlies perturbation theory (Sec. 1.5).

[107]Recall that $\int_{-\infty}^{\infty} dy\, e^{-\frac{a}{2}y^2} = \sqrt{2\pi/a}$ for $a > 0$.

1.E.2 *Complex variables*

Now, we consider the multidimensional integral

$$\begin{aligned} Z(J^*,J) &= \int_{-\infty}^{\infty} \prod_{i=1}^{n} \frac{dz_i^* dz_i}{2i\pi} e^{-\sum_{i,j} z_i^* A_{ij} z_j + \sum_i (J_i^* z_i + \text{c.c.})} \\ &= \int_{-\infty}^{\infty} \prod_{i=1}^{n} \frac{dz_i^* dz_i}{2i\pi} e^{-z^\dagger A z + (J^\dagger z + \text{c.c.})} \end{aligned} \tag{1.E.8}$$

over the complex variables z_i and z_i^* $(i = 1, \dots, n)$, where

$$\frac{dz_i^* dz_i}{2i\pi} = \frac{1}{\pi} d\Re(z_i) d\Im(z_i) \tag{1.E.9}$$

and $z^\dagger = (z_1^*, \dots, z_n^*)$. Let us first assume the complex matrix A to be positive definite Hermitian ($A^\dagger = A$). We can then proceed as in the real case. A can be diagonalized by a unitary transformation: $A = U^\dagger D U$ with U a unitary matrix ($UU^\dagger = U^\dagger U = 1$) and D a diagonal matrix with non-negative real elements ($D_{ii} > 0$). The change of variables

$$z' = U(z - A^{-1}J), \qquad z'^\dagger = (z^\dagger - J^\dagger A^{-1}) U^\dagger \tag{1.E.10}$$

has a Jacobian $|\det U|^2 = 1$, and leads to[108]

$$Z(J^*,J) = e^{J^\dagger A^{-1} J} \int_{-\infty}^{\infty} \prod_{i=1}^{n} \frac{dz_i'^* dz_i'}{2i\pi} e^{-z'^\dagger D z'} = e^{J^\dagger A^{-1} J} (\det A)^{-1} \tag{1.E.11}$$

(using $\det D = \det A$). It can be shown that (1.E.11) holds for any complex matrix with a positive definite Hermitian part $\frac{1}{2}(A + A^\dagger)$.[109]

1.E.2.1 *Wick's theorem*

The average value of a monomial can be obtained from the generating function $Z(J^*,J)$,

$$\begin{aligned} \langle z_{i_1} z_{i_1'}^* \cdots z_{i_k} z_{i_k'}^* \rangle &= \frac{\int_{-\infty}^{\infty} \prod_{i=1}^{n} \frac{dz_i^* dz_i}{2i\pi} z_{i_1} z_{i_1'}^* \cdots z_{i_k} z_{i_k'}^* e^{-z^\dagger A z}}{\int_{-\infty}^{\infty} \prod_{i=1}^{n} \frac{dz_i^* dz}{2i\pi} e^{-z^\dagger A z}} \\ &= \frac{1}{Z(0,0)} \frac{\partial}{\partial J_{i_1}^*} \frac{\partial}{\partial J_{i_1'}} \cdots \frac{\partial}{\partial J_{i_k}^*} \frac{\partial}{\partial J_{i_k'}} Z(J^*,J) \Big|_{J^*=J=0} \\ &= \frac{\partial}{\partial J_{i_1}^*} \frac{\partial}{\partial J_{i_1'}} \cdots \frac{\partial}{\partial J_{i_k}^*} \frac{\partial}{\partial J_{i_k'}} e^{J^\dagger A^{-1} J} \Big|_{J^*=J=0}. \end{aligned} \tag{1.E.12}$$

---

[108] We use $\int_{-\infty}^{\infty} \frac{dz^* dz}{2i\pi} e^{-a|z|^2} = a^{-1}$ for $a > 0$.

[109] This can be understood by noting that, contrary to the case of the real Gaussian integral, the result (1.E.11) is an algebraic function of the matrix elements A_{ij} and can therefore be analytically continued to any complex matrix with a positive definite Hermitian part [4].

Note that only monomials with equal number of z_i and $z^*_{i'}$ factors have a non-vanishing average. From (1.E.12), we obtain

$$\begin{aligned}\langle z_{i_1} z^*_{i'_1} \cdots z_{i_k} z^*_{i'_k} \rangle &= \prod_{j=1}^{k} \frac{\partial}{\partial J^*_{i_j}} \left(\sum_{l_1} J^*_{l_1} A^{-1}_{l_1 i'_1} \right) \cdots \left(\sum_{l_k} J^*_{l_k} A^{-1}_{l_k i'_k} \right) e^{J^\dagger A^{-1} J} \Big|_{J^*=J=0} \\ &= \sum_{P \in S_k} A^{-1}_{i_{P(1)} i'_1} \cdots A^{-1}_{i_{P(k)} i'_k}, \end{aligned} \tag{1.E.13}$$

where the sum is over all permutations P of $\{1, \ldots, k\}$.

1.E.3 *Grassmannian integrals*

In order to establish the analogous results for Grassmann variables, we first need to derive the law of change of variables in a Grassmannian integral.

1.E.3.1 *Change of variables*

Let us consider the integral

$$I = \int d\theta f(\theta) \tag{1.E.14}$$

with θ a Grassmann variable, and perform the change of variables $\theta = a\theta' + b$ with a a c-number and b a Grassmann number. Since integration and derivation are identical for Grassmann numbers (Sec. 1.3.2), we obtain

$$I = \frac{\partial}{\partial \theta} f(\theta) = \frac{\partial \theta'}{\partial \theta} \frac{\partial f}{\partial \theta'} = \frac{1}{a} \frac{\partial f}{\partial \theta'} = \frac{1}{a} \int d\theta' f(a\theta' + b) \tag{1.E.15}$$

using the chain rule for differentiation of Grassmann variables [Eq. (1.240)]. The Jacobian of the transformation is therefore a^{-1} instead of a for ordinary variables.

This result can be generalized to the multidimensional integral

$$\begin{aligned} I &= \int d\theta_1 \cdots d\theta_n \, f(\theta_1 \cdots \theta_n) \\ &= \prod_{i=1}^{n} \frac{\partial}{\partial \theta_i} f(\theta_1 \cdots \theta_n) \\ &= \prod_{i=1}^{n} \left(\sum_j \frac{\partial \theta'_j}{\partial \theta_i} \frac{\partial}{\partial \theta'_j} \right) f(\theta_1 \cdots \theta_n). \end{aligned} \tag{1.E.16}$$

For an affine transformation $\theta'_i = \sum_j M_{ij} \theta_j + B_i$, with M_{ij} a c-number and B_i a Grassmann number, we obtain

$$I = \prod_{i=1}^{n} \left(\sum_j M_{ji} \frac{\partial}{\partial \theta'_j} \right) f(\theta_1 \cdots \theta_n). \tag{1.E.17}$$

Since the differential operators $\partial/\partial\theta'_j$ anticommute, the differential operator acting on $f(\theta_1\cdots\theta_n)$ is proportional to $\partial_{\theta'_1}\cdots\partial_{\theta'_n}$, and the prefactor is easily seen to be the determinant of the matrix M,

$$I = \det\left(\frac{\partial\theta'_i}{\partial\theta_j}\right)\prod_{i=1}^{n}\frac{\partial}{\partial\theta'_i}f(\theta_1\cdots\theta_n) = \det\left(\frac{\partial\theta'_i}{\partial\theta_j}\right)\int\prod_{i=1}^{n}d\theta'_i\, f(\theta_1\cdots\theta_n), \tag{1.E.18}$$

i.e.

$$\prod_{i=1}^{n}d\theta_i = \det\left(\frac{\partial\theta'_i}{\partial\theta_j}\right)\prod_{i=1}^{n}d\theta'_i. \tag{1.E.19}$$

The Jacobian is the inverse of the determinant $\det(\partial\theta_i/\partial\theta'_j)$.

1.E.3.2 *Gaussian integrals*

The one-dimensional Gaussian integral over a pair of complex Grassmann variables reads

$$\int d\theta^* d\theta\, e^{-a\theta^*\theta} = \int d\theta^* d\theta(1 - a\theta^*\theta) = a \tag{1.E.20}$$

for any c-number a. This result is easily generalized to the multidimensional Gaussian integral

$$Z = \int\prod_{i=1}^{n}d\theta_i^* d\theta_i\, e^{-\sum_{i,j}\theta_i^* A_{ij}\theta_j}, \tag{1.E.21}$$

where A is an arbitrary complex matrix. With the linear transformation $\theta'_i = \sum_j A_{ij}\theta_j$, we obtain

$$Z = (\det A)\int\prod_{i=1}^{n}d\theta_i^* d\theta'_i\, e^{-\sum_i\theta_i^*\theta'_i} = (\det A)\prod_{i=1}^{n}\int d\theta_i^* d\theta'_i(1-\theta_i^*\theta'_i) = \det A. \tag{1.E.22}$$

Note that for ordinary complex variables, one would obtain $(\det A)^{-1}$.

The generalized Gaussian integral

$$Z(J^*, J) = \int\prod_{i=1}^{n}d\theta_i^* d\theta_i\, e^{-\theta^\dagger A\theta + (J^\dagger\theta + \text{c.c.})} = (\det A)\, e^{J^\dagger A^{-1}J} \tag{1.E.23}$$

(J_i^* and J_i are anticommuting numbers) is easily calculated using the affine change of variables $\theta = \theta' + A^{-1}J$ and $\theta^\dagger = \theta'^\dagger + J^\dagger A^{-1}$.

1.E.3.3 *Wick's theorem*

As with complex variables, the average value

$$\begin{aligned}\langle\theta_{i_1}\cdots\theta_{i_k}\theta^*_{i'_k}\cdots\theta^*_{i'_1}\rangle &= \frac{\int\Pi_{i=1}^n d\theta_i^* d\theta_i\,\theta_{i_1}\cdots\theta_{i_k}\theta^*_{i'_k}\cdots\theta^*_{i'_1}\, e^{-\theta^\dagger A\theta}}{\int\Pi_{i=1}^n d\theta_i^* d\theta_i\, e^{-\theta^\dagger A\theta}}\\ &= \frac{(-1)^k}{Z(0,0)}\frac{\partial}{\partial J^*_{i_1}}\cdots\frac{\partial}{\partial J^*_{i_k}}\frac{\partial}{\partial J_{i'_k}}\cdots\frac{\partial}{\partial J_{i'_1}}Z(J,J^*)\Big|_{J^*=J=0}\end{aligned} \tag{1.E.24}$$

can be obtained from the generating function (1.E.23). This yields

$$\langle \theta_{i_1} \cdots \theta_{i_k} \theta^*_{i'_k} \cdots \theta^*_{i'_1} \rangle = \frac{\partial}{\partial J^*_{i_1}} \cdots \frac{\partial}{\partial J^*_{i_k}} \left(\sum_{l_k} J^*_{l_k} A^{-1}_{l_k i'_k} \right) \cdots \left(\sum_{l_1} J^*_{l_1} A^{-1}_{l_1 i'_1} \right) e^{J^\dagger A^{-1} J} \Big|_{J^*=J=0}$$

$$= \sum_{P \in S_k} \epsilon_P A^{-1}_{i_{P(k)} i'_k} \cdots A^{-1}_{i_{P(1)} i'_1}, \tag{1.E.25}$$

where ϵ_P denotes the signature of the permutation P.

1.E.3.4 *"Real" Grassmannian Gaussian integrals*

We can also consider integrals of the type

$$Z = \int d\theta_1 \cdots d\theta_{2n} \, e^{-\frac{1}{2}\theta^T A \theta}, \tag{1.E.26}$$

where A is an antisymmetric matrix: $A_{ij} = -A_{ji}$.[110] Expanding the exponential in (1.E.26) and retaining the only non-vanishing terms, we obtain

$$Z = \frac{(-1)^n}{2^n n!} \int d\theta_1 \cdots d\theta_{2n} \left(\sum_{i,j=1}^{2n} \theta_i A_{ij} \theta_j \right)^n$$

$$= \frac{1}{2^n n!} \sum_{P \in S_{2n}} \epsilon_P A_{P(1),P(2)} \cdots A_{P(2n-1),P(2n)}. \tag{1.E.27}$$

The quantity in the rhs is called the Pfaffian of the matrix A,

$$Z = \mathrm{Pf}(A). \tag{1.E.28}$$

The Pfaffian is related in a simple way to the determinant. To see this, we consider the square of Z,

$$Z^2 = \int d\theta_1 \cdots d\theta_{2n} d\theta'_1 \cdots d\theta'_{2n} \, e^{-\frac{1}{2}\theta^T A \theta - \frac{1}{2}\theta'^T A \theta'}. \tag{1.E.29}$$

The change of variables

$$\eta_k = \frac{1}{\sqrt{2}}(\theta_k + i\theta'_k),$$
$$\eta^*_k = \frac{1}{\sqrt{2}}(\theta_k - i\theta'_k) \tag{1.E.30}$$

gives

$$d\theta_1 \cdots d\theta_{2n} d\theta'_1 \cdots d\theta'_{2n} = (-1)^n d\theta_1 d\theta'_1 \cdots d\theta_{2n} d\theta'_{2n}$$
$$= (-1)^n J d\eta^*_1 d\eta_1 \cdots d\eta^*_{2n} d\eta_{2n}, \tag{1.E.31}$$

[110]If A is not antisymmetric, only its antisymmetric part contributes since the product $\theta_i \theta_j$ is antisymmetric.

where J is the Jacobian of the transformation $\theta_k, \theta'_k \to \eta^*_k, \eta_k$. Since

$$\det\left(\frac{\partial(\eta^*_k, \eta_k)}{\partial(\theta_k, \theta'_k)}\right) = \det \frac{1}{\sqrt{2}} \begin{pmatrix} 1 & -i \\ 1 & i \end{pmatrix} = i, \tag{1.E.32}$$

we find $J = i^{2n} = (-1)^n$ and therefore

$$Z^2 = \int d\eta^*_1 d\eta_1 \cdots d\eta^*_{2n} d\eta_{2n}\, e^{-\eta^\dagger A \eta} = \det(A), \tag{1.E.33}$$

i.e.

$$\text{Pf}(A)^2 = \det A. \tag{1.E.34}$$

In most cases of interest, we can use $Z = \pm(\det A)^{1/2}$, since the sign of Z either does not matter (e.g. when calculating average values $\langle\theta_{i_1} \cdots \theta_{i_k}\rangle$) or can be easily determined.

The result for the generalized Gaussian integral

$$Z(J) = \int d\theta_1 \cdots d\theta_{2n}\, e^{-\frac{1}{2}\theta^T A\theta + J^T\theta} = \pm\sqrt{\det A}\, e^{-\frac{1}{2}J^T A^{-1} J} \tag{1.E.35}$$

(note the minus sign in the exponential) is derived from the change of variables $\theta = \theta' - A^{-1}J$. Equation (1.E.35) implies that Wick's theorem for real Grassmann variables reads

$$\langle\theta_{i_1} \cdots \theta_{i_{2k}}\rangle = \sum_{\substack{\text{all possible pairings} \\ P \text{ of } \{i_1,\cdots,i_{2k}\}}} \epsilon_P A^{-1}_{i_{P(1)} i_{P(2)}} \cdots A^{-1}_{i_{P(2k-1)} i_{P(2k)}}. \tag{1.E.36}$$

Except for the signature ϵ_P of the permutation, this result is similar to the one obtained for real variables [Eq. (1.E.7)].

1.E.4 *Gaussian functional integrals*

Since functional integrals can be defined as the continuum limit of ordinary integrals (Secs. 1.1 and 1.4), the preceding results generalize straightforwardly to Gaussian functional integrals, i.e.

$$\begin{aligned} Z[J] &= \int \mathcal{D}[\phi]\, e^{-\frac{1}{2}\int d^dx d^dy\, \phi(\mathbf{x})A(\mathbf{x},\mathbf{y})\phi(\mathbf{y}) + \int d^dx\, J(\mathbf{x})\phi(\mathbf{x})} \\ &= (\det A)^{-1/2}\, e^{\frac{1}{2}\int d^dx d^dy\, J(\mathbf{x})A^{-1}(\mathbf{x},\mathbf{y})J(\mathbf{y})} \end{aligned} \tag{1.E.37}$$

for a real field $\phi(\mathbf{x})$ coupled to a real external "source" $J(\mathbf{x})$, and

$$\begin{aligned} Z[J^*, J] &= \int \mathcal{D}[\psi^*, \psi]\, e^{-\int d^dx d^dy\, \psi^*(\mathbf{x})A(\mathbf{x},\mathbf{y})\psi(\mathbf{y}) + \int d^dx\, [J^*(\mathbf{x})\psi(\mathbf{x}) + \text{c.c.}]} \\ &= (\det A)^{-\zeta}\, e^{\int d^dx d^dy\, J^*(\mathbf{x})A^{-1}(\mathbf{x},\mathbf{y})J(\mathbf{y})} \end{aligned} \tag{1.E.38}$$

for a complex ($\zeta = 1$) or Grassmann ($\zeta = -1$) field which couples to an external source of the same type. For a (real) Grassmannian field $\psi(\mathbf{x})$ coupled to a Grassmannian source

$J(\mathbf{x})$, one finds

$$Z[J] = \int \mathcal{D}[\psi]\, e^{-\frac{1}{2}\int d^dx d^dy\, \psi(\mathbf{x})A(\mathbf{x},\mathbf{y})\psi(\mathbf{y}) + \int d^dx\, J(\mathbf{x})\psi(\mathbf{x})}$$
$$= \pm(\det A)^{1/2}\, e^{-\frac{1}{2}\int d^dx d^dy\, J(\mathbf{x})A^{-1}(\mathbf{x},\mathbf{y})J(\mathbf{y})} \tag{1.E.39}$$

(note the minus sign in the exponential). Equations (1.E.37)–(1.E.39) hold up to a multiplicative constant which depends on the precise definition of the functional integral measure $\mathcal{D}[\cdot]$.

The inverse A^{-1} of the operator A is defined by

$$\int d^dz\, A(\mathbf{x},\mathbf{z})A^{-1}(\mathbf{z},\mathbf{y}) = \int d^dz\, A^{-1}(\mathbf{x},\mathbf{z})A(\mathbf{z},\mathbf{y}) = \delta(\mathbf{x}-\mathbf{y}). \tag{1.E.40}$$

$\det A$ denotes the determinant of the operator A. A useful identity is

$$\det A = \exp(\ln\det A) = \exp(\operatorname{Tr}\ln A), \tag{1.E.41}$$

where $\operatorname{Tr} A$ denotes the trace of the operator A. When A can be written as $A = 1 + K$,

$$\operatorname{Tr}\ln A = \operatorname{Tr}\ln(1+K) = \sum_{n=1}^{\infty} \frac{(-1)^{n+1}}{n} \operatorname{Tr} K^n \tag{1.E.42}$$

(provided K^n exists for any n), where

$$\operatorname{Tr} K^n = \int d^dx_1 \cdots d^dx_n\, K(\mathbf{x}_1,\mathbf{x}_2)K(\mathbf{x}_2,\mathbf{x}_3)\cdots K(\mathbf{x}_n,\mathbf{x}_1). \tag{1.E.43}$$

1.E.4.1 *Wick's theorem*

The various forms of Wick's theorem derived in the previous sections translate into

$$\langle\phi(\mathbf{x}_1)\cdots\phi(\mathbf{x}_{2k})\rangle = \sum_{\substack{\text{all possible pairings}\\ P \text{ of } \{\mathbf{x}_1,\ldots,\mathbf{x}_{2k}\}}} A^{-1}\big(\mathbf{x}_{P(1)},\mathbf{x}_{P(2)}\big)\cdots A^{-1}\big(\mathbf{x}_{P(2k-1)},\mathbf{x}_{P(2k)}\big),$$
$$\langle\psi(\mathbf{x}_1)\cdots\psi(\mathbf{x}_k)\psi^*(\mathbf{x}'_k)\cdots\psi^*(\mathbf{x}'_1)\rangle = \sum_{P\in S_k} \zeta^P A^{-1}\big(\mathbf{x}_1,\mathbf{x}'_{P(1)}\big)\cdots A^{-1}\big(\mathbf{x}_k,\mathbf{x}'_{P(k)}\big),$$
$$\langle\psi(\mathbf{x}_1)\cdots\psi(\mathbf{x}_{2k})\rangle = \sum_{\substack{\text{all possible pairings}\\ P \text{ of } \{\mathbf{x}_1,\ldots,\mathbf{x}_{2k}\}}} \epsilon_P A^{-1}\big(\mathbf{x}_{P(1)},\mathbf{x}_{P(2)}\big)\cdots A^{-1}\big(\mathbf{x}_{P(2k-1)},\mathbf{x}_{P(2k)}\big),$$
$$\tag{1.E.44}$$

where the averages are obtained from the generating functionals (1.E.37), (1.E.38) and (1.E.39), respectively. ζ^P equals one for bosons and the signature of the permutation P for fermions. The last line in (1.E.44) holds for Grassmann variables.

Finally, we note that Wick's theorem for complex variables can be generalized from the result

$$Z[J^*,J] = \int \mathcal{D}[\psi^*,\psi]\, e^{-\frac{1}{2}\int d^dx d^dy (\psi^*(\mathbf{x}),\psi(\mathbf{x}))A(\mathbf{x},\mathbf{y})\binom{\psi(\mathbf{y})}{\psi^*(\mathbf{y})} + \int d^dx (J^*(\mathbf{x}),J(\mathbf{x}))\binom{\psi(\mathbf{x})}{\psi^*(\mathbf{x})}}$$
$$= \pm(\det A)^{-1/2} e^{\frac{1}{2}\int d^dx d^dy (J^*(\mathbf{x}),J(\mathbf{x}))A^{-1}(\mathbf{x},\mathbf{y})\binom{J(\mathbf{x})}{J^*(\mathbf{x})}}. \tag{1.E.45}$$

Contrary to the case previously discussed, the matrix $A(\mathbf{x},\mathbf{y})$ in (1.E.45) can connect $\psi(\mathbf{x})$ and $\psi(\mathbf{y})$, or $\psi^*(\mathbf{x})$ and $\psi^*(\mathbf{y})$. Thus, the averages $\langle\psi(\mathbf{x})\psi(\mathbf{y})\rangle$ and $\langle\psi^*(\mathbf{x})\psi^*(\mathbf{y})\rangle$ can be non-zero. Wick's theorem is otherwise unchanged: The average of a product of ψ and ψ^* fields is given by the sum of all possible pairings.

1.F Matsubara Frequency Sums

In this appendix, we describe the standard methods to carry out summations over Matsubara frequencies ω_n and discuss explicitly a few examples.

General method

We consider the Matsubara sum

$$S = \frac{1}{\beta}\sum_{\omega_n} f(i\omega_n), \tag{1.F.1}$$

where f is some function. To compute S, we consider the integral

$$I = \int_{\mathcal{C}} \frac{dz}{2i\pi} f(z) n_\zeta(z) \tag{1.F.2}$$

along some path $\mathcal{C}$ in the complex plane, where

$$n_\zeta(z) = \frac{1}{e^{\beta z}-\zeta} = \begin{cases} n_B(z) & \text{(bosons)}, \\ n_F(z) & \text{(fermions)}. \end{cases} \tag{1.F.3}$$

Since $n_\zeta(z)$ has simple poles at the imaginary frequencies $z = i\omega_n$ (with residue ζ/β), the integral I will be related to S if the integration path $\mathcal{C}$ is suitably chosen.

Example 1

As a first example, we consider the sum

$$S = \frac{1}{\beta}\sum_{\omega_n} \frac{e^{i\omega_n\eta}}{i\omega_n - \xi} \qquad (\eta \to 0^+). \tag{1.F.4}$$

S can be calculated from the integral

$$I = \oint_{\mathcal{C}} \frac{dz}{2i\pi} \frac{e^{z\eta}}{z-\xi} n_\zeta(z) \tag{1.F.5}$$

along a circle $\mathcal{C}$ of radius $R \to \infty$ around the origin $z = 0$ in the complex plane. In the limit $|z| \to \infty$, the function $e^{z\eta} n_\zeta(z)$ is exponentially suppressed[111] so that the integral I vanishes for $R \to \infty$. Now, we can also evaluate I by the residue theorem. Besides the poles of $n_\zeta(z)$ at the Matsubara frequencies ω_n, there is a pole at $z = \xi$, and

$$I = \sum_{\omega_n} \mathrm{Res}(i\omega_n) + \mathrm{Res}(\xi) = \zeta S + n_\zeta(\xi) = 0, \tag{1.F.6}$$

[111] Note that this property depends crucially on the factor $e^{z\eta}$ when $\Re(z) < 0$.

i.e.

$$\frac{1}{\beta}\sum_{\omega_n}\frac{e^{i\omega_n\eta}}{i\omega_n-\xi}=\begin{cases}-n_B(\xi) & \text{(bosons)},\\ n_F(\xi) & \text{(fermions)}.\end{cases} \tag{1.F.7}$$

Example 2

As a second example, we consider the grand potential of non-interacting bosons or fermions (Sec. 1.4.2),

$$\Omega_0=\frac{\zeta}{\beta}\sum_{\alpha,\omega_n}\ln(-i\omega_n+\xi_\alpha)e^{i\omega_n\eta}, \tag{1.F.8}$$

which can be calculated from the integral

$$I_\alpha=\oint_{\mathcal{C}}\frac{dz}{2i\pi}n_\zeta(z)\ln(-z+\xi_\alpha)e^{z\eta}. \tag{1.F.9}$$

The contour $\mathcal{C}$ is chosen such as to avoid the branch cut of the function $\ln(-z+\xi_\alpha)$ (Fig. 1.17). For non-interacting bosons, the chemical potential is always smaller than $\min_\alpha(\epsilon_\alpha)$. When $\mu\to\min_\alpha(\epsilon_\alpha)$, there is BEC. The condition $\xi_\alpha=\epsilon_\alpha-\mu>0$ implies that there is no overlap between the branch cut of $\ln(-z+\xi_\alpha)$ and the Matsubara frequency $\omega_{n=0}=0$. Again, the function $e^{z\eta}n_\zeta(z)$ ensures that the circular parts of $\mathcal{C}$ do not contribute to I_α in the limit $R\to\infty$. We therefore obtain

$$\begin{aligned}I_\alpha&=\int_{\xi_\alpha^-}^{\infty}\frac{d\epsilon}{2i\pi}n_\zeta(\epsilon)\ln(-\epsilon-i\eta+\xi_\alpha)+\int_{\infty}^{\xi_\alpha^-}\frac{d\epsilon}{2i\pi}n_\zeta(\epsilon)\ln(-\epsilon+i\eta+\xi_\alpha)\\&=\int_{\xi_\alpha^-}^{\infty}\frac{d\epsilon}{2i\pi}n_\zeta(\epsilon)[\ln(-\epsilon-i\eta+\xi_\alpha)-\text{c.c.}]\end{aligned} \tag{1.F.10}$$

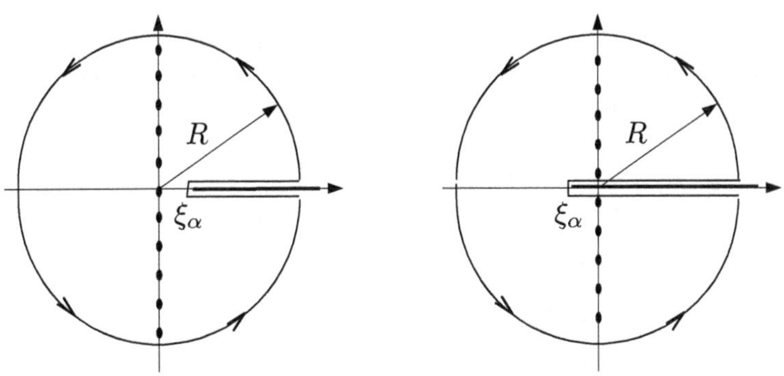

Fig. 1.17 Integration contour $\mathcal{C}$ in Eq. (1.F.9) for bosons (left) and fermions (right). The thick line on the real axis shows the branch cut of the function $\ln(-z+\xi_\alpha)$ and the black dots the positions of the imaginary frequencies $i\omega_n$. In the bosonic case, the branch cut is always located on the positive part of the real axis.

where $\xi_\alpha^- = \xi_\alpha - 0^+$.[112] Using

$$n_\zeta(\epsilon) = \frac{\zeta}{\beta}\frac{d}{d\epsilon}\ln\left|1-\zeta e^{-\beta\epsilon}\right| \tag{1.F.11}$$

and integrating by part,

$$\begin{aligned} I_\alpha &= -\frac{\zeta}{\beta}\int_{\xi_\alpha^-}^\infty \frac{d\epsilon}{2i\pi}\ln\left|1-\zeta e^{-\beta\epsilon}\right|\frac{d}{d\epsilon}[\ln(-\epsilon - i\eta + \xi_\alpha) - \text{c.c.}] \\ &= -\frac{\zeta}{\beta}\int_{\xi_\alpha^-}^\infty \frac{d\epsilon}{2i\pi}\ln\left|1-\zeta e^{-\beta\epsilon}\right|[-2i\pi\delta(\epsilon-\xi_\alpha)] \\ &= \frac{\zeta}{\beta}\ln\left(1-\zeta e^{-\beta\xi_\alpha}\right) \end{aligned} \tag{1.F.12}$$

since $\xi_\alpha > 0$ for bosons ($\zeta = 1$). On the other hand, the residue theorem gives

$$I_\alpha = \frac{\zeta}{\beta}\sum_{\omega_n}\ln(-i\omega_n + \xi_\alpha)e^{i\omega_n\eta}. \tag{1.F.13}$$

From (1.F.12) and (1.F.13), we finally deduce

$$\Omega_0 = \frac{\zeta}{\beta}\sum_\alpha \ln\left(1-\zeta e^{-\beta\xi_\alpha}\right), \tag{1.F.14}$$

which is the familiar result for non-interacting particles (Sec. 1.C.5).

Example 3

In some cases, it is possible to compute the Matsubara sum more directly. For instance, the particle–hole susceptibility (Lindhard function) encountered in Chapters 4 and 5 reads

$$\chi^0_{nn}(q) = -\frac{2}{\beta\mathcal{V}}\sum_k G_0(k)G_0(k+q), \tag{1.F.15}$$

where $G_0(k) = (i\omega_n - \xi_{\mathbf{k}})^{-1}$ is the non-interacting Green function and ω_n a fermionic Matsubara frequency. $q = (\mathbf{q}, i\omega_\nu)$ and ω_ν is a bosonic Matsubara frequency. Adding the factor $e^{i\omega_n\eta}$, which obviously does not change the result, we find

$$\begin{aligned} \chi^0_{nn}(q) &= -\frac{2}{\beta\mathcal{V}}\sum_k \frac{e^{i\omega_n\eta}}{i\omega_\nu + \xi_{\mathbf{k}} - \xi_{\mathbf{k}+\mathbf{q}}}\left(\frac{1}{i\omega_n - \xi_{\mathbf{k}}} - \frac{1}{i\omega_n + i\omega_\nu - \xi_{\mathbf{k}+\mathbf{q}}}\right) \\ &= -\int_{\mathbf{k}} \frac{n_F(\xi_{\mathbf{k}}) - n_F(\xi_{\mathbf{k}+\mathbf{q}})}{i\omega_\nu + \xi_{\mathbf{k}} - \xi_{\mathbf{k}+\mathbf{q}}}, \end{aligned} \tag{1.F.16}$$

using (1.F.7).

[112] It should be understood that the limit $\eta \to 0^+$ is taken at fixed $\xi_\alpha - \xi_\alpha^- > 0$. This ensures that the small vertical part of the integration contour does not contribute and the boundary term in (1.F.12) vanishes.

Similarly, the pairing susceptibility

$$\chi_0(q) = \frac{1}{\beta\mathcal{V}}\sum_k G_0(k)G_0(q-k) \tag{1.F.17}$$

which arises in the theory of superconductivity [Eq. (7.230)] is given by

$$\begin{aligned}\chi_0(q) &= \frac{1}{\beta\mathcal{V}}\sum_k \frac{e^{i\omega_n\eta}}{i\omega_\nu - \xi_{\mathbf{k}} - \xi_{\mathbf{q}-\mathbf{k}}}\left(\frac{1}{i\omega_n - \xi_{\mathbf{k}}} + \frac{1}{i\omega_\nu - i\omega_n - \xi_{\mathbf{q}-\mathbf{k}}}\right)\\ &= \int_{\mathbf{k}} \frac{n_F(\xi_{\mathbf{k}}) - n_F(-\xi_{\mathbf{q}-\mathbf{k}})}{i\omega_\nu - \xi_{\mathbf{k}} - \xi_{\mathbf{q}-\mathbf{k}}}\\ &= \int_{\mathbf{k}} \frac{n_F(\xi_{\mathbf{k}}) + n_F(\xi_{\mathbf{q}-\mathbf{k}}) - 1}{i\omega_\nu - \xi_{\mathbf{k}} - \xi_{\mathbf{q}-\mathbf{k}}}.\end{aligned} \tag{1.F.18}$$

Suppose that instead of $\chi^0_{nn}(q)$ [Eq. (1.F.15)], we wish to calculate

$$\tilde{\chi}^0_{nn}(q) = -\frac{2}{\beta\mathcal{V}}\sum_k G(k)G(k+q) \tag{1.F.19}$$

for a fermion system, where $G(k) = [i\omega_n - \xi_{\mathbf{k}} - \Sigma(k)]^{-1}$. The standard method starts from the integral

$$I = \oint_{\mathcal{C}} \frac{dz}{2i\pi} n_F(z) G(\mathbf{k}, z) G(\mathbf{k}+\mathbf{q}, z + i\omega_\nu). \tag{1.F.20}$$

Since $G(\mathbf{k}, z)$ can have branch cuts on the real axis $\Im(z) = 0$ (Sec. 3.5), the contour $\mathcal{C}$ is chosen as shown in Fig. 1.18. The circular parts of the contour do not contribute when the radius R of the circle is sent to infinity, and we find

$$\begin{aligned} I = &\int_{-\infty}^{\infty} \frac{d\epsilon}{2i\pi} n_F(\epsilon)\left[G^R(\mathbf{k},\epsilon) - G^A(\mathbf{k},\epsilon)\right] G(\mathbf{k}+\mathbf{q}, \epsilon + i\omega_\nu)\\ &+ \int_{-\infty}^{\infty} \frac{d\epsilon}{2i\pi} n_F(\epsilon)\left[G^R(\mathbf{k}+\mathbf{q},\epsilon) - G^A(\mathbf{k}+\mathbf{q},\epsilon)\right] G(\mathbf{k}, \epsilon - i\omega_\nu), \end{aligned} \tag{1.F.21}$$

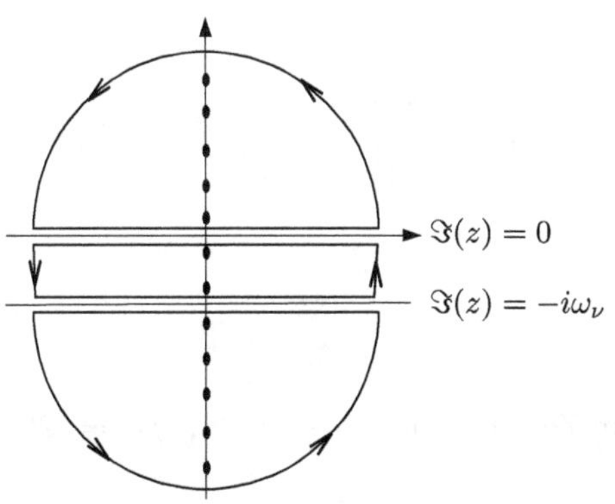

Fig. 1.18 Integration contour $\mathcal{C}$ in Eq. (1.F.20).

where $G^R(\mathbf{k},\epsilon) = G^A(\mathbf{k},\epsilon)^*$ denotes the retarded Green function $G(\mathbf{k}, i\omega_n \to \epsilon + i\eta)$. $G(\mathbf{k}, z)$ being analytic for $\Im(z) \neq 0$ (Sec. 3.5), the residue theorem gives

$$I = -\frac{1}{\beta}\sum_{\omega_n} G(k)G(k+q), \tag{1.F.22}$$

which leads to

$$\tilde{\chi}^0_{nn}(q) = -\int_{\mathbf{k}}\int_{-\infty}^{\infty} d\epsilon\, n_F(\epsilon)\left[A(\mathbf{k},\epsilon)G(\mathbf{k}+\mathbf{q},\epsilon+i\omega_\nu) + A(\mathbf{k}+\mathbf{q},\epsilon)G(\mathbf{k},\epsilon-i\omega_\nu)\right], \tag{1.F.23}$$

where $A(\mathbf{k},\epsilon) = -\frac{1}{\pi}\Im[G^R(\mathbf{k},\epsilon)]$ is the spectral function of the one-particle Green function (Sec. 3.5).

Alternatively, one can use the spectral representation (3.239) to obtain

$$\begin{aligned}\frac{1}{\beta}\sum_{\omega_n} G(k)G(k+q) &= \frac{1}{\beta}\sum_{\omega_n}\int_{-\infty}^{\infty} d\epsilon d\epsilon' \frac{A(\mathbf{k},\epsilon)A(\mathbf{k}+\mathbf{q},\epsilon')}{(i\omega_n-\epsilon)(i\omega_n+i\omega_\nu-\epsilon')} \\ &= \int_{-\infty}^{\infty} d\epsilon d\epsilon' A(\mathbf{k},\epsilon)A(\mathbf{k}+\mathbf{q},\epsilon')\frac{n_F(\epsilon)-n_F(\epsilon')}{i\omega_\nu+\epsilon-\epsilon'},\end{aligned} \tag{1.F.24}$$

where we have used (1.F.7). Using again the spectral representation, Eq. (1.F.24) can be rewritten as

$$\frac{1}{\beta}\sum_{\omega_n} G(k)G(k+q) = \int_{-\infty}^{\infty} d\epsilon\, n_F(\epsilon)[A(\mathbf{k},\epsilon)G(\mathbf{k}+\mathbf{q},\epsilon+i\omega_\nu) + A(\mathbf{k}+\mathbf{q},\epsilon)G(\mathbf{k},\epsilon-i\omega_\nu)], \tag{1.F.25}$$

which leads to (1.F.23).

Example 4

Finally, we note that it is sometimes possible to express Matsubara sums in terms of special functions. For example, the coefficient of the quartic term in the Ginzburg–Landau expansion of the free energy of a superconductor [Eq. (7.284)] depends on the Matsubara sum

$$S = \frac{1}{\beta}\sum_{\omega_n}\frac{1}{|\omega_n|^3} = \frac{1}{4\pi^3T^2}\sum_{n=0}^{\infty}\frac{1}{\left(n+\frac{1}{2}\right)^3} = \frac{1}{4\pi^3T^2}\zeta\left(3,\frac{1}{2}\right), \tag{1.F.26}$$

where

$$\zeta(z,q) = \sum_{n=0}^{\infty}\frac{1}{(n+q)^z} \tag{1.F.27}$$

is the generalized Riemann Zeta function. Using $\zeta(z,1/2) = (2^z-1)\zeta(z)$ where

$$\zeta(z) = \sum_{n=1}^{\infty}\frac{1}{n^z} \tag{1.F.28}$$

is the Riemann Zeta function, we finally obtain

$$S = \frac{7\zeta(3)}{4\pi^3 T^2}, \tag{1.F.29}$$

with $\zeta(3) \simeq 1.2021$.

1.G Perturbation Theory in the Operator Formalism

In this section, we briefly discuss perturbation theory in the operator formalism.[113] We consider a quantum system with a time-independent (grand canonical, i.e. including the term $-\mu\hat{N}$) Hamiltonian $\hat{H} = \hat{H}_0 + \hat{V}$ where $\hat{H}_0$ is the non-interacting part while $\hat{V}$ is due to the interactions between particles.

1.G.1 *Zero-temperature formalism*

1.G.1.1 *Interaction representation*

In the Schrödinger representation, the operators $\hat{O}_S$ are time independent (in the absence of time-dependent external fields) and the state vectors evolve in time according to $|\Psi_S(t)\rangle = \hat{U}_S(t,t')|\Psi_S(t')\rangle$ where $\hat{U}_S(t,t') = e^{-i\hat{H}(t-t')}$ is the unitary time-evolution operator and $\hat{H}$ the (time-independent) Hamiltonian. Alternatively, one can use the Heisenberg picture where the state vectors $|\Psi_H\rangle \equiv |\Psi(0)\rangle$ are time independent while the operators, defined by $\hat{O}_H(t) = e^{i\hat{H}t}\hat{O}_S e^{-i\hat{H}t}$, become time dependent (Appendix 1.B). In the interaction representation, both state vectors and operators are time dependent,

$$\begin{aligned} |\Psi_I(t)\rangle &= e^{i\hat{H}_0 t} e^{-i\hat{H}t}|\Psi_S(0)\rangle, \\ \ddot{O}_I(t) &= e^{i\hat{H}_0 t}\ddot{O}_S e^{-i\hat{H}_0 t}. \end{aligned} \tag{1.G.1}$$

These three representations coincide at time $t = 0$ and are equivalent since

$$\langle\hat{O}\rangle = \langle\Psi_S(t)|\hat{O}_S|\Psi_S(t)\rangle = \langle\Psi_H|\hat{O}_H(t)|\Psi_H\rangle = \langle\Psi_I(t)|\hat{O}_I(t)|\Psi_I(t)\rangle. \tag{1.G.2}$$

We may now define a time-evolution propagator

$$\hat{U}_I(t,t') = e^{i\hat{H}_0 t} e^{-i\hat{H}(t-t')} e^{-i\hat{H}_0 t'} \tag{1.G.3}$$

in the interaction representation such that $|\Psi_I(t)\rangle = \hat{U}_I(t,t')|\Psi_I(t')\rangle$. This propagator satisfies the equation

$$i\partial_t \hat{U}_I(t,t') = \hat{V}_I(t)\hat{U}_I(t,t') \quad \text{with} \quad \hat{U}_I(t,t) = 1. \tag{1.G.4}$$

This equation can be solved by iteration,

$$\begin{aligned} \hat{U}_I(t,t') &= \hat{U}_I(t',t') - i\int_{t'}^{t} dt_1 \hat{V}_I(t_1)\hat{U}_I(t_1,t') \\ &= 1 - i\int_{t'}^{t} dt_1 \hat{V}_I(t_1) + (-i)^2 \int_{t'}^{t} dt_1 \hat{V}_I(t_1) \int_{t'}^{t_1} dt_2 \hat{V}_I(t_2) + \cdots. \end{aligned} \tag{1.G.5}$$

[113] For more details, see Refs. [19–21].

Each term in this expansion can be rewritten in a simpler form using the time-ordering operator T, introduced in Sec. 1.2.3, which orders the operators from right to left in ascending time order and adds a factor $\zeta = \pm$ for each permutation of operators.[114] For instance, for the second-order term, one has

$$\begin{aligned}
&\int_{t'}^{t} dt_1 \hat{V}_I(t_1) \int_{t'}^{t_1} dt_2 \hat{V}_I(t_2) \\
&\quad = \frac{1}{2}\int_{t'}^{t} dt_1 \hat{V}_I(t_1) \int_{t'}^{t_1} dt_2 \hat{V}_I(t_2) + \frac{1}{2}\int_{t'}^{t_1} dt_2 \hat{V}_I(t_2) \int_{t'}^{t_2} dt_1 \hat{V}_I(t_1) \\
&\quad = \frac{1}{2}\int_{t'}^{t} dt_1 \int_{t'}^{t} dt_2 [\Theta(t_1 - t_2)\hat{V}_I(t_1)\hat{V}_I(t_2) + \Theta(t_2 - t_1)\hat{V}_I(t_2)\hat{V}_I(t_1)] \\
&\quad \equiv \frac{1}{2}\int_{t'}^{t} dt_1 \int_{t'}^{t} dt_2 \, T[\hat{V}_I(t_1)\hat{V}_I(t_2)].
\end{aligned} \tag{1.G.6}$$

This yields a compact expression of the propagator,

$$\hat{U}_I(t,t') = T\exp\left\{-i\int_{t'}^{t} dt'' \, \hat{V}_I(t'')\right\}, \tag{1.G.7}$$

which satisfies the following properties:

$$\hat{U}_I^\dagger(t,t') = \hat{U}_I(t',t), \qquad \hat{U}_I(t,t'')\hat{U}_I(t'',t') = \hat{U}_I(t,t'). \tag{1.G.8}$$

1.G.1.2 *Gell-Mann and Low theorem*

We now assume the interaction to be switched on adiabatically, i.e.

$$\hat{H}(t) = \hat{H}_0 + e^{-\epsilon|t|}\hat{V}, \tag{1.G.9}$$

where $\epsilon \to 0^+$. It is readily verified that Eq. (1.G.7) is still satisfied provided that we replace $\hat{V}_I(t)$ by $e^{-\epsilon|t|}\hat{V}_I(t)$. We shall denote the propagator by $\hat{U}_{I,\epsilon}(t,t')$ to emphasize that it depends on ϵ. At very large times, both in the past and in the future, the Hamiltonian reduces to $\hat{H}_0$, which we know how to solve. Suppose that in the distant past ($t \to -\infty$) the system is in the ground state $|\Phi_0\rangle$ of $\hat{H}_0$. Since the interaction is switched on adiabatically, one can expect the system at time t to be an eigenstate $|\Psi_0\rangle$ of the "instantaneous" Hamiltonian $\hat{H}(t)$.

The Gell-Mann and Low theorem [30] states that if the quantity

$$\lim_{\epsilon\to 0^+} \frac{\hat{U}_{I,\epsilon}(0,-\infty)|\Phi_0\rangle}{\langle\Phi_0|\hat{U}_{I,\epsilon}(0,-\infty)|\Phi_0\rangle} = \lim_{\epsilon\to 0^+} \frac{|\Psi_{0,\epsilon}\rangle}{\langle\Phi_0|\Psi_{0,\epsilon}\rangle} \tag{1.G.10}$$

exists to all orders in perturbation theory, then it is an eigenstate of $\hat{H}(0) \equiv \hat{H} = \hat{H}_0 + \hat{V}$,

$$\lim_{\epsilon\to 0^+} \hat{H}\frac{|\Psi_{0,\epsilon}\rangle}{\langle\Phi_0|\Psi_{0,\epsilon}\rangle} = \lim_{\epsilon\to 0^+} E\frac{|\Psi_{0,\epsilon}\rangle}{\langle\Phi_0|\Psi_{0,\epsilon}\rangle}. \tag{1.G.11}$$

[114] The interaction Hamiltonian $\hat{V}$ is necessarily a "bosonic" operator; for fermionic particles, it consists of a product of an even number of creation/annihilation operators.

An essential point of the theorem is that the limit $\epsilon \to 0^+$ does not separately exist for the numerator and the denominator in (1.G.10) and we cannot directly identify $\lim_{\epsilon\to 0^+} |\Psi_{0,\epsilon}\rangle$ with an eigenstate $|\Psi_0\rangle$ of $\hat{H}$ [20]. The denominator in (1.G.11) serves to cancel an infinite phase $\propto 1/\epsilon$ which appears in $|\Psi_{0,\epsilon}\rangle$. If $|\Phi_0\rangle$ is the ground state of $\hat{H}_0$, the corresponding eigenstate of $\hat{H}$ is usually the ground state but this is by no means necessary; in the following, we assume that the state that develops out from $|\Phi_0\rangle$ is non-degenerate and the ground state of $\hat{H}$.

The theorem equally applies to

$$\lim_{\epsilon\to 0^+} \frac{\hat{U}_{I,\epsilon}(0,\infty)|\Phi_0\rangle}{\langle\Phi_0|\hat{U}_{I,\epsilon}(0,\infty)|\Phi_0\rangle} = \lim_{\epsilon\to 0^+} \frac{|\tilde{\Psi}_{0,\epsilon}\rangle}{\langle\Phi_0|\tilde{\Psi}_{0,\epsilon}\rangle}. \tag{1.G.12}$$

Here, the system is in the state $|\Phi_0\rangle$ for $t \to \infty$ and evolves (backwards in time) to an eigenstate of $\hat{H}$ when $t = 0$. If the ground state of $\hat{H}$ is non-degenerate, the two states, obtained from $|\Psi(t=-\infty)\rangle = |\Phi_0\rangle$ and $|\Psi(t=\infty)\rangle = |\Phi_0\rangle$, must coincide at $t = 0$ up to a phase factor. But the equality

$$\langle\Phi_0|\frac{|\Psi_{0,\epsilon}\rangle}{\langle\Phi_0|\Psi_{0,\epsilon}\rangle} = \langle\Phi_0|\frac{|\tilde{\Psi}_{0,\epsilon}\rangle}{\langle\Phi_0|\tilde{\Psi}_{0,\epsilon}\rangle} = 1 \tag{1.G.13}$$

implies that the phase factor necessarily vanishes. Thus,

$$\lim_{\epsilon\to 0^+} \frac{|\Psi_{0,\epsilon}\rangle}{\langle\Phi_0|\Psi_{0,\epsilon}\rangle} = \lim_{\epsilon\to 0^+} \frac{|\tilde{\Psi}_{0,\epsilon}\rangle}{\langle\Phi_0|\tilde{\Psi}_{0,\epsilon}\rangle}. \tag{1.G.14}$$

The unitary operator $\hat{U}_{I,\epsilon}$ preserves the norm of the states. Thus, if $|\Phi_0\rangle$ is normalized, so are $|\Psi_{0,\epsilon}\rangle$ and $|\tilde{\Psi}_{0,\epsilon}\rangle$. Since these two state vectors become colinear with the interacting ground state $|\Psi_0\rangle$ in the limit $\epsilon \to 0^+$, they defer only by a (actually diverging) phase. This implies that the expectation value of an operator $\hat{O}$ in the interacting ground state can be written as

$$\begin{aligned}\lim_{\epsilon\to 0^+} \frac{\langle\tilde{\Psi}_{0,\epsilon}|\hat{O}|\Psi_{0,\epsilon}\rangle}{\langle\tilde{\Psi}_{0,\epsilon}|\Psi_{0,\epsilon}\rangle} &= \lim_{\epsilon\to 0^+} \frac{\langle\Phi_0|\hat{U}_{I,\epsilon}(\infty,0)\hat{O}_I(0)\hat{U}_{I,\epsilon}(0,-\infty)|\Phi_0\rangle}{\langle\Phi_0|\hat{U}_{I,\epsilon}(\infty,-\infty)|\Phi_0\rangle} \\ &= \lim_{\epsilon\to 0^+} \frac{\langle\Phi_0|T[\hat{U}_{I,\epsilon}(\infty,-\infty)\hat{O}_I(0)]|\Phi_0\rangle}{\langle\Phi_0|\hat{U}_{I,\epsilon}(\infty,-\infty)|\Phi_0\rangle},\end{aligned} \tag{1.G.15}$$

where we have used $\hat{O} = \hat{O}_I(0)$ and (1.G.8) as well as the definition of the time-ordering operator T. Since the expressions in (1.G.15) consist of a ratio, the divergent phase factors cancel and it is permissible to take the limit $\epsilon \to 0^+$.

These results allow us to write the energy of the interacting system as

$$\begin{aligned}E &= E_0 + \lim_{\epsilon\to 0^+} \frac{\langle\tilde{\Psi}_{0,\epsilon}|\hat{V}|\Psi_{0,\epsilon}\rangle}{\langle\tilde{\Psi}_{0,\epsilon}|\Psi_{0,\epsilon}\rangle} \\ &= E_0 + \lim_{\epsilon\to 0^+} \frac{\langle\Phi_0|T[\hat{U}_{I,\epsilon}(\infty,-\infty)\hat{V}_I(0)]|\Phi_0\rangle}{\langle\Phi_0|\hat{U}_{I,\epsilon}(\infty,-\infty)|\Phi_0\rangle}.\end{aligned} \tag{1.G.16}$$

Similarly, we can write the real-time one-particle propagator defined in Sec. 1.2.3 as

$$iG(\alpha t, \alpha' t') = \langle \Psi_0 | T[\hat{\psi}_{\alpha,H}(t)\hat{\psi}^\dagger_{\alpha',H}(t')] | \Psi_0 \rangle$$
$$= \lim_{\epsilon\to 0^+} \frac{\langle \Phi_0 | \hat{U}_{I,\epsilon}(\infty,0) T[\hat{\psi}_{\alpha,H}(t)\hat{\psi}^\dagger_{\alpha',H}(t')] \hat{U}_{I,\epsilon}(0,-\infty) | \Phi_0 \rangle}{\langle \Phi_0 | \hat{U}_{I,\epsilon}(\infty,-\infty) | \Phi_0 \rangle}, \tag{1.G.17}$$

where $\hat{\psi}^{(\dagger)}_{\alpha,H}(t) = e^{i\hat{H}t}\hat{\psi}^{(\dagger)}_\alpha e^{-i\hat{H}t}$ is defined in the Heisenberg representation. Using

$$\hat{\psi}_{\alpha,H}(t)\hat{\psi}^\dagger_{\alpha',H}(t') = \hat{U}_{I,\epsilon}(0,t)\hat{\psi}_{\alpha,I}(t)\hat{U}_{I,\epsilon}(t,t')\hat{\psi}^\dagger_{\alpha',I}(t')\hat{U}_{I,\epsilon}(t',0), \tag{1.G.18}$$

one finds

$$iG(\alpha t, \alpha' t') = \lim_{\epsilon\to 0^+} \frac{\langle \Phi_0 | T[\hat{\psi}_{\alpha,I}(t)\hat{\psi}^\dagger_{\alpha',I}(t')\hat{U}_{I,\epsilon}(\infty,-\infty)] | \Phi_0 \rangle}{\langle \Phi_0 | \hat{U}_{I,\epsilon}(\infty,-\infty) | \Phi_0 \rangle}. \tag{1.G.19}$$

1.G.1.3 *Wick's theorem*

This theorem states that the expectation value of a time-ordered product of operators in the non-interacting ground state is given by the sum of all complete contractions, i.e.

$$\langle \Phi_0 | T[\hat{\psi}_{\alpha_1,I}(t_1)\cdots\hat{\psi}_{\alpha_n,I}(t_n)\hat{\psi}^\dagger_{\alpha'_n,I}(t'_n)\cdots\hat{\psi}^\dagger_{\alpha'_1,I}(t'_1)] | \Phi_0 \rangle$$
$$= \sum_{P\in S_n} \zeta^P \langle \Phi_0 | T[\hat{\psi}_{\alpha_1,I}(t_1)\hat{\psi}^\dagger_{\alpha'_{P(1)},I}(t'_{P(1)})] | \Phi_0 \rangle \cdots$$
$$\cdots \langle \Phi_0 | T[\hat{\psi}_{\alpha_n,I}(t_n)\hat{\psi}^\dagger_{\alpha'_{P(n)},I}(t'_{P(n)})] | \Phi_0 \rangle, \tag{1.G.20}$$

where the sum is over all permutations P of $\{1,\dots,n\}$ and ζ^P is equal to one for bosons ($\zeta = 1$) and to the signature of P for fermions ($\zeta = -1$), and

$$\langle \Phi_0 | T[\hat{\psi}_{\alpha,I}(t)\hat{\psi}^\dagger_{\alpha',I}(t')] | \Phi_0 \rangle = iG_0(\alpha t, \alpha' t') \tag{1.G.21}$$

is the non-interacting one-particle propagator.

In the basis where the Hamiltonian $\hat{H}_0 = \sum_\alpha \xi_\alpha \hat{\psi}^\dagger_\alpha \hat{\psi}_\alpha$ is diagonal, one easily obtains $\hat{\psi}_{\alpha,I}(t) = e^{-i\xi_\alpha t}\hat{\psi}_\alpha$ and $\hat{\psi}^\dagger_{\alpha,I}(t) = e^{i\xi_\alpha t}\hat{\psi}^\dagger_\alpha$, so that

$$iG_0(\alpha t, \alpha' t') = \delta_{\alpha,\alpha'} e^{-i\xi_\alpha(t-t')}[\Theta(t-t')(1+\zeta n_\alpha) + \Theta(t'-t)\zeta n_\alpha], \tag{1.G.22}$$

where $n_\alpha = \langle \hat{\psi}^\dagger_\alpha \hat{\psi}_\alpha \rangle$ is given by the Fermi–Dirac function $n_F(\xi_F, T = 0)$ for fermions and the Bose–Einstein function $n_B(\xi_\alpha, T = 0)$ for bosons. In the latter case, $n_B(\xi_\alpha, T = 0) = \lim_{\beta\to\infty}(e^{\beta\xi_\alpha} - 1)^{-1}$ is defined only for a negative chemical potential μ (assuming $\min_\alpha \epsilon_\alpha = 0$ and that the ϵ_α's form a continuum), which corresponds to the vacuum state (no particles in the ground state). As discussed in Sec. 1.8.2, perturbation theory for finite-density boson systems requires to take into account the BEC of the particles in the minimum-energy one-particle state from the outset.

1.G.1.4 *Perturbation theory and Feynman diagrams*

Perturbation theory is obtained by expanding

$$\hat{U}_{I,\epsilon}(\infty,-\infty)=\sum_{n=0}^{\infty}\frac{(-i)^n}{n!}\int_{-\infty}^{\infty}dt_1\cdots dt_n\,e^{-\epsilon(|t_1|+\cdots+|t_n|)}T[\hat{V}_I(t_1)\cdots\hat{V}_I(t_n)] \qquad (1.G.23)$$

in series. From now on, the discussion is similar to that of Sec. 1.5. When computing a physical quantity, e.g. the ground state energy E or the one-particle propagator $G(\alpha t,\alpha' t')$, one uses Wick's theorem to represent each term in the expansion by Feynman diagrams where the non-interacting propagator is presented by a directed line and the interaction by a vertex with two incoming and two outgoing lines. Moreover, the sole role of $\langle\Phi_0|\hat{U}_I(\infty,-\infty)|\Phi_0\rangle$ in the denominators of Eqs. (1.G.16) and (1.G.19) is to remove the non-connected diagrams. Thus, both E and $G(\alpha t,\alpha' t')$ (as well as higher-order Green functions) are given by connected diagrams (the connected diagrams contributing to E are the so-called vacuum fluctuation diagrams). Once the precise diagrammatic rules are established (see Sec. 1.5), whether they were derived from the operator formalism or the functional integral one does not matter.

1.G.2 *Finite-temperature formalism*

In the finite-temperature formalism, the interaction picture is defined by

$$\begin{aligned}\hat{O}_I(\tau)&=e^{\tau\hat{H}_0}\hat{O}_S e^{-\tau\hat{H}_0},\\ \hat{U}_I(\tau,\tau')&=e^{\tau\hat{H}_0}e^{-(\tau-\tau')\hat{H}}e^{-\tau'\hat{H}_0},\end{aligned} \qquad (1.G.24)$$

where $\tau,\tau'\in[0,\beta]$. $\hat{U}_I$ satisfies the equation

$$\partial_\tau\hat{U}_I(\tau,\tau')=-\hat{V}_I(\tau)\hat{U}_I(\tau,\tau')\quad\text{with}\quad\hat{U}_I(\tau,\tau)=1 \qquad (1.G.25)$$

and can therefore be expressed as

$$\hat{U}_I(\tau,\tau')=T_\tau\exp\left\{-\int_{\tau'}^{\tau}d\tau''\,\hat{V}_I(\tau'')\right\}, \qquad (1.G.26)$$

where T_τ is the imaginary-time-ordering operator [Eq. (1.194)]. One therefore has

$$\begin{aligned}\hat{U}_I(\tau,\tau')&=\hat{U}_I(\tau-\tau'),\\ \hat{U}_I(\tau,\tau'')\hat{U}_I(\tau'',\tau')&=\hat{U}_I(\tau,\tau'),\\ \hat{U}_I(\tau,\tau')&=\hat{U}_I^{-1}(\tau',\tau).\end{aligned} \qquad (1.G.27)$$

All physical quantities can be expressed as a function of $\hat{U}_I$. The partition function Z is defined by

$$\frac{Z}{Z_0}=\frac{\mathrm{Tr}\,e^{-\beta\hat{H}}}{Z_0}=\frac{1}{Z_0}\mathrm{Tr}[e^{-\beta\hat{H}_0}\hat{U}_I(\beta,0)]\equiv\langle\hat{U}_I(\beta,0)\rangle_0, \qquad (1.G.28)$$

where $Z_0 = \operatorname{Tr} e^{-\beta \hat{H}_0}$ and $\langle \cdots \rangle_0 = Z_0^{-1} \operatorname{Tr}(e^{-\beta \hat{H}_0} \cdots)$. The one-particle Green function reads

$$
\begin{aligned}
G(\alpha\tau, \alpha'\tau') &= -\langle T_\tau[\hat{\psi}_{\alpha,H}(\tau)\hat{\psi}^\dagger_{\alpha',H}(\tau')]\rangle \\
&= -\frac{1}{Z}\operatorname{Tr}\{e^{-\beta\hat{H}}T_\tau[e^{\tau\hat{H}}\hat{\psi}_\alpha e^{-(\tau-\tau')\hat{H}}\hat{\psi}^\dagger_{\alpha'}e^{-\tau'\hat{H}}]\} \\
&= -\frac{\langle T_\tau[\hat{U}_I(\beta,0)\hat{\psi}_{\alpha,I}(\tau)\hat{\psi}^\dagger_{\alpha',I}(\tau')]\rangle_0}{\langle \hat{U}_I(\beta,0)\rangle_0},
\end{aligned}
\tag{1.G.29}
$$

where $\hat{\psi}^{(\dagger)}_{\alpha,H}(\tau) = e^{\tau\hat{H}}\hat{\psi}^{(\dagger)}_\alpha e^{-\tau\hat{H}}$ are creation and annihilation operators defined in the Heisenberg picture.

Perturbation theory is obtained by expanding

$$
\hat{U}_I(\beta,0) = \sum_{n=0}^{\infty} \frac{(-1)^n}{n!} \int_0^\beta d\tau_1 \cdots \int_0^\beta d\tau_n\, T_\tau[\hat{V}_I(\tau_1)\cdots\hat{V}_I(\tau_n)] \tag{1.G.30}
$$

and using Wick's theorem

$$
\begin{aligned}
&\langle T_\tau[\hat{\psi}_{\alpha_1}(\tau_1)\cdots\hat{\psi}_{\alpha_n}(\tau_n)\hat{\psi}^\dagger_{\alpha'_n}(\tau'_n)\cdots\hat{\psi}^\dagger_{\alpha'_1}(\tau'_1)]\rangle_0 \\
&\quad = \sum_{P\in S_n} \zeta^P \langle T_\tau[\hat{\psi}_{\alpha_1}(\tau_1)\hat{\psi}^\dagger_{\alpha'_{P(1)}}(\tau'_{P(1)})]\rangle_0 \cdots \langle T_\tau[\hat{\psi}_{\alpha_n}(\tau_n)\hat{\psi}^\dagger_{\alpha'_{P(n)}}(\tau'_{P(n)})]\rangle_0,
\end{aligned}
\tag{1.G.31}
$$

where

$$
-\langle T_\tau[\hat{\psi}_\alpha(\tau)\hat{\psi}^\dagger_{\alpha'}(\tau')]\rangle_0 = G_0(\alpha\tau, \alpha'\tau') \tag{1.G.32}
$$

is the non-interacting Green function in imaginary time. In the diagonal basis where $\hat{H}_0 = \sum_\alpha \xi_\alpha \hat{\psi}^\dagger_\alpha \hat{\psi}_\alpha$,

$$
G_0(\alpha\tau, \alpha'\tau') = -\delta_{\alpha,\alpha'} e^{-\xi_\alpha(\tau-\tau')}[\Theta(\tau-\tau')(1+\zeta n_\alpha) + \Theta(\tau'-\tau)\zeta n_\alpha]. \tag{1.G.33}
$$

From now on, the discussion is similar to that of Secs. 1.5 and 1.G.1. Both $\ln(Z/Z_0)$ (linked cluster theorem, Sec. 1.5.2) and $G(\alpha\tau, \alpha'\tau')$, as well as higher-order Green functions, are given by the sum of connected diagrams. We refer to Sec. 1.5 for a detailed discussion of the Feynman diagram rules.

Guide to the bibliography

- Many aspects of Feynman's path integral [1, 2] are reviewed in Schulman's book [3] (see also Refs. [4, 5]).
- Most textbooks on quantum field theory (e.g. Refs. [5–8]) discuss the functional integral formalism, perturbation theory, as well as the effective action (the generating functional of 1PI vertices).
- Functional integrals in non-relativistic systems and their applications to condensed-matter physics are discussed in Refs. [9, 10].
- The Luttinger–Ward functional (the generating functional of 2PI vertices) was introduced in Refs. [11–15] in condensed-matter physics and generalized to relativistic field theories in Ref. [16]. For a general introduction, see Ref. [17].
- The quantization of the electromagnetic field is presented in detail in Ref. [18].
- The "traditional" approach (based on the operator formalism) to non-relativistic quantum field theory is described in Refs. [19–21].

Bibliography

[1] R. P. Feynman, Space-time approach to non-relativistic quantum mechanics, *Rev. Mod. Phys.* **20**, 367 (1948).

[2] R. P. Feynman and A. R. Hibbs, *Quantum Mechanics and Path Integrals* (McGraw-Hill, New York, 1965).

[3] L. S. Schulman, *Techniques and Applications of Path Integration* (Dover, Mineola, New York, 2005).

[4] J. Zinn-Justin, *Intégrale de chemin en mécanique quantique: Introduction* (EDP Sciences, Les Ulis, 2003).

[5] H. Kleinert, *Path Integrals in Quantum Mechanics, Statistics, Polymer Physics, and Financial Markets* (World Scientific, Singapore, 2005).

[6] J. Zinn-Justin, *Quantum Field Theory and Critical Phenomena*, 3rd edition (Clarendon Press, Oxford, 1996).

[7] M. E. Peskin and D. V. Schroeder, *An Introduction to Quantum Field Theory* (Westview Press, Reading, Massachusetts, 1995).

[8] S. Weinberg, *The Theory of Quantum Fields*, Vols. I and II (Cambridge University Press, Cambridge, 1995).

[9] J. W. Negele and H. Orland, *Quantum Many-Particle Systems* (Addison-Wesley, Menlo Park, California, 1988).

[10] A. Altland and B. Simons, *Condensed-Matter Field Theory* (Cambridge University Press, Cambridge, 2006).

[11] J. M. Luttinger, Ground-state energy of a many-fermion system. II, *Phys. Rev.* **118**, 1417 (1960).

[12] G. Baym and L. P. Kadanoff, Conservation laws and correlation functions, *Phys. Rev.* **124**, 287 (1961).

[13] G. Baym, Self-consistent approximations in many-body systems, *Phys. Rev.* **127**, 1391 (1962).

[14] C. De Dominicis and P. C. Martin, Stationary entropy principle and renormalization in normal and superfluid systems. I. Algebraic formulation, *J. Math. Phys.* **5**, 14 (1964).

[15] C. De Dominicis and P. C. Martin, Stationary entropy principle and renormalization in normal and superfluid systems. II. Diagrammatic formulation, *J. Math. Phys.* **5**, 31 (1964).

[16] J. M. Cornwall, R. Jackiw, and E. Tomboulis, Effective action for composite operators, *Phys. Rev.* D **10**, 2428 (1974).

[17] G. Stefanucci and R. van Leeuwen, *Nonequilibrium Many-Body Theory of Quantum Systems: A Modern Introduction* (Cambridge University Press, Cambridge, 2013).

[18] C. Cohen-Tannoudji, J. Dupont-Roc, and G. Grynberg, *Photons and Atoms* (Wiley, New York, 1989).

[19] A. A. Abrikosov, L. P. Gor'kov, and I. E. Dzyaloshinski, *Methods of Quantum Field Theory in Statistical Physics* (Dover, Mineola, New York, 1975).

[20] A. L. Fetter and J. D. Walecka, *Quantum Theory of Many-Particle Systems* (Dover, Mineola, New York, 2003).

[21] G. D. Mahan, *Many-Particle Physics*, 2nd edition (Plenum Press, New York, 1990).

[22] H. Goldstein, *Classical Mechanics* (Addison-Wesley, Reading, 1950).

[23] T. Kibble, *Classical Mechanics* (Imperial College Press, London, 2004).

[24] L. I. Schiff, *Quantum Mechanics* (McGraw-Hill, 1955).

[25] A. Messiah, *Quantum Mechanics*, Vols. I and II (North-Holland, Amsterdam, 1962).

[26] C. Cohen-Tannoudji, B. Diu, and F. Laloë, *Quantum Mechanics*, Vols. I and II (Wiley and Herman, Paris, 1977).

[27] R. Shankar, *Principles of Quantum Mechanics*, 2nd edition (Plenum Press, New York, 1994).

[28] R. Balian, *From Microphysics to Macrophysics*, Vols. I and II (Springer-Verlag, Berlin, 1991).

[29] R. J. Glauber, Coherent and incoherent states of the radiation field, *Phys. Rev.* **131**, 2766 (1963).

[30] M. Gell-Mann and F. Low, Bound states in quantum field theory, *Phys. Rev.* **84**, 350 (1951).

Chapter 2

Symmetries

Contents

2.1 Symmetries, Invariances and Conservation Laws . . . 194
2.1.1 Symmetry transformations in quantum mechanics . . . 194
2.1.2 Wigner's theorem — Group structure . . . 194
2.1.3 Continuous symmetries . . . 195
2.1.4 Transformation of observables . . . 197
2.1.5 Invariance — Conservation laws . . . 198
2.2 Examples of Symmetry Transformations . . . 199
2.2.1 Parity . . . 199
2.2.2 Space translation . . . 201
2.2.3 Space rotation . . . 202
2.2.4 Time translation . . . 204
2.2.5 Galilean transformation . . . 204
2.2.6 Time reversal . . . 206
2.2.7 U(1) symmetry and gauge transformation . . . 212
2.3 Symmetries and Functional Integrals . . . 214
2.3.1 Classical field theory — Noether's theorem . . . 214
2.3.2 Ward identities . . . 219
2.3.3 Ward identities for the 1PI vertices . . . 222
2.3.4 Ward identities for the 2PI vertices . . . 226
Appendix 2.A Ward Identities in a Fermion System . . . 229
2.A.1 U(1) transformation . . . 230
2.A.2 Spin rotation . . . 230
2.A.3 Galilean transformation . . . 231
Bibliography . . . 233

This chapter is concerned with a discussion of symmetries in quantum many-body systems. After a general introduction to symmetries, invariances and conservation laws

(Sec. 2.1), we consider a number of standard symmetry transformations: geometric space transformations (e.g. parity or space rotation), time translation, Galilean transformation, time-reversal and U(1) transformation (Sec. 2.2). In Sec. 2.3, we briefly review Noether's theorem in classical field theory and show how the invariance of a quantum system in a symmetry transformation implies a set of relations, known as Ward identities, which must be satisfied by the Green functions or the 1PI/2PI vertices.

2.1 Symmetries, Invariances and Conservation Laws

2.1.1 *Symmetry transformations in quantum mechanics*

Physical states are represented by rays in the Hilbert space. A ray is a set of normalized state vectors which differ only by a phase factor: $|\Psi\rangle$ and $|\Psi'\rangle$ belong to the same ray if $|\Psi'\rangle = e^{i\alpha}|\Psi\rangle$ (α real). For a system in the state represented by the ray $\mathcal{R}$, the probability to find it in the state represented by the ray $\mathcal{R}_n$ is

$$P(\mathcal{R} \to \mathcal{R}_n) = |\langle \Psi_n | \Psi \rangle|^2, \tag{2.1}$$

where the vectors $|\Psi\rangle$ and $|\Psi_n\rangle$ belong to the rays $\mathcal{R}$ and $\mathcal{R}_n$, respectively.

Let us consider two observers $\mathcal{O}$ and $\mathcal{O}'$ who look at the same system.[1] They of course find the same result, in their own reference frame, for a given experiment. We shall say that $\mathcal{O}$ and $\mathcal{O}'$ are related by a symmetry transformation $\mathcal{T}$. When $\mathcal{O}$ sees the system in a state represented by the ray $\mathcal{R}$ ($\mathcal{R}_n$), the equivalent observer $\mathcal{O}'$ sees the system in a state represented by the ray $\mathcal{R}'$ ($\mathcal{R}'_n$). The two observers find the same probabilities $P(\mathcal{R} \to \mathcal{R}_n) = P(\mathcal{R}' \to \mathcal{R}'_n)$, i.e.

$$|\langle \Psi | \Psi_n \rangle|^2 = |\langle \Psi' | \Psi'_n \rangle|^2. \tag{2.2}$$

It is customary to call $\mathcal{T}$ a symmetry transformation regardless of whether the physical system itself possesses the symmetry. The conditions for the system to be invariant under a symmetry transformation will be discussed in Sec. 2.1.4. Note that a symmetry transformation can be time dependent.

2.1.2 *Wigner's theorem — Group structure*

The Wigner's theorem states that for any transformation $\mathcal{T} : \mathcal{R} \mapsto \mathcal{R}' = \mathcal{T}(\mathcal{R})$ which conserves the probability, $|\langle \Psi | \Phi \rangle| = |\langle \Psi' | \Phi' \rangle|$, there is an operator $\hat{U} \equiv \hat{U}(\mathcal{T})$ such that $|\Psi'\rangle = \hat{U}|\Psi\rangle$, which is either unitary or antiunitary.[2] $\hat{U}(\mathcal{T})$ is not uniquely defined: If $\hat{U}$ is an operator corresponding to the transformation $\mathcal{T}$, so is $e^{i\alpha}\hat{U}$ for any real constant α. The identity transformation being unitary, transformations that can be continuously connected to it are necessarily unitary; these are discussed in the following section. Except for time-reversal symmetry (Sec. 2.2.6), all symmetries discussed in this chapter will be associated with a unitary Wigner operator.

[1] This is the so-called passive point of view. Alternatively, one can adopt the active point of view with a single observer $\mathcal{O}$ and two systems $\mathcal{S}$ and $\mathcal{S}'$ [4].

[2] A linear operator $\hat{U}$ satisfies $\hat{U}(\alpha|\Psi\rangle + \beta|\Phi\rangle) = \alpha\hat{U}|\Psi\rangle + \beta\hat{U}|\Phi\rangle$. $\hat{U}$ is unitary if it is invertible with $\hat{U}^{-1} = \hat{U}^\dagger$. Antiunitary operators are defined in Sec. 2.2.6.

The set of symmetry transformations $\mathcal{T}$ has the properties of a group. The product of two symmetry transformations, $\mathcal{T}_1 : \mathcal{R} \mapsto \mathcal{T}_1(\mathcal{R})$ and $\mathcal{T}_2 : \mathcal{R} \mapsto \mathcal{T}_2(\mathcal{R})$, is defined by $\mathcal{T}_1\mathcal{T}_2 : \mathcal{R} \mapsto \mathcal{T}_1(\mathcal{T}_2(\mathcal{R}))$. The identity transformation leaves rays unchanged, and any transformation $\mathcal{T} : \mathcal{R} \mapsto \mathcal{R}'$ has an inverse defined by $\mathcal{T}^{-1} : \mathcal{R}' \mapsto \mathcal{R}$. The operators $\hat{U}(\mathcal{T})$ corresponding to the symmetry transformations $\mathcal{T}$ are expected to have a similar group structure. Since the $\hat{U}(\mathcal{T})$'s act on states of the Hilbert space, and not on rays, the requirement that $\hat{U}(\mathcal{T}_1\mathcal{T}_2)|\Psi\rangle$ and $\hat{U}(\mathcal{T}_1)\hat{U}(\mathcal{T}_2)|\Psi\rangle$ belong to the same ray leads to[3]

$$\hat{U}(\mathcal{T}_1)\hat{U}(\mathcal{T}_2) = e^{i\alpha(\mathcal{T}_1,\mathcal{T}_2)}\hat{U}(\mathcal{T}_1\mathcal{T}_2). \tag{2.3}$$

Thus, the operators $\hat{U}(\mathcal{T})$ provide a representation up to a phase (called a projective representation) of the group of symmetry transformations $\mathcal{T}$. It turns out that when the phase $\alpha(\mathcal{T}_1, \mathcal{T}_2) \neq 0$ cannot be eliminated by a (trivial) redefinition of $\hat{U}(\mathcal{T})$, the symmetry group can always be enlarged in such a way that its representations can all be defined as non-projective. Assuming that this has been done, we take $\alpha(\mathcal{T}_1, \mathcal{T}_2) = 0$ in the following.

The Wigner operator depends on time if the symmetry transformation does. Unless necessary to avoid confusion, we will not explicitly indicate the time dependence of the Wigner operator, the Hamiltonian, and the state vectors of the Hilbert space.

2.1.3 *Continuous symmetries*

In this section, we consider transformations described by a finite set $\theta = \{\theta_a\}$ of real parameters and continuously connected to the identity. Such transformations form a Lie group. The multiplication law takes the form

$$\mathcal{T}(\theta)\mathcal{T}(\theta') = \mathcal{T}(f(\theta, \theta')), \tag{2.4}$$

where $f_a(\theta, \theta')$ is a function of θ and θ'. If we choose $\theta_a = 0$ as the coordinates of the identity transformation, we must have

$$f_a(\theta, 0) = f_a(0, \theta) = \theta_a. \tag{2.5}$$

If the operators $\hat{U}(\theta) \equiv \hat{U}(\mathcal{T}(\theta))$ form a non-projective representation of the group of symmetry transformations, they satisfy

$$\hat{U}(\theta)\hat{U}(\theta') = \hat{U}(f(\theta, \theta')). \tag{2.6}$$

This equation leads to important consequences when expanded in powers of θ_a and θ'_a. In the neighborhood of the identity,

$$\hat{U}(\theta) = 1 + i\sum_a \theta_a \hat{t}_a + \frac{1}{2}\sum_{a,b} \theta_a\theta_b\hat{t}_{ab} + \cdots, \tag{2.7}$$

[3]This equation follows from $\hat{U}(\mathcal{T}_1)\hat{U}(\mathcal{T}_2)|\Psi\rangle = e^{i\alpha(\mathcal{T}_1,\mathcal{T}_2)}\hat{U}(\mathcal{T}_1\mathcal{T}_2)|\Psi\rangle$ where the phase $\alpha(\mathcal{T}_1, \mathcal{T}_2)$ can be shown to be independent of $|\Psi\rangle$ [4, 5].

where $\hat{t}_a$ and $\hat{t}_{ab}$ are Hermitian operators (as required by the unitarity of $\hat{U}(\theta)$). We have

$$f_a(\theta,\theta') = \theta_a + \theta'_a + \sum_{b,c} C_{abc}\theta_b\theta'_c + \cdots , \tag{2.8}$$

since terms proportional to $\theta_b\theta_c$ or $\theta'_b\theta'_c$ are forbidden by (2.5). By inserting (2.7) and (2.8) into (2.6), we conclude that the transformation law (2.6) holds at order $\theta_a\theta'_b$ only if

$$\hat{t}_{ab} = -\hat{t}_a\hat{t}_b - i\sum_c C_{cab}\hat{t}_c. \tag{2.9}$$

To derive Eq. (2.9), we have used the fact that $\hat{t}_{ab} = \hat{t}_{ba}$ is symmetric with respect to a and b, since it comes from the second-order derivative of $\hat{U}(\theta)$. The symmetry of the lhs of (2.9) with respect to a and b then yields

$$[\hat{t}_a, \hat{t}_b] = i\sum_c (C_{cab} - C_{cba})\hat{t}_c = i\sum_c f_{abc}\hat{t}_c. \tag{2.10}$$

The operators $\hat{t}_a$ are the generators of the group and the real constants $f_{abc} = -C_{cab} + C_{cba}$ the structure constants. Equation (2.9) means that for a given function $f(\theta,\theta')$, the second-order terms in the expansion of $\hat{U}(\theta)$ are completely determined by the first-order terms, i.e. the generators $\hat{t}_a$. The set of commutation relations (2.10) is known as a Lie algebra. It can be shown that these relations are sufficient to obtain the complete power series of $\hat{U}(\theta)$ and uniquely determine the operator $\hat{U}(\theta)$ (at least in the neighborhood of $\theta_a = 0$) [5]. When the function $f_a(\theta,\theta') = \theta_a + \theta'_a$ is simply additive, $f_{abc} = 0$ and the generators commute (Abelian group): $[\hat{t}_a, \hat{t}_b] = 0$.

Writing $\hat{U}(\theta)$ as an infinite number of infinitesimal transformations, we obtain[4]

$$\hat{U}(\theta) = \lim_{N\to\infty}\left[\hat{U}\left(\frac{\theta}{N}\right)\right]^N = \lim_{N\to\infty}\left(1 + i\sum_a \frac{\theta_a}{N}\hat{t}_a\right)^N = \exp\Big(i\sum_a \theta_a\hat{t}_a\Big). \tag{2.11}$$

This result will be illustrated by many examples in the following (see Table 2.1 and Sec. 2.2).

Table 2.1 Continuous symmetry transformations. (M is the total mass of the system and $\hat{\mathbf{R}}$ the position of its center of mass.)

Symmetry transformation	Wigner operator $\hat{U}$	Generator
Space translation	$e^{-i\mathbf{a}\cdot\hat{\mathbf{P}}}$	Momentum $\hat{\mathbf{P}}$
Space rotation	$e^{-i\theta\mathbf{n}\cdot\hat{\mathbf{J}}}$	Angular momentum $\hat{\mathbf{J}} = \hat{\mathbf{L}} + \hat{\mathbf{S}}$
Time translation	$\hat{T}(t+\tau,t)$	Hamiltonian $\hat{H}$
Galilean transformation	$e^{i(\hat{\mathbf{P}}t - M\hat{\mathbf{R}}).\mathbf{v}}$	$-\hat{\mathbf{P}}t + M\hat{\mathbf{R}}$
U(1) transformation	$e^{i\Lambda\hat{N}}$	Particle number $\hat{N}$

[4]The generators are sometimes defined with the opposite sign convention so that $\hat{U}(\theta) = e^{-i\sum_a \theta_a\hat{t}_a}$. This convention will sometimes be adopted in the following (e.g. in Table 2.1).

2.1.4 *Transformation of observables*

In order to ensure that the transformation $\mathcal{T}$ conserves the physical properties of the system, an operator $\hat{A}$ must transform as

$$\hat{A}' = \hat{U}\hat{A}\hat{U}^\dagger, \tag{2.12}$$

so that $\langle\Psi'|\hat{A}'|\Phi'\rangle = \langle\Psi|\hat{A}|\Phi\rangle$.

However, sometimes the operator $\hat{A}'$ is not used by the observer $\mathcal{O}'$ to make measurements on the system. Let us consider, for instance, the quantum mechanics of a single particle and call $\hat{\mathbf{r}}$ the operator for the distance between the particle and the origin of the reference frame used by $\mathcal{O}$. $\hat{U}\hat{\mathbf{r}}\hat{U}^\dagger$ is the operator used by $\mathcal{O}'$ for the same distance. But $\mathcal{O}'$ is rather interested in the position of the particle as measured with respect to its own reference frame and will therefore use the (non-transformed) operator $\hat{\mathbf{r}}$.[5]

Thus, when going from $\mathcal{O}$ to $\mathcal{O}'$, we shall not transform the operators corresponding to coordinates, momenta and spins of the particles. In some cases, however, it is necessary to transform the operator as in (2.12). Consider, for instance, the case of an external vector potential $\hat{\mathbf{A}}(\mathbf{r})$ in ordinary quantum mechanics. Assuming that $\mathcal{O}'$ uses a reference frame rotated with respect to that of $\mathcal{O}$, choosing $\hat{\mathbf{A}}'(\mathbf{r}) = \hat{\mathbf{A}}(\mathbf{r})$ amounts to having rotated the external field as well.

Let us go back to the case where the observable is not transformed. One can see the symmetry transformation as changing the operators rather than the states, since $\langle\Psi'|\hat{A}|\Phi'\rangle = \langle\Psi|\hat{U}^\dagger\hat{A}\hat{U}|\Phi\rangle$. In the following, we will sometimes take this point of view. A given symmetry transformation is then defined by the "transformation law" of a set of irreducible operators, e.g. the position, momentum and spin operators for a single particle.

Let us finally discuss the transformation law of the Hamiltonian. The transformed Hamiltonian $\hat{H}'$ determines the Schrödinger equation satisfied by the transformed vectors. From $i\partial_t|\Psi\rangle = \hat{H}|\Psi\rangle$ and

$$i\partial_t(\hat{U}|\Psi\rangle) = i(\partial_t\hat{U})|\Psi\rangle + i\hat{U}\partial_t|\Psi\rangle = [i(\partial_t\hat{U})\hat{U}^\dagger + \hat{U}\hat{H}\hat{U}^\dagger]\hat{U}|\Psi\rangle, \tag{2.13}$$

we deduce

$$\hat{H}' = \hat{U}\hat{H}\hat{U}^\dagger + i(\partial_t\hat{U})\hat{U}^\dagger \tag{2.14}$$

and

$$\langle\Psi'|\hat{H}'|\Psi'\rangle = \langle\Psi|\hat{H}|\Psi\rangle + i\langle\Psi|\hat{U}^\dagger(\partial_t\hat{U})|\Psi\rangle. \tag{2.15}$$

The expectation value of the energy is not the same for the two observers when the transformation is time dependent (see, e.g., Sec. 2.2.5).

[5]Thus, the same operator, $\hat{\mathbf{r}}$, is used by $\mathcal{O}$ and $\mathcal{O}'$ to describe the distance of a particle to the origin of their respective reference frames. Let us suppose that $\mathcal{O}'$ uses a reference frame translated by $-\mathbf{a}$ with respect to that used by $\mathcal{O}$. In the coordinate basis, $\Psi'(\mathbf{r}) = \Psi(\mathbf{r}-\mathbf{a})$ (see Sec. 2.2.2). The particle position as seen by $\mathcal{O}'$ is then $\langle\Psi'|\hat{\mathbf{r}}|\Psi'\rangle = \int d^dr|\Psi'(\mathbf{r})|^2\mathbf{r} = \int d^dr|\Psi(\mathbf{r})|^2(\mathbf{r}+\mathbf{a}) = \langle\Psi|\hat{\mathbf{r}}|\Psi\rangle + \mathbf{a}$.

2.1.5 *Invariance — Conservation laws*

A system is invariant under the symmetry transformation $\mathcal{T}$ if $|\Psi\rangle$ and $\hat{U}|\Psi\rangle$ satisfy the same Schrödinger equation, i.e. $\hat{H}' = \hat{H}$. Equation (2.14) then implies

$$[\hat{U}, \hat{H}] + i\partial_t \hat{U} = 0. \tag{2.16}$$

If the operator $\hat{U}$ is time independent, the invariance condition (2.16) reduces to $[\hat{U}, \hat{H}] = 0$. When $\hat{H}' = \hat{H}$, any vector $|\Psi\rangle$ solution of the Schrödinger equation $\hat{H}|\Psi\rangle = i\partial_t|\Psi\rangle$, and thus describing a possible physical state of the system as seen by $\mathcal{O}$, also corresponds to a possible physical state of the system as seen by $\mathcal{O}'$. The physical state described by $|\Psi\rangle$ looks the same subjectively to $\mathcal{O}$ and $\mathcal{O}'$, but obviously refers to objectively different physical situations for the system: The laws of physics are invariant under the symmetry transformation $\mathcal{O} \to \mathcal{O}'$.

The fact that $|\psi(t)\rangle$ and $|\psi'(t)\rangle = \hat{U}(t)|\psi(t)\rangle$ share the same time evolution can also be written as $|\psi'(t)\rangle = \hat{T}(t, t_0)|\psi'(t_0)\rangle$, where the evolution operator $\hat{T}$ is defined by $|\psi(t)\rangle = \hat{T}(t, t_0)|\psi(t_0)\rangle$. Since this is true for any state $|\psi(t_0)\rangle$, this implies

$$\hat{U}(t)\hat{T}(t, t_0) = \hat{T}(t, t_0)\hat{U}(t_0), \tag{2.17}$$

which is a mere reformulation of (2.16). Thus, to go from $|\psi(t_0)\rangle$ to $|\psi'(t)\rangle$, one can either first obtain $|\psi'(t_0)\rangle$ using $\hat{U}(t_0)$ and then let the state evolve in time with $\hat{T}(t, t_0)$, or first let the state $|\psi(t_0)\rangle$ evolve into $|\psi(t)\rangle$ and then obtain $|\psi'(t)\rangle$ with $\hat{U}(t)$ (Fig. 2.1).

So far, we have used the Schrödinger picture where the symmetry transformation affects the state vectors but not the operators such as position, momentum and spin. In the Heisenberg picture, the wavefunctions are time independent, taken to be equal to their values in the Schrödinger picture at an arbitrary time t_0, while the operators become time dependent: $\hat{A}_H(t) = \hat{T}^\dagger(t, t_0)\hat{A}\hat{T}(t, t_0)$. Since in general $\hat{T}$ and $\hat{T}'$ (the evolution operator used by $\mathcal{O}'$) differ, both the state vectors and the operators change when passing from $\mathcal{O}$ to $\mathcal{O}'$. However, when the system is invariant under the symmetry transformation, $\hat{H} = \hat{H}'$, and

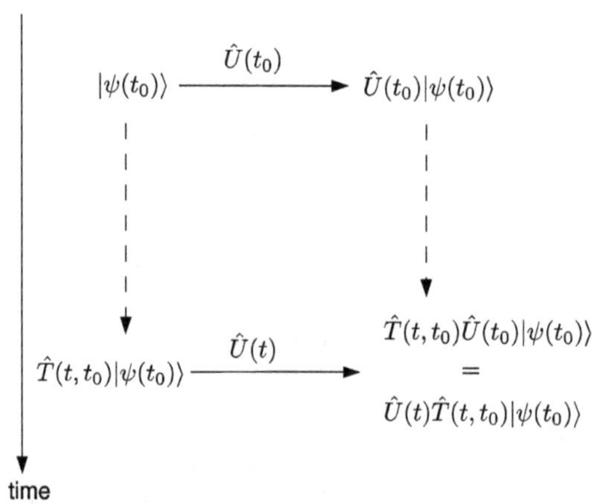

Fig. 2.1 Schematic representation of the invariance under the symmetry transformation $\hat{U}(t)$.

therefore, $\hat{T}(t, t_0) = \hat{T}'(t, t_0)$. It follows that the operator $\hat{A}_H(t)$ in the Heisenberg picture is the same for the two observers $\mathcal{O}$ and $\mathcal{O}'$.

Equations (2.16) and (2.17) take a very simple form in the Heisenberg picture,

$$\frac{d\hat{U}_H}{dt} = 0. \tag{2.18}$$

Thus, the system is invariant under the symmetry transformation $\mathcal{T}$ if the corresponding Wigner operator $\hat{U}_H$ is a constant of motion: Invariance implies conservation laws. For a continuous symmetry, a necessary and sufficient condition is that each generator of the group is a constant of motion. Considering (2.18) in the case of an infinitesimal transformation, $\hat{U}_H(\theta) \simeq 1 + i\sum_a \theta_a \hat{t}_{aH}$ (see Sec. 2.1.3), we obtain

$$\frac{d\hat{t}_{aH}}{dt} = 0, \tag{2.19}$$

where the $\hat{t}_{aH}$'s are the generators in the Heisenberg picture. If the Wigner operator is time independent, Eq. (2.19) reduces to $[\hat{H}, \hat{t}_a] = 0$.

The consequences of an invariance in a symmetry transformation are further discussed in Secs. 2.2 and 2.3, and in Chapters 3 and 9.

2.2 Examples of Symmetry Transformations

2.2.1 *Parity*

We consider the transformation $\mathcal{T}$ where the two observers use reference frames with opposite directions of the coordinate axis: $x' = -x$, $y' = -y$, etc. This is an example of a discrete (as opposed to continuous) transformation. Since two successive parity transformations reduce to the identity transformation, $\mathcal{T}$ and $\mathcal{T}^2 = 1$ form a group.

Let us first discuss the case of a single particle. By analogy with the parity transformation in classical mechanics, $\mathbf{r}' = -\mathbf{r}$ and $\mathbf{p}' = -\mathbf{p}$, we determine the Wigner operator $\hat{U}$ from

$$\begin{aligned} \langle\Psi|\hat{U}^\dagger\hat{\mathbf{r}}\hat{U}|\Psi\rangle &= -\langle\Psi|\hat{\mathbf{r}}|\Psi\rangle, \\ \langle\Psi|\hat{U}^\dagger\hat{\mathbf{p}}\hat{U}|\Psi\rangle &= -\langle\Psi|\hat{\mathbf{p}}|\Psi\rangle. \end{aligned} \tag{2.20}$$

Since these relations hold for any vector $|\Psi\rangle$, they imply

$$\begin{aligned} \hat{U}^\dagger\hat{\mathbf{r}}\hat{U} &= \mathcal{U}(\hat{\mathbf{r}}), \\ \hat{U}^\dagger\hat{\mathbf{p}}\hat{U} &= -\hat{\mathbf{p}}. \end{aligned} \tag{2.21}$$

The spin of the particle must transform as the orbital angular momentum $\hat{\mathbf{r}}\times\hat{\mathbf{p}}$ and is invariant under a parity transformation; we ignore it in the following. To make the discussion slightly more general and suitable to other space transformations, we have introduced the function $\mathcal{U}(\mathbf{r}) = -\mathbf{r}$. Equations (2.21) define $\hat{U}$ only up to a phase factor. For an eigenstate $|\mathbf{r}\rangle$ of the position operator $\hat{\mathbf{r}}$, we obtain

$$\hat{\mathbf{r}}\hat{U}|\mathbf{r}\rangle = \hat{U}\mathcal{U}(\hat{\mathbf{r}})|\mathbf{r}\rangle = \mathcal{U}(\mathbf{r})\hat{U}|\mathbf{r}\rangle. \tag{2.22}$$

$|\mathcal{U}(\mathbf{r})\rangle$ and $\hat{U}|\mathbf{r}\rangle$ being both eigenstates of $\hat{\mathbf{r}}$ with eigenvalue $\mathcal{U}(\mathbf{r})$, they should coincide up to a phase, $\hat{U}|\mathbf{r}\rangle = e^{i\alpha(\mathbf{r})}|\mathcal{U}(\mathbf{r})\rangle$. Choosing $\alpha(\mathbf{r}) = 0$, we obtain

$$\hat{U}|\mathbf{r}\rangle = |\mathcal{U}(\mathbf{r})\rangle. \tag{2.23}$$

Since $\hat{U}^\dagger = \hat{U}^{-1}$, we also have $\hat{U}^\dagger|\mathbf{r}\rangle = |\mathcal{U}^{-1}(\mathbf{r})\rangle$. Equation (2.23) completely determines $\hat{U}$. This definition ensures that $\{\hat{U}, \hat{U}^2 = 1\}$ forms a group and is compatible with the transformation law of $\mathbf{p}$ [Eq. (2.21)]:

$$\hat{U}^{(\dagger)}|\mathbf{p}\rangle = \int d^d r\, \hat{U}^{(\dagger)}|\mathbf{r}\rangle\langle\mathbf{r}|\mathbf{p}\rangle = \int d^d r\, |-\mathbf{r}\rangle\langle\mathbf{r}|\mathbf{p}\rangle = |-\mathbf{p}\rangle \tag{2.24}$$

(we have used $\langle\mathbf{r}|\mathbf{p}\rangle = \langle-\mathbf{r}|-\mathbf{p}\rangle$) and therefore $\hat{U}^\dagger\hat{\mathbf{p}}\hat{U} = -\hat{\mathbf{p}}$. The wavefunction $\Psi(\mathbf{r}) = \langle\mathbf{r}|\Psi\rangle$ transforms into

$$\Psi'(\mathbf{r}) = \langle\mathbf{r}|\hat{U}|\Psi\rangle = \int d^d r'\, \langle\mathbf{r}|\hat{U}|\mathbf{r}'\rangle\langle\mathbf{r}'|\Psi\rangle = \int d^d r'\, \langle\mathbf{r}|\mathcal{U}(\mathbf{r}')\rangle\Psi(\mathbf{r}') = \Psi(\mathcal{U}^{-1}(\mathbf{r})). \tag{2.25}$$

Invariance under a parity transformation means that the parity is a good quantum number: $[\hat{U}, \hat{H}] = 0$ and $d\hat{U}_H/dt = 0$.

2.2.1.1 *Many-particle system*

For a many-particle system, we define the Wigner operator of the parity transformation by

$$\hat{U}|\mathbf{r}_1, \dots, \mathbf{r}_N\} = |\mathcal{U}(\mathbf{r}_1), \dots, \mathcal{U}(\mathbf{r}_N)\}, \tag{2.26}$$

where $|\mathbf{r}_1, \dots, \mathbf{r}_N\} = \hat{\psi}^\dagger(\mathbf{r}_1)\cdots\hat{\psi}^\dagger(\mathbf{r}_N)|\text{vac}\rangle$, with $|\text{vac}\rangle$ the vacuum of particles (Sec. 1.2.1). By evaluating

$$\begin{aligned}\hat{U}^\dagger\hat{\psi}^\dagger(\mathbf{r})\hat{U}|\mathbf{r}_1, \dots, \mathbf{r}_N\} &= \hat{U}^\dagger\hat{\psi}^\dagger(\mathbf{r})|\mathcal{U}(\mathbf{r}_1), \dots, \mathcal{U}(\mathbf{r}_N)\} \\ &= \hat{U}^\dagger|\mathbf{r}, \mathcal{U}(\mathbf{r}_1), \dots, \mathcal{U}(\mathbf{r}_N)\} \\ &= |\mathcal{U}^{-1}(\mathbf{r}), \mathbf{r}_1, \dots, \mathbf{r}_N\},\end{aligned} \tag{2.27}$$

we conclude that the operators $\hat{\psi}^{(\dagger)}(\mathbf{r})$ transform as

$$\hat{U}^\dagger\hat{\psi}^{(\dagger)}(\mathbf{r})\hat{U} = \hat{\psi}^{(\dagger)}(\mathcal{U}^{-1}(\mathbf{r})). \tag{2.28}$$

Equations (2.25) and (2.28), which follow from (2.23), are general and do not depend on the space transformation $\mathcal{U}(\mathbf{r})$. The operators $\hat{\psi}(\mathbf{r})$ and the wavefunction $\Psi(\mathbf{r})$ transform analogously but in an opposite way with respect to the transformation of the eigenstates $|\mathbf{r}\rangle$ of the position operator $\hat{\mathbf{r}}$. The case $\mathcal{U}(\mathbf{r}) = -\mathbf{r}$ is special in this respect since $\mathcal{U}^{-1} = \mathcal{U}$.

Equation (2.28) has an important consequence for the parity transformation, $\mathcal{U}(\mathbf{r}) = -\mathbf{r}$. A local operator $\hat{A}(\mathbf{r})$, which can be written as a product of the fields $\hat{\psi}(\mathbf{r}), \hat{\psi}^\dagger(\mathbf{r})$ and their gradients, transforms as $\hat{U}^\dagger \hat{A}(\mathbf{r})\hat{U} = \epsilon_A^P \hat{A}(-\mathbf{r})$ where $\epsilon_A^P = \pm 1$ is the signature of the operator $\hat{A}$ under the parity transformation. $\epsilon_A^P = 1$ (-1) if $\hat{A}(\mathbf{r})$ involves an even (odd) number of derivative terms $\boldsymbol{\nabla}\hat{\psi}(\mathbf{r})$ or $\boldsymbol{\nabla}\hat{\psi}^\dagger(\mathbf{r})$. For instance, the particle density operator $\hat{n}(\mathbf{r}) = \hat{\psi}^\dagger(\mathbf{r})\hat{\psi}(\mathbf{r})$ is even under a parity transformation, whereas the current operator $\hat{\mathbf{j}}(\mathbf{r}) = -\frac{i}{2m}(\hat{\psi}^\dagger(\mathbf{r})\boldsymbol{\nabla}\hat{\psi}(\mathbf{r}) - \text{h.c.})$ is odd[6] :

$$\begin{aligned} \hat{U}^\dagger \hat{n}(\mathbf{r})\hat{U} &= \hat{n}(-\mathbf{r}) \qquad (\epsilon_n^P = 1), \\ \hat{U}^\dagger \hat{\mathbf{j}}(\mathbf{r})\hat{U} &= -\hat{\mathbf{j}}(-\mathbf{r}) \qquad (\epsilon_j^P = -1). \end{aligned} \tag{2.29}$$

From these results, we can deduce how correlation functions transform in a parity transformation. If the system is invariant under parity, $[\hat{H}, \hat{U}] = 0$ and $\hat{H}' = \hat{H}$, one finds

$$\hat{U}^\dagger \hat{\psi}(\mathbf{r}, t)\hat{U} = e^{i\hat{H}t}\hat{U}^\dagger \hat{\psi}(\mathbf{r})\hat{U}e^{-i\hat{H}t} = \hat{\psi}(-\mathbf{r}, t), \tag{2.30}$$

where $\hat{\psi}(\mathbf{r}, t) = e^{i\hat{H}t}\hat{\psi}(\mathbf{r})e^{-i\hat{H}t}$ is the field operator in the Heisenberg representation. Furthermore, since $\mathcal{O}$ and $\mathcal{O}'$ measure the same correlation functions in a parity-invariant system,

$$\begin{aligned} \langle 0|\hat{\psi}(\mathbf{r}, t)\hat{\psi}^\dagger(\mathbf{r}', t')|0\rangle &= \langle 0|\hat{U}^\dagger \hat{\psi}(\mathbf{r}, t)\hat{\psi}^\dagger(\mathbf{r}', t')\hat{U}|0\rangle \\ &= \langle 0|\hat{\psi}(-\mathbf{r}, t)\hat{\psi}^\dagger(-\mathbf{r}', t')|0\rangle, \end{aligned} \tag{2.31}$$

where $|0\rangle$ and $\hat{U}|0\rangle$ denote the ground states as seen by $\mathcal{O}$ and $\mathcal{O}'$, respectively. A similar result holds for the time-ordered correlation function (i.e. the one-particle Green function), both at zero and finite temperatures,

$$G(\mathbf{r}, t; \mathbf{r}', t') = G(-\mathbf{r}, t; -\mathbf{r}', t'). \tag{2.32}$$

Similarly, for two operators $\hat{A}$ and $\hat{B}$ with signatures ϵ_A^P and ϵ_B^P, one has

$$\begin{aligned} \chi_{AB}(\mathbf{r}, t; \mathbf{r}', t') &= \langle T\hat{A}(\mathbf{r}, t)\hat{B}(\mathbf{r}', t')\rangle \\ &= \chi_{AB}(-\mathbf{r}, t; -\mathbf{r}', t') = \epsilon_A^P \epsilon_B^P \chi_{AB}(\mathbf{r}, t; \mathbf{r}', t'). \end{aligned} \tag{2.33}$$

χ_{AB} vanishes if $\epsilon_A^P \neq \epsilon_B^P$. Equations (2.32) and (2.33) are also valid in imaginary time.

2.2.2 *Space translation*

As a second example, we consider the continuous transformations $\mathcal{T}(\mathbf{a})$ where the observer $\mathcal{O}'$ uses a reference frame translated by $-\mathbf{a}$ with respect to that used by $\mathcal{O}$. The set of transformations $\mathcal{T}(\mathbf{a})$ forms an Abelian group, $\mathcal{T}(\mathbf{a})\mathcal{T}(\mathbf{b}) = \mathcal{T}(\mathbf{b})\mathcal{T}(\mathbf{a}) = \mathcal{T}(\mathbf{a} + \mathbf{b})$.

[6]Here, we assume that the current operator has the same expression as in a free particle system. When the interaction part of the Hamiltonian is not invariant under a local U(1) transformation (see Sec. 2.3), there is an additional contribution due to the interactions.

By analogy with classical mechanics, we define the Wigner operator $\hat{U}(\mathbf{a}) \equiv \hat{U}(\mathcal{T}(\mathbf{a}))$ for a single particle by

$$\begin{aligned} \hat{U}^\dagger(\mathbf{a})\hat{\mathbf{r}}\hat{U}(\mathbf{a}) &= \mathcal{U}(\hat{\mathbf{r}}) = \hat{\mathbf{r}} + \mathbf{a}, \\ \hat{U}^\dagger(\mathbf{a})\hat{\mathbf{p}}\hat{U}(\mathbf{a}) &= \hat{\mathbf{p}}. \end{aligned} \tag{2.34}$$

The spin operator is invariant and is ignored in the following. Equations (2.34) define $\hat{U}$ only up to a phase. This ambiguity is lifted by specifying the action of $\hat{U}$ of the vectors $|\mathbf{r}\rangle$. A possible choice, compatible both with $\hat{U}^\dagger(\mathbf{a})\hat{\mathbf{p}}\hat{U}(\mathbf{a}) = \hat{\mathbf{p}}$ and the group structure of the transformations $\mathcal{T}(\mathbf{a})$, is

$$\hat{U}(\mathbf{a})|\mathbf{r}\rangle = |\mathbf{r} + \mathbf{a}\rangle. \tag{2.35}$$

The generator of the group is the momentum operator $\mathbf{p}$: Writing $\hat{U}(d\mathbf{a}) = 1 - id\mathbf{a}\cdot\hat{\mathbf{p}}$ (note the minus sign[4]) for an infinitesimal transformation, and using $[\hat{r}_\mu, \hat{p}_{\mu'}] = i\delta_{\mu,\mu'}$ ($\mu, \mu' = x, y, \ldots$), we verify that Eqs. (2.34) are satisfied. Proceeding as in Sec. 2.1.3 (see Eq. (2.11)), we then find

$$\hat{U}(\mathbf{a}) = e^{-i\mathbf{a}\cdot\hat{\mathbf{p}}}. \tag{2.36}$$

The wavefunction transforms as $\Psi'(\mathbf{r}) = \Psi(\mathbf{r} - \mathbf{a})$.

2.2.2.1 *Many-particle system*

The preceding results are readily extended to a many-particle system, where $\hat{U}(\mathbf{a}) = e^{-i\mathbf{a}\cdot\hat{\mathbf{P}}}$ with $\hat{\mathbf{P}}$ the total momentum operator. In second-quantized form, $\hat{\mathbf{P}} = \int d^dr\, \hat{\psi}^\dagger(\mathbf{r})(-i\boldsymbol{\nabla})\hat{\psi}(\mathbf{r})$. Translation invariance implies that the total momentum is conserved: $[\hat{\mathbf{P}}, \hat{H}] = 0$ and $d\hat{\mathbf{P}}_H/dt = 0$.

Equation (2.28) with $\mathcal{U}(\mathbf{r}) = \mathbf{r} + \mathbf{a}$ shows that $\hat{\psi}^{(\dagger)}(\mathbf{r})$ transforms as

$$\hat{U}^\dagger(\mathbf{a})\hat{\psi}^{(\dagger)}(\mathbf{r})\hat{U}(\mathbf{a}) = \hat{\psi}^{(\dagger)}(\mathbf{r} - \mathbf{a}). \tag{2.37}$$

Any local operator $\hat{A}(\mathbf{r})$ therefore transforms as $\hat{U}^\dagger(\mathbf{a})\hat{A}(\mathbf{r})\hat{U}(\mathbf{a}) = \hat{A}(\mathbf{r} - \mathbf{a})$. Thus, in a translation-invariant system, correlation functions satisfy $\chi_{AB}(\mathbf{r}, t; \mathbf{r}', t') = \chi_{AB}(\mathbf{r} - \mathbf{a}, t; \mathbf{r}' - \mathbf{a}, t')$.

2.2.3 *Space rotation*

In three dimensions, a space rotation is defined by a 3×3 orthogonal matrix $\mathcal{U} \in SO(3)$. We first consider a single spinless particle. The Wigner operator corresponding to an infinitesimal rotation of angle $d\theta$ about an axis $\mathbf{n}$ is defined by

$$\begin{aligned} \hat{U}^\dagger(d\theta, \mathbf{n})\hat{\mathbf{r}}\hat{U}(d\theta, \mathbf{n}) &= \hat{\mathbf{r}} + d\theta\mathbf{n} \times \hat{\mathbf{r}}, \\ \hat{U}^\dagger(d\theta, \mathbf{n})\hat{\mathbf{p}}\hat{U}(d\theta, \mathbf{n}) &= \hat{\mathbf{p}} + d\theta\mathbf{n} \times \hat{\mathbf{p}}. \end{aligned} \tag{2.38}$$

The generator of the group is the orbital angular momentum $\hat{\mathbf{L}} = \hat{\mathbf{r}} \times \hat{\mathbf{p}}$, i.e. $\hat{U}(d\theta, \mathbf{n}) = 1 - id\theta\mathbf{n}\cdot\hat{\mathbf{L}}$, as can be verified by direct substitution in (2.38) using the commutation relations $\hat{\mathbf{L}} \times \hat{\mathbf{L}} = i\hat{\mathbf{L}}$. Proceeding as in Sec. 2.1.3, we obtain

$$\hat{U}(\theta, \mathbf{n}) = e^{-i\theta\mathbf{n}\cdot\hat{\mathbf{L}}}. \tag{2.39}$$

By choosing the arbitrary phase factor in (2.39) equal to unity, we ensure that the operators $\hat{U}(\theta, \mathbf{n})$ constitute a representation of the group of rotations in ordinary space.

For a particle with spin, the Wigner operator can be written as

$$\hat{U}(\theta, \mathbf{n}) = \hat{U}_L(\theta, \mathbf{n})\hat{U}_S(\theta, \mathbf{n}), \tag{2.40}$$

where $\hat{U}_L(\theta, \mathbf{n})$ is given by (2.39) and $\hat{U}_S(\theta, \mathbf{n})$ acts on spin space. We determine $\hat{U}_S$ by requiring that the spin operator $\hat{\mathbf{S}}$ transforms as the orbital angular momentum operator $\hat{\mathbf{L}}$. Since $\hat{U}_L^\dagger(d\theta, \mathbf{n})\hat{\mathbf{L}}\hat{U}_L(d\theta, \mathbf{n}) = \hat{\mathbf{L}} + d\theta\mathbf{n} \times \hat{\mathbf{L}}$, we demand that

$$\hat{U}_S^\dagger(d\theta, \mathbf{n})\hat{\mathbf{S}}\hat{U}_S(d\theta, \mathbf{n}) = \hat{\mathbf{S}} + d\theta\mathbf{n} \times \hat{\mathbf{S}}. \tag{2.41}$$

Equation (2.41) gives $\hat{U}_S(d\theta, \mathbf{n}) = 1 - id\theta\mathbf{n}\cdot\hat{\mathbf{S}}$, as can be verified using the spin commutation relations $\hat{\mathbf{S}} \times \hat{\mathbf{S}} = i\hat{\mathbf{S}}$, and

$$\hat{U}_S(\theta, \mathbf{n}) = e^{-i\theta\mathbf{n}\cdot\hat{\mathbf{S}}}. \tag{2.42}$$

These matrices form a representation of the rotation group SU(2). We thus obtain

$$\hat{U}(\theta, \mathbf{n}) = \hat{U}_L(\theta, \mathbf{n})\hat{U}_S(\theta, \mathbf{n}) = e^{-i\theta\mathbf{n}\cdot\hat{\mathbf{J}}}, \tag{2.43}$$

where $\hat{\mathbf{J}} = \hat{\mathbf{L}} + \hat{\mathbf{S}}$ is the total angular momentum operator.

The wavefunction $\Psi_\sigma(\mathbf{r})$ transforms as

$$\begin{aligned} \Psi'_\sigma(\mathbf{r}) &= \langle \mathbf{r}, \sigma|\hat{U}(\theta, \mathbf{n})|\Psi\rangle \\ &= \sum_{\sigma'} \int d^d r' \langle \mathbf{r}, \sigma|\hat{U}(\theta, \mathbf{n})|\mathbf{r}', \sigma'\rangle\langle \mathbf{r}', \sigma'|\Psi\rangle \\ &= \sum_{\sigma'} \left(\hat{U}_S(\theta, \mathbf{n})\right)_{\sigma,\sigma'} \Psi_{\sigma'}(\mathcal{U}^{-1}(\mathbf{r})), \end{aligned} \tag{2.44}$$

where $\mathcal{U}(\mathbf{r})$ is the image of $\mathbf{r}$ by a rotation of angle θ about the $\mathbf{n}$ axis. We have introduced the eigenstates $|\sigma\rangle$ ($\sigma = -S, \ldots, S$) of the operator $\hat{S}^z$, where S is the spin of the particle.

2.2.3.1 *Many-particle system*

The Wigner operator is given by (2.43) where $\hat{\mathbf{J}}$ must now be understood as the total angular momentum operator of the many-particle system. The annihilation operator $\hat{\psi}(\mathbf{r})$ transforms as the wavefunction [Eq. (2.44)], i.e.

$$\hat{U}^\dagger(\theta, \mathbf{n})\hat{\psi}_\sigma^{(\dagger)}(\mathbf{r})\hat{U}(\theta, \mathbf{n}) = \sum_{\sigma'} \left(\hat{U}_S(\theta, \mathbf{n})\right)_{\sigma,\sigma'} \hat{\psi}_{\sigma'}^{(\dagger)}(\mathcal{U}^{-1}(\mathbf{r})). \tag{2.45}$$

Rotation invariance implies the conservation of the total angular momentum $\hat{\mathbf{J}}$.[7]

[7] If the system is invariant in a rotation in orbital and spin spaces, separately, then both $\hat{\mathbf{L}}$ and $\hat{\mathbf{S}}$ are conserved.

In a rotation-invariant system, the one-particle Green function satisfies

$$G_{\sigma\sigma'}(\mathbf{r},t;\mathbf{r}',t') = \sum_{\sigma_1,\sigma_1'} \left(\hat{U}_S(\theta,\mathbf{n})\right)_{\sigma,\sigma_1} \left(\hat{U}_S(\theta,\mathbf{n})\right)_{\sigma',\sigma_1'} G_{\sigma_1\sigma_1'}(\mathcal{U}^{-1}(\mathbf{r}),t;\mathcal{U}^{-1}(\mathbf{r}'),t'). \quad (2.46)$$

For two scalar operators $\hat{A}$ and $\hat{B}$, rotation invariance implies

$$\chi_{AB}(\mathbf{r},t;\mathbf{r}',t') = \chi_{AB}(\mathcal{U}^{-1}(\mathbf{r}),t;\mathcal{U}^{-1}(\mathbf{r}'),t'). \quad (2.47)$$

2.2.4 *Time translation*

When the clocks used by the two observers are shifted with respect to one another, i.e. $t(\mathcal{O}') = t(\mathcal{O}) - \tau$, the Wigner operator $\hat{U}_\tau(t)$ is the evolution operator $\hat{T}(t+\tau,t)$ (up to a phase factor that we take equal to unity),

$$|\Psi'(t)\rangle = |\Psi(t+\tau)\rangle = \hat{T}(t+\tau,t)|\Psi(t)\rangle. \quad (2.48)$$

The Wigner operators satisfy the group property $\hat{T}(t_1,t_2)\hat{T}(t_2,t_3) = \hat{T}(t_1,t_3)$. For an infinitesimal time translation, $|\Psi(t+d\tau)\rangle = |\Psi(t)\rangle + d\tau\partial_t|\Psi(t)\rangle = (1 - i\hat{H}(t)d\tau)|\Psi(t)\rangle$, so that

$$\hat{U}_{d\tau}(t) = 1 - i\hat{H}(t)d\tau. \quad (2.49)$$

The Hamiltonian is the generator of time translations. Given that

$$i\partial_t|\Psi'(t)\rangle = i\partial_t|\Psi(t+\tau)\rangle = \hat{H}(t+\tau)|\Psi(t+\tau)\rangle = \hat{H}(t+\tau)|\Psi'(t)\rangle, \quad (2.50)$$

we obtain

$$\hat{H}'(t) = \hat{H}(t+\tau). \quad (2.51)$$

Thus, time-translation invariance implies that the Hamiltonian $\hat{H}'(t) = \hat{H}(t+\tau) = \hat{H}(t)$ is time independent (the same conclusion can be reached by using (2.16)). In that case, the evolution operator is simply $\hat{U}_\tau(t) = \hat{T}(t+\tau,t) = e^{-i\hat{H}\tau}$.

2.2.5 *Galilean transformation*

Let us consider the case where the observer $\mathcal{O}'$ uses a reference frame which moves with velocity $\mathbf{v}$ with respect to that used by $\mathcal{O}$. The corresponding Wigner operator is defined by

$$\begin{aligned} \hat{U}^\dagger(\mathbf{v},t)\hat{\mathbf{r}}\hat{U}(\mathbf{v},t) &= \hat{\mathbf{r}} - \mathbf{v}t, \\ \hat{U}^\dagger(\mathbf{v},t)\hat{\mathbf{p}}\hat{U}(\mathbf{v},t) &= \hat{\mathbf{p}} - m\mathbf{v}, \end{aligned} \quad (2.52)$$

where m is the mass of the particle (we ignore the spin of the particle which is supposed to remain unaltered). We assume that $\hat{U}(\mathbf{v},t)$ can be written as the product $\hat{U}_p(\mathbf{v},t)\hat{U}_r(\mathbf{v},t)$ of two unitary operators where $\hat{U}_p$ ($\hat{U}_r$) is a function of $\hat{\mathbf{p}}$ ($\hat{\mathbf{r}}$) only. If we impose

$$\begin{aligned} \hat{U}_p^\dagger(\mathbf{v},t)\hat{\mathbf{r}}\hat{U}_p(\mathbf{v},t) &= \hat{\mathbf{r}} - \mathbf{v}t, \\ \hat{U}_r^\dagger(\mathbf{v},t)\hat{\mathbf{p}}\hat{U}_r(\mathbf{v},t) &= \hat{\mathbf{p}} - m\mathbf{v}, \end{aligned} \quad (2.53)$$

Eqs. (2.52) are automatically satisfied. $\hat{U}_p(\mathbf{v},t)$ is a translation operator and can be written as $\hat{U}_p(\mathbf{v},t) = e^{i\gamma_p(\mathbf{v},t)+it\mathbf{v}\cdot\hat{\mathbf{p}}}$ with $\gamma_p(\mathbf{v},t)$ an arbitrary phase (see Sec. 2.2.2). $\hat{U}_r(\mathbf{v},t)$ translates $\hat{\mathbf{p}}$ but leaves $\hat{\mathbf{r}}$ invariant. Following the analysis of Sec. 2.2.2, we find $\hat{U}_r(\mathbf{v},t) = e^{i\gamma_r(\mathbf{v},t)-im\mathbf{v}\cdot\hat{\mathbf{r}}}$ with $\gamma_r(\mathbf{v},t)$ an arbitrary phase. This gives

$$\begin{aligned}\hat{U}(\mathbf{v},t) &= e^{i\gamma_p(\mathbf{v},t)+i\gamma_r(\mathbf{v},t)}e^{it\mathbf{v}\cdot\hat{\mathbf{p}}}e^{-im\mathbf{v}\cdot\hat{\mathbf{r}}} \\ &= e^{i\gamma_p(\mathbf{v},t)+i\gamma_r(\mathbf{v},t)-\frac{i}{2}m\mathbf{v}^2t+i(t\hat{\mathbf{p}}-m\hat{\mathbf{r}})\cdot\mathbf{v}},\end{aligned} \tag{2.54}$$

where we have used the operator identity $e^{\hat{A}}e^{\hat{B}} = e^{\frac{1}{2}[\hat{A},\hat{B}]}e^{\hat{A}+\hat{B}}$ for two operators which commutes with their commutator. If we choose $\gamma_p(\mathbf{v},t)+\gamma_r(\mathbf{v},t) = \frac{1}{2}m\mathbf{v}^2t$, then the operators

$$\hat{U}(\mathbf{v},t) = e^{i(\hat{\mathbf{p}}t-m\hat{\mathbf{r}})\cdot\mathbf{v}} \tag{2.55}$$

form a group with the composition law $\hat{U}(\mathbf{v}_1)\hat{U}(\mathbf{v}_2) = \hat{U}(\mathbf{v}_1+\mathbf{v}_2)$. The generator of the group is $-\hat{\mathbf{p}}t+m\hat{\mathbf{r}}$. It turns out that the choice $\gamma(\mathbf{v},t) = \gamma_p(\mathbf{v},t)+\gamma_r(\mathbf{v},t) = \frac{1}{2}m\mathbf{v}^2t+\gamma_0(\mathbf{v})$, with $\gamma_0(\mathbf{v})$ a time-independent constant, is the only one compatible with invariance under a Galilean transformation. Indeed, for a free particle with Hamiltonian $\hat{H} = \hat{\mathbf{p}}^2/2m$, Galilean invariance [Eq. (2.16)] implies $d\gamma(\mathbf{v},t)/dt = \frac{1}{2}m\mathbf{v}^2$, i.e. $\gamma(\mathbf{v},t) = \frac{1}{2}m\mathbf{v}^2t+\gamma_0(\mathbf{v})$.

The wavefunction $\Psi(\mathbf{r},t)$ transforms as

$$\begin{aligned}\Psi'(\mathbf{r}',t) \equiv \langle\mathbf{r}'|\hat{U}(\mathbf{v},t)|\Psi(t)\rangle &= \int d^dr'' \sum_{\mathbf{p}} e^{i\gamma(\mathbf{v},t)}\langle\mathbf{r}'|e^{it\mathbf{v}\cdot\hat{\mathbf{p}}}|\mathbf{p}\rangle\langle\mathbf{p}|e^{-im\mathbf{v}\cdot\hat{\mathbf{r}}}|\mathbf{r}''\rangle\langle\mathbf{r}''|\Psi(t)\rangle \\ &= \int d^dr'' \int_{\mathbf{p}} e^{i\gamma(\mathbf{v},t)-im\mathbf{v}\cdot\mathbf{r}''+i\mathbf{p}\cdot(\mathbf{v}t+\mathbf{r}'-\mathbf{r}'')}\Psi(\mathbf{r}'',t) \\ &= \int d^dr''\, e^{i\gamma(\mathbf{v},t)-im\mathbf{v}\cdot\mathbf{r}''}\delta(\mathbf{v}t+\mathbf{r}'-\mathbf{r}'')\Psi(\mathbf{r}'',t) \\ &= e^{\frac{i}{2}m\mathbf{v}^2t-im\mathbf{v}\cdot\mathbf{r}}\Psi(\mathbf{r},t),\end{aligned} \tag{2.56}$$

where $\mathbf{r} = \mathbf{r}'+\mathbf{v}t$.

2.2.5.1 *Many-particle system*

For a many-particle system, the Wigner operator reads

$$\hat{U}(\mathbf{v},t) = e^{i(\hat{\mathbf{P}}t-M\hat{\mathbf{R}})\cdot\mathbf{v}}, \tag{2.57}$$

where $M = Nm$ is the total mass of the N particles, $\hat{\mathbf{R}}$ and $\hat{\mathbf{P}}$ the position and momentum operators of the center of mass. In second-quantized form, $M\hat{\mathbf{R}} = m\int d^dr\,\hat{\psi}^\dagger(\mathbf{r})\hat{\mathbf{r}}\hat{\psi}(\mathbf{r})$. The expression of $\hat{\mathbf{P}}$ is given in Sec. 2.2.2. Galilean invariance implies

$$\frac{d[\hat{\mathbf{P}}t-M\hat{\mathbf{R}}]_H}{dt} = 0. \tag{2.58}$$

If the system is translation invariant, $d\hat{\mathbf{P}}_H/dt = 0$, we obtain $\hat{\mathbf{P}}_H = Md\hat{\mathbf{R}}_H/dt$ and $\mathbf{R}_H(t) = t\hat{\mathbf{P}}_H/M+\mathbf{R}_H(0)$.

It is easy to verify that a system of N interacting particles with Hamiltonian $\hat{H} = \sum_i \hat{\mathbf{p}}_i^2/2m + \hat{V}$, where $\hat{V} \equiv \hat{V}(\{\mathbf{r}_i - \mathbf{r}_j\})$ depends only the relative particles' positions, is Galilean invariant. Indeed one finds[8]

$$\begin{aligned}\hat{H}' &= \hat{U}(\mathbf{v},t)\hat{H}\hat{U}^\dagger(\mathbf{v},t) + i(\partial_t\hat{U}(\mathbf{v},t))\hat{U}^\dagger(\mathbf{v},t) \\ &= \sum_i \frac{(\hat{\mathbf{p}}_i + m\mathbf{v})^2}{2m} + \hat{V} - \frac{M}{2}\mathbf{v}^2 - \mathbf{v}\cdot\hat{\mathbf{P}} = \hat{H},\end{aligned} \tag{2.59}$$

where $\hat{\mathbf{P}} = \sum_i \hat{\mathbf{p}}_i$. The mean value of the energy is however not the same for $\mathcal{O}$ and $\mathcal{O}'$, since

$$\langle \hat{H}' \rangle_{\mathcal{O}'} = \langle \hat{H} \rangle_{\mathcal{O}} - \mathbf{v}\cdot\langle \hat{\mathbf{P}} \rangle_{\mathcal{O}} + \frac{1}{2}Nm\mathbf{v}^2 \tag{2.60}$$

using (2.15). Equation (2.60) is the relation obtained in classical mechanics.

Since the operator $\hat{\psi}(\mathbf{r})$ transforms as the wavefunction, we have

$$\begin{aligned}\hat{\psi}'(\mathbf{r}',t) &= \hat{U}^\dagger(\mathbf{v},t)\hat{\psi}(\mathbf{r}')\hat{U}(\mathbf{v},t) = e^{\frac{i}{2}m\mathbf{v}^2 t - im\mathbf{v}\cdot\mathbf{r}}\hat{\psi}(\mathbf{r}), \\ \hat{\psi}'^\dagger(\mathbf{r}',t) &= \hat{U}^\dagger(\mathbf{v},t)\hat{\psi}^\dagger(\mathbf{r}')\hat{U}(\mathbf{v},t) = e^{-\frac{i}{2}m\mathbf{v}^2 t + im\mathbf{v}\cdot\mathbf{r}}\hat{\psi}^\dagger(\mathbf{r})\end{aligned} \tag{2.61}$$

(note that the Schrödinger-representation operator $\hat{\psi}'(\mathbf{r}',t)$ is time dependent). From (2.61), we obtain for the density and current operators,

$$\begin{aligned}\hat{n}'(\mathbf{r}') &= \hat{U}^\dagger(\mathbf{v},t)\hat{n}(\mathbf{r}')\hat{U}(\mathbf{v},t) = \hat{n}(\mathbf{r}), \\ \hat{\mathbf{j}}'(\mathbf{r}') &= \hat{U}^\dagger(\mathbf{v},t)\hat{\mathbf{j}}(\mathbf{r}')\hat{U}(\mathbf{v},t) = \hat{\mathbf{j}}(\mathbf{r}) - \hat{n}(\mathbf{r})\mathbf{v}.\end{aligned} \tag{2.62}$$

The transformation of correlation functions is easily deduced from (2.61).

2.2.6 *Time reversal*

2.2.6.1 *Antiunitary operators*

Before discussing time-reversal symmetry, we should recall the definition and properties of antilinear and antiunitary operators. An operator $\hat{A}$ is antilinear if

$$\hat{A}(\alpha|\Psi\rangle + \beta|\Phi\rangle) = \alpha^*\hat{A}|\Psi\rangle + \beta^*\hat{A}|\Phi\rangle, \tag{2.63}$$

where α and β are complex numbers. Antilinearity implies that $[\langle\Psi|(\hat{A}|\Phi\rangle)]^*$ is a linear function of $|\Phi\rangle$, i.e. $[\langle\Psi|(\hat{A}(\alpha_1|\Phi_1\rangle + \alpha_2|\Phi_2\rangle))]^* = \alpha_1[\langle\Psi|(\hat{A}|\Phi_1\rangle)]^* + \alpha_2[\langle\Psi|(\hat{A}|\Phi_2\rangle)]^*$. This defines a bra that we denote by $\langle\Psi|\hat{A}$:

$$(\langle\Psi|\hat{A})|\Phi\rangle = [\langle\Psi|(\hat{A}|\Phi\rangle)]^*. \tag{2.64}$$

The Dirac notation must be used with care with antilinear operators since $(\langle\Psi|\hat{A})|\Phi\rangle \neq \langle\Psi|(\hat{A}|\Phi\rangle)$; one should always specify whether the operator acts on the ket or the bra.

[8]We use the first expression of $\hat{U}(\mathbf{v},t)$ in (2.54), generalized to the case of N particles.

The adjoint $\hat{A}^\dagger$ is defined by $(|\Psi\rangle, \hat{A}|\Phi\rangle) = (\hat{A}^\dagger|\Psi\rangle, |\Phi\rangle)^*$, where $(\cdot,\cdot)$ denotes the scalar product.[9] In Dirac's notation,

$$\langle\Psi|(\hat{A}|\Phi\rangle) = \langle\Phi|(\hat{A}^\dagger|\Psi\rangle). \tag{2.65}$$

If each of the operators $\hat{A}_1$ and $\hat{A}_2$ is either linear or antilinear, $(\hat{A}_1\hat{A}_2)^\dagger = \hat{A}_2^\dagger\hat{A}_1^\dagger$. A product of N_1 linear and N_2 antilinear operators is linear (antilinear) if N_2 is even (odd).

An antilinear operator $\hat{A}$ is said to be antiunitary if $\hat{A}^\dagger\hat{A} = 1$. One then has

$$(\hat{A}|\Psi\rangle, \hat{A}|\Phi\rangle) = (\langle\Psi|\hat{A}^\dagger)(\hat{A}|\Phi\rangle) = \langle\Psi|\Phi\rangle^* = (|\Psi\rangle, |\Phi\rangle)^*. \tag{2.66}$$

This equation shows that an antiunitary operator conserves the modulus $|\langle\Psi|\Phi\rangle|$ of the scalar product and is therefore a possible candidate as a Wigner operator associated with a symmetry transformation.

2.2.6.2 *Wigner operator for a spin-zero particle*

After these technical preliminaries, we can turn back to the two observers $\mathcal{O}$ and $\mathcal{O}'$ and discuss time-reversal symmetry. In this section, we consider only time-independent Hamiltonians. We assume that the two observers use clocks moving opposite in time with respect to one another, i.e.

$$t(\mathcal{O}') = -t(\mathcal{O}) \tag{2.67}$$

if we take the two clocks to coincide at $t = 0$.

Let us start with classical mechanics. If $\mathcal{O}$ describes a classical particle by its position $\mathbf{r}(t)$ and momentum $\mathbf{p}(t)$, $\mathcal{O}'$ describes the same particle by $\mathbf{r}'(t) = \mathbf{r}(-t)$ and $\mathbf{p}'(t) = -\mathbf{p}(-t)$. Time-reversal invariance means that if $(\mathbf{r}(t), \mathbf{p}(t))$ is a possible trajectory, then $(\mathbf{r}'(t), \mathbf{p}'(t))$ is also a solution of the equations of motion $\dot{\mathbf{r}} = \partial H/\partial p$ and $\dot{\mathbf{p}} = -\partial H/\partial \mathbf{r}$. Clearly, this requires the Hamiltonian to be an even function of $\mathbf{p}$: $H(\mathbf{p}, \mathbf{r}) = H(-\mathbf{p}, \mathbf{r})$. The quantum Hamiltonian $H(-i\boldsymbol{\nabla}, \mathbf{r})$ is then real in the coordinate representation.[10]

In quantum mechanics, time-reversal symmetry defines a Wigner operator $\hat{U}(t)$ by

$$|\Psi'(t)\rangle = \hat{U}(t)|\Psi(t)\rangle, \tag{2.68}$$

where $|\Psi(t)\rangle$ and $|\Psi'(t)\rangle$ are the vectors assigned to the system by $\mathcal{O}$ and $\mathcal{O}'$ at the same subjective time t. It is convenient to introduce the operator $\hat{\Theta}$ relating the vectors $|\Psi\rangle$ and $|\Psi'\rangle$ at the same objective time (t for $\mathcal{O}'$ and $-t$ for $\mathcal{O}$),

$$|\Psi'(t)\rangle = \hat{\Theta}|\Psi(-t)\rangle. \tag{2.69}$$

---

[9] The usual definition of the adjoint of a linear operator, $(|\Psi\rangle, \hat{A}|\Phi\rangle) = (\hat{A}^\dagger|\Psi\rangle, |\Phi\rangle)$, cannot be used since the lhs is here linear in $|\Psi\rangle$ while the rhs would be antilinear if $\hat{A}$ were antilinear.

[10] The Hamiltonian $H(-i\boldsymbol{\nabla}, \mathbf{r}) = -\boldsymbol{\nabla}^2/2m + V(\mathbf{r})$ of a particle moving in a potential $V(\mathbf{r})$ is real. When the particle is subjected to a vector potential, the Hamiltonian is not real anymore and time-reversal invariance is broken.

We shall see that $\hat{\Theta}$ is time independent. $\hat{U}(t)$ and $\hat{\Theta}$ are related by $\hat{U}(t) = \hat{\Theta}e^{2i\hat{H}t} = e^{-2i\hat{H}'t}\hat{\Theta}$ where $\hat{H}'$ is the Hamiltonian used by $\mathcal{O}'$.[11] According to Wigner's theorem, $\hat{U}(t)$ and $\hat{\Theta}$ are both either unitary or antiunitary. In both cases, they satisfy $\hat{U}^\dagger(t)\hat{U}(t) = \hat{\Theta}^\dagger\hat{\Theta} = 1$.

By analogy with classical mechanics, we require

$$\begin{aligned}\langle\Psi'(t)|\hat{\mathbf{r}}|\Psi'(t)\rangle &= \langle\Psi(-t)|\hat{\mathbf{r}}|\Psi(-t)\rangle,\\ \langle\Psi'(t)|\hat{\mathbf{p}}|\Psi'(t)\rangle &= -\langle\Psi(-t)|\hat{\mathbf{p}}|\Psi(-t)\rangle.\end{aligned} \tag{2.70}$$

This implies

$$\begin{aligned}\hat{\Theta}^\dagger\hat{\mathbf{r}}\hat{\Theta} &= \hat{\mathbf{r}},\\ \hat{\Theta}^\dagger\hat{\mathbf{p}}\hat{\Theta} &= -\hat{\mathbf{p}}.\end{aligned} \tag{2.71}$$

The preceding equations define $\hat{\Theta}$ only up to a phase factor. We fix this phase factor by specifying the action of $\hat{\Theta}$ on the eigenvectors $|\mathbf{r}\rangle$ of the position operator,

$$\hat{\Theta}|\mathbf{r}\rangle = |\mathbf{r}\rangle. \tag{2.72}$$

It is easy to show that this equation implies $\hat{\Theta}|\mathbf{p}\rangle = |\mathbf{p}\rangle$ if $\hat{\Theta}$ is linear and $\hat{\Theta}|\mathbf{p}\rangle = |-\mathbf{p}\rangle$ if $\hat{\Theta}$ is antilinear.[12] Since Eqs. (2.71) imply $\hat{\Theta}|\mathbf{p}\rangle = |-\mathbf{p}\rangle$, we conclude that $\hat{\Theta}$ must be antilinear (and therefore antiunitary). The antilinearity of $\hat{\Theta}$ can also be seen by evaluating $\hat{\Theta}^\dagger i\hat{\Theta} = \hat{\Theta}^\dagger[\hat{r}_\mu, \hat{p}_\mu]\hat{\Theta} = -[\hat{r}_\mu, \hat{p}_\mu] = -i$.

The time reversed Hamiltonian $\hat{H}'$ is deduced from the equation of motion of the time reversed vector $|\Psi'(t)\rangle$,

$$\begin{aligned}i\partial_t|\Psi'(t)\rangle &= i\partial_t\hat{\Theta}|\Psi(-t)\rangle = -\hat{\Theta}i\partial_t|\Psi(-t)\rangle = \hat{\Theta}\hat{H}|\Psi(-t)\rangle = \hat{\Theta}\hat{H}\hat{\Theta}^\dagger|\Psi'(t)\rangle\\ &\equiv \hat{H}'|\Psi'(t)\rangle,\end{aligned} \tag{2.73}$$

i.e.

$$\hat{H}' = \hat{\Theta}\hat{H}\hat{\Theta}^\dagger. \tag{2.74}$$

Time-reversal invariance means that $|\Psi'(t)\rangle$ and $|\Psi(t)\rangle$ satisfy the same Schrödinger equation, $\hat{H}' = \hat{H}$, i.e.

$$[\hat{H}, \hat{\Theta}] = 0. \tag{2.75}$$

If $\hat{\Theta}$ were linear, we would obtain $\hat{H}' = -\hat{\Theta}\hat{H}\hat{\Theta}^\dagger$ and therefore $\hat{H}\hat{\Theta} + \hat{\Theta}\hat{H} = 0$ in a time-reversal-invariant system. From the latter equation, we deduce that if $|\Psi(t=0)\rangle$ is an eigenstate with energy E, then the time reversed vector $\hat{\Theta}|\Psi(t=0)\rangle$ is an eigenstate with energy $-E$: $\hat{H}\hat{\Theta}|\Psi\rangle = -\hat{\Theta}\hat{H}|\Psi\rangle = -E\hat{\Theta}|\Psi\rangle$. The elementary case of a free particle is

[11] We use the fact that the time-evolution operators are $\hat{T}(t+\tau, t) = e^{-i\hat{H}\tau}$ and $\hat{T}'(t+\tau, t) = e^{-i\hat{H}'\tau}$ for $\mathcal{O}$ and $\mathcal{O}'$, respectively.

[12] This follows from $\hat{\Theta}|\mathbf{p}\rangle = \hat{\Theta}\int d^dr|\mathbf{r}\rangle\langle\mathbf{r}|\mathbf{p}\rangle = \int d^dr\langle\mathbf{r}|\mathbf{p}\rangle^*\hat{\Theta}|\mathbf{r}\rangle = \int d^dr\langle\mathbf{r}|-\mathbf{p}\rangle|\mathbf{r}\rangle = |-\mathbf{p}\rangle$ when $\hat{\Theta}$ is antilinear.

sufficient to show that this result is meaningless, since the momentum eigenstate $|\mathbf{p}\rangle$ and $\hat{\Theta}|\mathbf{p}\rangle = |-\mathbf{p}\rangle$ are eigenstates of the Hamiltonian with the same energy $\mathbf{p}^2/2m$. Note that time-reversal invariance [Eq. (2.75)] does not imply a conservation law as for the other symmetry transformations. Since $\hat{\Theta}$ is antilinear, its equation of motion in the Heisenberg picture, $i\partial_t\hat{\Theta}_H = -\hat{H}\hat{\Theta}_H - \hat{\Theta}_H\hat{H}$, involves the anticommutator of $\hat{H}$ and $\hat{\Theta}$ instead of the commutator.

In order to find an explicit expression of the operator $\hat{\Theta}$, we consider the wavefunction in the coordinate representation,

$$\begin{aligned}\Psi'(\mathbf{r},t) &= \langle\mathbf{r}|\Psi'(t)\rangle = \langle\mathbf{r}|(\hat{\Theta}|\Psi(-t)\rangle) = \int d^dr'\,\langle\mathbf{r}|(\hat{\Theta}|\mathbf{r}'\rangle\langle\mathbf{r}'|\Psi(-t)\rangle) \\ &= \int d^dr'\,\Psi^*(\mathbf{r}',-t)\langle\mathbf{r}|(\hat{\Theta}|\mathbf{r}'\rangle) = \Psi^*(\mathbf{r},-t).\end{aligned} \tag{2.76}$$

This shows that for a spin-zero particle, $\hat{\Theta}$ is the complex conjugation operator $\hat{K}$ in the coordinate representation,

$$\Psi'(\mathbf{r},t) = \hat{K}\Psi(\mathbf{r},-t) = \Psi^*(\mathbf{r},-t). \tag{2.77}$$

One can verify that $\Psi^*(\mathbf{r},-t)$ is a solution of the Schrödinger equation when the Hamiltonian is real in the coordinate representation. By taking the complex conjugate of $i\partial_t\Psi(\mathbf{r},t) = \hat{H}\Psi(\mathbf{r},t)$, one indeed obtains

$$i\partial_t\Psi^*(\mathbf{r},-t) = \hat{H}\Psi^*(\mathbf{r},-t). \tag{2.78}$$

2.2.6.3 *Particle with spin*

So far, we have considered only a spin-zero particle. Since the orbital angular momentum transforms as $\hat{\Theta}^\dagger\hat{\mathbf{r}}\times\hat{\mathbf{p}}\hat{\Theta} = -\hat{\mathbf{r}}\times\hat{\mathbf{p}}$, we demand that $\hat{\Theta}$ reverses the direction of the spin,

$$\hat{\Theta}^\dagger\hat{\mathbf{S}}\hat{\Theta} = -\hat{\mathbf{S}}. \tag{2.79}$$

Let us now denote by $\hat{K}$ the complex conjugation operator in the $(\mathbf{r},S^z)$ representation. With the standard choice of the eigenstates $|\mathbf{r},\sigma\rangle$ $(\sigma = -S,\dots,S)$ of $\hat{S}^z$, the spin operators satisfy[13,14]

$$\begin{aligned}\hat{K}^\dagger\hat{S}^x\hat{K} &= (\hat{S}^x)^* = \hat{S}^x, \\ \hat{K}^\dagger\hat{S}^y\hat{K} &= (\hat{S}^y)^* = -\hat{S}^y, \\ \hat{K}^\dagger\hat{S}^z\hat{K} &= (\hat{S}^z)^* = \hat{S}^z.\end{aligned} \tag{2.80}$$

Thus, $\hat{K}$ performs the time-reversal transformation for S^y but not for S^x and S^z. It is clear that $\hat{\Theta}$ can be written as the product $\hat{Y}\hat{K}$ of $\hat{K}$ and the unitary operator $\hat{Y} = e^{-i\pi\hat{S}^y}$ that

---

[13]This comes from $\langle\mathbf{r},\sigma|(\hat{K}^\dagger\hat{S}^\nu\hat{K})|\mathbf{r}',\sigma'\rangle = \langle\mathbf{r},\sigma|(\hat{K}^\dagger\hat{S}^\nu|\mathbf{r}',\sigma'\rangle) = (\hat{S}^\nu_{\sigma,\sigma'})^*\langle\mathbf{r},\sigma|(\hat{K}^\dagger|\mathbf{r}',\sigma'\rangle) = (\hat{S}^\nu_{\sigma,\sigma'})^*\langle\mathbf{r}',\sigma'|\mathbf{r},\sigma\rangle = (\hat{S}^\nu_{\sigma,\sigma'})^*\langle\mathbf{r},\sigma|\mathbf{r}',\sigma'\rangle$ and the fact that $\hat{S}^\nu_{\sigma,\sigma'} = \langle\sigma|\hat{S}^\nu|\sigma'\rangle$ is real if $\nu = x,z$ and imaginary if $\nu = y$.

[14]Note that $\hat{K}^\dagger = \hat{K}$ and $\hat{K}^2 = 1$.

performs a rotation of angle π about the y axis. Thus, we obtain

$$\hat{\Theta} = \eta e^{-i\pi \hat{S}^y} \hat{K}, \tag{2.81}$$

where η is an arbitrary complex number of modulus unity. Equation (2.81) yields[14] $\hat{\Theta}^2 = e^{-2i\pi \hat{S}^y} = (-1)^{2S}$, i.e. $\hat{\Theta}^2 = 1$ for a boson and $\hat{\Theta}^2 = -1$ for a fermion, regardless of the value of the phase factor η ($\eta = 1$ in the following). Combined with $\hat{\Theta}^\dagger \hat{\Theta} = 1$, this implies $\hat{\Theta}^\dagger = \hat{\Theta}$ (boson) and $\hat{\Theta}^\dagger = -\hat{\Theta}$ (fermion). For N fermions, we would obtain $\hat{\Theta}^2 = (-1)^N$.[15]

For a spin-$\frac{1}{2}$ fermion, $\hat{\mathbf{S}} = \hat{\boldsymbol{\sigma}}/2$ and $\hat{\Theta} = -i\hat{\sigma}^y \hat{K}$. Using

$$\begin{aligned} \hat{\Theta}|\mathbf{r}, \sigma\rangle &= \sigma|\mathbf{r}, -\sigma\rangle, \\ \hat{\Theta}|\mathbf{p}, \sigma\rangle &= \sigma| - \mathbf{p}, -\sigma\rangle, \end{aligned} \tag{2.82}$$

we find that the wavefunction of the time-reversed state in the $(\mathbf{r}, S^z)$ representation reads

$$\begin{aligned} \Psi'_\sigma(\mathbf{r}, t) &= \langle \mathbf{r}, \sigma|(\hat{\Theta}|\Psi(-t)\rangle) \\ &= \int d^d r' \sum_{\sigma'} \langle \mathbf{r}, \sigma|(\hat{\Theta}|\mathbf{r}', \sigma'\rangle \langle \mathbf{r}', \sigma'|\Psi(-t)\rangle) \\ &= \int d^d r' \sum_{\sigma'} \sigma' \langle \mathbf{r}, \sigma|\mathbf{r}', -\sigma'\rangle \Psi^*_{\sigma'}(\mathbf{r}', -t) = -\sigma \Psi^*_{-\sigma}(\mathbf{r}, -t). \end{aligned} \tag{2.83}$$

2.2.6.4 *Many-particle system*

Equation (2.81) is readily generalized to a many-particle system, $\hat{\mathbf{S}}$ then denoting the total spin operator. One has

$$\begin{aligned} &\hat{\Theta}|\mathbf{r}_1, \ldots, \mathbf{r}_N\} = |\mathbf{r}_1, \ldots, \mathbf{r}_N\} \qquad \text{(spin-zero bosons)}, \\ &\hat{\Theta}|\mathbf{r}_1, \sigma_1; \ldots; \mathbf{r}_N, \sigma_N\} = \left(\prod_i \sigma_i\right) |\mathbf{r}_1, -\sigma_1; \ldots; \mathbf{r}_N, -\sigma_N\} \qquad \text{(spin-}\tfrac{1}{2}\text{ fermions)}. \end{aligned} \tag{2.84}$$

As can be verified by a direct calculation, the annihilation and creation operators transform as

$$\begin{aligned} \hat{\Theta}^\dagger \hat{\psi}^{(\dagger)}(\mathbf{r}) \hat{\Theta} &= \hat{\psi}^{(\dagger)}(\mathbf{r}), \\ \hat{\Theta}^\dagger \hat{\psi}^{(\dagger)}(\mathbf{r}, t) \hat{\Theta} &= \hat{\psi}^{(\dagger)}(\mathbf{r}, -t) \end{aligned} \qquad \text{(spin-zero bosons)} \tag{2.85}$$

and

$$\begin{aligned} \hat{\Theta}^\dagger \hat{\psi}^{(\dagger)}_\sigma(\mathbf{r}) \hat{\Theta} &= -\sigma \hat{\psi}^{(\dagger)}_{-\sigma}(\mathbf{r}), \\ \hat{\Theta}^\dagger \hat{\psi}^{(\dagger)}_\sigma(\mathbf{r}, t) \hat{\Theta} &= -\sigma \hat{\psi}^{(\dagger)}_{-\sigma}(\mathbf{r}, -t) \end{aligned} \qquad \text{(spin-}\tfrac{1}{2}\text{ fermions)}. \tag{2.86}$$

---

[15] This result is at the origin of Kramers' degeneracy in time-reversal-invariant systems with an odd number of fermions (i.e. a half-integer total spin). In that case, any eigenstate $|\Psi\rangle$ of the Hamiltonian is degenerate with $\hat{\Theta}|\Psi\rangle$ whereas $\hat{\Theta}^2 = -1$ implies $\langle \Psi|(\hat{\Theta}|\Psi\rangle) = 0$, so that every energy level is at least doubly degenerate.

For any operator $\hat{A}$, the matrix element between time-reversed states is given by

$$(\langle\Psi|\hat{\Theta}^\dagger)\hat{A}(\hat{\Theta}|\Phi\rangle) = \langle\Psi|(\hat{\Theta}^\dagger\hat{A}\hat{\Theta})|\Phi\rangle^* = \langle\Phi|(\hat{\Theta}^\dagger\hat{A}^\dagger\hat{\Theta})|\Psi\rangle. \tag{2.87}$$

For the operator $\hat{A}(t) = e^{i\hat{H}t}\hat{A}e^{-i\hat{H}t}$ defined in the Heisenberg representation,

$$(\langle\Psi|\hat{\Theta}^\dagger)\hat{A}(t)(\hat{\Theta}|\Phi\rangle) = \langle\Phi|e^{-i\hat{H}t}(\hat{\Theta}^\dagger\hat{A}^\dagger\hat{\Theta})e^{i\hat{H}t}|\Psi\rangle = \epsilon_A^T\langle\Phi|\hat{A}^\dagger(-t)|\Psi\rangle, \tag{2.88}$$

when the system is time-reversal invariant ($[\hat{H},\hat{\Theta}] = 0$). The last result in (2.88) holds if $\hat{A}$ is even or odd under time reversal: $\hat{\Theta}^\dagger\hat{A}\hat{\Theta} = \epsilon_A^T\hat{A}$ and $\hat{\Theta}^\dagger\hat{A}(t)\hat{\Theta} = \epsilon_A^T\hat{A}(-t)$ with $\epsilon_A^T = \pm$ the signature.[16,17] Let us now consider the zero-temperature correlation function $\chi_{AB}(t,t') = \langle 0|\hat{A}(t)\hat{B}(t')|0\rangle$, where $|0\rangle$ denotes the ground state. If the system is time-reversal invariant, $\hat{H}' = \hat{H}$, $\chi_{AB}(t,t')$ must satisfy

$$\chi_{AB}(t,t') = (\langle 0|\hat{\Theta}^\dagger)\hat{A}(t)\hat{B}(t')(\hat{\Theta}|0\rangle). \tag{2.89}$$

Using (2.87), one finds

$$\begin{aligned}\chi_{AB}(t,t') &= \langle 0|(\hat{\Theta}^\dagger\hat{B}^\dagger(t')\hat{A}^\dagger(t)\hat{\Theta})|0\rangle \\ &= \langle 0|(e^{-i\hat{H}t'}\hat{\Theta}^\dagger\hat{B}^\dagger\hat{\Theta}e^{i\hat{H}t'}e^{-i\hat{H}t}\hat{\Theta}\hat{A}^\dagger\hat{\Theta}e^{i\hat{H}t})|0\rangle \\ &= \chi_{\Theta^\dagger\hat{B}^\dagger\hat{\Theta},\Theta^\dagger\hat{A}^\dagger\Theta}(-t',-t),\end{aligned} \tag{2.90}$$

i.e.

$$\chi_{AB}(\omega) = \chi_{\Theta^\dagger\hat{B}^\dagger\hat{\Theta},\Theta^\dagger\hat{A}^\dagger\Theta}(\omega) = \epsilon_A^T\epsilon_B^T\chi_{B^\dagger A^\dagger}(\omega), \tag{2.91}$$

where the last result holds for two operators with signatures ϵ_A^T and ϵ_B^T. Equations (2.87) and (2.90) also hold for the imaginary-time correlation function,[18]

$$\begin{aligned}\chi_{AB}(\tau,\tau') &= \langle 0|\hat{A}(\tau)\hat{B}(\tau')|0\rangle = \chi_{\Theta^\dagger\hat{B}^\dagger\hat{\Theta},\Theta^\dagger\hat{A}^\dagger\Theta}(-\tau',-\tau), \\ \chi_{AB}(i\omega_n) &= \chi_{\Theta^\dagger\hat{B}^\dagger\hat{\Theta},\Theta^\dagger\hat{A}^\dagger\Theta}(i\omega_n) = \epsilon_A^T\epsilon_B^T\chi_{B^\dagger A^\dagger}(i\omega_n)\end{aligned} \tag{2.92}$$

(for two operators with signatures ϵ_A^T and ϵ_B^T). It is easy to show that Eqs. (2.90–2.92) also hold for the time-ordered correlation function $\chi_{AB}(t,t') = \langle 0|T\hat{A}(t)\hat{B}(t')|0\rangle$ and $\chi_{AB}(\tau,\tau') = \langle 0|T_\tau\hat{A}(\tau)\hat{B}(\tau')|0\rangle$ and at finite temperature where the ground state expectation value $\langle 0|\cdots|0\rangle$ is replaced by the thermal average $\langle\cdots\rangle = \frac{1}{Z}\mathrm{Tr}(e^{-\beta\hat{H}}\ldots)$.

---

[16] $\hat{A}$ and $\hat{A}^\dagger$ have obviously the same signature ϵ_A^T under time reversal.

[17] For example, the density operator $\hat{n}(\mathbf{r},t) = \sum_\sigma\hat{\psi}_\sigma^\dagger(\mathbf{r},t)\hat{\psi}_\sigma(\mathbf{r},t)$ is even under time reversal while the current operator $\hat{\mathbf{j}}(\mathbf{r},t) = -\frac{i}{2m}\sum_\sigma(\hat{\psi}_\sigma^\dagger(\mathbf{r},t)\boldsymbol{\nabla}\hat{\psi}_\sigma(\mathbf{r},t) - \text{c.c.})$ is odd: $\epsilon_n^T = 1$ and $\epsilon_j^T = -1$.

[18] This comes from $\hat{\Theta}^\dagger\hat{B}(\tau')^\dagger\hat{A}(\tau)^\dagger\hat{\Theta} = e^{-\tau'\hat{H}}\hat{\Theta}^\dagger\hat{B}\hat{\Theta}e^{(\tau'-\tau)\hat{H}}\hat{\Theta}^\dagger\hat{A}\hat{\Theta}e^{\tau\hat{H}}$, since $[\hat{A}(\tau)]^\dagger = \hat{A}^\dagger(-\tau)$ and $[\hat{B}(\tau)]^\dagger = \hat{B}^\dagger(-\tau)$.

At finite temperature, the correlation function $\chi_{AB}(t,t') = \mathrm{Tr}[\hat\rho\hat A(t)\hat B(t')]$ satisfies

$$\begin{aligned}\chi_{AB}(t,t') &= \sum_n (\langle n|\hat\Theta^\dagger)\hat\rho\hat A(t)\hat B(t')(\hat\Theta|n\rangle) \\ &= \sum_n \langle n|(\hat\Theta^\dagger\hat\rho\hat A(t)\hat B(t')\hat\Theta)|n\rangle^* \\ &= \mathrm{Tr}[\hat\rho\hat\Theta^\dagger\hat B^\dagger(t')\hat A^\dagger(t)\hat\Theta]\end{aligned} \tag{2.93}$$

in a time-reversal-invariant system, where the last line is obtained using $[\hat\rho,\hat\Theta] = 0$ in a time-reversal-invariant system and the cyclic invariance of the trace ($\hat\rho = e^{-\beta\hat H}/Z$). Hence, $\chi_{AB}(t,t') = \chi_{\Theta^\dagger\hat B^\dagger\hat\Theta,\Theta^\dagger\hat A^\dagger\Theta}(-t',-t)$, a property that can also be derived for the time-ordered correlation function $\chi_{AB}(t,t') = \mathrm{Tr}\{\hat\rho[T\hat A(t)\hat B(t')]\}$ and the imaginary-time correlation function $\chi_{AB}(\tau,\tau') = \mathrm{Tr}\{\hat\rho[T_\tau\hat A(\tau)\hat B(\tau')]\}$.

In a time-reversal-invariant system, the one-particle correlation function satisfies

$$G(\mathbf r,t;\mathbf r',t') = -i\langle T\hat\psi(\mathbf r,t)\hat\psi^\dagger(\mathbf r',t')\rangle = G(\mathbf r',-t';\mathbf r,-t) \tag{2.94}$$

for spin-zero bosons and

$$G_{\sigma\sigma'}(\mathbf r,t;\mathbf r',t') = -i\langle T\hat\psi_\sigma(\mathbf r,t)\hat\psi^\dagger_{\sigma}(\mathbf r',t')\rangle = \sigma\sigma' G_{\bar\sigma'\bar\sigma}(\mathbf r',-t';\mathbf r,-t) \tag{2.95}$$

for spin-$\frac12$ fermions (and analogous results in imaginary time). Similarly, the anomalous Green function in a superconductor (Sec. 7.5) satisfies

$$F_{\sigma\sigma'}(\mathbf r,t;\mathbf r',t') = -i\langle T\hat\psi_\sigma(\mathbf r,t)\hat\psi_{\sigma'}(\mathbf r',t')\rangle = \sigma\sigma' F^\dagger_{\bar\sigma'\bar\sigma}(\mathbf r',-t';\mathbf r,-t), \tag{2.96}$$

where $F^\dagger_{\sigma\sigma'}(\mathbf r,t;\mathbf r',t') = -i\langle T\hat\psi^\dagger_\sigma(\mathbf r,t)\hat\psi^\dagger_{\sigma'}(\mathbf r',t')\rangle$ (see also Sec. 3.5.4).

2.2.7 *U(1) symmetry and gauge transformation*

In the previous sections, we have considered spacetime symmetries, also called external symmetries, related to the structure of space and time. They are defined by a change of the spacetime coordinates $\mathbf r\to\mathbf r'$, $\mathbf p\to\mathbf p'$, $\mathbf S\to\mathbf S'$, $t\to t'$. The internal symmetries correspond to transformations that cannot be seen as arising from a change of the spacetime coordinates.

2.2.7.1 *Global U(1) transformation*

One of the simplest examples of internal symmetries is the global U(1) transformation, where the wavefunction changes by a mere phase factor,

$$\Psi'(\mathbf r,t) = e^{i\Lambda}\Psi(\mathbf r,t). \tag{2.97}$$

In a many-particle system, the Wigner operator is $\hat U = e^{i\Lambda\hat N}$, and the field operator transforms as

$$\begin{aligned}\hat U^\dagger\hat\psi(\mathbf r)\hat U &= e^{i\Lambda}\hat\psi(\mathbf r),\\ \hat U^\dagger\hat\psi^\dagger(\mathbf r)\hat U &= e^{-i\Lambda}\hat\psi^\dagger(\mathbf r)\end{aligned} \tag{2.98}$$

(we ignore the particle spin). Invariance of the Hamiltonian in the transformation (2.98) implies that the total number of particles $\hat{N}$ (the generator of the transformation) is conserved: $[\hat{N}, \hat{H}] = 0$ and $d\hat{N}_H/dt = 0$.

2.2.7.2 *Local U(1) transformation and gauge transformation*

In a local U(1) transformation, the field operator transforms as

$$\begin{aligned} \hat{U}^\dagger \hat{\psi}(\mathbf{r})\hat{U} &= e^{iq\Lambda(\mathbf{r},t)}\hat{\psi}(\mathbf{r}), \\ \hat{U}^\dagger \hat{\psi}^\dagger(\mathbf{r})\hat{U} &= e^{-iq\Lambda(\mathbf{r},t)}\hat{\psi}^\dagger(\mathbf{r}), \end{aligned} \tag{2.99}$$

where $\Lambda(\mathbf{r}, t)$ is now an arbitrary time- and space-dependent function. The Wigner operator is given by

$$\hat{U} = e^{iq\int d^d r\, \hat{n}(\mathbf{r})\Lambda(\mathbf{r},t)}, \tag{2.100}$$

where $\hat{n}(\mathbf{r})$ is the density of particles. The meaning of the parameter q will become clear in the following.

If the Hamiltonian conserves the number of particles, $[\hat{H}, \hat{N}] = 0$, it is invariant under a global U(1) transformation but not under a local one. One can however impose the invariance in the latter case but introducing a gauge field, i.e. a scalar potential ϕ and a vector potential $\mathbf{A}$, into the Hamiltonian

$$\hat{H} = \int d^d r \left\{ \frac{1}{2m}[(i\boldsymbol{\nabla} - q\mathbf{A})\hat{\psi}^\dagger(\mathbf{r})] \cdot (-i\boldsymbol{\nabla} - q\mathbf{A})\hat{\psi}(\mathbf{r}) + q\phi(\mathbf{r}, t)\hat{\psi}^\dagger(\mathbf{r})\hat{\psi}(\mathbf{r}) \right\} + \hat{H}_{\text{int}}. \tag{2.101}$$

In quantum field theory, the procedure which consists in including gauge fields in order to promote a global invariance to a local one is known as the gauge principle. The Hamiltonian (2.101) is invariant[19] in the gauge transformation defined by (2.99) and

$$\begin{aligned} \mathbf{A}'(\mathbf{r}, t) &= \mathbf{A}(\mathbf{r}, t) + \boldsymbol{\nabla}\Lambda(\mathbf{r}, t), \\ \phi'(\mathbf{r}, t) &= \phi(\mathbf{r}, t) - \partial_t \Lambda(\mathbf{r}, t), \end{aligned} \tag{2.102}$$

assuming that $\hat{H}_{\text{int}}$ is invariant in the local U(1) transformation.[20] $\hat{H}$ is nothing but the Hamiltonian of particles with electric charge q coupled to the electromagnetic field $\mathbf{E} = -\boldsymbol{\nabla}\phi - \partial_t \mathbf{A}$ and $\mathbf{B} = \boldsymbol{\nabla} \times \mathbf{A}$ (Sec. 1.9).

It should be noted that a gauge transformation, although it leaves the Hamiltonian invariant, is not a symmetry transformation. It is a mere mathematical manipulation which leaves the description of the system invariant and does not correspond to any physical transformation of the actual system.

---

[19]i.e. $\hat{U}\hat{H}[\hat{\psi}^\dagger, \hat{\psi}, \phi, \mathbf{A}]\hat{U}^\dagger + i(\partial_t \hat{U})\hat{U}^\dagger = \hat{H}[\hat{\psi}^\dagger, \hat{\psi}, \phi', \mathbf{A}']$ [see Eq. (2.14)].

[20]This is the case for a density–density interaction: $\hat{H}_{\text{int}} = \frac{1}{2}\int d^d r d^d r'\, v(\mathbf{r} - \mathbf{r}')\hat{n}(\mathbf{r})\hat{n}(\mathbf{r}')$.

2.3 Symmetries and Functional Integrals

In this section, we start with classical field theory and Noether's theorem before discussing symmetries in the functional integral formalism. In particular, we show that invariance in a continuous symmetry transformation implies a set of constraints for the Green functions and the vertex functions (Ward identities) and briefly discuss the consequences of spontaneous symmetry breaking (Goldstone's theorem).

2.3.1 *Classical field theory — Noether's theorem*

We consider a classical system with the action

$$S[\psi_j] = \int dt \int d^d r\, \mathcal{L}(\psi_j, \dot{\psi}_j, \boldsymbol{\nabla}\psi_j) \tag{2.103}$$

and the Lagrangian density $\mathcal{L}$, where the field $\psi_j(x)$ is a function of $x = (\mathbf{r}, t)$ and a discrete index j (which may include a spin index or refer to the real and imaginary parts of a complex field, etc.). The dynamics of the system is given by the Euler–Lagrange equations obtained from the minimization of the action,

$$\frac{d}{dt}\frac{\partial \mathcal{L}}{\partial \dot{\psi}_j(x)} = \frac{\partial \mathcal{L}}{\partial \psi_j(x)} - \boldsymbol{\nabla} \cdot \frac{\partial \mathcal{L}}{\partial \boldsymbol{\nabla}\psi_j(x)}. \tag{2.104}$$

In a symmetry transformation $\psi_j(x) \to \psi'_j(x)$, the Lagrangian density transforms into $\mathcal{L}'(\psi'_j, \dot{\psi}'_j, \boldsymbol{\nabla}\psi'_j)$. For instance, for a space translation, $\psi'_j(\mathbf{r}, t) = \psi_j(\mathbf{r} - \mathbf{a}, t)$. Since ψ_j and ψ'_j refer to the same system, one must have $S'[\psi'] = S[\psi]$; $\mathcal{L}(\psi_j, \dot{\psi}_j, \boldsymbol{\nabla}\psi_j)$ and $\mathcal{L}'(\psi'_j, \dot{\psi}'_j, \boldsymbol{\nabla}\psi'_j)$ then coincide up to a total time or space derivative. If $S'[\psi'] = S[\psi']$, the equation of motion of $\psi'_j(x)$ is the same as that of $\psi_j(x)$ and the system is invariant in the transformation $\psi_j(x) \to \psi'_j(x)$ (see Sec. 1.A.3).

2.3.1.1 *Noether's theorem*

Let us consider a continuous transformation, defined by

$$\psi_j(x) \to \psi_j(x) + i\epsilon F_j(x) \qquad (\epsilon \to 0) \tag{2.105}$$

in its infinitesimal version, which leaves the action invariant. Here, $F_j(x) \equiv F_j[x, \psi]$ is a function of x that depends functionally on ψ. When, as is often the case, $F_j(x)$ is a function of the field $\psi_{j'}(x)$ and its derivatives, Eq. (2.105) can be written as

$$\psi_j(x) \to \psi_j(x) - i\epsilon G \psi_j(x), \tag{2.106}$$

where G is the generator of the symmetry transformation (see Appendix 1.A.3). The Lagrangian density must be invariant up to a total derivative, i.e.

$$\mathcal{L}(\psi_j, \dot{\psi}_j, \boldsymbol{\nabla}\psi_j) \to \mathcal{L}(\psi_j, \dot{\psi}_j, \boldsymbol{\nabla}\psi_j) + \epsilon\partial_t \mathcal{F}_0(\psi_j) + \epsilon\boldsymbol{\nabla} \cdot \boldsymbol{\mathcal{F}}(\psi_j). \tag{2.107}$$

Noether's theorem states that for any continuous transformation, defined by (2.105) and (2.107), that leaves the action invariant, there is a density and a current,

$$\begin{aligned} j_0(x) &= -i\sum_j \frac{\partial \mathcal{L}}{\partial(\partial_t \psi_j(x))} F_j(x) + \mathcal{F}_0(x), \\ \mathbf{j}(x) &= -i\sum_j \frac{\partial \mathcal{L}}{\partial(\boldsymbol{\nabla} \psi_j(x))} F_j(x) + \boldsymbol{\mathcal{F}}(x) \end{aligned} \tag{2.108}$$

($\mathcal{F}_\mu(x) \equiv \mathcal{F}_\mu(\psi_j(x))$) that satisfy a continuity equation (assuming that the fields satisfy the Euler–Lagrange equations)

$$\partial_t j_0(x) + \boldsymbol{\nabla} \cdot \mathbf{j}(x) = 0. \tag{2.109}$$

This implies the existence of a conserved (time-independent) charge,

$$Q = \int d^d r\, j_0(x), \qquad \frac{d}{dt} Q = 0 \tag{2.110}$$

(see Sec. 1.A.3 for a more detailed discussion).

2.3.1.2 *What can we learn from Noether's theorem in quantum field theory?*

Let us consider an infinitesimal continuous transformation where the operator $\hat{\psi}_j(\mathbf{r})$ in the quantum theory transforms as

$$\hat{\psi}_j(\mathbf{r}) \to \hat{U}^\dagger \hat{\psi}_j(\mathbf{r}) \hat{U} \simeq \hat{\psi}_j(\mathbf{r}) + i\epsilon[\hat{t}, \hat{\psi}_j(\mathbf{r})] \equiv \hat{\psi}_j(\mathbf{r}) + i\epsilon \hat{F}_j(\mathbf{r}), \tag{2.111}$$

where $\hat{U} = e^{-i\epsilon\hat{t}}$ and $\hat{t}$ is the generator of the transformation.[21] For example, an infinitesimal space translation along the μ axis gives[22]

$$\begin{aligned} \hat{\psi}(\mathbf{r}) &\to \hat{\psi}(\mathbf{r}) - \epsilon \partial_\mu \hat{\psi}(\mathbf{r}), \\ \hat{\psi}^\dagger(\mathbf{r}) &\to \hat{\psi}^\dagger(\mathbf{r}) - \epsilon \partial_\mu \hat{\psi}^\dagger(\mathbf{r}) \end{aligned} \tag{2.112}$$

for a boson system, in agreement with the fact that the generator G of the translation is $P_\mu = -i\partial_\mu$ in the classical field theory (see Appendix 1.A.3). It is quite obvious that this result generalizes to other transformations: The operator $\hat{\psi}_j(x)$ in the quantum theory transforms as the classical field $\psi_j(x)$ in the Lagrangian formalism.

Let us now consider the Lagrangian formalism and an infinitesimal transformation defined by (2.105). The conserved classical current is given by (2.108). For simplicity, we assume that $\mathcal{F}_0 = \boldsymbol{\mathcal{F}} = 0$ and that for any canonical field, i.e. any field satisfying $\partial\mathcal{L}/\partial\dot{\psi}_j(\mathbf{r},t) \neq 0$, F_j does not depend on the conjugate momentum $\Pi_j(\mathbf{r},t) = \partial\mathcal{L}/\partial\dot{\psi}_j(\mathbf{r},t)$. We now quantize the theory and promote the classical (canonical) fields $\psi_j(\mathbf{r},t)$ and their conjugate momenta $\Pi_j(\mathbf{r},t)$ to operators satisfying the equal-time commutation relations

[21] See footnote 4 on page 196.

[22] We use $\hat{t} \equiv \hat{P}_\mu = \int d^d r\, \hat{\psi}^\dagger(\mathbf{r})(-i\partial_\mu)\hat{\psi}(\mathbf{r})$ (see Sec. 2.2.2).

$[\hat{\psi}_j(\mathbf{r}), \hat{\Pi}_{j'}(\mathbf{r}')] = i\delta_{j,j'}\delta(\mathbf{r}-\mathbf{r}')$ (we consider bosonic fields). From the conserved classical current, we obtain the density operator

$$\hat{j}_0(\mathbf{r}) = -i\sum_j \hat{\Pi}_j(\mathbf{r})\hat{F}_j(\mathbf{r}) \tag{2.113}$$

if we substitute the operators to the classical fields. Since by assumption $\hat{F}_j(\mathbf{r})$ does not depend on $\hat{\Pi}_j(\mathbf{r})$, we have

$$\begin{aligned} \int d^dr' [\hat{j}_0(\mathbf{r}'), \hat{\psi}_j(\mathbf{r})] &= -i\sum_{j'} \int d^dr' [\hat{\Pi}_{j'}(\mathbf{r}')\hat{F}_{j'}(\mathbf{r}'), \hat{\psi}_j(\mathbf{r})] \\ &= -i\sum_{j'} \int d^dr' [\hat{\Pi}_{j'}(\mathbf{r}'), \hat{\psi}_j(\mathbf{r})]\hat{F}_{j'}(\mathbf{r}') \\ &= -\hat{F}_j(\mathbf{r}) = -[\hat{t}, \hat{\psi}_j(\mathbf{r})], \end{aligned} \tag{2.114}$$

where the last result is obtained using (2.111). We conclude that[23]

$$\int d^dr\, \hat{j}_0(\mathbf{r}) = -\hat{t}. \tag{2.115}$$

The charge $\hat{Q} = \int d^dr\, \hat{j}_0(\mathbf{r})$, deduced from the classical field theory, is the generator of the symmetry transformation in the quantum theory. The invariance condition $d\hat{t}_H/dt = 0$ translates into

$$\frac{d}{dt}\int d^dr\, \hat{j}_0(\mathbf{r}, t) = 0, \tag{2.116}$$

where $\hat{j}_0(\mathbf{r}, t) = e^{i\hat{H}t}\hat{j}_0(\mathbf{r})e^{-i\hat{H}t}$ is the current operator in the Heisenberg picture. In most cases of interest, one can show, using the equation of motion $\partial_t\hat{j}_0(\mathbf{r}, t) = i[\hat{H}, \hat{j}_0(\mathbf{r}, t)]$, that the current operator satisfies the continuity equation

$$\partial_t\hat{j}_0(\mathbf{r}, t) + \boldsymbol{\nabla}\cdot\hat{\mathbf{j}}(\mathbf{r}, t) = 0, \tag{2.117}$$

directly obtained from its classical counterpart [Eq. (2.109)] by substituting the operators to the fields.

As an example, let us consider the Lagrangian density of a system of bosons with local interactions,

$$\mathcal{L} = i\psi^*\partial_t\psi - \frac{1}{2m}\boldsymbol{\nabla}\psi^*\cdot\boldsymbol{\nabla}\psi - \frac{g}{2}|\psi|^4, \tag{2.118}$$

[23]We should not worry about the minus sign in the rhs of (2.115). With the opposite sign convention for the generators (see footnote 4 on page 196), we would have obtained a plus sign. More generally, the Noether density and current are defined only up to a multiplication factor.

which is obtained in Sec. 1.8.2 from the derivation of the functional integral over coherent states starting from the quantum Hamiltonian. The action is invariant in the U(1) transformation

$$\begin{aligned}\psi(\mathbf{r},t) &\to \psi(\mathbf{r},t)e^{i\epsilon} = (1+i\epsilon)\psi(\mathbf{r},t) + \mathcal{O}(\epsilon^2),\\ \psi^*(\mathbf{r},t) &\to \psi^*(\mathbf{r},t)e^{-i\epsilon} = (1-i\epsilon)\psi(\mathbf{r},t) + \mathcal{O}(\epsilon^2).\end{aligned} \tag{2.119}$$

The Noether density and current are given by

$$\begin{aligned}j_0 &= -i\frac{\partial \mathcal{L}}{\partial(\partial_t \psi)}\psi + i\frac{\partial \mathcal{L}}{\partial(\partial_t \psi^*)}\psi^* = \psi^*\psi,\\ j_\mu &= -i\frac{\partial \mathcal{L}}{\partial(\partial_\mu \psi)}\psi + i\frac{\partial \mathcal{L}}{\partial(\partial_\mu \psi^*)}\psi^* = -\frac{i}{2m}(\psi^*\partial_\mu\psi - \text{c.c.}).\end{aligned} \tag{2.120}$$

If we substitute the operators $\hat\psi$ and $\hat\psi^\dagger$ to the classical fields ψ and ψ^*, we recover the usual expressions of the density $\hat\jmath_0$ and current $\hat{\mathbf{j}}$ operators. Using $i\partial_t\hat\jmath_0(\mathbf{r},t) = [\hat\jmath_0(\mathbf{r},t),\hat H]$ and the equal-time commutation relations $[\hat\psi(\mathbf{r},t),\hat\psi^\dagger(\mathbf{r}',t)] = \delta(\mathbf{r}-\mathbf{r}')$, it is straightforward to show that the continuity equation (2.117) is satisfied.

In the presence of an external gauge field $(\phi,\mathbf{A})$, the Lagrangian density is obtained by replacing the derivatives $\partial_t\psi$ and $\boldsymbol{\nabla}\psi$ by the covariant derivatives $D_t\psi = (\partial_t + iq\phi)\psi$ and $\mathbf{D}\psi = (\boldsymbol{\nabla} - iq\mathbf{A})\psi$, respectively, i.e.

$$\mathcal{L} = i\psi^*(\partial_t + iq\phi)\psi - \frac{1}{2m}(\boldsymbol{\nabla} + iq\mathbf{A})\psi^*\cdot(\boldsymbol{\nabla} - iq\mathbf{A})\psi - \frac{g}{2}|\psi|^4. \tag{2.121}$$

This procedure is referred to as minimal coupling and ensures that the Lagrangian is invariant in the gauge transformation defined by (2.102) and $\psi(\mathbf{r},t) \to \psi(\mathbf{r},t)e^{iq\Lambda(\mathbf{r},t)}$, $\psi^*(\mathbf{r},t) \to \psi^*(\mathbf{r},t)e^{-iq\Lambda(\mathbf{r},t)}$.[24] The action is still invariant in a global U(1) transformation (with the gauge field fixed). The Noether density $J_0 = j_0$ is unchanged, but the current density becomes

$$\mathbf{J} = \mathbf{j} - \frac{q}{m}\mathbf{A}\psi^*\psi \tag{2.122}$$

($\mathbf{j}$ is the paramagnetic part and the last term in the rhs the diamagnetic contribution), which again agrees with the expression of the current in the quantum theory once the fields have been replaced by operators. As a consequence of the minimal coupling, the density and current can be alternatively expressed as

$$\begin{aligned}qJ_0(x) &= -\frac{\partial \mathcal{L}}{\partial\phi(x)} = -\frac{\delta S}{\delta\phi(x)},\\ q\mathbf{J}(x) &= \frac{\partial \mathcal{L}}{\partial\mathbf{A}(x)} = \frac{\delta S}{\delta\mathbf{A}(x)}.\end{aligned} \tag{2.123}$$

[24]In the gauge transformation, the covariant derivatives transform as $D'_t\psi'(\mathbf{r},t) = e^{iq\Lambda(\mathbf{r},t)}D_t\psi(\mathbf{r},t)$ and $\mathbf{D}'\psi'(\mathbf{r},t) = e^{iq\Lambda(\mathbf{r},t)}\mathbf{D}\psi(\mathbf{r},t)$, where $D'_t = \partial_t + iq\phi'$ and $\mathbf{D}' = \boldsymbol{\nabla} - iq\mathbf{A}'$.

Since $\mathcal{L} \to -\mathcal{L}_E$ and $\phi \to -i\phi$ (Sec. 1.9.3) in the Wick rotation $t \to -i\tau$, in the Euclidean formalism

$$\begin{aligned} qJ_0(\mathbf{r},\tau) &= -i\frac{\delta S_E}{\delta\phi(\mathbf{r},\tau)} = \psi^*(\mathbf{r},\tau)\psi(\mathbf{r},\tau), \\ q\mathbf{J}(\mathbf{r},\tau) &= -\frac{\delta S_E}{\delta\mathbf{A}(\mathbf{r},\tau)} = -\frac{i}{2m}[\psi^*(\mathbf{r},\tau)\partial_\mu\psi(\mathbf{r},\tau) - \text{c.c.}] - \frac{q}{m}\mathbf{A}\psi^*(\mathbf{r},\tau)\psi(\mathbf{r},\tau). \end{aligned} \tag{2.124}$$

2.3.1.3 *Symmetries and invariances in the functional integral formalism*

In Sec. 2.2, we have seen how the invariance of a system in a symmetry transformation strongly constrains the correlation functions. This can also be seen in the functional integral formalism. Suppose we have identified a transformation $\psi_j(x) \to \psi'_j(x)$ which leaves the action invariant, $S[\psi'] = S[\psi]$, and has unit Jacobian.[25] One then has[26]

$$\begin{aligned} \langle \psi'_{j_1}(x_1)\cdots\psi'_{j_n}(x_n)\rangle_{S[\psi]} &= \frac{1}{Z}\int \mathcal{D}[\psi]\,\psi'_{j_1}(x_1)\cdots\psi'_{j_n}(x_n)e^{-S[\psi]} \\ &= \frac{1}{Z}\int \mathcal{D}[\psi']\,\psi'_{j_1}(x_1)\cdots\psi'_{j_n}(x_n)e^{-S[\psi']} \\ &= \langle \psi_{j_1}(x_1)\cdots\psi_{j_n}(x_n)\rangle_{S[\psi]}. \end{aligned} \tag{2.125}$$

The second line is obtained by a change of variable in the functional integral using $S[\psi] = S[\psi']$ and $\mathcal{D}[\psi] = \mathcal{D}[\psi']$, while the last result follows from a mere relabeling of the field. Invariance thus implies that the transformed field $\psi'_j(x)$ has the same correlation functions as the original field $\psi_j(x)$. For example, for a system which is invariant in a parity transformation, $\psi'_\sigma(\mathbf{r},\tau) = \psi_\sigma(-\mathbf{r},\tau)$, Eq. (2.125) implies

$$\langle \psi_\sigma(-\mathbf{r},\tau)\psi^*_{\sigma'}(-\mathbf{r}',\tau')\rangle = \langle \psi_\sigma(\mathbf{r},\tau)\psi^*_{\sigma'}(\mathbf{r}',\tau')\rangle \tag{2.126}$$

(σ, σ' are spin indices), in agreement with (the Euclidean version of) Eq. (2.32).

In the case of time reversal, the field $\psi^{(*)}(\mathbf{r},\tau)$ must transform like the one-particle wavefunction (and not like the field operators in (2.85, 2.86)), i.e.

$$\begin{aligned} \psi'(\mathbf{r},\tau) &= \psi^*(\mathbf{r},-\tau), \\ \psi'^*(\mathbf{r},\tau) &= \psi(\mathbf{r},-\tau), \end{aligned} \qquad \text{(bosons)} \tag{2.127}$$

and

$$\begin{aligned} \psi'_\sigma(\mathbf{r},\tau) &= -\sigma\psi^*_{-\sigma}(\mathbf{r},-\tau), \\ \psi'^*_\sigma(\mathbf{r},\tau) &= \sigma\psi_{-\sigma}(\mathbf{r},-\tau), \end{aligned} \qquad \text{(spin-}\tfrac{1}{2}\text{ fermions).} \tag{2.128}$$

The reason why ψ and ψ^* exchange roles under time reversal is that creation of a particle transforms into annihilation and *vice versa*. The different signs in the rhs of Eqs. (2.128)

[25]The case where the action is invariant but the Jacobian non-trivial is referred to as a (quantum) anomaly.
[26]From now on, we consider the Matsubara (imaginary-time) formalism where $x = (\mathbf{r},\tau)$. The field ψ_j can be a Grassmann variable.

come from the Grassmannian nature of the field.[27] The change $\tau \to -\tau$ in (2.127, 2.128) requires to extend the definition of $\psi^{(*)}(\tau)$ to the time interval $[-\beta, 0]$ with the property $\psi^{(*)}(\tau+\beta) = \zeta\psi^{(*)}(\tau)$.[28] From (2.125), (2.127) and (2.128), one easily checks that time-reversal invariance reproduces the previous results obtained within the operator formalism, e.g. Eqs. (2.94–2.96).

2.3.2 *Ward identities*

In this section and the following ones, we show that continuous transformations that leave the action invariant imply a set of relations, known as Ward identities, satisfied by correlation functions and 1PI/2PI vertices.

Let us consider the partition function $Z[J] = \int \mathcal{D}[\psi]\, e^{-S-S_J}$ in the presence of an external source,

$$\begin{aligned} S &= \int_0^\beta d\tau \int d^d r\, \mathcal{L}(\psi_j, \partial_\mu \psi_j), \\ S_J &= -\int_0^\beta d\tau \int d^d r \sum_j J_j(x)\psi_j(x), \end{aligned} \tag{2.129}$$

where $\mathcal{L}$ is now the Euclidean Lagrangian density. We assume that we have identified an infinitesimal transformation that leaves the measure of the functional integral and the action invariant [Eqs. (2.105) and (2.107)]. We then have[29]

$$\sum_j \left[iF_j(x) \frac{\partial \mathcal{L}}{\partial \psi_j(x)} + \sum_\mu i\partial_\mu F_j(x) \frac{\partial \mathcal{L}}{\partial(\partial_\mu \psi_j(x))} \right] = \sum_{\mu'} \partial_{\mu'} \mathcal{F}_{\mu'}(x), \tag{2.130}$$

where $\mu, \mu' = 0, x, y...$

We now generalize the infinitesimal transformation by allowing ϵ to depend on x,

$$\psi_j(x) \to \psi_j(x) + i\epsilon(x) F_j(x). \tag{2.131}$$

The change in the action is

$$\begin{aligned} \delta S &= \int_0^\beta d\tau \int d^d r \sum_j \left\{ i\epsilon(x) F_j(x) \frac{\partial \mathcal{L}}{\partial \psi_j(x)} + \sum_\mu i\partial_\mu[\epsilon(x) F_j(x)] \frac{\partial \mathcal{L}}{\partial(\partial_\mu \psi_j(x))} \right\} \\ &= \int_0^\beta d\tau \int d^d r \left\{ \epsilon(x) \sum_\mu \partial_\mu \mathcal{F}_\mu(x) + i \sum_{j,\mu} [\partial_\mu \epsilon(x)] F_j(x) \frac{\partial \mathcal{L}}{\partial(\partial_\mu \psi_j(x))} \right\}, \end{aligned} \tag{2.132}$$

---

[27] This is necessary to recover, for instance, $n'(\mathbf{r},\tau) = n(\mathbf{r},-\tau)$ and $\mathbf{j}'(\mathbf{r},\tau) = -\mathbf{j}(\mathbf{r},-\tau)$.

[28] Note also that the transformation (2.127, 2.128) preserves the proper time ordering in the action. For instance, the term $\psi^*(\tau)\psi(\tau) \equiv \psi^*(\tau+0^+)\psi(\tau)$ transforms into $\psi'^*(-\tau)\psi'(-\tau-0^+)$ in the case of bosons.

[29] Note the position of the field $F_j(x)$ in (2.130) which follows from the chain rule (1.240) for the derivation of Grassmann variables.

where we have used (2.130). Integrating by part the last term in (2.132), we obtain

$$\begin{aligned}\delta S &= -\int_0^\beta d\tau \int d^d r\, \epsilon(x) \sum_\mu \partial_\mu \left[i \sum_j F_j(x) \frac{\partial \mathcal{L}}{\partial(\partial_\mu \psi_j(x))} - \mathcal{F}_\mu(x) \right] \\ &= -\int_0^\beta d\tau \int d^d r\, [i\partial_\tau j_0(x) + \boldsymbol{\nabla}\cdot \mathbf{j}(x)]\, \epsilon(x),\end{aligned} \tag{2.133}$$

where the current densities are defined by[30]

$$\begin{aligned} j_0(x) &= \sum_j F_j(x) \frac{\partial \mathcal{L}}{\partial(\partial_\tau \psi_j(x))} + i\mathcal{F}_0(x), \\ j_\mu(x) &= i \sum_j F_j(x) \frac{\partial \mathcal{L}}{\partial(\partial_\mu \psi_j(x))} - \mathcal{F}_\mu(x) \qquad (\mu \neq 0). \end{aligned} \tag{2.134}$$

The transformation (2.131) is a change of variables which does not change the partition function $Z[J]$. If it leaves the measure invariant in the functional integral, we must have

$$\int \mathcal{D}[\psi]\, e^{-S-S_J} \int_0^\beta d\tau \int d^d r \Big[i\partial_\tau j_0(x) + \boldsymbol{\nabla}\cdot\mathbf{j}(x) + i \sum_j J_j(x) F_j(x) \Big] \epsilon(x) = 0 \tag{2.135}$$

to lowest order in $\epsilon(x)$. Since this equation holds for an arbitrary function $\epsilon(x)$, we deduce

$$\int \mathcal{D}[\psi]\, e^{-S-S_J} \Big[i\partial_\tau j_0(x) + \boldsymbol{\nabla}\cdot\mathbf{j}(x) + i \sum_j J_j(x) F_j(x) \Big] = 0. \tag{2.136}$$

Taking the nth-order functional derivative with respect to the source and setting $J_j(x) = 0$, we obtain

$$\begin{aligned} &\langle \psi_{j_n}(x_n) \cdots \psi_{j_1}(x_1) [i\partial_\tau j_0(x) + \boldsymbol{\nabla}\cdot\mathbf{j}(x)] \rangle \\ &\quad = -i \sum_{k=1}^n \langle \psi_{j_n}(x_n) \cdots \psi_{j_{k+1}}(x_{k+1}) F_{j_k}(x_k) \psi_{j_{k-1}}(x_{k-1}) \cdots \psi_{j_1}(x_1) \rangle \delta(x - x_k). \end{aligned} \tag{2.137}$$

Note that the time derivatives acting on the fields can be pulled outside the functional integral in (2.136) which can then be written as a time-ordered correlation function in the operator formalism. For instance, assuming that $j_0(x)$ does not contain any time derivative, one has $\langle \psi_j(x_1) \partial_\tau j_0(x) \rangle \equiv \partial_\tau \langle T_\tau \hat\psi_j(x_1) \hat j_0(x) \rangle$.[31] Equations (2.137) are exact relations

[30]The real-time current densities (2.108) can be deduced from their imaginary-time counterparts (2.134) by the usual Wick rotation $-S \to iS$ and $\tau = it$. (Note that $\mathcal{L}$ in (2.134) is the Euclidean Lagrangian density.)
[31]When one has a certain identity in the functional integral formalism, one must pull the time derivatives outside the functional integral before writing the identity in terms of time-ordered correlation functions in the operator formalism.

between the correlation functions. These relations must be fulfilled in any theory which satisfies the continuity equation associated with the conservation law considered.

It is often convenient to write the Ward identities (2.137) in a different form starting from

$$S = -\frac{1}{2}\sum_{j,j'}\int_0^\beta d\tau d\tau' \int d^d r d^d r' \, \psi_j(x) G^{-1}_{0,jj'}(x,x')\psi_{j'}(x') + S_{\text{int}}. \tag{2.138}$$

We assume that S_{int} is invariant under the local symmetry transformation (2.131), which means that the Noether current (2.134) depends only on S_0 (compare (2.139) with (2.133)). The change in the action is then

$$\delta S = \delta S_0 = -i\sum_{j,j'}\int_0^\beta d\tau d\tau' \int d^d r d^d r' \, \epsilon(x) F_j(x) G^{-1}_{0,jj'}(x,x')\psi_{j'}(x'), \tag{2.139}$$

where we have used $G_{0,jj'}(x,x') = \zeta G_{0,j'j}(x',x)$ with $\zeta = 1$ (-1) for bosons (fermions). Since $Z[J]$ is unaltered in the change of variable (2.131),

$$\int \mathcal{D}[\psi]\, e^{-S-S_J}\left[\int_0^\beta d\tau' \int d^d r' \sum_{j,j'} F_j(x) G^{-1}_{0,jj'}(x,x')\psi_{j'}(x') + \sum_j J_j(x) F_j(x)\right] = 0. \tag{2.140}$$

Taking the nth functional derivative with respect to the source and setting $J_j(x) = 0$, we obtain an alternative form of the Ward identities,

$$\begin{aligned} &\int_0^\beta d\tau' \int d^d r' \sum_{j,j'} \langle \psi_{j_n}(x_n)\cdots\psi_{j_1}(x_1) F_j(x)\psi_{j'}(x')\rangle G^{-1}_{0,jj'}(x,x') \\ &\quad = -\sum_{k=1}^{n} \langle \psi_{j_n}(x_n)\cdots\psi_{j_{k+1}}(x_{k+1}) F_{j_k}(x_k)\psi_{j_{k-1}}(x_{k-1})\cdots\psi_{j_1}(x_1)\rangle \delta(x-x_k). \end{aligned} \tag{2.141}$$

Unlike (2.137), Eq. (2.141) does not assume that the action can be written in terms of a Lagrangian density $\mathcal{L}(\psi_j, \partial_\mu \psi_j)$.

2.3.2.1 *Spontaneous symmetry breaking and Goldstone's theorem*

So far, we have assumed that the limit $J \to 0$ can be taken with no difficulty. This is not the case when the symmetry is spontaneously broken. To see this, let us consider a global symmetry transformation ($\epsilon(x) = \epsilon$) and assume that $F_j(x)$ is a linear function of $\psi_j(x)$,

$$F_j(x) = \int_0^\beta d\tau' \int d^d r' \sum_{j'} t_{jj'}(x,x')\psi_{j'}(x'). \tag{2.142}$$

Since $\delta S_0 = 0$ when $\epsilon(x) = \epsilon$,

$$\begin{aligned} 0 &= \int \mathcal{D}[\psi]\, e^{-S-S_J} \int_0^\beta d\tau \int d^d r \sum_j J_j(x) F_j(x) \\ &= \int_0^\beta d\tau d\tau' \int d^d r d^d r' \sum_{j,j'} J_j(x) t_{jj'}(x,x') \langle \psi_{j'}(x') \rangle \\ &= \int_0^\beta d\tau d\tau' \int d^d r d^d r' \sum_{j,j'} J_j(x) t_{jj'}(x,x') \frac{\delta \ln Z[J]}{\delta J_{j'}(x')}. \end{aligned} \tag{2.143}$$

Taking the functional derivative with respect to $J_k(y)$, we obtain

$$\begin{aligned} &\int_0^\beta d\tau' \int d^d r' \sum_{j'} t_{kj'}(y,x') \langle \psi_{j'}(x') \rangle \\ &\quad + \int_0^\beta d\tau d\tau' \int d^d r d^d r' \sum_{j,j'} J_j(x) t_{jj'}(x,x') \frac{\delta^2 \ln Z[J]}{\delta J_k(y) \delta J_{j'}(x')} = 0. \end{aligned} \tag{2.144}$$

Let us suppose that $\langle \psi_j(x) \rangle_{J=0} \neq 0$ (which is possible only if ψ is a bosonic field). If $\int_0^\beta d\tau' \int d^d r' \sum_{j'} t_{kj'}(y,x') \langle \psi_{j'}(x') \rangle$ is non-zero, the symmetry is spontaneously broken: Although the action is invariant in the transformation defined by (2.105) and (2.142), the state of the system is not since $\langle \psi_k(y) \rangle \to \langle \psi_k(y) \rangle + i\epsilon \int d^d r' \int_0^\beta d\tau' \sum_{j'} t_{kj'}(y,x') \langle \psi_{j'}(x') \rangle \neq \langle \psi_k(y) \rangle$. In that case, the second term in the lhs of (2.144) must remain finite as $J \to 0$. This is possible only if the correlation function

$$\int_0^\beta d\tau' \int d^d r' \sum_{j'} t_{jj'}(x,x') \frac{\delta^2 \ln Z[J]}{\delta J_k(y) \delta J_{j'}(x')} \bigg|_{J=0} \tag{2.145}$$

diverges. Such a divergence signals the existence of excitations with vanishing energies and is at the root of the Goldstone theorem (see Secs. 2.3.3 and 3.6.3).

2.3.3 *Ward identities for the 1PI vertices*

2.3.3.1 *Symmetries of the effective action* $\Gamma[\phi]$

The effective action $\Gamma[\phi]$ is defined as the Legendre transform

$$\begin{aligned} \Gamma[\phi] &= -\ln Z[J] + \int_0^\beta d\tau \int d^d r \sum_j J_j(x) \phi_j(x), \\ \phi_j[x;J] &= \langle \psi_j(x) \rangle = \frac{\delta \ln Z[J]}{\delta J_j(x)} \end{aligned} \tag{2.146}$$

of the generating functional $\ln Z[J]$ of connected Green functions (Sec. 1.6.2). In this section, we shall need two important properties of the effective action,[32]

$$\begin{aligned} \frac{\delta\Gamma[\phi]}{\delta\phi_j(x)} &= \zeta J_j(x), \\ \left.\frac{\delta^2\Gamma[\phi]}{\delta\phi_{j_1}(x_1)\delta\phi_{j_2}(x_2)}\right|_{\phi=\bar\phi} &= \zeta\Gamma^{(2)}_{j_1j_2}(x_1,x_2) = -\zeta G^{-1}_{j_1j_2}(x_1,x_2), \end{aligned} \tag{2.147}$$

where $G_{j_1j_2}(x_1,x_2) = -\langle\psi_{j_1}(x_1)\psi_{j_2}(x_2)\rangle + \langle\psi_{j_1}(x_1)\rangle\langle\psi_{j_2}(x_2)\rangle$ is the connected two-point Green function.[33] We denote by $\bar\phi_j(x) = \langle\psi_j(x)\rangle_{J=0}$ the expectation value of $\psi_j(x)$ in the absence of the external source ($\bar\phi_j$ always vanishes for a fermionic field ψ_j).

2.3.3.2 *Global Ward identities and Goldstone's theorem*

Let us first consider a global symmetry transformation ($\epsilon(x) = \epsilon$) defined by (2.131) and (2.142), which leaves the partition function invariant when $J=0$. Equation (2.143) can be rewritten as

$$\int_0^\beta d\tau d\tau' \int d^dr d^dr' \sum_{j,j'} t_{jj'}(x,x')\phi_{j'}(x')\frac{\delta\Gamma[\phi]}{\delta\phi_j(x)} = 0. \tag{2.148}$$

The effective action $\Gamma[\phi]$ is therefore invariant under the infinitesimal transformation

$$\phi_j(x) \to \phi_j(x) + i\epsilon\langle F_j(x)\rangle = \phi_j(x) + i\epsilon\int_0^\beta d\tau'\int d^dr'\sum_{j'} t_{jj'}(x,x')\phi_{j'}(x'). \tag{2.149}$$

We conclude that infinitesimal linear transformations that leave both the action $S[\psi]$ and the measure of the functional integral invariant also leave the effective action $\Gamma[\phi]$ invariant.

Taking the functional derivative with respect to $\phi_k(y)$ in (2.148) and setting $J=0$ (i.e. $\phi=\bar\phi$), we obtain

$$\int_0^\beta d\tau d\tau' \int d^dr d^dr' \sum_{j,j'} t_{jj'}(x,x')\bar\phi_{j'}(x')\Gamma^{(2)}_{kj}(y,x) = 0. \tag{2.150}$$

Similarly, by taking two functional derivatives, we obtain

$$\begin{aligned} \int_0^\beta d\tau \int d^dr \sum_j \Big\{ & t_{jk}(x,y)\Gamma^{(2)}_{jl}(x,z) + t_{jl}(x,z)\Gamma^{(2)}_{kj}(y,x) \\ & + \int_0^\beta d\tau'\int d^dr'\sum_{j'} t_{jj'}(x,x')\bar\phi_{j'}(x')\Gamma^{(3)}_{jkl}(x,y,z)\Big\} = 0. \end{aligned} \tag{2.151}$$

[32]Recall that the n-point 1PI vertices are defined by $\Gamma^{(n)}_{j_1\cdots j_n}(x_1,\dots,x_n) = \left.\frac{\delta^n\Gamma[\phi]}{\delta\phi_{j_n}(x_n)\cdots\delta\phi_{j_1}(x_1)}\right|_{\phi=\bar\phi}$ (see Sec. 1.6.2).

[33]Note that the sign convention of the Green function $G_{j_1j_2}(x_1,x_2)$ differs from that used in Sec. 1.6.2: $G(\tilde\alpha\tau,\tilde\alpha'\tau') = \langle\psi_{\tilde\alpha}(\tau)\psi_{\tilde\alpha'}(\tau')\rangle - \langle\psi_{\tilde\alpha}(\tau)\rangle\langle\psi_{\tilde\alpha'}(\tau')\rangle$.

(We have used the definition of the vertices given in footnote 32.) Equation (2.151), which relates $\Gamma^{(2)}$ and $\Gamma^{(3)}$, is an example of a (global) Ward identity.[34]

Equation (2.150) yields

$$\int_0^\beta d\tau d\tau' \int d^d r d^d r' \sum_{j,j'} G_{kj}^{-1}(y,x) t_{jj'}(x,x') \bar{\phi}_{j'}(x') = 0. \tag{2.152}$$

For simplicity, we assume that the system is translation invariant: $\bar{\phi}_j(x) = \bar{\phi}_j$, $t_{jj'}(x,x') = t_{jj'}(x-x')$ and $G_{jj'}(x,x') = G_{jj'}(x-x')$. Equation (2.152) then gives

$$\sum_{j,j'} G_{kj}^{-1}(q=0) t_{jj'}(q=0) \bar{\phi}_{j'} = 0, \tag{2.153}$$

where $q = (\mathbf{q}, i\omega_n)$ with ω_n a Matsubara frequency. This equation is trivially satisfied when $\sum_{j'} t_{jj'}(q=0)\bar{\phi}_{j'} = 0$. When $\sum_{j'} t_{jj'}(q=0)\bar{\phi}_{j'} \neq 0$, the state of the system is not invariant under the symmetry transformation although the action is; there is spontaneous symmetry breaking (see the discussion in Sec. 2.3.2.1). Equation (2.153) then shows that $\sum_{j'} t_{jj'}(q=0)\bar{\phi}_{j'}$ is an eigenvector of the matrix $G_{kj}^{-1}(q=0)$ with a vanishing eigenvalue. In other words, there is a zero momentum excitation with vanishing energy (which shows up as a pole of $G(q=0)$). This result is truly interesting only if it implies that there is an excitation mode with an energy $\omega_{\mathbf{q}}$ which vanishes in the limit $\mathbf{q} \to 0$. This turns out to be the case, provided that the interactions in the system are sufficiently short range. We shall come back to this issue in Sec. 3.6.3 where the Goldstone theorem will be discussed from a different point of view.

2.3.3.3 *Local Ward identities*

More generally, Ward identities for the 1PI vertices are derived by considering a local symmetry transformation. The invariance of the partition function $Z[J]$ in the infinitesimal transformation (2.131) implies (2.140), which can be rewritten as

$$\sum_{j_1} \Big\langle \sum_{1'} F(1) G_0^{-1}(1,1')\psi(1') + J(1)F(1) \Big\rangle = 0 \tag{2.154}$$

using the compact notation $1 = (j_1, \mathbf{r}_1, \tau_1)$ and $\sum_1 = \int_0^\beta d\tau_1 \int d^d r_1 \sum_{j_1}$, etc. For the linear transformation (2.142),

$$\begin{aligned} 0 &= \sum_{j_1,1''} t(1,1'') \Big\langle \sum_{1'} \psi(1'') G_0^{-1}(1,1')\psi(1') + J(1)\psi(1'') \Big\rangle \\ &= \sum_{j_1,1''} t(1,1'') \Big\{ \sum_{1'} G_0^{-1}(1,1') \left(\frac{\delta^2 \ln Z[J]}{\delta J(1'')\delta J(1')} + \frac{\delta \ln Z[J]}{\delta J(1'')} \frac{\delta \ln Z[J]}{\delta J(1')} \right) + J(1) \frac{\delta \ln Z[J]}{\delta J(1'')} \Big\}. \end{aligned} \tag{2.155}$$

[34]This Ward identity is global to the extent where it involves an integration over space (in Fourier space, it typically involves zero-momentum vertices).

Taking the functional derivative with respect to $\phi(2)$ and $\phi(3)$ and setting $J = 0$, we obtain

$$\sum_{j_1,1''} t(1,1'')\Big\{\sum_{1'} G_0^{-1}(1,1')\Big[-\sum_{2',3'}\Gamma^{(2)-1}(1'',2')\Gamma^{(4)}(3',2',2,3)\Gamma^{(2)-1}(3',1')$$
$$+\,\delta(1'',2)\delta(1',3)+\zeta\delta(1'',3)\delta(1',2)\Big]+\delta(1'',2)\Gamma^{(2)}(1,3)+\delta(1'',3)\Gamma^{(2)}(2,1)\Big\}=0 \tag{2.156}$$

assuming $\bar{\phi} = 0$. Using $\Gamma^{(2)} = -G^{-1}$ and $G^{-1} = G_0^{-1} - \Sigma$, we can rewrite (2.156) as

$$\sum_{j_1,1',1'',2',3'} t(1,1'')G_0^{-1}(1,1')G(1',2')\Gamma^{(4)}(3,2,2',3')G(3',1'')$$
$$= \sum_{j_1}[t(1,2)\Sigma(1,3)+t(1,3)\Sigma(2,1)]. \tag{2.157}$$

Equation (2.157) is a Ward identity relating the self-energy Σ to the two-particle vertex $\Gamma^{(4)}$. It is straightforward to extend the derivation to the case where $\bar{\phi} \neq 0$. Higher-order Ward identities can be obtained by considering additional functional derivatives with respect to ϕ before setting $J = 0$.

2.3.3.4 *Example 1: Superfluid boson system*

Let us consider an interacting boson system. Conservation of particles implies that the action is invariant under the infinitesimal U(1) transformation $\psi(x) \to \psi(x) + i\epsilon\psi(x)$, $\psi^*(x) \to \psi^*(x) - i\epsilon\psi^*(x)$.[35] This linear transformation also leaves the effective action $\Gamma[\phi^*, \phi]$ invariant. From (2.153), we deduce

$$G_c^{-1}(q=0)\begin{pmatrix}\bar{\phi}\\ -\bar{\psi}^*\end{pmatrix} = 0, \tag{2.158}$$

where $\bar{\phi} = \langle\psi(x)\rangle$, $\bar{\phi}^* = \langle\psi^*(x)\rangle$, and

$$G_c(x-x') = -\begin{pmatrix}\langle\psi(x)\psi(x')\rangle - \bar{\phi}^2 & \langle\psi(x)\psi^*(x')\rangle - |\bar{\phi}|^2\\ \langle\psi^*(x)\psi(x')\rangle - |\bar{\phi}|^2 & \langle\psi^*(x)\psi^*(x')\rangle - (\bar{\phi}^*)^2\end{pmatrix} \tag{2.159}$$

is the one-particle Green function (written here as a 2×2 matrix). If $|\bar{\phi}| \neq 0$, the U(1) symmetry is spontaneously broken. We shall see in Chapter 7 that this property is characteristic of superfluid systems. The Goldstone mode, whose existence can be inferred from (2.158) when the interactions are short range, is the Bogoliubov sound mode.

[35] Obviously, the U(1) transformation leaves the measure in the functional integral invariant.

2.3.3.5 *Example 2: Magnetism in localized spin systems*

Let us consider quantum spins located at the sites $\mathbf{r}$ of a three-dimensional cubic lattice. The partition function can be written as a spin coherent-state functional integral with the action $S[\boldsymbol{\Omega}]$, where $\boldsymbol{\Omega}_\mathbf{r}$ is a unit vector field ($\boldsymbol{\Omega}_\mathbf{r}^2 = 1$) (see Sec. 8.4.1). As usual, the effective action $\Gamma[\mathbf{M}]$ is defined as the Legendre transform of $\ln Z[\mathbf{h}]$ where $\mathbf{h}$ is an external field which couples to the spins. Assuming spin-rotation invariance in the absence of external field, $\Gamma[\mathbf{M}]$ is invariant under an infinitesimal rotation about an arbitrary axis $\mathbf{n}$, $\mathbf{M} \to \mathbf{M} + \epsilon \mathbf{n} \times \mathbf{M}$. Equation (2.152) gives

$$\int_0^\beta d\tau d\tau' \sum_{\substack{\mathbf{r},\mathbf{r}' \\ i,j,k}} G^{-1}_{li}(x,x') \epsilon_{ijk} n_j \bar{M}_k(x') = 0, \tag{2.160}$$

where $\bar{\mathbf{M}}(x) = \langle \boldsymbol{\Omega}(x) \rangle_{\mathbf{h}=0}$ is the zero-field magnetization, and

$$G_{li}(x,x') = \langle \Omega_l(x) \Omega_i(x') \rangle - \langle \Omega_l(x) \rangle \langle \Omega_i(x') \rangle \tag{2.161}$$

the spin–spin correlation function.

If the system is ferromagnetic, $\bar{M}_k(x) = \delta_{k,z} \bar{M}_0$ (assuming the magnetization along the z direction), we obtain

$$\sum_{i,j} G^{-1}_{li}(q=0) \epsilon_{ijz} n_j = 0. \tag{2.162}$$

This equation is trivially satisfied for a rotation about the z axis, since $\epsilon_{izz} = 0$. By considering rotations about the x or y axis, we conclude that $G^{-1}_{yy}(q=0)$ and $G^{-1}_{xx}(q=0)$ vanish.[36] The gapless excitations showing up as poles of $G_{xx}(q)$ and $G_{yy}(q)$ are (ferromagnetic) spin waves; they will be discussed in detail in Chapter 8.

2.3.4 *Ward identities for the 2PI vertices*

In the 2PI scheme (Sec. 1.6.3), the quantity to be considered is the functional

$$\begin{aligned} \Gamma[G] &= -\ln Z[J] - \frac{1}{2} \int d\tau_1 d\tau_2 \int d^d r_1 d^d r_2 \sum_{j_1,j_2} J_{j_1 j_2}(x_1,x_2) G_{j_1 j_2}(x_1,x_2) \\ &= -\frac{\zeta}{2} \mathrm{Tr} \ln(-G) + \frac{\zeta}{2} \mathrm{Tr}(G_0^{-1} G - 1) + \Phi[G]. \end{aligned} \tag{2.163}$$

Here, $J_{j_1 j_2}(x_1,x_2) = \zeta J_{j_2 j_1}(x_2,x_1)$ is a bilinear external source which couples to $\psi_{j_1}(x_1)\psi_{j_2}(x_2)$ and G is the one-particle Green function,[33]

$$\begin{aligned} G_{j_1 j_2}[x_1,x_2;J] &= -\langle \psi_{j_1}(x_1) \psi_{j_2}(x_2) \rangle = -\frac{\delta \ln Z[J]}{\delta J_{j_1 j_2}(x_1,x_2)}, \\ \frac{\delta \Gamma[G]}{\delta G_{j_1 j_2}(x_1,x_2)} &= -J_{j_1 j_2}(x_1,x_2). \end{aligned} \tag{2.164}$$

[36] For symmetry reasons, $G_{xy}(q=0)$ always vanishes.

The actual Green function $\bar{G}$, corresponding to a vanishing external source, is obtained from the stationary point of $\Gamma[G]$: $\delta\Gamma/\delta G|_{G=\bar{G}} = \Gamma^{(1)}[\bar{G}] = 0$. The Luttinger–Ward functional $\Phi[G]$ is the generating functional of the 2PI vertices $\Phi^{(n)}$.

2.3.4.1 *Global Ward identities and Goldstone's theorem*

Let us first consider a global linear transformation defined by (2.105) and (2.142),

$$\psi_j(x) \to \psi_j(x) + i\epsilon \int_0^\beta d\tau' \int d^d r' \sum_{j'} t_{jj'}(x')\psi_{j'}(x'), \tag{2.165}$$

which leaves the action $S_0 + S_{\text{int}}$ invariant. Proceeding as in Sec. 2.3.3 and using the same compact notation, one easily finds

$$\sum_{1,2,3} \frac{\delta\Gamma[G]}{\delta G(1,2)} t(1,3)G(3,2) = 0. \tag{2.166}$$

Equation (2.166) expresses the invariance of $\Gamma[G]$ in the transformation

$$G(1,2) \to G(1,2) + i\epsilon \sum_3 [t(1,3)G(3,2) + t(2,3)G(1,3)], \tag{2.167}$$

which follows from (2.165). Taking the functional derivative with respect to $G(1',2')$ in (2.166) and setting $J = 0$, we obtain the (global) Ward identity

$$\sum_{1,2,3} \left. \frac{\delta^2\Gamma[G]}{\delta G(1',2')\delta G(1,2)} \right|_{G=\bar{G}} [t(1,3)\bar{G}(3,2) + t(2,3)\bar{G}(1,3)] = 0. \tag{2.168}$$

If $\sum_3 [t(1,3)\bar{G}(3,2) + t(2,3)\bar{G}(1,3)] \neq 0$, the state of the system is not invariant under the symmetry transformation (2.167): The symmetry is spontaneously broken. $\delta^2\Gamma/\delta G\delta G|_{\bar{G}} = W^{(2)-1}$ must then have a vanishing eigenvalue.[37,38] We thus reproduce Goldstone's theorem. Additional Ward identities can be obtained from (2.166) by taking further functional derivatives.

2.3.4.2 *Symmetry properties of $\Phi[G]$ and local Ward identities*

We now consider a local symmetry transformation defined by (2.165) but with $\epsilon \to \epsilon(x)$, which leaves S_{int} invariant. The corresponding change in the Green function is

$$G(1,2) \to G(1,2) + i \sum_3 [\epsilon(x_1)t(1,3)G(3,2) + \epsilon(x_2)t(2,3)G(1,3)]. \tag{2.169}$$

The invariance of the partition function implies

$$\sum_{j_1,2,3} \left[G_0^{-1}(1,2) - \frac{\delta\Gamma[G]}{\delta G(1,2)} \right] t(1,3)G(3,2) = 0. \tag{2.170}$$

[37] $W^{(2)}(1,2;1',2') = \langle\psi(1)\psi(2)\psi(1')\psi(2')\rangle - \langle\psi(1)\psi(2)\rangle\langle\psi(1')\psi(2')\rangle$ (see Sec. 1.6.3).

[38] Note that the vanishing eigenvalue appears here in the two-particle propagator. For fermions, this is the lowest-order Green function where the Goldstone theorem can manifest itself since, as we shall see, order-parameter fields are always composite fields, e.g., $\psi^*_\sigma\psi_{\sigma'}$ or $\psi_\sigma\psi_{\sigma'}$.

From (2.169) and (2.170), we find that the functional $\Gamma[G]$ varies by[39]

$$\begin{aligned}\delta\Gamma[G] &= \frac{i}{2}\sum_{1,2,3}\frac{\delta\Gamma[G]}{\delta G(1,2)}[\epsilon(x_1)t(1,3)G(3,2)+\epsilon(x_2)t(2,3)G(1,3)] \\ &= \frac{i}{2}\sum_{1,2,3}G_0^{-1}(1,2)[\epsilon(x_1)t(1,3)G(3,2)+\epsilon(x_2)t(2,3)G(1,3)] \end{aligned} \tag{2.171}$$

to leading order in $\epsilon(x)$. Comparing this result with the definition (2.163) of $\Gamma[G]$, we see that the change in Γ is entirely due to $\frac{\zeta}{2}\mathrm{Tr}(G_0^{-1}G-1)$, which implies that $-\frac{\zeta}{2}\mathrm{Tr}\ln(-G)+\Phi[G]$ is invariant in the transformation (2.169). Now, since $\Phi[G]$ vanishes in the absence of interaction, $-\frac{\zeta}{2}\mathrm{Tr}\ln(-G)$ must be invariant. We conclude that the Luttinger–Ward functional is invariant in any infinitesimal transformation that leaves S_{int} invariant.

The invariance of $\Phi[G]$ can also be understood in a diagrammatic picture. As an example, let us consider a local U(1) transformation, $\psi(x)\to e^{i\epsilon(x)}\psi(x)$ and $\psi^*(x)\to e^{-i\epsilon(x)}\psi^*(x)$, whereby the Green function transforms as

$$G(x,x')\to e^{\pm i\epsilon(x)}G(x,x')e^{\pm i\epsilon(x')}. \tag{2.172}$$

For any skeleton diagram contributing to $\Phi[G]$, each vertex associated with

$$v(x_1,x_2,x_3,x_4)\psi^*(x_1)\psi^*(x_2)\psi(x_3)\psi(x_4) \tag{2.173}$$

picks up a factor $e^{i[-\epsilon(x_1)-\epsilon(x_2)+\epsilon(x_3)+\epsilon(x_4)]}$. But the invariance of S_{int} in the local U(1) transformation requires this phase factor to be unity, which implies the invariance of any diagram contributing to $\Phi[G]$. A similar reasoning can be made for other symmetry transformations that leave S_{int} invariant.

Local Ward identities for the 2PI vertices follow from the symmetry properties of the Luttinger–Ward functional. The invariance of $\Phi[G]$ in the transformation (2.169) implies

$$\sum_{j_1,2,3}\frac{\delta\Phi[G]}{\delta G(1,2)}t(1,3)G(3,2)=0. \tag{2.174}$$

Taking the functional derivative with respect to $G(1',2')$ and setting $J=0$, i.e. $G=\bar{G}$, we obtain

$$\sum_{j_1,2,3}\Phi^{(2)}(1',2';1,2)t(1,3)\bar{G}(3,2)=\sum_{j_1}[t(1,1')\Sigma(1,2')+t(1,2')\Sigma(1',1)], \tag{2.175}$$

where $\Sigma=-\Phi^{(1)}$. Equation (2.175) is a Ward identity relating the self-energy Σ to the 2PI two-particle vertex $\Phi^{(2)}$. Higher-order Ward identities can be obtained by taking additional functional derivatives.

[39]The reason for the factor 1/2 in (2.171) is explained in Sec. 1.6.3 (page 100).

Appendix 2.A Ward Identities in a Fermion System

In this appendix, we use the results of Sec. 2.3.3 to derive Ward identities in a spin-$\frac{1}{2}$ fermion system.[40] We start from Eq. (2.157) relating the self-energy to the 1PI two-particle vertex $\Gamma^{(4)}$. The index j includes a spin index $\sigma =\uparrow,\downarrow$ and a charge index $c = \pm$ such that $\psi_j = \psi_\sigma$ (ψ^*_σ) if $c = -$ (+). We assume that the system is not superconducting (the U(1) symmetry is not broken) and the propagators are diagonal in spin. The latter assumption does not eliminate a broken spin-rotation symmetry state but restricts the possible states of the system.[41] We further assume that

$$t_{jj'}(x,x') = \begin{cases} t^{(+)}_{\sigma\sigma'}(x,x') & \text{if } c = c' = +, \\ t^{(-)}_{\sigma\sigma'}(x,x') & \text{if } c = c' = -, \\ 0 & \text{if } c \neq c'. \end{cases} \tag{2.A.1}$$

Equation (2.157) then gives

$$\begin{aligned} &\int_0^\beta d\tau_{1'} d\tau_{1''} d\tau_{2'} d\tau_{3'} \int d^d r_{1'} d^d r_{1''} d^d r_{2'} d^d r_{3'} \sum_{\sigma_1,\sigma_{1''}} \Big[t^{(-)}_{\sigma_{1''}\sigma_1}(x_1,x_{1'}) G^{-1}_{0,\sigma_{1''}}(x_{1''},x_1) \\ &+ t^{(+)}_{\sigma_1\sigma_{1''}}(x_1,x_{1''}) G^{-1}_{0,\sigma_1}(x_1,x_{1'}) \Big] G_{\sigma_1}(x_{1'},x_{2'}) \Gamma^{(4)}_{\mathrm{ph},\sigma_3\sigma_2\sigma_1\sigma_{1''}}(x_3,x_2,x_{2'},x_{3'}) G_{\sigma_{1''}}(x_{3'},x_{1''}) \\ &\qquad = -t^{(-)}_{\sigma_3\sigma_2}(x_1,x_2)\Sigma_{\sigma_3}(x_3,x_1) - t^{(+)}_{\sigma_2,\sigma_3}(x_1,x_3)\Sigma_{\sigma_2}(x_1,x_2), \end{aligned} \tag{2.A.2}$$

where

$$\Gamma^{(4)}_{\mathrm{ph},\sigma_1\sigma_{1'}\sigma_2\sigma_{2'}}(x_1,x_{1'},x_2,x_{2'}) \equiv \Gamma^{(4)}_{\sigma_1\sigma_{1'}\sigma_2\sigma_{2'}}(x_1+,x_{1'}-,x_2+,x_{2'}-) \tag{2.A.3}$$

is the two-particle vertex in the particle-hole channel. Introducing the Fourier transform

$$t^{(\pm)}(k) = \int_0^\beta d\tau \int d^d r\, e^{\pm i(\mathbf{k}\cdot\mathbf{r} - \omega_n \tau)} t^{(\pm)}(x) \tag{2.A.4}$$

and multiplying (2.A.2) by $\frac{1}{\beta\mathcal{V}} \int_0^\beta d\tau_1 d\tau_2 d\tau_3 \int d^d r_1 d^d r_2 d^d r_3\, e^{iqx_1 + ikx_2 + ik_3 x_3}$, we obtain

$$\begin{aligned} &\frac{1}{\beta\mathcal{V}} \sum_{k',\sigma,\sigma'} \Big[t^{(-)}_{\sigma'\sigma}(k') G^{-1}_{0,\sigma'}(k'+q) + t^{(+)}_{\sigma\sigma'}(k'+q) G^{-1}_{0,\sigma}(k') \Big] G_\sigma(k') G_{\sigma'}(k'+q) \\ &\times \Gamma^{(4)}_{\mathrm{ph},\sigma_3\sigma_2\sigma\sigma'}(k+q,k,k',k'+q) = -t^{(-)}_{\sigma_3\sigma_2}(k)\Sigma_{\sigma_3}(k+q) - t^{(+)}_{\sigma_2\sigma_3}(k+q)\Sigma_{\sigma_2}(k). \end{aligned} \tag{2.A.5}$$

---

[40] These Ward identities play a crucial role in Fermi-liquid theory (Chapter 4).

[41] For instance, if the system is ferromagnetic or antiferromagnetic, the magnetization must be along the spin-quantization axis.

2.A.1 *U(1) transformation*

For a U(1) transformation,

$$\begin{aligned}\psi_\sigma(x) &\to \psi_\sigma(x)e^{i\epsilon(x)} = \psi_\sigma(x) + i\epsilon(x)\psi_\sigma(x) + \mathcal{O}(\epsilon^2),\\ \psi^*_\sigma(x) &\to \psi^*_\sigma(x)e^{-i\epsilon(x)} = \psi^*_\sigma(x) - i\epsilon(x)\psi^*_\sigma(x) + \mathcal{O}(\epsilon^2),\end{aligned}\tag{2.A.6}$$

we have $t^{(\pm)}_{\sigma\sigma'}(x,x') = \mp\delta_{\sigma,\sigma'}\delta(x-x')$. The corresponding Ward identity reads

$$\begin{aligned}\frac{1}{\beta\mathcal{V}}\sum_{k',\sigma}&\left[G^{-1}_{0,\sigma'}(k'+q) - G^{-1}_{0,\sigma'}(k')\right]G_{\sigma'}(k')G_{\sigma'}(k'+q)\\ &\times\Gamma^{(4)}_{\text{ph},\sigma\sigma\sigma'\sigma'}(k+q,k,k',k'+q) = -\Sigma_\sigma(k+q) + \Sigma_\sigma(k).\end{aligned}\tag{2.A.7}$$

2.A.2 *Spin rotation*

For a rotation of angle $\epsilon(x)$ about the $\mathbf{n}$ axis in spin space,

$$\begin{aligned}\psi(x) &\to e^{-\frac{i}{2}\epsilon(x)\sigma^{(n)}}\psi(x) = \psi(x) - \frac{i}{2}\epsilon(x)\sigma^{(n)}\psi(x) + \mathcal{O}(\epsilon^2),\\ \psi^\dagger(x) &\to \psi^\dagger(x)e^{\frac{i}{2}\epsilon(x)\sigma^{(n)}} = \psi^\dagger(x) + \frac{i}{2}\epsilon(x)\psi^\dagger(x)\sigma^{(n)} + \mathcal{O}(\epsilon^2),\end{aligned}\tag{2.A.8}$$

where $\psi(x) = (\psi_\uparrow(x),\psi_\downarrow(x))^T$, $\psi^\dagger(x) = (\psi^*_\uparrow(x),\psi^*_\downarrow(x))$ and $\sigma^{(n)} = \boldsymbol{\sigma}\cdot\mathbf{n}$. This gives $t^{(-)}_{\sigma\sigma'}(x,x') = -\frac{1}{2}\delta(x-x')\sigma^{(n)}_{\sigma\sigma'}$ and $t^{(+)}_{\sigma\sigma'}(x,x') = \frac{1}{2}\delta(x-x')\sigma^{(n)*}_{\sigma\sigma'}$, and in turn

$$\begin{aligned}\frac{1}{\beta\mathcal{V}}\sum_{k',\sigma,\sigma'}&\left[-\sigma^{(n)}_{\sigma'\sigma}G^{-1}_{0,\sigma'}(k'+q) + \sigma^{(n)*}_{\sigma\sigma'}G^{-1}_{0,\sigma'}(k')\right]G_\sigma(k')G_{\sigma'}(k'+q)\\ &\times\Gamma^{(4)}_{\text{ph},\sigma_2\sigma_1\sigma\sigma'}(k+q,k,k',k'+q) = \sigma^{(n)}_{\sigma_2\sigma_1}\Sigma_{\sigma_2}(k+q) - \sigma^{(n)*}_{\sigma_1\sigma_2}(k+q)\Sigma_{\sigma_1}(k).\end{aligned}\tag{2.A.9}$$

Considering $\mathbf{n}=\mathbf{z}$ and $\mathbf{n}=\mathbf{x}$, we deduce the Ward identities

$$\begin{aligned}\frac{1}{\beta\mathcal{V}}\sum_{k',\sigma'}\sigma'&\left[G^{-1}_{0,\sigma'}(k'+q) - G^{-1}_{0,\sigma'}(k')\right]G_{\sigma'}(k')G_{\sigma'}(k'+q)\\ &\times\Gamma^{(4)}_{\text{ph},\sigma\sigma\sigma'\sigma'}(k+q,k,k',k'+q) = -\sigma\left[\Sigma_\sigma(k+q) - \Sigma_\sigma(k)\right]\end{aligned}\tag{2.A.10}$$

and

$$\begin{aligned}\frac{1}{\beta\mathcal{V}}\sum_{k',\sigma'}&\left[G^{-1}_{0,\bar\sigma'}(k'+q) - G^{-1}_{0,\sigma'}(k')\right]G_{\sigma'}(k')G_{\bar\sigma'}(k'+q)\\ &\times\Gamma^{(4)}_{\text{ph},\bar\sigma\sigma\sigma'\bar\sigma'}(k+q,k,k',k'+q) = -\Sigma_{\bar\sigma}(k+q) + \Sigma_\sigma(k).\end{aligned}\tag{2.A.11}$$

For a spin-rotation-invariant system, the particle–hole interaction vertex can be written as

$$\Gamma^{(4)}_{\text{ph},\sigma_1\sigma_{1'}\sigma_2\sigma_{2'}} = \Gamma^{(4)}_{\text{ch}}\sigma^0_{\sigma_1\sigma_{1'}}\sigma^0_{\sigma_2\sigma_{2'}} + \Gamma^{(4)}_{\text{sp}}\boldsymbol{\sigma}_{\sigma_1\sigma_{1'}}\cdot\boldsymbol{\sigma}_{\sigma_2\sigma_{2'}}.\tag{2.A.12}$$

The three Ward identities (2.A.7, 2.A.10, 2.A.11) due to the U(1) and spin-rotation invariances then reduce to

$$\frac{2}{\beta\mathcal{V}}\sum_{k'}\left[G_0^{-1}(k'+q)-G_0^{-1}(k')\right]G(k')G(k'+q)$$
$$\times\,\Gamma^{(4)}_{\rm ch/sp}(k+q,k,k',k'+q)=-\Sigma(k+q)+\Sigma(k), \tag{2.A.13}$$

where the one-particle Green function and the self-energy are now spin independent.

2.A.3 *Galilean transformation*

Let us finally consider the "Galilean" transformation

$$\begin{aligned}\psi_\sigma(x)&\to\psi_\sigma(\mathbf{r}-\mathbf{R}(\tau),\tau)=\psi_\sigma(x)-\mathbf{R}(\tau)\cdot\boldsymbol{\nabla}\cdot\psi_\sigma(x)+\mathcal{O}(\mathbf{R}^2),\\ \psi^*_\sigma(x)&\to\psi^*_\sigma(\mathbf{r}-\mathbf{R}(\tau),\tau)=\psi^*_\sigma(x)-\mathbf{R}(\tau)\cdot\boldsymbol{\nabla}\cdot\psi^*_\sigma(x)+\mathcal{O}(\mathbf{R}^2),\end{aligned} \tag{2.A.14}$$

which amounts to viewing the system in a frame moving with a time-dependent velocity $\mathbf{v}(\tau)=id\mathbf{R}/d\tau$.[42] If the interaction $S_{\rm int}$ is local in time and translation invariant, it is invariant in this transformation. Equation (2.A.14) defines a linear transformation with $t^{(\pm)}_{\sigma\sigma'}(x,x')=i\delta_{\sigma,\sigma'}\hat{\boldsymbol{\epsilon}}\cdot\boldsymbol{\nabla}_{\mathbf{r}'}\delta(x-x')$, where we assume $\mathbf{R}(\tau)$ to be parallel to the unit vector $\hat{\boldsymbol{\epsilon}}$. Equation (2.157) then gives

$$\sum_{\sigma_1}\int dx_{1'}dx_{2'}dx_{3'}\Big[G^{-1}_{0,\sigma_1}(x_{1'},x_1)\boldsymbol{\nabla}_{\mathbf{r}_1}G_{\sigma_1}(x_1,x_{2'})G_{\sigma_1}(x_{3'},x_{1'})$$
$$+\,G^{-1}_{0,\sigma_1}(x_1,x_{1'})G_{\sigma_1}(x_{1'},x_{2'})\boldsymbol{\nabla}_{\mathbf{r}_1}G_{\sigma_1}(x_{3'},x_1)\Big]\Gamma^{(4)}_{{\rm ph},\sigma_2\sigma_2\sigma_1\sigma_1}(x_3,x_2,x_{2'},x_{3'})$$
$$=-\boldsymbol{\nabla}_{\mathbf{r}_2}\delta(x_1-x_2)\Sigma_{\sigma_2}(x_3-x_1)-\boldsymbol{\nabla}_{\mathbf{r}_3}\delta(x_1-x_3)\Sigma_{\sigma_2}(x_1-x_2). \tag{2.A.15}$$

In Fourier space, we obtain the Ward identity

$$\frac{1}{\beta\mathcal{V}}\sum_{k',\sigma'}\Big[\mathbf{k}'G^{-1}_{0,\sigma'}(k'+q)-(\mathbf{k}'+\mathbf{q})G^{-1}_{0,\sigma'}(k')\Big]G_{\sigma'}(k')G_{\sigma'}(k'+q)$$
$$\times\,\Gamma^{(4)}_{{\rm ph},\sigma\sigma\sigma'\sigma'}(k+q,k,k',k'+q)=-\mathbf{k}\Sigma_\sigma(k+q)+(\mathbf{k}+\mathbf{q})\Sigma_\sigma(k). \tag{2.A.16}$$

[42] In real time, $(\tau=it)$, $\mathbf{v}(t)=d\mathbf{R}/dt$.

Guide to the bibliography

Symmetries in quantum mechanics are discussed in a number of books. Here, I mention only those that were most useful when writing this chapter [1–5]. Sections 2.1 and 2.2 are mainly based on Ref. [4]. Symmetries in quantum field theory are discussed in Refs. [4, 5]. Ward identities were first obtained in quantum electrodynamics [6, 7]. For condensed-matter systems, they are discussed in Refs. [8–11]. Goldstone's theorem was first derived in Refs. [12–14]; to a large extent, our discussion follows Ref. [5].

Bibliography

[1] A. Messiah, *Quantum Mechanics*, Vol. II (North-Holland, Amsterdam, 1962).

[2] K. Gottfried, *Quantum Mechanics*, Vol. I: Fundamentals (Benjamin, New York, 1966).

[3] J. J. Sakurai, *Modern Quantum Mechanics* (Addison Wesley, Reading, 1994).

[4] L. Fonda and G. C. Ghirardi, *Symmetry Principles in Quantum Physics* (Marcel Dekker, Inc., New York, 1970).

[5] S. Weinberg, *The Theory of Quantum Fields*, Vols. I and II (Cambridge University Press, Cambridge, 1995).

[6] J. C. Ward, An identity in quantum electrodynamics, *Phys. Rev.* **78**, 182 (1950).

[7] Y. Takahashi, On the generalized Ward identity, *Nuovo Cimento* **6**, 371 (1957).

[8] A. A. Abrikosov, L. P. Gor'kov, and I. E. Dzyaloshinski, *Methods of Quantum Field Theory in Statistical Physics* (Dover, Mineola, New York, 1975).

[9] P. Nozières, *Theory of Interacting Fermi Systems* (Addison Wesley, Reading, Massachusetts, 1964).

[10] T. Toyoda, Nonperturbative canonical formulation and Ward-Takahashi relations for quantum many-body systems at finite temperatures, *Ann. Phys. (N.Y.)* **173**, 226 (1987).

[11] W. Metzner, C. Castellani, and C. Di Castro, Fermi systems with strong forward scattering, *Adv. Phys.* **47**, 317 (1998).

[12] J. Goldstone, Field theories with "Superconductor" solutions, *Nuovo Cimento* **19**, 154 (1961).

[13] Y. Nambu, Axial vector current conservation in weak interactions, *Phys. Rev. Lett.* **4**, 380 (1960).

[14] J. Goldstone, A. Salam, and S. Weinberg, Broken symmetries, *Phys. Rev.* **127**, 965 (1962).

Chapter 3

Correlation and Response Functions

Contents

3.1 Linear Response Theory . . . 236
- 3.1.1 Kubo's formula . . . 237
- 3.1.2 Periodic perturbations . . . 239

3.2 Properties of Correlation Functions . . . 240
- 3.2.1 Symmetry properties . . . 240
- 3.2.2 Dissipation . . . 243
- 3.2.3 Spectral representation . . . 244
- 3.2.4 Kramers–Kronig relations and causality . . . 248
- 3.2.5 Fluctuation–dissipation theorem . . . 249
- 3.2.6 Sum rules and high-frequency behavior . . . 250
- 3.2.7 Imaginary-time correlation function . . . 250

3.3 Density–Density Response Function . . . 258
- 3.3.1 Symmetry properties . . . 258
- 3.3.2 High-frequency behavior and f-sum rule . . . 259
- 3.3.3 Compressibility . . . 259

3.4 Electromagnetic Response Function . . . 261
- 3.4.1 Longitudinal dielectric function . . . 261
- 3.4.2 Transverse dielectric function . . . 267
- 3.4.3 Linear response to external fields . . . 270
- 3.4.4 Conductivity . . . 273
- 3.4.5 Superconductors . . . 280

3.5 One-Particle Green Function . . . 283
- 3.5.1 Spectral function . . . 284
- 3.5.2 Analyticity properties . . . 285
- 3.5.3 Sum rules and high-energy behavior . . . 285

3.5.4 Symmetry properties 287
3.5.5 Relation to observables 288
3.6 Spontaneous Symmetry Breaking and Goldstone's Theorem 290
3.6.1 The Bogoliubov inequality 291
3.6.2 Spontaneous symmetry breaking 291
3.6.3 Goldstone's theorem 296
3.6.4 Mermin–Wagner theorem 302
3.6.5 Quantum disorder at zero temperature 303
Bibliography 309

When studying a many-body system, we rarely try to compute the N-particle wavefunction $\Psi(\mathbf{r}_1, \ldots, \mathbf{r}_N)$. This calculation is usually hardly feasible, and we do not need all the information which is contained in the wavefunction. A reduced description in terms of a few variables is often sufficient if it allows one to explain the response of the system to an experimental probe. Correlation functions offer such a description. In Sec. 3.1, we show that the linear response of the system to an external field is given by a correlation function. In Sec. 3.2, we study the basic properties of correlation (or response) functions and show that they contain information about the excitations of the system. Furthermore, they are simply related to the imaginary-time Green functions introduced in Sec. 1.2.3 and can therefore be calculated perturbatively using the methods discussed in Chapter 1. In Secs. 3.3 and 3.4, we study the density–density and electromagnetic response functions. From the properties of the latter, we deduce a number of general features of electron systems (metals and superconductors). The one-particle Green function is just one type of correlation function and is studied in Sec. 3.5. Finally, in Sec. 3.6, we consider systems with a spontaneous broken symmetry and derive two important theorems due to Goldstone, and Mermin and Wagner.

3.1 Linear Response Theory

We consider a physical system in the presence of an external field $h_i(\mathbf{r}, t)$ which couples to observables of the system described by the operators $\hat{A}_i(\mathbf{r})$. The total Hamiltonian can be written as

$$\hat{H} + \hat{H}'(t) = \hat{H} - \int d^d r \sum_i h_i(\mathbf{r}, t) \hat{A}_i(\mathbf{r}), \tag{3.1}$$

where $\hat{H}$ is the Hamiltonian in the absence of the external field. In the grand canonical ensemble, $\hat{H}$ includes the term $-\mu\hat{N}$. The perturbation term $\hat{H}'(t)$ comes from the coupling to the external field which is assumed to be local in space. If we are dealing with a physical external field, e.g. an electromagnetic field, the operators $\hat{A}_i(\mathbf{r})$ correspond to observables of the system and are Hermitian, while $h_i(\mathbf{r}, t)$ is real. It is useful to consider a more general perturbation Hamiltonian $\hat{H}'(t)$ where the operator $\hat{A}_i(\mathbf{r})$ is bosonic but not necessarily Hermitian. In order to ensure that $\hat{H}'(t)$ is Hermitian, one has to include the complex conjugate part $h_i^*(\mathbf{r}, t)\hat{A}_i^\dagger(\mathbf{r})$ in $\hat{H}'(t)$. The functional integral formalism allows a further generalization where $\hat{A}_i$ is fermionic and the field h_i a Grassmann variable (Sec. 3.2.7).

3.1.1 *Kubo's formula*

We are interested in the response of the system in the limit of a vanishing external field, $h_i(\mathbf{r}, t) \to 0$. At finite temperature T, the mean value of the operator $\hat{A}_i(\mathbf{r})$ is given by[1]

$$\langle \hat{A}_i(\mathbf{r})\rangle(t) = \mathrm{Tr}[\hat{\rho}(t)\hat{A}_i(\mathbf{r})], \tag{3.2}$$

where $\hat{\rho}(t)$ is the density matrix of the whole system (including the external field) at time t. We assume that the external field is turned on adiabatically and vanishes for $t \to -\infty$. These boundary conditions ensure that the external perturbation remains small — provided the system is stable — and that nonlinear effects do not play an important role. As will become clear later on, they also ensure that the response of the system to the external field is causal.[2]

The time evolution of the density matrix is obtained under the assumption of adiabaticity: $\hat{\rho}(t) = \sum_n |\Psi_n(t)\rangle p_n \langle \Psi_n(t)|$ where $\{|\Psi_n(t = -\infty)\rangle, \epsilon_n\}$ is a complete set of eigenvectors of $\hat{H}$ and the probability occupations $p_n \propto e^{-\beta\epsilon_n}$ are time independent. This assumption of unitary evolution is justified for time evolutions that are fast on the scale of the thermal equilibrium time. When the time evolution is sufficiently slow, it is possible for the system to exchange energy (heat) with the reservoir, thus modifying the population of the energy levels.[3] Since $i\partial_t|\Psi_n(t)\rangle = (\hat{H} + \hat{H}'(t))|\Psi_n(t)\rangle$, the adiabaticity assumption leads to

$$i\frac{\partial \hat{\rho}(t)}{\partial t} = [\hat{H} + \hat{H}'(t), \hat{\rho}(t)], \tag{3.3}$$

with $\hat{\rho}(-\infty) = \hat{\rho}_{\text{eq}} = e^{-\beta\hat{H}}/\mathrm{Tr}(e^{-\beta\hat{H}})$ the density matrix of the system in thermal equilibrium. To linear order in the external fields, $\delta\hat{\rho}(t) = \hat{\rho}(t) - \hat{\rho}_{\text{eq}}$ satisfies the equation

$$i\frac{\partial \delta\hat{\rho}(t)}{\partial t} = [\hat{H}, \delta\hat{\rho}(t)] + [\hat{H}'(t), \hat{\rho}_{\text{eq}}], \tag{3.4}$$

the solution of which is given by

$$\delta\hat{\rho}(t) = -i\int_{-\infty}^{t} dt' e^{-i\hat{H}(t-t')}[\hat{H}'(t'), \hat{\rho}_{\text{eq}}]e^{i\hat{H}(t-t')}, \tag{3.5}$$

as can be verified by direct substitution into (3.4).

---

[1] We shall use the notation $\langle \hat{A}_i(\mathbf{r})\rangle(t)$ or $\langle A_i(\mathbf{r}, t)\rangle$ to denote the expectation value of the (time-independent) operator $\hat{A}_i(\mathbf{r})$ in the presence of a (time-dependent) external perturbation. The notation $\langle A_i(\mathbf{r}, t)\rangle$ is indeed natural when one computes $\langle A_i(\mathbf{r}, \tau)\rangle$ in the Euclidean formalism before performing the analytic continuation to real time (see Sec. 3.2.7).

[2] The relation between these boundary conditions and causality is made explicit by the analyticity properties of the retarded response function $\chi^R(\omega)$.

[3] See the discussion in Sec. 3.2.7.

This result can also be derived by writing the evolution operator (Sec. 2.2.4) in the interaction representation (Sec. 1.G.1.1):

$$|\Psi_n(t)\rangle = \hat{T}(t,t_0)|\Psi_n(t_0)\rangle, \qquad \hat{T}(t,t_0) = e^{-i\hat{H}t}\hat{T}_I(t,t_0)e^{i\hat{H}t_0}, \tag{3.6}$$

where $\hat{T}$ and $\hat{T}_I$ satisfy

$$\begin{aligned}\hat{T}^\dagger(t,t_0)\hat{T}(t,t_0) = \hat{T}_I^\dagger(t,t_0)\hat{T}_I(t,t_0) = 1,\\ i\partial_t\hat{T}(t,t_0) = [\hat{H}+\hat{H}'(t)]\hat{T}(t,t_0),\\ i\partial_t\hat{T}_I(t,t_0) = \hat{H}_I'(t)\hat{T}_I(t,t_0)\end{aligned} \tag{3.7}$$

with

$$\hat{H}_I'(t) = e^{i\hat{H}t}\hat{H}'(t)e^{-i\hat{H}t}. \tag{3.8}$$

To linear order in the external perturbation,

$$\hat{T}_I(t,t_0) = 1 - i\int_{t_0}^{t} dt'\,\hat{H}_I'(t'). \tag{3.9}$$

Equation (3.5) then follows from $\hat{\rho}(t) = \hat{T}(t,-\infty)\hat{\rho}_{\text{eq}}\hat{T}^\dagger(t,-\infty)$.

From the cyclic invariance of the trace, we then deduce

$$\begin{aligned}\delta\langle\hat{A}_i(\mathbf{r})\rangle(t) &= \text{Tr}[\delta\hat{\rho}(t)\hat{A}_i(\mathbf{r})]\\ &= -i\int_{-\infty}^{t} dt'\,\text{Tr}\Big\{e^{-i\hat{H}(t-t')}[\hat{H}'(t'),\hat{\rho}_{\text{eq}}]e^{i\hat{H}(t-t')}\hat{A}_i(\mathbf{r})\Big\}\\ &= -i\int_{-\infty}^{t} dt'\,\text{Tr}\Big\{\hat{\rho}_{\text{eq}}\Big[\hat{A}_i(\mathbf{r},t), e^{i\hat{H}t'}H'(t')e^{-i\hat{H}t'}\Big]\Big\}\\ &= i\int_{-\infty}^{t} dt'\int d^dr'\sum_j \text{Tr}\{\hat{\rho}_{\text{eq}}[\hat{A}_i(\mathbf{r},t),\hat{A}_j(\mathbf{r},t')]\}h_j(\mathbf{r}',t').\end{aligned} \tag{3.10}$$

Here, $\hat{A}_i(\mathbf{r},t) = e^{i\hat{H}t}\hat{A}_i(\mathbf{r})e^{-i\hat{H}t}$ denotes the operator $\hat{A}_i(\mathbf{r})$ in the Heisenberg representation (with respect to the Hamiltonian $\hat{H}$). Equation (3.10) can be written in terms of a "retarded" (causal) response function (or susceptibility) $\chi^R_{A_iA_j}$ as

$$\begin{aligned}\delta\langle\hat{A}_i(\mathbf{r})\rangle(t) &= \int_{-\infty}^{\infty} dt'\int d^dr'\sum_j \chi^R_{A_iA_j}(\mathbf{r},t;\mathbf{r}',t')h_j(\mathbf{r}',t'),\\ \chi^R_{A_iA_j}(\mathbf{r},t;\mathbf{r}',t') &= i\Theta(t-t')\langle[\hat{A}_i(\mathbf{r},t),\hat{A}_j(\mathbf{r}',t')]\rangle,\end{aligned} \tag{3.11}$$

where the average $\langle[.,.]\rangle$ is taken with the equilibrium density matrix $\hat{\rho}_{\text{eq}}$. Equation (3.11), which expresses the linear response function as an equilibrium correlation function, is known as Kubo's formula. It is convenient to introduce the correlation function $\chi''_{A_iA_j}$ defined by

$$\begin{aligned}\chi''_{A_iA_j}(\mathbf{r},t;\mathbf{r}',t') &= \frac{1}{2}\langle[\hat{A}_i(\mathbf{r},t),\hat{A}_j(\mathbf{r}',t')]\rangle,\\ \chi^R_{A_iA_j}(\mathbf{r},t;\mathbf{r}',t') &= 2i\Theta(t-t')\chi''_{A_iA_j}(\mathbf{r},t;\mathbf{r}',t').\end{aligned} \tag{3.12}$$

In the following, we include the $\mathbf{r}$ dependence of the operator $\hat{A}_i$ in the i index unless explicitly written. Accordingly, the response function is denoted by $\chi^R_{A_iA_j}(t,t') = 2i\Theta(t-t')\chi''_{A_iA_j}(t,t')$. We also write $\hat{\rho}$ instead of $\hat{\rho}_{\text{eq}}$ for the equilibrium density matrix $e^{-\beta\hat{H}}/\text{Tr}(e^{-\beta\hat{H}})$ as only equilibrium thermal averages will now be considered.

In addition to the retarded response function, we shall also discuss the "advanced" (anticausal) response function, defined by

$$\chi^A_{A_iA_j}(t,t') = -2i\Theta(-t+t')\chi''_{A_iA_j}(t,t'). \tag{3.13}$$

To make $\chi^A_{A_iA_j}$ well defined, one has to assume that the external field is switched off adiabatically and vanishes for $t\to\infty$.

3.1.2 *Periodic perturbations*

Both experimentally and theoretically, one often considers the response to a periodic perturbation. A field $h_i(t) = h_i(\omega)e^{-i\omega t}$ would violate the assumption that the perturbation is turned on — or switched off when considering the advanced response — adiabatically. This difficulty is avoided by choosing

$$h_i(t) = h_i(\omega)e^{-i\omega t+\eta t} \qquad (\eta\to 0^+) \tag{3.14}$$

when calculating the retarded response function. η^{-1} determines the adiabatic switching time of the external perturbation. The limit $\eta\to 0^+$ must be taken at the end of the calculation. A straightforward calculation then gives

$$\delta\langle\hat{A}_i\rangle(t) = \sum_j \chi^R_{A_iA_j}(\omega)h_j(\omega)e^{-i\omega t+\eta t}, \tag{3.15}$$

where the Fourier transform of χ^R is defined by

$$\chi^R_{A_iA_j}(\omega) = \int_{-\infty}^{\infty} dt\,\chi^R_{A_iA_j}(t)e^{i(\omega+i\eta)t}. \tag{3.16}$$

Similarly, we obtain

$$\chi^A_{A_iA_j}(\omega) = \int_{-\infty}^{\infty} dt\,\chi^A_{A_iA_j}(t)e^{i(\omega-i\eta)t} \tag{3.17}$$

for the advanced response function, with $h_i(t) = h_i(\omega)e^{-i\omega t-\eta t}$.

3.2 Properties of Correlation Functions

3.2.1 *Symmetry properties*

In this section, we discuss the symmetry properties of $\chi''(\omega)$. These imply analogous properties for χ^R and χ^A which are most easily derived from their spectral representation (Sec. 3.2.3). As the derivation does not raise any difficulty, we merely summarize the symmetry properties of χ^R and χ^A in Table 3.1.

3.2.1.1 *Properties following from the definition*

Some properties of χ'' directly follow from its definition (3.12). The cyclic invariance of the trace implies that $\chi''_{A_iA_j}(t,t') = \chi''_{A_iA_j}(t-t')$ depends only on the time difference $t-t'$, as expected for a system in thermal equilibrium. We denote its Fourier transform by $\chi''_{A_iA_j}(\omega)$. $\chi''_{A_iA_j}(t)$ being defined by a commutator, it satisfies $\chi''_{A_iA_j}(t) = -\chi''_{A_jA_i}(-t) = [\chi''_{A_j^\dagger A_i^\dagger}(-t)]^*$. In frequency space, this gives

$$\chi''_{A_iA_j}(\omega) = -\chi''_{A_jA_i}(-\omega) = [\chi''_{A_j^\dagger A_i^\dagger}(\omega)]^*. \tag{3.18}$$

3.2.1.2 *Properties following from symmetries of the system*

To each symmetry transformation, we can associate a unitary operator $\hat{U}$ (Chapter 2). Time-reversal symmetry, where the Wigner operator $\hat{U}$ is antiunitary, will be discussed in the following. If the system is invariant under the symmetry transformation, then the transformed Hamiltonian $\hat{H}' = \hat{U}\hat{H}\hat{U}^\dagger + i(\partial_t\hat{U})\hat{U}^\dagger$ is identical to the original Hamiltonian $\hat{H}$. In this section, except for the time-reversal symmetry discussed in the following, we restrict ourselves to time-independent Wigner operators.[4] The condition $\hat{H}' = \hat{H}$ then implies $[\hat{H},\hat{U}] = 0$ and, provided that the symmetry is not spontaneously broken, $[\hat{\rho},\hat{U}] = 0$. Let us denote by $\{|n\rangle, \epsilon_n\}$ a complete set of eigenvectors of $\hat{H}$. The vectors $\hat{U}|n\rangle$ are also eigenvectors of $\hat{H}$, $\hat{H}\hat{U}|n\rangle = \epsilon_n\hat{U}|n\rangle$, and can be used to compute χ'',

$$\begin{aligned}\chi''_{A_iA_j}(t) &= \frac{1}{2}\sum_n \langle n|\hat{U}^\dagger\hat{\rho}[\hat{A}_i(t),\hat{A}_j]\hat{U}|n\rangle \\ &= \frac{1}{2}\sum_n \langle n|\hat{\rho}[\hat{U}^\dagger\hat{A}_i(t)\hat{U},\hat{U}^\dagger\hat{A}_j\hat{U}]|n\rangle = \chi''_{U^\dagger A_iU,U^\dagger A_jU}(t),\end{aligned} \tag{3.19}$$

where we have used $[\hat{\rho},\hat{U}] = 0$, $\hat{U}\hat{U}^\dagger = 1$ and $\hat{U}^\dagger\hat{A}_i(t)\hat{U} = e^{i\hat{H}t}\hat{U}^\dagger\hat{A}_i\hat{U}e^{-i\hat{H}t}$. The correlation function $\chi''_{A_iA_j}(t)$ is thus invariant under the transformation $\hat{A}_i \to \hat{U}^\dagger\hat{A}_i\hat{U}$ and $\hat{A}_j \to \hat{U}^\dagger\hat{A}_j\hat{U}$.[5]

[4] Time-dependent symmetry transformations are more naturally handled in the functional integral formalism (see Sec. 3.2.7).

[5] In the language of Chapter 2, Eq. (3.19) means that the two observers $\mathcal{O}$ and $\mathcal{O}'$ measure the same correlation function $\chi''_{A_iA_j}$ in their respective reference frames when the system is invariant under the symmetry transformation.

Table 3.1 Symmetry properties of $\chi''_{A_iA_j}(\omega)$ and $\chi^{R,A}_{A_iA_j}(\omega)$. $\hat{A}_i$ and $\hat{A}_j$ are bosonic operators. ϵ^P_A and ϵ^T_A denote the signature of the operator $\hat{A}$ under parity and time reversal, respectively.

		$\chi''_{A_iA_j}(\omega)$	$\chi^{R,A}_{A_iA_j}(\omega)$
		$\chi''_{A_iA_j}(\omega) = -\chi''_{A_jA_i}(-\omega) = [\chi''_{A_j^\dagger A_i^\dagger}(\omega)]^*$	$\chi^R_{A_iA_j}(\omega) = [\chi^R_{A_i^\dagger A_j^\dagger}(-\omega)]^* = [\chi^A_{A_jA_i}(-\omega)]$
Space translation		$\chi''_{A_iA_j}(\mathbf{r},\mathbf{r}',\omega) = \chi''_{A_iA_j}(\mathbf{r}-\mathbf{r}',\omega)$	$\chi^R_{A_iA_j}(\mathbf{r},\mathbf{r}',\omega) = \chi^R_{A_iA_j}(\mathbf{r}-\mathbf{r}',\omega)$
Parity		$\chi''_{A_iA_j}(\mathbf{r},\mathbf{r}',\omega) = \epsilon^P_{A_i}\epsilon^P_{A_j}\chi''_{A_iA_j}(-\mathbf{r},-\mathbf{r}',\omega)$	$\chi^R_{A_iA_j}(\mathbf{r},\mathbf{r}',\omega) = \epsilon^P_{A_i}\epsilon^P_{A_j}\chi^R_{A_iA_j}(-\mathbf{r},-\mathbf{r}',\omega)$
Time reversal	$\epsilon^T_{A_i}\epsilon^T_{A_j} = 1$	$\chi''_{A_iA_j}(\omega) = [\chi''_{A_iA_j}(\omega)]^* = \Im[\chi^R_{A_iA_j}(\omega)]$	$\chi^R_{A_iA_j}(\omega) = \chi^R_{A_j^\dagger A_i^\dagger}(\omega) = [\chi^A_{A_iA_j}(\omega)]^*$
	$\epsilon^T_{A_i}\epsilon^T_{A_j} = -1$	$\chi''_{A_iA_j}(\omega) = -[\chi''_{A_iA_j}(\omega)]^* = -i\Re[\chi^R_{A_iA_j}(\omega)]$	$\chi^R_{A_iA_j}(\omega) = -\chi^R_{A_j^\dagger A_i^\dagger}(\omega) = -[\chi^A_{A_iA_j}(\omega)]^*$
$\hat{A}_i = \hat{A}_j^\dagger$		$\chi''_{A^\dagger A}(\omega) = \Im[\chi^R_{A^\dagger A}(\omega)]$	$\chi^R_{A^\dagger A}(\omega) = [\chi^A_{A^\dagger A}(\omega)]^*$
Stability		$\{\omega\chi''_{A_i^\dagger A_j}(\omega)\}$ positive matrix	$\{\chi^R_{A_i^\dagger A_j}(\omega=0)\}$ positive matrix
		$\mathrm{sgn}[\chi''_{A^\dagger A}(\omega)] = \mathrm{sgn}(\omega)$	$\chi^R_{A^\dagger A}(\omega=0) \geq 0$

As an example, let us consider a space translation, $\hat{U}^\dagger \hat{A}_i(\mathbf{r})\hat{U} = \hat{A}_i(\mathbf{r}-\mathbf{a})$, where $\mathbf{a}$ is an arbitrary vector (see Sec. 2.2.2). If the system is translation invariant, Eq. (3.19) implies $\chi''_{A_iA_j}(\mathbf{r},\mathbf{r}',t) = \chi''_{A_iA_j}(\mathbf{r}-\mathbf{a},\mathbf{r}'-\mathbf{a},t)$, i.e.

$$\chi''_{A_iA_j}(\mathbf{r},\mathbf{r}',t) = \chi''_{A_iA_j}(\mathbf{r}-\mathbf{r}',t). \tag{3.20}$$

Similarly, invariance under a parity transformation implies

$$\chi''_{A_iA_j}(\mathbf{r},\mathbf{r}',t) = \epsilon^P_{A_i}\epsilon^P_{A_j}\chi''_{A_iA_j}(-\mathbf{r},-\mathbf{r}',t) \tag{3.21}$$

for two operators with signatures $\epsilon^P_{A_i}$ and $\epsilon^P_{A_j}$: $\hat{U}^\dagger \hat{A}_i(\mathbf{r})\hat{U} = \epsilon^P_{A_i}\hat{A}_i(-\mathbf{r})$ and $\hat{U}^\dagger \hat{A}_j(\mathbf{r})\hat{U} = \epsilon^P_{A_j}\hat{A}_j(-\mathbf{r})$ (see Sec. 2.2.1).

In the case of time-reversal symmetry, the Wigner operator $\hat{U}(t)$ is antiunitary. χ'' can be evaluated using the time-reversed states $\hat{\Theta}|n\rangle$, where $\hat{\Theta} = \hat{U}(t)e^{-2i\hat{H}t}$ is the time-independent antiunitary operator introduced in Sec. 2.2.6. Time-reversal invariance implies $[\hat{H},\hat{\Theta}] = [\hat{\rho},\hat{\Theta}] = 0$. We then obtain[6]

$$\begin{aligned}\chi''_{A_iA_j}(t) &= \frac{1}{2}\sum_n ((\langle n|\hat{\Theta}^\dagger)\hat{\rho}[\hat{A}_i(t),\hat{A}_j](\hat{\Theta}|n\rangle)) = \frac{1}{2}\sum_n [\langle n|(\hat{\Theta}^\dagger\hat{\rho}[\hat{A}_i(t),\hat{A}_j]\hat{\Theta}|n\rangle)]^* \\ &= \frac{1}{2}\sum_n [\langle n|\hat{\rho}([\hat{\Theta}^\dagger\hat{A}_i(t)\hat{\Theta},\hat{\Theta}^\dagger\hat{A}_j\hat{\Theta}]|n\rangle)]^* = [\chi''_{\Theta^\dagger A_i\Theta,\Theta^\dagger A_j\Theta}(-t)]^* \\ &= \chi''_{\Theta^\dagger A_j^\dagger\Theta,\Theta^\dagger A_i^\dagger\Theta}(t),\end{aligned} \tag{3.22}$$

i.e.

$$\chi''_{A_iA_j}(\omega) = \chi''_{\Theta^\dagger A_j^\dagger\Theta,\Theta^\dagger A_i^\dagger\Theta}(\omega). \tag{3.23}$$

We have used $\hat{\Theta}\hat{\Theta}^\dagger = 1$ and $\Theta^\dagger\hat{A}(t)\hat{\Theta} = e^{-i\hat{H}t}\hat{\Theta}^\dagger\hat{A}\hat{\Theta}e^{i\hat{H}t}$ (Chapter 2). For two operators with signatures $\epsilon^T_{A_i}$ and $\epsilon^T_{A_j}$ under time reversal symmetry ($\Theta^\dagger\hat{A}_i(t)\hat{\Theta} = \epsilon^T_{A_i}\hat{A}_i(-t)$ and $\Theta^\dagger\hat{A}_j(t)\hat{\Theta} = \epsilon^T_{A_j}\hat{A}_j(-t)$), Eqs. (3.22) and (3.23) give

$$\begin{aligned}\chi''_{A_iA_j}(t) &= \epsilon^T_{A_i}\epsilon^T_{A_j}[\chi''_{A_iA_j}(-t)]^*, \\ \chi''_{A_iA_j}(\omega) &= \epsilon^T_{A_i}\epsilon^T_{A_j}[\chi''_{A_iA_j}(\omega)]^* = \epsilon^T_{A_i}\epsilon^T_{A_j}\chi''_{A_j^\dagger A_i^\dagger}(\omega).\end{aligned} \tag{3.24}$$

From Eq. (3.24), we conclude that $\chi''_{A_iA_j}(\omega)$ is real if $\epsilon^T_{A_i}\epsilon^T_{A_j} = 1$ and imaginary if $\epsilon^T_{A_i}\epsilon^T_{A_j} = -1$. Properties of real- and imaginary-time correlation functions, similar to (3.22) and (3.23), were obtained in Sec. 2.2.6.

[6]Recall that parentheses must be used to avoid the ambiguity of the Dirac notation when dealing with antilinear operators (Sec. 2.2.6).

3.2.2 Dissipation

Another property of $\chi''_{A_iA_j}$ can be obtained by considering the energy dissipation of the system coupled to an external perturbation. If the system is stable, it can only absorb energy from the external field over a long time interval. The energy transferred to the system per unit time is given by

$$\begin{aligned}\frac{dW}{dt} &= \frac{d}{dt}\mathrm{Tr}\{\hat{\rho}(t)[\hat{H}+\hat{H}'(t)]\}\\ &= \mathrm{Tr}\left\{\hat{\rho}(t)\frac{d\hat{H}'(t)}{dt}\right\} + \mathrm{Tr}\left\{\frac{d\hat{\rho}(t)}{dt}\left[\hat{H}+\hat{H}'(t)\right]\right\}. \end{aligned} \tag{3.25}$$

Using Eq. (3.3), the last term in (3.25) is found to vanish, and

$$\frac{dW}{dt} = -\sum_i \mathrm{Tr}[\hat{\rho}(t)\hat{A}_i]\dot{h}_i(t) = -\sum_i \{\mathrm{Tr}[\hat{\rho}\hat{A}_i] + \mathrm{Tr}[\delta\hat{\rho}(t)\hat{A}_i]\}\dot{h}_i(t), \tag{3.26}$$

where $\dot{h}_i(t) = \partial_t h_i(t)$ denotes the time derivative of $h_i(t)$. Here, $\mathrm{Tr}[\delta\hat{\rho}(t)\hat{A}_i] = \delta\langle\hat{A}_i(t)\rangle$ is the response of the system to the external fields.

Let us now consider a periodic perturbation of the form

$$h_i(t) = \frac{1}{2}\left[h_i(\omega)e^{-i\omega t+\eta t} + h_i(-\omega)e^{i\omega t+\eta t}\right], \tag{3.27}$$

with $h_{A_i^\dagger}(\omega) = h^*_{A_i}(-\omega)$ ($h_i(-\omega) = h_i^*(\omega)$ when $h_i(t)$ is real and $\hat{A}_i$ Hermitian). To lowest order in the field h_i, the mean power dissipated into the system reads

$$\begin{aligned}\overline{\frac{dW}{dt}} &= \frac{1}{T}\int_t^{t+T} dt\frac{dW}{dt}\\ &= -\frac{1}{4}\sum_{i,j}\left[\chi^R_{A_iA_j}(\omega)h_j(\omega)i\omega h_i(-\omega) - \chi^R_{A_iA_j}(-\omega)h_j(-\omega)i\omega h_i(\omega)\right]\\ &= -\frac{1}{4}\sum_{i,j} h_i(-\omega)\left[\chi^R_{A_iA_j}(\omega) - \chi^R_{A_jA_i}(-\omega)\right]i\omega h_j(\omega), \end{aligned} \tag{3.28}$$

where $T = 2\pi/\omega$. From the definition of $\chi^R_{A_iA_j}$, we easily deduce

$$\chi^R_{A_iA_j}(\omega) - \chi^R_{A_jA_i}(-\omega) = 2i\chi''_{A_iA_j}(\omega), \tag{3.29}$$

and therefore,

$$\overline{\frac{dW}{dt}} = \frac{1}{2}\sum_{i,j} h_i^*(\omega)\omega\chi''_{A_i^\dagger A_j}(\omega)h_j(\omega). \tag{3.30}$$

Here, we have changed $\hat{A}_i \to \hat{A}_i^\dagger$ in the sum over i. Since the stability of the system requires $\overline{dW/dt} \geq 0$ regardless of the external perturbation, we conclude that $\omega\chi''_{A_i^\dagger A_j}(\omega)$ is a positive matrix for any set of operators $\hat{A}_i$. In particular, for the minimal set $\{\hat{A}, \hat{A}^\dagger\}$, this implies

$$\omega\chi''_{A^\dagger A}(\omega) \geq 0. \tag{3.31}$$

Using Eq. (3.44), one then obtains the inequality $\chi^R_{A^\dagger A}(\omega = 0) \geq 0$ whose physical meaning will be discussed at the end of Sec. 3.2.7. Equation (3.31) can also be deduced from the expression of $\chi''_{A_iA_j}(\omega)$ given in (3.37).

3.2.3 *Spectral representation*

From the definition (3.16, 3.17) of the Fourier transforms of the retarded and advanced response functions, we obtain

$$\begin{aligned}\chi^{R,A}_{A_iA_j}(\omega) &= \pm 2i \int_{-\infty}^{\infty} dt\, e^{i(\omega \pm i\eta)t}\Theta(\pm t) \int_{-\infty}^{\infty} \frac{d\omega'}{2\pi} \chi''_{A_iA_j}(\omega') e^{-i\omega' t} \\ &= \int_{-\infty}^{\infty} \frac{d\omega'}{\pi} \frac{\chi''_{A_iA_j}(\omega')}{\omega' - (\omega \pm i\eta)}.\end{aligned} \tag{3.32}$$

Equation (3.32) allows one to generalize the definition of $\chi^{R,A}(\omega)$ to complex frequencies z by introducing

$$\chi_{A_iA_j}(z) = \int_{-\infty}^{\infty} \frac{d\omega}{\pi} \frac{\chi''_{A_iA_j}(\omega)}{\omega - z}. \tag{3.33}$$

χ^R, χ^A and χ are related by

$$\begin{aligned}\chi^R_{A_iA_j}(\omega) &= \chi_{A_iA_j}(z = \omega + i\eta), \\ \chi^A_{A_iA_j}(\omega) &= \chi_{A_iA_j}(z = \omega - i\eta).\end{aligned} \tag{3.34}$$

Since

$$\chi^R_{A_iA_j}(\omega) - \chi^A_{A_iA_j}(\omega) = 2i\chi''_{A_iA_j}(\omega), \tag{3.35}$$

$\chi_{A_iA_j}(z)$ is discontinuous across the real axis wherever $\chi''_{A_iA_j}(\omega)$ does not vanish. Equation (3.33) is known as a spectral or Lehmann representation. It relates $\chi_{A_iA_j}(z)$ in the entire complex plane to $\chi''_{A_iA_j}(\omega)$ along the real axis.

A more explicit definition of $\chi_{A_iA_j}(z)$ can be obtained by expressing $\chi''_{A_iA_j}(\omega)$ in terms of a complete set of eigenvectors $\{|n\rangle, \epsilon_n\}$ of the Hamiltonian $\hat{H}$. Using the closure relation $\sum_n |n\rangle\langle n| = 1$, we obtain

$$\begin{aligned}\chi''_{A_iA_j}(t) &= \frac{1}{2Z}\mathrm{Tr}\big[e^{-\beta\hat{H}}\big(e^{i\hat{H}t}\hat{A}_i e^{-i\hat{H}t}\hat{A}_j - \hat{A}_j e^{i\hat{H}t}\hat{A}_i e^{-i\hat{H}t}\big)\big] \\ &= \frac{1}{2Z}\sum_{n,m} e^{-\beta\epsilon_n}\left(e^{i\epsilon_n t}A_i^{nm}e^{-i\epsilon_m t}A_j^{mn} - A_j^{nm}e^{i\epsilon_m t}A_i^{mn}e^{-i\epsilon_n t}\right),\end{aligned} \tag{3.36}$$

where $A_i^{nm} = \langle n|\hat{A}_i|m\rangle$. In frequency space, this gives

$$\chi''_{A_iA_j}(\omega) = \frac{\pi}{Z}\sum_{n,m}\left(e^{-\beta\epsilon_n} - e^{-\beta\epsilon_m}\right)A_i^{nm}A_j^{mn}\delta(\omega + \epsilon_n - \epsilon_m). \tag{3.37}$$

Inserting (3.37) into (3.33), we obtain an alternative form of the Lehmann representation, in terms of the exact eigenstates of the system,

$$\chi_{A_iA_j}(z) = \frac{1}{Z} \sum_{\substack{n,m \\ (\epsilon_n \neq \epsilon_m)}} \frac{e^{-\beta\epsilon_n} - e^{-\beta\epsilon_m}}{\epsilon_m - \epsilon_n - z} A_i^{nm} A_j^{mn}. \tag{3.38}$$

We have excluded of the sum the terms with $\epsilon_n = \epsilon_m$ which do not contribute to $\chi''_{A_iA_j}(\omega)$ [Eq. (3.37)]. Since $\chi''_{A_iA_j}(\omega)$ is a sum of delta peaks at the transition energies $\epsilon_m - \epsilon_n$, it gives information about the excitation spectrum of the system. For this reason, it is often referred to as the spectral function.

$\chi_{A_iA_j}(z)$ is an analytic function of z for $\Im(z) \neq 0$. It follows that $\chi^R_{A_iA_j}(\omega)$, considered as a function of the complex variable ω, is analytic in the upper-half complex plane $\Im(\omega) \geq 0$, whereas $\chi^A_{A_iA_j}(\omega)$ is analytic in the lower-half complex plane. In a finite system, the singularities of $\chi^R_{A_iA_j}(\omega)$ ($\chi^A_{A_iA_j}(\omega)$) always consist of discrete poles located infinitesimally below (above) the real axis (Fig. 3.1). The thermodynamic limit (large-size systems), whereby the sum over the state index n may be partially replaced by a continuous integral, qualitatively changes this picture. An infinite number of poles of vanishing strength can merge and form a branch cut corresponding to a continuous excitation spectrum. Poles can also coalesce into an isolated pole located away from the real axis, i.e. at the complex frequency $\Re(\omega) + i\Im(\omega)$ with $\Im(\omega) < 0$ (> 0) for a pole of χ^R (χ^A). This gives rise to a strong enhancement of the response function χ^R at the real frequency $\Re(\omega)$. This suggests — as will be seen on many examples — that an isolated pole corresponds to a collective excitation (or collective mode) of the system, with an energy $\Re(\omega)$ and a lifetime $(-\Im(\omega))^{-1}$.[7] An isolated pole shows up as a sharp peak in the spectral function $\chi''_{A_iA_j}(\omega)$ with a width given by the inverse lifetime $-\Im(\omega)$, whereas a branch cut gives rise to a structureless continuous excitation background (Fig. 3.2).

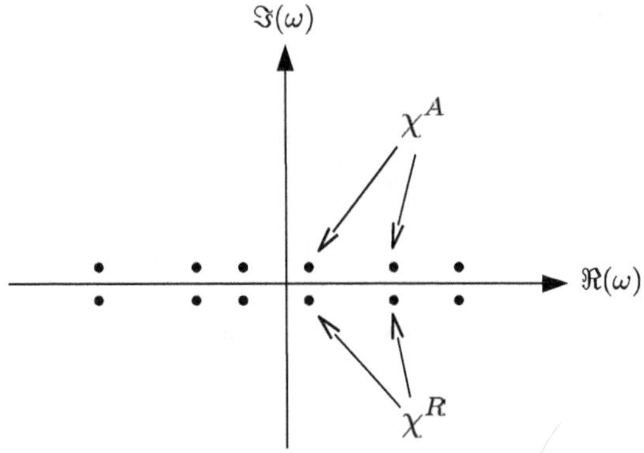

Fig. 3.1 Poles of $\chi^R(\omega)$ and $\chi^A(\omega)$ in the complex ω plane.

[7] An isolated pole can also correspond to a single-particle excitation when $\chi^R_{A_iA_j}$ is the one-particle Green function: $\hat{A}_i = \hat{\psi}$ and $\hat{A}_j = \hat{\psi}^\dagger$ (see Sec. 3.5).

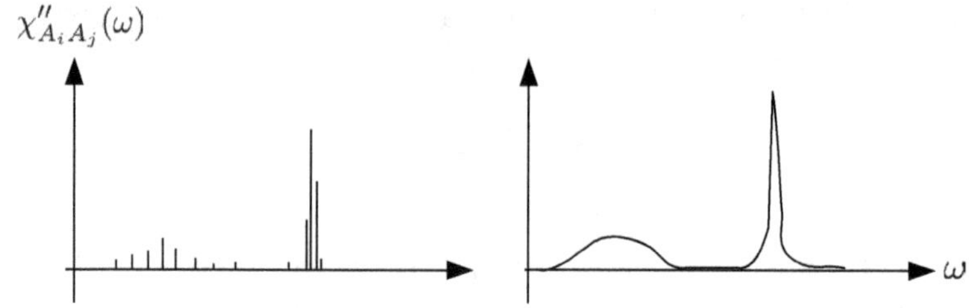

Fig. 3.2 Schematic representation of the spectral function $\chi''_{A_iA_j}(\omega)$ — here assumed real — in a finite (left) and infinite (right) system.

The spectral decomposition (3.32) of the retarded response function, which relates χ'' to χ^R and χ^A, can be written as

$$\chi^{R,A}_{A_iA_j}(\omega) = \mathcal{P}\int_{-\infty}^{\infty} \frac{d\omega'}{\pi}\frac{\chi''_{A_iA_j}(\omega')}{\omega'-\omega} \pm i\chi''_{A_iA_j}(\omega), \tag{3.39}$$

where $\mathcal{P}$ denotes the principal part. In a time-reversal-invariant system, $\chi''_{A_iA_j}(\omega)$ is real when the operators have the same signature ($\epsilon^T_{A_i} = \epsilon^T_{A_j}$). In the opposite case ($\epsilon^T_{A_i} = -\epsilon^T_{A_j}$), $\chi''_{A_iA_j}(\omega)$ is imaginary [Eq. (3.24)]. From (3.39), we then deduce

$$\begin{cases} \chi''_{A_iA_j}(\omega) = \Im[\chi^R_{A_iA_j}(\omega)] \\ \chi^A_{A_iA_j}(\omega) = [\chi^R_{A_iA_j}(\omega)]^* \end{cases} \quad \text{if} \quad \epsilon^T_{A_i} = \epsilon^T_{A_j} \tag{3.40}$$

and

$$\begin{cases} \chi''_{A_iA_j}(\omega) = -i\Re[\chi^R_{A_iA_j}(\omega)] \\ \chi^A_{A_iA_j}(\omega) = -[\chi^R_{A_iA_j}(\omega)]^* \end{cases} \quad \text{if} \quad \epsilon^T_{A_i} = -\epsilon^T_{A_j}. \tag{3.41}$$

When $\hat{A}_i = \hat{A}^\dagger_j$, $\chi''_{A_iA_j}(\omega)$ is real and has the sign of ω, regardless of the behavior of $\hat{A}_i$ under time-reversal symmetry [see Eq. (3.37)], so that

$$\chi''_{A^\dagger A}(\omega) = \Im[\chi^R_{A^\dagger A}(\omega)] \quad \text{and} \quad \chi^A_{A^\dagger A}(\omega) = [\chi^R_{A^\dagger A}(\omega)]^*. \tag{3.42}$$

In practice, most cases of interest correspond to (3.40), (3.41) or (3.42).

3.2.3.1 *The static limit*

In general, $\lim_{\mathbf{q},\omega\to 0}\chi^R_{A_iA_j}(\mathbf{q},\omega)$ depends on the order of the limits $\mathbf{q}\to 0$ and $\omega \to 0$ (we write explicitly the $\mathbf{q}$ dependence and assume translation invariance). The static limit $\omega\to 0$ determines the response to a time-independent external perturbation. In particular, the uniform susceptibility is obtained by taking $\omega\to 0$ and then $\mathbf{q}\to 0$, which amounts to taking the infinite volume (thermodynamic) limit ($\mathcal{V}\sim|\mathbf{q}|^{-3}\to\infty$) in the static case

$(\omega = 0)$,[8]

$$\chi_{A_iA_j} = \left.\frac{\partial\langle\hat{A}_i\rangle_h}{\partial h_j}\right|_{h=0} = \lim_{\mathbf{q}\to 0}\chi_{A_iA_j}(\mathbf{q}) \tag{3.43}$$

($\langle\hat{A}_i\rangle_h$ denotes the mean value of $\hat{A}_i$ in the presence of the external field h_j), where

$$\begin{aligned}\chi_{A_iA_j}(\mathbf{q}) &\equiv \chi^R_{A_iA_j}(\mathbf{q},\omega=0)\\ &= \mathcal{P}\int_{-\infty}^{\infty}\frac{d\omega'}{\pi}\frac{\chi''_{A_iA_j}(\mathbf{q},\omega')}{\omega'} + i\chi''_{A_iA_j}(\mathbf{q},0)\\ &= \int_{-\infty}^{\infty}\frac{d\omega'}{\pi}\frac{\chi''_{A_iA_j}(\mathbf{q},\omega')}{\omega'} \equiv \chi_{A_iA_j}(\mathbf{q},z=0).\end{aligned} \tag{3.44}$$

Here, we have used $\chi''_{A_iA_j}(\mathbf{q},0) = 0$ and the fact that the integral over ω' converges so that the principal part is not necessary.[9] The other limit, $\mathbf{q} = 0$ and $\omega \to 0$, describes transport properties as we shall see in Sec. 3.4.4.

Let us illustrate this point by considering the density–density response function χ_{nn}. The latter is introduced in Sec. 3.3 and will be encountered many times in the following chapters. For a propagating sound mode with velocity $c = 1/\sqrt{nm\kappa}$,

$$\begin{aligned}\chi_{nn}(\mathbf{q},z) &= \frac{n}{m}\frac{\mathbf{q}^2}{c^2\mathbf{q}^2 - z^2},\\ \chi''_{nn}(\mathbf{q},\omega) &= \frac{\pi n}{2mc}|\mathbf{q}|[\delta(\omega - c|\mathbf{q}|) - \delta(\omega + c|\mathbf{q}|)],\end{aligned} \tag{3.45}$$

while for a diffusive mode with diffusion constant D [1, 2]

$$\begin{aligned}\chi_{nn}(\mathbf{q},z) &= n^2\kappa\frac{D\mathbf{q}^2}{D\mathbf{q}^2 - iz\,\mathrm{sgn}(\Im z)},\\ \chi''_{nn}(\mathbf{q},\omega) &= n^2\kappa\frac{D\mathbf{q}^2\omega}{\omega^2 + D^2\mathbf{q}^4},\end{aligned} \tag{3.46}$$

where κ is the compressibility (Sec. 3.3). It is straightforward to verify that in both cases

$$n^2\kappa = \lim_{\mathbf{q}\to 0}\chi_{nn}(\mathbf{q}) = \lim_{\mathbf{q}\to 0}\mathcal{P}\int_{-\infty}^{\infty}\frac{d\omega}{\pi}\frac{\chi''_{nn}(\mathbf{q},\omega)}{\omega} = \int_{-\infty}^{\infty}\frac{d\omega}{\pi}\lim_{\mathbf{q}\to 0}\frac{\chi''_{nn}(\mathbf{q},\omega)}{\omega} \tag{3.47}$$

but

$$n^2\kappa \neq \mathcal{P}\int_{-\infty}^{\infty}\frac{d\omega}{\pi}\lim_{\mathbf{q}\to 0}\frac{\chi''_{nn}(\mathbf{q},\omega)}{\omega}. \tag{3.48}$$

[8]Whether $\chi_{A_iA_j}$ is the true thermodynamic susceptibility is discussed in Sec. 3.2.7.

[9]This can be seen from the fluctuation–dissipation theorem derived in Sec. 3.2.5. The structure factor $S_{A_iA_j}(t=0) = \langle\hat{A}_i\hat{A}_j\rangle_c = \int_{-\infty}^{\infty}(d\omega/2\pi)S_{A_iA_j}(\omega)$ is finite. For $|\omega| \ll T$, the integral behaves as $\int d\omega(2T/\omega)\chi''_{A_iA_j}(\omega)$ so that $T\int_{-T}^{T}(d\omega/\omega)\chi''_{A_iA_j}(\omega) < \infty$.

This last result shows that one must take the limit $\eta \to 0^+$ before $\mathbf{q} \to 0$. Physically, this means that the adiabatic turning-on time $1/\eta$ must be longer than the characteristic time scale set by $\mathbf{q}$ to allow the density to relax towards equilibrium. We also note that taking the limits $\mathbf{q} \to 0, \omega \to 0$ in reverse order does not yield the compressibility since $\chi_{nn}(0, z) = 0$ for $z \neq 0$, as follows from gauge invariance or particle number conservation (see Sec. 3.4.4).

3.2.4 *Kramers–Kronig relations and causality*

Kramers–Kronig relations are a consequence of the causality of the response function χ^R. The causality of

$$\chi^R(t)e^{-\eta t} = \int_{-\infty}^{\infty} \frac{d\omega}{2\pi} e^{-i\omega t} \chi^R(\omega) \tag{3.49}$$

is ensured by the analyticity of $\chi^R(\omega)$ in the upper-half complex plane. For $t < 0$, we can calculate $\chi^R(t)$ by closing the integration contour in the upper-half plane in Eq. (3.49). Since no pole or branch cut is enclosed, $\chi^R(t)$ vanishes for $t < 0$.

The Kramers–Kronig relations are derived by evaluating

$$\oint_{\mathcal{C}} \frac{d\omega'}{\pi} \frac{\chi^R(\omega')}{\omega' - \omega - i\eta} = 2i\chi^R(\omega + i\eta) = 2i\chi^R(\omega), \tag{3.50}$$

where $\mathcal{C}$ is the contour shown in Fig. 3.3. The result (3.50) relies only on the residue theorem and the analyticity of $\chi^R(\omega)$ in the upper-half complex plane. Assuming that $\chi^R(\omega') \to 0$ when $|\omega'| \to \infty$, there is no contribution from the part of the contour at infinity,[10] and Eq. (3.50) gives

$$\mathcal{P} \int_{-\infty}^{\infty} \frac{d\omega'}{\pi} \frac{\chi^R(\omega')}{\omega' - \omega} = i\chi^R(\omega). \tag{3.51}$$

Taking the real and imaginary parts of (3.51), we obtain the Kramers–Kronig relations

$$\begin{aligned} \Re[\chi^R(\omega)] &= \mathcal{P} \int_{-\infty}^{\infty} \frac{d\omega'}{\pi} \frac{\Im[\chi^R(\omega')]}{\omega' - \omega}, \\ \Im[\chi^R(\omega)] &= -\mathcal{P} \int_{-\infty}^{\infty} \frac{d\omega'}{\pi} \frac{\Re[\chi^R(\omega')]}{\omega' - \omega}. \end{aligned} \tag{3.52}$$

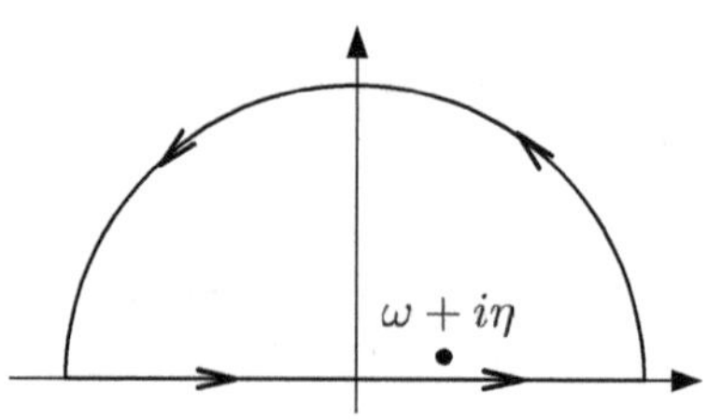

Fig. 3.3 Integration contour $\mathcal{C}$ in Eq. (3.50).

[10]When $\lim_{\omega\to\infty} \chi^R(\omega) = \text{const}$, the Kramers–Kronig relations (3.52) hold for $\chi^R(\omega) - \chi^R(\omega \to \infty)$.

3.2.5 *Fluctuation–dissipation theorem*

Let us consider the structure factor $S_{A_iA_j}$ defined as the correlation function

$$S_{A_iA_j}(t) = \langle\delta\hat{A}_i(t)\delta\hat{A}_j\rangle = \langle\hat{A}_i(t)\hat{A}_j\rangle - \langle\hat{A}_i\rangle\langle\hat{A}_j\rangle, \tag{3.53}$$

where $\delta\hat{A}_i(t) = \hat{A}_i(t) - \langle\hat{A}_i\rangle$. In frequency space, we easily obtain

$$S_{A_iA_j}(\omega) = \frac{2\pi}{Z}\sum_{n,m} e^{-\beta\epsilon_n} A_i^{nm} A_j^{mn}\delta(\omega + \epsilon_n - \epsilon_m) - 2\pi\delta(\omega)\langle\hat{A}_i\rangle\langle\hat{A}_j\rangle, \tag{3.54}$$

where $\{|n\rangle, \epsilon_n\}$ is a complete set of eigenstates of the Hamiltonian $\hat{H}$. The real function $S_{AA^\dagger}(\omega)$ gives the spectral density — or power spectrum — of the spontaneous fluctuations of the variable $\hat{A}$. Changing ω into $-\omega$ in (3.54), we obtain the relation

$$S_{A_iA_j}(\omega) = e^{\beta\omega} S_{A_jA_i}(-\omega) \tag{3.55}$$

between the absorption ($\omega = \epsilon_m - \epsilon_n > 0$) and emission ($\omega < 0$) parts of the spectrum. The emission spectrum vanishes when $T \to 0$.

$\chi''_{A_iA_j}(t)$ and $S_{A_iA_j}(t)$ are related by

$$\chi''_{A_iA_j}(t) = \frac{1}{2}[S_{A_iA_j}(t) - S_{A_jA_i}(-t)]. \tag{3.56}$$

From (3.55) and (3.56), we deduce

$$\chi''_{A_iA_j}(\omega) = \frac{1}{2}\left(1 - e^{-\beta\omega}\right) S_{A_iA_j}(\omega). \tag{3.57}$$

The fluctuation–dissipation theorem is usually written as

$$S_{A_iA_j}(\omega) = 2[n_B(\omega) + 1]\chi''_{A_iA_j}(\omega). \tag{3.58}$$

It relates the spontaneous fluctuations in the system, whose return to equilibrium is described by $S_{A_iA_j}$, to the dissipation induced by an external perturbation.

If $|\omega| \ll T$, Eq. (3.57) gives

$$\chi''_{A_iA_j}(\omega) = \frac{\beta\omega}{2} S_{A_iA_j}(\omega). \tag{3.59}$$

At sufficiently high temperatures, the condition $|\omega| \ll T$ holds for all excitation energies of the system, i.e. for all ω such that $\chi''_{A_iA_j}(\omega)$ is non-zero. Integrating (3.59), we then obtain the classical limit of the fluctuation–dissipation theorem,[11]

$$\chi_{A_iA_j}(t=0) = \beta S_{A_iA_j}(t=0) = \beta\left[\langle\hat{A}_i\hat{A}_j\rangle - \langle\hat{A}_i\rangle\langle\hat{A}_j\rangle\right]. \tag{3.60}$$

[11] Note that the classical fluctuation–dissipation theorem is a direct consequence of linear response theory for a classical system.

3.2.6 *Sum rules and high-frequency behavior*

3.2.6.1 *Thermodynamic sum rules*

The static uniform susceptibility is given by (3.43),

$$\chi_{A_iA_j} \equiv \frac{\partial\langle\hat{A}_i\rangle_h}{\partial h_j}\bigg|_{h=0} = \lim_{\mathbf{q}\to 0}\int_{-\infty}^{\infty}\frac{d\omega}{\pi}\frac{\chi''_{A_iA_j}(\mathbf{q},\omega)}{\omega}. \tag{3.61}$$

Equation (3.61) is a sum rule relating $\chi''_{A_iA_j}(\omega)$ to the susceptibility $\chi_{A_iA_j}$.

3.2.6.2 *Other sum rules*

Other sum rules are obtained by considering

$$\int_{-\infty}^{\infty}\frac{d\omega}{\pi}\omega^n\chi''_{A_iA_j}(\omega) = 2\big[(i\partial_t)^n\chi''_{A_iA_j}(t)\big]_{t=0} = \big[(i\partial_t)^n\langle[\hat{A}_i(t),\hat{A}_j]\rangle\big]_{t=0}. \tag{3.62}$$

Using $i\partial_t\hat{A}_i(t) = [\hat{A}_i(t),\hat{H}]$, we finally obtain

$$\int_{-\infty}^{\infty}\frac{d\omega}{\pi}\omega^n\chi''_{A_iA_j}(\omega) = \Big\langle\Big[[[\hat{A}_i,\hat{H}],\hat{H}],\dots,\hat{A}_j\Big]\Big\rangle. \tag{3.63}$$

3.2.6.3 *High-frequency behavior*

These sum rules determine the asymptotic behavior of $\chi_{A_iA_j}(z)$ at high frequency. Noting that the spectral function $\chi''_{A_iA_j}(\omega)$ is expected to be bounded in frequency in many cases of interest — in particular when i stands for a momentum variable — the spectral representation gives

$$\chi_{A_iA_j}(z) = -\frac{1}{z}\int_{-\infty}^{\infty}\frac{d\omega}{\pi}\chi''_{A_iA_j}(\omega) - \frac{1}{z^2}\int_{-\infty}^{\infty}\frac{d\omega}{\pi}\omega\chi''_{A_iA_j}(\omega) + \mathcal{O}\left(\frac{1}{z^3}\right). \tag{3.64}$$

In many cases, the leading terms in (3.64) can be exactly calculated from their expression in terms of commutators [Eq. (3.63)]. Examples are discussed in Secs. 3.3 and 3.4.

3.2.7 *Imaginary-time correlation function*

So far we have discussed the retarded and advanced response functions χ^R and χ^A, as well as the structure factor S (related to χ'' by the fluctuation–dissipation theorem). In this section, we show how one can deduce these real-time correlation functions from the knowledge of their imaginary-time counterparts.

Let us first rephrase the results of Sec. 3.1 in the framework of the imaginary-time functional integral. In presence of an external field $h_i(\tau)$, the partition function reads

$$Z[h] = \int\mathcal{D}[\psi]\exp\Big\{-S[\psi] + \int_0^\beta d\tau\sum_i h_i(\tau)A_i(\tau)\Big\}. \tag{3.65}$$

As in the preceding sections, the space dependence of the fields is included in the index i. The advantage of the functional integral formulation is that the field A_i need not be bosonic.

It can also be fermionic, in which case the external "field" (or source) h_i is a Grassmann variable. In order to calculate the linear response to the external perturbation, we integrate out the ψ field in (3.65),

$$Z[h] = Z[0] \exp\left\{ \int_0^\beta d\tau \sum_i h_i(\tau)\langle A_i(\tau)\rangle \right.$$
$$\left. + \frac{1}{2}\int_0^\beta d\tau d\tau' \sum_{i,j} h_i(\tau)\langle A_i(\tau)A_j(\tau')\rangle_c h_j(\tau') + \mathcal{O}(h^3)\right\}, \tag{3.66}$$

where the averages $\langle\cdots\rangle$ are taken with the action $S[\psi]$ and $\langle A_i(\tau)A_j(\tau')\rangle_c$ denotes the cumulant $\langle A_i(\tau)A_j(\tau')\rangle - \langle A_i(\tau)\rangle\langle A_j(\tau')\rangle$. Recall that $\langle A_i(\tau)A_j(\tau')\rangle$ corresponds to the time-ordered correlation function $\langle T_\tau \hat{A}_i(\tau)\hat{A}_j(\tau')\rangle$ in the operator formalism (Chapter 1). The mean value of the field A_i is then given by

$$\langle A_i(\tau)\rangle_h = \frac{\delta \ln Z[h]}{\delta h_i(\tau)}$$
$$= \langle A_i(\tau)\rangle_{h=0} + \int_0^\beta d\tau' \sum_j \langle A_i(\tau)A_j(\tau')\rangle_c h_j(\tau') + \mathcal{O}(h^2). \tag{3.67}$$

The linear response to the field h is

$$\delta\langle A_i(\tau)\rangle = \int_0^\beta d\tau' \sum_j \chi_{A_iA_j}(\tau,\tau')h_j(\tau'),$$
$$\chi_{A_iA_j}(\tau,\tau') = \left.\frac{\delta^2 \ln Z[h]}{\delta h_i(\tau)\delta h_j(\tau')}\right|_{h=0} = \langle A_i(\tau)A_j(\tau')\rangle_c. \tag{3.68}$$

The imaginary-time response function $\chi_{A_iA_j}(\tau,\tau')$ is given by a correlation function of the system at equilibrium ($h=0$). We shall show that it is related to the real-time correlation functions[12] (written here in operator form)

$$\chi^{R,A}_{A_iA_j}(t) = \pm 2i\Theta(\pm t)\chi''_{A_iA_j}(t), \quad \chi''_{A_iA_j}(t) = \frac{1}{2}\langle[\hat{A}_i(t),\hat{A}_j]_{-\zeta}\rangle,$$
$$\chi_{A_iA_j}(t) = i\langle T\hat{A}_i(t)\hat{A}_j\rangle. \tag{3.69}$$

For bosonic operators ($\zeta = 1$), χ^R and χ^A are the retarded and advanced response functions discussed in the preceding sections. For fermionic operators ($\zeta = -1$), the spectral function χ'' is defined by an anticommutator rather than a commutator. Equation (3.37) then takes the more general form

$$\chi''_{A_iA_j}(\omega) = \frac{\pi}{Z}\sum_{n,m} e^{-\beta\epsilon_n}\left(1-\zeta e^{-\beta\omega}\right) A_i^{nm}A_j^{mn}\delta(\omega+\epsilon_n-\epsilon_m), \tag{3.70}$$

[12]Note that although $\chi(\tau)$ and $\chi(t)$ are different functions, we use a unique symbol (χ) and distinguish them by their argument.

but the spectral representation remains unchanged [Eqs. (3.32–3.34)]. The real function $\chi''_{A^\dagger A}(\omega)$ is positive for fermionic operators and has the sign of ω for bosonic operators.

3.2.7.1 *Relation between $\chi_{A_iA_j}(i\omega_n)$ and $\chi^{R,A}_{A_iA_j}(\omega)$*

Proceeding as for the derivation of (3.70), we obtain

$$\chi_{A_iA_j}(\tau) = \langle T_\tau \hat{A}_i(\tau)\hat{A}_j\rangle - \langle \hat{A}_i\rangle\langle \hat{A}_j\rangle$$
$$= \frac{1}{Z}\sum_{n,m} e^{-(\beta-\tau)\epsilon_n - \tau\epsilon_m} A_i^{nm}A_j^{mn} - \langle \hat{A}_i\rangle\langle \hat{A}_j\rangle \qquad (\tau > 0). \tag{3.71}$$

Discarding first the static case $i\omega_n = 0$, we readily obtain

$$\chi_{A_iA_j}(i\omega_n) = \int_0^\beta d\tau\, e^{i\omega_n\tau}\chi_{A_iA_j}(\tau) = \frac{1}{Z}\sum_{n,m} A_i^{nm}A_j^{mn}\frac{e^{-\beta\epsilon_n} - \zeta e^{-\epsilon_m}}{\epsilon_m - \epsilon_n - i\omega_n}$$
$$= \int_{-\infty}^{\infty}\frac{d\omega}{\pi}\frac{\chi''_{A_iA_j}(\omega)}{\omega - i\omega_n} \qquad (i\omega_n \neq 0), \tag{3.72}$$

where ω_n is a bosonic ($\zeta = 1$) or fermionic ($\zeta = -1$) Matsubara frequency. Comparing with (3.33), we see that $\chi_{A_iA_j}(i\omega_n)$ and $\chi_{A_iA_j}(z = i\omega_n)$ are identical for $i\omega_n \neq 0$.

The static limit $i\omega_n = 0$ for boson operators requires some care. In that case, one should single out the terms with $\epsilon_n = \epsilon_m$ in (3.71), which yields

$$\chi_{A_iA_j}(i\omega_n = 0) = \int_{-\infty}^{\infty}\frac{d\omega}{\pi}\frac{\chi''_{A_iA_j}(\omega)}{\omega} + \frac{\beta}{Z}\sum_{\substack{n,m\\(\epsilon_n=\epsilon_m)}} e^{-\beta\epsilon_n}A_i^{nm}A_j^{mn} - \beta\langle \hat{A}_i\rangle\langle \hat{A}_j\rangle$$
$$= \chi_{A_iA_j}(z = 0) + \beta\big[\langle \hat{A}_i^{(0)}\hat{A}_j^{(0)}\rangle - \langle \hat{A}_i^{(0)}\rangle\langle \hat{A}_j^{(0)}\rangle\big]. \tag{3.73}$$

(Recall that for $\zeta = 1$, we can omit the terms with $\epsilon_n = \epsilon_m$ in (3.70).) Here, $\hat{A}_i^{(0)}$ denotes the energy diagonal part of $\hat{A}_i$: $A_i^{(0)nm} = A_i^{nm}$ if $\epsilon_n = \epsilon_m$ and $A_i^{(0)nm} = 0$ otherwise. Thus, $\chi_{A_iA_j}(i\omega_n = 0)$ may differ from $\chi_{A_iA_j}(z = 0) = \chi^R_{A_iA_j}(\omega = 0)$ as the last term in (3.73) does not necessarily vanish. $\chi_{A_iA_j}(i\omega_n = 0)$ and $\chi^R_{A_iA_j}(\omega = 0)$ carry different physical meanings. $\chi_{A_iA_j}(i\omega_n = 0)$ gives the static susceptibility as obtained from the equilibrium state of a system with the time-independent Hamiltonian $\hat{H} + \hat{H}'$. By contrast, $\chi^R_{A_iA_j}(\omega = 0)$ is obtained as the zero-frequency limit of the dynamic susceptibility calculated under the assumption of adiabatic evolution of the system subjected to the time-dependent Hamiltonian $\hat{H}'(t)$. That the occupation probabilities $p_n = e^{-\beta\epsilon_n}/Z$ remain constant in time as the external field is adiabatically switched on (unitary evolution) is guaranteed only when the system is isolated. For a system in thermal equilibrium with its environment, we do expect transitions between orthogonal states to occur when both the adiabatic switching time η^{-1} and the time scale $1/\omega$ of the perturbative field become longer and longer, thus allowing the system to reach the quasi-equilibrium state in which the density matrix takes the (grand) canonical form $e^{-\beta(\hat{H}+\hat{H}'(t))}/Z$.

Computing the static response function from the static limit $\chi^R(\omega = 0)$ of the dynamical susceptibility may therefore be dangerous. In Sec. 3.2.3, we argued that the static uniform susceptibility can be obtained from $\chi^R(\mathbf{q}, \omega)$ in the limit $\omega = 0$ and $\mathbf{q} \to 0$. This turns out to be correct since $A^{(0)nm}(\mathbf{q})$ usually vanishes for a non-zero vector $\mathbf{q}$. In other words, one has $\chi_{A_iA_j}(\mathbf{q} = 0, i\omega_n = 0) = \lim_{\mathbf{q}\to 0} \chi^R_{A_iA_j}(\mathbf{q}, 0)$.

It is instructive to illustrate this point with the calculation of the density–density response function of a free electron gas. We have

$$\begin{aligned}\chi_{nn}(\mathbf{q}, i\omega_\nu) &= \langle n(\mathbf{q}, i\omega_\nu) n(-\mathbf{q}, -i\omega_\nu)\rangle \\ &= -\frac{2}{\beta\mathcal{V}} \sum_{\mathbf{k},\omega_n} (i\omega_n - \xi_{\mathbf{k}})^{-1} (i\omega_n + i\omega_\nu - \xi_{\mathbf{k+q}})^{-1} \\ &= -2\int_{\mathbf{k}} \frac{n_F(\xi_{\mathbf{k}}) - n_F(\xi_{\mathbf{k+q}})}{i\omega_\nu + \xi_{\mathbf{k}} - \xi_{\mathbf{k+q}}}\end{aligned} \tag{3.74}$$

(see Appendix 1.F), where $n(\mathbf{q}, i\omega_\nu)$ is the Fourier transform of $n(\mathbf{r}, \tau) = \sum_\sigma \psi^*_\sigma(\mathbf{r}, \tau)\psi_\sigma(\mathbf{r}, \tau)$. Equation (3.74) gives

$$\chi^R_{nn}(\mathbf{q}, \omega) = -2\int_{\mathbf{k}} \frac{n_F(\xi_{\mathbf{k}}) - n_F(\xi_{\mathbf{k+q}})}{\omega + i\eta + \xi_{\mathbf{k}} - \xi_{\mathbf{k+q}}}. \tag{3.75}$$

Hence, we find

$$\lim_{\mathbf{q}\to 0} \chi^R_{nn}(\mathbf{q}, 0) = -2\int d\xi\, N(\xi) n'_F(\xi) = \frac{\partial n}{\partial \mu}, \tag{3.76}$$

where $n = 2\int d\xi N(\xi) n_F(\xi)$ is the electron density and $N(\xi) = \int_{\mathbf{k}} \delta(\xi - \xi_{\mathbf{k}})$ the density of state per spin. χ^R_{nn} gives the response to a scalar potential. In the static and uniform limit, this potential is equivalent to a change in the chemical potential, so that the charge susceptibility should be equal to $\partial n/\partial \mu$ as obtained in (3.76). Taking first the limit $\mathbf{q} \to 0$, one obtains $\chi^R_{nn}(0, \omega) = 0$. As shown in Sec. 3.4.4, this result follows from particle number conservation (or gauge invariance).

Let us now compute the imaginary-time correlation function $\chi_{nn}(\mathbf{q}, i\omega_\nu)$ for $\mathbf{q} = \omega_\nu = 0$,

$$\begin{aligned}\chi_{nn}(\mathbf{q} = 0, i\omega_\nu = 0) &= -\frac{2}{\beta\mathcal{V}} \sum_{\mathbf{k},\omega_n} (i\omega_n - \xi_{\mathbf{k}})^{-2} \\ &= \frac{i}{\pi} \sum_{n=0}^{\infty} \int d\xi\, N(\xi) \partial_\xi \left(\frac{1}{n + \frac{1}{2} + i\tilde{\xi}} - \text{c.c.} \right),\end{aligned} \tag{3.77}$$

where $\tilde{\xi} = \xi/(2\pi T)$. Using $\sum_{n\geq 0}(n + \frac{1}{2} + ix)^{-1} - \text{c.c.} = -\Psi(\frac{1}{2} + ix) - \text{c.c.} = -i\pi \tanh(\pi x)$ (Ψ is the digamma function), we finally obtain

$$\begin{aligned}\chi_{nn}(\mathbf{q} = 0, i\omega_\nu = 0) &= \int d\xi\, N(\xi) \partial_\xi \tanh\left(\frac{\xi}{2T}\right) \\ &= -2\int d\xi\, N(\xi) n'_F(\xi) \\ &= \frac{\partial n}{\partial \mu},\end{aligned} \tag{3.78}$$

in agreement with $\lim_{\mathbf{q}\to 0}\chi^R_{nn}(\mathbf{q},0)$. Note that if we take the limit $T\to 0$ before performing the Matsubara sum, we obtain

$$\chi_{nn}(\mathbf{q}=0, i\omega_\nu=0) \to -2\int_{\mathbf{k}}\int_{-\infty}^{\infty}\frac{d\omega}{2\pi}(i\omega-\xi_{\mathbf{k}})^{-2}=0. \qquad (3.79)$$

This shows that the limit $T\to 0$ must always be taken at the end of the calculation when computing static uniform response functions within the imaginary-time formalism.

Given that $\chi_{A_iA_j}(i\omega_n)=\chi_{A_iA_j}(z=i\omega_n)$ and $\chi_{A_iA_j}(z)$ is analytic everywhere in the complex plane except on the real axis, $\chi^{R,A}_{A_iA_j}(\omega)=\chi_{A_iA_j}(z=\omega\pm i\eta)$ can be obtained from $\chi_{A_iA_j}(i\omega_n)$ by analytic continuation (Fig. 3.4),

$$\begin{aligned}\chi^R_{A_iA_j}(\omega)&=\chi_{A_iA_j}(i\omega_n\to\omega+i\eta) \qquad (\omega_n>0),\\ \chi^A_{A_iA_j}(\omega)&=\chi_{A_iA_j}(i\omega_n\to\omega-i\eta) \qquad (\omega_n<0)\end{aligned} \qquad (3.80)$$

(keeping in mind the possible difference between $\chi^R_{A_iA_j}(\omega=0)$ and $\chi_{A_iA_j}(i\omega_n=0)$ discussed above).

In practical calculations, we do not make use of the expression (3.72) of $\chi_{A_iA_j}(i\omega_n)$ and the analytic continuation $i\omega_n\to z$ is not unambiguously defined. Suppose that we have (approximately) calculated $\chi_{A_iA_j}(i\omega_n)$, e.g. from perturbation theory. $\chi_{A_iA_j}(i\omega_n)$ is identical to an infinite number of analytic functions, e.g. $\chi_{A_iA_j}(i\omega_n)(1+\zeta e^{i\beta\omega_n})/2$. Which of these functions should be analytically continued to obtain $\chi_{A_iA_j}(z)$? We can pick up the correct function $\chi_{A_iA_j}(z)$ by verifying that its asymptotic behavior for $|z|\to\infty$ is in accordance with the spectral representation (3.33) [see Eq. (3.64)]. To be more precise, we use the following result from the theory of complex functions. Let $\chi(z)$ and $\chi(i\omega_n\to z)$

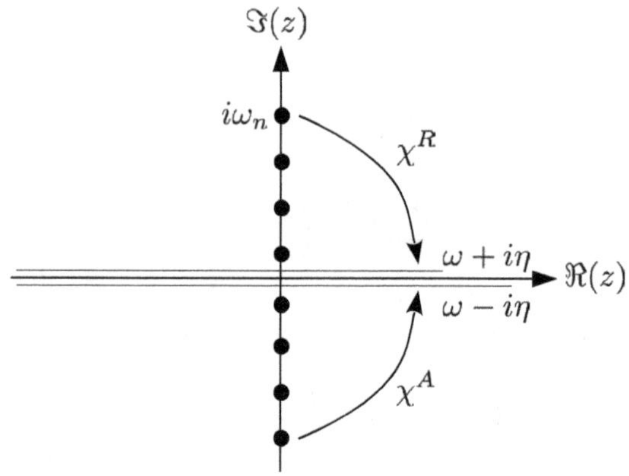

Fig. 3.4 Analytic continuation of the function $\chi(z)$ from the imaginary axis ($z=i\omega_n$) to points infinitesimally above ($z=\omega+i\eta$) and below ($z=\omega-i\eta$) the real axis where it gives $\chi^R(\omega)$ and $\chi^A(\omega)$, respectively.

be two analytic functions of the complex variable z in a certain domain of the complex plane. If

(i) $\chi(z = i\omega_n) = \chi(i\omega_n)$ for all Matsubara frequencies ω_n,
(ii) $\chi(z)$ and $\chi(i\omega_n \to z)$ exhibit the same asymptotic behavior for $|z| \to \infty$ and z in the analyticity domain,

then the functions $\chi(z)$ and $\chi(i\omega_n \to z)$ coincide in the domain of analyticity.

3.2.7.2 *Relation between* $\chi_{A_iA_j}(\omega)$ *and* $\chi^{R,A}_{A_iA_j}(\omega)$

Writing the real-time correlation function $\chi_{A_iA_j}(t) = i\langle T\hat{A}_i(t)\hat{A}_j\rangle$ as

$$\chi_{A_iA_j}(t) = \frac{i}{Z}\sum_{n,m} A_i^{nm} A_j^{mn} e^{i(\epsilon_n-\epsilon_m)t}[\Theta(t)e^{-\beta\epsilon_n} + \zeta\Theta(-t)e^{-\beta\epsilon_m}], \tag{3.81}$$

we obtain[13]

$$\chi_{A_iA_j}(\omega) = \frac{1}{Z}\sum_{n,m} A_i^{nm} A_j^{mn}\left(\frac{e^{-\beta\epsilon_n}}{\epsilon_m-\epsilon_n-\omega-i\eta} - \zeta\frac{e^{-\beta\epsilon_m}}{\epsilon_m-\epsilon_n-\omega+i\eta}\right). \tag{3.82}$$

When $\hat{A}_i$ and $\hat{A}_j$ have the same signature under time reversal, $\chi''_{A_iA_j}(\omega) = \Im[\chi^R_{A_iA_j}(\omega)]$ is real (see Sec. 3.2.3), we deduce

$$\begin{aligned}\Re[\chi_{A_iA_j}(\omega)] &= \Re[\chi^R_{A_iA_j}(\omega)] = \mathcal{P}\int_{-\infty}^{\infty}\frac{d\omega'}{\pi}\frac{\Im[\chi^R_{A_iA_j}(\omega')]}{\omega'-\omega},\\ \Im[\chi_{A_iA_j}(\omega)] &= \Im[\chi^R_{A_iA_j}(\omega)]\left[\tanh\left(\beta\frac{\omega}{2}\right)\right]^{-\zeta}.\end{aligned} \tag{3.83}$$

When they have opposite signatures, $\chi''_{A_iA_j}(\omega) = -i\Re[\chi^R_{A_iA_j}(\omega)]$ is imaginary, and $\chi_{A_iA_j}(\omega)$ and $\chi^R_{A_iA_j}(\omega)$ are related by

$$\begin{aligned}\Im[\chi_{A_iA_j}(\omega)] &= \Im[\chi^R_{A_iA_j}(\omega)] = -\mathcal{P}\int_{-\infty}^{\infty}\frac{d\omega'}{\pi}\frac{\Re[\chi^R_{A_iA_j}(\omega')]}{\omega'-\omega},\\ \Re[\chi_{A_iA_j}(\omega)] &= \Re[\chi^R_{A_iA_j}(\omega)]\left[\tanh\left(\beta\frac{\omega}{2}\right)\right]^{-\zeta}.\end{aligned} \tag{3.84}$$

Since $\chi^A_{A_iA_j}(\omega) = [\chi^R_{A_iA_j}(\omega)]^*$ in the first case ($\epsilon^T_{A_i} = \epsilon^T_{A_j}$) and $\chi^A_{A_iA_j}(\omega) = -[\chi^R_{A_iA_j}(\omega)]^*$ in the second one ($\epsilon^T_{A_i} = -\epsilon^T_{A_j}$), in both cases

$$\chi_{A_iA_j}(\omega) = [1+\zeta n_\zeta(\omega)]\chi^R_{A_iA_j}(\omega) - \zeta n_\zeta(\omega)\chi^A_{A_iA_j}(\omega), \tag{3.85}$$

---

[13] We use $\int_{-\infty}^{\infty} dt\, e^{i(\omega-\epsilon)t}\Theta(\pm t) = \pm\frac{i}{\omega\pm i\eta-\epsilon}$, which is easily deduced from the inverse relation $\int_{-\infty}^{\infty}\frac{d\omega}{2\pi}\frac{e^{-i\omega t}}{\omega\pm i\eta-\epsilon} = \mp i\Theta(\pm t)e^{-i\epsilon t}$.

where $n_\zeta(\omega) = (e^{\beta\omega} - \zeta)^{-1}$ is the Bose–Einstein ($\zeta = 1$) or Fermi–Dirac ($\zeta = -1$) distribution function. At zero temperature, one obtains

$$\chi_{A_iA_j}(\omega) = \Theta(\omega)\chi^R_{A_iA_j}(\omega) + \Theta(-\omega)\chi^A_{A_iA_j}(\omega) \qquad (T = 0), \tag{3.86}$$

showing that the poles of $\chi_{A_iA_j}(\omega)$ are in the upper-half (lower-half) complex plane if $\omega < 0$ ($\omega > 0$). It is easy to show that Eqs. (3.85) and (3.86) are verified when $\hat{A}_j = \hat{A}_i^\dagger$ regardless of the behavior of $\hat{A}_i$ under time reversal.

3.2.7.3 *Relation between $\chi_{A_iA_j}(\tau)$ and $\chi_{A_iA_j}(t)$*

It is possible to directly relate $\chi_{A_iA_j}(\tau)$ and $\chi_{A_iA_j}(t)$ without using the Fourier transforms. We write $\chi_{A_iA_j}(t)$ as

$$\begin{aligned}
\chi_{A_iA_j}(t) &= \Theta(t)\chi^>_{A_iA_j}(t) + \Theta(-t)\chi^<_{A_iA_j}(t), \\
\chi^>_{A_iA_j}(t) &= \frac{i}{Z}\mathrm{Tr}\left[e^{-(\beta-it)\hat{H}}\hat{A}_i e^{-i\hat{H}t}\hat{A}_j\right] = i\langle \hat{A}_i(t)\hat{A}_j\rangle, \\
\chi^<_{A_iA_j}(t) &= \zeta\frac{i}{Z}\mathrm{Tr}\left[e^{-(\beta+it)\hat{H}}\hat{A}_j e^{i\hat{H}t}\hat{A}_i\right] = i\zeta\langle \hat{A}_j(-t)\hat{A}_i\rangle,
\end{aligned} \tag{3.87}$$

and extend the definition of $\chi^>$ and $\chi^<$ to complex values of t. For the thermodynamic trace to converge in (3.87), the factors multiplying $\hat{H}$ in the exponentials must have negative real parts, since $\hat{H}$ must be positive for the partition function $Z = \mathrm{Tr}\, e^{-\beta\hat{H}}$ to exist.[14] Thus, $\chi^>(t)$ can be analytically continued if $\Re(it - \beta) \leq 0$ and $\Re(it) \geq 0$, i.e. $-\beta \leq \Im(t) \leq 0$, while the analogous condition for $\chi^<(t)$ is $0 \leq \Im(t) \leq \beta$. $\chi(\tau)$ can then be obtained by analytically continuing $\chi^>(t)$ and $\chi^<(t)$ to the imaginary axis within their respective analyticity domains,

$$\begin{aligned}
\chi^>_{A_iA_j}(t = -i\tau) &= i\chi_{A_iA_j}(\tau) \qquad (0 \leq \tau \leq \beta), \\
\chi^<_{A_iA_j}(t = -i\tau) &= i\chi_{A_iA_j}(\tau) \qquad (-\beta \leq \tau \leq 0).
\end{aligned} \tag{3.88}$$

This result can be used to recover the relation between $\chi_{A_iA_j}(i\omega_n)$ and $\chi^{R,A}_{A_iA_j}(\omega)$. Consider

$$\chi_{A_iA_j}(i\omega_n) = \int_0^\beta d\tau\, e^{i\omega_n\tau}\chi_{A_iA_j}(\tau) = -i\int_0^\beta d\tau\, e^{i\omega_n\tau}\chi^>_{A_iA_j}(t = -i\tau), \tag{3.89}$$

and allow ω_n to take any complex value with $\Re(\omega_n) > 0$. $\chi_{A_iA_j}(i\omega_n)$ can be calculated by deforming the integration contour into the analyticity domain of $\chi^>(t)$ as shown in Fig. 3.5,

$$\chi_{A_iA_j}(i\omega_n) = -i\int_{t=0}^{t=\infty} d(it)e^{-\omega_n t}\chi^>_{A_iA_j}(t) - i\int_{t=\infty}^{t=0} d(it)e^{-\omega_n(t-i\beta)}\chi^>_{A_iA_j}(t - i\beta). \tag{3.90}$$

The condition $\Re(\omega_n) > 0$ ensures that the small vertical segment of the contour does not contribute ($e^{-\omega_n t} \to 0$ for $\Re(t) \to \infty$). Using $e^{i\beta\omega_n} = \zeta$ and $\chi^>_{A_iA_j}(t - i\beta) = i\langle \hat{A}_j\hat{A}_i(t)\rangle$,

---

[14]More precisely, we can always shift the energies, $\hat{H} \to \hat{H} + \text{const}$, to make $\hat{H}$ positive.

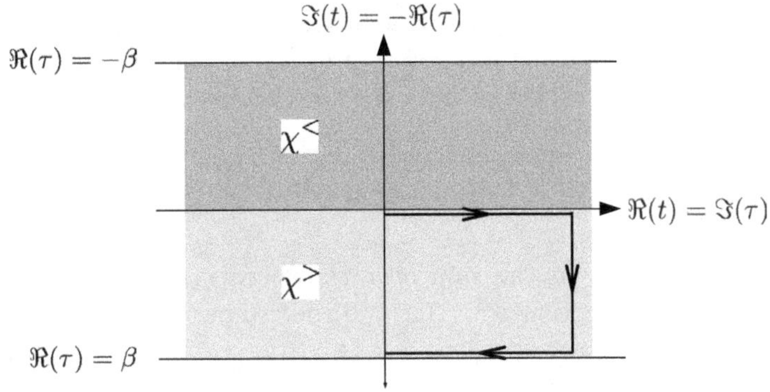

Fig. 3.5 Analyticity domains of $\chi^>(t)$ and $\chi^<(t)$ in the complex t plane ($\tau = it$) and integration contour used in the calculation of $\chi(i\omega_n)$ [Eqs. (3.89–3.90)].

a consequence of the cyclic invariance of the trace, we obtain

$$\chi_{A_iA_j}(i\omega_n) = i\int_0^\infty dt\, e^{-\omega_n t}\langle \hat{A}_i(t)\hat{A}_j - \zeta\hat{A}_j\hat{A}_i(t)\rangle = \int_{-\infty}^\infty dt\, e^{-\omega_n t}\chi^R_{A_iA_j}(t). \tag{3.91}$$

This equation holds for any frequency ω_n satisfying $\Re(\omega_n) > 0$. For $i\omega_n = \omega + i\eta$, it yields $\chi_{A_iA_j}(i\omega_n = \omega + i\eta) = \chi^R_{A_iA_j}(\omega)$ as derived previously. Similarly, $\chi_{A_iA_j}(i\omega_n = \omega - i\eta) = \chi^A_{A_iA_j}(\omega)$ can be obtained by deforming the integration contour in the region $\Re(t) < 0$.

3.2.7.4 *Stiffness and stability*

Let us consider the Legendre transform

$$\Gamma[M] = -\ln Z[h] + \int_0^\beta d\tau \sum_i h_i(\tau)M_i(\tau), \tag{3.92}$$

where

$$M_i(\tau) = \langle A_i(\tau)\rangle_h = \frac{\delta \ln Z[h]}{\delta h_i(\tau)} \tag{3.93}$$

(A_i is a bosonic field in this section). In Eq. (3.92), $h \equiv h[M]$ is obtained by inverting (3.93). $\Gamma[M]$ satisfies the equation of motion

$$\frac{\delta\Gamma[M]}{\delta M_i(\tau)} = h_i(\tau) \tag{3.94}$$

and is therefore stationary at equilibrium ($h = 0$). Expanding about the stationary point $M = \bar{M}$, we obtain

$$\begin{aligned}\Gamma[M] &= \Gamma[\bar{M}] + \frac{1}{2}\int_0^\beta d\tau d\tau' \sum_{i,j} \frac{\delta^2\Gamma[M]}{\delta M_i(\tau)M_j(\tau')}\bigg|_{M=\bar{M}} \delta M_i(\tau)\delta M_j(\tau') \\ &= \Gamma[\bar{M}] + \frac{1}{2}\int_0^\beta d\tau d\tau' \sum_{i,j} \chi^{-1}_{A_iA_j}(\tau - \tau')\delta M_i(\tau)\delta M_j(\tau'),\end{aligned} \tag{3.95}$$

where $\delta M_i(\tau) = M_i(\tau) - \bar{M}_i$ and $\chi_{A_iA_j}(\tau)$ is the susceptibility defined in (3.68). The last line in (3.95) follows from the usual properties of Legendre transforms (Chapter 1). When $M_i(\tau)$ is time independent, $\beta^{-1}\Gamma[M]$ defines the Gibbs free energy

$$G[M] = G[\bar{M}] + \frac{1}{2}\sum_{i,j} \chi^{-1}_{A_i^\dagger A_j}(z=0)\delta M_i^* \delta M_j \tag{3.96}$$

(we have changed $A_i \to A_i^*$ in the sum over i), where $\chi(z = 0)^{-1} = \chi^R(\omega = 0)^{-1}$. Thus, the static response function $\chi^R_{A_i^\dagger A_j}(\omega = 0)$ determines the stiffness (or rigidity) of the system with respect to a variation $\delta M_i = M_i - \bar{M}_i$ induced by a static external field $h_i = \delta G[M]/\delta M_i$. For the system to be stable, the stiffness matrix $\{\chi^R_{A_i^\dagger A_j}(\omega = 0)^{-1}\}$ and its inverse $\{\chi^R_{A_i^\dagger A_j}(\omega = 0)\}$ must be positive for any set of fields A_i. In particular,

$$\chi^R_{A^\dagger A}(\omega = 0) \geq 0. \tag{3.97}$$

As pointed out in Sec. 3.2.2, these results are a consequence of the positivity of the matrix $\omega\chi''_{A_i^\dagger A_j}(\omega)$.

3.3 Density–Density Response Function

In this section, we consider a homogeneous and isotropic bosonic or fermionic system. An external potential $\phi(\mathbf{r}, t)$ that couples to the density contributes to the Hamiltonian a term

$$\hat{H}'(t) = \int d^d r\, \phi(\mathbf{r}, t)\hat{n}(\mathbf{r}) \tag{3.98}$$

and induces a density modulation

$$\delta\langle\hat{n}(\mathbf{r})\rangle(t) = -\int_{-\infty}^{\infty} dt' \int d^d r'\, \chi^R_{nn}(\mathbf{r} - \mathbf{r}', t - t')\phi(\mathbf{r}', t') \tag{3.99}$$

to linear order. The density–density response is defined by

$$\begin{aligned} \chi^R_{nn}(\mathbf{r}, t) &= 2i\Theta(t)\chi''_{nn}(\mathbf{r}, t), \\ \chi''_{nn}(\mathbf{r}, t) &= \frac{1}{2}\langle[\hat{n}(\mathbf{r}, t), \hat{n}(0, 0)]\rangle. \end{aligned} \tag{3.100}$$

3.3.1 *Symmetry properties*

From parity invariance [Eq. (3.21)] and the general properties (3.18), one easily obtains

$$\chi''_{nn}(\mathbf{q}, \omega) = [\chi''_{nn}(\mathbf{q}, \omega)]^* = \chi''_{nn}(-\mathbf{q}, \omega) = -\chi''_{nn}(\mathbf{q}, -\omega). \tag{3.101}$$

$\chi''_{nn}(\mathbf{q}, \omega)$ is real, odd in ω and even in $\mathbf{q}$. It follows that $\chi^R_{nn}(\mathbf{q}, \omega)$ satisfies

$$\chi^R_{nn}(\mathbf{q}, \omega) = \chi^R_{nn}(-\mathbf{q}, \omega) = [\chi^R_{nn}(\mathbf{q}, -\omega)]^*. \tag{3.102}$$

3.3.2 *High-frequency behavior and f-sum rule*

$\chi''_{nn}(\mathbf{q},\omega)$ being odd in frequency, we obtain

$$\chi_{nn}(\mathbf{q},z) = -\frac{1}{z^2}\int_{-\infty}^{\infty}\frac{d\omega}{\pi}\omega\chi''_{nn}(\mathbf{q},\omega) + \mathcal{O}\left(\frac{1}{z^4}\right). \tag{3.103}$$

The coefficient of $1/z^2$ can be written as the mean value of an equal-time commutator (Sec. 3.2.6),

$$\int_{-\infty}^{\infty}\frac{d\omega}{\pi}\omega\chi''_{nn}(\mathbf{q},\omega) = \langle[i\partial_t\hat{n}(\mathbf{q},t),\hat{n}(-\mathbf{q},0)]\rangle_{t=0} = \mathbf{q}\cdot\langle[\hat{\mathbf{j}}(\mathbf{q}),\hat{n}(-\mathbf{q})]\rangle, \tag{3.104}$$

where the last result follows from the continuity equation $\partial_t\hat{n}(\mathbf{q},t) + i\mathbf{q}\cdot\hat{\mathbf{j}}(\mathbf{q},t) = 0$. By writing the operators $\hat{n}(\mathbf{q})$ and $\hat{\mathbf{j}}(\mathbf{q})$ in second-quantized form,[15] the equal-time commutator in (3.104) is easily calculated,

$$[\hat{\mathbf{j}}(\mathbf{q}),\hat{n}(-\mathbf{q})] = \frac{\mathbf{q}}{m\mathcal{V}}\hat{N}, \tag{3.105}$$

and we finally obtain

$$\int_{-\infty}^{\infty}\frac{d\omega}{\pi}\omega\chi''_{nn}(\mathbf{q},\omega) = \frac{n\mathbf{q}^2}{m}, \tag{3.106}$$

where $n = \langle\hat{N}\rangle/\mathcal{V}$ is the mean density of particles. This result is known as the f-sum rule. It holds for both fermions and bosons.

3.3.3 *Compressibility*

The isothermal compressibility κ is defined as the volume variation induced by a change in pressure,

$$\kappa = -\frac{1}{\mathcal{V}}\frac{\partial\mathcal{V}}{\partial P}\bigg|_{T,N}. \tag{3.107}$$

The free energy $F(T,\mathcal{V},N) \equiv E - TS$ being an extensive quantity, it can be written as $F(T,\mathcal{V},N) = \mathcal{V}f(T,n)$ where $n = N/\mathcal{V}$ is the density. Since $P = -\partial F/\partial\mathcal{V}|_{T,N}$, we can express the compressibility as

$$\frac{1}{\kappa} = n\frac{\partial P}{\partial n}\bigg|_T = n^2\frac{\partial^2 f}{\partial n^2}\bigg|_T. \tag{3.108}$$

Using

$$\mu = \frac{\partial F}{\partial N}\bigg|_{T,\mathcal{V}} = \frac{\partial f}{\partial n}\bigg|_T, \tag{3.109}$$

[15] $\hat{n}(\mathbf{q}) = \frac{1}{\mathcal{V}^{1/2}}\sum_{\mathbf{k},\sigma}\hat{\psi}^\dagger_\sigma(\mathbf{k})\hat{\psi}_\sigma(\mathbf{k}+\mathbf{q})$ and $\hat{\mathbf{j}}(\mathbf{q}) = \frac{1}{\mathcal{V}^{1/2}m}\sum_{\mathbf{k},\sigma}(\mathbf{k}+\frac{\mathbf{q}}{2})\hat{\psi}^\dagger_\sigma(\mathbf{k})\hat{\psi}_\sigma(\mathbf{k}+\mathbf{q})$. This expression of the current assumes that there is no contribution from the interacting part of the Hamiltonian, as in the case of a density–density interaction.

κ can be related to the variation of the density induced by a change in the chemical potential,

$$\kappa = \frac{1}{n^2}\frac{\partial n}{\partial \mu}\bigg|_T . \tag{3.110}$$

In the grand canonical ensemble, we can therefore obtain the compressibility from the response to a static uniform external potential $\phi \equiv \delta\mu$, i.e.

$$n^2\kappa = \frac{n}{mc_s^2} \equiv \frac{\partial n}{\partial \mu}\bigg|_T = \lim_{\mathbf{q}\to 0} \chi^R_{nn}(\mathbf{q},0) = \lim_{\mathbf{q}\to 0} \int_{-\infty}^{\infty} \frac{d\omega}{\pi} \frac{\chi''_{nn}(\mathbf{q},\omega)}{\omega}. \tag{3.111}$$

The compressibility sum rule (3.111) is an example of a thermodynamic sum rule [Eq. (3.61)]. It is often expressed in terms of the macroscopic (first) sound velocity c_s. In the following section, we shall see how the compressibility sum rule must be modified when the particles interact *via* the long-range Coulomb interaction.

The relation (3.111) between the compressibility and the macroscopic sound velocity c_s can be obtained from simple hydrodynamic arguments. Let us consider an infinitesimal change $\delta\langle n(\mathbf{r},t)\rangle$ in density and assume the density oscillations to be sufficiently slow so that the system remains locally in thermodynamic equilibrium with pressure $P_{\rm eq}+\delta P(\mathbf{r},t)$. This requires the characteristic time τ of return to equilibrium to be much smaller than the period $2\pi/\omega$ of the density wave, i.e. $\omega\tau \ll 1$ (collisional or hydrodynamic regime). From the continuity equation

$$\partial_t \delta\langle n(\mathbf{r},t)\rangle = -n\boldsymbol{\nabla}\cdot\langle \mathbf{v}(\mathbf{r},t)\rangle \tag{3.112}$$

and the Euler equation

$$mn\partial_t\langle \mathbf{v}(\mathbf{r},t)\rangle = -\boldsymbol{\nabla}\delta P(\mathbf{r},t) \tag{3.113}$$

[Eqs. (3.112, 3.113) are valid to lowest order in $\mathbf{v}$], we deduce

$$\partial_t^2 \delta\langle n(\mathbf{r},t)\rangle = m^{-1}\boldsymbol{\nabla}^2\delta P(\mathbf{r},t) = m^{-1}\left(\frac{\partial P}{\partial n}\right)_T \boldsymbol{\nabla}^2\delta\langle n(\mathbf{r},t)\rangle. \tag{3.114}$$

Here, we have used $P(\mathbf{r},t) = P(n+\delta\langle n(\mathbf{r},t)\rangle) \simeq P(n) + (\partial P/\partial n)\delta\langle n(\mathbf{r},t)\rangle$ for a system locally in equilibrium. This gives a density wave (first-sound mode) with linear dispersion $\omega = c_s|\mathbf{q}|$ and velocity $c_s^2 = m^{-1}(\partial P/\partial n)_T = 1/\kappa n m$. At finite temperature, a more correct hydrodynamic treatment would include the coupling between mass and energy density fluctuations. As a result, the sound velocity

$$c_s^2 = \frac{1}{m}\left(\frac{\partial P}{\partial n}\right)_S = \frac{1}{\kappa_S n m} \tag{3.115}$$

is related to the isentropic compressibility κ_S [1,2,5]. At low temperatures, the difference between the isothermal and isentropic sound velocities is in general weak and can be neglected.

3.4 Electromagnetic Response Function

In this section, we consider a homogeneous and isotropic electron system. The global electric neutrality is ensured by a uniform background of positive charge (jellium model, see Chapter 5). In the Coulomb gauge ($\nabla \cdot \mathbf{A} = 0$), the action of the particles and the electromagnetic field reads (see Chapter 1)

$$S[\psi, \phi, \mathbf{A}] = \int_0^\beta d\tau \int d^d r \biggl\{ \frac{\epsilon_0}{2} \Bigl[(\nabla\phi(x))^2 + (\partial_\tau \mathbf{A}_\perp(x))^2 + c_l^2 (\nabla \times \mathbf{A}_\perp(x))^2 \Bigr] \\ + i\phi(x)\rho(x) - \mathbf{j}(x)\cdot\mathbf{A}_\perp(x) + \frac{e}{2m}\rho(x)\mathbf{A}_\perp^2(x) \biggr\} + S'_m[\psi] \tag{3.116}$$

(c_l is the speed of light and $x = (\mathbf{r}, \tau)$), where $S'_m[\psi]$ is the action of the electrons without electromagnetic interactions. In the electron gas model (Chapter 5), it is nothing but the action of free electrons. In more realistic models of metals, it can include short-range (effective) interactions between electrons. $\rho(x) = en(x)$ is the charge density and $\mathbf{j}(x)$ the paramagnetic part of the current $\mathbf{J} = \mathbf{j} - (e/m)\rho\mathbf{A}_\perp$. The background of positive charge removes the $\mathbf{q} = 0$ component of the Coulomb interaction. This implies that the field $\phi(\mathbf{q}, \tau)$ is defined only for $\mathbf{q} \neq 0$.

Note that in the Euclidean formalism, $i\phi(x)$ plays the role of the scalar potential (Sec. 1.9.3). The mean values of the electric and magnetic fields are given by (1.636) and (1.637) and the propagators $D_{00}(q)$ and $D_{\mu\nu}(q)$ of the electromagnetic potentials are defined by (1.642). In the absence of coupling to the particles, they take the form (1.643).

In the following sections, we consider D_{00} and $D_{\mu\nu}$, as well as the longitudinal and transverse dielectric functions, when the electromagnetic potentials are coupled to the electron system.

3.4.1 *Longitudinal dielectric function*

To find the expression of the scalar potential propagator D_{00}, it is sufficient to consider the reduced action

$$S[\psi, \phi] = \int_0^\beta d\tau \int d^d r \biggl\{ \frac{\epsilon_0}{2} (\nabla\phi(x))^2 + [i\rho(x) - J(x)]\phi(x) \biggr\} + S'_m[\psi], \tag{3.117}$$

since the magnetic interactions can be neglected with respect to the Coulomb interaction in a non-relativistic system (Sec. 1.9.3). We have added an external source J to the action which allows us to obtain D_{00} from the functional derivatives of $\ln Z[J]$. Integrating out ϕ, we obtain

$$S[\psi] = -\frac{1}{2} \sum_q D_{00}^{(0)}(q)[J(-q)J(q) - 2iJ(-q)\rho(q)] + S_m[\psi], \tag{3.118}$$

where $D_{00}^{(0)}(q) = 1/\epsilon_0 \mathbf{q}^2$ and

$$S_m[\psi] = S'_m[\psi] + \frac{1}{2} \sum_q D_{00}^{(0)}(q)\rho(-q)\rho(q) \tag{3.119}$$

is the action of the electrons including the Coulomb interaction but not the magnetic interactions. Note that $S_m[\psi]$ is the usual action for non-relativistic charged particles. Integrating out the field ψ within a cumulant expansion, we finally obtain

$$\ln Z[J] = \frac{1}{2}\sum_q D_{00}^{(0)}(q)J(-q)J(q)[1 - D_{00}^{(0)}(q)\chi_{\rho\rho}(q)] + \mathcal{O}(J^3) \tag{3.120}$$

(the first-order cumulant $\sum_q D_{00}^{(0)}(q)J(-q)\langle\rho(q)\rangle = enD_{00}^{(0)}(q=0)J(q=0)$ vanishes), where

$$\chi_{\rho\rho}(q) = \langle\rho(q)\rho(-q)\rangle \tag{3.121}$$

is the density–density correlation function computed in the presence of the Coulomb interaction between particles, i.e. with the action $S_m[\psi]$. In the following, the expectation values $\langle\cdots\rangle$ will always be obtained with $S_m[\psi]$. From (3.120), we deduce

$$D_{00}(q) = \left.\frac{\delta^2 \ln Z[J]}{\delta J(-q)\delta J(q)}\right|_{J=0} = D_{00}^{(0)}(q)[1 - D_{00}^{(0)}(q)\chi_{\rho\rho}(q)]. \tag{3.122}$$

It is convenient to write $\chi_{\rho\rho}$ as

$$\chi_{\rho\rho}(q) = \frac{\Pi_{00}(q)}{1 + D_{00}^{(0)}(q)\Pi_{00}(q)}, \tag{3.123}$$

where Π_{00} is the "irreducible" part of $\chi_{\rho\rho}$, i.e. the part that cannot be split into two disconnected pieces by cutting a Coulomb line $D_{00}^{(0)}$. The relation (3.123) between $\chi_{\rho\rho}$ and Π_{00} follows from the diagrammatic rules discussed in Chapter 1. Equation (3.122) then gives

$$D_{00}^{-1}(q) = [D_{00}^{(0)}(q)]^{-1} + \Pi_{00}(q), \tag{3.124}$$

which shows that $\Pi_{00}(q)$ is the self-energy of the scalar potential. Equations (3.122–3.124) are shown diagrammatically in Fig. 3.6. In fact, one could have directly obtained D_{00} from its diagrammatic expansion. Note that the only approximation in (3.124) is the neglect of the magnetic interactions with respect to the Coulomb interaction in the irreducible density–density correlation function Π_{00}.

The longitudinal dielectric function $\epsilon_{\|}(q)$ is defined by

$$D_{00}(q) = \frac{1}{\epsilon_0\epsilon_{\|}(q)\mathbf{q}^2}, \tag{3.125}$$

i.e.

$$\frac{1}{\epsilon_{\|}(q)} = 1 - \frac{\chi_{\rho\rho}(q)}{\epsilon_0\mathbf{q}^2} \quad \text{or} \quad \epsilon_{\|}(q) = 1 + \frac{\Pi_{00}(q)}{\epsilon_0\mathbf{q}^2}. \tag{3.126}$$

$D_{00}(q)$ can be seen as an effective (screened) Coulomb interaction between electrons, the screening being encoded in the dielectric function $\epsilon_{\|}(q)$ or, equivalently, the density–density correlation function $\chi_{\rho\rho}(q)$. This interpretation is particularly natural when one considers the diagrams contributing to D_{00} (Fig. 3.6). We shall show below that $\epsilon_{\|}$ also determines

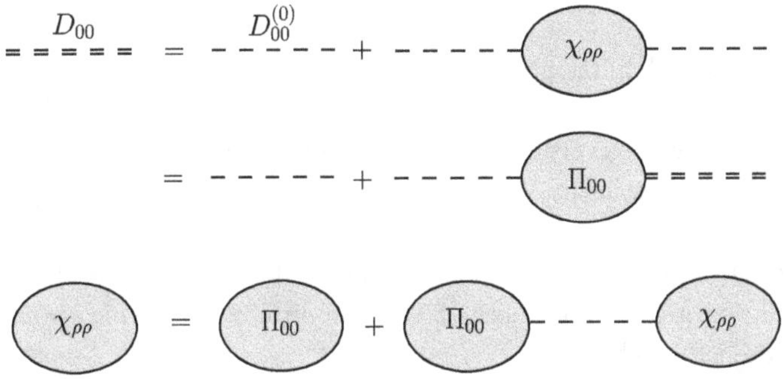

Fig. 3.6 Diagrammatic representation of D_{00} and $\chi_{\rho\rho}$ in terms of the irreducible part Π_{00} of the density–density correlation function.

the screening of an external charge distribution. The zeros of $\epsilon_{\parallel}^R(\mathbf{q},\omega)$, i.e. the poles of $\chi_{\rho\rho}^R(\mathbf{q},\omega)$, describe charge oscillations of the electron system and in particular the collective density mode (plasmon) discussed in the following.

The determination of the longitudinal dielectric function $\epsilon_{\parallel}$ reduces to the calculation of the irreducible part Π_{00} of the density–density correlation function $\chi_{\rho\rho}$. The simplest approximation, known as the RPA, amounts to approximating Π_{00} by its non-interacting value and will be discussed in Chapter 5. In the following, we show that certain properties of $\epsilon_{\parallel}$ can be obtained without relying on an explicit calculation.

3.4.1.1 *Analytic properties*

It follows from the definition (3.126) that $\epsilon_{\parallel}^{-1}(\mathbf{q},z) - 1 = -\chi_{\rho\rho}(\mathbf{q},z)/\epsilon_0\mathbf{q}^2$ shares all the analytic properties of a genuine response function. In particular, it admits the spectral representation

$$\frac{1}{\epsilon_{\parallel}(\mathbf{q},z)} - 1 = \int_{-\infty}^{\infty} \frac{d\omega}{\pi} \frac{1}{\omega - z} \Im\left[\frac{1}{\epsilon_{\parallel}^R(\mathbf{q},\omega)}\right], \tag{3.127}$$

and $1/\epsilon_{\parallel}^R(\mathbf{q},\omega) - 1$ satisfies the Kramers–Kronig relations. Note that neither $\epsilon_{\parallel}(\mathbf{q},z) - 1$ nor $\Pi_{00}(\mathbf{q},z)$ is *a priori* expected to satisfy these properties, since they do not describe the response of the system to an external probe.[16] However, the relation $\chi_{\rho\rho}^R(\mathbf{q},\omega) = \Pi_{00}^R(\mathbf{q},\omega)[1-\chi_{\rho\rho}^R(\mathbf{q},\omega)/(\epsilon_0\mathbf{q}^2)]$ together with the analytic properties of $\chi_{\rho\rho}^R(\mathbf{q},\omega)$ implies that a pole in $\Pi_{00}^R(\mathbf{q},\omega)$ at the complex frequency ω ($\Im(\omega) \geq 0$) is possible only if $\chi_{\rho\rho}^R(\mathbf{q},\omega) = \epsilon_0\mathbf{q}^2$ at the pole. We therefore expect $\Pi_{00}^R(\mathbf{q},\omega)$ and $\epsilon_{\parallel}^R(\mathbf{q},\omega)$ to be analytic in the upper-half complex plane in most cases. Possible non-analyticities of $\epsilon_{\parallel}^R(\mathbf{q},\omega)$ are discussed in Refs. [5,6].

[16] $\epsilon_{\parallel}(\mathbf{q},z) - 1$ and $\Pi_{00}(\mathbf{q},z)$ determine the response to the screened longitudinal field as can be seen from $\delta\langle\rho(\mathbf{q},\omega)\rangle = -\Pi_{00}^R(\mathbf{q},\omega)\langle\phi(\mathbf{q},\omega)\rangle = -\epsilon_0\mathbf{q}^2(\epsilon_{\parallel}^R(\mathbf{q},\omega) - 1)\langle\phi(\mathbf{q},\omega)\rangle$ (compare with $\delta\langle\rho(\mathbf{q},\omega)\rangle = -\chi_{\rho\rho}^R(\mathbf{q},\omega)\phi_{\text{ext}}(\mathbf{q},\omega) = \epsilon_0\mathbf{q}^2(1/\epsilon_{\parallel}^R(\mathbf{q},\omega) - 1)\phi_{\text{ext}}(\mathbf{q},\omega))$ (see Sec. 3.4.3).

3.4.1.2 *Compressibility*

In a charged system, the definitions (3.107) and (3.110) of the compressibility assume that the background of positive charge is compressed along with the electrons in order to maintain electric neutrality everywhere. By contrast, χ^R_{nn} determines the response of the system to a change in chemical potential at fixed background of positive charge. Consequently, the relation (3.111) between κ and χ^R_{nn} must be revised.

Let us consider the effect of a static longitudinal field $\langle \mathbf{D}_{\text{ext}}(\mathbf{r})\rangle$ created by a distribution $\rho_{\text{ext}}(\mathbf{r})$ of external charges.[17] The force $\langle \mathbf{F}(\mathbf{r})\rangle$ per unit volume due to the field induces a pressure variation $\delta P(\mathbf{r})$ such that

$$\boldsymbol{\nabla}\delta P(\mathbf{r}) = \langle \mathbf{F}(\mathbf{r})\rangle. \tag{3.128}$$

In the long wavelength limit, the pressure distribution $\delta P(\mathbf{r})$ is related to the particle density $\delta\langle n(\mathbf{r})\rangle$ by

$$\delta\langle n(\mathbf{r})\rangle = \kappa n \delta P(\mathbf{r}). \tag{3.129}$$

This equation is a direct consequence of the definition of the compressibility, $\kappa^{-1} = n\partial P/\partial n|_T$ (Sec. 3.3). From (3.128) and (3.129), we deduce

$$\boldsymbol{\nabla}\delta\langle n(\mathbf{r})\rangle = \kappa n \langle \mathbf{F}(\mathbf{r})\rangle. \tag{3.130}$$

This result can be used to calculate κ provided that $\langle \mathbf{F}(\mathbf{r})\rangle$ is defined carefully [5]. The electrons respond to the external field $\langle \mathbf{D}_{\text{ext}}(\mathbf{r})\rangle$ by producing a polarization field $\langle \mathbf{E}_P(\mathbf{r})\rangle$. The background of positive charge adjusts itself to maintain electric neutrality and thus produces a field $-\langle \mathbf{E}_P(\mathbf{r})\rangle$. The total field felt by the electrons is then $\langle \mathbf{D}(\mathbf{r})\rangle = \langle \mathbf{D}_{\text{ext}}(\mathbf{r})\rangle - \langle \mathbf{E}_P(\mathbf{r})\rangle$. When calculating the compressibility, we are interested in the forces that are external to the complete system consisting of the electrons and the background of positive charge, i.e. to the field $\langle \mathbf{D}_{\text{ext}}\rangle$. But $\langle \mathbf{D}_{\text{ext}}\rangle = \langle \mathbf{D}\rangle + \langle \mathbf{E}_P\rangle$ is nothing but the local electric field $-\boldsymbol{\nabla}\langle\phi(\mathbf{r})\rangle$, with $\phi(\mathbf{r})$ the scalar potential. We thus have $\langle \mathbf{F}(\mathbf{r})\rangle = -ne\boldsymbol{\nabla}\langle\phi(\mathbf{r})\rangle$ and

$$\boldsymbol{\nabla}\delta\langle n(\mathbf{r})\rangle = -e\kappa n^2 \boldsymbol{\nabla}\langle\phi(\mathbf{r})\rangle. \tag{3.131}$$

Using the relation between $\delta\langle n(\mathbf{r})\rangle$ and $\langle\phi(\mathbf{r})\rangle$ given in Eq. (3.174),

$$e\boldsymbol{\nabla}\delta\langle n(\mathbf{r})\rangle = -\lim_{\mathbf{q}\to 0}\Pi^R_{00}(\mathbf{q},0)\boldsymbol{\nabla}\langle\phi(\mathbf{r})\rangle = -\lim_{\mathbf{q}\to 0}\epsilon_0\mathbf{q}^2(\epsilon^R_{\parallel}(\mathbf{q},0)-1)\boldsymbol{\nabla}\langle\phi(\mathbf{r})\rangle, \tag{3.132}$$

we finally obtain

$$e^2 n^2 \kappa = \lim_{\mathbf{q}\to 0}\Pi^R_{00}(\mathbf{q},0) \tag{3.133}$$

or, equivalently,

$$\epsilon^R_{\parallel}(\mathbf{q},0) = 1 + \frac{n^2e^2\kappa}{\epsilon_0\mathbf{q}^2} = 1 + \frac{\omega_p^2}{c_s^2\mathbf{q}^2} \qquad (\mathbf{q}\to 0), \tag{3.134}$$

[17] The charge distribution $\rho_{\text{ext}}(\mathbf{r})$ is assumed classical (no fluctuations), but fluctuations of the electromagnetic field are taken into account (see Sec. 3.4.3).

where $c_s = (1/\kappa nm)^{1/2}$ is the sound velocity (Sec. 3.3) and $\omega_p^2 = ne^2/\epsilon_0 m$ (the physical meaning of ω_p is discussed in Sec. 3.4.1.4).

In a charged system, the compressibility is determined by the irreducible part $\Pi^R_{00}(\mathbf{q},\omega)$ of the density–density correlation function $\chi^R_{\rho\rho}(\mathbf{q},\omega)$ [compare (3.111) and (3.133)]. It should be noted here that the compressibility κ given by (3.134) is not the true compressibility of the system. Indeed, we have assumed that the background of positive charge adjusts itself to maintain electric neutrality at no energy cost. Since we neglect the stiffness of the background, κ can be negative even for a stable system [3].

3.4.1.3 *Sum rules and high-frequency behavior*

The f-sum rule (3.106) can be rewritten in terms of the longitudinal dielectric function,

$$\int_{-\infty}^{\infty} \frac{d\omega}{\pi} \omega \Im\left[\frac{1}{\epsilon^R_{\parallel}(\mathbf{q},\omega)}\right] = -\omega_p^2 \qquad (\mathbf{q} \neq 0). \tag{3.135}$$

The high-frequency limit of the density–density response function (Sec. 3.3.2), which follows from the f-sum rule, implies

$$\epsilon^R_{\parallel}(\mathbf{q},\omega) = 1 - \frac{\omega_p^2}{\omega^2} \qquad (\omega \to \infty). \tag{3.136}$$

Another sum rule can be derived from

$$\lim_{\mathbf{q}\to 0} \frac{1}{\epsilon^R_{\parallel}(\mathbf{q},0)} = \lim_{\mathbf{q}\to 0} \frac{c_s^2 \mathbf{q}^2}{\omega_p^2} = 0, \quad \text{i.e.} \quad \chi^R_{\rho\rho}(\mathbf{q},0) = \epsilon_0 \mathbf{q}^2 \quad (\mathbf{q} \to 0). \tag{3.137}$$

Equation (3.137), which follows from (3.134), means that the screening becomes perfect in the static ($\omega = 0$) long-wavelength ($\mathbf{q} \to 0$) limit. Using the spectral representation of $1/\epsilon^R_{\parallel}(\mathbf{q},\omega) - 1$ or the Kramers–Kronig relations, we obtain the perfect-screening sum rule

$$\lim_{\mathbf{q}\to 0} \int_{-\infty}^{\infty} \frac{d\omega}{\pi} \frac{1}{\omega} \Im\left[\frac{1}{\epsilon^R_{\parallel}(\mathbf{q},\omega)}\right] = -1. \tag{3.138}$$

When $\epsilon^R_{\parallel}(\mathbf{q},\omega)$ is analytic in the upper-half complex plane, it satisfies the Kramers–Kronig relation

$$\Re[\epsilon^R_{\parallel}(\mathbf{q},\omega)] - 1 = \mathcal{P}\int_{-\infty}^{\infty} \frac{d\omega'}{\pi} \frac{\Im[\epsilon^R_{\parallel}(\mathbf{q},\omega')]}{\omega' - \omega} \tag{3.139}$$

(note that $\lim_{\omega\to\infty} \epsilon^R_{\parallel}(\mathbf{q},\omega) - 1 = 0$). Taking the limit $\omega \to \infty$ and using the asymptotic behavior (3.136), we obtain

$$\int_{-\infty}^{\infty} \frac{d\omega}{\pi} \omega \Im[\epsilon^R_{\parallel}(\mathbf{q},\omega)] = \omega_p^2. \tag{3.140}$$

Equation (3.140) can be rewritten as a sum rule for the longitudinal conductivity and is known as the longitudinal-conductivity sum rule (see Sec. 3.4.3). Finally, we can derive the compressibility sum rule from (3.134) and (3.139),

$$\int_{-\infty}^{\infty} \frac{d\omega}{\pi} \frac{\Im[\epsilon^R_{\parallel}(\mathbf{q},\omega)]}{\omega} = \frac{\omega_p^2}{c_s^2 \mathbf{q}^2} = \frac{n^2 e^2 \kappa}{\epsilon_0 \mathbf{q}^2} \qquad (\mathbf{q} \to 0). \tag{3.141}$$

3.4.1.4 *Charge oscillations and plasmon mode*

The zeros of $\epsilon_{\parallel}^{R}(\mathbf{q},\omega)$ are given by the poles of $\chi_{\rho\rho}^{R}(\mathbf{q},\omega)$. They describe charge oscillations of the electron system in the presence of the long-range Coulomb interaction. Besides the continuum of particle–hole pair excitations $\omega = \epsilon_{\mathbf{k}+\mathbf{q}} - \epsilon_{\mathbf{k}}$, we also expect a collective density mode showing up as an isolated zero of $\epsilon_{\parallel}^{R}(\mathbf{q},\omega)$, i.e. an isolated pole of $\chi_{\rho\rho}^{R}(\mathbf{q},\omega)$. In the absence of long-range interaction, the collective density mode of the electron system is the zero-sound mode (Chapter 4) or, in the collisional (hydrodynamic) regime, the first-sound mode (Sec. 3.3). Both these modes have a linear dispersion $\omega \propto |\mathbf{q}|$ at low energies. Because of the Coulomb interaction, the collective density mode, known as the plasmon mode, is pushed up to high energies. To see this in a simple way, we consider the continuity equation

$$\partial_t \delta\langle n(\mathbf{r},t)\rangle + n\boldsymbol{\nabla}\cdot\langle \mathbf{v}(\mathbf{r},t)\rangle = 0 \tag{3.142}$$

($\delta\langle n(\mathbf{r},t)\rangle$ and $\langle \mathbf{v}(\mathbf{r},t)\rangle$ are assumed infinitesimal) and Newton's equation of motion

$$mn\partial_t\langle \mathbf{v}(\mathbf{r},t)\rangle = en\langle \mathbf{E}(\mathbf{r},t)\rangle. \tag{3.143}$$

Combining (3.142) and (3.143) with Poisson's equation

$$\boldsymbol{\nabla}\cdot\langle \mathbf{E}(\mathbf{r},t)\rangle = \epsilon_0^{-1} e\delta\langle n(\mathbf{r},t)\rangle, \tag{3.144}$$

we then obtain

$$\partial_t^2\delta\langle n(\mathbf{r},t)\rangle = -\frac{ne}{m}\boldsymbol{\nabla}\cdot\langle \mathbf{E}(\mathbf{r},t)\rangle = -\frac{ne^2}{\epsilon_0 m}\delta\langle n(\mathbf{r},t)\rangle \tag{3.145}$$

and, after Fourier transformation,

$$(\omega^2 - \omega_p^2)\delta\langle n(\mathbf{q},\omega)\rangle = 0 \tag{3.146}$$

in a three-dimensional system. The plasmon frequency $\omega_p = (ne^2/m\epsilon_0)^{1/2}$ lies above the continuum of particle–hole pair excitations ($\omega = \epsilon_{\mathbf{k}+\mathbf{q}} - \epsilon_{\mathbf{k}} \sim v_F|\mathbf{q}|$), which ensures that the plasmon mode remains undamped in the limit $\mathbf{q}\to 0$ (Fig. 3.7).[18] Note that in a typical metal, the plasmon energy is larger or of the order of the electronic bandwidth.

Let us assume that the particle–hole pair excitations do not contribute to $\Im[1/\epsilon_{\parallel}^{R}(\mathbf{q},\omega)]$ in the limit $\mathbf{q}\to 0$. This assumption can be verified within the RPA (Chapter 5) and more generally follows from phase space arguments (we now consider a three-dimensional system) [5]. We then expect

$$\lim_{\mathbf{q}\to 0}\Im\left[\frac{1}{\epsilon_{\parallel}^{R}(\mathbf{q},\omega)}\right] = A[\delta(\omega-\omega_0) - \delta(\omega+\omega_0)] \tag{3.147}$$

($\Im[1/\epsilon_{\parallel}^{R}(\mathbf{q},\omega)]$ is odd in frequency), where ω_0 is the frequency of the plasmon mode in the limit $\mathbf{q}\to 0$. The f-sum rule (3.135) and the perfect-screening sum rule (3.138) determine

[18] For electrons confined to a two-dimensional plane (but interacting *via* the three-dimensional Coulomb interaction), one finds the plasmon dispersion $\omega = (|\mathbf{q}|ne^2/\epsilon_0 m)^{1/2}$. It vanishes for $\mathbf{q}\to 0$, but the (undamped) plasmon mode lies nevertheless above the particle–hole pair excitation spectrum.

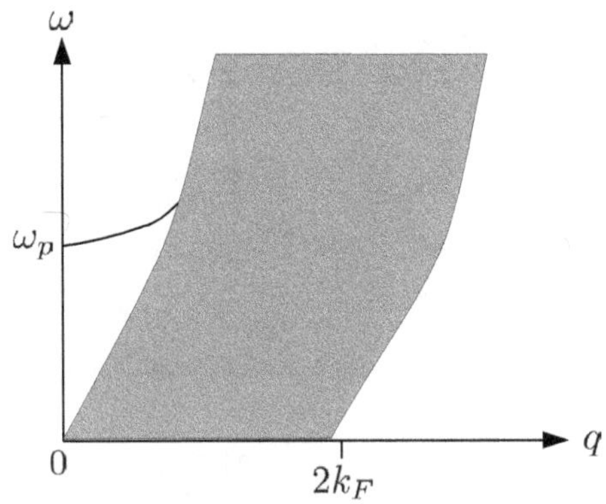

Fig. 3.7 Plasmon mode in a three-dimensional electron system. The shaded area shows the continuum of particle–hole pair excitations $\omega = \epsilon_{\mathbf{k+q}} - \epsilon_{\mathbf{k}}$.

both A and ω_0: $A = -\omega_p\pi/2$ and $\omega_0 = \omega_p$. Thus, ω_p is the exact value of the plasmon frequency in the $\mathbf{q} \to 0$ limit. Equivalently, Eq. (3.147) can be written as

$$\chi''_{\rho\rho}(\mathbf{q}, \omega) = \frac{\pi}{2}\omega_p\epsilon_0\mathbf{q}^2[\delta(\omega - \omega_p) - \delta(\omega + \omega_p)] \qquad (\mathbf{q} \to 0). \tag{3.148}$$

Using the spectral representation of $1/\epsilon^R_{\parallel}(\mathbf{q}, \omega) - 1$, we then obtain

$$\lim_{\mathbf{q}\to 0} \epsilon^R_{\parallel}(\mathbf{q}, \omega) = 1 - \frac{\omega_p^2}{(\omega + i\eta)^2}. \tag{3.149}$$

3.4.2 *Transverse dielectric function*

To calculate the propagator of the vector potential, we first integrate out ϕ and obtain

$$\begin{aligned} S[\psi, \mathbf{A}_\perp] = S_m[\psi] + \int_0^\beta d\tau \int d^d r \bigg\{ &\frac{\epsilon_0}{2}\left[\left(\partial_\tau \mathbf{A}_\perp(x)\right)^2 + c_l^2\left(\boldsymbol{\nabla} \times \mathbf{A}_\perp(x)\right)^2\right] \\ &- \mathbf{j}(x) \cdot \mathbf{A}_\perp(x) + \frac{e}{2m}\rho(x)\mathbf{A}_\perp^2(x)\bigg\}. \end{aligned} \tag{3.150}$$

The most direct way to determine $D_{\mu\nu}$ is to proceed diagrammatically as discussed in the preceding section. The Dyson equation satisfied by $D_{\mu\nu}$ is shown in Fig. 3.8. The self-energy $\Pi_{\mu\nu}$ is equal to

$$\Pi_{\mu\nu}(q) = \left(\delta_{\mu,\nu} - \frac{q_\mu q_\nu}{\mathbf{q}^2}\right)[-\epsilon_0\omega_p^2 + \chi_\perp(q)], \tag{3.151}$$

where $\chi_\perp(q)$ is the transverse part of the current–current correlation function

$$\chi_{j_\mu j_\nu}(q) = \langle j_\mu(q) j_\nu(-q)\rangle = \frac{q_\mu q_\nu}{\mathbf{q}^2}\chi_{\parallel}(q) + \left(\delta_{\mu,\nu} - \frac{q_\mu q_\nu}{\mathbf{q}^2}\right)\chi_\perp(q) \tag{3.152}$$

computed with the action $S_m[\psi]$. The first term in (3.151) comes from the diamagnetic term, while the second one is due to the coupling $\mathbf{j} \cdot \mathbf{A}_\perp = \mathbf{j}_\perp \cdot \mathbf{A}_\perp$ ($\mathbf{j}_\perp$ is the transverse

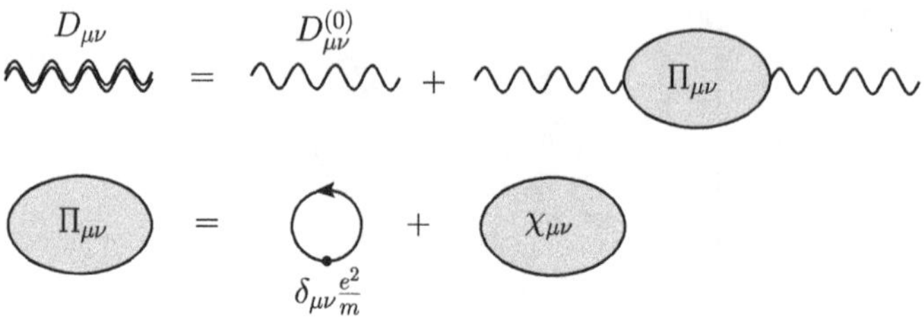

Fig. 3.8 Diagrammatic representation of $D_{\mu\nu}$ and $\Pi_{\mu\nu}$.

component of the current density). If one ignores the magnetic interactions with respect to the Coulomb interaction, Eq. (3.151) is exact. It yields

$$D_\perp^{-1}(q) = [D_\perp^{(0)}(q)]^{-1} - \Pi_\perp(q) = [D_\perp^{(0)}(q)]^{-1} + \epsilon_0\omega_p^2 - \chi_\perp(q), \tag{3.153}$$

where $\chi_\perp$ is the transverse part of the current–current correlation function.

The transverse dielectric function $\epsilon_\perp(q)$ is defined by

$$D_\perp(q) = \frac{1}{\epsilon_0(\epsilon_\perp(q)\omega_n^2 + c_l^2\mathbf{q}^2)}, \tag{3.154}$$

i.e.

$$\epsilon_\perp(q) = 1 + \frac{\omega_p^2}{\omega_n^2} - \frac{\chi_\perp(q)}{\epsilon_0\omega_n^2}. \tag{3.155}$$

It determines the propagating modes of the electromagnetic field coupled to the transverse oscillations of the electron system, i.e. the "screening" of the electromagnetic field (Fig. 3.8).

3.4.2.1 *Asymptotic behavior and analytic properties*

Since $\lim_{|z|\to\infty}\chi_\perp(\mathbf{q},z) = 0$,

$$\epsilon_\perp^R(\mathbf{q},\omega) = 1 - \frac{\omega_p^2}{\omega^2} \qquad (\omega\to\infty). \tag{3.156}$$

In a translation-invariant system where the total current $\hat{\mathbf{j}}(\mathbf{q}=0)$ is conserved, $\chi''_\perp(0,t)$ and $\chi^R_\perp(0,\omega)$ vanish,[19] and

$$\lim_{\mathbf{q}\to 0}\epsilon_\perp^R(\mathbf{q},\omega) = 1 - \frac{\omega_p^2}{(\omega+i\eta)^2}. \tag{3.157}$$

It readily follows from the definition (3.155) that $\epsilon_\perp^R(\mathbf{q},\omega)$ is analytic in the upper-half plane and satisfies the Kramers–Kronig relations. In particular,

$$\Re[\epsilon_\perp^R(\mathbf{q},\omega)] - 1 = \mathcal{P}\int_{-\infty}^{\infty}\frac{d\omega'}{\pi}\frac{\Im[\epsilon_\perp^R(\mathbf{q},\omega')]}{\omega'-\omega}. \tag{3.158}$$

---

[19] The vanishing of $\chi''_{j_\mu j_\nu}(0,\omega)$ and $\chi^R_{j_\mu j_\nu}(0,\omega)$ is a direct consequence of the definition (3.12) of $\chi''_{j_\mu j_\nu}(\mathbf{q}=0,t)$ when $\hat{\mathbf{j}}(\mathbf{q}=0,t)$ is independent of t.

The analyticity properties of $\epsilon_\perp^R(\mathbf{q},\omega)$ also imply that the integral $\int_{\mathcal{C}} d\omega\, \omega \epsilon_\perp^R(\mathbf{q},\omega)$, where $\mathcal{C}$ is the contour defined by the real axis and the half circle of radius $R \to \infty$ located in the upper-half complex plane, must vanish. Because of (3.156), the contribution of the half circle is $-i\pi\omega_p^2$, so that

$$\int_{-\infty}^{\infty} \frac{d\omega}{\pi} \omega \Im[\epsilon_\perp^R(\mathbf{q},\omega)] = \omega_p^2. \tag{3.159}$$

Equation (3.159) can be rewritten as a sum rule for the transverse conductivity (Sec. 3.4.4).

3.4.2.2 *Photon mode and transverse collective mode*

The dispersion of the transverse excitations of the electron system coupled to the electromagnetic field is obtained from the poles of $D_\perp^R(\mathbf{q},\omega)$,

$$\epsilon_\perp^R(\mathbf{q},\omega)\omega^2 = c_l^2 \mathbf{q}^2. \tag{3.160}$$

Here, we drop the infinitesimal $i\eta$ which does not play any role in the following discussion.

Let us first look for a high-frequency solution $\omega \gtrsim \omega_p$. Using the asymptotic behavior (3.156), we obtain

$$\omega^2 \simeq \omega_p^2 + c_l^2 \mathbf{q}^2. \tag{3.161}$$

For $c_l|\mathbf{q}| \gg \omega_p$, the photon mode (3.161) is weakly coupled to the system and its frequency $\omega \simeq c_l|\mathbf{q}|$ is only slightly modified by the electrons. When $c_l|\mathbf{q}| \lesssim \omega_p$, the coupling between the electromagnetic field and the electrons is strong and leads to an enormous modification of the natural frequency $\omega \sim c_l|\mathbf{q}|$ of the field. Note that this strong coupling is due only to the diamagnetic term and does not involve the dynamics of the electrons.

Besides the photon mode (3.161), there are also low-energy excitations arising from genuine transverse fluctuations of the electron system, i.e. from the (isolated) poles of $\chi_\perp^R(\mathbf{q},\omega)$. Let us rewrite (3.160) as

$$\epsilon_0^{-1} \chi_\perp^R(\mathbf{q},\omega) = \omega_p^2 + c_l^2 \mathbf{q}^2 - \omega^2. \tag{3.162}$$

Since transverse current fluctuations do not couple to the charge, $\chi_\perp^R$ does not contain the plasmon pole. The poles of $\chi_\perp^R(\mathbf{q},\omega)$ are therefore expected at low frequency $\omega \sim v_F|\mathbf{q}| \ll c_l|\mathbf{q}|$ in a metal. These poles satisfy (3.162), with only a slight shift due to the large but finite value of the rhs. These transverse excitations can exist only if they do not mix with the continuum of particle–hole pair excitations (the solutions of (3.160) would otherwise appear at complex frequencies). In Secs. 4.3.3 and 4.3.4, we shall study a transverse mode within a simple model based on Landau Fermi-liquid theory. The result suggests that a necessary condition for the propagation of transverse modes is that short-range interactions between electrons are sufficiently strong (thus making a transverse collective mode in real systems rather unlikely). When this condition is satisfied, one can show that the admixture of electromagnetic field into the transverse collective mode is negligible when $c_l|\mathbf{q}| \gg \omega_p$: The collective mode is essentially the same as in the neutral system. In the low-frequency regime $c_l|\mathbf{q}| \lesssim \omega_p$, the transverse collective mode is repelled by the photon mode (3.161) and disappears in the continuum of particle–hole pair excitations (Fig. 3.9) [5].

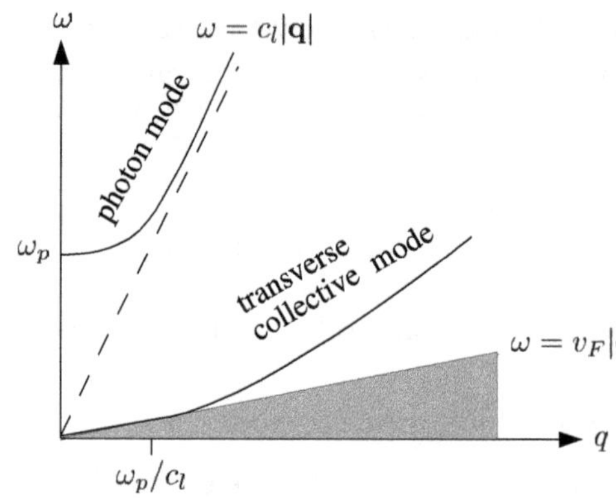

Fig. 3.9 Transverse excitation spectrum of an electron system coupled to the electromagnetic field in the case of strong short-range interactions. For $c_l|\mathbf{q}| \lesssim \omega_p$, the transverse mode disappears in the continuum of particle–hole pair excitations (shaded area). Reproduced from Ref. [5].

3.4.3 *Linear response to external fields*

In this section, we consider the effect of a field created by a charge ($\rho_{\rm ext}$) and current ($\mathbf{j}_{\rm ext}$) distribution external to the system. Fluctuations of $\rho_{\rm ext}$ and $\mathbf{j}_{\rm ext}$ are neglected. In the Coulomb gauge, the coupling between the external field and the system is described by the action

$$S_{\rm ext}[\phi, \mathbf{A}_\perp] = \int_0^\beta d\tau \int d^d r [i\phi(x)\rho_{\rm ext}(x) - \mathbf{j}_{\rm ext\perp}(x) \cdot \mathbf{A}_\perp(x)]. \tag{3.163}$$

3.4.3.1 *Screening of the external electromagnetic potentials*

To linear order in the external perturbation, the mean value of the electromagnetic potentials is given by

$$\begin{aligned} \langle i\phi(q)\rangle &= -\langle i\phi(q) i\phi(-q)\rangle \rho_{\rm ext}(q) = D_{00}(q)\rho_{\rm ext}(q), \\ \langle \mathbf{A}_\perp(q)\rangle &= \langle \mathbf{A}_{\perp\mu}(q)\mathbf{A}_{\perp\mu}(-q)\rangle \mathbf{j}_{\rm ext\perp}(q) = D_\perp(q)\mathbf{j}_{\rm ext\perp}(q), \end{aligned} \tag{3.164}$$

where D_{00} and $D_\perp$ are the propagators in the absence of the external perturbation. These equations can be rewritten in terms of the dielectric functions,

$$\begin{aligned} \epsilon_0 \epsilon_\parallel(q)\mathbf{q}^2 \langle i\phi(q)\rangle &= \rho_{\rm ext}(q), \\ \epsilon_0[\epsilon_\perp(q)\omega_n^2 + c_l^2\mathbf{q}^2]\langle \mathbf{A}_\perp(q)\rangle &= \mathbf{j}_{\rm ext\perp}(q). \end{aligned} \tag{3.165}$$

Real-time equations are obtained from the analytic continuation $i\omega_n \to \omega + i\eta$ with $i\phi(q) \to \phi(\mathbf{q},\omega)$ and $\mathbf{A}_\perp(q) \to \mathbf{A}_\perp(\mathbf{q},\omega)$ so that $\phi(\mathbf{q},\omega)$ is the scalar potential according to the usual definition (see the discussion at the beginning of Sec. 3.4):

$$\begin{aligned} \epsilon_0 \epsilon_\parallel^R(\mathbf{q},\omega)\mathbf{q}^2 \langle \phi(\mathbf{q},\omega)\rangle &= \rho_{\rm ext}(\mathbf{q},\omega), \\ \epsilon_0\left[-\epsilon_\perp^R(\mathbf{q},\omega)(\omega+i\eta)^2 + c_l^2\mathbf{q}^2\right]\langle \mathbf{A}_\perp(\mathbf{q},\omega)\rangle &= \mathbf{j}_{\rm ext\perp}(\mathbf{q},\omega). \end{aligned} \tag{3.166}$$

Thus, we find that the average value of the electromagnetic field satisfies the Maxwell equations,

$$\begin{aligned} \nabla \cdot \langle \mathbf{D}_{\|}(\mathbf{r},t)\rangle &= \rho_{\text{ext}}(\mathbf{r},t), \\ \mu_0^{-1}\nabla \times \langle \mathbf{B}(\mathbf{r},t)\rangle &= \frac{\partial \langle \mathbf{D}_{\perp}(\mathbf{r},t)\rangle}{\partial t} + \mathbf{j}_{\text{ext}\perp}(\mathbf{r},t) \end{aligned} \tag{3.167}$$

(recall that $\epsilon_0\mu_0 c_l^2 = 1$), where

$$\begin{aligned} \langle \mathbf{D}_{\|}(\mathbf{q},\omega)\rangle &= \epsilon_0 \epsilon_{\|}^R(\mathbf{q},\omega)\langle \mathbf{E}_{\|}(\mathbf{q},\omega)\rangle, \\ \langle \mathbf{D}_{\perp}(\mathbf{q},\omega)\rangle &= \epsilon_0 \epsilon_{\perp}^R(\mathbf{q},\omega)\langle \mathbf{E}_{\perp}(\mathbf{q},\omega)\rangle \end{aligned} \tag{3.168}$$

is the displacement field. The dielectric functions that we defined from the propagator of the electromagnetic field correspond to those introduced in classical electromagnetism.

In the absence of coupling to the electrons, the external charge and current distribution create an electromagnetic potential defined by

$$\begin{aligned} [D_{00}^{(0)}(q)]^{-1} i\phi_{\text{ext}}(q) &= \rho_{\text{ext}}(q), \\ [D_{\perp}^{(0)}(q)]^{-1} \mathbf{A}_{\text{ext}\perp}(q) &= \mathbf{j}_{\text{ext}\perp}(q). \end{aligned} \tag{3.169}$$

Combining Eqs. (3.164), (3.165) and (3.169), we obtain the relation between the local $(\phi, \mathbf{A}_\perp)$ and external $(\phi_{\text{ext}}, \mathbf{A}_{\text{ext}\perp})$ potentials,

$$\begin{aligned} \langle \phi(\mathbf{q},\omega)\rangle &= \frac{D_{00}^R(\mathbf{q},\omega)}{D_{00}^{(0)R}(\mathbf{q},\omega)} \phi_{\text{ext}}(\mathbf{q},\omega) = \frac{\phi_{\text{ext}}(\mathbf{q},\omega)}{\epsilon_{\|}^R(\mathbf{q},\omega)}, \\ \langle \mathbf{A}_{\perp}(\mathbf{q},\omega)\rangle &= \frac{D_{\perp}^R(\mathbf{q},\omega)}{D_{\perp}^{(0)R}(\mathbf{q},\omega)} \mathbf{A}_{\text{ext}\perp}(\mathbf{q},\omega) = \frac{(\omega+i\eta)^2 - c_l^2\mathbf{q}^2}{\epsilon_{\perp}^R(\mathbf{q},\omega)(\omega+i\eta)^2 - c_l^2\mathbf{q}^2} \mathbf{A}_{\text{ext}\perp}(\mathbf{q},\omega), \end{aligned} \tag{3.170}$$

after analytic continuation $i\omega_n \to \omega + i\eta$. These equations enable one to compute the local field as a function of the external field. In particular, the second one determines the screening of a static external magnetic field $\mathbf{B}_{\text{ext}\perp}(\mathbf{q},0) = i\mathbf{q} \times \mathbf{A}_{\text{ext}\perp}(\mathbf{q},0)$ and distinguishes between the weak orbital (Landau) diamagnetism of a normal metal and the perfect diamagnetism (Meissner effect) of a superconductor (Sec. 3.4.5).

3.4.3.2 *Screening of the external charge and current densities*

Similar relations can be obtained between the total and external charge and current densities. To linear order in the external perturbation

$$\delta\langle \rho(q)\rangle = -\langle \rho(q) i\phi(-q)\rangle \rho_{\text{ext}}(q) = -\Pi_{00}(q) D_{00}(q) \rho_{\text{ext}}(q) \tag{3.171}$$

(see the relevant diagrams in Fig. 3.10), and

$$\begin{aligned} \langle \mathbf{J}_{\perp}(q)\rangle &= \langle \mathbf{j}_{\perp}(q)\rangle - \frac{ne^2}{m}\langle \mathbf{A}_{\perp}(q)\rangle, \\ &= \Big[\langle j_{\perp\mu}(q) A_{\perp\mu}(-q)\rangle - \frac{ne^2}{m} D_{\perp}(q)\Big] \mathbf{j}_{\text{ext}\perp}(q) \\ &= \left[\chi_{\perp}(q) - \frac{ne^2}{m}\right] D_{\perp}(q) \mathbf{j}_{\text{ext}\perp}(q). \end{aligned} \tag{3.172}$$

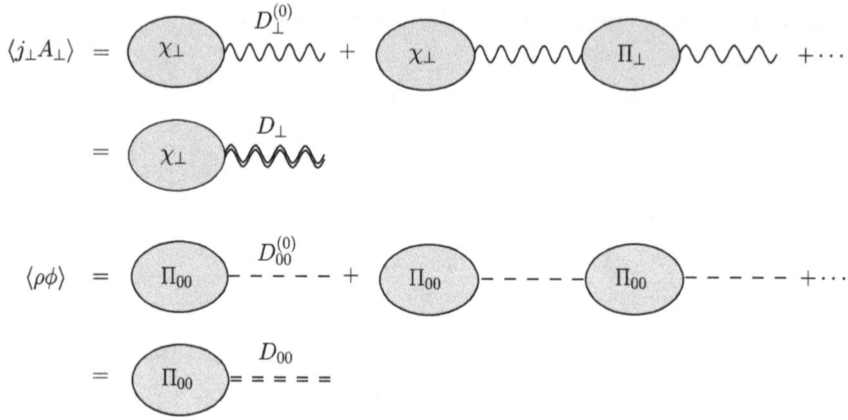

Fig. 3.10 Diagrammatic representation of $\langle j_{\perp\mu}(q)A_{\perp\mu}(-q)\rangle$ and $\langle\rho(q)\phi(-q)\rangle$.

This yields

$$\begin{aligned}
\delta\langle\rho(\mathbf{q},\omega)\rangle+\rho_{\text{ext}}(\mathbf{q},\omega)&=\frac{\rho_{\text{ext}}(\mathbf{q},\omega)}{\epsilon_{\parallel}^{R}(\mathbf{q},\omega)},\\
\langle\mathbf{J}_\perp(\mathbf{q},\omega)\rangle+\mathbf{j}_{\text{ext}\perp}(\mathbf{q},\omega)&=\frac{(\omega+i\eta)^2-c_l^2q^2}{\epsilon_\perp^R(\mathbf{q},\omega)(\omega+i\eta)^2-c_l^2q^2}\mathbf{j}_{\text{ext}\perp}(\mathbf{q},\omega).
\end{aligned}\tag{3.173}$$

3.4.3.3 *Conductivity*

Finally, we can relate the current density to the electromagnetic field and obtain the conductivity. Equations (3.164) and (3.171) give

$$\delta\langle\rho(\mathbf{q},\omega)\rangle=-\Pi_{00}^R(\mathbf{q},\omega)\langle\phi(\mathbf{q},\omega)\rangle.\tag{3.174}$$

Using the continuity equation $(\omega+i\eta)\delta\langle\rho(\mathbf{q},\omega)\rangle=\mathbf{q}\cdot\langle\mathbf{J}(\mathbf{q},\omega)\rangle$, we deduce $\langle\mathbf{J}_\parallel(\mathbf{q},\omega)\rangle=\sigma_\parallel(\mathbf{q},\omega)\langle\mathbf{E}_\parallel(\mathbf{q},\omega)\rangle$ with[20]

$$\sigma_\parallel(\mathbf{q},\omega)=\frac{\omega+i\eta}{i\mathbf{q}^2}\Pi_{00}^R(\mathbf{q},\omega).\tag{3.175}$$

From Eqs. (3.164, 3.172), we obtain

$$\langle\mathbf{J}_\perp(\mathbf{q},\omega)\rangle=\left[\chi_\perp^R(\mathbf{q},\omega)-\frac{ne^2}{m}\right]\langle\mathbf{A}_\perp(\mathbf{q},\omega)\rangle\tag{3.176}$$

and

$$\sigma_\perp(\mathbf{q},\omega)=\frac{1}{i(\omega+i\eta)}\left[\chi_\perp^R(\mathbf{q},\omega)-\frac{ne^2}{m}\right].\tag{3.177}$$

[20]The conductivity tensor can be written as $\sigma_{\mu\nu}(\mathbf{q},\omega)=(q_\mu q_\nu/\mathbf{q}^2)\sigma_\parallel(\mathbf{q},\omega)+(\delta_{\mu,\nu}-q_\mu q_\nu/\mathbf{q}^2)\sigma_\perp(\mathbf{q},\omega)$. This form follows from the homogeneity and isotropy of the system.

The conductivity is related to the dielectric function by

$$\epsilon^R_{\parallel,\perp}(\mathbf{q},\omega) = 1 + \frac{i}{\epsilon_0(\omega + i\eta)}\sigma_{\parallel,\perp}(\mathbf{q},\omega). \tag{3.178}$$

For each sum rule satisfied by $\epsilon^R_{\parallel,\perp}$, we can deduce a sum rule for $\sigma_{\parallel,\perp}$. Of particular interest are the longitudinal- and transverse-conductivity sum rules

$$\int_{-\infty}^{\infty} \frac{d\omega}{2\pi} \Re[\sigma_\parallel(\mathbf{q},\omega)] = \int_{-\infty}^{\infty} \frac{d\omega}{2\pi} \Re[\sigma_\perp(\mathbf{q},\omega)] = \frac{\epsilon_0}{2}\omega_p^2, \tag{3.179}$$

which follow from (3.140) and (3.159). Note that the longitudinal-conductivity sum rule assumes the analyticity of $\epsilon^R_\parallel(\mathbf{q},\omega)$ in the upper-half complex plane and may be violated. By contrast, $\epsilon^R_\perp(\mathbf{q},\omega)$ and $\sigma_\perp(\mathbf{q},\omega)$ are always analytic for $\Im(\omega) \geq 0$, and the transverse-conductivity sum rule is always valid.

3.4.4 *Conductivity*

The conductivity can also be calculated by considering the Hamiltonian

$$\hat{H} + \hat{H}'(t) = \hat{H} + \int d^d r \left[\phi(\mathbf{r},t)\hat{\rho}(\mathbf{r}) - \hat{\mathbf{j}}(\mathbf{r}) \cdot \mathbf{A}(\mathbf{r},t) + \frac{e}{2m}\hat{\rho}(\mathbf{r})\mathbf{A}^2(\mathbf{r},t)\right] \tag{3.180}$$

describing a system coupled to the electromagnetic potentials $(\phi, \mathbf{A})$. The Hamiltonian $\hat{H}$ includes the Coulomb interaction between particles but not the magnetic interactions. There is no necessity anymore to restrict ourselves to the Coulomb gauge (which was primarily dictated by the simple treatment of the electromagnetic field fluctuations, see Sec. 1.9).

The longitudinal potentials $(\phi, \mathbf{A}_\parallel)$ in (3.180) should be seen as external potentials. The polarization processes giving rise to screening will arise through a proper treatment of the Coulomb interaction included in the Hamiltonian (3.180). On the other hand, $\mathbf{A}_\perp$ should be seen as a local potential since, in the absence of the electromagnetic part $\hat{H}_{\rm em}$ of the Hamiltonian (describing the dynamics of the electromagnetic field), the screening of the transverse field cannot be taken into account. The absence of $\hat{H}_{\rm em}$ in (3.180) also implies that the magnetic interactions between electrons are not included but they can be safely ignored in a non-relativistic system.[21] In the following, we shall therefore consider the transverse response to the local potential $\mathbf{A}_\perp$ and the longitudinal response to the external potentials $(\phi, \mathbf{A}_\parallel)$.

3.4.4.1 *Kubo's formulae*

From Eq. (3.180), we readily obtain the linear response to the electromagnetic potentials,

$$\begin{aligned} \delta\langle\rho(\mathbf{q},\omega)\rangle &= -\chi^R_{\rho\rho}(\mathbf{q},\omega)\phi(\mathbf{q},\omega) + \sum_\mu \chi^R_{\rho j_\mu}(\mathbf{q},\omega)A_\mu(\mathbf{q},\omega), \\ \langle \mathbf{J}_\mu(\mathbf{q},\omega)\rangle &= \sum_\nu \left[\chi^R_{j_\mu j_\nu}(\mathbf{q},\omega) - \frac{ne^2}{m}\delta_{\mu,\nu}\right] A_\nu(\mathbf{q},\omega) - \chi^R_{j_\mu\rho}(\mathbf{q},\omega)\phi(\mathbf{q},\omega). \end{aligned} \tag{3.181}$$

[21] See the discussion at the end of Sec. 1.9.3.

Choosing the gauge $\phi = 0$, we deduce the expression of the conductivity tensor

$$\sigma_{\mu\nu}(\mathbf{q}, \omega) = \frac{1}{i(\omega + i\eta)} \left[\chi^R_{j_\mu j_\nu}(\mathbf{q}, \omega) - \frac{ne^2}{m} \delta_{\mu,\nu} \right], \tag{3.182}$$

i.e.

$$\sigma_{\parallel,\perp}(\mathbf{q}, \omega) = \frac{1}{i(\omega + i\eta)} \left[\chi^R_{\parallel,\perp}(\mathbf{q}, \omega) - \frac{ne^2}{m} \right]. \tag{3.183}$$

Note that $\sigma_\parallel(\mathbf{q}, \omega)$ differs from (3.175) which gives the response to the local field.[22] From now on, $\sigma_\parallel$ shall always denote the response to the external longitudinal electric field. We shall show below how Eq. (3.183) can be derived in an arbitrary gauge.

3.4.4.2 *Gauge invariance and particle number conservation*

When $\mathbf{A}_\parallel(\mathbf{r}, t) = \boldsymbol{\nabla}\Lambda(\mathbf{r}, t)$ and $\phi(\mathbf{r}, t) = -\partial_t \Lambda(\mathbf{r}, t)$, there is no longitudinal electric field, $\mathbf{E}_\parallel = -\boldsymbol{\nabla}\phi - \partial_t \mathbf{A}_\parallel = 0$, and the longitudinal response must therefore vanish for any function Λ. This gives

$$\begin{aligned} 0 &= (\omega + i\eta)\chi^R_{\rho\rho}(\mathbf{q}, \omega) - \sum_\mu q_\mu \chi^R_{\rho j_\mu}(\mathbf{q}, \omega), \\ 0 &= \sum_\nu \left[\chi^R_{j_\mu j_\nu}(\mathbf{q}, \omega) - \frac{ne^2}{m}\delta_{\mu,\nu} \right] q_\nu - (\omega + i\eta)\chi^R_{j_\mu \rho}(\mathbf{q}, \omega). \end{aligned} \tag{3.184}$$

Multiplying the last equation by q_μ and summing over μ gives

$$\begin{aligned} 0 &= \mathbf{q}^2 \left[\chi^R_\parallel(\mathbf{q}, \omega) - \frac{ne^2}{m} \right] - (\omega + i\eta) \sum_\mu q_\mu \chi^R_{j_\mu \rho}(\mathbf{q}, \omega) \\ &= \mathbf{q}^2 \left[\chi^R_\parallel(\mathbf{q}, \omega) - \frac{ne^2}{m} \right] - (\omega + i\eta)^2 \chi^R_{\rho\rho}(\mathbf{q}, \omega). \end{aligned} \tag{3.185}$$

The last result is obtained using $\chi^R_{\rho j_\mu}(\mathbf{q}, \omega) = \chi^R_{j_\mu \rho}(\mathbf{q}, \omega)$. The equality of these two correlation functions is easily shown using (3.18), as well as parity and time-reversal invariance (see Table 3.1):

$$\chi''_{j_\mu \rho}(\mathbf{q}, \omega) = [\chi''_{\rho j_\mu}(-\mathbf{q}, \omega)]^* = \epsilon^T_\rho \epsilon^T_{j_\mu} \epsilon^P_\rho \epsilon^P_{j_\mu} \chi''_{\rho j_\mu}(\mathbf{q}, \omega) = \chi''_{\rho j_\mu}(\mathbf{q}, \omega) \tag{3.186}$$

($\epsilon^T_\rho = \epsilon^P_\rho = 1$ and $\epsilon^T_{j_\mu} = \epsilon^P_{j_\mu} = -1$), where ϵ^P_A and ϵ^T_A denote the signatures of the operator $\hat{A}$ under parity and time reversal (Chapter 2).

These results can also be derived from the continuity equation $(\omega + i\eta)\delta\langle\rho(\mathbf{q}, \omega)\rangle - \mathbf{q} \cdot \langle \mathbf{J}_\parallel(\mathbf{q}, \omega)\rangle = 0$, where $\delta\langle\rho(\mathbf{q}, \omega)\rangle$ and $\langle \mathbf{J}_\parallel(\mathbf{q}, \omega)\rangle$ are given by (3.181) to linear order in the

[22]The expression $\sigma_\parallel(\mathbf{q}, \omega) = [(\omega + i\eta)/i\mathbf{q}^2]\chi^R_{\rho\rho}(\mathbf{q}, \omega)$ derived in Eq. (3.192) can be deduced from (3.175) and the relation (3.170) between $\langle\phi(\mathbf{q}, \omega)\rangle$ and $\phi_{\rm ext}(\mathbf{q}, \omega)$.

electromagnetic potentials. Since the continuity equation must be satisfied for any value of $\mathbf{A}_\parallel$ and ϕ, we reproduce (3.184) except for the exchange $\chi^R_{\rho j_\mu}(\mathbf{q},\omega) \leftrightarrow \chi^R_{j_\mu \rho}(\mathbf{q},\omega)$.

Equations (3.184) and (3.185) can be used to derive the expression of $\sigma_\parallel(\mathbf{q},\omega)$ in an arbitrary gauge. It is convenient to introduce the longitudinal and transverse parts of the current,

$$\hat{\mathbf{J}}_\parallel(\mathbf{q}) = \frac{\mathbf{q}}{\mathbf{q}^2}[\mathbf{q}\cdot\hat{\mathbf{J}}(\mathbf{q})] = \frac{\mathbf{q}}{|\mathbf{q}|}\hat{J}_\parallel(\mathbf{q}),$$
$$\hat{\mathbf{J}}_\perp(\mathbf{q}) = \hat{\mathbf{J}}(\mathbf{q}) - \hat{\mathbf{J}}_\parallel(\mathbf{q}) \tag{3.187}$$

(similar expressions hold for the vector potential $\mathbf{A}$) and the correlation functions

$$\chi_\parallel(\mathbf{q},\omega) = \sum_{\mu,\nu}\frac{q_\mu q_\nu}{\mathbf{q}^2}\chi_{j_\mu j_\nu}(\mathbf{q},\omega), \quad \chi_\perp(\mathbf{q},\omega) = \sum_{\mu,\nu}\left(\delta_{\mu,\nu} - \frac{q_\mu q_\nu}{\mathbf{q}^2}\right)\chi_{j_\mu j_\nu}(\mathbf{q},\omega),$$
$$\chi_{j_\parallel \rho}(\mathbf{q},\omega) = \sum_\mu \frac{q_\mu}{|\mathbf{q}|}\chi_{j_\mu \rho}(\mathbf{q},\omega), \quad \chi_{\rho j_\parallel}(\mathbf{q},\omega) = \sum_\mu \chi_{\rho j_\mu}(\mathbf{q},\omega)\frac{q_\mu}{|\mathbf{q}|}. \tag{3.188}$$

This allows us to rewrite Eqs. (3.181) as[23]

$$\delta\langle\rho(\mathbf{q},\omega)\rangle = -\chi^R_{\rho\rho}(\mathbf{q},\omega)\phi(\mathbf{q},\omega) + \chi^R_{\rho j_\parallel}(\mathbf{q},\omega)A_\parallel(\mathbf{q},\omega),$$
$$\langle J_\parallel(\mathbf{q},\omega)\rangle = \left[\chi^R_\parallel(\mathbf{q},\omega) - \frac{ne^2}{m}\right]A_\parallel(\mathbf{q},\omega) - \chi^R_{j_\parallel \rho}(\mathbf{q},\omega)\phi(\mathbf{q},\omega) \tag{3.189}$$

for the longitudinal part. Using Eqs. (3.184) and (3.185),

$$0 = (\omega + i\eta)\chi^R_{\rho\rho}(\mathbf{q},\omega) - |\mathbf{q}|\chi^R_{\rho j_\parallel}(\mathbf{q},\omega),$$
$$0 = |\mathbf{q}|\left[\chi^R_\parallel(\mathbf{q},\omega) - \frac{ne^2}{m}\right] - (\omega + i\eta)\chi^R_{j_\parallel \rho}(\mathbf{q},\omega) = 0, \tag{3.190}$$

we finally obtain

$$\delta\langle\rho(\mathbf{q},\omega)\rangle = \frac{1}{i|\mathbf{q}|}\chi^R_{\rho\rho}(\mathbf{q},\omega)E_\parallel(\mathbf{q},\omega),$$
$$\langle J_\parallel(\mathbf{q},\omega)\rangle = \frac{1}{i(\omega + i\eta)}\left[\chi^R_\parallel(\mathbf{q},\omega) - \frac{ne^2}{m}\right]E_\parallel(\mathbf{q},\omega), \tag{3.191}$$

where $E_\parallel(\mathbf{q},\omega) = (\mathbf{q}/|\mathbf{q}|)\cdot\mathbf{E}_\parallel(\mathbf{q},\omega) = -i|\mathbf{q}|\phi(\mathbf{q},\omega) + i(\omega + i\eta)A_\parallel(\mathbf{q},\omega)$. Equations (3.191) and (3.190) give the (gauge-invariant) longitudinal conductivity

$$\sigma_\parallel(\mathbf{q},\omega) = \frac{1}{i(\omega + i\eta)}\left[\chi^R_\parallel(\mathbf{q},\omega) - \frac{ne^2}{m}\right]$$
$$= \frac{1}{i|\mathbf{q}|}\chi^R_{j_\parallel \rho}(\mathbf{q},\omega) = \frac{\omega + i\eta}{i\mathbf{q}^2}\chi^R_{\rho\rho}(\mathbf{q},\omega). \tag{3.192}$$

[23]For symmetry reasons, the charge density does not couple to the transverse part of the current density.

Useful relations can be obtained from (3.184), (3.185) or (3.190) by considering the two special cases $\omega = 0, \mathbf{q} \neq 0$ and $\omega \neq 0, \mathbf{q} = 0$:

$$\chi^R_{\parallel}(\mathbf{q},0) - \frac{ne^2}{m} = 0 \quad \text{and} \quad \sum_\mu \frac{q_\mu}{|\mathbf{q}|}\chi^R_{\rho j_\mu}(\mathbf{q},0) = \chi^R_{\rho j_\parallel}(\mathbf{q},0) = 0 \quad \text{for} \quad \mathbf{q} \neq 0,$$
$$\chi^R_{\rho\rho}(0,\omega) = 0 \quad \text{and} \quad \chi^R_{j_\mu\rho}(0,\omega) = 0 \quad \text{for} \quad \omega \neq 0. \tag{3.193}$$

The first of these equations gives

$$\int_{-\infty}^{\infty} \frac{d\omega}{\pi} \frac{\chi''_{\parallel}(\mathbf{q},\omega)}{\omega} = \frac{ne^2}{m} \qquad (\mathbf{q} \neq 0), \tag{3.194}$$

which is nothing but the f-sum rule as can be seen by inserting the continuity equation for the Heisenberg operators, $\omega\hat{\rho}(\mathbf{q},\omega) - \mathbf{q}\cdot\hat{\mathbf{j}}_{\parallel}(\mathbf{q},\omega) = 0$, into Eq. (3.106).

It is easy to see that the second of Eqs. (3.193) also holds for the Matsubara correlation functions, i.e.

$$\chi_{\rho\rho}(0,i\omega_n) = 0 \quad \text{and} \quad \chi_{j_\mu\rho}(0,i\omega_n) = \chi_{\rho j_\mu}(0,-i\omega_n) = 0 \quad \text{for} \quad \omega_n \neq 0. \tag{3.195}$$

This follows from $\chi_{\rho\rho}(\mathbf{q},\tau) = \langle T_\tau\hat{\rho}(\mathbf{q},\tau)\hat{\rho}(-\mathbf{q},0)\rangle$ and $\chi_{\rho j_\mu}(\mathbf{q},\tau) = \langle T_\tau\hat{\rho}(\mathbf{q},\tau)\hat{\jmath}_\mu(-\mathbf{q},0)\rangle$ being independent of τ for $\mathbf{q} = 0$ since $\hat{\rho}(\mathbf{q}=0) = (e/\sqrt{\mathcal{V}})\hat{N}$ is nothing but the total particle number operator and therefore conserved: $[\hat{\rho}(\mathbf{q}=0),\hat{H}] = 0$ implies $\hat{\rho}(\mathbf{q}=0,\tau) = e^{\tau\hat{H}}\hat{\rho}(\mathbf{q}=0)e^{-\tau\hat{H}} = \hat{\rho}(\mathbf{q}=0)$.

3.4.4.3 *Kubo's formulae from the functional integral formalism*

It is instructive to derive the preceding results within the functional integral formalism, by considering the Euclidean action

$$S[\psi^*,\psi,A] = \int_0^\beta d\tau \int d^dr \sum_\sigma \psi^*_\sigma\Big[\partial_\tau - \mu - ieA_0 + \frac{1}{2m}(-i\boldsymbol{\nabla} - e\mathbf{A})^2\Big]\psi_\sigma + S_{\text{int}}[\psi^*,\psi], \tag{3.196}$$

where $A_0 = -\phi$. The current density is defined by

$$J_\mu(x) = -\frac{\delta S}{\delta A_\mu(x)} \qquad (\mu = 0, x, \dots) \tag{3.197}$$

so that $J_0(x) = j_0(x) = i\rho(x)$ whereas $J_\mu(x)$ corresponds to the usual definition when $\mu \neq 0$. We thus have

$$\langle J_\mu(x)\rangle = -\frac{1}{Z[A]}\int \mathcal{D}[\psi^*,\psi]\,\frac{\delta S}{\delta A_\mu(x)}\,e^{-S} = \frac{\delta \ln Z[A]}{\delta A_\mu(x)}, \tag{3.198}$$

where $Z[A]$ is the partition function. The linear response to the gauge field is therefore given by

$$\langle J_\mu(x)\rangle = \langle J_\mu(x)\rangle_{A=0} + \int_0^\beta d\tau' \int d^dr' \sum_\nu K_{\mu\nu}(x,x')A_\nu(x') + \mathcal{O}(A^2), \tag{3.199}$$

where

$$K_{\mu\nu}(x,x') = \frac{\delta^2 \ln Z[A]}{\delta A_\mu(x)\delta A_\nu(x')}\bigg|_{A=0}. \tag{3.200}$$

To obtain an explicit expression of the response function $K_{\mu\nu}$, we integrate out the field ψ in a cumulant expansion,

$$\begin{aligned} Z[A] &= \int \mathcal{D}[\psi^*,\psi]\, e^{-S[\psi^*,\psi,A=0]+\int_0^\beta d\tau \int d^d r \left(\sum_\mu j_\mu A_\mu + i\frac{ej_0}{2m}\mathbf{A}^2\right)} \\ &= Z[0]\, e^{\int_0^\beta d\tau \int d^d r \sum_\mu \langle j_\mu\rangle A_\mu + \frac{1}{2}\int_0^\beta d\tau d\tau' \int d^d r d^d r' \sum_{\mu,\nu} K_{\mu\nu}(x,x')A_\mu(x)A_\nu(x') + \mathcal{O}(A^3)}, \end{aligned} \tag{3.201}$$

where $K_{\mu\nu}(x,x')$ is defined by its Fourier transform

$$K_{\mu\nu}(q) = \chi_{j_\mu j_\nu}(q) - \frac{ne^2}{m}\delta_{\mu,\nu}(1-\delta_{\mu,0}) \tag{3.202}$$

and all expectation values are now taken with the action $S[\psi^*,\psi,A=0]$. Performing the analytic continuation to real frequencies with $i\phi = -iA_0 \to \phi$, and noting that $\chi_{j_0 j_0} = -\chi_{\rho\rho}$ and $\chi_{j_0 j_\mu} = i\chi_{\rho j_\mu}$ $(\mu \neq 0)$, we reproduce (3.181).

Since the action (3.196) is invariant in the gauge transformation

$$\begin{aligned} \psi_\sigma(x) \to e^{ie\Lambda(x)}\psi_\sigma(x), \qquad \psi^*_\sigma(x) \to e^{-ie\Lambda(x)}\psi^*_\sigma(x), \\ A_\mu(x) \to A_\mu(x) + \partial_\mu\Lambda(x), \end{aligned} \tag{3.203}$$

the response to a pure gauge field $A_\mu = \partial_\mu\Lambda$ must vanish, i.e.

$$\begin{aligned} 0 &= \int_0^\beta d\tau' \int d^d r' \sum_\nu K_{\mu\nu}(x,x')\partial_{x'_\nu}\Lambda(x') \\ &= -\int_0^\beta d\tau' \int d^d r' \sum_\nu \partial_{x'_\nu} K_{\mu\nu}(x,x')\Lambda(x'), \end{aligned} \tag{3.204}$$

which implies that the response function is transverse,

$$\sum_\nu \partial_{x'_\nu} K_{\mu\nu}(x,x') = 0. \tag{3.205}$$

In Fourier space, this gives

$$0 = -\omega_n K_{\mu 0}(q) + \sum_{\nu\neq 0} q_\nu K_{\mu\nu}(q). \tag{3.206}$$

The analytic continuation of this equation yields Eqs. (3.184).

Finally, we note that the continuity equation implies $\sum_\mu \partial_\mu \langle J_\mu(x)\rangle = 0$ and therefore

$$\sum_\mu \partial_\mu K_{\mu\nu}(x,x') = 0. \tag{3.207}$$

The equivalence between (3.205) and (3.207) follows from the property $K_{\mu\nu}(q) = K_{\nu\mu}(-q)$.

3.4.4.4 *Longitudinal-conductivity sum rule*

From Eq. (3.183), we deduce

$$\begin{aligned}\Re[\sigma_\parallel(\mathbf{q},\omega)] &= \Re\left[\mathcal{P}\frac{\chi_\parallel^R(\mathbf{q},\omega)}{i\omega}\right] - \pi\delta(\omega)\left\{\Re[\chi_\parallel^R(\mathbf{q},\omega)] - \frac{ne^2}{m}\right\} \\ &= \mathcal{P}\frac{\chi_\parallel''(\mathbf{q},\omega)}{\omega} - \pi\delta(\omega)\left\{\Re[\chi_\parallel^R(\mathbf{q},\omega)] - \frac{ne^2}{m}\right\}, \end{aligned} \tag{3.208}$$

where we have used the spectral representation (3.32) and the fact that $\chi_\parallel''(\mathbf{q},\omega) = \Im[\chi_\parallel^R(\mathbf{q},\omega)]$ is real [Eq. (3.42)]. The term $\delta(\omega)\chi_\parallel^R(\mathbf{q},\omega)$ in Eq. (3.208) has a different meaning whether $\mathbf{q} \neq 0$ or $\mathbf{q} = 0$. In the former case, it reduces to $\delta(\omega)\chi_\parallel^R(\mathbf{q},0)$, whereas in the latter case it should be interpreted as $\delta(\omega)\lim_{\omega\to 0}\chi^R(0,\omega)$ — as follows from taking the limit $\mathbf{q} = 0$ before $\eta \to 0^+$.

Equation (3.208) implies

$$\int_{-\infty}^{\infty}\frac{d\omega}{2\pi}\Re[\sigma_\parallel(\mathbf{q},\omega)] = \frac{ne^2}{2m} = \frac{\epsilon_0}{2}\omega_p^2. \tag{3.209}$$

For $\mathbf{q} \neq 0$, we have used the relation (3.44) between $\chi_\parallel^R(\mathbf{q},0)$ and $\chi_\parallel''(\mathbf{q},\omega)$ whereas for $\mathbf{q} = 0$, we note that $\chi_\parallel^R(0,\omega)$ and $\chi_\parallel''(0,\omega)$ vanish in a translation-invariant system where the total current $\hat{\mathbf{j}}(\mathbf{q} = 0)$ is conserved.[24] The longitudinal-conductivity sum rule (3.209), which involves the response to the external electric field, is identical to the sum rule (3.179) satisfied by the longitudinal response to the local field. However, the latter assumes the analyticity of $\epsilon_\parallel^R(\mathbf{q},\omega)$ in the upper-half complex plane, whereas the sum rule (3.209) is always valid.

3.4.4.5 *Zero-frequency conductivity and Drude weight*

For a uniform field ($\mathbf{q} = 0$), we cannot distinguish between longitudinal and transverse response functions, so that $\chi_\parallel^R(0,\omega) = \chi_\perp^R(0,\omega) \equiv \chi^R(0,\omega)$ and

$$\sigma_\parallel(0,\omega) = \sigma_\perp(0,\omega) \equiv \sigma(0,\omega). \tag{3.210}$$

The DC conductivity is defined by $\sigma_{\rm DC} = \lim_{\omega\to 0}\sigma(0,\omega)$. By taking the $\mathbf{q} \to 0$ limit first, we ensure that we are looking at an infinite volume, where energy levels can be arbitrarily close in energy, so that absorption remains possible even when $\omega \to 0$, the latter property being necessary to have a metallic (conducting) state. Recalling the definition of $\sigma_\parallel(\mathbf{q},\omega)$ and $\sigma_\perp(\mathbf{q},\omega)$ given at the beginning of Sec. 3.4.4, Eq. (3.210) implies that the local and external fields coincide, as expected when $\mathbf{q} = 0$.[25]

---

[24] See footnote 19 on page 268.

[25] The absence of screening of the longitudinal field, i.e. $\langle\phi(0,\omega)\rangle = \phi_{\rm ext}(0,\omega)$, comes from the vanishing of the $\mathbf{q} = 0$ component of the Coulomb interaction in the jellium model, so that $\epsilon_\parallel(0,\omega) = 1$ (note that $\epsilon_\parallel(0,\omega) \neq \lim_{\mathbf{q}\to 0}\epsilon_\parallel(\mathbf{q},\omega)$).

Equation (3.208) gives

$$\Re[\sigma(0,\omega)] = \mathcal{P}\frac{\chi''(0,\omega)}{\omega} + D\delta(\omega), \tag{3.211}$$

where the coefficient of the delta function,

$$D = \pi \lim_{\omega\to 0}\left\{\frac{ne^2}{m} - \Re[\chi^R(0,\omega)]\right\}, \tag{3.212}$$

is called the Drude weight. As pointed out above, $\chi^R(0,\omega)$ vanishes in translation-invariant systems and

$$\sigma(0,\omega) = \frac{D}{\pi}\frac{i}{\omega + i\eta} \quad \text{with} \quad D = \pi\frac{ne^2}{m}. \tag{3.213}$$

$\Re[\sigma(0,\omega)] = D\delta(\omega)$ and the Drude weight exhausts all the spectral weight in the conductivity sum rule (3.209) (perfect conductor). When translation invariance of the electron system is broken, for instance, by interactions with phonons or umklapp processes made possible by the crystal lattice, $\Re[\chi^R(0,\omega)]$ is finite and reduces the Drude weight. Part of the spectral weight is then pushed up to finite frequencies in agreement with the sum rule (3.209). A metal is defined by a finite Drude weight,[26] whereas an insulator corresponds to a vanishing Drude weight (Fig. 3.11), i.e.

$$\lim_{\omega\to 0}\Re[\chi^R(0,\omega)] = \frac{ne^2}{m}. \tag{3.214}$$

Together with the f-sum rule (3.193), this result means that

$$\lim_{\omega\to 0}\lim_{\mathbf{q}\to 0}\Re[\chi^R_\parallel(\mathbf{q},\omega)] = \lim_{\mathbf{q}\to 0}\lim_{\omega\to 0}\Re[\chi^R_\parallel(\mathbf{q},\omega)] \tag{3.215}$$

is independent of the order of the limits $\mathbf{q}\to 0$ and $\omega\to 0$. It is easy to show that a sufficient condition for (3.215) to hold is the presence of a gap in the longitudinal excitation

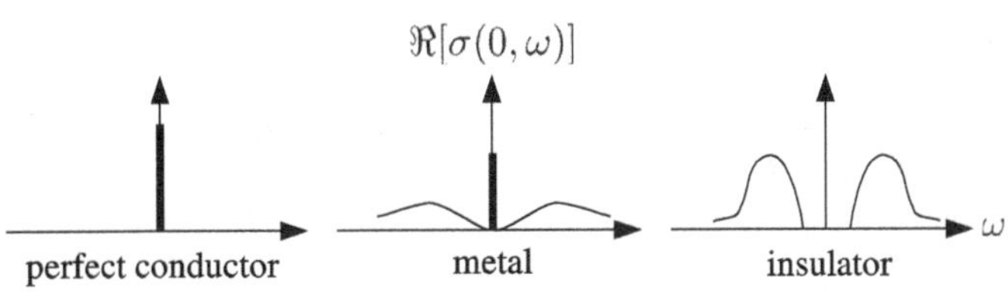

Fig. 3.11 Schematic representation of $\Re[\sigma(0,\omega)]$ in a perfect conductor ($D = \pi ne^2/m$), a metal ($0 < D < \pi ne^2/m$) and an insulator ($D = 0$).

[26] In a disordered conductor, the Drude peak $D\delta(\omega)$ broadens into $D\frac{\tau_{\rm tr}}{1-i\omega\tau_{\rm tr}}$, where $\tau_{\rm tr}$ is the transport scattering time.

spectrum, i.e. $\chi''_{\parallel}(\mathbf{q},\omega)=0$ for $|\omega|<\Delta$ (regardless of the value of $\mathbf{q}$). From the Kramers–Kronig relations, we obtain

$$\Re[\chi^R_{\parallel}(\mathbf{q},\omega)]=\mathcal{P}\int_{-\infty}^{\infty}\frac{d\omega'}{\pi}\frac{\chi''_{\parallel}(\mathbf{q},\omega')}{\omega'-\omega}. \tag{3.216}$$

For $|\omega|<\Delta$, there is no singularity in this integral, and the limits $\omega\to 0$ and $\mathbf{q}\to 0$ are interchangeable,

$$\lim_{\omega\to 0}\lim_{\mathbf{q}\to 0}\Re[\chi^R_{\parallel}(\mathbf{q},\omega)]=\lim_{\mathbf{q}\to 0}\lim_{\omega\to 0}\Re[\chi^R_{\parallel}(\mathbf{q},\omega)]=2\int_{\Delta}^{\infty}\frac{d\omega'}{\pi}\frac{\chi''_{\parallel}(0,\omega')}{\omega'}. \tag{3.217}$$

3.4.5 *Superconductors*

Superfluidity and superconductivity will be the subject of Chapter 7. Here, we give a first introduction to superconductivity.

3.4.5.1 *Landau diamagnetism in normal metals*

While gauge invariance implies the f-sum rule $\chi^R_{\parallel}(\mathbf{q},0)=ne^2/m$, there is *a priori* no constraint on $\chi^R_{\perp}(\mathbf{q},0)$. However, if the current–current correlation function is short-range, i.e. if it decays fast enough in real space, the $\mathbf{q}\to 0$ limit should not be singular:

$$\lim_{\mathbf{q}\to 0}\chi^R_{\perp}(\mathbf{q},0)=\lim_{q_y\to 0}\lim_{q_x\to 0}\chi^R_{j_xj_x}(\mathbf{q},0)=\lim_{q_x\to 0}\lim_{q_y\to 0}\chi^R_{j_xj_x}(\mathbf{q},0)=\frac{ne^2}{m} \tag{3.218}$$

($q_z,\dots=0$ in this equation), where the last result follows from the f-sum rule. In the long wavelength limit, we thus have $\epsilon_0^{-1}\chi^R_{\perp}(\mathbf{q},0)=\omega_p^2-\alpha\mathbf{q}^2+\mathcal{O}(|\mathbf{q}|^4)$. The screening of a slowly varying external static magnetic field is determined by (3.170),

$$\langle\mathbf{A}_{\perp}(\mathbf{q},0)\rangle=\frac{c_l^2\mathbf{q}^2}{c_l^2\mathbf{q}^2+\omega_p^2-\epsilon_0^{-1}\chi^R_{\perp}(\mathbf{q},0)}\mathbf{A}_{\text{ext}\perp}(\mathbf{q},0)\simeq\frac{c_l^2}{c_l^2+\alpha}\mathbf{A}_{\text{ext}\perp}(\mathbf{q},0). \tag{3.219}$$

The current $\langle\mathbf{J}_{\perp}\rangle$ produced by a static magnetic field is related to the orbital magnetization density $\langle\mathbf{M}\rangle$ by $\langle\mathbf{J}_{\perp}\rangle=\boldsymbol{\nabla}\times\langle\mathbf{M}\rangle$, i.e.

$$\left[\chi^R_{\perp}(\mathbf{q},0)-\frac{ne^2}{m}\right]\langle\mathbf{A}_{\perp}(\mathbf{q},0)\rangle=i\mathbf{q}\times\langle\mathbf{M}(\mathbf{q},0)\rangle. \tag{3.220}$$

Using (3.219) and taking the curl, one finds

$$\langle\mathbf{M}(\mathbf{q},0)\rangle=-\frac{\alpha\epsilon_0c_l^2}{c_l^2+\alpha}\mathbf{B}_{\text{ext}}(\mathbf{q},0)\simeq-\alpha\epsilon_0\mathbf{B}_{\text{ext}}(\mathbf{q},0)\qquad(\mathbf{q}\to 0), \tag{3.221}$$

and the orbital susceptibility

$$\chi_{\text{orb}}=-\alpha\epsilon_0=\lim_{\mathbf{q}\to 0}\frac{1}{\mathbf{q}^2}\left[\chi^R_{\perp}(\mathbf{q},0)-\frac{ne^2}{m}\right]. \tag{3.222}$$

In normal metals, χ_{orb} is negative (Landau diamagnetism).[27]

[27] Note that a negative orbital susceptibility is not in contradiction with the stability of the system [6], which requires $\Re[\sigma_{\perp}(\mathbf{q},\omega)]\geq 0$ (see Sec. 3.4.5.3).

For free electrons, $\chi_\perp^R(\mathbf{q},0)=\chi_\perp(\mathbf{q},i\omega_\nu=0)$ is given by

$$\begin{aligned}\chi_\perp^R(\mathbf{q},0) &= \langle j_x(\mathbf{q},i\omega_\nu=0)j_x(-\mathbf{q},i\omega_\nu=0)\rangle \\ &= -\frac{2e^2}{\beta V m^2}\sum_{\mathbf{k},\omega_n} k_x^2 (i\omega_n-\xi_\mathbf{k})^{-1}(i\omega_n-\xi_{\mathbf{k}+\mathbf{q}})^{-1} \\ &= -\frac{2e^2}{m^2}\int_\mathbf{k} k_x^2 \frac{n_F(\xi_\mathbf{k})-n_F(\xi_{\mathbf{k}+\mathbf{q}})}{\xi_\mathbf{k}-\xi_{\mathbf{k}+\mathbf{q}}}, \end{aligned} \tag{3.223}$$

where we have assumed $\mathbf{q}\perp\hat{\mathbf{x}}$ and used the expression of the current operator given in footnote 15 on page 259. An elementary calculation gives

$$\chi_\perp^R(\mathbf{q},0)=\frac{ne^2}{m}\left(1-\frac{\mathbf{q}^2}{4k_F^2}\right)+\mathcal{O}(\mathbf{q}^4), \tag{3.224}$$

for $T\to 0$ and $d=3$, and

$$\chi_{\rm orb}=-\frac{ne^2}{4mk_F^2}=-\frac{1}{3}\chi_P. \tag{3.225}$$

$\chi_P=(g\mu_B/2)^2 2N(0)$ is the (paramagnetic) Pauli susceptibility (with $g\simeq 2$) and $\mu_B=e/2m$ the Bohr magneton. The calculation of $\chi_P=(g\mu_B)^2\chi_{S^zS^z}(\mathbf{q}=0, i\omega_\nu=0)$ is analogous to that of the charge susceptibility discussed in Sec. 3.2.7. ($S^z(\mathbf{r})=\frac{1}{2}\sum_\sigma \psi_\sigma^*(\mathbf{r})\sigma\psi_\sigma(\mathbf{r})$ is the z-component of the spin density.)

3.4.5.2 *Long-range current–current correlations and Meissner effect in superconductors*

When the current–current correlations are long range, Eq. (3.218) can be violated, i.e.

$$\lim_{\mathbf{q}\to 0}\chi_\perp^R(\mathbf{q},0)=\frac{(n-n_s)e^2}{m}. \tag{3.226}$$

Equation (3.226) yields the London equation

$$\langle \mathbf{J}_\perp(\mathbf{q},0)\rangle = -\frac{n_s e^2}{m}\langle \mathbf{A}_\perp(\mathbf{q},0)\rangle \tag{3.227}$$

if we approximate $\chi_\perp^R(\mathbf{q},0)$ by its $\mathbf{q}\to 0$ limit. We shall see in Chapter 7 that the "superfluid density" n_s can be interpreted as the density of superfluid particles. We show in the following that $0\le n_s\le n$.

The screening of a slowly varying static external magnetic field is then derived from[28]

$$\langle \mathbf{A}_\perp(\mathbf{q},0)\rangle = \frac{\mathbf{q}^2}{\mathbf{q}^2+\lambda_L^{-2}}\langle \mathbf{A}_{\perp\rm ext}(\mathbf{q},0)\rangle \tag{3.228}$$

[28]Equation (3.228) is valid in the London limit where the superconducting coherence length ξ is much smaller than λ_L. In the (non-local) Pippard limit $\xi\gg\lambda_L$, the $\mathbf{q}$ dependence of $\chi_\perp^R(\mathbf{q},0)$ cannot be neglected (Chapter 7).

[see Eq. (3.170)], where

$$\lambda_L = c\left(\frac{\epsilon_0 m}{n_s e^2}\right)^{1/2} = \frac{c}{\omega_p}\sqrt{\frac{n}{n_s}} \tag{3.229}$$

is the London penetration depth. In real space, (3.228) becomes $\nabla \times (\nabla \times \langle \mathbf{A}_\perp \rangle) + \lambda_L^{-2}\langle \mathbf{A}_\perp \rangle = \nabla \times (\nabla \times \langle \mathbf{A}_{\perp \rm ext} \rangle)$. Taking the curl and using $\nabla \times (\nabla \times \langle \mathbf{B} \rangle) = -\nabla^2 \langle \mathbf{B} \rangle$, we finally obtain[29]

$$\nabla^2 \langle \mathbf{B}(\mathbf{r}) \rangle = \lambda_L^{-2} \langle \mathbf{B}(\mathbf{r}) \rangle \tag{3.230}$$

for a uniform external field $\mathbf{B}_{\rm ext}$. Let us assume the superconductor to occupy the half space $z > 0$ and the external field $\mathbf{B}_{\rm ext}$ to be parallel to the interface between the superconductor and the vacuum. From (3.230) and the boundary condition $\langle \mathbf{B}(z=0) \rangle = \mathbf{B}_{\rm ext}$ ($\langle \mathbf{B} \rangle = \langle \mathbf{B}_{\rm ext} \rangle$ outside the superconductor), we find that the external field is screened in the superconductor ($z > 0$) over a characteristic length λ_L,

$$\langle \mathbf{B}(z) \rangle = \mathbf{B}_{\rm ext}\, e^{-z/\lambda_L}. \tag{3.231}$$

From the Maxwell equation $\nabla \times \langle \mathbf{B} \rangle = \mu_0 \langle \mathbf{J}_\perp \rangle$, we deduce the existence of a current in a layer of width λ_L in the superconductor,

$$\langle \mathbf{J}_\perp(z) \rangle = -\frac{1}{\mu_0 \lambda_L} \hat{\mathbf{z}} \times \mathbf{B}_{\rm ext}\, e^{-z/\lambda_L}. \tag{3.232}$$

In the bulk of the superconductor ($z \gg \lambda_L$), the magnetic field is completely expelled. This perfect diamagnetism (Meissner effect) is the hallmark of a superconductor. We will see in Chapter 7 that long-range current–current correlations are at the root of superfluidity defined as the ability of a system to sustain a (dissipationless) superflow.

3.4.5.3 *Transverse conductivity and superfluid weight*

The analog of (3.208) in the transverse case gives

$$\lim_{\mathbf{q}\to 0} \Re[\sigma_\perp(\mathbf{q},\omega)] = \lim_{\mathbf{q}\to 0} \mathcal{P}\frac{\chi''_\perp(\mathbf{q},\omega)}{\omega} + D_s \delta(\omega), \tag{3.233}$$

where

$$D_s = \pi \lim_{\mathbf{q}\to 0}\left[\frac{ne^2}{m} - \chi^R_\perp(\mathbf{q},0)\right] = \frac{\pi n_s e^2}{m} \tag{3.234}$$

is the superfluid weight. A superconductor is therefore characterized by a non-zero weight D_s of the zero-frequency delta peak in the transverse conductivity $\lim_{\mathbf{q}\to 0} \Re[\sigma_\perp(\mathbf{q},\omega)]$. Note

[29] Equation (3.230) can also be obtained from the London equation (3.229) and the Maxwell equation $\nabla \times \langle \mathbf{B} \rangle = \mu_0 \langle \mathbf{J}_\perp \rangle$.

Table 3.2 Drude weight D and superfluid weight D_s in a metal, an insulator and a superconductor.

	D	D_s
Metal	D	0
Insulator	0	0
Superconductor	D	D_s

the different order of the limits $\mathbf{q} \to 0$ and $\omega \to 0$ in the respective definitions of D and D_s. Equations (3.226) and (3.233) yield

$$\lim_{\mathbf{q}\to 0} \int_{-\infty}^{\infty} \frac{d\omega}{2\pi} \Re[\sigma_\perp(\mathbf{q}, \omega)] = \frac{1}{2} \lim_{\mathbf{q}\to 0} \chi_\perp^R(\mathbf{q}, 0) + \frac{D_s}{2\pi} = \frac{\epsilon_0}{2} \omega_p^2, \tag{3.235}$$

which is the $\mathbf{q} \to 0$ limit of the transverse-conductivity sum rule (3.179). The stability of the system implies $\Re[\sigma_\perp(\mathbf{q}, \omega)] \geq 0$; the system can only absorb energy from the external field.[30] In particular, $D_s \geq 0$ and therefore $n_s \geq 0$. Furthermore, the constraint $\chi_\perp^R(\mathbf{q}, 0) \geq 0$ together with the sum rule (3.235) gives $n_s \leq n$. In Chapter 7, we shall see that $n_s(T = 0) = n$ in Galilean-invariant systems.

One can show that when there is a gap in the transverse excitation spectrum, D and D_s are equal. This is the case, for instance, in an s-wave superconductor where quasi-particle excitations are gapped while the Anderson–Bogoliubov mode appears in the longitudinal channel (Chapter 7). The same argument as that used to derive (3.217) gives

$$\lim_{\mathbf{q}\to 0} \lim_{\omega\to 0} \Re[\chi_\perp^R(\mathbf{q}, \omega)] = \lim_{\omega\to 0} \lim_{\mathbf{q}\to 0} \Re[\chi_\perp^R(\mathbf{q}, \omega)] = \lim_{\omega\to 0} \Re[\chi^R(0, \omega)] \tag{3.236}$$

and in turn $D_s = D$. The differences between a metal, an insulator and a superconductor are summarized in Table 3.2.

3.5 One-Particle Green Function

Apart from the correlation functions determining the linear response to external fields, such as the density–density or electromagnetic response functions discussed in Secs. 3.3 and 3.4, the one-particle Green function is the most important correlation function in a many-particle quantum system. On the one hand, it is related to many observables (particle and current densities, mean kinetic energy, etc.). On the other hand, it provides us with information

[30]When the external perturbation is an electromagnetic field, the calculation of the dissipation in Sec. 3.2.2 must be modified to include the diamagnetic term (which is quadratic in the vector potential). It is not difficult to show that the dissipation is related to the real part of the conductivity and the stability of the system requires $\Re[\sigma_{\parallel,\perp}(\mathbf{q}, \omega)] \geq 0$. On the other hand, the properties $\chi_{\parallel,\perp}^R(\mathbf{q}, 0) \geq 0$ follow from the definition of $\chi''_{\parallel,\perp}(\mathbf{q}, \omega)$ (see Sec. 3.2.3).

about the propagation of an additional particle in the system and the possible existence of quasi-particle excitations, and can actually be measured in experiments where a single particle is injected into or extracted from the system. While the existence of quasi-particles in the case of fermions is discussed at length in Chapter 4 on Fermi-liquid theory, in the following, we focus on the formal properties of the one-particle Green function.

3.5.1 *Spectral function*

The one-particle Green function is defined by

$$\begin{aligned} G_{\sigma\sigma'}(\mathbf{r},\tau;\mathbf{r}',\tau') &= -\langle T_\tau \hat\psi_\sigma(\mathbf{r},\tau)\hat\psi^\dagger_{\sigma'}(\mathbf{r}',\tau')\rangle, \\ F_{\sigma\sigma'}(\mathbf{r},\tau;\mathbf{r}',\tau') &= -\langle T_\tau \hat\psi_\sigma(\mathbf{r},\tau)\hat\psi_{\sigma'}(\mathbf{r}',\tau')\rangle, \\ F^\dagger_{\sigma\sigma'}(\mathbf{r},\tau;\mathbf{r}',\tau') &= -\langle T_\tau \hat\psi^\dagger_\sigma(\mathbf{r},\tau)\hat\psi^\dagger_{\sigma'}(\mathbf{r}',\tau')\rangle, \end{aligned} \tag{3.237}$$

where the anomalous Green functions F and $F^\dagger$ are non-zero if the global U(1) symmetry is broken, i.e. if the system is superfluid (Chapter 7). In the following, we focus on the normal (non-superfluid) component G, assumed to be translation invariant and diagonal in the spin indices. We thus consider

$$G_\sigma(\mathbf{k},i\omega_n) = \int_0^\beta d\tau \int d^dr\, e^{-i(\mathbf{k}\cdot\mathbf{r}-\omega_n\tau)} G_\sigma(\mathbf{r}-\mathbf{r}',\tau-\tau'). \tag{3.238}$$

The one-particle Green function $G_\sigma(\mathbf{k},z)$, considered as a function of the complex frequency z, admits the spectral representation (3.33), i.e.

$$G_\sigma(\mathbf{k},z) = \int_{-\infty}^{\infty} d\omega \frac{A_\sigma(\mathbf{k},\omega)}{z-\omega}, \tag{3.239}$$

where the spectral function is defined by

$$\begin{aligned} A_\sigma(\mathbf{k},t) &= \frac{1}{2\pi}\langle[\hat\psi_\sigma(\mathbf{k},t),\hat\psi^\dagger_\sigma(\mathbf{k},0)]_{-\zeta}\rangle, \\ A_\sigma(\mathbf{k},\omega) &= -\frac{1}{\pi}\Im[G^R_\sigma(\mathbf{k},\omega)]. \end{aligned} \tag{3.240}$$

The Matsubara Green function $G_\sigma(\mathbf{k},i\omega_n)$ corresponds to $z=i\omega_n$ while the retarded and advanced correlation functions $G^{R,A}_\sigma(\mathbf{k},\omega)$ are obtained for $z=\omega\pm i\eta$ ($\eta\to 0^+$). The spectral function A_σ of the one-particle propagator is usually defined with a prefactor $-1/\pi$ as in (3.240) which explains why (3.239) slightly differs from (3.33). A_σ can be expressed in terms of a complete set $\{|n\rangle,\epsilon_n\}$ of eigenvectors of the (grand-canonical) Hamiltonian [see, e.g., Eq. (3.37) or (3.70)],

$$A_\sigma(\mathbf{k},\omega) = \frac{1}{Z}\sum_{n,m} e^{-\beta\epsilon_n}(1-\zeta e^{-\beta\omega})|\langle n|\hat\psi_\sigma(\mathbf{k})|m\rangle|^2\delta(\omega+\epsilon_n-\epsilon_m). \tag{3.241}$$

$A_\sigma(\mathbf{k},\omega)$ is positive for fermions ($\zeta=-1$) and has the sign of ω for bosons ($\zeta=1$).[31]

[31] The real-time Green function $G_\sigma(\mathbf{k},\omega) = -i\langle T\hat\psi_\sigma(\mathbf{k},t)\hat\psi^\dagger_\sigma(\mathbf{k},0)\rangle$ (that we do not discuss here) is given by $G_\sigma(\mathbf{k},\omega) = [1+\zeta n_\zeta(\omega)]G^R_\sigma(\mathbf{k},\omega) - \zeta n_\zeta(\omega)G^A(\mathbf{k},\omega)$ (see Sec. 3.2.7.2).

For a non-interacting system, one simply has $G_0(\mathbf{k}, i\omega_n) = (i\omega_n - \xi_\mathbf{k})^{-1}$ (assuming the one-particle energy $\xi_\mathbf{k} = \epsilon_\mathbf{k} - \mu$ to be spin independent) and the spectral function $A_0(\mathbf{k}, \omega) = \delta(\omega - \xi_\mathbf{k})$ reduces to a Dirac function. In an interacting normal system, where the global U(1) symmetry is not broken ($F = F^\dagger = 0$), one usually expresses the Green function $G_\sigma(\mathbf{k}, z) = [z - \xi_\mathbf{k} - \Sigma_\sigma(\mathbf{k}, z)]^{-1}$ by introducing the self-energy $\Sigma_\sigma(\mathbf{k}, z)$ (Sec. 1.6.2) and the spectral function takes the form

$$A_\sigma(\mathbf{k}, \omega) = -\frac{1}{\pi} \frac{\Im[\Sigma_\sigma^R(\mathbf{k}, \omega)]}{(\omega - \xi_\mathbf{k} - \Re[\Sigma_\sigma^R(\mathbf{k}, \omega)])^2 + (\Im[\Sigma_\sigma^R(\mathbf{k}, \omega)])^2}, \tag{3.242}$$

where $\Sigma_\sigma^R(\mathbf{k}, \omega) = \Sigma_\sigma(\mathbf{k}, z = \omega + i\eta)$. The consequences of this equation for fermion systems are discussed at length in Chapter 4 in the context of Fermi-liquid theory. In superfluid systems, the self-energy has both normal and anomalous parts [see Eq. (3.252)] and the expression of the spectral function is more complicated (Chapters 7 and 13).

3.5.2 *Analyticity properties*

It follows from the spectral representation (3.239) that $G_\sigma(\mathbf{k}, z)$ is an analytic function of z when $\Im(z) \neq 0$ whereas $G_\sigma^R(\mathbf{k}, \omega)$ and $G_\sigma^A(\mathbf{k}, \omega)$ are analytic in the upper- and lower-half complex planes, respectively. $G_\sigma^R(\mathbf{k}, \omega)$ therefore satisfies the Kramers–Kronig relations (3.52). $G_\sigma(\mathbf{k}, z)$ is discontinuous across the real axis whenever the spectral function $A_\sigma(\mathbf{k}, \omega) = -[G_\sigma(\mathbf{k}, \omega + i\eta) - G_\sigma(\mathbf{k}, \omega - i\eta)]/(2i\pi)$ is non-zero.

The self-energy $\Sigma_\sigma(\mathbf{k}, z)$ inherits the analyticity properties of the Green function: It is analytic for $\Im(z) \neq 0$, the retarded part $\Sigma_\sigma^R(\mathbf{k}, \omega)$ is analytic in the upper-half complex plane and satisfies the Kramers–Kronig relations[32]

$$\begin{aligned} \Re[\Sigma_\sigma^R(\mathbf{k}, \omega) - \Sigma_\sigma^R(\mathbf{k}, \omega \to \infty)] &= \mathcal{P} \int_{-\infty}^{\infty} \frac{d\omega'}{\pi} \frac{\Im[\Sigma_\sigma^R(\mathbf{k}, \omega')]}{\omega' - \omega}, \\ \Im[\Sigma_\sigma^R(\mathbf{k}, \omega)] &= -\mathcal{P} \int_{-\infty}^{\infty} \frac{d\omega'}{\pi} \frac{\Re[\Sigma_\sigma^R(\mathbf{k}, \omega') - \Sigma_\sigma^R(\mathbf{k}, \omega \to \infty)]}{\omega' - \omega}. \end{aligned} \tag{3.243}$$

Furthermore, the sign of the spectral function $A_\sigma(\mathbf{k}, \omega)$ implies $\Im[\Sigma_\sigma^R(\mathbf{k}, \omega)] \leq 0$ for fermions and $\text{sgn}(\Im[\Sigma_\sigma^R(\mathbf{k}, \omega)]) = -\text{sgn}(\omega)$ for bosons.

3.5.3 *Sum rules and high-energy behavior*

Sum rules are given by Eq. (3.63), i.e.

$$\int_{-\infty}^{\infty} d\omega\, \omega^n A_\sigma(\mathbf{k}, \omega) = \left\langle \left[[[\hat\psi_\sigma(\mathbf{k}), \hat H], \hat H], \cdots, \hat\psi_\sigma^\dagger(\mathbf{k}) \right]_{-\zeta} \right\rangle \tag{3.244}$$

[32] $\Sigma_\sigma^R(\mathbf{k}, \omega \to \infty)$ is real (see Sec. 3.5.3).

($\hat{H}$ appears n times). For $n = 0$, one concludes that the spectral function is normalized to unity,

$$\int_{-\infty}^{\infty} d\omega\, A_\sigma(\mathbf{k}, \omega) = \langle [\hat{\psi}_\sigma(\mathbf{k}), \hat{\psi}_\sigma^\dagger(\mathbf{k})]_{-\varsigma} \rangle = 1. \tag{3.245}$$

For $n = 1$, Eq. (3.244) gives

$$\int_{-\infty}^{\infty} d\omega\, \omega A_\sigma(\mathbf{k}, \omega) = \langle \big[[\hat{\psi}_\sigma(\mathbf{k}), \hat{H}], \hat{\psi}_\sigma^\dagger(\mathbf{k})\big]_{-\varsigma} \rangle. \tag{3.246}$$

We thus deduce the high-frequency behavior [see Eq. (3.64)]

$$G_\sigma(\mathbf{k}, z) = \frac{1}{z} + \frac{1}{z^2} \langle \big[[\hat{\psi}_\sigma(\mathbf{k}), \hat{H}], \hat{\psi}_\sigma^\dagger(\mathbf{k})\big]_{-\varsigma} \rangle + \mathcal{O}\left(\frac{1}{z^3}\right). \tag{3.247}$$

For a normal (non-superfluid) system, this leads in turn to the high-frequency limit of the self-energy,

$$\Sigma_\sigma(\mathbf{k}, z) = -\xi_\mathbf{k} + \langle \big[[\hat{\psi}_\sigma(\mathbf{k}), \hat{H}], \hat{\psi}_\sigma^\dagger(\mathbf{k})\big]_{-\varsigma} \rangle + \mathcal{O}\left(\frac{1}{z}\right). \tag{3.248}$$

As an example, consider a spin-$\frac{1}{2}$ fermion system with Hamiltonian

$$\hat{H} = \sum_{\mathbf{k},\sigma} \xi_\mathbf{k} \hat{\psi}_\sigma^\dagger(\mathbf{k}) \hat{\psi}_\sigma(\mathbf{k}) + \frac{g}{2\mathcal{V}} \sum_{\substack{\mathbf{k},\mathbf{k}',\mathbf{q} \\ \sigma,\sigma'}} \hat{\psi}_\sigma^\dagger(\mathbf{k}+\mathbf{q}) \hat{\psi}_{\sigma'}^\dagger(\mathbf{k}'-\mathbf{q}) \hat{\psi}_{\sigma'}(\mathbf{k}') \hat{\psi}_\sigma(\mathbf{k}). \tag{3.249}$$

One easily obtains[33]

$$\lim_{|z|\to\infty} \Sigma_\sigma(\mathbf{k}, z) = g \langle \hat{\psi}_{\bar{\sigma}}^\dagger(\mathbf{r}) \hat{\psi}_{\bar{\sigma}}(\mathbf{r}) \rangle = \frac{gn}{2} \tag{3.250}$$

(with the notation $\bar{\sigma} = -\sigma$), where the last result holds for a spin-rotation-invariant system (with n the mean particle density). For spin-zero bosons, one finds a similar expression,

$$\lim_{|z|\to\infty} \Sigma(\mathbf{k}, z) = 2g \langle \hat{\psi}^\dagger(\mathbf{r}) \hat{\psi}(\mathbf{r}) \rangle = 2gn. \tag{3.251}$$

In both cases, spin-$\frac{1}{2}$ fermions and spin-zero bosons, the high-frequency limit of the self-energy is given by the Hartree–Fock result (i.e. the first-order self-consistent result). This result extends to the superfluid phase with broken global U(1) symmetry. For instance, for a superfluid boson system, the propagator is given by

$$\mathcal{G}^{-1}(\mathbf{k}, i\omega_n) = \begin{pmatrix} i\omega + \mu - \epsilon_\mathbf{k} - \Sigma_\mathrm{n}(\mathbf{k}, i\omega_n) & -\Sigma_\mathrm{an}(\mathbf{k}, i\omega_n) \\ -\Sigma_\mathrm{an}^*(\mathbf{k}, i\omega_n) & -i\omega + \mu - \epsilon_\mathbf{k} - \Sigma_\mathrm{n}(-\mathbf{k}, -i\omega_n) \end{pmatrix}, \tag{3.252}$$

where $\mathcal{G}_{11}(\mathbf{k}, i\omega_n) = -\langle \psi(\mathbf{k}, i\omega_n) \psi^*(\mathbf{k}, i\omega_n) \rangle$ and $\mathcal{G}_{12}(\mathbf{k}, i\omega_n) = -\langle \psi(\mathbf{k}, i\omega_n) \psi (-\mathbf{k}, -i\omega_n) \rangle$ (Chapter 13). The high-frequency limit of the normal and anomalous

[33] In the more general case of a momentum-dependent interaction $v(\mathbf{q})$: $\lim_{|z|\to\infty} \Sigma_\sigma(\mathbf{k}, z) = v(\mathbf{q}=0)n - \int_\mathbf{q} v(\mathbf{q}) \langle \hat{\psi}_\sigma^\dagger(\mathbf{k}+\mathbf{q}) \hat{\psi}_\sigma(\mathbf{k}+\mathbf{q}) \rangle$.

self-energies $\Sigma_{\rm n}$ and $\Sigma_{\rm an}$ is given by [29]

$$\lim_{|z|\to\infty} \Sigma_{\rm n}(\mathbf{k},z) = 2g\langle\hat\psi^\dagger(\mathbf{r})\hat\psi(\mathbf{r})\rangle, \qquad \lim_{|z|\to\infty} \Sigma_{\rm an}(\mathbf{k},z) = g\langle\hat\psi(\mathbf{r})^2\rangle. \tag{3.253}$$

3.5.4 *Symmetry properties*

The discussion of Sec. 3.2.1 about the symmetry properties of correlation functions applies to the one-particle Green function. As an example, let us consider an s-wave superconductor (spin-$\frac{1}{2}$ fermion system) and the three Green functions defined in (3.237).

Invariance under parity implies

$$\begin{aligned} G_{\sigma\sigma'}(\mathbf{r},\tau;\mathbf{r}',\tau';\Delta) &= G_{\sigma\sigma'}(-\mathbf{r},\tau;-\mathbf{r}',\tau';\Delta), \\ \text{i.e.}\quad G_{\sigma\sigma'}(\mathbf{k},i\omega_n;\Delta) &= G_{\sigma\sigma'}(-\mathbf{k},i\omega_n;\Delta) \end{aligned} \tag{3.254}$$

in a translation-invariant system, and similar relations for F and $F^\dagger$. The notation emphasizes the fact that the Green function depends on the order parameter $\Delta \propto \langle\hat\psi_\downarrow(\mathbf{r})\hat\psi_\uparrow(\mathbf{r})\rangle$ (see Sec. 7.5.2), which is assumed here uniform and invariant under parity. Time-reversal invariance leads to (see Sec. 2.2.6)

$$\begin{aligned} G_{\sigma\sigma'}(\mathbf{r},\tau;\mathbf{r}',\tau';\Delta) &= \sigma\sigma' G_{\bar\sigma'\bar\sigma}(\mathbf{r}',-\tau';\mathbf{r},-\tau;\Delta^*), \\ F_{\sigma\sigma'}(\mathbf{r},\tau;\mathbf{r}',\tau';\Delta) &= \sigma\sigma' F^\dagger_{\bar\sigma'\bar\sigma}(\mathbf{r}',-\tau';\mathbf{r},-\tau;\Delta^*), \end{aligned} \tag{3.255}$$

i.e.

$$\begin{aligned} G_{\sigma\sigma'}(\mathbf{k},i\omega_n;\Delta) &= \sigma\sigma' G_{\bar\sigma'\bar\sigma}(-\mathbf{k},i\omega_n;\Delta^*), \\ F_{\sigma\sigma'}(\mathbf{k},i\omega_n;\Delta) &= \sigma\sigma' F^\dagger_{\bar\sigma'\bar\sigma}(-\mathbf{k},i\omega_n;\Delta^*). \end{aligned} \tag{3.256}$$

We have used the fact that $\Delta \to \Delta^*$ under time reversal.[34,35]

In systems that are spin-rotation invariant, it is often convenient to write the one-particle Green function

$$\begin{aligned} G_{\sigma\sigma'}(\mathbf{r},\tau;\mathbf{r}',\tau') &= \frac{1}{2}\sigma^0_{\sigma\sigma'} G_{\rm ch}(\mathbf{r},\tau;\mathbf{r}',\tau') + \frac{1}{2}\boldsymbol{\sigma}_{\sigma\sigma'}\cdot \mathbf{G}_{\rm sp}(\mathbf{r},\tau;\mathbf{r}',\tau'), \\ F_{\sigma\sigma'}(\mathbf{r},\tau;\mathbf{r}',\tau') &= \frac{1}{2}(i\sigma^y)_{\sigma\sigma'} F_{\rm s}(\mathbf{r},\tau;\mathbf{r}',\tau') + \frac{1}{2}(i\boldsymbol{\sigma}\sigma^y)_{\sigma\sigma'}\cdot \mathbf{F}_{\rm t}(\mathbf{r},\tau;\mathbf{r}',\tau'), \\ F^\dagger_{\sigma\sigma'}(\mathbf{r},\tau;\mathbf{r}',\tau') &= \frac{1}{2}(i\sigma^y)^\dagger_{\sigma\sigma'} F^\dagger_{\rm s}(\mathbf{r},\tau;\mathbf{r}',\tau') + \frac{1}{2}(i\boldsymbol{\sigma}\sigma^y)^\dagger_{\sigma\sigma'}\cdot \mathbf{F}^\dagger_{\rm t}(\mathbf{r},\tau;\mathbf{r}',\tau') \end{aligned} \tag{3.257}$$

[σ^0 denotes the unit 2×2 matrix and $\boldsymbol{\sigma} = (\sigma^x,\sigma^y,\sigma^z)$ stands for the Pauli matrices] in terms of charge/spin and singlet/triplet components [$\mathbf{G}_{\rm sp} = (G^x_{\rm sp}, G^y_{\rm sp}, G^z_{\rm sp})$ and $\mathbf{F}^{(\dagger)}_{\rm t} = (F^{(\dagger)x}_{\rm t}, F^{(\dagger)y}_{\rm t}, F^{(\dagger)z}_{\rm t})$].

---

[34]This follows from $(\langle\Psi|\hat\Theta^\dagger)\hat\psi_\downarrow(\mathbf{r})\hat\psi_\uparrow(\mathbf{r})(\hat\Theta|\Psi\rangle) = \langle\Psi|\hat\psi_\downarrow(\mathbf{r})\hat\psi_\uparrow(\mathbf{r})|\Psi\rangle^*$ for any ket $|\Psi\rangle$, where $\hat\Theta$ is the time-reversal operator introduced in Sec. 2.2.6.

[35]Equations (3.256) are satisfied by the BCS correlation functions (7.244).

3.5.5 *Relation to observables*

Many physical observables can be related to the one-particle Green function.

3.5.5.1 *One-body observable*

Let us first consider a one-body operator

$$\hat{V}^{(1)} = \int d^d r \int d^d r' \sum_{\sigma,\sigma'} \langle \mathbf{r}, \sigma | \hat{V} | \mathbf{r}', \sigma' \rangle \hat{\psi}^\dagger_\sigma(\mathbf{r}) \hat{\psi}_{\sigma'}(\mathbf{r}') \tag{3.258}$$

in second-quantized form (Sec. 1.2.2). Its expectation value can be written as

$$\begin{aligned} \langle \hat{V}^{(1)} \rangle &= -\zeta \int d^d r \int d^d r' \sum_{\sigma,\sigma'} \langle \mathbf{r}, \sigma | \hat{V} | \mathbf{r}', \sigma' \rangle G_{\sigma'\sigma}(\mathbf{r}', \tau; \mathbf{r}, \tau^+) \\ &= -\frac{\zeta}{\beta} \sum_{\omega_n} \int d^d r \int d^d r' \sum_{\sigma,\sigma'} \langle \mathbf{r}, \sigma | \hat{V} | \mathbf{r}', \sigma' \rangle G_{\sigma'\sigma}(\mathbf{r}', \mathbf{r}; i\omega_n) e^{i\omega_n \eta}. \end{aligned} \tag{3.259}$$

For the density operator

$$\hat{\jmath}^\nu_0(\mathbf{r}) = \sum_{\sigma,\sigma'} \hat{\psi}^\dagger_\sigma(\mathbf{r}) S^\nu_{\sigma\sigma'} \hat{\psi}_{\sigma'}(\mathbf{r}) \qquad (\nu = 0, x, y, \ldots), \tag{3.260}$$

one obtains

$$\langle \hat{\jmath}^\nu_0(\mathbf{r}) \rangle = -\zeta \sum_{\sigma,\sigma'} S^\nu_{\sigma\sigma'} G_{\sigma'\sigma}(\mathbf{r}, \tau; \mathbf{r}, \tau^+). \tag{3.261}$$

Here, S^ν is the unit matrix if $\nu = 0$ and a spin matrix if $\nu \neq 0$ (e.g. for spin-$\frac{1}{2}$ fermions, $S^\nu = \frac{1}{2}\sigma^\nu$ with $\{\sigma^\nu\}$ the Pauli matrices). Similarly, for the current-density operator

$$\hat{\jmath}^\nu_\mu(\mathbf{r}) = -\frac{i}{2m} \sum_{\sigma,\sigma'} [\hat{\psi}^\dagger_\sigma(\mathbf{r}) S^\nu_{\sigma\sigma'} \partial_\mu \hat{\psi}_{\sigma'}(\mathbf{r}) - \text{h.c.}] \qquad (\mu = x, y, \ldots), \tag{3.262}$$

one finds

$$\langle \hat{\jmath}^\nu_\mu(\mathbf{r}) \rangle = -\frac{\zeta}{m} \sum_{\sigma,\sigma'} S^\nu_{\sigma\sigma'} \lim_{\mathbf{r}' \to \mathbf{r}} \partial_\mu \Im[G_{\sigma'\sigma}(\mathbf{r}, \tau; \mathbf{r}', \tau^+)]. \tag{3.263}$$

Finally, the expectation value of the kinetic energy operator

$$\hat{T} = -\frac{1}{2m} \int d^d r \sum_\sigma \hat{\psi}^\dagger_\sigma(\mathbf{r}) \boldsymbol{\nabla}^2 \hat{\psi}_\sigma(\mathbf{r}) \tag{3.264}$$

reads

$$\begin{aligned} \langle \hat{T} \rangle &= \frac{\zeta}{2m} \int d^d r \lim_{\mathbf{r}' \to \mathbf{r}} \sum_\sigma \boldsymbol{\nabla}^2 G_{\sigma\sigma}(\mathbf{r}, \tau; \mathbf{r}', \tau^+) \\ &= -\frac{\zeta}{\beta \mathcal{V}} \sum_{\mathbf{k}, \omega_n, \sigma} \epsilon_{\mathbf{k}} G_{\sigma\sigma}(\mathbf{k}, i\omega_n) e^{i\omega_n \eta}, \end{aligned} \tag{3.265}$$

where $\epsilon_{\mathbf{k}} = \mathbf{k}^2/2m$.

3.5.5.2 Energy and thermodynamic potential

The expectation value of a two-body operator can be related to the two-particle Green function. In the case where the Hamiltonian $\hat{H} = \hat{H}_0 + \hat{V}$ consists of the sum of the kinetic energy and a two-body interaction term,

$$\hat{V} = \frac{1}{2} \int d^d r \int d^d r' \sum_{\sigma,\sigma'} v(\mathbf{r} - \mathbf{r}') \hat{\psi}^\dagger_\sigma(\mathbf{r}) \hat{\psi}^\dagger_{\sigma'}(\mathbf{r}') \hat{\psi}_{\sigma'}(\mathbf{r}') \hat{\psi}_\sigma(\mathbf{r}), \tag{3.266}$$

the expectation value $\langle \hat{V} \rangle$ can be expressed solely in terms of the one-particle Green function G. The equation of motion $\partial_\tau \hat{\psi}_\sigma(\mathbf{r}, \tau) = [\hat{H}, \hat{\psi}_\sigma(\mathbf{r}, \tau)]$ gives

$$\begin{aligned} \partial_\tau \hat{\psi}_\sigma(\mathbf{r}, \tau) &= \left(\frac{\boldsymbol{\nabla}^2}{2m} + \mu \right) \hat{\psi}_\sigma(\mathbf{r}, \tau) \\ &\quad - \int d^d r'' \sum_{\sigma''} v(\mathbf{r} - \mathbf{r}'') \hat{\psi}^\dagger_{\sigma''}(\mathbf{r}'', \tau) \hat{\psi}_{\sigma''}(\mathbf{r}'', \tau) \hat{\psi}_\sigma(\mathbf{r}, \tau). \end{aligned} \tag{3.267}$$

Multiplying from the left by $\hat{\psi}^\dagger_\sigma(\mathbf{r}, \tau^+)$ and taking the expectation value, we obtain[36]

$$\begin{aligned} -\zeta \lim_{\tau' \to \tau^+} \partial_\tau G_{\sigma\sigma}(\mathbf{r}, \tau; \mathbf{r}, \tau') &= -\zeta \lim_{\mathbf{r}' \to \mathbf{r}} \left(\frac{\boldsymbol{\nabla}^2}{2m} + \mu \right) G_{\sigma\sigma}(\mathbf{r}, \tau; \mathbf{r}', \tau^+) \\ &\quad - \int d^d r'' \sum_{\sigma''} v(\mathbf{r} - \mathbf{r}'') \langle \hat{\psi}^\dagger_\sigma(\mathbf{r}, \tau) \hat{\psi}^\dagger_{\sigma''}(\mathbf{r}'', \tau) \hat{\psi}_{\sigma''}(\mathbf{r}'', \tau) \hat{\psi}_\sigma(\mathbf{r}, \tau) \rangle, \end{aligned} \tag{3.268}$$

where $G_{\sigma\sigma'} = \delta_{\sigma,\sigma'} G$ since the Hamiltonian is spin-rotation invariant. The last term in the rhs gives $2\langle \hat{V} \rangle$ (after summation over σ and $\mathbf{r}$) and we conclude that

$$\langle \hat{V} \rangle = -\zeta \int d^d r \lim_{\substack{\mathbf{r}' \to \mathbf{r} \\ \tau' \to \tau^+}} \Big(-\partial_\tau + \frac{\boldsymbol{\nabla}^2}{2m} + \mu \Big) G(\mathbf{r}, \tau; \mathbf{r}', \tau'). \tag{3.269}$$

The internal energy $E = \langle \hat{H}_0 + \mu \hat{N} + \hat{V} \rangle$ is therefore[37]

$$\begin{aligned} E &= -\zeta \int d^d r \lim_{\substack{\mathbf{r}' \to \mathbf{r} \\ \tau' \to \tau^+}} \Big(-\partial_\tau - \frac{\boldsymbol{\nabla}^2}{2m} + \mu \Big) G(\mathbf{r}, \tau; \mathbf{r}', \tau') \\ &= -\frac{\zeta}{\beta} \sum_{\mathbf{k}, \omega_n} (i\omega_n + \mu + \epsilon_{\mathbf{k}}) G(\mathbf{k}, i\omega_n) e^{i\omega_n \eta}. \end{aligned} \tag{3.270}$$

The grand potential can now be obtained by means of an integration over a variable coupling constant considering the Hamiltonian

$$\hat{H}_\lambda = \hat{H}_0 + \lambda \hat{V}. \tag{3.271}$$

[36] We use $\langle \hat{\psi}^\dagger_\sigma(\mathbf{r}', \tau^+) \partial_\tau \hat{\psi}_\sigma(\mathbf{r}, \tau) \rangle = -\zeta \lim_{\tau' \to \tau^+} \partial_\tau G(\mathbf{r}, \tau; \mathbf{r}', \tau')$, etc.

[37] Recall that $\hat{H}$ is a grand-canonical Hamiltonian and therefore includes the term $-\mu \hat{N}$.

To compute the thermodynamic potential $\Omega_\lambda = -\frac{1}{\beta}\ln Z_\lambda$, we consider the equation

$$\frac{\partial \Omega_\lambda}{\partial \lambda} = -\frac{1}{\beta Z_\lambda}\frac{\partial Z_\lambda}{\partial \lambda} \quad \text{where} \quad Z_\lambda = \mathrm{Tr}\big(e^{-\beta \hat{H}_\lambda}\big). \tag{3.272}$$

By expanding the exponential, one obtains

$$\frac{\partial Z_\lambda}{\partial \lambda} = \sum_{n=1}^{\infty} \frac{(-\beta)^n}{n!}\frac{\partial}{\partial \lambda}\mathrm{Tr}(\hat{H}_0 + \lambda \hat{V})^n = -\beta \mathrm{Tr}\big(e^{-\beta \hat{H}_\lambda}\hat{V}\big), \tag{3.273}$$

using

$$\frac{\partial}{\partial \lambda}\mathrm{Tr}(\hat{H}_0 + \hat{V})^n = n\mathrm{Tr}[\hat{V}(\hat{H}_0 + \lambda \hat{V})^{n-1}] \tag{3.274}$$

which follows from the cyclic property of the trace. We therefore have

$$\frac{\partial \Omega_\lambda}{\partial \lambda} = \frac{1}{\lambda}\langle \lambda \hat{V}\rangle_\lambda, \tag{3.275}$$

where $\langle \cdots \rangle_\lambda$ denotes the ensemble average with the Hamiltonian $\hat{H}_\lambda$. Integrating (3.275) from $\lambda = 0$ to $\lambda = 1$ gives the thermodynamic potential $\Omega \equiv \Omega_{\lambda=1}$,

$$\Omega = \Omega_0 + \int_0^1 \frac{d\lambda}{\lambda}\langle \lambda \hat{V}\rangle_\lambda, \tag{3.276}$$

where Ω_0 is the thermodynamic potential for the Hamiltonian $\hat{H}_0$. From (3.269), we finally express

$$\begin{aligned}
\Omega &= \Omega_0 - \zeta \int_0^1 \frac{d\lambda}{\lambda}\int d^d r \lim_{\substack{\mathbf{r}'\to\mathbf{r} \\ \tau'\to\tau+}} \Big(-\partial_\tau + \frac{\boldsymbol{\nabla}^2}{2m} + \mu\Big) G_\lambda(\mathbf{r},\tau;\mathbf{r}',\tau') \\
&= \Omega_0 - \zeta \int_0^1 \frac{d\lambda}{\lambda}\frac{1}{\beta}\sum_{\mathbf{k},\omega_n} G_0^{-1}(\mathbf{k}, i\omega_n) G_\lambda(\mathbf{k}, i\omega_n) e^{i\omega_n \eta},
\end{aligned} \tag{3.277}$$

in terms of the Green function G_λ for a system with Hamiltonian $\hat{H}_\lambda$.

3.6 Spontaneous Symmetry Breaking and Goldstone's Theorem

Spontaneous symmetry breaking and the Goldstone theorem were discussed in Chapter 2. In this section, we reconsider these issues, but from a less formal point of view, using the general properties of correlation functions.[38] We first derive the Bogoliubov inequality which will be used in the subsequent sections.

[38] Spontaneous symmetry breaking will also be discussed in Chapter 10.

3.6.1 *The Bogoliubov inequality*

We define the scalar product

$$\langle A|B\rangle = \chi_{A^\dagger B}(z=0) = \int_{-\infty}^{\infty} \frac{d\omega}{\pi} \frac{\chi''_{A^\dagger B}(\omega)}{\omega}$$

$$= \frac{1}{Z} \sum_{\substack{n,m \\ (\epsilon_n \neq \epsilon_m)}} \frac{e^{-\beta\epsilon_n} - e^{-\beta\epsilon_m}}{\epsilon_m - \epsilon_n} \langle n|\hat{A}^\dagger|m\rangle\langle m|\hat{B}|n\rangle \tag{3.278}$$

for two operators $\hat{A}$ and $\hat{B}$ (not necessarily hermitian). The last line of (3.278) is the Lehmann representation (3.38) of $\chi_{A^\dagger B}(z=0)$. One readily verifies that this definition satisfies the basic properties of a scalar product: $\langle A|A\rangle \geq 0$, $\langle A|bB + cC\rangle = b\langle A|B\rangle + c\langle A|C\rangle$ (b, c complex numbers), and $\langle B|A\rangle = \langle A|B\rangle^*$. Therefore, the Schwarz inequality holds,

$$\langle A|A\rangle\langle B|B\rangle \geq |\langle A|B\rangle|^2. \tag{3.279}$$

More explicitly, Eq. (3.279) reads

$$\int_{-\infty}^{\infty} \frac{d\omega}{\pi} \frac{\chi''_{A^\dagger A}(\omega)}{\omega} \int_{-\infty}^{\infty} \frac{d\omega}{\pi} \frac{\chi''_{B^\dagger B}(\omega)}{\omega} \geq \left| \int_{-\infty}^{\infty} \frac{d\omega}{\pi} \frac{\chi''_{A^\dagger B}(\omega)}{\omega} \right|^2, \tag{3.280}$$

which is one form of the Bogoliubov inequality.

3.6.2 *Spontaneous symmetry breaking*

3.6.2.1 *Example: The Heisenberg ferromagnet*

Let us start with a simple example of broken symmetry, namely a three-dimensional isotropic Heisenberg ferromagnet with the Hamiltonian[39]

$$\hat{H} = -\frac{1}{2} \sum_{\mathbf{r},\mathbf{r}'} J_{\mathbf{r},\mathbf{r}'} \hat{\mathbf{S}}_{\mathbf{r}} \cdot \hat{\mathbf{S}}_{\mathbf{r}'}, \tag{3.281}$$

where $\mathbf{r}$ labels the sites of a lattice and $J_{\mathbf{r},\mathbf{r}'} \geq 0$ denotes the exchange interaction between spins located at sites $\mathbf{r}$ and $\mathbf{r}'$. The spin operators satisfy the commutation relations

$$[\hat{\mathbf{S}}^\alpha_{\mathbf{r}}, \hat{\mathbf{S}}^\beta_{\mathbf{r}'}] = i\delta_{\mathbf{r},\mathbf{r}'}\epsilon_{\alpha\beta\gamma}\hat{\mathbf{S}}^\gamma_{\mathbf{r}} \qquad (\alpha, \beta, \gamma = x, y, z). \tag{3.282}$$

The SU(2) spin-rotation invariance implies that the generator of spin rotations, $\hat{\mathbf{S}}_{\text{tot}} = \sum_{\mathbf{r}} \hat{\mathbf{S}}_{\mathbf{r}}$, commutes with the Hamiltonian: $[\hat{U}, \hat{H}] = 0$ where $\hat{U} = e^{-i\theta\hat{\mathbf{S}}_{\text{tot}}\cdot\mathbf{n}}$ with θ an arbitrary angle and $\mathbf{n}$ an arbitrary unit vector (see Sec. 2.2.3).

In the ferromagnetic state, spin-rotation invariance is spontaneously broken and the system exhibits a finite magnetization (assumed here to point in the z direction),

$$\langle \hat{\mathbf{S}}_{\text{tot}} \rangle = NM_0\hat{\mathbf{z}} \qquad (M_0 \neq 0), \tag{3.283}$$

where N is the number of lattice sites. Equation (3.283) implies that the density matrix cannot commute with $\hat{\mathbf{S}}_{\text{tot}}$. Indeed, using $[\hat{S}^x_{\text{tot}}, \hat{S}^y_{\text{tot}}] = i\hat{S}^z_{\text{tot}}$ and the cyclic invariance of the

[39]The Heisenberg (anti)ferromagnet is discussed in detail in Chapter 8.

trace, we find

$$\langle \hat{S}^z_{\rm tot}\rangle = -i{\rm Tr}(\hat{\rho}[\hat{S}^x_{\rm tot}, \hat{S}^y_{\rm tot}]) = i{\rm Tr}([\hat{\rho}, \hat{S}^y_{\rm tot}]\hat{S}^x_{\rm tot}) = -i{\rm Tr}([\hat{\rho}, \hat{S}^x_{\rm tot}]\hat{S}^y_{\rm tot}), \tag{3.284}$$

and therefore

$$[\hat{\rho}, \hat{S}^x_{\rm tot}] \neq 0, \qquad [\hat{\rho}, \hat{S}^y_{\rm tot}] \neq 0 \tag{3.285}$$

when $\langle \hat{S}^z_{\rm tot}\rangle \neq 0$. The canonical ensemble, where $\hat{\rho} \propto e^{-\beta \hat{H}}$, is no longer a good ensemble when the symmetry is spontaneously broken. This ensemble implies an average over all possible directions of the magnetization and inevitably gives $\langle \hat{\mathbf{S}}_{\rm tot}\rangle = 0$, in contradiction with the assumption that the symmetry is spontaneously broken [Eq. (3.283)]. A possible way out of this difficulty is to assume the presence of an infinitesimal external magnetic field, i.e.

$$\hat{\rho}_h = \frac{e^{-\beta(\hat{H} - h\hat{S}^z_{\rm tot})}}{{\rm Tr}\, e^{-\beta(\hat{H} - h\hat{S}^z_{\rm tot})}}. \tag{3.286}$$

In the ferromagnetic state, one then has

$$\lim_{h\to 0} \langle \hat{S}^z_{\rm tot}\rangle = \lim_{h\to 0} {\rm Tr}(\hat{\rho}_h \hat{S}^z_{\rm tot}) \neq 0, \tag{3.287}$$

although ${\rm Tr}(\hat{\rho}_{h=0}\hat{S}^z_{\rm tot}) = 0$. As we shall see from many examples, it is rarely necessary to explicitly introduce a symmetry-breaking field. In general, the symmetry can be broken "by hand", e.g. by assuming a properly chosen operator to take a finite average value.

The expectation value $\langle \hat{\mathbf{S}}_{\rm tot}\rangle$ (or, equivalently, $\langle \hat{\mathbf{S}}_{\mathbf{r}}\rangle$) allows one to distinguish the ferromagnetic phase from the paramagnetic phase (where $\langle \hat{\mathbf{S}}_{\mathbf{r}}\rangle = 0$) and is referred to as the order parameter. A non-zero order parameter $\langle \hat{\mathbf{S}}_{\mathbf{r}}\rangle$ implies the existence of long-range order: The spin–spin correlation function

$$C(\mathbf{r}, \mathbf{r}') = \langle \hat{\mathbf{S}}_{\mathbf{r}} \cdot \hat{\mathbf{S}}_{\mathbf{r}'}\rangle \to \langle \hat{\mathbf{S}}_{\mathbf{r}}\rangle \cdot \langle \hat{\mathbf{S}}_{\mathbf{r}'}\rangle \neq 0 \tag{3.288}$$

does not vanish in the limit $|\mathbf{r} - \mathbf{r}'| \to \infty$. By contrast, in the paramagnetic phase, we expect the spin–spin correlation function $\langle \hat{\mathbf{S}}_{\mathbf{r}} \cdot \hat{\mathbf{S}}_{\mathbf{r}'}\rangle \sim e^{-|\mathbf{r}-\mathbf{r}'|/\xi}$ to decay exponentially with distance (ξ is the correlation length).[40]

Let us now adiabatically apply a magnetic field in the x direction. This contributes to the Hamiltonian a term

$$\hat{H}'(t) = -\sum_{\mathbf{r}} h^x_{\mathbf{r}} \hat{S}^x_{\mathbf{r}} e^{\eta t} = -\sum_{\mathbf{q}} h^x_{-\mathbf{q}} \hat{S}^x_{\mathbf{q}} e^{\eta t} \qquad (\eta \to 0^+). \tag{3.289}$$

The magnetization is taken along the z axis for $t \to -\infty$. For a uniform field $h^x_{\mathbf{r}} = h^x$, we expect the magnetization $\langle \hat{\mathbf{S}}_{\rm tot}\rangle$ at time $t = 0$ to be in the x direction whatever the strength

[40] In low-dimensional systems, it is possible for the correlation function to decay as a power law: $C(\mathbf{r}, \mathbf{r}') \sim 1/|\mathbf{r} - \mathbf{r}'|^a$ with some exponent a. This is called algebraic long-range order or quasi-long-range order and is realized, e.g. in the low-temperature phase of the two-dimensional XY model (Chapter 10).

of the field, since all the ferromagnetic states with $|\langle \hat{\mathbf{S}}_{\rm tot}\rangle| = NM_0$ are degenerate. The linear response theory gives

$$\langle \hat{S}^x_{\mathbf{q}}(t=0)\rangle = \int_{-\infty}^{\infty} dt' \chi^R_{xx}(\mathbf{q}, -t') e^{\eta t'} h^x_{\mathbf{q}} = \chi_{xx}(\mathbf{q}) h^x_{\mathbf{q}}, \tag{3.290}$$

where $\chi_{xx}(\mathbf{q}) \equiv \chi_{S^x S^x}(\mathbf{q})$ is the static limit of the retarded response function [Eq. (3.44)]. Since $\lim_{\mathbf{q}\to 0}\langle \hat{S}^x_{\mathbf{q}}(t=0)\rangle = NM_0$ even when $h^x_{\mathbf{q}} \to 0$, (3.290) implies that the limit $\mathbf{q}\to 0$ of $\chi_{xx}(\mathbf{q})$ is singular,

$$\lim_{\mathbf{q}\to 0} \chi_{xx}(\mathbf{q}) = \infty. \tag{3.291}$$

A similar result holds for $\chi_{yy}(\mathbf{q})$. Both operators $\hat{S}^x$ and $\hat{S}^y$ being even under a parity transformation (Chapter 2), $\chi_{xx}(\mathbf{q})$ and $\chi_{yy}(\mathbf{q})$ are even in $\mathbf{q}$, and we expect $\chi_{xx}(\mathbf{q}), \chi_{yy}(\mathbf{q}) \sim 1/\mathbf{q}^2$ if we assume the divergence to be minimal.

The divergence of the transverse response functions $\chi_{xx}(\mathbf{q})$ and $\chi_{yy}(\mathbf{q})$ can also be derived using the Bogoliubov inequality. Taking the expectation value of $[\hat{S}^x_{\mathbf{r}}, \hat{S}^y_{\mathbf{r}'}] = i\delta_{\mathbf{r},\mathbf{r}'}\hat{S}^z_{\mathbf{r}}$ and summing over $\mathbf{r}$, we obtain

$$\chi''_{xy}(\mathbf{q}=0, t=0) = \frac{i}{2}M_0, \quad \text{i.e.} \quad \int_{-\infty}^{\infty}\frac{d\omega}{\pi}\chi''_{xy}(\mathbf{q}=0,\omega) = iM_0. \tag{3.292}$$

The Bogoliubov inequality (3.280) with $\hat{A}_{\mathbf{r}}(t) = \partial_t \hat{S}^x_{\mathbf{r}}(t)$ and $\hat{B}_{\mathbf{r}}(t) = \hat{S}^y_{\mathbf{r}}(t)$ gives

$$\int_{-\infty}^{\infty}\frac{d\omega}{\pi}\omega\chi''_{xx}(\mathbf{q},\omega)\int_{-\infty}^{\infty}\frac{d\omega}{\pi}\frac{\chi''_{yy}(\mathbf{q},\omega)}{\omega} \geq \left|\int_{-\infty}^{\infty}\frac{d\omega}{\pi}\chi''_{xy}(\mathbf{q},\omega)\right|^2, \tag{3.293}$$

i.e.

$$\lim_{\mathbf{q}\to 0}\int_{-\infty}^{\infty}\frac{d\omega}{\pi}\omega\chi''_{xx}(\mathbf{q},\omega)\chi_{yy}(\mathbf{q}) \geq M_0^2 \tag{3.294}$$

using (3.292). Since the total spin is conserved ($[\hat{\mathbf{S}}_{\rm tot}, \hat{H}] = 0$), the spin density obeys a continuity equation. In the long-wavelength limit, the lattice does not matter and the continuity equation takes the usual form

$$\partial_t \hat{S}^\nu_{\mathbf{r}}(t) + \boldsymbol{\nabla}\cdot\hat{\mathbf{j}}^\nu_{\mathbf{r}}(t) = 0 \tag{3.295}$$

(in the Heisenberg representation), where $\hat{\mathbf{j}}^\nu_{\mathbf{r}}$ is the spin current. From (3.295), we deduce

$$\omega^2\chi''_{\nu\nu}(\mathbf{q},\omega) = \sum_{\mu,\mu'} q_\mu q_{\mu'}\chi''_{j^\nu_\mu j^\nu_{\mu'}}(\mathbf{q},\omega) = \mathcal{O}(\mathbf{q}^2), \tag{3.296}$$

and therefore

$$\int_{-\infty}^{\infty}\frac{d\omega}{\pi}\omega\chi''_{xx}(\mathbf{q},\omega) = \mathcal{O}(\mathbf{q}^2). \tag{3.297}$$

Equation (3.294) then implies that $\chi_{yy}(\mathbf{q})$ diverges at least as $1/\mathbf{q}^2$ in the limit $\mathbf{q} \to 0$. If we assume the divergence to be minimal, i.e. $\chi_{yy}(\mathbf{q}) \propto 1/\mathbf{q}^2$, we eventually obtain

$$\chi_{yy}(\mathbf{q}) = \chi_{xx}(\mathbf{q}) = \frac{M_0^2}{R\mathbf{q}^2} \qquad (\mathbf{q} \to 0), \tag{3.298}$$

where the "stiffness" (this terminology is explained below) R is bounded by

$$R \leq \lim_{\mathbf{q}\to 0} \frac{1}{\mathbf{q}^2} \int_{-\infty}^{\infty} \frac{d\omega}{\pi} \omega \chi''_{xx}(\mathbf{q}, \omega). \tag{3.299}$$

The bound in (3.299) can be shown to be finite if the exchange interactions are short-range, i.e. if $\sum_{\mathbf{r}'} J_{\mathbf{r},\mathbf{r}'} < \infty$ [19].

3.6.2.2 *Spontaneous symmetry breaking*

Although the ferromagnetic state exemplifies some aspects of spontaneous symmetry breaking — the existence of long-range order, the divergence of the transverse response function as well as the Goldstone theorem derived in Sec. 3.6.3 — it is a particular case to the extent where the order parameter operator $\hat{\mathbf{S}}_{\text{tot}} = \sum_{\mathbf{r}} \hat{\mathbf{S}}_{\mathbf{r}}$ commutes with the Hamiltonian. Any fully polarized state with all spins simultaneously pointing in the same direction is an eigenstate.[39] Even for a finite-size system, the ground state is therefore not unique. The actual ground state is selected by the infinitesimal applied field $\mathbf{h}$.

In the generic case however, the order parameter operator may not commute with the Hamiltonian and thus not be conserved. The ground state of a finite-size system is in general symmetric (i.e. does not break the symmetry) and unique. In the large-size limit, there are low-energy states that are symmetric and quasi-degenerate ($\Delta E \sim 1/\mathcal{V}$): They form a so-called tower of states. By taking a linear combination of these states, it is possible to construct non-symmetric states (i.e. exhibiting spontaneous symmetry breaking). For a finite volume $\mathcal{V}$, none of these non-symmetric states is an eigenstate of the Hamiltonian but their lifetimes diverge with $\mathcal{V}$. Thus, in the thermodynamic limit, the spontaneous-symmetry-breaking picture becomes valid.

The Heisenberg antiferromagnet falls in that category. It corresponds to the Hamiltonian (3.281) defined on a bipartite lattice with the exchange integrals $J_{\mathbf{r},\mathbf{r}'}$ being negative when $\mathbf{r}, \mathbf{r}'$ are nearest-neighbor sites and vanishing otherwise. The classical ground state is the Néel state with all spins on one sublattice pointing in a direction $\mathbf{n}$ and all spins on the other sublattice pointing in the opposite direction. The order parameter $\sum_{\mathbf{r}} (-1)^{\mathbf{r}} \langle \hat{\mathbf{S}}_{\mathbf{r}} \rangle$ corresponds to a staggered magnetization. Contrary to the ferromagnetic case, the classical state is not an eigenstate of the Hamiltonian.[39] Moreover, the ground state for an even finite number of sites N is unique and a spin singlet ($S_{\text{tot}} = 0$), i.e. rotationally invariant, which rules out symmetry breaking (Marshall's theorem). In the large-size limit, there is nevertheless a tower of symmetric states with $S_{\text{tot}} = 1, 2, ...$ The degeneracy of these states in the limit $N \to \infty$ can be used to construct a new set of eigenstates with a non-zero value of the staggered magnetization and differing from one another by the direction of the latter.[41] As in the ferromagnetic case, the broken-symmetry picture requires a statistical ensemble where the density matrix does not commute with $\hat{\mathbf{S}}_{\text{tot}}$. Such a statistical ensemble can be defined by considering an infinitesimal staggered field $h_{\mathbf{r}} = (-1)^{\mathbf{r}} h$ (e.g. parallel to the z axis) that couples to the spins. Spontaneous symmetry breaking occurs if

$$\lim_{h\to 0} \lim_{N\to\infty} (-1)^{\mathbf{r}} \langle \hat{\mathbf{S}}_{\mathbf{r}} \rangle \neq 0. \tag{3.300}$$

We stress however that the classical Néel state is not the ground state because of quantum fluctuations that reduce the amplitude of the staggered magnetization.

As we shall see in the following chapters, spontaneous symmetry breaking includes many other examples: superfluids and superconductors, charge and magnetic instabilities in itinerant fermion systems, etc.

Long-range order without spontaneous symmetry breaking. In the thermodynamic limit, long-range order is a consequence of spontaneous symmetry breaking. In a finite system with a unique ground state, there cannot be spontaneous symmetry breaking for reasons discussed above but there can be long-range order, e.g.

$$\lim_{|\mathbf{r}-\mathbf{r}'|\to\infty} \langle \hat{\mathbf{S}}_{\mathbf{r}} \cdot \hat{\mathbf{S}}_{\mathbf{r}'} \rangle \neq 0 \quad \text{and} \quad \langle \hat{\mathbf{S}}_{\mathbf{r}} \rangle = 0 \tag{3.301}$$

in the Heisenberg antiferromagnet. Such a state is however quite unnatural since it exhibits very large fluctuations of the Néel order parameter $\sum_{\mathbf{r}} (-1)^{\mathbf{r}} \langle \hat{\mathbf{S}}_{\mathbf{r}} \rangle$ (which is an extensive quantity) and therefore does not qualify as a macroscopic stable state. It also violates the cluster decomposition property which requires that

$$C_{ab}(\mathbf{r},\mathbf{r}') = \langle \hat{a}(\mathbf{r}) \hat{b}(\mathbf{r}') \rangle - \langle \hat{a}(\mathbf{r}) \rangle \langle \hat{b}(\mathbf{r}') \rangle \to 0 \quad \text{when} \quad |\mathbf{r}-\mathbf{r}'| \to \infty \tag{3.302}$$

for all local observables $\hat{a}(\mathbf{r})$ and $\hat{b}(\mathbf{r})$. In magnetic systems, states with long-range order without spontaneous symmetry breaking are never observed experimentally. A physically natural state with both long-range order and spontaneous symmetry breaking can be obtained, as explained above, from a linear combination of the quasi-degenerate states forming the low-energy tower of states. Although such a state is not an eigenstate of the Hamiltonian, it is expected to be stabilized by small symmetry-breaking perturbations that are inevitably present experimentally. This also explains why spontaneous symmetry breaking can occur in finite-size systems.

3.6.2.3 *Classification of broken-symmetry states*

A symmetry transformation g is represented by a unitary or antiunitary operator $\hat{U}(g)$ acting on the Hilbert space of state vectors (Chapter 2).[42] The set of all symmetry transformations forms a group G, e.g. the group SU(2) associated with spin rotations in the Heisenberg model. For simplicity, we assume that $\hat{U}(g)$ is unitary and time independent and consider the zero-temperature limit.[43] Invariance of the system under the group G implies that $[\hat{H}, \hat{U}(g)] = 0$ for all elements $g \in G$. Spontaneous symmetry breaking occurs when the

[41]See also the discussion of spontaneous U(1) symmetry breaking in superfluids in Sec. 7.2.1.2.

[42]Note that we discuss here only global symmetries. Spontaneous breaking of a local symmetry is impossible (Elitzur's theorem [31]).

[43]We thus discard time-reversal symmetry and Galilean invariance.

ground state $|\psi\rangle$ of the system breaks some of the symmetries of the group G but is invariant under others: $|\psi\rangle$ and $\hat{U}(g)|\psi\rangle = e^{i\alpha}|\psi\rangle$ are equal (up to a phase) when g is unbroken but are distinct, inequivalent, states when g is broken. The set of all unbroken transformations forms a subgroup $H \subset G$ called the residual symmetry group.

Consider now the set of all states obtained from the ground state by applying the symmetry transformations of the group G. Two of these states, e.g. $\hat{U}(g_i)|\psi\rangle$ and $\hat{U}(g_j)|\psi\rangle$, are equivalent if there is an element h of H such that $g_i = g_j h$. Conversely, if $g_i \neq g_j h$ for all $h \in H$, $\hat{U}(g_i)|\psi\rangle$ and $\hat{U}(g_j)|\psi\rangle$ are distinct, inequivalent, states. This identifies the quotient set G/H as the set of inequivalent broken-symmetry states.

For the (anti)ferromagnetic Heisenberg model, the relevant symmetry transformations are spin rotations, i.e. $G = SU(2)$, and the residual symmetry group $H = U(1)$ corresponds to rotations about the axis $\mathbf{n}$ defining the direction of the (staggered) magnetization in the broken-symmetry state: $\hat{U}(\theta) = e^{-i\theta\hat{\mathbf{S}}\cdot\mathbf{n}}$. The quotient set $SU(2)/U(1)$ is isomorphic to the two-sphere S^2, which corresponds to the set of points on the surface of a sphere. These points indicate the direction of the residual rotation axis, i.e. the direction of the magnetization $\langle \mathbf{S}_\mathbf{r}\rangle$ (or $(-1)^\mathbf{r}\langle \mathbf{S}_\mathbf{r}\rangle$ in the antiferromagnetic case). A superfluid breaks the U(1) symmetry of the Hamiltonian and is characterized by a non-zero expectation value of the field operator $\hat{\psi}(\mathbf{r})$ (Chapter 7). Different ground states differ by the phase of the order parameter $\langle\hat{\psi}(\mathbf{r})\rangle$, but the modulus is the same. The residual symmetry group H is therefore the trivial group containing only the identity operator and the quotient space $G/H = U(1)$ is in one-to-one correspondence with the order parameter. These two examples exemplify a general result: Inequivalent broken-symmetry states can also be classified by the order parameter.[44]

For continuous groups, the symmetry operators $\hat{U}(g) = e^{i\sum_a \theta_a \hat{Q}_a}$ can be written in terms of the generators $\hat{Q}_a$. The system is invariant under the Lie group G if the Hamiltonian commutes with all generators: $[\hat{H}, \hat{Q}_a] = 0$. Symmetric states can be chosen to be also eigenstates of the generators. If the broken-symmetry state remains an eigenstate of $\hat{Q}_a$, it is invariant under the transformation $\hat{U}_a(\theta) = e^{i\theta\hat{Q}_a}$ and the generator is said to be unbroken. Generators that do not leave the state invariant are said to be broken. The dimension of the quotient space G/H is equal to the number of broken generators. In the (anti)ferromagnetic phase of the Heisenberg model, one finds an unbroken generator, namely $\hat{S}^z_{\rm tot}$ if the (staggered) magnetization points in the z direction, and two broken generators: $\hat{S}^x_{\rm tot}$ and $\hat{S}^y_{\rm tot}$.

3.6.3 *Goldstone's theorem*

Let us consider a system invariant under the transformations of a continuous group and denote by $\hat{Q}$ one of the generators with $\hat{\jmath}_0(\mathbf{r})$ the associated local density,

$$\hat{Q} = \int d^d r\, \hat{\jmath}_0(\mathbf{r}). \tag{3.303}$$

[44]The order parameter does not necessarily distinguish between the different broken-symmetry states. There is however some freedom in its definition and it turns out that it is always possible to find an order parameter which is in one-to-one correspondence with the quotient space G/H.

Since $\hat{Q}$ commutes with the Hamiltonian, the density $\hat{\jmath}_0(\mathbf{r},t)$ in the Heisenberg representation satisfies a continuity equation,

$$\partial_t \hat{\jmath}_0(\mathbf{r},t) + \boldsymbol{\nabla}\cdot\hat{\mathbf{j}}(\mathbf{r},t) = 0. \tag{3.304}$$

Let us now suppose that we have identified two local observables[45] $\hat{A}(\mathbf{r})$ and $\hat{B}(\mathbf{r})$ such that $\hat{A}(\mathbf{r})$ transforms into $\hat{B}(\mathbf{r})$ under the symmetry transformation generated by $\hat{Q}$,[46]

$$\hat{B}(\mathbf{r}) = -i[\hat{Q},\hat{A}(\mathbf{r})]. \tag{3.305}$$

This equation implies

$$\langle\hat{B}(\mathbf{r})\rangle = -i\mathrm{Tr}\{\hat{\rho}[\hat{Q},\hat{A}(\mathbf{r})]\} = -i\mathrm{Tr}\{[\hat{\rho},\hat{Q}]\hat{A}(\mathbf{r})\}. \tag{3.306}$$

If $\langle\hat{B}(\mathbf{r})\rangle \neq 0$, then $[\hat{\rho},\hat{Q}]\neq 0$: The symmetry is spontaneously broken and $\langle\hat{B}(\mathbf{r})\rangle$ is nothing but the order parameter. $\hat{A}(\mathbf{r})$ is referred to as the interpolating operator (or field) or the symmetry-restoring variable. From Eqs. (3.303) and (3.305), we then deduce

$$-i\int d^dr'\langle[\hat{\jmath}_0(\mathbf{r}',t),\hat{A}(\mathbf{r})]\rangle = \langle\hat{B}(\mathbf{r})\rangle \equiv \langle\hat{B}\rangle, \tag{3.307}$$

where for simplicity we assume that $\langle\hat{B}(\mathbf{r})\rangle\equiv\langle\hat{B}\rangle$ is uniform and does not break translation invariance.[47] The time t in (3.307) is arbitrary, since $\hat{Q}(t)=\hat{Q}$ is conserved. We thus obtain

$$\chi''_{j_0A}(\mathbf{q}=0,\omega) = i\pi\langle\hat{B}\rangle\delta(\omega). \tag{3.308}$$

The apparent conflict between (3.308) and the general property $\chi''_{A_iA_j}(\omega=0)=0$ derived in Sec. 3.2.3 is resolved by noting that the latter holds in the (grand) canonical ensemble which is incompatible with spontaneous symmetry breaking. In the example of the ferromagnet discussed above, the density matrix ρ_h would lead to an equation similar to (3.308) but with $\delta(\omega)$ replaced by $\delta(\omega-\omega_h)$ and ω_h vanishing for $h\to 0$.

The spectral decomposition (3.37) shows that χ'' is related to the density of excitation energies of the physical system. Equation (3.308) can then be understood as showing that there are $\mathbf{q}=\omega=0$ "excitations". But these "excitations" are just other equivalent (degenerate) states, so that it appears that (3.308) does not carry much information. However, if we assume that the $\mathbf{q}\to 0$ limit is regular, we obtain

$$\lim_{\mathbf{q}\to 0}\chi''_{j_0A}(\mathbf{q},\omega) = i\pi\langle\hat{B}\rangle\delta(\omega), \tag{3.309}$$

which implies the existence of excitations with energies $\omega_{\mathbf{q}}$ satisfying $\lim_{\mathbf{q}\to 0}\omega_{\mathbf{q}}=0$. This is the Goldstone theorem, first encountered in Chapter 2: The spontaneous breaking of a

---

[45] Although we now assume $\hat{A}(\mathbf{r})$ and $\hat{B}(\mathbf{r})$ to be Hermitian, this condition can easily be relaxed.

[46] Equation (3.305) simply follows from $\hat{A}(\mathbf{r}) \to \hat{U}^\dagger(\theta)\hat{A}(\mathbf{r})\hat{U}(\theta) = \hat{A}(\mathbf{r}) - i[\hat{Q},\hat{A}(\mathbf{r})]\theta + \mathcal{O}(\theta^2)$ with $\hat{U}(\theta)=e^{i\theta\hat{Q}}$.

[47] This includes the case where the system is invariant under a set of discrete translations.

continuous symmetry implies the existence of a collective mode (Goldstone mode) whose energy $\omega_{\mathbf{q}}$ goes to zero when $\mathbf{q} \to 0$.[48,49]

It is possible to show that Eqs. (3.308) and (3.309) are equivalent when the interactions are short-range. Equation (3.308) is equivalent to

$$\int_{-\infty}^{\infty} \frac{d\omega}{\pi} \omega^n \chi''_{j_0 A}(\mathbf{q}=0,\omega) = i\langle \hat{B} \rangle \delta_{n,0} \quad \text{for} \quad n \geq 0, \tag{3.310}$$

so that what we need to show is

$$\lim_{\mathbf{q}\to 0} \int_{-\infty}^{\infty} \frac{d\omega}{\pi} \omega^n \chi''_{j_0 A}(\mathbf{q},\omega) = \int_{-\infty}^{\infty} \frac{d\omega}{\pi} \omega^n \chi''_{j_0 A}(\mathbf{q}=0,\omega) \quad \text{for} \quad n \geq 0. \tag{3.311}$$

When $\hat{\jmath}_0(\mathbf{r})$ and $\hat{A}(\mathbf{r})$ are local operators which depend on degrees of freedom located in the vicinity of $\mathbf{r}$, the commutator $[\hat{\jmath}_0(\mathbf{r}), \hat{A}(\mathbf{r}')]$ should vanish when $\mathbf{r}$ and $\mathbf{r}'$ are sufficiently far away. We then expect $\chi''_{j_0 A}(\mathbf{r},t;\mathbf{r}',t)$ to rapidly decay as $|\mathbf{r}-\mathbf{r}'| \to \infty$ and its Fourier transform $\chi''_{j_0 A}(\mathbf{q},t=0)$ to be regular in the limit $\mathbf{q} \to 0$. This proves (3.311) for $n=0$. For $n=1$, we have

$$\begin{aligned}\int_{-\infty}^{\infty} \frac{d\omega}{\pi} \omega \chi''_{j_0 A}(\mathbf{r}-\mathbf{r}',\omega) &= \big\langle \big[[\hat{\jmath}_0(\mathbf{r}), \hat{H}], \hat{A}(\mathbf{r}')\big]\big\rangle \\ &= \int d^d r'' \big\langle \big[[\hat{\jmath}_0(\mathbf{r}), \hat{\mathcal{H}}(\mathbf{r}'')], \hat{A}(\mathbf{r}')\big]\big\rangle,\end{aligned} \tag{3.312}$$

where we have written the Hamiltonian $\hat{H}$ in terms of an energy density operator $\hat{\mathcal{H}}(\mathbf{r})$. If all interactions are short-range, then $\hat{\mathcal{H}}(\mathbf{r})$ is a local operator and the rhs of (3.312) is expected to rapidly vanish as $|\mathbf{r}-\mathbf{r}'| \to \infty$. This ensures that Eq. (3.311) is satisfied for $n=1$. This argument is easily extended to $n \geq 2$.

The Goldstone theorem breaks down when interactions are long-range. For instance, a ferromagnetic spin system with sufficiently long-range interactions $J_{\mathbf{r},\mathbf{r}'} \geq 0$ [Eq. (3.281)] would not exhibit a gapless spin-wave mode.[50] Another example is provided by a charged superfluid where the long-range Coulomb interaction obliterates the Goldstone mode (see Chapter 7).

3.6.3.1 *Number of Goldstone modes*

One could naively expect the number of Goldstone modes to be equal to the number of broken generators. The example of the ferromagnet discussed above shows that this is not the case. Although there are two broken generators (e.g. $\hat{S}^x_{\text{tot}}$ and $\hat{S}^x_{\text{tot}}$ if the magnetization points in the z direction), the Heisenberg ferromagnet is known to have a single Goldstone

---

[48] In the antiferromagnet, where $\langle \hat{B}_{\mathbf{r}} \rangle \equiv \langle \hat{S}^z_{\mathbf{r}} \rangle$ is not uniform, the energy of the collective mode vanishes for $\mathbf{q} \to \pm\mathbf{Q}$ as well as $\mathbf{q} \to 0$ (see Chapter 8).

[49] In high-energy physics, a mode whose energy vanishes as $\mathbf{q} \to 0$ is referred to as a massless mode since it corresponds to a massless particle. In statistical physics, such a mode is sometimes referred to as a soft mode.

[50] A thorough derivation of the Goldstone theorem for Heisenberg spins is given by Lange [19].

mode with a quadratic dispersion law $\omega \sim \mathbf{q}^2$ (Chapter 8). By contrast, the Heisenberg antiferromagnet has two Goldstone modes with a linear dispersion $\omega \sim |\mathbf{q}|$. In condensed-matter physics, the situation is therefore more diverse than in relativistic systems where the Lorentz invariance ensures the existence of a Goldstone mode with a linear dispersion for each broken generator.

In the general case, for internal symmetries,[51] the number of Goldstone modes is given by

$$n_G = n_{BG} - \frac{1}{2}\,\mathrm{rank}\,M, \tag{3.313}$$

where

$$n_{BG} = \dim(G/H) = \dim(G) - \dim(H) \tag{3.314}$$

is the number of broken generators and

$$M_{ab} = -\frac{i}{\mathcal{V}}\langle[\hat{Q}_a, \hat{Q}_b]\rangle = \frac{1}{\mathcal{V}}\sum_c f_{abc}\langle\hat{Q}_c\rangle \tag{3.315}$$

is the Watanabe–Brauner matrix (f_{abc} denotes the structure constants of the Lie group G). Here, a and b label all broken symmetry generators so that M is an $n_{BG} \times n_{BG}$ matrix. When the state of the system is translation invariant, the matrix element (3.315) can be written as[52]

$$M_{ab} = -i\langle[\hat{Q}_a, \hat{j}_{b,0}(\mathbf{r})]\rangle, \tag{3.316}$$

where $\hat{j}_{b,0}$ is the local density associated with the generator $\hat{Q}_b$. The real antisymmetric matrix M can always be diagonalized by an orthogonal matrix into the following form:

$$\begin{pmatrix} \bar{M}_1 & & & & & \\ & \ddots & & & & \\ & & \bar{M}_m & & & \\ & & & 0 & & \\ & & & & \ddots & \\ & & & & & 0 \end{pmatrix} \quad \text{with} \quad \bar{M}_i = \begin{pmatrix} 0 & \lambda_i \\ -\lambda_i & 0 \end{pmatrix}, \tag{3.317}$$

where all blank entries are zero and the λ_i's are non-zero. We thus get $\mathrm{rank}\,M = 2m$ and the rhs of (3.313) is an integer. The form (3.317) of the matrix M suggests that we can distinguish two types of Goldstone modes. To each pair $(\hat{Q}_{2l-1}, \hat{Q}_{2l})$ $(l = 1, 2, \ldots, m)$ of broken

---

[51] We briefly comment on the case of spacetime symmetries in the following.
[52] Equation (3.316) does not hold for the antiferromagnet.

generators, we associate a type-B mode. The remaining broken generators correspond to type-A modes. Equation (3.313) is equivalent to

$$n_G = n_A + n_B \quad \text{with} \quad \begin{cases} n_B = \frac{1}{2}\,\text{rank}\, M, \\ n_A = n_{BG} - 2n_B. \end{cases} \tag{3.318}$$

To understand why the pair $(\hat{Q}_{2l-1}, \hat{Q}_{2l})$ $(l = 1, 2, \dots, m)$ leads to a single (type-B) Goldstone mode, let us consider the explicit form of the spectral function

$$\chi''_{j_{a,0}A}(\mathbf{q}, \omega) = \pi \sum_{n \neq 0} \big[\langle \psi | \hat{\jmath}_{a,0}(\mathbf{q}) | n \rangle \langle n | \hat{A}(-\mathbf{q}) | \psi \rangle \delta(\omega + \epsilon_0 - \epsilon_n)$$
$$- \langle \psi | \hat{A}(\mathbf{q}) | n \rangle \langle n | \hat{\jmath}_{a,0}(-\mathbf{q}) | \psi \rangle \delta(\omega + \epsilon_n - \epsilon_0) \big] \tag{3.319}$$

in the zero-temperature limit, where $\{|n\rangle, \epsilon_n\}$ denotes a complete set of eigenstates of the Hamiltonian with $|0\rangle \equiv |\psi\rangle$ the ground state and $\hat{A}(\mathbf{r})$ is the interpolating operator. Since $\chi''_{j_{a,0}A}(\mathbf{q}, \omega) \sim \delta(\omega \pm \omega_{\mathbf{q}})$ with $\omega_{\mathbf{q}}$ the energy of the Goldstone mode, we see that the latter can be excited from the ground state by the operator $\hat{\jmath}_{a,0}$ or $\hat{A}$. A special case occurs when $\hat{A}(\mathbf{r})$ can be chosen as the density $\hat{\jmath}_{b,0}(\mathbf{r})$ associated with the broken generator $\hat{Q}_b$. Using (3.307), one then finds

$$-\frac{i}{\mathcal{V}} \langle [\hat{Q}_a, \hat{Q}_b] \rangle = -i \langle [\hat{Q}_a, \hat{\jmath}_{b,0}(\mathbf{r})] \rangle = \langle \hat{B} \rangle, \tag{3.320}$$

so that $\hat{Q}_a$ and $\hat{Q}_b$ form a pair of type-B generators. We therefore conclude that the pair $(\hat{Q}_{2l-1}, \hat{Q}_{2l})$ of generators excites the same (type-B) Goldstone mode, which leads to $n_B = m = \frac{1}{2}\,\text{rank}\, M$.

A rigorous derivation of (3.313) is beyond the scope of our discussion, but further insight can be gained by examining the most general action allowed by the symmetries in terms of the field $\Pi_a(\mathbf{r}, t)$ associated with the broken generators $\hat{Q}_a$.[53] To second order in a derivative expansion, the Lagrangian density reads

$$\mathcal{L}_{\text{eff}} = \sum_{a,b} \left\{ \frac{1}{2} m_{ab} (\Pi_a \partial_t \Pi_b - \Pi_b \partial_t \Pi_a) + \frac{1}{2} \bar{g}_{ab} \partial_t \Pi_a \partial_t \Pi_b - \frac{1}{2} g_{ab} \boldsymbol{\nabla} \Pi_a \cdot \boldsymbol{\nabla} \Pi_b \right\} \tag{3.321}$$

assuming space to be isotropic (which forbids linear terms in gradient). The key point of the derivation of (3.313) is that the coefficients m_{ab} coincide with the matrix elements M_{ab} of the Watanabe–Brauner matrix (3.315) [24]. It is now easy to find the collective modes from the equations of motion derived from $\mathcal{L}_{\text{eff}}$. When $m_{ab} \neq 0$, the first-order time-derivative term in (3.321) dominates over the second-order one in the low-energy limit. The two fields Π_a and Π_b can then be considered as conjugate fields (in the low-energy sector) and describe the same degrees of freedom; as a result, they produce a single (type-B) Goldstone mode with quadratic dispersion $\omega \sim \mathbf{q}^2$. When both m_{ab} and $\bar{g}_{ab}$ are non-zero, a gapped

[53] The fields $\Pi_a(\mathbf{r}, t)$ can be defined by writing the Wigner operator of a global symmetry transformation as $\hat{U}(\{\Pi_a\}) = e^{i \sum_a \Pi_a \hat{Q}_a}$ and promoting the real variables Π_a to space- and time-dependent fields.

partner to the type-B Goldstone mode may exist with a mass of the order $m_{ab}/\sqrt{\bar{g}_{aa}\bar{g}_{bb}}$ (this conclusion may however be beyond the validity of the derivative expansion). The remaining $n_A = n_{BG} - n_B$ fields produce n_A modes with a linear dispersion $\omega \sim |\mathbf{q}|$.[54]

We thus reproduce the known results in the Heisenberg model mentioned above. In the ferromagnet, the two broken generators are $\hat{S}^x_{\rm tot}$ and $\hat{S}^y_{\rm tot}$. Since $\langle[\hat{S}^x_{\rm tot}, \hat{S}^y_{\rm tot}]\rangle = i\langle\hat{S}^z_{\rm tot}\rangle$, the off-diagonal elements of the 2×2 Watanabe–Brauner matrix do not vanish. There is a single, type-B, Goldstone mode with quadratic dispersion ($n_B = 1$ and $n_A = 0$). In the antiferromagnet, $\langle\hat{S}^z_{\rm tot}\rangle = 0$ and $M_{ab} = 0$, so that one has two type-A Goldstone modes with linear dispersion ($n_A = 2$ and $n_B = 0$).

In a superfluid, the generator of the U(1) symmetry is the number of particles $\hat{N} = \int d^d r\, \hat{\psi}^\dagger(\mathbf{r})\hat{\psi}(\mathbf{r})$ and the field operator $\hat{\psi}(\mathbf{r})$ acts as the order parameter.[55] Since there is a single generator, one has a type-A Goldstone mode with linear dispersion. The interpolating operator is the field operator.[56] The Goldstone mode can be seen as a phase mode or a density mode (phonon mode) since phase and density are conjugate variables (Chapter 7).

A crystal in d dimensions breaks the symmetries of space. There are d generators for translation and $\frac{1}{2}d(d-1)$ generators for rotations. The translation group being Abelian, the associated Goldstone modes (phonons or sound waves) are all type-A. Local rotations and local translations are not independent so that broken rotations do not lead to independent Goldstone modes in the crystal [23, 25].

The apparent absence of Goldstone modes due to rotation symmetry breaking in a crystal is a particular case of a general result for spacetime symmetries. When the local densities $\hat{\jmath}_{a,0}(\mathbf{r})$ and $\hat{\jmath}_{b,0}(\mathbf{r})$ associated with the broken generators $\hat{Q}_a$ and $\hat{Q}_b$ are not independent, the gapless modes due to these two broken generators are not independent either and Eq. (3.313) does not give the correct number of Goldstone modes [25]. This also explains why in a superfluid the Goldstone mode associated with the spontaneous breaking of Galilean symmetry is redundant with phonons.

3.6.3.2 *Stiffness*

It is straightforward to extend Eq. (3.298) to the general case by using (3.309), the continuity equation (3.304), and the Bogoliubov inequality with the operators $\hat{A}(\mathbf{r},t)$ and $\partial_t\hat{\jmath}_0(\mathbf{r},t)$. Assuming that the divergence of $\chi_{AA}(\mathbf{q})$ is minimal ($\propto 1/\mathbf{q}^2$), this gives

$$\chi_{AA}(\mathbf{q}) = \frac{\langle\hat{B}\rangle^2}{R\mathbf{q}^2} \quad (\mathbf{q}\to 0), \qquad R \leq \lim_{\mathbf{q}\to 0}\frac{1}{\mathbf{q}^2}\int_{-\infty}^{\infty}\frac{d\omega}{\pi}\omega\chi''_{j_0 j_0}(\mathbf{q},\omega). \tag{3.322}$$

One can generally show that the bound (3.322) is finite for systems with sufficiently short-range interactions.

R measures the stiffness of the system with respect to spatial variation of the observable $\langle\hat{A}(\mathbf{r})\rangle$. To see this, let us consider the Gibbs free energy $G[\langle\hat{A}\rangle]$. Using the results of

[54] If for some reason $g_{ab} = 0$, one must include fourth-order space-derivative terms in the effective Lagrangian. One then finds the dispersion $\omega \sim |\mathbf{q}|^4$ for type-B modes and $\omega \sim \mathbf{q}^2$ for type-A modes.

[55] A similar analysis holds for superconductors where the order parameter is $\langle\hat{\psi}_{\mathbf{r}\downarrow}\hat{\psi}_{\mathbf{r}\uparrow}\rangle$ (see Sec. 7.5).

[56] $[\hat{N}, \hat{\psi}(\mathbf{r})] = \int d^d r' [\hat{\psi}^\dagger(\mathbf{r}')\hat{\psi}(\mathbf{r}'), \hat{\psi}(\mathbf{r})] = -\hat{\psi}(\mathbf{r})$.

Sec. 3.2.7, its variation can be written as

$$\Delta G[\langle\hat{A}\rangle] = \frac{1}{2}\sum_{\mathbf{q}}\langle\hat{A}(-\mathbf{q})\rangle\chi_{AA}^{-1}(\mathbf{q})\langle\hat{A}(\mathbf{q})\rangle = \frac{R}{2\langle\hat{B}\rangle^2}\int d^dr[\boldsymbol{\nabla}\langle\hat{A}(\mathbf{r})\rangle]^2 \tag{3.323}$$

to lowest order in $\boldsymbol{\nabla}\langle\hat{A}(\mathbf{r})\rangle$. The external field required to create the configuration $\langle\hat{A}(\mathbf{r})\rangle$ is given by $\delta G[\langle\hat{A}\rangle]/\delta\langle\hat{A}(\mathbf{r})\rangle$. Local variations of the expectation value of the symmetry-restoring variable $\langle\hat{A}(\mathbf{r})\rangle$ yield an elastic contribution to the Gibbs free energy with a stiffness constant $R/\langle\hat{B}\rangle^2$.

It can also be argued in many cases that R determines the stiffness with respect to both external perturbations and spontaneous fluctuations. Suppose we are able to integrate out all degrees of freedom but those related to the symmetry-restoring field. The simplest effective action describing (small) fluctuations of A about its equilibrium value reads

$$S_{\text{eff}}[A] = \frac{\rho}{2}\int_0^\beta d\tau\int d^dr\left\{[\boldsymbol{\nabla}A(\mathbf{r},\tau)]^2 + \frac{[\partial_\tau A(\mathbf{r},\tau)]^2}{c^2}\right\} \tag{3.324}$$

when A corresponds to a type-A Goldstone mode. This action gives the imaginary-time response function

$$\chi_{AA}(\mathbf{q},i\omega_n) = \frac{\rho^{-1}}{\mathbf{q}^2+\omega_n^2/c^2}, \tag{3.325}$$

from which we deduce the existence of a gapless mode $\omega = c|\mathbf{q}|$. The parameter ρ in (3.324) determines the stiffness of the system with respect to spontaneous fluctuations. Comparing $\chi_{AA}(\mathbf{q},i\omega_n)$ [Eq. (3.325)] with $\chi_{AA}(\mathbf{q}) = \chi_{AA}(\mathbf{q},i\omega_n=0)$ [Eq. (3.322)], we deduce $\rho = R/\langle\hat{B}\rangle^2$.

3.6.4 *Mermin–Wagner theorem*

The Mermin–Wagner theorem states that in one and two dimensions a continuous symmetry cannot be spontaneously broken at finite temperature in systems with sufficiently short-range interactions. Let us consider the local fluctuations

$$\left\langle\left(\hat{A}(\mathbf{r})-\langle\hat{A}(\mathbf{r})\rangle\right)^2\right\rangle = \int_{\mathbf{q}}\int_{-\infty}^{\infty}\frac{d\omega}{2\pi}S_{AA}(\mathbf{q},\omega) \tag{3.326}$$

of the symmetry-restoring variable $\hat{A}(\mathbf{r})$. $S_{AA}(\mathbf{q},\omega)$ is the structure factor defined in (3.53). Using the fluctuation–dissipation theorem, we obtain

$$\begin{aligned}\left\langle\left(\hat{A}(\mathbf{r})-\langle\hat{A}(\mathbf{r})\rangle\right)^2\right\rangle &= \int_{\mathbf{q}}\int_{-\infty}^{\infty}\frac{d\omega}{2\pi}\left[\coth\left(\frac{\beta\omega}{2}\right)+1\right]\chi''_{AA}(\mathbf{q},\omega)\\ &\geq 2T\int_{\mathbf{q}}\chi_{AA}(\mathbf{q}).\end{aligned} \tag{3.327}$$

The last inequality follows from $|x| \coth |x| \geq 1$ and $\chi''_{AA}(\mathbf{q}, -\omega) = -\chi''_{AA}(\mathbf{q}, \omega)$. Using Eq. (3.322), we see that the integral in (3.327) behaves as

$$\frac{\langle \hat{B} \rangle^2}{R} \int_0 |\mathbf{q}|^{d-3} d|\mathbf{q}| \tag{3.328}$$

for small $\mathbf{q}$. While the integral (3.328) is finite for $d > 2$, it diverges logarithmically when $d = 2$ and more strongly for $d < 2$. Since the fluctuations of $\hat{A}(\mathbf{r})$ are necessarily bounded, we conclude that we must have $\langle \hat{B} \rangle = 0$ for $d \leq 2$ and $T > 0$. This conclusion holds only if R is finite, which is the case for sufficiently short-range interactions.

This result can also be obtained from a classical model starting from the free energy functional $F[A] = \frac{\rho}{2} \int d^d r [\boldsymbol{\nabla} A(\mathbf{r})]^2 = \frac{\rho}{2} \sum_{\mathbf{q}} \mathbf{q}^2 A(-\mathbf{q}) A(\mathbf{q})$ describing energy fluctuations due to deviations from the perfectly ordered state. By the equipartition theorem $\rho \mathbf{q}^2 \langle A(-\mathbf{q}) A(\mathbf{q}) \rangle \sim T$, so that the local fluctuations

$$\langle A(\mathbf{r})^2 \rangle = \int_{\mathbf{q}} \langle A(-\mathbf{q}) A(\mathbf{q}) \rangle \sim \int_{\mathbf{q}} \frac{T}{\rho \mathbf{q}^2} \tag{3.329}$$

diverge in dimension $d \leq 2$.

3.6.5 *Quantum disorder at zero temperature*

The Mermin–Wagner theorem emphasizes the role of classical fluctuations at finite temperatures. In this section, we show that fluctuations can also suppress long-range order in the ground state of a quantum system in low dimensions. For two operators $\hat{A}$ and $\hat{B}$, we define the scalar product

$$\langle A | B \rangle = \langle 0 | [\hat{A}^\dagger, \hat{B}]_+ | 0 \rangle, \tag{3.330}$$

where $|0\rangle$ is the ground state (assumed unique) of the system. The Schwarz inequality implies

$$\langle 0 | [\hat{A}^\dagger, \hat{A}]_+ | 0 \rangle \langle 0 | [\hat{B}^\dagger, \hat{B}]_+ | 0 \rangle \geq |\langle 0 | [\hat{A}^\dagger, \hat{B}]_+ | 0 \rangle|^2. \tag{3.331}$$

We now introduce the auxiliary operator $\hat{\hat{B}}$ defined by

$$\langle n | \hat{\hat{B}} | 0 \rangle = \langle n | \hat{B} | 0 \rangle, \qquad \langle 0 | \hat{\hat{B}} | n \rangle = -\langle 0 | \hat{B} | n \rangle, \tag{3.332}$$

and therefore,

$$\langle n | \hat{\hat{B}}^\dagger | 0 \rangle = -\langle n | \hat{B}^\dagger | 0 \rangle, \qquad \langle 0 | \hat{\hat{B}}^\dagger | n \rangle = \langle 0 | \hat{B}^\dagger | n \rangle, \tag{3.333}$$

where $\{|n\rangle\}$ denotes a complete set of states. Using

$$\langle 0 | [\hat{A}^\dagger, \hat{B}]_+ | 0 \rangle = \sum_n \left(\langle 0 | \hat{A}^\dagger | n \rangle \langle n | \hat{B} | 0 \rangle + \langle 0 | \hat{B} | n \rangle \langle n | \hat{A}^\dagger | 0 \rangle \right), \tag{3.334}$$

one straightforwardly finds

$$\begin{aligned} \langle 0 | [\hat{A}^\dagger, \hat{\hat{B}}]_+ | 0 \rangle &= \langle 0 | [\hat{A}^\dagger, \hat{B}] | 0 \rangle, \\ \langle 0 | [\hat{\hat{B}}^\dagger, \hat{\hat{B}}]_+ | 0 \rangle &= \langle 0 | [\hat{B}^\dagger, \hat{B}]_+ | 0 \rangle. \end{aligned} \tag{3.335}$$

Applying Eq. (3.331) to $\hat{A}$ and $\hat{\tilde{B}}$ (instead of $\hat{A}$ and $\hat{B}$), thus gives

$$\langle 0|[\hat{A}^\dagger, \hat{A}]_+|0\rangle\langle 0|[\hat{B}^\dagger, \hat{B}]_+|0\rangle \geq |\langle 0|[\hat{A}^\dagger, \hat{B}]|0\rangle|^2 \tag{3.336}$$

(note the commutator, instead of the anticommutator, in the rhs). Equation (3.336) can be used to show the absence of long-range order in the ground state of some one-dimensional quantum systems.[57]

3.6.5.1 *Bose superfluid*

As an example, let us consider an interacting boson system. In three and two dimensions, a superfluid ground state is characterized by broken U(1) invariance and a non-vanishing value of $\langle\hat{\psi}(\mathbf{r})\rangle$, where $\hat{\psi}(\mathbf{r})$ is the annihilation boson operator.[58] With the operators

$$\begin{aligned} \hat{A} &= \hat{\psi}^\dagger(\mathbf{q}), \\ \hat{B} &= \hat{n}(-\mathbf{q}) = \frac{1}{\sqrt{\mathcal{V}}}\sum_{\mathbf{q}'}\hat{\psi}^\dagger(\mathbf{q}')\hat{\psi}(\mathbf{q}'-\mathbf{q}), \end{aligned} \tag{3.337}$$

Eq. (3.336) gives

$$2(2n_{\mathbf{q}}+1)\mathcal{S}(\mathbf{q}) \geq n_0 \quad \text{i.e.} \quad n_{\mathbf{q}} \geq \frac{n_0}{4\mathcal{S}(\mathbf{q})} - \frac{1}{2}, \tag{3.338}$$

where $n_{\mathbf{q}} = \langle\hat{\psi}^\dagger(\mathbf{q})\hat{\psi}(\mathbf{q})\rangle$ is the mean number of bosons with momentum $\mathbf{q}$, $n_0 = |\langle\hat{\psi}(\mathbf{r})\rangle|^2$ the condensate density[58] and $\mathcal{S}(\mathbf{q}) = \langle\hat{n}(\mathbf{q})\hat{n}(-\mathbf{q})\rangle$ the equal-time density–density correlation function (structure factor).[59]

Now, we note that

$$\mathcal{S}(\mathbf{q}) = \int_{-\infty}^{\infty}\frac{d\omega}{2\pi}\mathcal{S}(\mathbf{q},\omega) \leq \frac{1}{2\pi}\left[\int_0^\infty d\omega\,\omega\mathcal{S}(\mathbf{q},\omega)\int_0^\infty d\omega\frac{\mathcal{S}(\mathbf{q},\omega)}{\omega}\right]^{1/2}, \tag{3.339}$$

which follows from the fact that at zero temperature $\mathcal{S}(\mathbf{q},\omega)$ is non-zero only when $\omega \geq 0$ and is then positive.[60] The first integral in the rhs of (3.339) can be calculated using the fluctuation–dissipation theorem (3.58) and the f-sum rule (3.106),[61]

$$\int_0^\infty d\omega\,\omega\mathcal{S}(\mathbf{q},\omega) = \int_{-\infty}^{\infty} d\omega\,\omega\chi''_{nn}(\mathbf{q},\omega) = \frac{\pi n\mathbf{q}^2}{m}, \tag{3.340}$$

[57] The ground state of some one-dimensional systems may exhibit long-range order. For example, the non-interacting Bose gas exhibits (BEC) (see the following) and the ground state of a one-dimensional Bose fluid with sufficiently long-range interactions is a Wigner crystal with spontaneously broken translation invariance [28].

[58] See Chapters 7 and 13 for a discussion of superfluidity.

[59] We have used $\langle[\hat{\psi}(\mathbf{q}),\hat{n}(-\mathbf{q})]\rangle = \frac{1}{\sqrt{L}}\langle\hat{\psi}(\mathbf{q}=0)\rangle = \langle\hat{\psi}(\mathbf{r})\rangle$.

[60] This can be deduced from (3.55) or (3.58).

[61] We use $\chi''_{nn}(\mathbf{q},-\omega) = -\chi''_{nn}(\mathbf{q},\omega)$ [Eq. (3.101)].

with n the density of bosons. As for the second integral, we obtain

$$\int_0^\infty d\omega \frac{\mathcal{S}(\mathbf{q},\omega)}{\omega} = \int_{-\infty}^\infty d\omega \frac{\chi''_{nn}(\mathbf{q},\omega)}{\omega} \equiv \pi n^2 \kappa(\mathbf{q}), \tag{3.341}$$

where $\lim_{\mathbf{q}\to 0}\kappa(\mathbf{q})$ is nothing but the compressibility $\kappa = 1/nmc_s^2$ with c_s the sound velocity [Eq. (3.111)]. Thus, in the limit $\mathbf{q}\to 0$, Eq. (3.338) becomes

$$n_{\mathbf{q}} \geq \frac{n_0 m c_s}{2n|\mathbf{q}|} \qquad (\mathbf{q}\to 0). \tag{3.342}$$

Since the integral $\int_{\mathbf{q}} n_{\mathbf{q}} = n$ must be convergent, we conclude that BEC, i.e. $n_0 \neq 0$, is not possible in a one-dimensional Bose fluid. This conclusion does not hold for the ideal Bose gas since in this case the compressibility is infinite and c_s vanishes. Superfluidity in an interacting one-dimensional Bose fluid is nevertheless possible but without BEC (see Sec. 7.2.2.1 and Chapter 13).

3.6.5.2 *Quantum antiferromagnet*

As a second example, we consider a quantum antiferromagnet defined on a d-dimensional hypercubic lattice with order parameter

$$M_0 = \frac{1}{N}\sum_{\mathbf{r}} e^{i\mathbf{Q}\cdot\mathbf{r}}\langle \hat{S}^z_{\mathbf{r}}\rangle, \tag{3.343}$$

where $\mathbf{Q}=(\pi,\dots,\pi)$, N is the number of lattice sites, and $\hat{\mathbf{S}}^2_{\mathbf{r}} = S(S+1)$. We consider the inequality (3.336) with operators

$$\begin{aligned} \hat{A} &= \hat{S}^+_{\mathbf{q}+\mathbf{Q}} = \frac{1}{\sqrt{N}}\sum_{\mathbf{r}} e^{-i(\mathbf{q}+\mathbf{Q})\cdot\mathbf{r}}\hat{S}^+_{\mathbf{r}}, \\ \hat{B} &= \hat{S}^+_{\mathbf{q}} = \frac{1}{\sqrt{N}}\sum_{\mathbf{r}} e^{-i\mathbf{q}\cdot\mathbf{r}}\hat{S}^+_{\mathbf{r}}, \end{aligned} \tag{3.344}$$

where $\hat{S}^\pm_{\mathbf{r}} = \hat{S}^x_{\mathbf{r}} \pm \hat{S}^y_{\mathbf{r}} = (\hat{S}^\mp_{\mathbf{r}})^\dagger$. Using

$$\begin{aligned} \langle[\hat{S}^-_{-\mathbf{q}-\mathbf{Q}},\hat{S}^+_{\mathbf{q}}]\rangle &= \frac{1}{N}\sum_{\mathbf{r}} e^{i\mathbf{Q}\cdot\mathbf{r}}\langle[\hat{S}^-_{\mathbf{r}},\hat{S}^+_{\mathbf{r}}\rangle = -2M_0, \\ \langle[\hat{S}^-_{-\mathbf{q}},\hat{S}^+_{\mathbf{q}}]_+\rangle &= 2\langle \hat{S}^x_{-\mathbf{q}},\hat{S}^x_{\mathbf{q}} + \hat{S}^y_{-\mathbf{q}},\hat{S}^y_{\mathbf{q}}\rangle = 2\mathcal{S}_\perp(\mathbf{q}), \end{aligned} \tag{3.345}$$

we obtain

$$\mathcal{S}_\perp(\mathbf{q}+\mathbf{Q})\mathcal{S}_\perp(\mathbf{q}) \geq M_0^2, \tag{3.346}$$

where $\mathcal{S}_\perp(\mathbf{q})$ is the transverse (equal-time) structure factor.

We can now use (3.339) with

$$\int_0^\infty d\omega \frac{\mathcal{S}_\perp(\mathbf{q},\omega)}{\omega} = \int_{-\infty}^\infty d\omega \frac{\chi''_\perp(\mathbf{q},\omega)}{\omega} = \pi\chi_\perp(\mathbf{q}),$$
$$\int_0^\infty d\omega\, \omega \mathcal{S}_\perp(\mathbf{q},\omega) = c^2_{\text{eff}}(\mathbf{q})\mathbf{q}^2 \int_0^\infty d\omega \frac{\mathcal{S}_\perp(\mathbf{q},\omega)}{\omega}, \tag{3.347}$$

where $c_{\text{eff}}(\mathbf{q})$ is an effective spin-wave velocity and $\chi_\perp$ the transverse spin susceptibility,[62] which leads to

$$\mathcal{S}_\perp(\mathbf{q}+\mathbf{Q}) \geq \frac{2M_0^2}{\chi_\perp(\mathbf{q})c_{\text{eff}}(\mathbf{q})|\mathbf{q}|}. \tag{3.348}$$

Since the integral ($N \to \infty$)

$$\int_{\mathbf{q}} \mathcal{S}_\perp(\mathbf{q}) \equiv \frac{1}{N}\sum_{\mathbf{q}} \mathcal{S}_\perp(\mathbf{q}) = \frac{1}{N}\sum_{\mathbf{r}} \langle \hat{S}^{x2}_{\mathbf{r}} + \hat{S}^{y2}_{\mathbf{r}} \rangle \leq S(S+1) \tag{3.349}$$

over the first Brillouin zone is finite, we conclude that long-range antiferromagnetic order, i.e. $M_0 \neq 0$, is not possible in one dimension if $\chi_\perp(\mathbf{q})$ and $c_{\text{eff}}(\mathbf{q})$ are finite in the limit $\mathbf{q} \to 0$. There is no reason for $\lim_{\mathbf{q}\to 0} \chi_\perp(\mathbf{q})$ to be infinite, since a uniform transverse field is expected to produce a non-zero restoring force in an antiferromagnet. In the Heisenberg model, one can obtain an explicit expression of $\int_0^\infty d\omega\, \omega \mathcal{S}_\perp(\mathbf{q},\omega) = \pi\chi_\perp(\mathbf{q})c^2_{\text{eff}}(\mathbf{q})\mathbf{q}^2$, which leads to a finite value of $c_{\text{eff}}(\mathbf{q})$ in the limit $\mathbf{q} \to 0$ [27].

62 For a system with a spin-wave mode (with velocity c) as sole transverse mode at low energies, one would have $\chi''_\perp(\mathbf{q},\omega) \propto \delta(\omega - c|\mathbf{q}|) - \delta(\omega + c|\mathbf{q}|)$ and therefore $c_{\text{eff}}(\mathbf{q}) = c$.

Guide to the bibliography

- Linear response theory is discussed in many textbooks [1–4].
- For the electromagnetic response of fermion systems, see Refs. [3–9]
- Properties of the one-particle Green function are discussed in Refs. [3, 9, 10].
- Spontaneous symmetry breaking and its consequences (Goldstone's theorem [20–22], etc.) are reviewed in Refs. [1, 14, 15]. The Mermin–Wagner theorem, also known as the Mermin–Wagner–Hohenberg theorem [11] or the Coleman–Mermin–Wagner theorem [12], is proven in Ref. [13].
- For the counting of Goldstone modes, see Refs. [23–26].
- Quantum disorder at zero temperature in one-dimensional systems is discussed in Ref. [27].

Bibliography

[1] D. Forster, *Hydrodynamic Fluctuations, Broken Symmetry, and Correlation Functions* (Addison-Wesley, Reading, Massachusetts, 1975).

[2] P. M. Chaikin and T. C. Lubensky, *Principles of Condensed Matter Physics* (Cambridge University Press, Cambridge, 1995).

[3] G. F. Giuliani and G. Vignale, *Quantum Theory of the Electron Liquid*, Chapter 3 (Cambridge University Press, Cambridge, 2005).

[4] A.-M. Tremblay, *Quantum Material's Theory, from perturbation theory to dynamical-mean field theory*, https://www.physique.usherbrooke.ca/pages/tremblay.

[5] D. Pines and P. Nozières, *The Theory of Quantum Liquids* (Addison-Wesley, Reading, Massachusetts, 1968).

[6] P. C. Martin, Sum rules, Kramers-Kronig relations, and transport coefficients in charged systems, *Phys. Rev.* **161**, 143 (1967).

[7] W. Kohn, Theory of the insulating state, *Phys. Rev.* **133**, A171 (1964).

[8] D. J. Scalapino, S. R. White, and S. Zhang, Insulator, metal, or superconductor: The criteria, *Phys. Rev.* B **47**, 7995 (1993).

[9] A. A. Abrikosov, L. P. Gor'kov, and I. E. Dzyaloshinski, *Methods of Quantum Field Theory in Statistical Physics*, Chapter 6 (Dover, Mineola, New York, 1975).

[10] A. L. Fetter and J. D. Walecka, *Quantum Theory of Many-Particle Systems* (Dover, Mineola, New York, 2003).

[11] P. C. Hohenberg, Existence of long-range order in one and two dimensions, *Phys. Rev.* **158**, 383 (1967).

[12] S. Coleman, There are no Goldstone bosons in two dimensions, *Comm. Math. Phys.* **31**, 259 (1973).

[13] N. D. Mermin and H. Wagner, Absence of ferromagnetism or antiferromagnetism in one- or two-dimensional isotropic Heisenberg models, *Phys. Rev. Lett.* **17**, 1133 (1966).

[14] P. W. Anderson, *Basic Notions of Condensed Matter Physics* (Advanced Book Classics, Menlo Park, California, 1984).

[15] A. J. Beekman, L. Rademaker, and J. van Wezel, An introduction to spontaneous symmetry breaking, *SciPost Phys. Lect. Notes 11* (2019).

[16] L. P. Kadanoff and P. C. Martin, Hydrodynamic equations and correlation functions, *Ann. Phys.* **24**, 419 (1963).

[17] P. W. Anderson, An approximate quantum theory of the antiferromagnetic ground state, *Phys. Rev.* **86**, 694 (1952).

[18] W. Marshall, Antiferromagnetism, *Proc. R. Soc. Lond.* A **232**, 48 (1955).

[19] R. V. Lange, Nonrelativistic theorem analogous to the Goldstone theorem, *Phys. Rev.* **146**, 301 (1966).

[20] J. Goldstone, Field theories with "superconductor" solutions, *Nuovo Cimento* **19**, 154 (1961).

[21] Y. Nambu, Axial vector current conservation in weak interactions, *Phys. Rev. Lett.* **4**, 380 (1960).

[22] J. Goldstone, A. Salam, and S. Weinberg, Broken symmetries, *Phys. Rev.* **127**, 965 (1962).

[23] I. Low and A. V. Manohar, Spontaneously broken spacetime symmetries and Goldstone's theorem, *Phys. Rev. Lett.* **88**, 101602 (2002).

[24] H. Watanabe and H. Murayama, Unified description of Nambu-Goldstone bosons without Lorentz invariance, *Phys. Rev. Lett.* **108**, 251602 (2012).

[25] H. Watanabe and H. Murayama, Redundancies in Nambu-Goldstone bosons, *Phys. Rev. Lett.* **110**, 181601 (2013).

[26] H. Watanabe, Counting Rules of Nambu-Goldstone Modes, *Ann. Rev. Cond. Mat. Phys.* **11**, 169 (2020).

[27] L. Pitaevskii and S. Stringari, Uncertaintly principle, quantum fluctuations, and broken symmetries, *J. Low. Temp. Phys.* **85**, 377 (1991).

[28] R. Daviet and N. Dupuis, Mott-glass phase of a one-dimensional quantum fluid with long-range interactions, *Phys. Rev. Lett.* **125**, 235301 (2020).

[29] N. Dupuis, Infrared behavior and spectral function of a Bose superfluid at zero temperature, *Phys. Rev.* A **80**, 043627 (2009).

[30] H. Tasaki, Long-range order, "tower" of states, and symmetry breaking in lattice quantum systems, *J. Stat. Phys.* **174**, 735 (2019).

[31] S. Elitzur, Impossibility of spontaneously breaking local symmetries, *Phys. Rev.* D **12**, 3978 (1975).

Part 2

Quantum Many-Body Systems (I)

Chapter 4

Fermi-Liquid Theory

Contents

4.1 The Quasi-Particle Concept . . . 314
- 4.1.1 The ideal Fermi gas . . . 315
- 4.1.2 The interacting Fermi liquid . . . 316
- 4.1.3 Landau energy functional $E[n]$ . . . 318
- 4.1.4 Stability of the ground state . . . 320
- 4.1.5 Effective mass . . . 321

4.2 Thermodynamics . . . 322
- 4.2.1 Specific heat . . . 322
- 4.2.2 Compressibility . . . 322
- 4.2.3 Spin susceptibility . . . 323

4.3 Non-Equilibrium Properties . . . 324
- 4.3.1 Kinetic equation . . . 325
- 4.3.2 Conservation laws . . . 325
- 4.3.3 Collective modes . . . 327
- 4.3.4 Response functions . . . 332
- 4.3.5 Multi-pair excitations . . . 338

4.4 Microscopic Basis of Fermi-Liquid Theory . . . 339
- 4.4.1 Quasi-particles . . . 340
- 4.4.2 Thermodynamic potential $\Omega[n]$ . . . 347
- 4.4.3 Quantum Boltzmann equation . . . 353
- 4.4.4 Ward identities for the Fermi liquid . . . 355
- 4.4.5 Response to external fields . . . 357
- 4.4.6 Luttinger theorem . . . 360

4.5 Fermi Liquids in Low Dimensions . . . 362

Appendix 4.A Second-Order Self-energy . . . 363

Bibliography . . . 367

One of the main goals of this book is to introduce the theoretical methods necessary to study (strongly) correlated quantum systems. This chapter is devoted to a class of fermion systems, known as Landau Fermi liquids or merely Fermi liquids, that can be understood without resorting to sophisticated many-body techniques. The Fermi-liquid paradigm describes not only the low-energy behavior of electrons in metals but also Fermi systems, such as ^{3}He, nuclear matter, or ultracold fermion gases.

In a Fermi liquid, the elementary excitations (quasi-particles and quasi-holes) are in direct correspondence with the (particle or hole) excitations of the ideal Fermi gas; they carry the same quantum numbers and satisfy the Fermi–Dirac statistics.[1,2] This correspondence can be made explicit by means of an adiabatic switching-on of the interactions. The quasi-particles and quasi-holes determine both the low-temperature thermodynamics and the response of the system to macroscopic perturbations. In a very elegant phenomenological theory [1–3], Landau has shown that the low-energy behavior of the system can be expressed in terms of a few unknown parameters (the Landau parameters) that depend on the interactions between quasi-particles.

In the first part of the chapter (Secs. 4.1, 4.2 and 4.3), we review the main aspects of Landau's Fermi-liquid theory starting from the quasi-particle concept. We mainly consider neutral Fermi liquids.[3] In the second part (Sec. 4.4), we discuss the microscopic underpinning of Fermi-liquid theory. Except in Secs. 4.4.6 and 4.5, we consider only isotropic three-dimensional systems.

4.1 The Quasi-Particle Concept

Landau's Fermi-liquid theory relies on the assumption that the low-lying eigenstates of the ideal Fermi gas continuously evolve into eigenstates of the real system as the interaction is adiabatically switched on. The quasi-particle concept, which is the starting point of Fermi-liquid theory, is a direct consequence of this assumption.

Before discussing this concept in detail, it should be noted that the adiabatic continuity assumption is quite restrictive and there are a number of cases where it is obviously violated. For instance, in a superconductor — and more generally whenever an instability of the Fermi surface leads to a broken symmetry state — the ground state is not related in any direct way to any one state of the free Fermi gas but rather to a coherent superposition of a large number of states. Fermi-liquid theory can also break down without the occurrence of a broken symmetry state as in the one-dimensional interacting fermion gas (Chapter 15).

[1]Generally, the term "quasi-particles" refers to the elementary excitations whatever their relation to the bare particles. In the Fermi-liquid theory context, it has a narrower sense as explained in Sec. 4.1.

[2]The Fermi-liquid theory elucidates the success of the ideal Fermi gas model in explaining some physical properties of electrons in metals despite the importance of the Coulomb interaction at metallic densities (this latter point is discussed in detail in Chapter 5).

[3]The electromagnetic response of charged systems was the subject of Sec. 3.4. The electron gas will be studied in Chapter 5 within the framework of the RPA.

4.1.1 *The ideal Fermi gas*

Let us start with an ideal gas of spin-$\frac{1}{2}$ fermions. The eigenstates are antisymmetric combinations of plane waves, and a state of the system is fully determined by the momentum distribution function $n_{\mathbf{k}\sigma}$ giving the number of particles with momentum $\mathbf{k}$ and spin σ. The ground state corresponds to the distribution function $n^0_{\mathbf{k}} = \Theta(k_F - |\mathbf{k}|)$ where the Fermi momentum k_F is related to the mean particle density

$$n = \frac{1}{\mathcal{V}} \sum_{\mathbf{k},\sigma} \Theta(k_F - |\mathbf{k}|) = 2 \int_{\mathbf{k}} \Theta(k_F - |\mathbf{k}|) = \frac{k_F^3}{3\pi^2} \tag{4.1}$$

(the sum over σ gives a factor 2). The ground state energy is given by

$$E_0 = \sum_{\substack{\mathbf{k},\sigma \\ |\mathbf{k}| \leq k_F}} \epsilon^0_{\mathbf{k}} = \frac{3}{5} n \epsilon^0_F \mathcal{V}, \tag{4.2}$$

where $\epsilon^0_{\mathbf{k}} = \mathbf{k}^2/2m$ is the free fermion dispersion and $\epsilon^0_F = k_F^2/2m = \mu(T=0)$ the Fermi energy.

Low-lying excited states are defined by their distribution function

$$n_{\mathbf{k}\sigma} = n^0_{\mathbf{k}} + \delta n_{\mathbf{k}\sigma}. \tag{4.3}$$

The change in the total energy corresponding to $\delta n_{\mathbf{k}\sigma}$ is

$$\delta E[\delta n] = E[n] - E_0 = \sum_{\mathbf{k},\sigma} \epsilon^0_{\mathbf{k}} \delta n_{\mathbf{k}\sigma}. \tag{4.4}$$

We denote by $n \equiv \{n_{\mathbf{k}\sigma}\}$ the momentum distribution function (not to be confused with the mean particle density). The particle energy can be defined as the functional derivative of the total energy with respect to the momentum distribution function $n_{\mathbf{k}\sigma}$,

$$\epsilon^0_{\mathbf{k}} = \frac{\delta E[n]}{\delta n_{\mathbf{k}\sigma}}. \tag{4.5}$$

The particle group velocity and the Fermi velocity are obtained from

$$\mathbf{v}_{\mathbf{k}} = \boldsymbol{\nabla}_{\mathbf{k}} \epsilon^0_{\mathbf{k}}, \qquad v_F = |\mathbf{v}_{\mathbf{k}}|_{|\mathbf{k}|=k_F} = \left. \frac{\partial \epsilon^0_{\mathbf{k}}}{\partial |\mathbf{k}|} \right|_{|\mathbf{k}|=k_F} = \frac{k_F}{m}, \tag{4.6}$$

respectively. Near the Fermi surface $|\mathbf{k}| = k_F$, one can therefore write the dispersion law as

$$\epsilon^0_{\mathbf{k}} = \epsilon^0_F + v_F(|\mathbf{k}| - k_F) + \mathcal{O}\left((|\mathbf{k}| - k_F)^2\right). \tag{4.7}$$

An elementary excitation corresponds to a particle added to or removed from the ground state. Any excited eigenstate of the system can be constructed by creating a certain number of these elementary particle or hole excitations. Since the latter are non-interacting, the total energy δE of the excited state is simply the sum of the particle and hole excitation energies [Eq. (4.4)].

4.1.2 *The interacting Fermi liquid*

According to the central hypothesis of Landau's Fermi-liquid theory, any state of the ideal Fermi gas, characterized by a momentum distribution function $n_{\mathbf{k}\sigma} = n^0_{\mathbf{k}} + \delta n_{\mathbf{k}\sigma}$, generates an eigenstate of the interacting system as the interactions are switched on. This eigenstate can therefore be labeled by the distribution function $n_{\mathbf{k}\sigma}$. For reasons given below, this distribution function is referred to as the quasi-particle distribution function of the interacting Fermi liquid. As we shall see in Sec. 4.4.1, it differs from — and should not be confused with — the momentum distribution $\langle \hat{\psi}^\dagger_\sigma(\mathbf{k})\hat{\psi}_\sigma(\mathbf{k})\rangle$ of the interacting system.

For reasons of symmetry, the Fermi surface of the interacting (isotropic) system is spherical.[4] Basic to Fermi-liquid theory is the fact that the volume of the Fermi surface is not changed by interactions (Luttinger theorem, derived in Sec. 4.4.6) so that the Fermi momentum k_F of the interacting system is the same as that of the ideal gas. The ground state of the interacting system is then generated adiabatically from that of the ideal gas.[5]

Let us now add a particle with momentum $\mathbf{k}$ ($|\mathbf{k}| > k_F$) and spin σ to the ground state of the ideal gas. According to the adiabatic continuity assumption, as the interaction is slowly turned on we generate an (excited) eigenstate of the interacting system. However, because of the interactions the state under study is damped and acquires a finite lifetime. Central to Fermi-liquid theory is the fact that the lifetime becomes larger and larger at low energies ($|\mathbf{k}| \to k_F$). This property is a consequence of the Pauli principle which makes interactions ineffective near the Fermi surface (Sec. 4.4.1).[6] Thus, the state obtained by adding a low-lying ($|\mathbf{k}| \gtrsim k_F$) particle to the non-interacting Fermi sea evolves into a quasi-eigenstate of the interacting system, which is referred to as a quasi-particle. Similarly, one can define a quasi-hole by removing a particle with momentum $|\mathbf{k}| \lesssim k_F$ from the non-interacting Fermi sea and switching on the interactions adiabatically. Since the total momentum and spin are conserved, quasi-particles and quasi-holes can be labeled by the same quantum numbers as in the non-interacting case, namely the momentum $\mathbf{k}$ and the spin projection σ along a given axis.[7]

Near the Fermi surface, the quasi-particle dispersion can be expanded as

$$\epsilon_{\mathbf{k}} = \epsilon_{k_F} + v^*_F(|\mathbf{k}| - k_F) + \mathcal{O}\left((|\mathbf{k}| - k_F)^2\right), \qquad v^*_F = \frac{k_F}{m^*}, \tag{4.8}$$

[4]Note that at this stage we have not rigorously defined the Fermi surface of an interacting system; this will be done in Sec. 4.4.1.

[5]This is not true in anisotropic systems, where in general interactions deform the Fermi surface. The ground state then follows adiabatically from an excited state of the non-interacting system.

[6]This result can be obtained from a simple phase space argument. Consider the process where a particle above the Fermi sea ($|\mathbf{k}| > k_F$) is scattered into the state $\mathbf{k}+\mathbf{q}$ ($|\mathbf{k}+\mathbf{q}| > k_F$) by creating a particle–hole pair $(\mathbf{k}', \mathbf{k}'-\mathbf{q})$ ($|\mathbf{k}'| < k_F$ and $|\mathbf{k}'-\mathbf{q}| > k_F$). Because of energy conservation, $\omega = \epsilon_{\mathbf{k}+\mathbf{q}} - \epsilon_{\mathbf{k}} = -\epsilon_{\mathbf{k}'-\mathbf{q}} + \epsilon_{\mathbf{k}'} < 0$, the phase phase available for this scattering process is proportional to $(|\mathbf{k}| - k_F)^2$. This can be seen by evaluating the integral $\int d^3k' d^3q\, \delta(\epsilon_{\mathbf{k}+\mathbf{q}} + \epsilon_{\mathbf{k}'-\mathbf{q}} - \epsilon_{\mathbf{k}} + \epsilon_{\mathbf{k}'})$ where the momenta satisfied the above-mentioned constraints. Higher-order processes, involving multi-pair excitations, are more strongly suppressed as the corresponding phase space is smaller.

[7]In a charged Fermi liquid, the quasi-particles (holes) carry charge e ($-e$) where e is the charge of the bare particles.

which defines the (renormalized) Fermi velocity v_F^* and the effective mass m^*. $\epsilon_F = \mu(T=0)$ is the Fermi energy. The quasi-particle group velocity is defined by

$$\mathbf{v}_{\mathbf{k}}^* = \boldsymbol{\nabla}_{\mathbf{k}} \epsilon_{\mathbf{k}} \to v_F^* \hat{\mathbf{k}} \quad \text{for} \quad |\mathbf{k}| \to k_F. \tag{4.9}$$

At the Fermi level, the density of quasi-particle states (per spin) $N^*(\xi) = \int_{\mathbf{k}} \delta(\xi - \xi_{\mathbf{k}})$ takes the value

$$N^*(0) = \frac{m^* k_F}{2\pi^2}. \tag{4.10}$$

The only difference with the case of the ideal Fermi gas is that the mass is replaced by the effective mass.

A generic low-lying excited state of the ideal Fermi gas, defined by its momentum distribution function $n_{\mathbf{k}\sigma} \neq n_{\mathbf{k}}^0$, can be constructed by combining particle and hole excitations. As the interaction is switched on, it evolves into a quasi-eigenstate of the interacting system characterized by the distribution function $n_{\mathbf{k}\sigma}$.[8,9] Because of the one-to-one correspondence between particle (or hole) excitations in the ideal Fermi gas and quasi-particle excitations in the Fermi liquid, quasi-particles follow the Fermi–Dirac statistics. Since the concept of quasi-particles refers only to low-lying excited states, $\delta n_{\mathbf{k}\sigma} = n_{\mathbf{k}\sigma} - n_{\mathbf{k}}^0$ should be appreciable only in the vicinity of the Fermi surface.

More precisely, for the notion of quasi-particles (or quasi-holes) to make sense, their lifetime $\tau_{\mathbf{k}}$ and excitation energy $\xi_{\mathbf{k}} = \epsilon_{\mathbf{k}} - \mu$ should satisfy

$$\frac{1}{\tau_{\mathbf{k}}} \ll |\xi_{\mathbf{k}}|, \tag{4.11}$$

since $1/|\xi_{\mathbf{k}}|$ is the minimum time required to observe (or create with an external field) the quasi-particle. We shall see in Sec. 4.4.1 that in a three-dimensional Fermi liquid[6]

$$\frac{1}{\tau_{\mathbf{k}}} = \mathcal{O}\left((|\mathbf{k}| - k_F)^2\right) \tag{4.12}$$

at zero temperature, so that the condition (4.11) is satisfied in the vicinity of the Fermi surface (recall that $\xi_{\mathbf{k}} = \mathcal{O}(|\mathbf{k}| - k_F)$). Suppose that the interaction is switched on within a characteristic time η^{-1}: $\hat{H}_{\text{int}}(t) = \hat{H}_{\text{int}}(t=0)e^{\eta t}$. Quasi-particles can be observed if their lifetime is larger than η^{-1} and $1/|\xi_{\mathbf{k}}|$ smaller than η^{-1}, i.e.

$$\frac{1}{\tau_{\mathbf{k}}} \ll \eta \ll |\xi_{\mathbf{k}}|. \tag{4.13}$$

When condition (4.11) is fulfilled, it is possible to satisfy the inequality (4.13).

It should be emphasized that quasi-particle and quasi-hole excitations are not necessarily the only elementary excitations in the interacting system. The adiabatic continuity hypothesis does not exclude the possibility of other elementary excitations of the real system

[8] Note that the term "quasi-particle" often refers to both quasi-particles and quasi-holes. We shall explicitly distinguish between quasi-particles ($|\mathbf{k}| > k_F$) and quasi-holes ($|\mathbf{k}| < k_F$) only when necessary.

[9] If the system is anisotropic in spin space and the spin projection not a good quantum number, the quasi-particle distribution function $n_{\mathbf{k}\sigma\sigma'}$ becomes a matrix in spin space.

which disappear when the interaction is reduced to zero. These states correspond to collective excitations and emerge naturally in Landau Fermi-liquid theory (Sec. 4.3).

4.1.3 *Landau energy functional $E[n]$*

In the interacting Fermi liquid, the change in energy due a change $\delta n = n - n^0$ in the distribution function reads

$$\delta E[\delta n] = \sum_{\mathbf{k},\sigma} \epsilon_{\mathbf{k}} \delta n_{\mathbf{k}\sigma} \tag{4.14}$$

to first order in δn. Here, $\epsilon_{\mathbf{k}}$ is the energy of a single quasi-particle added to the ground state of the system (as defined in the preceding section by (4.8)). According to (4.14), there is no interaction between quasi-particles or quasi-holes, since the total energy is simply additive. This suggests to push (4.14) one step further,

$$\delta E[\delta n] = \sum_{\mathbf{k},\sigma} \epsilon_{\mathbf{k}} \delta n_{\mathbf{k}\sigma} + \frac{1}{2\mathcal{V}} \sum_{\mathbf{k},\mathbf{k}',\sigma,\sigma'} f_{\sigma\sigma'}(\mathbf{k},\mathbf{k}') \delta n_{\mathbf{k}\sigma} \delta n_{\mathbf{k}'\sigma'}. \tag{4.15}$$

The quadratic term in (4.15) is due to the interactions between quasi-particles. The Landau function f is defined as the second-order functional derivative of the total energy,

$$\frac{1}{\mathcal{V}} f_{\sigma\sigma'}(\mathbf{k},\mathbf{k}') = \left. \frac{\delta^2 E[n]}{\delta n_{\mathbf{k}\sigma} \delta n_{\mathbf{k}'\sigma'}} \right|_{n=n^0}, \tag{4.16}$$

and is therefore symmetric under the exchange $(\mathbf{k},\sigma) \leftrightarrow (\mathbf{k}',\sigma')$.

In the grand canonical ensemble at $T = 0$, the relevant thermodynamic potential is $\Omega(T = 0) = E - \mu N$, where $N = \sum_{\mathbf{k},\sigma} n_{\mathbf{k}\sigma}$ is the total number of quasi-particles (see in the following). Its variation is given by

$$\delta E[\delta n] - \mu \delta N[\delta n] = \sum_{\mathbf{k},\sigma} (\epsilon_{\mathbf{k}} - \mu) \delta n_{\mathbf{k}\sigma} + \frac{1}{2\mathcal{V}} \sum_{\mathbf{k},\mathbf{k}',\sigma,\sigma'} f_{\sigma\sigma'}(\mathbf{k},\mathbf{k}') \delta n_{\mathbf{k}\sigma} \delta n_{\mathbf{k}'\sigma'}. \tag{4.17}$$

Suppose that $\delta n_{\mathbf{k}\sigma}$ is appreciable only for $||\mathbf{k}| - k_F| \lesssim \delta$. Then, both terms in (4.17) are of order δ^2, which shows the need to push the expansion in δn to second order.

The energy $\tilde{\epsilon}_{\mathbf{k}}$ of a quasi-particle is defined as the variation of the total energy of the system due to the introduction of this quasi-particle. Mathematically, this means that $\tilde{\epsilon}_{\mathbf{k}}$ is given by the functional derivative of $E[n]$ with respect to the distribution function $n_{\mathbf{k}\sigma}$,

$$\tilde{\epsilon}_{\mathbf{k}} = \frac{\delta E[n]}{\delta n_{\mathbf{k}\sigma}} = \epsilon_{\mathbf{k}} + \int_{\mathbf{k}'} \sum_{\sigma'} f_{\sigma\sigma'}(\mathbf{k},\mathbf{k}') \delta n_{\mathbf{k}'\sigma'}. \tag{4.18}$$

Thus, the quasi-particle energy $\tilde{\epsilon}_{\mathbf{k}} \equiv \tilde{\epsilon}_{\mathbf{k}}[\delta n]$ depends on the distribution $\delta n_{\mathbf{k}\sigma} = n_{\mathbf{k}\sigma} - n^0_{\mathbf{k}}$ of quasi-particles present in the system; it coincides with $\epsilon_{\mathbf{k}}$ only when $\delta n = 0$. In an isotropic liquid, spin-rotation invariance ensures that $\tilde{\epsilon}_{\mathbf{k}}$ is independent of σ. Equation (4.18) gives the quasi-particle energy change coming from the average field due to the other quasi-particles. This mean-field-like description is characteristic of Landau's Fermi-liquid theory. It also shows up in the RPA form of the response functions (Sec. 4.3.4).

4.1.3.1 *Landau parameters*

The Landau function f plays a crucial role in Fermi-liquid theory. Spin-rotation invariance implies that it can be written in terms of a spin symmetric and a spin antisymmetric part,

$$f_{\sigma\sigma'}(\mathbf{k},\mathbf{k}') = f^s(\mathbf{k},\mathbf{k}') + \sigma\sigma' f^a(\mathbf{k},\mathbf{k}'). \tag{4.19}$$

Furthermore, for states near the Fermi surface one can set $|\mathbf{k}| = |\mathbf{k}'| = k_F$ so that $f_{\sigma\sigma'}(\mathbf{k},\mathbf{k}')$ depends only on the angle θ between $\mathbf{k}_F = k_F\hat{\mathbf{k}}$ and $\mathbf{k}'_F = k_F\hat{\mathbf{k}}'$,

$$f_{\sigma\sigma'}(\mathbf{k},\mathbf{k}') = f_{\sigma\sigma'}(\mathbf{k}_F,\mathbf{k}'_F) = f^s(\theta) + \sigma\sigma' f^a(\theta), \tag{4.20}$$

where the functions $f^s(\theta)$ and $f^a(\theta)$ can be expanded in Legendre polynomials,

$$f^{s,a}(\theta) = \sum_{l=0}^{\infty} f_l^{s,a} P_l(\cos\theta), \qquad f_l^{s,a} = (2l+1)\int_0^{\pi} \frac{d\Omega}{4\pi} f^{s,a}(\theta)P_l(\cos\theta) \tag{4.21}$$

($d\Omega = d\varphi d\theta \sin\theta$ denotes the elementary solid angle in the direction (φ,θ)). It is convenient to introduce dimensionless parameters — the Landau parameters — by multiplying $f_l^{s,a}$ by the density of states at the Fermi level,

$$F_l^{s,a} = 2N^*(0) f_l^{s,a} \tag{4.22}$$

(recall that $N^*(\xi)$ is the quasi-particle density of states per spin and $2N^*(\xi)$ the total density of states).

4.1.3.2 *Entropy and thermodynamic potential*

Since quasi-particles obey the Fermi–Dirac statistics, their entropy takes the form

$$S[n] = -\sum_{\mathbf{k},\sigma} \left[n_{\mathbf{k}\sigma} \ln n_{\mathbf{k}\sigma} + (1 - n_{\mathbf{k}\sigma}) \ln(1 - n_{\mathbf{k}\sigma}) \right]. \tag{4.23}$$

The thermodynamic potential is given by

$$\Omega[n] = E[n] - \mu N[n] - TS[n], \tag{4.24}$$

where $N[n] = \sum_{\mathbf{k},\sigma} n_{\mathbf{k}\sigma}$ is the total quasi-particle number. The equilibrium distribution function $\bar{n} \equiv \{\bar{n}_{\mathbf{k}\sigma}\}$ is obtained from the stationarity condition $\delta\Omega[n]/\delta n_{\mathbf{k}\sigma} = 0$,

$$\bar{n}_{\mathbf{k}\sigma} = n_F(\tilde{\xi}_{\mathbf{k}}), \tag{4.25}$$

where

$$\tilde{\epsilon}_{\mathbf{k}} = \left. \frac{\delta E[n]}{\delta n_{\mathbf{k}\sigma}} \right|_{n=\bar{n}} = \epsilon_{\mathbf{k}} + \int_{\mathbf{k}'} \sum_{\sigma'} f_{\sigma\sigma'}(\mathbf{k},\mathbf{k}') \left(\bar{n}_{\mathbf{k}'\sigma'} - n^0_{\mathbf{k}'\sigma'} \right) \tag{4.26}$$

is the quasi-particle energy corresponding to the equilibrium distribution $\bar{n}$. If we expand $\Omega[n]$ about its equilibrium value, we obtain[10]

$$\Omega[\bar{n} + \delta n] - \Omega[\bar{n}] = \frac{1}{2} \sum_{\mathbf{k},\mathbf{k}',\sigma,\sigma'} \left[-\frac{\delta_{\sigma,\sigma'}\delta_{\mathbf{k},\mathbf{k}'}}{n'_F(\tilde{\xi}_{\mathbf{k}})} + \frac{1}{\mathcal{V}} f_{\sigma\sigma'}(\mathbf{k},\mathbf{k}') \right] \delta n_{\mathbf{k}\sigma} \delta n_{\mathbf{k}'\sigma'} \tag{4.27}$$

[10]The first term in the rhs of (4.27) comes from $\left. \frac{\delta^2 S[n]}{\delta n_{\mathbf{k}\sigma}\delta n_{\mathbf{k}'\sigma'}} \right|_{n=\bar{n}} = -\frac{\delta_{\mathbf{k},\mathbf{k}'}\delta_{\sigma,\sigma'}}{\bar{n}_{\mathbf{k}\sigma}(1-\bar{n}_{\mathbf{k}\sigma})} = \delta_{\mathbf{k},\mathbf{k}'}\delta_{\sigma,\sigma'}\beta/n'_F(\tilde{\xi}_{\mathbf{k}})$.

to lowest order in δn. There is no linear term since $\Omega[n]$ is stationary for $n = \bar{n}$. Equation (4.27) shows that the f function can also be defined from the thermodynamic potential,

$$\frac{1}{\mathcal{V}} f_{\sigma\sigma'}(\mathbf{k},\mathbf{k}') = \frac{\delta_{\sigma,\sigma'}\delta_{\mathbf{k},\mathbf{k}'}}{n'_F(\tilde{\xi}_\mathbf{k})} + \left.\frac{\delta^2\Omega[n]}{\delta n_{\mathbf{k}\sigma}\delta n_{\mathbf{k}'\sigma'}}\right|_{n=\bar{n}}. \tag{4.28}$$

This relation will be used in Sec. 4.4.2 to obtain a microscopic definition of the Landau function.

4.1.4 *Stability of the ground state*

Because of the thermal factor $1/n'_F(\tilde{\xi}_\mathbf{k})$ in (4.27), small variations of the thermodynamic potential are due to quasi-particle excitations lying in the thermal broadening of the Fermi surface ($|\tilde{\xi}_\mathbf{k}| \lesssim T$). When $T \to 0$, these excitations have vanishing energies and can be viewed as resulting from a displacement (that can depend on spin) of the Fermi surface. Suppose that in the direction $\hat{\mathbf{k}}$, the Fermi momentum k_F varies by an infinitesimal amount $u_\sigma(\hat{\mathbf{k}})$ for spin-σ particles. This induces a change

$$\begin{aligned}\delta n_{\mathbf{k}\sigma} &= n_F[\tilde{\xi}_\mathbf{k} - v_F^* u_\sigma(\hat{\mathbf{k}})] - n_F(\tilde{\xi}_\mathbf{k}) \\ &= -v_F^* n'_F(\tilde{\xi}_\mathbf{k}) u_\sigma(\hat{\mathbf{k}}) = v_F^*\delta(\xi_\mathbf{k}) u_\sigma(\hat{\mathbf{k}}) \qquad (T\to 0)\end{aligned} \tag{4.29}$$

in the distribution function. We have used $\tilde{\epsilon}_\mathbf{k} \to \epsilon_\mathbf{k}$ and $n'_F(x) \to -\delta(x)$ when $T \to 0$. The corresponding variation of the thermodynamic potential reads

$$\begin{aligned}\delta\Omega[u] = \mathcal{V}\frac{v_F^{*\,2} N^*(0)}{2} \sum_{\sigma,\sigma'} \biggl\{ & \delta_{\sigma,\sigma'} \int \frac{d\Omega_{\hat{\mathbf{k}}}}{4\pi} u_\sigma^2(\hat{\mathbf{k}}) \\ & + \frac{1}{2}\int \frac{d\Omega_{\hat{\mathbf{k}}}}{4\pi}\frac{d\Omega_{\hat{\mathbf{k}}'}}{4\pi} F_{\sigma\sigma'}(\mathbf{k}_F,\mathbf{k}'_F) u_\sigma(\hat{\mathbf{k}}) u_{\sigma'}(\hat{\mathbf{k}}')\biggr\}\end{aligned} \tag{4.30}$$

in the limit $T \to 0$. $d\Omega_{\hat{\mathbf{k}}}$ denotes the elementary solid angle in the direction of $\hat{\mathbf{k}}$. To proceed further, we expand $u_\sigma(\hat{\mathbf{k}})$ in spherical harmonics,

$$u_\sigma(\hat{\mathbf{k}}) = u^s(\hat{\mathbf{k}}) + \sigma u^a(\hat{\mathbf{k}}) = \sum_{l=0}^{\infty}\sum_{m=-l}^{l} (u^s_{lm} + \sigma u^a_{lm})\, Y_l^m(\hat{\mathbf{k}}), \tag{4.31}$$

where $u^{s,a}_{l,-m} = (-1)^m u^{s,a\,*}_{lm}$, since $u_\sigma(\hat{\mathbf{k}})$ is real. Using the addition theorem and other standard properties of spherical harmonics, we obtain

$$\delta\Omega[u] = \mathcal{V}\frac{v_F^{*\,2} N^*(0)}{4\pi} \sum_{l=0}^{\infty}\sum_{m=-l}^{l} \left[|u^s_{lm}|^2 \left(1 + \frac{F_l^s}{2l+1}\right) + |u^a_{lm}|^2 \left(1 + \frac{F_l^a}{2l+1}\right)\right]. \tag{4.32}$$

The stability of the spherical Fermi surface requires $\delta\Omega[u]$ to be positive for any deformation $u_\sigma(\hat{\mathbf{k}})$, i.e.

$$F_l^s > -2l - 1, \qquad F_l^a > -2l - 1. \tag{4.33}$$

The instabilities occurring when the conditions (4.33) are violated are known as Pomeranchuk instabilities.

4.1.5 *Effective mass*

The current carried by a quasi-particle, as well as its effective mass, can be obtained by considering the system as seen from a reference frame moving at a velocity $\mathbf{v} = \mathbf{q}/m$ with respect to the laboratory frame. In the moving frame, the energy is given by

$$E(\mathbf{q}) = E(0) - \mathbf{P} \cdot \mathbf{v} + \frac{1}{2} M \mathbf{v}^2, \tag{4.34}$$

where M is the total mass, and $\mathbf{P}$ the expectation of the total momentum measured with respect to the laboratory frame (see Eq. (2.60)). Since the momentum $\mathbf{P} = m\mathbf{J}$ coincides with the current in a translation-invariant system,

$$\mathbf{J} = \frac{\mathbf{P}}{m} = -\frac{1}{m} \frac{\partial E}{\partial \mathbf{v}}\bigg|_{\mathbf{v}=0} = -\frac{\partial E}{\partial \mathbf{q}}\bigg|_{\mathbf{q}=0}. \tag{4.35}$$

Let us consider the state corresponding to a quasi-particle of momentum $\mathbf{k}$ (in the laboratory frame) added to the ground state. The current in that state is simply

$$\mathbf{j}_{\mathbf{k}} = \frac{\mathbf{k}}{m} = \mathbf{v}_{\mathbf{k}}, \tag{4.36}$$

since the adiabatic switching of the interactions does not modify the total momentum. The current can also be calculated by considering the same physical state in the moving frame, where the quasi-particle has momentum $\mathbf{k} - \mathbf{q}$ and the ground state is a shifted Fermi sea: $n_{\mathbf{k}} = n^0_{\mathbf{k}+\mathbf{q}}$. Since Eq. (4.15) is valid in any inertial frame, the quasi-particle energy in the moving frame reads[11]

$$\tilde{\epsilon}_{\mathbf{k}-\mathbf{q}} = \epsilon_{\mathbf{k}-\mathbf{q}} + \int_{\mathbf{k}'} \sum_{\sigma'} f_{\sigma\sigma'}(\mathbf{k}, \mathbf{k}')[n^0_{\mathbf{k}'+\mathbf{q}} - n^0_{\mathbf{k}'}], \tag{4.37}$$

so that the current $\mathbf{j}_{\mathbf{k}}$ — measured in the laboratory frame — carried by a quasi-particle of momentum $\mathbf{k}$ is

$$\begin{aligned} \mathbf{v}_{\mathbf{k}} &= -\frac{\partial \tilde{\epsilon}_{\mathbf{k}-\mathbf{q}}}{\partial \mathbf{q}}\bigg|_{\mathbf{q}=0} = \boldsymbol{\nabla}_{\mathbf{k}} \epsilon_{\mathbf{k}} - \int_{\mathbf{k}'} \sum_{\sigma'} f_{\sigma\sigma'}(\mathbf{k}, \mathbf{k}') \boldsymbol{\nabla}_{\mathbf{k}'} n^0_{\mathbf{k}'} \\ &= \mathbf{v}^*_{\mathbf{k}} + \int_{\mathbf{k}'} \sum_{\sigma'} f_{\sigma\sigma'}(\mathbf{k}, \mathbf{k}') \mathbf{v}^*_{\mathbf{k}'} \delta(\xi_{\mathbf{k}'}). \end{aligned} \tag{4.38}$$

The first term in the rhs of (4.38) corresponds to the group velocity of the quasi-particle. The second one is a drag current that comes from the interaction of the moving quasi-particle with the surrounding fluid.

Near the Fermi surface, where $\xi_{\mathbf{k}} \simeq v^*_F(|\mathbf{k}| - k_F)$ and $\mathbf{v}^*_{\mathbf{k}} \simeq v^*_F \hat{\mathbf{k}}$, Eq. (4.38) gives

$$\mathbf{v}_{\mathbf{k}} = v^*_F \hat{\mathbf{k}} \left[1 + 2N^*(0) \int \frac{d\Omega_{\mathbf{k}'}}{4\pi} f^s(\theta) \cos\theta \right] = v^*_F \hat{\mathbf{k}} \left(1 + \frac{F^s_1}{3}\right). \tag{4.39}$$

[11] Note that $\mathbf{k}$ must not be shifted in the Landau function, since the interaction between particles is velocity independent.

From (4.36) and (4.39), we obtain

$$\frac{m^*}{m} = 1 + \frac{F_1^s}{3}. \tag{4.40}$$

Depending on the sign of F_1^s, the effective mass can be larger or smaller than the bare mass. When $F_1^s < -3$, the effective mass is negative and the system unstable since quasi-particle excitations across the Fermi surface are energetically favorable. This is a special case of Pomeranchuk instabilities [Eqs. (4.33)].

4.2 Thermodynamics

In this section, we show how to obtain the thermodynamic quantities from the Landau energy functional $E[n]$.

4.2.1 *Specific heat*

The specific heat is defined by

$$C_V = \left.\frac{\partial E}{\partial T}\right|_{\mathcal{V},N}, \tag{4.41}$$

where

$$E = E(T=0) + \sum_{\mathbf{k},\sigma} \epsilon_{\mathbf{k}} \delta n_{\mathbf{k}\sigma} + \frac{1}{2\mathcal{V}} \sum_{\substack{\mathbf{k},\mathbf{k}' \\ \sigma,\sigma'}} f_{\sigma\sigma'}(\mathbf{k},\mathbf{k}') \delta n_{\mathbf{k}\sigma} \delta n_{\mathbf{k}'\sigma'} \tag{4.42}$$

and $\delta n_{\mathbf{k}\sigma} = n_F(\tilde\epsilon_{\mathbf{k}}) - n^0_{\mathbf{k}}$. If we neglect the interactions between quasi-particles, then

$$C_V = \mathcal{V}\frac{2\pi^2}{3} N^*(0) T = \mathcal{V}\frac{m^* k_F}{3} T \qquad (T \to 0) \tag{4.43}$$

is simply the specific heat of non-interacting fermions with mass m^*. The interaction term in (4.42) is $\mathcal{O}(T^4)$, since $\int d|\mathbf{k}|\mathbf{k}^2 \delta n_{\mathbf{k}\sigma} = \mathcal{O}(T^2)$ in the grand canonical ensemble, and can be neglected. Clearly, this conclusion will not change if we consider the $\mathcal{O}(T^2)$ shift of the chemical potential necessary to keep the total number of particles constant.

4.2.2 *Compressibility*

The isothermal compressibility is defined as

$$\kappa = -\frac{1}{\mathcal{V}} \left.\frac{\partial \mathcal{V}}{\partial P}\right|_{T,N} = \frac{1}{n^2} \left.\frac{\partial n}{\partial \mu}\right|_{T} \tag{4.44}$$

(Sec. 3.3) so that what we need to calculate is $\partial n/\partial \mu$. A variation in the density $n = k_F^3/3\pi^2$ is equivalent to a variation of the Fermi momentum k_F: $\partial k_F/\partial n = \pi^2/k_F^2$. When k_F varies, the quasi-particle distribution also varies, so that the change in the chemical potential $\mu = \epsilon_{k_F}$ is given by

$$\frac{\partial \mu}{\partial n} = \frac{\partial \epsilon_{k_F}}{\partial k_F}\frac{\partial k_F}{\partial n} + \int_{\mathbf{k}'} \sum_{\sigma'} f_{\sigma\sigma'}(\mathbf{k}_F, \mathbf{k}') \frac{\partial n_{\mathbf{k}'\sigma'}}{\partial k_F}\frac{\partial k_F}{\partial n}. \tag{4.45}$$

Using $\partial \epsilon_{k_F}/\partial k_F = v_F^* = k_F/m^*$ and $\partial n_{\mathbf{k}\sigma}/\partial k_F = \delta(k_F - |\mathbf{k}|)$ $(T = 0)$, we obtain

$$\begin{aligned}\frac{\partial \mu}{\partial n} &= \frac{\pi^2}{k_F^2}\left[v_F^* + \sum_{\sigma'}\int_{\mathbf{k}'} f_{\sigma\sigma'}(\mathbf{k}_F, \mathbf{k}')\delta(k_F - |\mathbf{k}'|)\right] \\ &= \frac{\pi^2}{k_F^2}\left[v_F^* + 2N^*(0)v_F^* f_0^s\right] = \frac{1 + F_0^s}{2N^*(0)},\end{aligned} \tag{4.46}$$

and in turn

$$n^2 \kappa = \frac{2N^*(0)}{1 + F_0^s}. \tag{4.47}$$

The interactions between quasi-particles lead to a renormalization by a factor $1/(1 + F_0^s)$ of the naive result $2N^*(0)$ obtained from the compressibility of the ideal Fermi gas by the mere replacement $m \to m^*$. Again, we note that the stability of the system requires $F_0^s > -1$ in agreement with (4.33).

Equation (4.47) yields the macroscopic sound velocity

$$c_s = \frac{1}{\sqrt{\kappa n m}} = \frac{v_F}{\sqrt{3}}\left(\frac{1 + F_0^s}{1 + F_1^s/3}\right)^{1/2} = \frac{v_F^*}{\sqrt{3}}\left[\left(1 + F_0^s\right)\left(1 + \frac{F_1^s}{3}\right)\right]^{1/2} \tag{4.48}$$

[see Eq. (3.111)]. In the absence of interaction, we recover the sound velocity $v_F/\sqrt{3}$ of the ideal Fermi gas.

4.2.3 *Spin susceptibility*

In the presence of a magnetic field $\mathbf{B} = B\hat{\mathbf{z}}$, the energy $\epsilon_{\mathbf{k}\sigma}$ is shifted by $\frac{\sigma}{2} g\mu_B B$, where $g \simeq 2$ is the Landé factor and μ_B is the Bohr magneton. Since the field displaces the Fermi surface and changes the quasi-particle distribution, the quasi-particle energy becomes

$$\tilde{\epsilon}_{\mathbf{k}\sigma} = \epsilon_{\mathbf{k}} + \frac{\sigma}{2} g\mu_B B + \int_{\mathbf{k}'} \sum_{\sigma'} f_{\sigma\sigma'}(\mathbf{k}, \mathbf{k}')\delta n_{\mathbf{k}'\sigma'}. \tag{4.49}$$

The new (spin-dependent) Fermi surface is defined by $\tilde{\epsilon}_{k_{F\uparrow},\uparrow} = \tilde{\epsilon}_{k_{F\downarrow},\downarrow} = \mu$,[12] i.e.

$$\begin{aligned}&\epsilon_{k_{F\uparrow}} + \frac{g\mu_B B}{2} + \int_{\mathbf{k}'} \sum_{\sigma'} f_{\uparrow\sigma'}(\mathbf{k}_F, \mathbf{k}')\delta n_{\mathbf{k}'\sigma'} \\ &= \epsilon_{k_{F\downarrow}} - \frac{g\mu_B B}{2} + \int_{\mathbf{k}'} \sum_{\sigma'} f_{\downarrow\sigma'}(\mathbf{k}_F, \mathbf{k}')\delta n_{\mathbf{k}'\sigma'},\end{aligned} \tag{4.50}$$

[12]Since $\delta\mu$ cannot depend on the direction of $\mathbf{B}$, it is at least of order B^2 and the chemical potential is constant to leading order in B.

where

$$\delta n_{\mathbf{k}\sigma} = \Theta(k_F + \delta k_{F\sigma} - |\mathbf{k}|) - \Theta(k_F - |\mathbf{k}|) = \delta k_{F\sigma} v_F^* \delta(\xi_{\mathbf{k}}) \tag{4.51}$$

($\delta k_{F\sigma} = k_{F\sigma} - k_F$ and $T \to 0$) to leading order in B. Using $\epsilon_{k_{F\sigma}} = \epsilon_{k_F} + v_F^* \delta k_{F\sigma}$, Eq. (4.50) gives

$$v_F^*(\delta k_{F\uparrow} - \delta k_{F\downarrow}) = -\frac{g\mu_B B}{1+F_0^a}. \tag{4.52}$$

The magnetization per unit volume is given by

$$\begin{aligned} M &= -\frac{g\mu_B}{2}\int_{\mathbf{k}} (\delta n_{\mathbf{k}\uparrow} - \delta n_{\mathbf{k}\downarrow}) = -\frac{1}{2} g\mu_B N^*(0) v_F^*(\delta k_{F\uparrow} - \delta k_{F\downarrow}) \\ &= \left(\frac{g\mu_B}{2}\right)^2 \frac{2N^*(0)}{1+F_0^a} B. \end{aligned} \tag{4.53}$$

This yields the spin (paramagnetic) susceptibility

$$\chi = \frac{\partial M}{\partial B} = \left(\frac{g\mu_B}{2}\right)^2 \frac{2N^*(0)}{1+F_0^a}. \tag{4.54}$$

Stability against ferromagnetism requires $F_0^a > -1$.

4.3 Non-Equilibrium Properties

The Landau energy function $\delta E[\delta n]$ enables to compute the thermodynamic properties of the Fermi liquid but does not contain any information about the quasi-particle dynamics. To study the latter, one has to extend the definition of δE to non-equilibrium states. When physical properties vary only on macroscopic scales ($\gg k_F^{-1}$), one can adopt a semiclassical description and define a local quasi-particle distribution function $n_{\mathbf{k}\sigma}(\mathbf{r},t)$ giving the density of quasi-particles with momentum $\mathbf{k}$ and spin σ in the vicinity of point $\mathbf{r}$ at time t. By analogy with (4.15), we define the time-dependent functional

$$\begin{aligned} \delta E[\delta n, t] &= \sum_{\mathbf{k},\sigma} \int d^3r\, \epsilon_{\mathbf{k}} \delta n_{\mathbf{k}\sigma}(\mathbf{r},t) \\ &+ \frac{1}{2\mathcal{V}} \sum_{\mathbf{k},\mathbf{k}',\sigma,\sigma'} \int d^3r d^3r'\, f_{\sigma\sigma'}(\mathbf{k},\mathbf{k}';\mathbf{r}-\mathbf{r}') \delta n_{\mathbf{k}\sigma}(\mathbf{r},t) \delta n_{\mathbf{k}'\sigma'}(\mathbf{r}',t). \end{aligned} \tag{4.55}$$

For a homogeneous system, $\epsilon_{\mathbf{k}}$ is independent of the spin σ and position $\mathbf{r}$ of the quasi-particle. The interaction is assumed to be instantaneous in time and short-range in space. $f_{\sigma\sigma'}(\mathbf{k},\mathbf{k}';\mathbf{r}-\mathbf{r}')$ then decreases rapidly in space and we can approximate $\delta n_{\mathbf{k}'\sigma'}(\mathbf{r}',t)$ by $\delta n_{\mathbf{k}'\sigma'}(\mathbf{r},t)$ in (4.55). This leads to

$$\delta E[\delta n, t] = \sum_{\mathbf{k},\sigma} \int d^3r\, \epsilon_{\mathbf{k}} \delta n_{\mathbf{k}\sigma}(x) + \frac{1}{2\mathcal{V}} \sum_{\mathbf{k},\mathbf{k}',\sigma,\sigma'} \int d^3r\, f_{\sigma\sigma'}(\mathbf{k},\mathbf{k}') \delta n_{\mathbf{k}\sigma}(x) \delta n_{\mathbf{k}'\sigma'}(x), \tag{4.56}$$

where $f_{\sigma\sigma'}(\mathbf{k},\mathbf{k}') = \int d^3r'\, f_{\sigma\sigma'}(\mathbf{k},\mathbf{k}';\mathbf{r}-\mathbf{r}')$ and $x = (\mathbf{r},t)$.

4.3.1 *Kinetic equation*

To obtain the equation governing the quasi-particle dynamics, we consider the time-dependent quasi-particle energy

$$\tilde{\epsilon}_{\mathbf{k}\sigma}(x) = \frac{\delta E[n,t]}{\delta n_{\mathbf{k}\sigma}(x)} = \epsilon_{\mathbf{k}} + \int_{\mathbf{k}'} \sum_{\sigma'} f_{\sigma\sigma'}(\mathbf{k},\mathbf{k}')\delta n_{\mathbf{k}'\sigma'}(x) \tag{4.57}$$

as a quasi-classical Hamiltonian. This assumption leads to the equations of motion

$$\partial_t \mathbf{r} = \boldsymbol{\nabla}_{\mathbf{k}} \tilde{\epsilon}_{\mathbf{k}\sigma}(x), \qquad \partial_t \mathbf{k} = -\boldsymbol{\nabla}_{\mathbf{r}} \tilde{\epsilon}_{\mathbf{k}\sigma}(x). \tag{4.58}$$

The time evolution of the quasi-particle distribution is then governed by the usual Boltzmann equation [9],

$$\frac{dn_{\mathbf{k}\sigma}(x)}{dt} = \partial_t n_{\mathbf{k}\sigma}(x) - \boldsymbol{\nabla}_{\mathbf{k}} n_{\mathbf{k}\sigma}(x) \cdot \boldsymbol{\nabla}_{\mathbf{r}} \tilde{\epsilon}_{\mathbf{k}\sigma}(x) + \boldsymbol{\nabla}_{\mathbf{r}} n_{\mathbf{k}\sigma}(x) \cdot \boldsymbol{\nabla}_{\mathbf{k}} \tilde{\epsilon}_{\mathbf{k}\sigma}(x) = I[n_{\mathbf{k}\sigma}(x)], \tag{4.59}$$

where the "collision integral" $I[n_{\mathbf{k}\sigma}(x)] = \partial_t n_{\mathbf{k}\sigma}(x)|_{\text{coll}}$ takes into account the collisions between particles. To first order in $\delta n_{\mathbf{k}\sigma}(x) = n_{\mathbf{k}\sigma}(x) - n^0_{\mathbf{k}}$,

$$\partial_t \delta n_{\mathbf{k}\sigma}(x) - \boldsymbol{\nabla}_{\mathbf{k}} n^0_{\mathbf{k}} \cdot \boldsymbol{\nabla}_{\mathbf{r}} \tilde{\epsilon}_{\mathbf{k}\sigma}(x) + \boldsymbol{\nabla}_{\mathbf{r}} \delta n_{\mathbf{k}\sigma}(x) \cdot \boldsymbol{\nabla}_{\mathbf{k}} \epsilon_{\mathbf{k}} = I[n_{\mathbf{k}\sigma}(x)] \tag{4.60}$$

so that we finally obtain

$$\partial_t \delta n_{\mathbf{k}\sigma}(x) + \mathbf{v}^*_{\mathbf{k}} \cdot \boldsymbol{\nabla}_{\mathbf{r}} \delta n_{\mathbf{k}\sigma}(x) + \int_{\mathbf{k}'} \sum_{\sigma'} f_{\sigma\sigma'}(\mathbf{k},\mathbf{k}') \boldsymbol{\nabla}_{\mathbf{r}} \delta n_{\mathbf{k}'\sigma'}(x) \cdot \mathbf{v}^*_{\mathbf{k}} \delta(\xi_{\mathbf{k}}) = I[n_{\mathbf{k}\sigma}(x)] \tag{4.61}$$

for $T \to 0$. Note that this equation involves only states near the Fermi surface where the quasi-particle concept is valid.

4.3.2 *Conservation laws*

4.3.2.1 *Particle number conservation*

Since the collisions conserve the total number of particles,

$$\sum_{\mathbf{k},\sigma} \frac{dn_{\mathbf{k}\sigma}(x)}{dt} = \sum_{\mathbf{k},\sigma} I[n_{\mathbf{k}\sigma}(x)] = 0. \tag{4.62}$$

Using (4.59), this equation can be written as the continuity equation $\partial_t n(x) + \boldsymbol{\nabla} \cdot \mathbf{j}(x) = 0$, where

$$n(x) = \int_{\mathbf{k}} \sum_{\sigma} n_{\mathbf{k}\sigma}(x) \tag{4.63}$$

is the particle density at point $\mathbf{r}$ and time t and

$$\mathbf{j}(x) = \int_{\mathbf{k}} \sum_{\sigma} n_{\mathbf{k}\sigma}(x) \boldsymbol{\nabla}_{\mathbf{k}} \tilde{\epsilon}_{\mathbf{k}\sigma}(x) \tag{4.64}$$

the current density. To linear order in δn, we obtain

$$\begin{aligned}\mathbf{j}(x) &= \int_{\mathbf{k}}\sum_{\sigma}\delta n_{\mathbf{k}\sigma}(x)\Big[\mathbf{v}^*_{\mathbf{k}} + \int_{\mathbf{k}'}\sum_{\sigma'} f_{\sigma\sigma'}(\mathbf{k},\mathbf{k}')\mathbf{v}^*_{\mathbf{k}'}\delta(\xi_{\mathbf{k}'})\Big] \\ &= \int_{\mathbf{k}}\sum_{\sigma}\delta n_{\mathbf{k}\sigma}(x)\mathbf{j}_{\mathbf{k}},\end{aligned} \tag{4.65}$$

where $\mathbf{j}_{\mathbf{k}}$ is the current carried by a quasi-particle of momentum $\mathbf{k}$ [Eqs. (4.36, 4.38)].

4.3.2.2 *Momentum conservation*

Similarly, by multiplying (4.59) by k_i and summing over $\mathbf{k}$ and σ, we obtain

$$\partial_t g_i(x) + \int_{\mathbf{k}}\sum_{\sigma,j} k_i\left[\frac{\partial}{\partial r_j}\left(n_{\mathbf{k}\sigma}(x)\frac{\partial\tilde\epsilon_{\mathbf{k}\sigma}(x)}{\partial k_j}\right) - \frac{\partial}{\partial k_j}\left(n_{\mathbf{k}\sigma}(x)\frac{\partial\tilde\epsilon_{\mathbf{k}\sigma}(x)}{\partial r_j}\right)\right] = 0, \tag{4.66}$$

where

$$\mathbf{g}(x) = \int_{\mathbf{k}}\sum_{\sigma}\mathbf{k}n_{\mathbf{k}\sigma}(x) \tag{4.67}$$

is the momentum density. Again, from momentum conservation in collisions, the collision term does not appear in (4.66). By integrating by part, we rewrite the last term of (4.66) as

$$\begin{aligned}\int_{\mathbf{k}}\sum_{\sigma} n_{\mathbf{k}\sigma}(x)\frac{\partial\tilde\epsilon_{\mathbf{k}\sigma}(x)}{\partial r_i} &= \int_{\mathbf{k}}\sum_{\sigma}\left[\frac{\partial}{\partial r_i}\left(n_{\mathbf{k}\sigma}(x)\tilde\epsilon_{\mathbf{k}\sigma}(x)\right) - \frac{\partial n_{\mathbf{k}\sigma}(x)}{\partial r_i}\tilde\epsilon_{\mathbf{k}\sigma}(x)\right] \\ &= \int_{\mathbf{k}}\sum_{\sigma}\frac{\partial}{\partial r_i}\left[n_{\mathbf{k}\sigma}(x)\tilde\epsilon_{\mathbf{k}\sigma}(x)\right] - \frac{1}{\mathcal{V}}\frac{\partial E}{\partial r_i},\end{aligned} \tag{4.68}$$

where we have used (4.18). From (4.66, 4.68), we deduce the equation

$$\partial_t g_i(x) + \sum_j \nabla_{r_j}\Pi_{ij}(x) = 0, \tag{4.69}$$

where

$$\Pi_{ij}(x) = \int_{\mathbf{k}}\sum_{\sigma} k_i n_{\mathbf{k}\sigma}(x)\frac{\partial\tilde\epsilon_{\mathbf{k}\sigma}(x)}{\partial k_j} + \delta_{i,j}\left[\int_{\mathbf{k}}\sum_{\sigma} n_{\mathbf{k}\sigma}(x)\tilde\epsilon_{\mathbf{k}\sigma}(x) - \frac{E}{\mathcal{V}}\right] \tag{4.70}$$

is the spatial part of the energy-momentum tensor. To linear order in δn,

$$\Pi_{ij}(x) = \int_{\mathbf{k}}\sum_{\sigma} k_i\Big[\delta n_{\mathbf{k}\sigma}(x) + \delta(\xi_{\mathbf{k}})\int_{\mathbf{k}'}\sum_{\sigma'} f_{\sigma\sigma'}(\mathbf{k},\mathbf{k}')\delta n_{\mathbf{k}'\sigma'}(x)\Big]v^*_{\mathbf{k}j}. \tag{4.71}$$

4.3.2.3 *Energy conservation*

Last, we obtain the expression of the energy current by multiplying (4.59) by $\tilde{\epsilon}_{\mathbf{k}\sigma}$ and summing over $\mathbf{k}$ and σ,

$$\partial_t E + \int_{\mathbf{k}} \sum_{\sigma,i} \tilde{\epsilon}_{\mathbf{k}\sigma}(x) \left[\frac{\partial}{\partial r_i} \left(n_{\mathbf{k}\sigma}(x) \frac{\partial \tilde{\epsilon}_{\mathbf{k}\sigma}(x)}{\partial k_i} \right) - \frac{\partial}{\partial k_i} \left(n_{\mathbf{k}\sigma}(x) \frac{\partial \tilde{\epsilon}_{\mathbf{k}\sigma}(x)}{\partial r_i} \right) \right] = 0, \qquad (4.72)$$

where

$$\partial_t E = \int_{\mathbf{k}} \sum_{\sigma} \frac{\delta E}{\delta n_{\mathbf{k}\sigma}(x)} \partial_t n_{\mathbf{k}\sigma}(x) = \int_{\mathbf{k}} \sum_{\sigma} \tilde{\epsilon}_{\mathbf{k}\sigma}(x) \partial_t n_{\mathbf{k}\sigma}(x) \qquad (4.73)$$

is the time derivative of the energy. Integrating by part the last term in (4.72), we obtain

$$\partial_t E + \boldsymbol{\nabla} \cdot \mathbf{j}_E(x) = 0, \qquad (4.74)$$

where the energy current is given by

$$\mathbf{j}_E(x) = \int_{\mathbf{k}} \sum_{\sigma} \tilde{\epsilon}_{\mathbf{k}\sigma}(x) n_{\mathbf{k}\sigma}(x) \boldsymbol{\nabla}_{\mathbf{k}} \tilde{\epsilon}_{\mathbf{k}\sigma}(x). \qquad (4.75)$$

To linear order in δn,

$$\mathbf{j}_E(x) = \int_{\mathbf{k}} \sum_{\sigma} \epsilon_{\mathbf{k}} \left[\delta n_{\mathbf{k}\sigma}(x) + \delta(\xi_{\mathbf{k}}) \int_{\mathbf{k}'} \sum_{\sigma'} f_{\sigma\sigma'}(\mathbf{k}, \mathbf{k}') \delta n_{\mathbf{k}'\sigma'}(x) \right] \mathbf{v}^*_{\mathbf{k}}. \qquad (4.76)$$

4.3.3 *Collective modes*

A collective mode with momentum $\mathbf{q}$ and frequency ω is a coherent superposition of quasi-particle–quasi-hole pair excitations. When $\mathbf{q} \to 0$, the quasi-particles (holes) excitations have vanishing energy and the collective mode can be seen as a time-dependent displacement $u_\sigma(\hat{\mathbf{k}}) e^{i(\mathbf{q}\cdot\mathbf{r} - \omega t)}$ + c.c. of the Fermi surface. Generalizing (4.29), we therefore consider

$$\delta n_{\mathbf{k}\sigma}(x) = v^*_F \delta(\xi_{\mathbf{k}}) u_\sigma(\hat{\mathbf{k}}) e^{i(\mathbf{q}\cdot\mathbf{r} - \omega t)} + \text{c.c.} \qquad (T \to 0) \qquad (4.77)$$

(note that $u_\sigma(\hat{\mathbf{k}})$ is now complex). As before, we expand $u_\sigma(\hat{\mathbf{k}})$ in spherical harmonics [Eq. (4.31)] and choose $\mathbf{q}$ as the polar axis. Since $\delta n_{\mathbf{k}\sigma}(x)$ satisfies the kinetic equation (4.61), we have

$$(\cos\theta - s) u^\nu(\hat{\mathbf{k}}) + \cos\theta \int \frac{d\Omega_{\hat{\mathbf{k}}'}}{4\pi} F^\nu(\mathbf{k}, \mathbf{k}') u^\nu(\hat{\mathbf{k}}') = I[u] \qquad (4.78)$$

($\nu = s, a$), where $s = \omega / v^*_F |\mathbf{q}|$ and θ is the angle between $\mathbf{k}$ and $\mathbf{q}$. Using standard properties of the spherical harmonics, we obtain

$$(\cos\theta - s) \sum_{l=0}^{\infty} \sum_{m=-l}^{l} u^\nu_{lm} Y^m_l(\hat{\mathbf{k}}) + \cos\theta \sum_{l=0}^{\infty} \frac{F^\nu_l}{2l+1} \sum_{m=-l}^{l} u^\nu_{lm} Y^m_l(\hat{\mathbf{k}}) = I[u]. \qquad (4.79)$$

A set of equations for the u^ν_{lm}'s can be obtained multiplying (4.79) by $\int d\Omega_{\hat{\mathbf{k}}} Y^m_l(\hat{\mathbf{k}})^*$. One readily sees that m is a good quantum number (but l is not) if one ignores the collision term.

The "longitudinal" mode $m = 0$ is particular as it is the only one to involve density fluctuations. Indeed, we have

$$\begin{aligned}\delta n(x) &= 2N^*(0)v_F^* \int \frac{d\Omega_{\hat{\mathbf{k}}}}{4\pi} u^s(\hat{\mathbf{k}}) e^{i(\mathbf{q}\cdot\mathbf{r}-\omega t)} + \text{c.c.} \\ &= N^*(0)v_F^* \frac{u_{00}^s}{\sqrt{\pi}} e^{i(\mathbf{q}\cdot\mathbf{r}-\omega t)} + \text{c.c.} \\ &= \frac{k_F^2}{2\pi^{5/2}} u_{00}^s e^{i(\mathbf{q}\cdot\mathbf{r}-\omega t)} + \text{c.c.} \end{aligned} \tag{4.80}$$

After straightforward manipulations, one finds that in the mode $m = 0$ the current is longitudinal and takes the form

$$\begin{aligned}\mathbf{j}(x) &= \hat{\mathbf{q}} N^*(0) v_F^{*\,2} \frac{u_{10}^s}{\sqrt{3\pi}} \left(1 + \frac{F_1^s}{3}\right) e^{i(\mathbf{q}\cdot\mathbf{r}-\omega t)} + \text{c.c.} \\ &= \hat{\mathbf{q}} \frac{k_F^3}{2\sqrt{3}\pi^{5/2} m} u_{10}^s e^{i(\mathbf{q}\cdot\mathbf{r}-\omega t)} + \text{c.c.} \end{aligned} \tag{4.81}$$

4.3.3.1 *Zero sound*

Let us consider the longitudinal mode $m = 0$ in the frequency range $\omega\tau \gg 1$ where the collision term $I[n_{\mathbf{k}\sigma}] \sim -\delta n_{\mathbf{k}\sigma}/\tau$ can be neglected with respect to $\partial_t \delta n_{\mathbf{k}\sigma}(x)$. τ is a characteristic quasi-particle collision time. We shall see later that $\tau \sim 1/T^2$ at low temperatures (Sec. 4.4.1). We further assume that $f_{\sigma\sigma'}(\mathbf{k},\mathbf{k}') = f_0^s$. For a solution where both spin states oscillate in phase (sound mode: $u^a = 0$), Eq. (4.78) gives

$$(\cos\theta - s)u^s(\hat{\mathbf{k}}) + F_0^s \cos\theta \int \frac{d\Omega_{\hat{\mathbf{k}}'}}{4\pi} u^s(\hat{\mathbf{k}}') = 0, \tag{4.82}$$

the solution of which is

$$\begin{aligned} u^s(\hat{\mathbf{k}}) &= \text{const} \times \frac{\cos\theta}{s + i\eta - \cos\theta}, \\ \frac{1}{F_0^s} &= \int \frac{d\Omega}{4\pi} \frac{\cos\theta}{s + i\eta - \cos\theta} = -1 + \frac{s}{2} \ln\left(\frac{s + i\eta + 1}{s + i\eta - 1}\right). \end{aligned} \tag{4.83}$$

We have added to the real frequency s an infinitesimal imaginary part $i\eta = i0^+$, which amounts to switching the collective fluctuations adiabatically. This makes the logarithm in (4.83) well defined even when $|s| \leq 1$. As when considering a retarded response function (Sec. 3.2.3), one can allow s to take complex values and interpret the imaginary part of ω as the inverse lifetime of the collective mode.

For a repulsive interaction $F_0^s > 0$, s is real and larger than unity. The limiting cases are

$$\begin{aligned} s &\to 1 + 2e^{-2/F_0^s - 2} \quad \text{for} \quad F_0^s \to 0, \\ s &\to (F_0^s/3)^{1/2} \quad \text{for} \quad F_0^s \to \infty. \end{aligned} \tag{4.84}$$

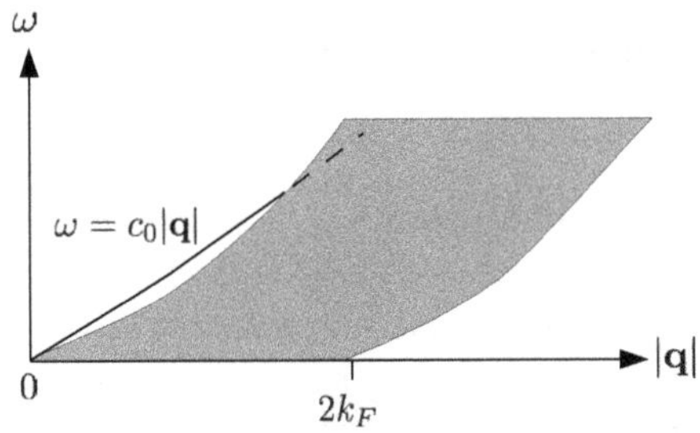

Fig. 4.1 Dispersion $\omega = c_0|\mathbf{q}|$ of an undamped zero-sound mode ($c_0 > v_F^*$). The shaded area shows the continuum of particle–hole pair excitations.

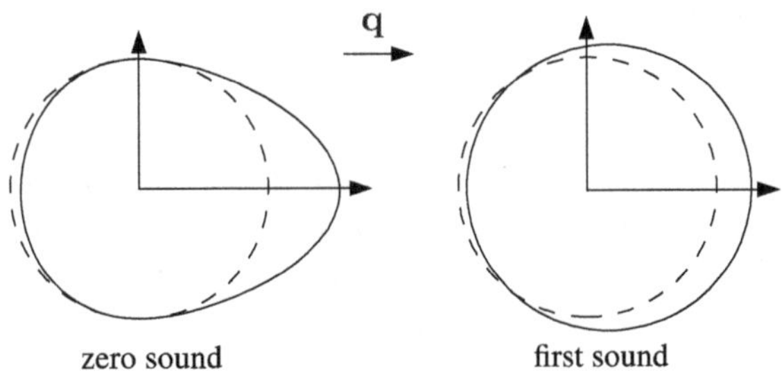

Fig. 4.2 Fermi surface deformations in the zero-sound and first-sound modes for a momentum-independent interaction $f_{\sigma\sigma'}(\mathbf{k},\mathbf{k}') = f_0^s$. The dashed line shows the equilibrium Fermi surface. In the first-sound mode, the Fermi surface keeps its spherical shape.

The solution corresponds to an undamped mode — known as the zero-sound mode — propagating at the velocity $c_0 = \omega/|\mathbf{q}| = sv_F^*$ larger than v_F^* (Fig. 4.1). The Fermi surface deformation corresponding to a zero- and first-sound mode is shown in Fig. 4.2. In practice, the collisions between particles give a finite lifetime to the zero-sound mode. In the limit $\omega\tau \gg 1$, this effect is however negligible and the main source of damping comes from multi-pair excitations (Sec. 4.3.5).

For moderate attractive interactions, $-1 < F_0^s < 0$, one can numerically verify that s is complex and satisfies $|\Re(s)| < 1$ and $\Im(s) < 0$, corresponding to a damped zero-sound mode. From (4.83), it is clear that the imaginary part of s is due to the interaction of the collective mode with quasi-hole–quasi-particle pair excitations. When $\omega = \epsilon_{\mathbf{k}+\mathbf{q}} - \epsilon_{\mathbf{k}} \simeq v_F^*|\mathbf{q}|\cos\theta$, the interaction is resonant and gives rise to a damping of the collective mode (known as Landau damping). The collective mode has then a short lifetime and does not represent a well-defined excitation of the system.

Last, for $F_0^s < -1$, there are two purely imaginary solutions. Substituting $s = i\alpha$ into (4.83), one finds

$$\frac{1}{F_0^s} = -1 + \frac{i}{2}\alpha \ln\left(\frac{1+i\alpha}{i\alpha - 1}\right) = -1 - \alpha\left(\gamma - \frac{\pi}{2}\right), \tag{4.85}$$

where $\gamma \in]-\pi, \pi]$ is defined by $1 + i\alpha = \sqrt{1+\alpha^2}e^{i\gamma}$ and $-1 + i\alpha = \sqrt{1+\alpha^2}e^{i(\pi-\gamma)}$. Since $\tan\gamma = \alpha$, we eventually obtain

$$\frac{1}{F_0^s} = -1 - \alpha\left(\arctan\alpha - \frac{\pi}{2}\right) = -1 + \alpha\arctan\left(\frac{1}{\alpha}\right). \tag{4.86}$$

For $F_0^s < -1$, this equation possesses two real solutions of opposite signs. One of these ($\alpha > 0$, i.e. $\Im(\omega) > 0$) corresponds to an unstable collective mode. This instability, characterized by divergent density fluctuations, also shows up in the negative compressibility [Eq. (4.47)].

It should be noted that the zero-sound mode exists only in neutral Fermi liquids. In a charged system, it is replaced by a plasmon mode as discussed in Sec. 3.4.1. We shall come back to this point in Sec. 4.3.4.

4.3.3.2 *First sound*

Since the Fermi surface relaxes towards its equilibrium position within a characteristic time τ, in the hydrodynamic regime $\omega\tau \ll 1$ the displacement $u_\sigma(\hat{\mathbf{k}})$ of the Fermi surface is expected to be negligible. This is actually true of all components u_{lm}^s except u_{00}^s and u_{10}^s whose fluctuations (and return to equilibrium) are constrained by the conservation of particle number [Eqs. (4.80, 4.81)]. Thus, the sound propagation in the hydrodynamic regime can be studied by retaining only the hydrodynamic components u_{00}^s and u_{10}^s (which are not affected by the collisions).

With $Y_0^0(\hat{\mathbf{k}}) = 1/2\sqrt{\pi}$ and $Y_1^0(\hat{\mathbf{k}}) = \cos\theta\sqrt{3}Y_0^0(\hat{\mathbf{k}})$, the kinetic equation then becomes

$$(\cos\theta - s)\left(u_{00}^s + u_{10}^s\sqrt{3}\cos\theta\right) + \cos\theta\left(F_0^s u_{00}^s + \frac{F_1^s}{\sqrt{3}}u_{10}^s\cos\theta\right) = 0. \tag{4.87}$$

Multiplying this equation by $\int d\Omega_{\hat{\mathbf{k}}} Y_0^0(\hat{\mathbf{k}})$ and $\int d\Omega_{\hat{\mathbf{k}}} Y_1^0(\hat{\mathbf{k}})$, we deduce

$$\begin{aligned} s u_{00}^s - \frac{u_{10}^s}{\sqrt{3}}\left(1 + \frac{F_1^s}{3}\right) &= 0, \\ u_{00}^s\,(1 + F_0^s) - s\sqrt{3}u_{10}^s &= 0. \end{aligned} \tag{4.88}$$

The first of these equations is nothing else but the continuity equation $\partial_t n(x) + \boldsymbol{\nabla}\cdot\mathbf{j}(x) = 0$ in the longitudinal mode $m = 0$ [Eqs. (4.80, 4.81)]. The second one can be identified with (4.69).[13] They admit a solution if

$$s^2 = \frac{\omega^2}{(v_F^*\mathbf{q})^2} = \frac{1}{3}(1 + F_0^s)\left(1 + \frac{F_1^s}{3}\right)^{1/2}, \tag{4.89}$$

[13] In the hydrodynamic mode $m = 0$ (where only u_{00}^s and u_{10}^s need to be considered), the energy-momentum tensor $\Pi_{ij} = \delta_{i,j}\Pi$ is diagonal with $\Pi(x) = \frac{1}{3\sqrt{\pi}}N^*(0)k_F {v_F^*}^2 u_{00}^s(1 + F_0^s)e^{i(\mathbf{q}\cdot\mathbf{r}-\omega t)}$ + c.c. The second of Eqs. (4.88) can be rewritten as $m\partial_t\mathbf{j}(x) + \boldsymbol{\nabla}\Pi(x) = 0$.

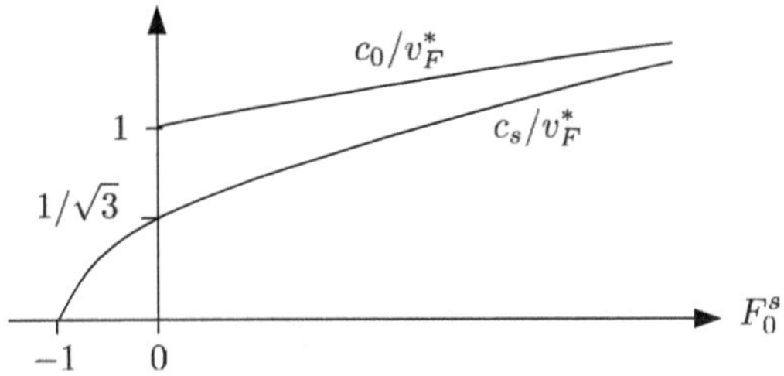

Fig. 4.3 Velocities of the zero- and first-sound modes when $f_{\sigma\sigma'}(\mathbf{k},\mathbf{k}') = f_0^s$.[14] For $F_0^s \to \infty$, $c_s \simeq c_0 \simeq v_F^*(F_0^s/3)^{1/2}$.

which agrees with the macroscopic sound velocity c_s obtained from the compressibility by the usual hydrodynamic arguments [Eq. (4.48)]. The solution reads

$$u^s(\hat{\mathbf{k}}) = \frac{u_{00}^s}{2\sqrt{\pi}} \left[1 + \sqrt{3} \left(\frac{1+F_0^s}{1+F_1^s/3} \right)^{1/2} \cos\theta \right]. \tag{4.90}$$

The Fermi surface keeps its spherical shape but its center oscillates about the origin in momentum space (hence the $\cos\theta$ term in (4.90)).

It is instructive to compare the zero- and first-sound modes within the simple model where $f_{\sigma\sigma'}(\mathbf{k},\mathbf{k}') = f_0^s$.[14] In the limit $F_0^s \to \infty$, the velocity of the two modes tends to the same value $v_F^*(F_0^s/3)^{1/2}$ (Fig. 4.3), while the Fermi surface deformation takes the simple form $u^s(\hat{\mathbf{k}}) \propto \cos\theta$.

4.3.3.3 *An example of transverse mode*

Let us consider the transverse mode $m = 1$ where

$$u_\sigma(\hat{\mathbf{k}}) = \sum_{l=1}^{\infty} u_{l,m=1}^s Y_l^1(\hat{\mathbf{k}}) \equiv u^s(\theta) e^{i\varphi}. \tag{4.91}$$

There are no density fluctuations in this mode. Using (4.65), one easily finds that the current is transverse (recall that $\mathbf{q} \parallel \hat{\mathbf{z}}$ defines the polar axis) and circularly polarized,

$$\mathbf{j}(x) = j_0[\hat{\mathbf{x}}\cos(\mathbf{q}\cdot\mathbf{r} - \omega t) - \hat{\mathbf{y}}\sin(\mathbf{q}\cdot\mathbf{r} - \omega t)], \tag{4.92}$$

with j_0 a constant depending on $u^s(\theta)$ and $F^s(\theta)$.

In the collisionless regime, the kinetic equation (4.78) then gives

$$(\cos\theta - s)u^s(\theta)e^{i\varphi} + \cos\theta \int \frac{d\Omega'}{4\pi} F^s(\Omega,\Omega') u^s(\theta') e^{i\varphi'} = 0. \tag{4.93}$$

In the following, we assume that only F_0^s and F_1^s are non-zero, i.e. $F^s(\alpha) = F_0^s + F_1^s \cos\alpha$ (α is the angle between Ω and Ω'). Using $\cos\alpha = \cos\theta\cos\theta' + \sin\theta\sin\theta'\cos(\varphi - \varphi')$,

[14] Note that $m^* = m$ and $v_F^* = v_F$ when $f_{\sigma\sigma'}(\mathbf{k},\mathbf{k}') = f_0^s$.

we obtain

$$(\cos\theta - s)u^s(\theta) + \frac{F_1^s}{4}\cos\theta\sin\theta\int_0^\pi d\theta'\sin^2(\theta')u^s(\theta') = 0 \tag{4.94}$$

so that

$$u^s(\theta,\varphi) = \text{const} \times \frac{\cos\theta\sin\theta}{s + i\eta - \cos\theta}e^{i\varphi}, \tag{4.95}$$

$$\frac{4}{F_1^s} = \int_0^\pi d\theta \frac{\sin^3\theta\cos\theta}{s + i\eta - \cos\theta} = -\frac{4}{3} + 2s^2 + s(1 - s^2)\ln\left(\frac{s + i\eta + 1}{s + i\eta - 1}\right). \tag{4.96}$$

As previously, we have added an infinitesimal imaginary part to the frequency ω. A real solution must satisfy $|s| > 1$. Since the rhs of (4.96) is maximum at $|s| = 1$ where it takes the value 2/3, a real solution is possible only if $F_1^s > 6$. The interaction between quasi-particles must therefore be quite strong for the mode $m = 1$ to propagate. For $F_1^s \leq 6$, the collective mode mixes with quasi-particle–quasi-hole pair excitations and is not a well-defined excitation of the system anymore.

4.3.3.4 *Spin-wave modes*

Until now, we have only considered spin symmetric solutions ($u^a = 0$) corresponding to density oscillations (sound modes). We could repeat the same discussion for spin antisymmetric solutions ($u^a \neq 0$) corresponding to spin density oscillations. The spin collective modes are similar to their charge counterparts but involve the spin antisymmetric Landau parameters F^a instead of F^s. For instance, for $F_0^a > 0$, one finds a "spin zero-sound" mode analogous to the (charge) zero-sound mode.

4.3.4 *Response functions*

In this section, we compute the density–density and current–current response functions in the long-wavelength low-energy limit. To this end, we consider the quasi-classical Hamiltonian

$$\tilde{\epsilon}_{\mathbf{k}\sigma}(x) + \phi(x) - \frac{\mathbf{k}}{m}\cdot\mathbf{A}(x). \tag{4.97}$$

For a neutral Fermi liquid, ϕ and $\mathbf{A}$ should be considered as fictitious external fields introduced in order to derive the response functions; in a charged system, they would correspond to the usual electromagnetic potentials. The vector potential couples to the quasi-particle current $\mathbf{j}_\mathbf{k} = \mathbf{v}_\mathbf{k}$ and not to the group velocity $\mathbf{v}^*_\mathbf{k}$. The linearized kinetic equation now reads

$$\begin{aligned}\partial_t\delta n_{\mathbf{k}\sigma}(x) + \mathbf{v}^*_\mathbf{k}\cdot\boldsymbol{\nabla}_\mathbf{r}\delta n_{\mathbf{k}\sigma}(x) + \int_{\mathbf{k}'}\sum_{\sigma'} f_{\sigma\sigma'}(\mathbf{k},\mathbf{k}')\boldsymbol{\nabla}_\mathbf{r}\delta n_{\mathbf{k}'\sigma'}(x)\cdot\mathbf{v}^*_\mathbf{k}\delta(\xi_\mathbf{k}) \\ + \delta(\xi_\mathbf{k})\mathbf{v}^*_\mathbf{k}\cdot\boldsymbol{\nabla}\left[\phi(x) - \frac{e}{m}\mathbf{k}\cdot\mathbf{A}(x)\right] = I[n_{\mathbf{k}\sigma}(x)].\end{aligned} \tag{4.98}$$

Following the analysis of Sec. 4.3.2, we can verify that the conservation of particle number implies the continuity equation $\partial_t n(x) + \boldsymbol{\nabla} \cdot \mathbf{J}(x) = 0$, where the current

$$\mathbf{J}(x) = \mathbf{j}(x) - \frac{ne}{m}\mathbf{A}(x), \qquad \mathbf{j}(x) = \int_{\mathbf{k}} \sum_{\sigma} \frac{\mathbf{k}}{m} \delta n_{\mathbf{k}\sigma}(x) \tag{4.99}$$

includes the usual diamagnetic part.

4.3.4.1 *Density–density response function in the collisionless regime*

We first consider the response to a scalar potential

$$\phi(x) = \phi(\mathbf{q}, \omega) e^{i[\mathbf{q}\cdot\mathbf{r} - (\omega + i\eta)t]} + \text{c.c.} \tag{4.100}$$

in the regime $\omega\tau \gg 1$ where collisions can be neglected. We use (4.77), valid for $\mathbf{q} \to 0$, and write the induced density as

$$\begin{aligned} \delta n(x) &= \int_{\mathbf{k}} \sum_{\sigma} \delta n_{\mathbf{k}\sigma}(x) = \delta n(\mathbf{q}, \omega) e^{i[\mathbf{q}\cdot\mathbf{r} - (\omega + i\eta)t]} + \text{c.c.}, \\ \delta n(\mathbf{q}, \omega) &= \int_{\mathbf{k}} \sum_{\sigma} v_F^* \delta(\xi_{\mathbf{k}}) u_\sigma(\hat{\mathbf{k}}) = 2N^*(0) v_F^* \int \frac{d\Omega_{\hat{\mathbf{k}}}}{4\pi} u^s(\hat{\mathbf{k}}). \end{aligned} \tag{4.101}$$

As shown in Sec. 3.3, the linear response to the external field is determined by the retarded density–density response function,

$$\delta n(\mathbf{q}, \omega) = -\chi^R_{nn}(\mathbf{q}, \omega) \phi(\mathbf{q}, \omega) \tag{4.102}$$

(note the minus sign).

Without the collision terms, the kinetic equation reads

$$(\omega + i\eta - \mathbf{v}^*_{\mathbf{k}} \cdot \mathbf{q}) u_\sigma(\hat{\mathbf{k}}) - \mathbf{v}^*_{\mathbf{k}} \cdot \mathbf{q} \int_{\mathbf{k}'} \sum_{\sigma'} f_{\sigma\sigma'}(\mathbf{k}, \mathbf{k}') \delta(\xi_{\mathbf{k}'}) u_{\sigma'}(\hat{\mathbf{k}}') = \mathbf{v}^*_{\mathbf{k}} \cdot \mathbf{q} \frac{\phi(\mathbf{q}, \omega)}{v_F^*}. \tag{4.103}$$

In the absence of interactions, Eq. (4.103) gives $\delta n(\mathbf{q}, \omega) = -\chi^{0R}_{nn}(\mathbf{q}, \omega)\phi(\mathbf{q}, \omega)$, where

$$\begin{aligned} \chi^{0R}_{nn}(\mathbf{q}, \omega) &= -2 \int_{\mathbf{k}} \frac{\mathbf{v}^*_{\mathbf{k}} \cdot \mathbf{q}}{\omega + i\eta - \mathbf{v}^*_{\mathbf{k}} \cdot \mathbf{q}} \delta(\xi_{\mathbf{k}}) \\ &= 2N^*(0) \left[1 - \frac{s}{2} \ln \left(\frac{s + i\eta + 1}{s + i\eta - 1} \right) \right] \end{aligned} \tag{4.104}$$

($s = \omega / v_F^* |\mathbf{q}|$) is the non-interacting density–density response function. Except for the effective mass correction, Eq. (4.104) agrees with the direct evaluation of the density–density correlation function of the ideal Fermi gas in the limit $\mathbf{q} \to 0$ [Eq. (3.74)].

It is difficult to solve (4.103) in the general case. Assuming that $f_{\sigma\sigma'}(\mathbf{k}, \mathbf{k}') = f_0^s$, we find

$$\chi^R_{nn}(\mathbf{q}, \omega) = \frac{\chi^{0R}_{nn}(\mathbf{q}, \omega)}{1 + f_0^s \chi^{0R}_{nn}(\mathbf{q}, \omega)}. \tag{4.105}$$

The density–density response function possesses a pole for $1 + f_0^s \chi^{0R}_{nn}(\mathbf{q}, \omega) = 0$, which is precisely the equation determining the zero-sound frequency obtained previously [Eq. (4.83)].

Such a pole (for ω real) exists only for a repulsive interaction ($f_0^s > 0$). More generally, the density excitation spectrum can be obtained from the imaginary part of the response function or, equivalently, the structure factor $S_{nn}(\mathbf{q}, \omega; T = 0) = 2\Theta(\omega)\Im[\chi_{nn}^R(\mathbf{q}, \omega)] = 2\Theta(\omega)\chi''_{nn}(\mathbf{q}, \omega)$ (Sec. 3.2.5). In the non-interacting case,

$$S_{nn}^0(\mathbf{q}, \omega) = \begin{cases} 2\pi N(0)\dfrac{\omega}{v_F|\mathbf{q}|} = \dfrac{m^2}{\pi}\dfrac{\omega}{|\mathbf{q}|} & \text{if} \quad 0 \le \omega \le v_F|\mathbf{q}|, \\ 0 & \text{otherwise.} \end{cases} \tag{4.106}$$

In the interacting case, there are two contributions to the structure factor

$$S_{nn}(\mathbf{q}, \omega) = 2\Theta(\omega)\frac{\chi_{nn}^{0\,\prime\prime}(\mathbf{q}, \omega)}{\left(1 + f_0^s\Re[\chi_{nn}^{0R}(\mathbf{q}, \omega)]\right)^2 + \left(f_0^s\chi_{nn}^{0\,\prime\prime}(\mathbf{q}, \omega)\right)^2}. \tag{4.107}$$

The first one is due to a non-vanishing $\chi_{nn}^{0\,\prime\prime}$ and comes from the quasi-particle–quasi-hole pair excitations. These are quite similar to the particle–hole pair excitations of the non-interacting system and give a continuous excitation spectrum for $|\omega| \le v_F^*|\mathbf{q}|$. When $f_0^s > 0$, there is a second contribution due to the pole of $\chi_{nn}^R(\mathbf{q}, \omega)$ at the (real) zero-sound frequency $\omega = c_0|\mathbf{q}|$. For ω near $c_0|\mathbf{q}|$,

$$\begin{aligned} \chi_{nn}^R(\mathbf{q}, \omega) &\simeq \frac{1}{\omega + i\eta - c_0|\mathbf{q}|}\frac{1}{f_0^{s2}\partial_\omega\chi_{nn}^{0R}(\mathbf{q}, \omega)|_{\omega=c_0|\mathbf{q}|}}, \\ \chi''_{nn}(\mathbf{q}, \omega) &= -\frac{\pi}{f_0^{s2}\partial_\omega\chi_{nn}^{0R}(\mathbf{q}, \omega)|_{\omega=c_0|\mathbf{q}|}}\delta(\omega - c_0|\mathbf{q}|). \end{aligned} \tag{4.108}$$

Thus, the zero-sound mode gives a delta peak contribution to the structure factor $S_{nn}(\mathbf{q}, \omega)$ (Fig. 4.4). When $-1 < F_0^s < 0$, the zero-sound mode strongly couples to the quasi-particle–quasi-hole pair excitations (Landau damping) and is not a well-defined excitation of the interacting Fermi liquid; it appears as a pole of $\chi_{nn}^R(\mathbf{q}, \omega)$ at a complex frequency. In the structure factor $S_{nn}(\mathbf{q}, \omega)$, it manifests itself as a broad resonance, characteristic of a damped collective mode, located at the frequency ω defined by $1 + f_0^s\Re[\chi_{nn}^{0R}(\mathbf{q}, \omega)] = 0$ ($\omega < v_F^*|\mathbf{q}|$).

We have already pointed out that the zero-sound mode exists only in a neutral Fermi liquid. In the presence of long-range Coulomb interactions, it is convenient to write the

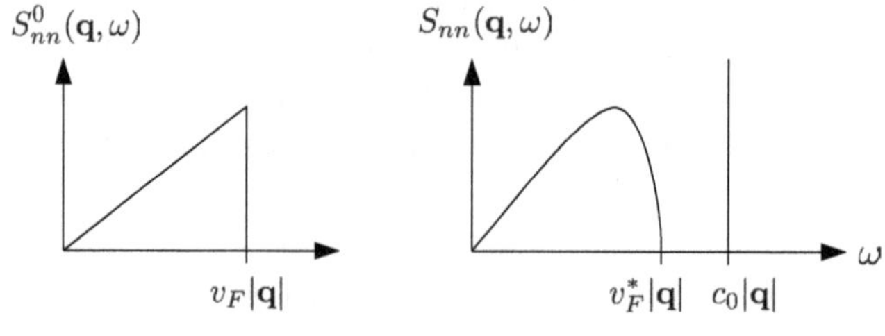

Fig. 4.4 Structure factors $S_{nn}^0(\mathbf{q}, \omega)$ and $S_{nn}(\mathbf{q}, \omega)$ for $|\mathbf{q}| \ll k_F$ obtained within the Landau Fermi-liquid theory when $f^s(\mathbf{k}, \mathbf{k}') = f_0^s > 0$.

density–density response function as in (3.123). Its poles then appear as zeros of the longitudinal dielectric function

$$\epsilon_{\parallel}^R(\mathbf{q},\omega) = 1 + \frac{e^2}{\epsilon_0 \mathbf{q}^2}\Pi_{nn}^R(\mathbf{q},\omega), \tag{4.109}$$

where Π_{nn}^R is the irreducible part of the density–density response function (Sec. 3.4.1). The simplest approximation amounts to approximating Π_{nn} by the density–density response function of the neutral (i.e. without the long-range Coulomb interaction) system. Using (4.105), we then obtain

$$\Pi_{nn}^R(\mathbf{q},\omega) \simeq \Pi_{nn}^{0R}(\mathbf{q},\omega) \simeq -\frac{2N^*(0){v_F^*}^2\mathbf{q}^2}{3\omega^2} = -\frac{n\mathbf{q}^2}{m\omega^2} \quad \text{for} \quad |\omega| \gg v_F^*|\mathbf{q}| \tag{4.110}$$

and

$$\epsilon_{\parallel}^R(\mathbf{q},\omega) \simeq 1 - \frac{\omega_p^2}{\omega^2} \quad \text{for} \quad |\omega| \gg v_F^*|\mathbf{q}|, \tag{4.111}$$

where $\omega_p = (ne^2/\epsilon_0 m)^{1/2}$ is the plasma frequency (Sec. 3.4). $\epsilon_{\parallel}^R(\mathbf{q},\omega)$ possesses a pole at $\omega = \pm\omega_p$. We conclude that the zero-sound mode of the neutral Fermi liquid has been replaced by the plasmon mode of the charged Fermi liquid.

4.3.4.2 *Density–density response function in the hydrodynamic regime*

We have pointed out earlier that in the regime $\omega\tau \ll 1$, the effect of collisions is to suppress all components u_{lm}^s except the "hydrodynamic" variables u_{00}^s and u_{10}^s. The kinetic equation (4.103) then reduces to

$$\begin{aligned}(\omega + i\eta - \mathbf{v}_{\mathbf{k}}^* \cdot \mathbf{q}) &\left[u_{00}^s Y_0^0(\hat{\mathbf{k}}) + u_{10}^s Y_1^0(\hat{\mathbf{k}})\right] \\ &- \mathbf{v}_{\mathbf{k}}^* \cdot \mathbf{q}\left[F_0^s u_{00}^s Y_0^0(\hat{\mathbf{k}}) + \frac{F_1^s}{3} Y_1^0(\hat{\mathbf{k}})\right] = \frac{\mathbf{v}_{\mathbf{k}}^* \cdot \mathbf{q}}{v_F^*}\phi(\mathbf{q},\omega).\end{aligned} \tag{4.112}$$

Multiplying this equation by $\int d\Omega_{\hat{\mathbf{k}}} Y_0^0(\hat{\mathbf{k}})$ and $\int d\Omega_{\hat{\mathbf{k}}} Y_1^0(\hat{\mathbf{k}})$, we obtain

$$\begin{aligned}&(\omega + i\eta)u_{00}^s - v_F^*|\mathbf{q}|\frac{u_{10}^s}{\sqrt{3}}\left(1 + \frac{F_1^s}{3}\right) = 0, \\ &v_F^*|\mathbf{q}|u_{00}^s(1 + F_0^s) - (\omega + i\eta)\sqrt{3}u_{10}^s = -2\sqrt{\pi}|\mathbf{q}|\phi(\mathbf{q},\omega).\end{aligned} \tag{4.113}$$

From (4.101), $\delta n(\mathbf{q},\omega) = \pi^{-1/2}N^*(0)v_F^* u_{00}^s$, we deduce

$$\chi_{nn}^R(\mathbf{q},\omega) = -\frac{n\mathbf{q}^2/m}{(\omega + i\eta)^2 - c_s^2\mathbf{q}^2} \tag{4.114}$$

and

$$S_{nn}(\mathbf{q},\omega) = \Theta(\omega)\frac{\pi n}{mc_s}|\mathbf{q}|\delta(\omega - c_s|\mathbf{q}|). \tag{4.115}$$

The structure factor exhibits a single delta peak at the first-sound frequency $c_s|\mathbf{q}|$. The quasi-particle–quasi-hole pair excitations have been washed out by collisions. Remarkably, the structure factor (4.115) satisfies the f-sum rule (3.106) and the compressibility sum rule (3.111), namely

$$\begin{aligned}\int_0^\infty \frac{d\omega}{\pi}\omega S_{nn}(\mathbf{q},\omega) &= \frac{n\mathbf{q}^2}{m},\\ \int_0^\infty \frac{d\omega}{\pi}\frac{S_{nn}(\mathbf{q},\omega)}{\omega} &= \frac{n}{mc_s^2}.\end{aligned} \tag{4.116}$$

This shows that the Landau theory, in spite of being a low-energy theory, describes all excitations of the system in the hydrodynamic regime.

4.3.4.3 *Current–current response function*

In this section, we compute the transverse current–current response function and the conductivity in the collisionless regime. We take

$$\mathbf{A}(x) = \mathbf{A}(\mathbf{q},\omega)e^{i[\mathbf{q}\cdot\mathbf{r}-(\omega+i\eta)t]} + \text{c.c.} \tag{4.117}$$

and $\mathbf{q} \parallel \hat{\mathbf{z}}$. We write the induced paramagnetic current as

$$\begin{aligned}\mathbf{j}(x) &= \int_{\mathbf{k}}\sum_\sigma \frac{\mathbf{k}}{m}\delta n_{\mathbf{k}\sigma}(x) = \mathbf{j}(\mathbf{q},\omega)e^{i[\mathbf{q}\cdot\mathbf{r}-(\omega+i\eta)t]} + \text{c.c.}\\ \mathbf{j}(\mathbf{q},\omega) &= \int_{\mathbf{k}}\sum_\sigma \frac{\mathbf{k}}{m}v_F^*\delta(\xi_{\mathbf{k}})u_\sigma(\hat{\mathbf{k}}) = \frac{3n}{m}\int\frac{d\Omega_{\hat{\mathbf{k}}}}{4\pi}\hat{\mathbf{k}}u^s(\hat{\mathbf{k}}).\end{aligned} \tag{4.118}$$

The linear response to the potential is given by the current–current response function,

$$J_\mu(\mathbf{q},\omega) = \sum_\nu \left[\chi^R_{j_\mu j_\nu}(\mathbf{q},\omega) - \frac{n}{m}\delta_{\mu,\nu}\right]A_\nu(\mathbf{q},\omega) \tag{4.119}$$

(see Sec. 3.4.4).

Without the collision term, the kinetic equation gives

$$(\omega + i\eta - \mathbf{v}^*_{\mathbf{k}}\cdot\mathbf{q})u_\sigma(\hat{\mathbf{k}}) - \mathbf{v}^*_{\mathbf{k}}\cdot\mathbf{q}\int_{\mathbf{k}'}\sum_{\sigma'} f_{\sigma\sigma'}(\mathbf{k},\mathbf{k}')\delta(\xi_{\mathbf{k}'})u_{\sigma'}(\hat{\mathbf{k}}') \tag{4.120}$$

$$= -\frac{1}{mv_F^*}(\mathbf{v}^*_{\mathbf{k}}\cdot\mathbf{q})\mathbf{k}\cdot\mathbf{A}(\mathbf{q},\omega). \tag{4.121}$$

Setting $f_{\sigma\sigma'}(\mathbf{k},\mathbf{k}') = 0$ in Eq. (4.121) gives $j_\mu(\mathbf{q},\omega) = \sum_\nu \chi^{0R}_{j_\mu j_\nu}(\mathbf{q},\omega)A_\nu(\mathbf{q},\omega)$, where

$$\chi^{0R}_{j_\mu j_\nu}(\mathbf{q},\omega) = -2\int_{\mathbf{k}}\delta(\xi_{\mathbf{k}})\frac{\mathbf{v}^*_{\mathbf{k}}\cdot\mathbf{q}}{\omega+i\eta-\mathbf{v}^*_{\mathbf{k}}\cdot\mathbf{q}}\frac{k_\mu k_\nu}{m^2}. \tag{4.122}$$

Apart from the effective mass correction, $\chi^{0R}_{j_\mu j_\nu}$ is the non-interacting (paramagnetic) current–current response function. To evaluate the transverse response, we take $\mathbf{q} \parallel \hat{\mathbf{z}}$ and

$\mu = \nu = x$,

$$\chi_\perp^{0R}(\mathbf{q},\omega) = -2\int_{\mathbf{k}} \delta(\xi_{\mathbf{k}}) \frac{\mathbf{v}_{\mathbf{k}}^* \cdot \mathbf{q}}{\omega + i\eta - \mathbf{v}_{\mathbf{k}}^* \cdot \mathbf{q}} \frac{k_x^2}{m^2} = -\frac{3}{4}\frac{nm^*}{m^2} I(s+i\eta), \tag{4.123}$$

where $s = \omega / v_F^* |\mathbf{q}|$ and

$$I(x) = \int_0^\pi d\theta \frac{\sin^3\theta\cos\theta}{x - \cos\theta} = -\frac{4}{3} + 2x^2 + x(1-x^2)\ln\left(\frac{x+1}{x-1}\right). \tag{4.124}$$

In order to solve (4.121) in the interacting case, we assume that $f_{\sigma\sigma}(\mathbf{k},\mathbf{k}') = f_0^s + f_1^s\cos(\alpha)$ (α is the angle between $\mathbf{k}$ and $\mathbf{k}'$). After some algebra, one finds

$$\begin{aligned}\chi_\perp^R(\mathbf{q},\omega) &= \frac{\chi_\perp^{0R}(\mathbf{q},\omega)}{1 + \frac{F_1^s}{3+F_1^s}\frac{m}{n}\chi_\perp^{0R}(\mathbf{q},\omega)} \\ &= -\frac{3}{4}\frac{nm^*}{m^2}\frac{I(s+i\eta)}{1 - \frac{F_1^s}{4}I(s+i\eta)}.\end{aligned} \tag{4.125}$$

The transverse current–current correlation function possesses a pole for $1 - \frac{F_1^s}{4}I(s+i\eta) = 0$. We recover Eq. (4.96) determining the frequency of the transverse mode $m = 1$.

The preceding results enable us to calculate the response to the (local) electromagnetic field in a charged Fermi liquid.[15] The transverse conductivity is defined by

$$\sigma_\perp(\mathbf{q},\omega) = \frac{e^2}{i(\omega+i\eta)}\left[\chi_\perp^R(\mathbf{q},\omega) - \frac{n}{m}\right] \tag{4.126}$$

(see Sec. 3.4.4). Using $\chi_\perp^{0R}(0,\omega) = 0$, we obtain

$$\sigma_\perp(0,\omega) \equiv \sigma(0,\omega) = \frac{i}{\omega+i\eta}\frac{ne^2}{m}, \tag{4.127}$$

which is the expected result for a translation-invariant system (Sec. 3.4.4). With $I(x) = -4/3 - i\pi x + \mathcal{O}(x^2)$, we obtain the static transverse conductivity

$$\sigma_\perp(\mathbf{q},0) = \frac{3\pi}{4}\frac{ne^2}{k_F|\mathbf{q}|}, \tag{4.128}$$

a result that does not depend on the mass of the particles.

The transverse mode $m = 1$ which appears as a pole of $\chi_\perp^R(\mathbf{q},\omega)$ when $F_1^s > 6$ [Eq. (4.96)] is modified by the coupling to the electromagnetic field. The dispersion of the transverse modes of the fermion system coupled to the electromagnetic field is obtained from (3.162). Solving this equation together with (4.125), one finds that the transverse excitations of the neutral system — that appear as a pole of $\chi_\perp^R$ — are little affected by the coupling to the electromagnetic field for $|\mathbf{q}| \gg \omega_p/c_l$ (c_l is the velocity of light). But at low frequency, when $|\mathbf{q}| \lesssim \omega_p/c_l$, the transverse mode disappears in the continuum of particle–hole pair excitations (see Fig. 3.9 and the discussion in Sec. 3.4.2).

[15]The Landau theory gives a transverse current–current response function $\chi_\perp^R(\mathbf{q},0) = n/m$ which is independent of $\mathbf{q}$. As a result, it does not describe the orbital diamagnetism of the charged Fermi liquid (Sec. 3.4.5).

4.3.5 *Multi-pair excitations*

The preceding study of the response functions $\chi^R_{nn}(\mathbf{q},\omega)$ and $\chi^R_\perp(\mathbf{q},\omega)$ shows that the Landau theory describes single-pair excitations and collective modes but does not take into account multi-pair excitations. In an interacting Fermi liquid, a single quasi-particle–quasi-hole pair excitation can decay into multiple excited pairs. In other words — focusing on the density–density response function — the density operator $\hat{n}(\mathbf{q})$ couples the ground state to excited states with an arbitrary number of quasi-particle–quasi-hole pairs. In this section, we briefly discuss to what extent the multi-pair excitations are expected to affect the structure factor $S_{nn}(\mathbf{q},\omega)$. For a thorough analysis, we refer to Refs. [4, 5].

When $|\mathbf{q}| \ll k_F$, the excitation energy $\omega = \epsilon_{\mathbf{k}+\mathbf{q}} - \epsilon_{\mathbf{k}}$ of a single pair is necessarily small since the Pauli principle requires $|\mathbf{k}| < k_F$ and $|\mathbf{k}+\mathbf{q}| > k_F$ (Fig. 4.5). Energy conservation imposes $\epsilon_{\mathbf{k}}$ and $\epsilon_{\mathbf{k}+\mathbf{q}}$ to be within ω of the Fermi surface. Since the single-particle density of states $N^*(\xi) \simeq N^*(0)$ is nearly constant near the Fermi surface, the density per unit energy $\rho^{(1)}(\omega)$ of the single-pair excitations is proportional to ω at small energy. In the ideal gas, $S^0_{nn}(\mathbf{q},\omega)$ is directly given by $\rho^{(1)}(\omega)$ since the matrix element $\langle 0|\hat{n}(\mathbf{q})|m\rangle$ is equal to unity when the transition is allowed by momentum conservation.[16]

The case of multi-pair excitations is different. Since only the total momentum $\mathbf{q}$ is fixed, the excitation energy can extend up to infinity. An example of a two-pair excitation is shown in Fig. 4.5. Multi-pair excitations are therefore expected to contribute a broad spectrum to the structure factor $S_{nn}(\mathbf{q},\omega)$. For an n-pair excitation, the excitation energy is determined by $\omega = \sum_{i=1}^n (\epsilon_{\mathbf{k}_i+\Delta\mathbf{k}_i} - \epsilon_{\mathbf{k}_i})$ with $|\mathbf{k}_i| < k_F$, $|\mathbf{k}_i + \Delta\mathbf{k}_i| > k_F$ and $\mathbf{q} = \sum_{i=1}^n \Delta\mathbf{k}_i$. Energy conservation requires that the $2n$ quasi-particles and quasi-holes lie within ω of the Fermi surface. The density per unit energy $\rho^{(n)}(\omega)$ of the n-pair excitations is therefore of order ω^{2n-1} for $\omega \to 0$. At low energies, multi-pair excitations are therefore negligible with respect to single-pair excitations. Their main effect is to produce a continuous excitation spectrum extended up to very high energies and leading to a small damping of the zero-sound mode (Fig. 4.6).

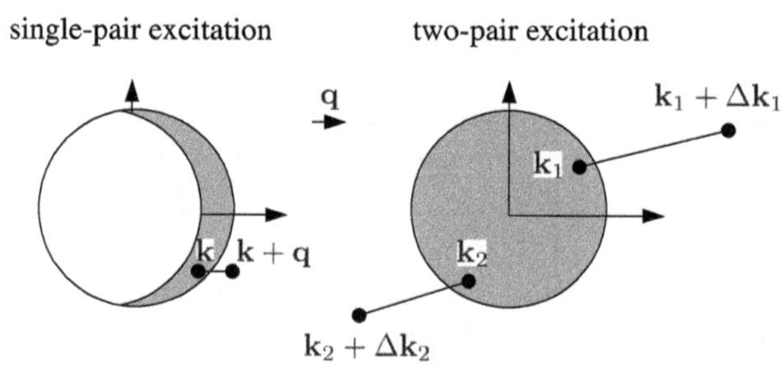

Fig. 4.5 Single-pair *vs* two-pair excitations ($\Delta\mathbf{k}_1 + \Delta\mathbf{k}_2 = \mathbf{q}$). The shaded areas show the allowed regions for the hole wavevectors $\mathbf{k}_i$.

[16] Recall that the $T=0$ structure factor reads $S_{nn}(\mathbf{q},\omega) = 2\pi\sum_{m\neq 0}|\langle 0|\hat{n}(\mathbf{q})|m\rangle|^2\delta(\omega+\epsilon_0-\epsilon_m)$ [see Eq. (3.37)].

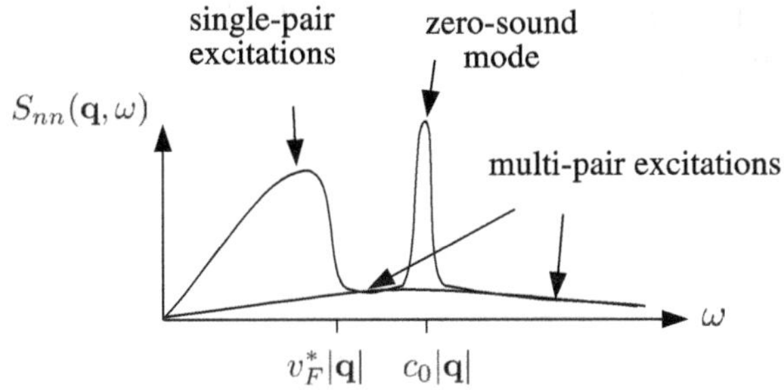

Fig. 4.6 Structure factor $S_{nn}(\mathbf{q}, \omega)$ of a Fermi liquid when multi-pair excitations are taken into account. The zero-sound mode is broadened due to the coupling to multi-pair excitations. Reproduced from Ref. [4].

It can also be argued that multi-pair excitations are negligible in the limit $\mathbf{q} \to 0$ regardless of the value of the energy ω [4]. The reason is that the matrix element $\langle m|\hat{n}(\mathbf{q})|0\rangle$ is of order $\mathbf{q}^2$ (in a translation-invariant system) for a multi-pair excited state $|m\rangle$. ($|0\rangle$ denotes the ground state.) By contrast, $\langle m|\hat{n}(\mathbf{q})|0\rangle$ is $\mathcal{O}(1)$ for single-pair excitations (as in the ideal Fermi gas) and $\mathcal{O}(\sqrt{|\mathbf{q}|})$ for the zero-sound mode. The suppression of multi-pair excitations in the long-wavelength limit is a direct consequence of translation invariance. The latter implies $[\hat{H}, \hat{\mathbf{j}}(\mathbf{q} = 0)] = 0$ so that we expect $[\hat{H}, \hat{\mathbf{j}}(\mathbf{q})] = \mathcal{O}(\mathbf{q})$. Let us now consider the continuity equation

$$\partial_t \hat{n}(\mathbf{q}, t) + i\mathbf{q} \cdot \hat{\mathbf{j}}(\mathbf{q}, t) = i[\hat{H}, \hat{n}(\mathbf{q}, t)] + i\mathbf{q} \cdot \hat{\mathbf{j}}(\mathbf{q}, t) = 0. \tag{4.129}$$

It implies

$$\begin{aligned} 0 &= \langle m|[\hat{H}, \hat{n}(\mathbf{q})]|0\rangle + \mathbf{q} \cdot \langle m|\hat{\mathbf{j}}(\mathbf{q})|0\rangle \\ &= (\epsilon_m - \epsilon_0)\langle m|\hat{n}(\mathbf{q})|0\rangle + \mathbf{q} \cdot \langle m|\hat{\mathbf{j}}(\mathbf{q})|0\rangle. \end{aligned} \tag{4.130}$$

Since the typical multi-pair excitation energy $\epsilon_m - \epsilon_0$ remains finite as $\mathbf{q} \to 0$, $\langle m|\hat{n}(\mathbf{q})|0\rangle = \mathcal{O}(\mathbf{q}^2)$.

4.4 Microscopic Basis of Fermi-Liquid Theory

The main goal of a microscopic approach to Fermi-liquid theory is to show how the quasi-particle concept emerges from the one-particle Green function. But it should also give a microscopic interpretation of the Landau functional $E[n]$ and the Landau parameters, reproduce the collective modes and response to macroscopic perturbations as obtained in Landau's theory, and prove Luttinger theorem. Before discussing these points in more detail, let us summarize the characteristic features of a Fermi liquid that emerge from a microscopic theory:

- The central (and often taken as the most fundamental) property of a Fermi liquid is that the self-energy satisfies

$$\Im[\Sigma^R(\mathbf{k}, \omega)] \propto -(\omega^2 + \pi^2 T^2) \tag{4.131}$$

at low energies and temperatures. Equation (4.131) implies that the $T = 0$ scattering rate $1/\tau_{\mathbf{k}} \sim -\Im[\Sigma^R(\mathbf{k}, \xi_{\mathbf{k}})] \propto (|\mathbf{k}| - k_F)^2$ vanishes faster than $\xi_{\mathbf{k}} \simeq v_F^*(|\mathbf{k}| - k_F)$ as one approaches the Fermi surface, a necessary condition for the existence of quasi-particles.

- Another fundamental property is that

$$\frac{\partial}{\partial\omega}\Re[\Sigma^R(\mathbf{k},\omega)]\Big|_{\omega=\xi_{\mathbf{k}}} \leq 0. \tag{4.132}$$

  This equation implies that the "quasi-particle weight"

$$z_{\mathbf{k}} = \frac{1}{1 - \partial_\omega \Re[\Sigma^R(\mathbf{k},\omega)]}\Bigg|_{\omega=\xi_{\mathbf{k}}}, \tag{4.133}$$

  which measures the overlap between a particle excitation $\hat{\psi}_\sigma^\dagger(\mathbf{k})|0\rangle$ and a quasi-particle state is necessarily finite and satisfies $0 < z_{\mathbf{k}} \leq 1$. $z_{\mathbf{k}}$ also determines the discontinuity at k_F in the momentum distribution $\langle \hat{\psi}_\sigma^\dagger(\mathbf{k})\hat{\psi}_\sigma(\mathbf{k})\rangle$. The existence of such a discontinuity is a characteristic property of a Fermi liquid.
- The Landau energy functional $E[n]$ — or, equivalently, the thermodynamic potential $\Omega[n]$ — can be obtained as a suitably defined Legendre transform of the grand potential $-T\ln Z$. The Landau function f is related in a simple way to the particle–hole interaction vertex Γ_{ph}.
- The volume of the Fermi surface is independent of the interactions (Luttinger theorem).

4.4.1 *Quasi-particles*

4.4.1.1 *Spectral function of the ideal Fermi gas*

We start by considering an ideal Fermi gas with (grand canonical) Hamiltonian

$$\hat{H}_0 = \sum_{\mathbf{k},\sigma} \xi_{\mathbf{k}}^0 \hat{\psi}_\sigma^\dagger(\mathbf{k})\hat{\psi}_\sigma(\mathbf{k}), \qquad \xi_{\mathbf{k}}^0 = \epsilon_{\mathbf{k}}^0 - \mu, \tag{4.134}$$

where μ is the chemical potential $[\mu(T=0) = \epsilon_F^0]$. The ground state reads

$$|0\rangle = \prod_{\substack{\mathbf{k},\sigma \\ |\mathbf{k}|\leq k_F}} \hat{\psi}_\sigma^\dagger(\mathbf{k})|\text{vac}\rangle, \tag{4.135}$$

and its (grand canonical) energy is $E_0 - \mu N$ [E_0 is given by (4.2)].

The state $\hat{\psi}_\sigma^\dagger(\mathbf{k})|0\rangle$ with an additional particle of momentum $\mathbf{k}$ ($|\mathbf{k}| > k_F$) and spin σ has an energy $E_0 - \mu N + \xi_{\mathbf{k}}^0$ and evolves in time according to

$$e^{-i\hat{H}_0 t}\hat{\psi}_\sigma^\dagger(\mathbf{k})|0\rangle = e^{-i(E_0 - \mu N + \xi_{\mathbf{k}}^0)t}\hat{\psi}_\sigma^\dagger(\mathbf{k})|0\rangle. \tag{4.136}$$

The purely oscillating time dependence is due to the fact that $\hat{\psi}_\sigma^\dagger(\mathbf{k})|0\rangle$ is an exact eigenstate. This property also shows up in the retarded Green function (which will turn out to be the quantity of interest in the interacting case)

$$G_0^R(\mathbf{k},t) = \int_{-\infty}^{\infty} \frac{d\omega}{2\pi} \frac{e^{-i\omega t}}{\omega + i\eta - \xi_{\mathbf{k}}^0} = -i\Theta(t)e^{-i\xi_{\mathbf{k}}^0 t} \tag{4.137}$$

and the spectral function

$$A_0(\mathbf{k},\omega) = -\frac{1}{\pi}\Im[G_0^R(\mathbf{k},\omega)] = \delta(\omega - \xi_{\mathbf{k}}^0), \tag{4.138}$$

where the exact excited state appears as a Dirac peak. A similar reasoning can be made for a hole excitation ($\hat{\psi}_\sigma(\mathbf{k})|0\rangle$ with $|\mathbf{k}| < k_F$).

4.4.1.2 *Spectral function of the interacting Fermi liquid*

In the interacting system, the retarded Green function and the spectral function,

$$\begin{aligned} G^R(\mathbf{k},\omega) &= \frac{1}{\omega + i\eta - \xi_{\mathbf{k}}^0 - \Sigma^R(\mathbf{k},\omega)}, \\ A(\mathbf{k},\omega) &= -\frac{1}{\pi} \frac{\Im[\Sigma^R(\mathbf{k},\omega)]}{\left(\omega - \xi_{\mathbf{k}}^0 - \Re[\Sigma^R(\mathbf{k},\omega)]\right)^2 + \left(\Im[\Sigma^R(\mathbf{k},\omega)]\right)^2} \end{aligned} \tag{4.139}$$

can be expressed in terms of the retarded self-energy $\Sigma^R(\mathbf{k},\omega)$ (Sec. 3.5.1). Spin-rotation invariance ensures that Σ^R, G^R and A are spin independent. Equation (4.139) holds for $\Im[\Sigma^R(\mathbf{k},\omega)] \neq 0$; when $\Sigma^R(\mathbf{k},\omega)$ is real $A(\mathbf{k},\omega) = \delta(\omega - \xi_{\mathbf{k}}^0 - \Sigma^R(\mathbf{k},\omega))$. A Fermi liquid is defined by a spectral function $A(\mathbf{k},\omega)$ which, for $\mathbf{k}$ in the vicinity of the Fermi surface, exhibits a sharp peak at an energy $\xi_{\mathbf{k}}$ with a width $1/2\tau_{\mathbf{k}}$ which goes to zero faster than $\xi_{\mathbf{k}}$ when $|\mathbf{k}| \to k_F$ (Fig. 4.7). As we shall see, such a peak is the signature of a quasi-particle ($|\mathbf{k}| > k_F$) or quasi-hole ($|\mathbf{k}| < k_F$) excitation.

If $\Im[\Sigma^R(\mathbf{k},\omega)]$ varies weakly for $\omega \approx \xi_{\mathbf{k}}$, then the position of the maximum is determined by

$$\xi_{\mathbf{k}} - \xi_{\mathbf{k}}^0 - \Re[\Sigma^R(\mathbf{k},\xi_{\mathbf{k}})] = 0 \tag{4.140}$$

(using $\partial_\omega \Im[\Sigma^R(\mathbf{k},\omega)] = 0$ for $\omega \simeq \xi_{\mathbf{k}}$). This equation determines the energy $\xi_{\mathbf{k}}$ of a quasi-particle with momentum $\mathbf{k}$. In particular, the Fermi momentum k_F is obtained from

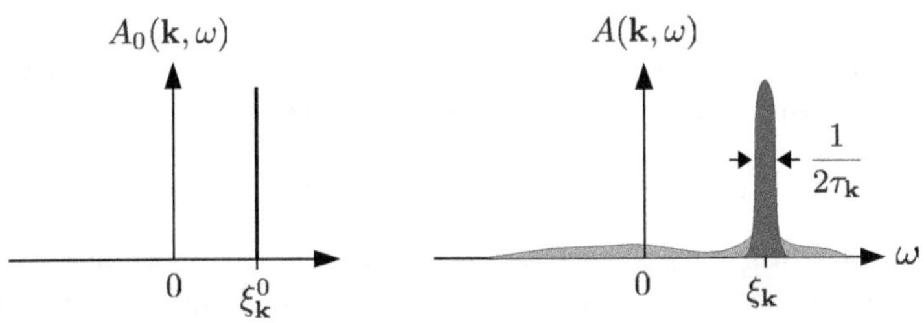

Fig. 4.7 Spectral function in an ideal Fermi gas and in a Fermi liquid. The dark shaded area shows the quasi-particle peak (spectral weight $z_{\mathbf{k}}$) and the light shaded one the incoherent part of the spectrum (spectral weight $1 - z_{\mathbf{k}}$). The width $1/\tau_{\mathbf{k}}$ of the quasi-particle peak goes to zero faster than $|\xi_{\mathbf{k}}|$ when $|\mathbf{k}| \to k_F$.

$\xi_{k_F} = 0$, i.e.

$$\xi^0_{k_F} + \Sigma^R(k_F, 0) = 0. \tag{4.141}$$

Here, we have used the fact that $\Sigma^R(\mathbf{k}_F, 0)$ is real at zero temperature [see Eq. (4.160) below]. In general, μ differs from $\epsilon^0_F = k_F^2/2m$ in the interacting system so that $\xi^0_{k_F} \neq 0$. We shall see in Sec. 4.4.6 that k_F (for a given density n) is not changed by the interactions (Luttinger theorem) and therefore given by (4.1). For ω near $\xi_{\mathbf{k}}$,

$$\begin{aligned}\omega - \xi^0_{\mathbf{k}} - \Re[\Sigma^R(\mathbf{k}, \omega)] &\simeq \omega - \xi^0_{\mathbf{k}} - \Re[\Sigma^R(\mathbf{k}, \xi_{\mathbf{k}})] - (\omega - \xi_{\mathbf{k}})\partial_\omega \Re[\Sigma^R(\mathbf{k}, \omega)]\big|_{\omega=\xi_{\mathbf{k}}} \\ &= \frac{\omega - \xi_{\mathbf{k}}}{z_{\mathbf{k}}},\end{aligned} \tag{4.142}$$

where $z_{\mathbf{k}}$, defined in (4.133), is referred to as the quasi-particle weight (for reasons that will be explained below). We deduce that the spectral function can be written as

$$A(\mathbf{k}, \omega) = \frac{z_{\mathbf{k}}}{\pi} \frac{1/2\tau_{\mathbf{k}}}{(\omega - \xi_{\mathbf{k}})^2 + (1/2\tau_{\mathbf{k}})^2} + A_{\text{inc}}(\mathbf{k}, \omega), \tag{4.143}$$

where

$$\frac{1}{\tau_{\mathbf{k}}} = -2 z_{\mathbf{k}} \Im[\Sigma^R(\mathbf{k}, \xi_{\mathbf{k}})] \tag{4.144}$$

is, as we shall see later, the inverse quasi-particle lifetime. The first term in the rhs of (4.143), the "quasi-particle peak", follows from (4.142) and determines $A(\mathbf{k}, \omega)$ near the maximum at $\omega = \xi_{\mathbf{k}}$. It corresponds to a Lorentzian peak of width $\sim 1/\tau_{\mathbf{k}}$ and spectral weight (defined as the area under the peak) $z_{\mathbf{k}}$. Since $A(\mathbf{k}, \omega) \geq 0$ and $\int_{-\infty}^{\infty} d\omega A(\mathbf{k}, \omega) = 1$, one has $0 \leq z_{\mathbf{k}} \leq 1$ and in turn (4.132). $A_{\text{inc}}(\mathbf{k}, \omega)$ denotes the "incoherent" part of the spectral function. It typically corresponds to a broad (featureless) excitation spectrum extending up to very high energies. In order for the total spectral weight to be unity, its weight must be equal to $1 - z_{\mathbf{k}}$. From (4.143) and the spectral representation (3.239), we deduce that the retarded Green function takes the form

$$G^R(\mathbf{k}, \omega) = \int_{-\infty}^{\infty} d\omega' \frac{A(\mathbf{k}, \omega')}{\omega + i\eta - \omega'} = \frac{z_{\mathbf{k}}}{\omega - \xi_{\mathbf{k}} + \frac{i}{2\tau_{\mathbf{k}}}} + G^R_{\text{inc}}(\mathbf{k}, \omega). \tag{4.145}$$

The quasi-particle peak gives rise to a pole at the complex energy $\xi_{\mathbf{k}} - i/2\tau_{\mathbf{k}}$ with a residue determined by the quasi-particle weight $z_{\mathbf{k}}$. In time space, Eq. (4.145) gives

$$G^R(\mathbf{k}, t) = -i z_{\mathbf{k}} \Theta(t) e^{-i\xi_{\mathbf{k}} t - t/2\tau_{\mathbf{k}}} + G^R_{\text{inc}}(\mathbf{k}, t). \tag{4.146}$$

Because $A_{\text{inc}}(\mathbf{k}, \omega)$ has no sharp structure in the variable ω, the incoherent part $G^R_{\text{inc}}(\mathbf{k}, t)$ of the Green function decays quickly in time, and the long-time behavior of $G^R(\mathbf{k}, t)$ is dominated by the quasi-particle pole. Thus, for $1/|\xi_{\mathbf{k}}| \ll t \ll \tau_{\mathbf{k}}$, which requires $1/\tau_{\mathbf{k}} \ll |\xi_{\mathbf{k}}|$, one observes the oscillating behavior characteristic of an eigenstate of the Hamiltonian. Equation (4.146) confirms the interpretation of $\tau_{\mathbf{k}}$ as the lifetime of the quasi-particle. The oscillating part of the Green function (4.146) is reduced by a factor $z_{\mathbf{k}}$ in the interacting system. Thus, $z_{\mathbf{k}}$ is a measure of the overlap between the state $\hat{\psi}_\sigma^\dagger(\mathbf{k})|0\rangle$ and the quasi-particle state with momentum $\mathbf{k}$ and spin σ; it can be seen as the fraction of bare particle contained in the quasi-particle.[17]

4.4.1.3 *Momentum distribution*

The existence of quasi-particles with reduced spectral weight has an important consequence for the momentum distribution function of the bare particles,

$$\begin{aligned}\langle\hat{\psi}_\sigma^\dagger(\mathbf{k})\hat{\psi}_\sigma(\mathbf{k})\rangle &= G(\mathbf{k},\tau=0^-) = \frac{1}{\beta}\sum_{\omega_n} e^{i\omega_n\eta}G(\mathbf{k},i\omega_n)\\ &= \frac{1}{\beta}\sum_{\omega_n} e^{i\omega_n\eta}\int_{-\infty}^{\infty} d\omega\frac{A(\mathbf{k},\omega)}{i\omega_n-\omega} = \int_{-\infty}^{\infty} d\omega\, n_F(\omega)A(\mathbf{k},\omega)\\ &= \int_{-\infty}^{0} d\omega\, A(\mathbf{k},\omega)\end{aligned} \tag{4.147}$$

($\eta\to 0^+$), where the last result holds at zero temperature. Here, $G(\mathbf{k},\tau)$ is the imaginary-time Green function. When $1/\tau_{\mathbf{k}} \ll |\xi_{\mathbf{k}}|$, the quasi-particle peak in $A(\mathbf{k},\omega)$ becomes sharper and sharper as we approach the Fermi surface and tends to $z_{\mathbf{k}}\delta(\omega-\xi_{\mathbf{k}})$ for $\xi_{\mathbf{k}}\to 0$. Since the incoherent part of the spectral function varies smoothly with $\mathbf{k}$, it is continuous across the Fermi level $\xi_{\mathbf{k}}=0$. We then deduce

$$\left[\lim_{|\mathbf{k}|\to k_F^+} - \lim_{|\mathbf{k}|\to k_F^-}\right]\langle\hat{\psi}_\sigma^\dagger(\mathbf{k})\hat{\psi}_\sigma(\mathbf{k})\rangle = \left[\lim_{|\mathbf{k}|\to k_F^+} - \lim_{|\mathbf{k}|\to k_F^-}\right]\int_{-\infty}^{0} d\omega\, z_{\mathbf{k}}\delta(\omega-\xi_{\mathbf{k}}), \tag{4.148}$$

where $k_F^\pm = k_F \pm 0^+$. Since $\xi_{k_F^+} > 0$ whereas $\xi_{k_F^-} < 0$, we conclude that the momentum distribution function exhibits a jump

$$\left[\lim_{|\mathbf{k}|\to k_F^+} - \lim_{|\mathbf{k}|\to k_F^-}\right]\langle\hat{\psi}_\sigma^\dagger(\mathbf{k})\hat{\psi}_\sigma(\mathbf{k})\rangle = -z_{\mathbf{k}}\big|_{|\mathbf{k}|=k_F} \equiv -z_{k_F} \tag{4.149}$$

across the Fermi level (Fig. 4.8). The existence of quasi-particles requiring $z_{k_F} > 0$, the discontinuity in the momentum distribution function $\langle\hat{\psi}_\sigma^\dagger(\mathbf{k})\hat{\psi}_\sigma(\mathbf{k})\rangle$ of the bare particles is an important characteristic of a Fermi liquid. This momentum distribution should not be confused with the quasi-particle momentum distribution $n_{\mathbf{k}}^0 = \Theta(k_F - |\mathbf{k}|)$ introduced in Sec. 4.1.

4.4.1.4 *Effective mass*

The quasi-particle group velocity is defined by

$$\mathbf{v}_{\mathbf{k}}^* = \boldsymbol{\nabla}\xi_{\mathbf{k}}. \tag{4.150}$$

[17]Recall that the $G^R(\mathbf{k},t)$ essentially measures the probability amplitude for a bare particle (or hole) with momentum $\mathbf{k}$ created at $t=0$ to be in the same quantum state at time t.

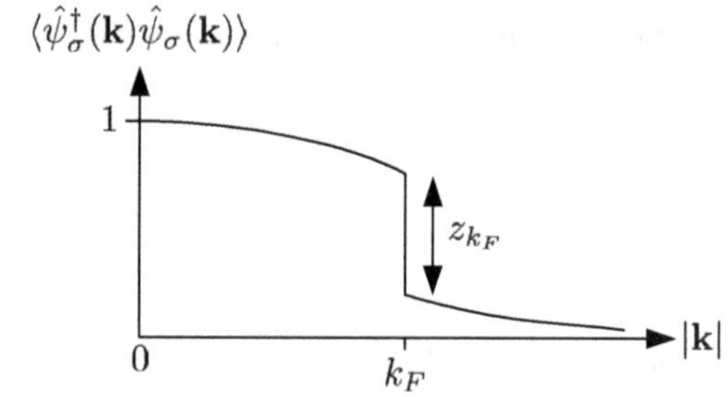

Fig. 4.8 Momentum distribution $\langle\hat\psi^\dagger_\sigma(\mathbf{k})\hat\psi_\sigma(\mathbf{k})\rangle$ in a Fermi liquid.

Using (4.140), we obtain

$$\begin{aligned}\mathbf{v}^*_\mathbf{k} &= \boldsymbol{\nabla}\{\xi^0_\mathbf{k} + \Re[\Sigma^R(\mathbf{k},\xi_\mathbf{k})]\}\\ &= \mathbf{v}_\mathbf{k} + \left(\frac{\partial}{\partial\mathbf{k}}\Re[\Sigma^R(\mathbf{k},\omega)] + \frac{\partial}{\partial\omega}\Re[\Sigma^R(\mathbf{k},\omega)]\boldsymbol{\nabla}\xi_\mathbf{k}\right)_{\omega=\xi_\mathbf{k}}\\ &= z_\mathbf{k}\left(\mathbf{v}_\mathbf{k} + \frac{\partial}{\partial\mathbf{k}}\Re[\Sigma^R(\mathbf{k},\omega)]\Big|_{\omega=\xi_\mathbf{k}}\right).\end{aligned} \tag{4.151}$$

For symmetry reasons, $\mathbf{v}^*_\mathbf{k} = v^*_\mathbf{k}\hat{\mathbf{k}}$ and $\Sigma^R(\mathbf{k},\omega)$ is a function of $|\mathbf{k}|$. From (4.151), we then obtain

$$v^*_\mathbf{k} = z_\mathbf{k}\left(\frac{|\mathbf{k}|}{m} + \frac{\partial}{\partial|\mathbf{k}|}\Re[\Sigma^R(\mathbf{k},\omega)]\Big|_{\omega=\xi_\mathbf{k}}\right). \tag{4.152}$$

From the definition (4.8) of the effective mass, we finally deduce

$$\frac{m}{m^*} = z_{k_F}\left(1 + \frac{m}{k_F}\frac{\partial}{\partial|\mathbf{k}|}\Re[\Sigma^R(\mathbf{k},\omega)]\Big|_{|\mathbf{k}|=k_F,\omega=0}\right). \tag{4.153}$$

When the self-energy is momentum independent, the effective mass is simply determined by $m^* = m/z_{k_F}$ and is larger than the bare mass. More generally, however, $\partial_{|\mathbf{k}|}\Re[\Sigma^R(\mathbf{k},\omega)]|_{|\mathbf{k}|=k_F,\omega=0}$ can have either sign and the effective mass can be larger or smaller than the bare mass. The same conclusion was reached in the phenomenological approach (Sec. 4.1.5). In Sec. 4.4.4, we shall show, using the Ward identities and the microscopic definition of the Landau function f, that Eq. (4.153) agrees with (4.40).

4.4.1.5 *Quasi-particle operators*

We can formally define quasi-particle operators (or fields) as follows [8]. In a first step, one eliminates the incoherent part of the spectral function $A(\mathbf{k},\omega)$ by filtering out $A_{\rm inc}(\mathbf{k},\omega)$. The retarded Green function then exhibits a purely propagating behavior $-iz_\mathbf{k}\Theta(t)e^{-i\xi_\mathbf{k}t}$ for $t \ll \tau_\mathbf{k}$. Let us suppose that this step can be seen as a change $\hat\psi_\sigma(\mathbf{k}) \to \hat\psi'_\sigma(\mathbf{k})$ of the fermion operator. In a second step, one introduces rescaled operators $\hat{\hat\psi}_\sigma(\mathbf{k}) = z_\mathbf{k}^{-1/2}\hat\psi'_\sigma(\mathbf{k})$

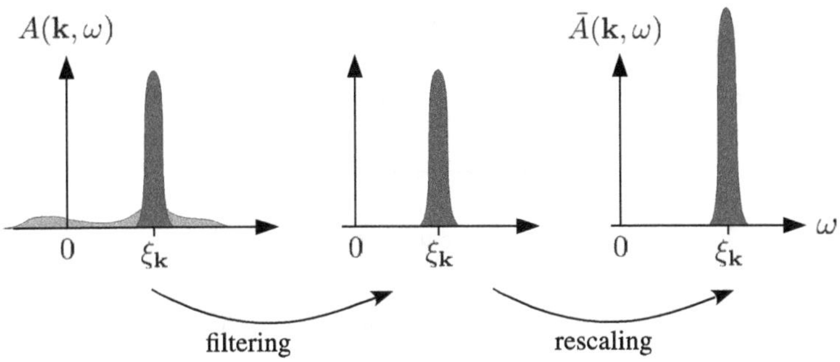

Fig. 4.9 Quasi-particle spectral function $\bar{A}(\mathbf{k},\omega)$ obtained by filtering out the incoherent part of $A(\mathbf{k},\omega)$ and rescaling in order to have a total spectral weight normalized to unity.

in order to recover a spectral function normalized to unity,

$$\bar{A}(\mathbf{k},\omega) = \frac{1}{\pi}\frac{1/2\tau_\mathbf{k}}{(\omega-\xi_\mathbf{k})^2+(1/2\tau_\mathbf{k})^2}. \tag{4.154}$$

Note that this program, filtering followed by rescaling (Fig. 4.9), is naturally realized in the RG framework (see the bibliography at the end of the chapter).

The retarded quasi-particle Green function then reads

$$\bar{G}^R(\mathbf{k},\omega) = \frac{1}{\omega-\xi_\mathbf{k}+\frac{i}{2\tau_\mathbf{k}}} \tag{4.155}$$

and is related to the fermion Green function by

$$G^R(\mathbf{k},\omega) = z_\mathbf{k}\bar{G}^R(\mathbf{k},\omega) + G_{\rm inc}(\mathbf{k},\omega). \tag{4.156}$$

The corresponding distribution function $\langle\hat{\bar{\psi}}^\dagger_\sigma(\mathbf{k})\hat{\bar{\psi}}_\sigma(\mathbf{k})\rangle = \bar{G}^R(\mathbf{k},\tau=0^-)$ takes the quasi-particle form $n^0_\mathbf{k} = \Theta(k_F-|\mathbf{k}|)$.

4.4.1.6 *Quasi-particle lifetime*

The divergence of the quasi-particle lifetime near the Fermi surface is a consequence of the reduced phase space available for the decay of an incoming particle induced by the excitation of a single or several quasi-particle–quasi-hole pairs.[18] This essential property of the Fermi liquid does not depend on the strength of the interactions.

In this section, we want to substantiate the phase space argument by considering the second-order self-energy

$$\Sigma(k) = \frac{1}{\beta\mathcal{V}}\sum_q v^2_\mathbf{q}\chi_0(q)G_0(k+q) \tag{4.157}$$

[18]See footnote 6 on page 316.

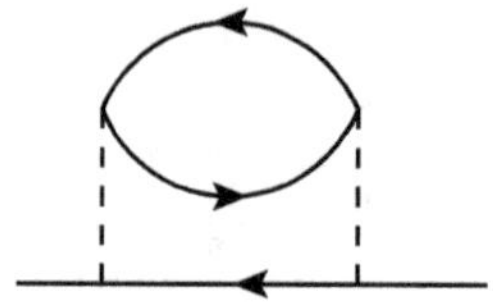

Fig. 4.10 Second-order contribution to the self-energy.

[with the notation $k = (\mathbf{k}, i\omega_n)$ and $q = (\mathbf{q}, i\omega_\nu)$] shown in Fig. 4.10, where

$$\chi_0(q) = -\frac{2}{\beta\mathcal{V}}\sum_k G_0(k)G_0(k+q) \tag{4.158}$$

is the bare particle–hole response function and $v_\mathbf{q}$ is the Fourier transform of the interaction. The calculation gives

$$\Im[\Sigma^R(\mathbf{k},\omega)] = -\frac{m^3}{16\pi^3 k_F}\left(\omega^2+\pi^2T^2\right)\int_0^{2k_F} d|\mathbf{q}|v_\mathbf{q}^2 \qquad (|\omega|, |\xi^0_\mathbf{k}|, T \ll \epsilon^0_F) \tag{4.159}$$

(see Appendix 4.A). At zero temperature, the self-energy $\Sigma(\mathbf{k},0) = \Sigma^R(\mathbf{k},0) = \Sigma^A(\mathbf{k},0)$ is real. Equation (4.159) yields the low-energy scattering rate

$$\frac{1}{\tau_\mathbf{k}} = -2z_\mathbf{k}\Im[\Sigma^R(\mathbf{k},\xi_\mathbf{k})] = z_\mathbf{k}\frac{m^3}{8\pi^3 k_F}\left(\xi_\mathbf{k}^2+\pi^2T^2\right)\int_0^{2k_F} d|\mathbf{q}|v_\mathbf{q}^2. \tag{4.160}$$

At zero temperature, $1/\tau_\mathbf{k} \sim (|\mathbf{k}|-k_F)^2$ vanishes faster than $\xi_\mathbf{k}$ when approaching the Fermi surface.

Similarly, one also could calculate higher-order self-energy diagrams. One would find that they give weaker contributions $\sim \omega^n, T^n$ (with $n > 2$) than the second-order one. The reason is that the phase space available for multi-pair excitations (i.e. the density per unit energy of the multi-pair excitations) is very small at low energies (Sec. 4.3.5).

4.4.1.7 *The quasi-particle picture in the Euclidean formalism*

In the zero-temperature Euclidean formalism, the quasi-particle propagator $\bar{G}(\mathbf{k}, i\omega)$ naturally emerges if one expands the self-energy $\Sigma(\mathbf{k}, i\omega)$ about $\omega = 0$. The discontinuity of the self-energy $\Sigma(\mathbf{k}, z)$ across the real axis comes from the imaginary part of the retarded self-energy: $\Sigma(\mathbf{k}, \omega + i0^+) - \Sigma(\mathbf{k}, \omega - i0^+) = 2i\Im[\Sigma^R(\mathbf{k},\omega)]$. Equation (4.159) shows that when $T \to 0$, this discontinuity originates from a contribution $-\gamma z^2 \operatorname{sgn}(\Im z)$ to $\Sigma(\mathbf{k}, z)$ when $|\Im z|, |\xi_\mathbf{k}| \ll \epsilon^0_F$, with γ a positive constant. $\Sigma(\mathbf{k}, i\omega)$ (ω real) is therefore continuous and differentiable at $\omega = 0$,[19]

$$\Sigma(\mathbf{k}, i\omega) = \Sigma(\mathbf{k}, 0) + i\omega\frac{\partial\Sigma(\mathbf{k}, i\omega)}{\partial i\omega}\bigg|_{\omega=0} + \mathcal{O}(\omega^2), \tag{4.161}$$

[19]We denote the Matsubara frequency by ω (and not ω_n) since it becomes a continuous variable when $T \to 0$.

where $\Sigma(\mathbf{k},0)$ is real. Equation (4.161) allows us to express the Green function as

$$\begin{aligned} G(\mathbf{k},i\omega) &= \frac{z_{\mathbf{k}}}{i\omega-\xi_{\mathbf{k}}} + G_{\text{inc}}(\mathbf{k},i\omega) \\ &\equiv z_{\mathbf{k}}\bar{G}(\mathbf{k},i\omega) + G_{\text{inc}}(\mathbf{k},i\omega), \end{aligned} \tag{4.162}$$

where $G_{\text{inc}}(\mathbf{k},i\omega)$ denotes the incoherent part of the propagator and

$$\frac{1}{z_{\mathbf{k}}} = 1 - \left.\frac{\partial\Sigma(\mathbf{k},i\omega)}{\partial i\omega}\right|_{\omega=0}, \qquad \xi_{\mathbf{k}} = z_{\mathbf{k}}[\xi^0_{\mathbf{k}} + \Sigma(\mathbf{k},0)]. \tag{4.163}$$

These two expressions can be deduced from (4.133) and (4.140) by expanding $\Sigma^R(\mathbf{k},\xi_{\mathbf{k}})$ about $\xi_{\mathbf{k}} = 0$.[20] Being the residue of the propagator at the pole $i\omega = \xi_{\mathbf{k}}$, $z_{\mathbf{k}}$ is real by virtue of the spectral representation (3.239). The finite quasi-particle lifetime is not taken into account in the expansion (4.161) but can be safely ignored in the low-energy limit since it does not contribute to leading order. In the following, we will often approximate the quasi-particle weight $z_{\mathbf{k}}$ by its value at the Fermi level $z_{k_F} \equiv z$.

4.4.2 *Thermodynamic potential* $\Omega[n]$

In this section, we derive the thermodynamic potential $\Omega[n]$ introduced in Landau's Fermi-liquid theory (Sec. 4.1.3) and obtain a microscopic definition of the Landau function f. We then relate f to the particle–hole interaction vertex Γ_{ph}.

4.4.2.1 *Microscopic definitions of* $\Omega[n]$ *and the Landau function* f

We consider the partition function

$$Z[h] = \int \mathcal{D}[\psi^*,\psi] \exp\Big\{-S[\psi^*,\psi] + \sum_{\mathbf{k},\sigma} h_{\mathbf{k}\sigma} \int_0^\beta d\tau\, \hat{n}_{\mathbf{k}\sigma}(\tau)\Big\} \tag{4.164}$$

(ψ and ψ^* are Grassmann variables) in the presence of a static external field that couples to the quasi-particle number operator

$$\hat{n}_{\mathbf{k}\sigma}(\tau) = \bar{\psi}^*_\sigma(\mathbf{k},\tau)\bar{\psi}_\sigma(\mathbf{k},\tau). \tag{4.165}$$

Note that $\hat{n}_{\mathbf{k}\sigma}$ is defined as a function of the quasi-particle field $\bar{\psi}^{(*)}$ which differs from the (bare) fermion field $\psi^{(*)}$ (Sec. 4.4.1.5). The quasi-particle occupation number is then obtained from a functional derivative of the partition function,

$$n_{\mathbf{k}\sigma} = \langle \hat{n}_{\mathbf{k}\sigma}(\tau)\rangle = \frac{1}{\beta}\frac{\delta \ln Z[h]}{\delta h_{\mathbf{k}\sigma}}. \tag{4.166}$$

[20]Equation (4.133) gives $z_{\mathbf{k}}^{-1} \simeq 1 - \partial_\omega \Re[\Sigma^R(\mathbf{k},\omega)]|_{\omega=0} = 1 - \partial_{i\omega}\Sigma(\mathbf{k},i\omega)|_{\omega=0}$ using $\Im[\Sigma^R(\mathbf{k},\omega)] = \mathcal{O}(\omega^2)$ and $\partial_\omega\Sigma^R(\mathbf{k},\omega)|_{\omega=0} = \partial_{i\omega}\Sigma(\mathbf{k},i\omega)|_{\omega=0}$. Equation (4.140) gives $\xi_{\mathbf{k}} \simeq \xi^0_{\mathbf{k}} + \Sigma^R(\mathbf{k},0) + \xi_{\mathbf{k}}\partial_\omega\Re[\Sigma^R(\mathbf{k},\omega)]|_{\omega=0} = \xi^0_{\mathbf{k}} + \Sigma(\mathbf{k},0) + \xi_{\mathbf{k}}(1-1/z_{\mathbf{k}})$.

In order to write the grand potential $\Omega[n]$ as a function of the quasi-particle distribution function $n \equiv \{n_{\mathbf{k}\sigma}\}$, we consider the Legendre transform

$$\Omega[n] = -\frac{1}{\beta} \ln Z[h] + \sum_{\mathbf{k},\sigma} h_{\mathbf{k}\sigma} n_{\mathbf{k}\sigma}, \tag{4.167}$$

where $h_{\mathbf{k}\sigma} \equiv h_{\mathbf{k}\sigma}[n]$ is obtained by inverting (4.166). $\Omega[n]$ satisfies the equation of state

$$\frac{\delta\Omega[n]}{\delta n_{\mathbf{k}\sigma}} = h_{\mathbf{k}\sigma}. \tag{4.168}$$

At equilibrium ($h = 0$), it is stationary with respect to variations of the quasi-particle distribution.

For a non-interaction system, the calculation of $\Omega[n]$ is straightforward. In that case, the quasi-particles coincide with the bare fermions ($\bar{\psi}^{(*)} = \psi^{(*)}$) so that

$$\begin{aligned} Z[h] &= \int \mathcal{D}[\psi^*, \psi] \exp\Biggl\{ \sum_{\mathbf{k},\sigma,\omega_n} \psi^*_\sigma(\mathbf{k}, i\omega_n)(i\omega_n - \xi_{\mathbf{k}} + h_{\mathbf{k}\sigma})\psi_\sigma(\mathbf{k}, i\omega_n) \Biggr\} \\ &= \prod_{\mathbf{k},\sigma} \left[1 + e^{-\beta(\xi_{\mathbf{k}} - h_{\mathbf{k}\sigma})} \right]. \end{aligned} \tag{4.169}$$

This gives $n_{\mathbf{k}\sigma} = n_F(\xi_{\mathbf{k}} - h_{\mathbf{k}\sigma})$ and

$$\Omega[n] = \sum_{\mathbf{k},\sigma} \xi_{\mathbf{k}} n_{\mathbf{k}\sigma} + \frac{1}{\beta} \sum_{\mathbf{k},\sigma} \left[n_{\mathbf{k}\sigma} \ln n_{\mathbf{k}\sigma} + (1 - n_{\mathbf{k}\sigma}) \ln(1 - n_{\mathbf{k}\sigma}) \right], \tag{4.170}$$

which is the expected result for non-interacting fermions.

For interacting fermions, it is not possible to calculate exactly the grand potential. However, we do not require the whole knowledge of $\Omega[n]$, but only its variation $\delta\Omega$ when the quasi-particle distribution n varies from its equilibrium value $\bar{n} = n|_{h=0}$ by δn. For $T \to 0$, the case we are interested in, $\bar{n}_{\mathbf{k}\sigma} = \Theta(k_F - |\mathbf{k}|)$. Expanding $\delta\Omega[\delta n] = \Omega[n + \delta n] - \Omega[n]$ to second order in δn, we obtain

$$\delta\Omega[\delta n] = \frac{1}{2} \sum_{\mathbf{k},\mathbf{k}',\sigma,\sigma'} \left. \frac{\delta^2\Omega[n]}{\delta n_{\mathbf{k}\sigma} \delta n_{\mathbf{k}'\sigma'}} \right|_{n=\bar{n}} \delta n_{\mathbf{k}\sigma} \delta n_{\mathbf{k}'\sigma'}. \tag{4.171}$$

There is no linear term since we expand about the stationary (equilibrium) state. Taking the functional derivative of the equation of state (4.168), one easily obtains

$$\frac{1}{\beta} \sum_{\mathbf{k}_3,\sigma_3} \frac{\delta^2\Omega[n]}{\delta n_{\mathbf{k}_1\sigma_1} \delta n_{\mathbf{k}_3\sigma_3}} \frac{\delta^2 \ln Z[h]}{\delta h_{\mathbf{k}_3\sigma_3} \delta h_{\mathbf{k}_2\sigma_2}} = \delta_{\mathbf{k}_1,\mathbf{k}_2} \delta_{\sigma_1,\sigma_2}. \tag{4.172}$$

This allows us to rewrite $\delta\Omega$ as

$$\delta\Omega[\delta n] = \frac{1}{2} \sum_{\mathbf{k},\mathbf{k}',\sigma,\sigma'} \bar{\chi}^{-1}_{\sigma\sigma'}(\mathbf{k}, \mathbf{k}') \delta n_{\mathbf{k}\sigma} \delta n_{\mathbf{k}'\sigma'}, \tag{4.173}$$

where $\bar{\chi}^{-1}$ is the inverse (in a matrix sense) of the correlation function

$$\bar{\chi}_{\sigma\sigma'}(\mathbf{k},\mathbf{k}') = \frac{1}{\beta}\frac{\delta^2 \ln Z[h]}{\delta h_{\mathbf{k}\sigma}\delta h_{\mathbf{k}'\sigma'}}\bigg|_{h=0} = \frac{1}{\beta}\int_0^\beta d\tau d\tau' \langle \hat{n}_{\mathbf{k}\sigma}(\tau)\hat{n}_{\mathbf{k}'\sigma'}(\tau')\rangle. \tag{4.174}$$

Note that $\bar{\chi}$ is nothing but the linear response function to the external field h. Comparing (4.173) and (4.28), we obtain the following microscopic definition of the Landau function f:

$$\frac{1}{\mathcal{V}} f_{\sigma\sigma'}(\mathbf{k},\mathbf{k}') = \frac{\delta_{\sigma,\sigma'}\delta_{\mathbf{k},\mathbf{k}'}}{n_F'(\xi_{\mathbf{k}})} + \bar{\chi}^{-1}_{\sigma\sigma'}(\mathbf{k},\mathbf{k}'). \tag{4.175}$$

Since we are interested in the limit $T \to 0$, we have taken $\tilde{\xi}_{\mathbf{k}} = \xi_{\mathbf{k}}$ in (4.28).

Thus, the calculation of the Landau function f reduces to that of the response function $\bar{\chi}$. In the following, we show that f can be identified with the irreducible (2PI) quasi-particle–quasi-hole interaction vertex. This will enable us to relate f to the particle–hole interaction vertex.

4.4.2.2 *The Landau function f vs the particle–hole interaction vertex $\Gamma_{\rm ph}$*

Let us introduce the correlation functions

$$\begin{aligned}\bar{\chi}_{\sigma\sigma'}(k,k';q) &= \langle \bar{\psi}^*_\sigma(k_-)\bar{\psi}_\sigma(k_+)\bar{\psi}^*_{\sigma'}(k'_+)\bar{\psi}_{\sigma'}(k'_-)\rangle \\ &\quad - \langle \bar{\psi}^*_\sigma(k_-)\bar{\psi}_\sigma(k_+)\rangle\langle \bar{\psi}^*_{\sigma'}(k'_+)\bar{\psi}_{\sigma'}(k'_-)\rangle\end{aligned} \tag{4.176}$$

and

$$\bar{\chi}_{\sigma\sigma'}(\mathbf{k},\mathbf{k}';q) = \frac{1}{\beta}\sum_{\omega_n,\omega_{n'}} \bar{\chi}_{\sigma\sigma'}(k,k';q), \tag{4.177}$$

where $k = (\mathbf{k}, i\omega_n)$, $q = (\mathbf{q}, i\omega_\nu)$, $k_+ = (\mathbf{k}+\mathbf{q}/2, i\omega_n + i\omega_\nu)$, $k_- = (\mathbf{k}-\mathbf{q}/2, i\omega_n)$ (ω_ν is a bosonic Matsubara frequency). The function $\bar{\chi}_{\sigma\sigma'}(\mathbf{k},\mathbf{k}')$, which is related to the Landau function f by (4.175), corresponds to the limit $q \to 0$ of $\bar{\chi}_{\upsilon\upsilon'}(\mathbf{k},\mathbf{k}';q)$. We shall see in the following that this limit is ill-defined, since the limits $\mathbf{q} \to 0$ and $i\omega_\nu \to 0$ do not commute. At zero temperature, it is not possible to create a quasi-particle–quasi-hole pair with a finite total energy and a vanishing total momentum because of the Pauli principle.[21] Thus, for the external field $h_{\mathbf{k}\sigma}$ to modify the quasi-particle distribution function, it should be understood as

$$\begin{aligned}h_{\mathbf{k}\sigma}\int_0^\beta d\tau\, \bar{\psi}^*_\sigma(\mathbf{k},\tau)\bar{\psi}_\sigma(\mathbf{k},\tau) &= h_{\mathbf{k}\sigma}\sum_{\omega_n} \bar{\psi}^*_\sigma(k)\bar{\psi}_\sigma(k) \\ &\equiv h_{\mathbf{k}\sigma}\lim_{\mathbf{q}\to 0}\left[\lim_{\omega_\nu\to 0}\sum_{\omega_n}\bar{\psi}^*_\sigma(k_-)\bar{\psi}_\sigma(k_+)\right].\end{aligned} \tag{4.178}$$

[21]This means that the equation $\omega = \epsilon_{\mathbf{k}+\mathbf{q}} - \epsilon_{\mathbf{k}}$, together with the constraints $|\mathbf{k}| < k_F$ and $|\mathbf{k}+\mathbf{q}| > k_F$, has no solution for ω finite and $\mathbf{q} \to 0$.

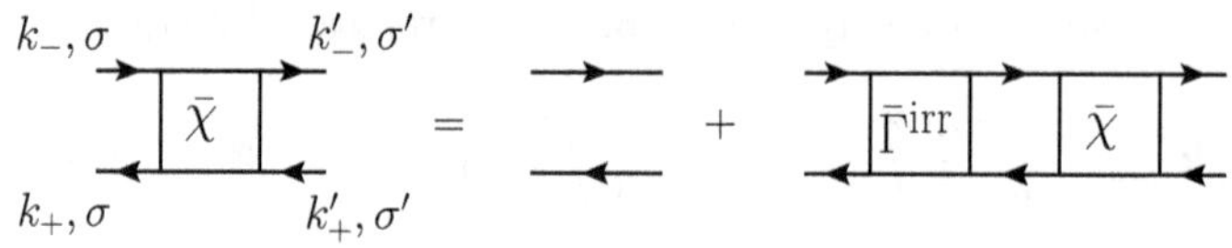

Fig. 4.11 Diagrammatic representation of the Bethe–Salpeter equation satisfied by $\bar\chi_{\sigma\sigma'}(k, k'; q)$.

This leads us to define $\bar\chi_{\sigma\sigma'}(\mathbf{k}, \mathbf{k}')$ as

$$\bar\chi_{\sigma\sigma'}(\mathbf{k}, \mathbf{k}') = \lim_{\mathbf{q}\to 0}\Big[\lim_{\omega_\nu\to 0} \bar\chi_{\sigma\sigma'}(\mathbf{k}, \mathbf{k}'; q)\Big]. \tag{4.179}$$

It is customary to refer to the limits $(i\omega_\nu \to 0, \mathbf{q} = 0)$ and $(i\omega_\nu = 0, \mathbf{q} \to 0)$ as the ω- and $\mathbf{q}$-limits, respectively.

The two-particle Green function $\bar\chi_{\sigma\sigma'}(k, k'; q)$ satisfies the Bethe–Salpeter equation (Fig. 4.11)

$$\begin{aligned}\bar\chi_{\sigma\sigma'}(k, k'; q) &= \bar\Pi_{\sigma\sigma'}(k, k'; q)\\ &\quad - \frac{1}{\beta\mathcal{V}} \sum_{k_1, k_2, \sigma_1, \sigma_2} \bar\Pi_{\sigma\sigma_1}(k, k_1; q)\bar\Gamma^{\rm irr}_{\sigma_1\sigma_2}(k_1, k_2; q)\bar\chi_{\sigma_2\sigma'}(k_2, k'; q)\end{aligned} \tag{4.180}$$

where

$$\bar\Gamma^{\rm irr}_{\sigma\sigma'}(k, k'; q) = \bar\Gamma^{\rm irr}_{{\rm ph},\sigma\sigma\sigma'\sigma'}(k_+, k_-; k'_-, k'_+) \tag{4.181}$$

is the 2PI vertex in the particle–hole channel, and

$$\bar\Pi_{\sigma\sigma'}(k, k'; q) = -\delta_{\sigma,\sigma'}\delta_{k,k'}\bar G(k_-)\bar G(k_+) \tag{4.182}$$

the quasi-particle–quasi-hole pair propagator. The quasi-particle propagator $\bar G$ is defined by (4.162). In the following, we approximate $z_{\mathbf{k}}$ by its value $z_{k_F} \equiv z$ at the Fermi level.

The quasi-particle–quasi-hole pair propagator $\bar\Pi$ is singular in the limit $q \to 0$. To see this, we consider

$$\begin{aligned}\bar\Pi_{\sigma\sigma'}(\mathbf{k}, \mathbf{k}'; q) &= \frac{1}{\beta}\sum_{\omega_n, \omega_{n'}} \bar\Pi_{\sigma\sigma'}(k, k'; q) = -\delta_{\sigma,\sigma'}\delta_{\mathbf{k},\mathbf{k}'}\frac{1}{\beta}\sum_{\omega_n} \bar G(k_-)\bar G(k_+)\\ &= \delta_{\sigma,\sigma'}\delta_{\mathbf{k},\mathbf{k}'}\frac{n_F(\xi_{\mathbf{k}+\mathbf{q}/2}) - n_F(\xi_{\mathbf{k}} - \mathbf{q}/2)}{i\omega_\nu - \xi_{\mathbf{k}+\mathbf{q}/2} + \xi_{\mathbf{k}-\mathbf{q}/2}}\\ &= \delta_{\sigma,\sigma'}\delta_{\mathbf{k},\mathbf{k}'}\frac{\mathbf{v}^*_{\mathbf{k}}\cdot\mathbf{q}}{i\omega_\nu - \mathbf{v}^*_{\mathbf{k}}\cdot\mathbf{q}} n'_F(\xi_{\mathbf{k}}) \qquad (\mathbf{q}\to 0).\end{aligned} \tag{4.183}$$

Thus, for $q \to 0$, we obtain 0 if $|\mathbf{q}|/\omega_\nu \to 0$, and $-\delta_{\sigma,\sigma'}\delta_{\mathbf{k},\mathbf{k}'}n'_F(\xi_{\mathbf{k}})$ if $\omega_\nu/|\mathbf{q}| \to 0$. In Fermi-liquid theory, one assumes that only the quasi-particle–quasi-hole pair propagator $\bar\Pi$ leads to singularities when $q \to 0$. This can be checked in perturbation theory, at least to lowest orders. The one-loop corrections to $\bar\chi$ (or $\bar\Gamma$) are shown in Fig. 4.12. The Feynman diagrams are labeled according to the type of fluctuations they describe. The zero-sound (ZS) channel

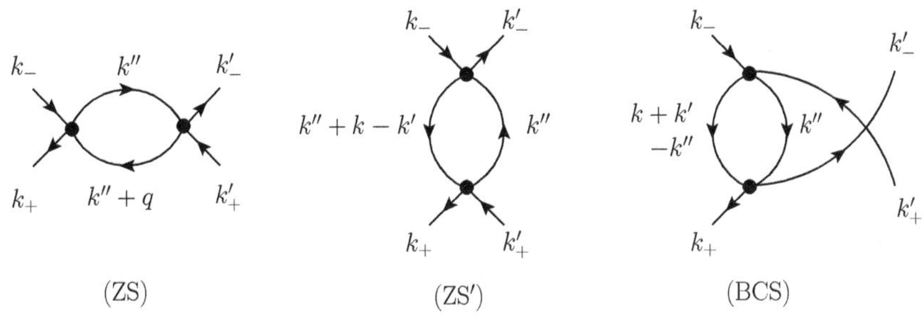

Fig. 4.12 First-order (one-loop) corrections to the response function $\bar\chi$ (or, without the external legs, the particle–hole interaction vertex $\bar\Gamma$). The line denotes the quasi-particle propagator $\bar G$ and the dot the interaction.

corresponds to propagation of particle–hole pairs with a small total frequency–momentum. The crossed particle–hole channel is referred to as ZS′, while the BCS channel involves particle–particle pair propagation. One readily sees that only the zero-sound channel leads to a singularity in the $q \to 0$ limit; the ZS′ and BCS loops (after summation over the internal frequency–momentum) are not singular in that limit.[22]

Since the irreducible vertex $\bar\Gamma^{\rm irr}$ does not contain the ZS loop (the latter being two-particle reducible), it has a well-defined limit when $q \to 0$. We can therefore set $q = 0$ in $\bar\Gamma^{\rm irr}$. Furthermore, since the singularity of $\bar\Pi$ is due to states near the Fermi surface [Eq. (4.183)], we can ignore the ω_n dependence of $\bar\Gamma^{\rm irr}$ which then becomes a function $\bar\Gamma^{\rm irr}_{\sigma\sigma'}(\mathbf{k},\mathbf{k}')$ of the momenta $\mathbf{k}$ and $\mathbf{k}'$. This enables us to carry out the frequency sum in Eq. (4.180) satisfied by $\bar\chi_{\sigma\sigma'}(k,k';q)$, which gives

$$\begin{aligned}\bar\chi_{\sigma\sigma'}(\mathbf{k},\mathbf{k}';q) &= \bar\Pi_{\sigma\sigma'}(\mathbf{k},\mathbf{k}';q) \\ &\quad - \frac{1}{\mathcal{V}} \sum_{\mathbf{k}_1,\mathbf{k}_2,\sigma_1,\sigma_2} \bar\Pi_{\sigma\sigma_1}(\mathbf{k},\mathbf{k}_1;q)\bar\Gamma^{\rm irr}_{\sigma_1\sigma_2}(\mathbf{k}_1,\mathbf{k}_2)\bar\chi_{\sigma_2\sigma'}(\mathbf{k}_2,\mathbf{k}';q). \end{aligned} \tag{4.184}$$

This equation can be rewritten in matrix form as

$$\bar\chi = \bar\Pi - \bar\Pi\bar\Gamma^{\rm irr}\bar\chi \quad \text{i.e.} \quad \bar\chi^{-1} = \bar\Pi^{-1} + \bar\Gamma^{\rm irr}. \tag{4.185}$$

In the $\mathbf{q}$-limit, this gives

$$\bar\chi^{-1}_{\sigma\sigma'}(\mathbf{k},\mathbf{k}') = -\frac{\delta_{\sigma,\sigma'}\delta_{\mathbf{k},\mathbf{k}'}}{n'_F(\xi_{\mathbf{k}})} + \frac{1}{\mathcal{V}}\bar\Gamma^{\rm irr}_{\sigma\sigma'}(\mathbf{k},\mathbf{k}'). \tag{4.186}$$

Comparing this result with (4.175), we deduce

$$f_{\sigma\sigma'}(\mathbf{k},\mathbf{k}') = \bar\Gamma^{\rm irr}_{\sigma\sigma'}(\mathbf{k},\mathbf{k}'). \tag{4.187}$$

[22]For $k \to k'$, the ZS′ channel also becomes singular since the total frequency–momentum $k - k'$ in the ZS′ loop vanishes (Fig. 4.12). For $k \to k'$, both the ZS and ZS′ channels should therefore be taken into account. In particular, this is necessary for a correct description of the Pauli principle. But for most physical properties, the ZS′ channel can be safely discarded. This issue is discussed in detail in Ref. [14].

Since $\bar{\Gamma}^{\rm irr}_{\sigma\sigma'}(\mathbf{k},\mathbf{k}')$ does not contain the ZS loop, it is a non-singular function of $\mathbf{k}$ and $\mathbf{k}'$ and one can approximate the latter by $\mathbf{k}_F$ and $\mathbf{k}'_F$ at low energies, which amounts to approximating $f_{\sigma\sigma'}(\mathbf{k},\mathbf{k}')$ by $f_{\sigma\sigma'}(\mathbf{k}_F,\mathbf{k}'_F)$.

One can proceed one step further by relating the 2PI vertex $\bar{\Gamma}^{\rm irr}$ to the full (1PI) vertex $\bar{\Gamma} \equiv \bar{\Gamma}_{\rm ph}$. The latter satisfies the Bethe–Salpeter equation

$$\begin{aligned}\bar{\Gamma}_{\sigma\sigma'}(k,k';q) &= \bar{\Gamma}^{\rm irr}_{\sigma\sigma'}(k,k') \\ &\quad - \frac{1}{\beta\mathcal{V}} \sum_{k_1,k_2,\sigma_1,\sigma_2} \bar{\Gamma}^{\rm irr}_{\sigma\sigma_1}(k,k_1)\bar{\Pi}_{\sigma_1\sigma_2}(k_1,k_2;q)\bar{\Gamma}_{\sigma_2\sigma'}(k_2,k';q), \end{aligned} \tag{4.188}$$

where again we set $q=0$ in $\bar{\Gamma}^{\rm irr}$. Since the frequency dependence of $\bar{\Gamma}^{\rm irr}$ can be neglected, we can carry out the frequency sums,

$$\begin{aligned}\bar{\Gamma}_{\sigma\sigma'}(\mathbf{k},\mathbf{k}';q) &= \bar{\Gamma}^{\rm irr}_{\sigma\sigma'}(\mathbf{k},\mathbf{k}') \\ &\quad - \frac{1}{\mathcal{V}} \sum_{\mathbf{k}_1,\mathbf{k}_2,\sigma_1,\sigma_2} \bar{\Gamma}^{\rm irr}_{\sigma\sigma_1}(\mathbf{k},\mathbf{k}_1)\bar{\Pi}_{\sigma_1\sigma_2}(\mathbf{k}_1,\mathbf{k}_2;q)\bar{\Gamma}_{\sigma_2\sigma'}(\mathbf{k}_2,\mathbf{k}';q). \end{aligned} \tag{4.189}$$

$\bar{\Pi}$ vanishing in the ω-limit,

$$\bar{\Gamma}^{\rm irr}_{\sigma\sigma'}(\mathbf{k},\mathbf{k}') = \lim_{\omega_\nu\to 0}\left[\lim_{\mathbf{q}\to 0}\bar{\Gamma}_{\sigma\sigma'}(\mathbf{k},\mathbf{k}';q)\right] \equiv \bar{\Gamma}^{\omega}_{\sigma\sigma'}(\mathbf{k},\mathbf{k}'). \tag{4.190}$$

Equations (4.187) and (4.190) relate the Landau function f to $\bar{\Gamma}^{\omega}$.

The final step is to relate the quasi-particle–quasi-hole interaction vertex $\bar{\Gamma}^{\omega}$ to the particle–hole interaction vertex Γ. Let us introduce the correlation function

$$\begin{aligned}\chi_{\sigma\sigma'}(k,k';q) &= \langle \psi^*_\sigma(k_-)\psi_\sigma(k_+)\psi^*_{\sigma'}(k'_+)\psi_{\sigma'}(k'_-)\rangle \\ &\quad - \langle \psi^*_\sigma(k_-)\psi_\sigma(k_+)\rangle\langle \psi^*_{\sigma'}(k'_+)\psi_{\sigma'}(k'_-)\rangle. \end{aligned} \tag{4.191}$$

χ satisfies the equation

$$\begin{aligned}\chi_{\sigma\sigma'}(k,k';q) &= \Pi_{\sigma\sigma'}(k,k';q) \\ &\quad - \frac{1}{\beta\mathcal{V}} \sum_{k_1,k_2,\sigma_1,\sigma_2} \Pi_{\sigma\sigma_1}(k,k_1;q)\Gamma_{\sigma_1\sigma_2}(k_1,k_2;q)\Pi_{\sigma_2\sigma'}(k_2,k';q), \end{aligned} \tag{4.192}$$

where $\Gamma_{\sigma\sigma'}(k,k';q) = \Gamma_{{\rm ph},\sigma\sigma\sigma'\sigma'}(k_+,k_-;k'_-,k'_+)$ and

$$\begin{aligned}\Pi_{\sigma\sigma'}(k,k';q) &= -\delta_{\sigma,\sigma'}\delta_{k,k'}G(k_-)G(k_+) \\ &= z^2\bar{\Pi}_{\sigma\sigma'}(k,k';q) + \delta_{\sigma,\sigma'}\delta_{k,k'}\varphi(k) \end{aligned} \tag{4.193}$$

is the particle–hole pair propagator. We have separated the coherent part $z^2\bar{\Pi} \equiv z^2\bar{G}\bar{G}$ from the incoherent one (φ) using (4.162). φ being non-singular in the limit $q \to 0$, it is evaluated at $q=0$. Retaining only the quasi-particle (coherent) part in (4.192), we obtain

$$\chi\big|_{\rm coh} = z^2\bar{\Pi} - z^2\bar{\Pi}\Gamma z^2\bar{\Pi} \tag{4.194}$$

(in matrix notation). Since $\bar{\chi} = \bar{\Pi} - \bar{\Pi}\bar{\Gamma}\bar{\Pi}$, we conclude that $\chi\big|_{\rm coh} = z^2\bar{\chi}$ and $z^2\Gamma = \bar{\Gamma}$. This yields the final expression of the Landau function f,

$$f_{\sigma\sigma'}(\mathbf{k}, \mathbf{k}') = z^2\Gamma^{\omega}_{\sigma\sigma'}(\mathbf{k}, \mathbf{k}'). \tag{4.195}$$

4.4.3 *Quantum Boltzmann equation*

The preceding analysis can easily be extended to the non-equilibrium case. Instead of the quasi-particle distribution function, we must consider the quasi-particle Wigner distribution function

$$n_{\mathbf{k}\sigma}(\mathbf{r}, \tau) = \int d^3r'\, e^{-i\mathbf{k}\cdot\mathbf{r}'} \left\langle \bar{\psi}^*_\sigma\left(\mathbf{r} - \frac{\mathbf{r}'}{2}, \tau\right)\bar{\psi}_\sigma\left(\mathbf{r} + \frac{\mathbf{r}'}{2}, \tau\right)\right\rangle \tag{4.196}$$

and its Fourier transform

$$\begin{aligned} n_{\mathbf{k}\sigma}(q) &= \frac{1}{\beta\mathcal{V}} \int_0^\beta d\tau \int d^3r\, e^{-i(\mathbf{q}\cdot\mathbf{r} - \omega_\nu\tau)} n_{\mathbf{k}\sigma}(\mathbf{r}, \tau) \\ &= \frac{1}{\beta}\sum_{\omega_n} \langle \bar{\psi}^*_\sigma(k_-)\bar{\psi}_\sigma(k_+)\rangle. \end{aligned} \tag{4.197}$$

The Wigner distribution function is the quantum analog of the semiclassical distribution function considered in the phenomenological Fermi-liquid theory (Sec. 4.3). Although it is not positive definite and therefore not a true distribution function, as far as its moments are concerned it generally behaves similarly to a distribution function [22, 23].

We consider the system in the presence of a source field $h_{\mathbf{k}\sigma}(q) = h^*_{\mathbf{k}\sigma}(-q)$ that couples to the quasi-particle operator

$$\hat{n}_{\mathbf{k}\sigma}(q) = \frac{1}{\beta}\sum_{\omega_n} \bar{\psi}^*_\sigma(k_-)\bar{\psi}_\sigma(k_+). \tag{4.198}$$

The source term in the action reads

$$S_h = -\beta \sum_{\mathbf{k},\sigma,q} h_{\mathbf{k}\sigma}(-q)\hat{n}_{\mathbf{k}\sigma}(q) \tag{4.199}$$

and the Wigner distribution function is given by

$$n_{\mathbf{k}\sigma}(q) = \langle \hat{n}_{\mathbf{k}\sigma}(q)\rangle = \frac{1}{\beta}\frac{\delta \ln Z[h]}{\delta h_{\mathbf{k}\sigma}(-q)}. \tag{4.200}$$

We are now in a position to introduce a functional of the Wigner distribution function — analogous to the grand potential $\Omega[n]$ in the equilibrium case — by means of a Legendre transform,

$$\Omega[n] = -\frac{1}{\beta}\ln Z[h] + \sum_{\mathbf{k},\sigma,q} h_{\mathbf{k}\sigma}(-q)n_{\mathbf{k}\sigma}(q). \tag{4.201}$$

The "equation of state" reads

$$\frac{\delta\Omega[n]}{\delta n_{\mathbf{k}\sigma}(q)} = h_{\mathbf{k}\sigma}(-q). \tag{4.202}$$

Even for non-interacting fermions, $\Omega[n]$ cannot be calculated exactly. We shall therefore consider only small fluctuations about the equilibrium state,

$$n_{\mathbf{k}\sigma}(q) = \delta_{q,0}\bar{n}_{\mathbf{k}} + \delta n_{\mathbf{k}\sigma}(q), \tag{4.203}$$

where $\delta_{q,0}\bar{n}_{\mathbf{k}}$ is solution of (4.202) when $h_{\mathbf{k}\sigma}(q) = 0$ [$\bar{n}_{\mathbf{k}} = \Theta(k_F - |\mathbf{k}|)$ when $T = 0$]. To lowest order in δn,

$$\delta\Omega[\delta n] = \frac{1}{2} \sum_{\mathbf{k},\mathbf{k}',\sigma,\sigma',q,q'} \left.\frac{\delta^2\Omega[n]}{\delta n_{\mathbf{k}\sigma}(-q)\delta n_{\mathbf{k}'\sigma'}(q')}\right|_{n=\bar{n}} \delta n_{\mathbf{k}\sigma}(-q)\delta n_{\mathbf{k}'\sigma'}(q'). \tag{4.204}$$

Equation (4.172) can easily be generalized into

$$\left.\frac{\delta^2\Omega[n]}{\delta n_{\mathbf{k}\sigma}(-q)\delta n_{\mathbf{k}'\sigma'}(q')}\right|_{n=\bar{n}} = \delta_{q,q'}\bar{\chi}^{-1}_{\sigma\sigma'}(\mathbf{k},\mathbf{k}';q), \tag{4.205}$$

where $\bar{\chi}_{\sigma\sigma'}(\mathbf{k},\mathbf{k}';q)$ is the correlation function defined in (4.177). The Kronecker symbol in (4.205) results from translation invariance. Using (4.183), (4.185) and (4.187), we find

$$\begin{aligned}\delta\Omega[\delta n] &= \frac{1}{2} \sum_{\mathbf{k},\mathbf{k}',\sigma,\sigma',q} \bar{\chi}^{-1}_{\sigma\sigma'}(\mathbf{k},\mathbf{k}';q)\delta n_{\mathbf{k}\sigma}(-q)\delta n_{\mathbf{k}'\sigma'}(q) \\ &= \frac{1}{2} \sum_{\mathbf{k},\mathbf{k}',\sigma,\sigma',q} \left\{ \frac{\delta_{\sigma,\sigma'}\delta_{\mathbf{k},\mathbf{k}'}}{n_F'(\xi_{\mathbf{k}})} \frac{i\omega_\nu - \mathbf{v}^*_{\mathbf{k}}\cdot\mathbf{q}}{\mathbf{v}^*_{\mathbf{k}}\cdot\mathbf{q}} + \frac{1}{\mathcal{V}} f_{\sigma\sigma'}(\mathbf{k},\mathbf{k}') \right\} \delta n_{\mathbf{k}\sigma}(-q)\delta n_{\mathbf{k}'\sigma'}(q). \end{aligned} \tag{4.206}$$

In the absence of a source field, the stationary condition (4.202) gives the quantum Boltzmann equation

$$(i\omega_\nu - \mathbf{v}^*_{\mathbf{k}}\cdot\mathbf{q})\delta n_{\mathbf{k}\sigma}(q) + n_F'(\xi_{\mathbf{k}})\mathbf{v}^*_{\mathbf{k}}\cdot\mathbf{q} \int_{\mathbf{k}'} \sum_{\sigma'} f_{\sigma\sigma'}(\mathbf{k},\mathbf{k}')\delta n_{\mathbf{k}'\sigma'}(q) = 0 \tag{4.207}$$

satisfied by the Wigner distribution function $n_{\mathbf{k}\sigma}(q)$; as usual, the real-time formalism can be recovered from the analytic continuation $i\omega_\nu \to \omega + i\eta$. This equation is identical to the semiclassical kinetic equation obtained in Sec. 4.3. Its solution can be written as

$$\delta n_{\mathbf{k}\sigma}(q) = v_F^*\delta(\xi_{\mathbf{k}})u_\sigma(\hat{\mathbf{k}},q) \qquad (T \to 0) \tag{4.208}$$

where $u_\sigma(\hat{\mathbf{k}},q)$ is the displacement of the Fermi surface in the direction $\hat{\mathbf{k}}$. The functional $\delta\Omega$ can be expressed in terms of u,

$$\begin{aligned}\delta\Omega[u] = \frac{\mathcal{V}}{2}N^*(0){v_F^*}^2 \sum_{q,\sigma,\sigma'} \Bigg\{ &\delta_{\sigma,\sigma'} \int \frac{d\Omega_{\hat{\mathbf{k}}}}{4\pi} \frac{v_F^*\hat{\mathbf{k}}\cdot\mathbf{q}}{v_F^*\hat{\mathbf{k}}\cdot\mathbf{q} - i\omega_\nu} u_\sigma(\hat{\mathbf{k}},-q)u_\sigma(\hat{\mathbf{k}},q) \\ &+ \frac{1}{2} \int \frac{d\Omega_{\hat{\mathbf{k}}}}{4\pi}\frac{d\Omega_{\hat{\mathbf{k}}'}}{4\pi} F_{\sigma\sigma'}(\mathbf{k}_F,\mathbf{k}_F')u_\sigma(\hat{\mathbf{k}},-q)u_{\sigma'}(\hat{\mathbf{k}}',q) \Bigg\}. \end{aligned} \tag{4.209}$$

This equation generalizes (4.30) — to which it reduces in the **q**-limit — to dynamic fluctuations of the Fermi surface. It can be used as the starting point for the calculation of both the static (thermodynamics) and dynamic (collective modes, response functions) properties of the Fermi liquid.

4.4.4 *Ward identities for the Fermi liquid*

In Sec. 4.4.1, we have seen that the quasi-particle properties — velocity, effective mass, lifetime — are related to the self-energy of the one-particle Green function. In Sec. 4.4.2, we have obtained a relation between the Landau function f and the particle–hole interaction vertex Γ. To complete the microscopic description of the Fermi liquid, we should also consider the relations between the self-energy and the particle–hole interaction vertex that follow from the symmetries of the physical system.

Symmetries and their consequences were discussed in Chapter 2. Any continuous symmetry of the action implies a set of relations (Ward identities) between vertices. U(1) and Galilean invariances imply[23]

$$\Sigma(k_-) - \Sigma(k_+) = \frac{1}{\beta\mathcal{V}} \sum_{k',\sigma'} [G_0^{-1}(k'_+) - G_0^{-1}(k'_-)] G(k'_-) G(k'_+) \Gamma_{\sigma\sigma'}(k, k'; q) \tag{4.210}$$

and

$$\begin{aligned}\mathbf{k}_+\Sigma(k_-) - \mathbf{k}_-\Sigma(k_+) &= \frac{1}{\beta\mathcal{V}} \sum_{k',\sigma'} [\mathbf{k}'_- G_0^{-1}(k'_+) - \mathbf{k}'_+ G_0^{-1}(k'_-)] \\ &\quad \times G(k'_-) G(k'_+) \Gamma_{\sigma\sigma'}(k, k'; q)\end{aligned} \tag{4.211}$$

[see Eqs. (2.A.7) and (2.A.16)]. Considering the first identity both in the ω- and **q**-limits and the second one in the ω-limit, we obtain

$$\frac{\partial\Sigma(k)}{\partial i\omega} = -\frac{1}{\beta\mathcal{V}} \sum_{k',\sigma'} \left\{G(k')^2\right\}_\omega \Gamma^\omega_{\sigma\sigma'}(k, k'), \tag{4.212}$$

$$\boldsymbol{\nabla}_\mathbf{k}\Sigma(k) = \frac{1}{\beta\mathcal{V}} \sum_{k',\sigma'} \boldsymbol{\nabla}_{\mathbf{k}'}\xi^0_{\mathbf{k}'} \left\{G(k')^2\right\}_\mathbf{q} \Gamma^\mathbf{q}_{\sigma\sigma'}(k, k'), \tag{4.213}$$

$$\mathbf{k}\frac{\partial\Sigma(k)}{\partial i\omega} = -\frac{1}{\beta\mathcal{V}} \sum_{k',\sigma'} \mathbf{k}' \left\{G(k')^2\right\}_\omega \Gamma^\omega_{\sigma\sigma'}(k, k'), \tag{4.214}$$

where

$$\begin{aligned}\left\{G(k)^2\right\}_\omega &= \lim_{\omega_\nu\to 0}\left[\lim_{\mathbf{q}\to 0} G(k_-)G(k_+)\right], \\ \left\{G(k)^2\right\}_\mathbf{q} &= \lim_{\mathbf{q}\to 0}\left[\lim_{\omega_\nu\to 0} G(k_-)G(k_+)\right].\end{aligned} \tag{4.215}$$

We consider the $T \to 0$ limit where the Matsubara frequency $i\omega_n \equiv i\omega$ becomes a continuous variable and $\frac{1}{\beta}\sum_{\omega_n} \equiv \int \frac{d\omega}{2\pi}$.

[23] $\Gamma_{\sigma\sigma'}(k, k'; q)$ is defined after (4.192).

These three Ward identities must be supplemented by the relation between Γ^ω and $\Gamma^{\mathbf{q}}$. The particle–hole interaction vertex satisfies the Bethe–Salpeter equation

$$\Gamma = \Gamma^{\text{irr}} - \Gamma^{\text{irr}}\Pi\Gamma \tag{4.216}$$

in matrix form. Writing the particle–hole pair propagator as in (4.193),

$$\Pi = z^2\bar{\Pi} + \varphi, \tag{4.217}$$

we have

$$\Gamma = \Gamma^{\text{irr}} - \Gamma^{\text{irr}}(z^2\bar{\Pi} + \varphi)\Gamma, \tag{4.218}$$

where $\Gamma^{\text{irr}}_{\sigma\sigma'}(k,k';q)$ can be approximated by $\Gamma^{\text{irr}}_{\sigma\sigma'}(k,k';q=0) \equiv \Gamma^{\text{irr}}_{\sigma\sigma'}(k,k')$ (see Sec. 4.4.2.2). Since the coherent part $z^2\bar{\Pi}$ of the pair propagator does not contribute in the ω limit,

$$\Gamma^\omega = \Gamma^{\text{irr}} - \Gamma^{\text{irr}}\varphi\Gamma^\omega \quad \text{i.e.} \quad \Gamma^\omega = (1+\Gamma^{\text{irr}}\varphi)^{-1}\Gamma^{\text{irr}}. \tag{4.219}$$

From (4.216) and (4.219), we deduce $\Gamma = \Gamma^\omega - z^2\Gamma^\omega\bar{\Pi}\Gamma$, i.e.

$$\begin{aligned}\Gamma_{\sigma\sigma'}(k,k';q) &= \Gamma^\omega_{\sigma\sigma'}(k,k') \\ &\quad - \frac{1}{\beta\mathcal{V}}\sum_{k_1,k_2,\sigma_1,\sigma_2} z^2\Gamma^\omega_{\sigma\sigma_1}(k,k_1)\bar{\Pi}_{\sigma_1\sigma_2}(k_1,k_2;q)\Gamma_{\sigma_2\sigma'}(k_2,k';q)\end{aligned} \tag{4.220}$$

and, taking the $\mathbf{q}$-limit,

$$\begin{aligned}\Gamma^{\mathbf{q}}_{\sigma\sigma'}(k,k') &= \Gamma^\omega_{\sigma\sigma'}(k,k') + \frac{1}{\beta\mathcal{V}}\sum_{k'',\sigma''} z^2\Gamma^\omega_{\sigma\sigma''}(k,k'')\left\{\bar{G}(k'')^2\right\}_{\mathbf{q}}\Gamma^{\mathbf{q}}_{\sigma''\sigma'}(k'',k') \\ &= \Gamma^\omega_{\sigma\sigma'}(k,k') - z^2N^*(0)\sum_{\sigma''}\int\frac{d\Omega_{\hat{\mathbf{k}}''}}{4\pi}\Gamma^\omega_{\sigma\sigma''}(k,\mathbf{k}''_F)\Gamma^{\mathbf{q}}_{\sigma''\sigma'}(\mathbf{k}''_F,k').\end{aligned} \tag{4.221}$$

To obtain the last line, we have ignored the dependence of $\Gamma^\omega_{\sigma\sigma''}(k,k'')$ and $\Gamma^{\mathbf{q}}_{\sigma''\sigma'}(k'',k')$ on $\omega_{n''}$ and used

$$\frac{1}{\beta}\sum_{\omega_{n''}}\left\{\bar{G}(k'')^2\right\}_{\mathbf{q}} = -\delta(\xi_{\mathbf{k}''}). \tag{4.222}$$

4.4.4.1 *Quasi-particle current and effective mass*

We are now in a position to compute the current carried by a quasi-particle and deduce its effective mass. From (4.213) and (4.221), we obtain

$$\begin{aligned}\nabla_{\mathbf{k}}\Sigma(k) &= \frac{1}{\beta\mathcal{V}}\sum_{k',\sigma'}\mathbf{v}_{\mathbf{k}'}\left\{G(k')^2\right\}_{\mathbf{q}}\Gamma^\omega_{\sigma\sigma'}(k,k') \\ &\quad - \frac{z^2N^*(0)}{\beta\mathcal{V}}\sum_{k',\sigma'}\mathbf{v}_{\mathbf{k}'}\left\{G(k')^2\right\}_{\mathbf{q}}\sum_{\sigma''}\int\frac{d\Omega_{\hat{\mathbf{k}}''}}{4\pi}\Gamma^\omega_{\sigma\sigma''}(k,\mathbf{k}''_F)\Gamma^{\mathbf{q}}_{\sigma''\sigma'}(\mathbf{k}''_F,k'),\end{aligned} \tag{4.223}$$

where $\mathbf{v_k} = \boldsymbol{\nabla}_\mathbf{k}\xi^0_\mathbf{k} = \mathbf{k}/m$. The second line can be simplified using again (4.193) and (4.217),

$$\begin{aligned}\boldsymbol{\nabla}_\mathbf{k}\Sigma(k) &= \frac{1}{\beta\mathcal{V}}\sum_{k',\sigma'}\mathbf{v}_{\mathbf{k}'}\left[\left\{G(k')^2\right\}_\omega + z^2\left\{\bar{G}(k')^2\right\}_\mathbf{q}\right]\Gamma^\omega_{\sigma\sigma'}(k,k') \\ &\quad - z^2N^*(0)\sum_{\sigma'}\int\frac{d\Omega_{\hat{\mathbf{k}}'}}{4\pi}\Gamma^\omega_{\sigma\sigma'}(k,\mathbf{k}'_F)\nabla_{\mathbf{k}'}\Sigma(k')|_{|\mathbf{k}'|=k_F}. \end{aligned}\tag{4.224}$$

The first term in the rhs simplifies using the Ward identity (4.214), which leads to

$$\begin{aligned}\boldsymbol{\nabla}_\mathbf{k}\Sigma(k) = &-\frac{\mathbf{k}}{m}\frac{\partial\Sigma(k)}{\partial i\omega} \\ &- z^2N^*(0)\sum_{\sigma'}\int\frac{d\Omega_{\hat{\mathbf{k}}'}}{4\pi}\left[\mathbf{v}_{\mathbf{k}'_F} + \boldsymbol{\nabla}_{\mathbf{k}'}\Sigma(k')|_{|\mathbf{k}'|=\mathbf{k}'_F}\right]\Gamma^\omega_{\sigma\sigma'}(k,\mathbf{k}'_F). \end{aligned}\tag{4.225}$$

For $i\omega\to 0$ and $|\mathbf{k}|\to k_F$, using

$$\frac{\partial\Sigma(k)}{\partial i\omega}\to 1-\frac{1}{z},\qquad \boldsymbol{\nabla}_\mathbf{k}\Sigma(k)\to\frac{\mathbf{v}^*_{\mathbf{k}_F}}{z}-\mathbf{v}_{\mathbf{k}_F},\tag{4.226}$$

we finally obtain

$$\begin{aligned}\mathbf{v}_{\mathbf{k}_F} &= \mathbf{v}^*_{\mathbf{k}_F} + z^2N^*(0)\sum_{\sigma'}\int\frac{d\Omega_{\hat{\mathbf{k}}'}}{4\pi}\mathbf{v}^*_{\mathbf{k}'_F}\Gamma^\omega_{\sigma\sigma'}(\mathbf{k}_F,\mathbf{k}'_F) \\ &= \mathbf{v}^*_{\mathbf{k}_F} + N^*(0)\sum_{\sigma'}\int\frac{d\Omega_{\hat{\mathbf{k}}'}}{4\pi}\mathbf{v}^*_{\mathbf{k}'_F}f_{\sigma\sigma'}(\mathbf{k}_F,\mathbf{k}'_F) \\ &= v^*_F\hat{\mathbf{k}}\left(1+\frac{F^s_1}{3}\right). \end{aligned}\tag{4.227}$$

This agrees with the expression of the quasi-particle current $\mathbf{j_k} = \mathbf{v}_k$ obtained within Landau's Fermi-liquid theory [Eq. (4.38)]. Equation (4.227) implies that the quasi-particle effective mass is determined by (4.40).

4.4.5 *Response to external fields*

To compute the response functions to an external field, we must add to the action the term

$$\begin{aligned}S_\text{ext}[\psi^*,\psi] &= \int_0^\beta d\tau\int d^3r\left[\phi(\mathbf{r},\tau)n(\mathbf{r},\tau) - \mathbf{j}(\mathbf{r},\tau)\cdot\mathbf{A}(\mathbf{r},\tau)\right] \\ &= -\frac{1}{\sqrt{\beta\mathcal{V}}}\sum_{\mu=0,x,y,z}\sum_{k,q,\sigma}A_\mu(-q)\lambda_\mu(\mathbf{k})\psi^*_\sigma(k_-)\psi_\sigma(k_+), \end{aligned}\tag{4.228}$$

where ϕ and $\mathbf{A}$ are the scalar and vector potentials, respectively. In the second line of (4.228), we have introduced a compact notation with $A_0 = -\phi$ and

$$\lambda_0(\mathbf{k}) = 1,\qquad \lambda_{\mu\neq 0}(\mathbf{k}) = \frac{k_\mu}{m}.\tag{4.229}$$

The linear response

$$\langle j_\mu(q)\rangle = \sum_\nu \chi_{\mu\nu}(q) A_\nu(q) \tag{4.230}$$

to the external field is given by the correlation function

$$\begin{aligned}\chi_{\mu\nu}(q) &= \left.\frac{\delta^2 \ln Z[A]}{\delta A_\mu(-q)\delta A_\nu(q)}\right|_{A=0} \\ &= \frac{1}{\beta\mathcal{V}} \sum_{k,k',\sigma,\sigma'} \lambda_\mu(\mathbf{k})\chi_{\sigma\sigma'}(k,k';q)\lambda_\nu(\mathbf{k}'),\end{aligned} \tag{4.231}$$

where χ is defined by (4.191). Since the response is dominated by the quasi-particles in the low-energy limit, it is tempting to approximate χ by its coherent part $\chi|_{\rm coh} = z^2\bar\chi$ [Eq. (4.194)]. This however is not sufficient. One must also determine how the external field couples to the quasi-particles. The mere replacement of $\psi^{(*)}$ by $\sqrt{z}\bar\psi^{(*)}$ in $S_{\rm ext}[\psi^*,\psi]$, which would indeed lead to $\bar\chi$ appearing in (4.231) instead of χ, is not correct.

To understand this issue in more detail, let us consider the coupling between the field and the fermions shown in Fig. 4.13. The renormalized vertex satisfies the equation

$$\begin{aligned}\Lambda_\mu(\mathbf{k},\mathbf{q}) &= \lambda_\mu(\mathbf{k}) + \frac{1}{\beta\mathcal{V}}\sum_{k',\sigma'} \Lambda_\mu(\mathbf{k}',\mathbf{q}) G(k'_-)G(k'_+)\Gamma^{\rm irr}_{\sigma'\sigma}(k',k;q) \\ &= \lambda_\mu(\mathbf{k}) + \frac{1}{\beta\mathcal{V}}\sum_{k',\sigma'} \Lambda_\mu(\mathbf{k}',\mathbf{q})[z^2\bar G(k'_-)\bar G(k'_+) - \varphi(k')]\Gamma^{\rm irr}_{\sigma'\sigma}(k',k),\end{aligned} \tag{4.232}$$

where $\varphi(k') \equiv -\left\{G(k')^2\right\}_\omega$ denotes the incoherent part of the particle–hole pair propagator [Eq. (4.193)] and we have set $q = 0$ in the non-singular vertex $\Gamma^{\rm irr}$. We can now understand why it is not possible to entirely ignore the incoherent part of the particle–hole pair propagator. The latter leads to a renormalization of the coupling between the field and the quasi-particles (Fig. 4.14),[24]

$$\begin{aligned}\Lambda^R_\mu(\mathbf{k}) &= \lambda_\mu(\mathbf{k}) + \frac{1}{\beta\mathcal{V}}\sum_{k',\sigma'} \Lambda^R_\mu(\mathbf{k}')\left\{G(k')^2\right\}_\omega \Gamma^{\rm irr}_{\sigma'\sigma}(k',k) \\ &= \lambda_\mu(\mathbf{k}) + \frac{1}{\beta\mathcal{V}}\sum_{k',\sigma'} \lambda_\mu(\mathbf{k}')\left\{G(k')^2\right\}_\omega \Gamma^{\omega}_{\sigma'\sigma}(k',k),\end{aligned} \tag{4.233}$$

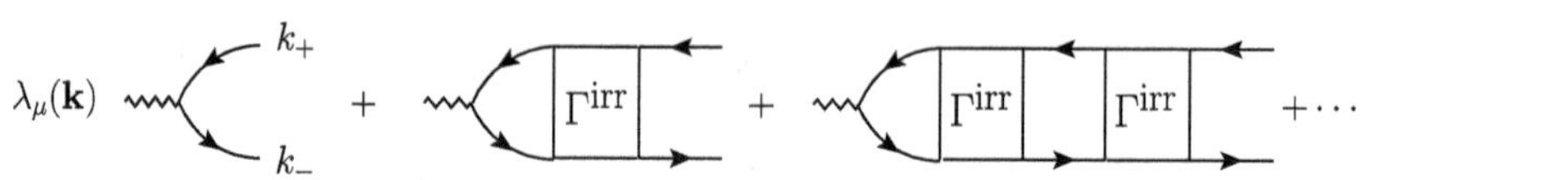

Fig. 4.13 Diagrammatic representation of the renormalized interaction vertex $\Lambda_\mu(\mathbf{k},\mathbf{q})$ between the fermions and the external field.

[24] Λ^R_μ can be seen as an effective (bare) vertex between the external field and the quasi-particles obtained by "integrating out" the incoherent part of the particle–hole pair propagator. Note that it is not the full renormalized vertex for the quasi-particles that will be obtained by considering the coherent part $z^2\bar G\bar G$ of the particle–hole pair propagator in (4.232).

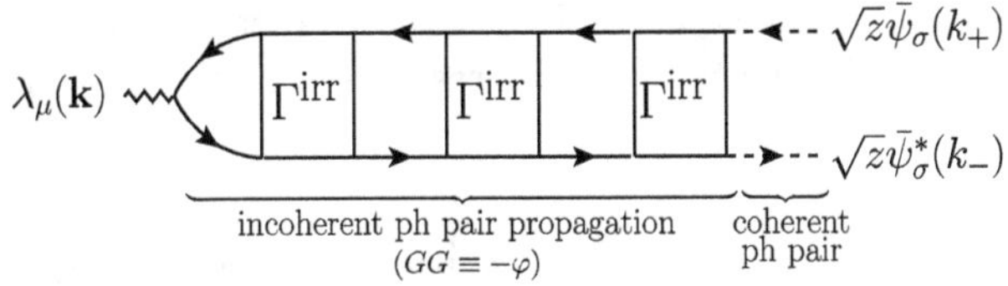

Fig. 4.14 Diagrammatic representation of the renormalized interaction vertex $\Lambda^R_\mu(\mathbf{k})$ between the external field and the quasi-particles due to the incoherent degrees of freedom. The solid lines represent the incoherent part φ of the particle–hole (ph) pair propagator.

where the last result follows from the relation (4.219) between $\Gamma^{\rm irr}$ and Γ^ω. Using the Ward identities (4.212) and (4.214), we obtain the very simple result[25]

$$\Lambda^R_\mu(\mathbf{k}) = \frac{\lambda_\mu(\mathbf{k})}{z}. \tag{4.234}$$

The effective (bare) coupling between the external field and the quasi-particles is therefore described by the action

$$\begin{aligned} S_{\rm ext,eff}[\bar\psi^*,\bar\psi] &= -\frac{1}{\sqrt{\beta\mathcal{V}}}\sum_{\mu=0,x,y,z}\sum_{k,q,\sigma} A_\mu(-q)\Lambda^R_\mu(\mathbf{k})\sqrt{z}\bar\psi^*_\sigma(k_-)\sqrt{z}\bar\psi_\sigma(k_+) \\ &= -\frac{1}{\sqrt{\beta\mathcal{V}}}\sum_{\mu=0,x,y,z}\sum_{k,q,\sigma} A_\mu(-q)\lambda_\mu(\mathbf{k})\bar\psi^*_\sigma(k_-)\bar\psi_\sigma(k_+). \end{aligned} \tag{4.235}$$

The renormalization of the vertex $\lambda_\mu(\mathbf{k})$ cancels the rescaling introduced in the definition of the quasi-particle field $\bar\psi$. This justifies the fact that in the phenomenological Fermi-liquid theory, one calculates the quasi-particle response to an external field assuming that the latter couples to the quasi-particles in the standard way [see Eq. (4.97)]. This also explains why the response functions do not depend on the quasi-particle weight z. In order for the Fermi-liquid theory to be valid, we need z to be finite, but its precise value does not influence the physical properties of the system that can be measured experimentally.[26]

From (4.235), one obtains the quasi-particle (leading) contribution to the response function,

$$\begin{aligned} \chi_{\mu\nu}(q) &= \frac{1}{\beta\mathcal{V}}\sum_{k,k',\sigma,\sigma'} \lambda_\mu(\mathbf{k})\bar\chi_{\sigma\sigma'}(k,k';q)\lambda_\nu(\mathbf{k}') \\ &= \frac{1}{\mathcal{V}}\sum_{\mathbf{k},\mathbf{k}',\sigma,\sigma'} \lambda_\mu(\mathbf{k})\bar\chi_{\sigma\sigma'}(\mathbf{k},\mathbf{k}';q)\lambda_\nu(\mathbf{k}'), \end{aligned} \tag{4.236}$$

[25] The renormalization of the vertex, Eq. (4.234), appears naturally in the RG framework (see the bibliography).

[26] From a diagrammatic point of view, this illustrates the importance of vertex corrections when calculating response functions. A mere renormalization of the one-particle propagator ($G_0 \to z\bar G$) in the perturbation expansion would violate the Ward identity (4.234) and give response functions that depend on z.

where $\bar{\chi}_{\sigma\sigma'}(\mathbf{k},\mathbf{k}';q)$ is defined by (4.177). The Bethe–Salpeter equation (4.184) satisfied by $\bar{\chi}_{\sigma\sigma'}(\mathbf{k},\mathbf{k}';q)$ implies that the density–density response function $\chi_{nn}(q) \equiv \chi_{00}(q)$ is determined by

$$\chi_{nn}(q) = \bar{\Pi}(q) - \frac{1}{\mathcal{V}^2} \sum_{k,k',\mathbf{k}_1,\mathbf{k}_2} \sum_{\sigma,\sigma',\sigma_1,\sigma_2} \bar{\Pi}_{\sigma\sigma_1}(\mathbf{k},\mathbf{k}_1;q) f_{\sigma_1\sigma_2}(\mathbf{k}_1,\mathbf{k}_2) \bar{\chi}_{\sigma_2\sigma'}(\mathbf{k}_2,\mathbf{k}';q), \quad (4.237)$$

where

$$\bar{\Pi}(q) = \frac{1}{\mathcal{V}} \sum_{\mathbf{k},\mathbf{k}',\sigma,\sigma'} \bar{\Pi}_{\sigma\sigma'}(\mathbf{k},\mathbf{k}';q) = \chi^0_{nn}(q) \quad (4.238)$$

[see Eq. (4.104)]. If we assume $f_{\sigma\sigma'}(\mathbf{k},\mathbf{k}') = f_0^s$, this equation is easily solved,

$$\chi_{nn}(q) = \chi^0_{nn}(q) - f_0^s \chi^0_{nn}(q) \chi_{nn}(q) = \frac{\chi^0_{nn}(q)}{1 + f_0^s \chi^0_{nn}(q)}, \quad (4.239)$$

and we recover the result obtained in the phenomenological approach [Eq. (4.105)]. Similarly, one can recover from (4.236) the expression of the current–current correlation function derived in Sec. 4.3.4.3 in the case where the Landau function assumes the simple form $f_{\sigma\sigma}(\mathbf{k},\mathbf{k}') = f_0^s + f_1^s \cos(\alpha)$ (with α the angle between $\mathbf{k}$ and $\mathbf{k}'$).

4.4.6 *Luttinger theorem*

In this section, we do not assume that the system is isotropic and consider a three-dimensional Fermi liquid with an arbitrary dispersion $\xi^0_{\mathbf{k}}$. At zero temperature, the density $n = N/\mathcal{V}$ is given by

$$\begin{aligned} n &= 2 \int_{\mathbf{k}} \int_{-\infty}^{\infty} \frac{d\omega}{2\pi} e^{i\omega\eta} G(\mathbf{k}, i\omega) \\ &= 2 \int_{\mathbf{k}} \int_{-\infty}^{\infty} \frac{d\omega}{2\pi} e^{i\omega\eta} \left\{ \frac{\partial}{\partial i\omega} \ln[-G(\mathbf{k},i\omega)^{-1}] + G(\mathbf{k},i\omega) \frac{\partial \Sigma(\mathbf{k},i\omega)}{\partial i\omega} \right\}, \end{aligned} \quad (4.240)$$

where $G(\mathbf{k},i\omega)^{-1} = i\omega_n - \xi^0_{\mathbf{k}} - \Sigma(\mathbf{k},i\omega)$. In (4.240), it is understood that we actually consider the principal part of the integrand in order to deal with a possible non-analyticity (pole) of $G(\mathbf{k},i\omega)$ at $\omega = 0$.

Let us first show that the last term in the rhs of (4.240) vanishes. Integrating by part, we find

$$\int_{-\infty}^{\infty} \frac{d\omega}{2\pi} G(\mathbf{k},i\omega) \frac{\partial \Sigma(\mathbf{k},i\omega)}{\partial i\omega} = - \int_{-\infty}^{\infty} \frac{d\omega}{2\pi} \Sigma(\mathbf{k},i\omega) \frac{\partial G(\mathbf{k},i\omega)}{\partial i\omega}. \quad (4.241)$$

Here, we have used $\lim_{|\omega|\to\infty} G(\mathbf{k},i\omega) = 1/i\omega$ and $\lim_{|\omega|\to\infty} \Sigma(\mathbf{k},i\omega) = \text{const}$ (see Sec. 3.5.3). The self-energy $\Sigma(\mathbf{k},i\omega)$ can be expressed as the functional derivative of the Luttinger–Ward functional $\Phi[G]$ with respect to the Green function $G(\mathbf{k},i\omega)$ (Chapter 1). This functional is given by the sum of skeleton diagrams; it is clearly invariant if we shift all Matsubara frequencies $i\omega$ in the propagators of these diagrams by an infinitesimal amount $i\epsilon$,

$$\begin{aligned} 0 &= \int_{\mathbf{k}} \int_{-\infty}^{\infty} \frac{d\omega}{2\pi} \frac{\delta\Phi[G]}{\delta G(\mathbf{k}, i\omega)} \frac{\partial G(\mathbf{k}, i\omega)}{\partial i\omega} \\ &= \int_{\mathbf{k}} \int_{-\infty}^{\infty} \frac{d\omega}{2\pi} \Sigma(\mathbf{k}, i\omega) \frac{\partial G(\mathbf{k}, i\omega)}{\partial i\omega}. \end{aligned} \tag{4.242}$$

Equations (4.241) and (4.242) prove our assertion.

We therefore have

$$n = -2 \int_{\mathbf{k}} \int_{-i\infty}^{i\infty} \frac{dz}{2i\pi} e^{z\eta} \frac{\partial}{\partial z} \ln[-G(\mathbf{k}, z)]. \tag{4.243}$$

Since $G(\mathbf{k}, z)$ is an analytic function of the complex variable z, except possibly on the real axis, the convergence factor $e^{z\eta}$ enables us to change the contour of integration as shown in Fig. 4.15,

$$\begin{aligned} n &= -2 \int_{\mathbf{k}} \left\{ \int_{-\infty}^{0} \frac{d\omega}{2i\pi} e^{\omega\eta} \frac{\partial}{\partial\omega} \ln[-G(\mathbf{k}, \omega - i\eta)] + \int_{0}^{-\infty} \frac{d\omega}{2i\pi} e^{\omega\eta} \frac{\partial}{\partial\omega} \ln[-G(\mathbf{k}, \omega + i\eta)] \right\} \\ &= -\frac{i}{\pi} \int_{\mathbf{k}} \int_{-\infty}^{0} d\omega \frac{\partial}{\partial\omega} \ln\left(\frac{G^R(\mathbf{k}, \omega)}{G^A(\mathbf{k}, \omega)} \right) \end{aligned} \tag{4.244}$$

(we have dropped the convergence factor which is not necessary anymore). Denoting the phase of $G^R(\mathbf{k}, \omega) = [\omega + i\eta - \xi^0_{\mathbf{k}} - \Sigma^R(\mathbf{k}, \omega)]^{-1}$ by $\varphi(\mathbf{k}, \omega)$, we obtain

$$n = \frac{2}{\pi} \int_{\mathbf{k}} \int_{-\infty}^{0} d\omega \frac{\partial}{\partial\omega} \varphi(\mathbf{k}, \omega). \tag{4.245}$$

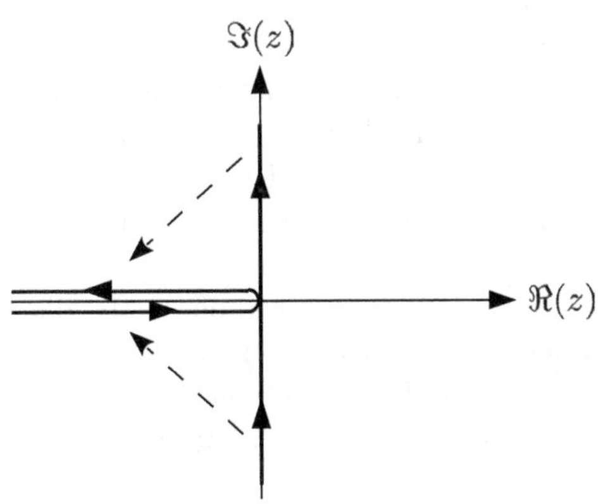

Fig. 4.15 Integration contour used in Eq. (4.244).

$G^R(\mathbf{k}, -\infty)$ being real and negative (with an infinitesimal negative imaginary part), $\varphi(\mathbf{k}, -\infty) = -\pi$. Since $\Im[\Sigma^R(\mathbf{k}, \omega)] \leq 0$ (Sec. 3.5.2), $G^R(\mathbf{k}, \omega)$ remains in the lower complex plane and its phase $\varphi(\mathbf{k}, \omega)$ can only vary between $-\pi$ and 0. There is therefore no winding of the phase $\varphi(\mathbf{k}, \omega)$ as ω varies between $-\infty$ and 0, and

$$n = \frac{2}{\pi} \int_{\mathbf{k}} [\varphi(\mathbf{k}, 0) - \varphi(\mathbf{k}, -\infty)] = \frac{2}{\pi} \int_{\mathbf{k}} [\varphi(\mathbf{k}, 0) + \pi]. \tag{4.246}$$

Given that $\Im[\Sigma^R(\mathbf{k}, 0)] = 0$ in a Fermi liquid (Sec. 4.4.1.6), the phase $\varphi(\mathbf{k}, 0)$ is either 0 or $-\pi$ depending on the sign of $\Re[G^R(\mathbf{k}, 0)]$. We finally obtain

$$n = 2 \int_{\mathbf{k}} \Theta(\Re[G^R(\mathbf{k}, 0)]). \tag{4.247}$$

The region $\Re[G^R(\mathbf{k}, 0)] > 0$, i.e. $\xi^0_{\mathbf{k}} + \Sigma^R(\mathbf{k}, 0) < 0$, is bounded by the Fermi surface defined by $\xi^0_{\mathbf{k}} + \Sigma^R(\mathbf{k}, 0) = 0$. Equation (4.247) then states that for a given density of particles, the volume of the Fermi surface in $\mathbf{k}$ space is the same as that of the non-interacting Fermi system (Luttinger theorem). In the case of an isotropic system, where the interactions do not change the spherical shape of the Fermi surface, the Fermi momentum k_F coincides with that of the non-interacting system.

4.5 Fermi Liquids in Low Dimensions

So far, we have discussed only three-dimensional systems for which Fermi-liquid theory is on firm grounds. In Sec. 4.1, we pointed out that there are nevertheless many cases where the adiabatic continuity assumption — which underlies Fermi-liquid theory — does not apply, for instance, when there is an instability of the Fermi surface. But even in that case, one expects Fermi-liquid theory to be valid above the transition temperature T_c to the broken-symmetry state and therefore provide us with a reliable starting point to understand the instability of the metallic state.[27]

In two dimensions, the situation is different. Although there is little doubt that Landau's theory is valid in isotropic systems with a circular Fermi surface and repulsive interactions (at least away from Fermi-surface instabilities), Fermi-liquid theory may break down in some cases. In Chapter 6, we shall see that a Fermi surface with nesting properties gives rise to strong antiferromagnetic fluctuations which invalidate Fermi-liquid theory in a broad temperature range.

In one dimension, Fermi-liquid theory never applies. Metallic states are known as Luttinger liquids and are characterized by a number of properties that markedly differ from those of the Fermi liquid, most notably the absence of quasi-particles. One-dimensional interacting fermion systems will be discussed in Chapter 15.

[27] Fermi-liquid behavior in three-dimensional systems is destroyed only in a very narrow temperature range above T_c [24].

Appendix 4.A Second-Order Self-energy

In this appendix, we compute the second-order self-energy defined by (4.157). To carry out the sum over the bosonic Matsubara frequencies ω_ν, we consider the integral

$$\oint_{\mathcal{C}} \frac{dz}{2i\pi} n_B(z)\chi_0(\mathbf{q}, z)G_0(\mathbf{k}+\mathbf{q}, i\omega_n + z) \tag{4.A.1}$$

where $\mathcal{C}$ is the contour shown in Fig. 4.16. Using the residue theorem and noting that the part of the contour at infinity does not contribute, we obtain

$$\begin{aligned}
&\frac{1}{\beta}\sum_{\omega_\nu\neq 0} \chi_0(q)G_0(k+q) + n_B(-i\omega_n + \xi^0_{\mathbf{k}})\chi_0(\mathbf{q}, \xi^0_{\mathbf{k}} - i\omega_n) \\
&= \mathcal{P}\int_{-\infty}^{\infty} \frac{d\omega}{2i\pi} n_B(\omega) \frac{\chi_0(\mathbf{q}, \omega + i\eta) - \chi_0(\mathbf{q}, \omega - i\eta)}{i\omega_n + \omega - \xi^0_{\mathbf{k}}} - \frac{1}{\beta}\chi_0(\mathbf{q}, 0)G_0(\mathbf{k}+\mathbf{q}, i\omega_n),
\end{aligned} \tag{4.A.2}$$

where the last term comes from the part of the contour near the origin. Thus, we have

$$\begin{aligned}
\frac{1}{\beta}\sum_{\omega_\nu} \chi_0(q)G_0(k+q) &= n_F(\xi^0_{\mathbf{k}})\chi_0(\mathbf{q}, \xi^0_{\mathbf{k}} - i\omega_n) + \mathcal{P}\int_{-\infty}^{\infty} \frac{d\omega}{\pi} n_B(\omega) \frac{\chi''_0(\mathbf{q}, \omega)}{i\omega_n + \omega - \xi^0_{\mathbf{k}}} \\
&= \mathcal{P}\int_{-\infty}^{\infty} \frac{d\omega}{\pi} \frac{\chi''_0(\mathbf{q}, \omega)}{i\omega_n + \omega - \xi^0_{\mathbf{k}}} \left[n_B(\omega) + n_F(\xi^0_{\mathbf{k}})\right]
\end{aligned} \tag{4.A.3}$$

($\chi''_0(\mathbf{q}, \omega) = \Im[\chi_0(\mathbf{q}, \omega + i\eta)]$) using the spectral representation (3.33) of $\chi_0(\mathbf{q}, \xi^0_{\mathbf{k}} - i\omega_n)$. This gives

$$\Sigma(k) = \int_{\mathbf{q}} v^2_{\mathbf{q}} \mathcal{P}\int_{-\infty}^{\infty} \frac{d\omega}{\pi} \frac{\chi''_0(\mathbf{q}, \omega)}{i\omega_n + \omega - \xi^0_{\mathbf{k}+\mathbf{q}}} \left[n_B(\omega) + n_F(\xi^0_{\mathbf{k}+\mathbf{q}})\right]. \tag{4.A.4}$$

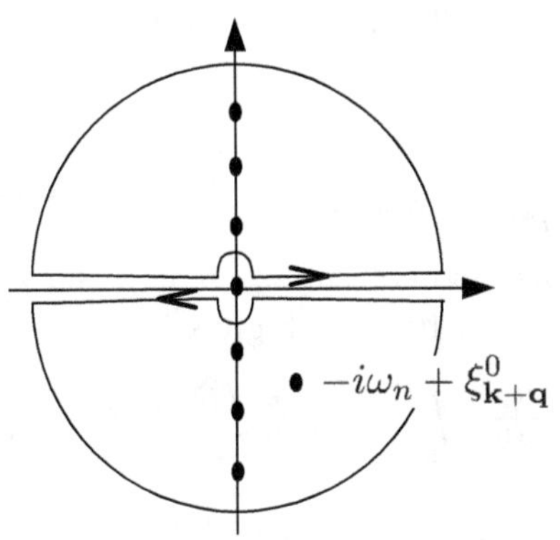

Fig. 4.16 Contour $\mathcal{C}$ used in Eq. (4.157). The black dots indicate the position of the bosonic Matsubara frequencies $i\omega_\nu$ as well as $-i\omega_n + \xi^0_{\mathbf{k}}$.

To obtain the quasi-particle lifetime, we need to calculate the imaginary part of the retarded self-energy,

$$\begin{aligned}\Im[\Sigma^R(\mathbf{k},\omega)] &= -\int_{\mathbf{q}} v_{\mathbf{q}}^2 \int_{-\infty}^{\infty} d\omega' \chi_0''(\mathbf{q},\omega')\delta(\omega+\omega'-\xi_{\mathbf{k}+\mathbf{q}}^0)\left[n_B(\omega')+n_F(\xi_{\mathbf{k}+\mathbf{q}}^0)\right] \\ &= -\int_0^{\infty} \frac{d|\mathbf{q}|}{2\pi^2}\mathbf{q}^2 v_{\mathbf{q}}^2 \int \frac{d\Omega_{\hat{\mathbf{q}}}}{4\pi}\int_{-\infty}^{\infty} d\omega' \chi_0''(\mathbf{q},\omega')\delta(\omega+\omega'-\xi_{\mathbf{k}+\mathbf{q}}^0) \\ &\qquad \times \left[n_B(\omega')+n_F(\omega+\omega')\right]. \end{aligned} \tag{4.A.5}$$

Since $\chi_0''(\mathbf{q},\omega) = \chi_0''(|\mathbf{q}|,\omega)$, we can carry out the angular integration,

$$\begin{aligned}\int \frac{d\Omega_{\hat{\mathbf{q}}}}{4\pi}&\delta\left(\omega+\omega'-\xi_{\mathbf{k}}^0-\frac{\mathbf{q}^2}{2m}-\frac{\mathbf{q}\cdot\mathbf{k}}{m}\right) \\ &= \frac{1}{2}\int_0^{\pi} d\theta \sin\theta\,\delta\left(\omega+\omega'-\xi_{\mathbf{k}}^0-\frac{\mathbf{q}^2}{2m}-\frac{|\mathbf{q}||\mathbf{k}|\cos\theta}{m}\right) \\ &= \frac{m}{2|\mathbf{k}||\mathbf{q}|}\Theta\left(1-\frac{|\omega+\omega'-\xi_{\mathbf{k}}^0-\mathbf{q}^2/2m|}{|\mathbf{k}||\mathbf{q}|/m}\right). \end{aligned} \tag{4.A.6}$$

From (4.A.5) and (4.A.6), we deduce

$$\begin{aligned}\Im[\Sigma^R(\mathbf{k},\omega)] = &-\frac{m}{2|\mathbf{k}|}\int_0^{\infty}\frac{d|\mathbf{q}|}{2\pi^2}|\mathbf{q}|\int_{-\infty}^{\infty} d\omega' \chi_0''(\mathbf{q},\omega')v_{\mathbf{q}}^2\left[n_B(\omega')+n_F(\omega+\omega')\right] \\ &\times\Theta\left(1-\frac{|\omega+\omega'-\xi_{\mathbf{k}}^0-\mathbf{q}^2/2m|}{|\mathbf{k}||\mathbf{q}|/m}\right). \end{aligned} \tag{4.A.7}$$

Because of the Bose and Fermi functions, the relevant part of the integration over ω' corresponds to $|\omega'| \lesssim \max(|\omega|, T)$. Since we are interested in the low-energy behavior of the self-energy where $|\omega|, |\xi_{\mathbf{k}}^0|, T \ll \epsilon_F^0$, we have $|\omega'| \ll \epsilon_F^0$,

$$\Theta\left(1-\frac{|\omega+\omega'-\xi_{\mathbf{k}}^0-\mathbf{q}^2/2m|}{|\mathbf{k}||\mathbf{q}|/m}\right) \simeq \Theta\left(1-\frac{|\mathbf{q}|}{2k_F}\right) \tag{4.A.8}$$

and

$$\Im[\Sigma^R(\mathbf{k},\omega)] = -\frac{m}{4\pi^2 k_F}\int_0^{2k_F} v_{\mathbf{q}}^2|\mathbf{q}|d|\mathbf{q}|\int_{-\infty}^{\infty} d\omega' \chi_0''(\mathbf{q},\omega')\left[n_B(\omega')+n_F(\omega'+\omega)\right]. \tag{4.A.9}$$

For $|\mathbf{q}| \le 2k_F$, $\chi_0''(\mathbf{q},\omega)$ is given by

$$\chi_0''(\mathbf{q},\omega) = \pi N(0)\frac{\omega}{v_F|\mathbf{q}|} = \frac{m^2\omega}{2\pi|\mathbf{q}|} \quad \text{for} \quad |\omega| \le \omega_- = v_F|\mathbf{q}| - \frac{\mathbf{q}^2}{2m} \tag{4.A.10}$$

(the function $\chi_0''(\mathbf{q},\omega)$ is studied in detail in Sec. 5.3.1). Since $|\omega'| \ll \epsilon_F^0$ in (4.A.9), we can use the low-energy expression (4.A.10), which gives[28]

$$\begin{aligned}\Im[\Sigma^R(\mathbf{k},\omega)] &= -\frac{m^3}{8\pi^3 k_F}\int_0^{2k_F} d|\mathbf{q}|v_{\mathbf{q}}^2\int_{-\infty}^{\infty} d\omega'\omega'\left[n_B(\omega')+n_F(\omega'+\omega)\right] \\ &= -\frac{m^3}{16\pi^3 k_F}\left(\omega^2+\pi^2T^2\right)\int_0^{2k_F} d|\mathbf{q}|v_{\mathbf{q}}^2. \end{aligned} \tag{4.A.11}$$

[28] The frequency integral is done using $\int_{-\infty}^{\infty} dy \frac{y-x}{(1-e^{y-x})(1+e^{-y})} = \frac{1}{2}\frac{x^2+\pi^2}{1+e^{-x}}$.

Guide to the bibliography

- Besides Landau's original papers [1–3], there are excellent textbooks on Landau's Fermi-liquid theory [4–7]. Sections 4.1, 4.2 and 4.3 rely heavily on Refs. [4,5].
- The microscopic foundations of Fermi-liquid theory are discussed in Refs. [3,6–8].
- The derivation of the thermodynamic potential $\Omega[n]$ and the quantum Boltzmann equation follows Ref. [10]. Similar ideas were discussed in the context of the so-called statistical Fermi-liquid theory [11–13]. A direct derivation of $E[n]$ can be found in Ref. [7].
- Ward identities for the Fermi liquid are discussed in Refs. [6,8,15].
- The Luttinger theorem [16,17] is considered in Refs. [6,7]. For a discussion of the validity of possible extensions of this theorem, see Refs. [18–21].
- Fermi-liquid theory was also discussed in the framework of the RG [10,15,27–30].[29]

[29] See Chapter 15 for an introduction to the RG approach in fermion systems.

Bibliography

[1] L. D. Landau, The theory of a Fermi liquid, *JETP* **3**, 920 (1957).

[2] L. D. Landau, Oscillations in a Fermi liquid, *JETP* **5**, 101 (1957).

[3] L. D. Landau, On the theory of the Fermi liquid, *JETP* **8**, 70 (1959).

[4] D. Pines and P. Nozières, *The Theory of Quantum Liquids* (Addison-Wesley, Reading, Massachusetts, 1968).

[5] G. Baym and C. Pethick, *Landau Fermi-Liquid Theory* (Wiley, New York, 1991).

[6] A. A. Abrikosov, L. P. Gor'kov, and I. E. Dzyaloshinski, *Methods of Quantum Field Theory in Statistical Physics*, Chapters 1 and 4 (Dover, Mineola, New York, 1975).

[7] G. F. Giuliani and G. Vignale, *Quantum Theory of the Electron Liquid*, Chapter 8 (Cambridge University Press, Cambridge, 2005).

[8] P. Nozières, *Theory of Interacting Fermi Systems* (Addison Wesley, Reading, Massachusetts, 1964).

[9] N. W. Ashcroft and N. D. Mermin, *Solid State Physics* (Saunders College Publishing, New York, 1976).

[10] N. Dupuis, A unified description of static and dynamic properties of Fermi liquids, *Int. J. Mod. Phys.* B **14**, 379 (2000).

[11] R. Balian, C. Bloch, and C. De Dominicis, Formulation de la mécanique statistique en termes de nombres d'occupation (I), *Nucl. Phys.* **25**, 529 (1961); Formulation de la mécanique statistique en termes de nombres d'occupation (II), *Nucl. Phys.* **27**, 294 (1961).

[12] R. Balian and C. De Dominicis, Quasi-particle formulation of quantum statistics, *Physica* **30**, 1927 (1964); Microscopic approach to the Landau theory of the Bose liquid, *Physica* **30**, 1933 (1964); Statistical quasiparticles, effective Green's functions, and Landau descriptions of quantum liquids I. Impurity systems at equilibrium, *Ann. Phys. (N.Y.)* **62**, 229 (1971).

[13] J. M. Luttinger, Mean-occupation-number formalism in statistical mechanics, *Phys. Rev.* **174**, 263 (1968).

[14] G. Y. Chitov and D. Sénéchal, Fermi liquid as a renormalization-group fixed point: The role of interference in the Landau channel, *Phys. Rev.* B **57**, 1444 (1998).

[15] W. Metzner, C. Castellani, and C. D. Castro, Fermi systems with strong forward scattering, *Adv. in Phys.* **47**, 317 (1998).

[16] J. M. Luttinger and J. C. Ward, Ground-state energy of a many-fermion system. II, *Phys. Rev.* **118**, 1417 (1960).

[17] J. M. Luttinger, Fermi surface and some simple equilibrium properties of a system of interacting fermions, *Phys. Rev.* **119**, 1153 (1960).

[18] I. Dzyaloshinskii, Some consequences of the Luttinger theorem: The Luttinger surfaces in non-Fermi liquids and Mott insulators, *Phys. Rev.* B **68**, 085113 (2003).

[19] A. Rosch, Breakdown of Luttinger's theorem in two-orbital Mott insulators, *Eur. Phys. J.* B **59**, 495 (2007).

[20] B. Farid, On the Luttinger theorem concerning number of particles in the ground states of systems of interacting fermions, arXiv:0711.0952, see also the comment by A. Rosch (arXiv:0711.3093) and Farid's reply (arXiv:0711.3195).

[21] J. T. Heath and K. S. Bedell, Necessary and sufficient conditions for the validity of Luttinger's theorem, *New J. Phys.* **22**, 063011 (2020).

[22] E. P. Wigner, On the quantum correction for thermodynamic equilibrium, *Phys. Rev.* **40**, 749 (1932); M. Hillery, R. F. O'Connell, M. O. Scully, and E. P. Wigner, Distribution functions in physics: Fundamentals, *Phys. Rep.* **106**, 121 (1984).

[23] G. D. Mahan, *Many-Particle Physics*, Chapter 3 (Plenum Press, New York, 1990).

[24] Y. Vilk and A.-M. S. Tremblay, Non-perturbative many-body approach to the Hubbard model and single-particle pseudogap, *J. Phys. I France* **7**, 1309 (1997).

[25] P. Bloom, Two-dimensional Fermi gas, *Phys. Rev.* B **12**, 125 (1975).

[26] C. Hodges, H. Smith, and J. W. Wilkins, Effect of Fermi surface geometry on electron-electron scattering, *Phys. Rev.* B **4**, 302 (1971).

[27] R. Shankar, Renormalization-group approach to interacting fermions, *Rev. Mod. Phys.* **66**, 129 (1994).

[28] G. Y. Chitov and D. Sénéchal, Renormalization-group study of interacting electrons, *Phys. Rev.* B **52**, 13487 (1995).

[29] N. Dupuis and G. Y. Chitov, Renormalization-group approach to Fermi-liquid theory, *Phys. Rev.* B **54**, 3040 (1996).

[30] N. Dupuis, Fermi liquid theory: A renormalization group approach, *Eur. Phys. J.* B **3**, 315 (1998).

Chapter 5

The Electron Gas

Contents

5.1 The Jellium Model . . . 370
5.2 The Random-Phase Approximation . . . 371
5.2.1 Dielectric function and density–density response function . . . 371
5.2.2 Mean-field character of the RPA . . . 373
5.2.3 The RPA in the functional integral formalism . . . 375
5.3 Physical Properties of the RPA . . . 378
5.3.1 The Lindhard function . . . 378
5.3.2 Density fluctuation spectrum . . . 380
5.3.3 Screening . . . 383
5.3.4 Ground-state energy and compressibility . . . 385
5.4 One-Particle Properties . . . 387
5.4.1 Hartree–Fock theory . . . 388
5.4.2 RPA self-energy . . . 389
5.5 Beyond the RPA . . . 393
5.6 Wigner Crystallization . . . 395
Bibliography . . . 397

The aim of this chapter is twofold. On the one hand, we wish to give a short (and incomplete) description of the three-dimensional electron gas at zero temperature. The analysis is based on the jellium model where the electrons interact *via* the Coulomb interaction and the neutrality of the system is maintained by a uniform background of positive charge with infinite inertia. We discuss the results obtained within the RPA: plasmon mode, screening, Fermi-liquid behavior, etc. In this respect, this chapter is complementary to the discussion of the neutral Fermi liquid in Chapter 4 as well as the general considerations of Sec. 3.4. On the other hand, the electron gas provides an opportunity to introduce standard many-body approaches (RPA, (time-dependent) Hartree–Fock theory) that come up in different contexts and discuss in particular how they are formulated within diagrammatic or functional integral techniques. Except in Sec. 5.6, where we briefly discuss the Wigner crystallization of the electron gas, we deal only with a three-dimensional system.

5.1 The Jellium Model

In the jellium model, one assumes that the global neutrality of the system is maintained by the presence of a uniform background of positive charge. This can be seen as the simplest approximation of a metallic system. In a real crystal, the positive charge (ions) is not uniform and can also have its own dynamics (lattice vibrations).

The Hamiltonian reads $\hat{H} = \hat{H}_0 + \hat{H}_{\text{int}} + \hat{H}_{e-b} + \hat{H}_{b-b}$, where

$$\begin{aligned} \hat{H}_0 &= \sum_{i=1}^{N} \frac{\hat{p}_i^2}{2m}, \\ \hat{H}_{\text{int}} &= \frac{1}{2} \sum_{\substack{i,j=1 \\ i\neq j}}^{N} \frac{e^2}{4\pi\epsilon_0 |\hat{\mathbf{r}}_i - \hat{\mathbf{r}}_j|} \end{aligned} \tag{5.1}$$

are the kinetic and interaction Hamiltonians of the electrons. In second-quantized form, one obtains

$$\hat{H}_0 = \sum_{\mathbf{k},\sigma} \xi_{\mathbf{k}} \hat{\psi}_\sigma^\dagger(\mathbf{k}) \hat{\psi}_\sigma(\mathbf{k}) \tag{5.2}$$

and

$$\begin{aligned} \hat{H}_{\text{int}} &= \frac{1}{2\mathcal{V}} \sum_{\substack{\mathbf{k},\mathbf{k}',\mathbf{q} \\ \sigma,\sigma'}} v(\mathbf{q}) \hat{\psi}_\sigma^\dagger(\mathbf{k}+\mathbf{q}) \hat{\psi}_{\sigma'}^\dagger(\mathbf{k}'-\mathbf{q}) \hat{\psi}_{\sigma'}(\mathbf{k}') \hat{\psi}_\sigma(\mathbf{k}) \\ &= \frac{1}{2} \sum_{\mathbf{q}} v(\mathbf{q}) \left[\hat{n}(-\mathbf{q}) \hat{n}(\mathbf{q}) - \frac{\hat{N}}{\mathcal{V}} \right], \end{aligned} \tag{5.3}$$

where

$$\hat{n}(\mathbf{q}) = \frac{1}{\sqrt{\mathcal{V}}} \int d^3r\, e^{-i\mathbf{q}\cdot\mathbf{r}} \hat{n}(\mathbf{r}) = \frac{1}{\sqrt{\mathcal{V}}} \sum_{\mathbf{k},\sigma} \hat{\psi}_\sigma^\dagger(\mathbf{k}) \hat{\psi}_\sigma(\mathbf{k}+\mathbf{q}) \tag{5.4}$$

is the Fourier transform of the density operator $\hat{n}(\mathbf{r}) = \sum_\sigma \hat{\psi}_\sigma^\dagger(\mathbf{r}) \hat{\psi}_\sigma(\mathbf{r})$. $v(\mathbf{q}) = e^2/\epsilon_0 \mathbf{q}^2$ is the Fourier transform of the Coulomb potential $e^2/4\pi\epsilon_0|\mathbf{r}|$. We now use the grand canonical ensemble and include the term $-\mu\hat{N}$ into the Hamiltonian $\hat{H}_0$ ($\xi_{\mathbf{k}} = \mathbf{k}^2/2m - \mu$). The background of positive charge gives rise to the two additional terms:

$$\begin{aligned} \hat{H}_{e-b} &= -e^2 \int d^3r d^3r' \frac{n\hat{n}(\mathbf{r})}{4\pi\epsilon_0 |\mathbf{r}-\mathbf{r}'|} = -v(0) n\hat{N}, \\ \hat{H}_{b-b} &= \frac{e^2}{2} \int d^3r d^3r' \frac{n^2}{4\pi\epsilon_0 |\mathbf{r}-\mathbf{r}'|} = \frac{1}{2} n^2 \mathcal{V} v(0), \end{aligned} \tag{5.5}$$

where $v(0) \equiv v(\mathbf{q}=0)$. $\hat{H}_{b-b}$ gives the electrostatic energy of the uniform background of charge density $-en$ ($n = N/\mathcal{V}$ is the electron density) and $\hat{H}_{e-b}$ the interaction energy between the positive background and the electrons.

Due to the long-range nature of the Coulomb interaction, the integrals appearing in (5.5) are not convergent and $v(0)$ is not defined. Since the system is neutral, the various infinities should cancel and the energy of the system be finite. To deal with this difficulty, one regularizes the Coulomb interaction by choosing

$$v(\mathbf{r}) = \frac{e^2}{4\pi\epsilon_0}\frac{e^{-\kappa|\mathbf{r}|}}{|\mathbf{r}|}, \quad \text{i.e.} \quad v(\mathbf{q}) = \frac{e^2}{\epsilon_0(\mathbf{q}^2+\kappa^2)}. \tag{5.6}$$

All terms in (5.5) are now well defined. Physical quantities are obtained by taking first the thermodynamic limit $\mathcal{V}\to\infty$ and then $\kappa\to 0$. This procedure can be justified *a posteriori* by noting that the final results are independent of κ.

The part of the Hamiltonian $\hat{H}$ involving $v(0) = v(\mathbf{q}=0)$ reads

$$\frac{1}{2\mathcal{V}}v(0)(\hat{N}^2-\hat{N}) - v(0)n\hat{N} + \frac{1}{2}\mathcal{V}v(0)n^2 \simeq -\frac{1}{2}v(0)n, \tag{5.7}$$

since the fluctuations of the operator $\hat{N}$ becomes negligible in the thermodynamic limit.[1] The energy $\langle\hat{H}\rangle$ being an extensive quantity, the part involving $v(0)$ is of relative order

$$\frac{v(0)n}{\mathcal{V}} = \frac{ne^2}{\mathcal{V}\epsilon_0\kappa^2} \tag{5.8}$$

and vanishes in the thermodynamic limit ($\mathcal{V}\to\infty$ with κ fixed). Thus, we see that $\hat{H}_{e-b}+\hat{H}_{b-b}$ cancel the $\mathbf{q}=0$ term in $\hat{H}_{\rm int}$. The Hamiltonian in the jellium model can therefore be written as $\hat{H}_0+\hat{H}_{\rm int}$ without the $\mathbf{q}=0$ term in $\hat{H}_{\rm int}$. It is convenient to define $v(\mathbf{q})$ as $(1-\delta_{\mathbf{q},0})e^2/\epsilon_0\mathbf{q}^2$ so that $v(0)=0$.

5.2 The Random-Phase Approximation

The RPA is one of the simplest approximations to the density–density response function and the dynamic longitudinal dielectric function of the electron gas. In this section, we derive the RPA from several points of view. In particular, we show that the RPA arises very naturally in the functional integral formalism.

5.2.1 *Dielectric function and density–density response function*

In Sec. 3.4.1, we have shown that the longitudinal dielectric function $\epsilon_\parallel(q)$ is related to the density–density response function $\chi_{nn}(q)$,

$$\frac{1}{\epsilon_\parallel(q)} = 1 - v(\mathbf{q})\chi_{nn}(q) \quad \text{or} \quad \epsilon_\parallel(q) = 1 + v(\mathbf{q})\Pi_{nn}(q), \tag{5.9}$$

where $\Pi_{nn}(q)$ is the "irreducible" part of $\chi_{nn}(q)$, i.e. the part that cannot be split into two disconnected pieces by cutting a single Coulomb line $v(\mathbf{q})$. The general properties of χ_{nn} and $\epsilon_\parallel$ are discussed in Sec. 3.4.1 and the sum rules satisfied by $\epsilon_\parallel^R(\mathbf{q},\omega)$ are summarized

[1] $\frac{\sqrt{\langle\hat{N}^2\rangle-\langle\hat{N}\rangle^2}}{\langle\hat{N}\rangle} \sim \frac{1}{\sqrt{\langle\hat{N}\rangle}}$ when $\mathcal{V}\to\infty$.

Table 5.1 Sum rules satisfied by the longitudinal dielectric function $\epsilon_{\parallel}^R(\mathbf{q},\omega)$ (see Sec. 3.4.1). κ denotes the compressibility and ω_p the plasma frequency.

f-sum rule	$\int_{-\infty}^{\infty}\frac{d\omega}{\pi}\omega\Im\left[\frac{1}{\epsilon_{\parallel}^R(\mathbf{q},\omega)}\right]=-\omega_p^2$
perfect-screening sum rule	$\lim_{\mathbf{q}\to 0}\int_{-\infty}^{\infty}\frac{d\omega}{\pi}\frac{1}{\omega}\Im\left[\frac{1}{\epsilon_{\parallel}^R(\mathbf{q},\omega)}\right]=-1$
longitudinal-conductivity sum rule	$\int_{-\infty}^{\infty}\frac{d\omega}{\pi}\omega\Im\left[\epsilon_{\parallel}^R(\mathbf{q},\omega)\right]=\omega_p^2$
compressibility sum rule	$\int_{-\infty}^{\infty}\frac{d\omega}{\pi}\frac{1}{\omega}\Im\left[\epsilon_{\parallel}^R(\mathbf{q},\omega)\right]=\frac{e^2n^2\kappa}{\epsilon_0\mathbf{q}^2}\quad(\mathbf{q}\to 0)$

in Table 5.1. These sum rules, together with the fact that the particle–hole pair excitations do not contribute to the spectral function $\Im[1/\epsilon_{\parallel}^R(\mathbf{q},\omega)]$ for $\mathbf{q}\to 0$, imply that the electron gas possesses a (collective) plasmon mode with energy $\omega_p=(ne^2/\epsilon_0 m)^{1/2}$ in the long wavelength limit.

An explicit expression of $\epsilon_{\parallel}(q)$ and $\chi_{nn}(q)$ requires to compute, at least approximately, the irreducible part $\Pi_{nn}(q)$.[2] In the following, we consider the zeroth-order approximation — known as the RPA — where Π_{nn} is identified with the non-interacting susceptibility (Fig. 5.1),

$$\Pi_{nn}^{\rm RPA}(q)=\chi_{nn}^0(q)=-\frac{2}{\beta\mathcal{V}}\sum_k G_0(k)G_0(k+q). \tag{5.10}$$

This gives

$$\chi_{nn}^{\rm RPA}(q)=\frac{\chi_{nn}^0(q)}{1+v(\mathbf{q})\chi_{nn}^0(q)}\quad\text{and}\quad\epsilon_{\parallel}^{\rm RPA}(q)=1+v(\mathbf{q})\chi_{nn}^0(q). \tag{5.11}$$

The density–density response function $\chi_{nn}^{\rm RPA}$ can be represented by the sum of bubble diagrams shown in Fig. 5.2.

To estimate the domain of validity of a perturbative treatment of the Coulomb interaction, we consider the ratio

$$\frac{\text{potential energy}}{\text{kinetic energy}}\sim\frac{\frac{e^2}{\epsilon_0 n^{-1/3}}}{\frac{k_F^2}{2m}}\sim\frac{e^2m}{\epsilon_0 k_F}\sim\frac{1}{n^{1/3}a_0}, \tag{5.12}$$

where $a_0=h^2\epsilon_0/\pi me^2$ is the Bohr radius. It is customary to write the volume per particle as

$$\frac{1}{n}=\frac{3\pi^2}{k_F^3}=\frac{4}{3}\pi(r_s a_0)^3 \tag{5.13}$$

[2]Note that Π_{nn} is the right quantity to consider in order to obtain satisfying expressions for χ_{nn} and $\epsilon_{\parallel}$. Any approximate Π_{nn} will give a result for χ_{nn} and $\epsilon_{\parallel}$ to infinite order in the Coulomb interaction. By contrast, a finite-order approximation to χ_{nn} is likely to violate basic properties such as causality. This is the case, for instance, if one retains only the first two diagrams in Fig. 5.2. Note that a similar observation was made when discussing the one-particle Green function G (Sec. 3.5): The self-energy Σ (and not G) is the right quantity to consider in perturbation theory.

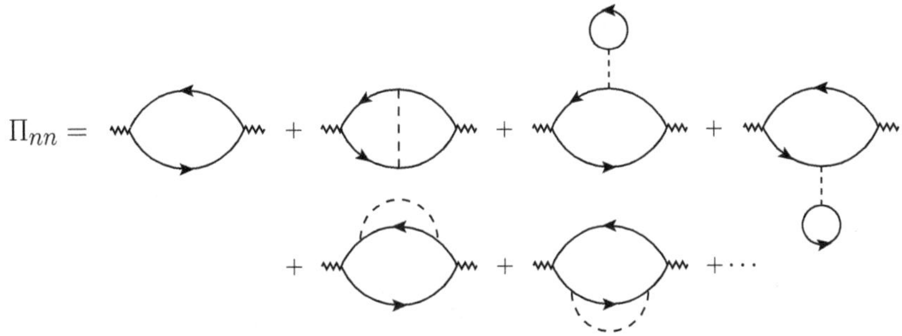

Fig. 5.1 Lowest-order contributions to Π_{nn}. The RPA retains only the zeroth-order term.

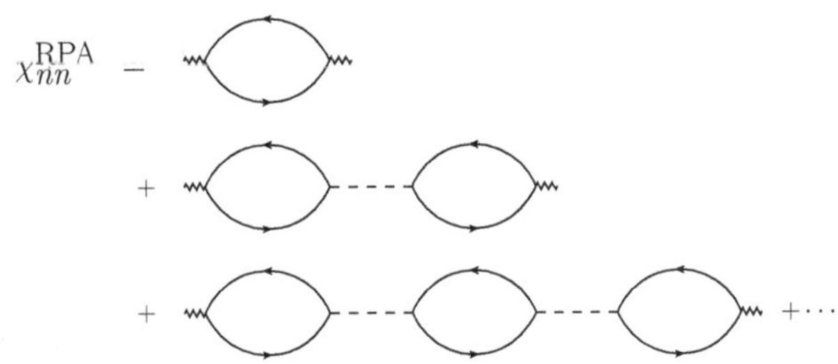

Fig. 5.2 Diagrammatic representation of the density–density response function χ_{nn} in the RPA.

so that

$$\frac{\text{potential energy}}{\text{kinetic energy}} \sim r_s. \tag{5.14}$$

Thus, the RPA is expected to be justified at high density when $r_s \ll 1$. Since typical metals correspond to a range $r_s \sim 2$–5, the RPA results have limited applicability even for conventional metals. In Sec. 5.5, we briefly discuss how one can go beyond the RPA. In the opposite limit $r_s \gg 1$, the electrons are expected to localize and form a Wigner crystal since at very low density, the zero-point kinetic energy associated with localizing the electrons eventually becomes negligible in comparison with the electrostatic energy of a classical lattice (Sec. 5.6).

5.2.2 *Mean-field character of the RPA*

To see the mean-field character of the RPA, let us rederive it within an elementary approach. We consider the system in the presence of an external potential

$$\phi_{\text{ext}}(\mathbf{r}, t) = \frac{1}{\sqrt{\mathcal{V}}} \sum_{\mathbf{q}} \phi_{\text{ext}}(\mathbf{q}, \omega) e^{i[\mathbf{q}\cdot\mathbf{r} - (\omega + i\eta)t]} + \text{c.c.} \qquad (\eta \to 0^+). \tag{5.15}$$

Keeping only the part that oscillates with $e^{i[\mathbf{q}\cdot\mathbf{r}-(\omega+i\eta)t]}$ (we are interested only in the linear response to ϕ_{ext}), the Hamiltonian reads

$$\hat{H} = \hat{H}_0 + \frac{1}{2}\sum_{\mathbf{q}} v(\mathbf{q})\left[\hat{n}(-\mathbf{q})\hat{n}(\mathbf{q}) - \frac{\hat{N}}{\mathcal{V}}\right] + \sum_{\mathbf{q}} \phi_{\text{ext}}(\mathbf{q},\omega)e^{-i(\omega+i\eta)t}\hat{n}(-\mathbf{q}). \tag{5.16}$$

The term $\hat{N}/\mathcal{V}$ can be ignored since it does not play any role in the following argument. In a mean-field approximation, the interacting Hamiltonian $\hat{H}_{\text{int}}$ is linearized with respect to the fluctuations $\delta\hat{n}(\mathbf{q}) = \hat{n}(\mathbf{q}) - \langle\hat{n}(\mathbf{q})\rangle$ of the density operator,

$$\begin{aligned}
\hat{n}(-\mathbf{q})\hat{n}(\mathbf{q}) &= [\delta\hat{n}(-\mathbf{q}) + \langle\hat{n}(-\mathbf{q})\rangle][\delta\hat{n}(\mathbf{q}) + \langle\hat{n}(\mathbf{q})\rangle] \\
&= \delta\hat{n}(-\mathbf{q})\langle\hat{n}(\mathbf{q})\rangle + \langle\hat{n}(-\mathbf{q})\rangle\delta\hat{n}(\mathbf{q}) + \langle\hat{n}(-\mathbf{q})\rangle\langle\hat{n}(\mathbf{q})\rangle + \mathcal{O}\left(\delta\hat{n}^2\right) \\
&= \hat{n}(-\mathbf{q})\langle\hat{n}(\mathbf{q})\rangle + \langle\hat{n}(-\mathbf{q})\rangle\hat{n}(\mathbf{q}) - \langle\hat{n}(-\mathbf{q})\rangle\langle\hat{n}(\mathbf{q})\rangle + \mathcal{O}\left(\delta\hat{n}^2\right),
\end{aligned} \tag{5.17}$$

so that

$$\hat{H}_{\text{MF}} = \hat{H}_0 + \sum_{\mathbf{q}}\left[\phi_{\text{ext}}(\mathbf{q},\omega)e^{-i(\omega+i\eta)t} + v(\mathbf{q})\langle\hat{n}(\mathbf{q})\rangle\right]\hat{n}(-\mathbf{q}) \tag{5.18}$$

(up to an irrelevant constant), where the expectation value $\langle\hat{n}(\mathbf{q})\rangle \equiv \langle\hat{n}(\mathbf{q})\rangle(t)$ is time dependent. We are therefore left with a system of free electrons subject to the effective potential

$$\phi_{\text{eff}}(\mathbf{q},t) = \phi_{\text{ext}}(\mathbf{q},\omega)e^{-i(\omega+i\eta)t} + v(\mathbf{q})\langle\hat{n}(\mathbf{q})\rangle. \tag{5.19}$$

The linear response to a scalar potential is given by the density–density response function (Sec. 3.3). We thus have

$$\begin{aligned}
\langle\hat{n}(\mathbf{q})\rangle &= -\chi_{nn}^{0R}(\mathbf{q},\omega)\phi_{\text{eff}}(\mathbf{q},\omega)e^{-i(\omega+i\eta)t} \\
&= -\chi_{nn}^{R}(\mathbf{q},\omega)\phi_{\text{ext}}(\mathbf{q},\omega)e^{-i(\omega+i\eta)t},
\end{aligned} \tag{5.20}$$

and therefore,

$$\chi_{nn}^{R}(\mathbf{q},\omega) = \frac{\chi_{nn}^{0R}(\mathbf{q},\omega)}{1 + v(\mathbf{q})\chi_{nn}^{0R}(\mathbf{q},\omega)}, \tag{5.21}$$

which is the RPA result for the retarded density–density correlation function.

5.2.3 *The RPA in the functional integral formalism*

5.2.3.1 *Hubbard–Stratonovich transformation*

We consider the action $S = S_0 + S_{\rm int} + S_J$ in the presence of external sources that couple to the density,

$$\begin{aligned} S_{\rm int} &= \frac{1}{2\beta\mathcal{V}} \sum_{\substack{k,k',q \\ \sigma,\sigma'}} v(\mathbf{q})\psi^*_\sigma(k+q)\psi^*_{\sigma'}(k'-q)\psi_{\sigma'}(k')\psi_\sigma(k) \\ &= \frac{1}{2}\sum_q v(\mathbf{q})n(-q)n(q), \\ S_J &= -\sum_q J(-q)n(q), \end{aligned} \tag{5.22}$$

where

$$n(q) = \frac{1}{\sqrt{\beta\mathcal{V}}} \sum_{k,\sigma} \psi^*_\sigma(k)\psi_\sigma(k+q). \tag{5.23}$$

We rewrite $S_{\rm int}$ by means of a Gaussian functional integral over an auxiliary real bosonic field $\phi(x)$,

$$e^{-S_{\rm int}} = \mathcal{N}\int \mathcal{D}[\phi]\, e^{-\frac{1}{2}\sum_q \phi(-q)v(\mathbf{q})^{-1}\phi(q) + i\sum_q \phi(-q)n(q)}. \tag{5.24}$$

The normalization constant $\mathcal{N} = \det(v^{-1/2})$ will be ignored in the following. This kind of transformation — known as a Hubbard–Stratonovich transformation — will be used repeatedly in this book.[3] It transforms a system of interacting particles into a system of "free" particles interacting with a bosonic field. The action now reads

$$S[\psi^*,\psi,\phi] = S_0[\psi^*,\psi] + \frac{1}{2}\sum_q \Big\{\phi(-q)v(\mathbf{q})^{-1}\phi(q) - 2[i\phi(-q) + J(-q)]n(q)\Big\}, \tag{5.25}$$

and the partition function involves a functional integral over both the fermionic field ψ and the bosonic field ϕ. Note that the Hubbard–Stratonovich field ϕ is nothing but the scalar potential of the electromagnetic field (Sec. 1.9.3).[4] Its propagator can therefore be expressed in terms of the dielectric function,

$$D_{00}(q) = \langle \phi(q)\phi(-q)\rangle = \frac{v(\mathbf{q})}{\epsilon_\parallel(q)}. \tag{5.26}$$

[3]See Sec. 6.3.3.1 for a discussion of the limitations of Hubbard–Stratonovich transformations.

[4]ϕ differs from the scalar potential by a factor e (the electron charge). Recall also that $i\phi$ plays the role of the scalar potential in the imaginary-time formalism.

5.2.3.2 *Saddle-point approximation: Hartree theory*

The Hubbard–Stratonovich transformation suggests various approximations to treat the electron–electron interactions (now encoded in the dynamics of the bosonic field ϕ). The simplest one amounts to performing a saddle-point (or mean-field) approximation for the functional integral over ϕ whereby

$$Z = \int \mathcal{D}[\psi^*, \psi, \phi]\, e^{-S[\psi^*,\psi,\phi]} \to Z_{\rm MF} = \int \mathcal{D}[\psi^*, \psi]\, e^{-S_{\rm MF}[\psi^*,\psi]}. \tag{5.27}$$

The mean field $\phi(q) = \sqrt{\beta\mathcal{V}}\delta_{q,0}\phi_0$ (i.e. $\phi(\mathbf{r},\tau) = \phi_0$) is chosen static and uniform. For $J = 0$, the mean-field action then reads

$$S_{\rm MF}[\psi^*, \psi] = \frac{\beta\mathcal{V}}{2} v(0)^{-1}\phi_0^2 - \sum_{k,\sigma} \psi^*_\sigma(k)(i\omega_n - \xi_{\mathbf{k}} + i\phi_0)\psi_\sigma(k). \tag{5.28}$$

The value of ϕ_0 is determined by requiring the thermodynamic potential to be extremum: $\partial\Omega_{\rm MF}/\partial\phi_0 = 0$ or, equivalently, $\partial Z_{\rm MF}/\partial\phi_0 = 0$. This gives

$$\int \mathcal{D}[\psi^*, \psi] \Big[\beta\mathcal{V} v(0)^{-1}\phi_0 - i\sum_{k,\sigma} \psi^*_\sigma(k)\psi_\sigma(k)\Big]\, e^{-S_{\rm MF}[\psi^*,\psi]} = 0, \tag{5.29}$$

i.e.

$$i\phi_0 = -v(0)\frac{1}{\beta\mathcal{V}}\sum_{k,\sigma}\langle\psi^*_\sigma(k)\psi_\sigma(k)\rangle e^{i\omega_n\eta} = -v(0)n \tag{5.30}$$

(we have added the usual convergence factor $e^{i\omega_n\eta}$). Note that ϕ_0 is purely imaginary, which ensures that the mean-field action is real.[5] The result (5.30) has a simple interpretation in terms of Feynman diagrams. It corresponds to the first-order self-consistent self-energy shown in Fig. 5.3,

$$\Sigma_{\rm H}(k) = v(0)\frac{1}{\beta\mathcal{V}}\sum_{k,\sigma} G_\sigma(k)e^{i\omega_n\eta} = v(0)n. \tag{5.31}$$

In the jellium model, $v(0) = 0$ and the Hartree self-energy (5.31) vanishes ($\phi_0 = 0$). In short-range interaction models however, $\Sigma_{\rm H}$ is in general finite. Being independent of k, it merely renormalizes the chemical potential.

5.2.3.3 *Gaussian fluctuations: RPA*

To proceed beyond the saddle-point approximation, we derive the effective action of the bosonic field to second order in $\phi(x) - \phi_0 \equiv \phi(x)$. To do so, one has to integrate out the

[5]Recall that mathematically there is nothing that prevents the saddle point ϕ_0 to be imaginary even though the auxiliary bosonic field $\phi(x)$ is real. One can deform the integration contour into the complex plane when looking for a saddle point; see Sec. 1.7.1.

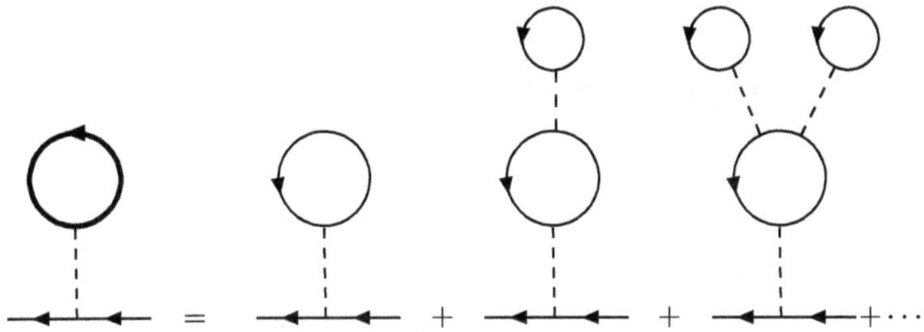

Fig. 5.3 Hartree self-energy $\Sigma_{\rm H}$. The thick solid line denotes the Green function $G = (G_0 - \Sigma_{\rm H})^{-1}$.

ψ field retaining only the second-order cumulant,[6]

$$\begin{aligned} Z[J] &= \int \mathcal{D}[\psi^*, \psi, \phi]\, e^{-S_0[\psi^*,\psi] - \frac{1}{2}\sum_q [\phi(-q)v(\mathbf{q})^{-1}\phi(q) - 2[i\phi(-q)+J(-q)]n(q)]} \\ &= Z_0 \int \mathcal{D}[\phi]\, e^{-S[\phi]}, \end{aligned} \tag{5.32}$$

where

$$S[\phi] = \frac{1}{2}\sum_q \Big\{\phi(-q)v(\mathbf{q})^{-1}\phi(q) - [i\phi(-q) + J(-q)]\chi^0_{nn}(q)[i\phi(q) + J(q)]\Big\}. \tag{5.33}$$

For $J = 0$, one obtains the propagator of the auxiliary field,

$$D_{00}(q) = \langle \phi(q)\phi(-q)\rangle = \frac{1}{v(\mathbf{q})^{-1} + \chi^0_{nn}(q)} = \frac{v(\mathbf{q})}{1 + v(\mathbf{q})\chi^0_{nn}(q)}, \tag{5.34}$$

which reproduces the RPA result (5.11) for the longitudinal dielectric function (compare (5.26) and (5.34)). From the action (5.33), one can also obtain the partition function by integrating out the ϕ field,

$$Z[J] = Z_0 \exp\bigg\{\frac{1}{2}\sum_q J(-q)\bigg[\chi^0_{nn}(q) - \frac{v(\mathbf{q})[\chi^0_{nn}(q)]^2}{1 + v(\mathbf{q})\chi^0_{nn}(q)}\bigg]J(q)\bigg\}, \tag{5.35}$$

and deduce the density–density correlation function

$$\chi_{nn}(q) = \frac{\delta^2 \ln Z[J]}{\delta J(-q)\delta J(q)}\bigg|_{J=0} = \frac{\chi^0_{nn}(q)}{1 + v(\mathbf{q})\chi^0_{nn}(q)}. \tag{5.36}$$

Thus, we see that by taking into account the Gaussian fluctuations of ϕ about the saddle point ϕ_0, we reproduce the RPA.

[6]The first-order cumulant gives a trivial contribution to the action: $\sum_q [i\phi(-q) + J(-q)]\langle n(q)\rangle = J(q = 0)\sqrt{\beta \mathcal{V}}n$ since $\langle n(q)\rangle = \delta_{q,0}\sqrt{\beta\mathcal{V}}n$.

5.3 Physical Properties of the RPA

In this section, we discuss some basic properties of the electron gas in the RPA: density fluctuation spectrum and plasmon mode, screening, ground-state energy and compressibility. We start with a calculation of the Lindhard function χ_{nn}^{0R}.

5.3.1 *The Lindhard function*

After summation over the Matsubara frequency in (5.10) and the analytic continuation $q = (\mathbf{q}, i\omega_\nu) \to (\mathbf{q}, \omega + i\eta)$,[7] one obtains the retarded response function (or Lindhard function)

$$\chi_{nn}^{0R}(\mathbf{q},\omega) = -2\int_{\mathbf{k}} \frac{n_F(\xi_{\mathbf{k}}) - n_F(\xi_{\mathbf{k+q}})}{\omega + i\eta + \xi_{\mathbf{k}} - \xi_{\mathbf{k+q}}}. \tag{5.37}$$

An elementary change of variables gives

$$\chi_{nn}^{0R}(\mathbf{q},\omega) = -2\int_{\mathbf{k}} \frac{n_F(\xi_{\mathbf{k}})}{\omega + i\eta + \xi_{\mathbf{k}} - \xi_{\mathbf{k+q}}} + (\omega + i\eta \to -\omega - i\eta), \tag{5.38}$$

with $n_F(\xi_{\mathbf{k}}) = \Theta(k_F - |\mathbf{k}|)$ at zero temperature. Introducing the dimensionless variables $x = |\mathbf{k}|/k_F$, $\tilde{q} = |\mathbf{q}|/k_F$ and $\tilde{\omega}^+ = (\omega + i\eta)/v_F|\mathbf{q}|$, we have

$$\begin{aligned}
\int_{\mathbf{k}} \frac{\Theta(k_F - |\mathbf{k}|)}{\omega + i\eta + \xi_{\mathbf{k}} - \xi_{\mathbf{k+q}}} &= \frac{1}{4\pi^2}\int_0^{k_F} k^2 dk \int_0^{\pi} d\theta \frac{\sin\theta}{\omega + i\eta - \frac{|\mathbf{k}||\mathbf{q}|}{m}\cos\theta - \frac{\mathbf{q}^2}{2m}} \\
&= \frac{mk_F^2}{4\pi^2|\mathbf{q}|}\int_0^1 dx\, x^2 \int_0^{\pi} \frac{\sin\theta}{\tilde{\omega}^+ - x\cos\theta - \frac{\tilde{q}}{2}} \\
&= \frac{N(0)}{2\tilde{q}}\int_0^1 dx\, x \ln\left(\frac{\tilde{\omega}^+ + x - \tilde{q}/2}{\tilde{\omega}^+ - x - \tilde{q}/2}\right) \\
&= \frac{N(0)}{2\tilde{q}}\left[a - \frac{a^2-1}{2}\ln\left(\frac{a+1}{a-1}\right)\right],
\end{aligned} \tag{5.39}$$

where $a = \tilde{\omega}^+ - \tilde{q}/2$. $N(0) \equiv N(\xi = 0) = mk_F/2\pi^2$ is the non-interacting density of states per spin at the Fermi level. From (5.38) and (5.39), we finally obtain

$$\begin{aligned}
\chi_{nn}^{0R}(\mathbf{q},\omega) = N(0)\Bigg\{1 &- \frac{1}{2\tilde{q}}\left[1 - \left(\tilde{\omega}^+ - \frac{\tilde{q}}{2}\right)^2\right]\ln\left(\frac{\tilde{\omega}^+ + 1 - \tilde{q}/2}{\tilde{\omega}^+ - 1 - \tilde{q}/2}\right) \\
&- \frac{1}{2\tilde{q}}\left[1 - \left(\tilde{\omega}^+ + \frac{\tilde{q}}{2}\right)^2\right]\ln\left(\frac{\tilde{\omega}^+ - 1 + \tilde{q}/2}{\tilde{\omega}^+ + 1 + \tilde{q}/2}\right)\Bigg\}.
\end{aligned} \tag{5.40}$$

[7]The Matsubara sum is done explicitly in Appendix 1.F.

The real and imaginary parts of the Lindhard function read

$$\begin{aligned}\Re[\chi^{0R}_{nn}(\mathbf{q},\omega)] = N(0)\Bigg\{1 &- \frac{1}{2\tilde{q}}\left[1-\left(\tilde{\omega}-\frac{\tilde{q}}{2}\right)^2\right]\ln\left|\frac{\tilde{\omega}+1-\tilde{q}/2}{\tilde{\omega}-1-\tilde{q}/2}\right| \\ &- \frac{1}{2\tilde{q}}\left[1-\left(\tilde{\omega}+\frac{\tilde{q}}{2}\right)^2\right]\ln\left|\frac{\tilde{\omega}-1+\tilde{q}/2}{\tilde{\omega}+1+\tilde{q}/2}\right|\Bigg\}\end{aligned} \tag{5.41}$$

and

$$\begin{aligned}\Im[\chi^{0R}_{nn}(\mathbf{q},\omega)] = N(0)\frac{\pi}{2\tilde{q}}\Bigg\{&\left[1-\left(\tilde{\omega}-\frac{\tilde{q}}{2}\right)^2\right]\Theta\left[1-\left(\tilde{\omega}-\frac{\tilde{q}}{2}\right)^2\right] \\ &-\left[1-\left(\tilde{\omega}+\frac{\tilde{q}}{2}\right)^2\right]\Theta\left[1-\left(\tilde{\omega}+\frac{\tilde{q}}{2}\right)^2\right]\Bigg\},\end{aligned} \tag{5.42}$$

where $\tilde{\omega} = \omega/v_F|\mathbf{q}|$.

For $\omega = 0$, the Lindhard function is real and given by

$$\chi^{0R}_{nn}(\mathbf{q},0) = N(0)\left[1-\frac{1}{\tilde{q}}\left(1-\frac{\tilde{q}^2}{4}\right)\ln\left|\frac{\tilde{q}-2}{\tilde{q}+2}\right|\right]. \tag{5.43}$$

In Sec. 5.3.3, we shall see the non-analyticity at $\tilde{q} = 2$ has important consequences. For $|\mathbf{q}| \to 0$ and $\tilde{\omega} = \omega/v_F|\mathbf{q}|$ fixed, χ^{0R}_{nn} takes the form

$$\lim_{\substack{|\mathbf{q}|\to 0 \\ \tilde{\omega}\text{ fixed}}} \chi^{0R}_{nn}(\mathbf{q},\omega) = 2N(0)\left[1-\frac{\tilde{\omega}^+}{2}\left(\frac{\tilde{\omega}^+ +1}{\tilde{\omega}^+ -1}\right)\right], \tag{5.44}$$

a result previously obtained in Fermi-liquid theory (Sec. 4.3.4).

5.3.1.1 *Spectral function* $\chi^{0\,\prime\prime}_{nn}(\mathbf{q},\omega) = \Im[\chi^{0R}_{nn}(\mathbf{q},\omega)]$

For $\tilde{q} \le 2$, the imaginary part of the Lindhard function has the form shown in Fig. 5.4 (left panel). It is linear at low energies and corresponds to an arc of parabola at higher energy,

$$\chi^{0\,\prime\prime}_{nn}(\mathbf{q},\omega) = \begin{cases} \pi N(0)\dfrac{\omega}{v_F|\mathbf{q}|} & \text{for } 0 \le \omega \le |\omega^-_{\mathbf{q}}|, \\ \dfrac{\pi N(0)}{2\tilde{q}}\left[1-(\tilde{\omega}-\frac{\tilde{q}}{2})^2\right] & \text{for } |\omega^-_{\mathbf{q}}| \le \omega \le \omega^+_{\mathbf{q}}, \end{cases} \tag{5.45}$$

where

$$\omega^{\pm}_{\mathbf{q}} = \pm v_F|\mathbf{q}| + \frac{\mathbf{q}^2}{2m}. \tag{5.46}$$

The negative energy part can be deduced from (5.45) by noting that $\chi^{0\,\prime\prime}_{nn}(\mathbf{q},\omega) = -\chi^{0\,\prime\prime}_{nn}(\mathbf{q},-\omega)$.

For $\tilde{q} \ge 2$, the linear part is not present anymore and the Lindhard function reduces to the arc of parabola (Fig. 5.4, right panel),

$$\chi^{0\,\prime\prime}_{nn}(\mathbf{q},\omega) = \frac{\pi N(0)}{2\tilde{q}}\left[1-\left(\tilde{\omega}-\frac{\tilde{q}}{2}\right)^2\right] \quad \text{for } \omega^-_{\mathbf{q}} \le \omega \le \omega^+_{\mathbf{q}}. \tag{5.47}$$

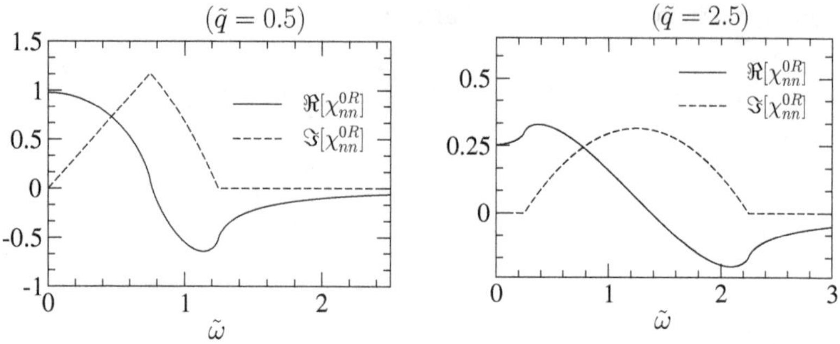

Fig. 5.4 Real and imaginary parts of the Lindhard function $\chi_{nn}^{0R}(\mathbf{q},\omega)/2N(0)$ for $|\mathbf{q}| = 0.5k_F$ and $|\mathbf{q}| = 2.5k_F$.

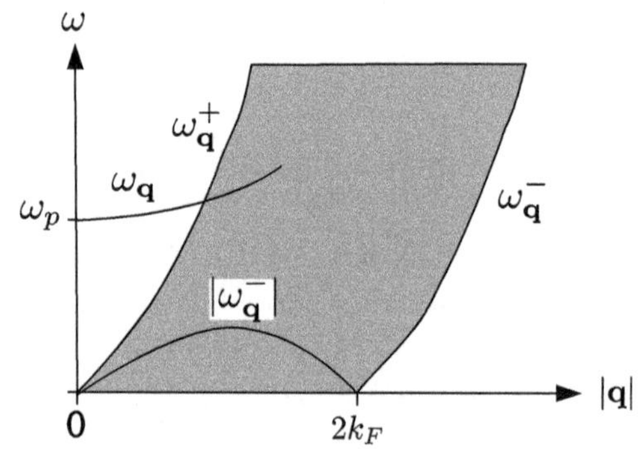

Fig. 5.5 Plasmon mode dispersion $\omega_{\mathbf{q}}$ in the electron gas. The shaded area shows the particle–hole excitation continuum. In the RPA, the latter has the same boundaries as that of the non-interacting electron gas.

5.3.1.2 *Particle–hole pair excitation continuum*

In the non-interacting system, the particle–hole pair excitation spectrum is defined by $\chi_{nn}^{0\,\prime\prime}(\mathbf{q},\omega) \neq 0$. It gives the values of $(\mathbf{q},\omega)$ for which the equation $\omega = \epsilon_{\mathbf{k}+\mathbf{q}} - \epsilon_{\mathbf{k}}$ possesses at least a solution with $\xi_{\mathbf{k}} < 0$ and $\xi_{\mathbf{k}+\mathbf{q}} > 0$. For $|\mathbf{q}| \leq 2k_F$, the particle–hole pair continuum extends from $\omega = 0$ up to $\omega = \omega_{\mathbf{q}}^{+}$. For $|\mathbf{q}| > 2k_F$, there is no excitation at low energy and the particle–hole pair continuum extends from $\omega_{\mathbf{q}}^{-}$ to $\omega_{\mathbf{q}}^{+}$ (Fig. 5.5).

5.3.2 *Density fluctuation spectrum*

The spectrum of the density fluctuations can be obtained from the spectral function $\chi_{nn}''(\mathbf{q},\omega) = \Im[\chi_{nn}^{R}(\mathbf{q},\omega)]$ or the structure factor $S_{nn}(\mathbf{q},\omega;T=0) = 2\Theta(\omega)\chi_{nn}''(\mathbf{q},\omega)$.[8] In the RPA, there are two contributions to

$$S_{nn}^{\mathrm{RPA}}(\mathbf{q},\omega) = 2\Theta(\omega)\frac{\chi_{nn}^{0\,\prime\prime}(\mathbf{q},\omega)}{\left(1+v(\mathbf{q})\Re[\chi_{nn}^{0R}(\mathbf{q},\omega)]\right)^2 + \left(v(\mathbf{q})\chi_{nn}^{0\,\prime\prime}(\mathbf{q},\omega)\right)^2}. \tag{5.48}$$

[8]The following discussion is similar to that of Sec. 4.3.4 on the structure factor of the neutral Fermi liquid.

The first one is due to a non-vanishing $\chi''^{0}_{nn}(\mathbf{q},\omega)$ and comes from the particle–hole pair excitations. These are quite similar to the particle–hole pair excitations of the non-interacting electron gas and gives a continuous excitation spectrum (Fig. 5.5). The second contribution to the structure factor comes from the (collective) plasmon mode at the frequency $\omega_{\mathbf{q}}$ defined by

$$1 + v(\mathbf{q})\Re[\chi^{0R}_{nn}(\mathbf{q},\omega_{\mathbf{q}})] = 0. \tag{5.49}$$

Such an excitation is well defined if its lifetime — determined by $1/\chi''^{0}_{nn}(\mathbf{q},\omega_{\mathbf{q}})$ — is long enough; only in this case does it show up as a sharp peak in $S_{nn}(\mathbf{q},\omega)$ (Fig. 5.6). In the RPA, the plasmon has an infinite lifetime as long as $\chi''^{0}_{nn}(\mathbf{q},\omega) = 0$, i.e. as long as it does not overlap with the particle-hole pair continuum. When its energy satisfies $\omega = \epsilon_{\mathbf{k}+\mathbf{q}} - \epsilon_{\mathbf{k}}$ (with $|\mathbf{k}| < k_F$ and $|\mathbf{k}+\mathbf{q}| > k_F$), the plasmon becomes strongly damped (Landau damping) and is not a well-defined excitation of the system anymore. The absence of damping at small $\mathbf{q}$ is an artifact of the RPA. Multi-pair excitations — not taken into account in the RPA — provides the main damping mechanism when Landau damping is ineffective. These issues have been discussed in detail in connection with the zero-sound mode of the neutral Fermi liquid (Secs. 4.3.3 and 4.3.5).

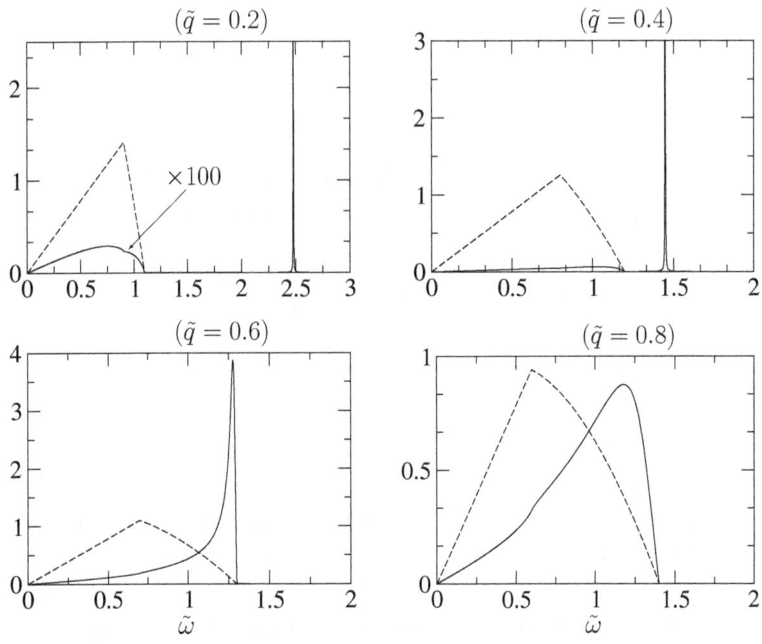

Fig. 5.6 Structure factor $S_{nn}(\mathbf{q},\omega)/4N(0)$ of the electron gas in the RPA for different values of $\tilde q = |\mathbf{q}|/k_F$ and $r_s = 1$ ($\tilde\omega = \omega/v_F|\mathbf{q}|$). The dashed line shows the results for the non-interacting electron gas. The plots are obtained with a small but finite value of η (hence the non-zero width of the plasmon peak). For small values of $\tilde q$, the plasmon peak is clearly visible — and gives the main contribution to the spectral weight $\int_0^\infty d\omega\, S_{nn}(\mathbf{q},\omega)$ — while for larger values, it merges with the continuum of particle–hole pair excitations.

5.3.2.1 *Plasmon mode and sum rules*

For $\mathbf{q} \to 0$ with ω fixed, $\chi^{0R}_{nn}(\mathbf{q},\omega) \to -n\mathbf{q}^2/\omega^2 m$, i.e.

$$\epsilon^R_{\parallel}(\mathbf{q},\omega) = 1 - \frac{\omega_p^2}{\omega^2} \qquad (\mathbf{q} \to 0), \tag{5.50}$$

and we recover $\omega_p = (ne^2/\epsilon_0 m)^{1/2}$ as the excitation energy of the plasmon mode in the long-wavelength limit. Pushing the expansion of $\chi^{0R}_{nn}(\mathbf{q},\omega)$ to higher order in $\mathbf{q}^2$ yields the long-wavelength plasmon dispersion

$$\omega_{\mathbf{q}}^2 = \omega_p^2 + \frac{3}{5} v_F^2 \mathbf{q}^2 + \mathcal{O}(|\mathbf{q}|^4). \tag{5.51}$$

For $\mathbf{q} \to 0$ and $\omega \simeq \omega_p$, one has

$$\begin{aligned} 1 + v(\mathbf{q})\chi^{0R}_{nn}(\mathbf{q},\omega) &\simeq (\omega + i\eta - \omega_p) v(\mathbf{q}) \partial_\omega \chi^{0R}_{nn}(\mathbf{q},\omega)\Big|_{\omega=\omega_p} \\ &= (\omega + i\eta - \omega_p)\frac{2}{\omega_p}, \end{aligned} \tag{5.52}$$

and therefore,

$$\left. \begin{aligned} \chi^R_{nn}(\mathbf{q},\omega) &= -\frac{\omega_p}{2v(\mathbf{q})} \frac{1}{\omega + i\eta - \omega_p}, \\ \frac{1}{\epsilon^R_{\parallel}(\mathbf{q},\omega)} &= \frac{\omega_p}{2} \frac{1}{\omega + i\eta - \omega_p}, \end{aligned} \right. \qquad (\mathbf{q} \to 0 \text{ and } \omega \to \omega_p). \tag{5.53}$$

For $\mathbf{q} \to 0$, the spectral function $\chi''_{nn}(\mathbf{q},\omega)$ exhibits a delta peak $(\pi\omega_p/2v(\mathbf{q}))\delta(\omega - \omega_p)$ and another one at $\omega = -\omega_p$ with negative weight. Using (5.53), one obtains

$$\begin{aligned} \lim_{\mathbf{q}\to 0} \int_{-\infty}^{\infty} \frac{d\omega}{\pi} \omega \Im\left[\frac{1}{\epsilon^R_{\parallel}(\mathbf{q},\omega)}\right]_{\text{plasmon}} &= -\omega_p^2, \\ \lim_{\mathbf{q}\to 0} \int_{-\infty}^{\infty} \frac{d\omega}{\pi} \frac{1}{\omega} \Im\left[\frac{1}{\epsilon^R_{\parallel}(\mathbf{q},\omega)}\right]_{\text{plasmon}} &= -1. \end{aligned} \tag{5.54}$$

We conclude that the RPA satisfies the f-sum rule and the perfect-screening sum rule in the $\mathbf{q} \to 0$ limit, the plasmon mode exhausting the spectral weight. Equations (5.54) are exact. They are a consequence of the vanishing contribution of particle–hole pair excitations to the spectral function $\Im[1/\epsilon^R_{\parallel}(\mathbf{q},\omega)]$ when $\mathbf{q} \to 0$.[9]

With $\epsilon^R_{\parallel}(\mathbf{q},\omega) = 1 + v(\mathbf{q})\chi^{0R}_{nn}(\mathbf{q},\omega)$, the longitudinal-conductivity sum rule reduces to the f-sum rule (3.106) for non-interacting electrons and is therefore trivially satisfied. Similarly, if we define the compressibility κ of the charged system by $n^2\kappa = \lim_{\mathbf{q}\to 0} \Pi_{nn}(\mathbf{q}, 0)$, as discussed in Sec. 3.4.1.2, the compressibility sum rule is also trivially satisfied since it follows

[9]In Sec. 3.4.1, we have shown that the exact result $\lim_{\mathbf{q}\to 0} \omega_{\mathbf{q}} = \omega_p = (ne^2/\epsilon_0 m)^{1/2}$ can be deduced from the sum rules (5.54).

from that definition.[10] Furthermore, κ is independent of the Coulomb interaction since $\Pi^{\rm RPA}_{nn} = \chi^0_{nn}$. The compressibility does depend on the interactions if it is calculated from the density dependence of the ground-state energy, as in Sec. 5.3.4.2, but in that case, the compressibility sum rule is violated.

5.3.3 *Screening*

5.3.3.1 *Thomas–Fermi theory*

Let us start with a simple theory of screening. We consider the response of an electron gas to a static impurity charge density $n_i(\mathbf{r})$. If the induced charge density varies slowly with respect to k_F, we expect a semiclassical description to be justified. The scalar potential $\phi(\mathbf{r})$ satisfies Poisson's equation[11]

$$\boldsymbol{\nabla}^2\phi(\mathbf{r}) + \frac{e^2}{\epsilon_0}[\delta n(\mathbf{r}) + n_i(\mathbf{r})] = 0, \tag{5.55}$$

where $\delta n(\mathbf{r}) = \langle \hat n(\mathbf{r})\rangle - n$ is the induced charge density. In a semiclassical picture, $\mu - \phi(\mathbf{r})$ acts as a local chemical potential and

$$\delta n(\mathbf{r}) = -\phi(\mathbf{r})\frac{\partial n}{\partial \mu} + \mathcal{O}(\phi^2) = -n^2\kappa\phi(\mathbf{r}) + \mathcal{O}(\phi^2), \tag{5.56}$$

where $\kappa = n^{-2}\partial n/\partial\mu = 3/2n\epsilon_F$ is the compressibility of the free electron gas. To first order in $\phi(\mathbf{r})$, Poisson's equation then gives

$$\boldsymbol{\nabla}^2\phi(\mathbf{r}) - \frac{e^2}{\epsilon_0}n^2\kappa\phi(\mathbf{r}) = -\frac{e^2}{\epsilon_0}n_i(\mathbf{r}). \tag{5.57}$$

In Fourier space, we find

$$\phi(\mathbf{q}) = \frac{e^2}{\epsilon_0}\frac{n_i(\mathbf{q})}{\mathbf{q}^2 + q_{\rm TF}^2}, \tag{5.58}$$

where

$$q_{\rm TF}^2 = \frac{e^2}{\epsilon_0}n^2\kappa = \frac{3e^2 n}{2\epsilon_0\epsilon_F} = r_s k_F^2\frac{4}{\pi}\left(\frac{4}{9\pi}\right)^{1/3} \tag{5.59}$$

is the (square of the) Thomas–Fermi screening wavevector. From the standard definition of the longitudinal dielectric function,

$$\phi(\mathbf{q}) = \frac{e^2 n_i(\mathbf{q})}{\epsilon_0\epsilon_\parallel(\mathbf{q})\mathbf{q}^2}, \tag{5.60}$$

we obtain

$$\epsilon_{\rm TF}(\mathbf{q}) = 1 + \frac{q_{\rm TF}^2}{\mathbf{q}^2}. \tag{5.61}$$

[10] There is no contribution of the $\mathbf{q}\to 0$ plasmon to the longitudinal-conductivity and compressibility sum rules since $\epsilon^R_\parallel(\mathbf{q}\to 0,\omega)$ is real for $\omega \simeq \omega_p$.

[11] See footnote 4 on page 375 for the explanation of e^2 (and not e) appearing in (5.55).

At small distances ($|\mathbf{q}| \gg q_{\rm TF}$), the impurity charge is unscreened since $\epsilon_{\rm TF}(\mathbf{q}) \simeq 1$. On the contrary, at large distances, the screening is very effective. For a point-like impurity $n_i(\mathbf{r}) = \delta(\mathbf{r})$, the screened potential has the form of a Yukawa potential,

$$\phi(\mathbf{r}) = \frac{e^2}{4\pi\epsilon_0|\mathbf{r}|}e^{-q_{\rm TF}|\mathbf{r}|}, \tag{5.62}$$

and is exponentially suppressed at length scales larger than the screening length $q_{\rm TF}^{-1}$. Away from the impurity ($\mathbf{r} \neq 0$), the induced charge density is given by

$$\delta n(\mathbf{r}) = -\frac{\epsilon_0}{e^2}\boldsymbol{\nabla}^2\phi(\mathbf{r}) = -q_{\rm TF}^2\frac{e^{-q_{\rm TF}|\mathbf{r}|}}{4\pi|\mathbf{r}|}. \tag{5.63}$$

A necessary condition for the (semiclassical) Thomas–Fermi theory to be justified is $q_{\rm TF} \ll k_F$, i.e. $r_s \ll 1$. We shall see in the following that even when this condition is satisfied, the Thomas–Fermi theory fails to provide an accurate description of the response of the electron gas to a static impurity charge density.

5.3.3.2 *Static RPA dielectric function*

In the RPA, the screening of a static impurity charge is determined by the dielectric function

$$\epsilon_{\rm RPA}(\mathbf{q}) = 1 + \frac{q_{\rm TF}^2}{2\mathbf{q}^2}\left\{1 - \frac{1}{\tilde{q}}\left(1 - \frac{\tilde{q}^2}{4}\right)\ln\left|\frac{\tilde{q}-2}{\tilde{q}+2}\right|\right\}. \tag{5.64}$$

For $\mathbf{q} \to 0$, one can approximate $\epsilon_{\rm RPA}(\mathbf{q})$ by

$$\epsilon_{\rm RPA}(\mathbf{q}) \simeq 1 + v(\mathbf{q})\lim_{\mathbf{q}\to 0}\chi^0_{nn}(\mathbf{q}, i\omega_\nu = 0) = 1 + \frac{e^2 n^2\kappa}{\epsilon_0\, q^2} = \epsilon_{\rm TF}(\mathbf{q}). \tag{5.65}$$

It is quite tempting to conclude that at long distances, the RPA predicts results similar to those of the Thomas–Fermi theory. This turns out to be wrong because of the non-analyticity of $\epsilon_{\rm RPA}(\mathbf{q})$ at $|\mathbf{q}| = 2k_F$ (Fig. 5.7). This non-analyticity, which originates in the

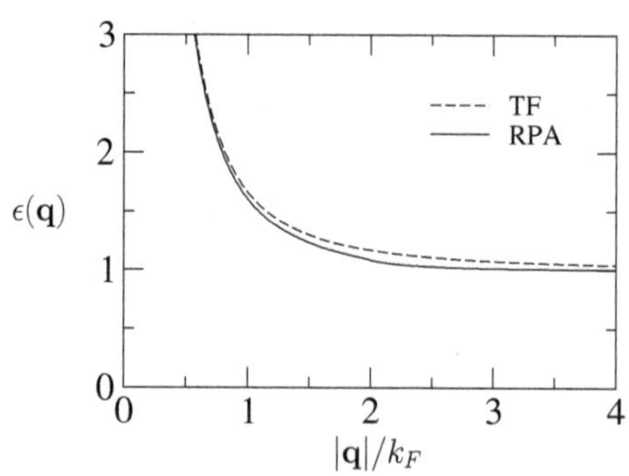

Fig. 5.7 Static dielectric functions obtained from the Thomas–Fermi theory and the RPA. The non-analyticity of $\epsilon_{\rm RPA}(\mathbf{q})$ at $|\mathbf{q}| = 2k_F$ gives rise to Friedel oscillations and a power-law decrease of the charge induced by a point-like impurity [Eq. (5.66)].

Fermi surface of the electron gas, prevents the mere replacement $\epsilon_{\text{RPA}}(\mathbf{q}) \to \epsilon_{\text{TF}}(\mathbf{q})$ when computing the long-distance screening of an external potential. A careful analysis shows that the charge induced by a point-like impurity decays as a power law and exhibits Friedel oscillations [1, 3],

$$\delta n(\mathbf{r}) \propto \frac{\cos(2k_F|\mathbf{r}|)}{|\mathbf{r}|^3}, \tag{5.66}$$

in marked contrast with the results of the Thomas–Fermi theory [Eq. (5.63)].

5.3.4 *Ground-state energy and compressibility*

5.3.4.1 *Ground-state energy*

The thermodynamic potential can be expressed as a function of the interaction energy by means of a coupling constant integration. Let us write the Hamiltonian as $\hat{H} = \hat{H}_0 + \lambda \hat{H}_{\text{int}}$ and use (3.275),

$$\frac{\partial \Omega}{\partial \lambda} = \frac{1}{\lambda}\langle \lambda \hat{H}_{\text{int}} \rangle_\lambda, \tag{5.67}$$

where the expectation value $\langle \cdots \rangle_\lambda$ is taken with the Hamiltonian $\hat{H}_0 + \lambda \hat{H}_{\text{int}}$. The potential energy can be expressed as a function of the density–density response function,

$$\begin{aligned} \langle \hat{H}_{\text{int}} \rangle_\lambda &= \frac{1}{2}\sum_{\mathbf{q}} v(\mathbf{q}) \left\langle \hat{n}(-\mathbf{q})\hat{n}(\mathbf{q}) - \frac{\hat{N}}{\mathcal{V}} \right\rangle_\lambda \\ &= \frac{1}{2}\sum_{\mathbf{q}} v(\mathbf{q}) \left[\frac{1}{\beta}\sum_{\omega_\nu} \chi^\lambda_{nn}(q) e^{i\omega_\nu \eta} - n^\lambda \right], \end{aligned} \tag{5.68}$$

where both $\chi^\lambda_{nn}(q)$ and $n^\lambda = \langle \hat{N} \rangle_\lambda / \mathcal{V}$ *a priori* depend on λ.

In the RPA, the one-particle propagator is approximated by the Hartree (non-interacting) propagator. This ensures that the RPA is a conserving approximation that satisfies the conservation laws.[12] $n^\lambda \equiv n$ is thus independent of the interaction $\lambda v(\mathbf{q})$ and equal to the density of the free gas (for a fixed value of the chemical potential); working at fixed density or fixed chemical potential in the RPA is the same. From (5.67) and (5.68), we then obtain

$$\Omega = \Omega_0 + \frac{1}{2}\int_0^1 \frac{d\lambda}{\lambda} \sum_{\mathbf{q}} \lambda v(\mathbf{q}) \left[\frac{1}{\beta}\sum_{\omega_\nu} \chi^\lambda_{nn}(q) e^{i\omega_\nu \eta} - n \right], \tag{5.69}$$

where

$$\chi^\lambda_{nn}(q) = \frac{\chi^0_{nn}(q)}{1 + \lambda v(\mathbf{q})\chi^0_{nn}(q)} = \chi^0_{nn}(q) - \lambda v(\mathbf{q}) \frac{[\chi^0_{nn}(q)]^2}{1 + \lambda v(\mathbf{q})\chi^0_{nn}(q)} \tag{5.70}$$

[12] Conserving approximations are discussed in Chapter 9; see in particular the discussion of the electron gas in Sec. 9.1.3.1.

and $\Omega_0 \equiv \Omega(\lambda = 0)$ is the thermodynamic potential of the non-interacting electron gas. Inserting (5.70) into (5.69) and integrating over λ, one obtains the ground-state energy

$$E_{\text{RPA}} = \Omega(T=0) + \mu N = E_0 + E_x + E_{\text{corr}}^{\text{RPA}}, \tag{5.71}$$

where ($\beta \to \infty$)

$$\begin{aligned} E_x &= \frac{1}{2}\sum_{\mathbf{q}} v(\mathbf{q})\left[\frac{1}{\beta}\sum_{\omega_\nu} \chi_{nn}^0(q)e^{i\omega_\nu \eta} - n\right], \\ E_{\text{corr}}^{\text{RPA}} &= \frac{1}{2\beta}\sum_q \left\{-v(\mathbf{q})\chi_{nn}^0(q) + \ln\left[1 + v(\mathbf{q})\chi_{nn}^0(q)\right]\right\}. \end{aligned} \tag{5.72}$$

The energy of the non-interacting gas per electron is given by

$$\frac{E_0}{N} = \frac{3}{5}\epsilon_F = \frac{3}{5}\left(\frac{9\pi}{4}\right)^{2/3}\frac{1}{r_s^2}\ \text{Ry}, \tag{5.73}$$

where the Rydberg $\text{Ry} = e^2/8\pi\epsilon_0 a_0 = \hbar^2/2ma_0^2$.

An elementary calculation gives

$$E_x = -\frac{1}{\mathcal{V}}\sum_{\mathbf{k},\mathbf{q}} v(\mathbf{q}) n_{\mathbf{k}+\mathbf{q}}^0 \left(n_{\mathbf{k}}^0 - 1\right) - \frac{1}{2}\sum_{\mathbf{q}} v(\mathbf{q}) n = -\frac{1}{\mathcal{V}}\sum_{\mathbf{k},\mathbf{q}} v(\mathbf{q}) n_{\mathbf{k}+\mathbf{q}}^0 n_{\mathbf{k}}^0, \tag{5.74}$$

where $n_{\mathbf{k}}^0 = \Theta(k_F - |\mathbf{k}|)$. E_x is called the exchange energy. It is nothing but the first-order correction $\langle \hat{H}_{\text{int}} \rangle$ to the energy E_0 of the non-interacting electron gas and arises from the antisymmetry of the ground state with respect to the exchange of two particles (Fig. 5.8(a)).[13] The exchange energy per electron reads

$$\begin{aligned} \frac{E_x}{N} &= -\frac{1}{n}\int_{\mathbf{q},\mathbf{k}} \frac{e^2}{\epsilon_0 |\mathbf{q} - \mathbf{k}|^2}\Theta(k_F - |\mathbf{k}|)\Theta(k_F - |\mathbf{q}|) \\ &= -\frac{e^2 k_F}{2\pi^2 \epsilon_0 n}\int_{\mathbf{k}} \Theta(k_F - |\mathbf{k}|)\left[\frac{1}{2} + \frac{1 - \tilde{k}^2}{4\tilde{k}}\ln\left|\frac{1 + \tilde{k}}{1 - \tilde{k}}\right|\right] \\ &= -\frac{3e^2 k_F}{16\pi^2 \epsilon_0} \\ &= -\frac{3}{2\pi}\left(\frac{9\pi}{4}\right)^{1/3}\frac{1}{r_s}\ \text{Ry}. \end{aligned} \tag{5.75}$$

The difference $E - E_0 - E_x$ is referred to as the correlation energy. In the RPA, it is given by (5.72) and corresponds to the bubble (or ring) diagrams shown in Fig. 5.8(b). As the full calculation is not possible analytically, we only quote the final result [3],

$$\frac{E_{\text{corr}}^{\text{RPA}}}{N} = \frac{2}{\pi^2}(1 - \ln 2)\ln r_s - 0.141\ \ \text{Ry} \simeq 0.0622 \ln r_s - 0.141\ \ \text{Ry}. \tag{5.76}$$

[13] Note that the other first-order correction, the so-called direct term, involves $v(0)$ and vanishes in the jellium model.

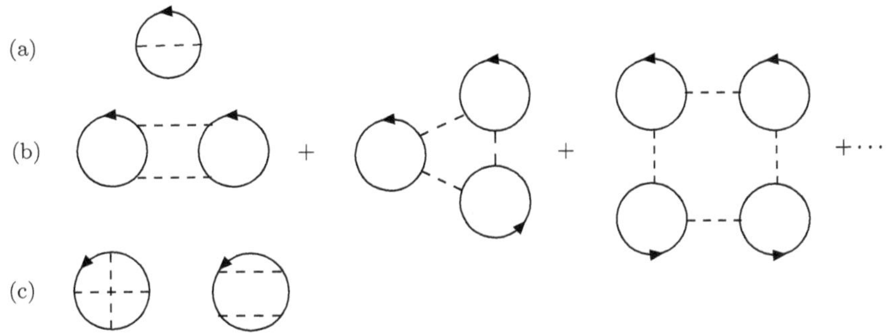

Fig. 5.8 Feynman diagrams for the ground-state energy $E = \Omega(T=0) + \mu N$. (a) Exchange energy E_x. (b) Correlation energy $E_{\rm corr}^{\rm RPA}$ in the RPA. (c) Second-order contributions not included in $E_{\rm corr}^{\rm RPA}$.

The validity of the RPA can be improved by adding the second-order contributions shown in Fig. 5.8(c). One then obtains the following expansion of the correlation energy in the high-density limit [3]:

$$\frac{E_{\rm corr}}{N} = 0.0622 \ln r_s - 0.094 + \mathcal{O}(r_s \ln r_s) \ {\rm Ry}. \tag{5.77}$$

5.3.4.2 *Compressibility*

Let us introduce the notation $\epsilon(r_s) = (E_0 + E_x + E_{\rm corr})/N$ for the energy per electron,

$$\epsilon(r_s) = \frac{3}{5}\left(\frac{9\pi}{4}\right)^{2/3}\frac{1}{r_s^2} - \frac{3}{2\pi}\left(\frac{9\pi}{4}\right)^{1/3}\frac{1}{r_s} + 0.0622\ln r_s - 0.094 + \mathcal{O}(r_s \ln r_s) \ {\rm Ry}. \tag{5.78}$$

We deduce the pressure

$$P = -\left.\frac{\partial E}{\partial \mathcal{V}}\right|_N = n^2 \frac{\partial \epsilon}{\partial n} = -\frac{n r_s}{3}\frac{\partial \epsilon}{\partial r_s} \ {\rm Ry} \tag{5.79}$$

and the inverse compressibility

$$\frac{1}{\kappa} = -\mathcal{V}\left.\frac{\partial P}{\partial \mathcal{V}}\right|_N = n\frac{\partial P}{\partial n} = \frac{n r_s}{3}\left[-\frac{2}{3}\frac{\partial \epsilon}{\partial r_s} + \frac{r_s}{3}\frac{\partial^2 \epsilon}{\partial r_s^2}\right] \ {\rm Ry}. \tag{5.80}$$

The compressibility becomes negative for $r_s > 4.84$ (Fig. 5.9). As noted in Sec. 3.4.1, the condition $\kappa < 0$ does not necessarily imply an instability of the system as κ does not include the stiffness of the background of positive charge with respect to a change of volume.[14]

5.4 One-Particle Properties

The RPA seems to give a reliable description of the density fluctuations: It provides us with a description of screening and the plasmon mode and satisfies the sum rules of Table 5.1

[14] κ in (5.80) is calculated by assuming that the background of positive charge adjusts itself to maintain neutrality at no energy cost (Sec. 3.4.1).

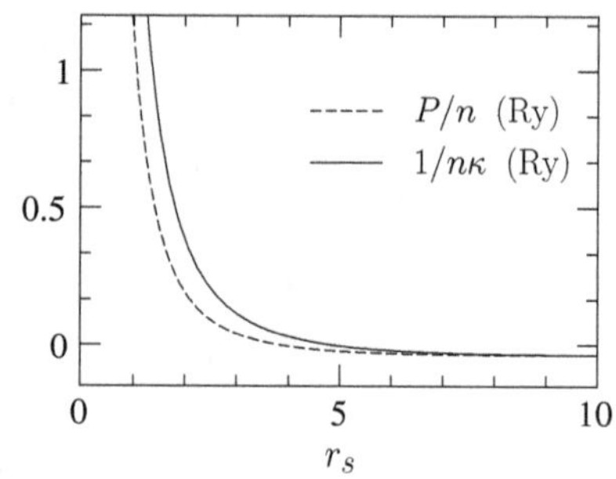

Fig. 5.9 Pressure P and compressibility κ of the electron gas obtained from (5.78).

Fig. 5.10 Diagrammatic representation of the exchange (or Fock) self-energy Σ_x. The thick solid line denotes the Green function $G = (G_0^{-1} - \Sigma_x)^{-1}$.

(except the one related to the compressibility, see Sec. 5.3.2). On the other hand, the RPA does not say much about the one-particle properties since it approximates the one-particle propagator by the non-interacting (Hartree) propagator. In particular, one would like to know whether the electron gas is a Fermi liquid (in the sense discussed in Chapter 4).

5.4.1 *Hartree–Fock theory*

The simplest, and the most natural, way to improve the Hartree theory is to consider the exchange (or Fock) term shown in Fig. 5.10:

$$\begin{aligned}\Sigma_x(\mathbf{k}) &= -\frac{1}{\beta\mathcal{V}}\sum_{k'} v(\mathbf{k}'-\mathbf{k})G(\mathbf{k}')e^{i\omega'_n\eta} \\ &= -\frac{1}{\beta\mathcal{V}}\sum_{k'} v(\mathbf{k}'-\mathbf{k})\frac{e^{i\omega'_n\eta}}{i\omega'_n - \xi_{\mathbf{k}'} - \Sigma_x(\mathbf{k}')} \\ &= -\int_{\mathbf{k}'} v(\mathbf{k}'-\mathbf{k})n_F[\xi_{\mathbf{k}'} + \Sigma_x(\mathbf{k}')].\end{aligned} \tag{5.81}$$

At zero temperature, one therefore obtains

$$\Sigma_x(\mathbf{k}) = -\int_{\mathbf{k}'} v(\mathbf{k}'-\mathbf{k})\Theta(k_F - |\mathbf{k}'|), \tag{5.82}$$

where the Fermi wavevector is defined by $\xi_{k_F} + \Sigma_x(k_F) = 0$ and

$$n = \frac{2}{\beta\mathcal{V}}\sum_k G(k)e^{i\omega_n\eta} = 2\int_{\mathbf{k}} n_F[\xi_{\mathbf{k}} + \Sigma_x(\mathbf{k})] = 2\int_{\mathbf{k}} \Theta(k_F - |\mathbf{k}|). \tag{5.83}$$

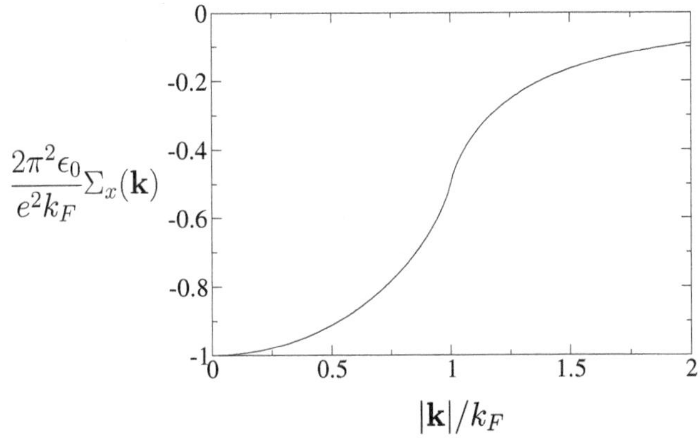

Fig. 5.11 Exchange self-energy $\Sigma_x(\mathbf{k})$ *vs* $\tilde{k} = |\mathbf{k}|/k_F$.

Thus, k_F is the same as in the non-interacting electron gas (for a given density). This is nothing but the Luttinger theorem: The interactions do not change the volume of the Fermi surface (Sec. 4.4.6). For a frequency-independent self-energy, the theorem is trivially satisfied. Note that $\Sigma_x(\mathbf{k})$ is actually first order in the Coulomb interaction: Only the first-order diagram contributes, all higher-order ones vanishing as can be verified by a direct calculation.

From (5.82), we thus obtain

$$\Sigma_x(\mathbf{k}) = -\frac{e^2k_F}{2\pi^2\epsilon_0}\left[\frac{1}{2} + \frac{1-\tilde{k}^2}{4\tilde{k}}\ln\left|\frac{1+\tilde{k}}{1-\tilde{k}}\right|\right] \tag{5.84}$$

(Fig. 5.11), where $\tilde{k} = |\mathbf{k}|/k_F$. Since

$$\frac{\Sigma_x(k_F)}{\epsilon_F} = -\frac{e^2m}{2\pi^2\epsilon_0 k_F} = -\frac{2}{\pi k_F a_0} \sim -r_s, \tag{5.85}$$

the exchange self-energy is small in the high-density limit $r_s \ll 1$, where a perturbative treatment of the Coulomb interaction is expected to be valid. Yet, the Hartree–Fock theory is not satisfactory for the electron gas. The reason is that the non-analyticity of $\Sigma_x(\mathbf{k})$ at $|\mathbf{k}| = k_F$ gives rise to a vanishing effective mass at the Fermi level,

$$\frac{m}{m^*} = 1 + \frac{m}{k_F}\frac{\partial}{\partial|\mathbf{k}|}\Sigma_x(\mathbf{k})\bigg|_{|\mathbf{k}|=k_F} = \lim_{\tilde{k}\to 1}\left[1 + \frac{1}{\pi k_F a_0}\ln\left|\frac{2}{1-\tilde{k}}\right|\right] = \infty. \tag{5.86}$$

5.4.2 *RPA self-energy*

The failure of the Hartree–Fock approximation is not surprising since the Fock self-energy involves the bare, unscreened, Coulomb interaction $v(\mathbf{q})$. The simplest improvement, consistent with screening, is obtained by replacing $v(\mathbf{q})$ by $v(\mathbf{q})/\epsilon_{\parallel}^{\rm RPA}(q)$ in Σ_x and neglecting

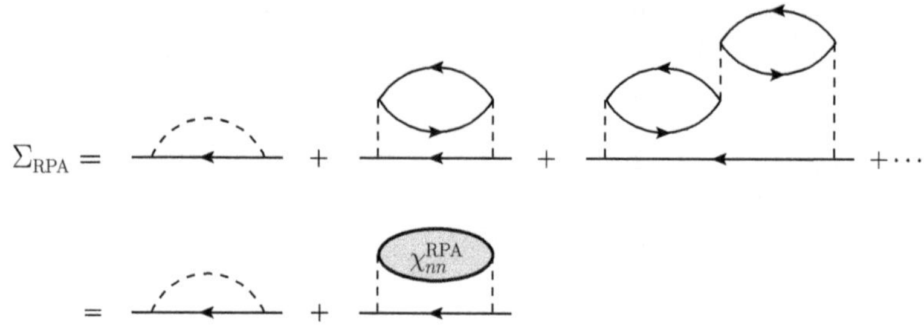

Fig. 5.12 Diagrammatic representation of the RPA self-energy.

the self-consistency, i.e. replacing the dressed propagator G by G_0. This gives the RPA self-energy

$$\Sigma_{\rm RPA}(k) = -\frac{1}{\beta\mathcal{V}}\sum_q \frac{v(\mathbf{q})}{\epsilon_{\|}^{\rm RPA}(q)}G_0(k+q)$$
$$= -\frac{1}{\beta\mathcal{V}}\sum_q \frac{v(\mathbf{q})}{1+v(\mathbf{q})\chi_{nn}^0(q)}G_0(k+q), \tag{5.87}$$

which can be rewritten as

$$\Sigma_{\rm RPA}(k) = \Sigma_x(\mathbf{k}) + \frac{1}{\beta\mathcal{V}}\sum_q v(\mathbf{q})^2\chi_{nn}^{\rm RPA}(q)G_0(k+q), \tag{5.88}$$

as shown diagrammatically in Fig. 5.12.

The RPA self-energy can also be justified as follows. The self-energy is related to the density–density response function by (1.439) or (1.440), i.e.

$$\frac{1}{\beta\mathcal{V}}\sum_k \Sigma(k)G(k)e^{i\omega_n\eta} = \frac{1}{2}\int_{\mathbf{q}} v(\mathbf{q})\left[\frac{1}{\beta}\sum_{\omega_\nu}\chi_{nn}(q) - n\right]. \tag{5.89}$$

One can therefore try to find an expression for the self-energy that satisfies (5.89) when χ_{nn} is approximated by $\chi_{nn}^{\rm RPA}$. Given that $\chi_{nn}^{\rm RPA}$ is expressed in terms of G_0, the RPA self-energy can be the solution only if we replace G by G_0 in the lhs. In that case, both sides of (5.89) can be represented by the sum of the bubble diagrams:

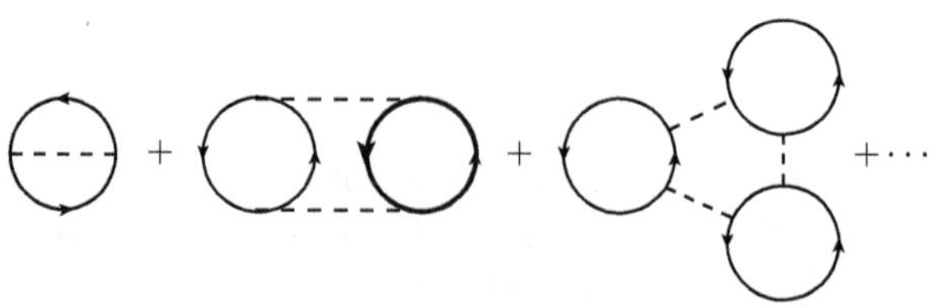

To show the validity of (5.89) more rigorously (and the role of n in the rhs), we consider

$$\frac{1}{\beta\mathcal{V}}\sum_k \Sigma_{\text{RPA}}(k)G_0(k)e^{i\omega_n\eta} = \frac{1}{\beta\mathcal{V}}\sum_k [\Sigma_{\text{RPA}}(k) - \Sigma_x(\mathbf{k})]G_0(k) + \frac{1}{\beta\mathcal{V}}\sum_k \Sigma_x(\mathbf{k})G_0(k)e^{i\omega_n\eta}. \tag{5.90}$$

Since

$$\Sigma_x(\mathbf{k}) = \lim_{|\omega_n|\to\infty} \Sigma_{\text{RPA}}(k), \tag{5.91}$$

the Matsubara sum in the first term in the rhs of (5.90) is convergent and the factor $e^{i\omega_n\eta}$ is not necessary anymore. One easily finds

$$\begin{aligned}\frac{1}{\beta\mathcal{V}}\sum_k [\Sigma_{\text{RPA}}(k) - \Sigma_x(\mathbf{k})]G_0(k) &= \frac{1}{(\beta\mathcal{V})^2}\sum_{k,q} v(\mathbf{q})^2\chi_{nn}^{\text{RPA}}(q)G_0(k)G_0(k+q) \\ &= \frac{1}{2\beta\mathcal{V}}\sum_q v(\mathbf{q})[\chi_{nn}^{\text{RPA}}(q) - \chi_{nn}^0(q)]. \end{aligned}\tag{5.92}$$

The second term in the rhs of (5.90) gives

$$\begin{aligned}\frac{1}{\beta\mathcal{V}}\sum_k \Sigma_x(\mathbf{k})G_0(k)e^{i\omega_n\eta} &= -\frac{1}{(\beta\mathcal{V})^2}\sum_{k,q} v(\mathbf{q})G_0(k+q)e^{i(\omega_n+\omega_\nu)\eta}G_0(k)e^{i\omega_n\eta} \\ &= -\frac{1}{\mathcal{V}^2}\sum_{\mathbf{k},\mathbf{q}} v(\mathbf{q})n_F(\xi_{\mathbf{k}+\mathbf{q}})n_F(\xi_{\mathbf{k}}). \end{aligned}\tag{5.93}$$

This should be compared with[15]

$$\begin{aligned}\frac{1}{\beta\mathcal{V}}\sum_q v(\mathbf{q})\chi_{nn}^0(q) &= -\frac{2}{\beta\mathcal{V}^2}\sum_{q,\mathbf{k}} v(\mathbf{q})\frac{n_F(\xi_{\mathbf{k}}) - n_F(\xi_{\mathbf{k}+\mathbf{q}})}{i\omega_\nu + \xi_{\mathbf{k}} - \xi_{\mathbf{k}+\mathbf{q}}}e^{i\omega_\nu\eta} \\ &= \frac{2}{\mathcal{V}^2}\sum_{\mathbf{q},\mathbf{k}} v(\mathbf{q})[n_F(\xi_{\mathbf{k}}) - n_F(\xi_{\mathbf{k}+\mathbf{q}})]n_B(\xi_{\mathbf{k}+\mathbf{q}} - \xi_{\mathbf{k}}). \end{aligned}\tag{5.94}$$

Using $[n_F(\xi_{\mathbf{k}}) - n_F(\xi_{\mathbf{k}+\mathbf{q}})]n_B(\xi_{\mathbf{k}+\mathbf{q}} - \xi_{\mathbf{k}}) = n_F(\xi_{\mathbf{k}+\mathbf{q}})[1 - n_F(\xi_{\mathbf{k}})]$, this gives

$$\frac{1}{\beta\mathcal{V}}\sum_q v(\mathbf{q})\chi_{nn}^0(q) = -\frac{2}{\mathcal{V}^2}\sum_{\mathbf{k},\mathbf{q}} v(\mathbf{q})n_F(\xi_{\mathbf{k}})n_F(\xi_{\mathbf{k}+\mathbf{q}}) + \frac{1}{\mathcal{V}}\sum_{\mathbf{q}} v(\mathbf{q})n_0, \tag{5.95}$$

and therefore,

$$\frac{1}{\beta\mathcal{V}}\sum_k \Sigma_x(\mathbf{k})G_0(k)e^{i\omega_n\eta} = \frac{1}{2\beta\mathcal{V}}\sum_q v(\mathbf{q})\chi_{nn}^0(q) - \frac{1}{2\mathcal{V}}\sum_{\mathbf{q}} v(\mathbf{q})n_0, \tag{5.96}$$

[15] Since the Matsubara sum in (5.94) is convergent, we may add the convergence factor $e^{i\omega_\nu\eta}$.

where $n_0 \equiv n_0(\mu) = 2\sum_{\mathbf{k}} n_F(\xi_{\mathbf{k}})$ is the average density in the non-interacting limit (for a given chemical potential). From (5.92) and (5.96), we finally obtain

$$\frac{1}{\beta\mathcal{V}}\sum_k \Sigma_{\rm RPA}(k)G_0(k)e^{i\omega_n\eta} = \frac{1}{2}\int_{\mathbf{q}} v(\mathbf{q})\left[\frac{1}{\beta}\sum_{\omega_\nu}\chi_{nn}^{\rm RPA}(q) - n_0\right], \tag{5.97}$$

which is the expected result except for $n_0 = (2/\beta\mathcal{V})\sum_k G_0(k)e^{i\omega_n\eta}$ appearing in the rhs instead of the actual density $n = (2/\beta\mathcal{V})\sum_k G_{\rm RPA}(k)e^{i\omega_n\eta}$ (with $G_{\rm RPA} = (G_0 - \Sigma_{\rm RPA})^{-1}$).

5.4.2.1 *Fermi-liquid results*

In order to find out whether the RPA self-energy describes a Fermi liquid, one must calculate the imaginary part $\Im[\Sigma^R_{\rm RPA}(\mathbf{k},\omega)]$ at low energies and temperatures (Sec. 4.4). The first term in the rhs of (5.88), Σ_x, does not contribute to $\Im[\Sigma^R_{\rm RPA}(\mathbf{k},\omega)]$. The second term is similar to the second-order contribution to the self-energy studied in Sec. 4.4.1.6 and Appendix 4.A with the bare charge susceptibility χ^0_{nn} replaced by $\chi^{\rm RPA}_{nn}$. We thus deduce from (4.A.9)

$$\begin{aligned}\Im[\Sigma^R_{\rm RPA}(\mathbf{k},\omega)] =& -\frac{m}{4\pi^2 k_F}\int_0^{2k_F} v(\mathbf{q})^2|\mathbf{q}|d|\mathbf{q}| \\ &\times \int_{-\infty}^{\infty} d\omega' \chi^{\rm RPA\prime\prime}_{nn}(\mathbf{q},\omega')\left[n_B(\omega') + n_F(\omega'+\omega)\right].\end{aligned} \tag{5.98}$$

Due to the Bose and Fermi factors, the relevant part of the sum over ω' corresponds to $|\omega'| \leq \max(|\omega|, T)$. For $|\omega|, T \to 0$, we can therefore use the low-energy limit of $\chi^{\rm RPA\prime\prime}_{nn}(\mathbf{q},\omega')$ (with $|\mathbf{q}| \leq 2k_F$),

$$\begin{aligned}\chi^{\rm RPA\prime\prime}_{nn}(\mathbf{q},\omega) &= \frac{\chi^{0\prime\prime}_{nn}(\mathbf{q},\omega)}{\left(1 + v(\mathbf{q})\Re[\chi^{0R}_{nn}(\mathbf{q},\omega)]\right)^2 + \left(v(\mathbf{q})\chi^{0\prime\prime}_{nn}(\mathbf{q},\omega)\right)^2} \\ &\simeq \frac{\chi^{0\prime\prime}_{nn}(\mathbf{q},\omega)}{[1 + v(\mathbf{q})\chi^{0R}_{nn}(\mathbf{q},0)]^2} \\ &\simeq \frac{\pi N(0)\omega}{\left[\epsilon^{\rm RPA}_{\parallel}(\mathbf{q},0)\right]^2 v_F|\mathbf{q}|} \qquad (\omega \to 0).\end{aligned} \tag{5.99}$$

Since $\chi^{\rm RPA\prime\prime}_{nn}(\mathbf{q},\omega)$ is linear in ω for $\omega \to 0$, we can proceed as in Appendix 4.A to obtain

$$\Im[\Sigma^R_{\rm RPA}(\mathbf{k},\omega)] = -\frac{m^3}{16\pi^3 k_F}(\omega^2 + \pi^2T^2)\int_0^{2k_F} d|\mathbf{q}|\frac{v(\mathbf{q})^2}{\left[\epsilon^{\rm RPA}_{\parallel}(\mathbf{q},0)\right]^2}, \tag{5.100}$$

which is the standard Fermi-liquid result. Approximating $\epsilon^{\rm RPA}_{\parallel}(\mathbf{q},0)$ by the Thomas–Fermi result $\epsilon_{\rm TF}(\mathbf{q}) = 1 + q^2_{\rm TF}/\mathbf{q}^2$, we find

$$\int_0^{2k_F} d|\mathbf{q}|\frac{v(\mathbf{q})^2}{\left[\epsilon^{\rm RPA}_{\parallel}(\mathbf{q},0)\right]^2} = \frac{2\pi^4}{m^2k_F}\xi(r_s), \tag{5.101}$$

Table 5.2 Quasi-particle weight z_{k_F} [7] and effective mass m^* (reproduced from Ref. [1]) in the RPA.

r_s	0	1	2	3	4	5	6
$z_{k_F}^{\rm RPA}$	1	0.859	0.768	0.700	0.646	0.602	0.566
$(m^*/m)_{\rm RPA}$	1	0.97	0.99	1.02	1.04	1.06	1.08

where

$$\xi(r_s) = \int_0^1 dx \left[1 + \frac{\pi}{r_s}\left(\frac{9\pi}{4}\right)^{1/3} x^2\right]^{-2}. \tag{5.102}$$

We deduce that the quasi-particle lifetime $\tau_{\mathbf{k}}$ is given by

$$\frac{1}{\tau_{\mathbf{k}}} = -2z_{\mathbf{k}}\Im[\Sigma^R(\mathbf{k},\tilde{\xi}_{\mathbf{k}})] = z_{\mathbf{k}}\frac{\pi}{8}\xi(r_s)\frac{\tilde{\xi}_{\mathbf{k}}^2 + \pi^2T^2}{\epsilon_F}, \tag{5.103}$$

where $z_{\mathbf{k}}$ is the quasi-particle weight and $\tilde{\xi}_{\mathbf{k}}$ the quasi-particle dispersion defined by $\tilde{\xi}_{\mathbf{k}} = \xi_{\mathbf{k}} + \Re[\Sigma^R(\mathbf{k},\tilde{\xi}_{\mathbf{k}})]$ (Sec. 4.4). In the high-density limit $r_s \to 0$, $\xi(r_s) \sim \sqrt{r_s}$ and (5.103) gives

$$\frac{1}{\tau_{\mathbf{k}}} \sim z_{\mathbf{k}}\sqrt{r_s}\frac{\tilde{\xi}_{\mathbf{k}}^2}{\epsilon_F} \sim z_{\mathbf{k}}\omega_p\frac{\tilde{\xi}_{\mathbf{k}}^2}{\epsilon_F^2} \qquad (T \to 0). \tag{5.104}$$

The quasi-particle scattering rate is proportional to the plasma frequency but reduced by the factor $\tilde{\xi}_{\mathbf{k}}^2/\epsilon_F^2$. At metallic densities, $r_s \sim 1$, $\xi(r_s) \sim 1$ and $1/\tau_{\mathbf{k}} \sim z_{\mathbf{k}}\tilde{\xi}_{\mathbf{k}}^2/\epsilon_F$. In both cases, $r_s \to 0$ and $r_s \sim 1$, the quasi-particle scattering rate $1/\tau_{\mathbf{k}} \sim (|\mathbf{k}| - k_F)^2$ diverges near the Fermi surface as in a Fermi liquid.

The calculation of $z_{\mathbf{k}}$ — as well as the effective mass m^* — is quite cumbersome and we therefore only quote the results in Table 5.2. As expected, z_{k_F} decreases monotonically with r_s but remains finite, confirming the Fermi-liquid behavior of the electron gas up to rather large values of r_s. The effective mass is smaller than the bare mass when $r_s \lesssim 2$, but becomes larger at sufficiently low density.[16]

5.5 Beyond the RPA

The main advantage and appeal of the RPA is its simplicity. Not only does it account for important physical properties of a metal (plasmon mode, screening, etc.), but it also satisfies a number of exact results such as the sum rules discussed in Sec. 5.3.2 or the value of the plasmon energy in the long wavelength limit. However, the RPA is based on the one-particle Hartree (non-interacting) propagator and the correlations between a given electron and the other particles are not taken into account. The Pauli principle reduces the probability of finding two electrons with the same spin orientation nearby. This "exchange hole" is not

[16]This suggests that the Landau parameter F_1^s is negative in the high-density limit of the three-dimensional electron gas ($m^*/m = 1 + F_1^s/3$ in a Galilean-invariant system (Sec. 4.1.5)). This agrees with predictions of local field factor approaches (Sec. 5.5) [1].

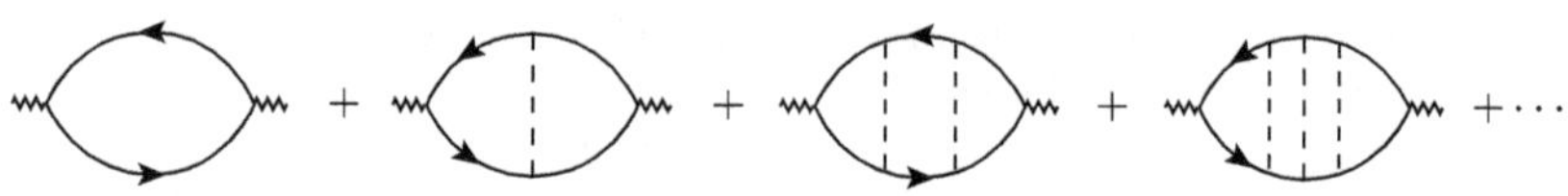

Fig. 5.13 Π_{nn} in the time-dependent Hartree–Fock approximation. The one-particle propagator includes the exchange self-energy Σ_x.

accounted for in the Hartree propagator. The probability of finding two electrons nearby, irrespective of their spin orientations, is further altered by the Coulomb interaction. This "correlation hole" is also neglected.

We have shown in Sec. 5.4.2 how the RPA can nevertheless be used to build a non-trivial approximation for the self-energy ($\Sigma_{\rm RPA}$). It would be tempting to include the RPA self-energy into the irreducible density–density response function Π_{nn}. This would however violate conservation laws, i.e. make the approximation non-conserving.[17] The simplest conserving approximation beyond the RPA includes both the Hartree and Fock self-energies in the one-particle propagator; it is sometimes referred to as the time-dependent Hartree–Fock approximation (the RPA corresponding to the time-dependent Hartree approximation). In order to satisfy conservation laws and other exact results, one must then include all ladder diagrams in the calculation of Π_{nn} (Fig. 5.13). The justification of this statement is given in Chapter 9 where a general method — based on the Luttinger–Ward functional — to generate diagrammatic approximations that satisfy conservation laws is described. The time-dependent Hartree–Fock approximation requires a numerical solution. It does not yield a significant improvement over the RPA and, as discussed in Sec. 5.4.1, the Fock self-energy is itself pathological. Thus, the time-dependent Hartree–Fock approximation has its own limitations and difficulties. A detailed discussion of the results is beyond the scope of this chapter; we refer the interested reader to Ref. [1].

It is possible to treat exchange and correlation effects in a non-diagrammatic approximation that preserves the mathematical structure of the RPA.[18] In the latter, the electrons are subjected to an effective potential

$$\phi_{\rm ext}(\mathbf{q},\omega) + v(\mathbf{q})\langle\hat{n}(\mathbf{q})\rangle(\omega) \tag{5.105}$$

(Sec. 5.2.2). In order to account for the correlations neglected in the RPA, one assumes that the effective potential seen by the spin-σ electrons can be written as

$$\phi_{{\rm eff},\sigma}(\mathbf{q},\omega) = \phi_{\rm ext}(\mathbf{q},\omega) + v(\mathbf{q})\sum_{\sigma'}\langle\hat{n}_{\sigma'}(\mathbf{q})\rangle(\omega) - \sum_{\sigma'} v(\mathbf{q})G_{\sigma\sigma'}(\mathbf{q},\omega)\langle\hat{n}_{\sigma'}(\mathbf{q})\rangle(\omega). \tag{5.106}$$

The local field factor $G_{\sigma\sigma'}(\mathbf{q},\omega)$ contains the correction to the effective potential stemming from the exchange and correlation effects. The density–density response function then reads

$$\chi_{nn}(q) = \frac{\chi^0_{nn}(q)}{1 + v(\mathbf{q})[1 - G_+(q)]\chi^0_{nn}(q)}, \tag{5.107}$$

[17] See also the discussion at the end of Sec. 9.1.3.1.

[18] The idea to improve the RPA while preserving its simple mathematical structure is also at the heart of the two-particle self-consistent (TPSC) theory of the Hubbard model (Sec. 6.4.5).

where $G_+(q) = \frac{1}{2}[G_{\uparrow\uparrow}(q)+G_{\uparrow\downarrow}(q)]$. Thus, $G_{\sigma\sigma'}$ can be seen as an effective vertex correction. The local field factor approach has proven to be a successful method to calculate many quantities of interest (effective mass, compressibility, spin susceptibility, Landau parameters, etc.) while treating exchange and correlation effects beyond the RPA. Again, we refer to Ref. [1] for a detailed discussion.

5.6 Wigner Crystallization

The RPA is valid only in the high-density limit where the kinetic energy dominates. In the opposite limit, the interaction energy prevails. When $r_s \to 0$, the system behaves classically and one expects the charged particles to form a crystalline structure, the so-called Wigner crystal. In three dimensions, the most stable structure is body-centered-cubic (bcc). Quantum-mechanical corrections — zero-point motion of the electrons about their classical equilibrium position — to the classical electrostatic energy modify the ground-state energy but do not destroy the crystalline order. The lowest-lying excitations of the Wigner crystal are phonons and collective spin excitations corresponding to fluctuations in the relative spin orientations on different lattice sites.

Thus, as the density is varied, one expects a phase transition between a homogeneous fluid and a crystal. Other phases are actually possible, in particular a spin-polarized fluid. Considering only these three possible ground states, one finds a paramagnetic fluid for $r_s \lesssim 75$, a ferromagnetic fluid for $75 \lesssim r_s \lesssim 100$ and a Wigner crystal for $r_s \gtrsim 100$ [1].

In two dimensions, the ground state is a metallic Fermi liquid at high densities and a Wigner crystal with a hexagonal lattice structure at low densities. The determination of the ground state in the intermediate range $20 \lesssim r_s \lesssim 34$ is very hard and whether or not the fluid is polarized remains an open issue [1].

In one dimension, the concept of Fermi liquid does not apply. A fermion system with short-range repulsive interactions is a Luttinger liquid (Chapters 4 and 15). The ground state of an electron system with a long-range Coulomb interaction potential $v(x) \sim e^2/|x|$ (i.e. $v(q) \sim -e^2 \ln |q|$) is a Wigner crystal [8] (see Chapter 15).

Guide to the bibliography

The electron gas in the RPA is discussed in Refs. [1–6]. For a discussion which goes beyond the RPA, see Refs. [1, 6]. Wigner crystallization is described in Refs. [1, 6].

Bibliography

[1] G. F. Giuliani and G. Vignale, *Quantum Theory of the Electron Liquid* (Cambridge University Press, Cambridge, 2005).

[2] G. D. Mahan, *Many-Particle Physics*, 2nd edition, Chapter 5 (Plenum Press, New York, 1990).

[3] A. L. Fetter and J. D. Walecka, *Quantum Theory of Many-Particle Systems* (Dover, 2003).

[4] D. Pines and P. Nozières, *The Theory of Quantum Liquids* (Addison-Wesley, Reading, Massachusetts, 1968).

[5] A.-M. Tremblay, *Quantum Material's Theory, From Perturbation Theory to Dynamical-Mean Field Theory*, Chapter 5, https://www.physique.usherbrooke.ca/pages/tremblay.

[6] R. M. Martin, L. Reining, and D. M. Ceperley, *Interacting Electrons: Theory and Computational Approaches* (Cambridge University Press, Cambridge, 2016).

[7] L. Hedin, New method for calculating the one-particle Green's function with application to the electron-gas problem, *Phys. Rev.* **139**, A796 (1965).

[8] H. J. Schulz, Wigner crystal in one dimension, *Phys. Rev. Lett.* **71**, 1864 (1993).

Chapter 6

Magnetism in Lattice Fermion Systems

Contents

6.1 The Hubbard Model 400
- 6.1.1 Symmetries 400
- 6.1.2 The non-interacting limit $U = 0$ 401
- 6.1.3 The local limit $t = 0$ 403
- 6.1.4 Ground states of the Hubbard model 405

6.2 Charge and Magnetic Instabilities 405
- 6.2.1 Non-interacting susceptibilities 406
- 6.2.2 Susceptibilities in the RPA 407

6.3 Stoner's Theory of Ferromagnetism 409
- 6.3.1 Hartree–Fock theory 409
- 6.3.2 Charge and spin collective modes 412
- 6.3.3 Functional integral formalism 415

6.4 Slater's Theory of Antiferromagnetism 421
- 6.4.1 Hartree–Fock theory 422
- 6.4.2 Transverse spin-wave modes 424
- 6.4.3 Beyond spin waves: The quantum nonlinear σ model 426
- 6.4.4 Large-U limit of the half-filled Hubbard model 434
- 6.4.5 Two-particle self-consistent theory 436

Appendix 6.A Hartree–Fock Correlation Functions 447
- 6.A.1 Ferromagnet 447
- 6.A.2 Antiferromagnet 448

Appendix 6.B SU(2) Gauge Fields 450

Bibliography 453

In this chapter, we discuss the magnetic instabilities of a spin-$\frac{1}{2}$ fermion system in the framework of the Hubbard model. The latter describes fermions moving on a lattice and subject to a local (onsite) repulsion (Sec. 6.1). Our aim is not to present an exhaustive description of the Hubbard model but rather to consider limiting cases within simple (and therefore questionable) theoretical approaches. In Sec. 6.2, we discuss the magnetic (and charge) instabilities in the RPA starting from the non-interacting limit. Section 6.3 is devoted to the Stoner theory of ferromagnetism. The ground state is obtained from the Hartree–Fock theory and the collective modes, in particular the transverse spin-wave modes, from the RPA. We use both diagrammatic and functional integral approaches. The spin-density-wave (antiferromagnetic) instability in the two-dimensional half-filled Hubbard model is discussed in Sec. 6.4 in the framework of the Slater theory. We show that the low-energy effective action describing magnetic fluctuations beyond the Slater theory is a nonlinear σ model (NLσM) (Sec. 6.4.3). In Sec. 6.4.5, we finally discuss the two-particle self-consistent (TPSC) theory, an improvement of the RPA valid at intermediate coupling and not too low temperatures. The equivalence between the Hubbard model in the strong-coupling limit and the Heisenberg model is derived in Sec. 6.4.4 and the attractive Hubbard model is briefly studied in Sec. 6.4.3.[1]

6.1 The Hubbard Model

The Hubbard model is defined by the Hamiltonian

$$\hat{H} = -t \sum_{\langle \mathbf{r},\mathbf{r}' \rangle,\sigma} \left(\hat{\psi}^\dagger_{\mathbf{r}\sigma} \hat{\psi}_{\mathbf{r}'\sigma} + \text{h.c.} \right) - \mu \sum_{\mathbf{r},\sigma} \hat{\psi}^\dagger_{\mathbf{r}\sigma} \hat{\psi}_{\mathbf{r}\sigma} + U \sum_{\mathbf{r}} \hat{\psi}^\dagger_{\mathbf{r}\uparrow} \hat{\psi}^\dagger_{\mathbf{r}\downarrow} \hat{\psi}_{\mathbf{r}\downarrow} \hat{\psi}_{\mathbf{r}\uparrow}, \tag{6.1}$$

where $\{\mathbf{r}\}$ denotes the N sites of a lattice and $\langle \mathbf{r}, \mathbf{r}' \rangle$ nearest-neighbor sites. $\psi^{(\dagger)}_{\mathbf{r}\sigma}$ is an annihilation (creation) operator of a fermion at site $\mathbf{r}$ with spin $\sigma = \uparrow, \downarrow$. t is the hopping amplitude, U the onsite interaction and μ the chemical potential. The interaction being local (onsite), only opposite-spin fermions can interact. In the following, we mainly discuss the repulsive Hubbard model ($U > 0$). We consider a d-dimensional hypercubic lattice and take the lattice spacing as the unit length.

In this section, we first consider the symmetries of the Hubbard model and two limiting cases — the non-interacting limit $U = 0$ and the local limit $t = 0$ — before discussing briefly the possible ground states.

6.1.1 *Symmetries*

In addition to invariance under a translation by a lattice vector or a spin rotation, the Hubbard model exhibits a number of symmetries. Here, we discuss two of them that will be useful for our discussion.

Under the particle–hole transformation

$$\hat{\psi}_{\mathbf{r}\sigma} \to (-1)^{\mathbf{r}} \hat{\psi}^\dagger_{\mathbf{r}\sigma}, \qquad \hat{\psi}^\dagger_{\mathbf{r}\sigma} \to (-1)^{\mathbf{r}} \hat{\psi}_{\mathbf{r}\sigma}, \tag{6.2}$$

[1] The Hubbard model, and in particular the Mott transition in the framework of the dynamical mean-field theory (DMFT), is further discussed in Chapter 9.

the (grand canonical) Hamiltonian transforms as

$$\hat{H}(t,U,\mu) \to \hat{H}(t,U,U-\mu) - (U-2\mu)N. \tag{6.3}$$

This implies that the particle density (i.e. the mean number of particles per site) satisfies

$$\begin{aligned} n(\mu,T) &= \frac{1}{\beta N}\frac{\partial}{\partial\mu}\ln\left\{\mathrm{Tr}\, e^{-\beta\hat{H}(t,U,\mu)}\right\} \\ &= \frac{1}{\beta N}\frac{\partial}{\partial\mu}\ln\left\{\mathrm{Tr}\, e^{-\beta[\hat{H}(t,U,U-\mu)-(U-2\mu)N]}\right\} = -n(U-\mu,T)+2. \end{aligned} \tag{6.4}$$

We deduce that at half-filling, where $n(\mu,T)=1$, $\mu=U/2$ for any temperature.

6.1.1.1 *Attractive vs repulsive Hubbard models*

If we perform the particle–hole transformation (6.2) only for the spin-down operator, i.e.

$$\hat{\psi}_{\mathbf{r}\downarrow} \to (-1)^{\mathbf{r}}\hat{\psi}^\dagger_{\mathbf{r}\downarrow}, \qquad \hat{\psi}^\dagger_{\mathbf{r}\downarrow} \to (-1)^{\mathbf{r}}\hat{\psi}_{\mathbf{r}\downarrow}, \tag{6.5}$$

the Hamiltonian transforms as

$$\hat{H}(t,U,\mu) \to \hat{H}(t,-U,-U/2) - (\mu-U/2)\sum_{\mathbf{r}}(\hat{\psi}^\dagger_{\mathbf{r}\uparrow}\hat{\psi}_{\mathbf{r}\uparrow} - \hat{\psi}^\dagger_{\mathbf{r}\downarrow}\hat{\psi}_{\mathbf{r}\downarrow}) - \mu N. \tag{6.6}$$

If we start from the attractive Hubbard model ($U<0$), the transformed Hamiltonian corresponds to the half-filled repulsive model with chemical potential $\mu=|U|/2$ and interaction $|U|$ in the presence of a magnetic field $\mu+|U|/2$ along the z axis that couples to the fermion spins. At half-filling, where $\mu=U/2$, the magnetic field vanishes and the attractive model maps onto the repulsive one. This mapping will be used in Sec. 6.4.3.

6.1.2 The non-interacting limit $U=0$

In the limit $U=0$, the (grand canonical) Hamiltonian reduces to the kinetic energy term

$$\hat{H}_0 = \sum_{\mathbf{k},\sigma}\xi_{\mathbf{k}}\hat{\psi}^\dagger_\sigma(\mathbf{k})\hat{\psi}_\sigma(\mathbf{k}) \tag{6.7}$$

where $\hat{\psi}^{(\dagger)}_\sigma(\mathbf{k})$ is the Fourier transform of $\hat{\psi}^{(\dagger)}_{\mathbf{r}\sigma}$ and

$$\xi_{\mathbf{k}} = \epsilon_{\mathbf{k}} - \mu = -2t\sum_{\nu=1}^{d}\cos k_\nu - \mu \tag{6.8}$$

is the dispersion of the free fermions moving on the lattice. The density of states reads[2,3]

$$N_d(\xi) = \int_{\mathbf{k}}\delta(\xi+\mu-\epsilon_{\mathbf{k}}), \tag{6.9}$$

---

[2]To be consistent with the notation of previous chapters (in particular Chapter 4), we consider $N(\xi)$ as a function of ξ rather than $\epsilon=\xi+\mu$.

[3]In the thermodynamic limit $N\to\infty$, $\frac{1}{N}\sum_{\mathbf{k}} \to \int\frac{d^dk}{(2\pi)^d} \equiv \int_{\mathbf{k}}$ where the integration is restricted to the first Brillouin zone $]-\pi,\pi]^d$.

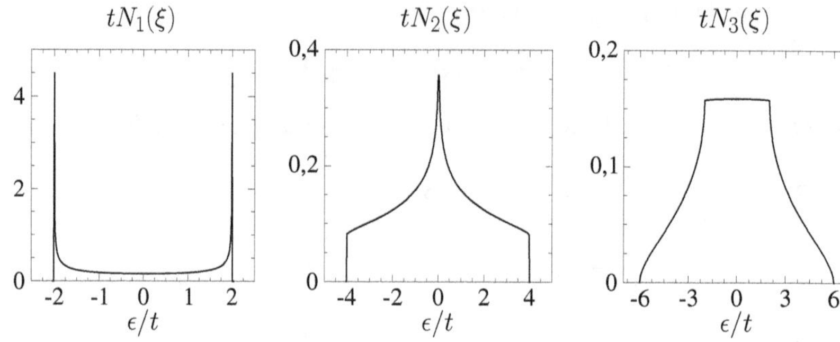

Fig. 6.1 d-dimensional density of states $N_d(\xi)$ *vs* $\epsilon = \xi + \mu$.

with $\epsilon_{\mathbf{k}} \in [-2dt, 2dt]$. It strongly depends on the space dimension:

$$\begin{aligned} N_1(\xi) &= \frac{\Theta(2t-|\epsilon|)}{2\pi t}\left(1-\frac{\epsilon^2}{4t^2}\right)^{-1/2}, \\ N_2(\xi) &= \frac{\Theta(4t-|\epsilon|)}{2\pi^2 t} K\left[\left(1-\frac{\epsilon^2}{16t^2}\right)^{1/2}\right], \\ N_3(\xi) &= \int_0^\pi \frac{du}{\pi} N_2(\xi + 2t\cos u), \end{aligned} \tag{6.10}$$

where K is the complete elliptic integral of the first kind.[4] The two-dimensional density of states exhibits a logarithmic divergence (Van Hove singularity) at $\epsilon = 0$ ($\xi = -\mu$).[5] In three dimensions, the Van Hove singularities are located at $\epsilon = \pm 2t$ (Fig. 6.1).

When $|\mathbf{k}| \ll \pi$, we can approximate $\epsilon_{\mathbf{k}}$ by a quadratic form,

$$\xi_{\mathbf{k}} \simeq -2dt + \frac{\mathbf{k}^2}{2m^*} - \mu \qquad (|\mathbf{k}| \ll \pi) \tag{6.11}$$

with an effective mass $m^* = 1/2t$, which gives a density of states $N_d(\xi) \sim (\epsilon + 2dt)^{d/2-1}$ for $0 \leq \epsilon + 2dt \ll 4dt$ as in a free fermion system in the absence of a lattice. When the mean density n is much smaller than unity, the chemical potential μ lies near the bottom $-2dt$ of the dispersion. In that case, the low-energy physics can be understood in terms of fermions with the isotropic dispersion (6.11). This is no longer true at higher densities. When n is of order one, the lattice plays an essential role and the Fermi surface differs significantly from a $(d-1)$-dimensional hypersphere. For example, in the two-dimensional half-filled model ($n = 1$) the Fermi surface $\epsilon_{\mathbf{k}} = \mu = 0$ is a square with vertices located at $(\pm\pi, 0)$ and $(0, \pm\pi)$ and the density of states at the Fermi level diverges due to the Van Hove singularity (Fig. 6.2).

[4] $K(k) = \int_0^{\pi/2} \frac{d\theta}{\sqrt{1-k^2\sin^2\theta}}$.

[5] $N_2(\xi) \simeq \frac{1}{2\pi^2 t}\ln(16t/|\epsilon|)$ for $|\epsilon| \ll t$. The logarithmic divergence at $\epsilon = 0$ is due to the fermion states $\mathbf{k}$ located near $(\pm\pi, 0)$ and $(0, \pm\pi)$.

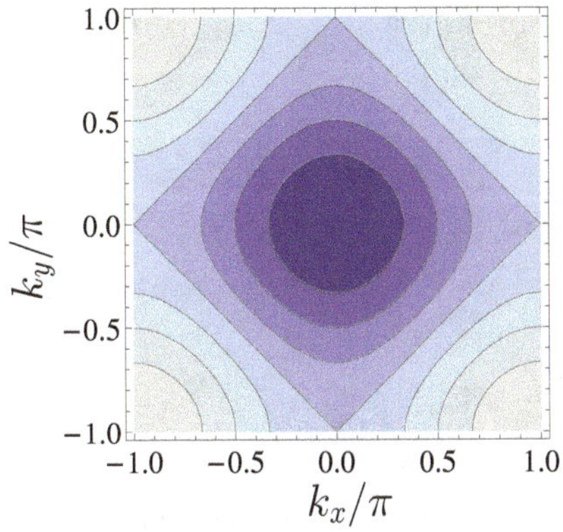

Fig. 6.2 Isoenergy curves $\epsilon_{\mathbf{k}} = \mu$ in two dimensions. At low density (μ near $-4t$), the Fermi surface can be approximated by a circle of radius $[2m^*(\mu + 4t)]^{1/2}$ [Eq. (6.11)]. At half-filling ($\mu = 0$), the Fermi surface is a square with vertices located at $(\pm\pi, 0)$ and $(0, \pm\pi)$.

In the non-interacting limit, the one-particle Green function

$$G_0(\mathbf{k}, i\omega_n) = (i\omega_n - \xi_{\mathbf{k}})^{-1} \tag{6.12}$$

exhibits a pole and the system is trivially a Fermi liquid for any non-zero density except for $n = 2$ where it becomes a band insulator. In the presence of a repulsive interaction $U > 0$, the Fermi liquid may be unstable. At low densities, we expect the standard instabilities of the Fermi liquid, e.g. the ferromagnetic instability (Chapter 4). For $n \simeq 1$, we shall see that the Fermi liquid is unstable against a spin-density-wave phase where the spin density $\langle \hat{\mathbf{S}}_{\mathbf{r}} \rangle$ oscillates in space with wavevector $\mathbf{Q} = (\pi, \dots, \pi)$.[6]

6.1.3 *The local limit $t = 0$*

For vanishing hopping amplitude $t = 0$, the Hamiltonian reduces to

$$\hat{H}_{\text{int}} = \sum_{\mathbf{r}} \big[U \hat{n}_{\mathbf{r}\uparrow} \hat{n}_{\mathbf{r}\downarrow} - \mu(\hat{n}_{\mathbf{r}\uparrow} + \hat{n}_{\mathbf{r}\downarrow}) \big], \tag{6.13}$$

where $\hat{n}_{\mathbf{r}\sigma} = \hat{\psi}^\dagger_{\mathbf{r}\sigma} \hat{\psi}_{\mathbf{r}\sigma}$ is the number operator of spin-σ fermions at site $\mathbf{r}$. Because of the Pauli principle, there are only four states in the Hilbert space of the single-site problem: the empty state $|\text{vac}\rangle$ (energy $\epsilon_0 = 0$), the singly occupied states $|\sigma\rangle = \hat{\psi}^\dagger_\sigma |\text{vac}\rangle$ with spin σ (energy $\epsilon_\sigma = -\mu$) and the doubly occupied site $|\uparrow\downarrow\rangle = \hat{\psi}^\dagger_\uparrow \hat{\psi}^\dagger_\downarrow |\text{vac}\rangle$ (energy $\epsilon_{\uparrow\downarrow} = U - 2\mu$). We deduce the partition function of the N-site problem

$$Z = Z_{\text{loc}}^N, \qquad Z_{\text{loc}} = 1 + 2e^{\beta\mu} + e^{-\beta(U - 2\mu)}. \tag{6.14}$$

[6] The spin-density operator $\hat{\mathbf{S}}_{\mathbf{r}}$ is defined in (6.20).

The one-particle propagator $G_\sigma(\tau) = -\langle T_\tau \hat{\psi}_{\mathbf{r}\sigma}(\tau)\hat{\psi}^\dagger_{\mathbf{r}\sigma}(0)\rangle$, which is obviously local in space, can be straightforwardly computed. For $\tau > 0$,

$$\begin{aligned} G_\sigma(\tau) &= -\frac{1}{Z_{\rm loc}} \sum_{m,m'} \langle m|e^{-(\beta-\tau)\hat{H}_{\rm loc}}\hat{\psi}_{\mathbf{r}\sigma}|m'\rangle\langle m'|e^{-\tau\hat{H}_{\rm loc}}\hat{\psi}^\dagger_{\mathbf{r}\sigma}|m\rangle \\ &= -\frac{1}{Z_{\rm loc}} \sum_{m,m'} e^{-(\beta-\tau)\epsilon_m - \tau\epsilon_{m'}} |\langle m'|\hat{\psi}^\dagger_{\mathbf{r}\sigma}|m\rangle|^2, \\ &= -\frac{1}{Z_{\rm loc}} \left[e^{-\beta\epsilon_0+\tau(\epsilon_0-\epsilon_\sigma)} + e^{-\beta\epsilon_{\bar\sigma}+\tau(\epsilon_{\bar\sigma}-\epsilon_{\uparrow\downarrow})}\right], \end{aligned} \tag{6.15}$$

where $\{|m\rangle\}$ denotes the four-state basis of the single-site problem and $\hat{H}_{\rm loc}$ the corresponding Hamiltonian. This yields[7]

$$G_\sigma(i\omega_n) = \frac{1-n_{\bar\sigma}}{i\omega_n + \mu} + \frac{n_{\bar\sigma}}{i\omega_n + \mu - U}, \tag{6.16}$$

where

$$n_\sigma = \frac{1}{Z_{\rm loc}}(e^{-\beta\epsilon_\sigma} + \epsilon^{-\beta\epsilon_{\uparrow\downarrow}}) = \frac{1}{Z_{\rm loc}}(e^{\beta\mu} + \epsilon^{-\beta(U-2\mu)}) \tag{6.17}$$

is the mean number of spin-σ fermions per site.

The two-pole structure of the local propagator has a simple physical interpretation. The pole at $\omega = U - \mu$ in the retarded propagator $G^R_\sigma(\omega) = G_\sigma(i\omega_n \to \omega + i0^+)$ corresponds to the excitation energy of the transitions $|\bar\sigma\rangle \to |\uparrow\downarrow\rangle$ and $|\uparrow\downarrow\rangle \to |\bar\sigma\rangle$;[8] the residue $n_{\bar\sigma}$ gives the probability than the site is either occupied with a spin-$\bar\sigma$ fermion or doubly occupied [Eq. (6.17)]. The pole at $\omega = -\mu$ corresponds to the transitions $|\text{vac}\rangle \to |\sigma\rangle$ and $|\sigma\rangle \to |\text{vac}\rangle$, with the residue $1 - n_{\bar\sigma}$ giving the probability that the site is either empty or occupied with a spin-σ fermion.

The structure of $G_\sigma(i\omega_n)$ is thus very different from a quasi-particle pole: There is no pole at $\omega = 0$ unless $\mu = 0$ or $\mu = U$.[9] For example, at half-filling where $n_\sigma = 1/2$ and $\mu = U/2$, Eq. (6.16) can be written in the form $G_\sigma(i\omega_n) = [i\omega_n + \mu - \Sigma(i\omega_n)]^{-1}$ with a self-energy

$$\Sigma(i\omega_n) = \frac{U}{2} + \frac{U^2}{4i\omega_n} \tag{6.18}$$

which is singular at low frequency for $U \neq 0$. The spectral function $A(\omega) = -\frac{1}{\pi}\Im[G^R_\sigma(\omega)]$ exhibits two delta peaks at $\omega = U/2$ and $\omega = -U/2$ separated by an energy gap equal to U. When t is non-zero but small ($t/U \ll 1$), we expect the two peaks to broaden but without suppressing the gap at zero frequency. This state, where the gap (and the insulating

[7] $G_\uparrow(i\omega_n) = G_\downarrow(i\omega_n)$, since $n_\uparrow = n_\downarrow$, in agreement with spin-rotation invariance.

[8] Recall that the pole of the retarded propagator gives the (opposite of the) energy excitation for a particle-like (hole-like) excitation.

[9] For $\mu = 0$, the states $|\text{vac}\rangle$ and $|\sigma\rangle$ are degenerate; for $\mu = U$, degeneracy occurs for $|\sigma\rangle$ and $|\uparrow\downarrow\rangle$.

behavior) is due to the strong onsite repulsion, is known as a Mott (or Mott–Hubbard) insulator.[1]

In the case of a hypercubic lattice, the Mott-insulating ground state is always accompanied by antiferromagnetism when $d \geq 2$. In the Mott–Hubbard insulator, each fermion is localized on a site and behaves as a (localized) spin-$\frac{1}{2}$. Second-order perturbation theory around the half-filled ground state yields an effective antiferromagnetic coupling $J = 4t^2/U$ between neighboring sites and the system is well described by the Heisenberg model (Sec. 6.4.4). The local magnetic moments can therefore order at low temperatures, giving rise to a Mott–Heisenberg antiferromagnetic insulator below the Néel temperature $T_{\rm N}$ (Chapter 8).[10]

6.1.4 *Ground states of the Hubbard model*

Although the (repulsive) Hubbard model looks simple, its solution is not known for arbitrary values of U/t and μ. From the discussion of Secs. 6.1.2 and 6.1.3, it is clear that the model has various ground states depending on the strength of the interaction and the density of fermions: metal, Mott insulator, broken-symmetry states, etc. The aim of the following sections is to study the magnetic instabilities in the repulsive Hubbard model, where the SU(2) spin-rotation symmetry is spontaneously broken, using mainly the Hartree–Fock theory and the RPA. The magnetic states discussed in this chapter are not the only possible broken-symmetry states in the repulsive Hubbard model. For example, the ground state can be superconducting (see Chapter 15).

6.2 Charge and Magnetic Instabilities

In order to determine the possible magnetic and charge instabilities, we must compute the corresponding susceptibilities,

$$\begin{aligned} \chi_{nn}(\mathbf{r},\tau;\mathbf{r}',\tau') &= \langle T_\tau \hat{n}(\mathbf{r},\tau)\hat{n}(\mathbf{r}',\tau')\rangle - \langle \hat{n}(\mathbf{r},\tau)\rangle\langle \hat{n}(\mathbf{r}',\tau')\rangle, \\ \chi^{\nu\nu'}(\mathbf{r},\tau;\mathbf{r}',\tau') &= \langle T_\tau \hat{S}^\nu(\mathbf{r},\tau)\hat{S}^{\nu'}(\mathbf{r}',\tau')\rangle, \end{aligned} \tag{6.19}$$

where

$$\hat{n}_{\mathbf{r}} = \sum_\sigma \hat{\psi}^\dagger_{\mathbf{r}\sigma}\hat{\psi}_{\mathbf{r}\sigma}, \qquad \hat{S}^\nu_{\mathbf{r}} = \sum_{\sigma,\sigma'} \hat{\psi}^\dagger_{\mathbf{r}\sigma}\sigma^\nu_{\sigma\sigma'}\hat{\psi}_{\mathbf{r}\sigma'} \tag{6.20}$$

are the particle-density and spin-density operators. It is often convenient to introduce the "transverse" spin susceptibilities

$$\begin{aligned} \chi^{+-} &= \langle T_\tau S^+ S^-\rangle = \frac{1}{4}(\chi^{xx} - i\chi^{xy} + i\chi^{yx} + \chi^{yy}), \\ \chi^{-+} &= \langle T_\tau S^- S^+\rangle = \frac{1}{4}(\chi^{xx} + i\chi^{xy} - i\chi^{yx} + \chi^{yy}), \end{aligned} \tag{6.21}$$

[10] In other types of lattices (e.g. the triangular lattice), the antiferromagnetic instability may be suppressed and the ground state is then a paramagnetic Mott insulator.

where $S^{\pm} = \frac{1}{2}(S^x \pm iS^y)$. When spin-rotation invariance is satisfied, $\chi^{xy} = \chi^{yx} = 0$ and

$$\chi^{+-} = \chi^{-+} = \frac{1}{2}\chi^{\nu\nu}. \tag{6.22}$$

6.2.1 *Non-interacting susceptibilities*

In the absence of interaction, the susceptibilities can be computed using Wick's theorem. An elementary calculation gives $\chi^{\nu\nu'}(q) = \delta_{\nu,\nu'}\chi_0(q)$ and $\chi_{nn}(q) = \chi_0(q)$, where

$$\chi_0(q) = -\frac{2}{\beta N}\sum_k G_0(k)G_0(k+q) = -2\int_{\mathbf{k}} \frac{n_F(\xi_{\mathbf{k}}) - n_F(\xi_{\mathbf{k+q}})}{i\omega_\nu + \xi_{\mathbf{k}} - \xi_{\mathbf{k+q}}} \tag{6.23}$$

with $G_0(k)$ the non-interacting propagator (6.12).

Figure 6.3 shows $\chi_0(\mathbf{q})$ in the two-dimensional Hubbard model at low temperatures and for two values of the chemical potential: $\mu = -3.75t$ (i.e. near the bottom of the non-interacting dispersion $\epsilon_{\mathbf{k}}$) and $\mu = 0$ (corresponding to the half-filled non-interacting system).[11] In the low-density limit $n \ll 1$, the susceptibility $\chi_0(\mathbf{q})$ is maximum for $\mathbf{q} = 0$. At half-filling, $n = 1$, the maximum is reached for $\mathbf{q} = \mathbf{Q} = (\pi, \dots, \pi)$ (Fig. 6.3).[12] This result follows from the "nesting" property of the Fermi surface of the half-filled system. If we translate the Fermi surface (defined by $\epsilon_{\mathbf{k}} = \mu = 0$) by $\mathbf{Q}$, it reproduces itself since

$$\epsilon_{\mathbf{k+Q}} = -\epsilon_{\mathbf{k}}. \tag{6.24}$$

This property leads to

$$\chi_0(\mathbf{Q}) = \int_{-2dt}^{2dt} \frac{N(\epsilon)}{\epsilon} \tanh\left(\frac{\beta\epsilon}{2}\right), \tag{6.25}$$

where $N(\epsilon) \equiv N_d(\epsilon)$ is the d-dimensional density of states. When $N(\epsilon)$ has no particular structure for small ϵ, $\chi_0(\mathbf{Q}) \simeq 2N(0)\ln(E_0/T)$ diverges logarithmically at low temperatures

Fig. 6.3 Non-interacting susceptibility $\chi_0(\mathbf{q})$ of the two-dimensional Hubbard model for $T = 0.05t$. Left: $\mu = -3.75t$ (low-density limit). Right: $\mu = 0$ (half-filling).

---

[11]We follow the notations of Chapter 3: $\chi(\mathbf{q}) \equiv \chi^R(\mathbf{q}, \omega = 0) = \chi(\mathbf{q}, i\omega_\nu = 0)$.

[12]Near half-filling and at sufficiently low temperatures, the maximum of $\chi_0(\mathbf{q})$ occurs for an incommensurate vector $\mathbf{q}$ close to (but different from) $\mathbf{Q}$. This suggests that the ground state of the nearly half-filled Hubbard model could be an incommensurate spin-density wave; see footnote 14 on page 408.

($E_0 \sim 2dt$ is the order of the half bandwidth). In the two-dimensional Hubbard model, the Van Hove singularity of the density of states leads to $\chi_0(\mathbf{Q}) \sim E_0^{-1} \ln^2(E_0/T)$ while $\chi_0(\mathbf{q}=0) \sim E_0^{-1} \ln(E_0/T)$. Thus, $\chi_0(\mathbf{Q})$ is larger than $\chi_0(\mathbf{q}=0)$ and antiferromagnetic fluctuations dominate over ferromagnetic ones at half-filling.

6.2.2 *Susceptibilities in the RPA*

In Sec. 5.2, we have seen that the RPA for the density–density correlation function χ_{nn} amounts to summing bubble diagrams (Fig. 5.2). A similar approximation can be made for the Hubbard model. The diagrams contributing to χ_{nn} and $\chi^{\nu\nu}$ are shown in Fig. 6.4. One can define the RPA as the sum of all ladder and bubble diagrams. Because of the local interaction (and the Pauli principle), only a subclass of these diagrams contribute to χ_{nn}, χ^{zz} or $\chi^{\pm}$. The physical content of the RPA is that interactions between collective spin/charge fluctuations are neglected. This interpretation is clear in the functional integral formalism where the RPA is formulated as a Gaussian approximation for the bosonic fields describing charge and spin fluctuations (Sec. 5.2.3; see also Sec. 6.3.3). While the RPA is *a priori* a weak-coupling approximation, in some cases it also captures the physics at strong coupling (see, e.g., Sec. 6.4.2).

Only bubble diagrams contribute to χ_{nn} (Fig. 6.4(a)) and one finds

$$\chi_{nn}(q) = \chi_0(q) - \frac{U}{2}\chi_0(q)^2 + \frac{U^2}{4}\chi_0(q)^3 + \cdots = \frac{\chi_0(q)}{1+\frac{U}{2}\chi_0(q)}. \tag{6.26}$$

The diagrams contributing to the spin–spin correlation function χ^{zz} are the same as those contributing to χ_{nn} (Fig. 6.4(a)) but for each diagram the two external bubbles must be multiplied by the spin index of the propagators. Thus, the diagrams with an even number of bubbles are multiplied by $\sigma\bar{\sigma} = -1$ while those with an odd number are multiplied by

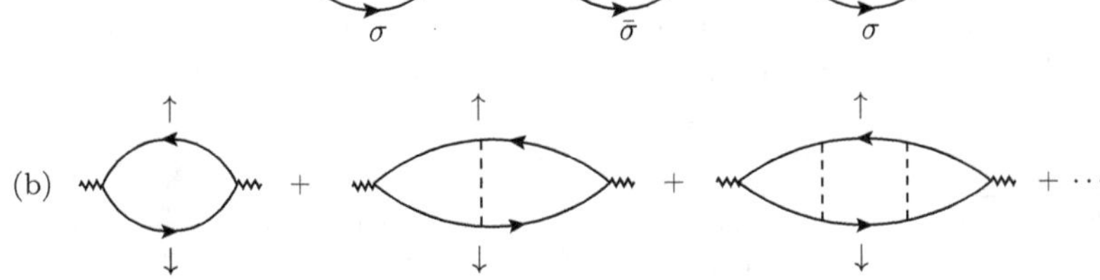

Fig. 6.4 Diagrammatic representation of the density–density (a) and transverse spin–spin (b) correlation functions in the RPA (a sum over the spin σ is implied). The dashed line stands for the interaction U. Signs and symmetry factors are not shown.

$\sigma^2 = 1$. It follows that

$$\chi^{zz}(q) = \chi_0(q) + \frac{U}{2}\chi_0(q)^2 + \frac{U^2}{4}\chi_0(q)^3 + \cdots = \frac{\chi_0(q)}{1 - \frac{U}{2}\chi_0(q)}. \tag{6.27}$$

Alternatively, one can calculate the transverse susceptibility $\chi^{+-}(q)$ [Eq. (6.21)]. In the absence of interaction, $\chi^{+-}(q) = \frac{1}{2}\chi_0(q)$. Only ladder diagrams contribute to χ^{+-} (Fig. 6.4(b)) and one obtains

$$\chi^{+-}(q) = \frac{1}{2}\chi_0(q) + \frac{U}{4}\chi_0(q)^2 + \frac{U^2}{8}\chi_0(q)^3 + \cdots = \frac{1}{2}\frac{\chi_0(q)}{1 - \frac{U}{2}\chi_0(q)}, \tag{6.28}$$

which agrees with (6.27) for a spin-rotation-invariant system [Eq. (6.22)].

The spin susceptibility $\chi^{\nu\nu}(q)$ [Eqs. (6.27, 6.28)] differs from the charge susceptibility (6.26) by the sign in the denominator. This difference is crucial as it implies that a magnetic instability is possible in the (repulsive) Hubbard model. To see this, let us consider the spectral function

$$\chi_0''(\mathbf{q}, \omega) = \Im[\chi_0^R(\mathbf{q}, \omega)] = 2\int_{\mathbf{k}} [n_F(\xi_{\mathbf{k}}) - n_F(\xi_{\mathbf{k}+\mathbf{q}})]\delta(\omega + \xi_{\mathbf{k}} - \xi_{\mathbf{k}+\mathbf{q}}), \tag{6.29}$$

where $\chi_0^R(\mathbf{q}, \omega) = \chi_0(\mathbf{q}, i\omega_\nu \to \omega + i0^+)$ is the retarded bare susceptibility. $\chi_0''(\mathbf{q}, \omega)$ is an odd function of ω. Using the spectral representation (3.32) of $\chi_0(q)$, one deduces that

$$\chi_0(q) = \int_{-\infty}^{\infty} \frac{d\omega}{\pi} \frac{\chi_0''(\mathbf{q}, \omega)}{\omega - i\omega_\nu} = \int_{-\infty}^{\infty} \frac{d\omega}{\pi} \chi_0''(\mathbf{q}, \omega) \frac{\omega}{\omega^2 + \omega_\nu^2} \tag{6.30}$$

is positive since $\omega\chi_0''(\mathbf{q}, \omega) \geq 0$ (Sec. 3.2.2). It follows that $\chi^{+-}(\mathbf{q})$, contrary to $\chi_{nn}(\mathbf{q})$, may diverge. A magnetic instability at wave vector $\mathbf{q}$ is signaled by the Stoner criterion

$$1 - \frac{U}{2}\chi_0(\mathbf{q}) = 0. \tag{6.31}$$

In the case of ferromagnetism ($\mathbf{q} = 0$), $\chi_0(\mathbf{q} = 0) = 2N(0)$ for $T = 0$ and the Stoner criterion becomes $1 - UN(0) = 0$, where $N(0)$ is the density of states per spin at the Fermi level. If the Stoner criterion is satisfied for a non-zero value of $\mathbf{q}$, the ground state is a spin-density wave with a non-uniform order parameter $\langle \hat{\mathbf{S}}_{\mathbf{r}} \rangle = \frac{1}{\sqrt{N}} \langle \hat{\mathbf{S}}(\mathbf{q}) \rangle e^{i\mathbf{q}\cdot\mathbf{r}} + \text{c.c.}$[13]

At low density, where $\chi_0(\mathbf{q})$ is maximum for $\mathbf{q} = 0$ (Sec. 6.2.1), we therefore expect a ferromagnetic instability. On the other hand, at half-filling, $\chi_0(\mathbf{q})$ being maximum for $\mathbf{Q} = (\pi, \ldots, \pi)$, the ground state should be antiferromagnetic (i.e. a spin-density wave with wavevector $\mathbf{Q}$). Our aim is not to make an exhaustive study of all possible instabilities as a function of the band filling. Instead, we will focus on ferromagnetism (Sec. 6.3) and, at half-filling, antiferromagnetism (Sec. 6.4).[14]

[13]Note that the divergence of the spin susceptibility (given by ladder diagrams in the particle–hole channel in the RPA) can be interpreted as the appearance of bound particle–hole pairs. The latter are analogous to the Cooper pairs in the BCS theory of superconductivity (Sec. 7.5).

[14]The fate of the antiferromagnetic state as one dopes the system away from half-filling is still under debate. Does one obtain an incommensurate spin-density wave (with wavevector $\mathbf{q} \neq \mathbf{Q}$) or a more exotic state? See references at the end of the chapter.

6.2.2.1 *Charge and spin fluctuations in the small density limit*

The spectrum of charge and spin fluctuations is obtained from the spectral functions

$$\begin{aligned}\chi''_{nn}(\mathbf{q},\omega) &= \frac{\chi''_0(\mathbf{q},\omega)}{\left[1+\frac{U}{2}\chi'_0(\mathbf{q},\omega)\right]^2+\frac{U^2}{4}\chi''_0(\mathbf{q},\omega)^2},\\ \chi^{+-\prime\prime}(\mathbf{q},\omega) &= \frac{1}{2}\frac{\chi''_0(\mathbf{q},\omega)}{\left[1-\frac{U}{2}\chi'_0(\mathbf{q},\omega)\right]^2+\frac{U^2}{4}\chi''_0(\mathbf{q},\omega)^2},\end{aligned} \tag{6.32}$$

where $\chi'_0(\mathbf{q},\omega)=\Re[\chi^R_0(\mathbf{q},\omega)]$. For a small density, we can approximate $\xi_\mathbf{k}\simeq\frac{\mathbf{k}^2}{2m^*}-\mu$ by a quadratic dispersion with an effective mass m^* (see Eq. (6.11), the constant $-2dt$ can be eliminated by redefining the chemical potential). For small $\mathbf{q}$ and $T=0$, we then have

$$\chi^R_0(\mathbf{q},\omega)=-2\int_\mathbf{k}\frac{n_F(\xi_\mathbf{k})-n_F(\xi_{\mathbf{k}+\mathbf{q}})}{\omega+i0^++\xi_\mathbf{k}-\xi_{\mathbf{k}+\mathbf{q}}}\simeq-2\int_\mathbf{k}\frac{\mathbf{v}^*_\mathbf{k}\cdot\mathbf{q}}{\omega+i0^+-\mathbf{v}^*_\mathbf{k}\cdot\mathbf{q}}\delta(\xi_\mathbf{k}), \tag{6.33}$$

where $\mathbf{v}^*_\mathbf{k}=\boldsymbol{\nabla}_\mathbf{k}\xi_\mathbf{k}=\mathbf{k}/m^*$, which is the form used in Fermi-liquid theory. Equations (6.32) are thus similar to the Fermi liquid results with Landau parameters $F^s_0=2N(0)f^s_0=UN(0)$ and $F^a_0=-2N(0)f^a_0=-UN(0)$.[15] In three dimensions, a positive F^s_0 implies the existence of a well-defined zero-sound mode with a linear dispersion $\omega=c_0|\mathbf{q}|$. On the other hand, a negative F^a_0 implies that the collective spin-wave mode strongly couples to particle–hole pair excitations and is not a well-defined (long-lived) mode. This damped mode is known as a paramagnon. The Fermi liquid becomes unstable when $F^a_0<-1$ (Pomeranchuk instability, see Sec. 4.1.4), which is nothing but the Stoner criterion for the ferromagnetic instability.

6.3 Stoner's Theory of Ferromagnetism

In this section, we discuss the Stoner theory of ferromagnetism: the Hartree–Fock theory for the equilibrium state and the RPA for the collective modes.

6.3.1 *Hartree–Fock theory*

To study the ferromagnetic phase, we use the Hartree–Fock approximation where the self-energy is given by[16] (Fig. 6.5)

$$\Sigma_\sigma(k)=\frac{U}{\beta N}\sum_{k'}G_{\bar\sigma}(k')e^{i\omega_{n'}0^+}=Un_{\bar\sigma}, \tag{6.34}$$

where

$$n_\sigma=\langle\hat n_{\mathbf{r}\sigma}\rangle=\int_\mathbf{k}\langle\hat\psi^\dagger_\sigma(\mathbf{k})\hat\psi_\sigma(\mathbf{k})\rangle \tag{6.35}$$

[15] See Eq. (4.107) for the $T=0$ structure factor $S_{nn}(\mathbf{q},\omega)=2\Theta(\omega)\chi''_{nn}(\mathbf{q},\omega)$ in a Fermi liquid.
[16] The Fock diagram vanishes for the Hubbard model because of the Pauli principle.

Fig. 6.5 Self-energy Σ_σ in the Hartree–Fock approximation (the propagator in the closed loop carries a spin $\bar\sigma$).

denotes the mean density of particles with spin σ. The propagator then takes the simple form

$$G_\sigma(k) = (i\omega_n - E_{\mathbf{k}\sigma})^{-1}, \tag{6.36}$$

with the Hartree–Fock energy[17]

$$E_{\mathbf{k}\sigma} = \xi_{\mathbf{k}} + U n_{\bar\sigma}. \tag{6.37}$$

It is convenient to introduce the particle density $n = n_\uparrow + n_\downarrow$ and the magnetization per site $m = n_\uparrow - n_\downarrow$. n and m are determined by the self-consistent equations

$$\begin{aligned} n &= \frac{1}{\beta N}\sum_{k,\sigma} G_\sigma(k) e^{i\omega_n 0^+} = \int_{\mathbf{k}} [n_F(E_{\mathbf{k}\uparrow}) + n_F(E_{\mathbf{k}\downarrow})], \\ m &= \frac{1}{\beta N}\sum_{k,\sigma} \sigma G_\sigma(k) e^{i\omega_n 0^+} = \int_{\mathbf{k}} [n_F(E_{\mathbf{k}\uparrow}) - n_F(E_{\mathbf{k}\downarrow})]. \end{aligned} \tag{6.38}$$

m vanishes in the metallic phase and takes a non-zero value in the ferromagnetic phase. Near the transition, m is small and we can expand the self-consistent equation it obeys,[18]

$$m = -Um \int_{\mathbf{k}} n_F'\left(\xi_{\mathbf{k}} + \frac{U}{2}n\right) + \mathcal{O}(m^3), \tag{6.39}$$

i.e.

$$m\left(1 - \frac{U}{2}\chi_0(\mathbf{q}=0)\right) + \mathcal{O}(m^3) = 0, \tag{6.40}$$

where the bare susceptibility χ_0 includes the Hartree shift $Un/2$ of the band energy $\epsilon_{\mathbf{k}}$ (which can be absorbed into a renormalization of the chemical potential). Equivalently,

[17] In the chapter on the electron gas, we noted that the Hartree–Fock theory is equivalent to a mean-field theory (Sec. 5.2.2). The same is true for the Hubbard model and Eq. (6.36) can also be obtained from a mean-field decoupling of the interaction term in the Hamiltonian: $U\hat n_\uparrow \hat n_\downarrow \to U(n_\uparrow \hat n_\downarrow + \hat n_\uparrow n_\downarrow - n_\uparrow n_\downarrow)$ (see also Sec. 6.3.3).

[18] Note that μ depends on m if we fix the density n, but this dependence does not matter when we compute the $\mathcal{O}(m)$ term in (6.39).

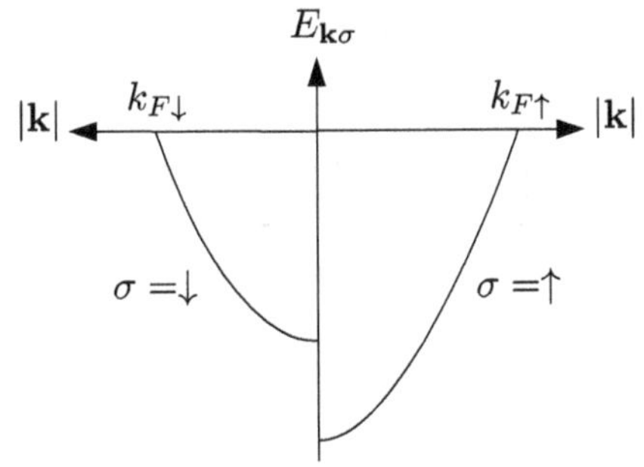

Fig. 6.6 Excitation energy $E_{\mathbf{k}\sigma}$ in the ferromagnetic ground state [Eq. (6.36)] in the low-density limit where $\epsilon_{\mathbf{k}} \simeq k^2/2m^*$ [Eq. (6.11)].

Eq. (6.40) can be derived by computing the thermodynamic potential $\Omega(m) = Am^2 + Bm^4 + \mathcal{O}(m^6)$ as a function of m and determining the value of m by requiring $\Omega(m)$ to be minimum: $Am + 2Bm^3 = 0$.[19] This equation admits a non-trivial solution $m \neq 0$ when the coefficient $A \sim 1 - (U/2)\chi_0(\mathbf{q} = 0)$ of the linear term is negative, and we recover the Stoner criterion (6.31) for the ferromagnetic instability.[20]

In the ground state ($T = 0$), the occupation factors of the one-particle state are given by $n_F(E_{\mathbf{k}\sigma}) = \Theta(-E_{\mathbf{k}\sigma})$ and the spin-dependent Fermi surfaces are defined by $E_{\mathbf{k}_{F\sigma}} = 0$, i.e.

$$\epsilon_{\mathbf{k}_{F\uparrow}} - \mu + \frac{U}{2}(n - m) = 0, \qquad \epsilon_{\mathbf{k}_{F\downarrow}} - \mu + \frac{U}{2}(n + m) = 0. \tag{6.41}$$

In the low-density limit where $\epsilon_{\mathbf{k}} \simeq \mathbf{k}^2/2m^*$ (absorbing the constant $-2dt$ in the chemical potential), we obtain two circular Fermi surfaces as shown in Fig. 6.6.[21]

6.3.1.1 *Limitations of the Stoner theory*

In the low-density limit, Kanamori and Brückner showed that the cross section for fermion–fermion scattering should be calculated beyond the lowest-order (Born) approximation. This amounts to replacing the bare interaction U in the calculation of the normal-phase

[19] This would be nothing but a microscopic derivation of the Landau free energy for the ferromagnetic transition (see Chapter 10).

[20] Stability requires that the coefficient B of the Landau expansion of the thermodynamic potential $\Omega(m)$ (i.e. the coefficient of the $\mathcal{O}(m^3)$ term in (6.40)) be positive if we expand $\Omega(m)$ only up to fourth order.

[21] At half-filling and for $U \gg t$, the self-consistent Eqs. (6.38) give $E_{\mathbf{k}\uparrow} < 0$ and $E_{\mathbf{k}\downarrow} > 0$, and therefore $n_\uparrow = 1$ and $n_\downarrow = 0$, which corresponds to saturated ferromagnetism. Due to the large local repulsion, fermions behave as localized spins-$\frac{1}{2}$ (Sec. 6.1.3). However, the ground state is not ferromagnetic but antiferromagnetic. The instability of the ferromagnetic state manifests itself in the collective modes: The constant D introduced in Sec. 6.3.2 turns out to be negative, $D \to -J/2$ with $J = 4t^2/U$ in contradiction with the stability requirement $D > 0$.

RPA susceptibilities by the effective interaction[22]

$U_{\text{eff}}(q) =$

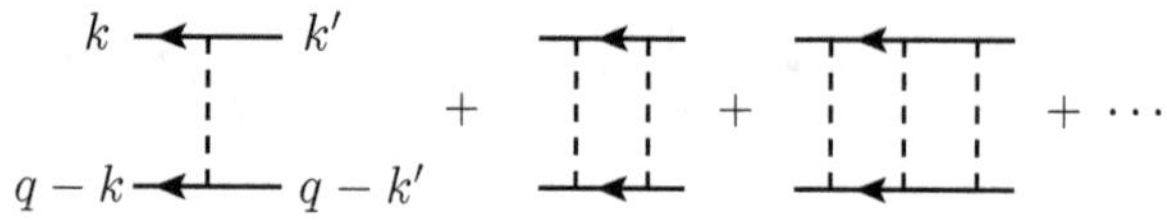

$$= \frac{U}{1 + U\Pi(q)}, \tag{6.42}$$

where

$$\Pi(q) = \frac{1}{\beta N} \sum_k G_0(k) G_0(q - k). \tag{6.43}$$

To recover a frequency- and momentum-independent interaction U_{eff}, one can set $q = (\mathbf{q}, 0)$ and average (6.42) over $\mathbf{q}$. As a result, when $U \to \infty$, U_{eff} saturates to the average of $1/\Pi(\mathbf{q}, 0)$. The latter is of the order of the bandwidth $W = 4dt$. Since the density of states $N(0)$ at the Fermi level is of the order of $1/W$, we see that the Stoner criterion $1 - UN(0) = 0$ might not be satisfied even in the large-U limit. Thus, Stoner's theory clearly overestimates the tendency to ferromagnetism. More refined calculations indicate that ferromagnetism does not in general occur in the Hubbard model except perhaps in very special cases when there is a Van Hove singularity in the density of states that is not located at half-filling (otherwise antiferromagnetism dominates).[23,24]

6.3.2 *Charge and spin collective modes*

Charge and spin collective modes in the ferromagnetic phase can be obtained from the correlation functions χ_{nn} and $\chi^{\nu\nu'}$ in the RPA (Sec. 6.2).

Let us first consider the transverse susceptibility χ^{+-}. It is given by the sum of the ladder diagrams (Fig. 6.4(b)), i.e.

$$\chi^{+-}(q) = \frac{\chi_0^{+-}(q)}{1 - U\chi_0^{+-}(q)}, \tag{6.44}$$

where

$$\chi_0^{+-}(q) = -\frac{1}{\beta N} \sum_k G_\downarrow(k) G_\uparrow(k - q) = -\int_{\mathbf{k}} \frac{n_F(E_{\mathbf{k}-\mathbf{q}\uparrow}) - n_F(E_{\mathbf{k}\downarrow})}{i\omega_\nu + E_{\mathbf{k}-\mathbf{q}\uparrow} - E_{\mathbf{k}\downarrow}} \tag{6.45}$$

must be computed with the Hartree–Fock propagator (6.36). The Hartree–Fock energy is defined by $E_{\mathbf{k}\sigma} = \xi_{\mathbf{k}} - \sigma\Delta$, with $\Delta = Um/2$, if we absorb the Hartree shift $Un/2$ into the

---

[22] In vacuum (vanishing density of particles), Eq. (6.42) yields the exact cross section for the fermion–fermion scattering.

[23] See, e.g., Ref. [24].

[24] There are other cases where a ferromagnetic ground state can arise. For example, the ground state of the half-filled infinite-U Hubbard model with a single hole is ferromagnetic (Nagaoka's theorem [25]). For certain special lattices (e.g. the Kagome lattice) with "flat bands", the ground state can also be ferromagnetic [8,9].

chemical potential. To order $\mathcal{O}(\omega_\nu, \mathbf{q}^2)$, one finds

$$\chi_0^{+-}(q) = \frac{m}{2\Delta}\left(1 + \frac{i\omega_\nu}{2\Delta}\right) - \frac{D}{2U\Delta}\mathbf{q}^2, \tag{6.46}$$

with D a constant that we shall not calculate explicitly (see Sec. 6.3.3 for a general expression in terms of Hartree–Fock current–current correlation functions). Equation (6.46) implies

$$1 - U\chi_0^{+-}(q) = -\frac{i\omega_\nu - D\mathbf{q}^2}{2\Delta} \tag{6.47}$$

and in turn

$$\chi^{+-}(q) = -\frac{2\Delta/U}{i\omega_\nu - D\mathbf{q}^2} \qquad (\mathbf{q}, \omega_\nu \to 0). \tag{6.48}$$

Since $\chi^{-+}(q) = \chi^{+-}(-q)$, we find

$$\chi^{xx}(q) = \chi^{yy}(q) = \chi^{+-}(q) + \chi^{-+}(q) = \frac{4\Delta}{U}\frac{D\mathbf{q}^2}{\omega_\nu^2 + (D\mathbf{q}^2)^2}. \tag{6.49}$$

From the retarded correlation functions $\chi^{xx}(\mathbf{q}, \omega + i0^+) = \chi^{yy}(\mathbf{q}, \omega + i0^+)$, we find a transverse spin-wave mode $\omega = \pm D\mathbf{q}^2$, which is the Goldstone mode associated with the spontaneous breaking of rotation symmetry in the ferromagnetic state (Sec. 3.6). This mode gives a delta peak contribution to the spectral function

$$\chi^{xx\prime\prime}(\mathbf{q}, \omega) = \Im[\chi^{xx}(\mathbf{q}, \omega + i0^+)] = 2\pi\frac{\Delta}{U}[\delta(\omega - D\mathbf{q}^2) - \delta(\omega + D\mathbf{q}^2)]. \tag{6.50}$$

Stability requires the constant D to be positive, otherwise the condition $\omega\chi^{xx\prime\prime}(\mathbf{q}, \omega) \geq 0$ for positivity of dissipation is violated (Sec. 3.2.2). According to (6.49), the static transverse susceptibility

$$\chi^{xx}(\mathbf{q}) = \chi^{yy}(\mathbf{q}) = \chi^{+-}(\mathbf{q}) + \chi^{-+}(\mathbf{q}) = \frac{4\Delta}{UD\mathbf{q}^2} \tag{6.51}$$

diverges for $\mathbf{q} \to 0$. This divergence is a consequence of the spontaneous breaking of spin-rotation symmetry as discussed in Sec. 3.6.2.

Besides the Goldstone mode, particle–hole pair excitations also contribute to the spectral function $\chi^{xx\prime\prime}(\mathbf{q}, \omega)$. The latter correspond to a non-vanishing spectral density $\chi_0^{+-\prime\prime}(\mathbf{q}, \omega)$ or $\chi_0^{-+\prime\prime}(\mathbf{q}, \omega)$. At zero temperature, the excitation energies obtained from $\chi_0^{+-\prime\prime}(\mathbf{q}, \omega)$ are given by $\omega = E_{\mathbf{k}\downarrow} - E_{\mathbf{k}-\mathbf{q}\uparrow}$ with the state $(\mathbf{k} - \mathbf{q}, \uparrow)$ occupied and the state $(\mathbf{k}, \downarrow)$ empty, i.e.

$$\omega = 2\Delta + \epsilon_{\mathbf{k}} - \epsilon_{\mathbf{k}-\mathbf{q}} \tag{6.52}$$

with $E_{\mathbf{k}\downarrow} > 0$ and $E_{\mathbf{k}-\mathbf{q}\uparrow} < 0$. For $\mathbf{q} \to 0$, these excitations have a gap 2Δ. Thus, for sufficiently small $\mathbf{q}$, the Goldstone mode $\omega = D\mathbf{q}^2$ appears in the particle–hole excitation gap and is a propagating mode (Fig. 6.7). (By considering $\chi_0^{-+\prime\prime}(\mathbf{q}, \omega)$, we obtain the negative part of the spectrum.)

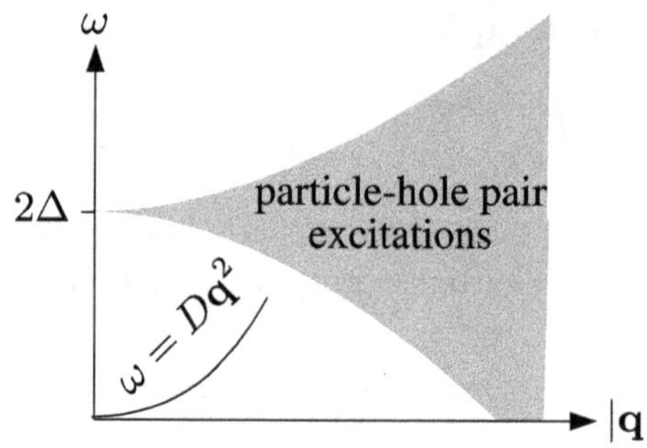

Fig. 6.7 Transverse excitation spectrum in the ferromagnetic ground state: spin-wave mode ($\omega = D\mathbf{q}^2$) and continuum of particle–hole pair excitations ($\omega = E_{\mathbf{k}\downarrow} - E_{\mathbf{k}-\mathbf{q}\uparrow}$).

Longitudinal spin-wave modes are obtained from the spectral function $\chi^{zz\prime\prime}$ of the spin susceptibility

$$\chi^{zz}(q) = \frac{\chi_0^{zz}(q)}{1 - \frac{U}{2}\chi_0^{zz}(q)}, \tag{6.53}$$

where

$$\chi_0^{zz}(q) = -\frac{1}{\beta N}\sum_{k,\sigma} G_\sigma(k)G_\sigma(k+q) = -\sum_\sigma \int_{\mathbf{k}} \frac{n_F(E_{\mathbf{k}\sigma}) - n_F(E_{\mathbf{k}+\mathbf{q}\sigma})}{i\omega_\nu + E_{\mathbf{k}\sigma} - E_{\mathbf{k}+\mathbf{q}\sigma}}, \tag{6.54}$$

with G_σ the Hartree–Fock propagator (6.36). If we include the Hartree shift $Un/2$ into the chemical potential, we obtain

$$\begin{aligned}\chi_0^{zz}(q) &= -\sum_\sigma \int_{\mathbf{k}} \frac{n_F(\xi_{\mathbf{k}} - \sigma\Delta) - n_F(\xi_{\mathbf{k}+\mathbf{q}} - \sigma\Delta)}{i\omega_\nu + \xi_{\mathbf{k}} - \xi_{\mathbf{k}+\mathbf{q}}} \\ &\simeq \sum_\sigma \int_{\mathbf{k}} \frac{\mathbf{v}_{\mathbf{k}} \cdot \mathbf{q}}{i\omega_\nu - \mathbf{v}_{\mathbf{k}} \cdot \mathbf{q}} \delta(\xi_{\mathbf{k}} - \sigma\Delta)\end{aligned} \tag{6.55}$$

for small $\mathbf{q}$ ($\mathbf{v}_{\mathbf{k}} = \boldsymbol{\nabla}_{\mathbf{k}}\xi_{\mathbf{k}}$), which is similar to the result (6.33) in the paramagnetic phase, except for a change in the (now spin-dependent) chemical potential $\mu_\sigma = \mu - \sigma\Delta$. This implies that the density of states at the Fermi level $N(\xi = 0)$ will be replaced by $\frac{1}{2}[N(\Delta) + N(-\Delta)]$, but the spectral function $\chi^{zz\prime\prime}(\mathbf{q}, \omega)$ will be otherwise unchanged. Thus, we expect (at least in the low-density limit) a damped longitudinal spin-wave mode coupled to the particle–hole pair continuum $\omega = E_{\mathbf{k}+\mathbf{q}\sigma} - E_{\mathbf{k}\sigma} = \epsilon_{\mathbf{k}+\mathbf{q}} - \epsilon_{\mathbf{k}}$ (with the state $(\mathbf{k}, \sigma)$ occupied and the state $(\mathbf{k} + \mathbf{q}, \sigma)$ empty); see Sec. 6.2.2.

The density–density correlation function takes the form

$$\chi_{nn}(q) = \frac{\chi_{nn}^0(q)}{1 + \frac{U}{2}\chi_{nn}^0(q)}, \tag{6.56}$$

where $\chi_{nn}^0(q) = \chi_0^{zz}(q)$. It differs from $\chi^{zz}(q)$ only by the sign in the denominator. Thus, the spectral function $\chi_{nn}''(\mathbf{q}, \omega)$ in the ferromagnetic phase is similar to the one in the paramagnetic phase (except for a renormalization of the density of states; see the above discussion). As a result, we expect (at least in the low-density limit) a well-defined zero-sound mode lying above the continuum of particle–hole pair excitations $\omega = \epsilon_{\mathbf{k}+\mathbf{q}} - \epsilon_{\mathbf{k}}$.

6.3.3 *Functional integral formalism*

In this section, we recover the previous results using the functional integral formalism. The action of the Hubbard model reads $S_0 + S_{\rm int}$,

$$\begin{aligned} S_0[\psi^*, \psi] &= \int_0^\beta d\tau \biggl\{ \sum_{\mathbf{r},\sigma} \psi^*_{\mathbf{r}\sigma}(\partial_\tau - \mu)\psi_{\mathbf{r}\sigma} - t \sum_{\langle \mathbf{r},\mathbf{r}'\rangle,\sigma} (\psi^*_{\mathbf{r}\sigma}\psi_{\mathbf{r}'\sigma} + \text{c.c.}) \biggr\}, \\ S_{\rm int}[\psi^*, \psi] &= U \int_0^\beta d\tau \sum_{\mathbf{r}} \psi^*_{\mathbf{r}\uparrow}\psi^*_{\mathbf{r}\downarrow}\psi_{\mathbf{r}\downarrow}\psi_{\mathbf{r}\uparrow}, \end{aligned} \tag{6.57}$$

where ψ is a Grassmannian field. For simplicity, we now consider a two-dimensional square lattice.

6.3.3.1 *Hubbard–Stratonovich transformations*

The first step in the functional integral formalism is to perform a Hubbard–Stratonovich transformation in order to introduce one or several auxiliary fields describing the collective fluctuations we are interested in. To describe charge and spin fluctuations, one writes

$$S_{\rm int}[\psi^*, \psi] = \frac{U}{4}\int_0^\beta d\tau \sum_{\mathbf{r}} [(\Psi^\dagger_{\mathbf{r}}\Psi_{\mathbf{r}})^2 - (\Psi^\dagger_{\mathbf{r}}\sigma^z\Psi_{\mathbf{r}})^2], \tag{6.58}$$

where

$$\Psi^\dagger_{\mathbf{r}} = (\psi^*_{\mathbf{r}\uparrow}, \psi^*_{\mathbf{r}\downarrow}), \qquad \Psi_{\mathbf{r}} = \begin{pmatrix} \psi_{\mathbf{r}\uparrow} \\ \psi_{\mathbf{r}\downarrow} \end{pmatrix} \tag{6.59}$$

is a two-component spinor. Introducing then two real auxiliary fields, we rewrite the interacting action as

$$e^{-S_{\rm int}} = \int \mathcal{D}[\Delta_c, \Delta_s] \exp\biggl\{ -\int_0^\beta d\tau \sum_{\mathbf{r}} \biggl[\frac{1}{U}(\Delta^2_{c\mathbf{r}} + \Delta^2_{s\mathbf{r}}) - \Psi^\dagger_{\mathbf{r}}(i\Delta_{c\mathbf{r}} + \Delta_{s\mathbf{r}}\sigma^z)\Psi_{\mathbf{r}} \biggr] \biggr\}. \tag{6.60}$$

As we shall see in the following, a saddle-point approximation for the functional integral over $\Delta_{c\mathbf{r}}$ and $\Delta_{s\mathbf{r}}$ reproduces the Hartree–Fock approximation. There is however an obvious shortcoming of (6.60). Although this equation is formally exact, spin-rotation invariance is not explicit anymore and is likely to be violated unless the functional integral over $\Delta_{c\mathbf{r}}$ and $\Delta_{s\mathbf{r}}$ is carried out exactly. In particular, it is not possible to recover the gapless spin-wave mode within a Gaussian approximation about the ferromagnetic saddle-point solution.

Alternatively, one could write $S_{\rm int}$ in an explicitly spin-rotation-invariant form, e.g.

$$S_{\rm int}[\psi^*, \psi] = -\frac{U}{6}\int_0^\beta d\tau \sum_{\mathbf{r}} (\Psi_{\mathbf{r}}^\dagger \boldsymbol{\sigma} \Psi_{\mathbf{r}})^2 \tag{6.61}$$

[with $\boldsymbol{\sigma} = (\sigma^x, \sigma^y, \sigma^z)$] and use a Hubbard–Stratonovich vector field. However, Eq. (6.61) does not reproduce the Hartree–Fock approximation at the saddle-point level. We encounter here one of the main limitations of Hubbard–Stratonovich transformations. There are many possible transformations[25] (this is known as the Fierz ambiguity) and most of them do not reproduce the Hartree–Fock approximation at the saddle-point level.

This difficulty can be circumvented by using the decomposition

$$S_{\rm int}[\psi^*, \psi] = \frac{U}{4}\int_0^\beta d\tau \sum_{\mathbf{r}} [(\Psi_{\mathbf{r}}^\dagger \Psi_{\mathbf{r}})^2 - (\Psi_{\mathbf{r}}^\dagger \boldsymbol{\sigma}\cdot\boldsymbol{\Omega}_{\mathbf{r}} \Psi_{\mathbf{r}})^2], \tag{6.62}$$

where $\boldsymbol{\Omega}_{\mathbf{r}}(\tau)$ is a site- and time-dependent unit vector. Spin-rotation invariance can then be made explicit by performing an angular integration over $\boldsymbol{\Omega}_{\mathbf{r}}$ at each site and time (with a measure normalized to unity) [26, 27]. The Hubbard–Stratonovich transformation reads

$$\begin{aligned} e^{-S_{\rm int}} = \int \mathcal{D}[\Delta_c, \Delta_s, \boldsymbol{\Omega}] \exp\biggl\{ -\int_0^\beta d\tau \sum_{\mathbf{r}} \biggl[\frac{1}{U}(\Delta_{c\mathbf{r}}^2 + \Delta_{s\mathbf{r}}^2) \\ - \Psi_{\mathbf{r}}^\dagger (i\Delta_{c\mathbf{r}} + \Delta_{s\mathbf{r}} \boldsymbol{\sigma}\cdot\boldsymbol{\Omega}_{\mathbf{r}}) \Psi_{\mathbf{r}} \biggr] \biggr\} \end{aligned} \tag{6.63}$$

and the partition function becomes a functional integral over the fermion field ψ and the real fields Δ_c, Δ_s, $\boldsymbol{\Omega}$ with an action $S[\psi^*, \psi, \Delta_c, \Delta_s, \boldsymbol{\Omega}]$ which can be read off from Eqs. (6.57) and (6.63). Equation (6.63) corresponds to an "amplitude-direction" representation where the magnetic order-parameter field is given by $\Delta_{s\mathbf{r}}\boldsymbol{\Omega}_{\mathbf{r}}$. As we shall see in the following, the Hartree–Fock theory is reproduced from a saddle-point approximation over the auxiliary fields $\Delta_{c\mathbf{r}}$, $\Delta_{s\mathbf{r}}$ and $\boldsymbol{\Omega}_{\mathbf{r}}$, and the transverse spin-wave excitations are obtained by considering small fluctuations of the field $\boldsymbol{\Omega}_{\mathbf{r}}$ about its saddle-point value.

6.3.3.2 *Hartree–Fock theory as a saddle-point approximation*

Assuming ferromagnetic order, i.e. taking uniform and time-independent auxiliary fields $\Delta_{c\mathbf{r}} = \Delta_c$, $\Delta_{s\mathbf{r}} = \Delta_s$ and $\boldsymbol{\Omega}_{\mathbf{r}} = \mathbf{u}_x$,[26] one obtains the (Hartree–Fock) saddle-point action

$$\begin{aligned} S_{\rm HF}[\Phi^\dagger, \Phi] = \int_0^\beta d\tau \biggl\{ \sum_{\mathbf{r}} \biggl[\frac{1}{U}(\Delta_c^2 + \Delta_s^2) + \Phi_{\mathbf{r}}^\dagger (\partial_\tau - \mu - i\Delta_c - \Delta_s \sigma^z) \Phi_{\mathbf{r}} \biggr] \\ - t \sum_{\langle \mathbf{r}, \mathbf{r}' \rangle} (\Phi_{\mathbf{r}}^\dagger \Phi_{\mathbf{r}'} + \text{c.c.}) \biggr\}, \end{aligned} \tag{6.64}$$

---

[25] There is in fact an infinite number of possible transformations since one can write $\psi_{\mathbf{r}\uparrow}^* \psi_{\mathbf{r}\downarrow}^* \psi_{\mathbf{r}\downarrow} \psi_{\mathbf{r}\uparrow} = \frac{a}{2}(\Psi_{\mathbf{r}}^\dagger \Psi_{\mathbf{r}})^2 - \frac{b}{2}(\Psi_{\mathbf{r}}^\dagger \sigma^z \Psi_{\mathbf{r}})^2 - \frac{c}{6}(\Psi_{\mathbf{r}}^\dagger \boldsymbol{\sigma} \Psi_{\mathbf{r}})^2$, with $a + b + c = 1$, and decouple each term in the rhs with an auxiliary field. To describe collective modes in the particle–particle channel, one could also consider an auxiliary field that couples to $\psi_{\mathbf{r}\downarrow}\psi_{\mathbf{r}\uparrow}$ as done in the BCS theory of superconductivity (Sec. 7.5).

[26] We denote by $(\mathbf{u}_x, \mathbf{u}_y, \mathbf{u}_z)$ an orthonormal basis in spin space.

where $\Phi_{\mathbf{r}} = R^\dagger \Psi_{\mathbf{r}}$ is a rotated fermion field such that $\Psi^\dagger_{\mathbf{r}} \sigma^x \Psi_{\mathbf{r}} = \Phi^\dagger_{\mathbf{r}} \sigma^z \Phi_{\mathbf{r}}$ [$R \in$SU(2), see Eqs. (6.69, 6.70)]. The saddle-point equations[27]

$$\frac{2}{U}\Delta_c = i\langle \Phi^\dagger_{\mathbf{r}} \Phi_{\mathbf{r}} \rangle, \qquad \frac{2}{U}\Delta_s = \langle \Phi^\dagger_{\mathbf{r}} \sigma^z \Phi_{\mathbf{r}} \rangle \tag{6.65}$$

are deduced from $0 = \partial Z_{\rm HF}/\partial \Delta_c = \partial Z_{\rm HF}/\partial \Delta_s$, where $Z_{\rm HF}$ is the partition function obtained from $S_{\rm HF}$. From (6.64), one obtains the fermion self-energy

$$\Sigma_\sigma(k) = -i\Delta_c - \sigma\Delta_s = \frac{U}{2}(n - \sigma m) = U n_{\bar{\sigma}}, \tag{6.66}$$

where $n = n_\uparrow + n_\downarrow = \langle \Phi^\dagger_{\mathbf{r}} \Phi_{\mathbf{r}} \rangle$ and $m = n_\uparrow - n_\downarrow = \langle \Phi^\dagger_{\mathbf{r}} \sigma^z \Phi_{\mathbf{r}} \rangle$, in agreement with the Hartree–Fock self-energy (6.34). Equations (6.65) reproduce the self-consistent equations (6.38).

6.3.3.3 *Transverse spin-wave modes*

In the following, we focus on the transverse spin-wave modes corresponding to fluctuations of the unit vector field $\boldsymbol{\Omega}_{\mathbf{r}}$, and ignore the fluctuations of $\Delta_{c\mathbf{r}}$ and $\Delta_{s\mathbf{r}}$ about their saddle-point values.[28] To simplify the notations, we absorb $i\Delta_c = Un/2$ in the chemical potential and denote Δ_s merely by Δ. We parameterize

$$\boldsymbol{\Omega}_{\mathbf{r}} = \begin{pmatrix} \sin\theta_{\mathbf{r}} \cos\varphi_{\mathbf{r}} \\ \sin\theta_{\mathbf{r}} \sin\varphi_{\mathbf{r}} \\ \cos\theta_{\mathbf{r}} \end{pmatrix} \tag{6.67}$$

in the basis $(\mathbf{u}_x, \mathbf{u}_y, \mathbf{u}_z)$ by the polar and azimuthal angles $\theta_{\mathbf{r}}$ and $\varphi_{\mathbf{r}}$ so that the Hartree–Fock ground state corresponds to the classical solution $\theta^{\rm cl}_{\mathbf{r}} = \pi/2$ and $\varphi^{\rm cl}_{\mathbf{r}} = 0$.[29]

Fluctuations of $\boldsymbol{\Omega}_{\mathbf{r}}$ are parameterized by[30]

$$P_{\mathbf{r}} = \frac{1}{2}\delta\boldsymbol{\Omega}_{\mathbf{r}} \cdot \mathbf{u}_z = \frac{1}{2}\cos\theta_{\mathbf{r}}, \qquad Q_{\mathbf{r}} = \delta\boldsymbol{\Omega}_{\mathbf{r}} \cdot \mathbf{u}_y = \sin\theta_{\mathbf{r}} \sin\varphi_{\mathbf{r}} \simeq \varphi_{\mathbf{r}}. \tag{6.68}$$

To obtain the transverse spin-wave excitations, we must derive the effective action $S[P, Q]$ governing the dynamics of P and Q by integrating out the fermions. It is convenient to introduce a new fermion field defined by

$$\Psi_{\mathbf{r}} = R_{\mathbf{r}} \Phi_{\mathbf{r}}, \tag{6.69}$$

where $R_{\mathbf{r}}$ is a site- and time-dependent SU(2)/U(1) rotation matrix satisfying

$$\boldsymbol{\sigma} \cdot \boldsymbol{\Omega}_{\mathbf{r}} = R_{\mathbf{r}} \sigma^z R^\dagger_{\mathbf{r}}. \tag{6.70}$$

---

[27] Δ_c is purely imaginary at the saddle point. This is allowed by the possibility to deform the integration contour into the complex plane (see footnote 5 on page 376).

[28] Fluctuations in $\Delta_{c\mathbf{r}}$ and $\Delta_{s\mathbf{r}}$ correspond to density and longitudinal spin fluctuations, respectively (see Sec. 6.3.2).

[29] Here, we use the term "classical" to emphasize the absence of quantum fluctuations of $\boldsymbol{\Omega}_{\mathbf{r}}$ in the Hartree–Fock ground state.

[30] The definition of $P_{\mathbf{r}}$ differs by a minus sign from that used in Sec. 8.4 [Eq. (8.111)].

This definition means that $\mathcal{R}_\mathbf{r}$, the SO(3) element associated with $R_\mathbf{r}$, maps $\mathbf{u}_z$ onto $\boldsymbol{\Omega}_\mathbf{r}$: $\mathcal{R}_\mathbf{r}(\mathbf{u}_z) = \boldsymbol{\Omega}_\mathbf{r}$. The spin of the fermions $\Phi_\mathbf{r}$ is quantized along the axis $\boldsymbol{\Omega}_\mathbf{r}$. The U(1) gauge freedom is due to rotations about $\mathbf{u}_z$ which do not change $\boldsymbol{\Omega}_\mathbf{r}$. The unitary transformation (6.69) has unit Jacobian but the action becomes

$$S[\Phi^\dagger, \Phi, P, Q] = S_{\rm HF}[\Phi^\dagger, \Phi] + S_1[\Phi^\dagger, \Phi, P, Q] + S_2[\Phi^\dagger, \Phi, P, Q], \tag{6.71}$$

where

$$\begin{aligned} S_1[\Phi^\dagger, \Phi, P, Q] &= \int_0^\beta d\tau \sum_\mathbf{r} \Phi_\mathbf{r}^\dagger R_\mathbf{r}^\dagger \dot{R}_\mathbf{r} \Phi_\mathbf{r}, \\ S_2[\Phi^\dagger, \Phi, P, Q] &= -\int_0^\beta d\tau \sum_{\mathbf{r},\mathbf{r}'} t_{\mathbf{r}\mathbf{r}'} \Phi_\mathbf{r}^\dagger (R_\mathbf{r}^\dagger R_{\mathbf{r}'} - 1)\Phi_{\mathbf{r}'} \end{aligned} \tag{6.72}$$

($\dot{R}_\mathbf{r} = \partial_\tau R_\mathbf{r}$). $t_{\mathbf{r}\mathbf{r}'}$ is equal to t if $\mathbf{r}$ and $\mathbf{r}'$ are first neighbors and vanishes otherwise. The action of the transverse spin waves is then defined by

$$e^{-S[P,Q]} = \int \mathcal{D}[\Phi^\dagger, \Phi]\, e^{-S_{\rm HF}[\Phi^\dagger, \Phi] - S_1[\Phi^\dagger, \Phi, P, Q] - S_2[\Phi^\dagger, \Phi, P, Q]}. \tag{6.73}$$

The functional integral over the field Φ can be carried out within a cumulant expansion.[31] To obtain the action $S[P, Q]$ to quadratic order in P and Q, it is sufficient to retain only the first- and second-order cumulants, i.e.

$$\begin{aligned} S[P, Q] = {}& \langle S_1[\Phi^\dagger, \Phi, P, Q] + S_2[\Phi^\dagger, \Phi, P, Q]\rangle_c \\ & - \frac{1}{2}\langle (S_1[\Phi^\dagger, \Phi, P, Q] + S_2[\Phi^\dagger, \Phi, P, Q])^2\rangle_c, \end{aligned} \tag{6.74}$$

where the cumulants $\langle \cdots \rangle_c$ are to be calculated with the Hartree–Fock action $S_{\rm HF}$.

The matrix $R_\mathbf{r}$ can be explicitly written as

$$R_\mathbf{r} = e^{-\frac{i}{2}\varphi_\mathbf{r}\sigma^z} e^{-\frac{i}{2}\theta_\mathbf{r}\sigma^y} e^{-\frac{i}{2}\psi_\mathbf{r}\sigma^z} = \begin{pmatrix} \cos\left(\frac{\theta_\mathbf{r}}{2}\right) e^{-\frac{i}{2}\varphi_\mathbf{r}} & -\sin\left(\frac{\theta_\mathbf{r}}{2}\right) e^{-\frac{i}{2}\varphi_\mathbf{r}} \\ \sin\left(\frac{\theta_\mathbf{r}}{2}\right) e^{\frac{i}{2}\varphi_\mathbf{r}} & \cos\left(\frac{\theta_\mathbf{r}}{2}\right) e^{\frac{i}{2}\varphi_\mathbf{r}} \end{pmatrix}, \tag{6.75}$$

where the last expression holds for the gauge choice $\psi_\mathbf{r} = 0$. In this gauge, one easily obtains

$$A_{0\mathbf{r}} = R_\mathbf{r}^\dagger \dot{R}_\mathbf{r} = \sum_{\nu=x,y,z} A_{0\mathbf{r}}^\nu \sigma^\nu, \tag{6.76}$$

with

$$A_{0\mathbf{r}}^x = \frac{i}{2}\dot{Q}_\mathbf{r}, \qquad A_{0\mathbf{r}}^y = i\dot{P}_\mathbf{r}, \qquad A_{0\mathbf{r}}^z = -iP_\mathbf{r}\dot{Q}_\mathbf{r}. \tag{6.77}$$

Similarly, one obtains

$$A_{\mu\mathbf{r}} = R_\mathbf{r}^\dagger R_{\mathbf{r}+\mu} - 1 = \sum_{\nu=0,x,y,z} A_{\mu\mathbf{r}}^\nu \sigma^\nu, \tag{6.78}$$

[31]The cumulant expansion is discussed in Sec. 1.6.1.

where

$$A^0_{\mu\mathbf{r}} = -\frac{1}{2}(P_\mathbf{r} - P_{\mathbf{r}+\boldsymbol{\mu}})^2 - \frac{1}{8}(Q_\mathbf{r} - Q_{\mathbf{r}+\boldsymbol{\mu}})^2, \quad A^z_{\mu\mathbf{r}} = \frac{i}{2}(P_\mathbf{r} + P_{\mathbf{r}+\boldsymbol{\mu}})(Q_\mathbf{r} - Q_{\mathbf{r}+\boldsymbol{\mu}}),$$
$$A^x_{\mu\mathbf{r}} = -\frac{i}{2}(Q_\mathbf{r} - Q_{\mathbf{r}+\boldsymbol{\mu}}), \quad A^y_{\mu\mathbf{r}} = -i(P_\mathbf{r} - P_{\mathbf{r}+\boldsymbol{\mu}}). \tag{6.79}$$

$\boldsymbol{\mu}$ denotes a unit vector (with the lattice spacing taking as the unit length) in the direction $\mu = x, y$. In (6.77) and (6.79), we have expanded $A^x_{\mu\mathbf{r}}$ and $A^y_{\mu\mathbf{r}}$ to linear order in $P_\mathbf{r}$, $Q_\mathbf{r}$ and $A^z_{\mu\mathbf{r}}$ to quadratic order, since this is sufficient to obtain $S[P,Q]$ to quadratic order. We introduce current (and density) fields,

$$j^\nu_{0\mathbf{r}} = \Phi^\dagger_\mathbf{r}\sigma^\nu\Phi_\mathbf{r}, \qquad j^\nu_{\mu\mathbf{r}} = -it(\Phi^\dagger_\mathbf{r}\sigma^\nu\Phi_{\mathbf{r}+\boldsymbol{\mu}} \pm \Phi^\dagger_{\mathbf{r}+\boldsymbol{\mu}}\sigma^\nu\Phi_\mathbf{r}), \tag{6.80}$$

where the + (−) sign corresponds to $\nu = 0$ ($\nu \neq 0$). This allows us to write S_1 and S_2 as

$$S_1 = \int_0^\beta d\tau \sum_{\mathbf{r},\nu} A^\nu_{0\mathbf{r}} j^\nu_{0\mathbf{r}}, \qquad S_2 = -i\int_0^\beta d\tau \sum_{\mathbf{r},\mu,\nu} A^\nu_{\mathbf{r}\mu} j^\nu_{\mu\mathbf{r}} \tag{6.81}$$

(we sum over $\nu = 0, x, y, z$ and set $A^0_{0\mathbf{r}} = 0$). We are now in a position to compute the first- and second-order cumulants.

The first-order cumulant $\langle S_1\rangle$ is given by

$$\langle S_1\rangle = \int_0^\beta d\tau \sum_\mathbf{r} A^z_{0\mathbf{r}}\langle j^z_{0\mathbf{r}}\rangle = -i\frac{\Delta}{U}\int_0^\beta d\tau \sum_\mathbf{r}(P_\mathbf{r}\dot{Q}_\mathbf{r} - \dot{P}_\mathbf{r}Q_\mathbf{r}), \tag{6.82}$$

where we have used (6.65) and written the last result in a symmetric form with respect to P and Q. As for $\langle S_2\rangle$, we obtain

$$\langle S_2\rangle = -i\int_0^\beta d\tau \sum_{\mathbf{r},\mu} A^0_{\mu\mathbf{r}}\langle j^0_{\mu\mathbf{r}}\rangle = -\frac{K}{4}\int_0^\beta d\tau \sum_{\langle\mathbf{r},\mathbf{r}'\rangle}\Big[(P_\mathbf{r} - P_{\mathbf{r}'})^2 + \frac{1}{4}(Q_\mathbf{r} - Q_{\mathbf{r}'})^2\Big], \tag{6.83}$$

where

$$K = -2i\langle j^0_{\mu\mathbf{r}}\rangle = \frac{1}{\beta N}\sum_{k,\sigma}\epsilon_\mathbf{k} G_\sigma(k)e^{i\omega_n 0^+} = \int_\mathbf{k}\sum_\sigma \epsilon_\mathbf{k} n_F(\xi_\mathbf{k} - \sigma\Delta) \tag{6.84}$$

is the mean value of the kinetic energy per site.

The second-order cumulant

$$\begin{aligned}\langle(S_1+S_2)^2\rangle_c = \int_0^\beta d\tau d\tau' \sum_{\mathbf{r},\mathbf{r}'}\sum_{\nu,\nu'}\Big[&A^\nu_{0\mathbf{r}}(\tau)\Pi^{\nu\nu'}_{00}(\mathbf{r}\tau,\mathbf{r}'\tau')A^{\nu'}_{0\mathbf{r}'}(\tau')\\ &- 2i\sum_\mu A^\nu_{\mu\mathbf{r}}(\tau)\Pi^{\nu\nu'}_{\mu 0}(\mathbf{r}\tau,\mathbf{r}'\tau')A^{\nu'}_{0\mathbf{r}'}(\tau') - \sum_{\mu,\mu'} A^\nu_{\mu\mathbf{r}}(\tau)\Pi^{\nu\nu'}_{\mu\mu'}(\mathbf{r}\tau,\mathbf{r}'\tau')A^{\nu'}_{\mu'\mathbf{r}'}(\tau')\Big],\end{aligned} \tag{6.85}$$

is expressed in terms of the Hartree–Fock current–current correlation function

$$\Pi^{\nu\nu'}_{\mu\mu'}(\mathbf{r}\tau,\mathbf{r}'\tau') = \langle j^{\nu}_{\mu\mathbf{r}}(\tau) j^{\nu'}_{\mu'\mathbf{r}'}(\tau')\rangle_c \qquad (\mu = 0, x, y). \tag{6.86}$$

To linear order in P and Q,

$$\begin{aligned} A^x_0(q) &= \frac{1}{2} z_0(q) Q_q, \qquad A^y_0(q) = z_0(q) P_q, \qquad A^z_0(q) = 0, \\ A^x_\mu(q) &= \frac{i}{2} z_\mu(q) Q_q, \qquad A^y_\mu(q) = i z_\mu(q) P_q, \qquad A^z_\mu(q) = 0 \end{aligned} \tag{6.87}$$

($\mu \neq 0$), where $z_0(q)$ and $z_\mu(q)$ are defined by

$$z_0(q) = \omega_\nu, \qquad z_\mu(q) = e^{iq_\mu} - 1. \tag{6.88}$$

From (6.82), (6.83) and (6.85), we finally deduce the action

$$S[P,Q] = \frac{1}{2}\sum_q (P_{-q}, Q_{-q}) \mathcal{D}(q) \begin{pmatrix} P_q \\ Q_q \end{pmatrix}, \tag{6.89}$$

where

$$\mathcal{D}(q) = \begin{pmatrix} -\frac{K}{2}\gamma_{\mathbf{q}} - \tilde\Pi^{yy}(q) & -2\frac{\Delta}{U}\omega_\nu - \frac{1}{2}\tilde\Pi^{yx}(q) \\ 2\frac{\Delta}{U}\omega_\nu - \frac{1}{2}\tilde\Pi^{xy}(q) & -\frac{K}{8}\gamma_{\mathbf{q}} - \frac{1}{4}\tilde\Pi^{xx}(q) \end{pmatrix} \tag{6.90}$$

with $\gamma_{\mathbf{q}} = 4 - 2[\cos(q_x) + \cos(q_y)]$. We have introduced

$$\tilde\Pi^{\nu\nu'}(q) = \sum_{\mu,\mu'=0,x,y} z_\mu(-q) z_{\mu'}(q) \Pi^{\nu\nu'}_{\mu\mu'}(q), \tag{6.91}$$

where $\Pi^{\nu\nu'}_{\mu\mu'}(q)$ is the Fourier transform of $\Pi^{\nu\nu'}_{\mu\mu'}(\mathbf{r}\tau,\mathbf{r}'\tau')$. $\Pi^{\nu\nu'}_{\mu\mu'}(q)$ and $\tilde\Pi^{\nu\nu'}(q)$ are further discussed in Appendix 6.A.1. The transverse spin-wave excitation spectrum can be obtained from the vanishing of $\det\mathcal{D}(q)$ after analytic continuation $i\omega_\nu \to \omega + i0^+$,

$$\det\mathcal{D}(q) = \frac{1}{4}\left[\frac{K}{2}\gamma_{\mathbf{q}} + \tilde\Pi^{xx}(q)\right]^2 + 4\frac{\Delta^2}{U^2}\omega_\nu^2 - \frac{1}{4}\tilde\Pi^{xy}(q)^2. \tag{6.92}$$

Since $z_\mu(q)$ and $\tilde\Pi^{\nu\nu'}(q)$ vanish for $q = 0$, $\det\mathcal{D}(q=0) = 0$, which agrees with the existence of a gapless mode. To obtain the dispersion in the low-energy limit, we expand $\mathcal{D}(q)$ about $q = 0$ using $\gamma_{\mathbf{q}} = \mathbf{q}^2$, $z_0(q) = \omega_\nu$ and $z_\mu(q) = iq_\mu$ to leading order in q. This gives (see Appendix 6.A.1)

$$\det\mathcal{D}(q) = \frac{1}{4}\left[\frac{K}{2} + \Pi^{xx}_{xx}(0)\right]^2 \mathbf{q}^4 + 4\frac{\Delta^2}{U^2}\omega_\nu^2 \tag{6.93}$$

retaining terms up to $\mathbf{q}^4$ and ω_ν^2 (since the dispersion is quadratic $\omega_\nu = \mathcal{O}(\mathbf{q}^2)$). We deduce the existence of a transverse spin-wave mode with quadratic dispersion $\omega = \pm D\mathbf{q}^2$, in agreement with the conclusion of Sec. 6.3.2, where

$$D = \frac{U}{4\Delta}\left[\frac{K}{2} + \Pi^{xx}_{xx}(0)\right] \tag{6.94}$$

is defined so as to be positive.[32]

6.4 Slater's Theory of Antiferromagnetism

In this section, we discuss the Slater theory of the (commensurate) spin-density-wave state in the half-filled Hubbard model. This theory is similar to the Stoner theory of ferromagnetism except that the magnetic instability occurs at wavevector $\mathbf{Q} = (\pi, \ldots, \pi)$ rather than $\mathbf{q} = 0$.

For fermions on a bipartite lattice, the origin of the antiferromagnetic ground state can be easily understood. Since spin-up and spin-down fermions interact repulsively, they tend to arrange themselves on alternating lattice sites to minimize their interaction and thus form a spin-density wave with wavevector $\mathbf{Q}$. The decrease of the interaction energy is balanced by an increase in kinetic energy due to the restricted motion. In the weak-coupling limit $U \ll t$, the amplitude of the spin-density wave is small with respect to unity, i.e. $|\langle \hat{\mathbf{S}}_\mathbf{r} \rangle| \ll 1$. The state of the system is referred to as a Slater antiferromagnet. In the strong-coupling limit, each fermion is localized on a lattice site and behaves as a (localized) spin-$\frac{1}{2}$ (Mott localization, see Sec. 6.1.3). The antiferromagnetic state is due to the effective antiferromagnetic coupling $J = 4t^2/U$ between first-neighbor sites. This limit is well described by the Heisenberg model (see Secs. 6.1.3 and 6.4.4). $|\langle \hat{\mathbf{S}}_\mathbf{r} \rangle|$ is of the order of unity and the state of the system is called a Mott–Heisenberg antiferromagnet. At zero temperature, the Hartree–Fock-RPA theory turns out to be sufficient to understand the crossover between the (weak-coupling) Slater antiferromagnet and the (strong-coupling) Mott–Heisenberg antiferromagnet.

In the RPA (Sec. 6.2.2), an instability of the metallic phase towards an antiferromagnetic phase is obtained from the criterion $1 - (U/2)\chi_0(\mathbf{Q}) = 0$, where $\chi_0(\mathbf{Q})$ is given by (6.25).[33] In three dimensions, the susceptibility $\chi_0(\mathbf{Q}) \simeq 2N(0)\ln(E_0/T)$ exhibits a logarithmic divergence at low temperatures $T \ll E_0 \sim 6t$ (Sec. 6.2.1) and one finds the weak-coupling Néel temperature[34]

$$T_\mathrm{N}^\mathrm{HF} \sim E_0 e^{-1/UN(0)}. \tag{6.95}$$

Note that contrary to the ferromagnetic ground state at low filling (which requires U to be larger than $1/N(0)$), the RPA predicts the ground state at half-filling to be antiferromagnetic for any non-zero value of the interaction U. In two dimensions, because of the Van Hove

32 The sign of D can be obtained by considering the limit $U \to \infty$ where $K \to 0$ and $\Pi^{xx}_{xx}(0) \to 4t^2/U$.

33 Which value of the chemical potential should be used in $\chi_0(\mathbf{Q})$? It is natural to set $\mu = 0$, since this ensures that we consider the true Fermi surface of the (non-interacting) half-filled system. However, this seems to contradict the fact that $\mu = U/2$ in the interacting system. The way out of this apparent contradiction is to include in the non-interacting propagator the Hartree–Fock self-energy, which exactly cancels the chemical potential $\mu = U/2$ (see Sec. 6.4.1).

34 By analogy with the Néel state in localized-spin systems (Chapter 8), we refer to T_N^HF as the Néel temperature (in the Hartree–Fock approximation).

singularity at $\epsilon = 0$ in the density of states, $\chi_0(\mathbf{Q}) \simeq \frac{1}{2\pi^2 t} \ln^2(At/T)$ for $U \ll t$ (with A of order unity) and

$$T_{\rm N}^{\rm HF} \sim te^{-2\pi\sqrt{t/U}} \tag{6.96}$$

in the weak-coupling limit.

In the strong-coupling limit $U \gg t$, the transition temperature $T_{\rm N}^{\rm HF}$ is given by $U/4$, which is obviously not the true Néel temperature $T_{\rm N}$. We expect $T_{\rm N}$ to be of the order of J in three dimensions (see Chapter 8) and to vanish in two dimensions because of the Mermin–Wagner theorem. Nevertheless, the Hartree–Fock transition temperature has a physical meaning; it can be interpreted as the temperature below which local moments form as a consequence of the strong onsite repulsion. These local moments order only below the Néel temperature $T_{\rm N}$. Taking into account strong correlations beyond the Hartree–Fock approximation is therefore essential to understand the metallic phase in the temperature range $T_{\rm N} \leq T \lesssim T_{\rm N}^{\rm HF}$. Nevertheless, the Hartree–Fock theory remains a good starting point at zero temperature to describe the crossover from the Slater to the Mott–Heisenberg antiferromagnet as U increases. In Sec. 6.4.2, we shall see that the collective modes, both in the weak- and strong-coupling limits, can be obtained by including fluctuations of the spin density about its Hartree–Fock value (RPA theory). In the following sections, we discuss two approaches which go beyond the Hartree–Fock-RPA theory and remain valid at low temperatures even in two dimensions where the Mermin–Wagner theorem forbids antiferromagnetic long-range order: an NLσM approach (Sec. 6.4.3) and the TPSC theory (Sec. 6.4.5).

6.4.1 *Hartree–Fock theory*

In the antiferromagnetic state, $n = n_{\mathbf{r}\uparrow} + n_{\mathbf{r}\downarrow} = 1$ and $m = (-1)^{\mathbf{r}}(n_{\mathbf{r}\uparrow} - n_{\mathbf{r}\downarrow})$ (staggered magnetization) so that the Hartree–Fock self-energy (Fig. 6.5) is given by

$$\begin{aligned}\Sigma_\sigma(\mathbf{r},\tau;\mathbf{r}',\tau') &= \delta(\mathbf{r}-\mathbf{r}')\delta(\tau-\tau')Un_{\mathbf{r}\bar\sigma} \\ &= \delta(\mathbf{r}-\mathbf{r}')\delta(\tau-\tau')\frac{U}{2}[1-\sigma m(-1)^{\mathbf{r}}],\end{aligned} \tag{6.97}$$

i.e.

$$\Sigma_\sigma(k,k') = \delta_{k,k'}\frac{U}{2} - \delta_{k+Q,k'}\sigma\Delta \tag{6.98}$$

with $Q = (\mathbf{Q}, 0)$ and, with no loss of generality, assuming $\Delta = Um/2 \geq 0$. In the following, we absorb the diagonal part of the self-energy into the chemical potential, which amounts to setting $\mu = 0$ since $\mu = U/2$ at half-filling, so that $\xi_{\mathbf{k}} = \epsilon_{\mathbf{k}}$ (see Sec. 6.1.1).

Since the self-energy is non-diagonal in Fourier space, the one-particle propagator has both diagonal and non-diagonal components. The Dyson equation

$$G_\sigma(k,k') = \delta_{k,k'}G_0(k) + G_0(k)\Sigma_\sigma(k,k+Q)G_\sigma(k+Q,k') \tag{6.99}$$

gives

$$\begin{aligned} G_\sigma(k,k) &= G_0(k) - G_0(k)\sigma\Delta G_\sigma(k+Q,k), \\ G_\sigma(k,k+Q) &= -G_0(k)\sigma\Delta G_\sigma(k+Q,k+Q). \end{aligned} \tag{6.100}$$

Solving (6.100), we find the propagators

$$\begin{aligned} G_\sigma(k,k) &= -\langle\psi_\sigma(k)\psi^*_\sigma(k)\rangle = \frac{-i\omega_n - \epsilon_{\mathbf{k}}}{\omega_n^2 + E_{\mathbf{k}}^2} \equiv G(k), \\ G_\sigma(k,k+Q) &= -\langle\psi_\sigma(k)\psi^*_\sigma(k+Q)\rangle = \frac{\sigma\Delta}{\omega_n^2 + E_{\mathbf{k}}^2} \equiv \sigma F(k), \end{aligned} \tag{6.101}$$

where

$$E_{\mathbf{k}} = \sqrt{\epsilon_{\mathbf{k}}^2 + \Delta^2} \tag{6.102}$$

is the Hartree–Fock energy. The one-particle propagator can be written as

$$\begin{aligned} G(k) &= \frac{u_{\mathbf{k}}^2}{i\omega_n - E_{\mathbf{k}}} + \frac{v_{\mathbf{k}}^2}{i\omega_n + E_{\mathbf{k}}}, \\ F(k) &= -u_{\mathbf{k}}v_{\mathbf{k}}\left(\frac{1}{i\omega_n - E_{\mathbf{k}}} - \frac{1}{i\omega_n + E_{\mathbf{k}}}\right), \end{aligned} \tag{6.103}$$

with

$$u_{\mathbf{k}}^2 = \frac{1}{2}\left(1 + \frac{\epsilon_{\mathbf{k}}}{E_{\mathbf{k}}}\right), \qquad v_{\mathbf{k}}^2 = \frac{1}{2}\left(1 - \frac{\epsilon_{\mathbf{k}}}{E_{\mathbf{k}}}\right), \qquad u_{\mathbf{k}}v_{\mathbf{k}} = \frac{\Delta}{2E_{\mathbf{k}}}. \tag{6.104}$$

With these notations, the spectral function reads

$$A(\mathbf{k},\omega) = -\frac{1}{\pi}\Im[G(\mathbf{k},\omega + i0^+)] = u_{\mathbf{k}}^2\delta(\omega - E_{\mathbf{k}}) + v_{\mathbf{k}}^2\delta(\omega + E_{\mathbf{k}}). \tag{6.105}$$

For $\Delta = 0$, we recover the spectral function $A(\mathbf{k},\omega) = \delta(\omega - \epsilon_{\mathbf{k}})$ of the non-interacting system. In the antiferromagnetic state, the spectrum exhibits two bands $\pm E_{\mathbf{k}}$ separated by a gap $2\Delta = 2\min_{\mathbf{k}} E_{\mathbf{k}}$, which shows that the system is insulating.[35] The origin of this insulating behavior is the spin-density modulation which acts as periodic potential for the fermions and thus opens a gap.[36]

[35] Note that a gap in the one-particle spectrum is not always sufficient to make the system insulating (one-particle excitations are also gapped in a superconductor). To definitely conclude that the antiferromagnetic state is insulating, one should in principle compute the conductivity.

[36] In the Hartree–Fock (mean-field) theory, where $U\hat n_{\mathbf{r}\uparrow}\hat n_{\mathbf{r}\downarrow} \to U(n_{\mathbf{r}\uparrow}\hat n_{\mathbf{r}\downarrow} + \hat n_{\mathbf{r}\uparrow}n_{\mathbf{r}\downarrow} - n_{\mathbf{r}\uparrow}n_{\mathbf{r}\downarrow})$, $Un_{\mathbf{r}\sigma} = U\langle\hat n_{\mathbf{r}\sigma}\rangle$ acts as a periodic potential for the spin-$\bar\sigma$ fermions.

The self-consistency equation for the gap $\Delta = Um/2 = (U/2)(-1)^{\mathbf{r}}\langle \hat{n}_{\mathbf{r}\uparrow} - \hat{n}_{\mathbf{r}\downarrow}\rangle$ reads

$$\frac{2}{U}\Delta = (-1)^{\mathbf{r}} \sum_{\sigma} \sigma G_\sigma(\mathbf{r},\tau;\mathbf{r},\tau^+). \tag{6.106}$$

Using (6.101), we find

$$\frac{2}{U}\Delta = \frac{2}{\beta N}\sum_k F(k) = \Delta \int_{\mathbf{k}} \frac{\tanh(\beta E_{\mathbf{k}}/2)}{E_{\mathbf{k}}}. \tag{6.107}$$

Solving this equation for $\Delta \to 0^+$, we recover the equation $1 - (U/2)\chi_0(\mathbf{Q}) = 0$ which determines the transition temperature $T_{\rm N}^{\rm HF}$. At zero temperature, $\Delta \simeq U/2$ in the strong-coupling limit $U \gg t$ and

$$\begin{aligned} &\Delta \sim E_0 e^{-1/UN(0)} \qquad (d=3), \\ &\Delta \simeq 32te^{-2\pi\sqrt{t/U}} \qquad (d=2), \end{aligned} \tag{6.108}$$

in the weak-coupling limit $U \ll t$. While $\Delta \ll t$ and $|\langle \hat{n}_{\mathbf{r}\uparrow} - \hat{n}_{\mathbf{r}\downarrow}\rangle| \ll 1$ when $U \ll t$, we find $|\langle \hat{n}_{\mathbf{r}\uparrow} - \hat{n}_{\mathbf{r}\downarrow}\rangle| \simeq 1$ when $U \gg t$.

6.4.2 *Transverse spin-wave modes*

In this section, we study transverse spin-wave modes in the antiferromagnetic phase at half-filling and zero temperature using the functional integral formalism. We start from the action $S[\psi^*,\psi,\Delta_c,\Delta_s,\mathbf{\Omega}]$ defined by (6.57) and (6.63) and, for simplicity, consider a two-dimensional square lattice.

The Hartree–Fock theory of the antiferromagnetic phase is obtained from a saddle-point approximation with $\Delta_{c\mathbf{r}} = \Delta_c$, $\Delta_{s\mathbf{r}} = (-1)^{\mathbf{r}}\Delta$ and $\mathbf{\Omega}_{\mathbf{r}}^{\rm cl} = \mathbf{u}_x$.[26] From the saddle-point action

$$\begin{aligned} S_{\rm HF}[\Phi^\dagger,\Phi] = \int_0^\beta d\tau \biggl\{ &\sum_{\mathbf{r}} \left[\frac{1}{U}(\Delta_c^2 + \Delta^2) + \Phi_{\mathbf{r}}^\dagger(\partial_\tau - \mu - i\Delta_c - \Delta(-1)^{\mathbf{r}}\sigma^z)\Phi_{\mathbf{r}} \right] \\ &- t\sum_{\langle \mathbf{r},\mathbf{r}'\rangle} (\Phi_{\mathbf{r}}^\dagger \Phi_{\mathbf{r}'} + \text{c.c.}) \biggr\}, \end{aligned} \tag{6.109}$$

where $\Phi_{\mathbf{r}} = R^\dagger \Psi_{\mathbf{r}}$ is a rotated fermion field with $\mathbf{u}_x$ as spin-quantization axis (see Sec. 6.3.3.2), we obtain the self-consistent equations

$$\frac{2}{U}\Delta_c = i\langle \Phi_{\mathbf{r}}^\dagger \Phi_{\mathbf{r}}\rangle, \qquad \frac{2}{U}\Delta = (-1)^{\mathbf{r}}\langle \Phi_{\mathbf{r}}^\dagger \sigma^z \Phi_{\mathbf{r}}\rangle. \tag{6.110}$$

The equation for Δ is identical to (6.106). At half-filling, $i\Delta_c = -U/2$. Since $\mu = U/2$, we take into account Δ_c by setting the chemical potential μ to zero. In Fourier space, the action is not diagonal,

$$S_{\rm HF}[\Phi^*,\Phi] = -\frac{1}{2}\sum_{k,\sigma} (\phi_\sigma^*(k), \phi_\sigma^*(k+Q)) \begin{pmatrix} i\omega_n - \epsilon_{\mathbf{k}} & \sigma\Delta \\ \sigma\Delta & i\omega_n - \epsilon_{\mathbf{k}+\mathbf{Q}} \end{pmatrix} \begin{pmatrix} \phi_\sigma(k) \\ \phi_\sigma(k+Q) \end{pmatrix} \tag{6.111}$$

but can be straightforwardly diagonalized, and we recover the propagators (6.101) obtained in the Hartree–Fock theory.

The calculation of the collective modes in the antiferromagnetic phase follows the method used in Sec. 6.3.3. We leave it to the reader to show that the fluctuations of the fields $\Delta_{c\mathbf{r}}$ and $\Delta_{s\mathbf{r}}$ are gapped and focus on the transverse spin-wave modes. We parameterize the fluctuations of the unit vector field $\mathbf{\Omega}_\mathbf{r}$ about its classical value $\mathbf{\Omega}_\mathbf{r}^{\rm cl} = \mathbf{u}_x$ by $P_\mathbf{r}$ and $Q_\mathbf{r}$ [Eq. (6.68)]. The action then becomes $S = S_{\rm HF} + S_1 + S_2$ with S_1 and S_2 defined by (6.72) or (6.81). Integrating out the fermion field, we obtain the effective action $S[P,Q]$ defined in (6.73). The only difference with the ferromagnetic case is that the cumulants must now be calculated with the antiferromagnetic Hartree–Fock action (6.109).

The first-order cumulant $\langle S_1 + S_2\rangle$ is given by

$$\begin{aligned} \langle S_1\rangle &= -i\frac{\Delta}{U}\int_0^\beta d\tau \sum_\mathbf{r} (-1)^\mathbf{r}(P_\mathbf{r}\dot{Q}_\mathbf{r} - \dot{P}_\mathbf{r}Q_\mathbf{r}), \\ \langle S_2\rangle &= -\frac{K}{4}\int_0^\beta d\tau \sum_{\langle\mathbf{r},\mathbf{r}'\rangle}\left[(P_\mathbf{r} - P_{\mathbf{r}'})^2 + \frac{1}{4}(Q_\mathbf{r} - Q_{\mathbf{r}'})^2\right], \end{aligned} \tag{6.112}$$

where

$$K = \frac{2}{\beta N}\sum_k \epsilon_\mathbf{k} G(k) e^{i\omega_n 0^+} = 2\int_\mathbf{k} \epsilon_\mathbf{k}[u_\mathbf{k}^2 n_F(E_\mathbf{k}) + v_\mathbf{k}^2 n_F(-E_\mathbf{k})] \tag{6.113}$$

is the mean value of the kinetic energy per site.

The second-order cumulant $\langle (S_1 + S_2)^2\rangle$ can be expressed in terms of the Hartree–Fock current–current correlation functions $\Pi_{\mu\mu'}^{\nu\nu'}$ [Eq. (6.86)]. Contrary to the ferromagnetic phase, $\Pi_{\mu\mu'}^{\nu\nu'}(q,q')$ has both diagonal $q' = q$ and non-diagonal $q' = q + Q$ components. Retaining only the non-zero correlation functions (Appendix 6.A.2), we find[37]

$$S[P,Q] = \frac{1}{2}\sum_q (P_{-q}, Q_{-q-Q})\mathcal{D}(q)\begin{pmatrix} P_q \\ Q_{q+Q}\end{pmatrix}, \tag{6.114}$$

where

$$\mathcal{D}(q) = \begin{pmatrix} -\frac{K}{2}\gamma_\mathbf{q} - \tilde{\Pi}^{yy}(q) & -2\frac{\Delta}{U}\omega_\nu - \frac{1}{2}\tilde{\Pi}^{yx}(q,q+Q) \\ 2\frac{\Delta}{U}\omega_\nu - \frac{1}{2}\tilde{\Pi}^{xy}(q+Q,q) & -\frac{K}{8}\gamma_{\mathbf{q}+\mathbf{Q}} - \frac{1}{4}\tilde{\Pi}^{xx}(q+Q) \end{pmatrix}. \tag{6.115}$$

The spectrum of the transverse spin-wave excitations can be deduced from the vanishing of the determinant

$$\begin{aligned} \det\mathcal{D}(q) &= \frac{1}{4}\left[\frac{K}{2}\gamma_\mathbf{q} + \tilde{\Pi}^{yy}(q)\right]\left[\frac{K}{2}\gamma_{\mathbf{q}+\mathbf{Q}} + \tilde{\Pi}^{xx}(q+Q)\right] \\ &\quad + \frac{1}{4}\left[4\frac{\Delta}{U}\omega_\nu - \tilde{\Pi}^{xy}(q+Q,q)\right]\left[4\frac{\Delta}{U}\omega_\nu - \tilde{\Pi}^{yx}(q,q+Q)\right] \end{aligned} \tag{6.116}$$

[37] We use the notation Q both for the momentum–frequency $Q = (\mathbf{Q}, 0)$ and for the angle fluctuation $\delta\mathbf{\Omega}_\mathbf{r}\cdot\mathbf{u}_y$ as there is in general no ambiguity in the meaning of this variable.

after analytic continuation $i\omega_\nu \to \omega + i0^+$. Since $\gamma_{\mathbf{q}=0} = 0$, $\tilde{\Pi}^{xx}(0) = \tilde{\Pi}^{yy}(0) = 0$ and $\tilde{\Pi}^{xy}(0,Q) = \tilde{\Pi}^{xy}(0,Q) = 0$ (for $T = 0$), the determinant vanishes for $q = 0$ and $q = Q$. We thus obtain the two Goldstone modes expected in the antiferromagnetic phase (Sec. 3.6.3.1). To obtain the dispersion of these modes in the low-energy limit, we expand $\det \mathcal{D}(q)$ about $q = 0$ and $q = Q$. Using the results of Appendix 6.A.2, one finds

$$\det \mathcal{D}(q) = \frac{K}{8}[4K + \tilde{\Pi}^{xx}(Q)]\mathbf{q}^2 + \frac{1}{4}\left[16\left(\frac{\Delta}{U} + \Pi^{xy}_{x0}(Q,0)\right)^2 - \Pi^{yy}_{00}(0)[4K + \tilde{\Pi}^{xx}(Q)]\right]\omega_\nu^2, \tag{6.117}$$

to order $\mathcal{O}(\mathbf{q}^2, \omega_\nu^2)$. We deduce the existence of a mode with linear dispersion $\omega = \pm c|\mathbf{q}|$. With the help of Appendix 6.A.2, we can obtain the final form of the velocity,[38]

$$c = 2t\sqrt{\frac{Uyz}{x}}, \tag{6.118}$$

where

$$x = \int_{\mathbf{k}} \frac{1}{2E^3_{\mathbf{k}}}, \qquad y = \int_{\mathbf{k}} \frac{\sin(k_x)^2}{2E^3_{\mathbf{k}}}, \qquad z = \int_{\mathbf{k}} \frac{\epsilon^2_{\mathbf{k}}}{2E^3_{\mathbf{k}}}. \tag{6.119}$$

Expanding $\det \mathcal{D}(q)$ about $q = Q$, we find the dispersion $\omega = \pm c|\mathbf{Q} - \mathbf{q}|$. In the strong-coupling limit $U \gg t$, using $E_{\mathbf{k}} \simeq \Delta \simeq U/2$ we find $c = \sqrt{2}J$ where $J = 4t^2/U$: For the Mott–Heisenberg antiferromagnet, we recover the velocity obtained in the two-dimensional Heisenberg model in the spin-wave approximation (Sec. 8.2.1). In the weak-coupling limit $U \ll t$, the momentum integrals in (6.119) are dominated by the Van Hove singularities[39] and $c \sim t(t/U)^{1/4}$. The factor $(t/U)^{1/4}$ is a consequence of the Van Hove singularities.

6.4.3 *Beyond spin waves: The quantum nonlinear σ model*[40]

The discussion of the transverse spin-wave excitations in Sec. 6.4.2 is valid only when there is long-range antiferromagnetic order. In a two-dimensional system at finite temperature, a spontaneous symmetry breaking is forbidden by the Mermin–Wagner theorem (Sec. 3.6.4). As already pointed out, the Hartree–Fock temperature $T_{\mathrm{N}}^{\mathrm{HF}}$ nevertheless bears a physical meaning as a crossover temperature below which the amplitude of the antiferromagnetic order parameter takes a non-zero value of order Δ/U (with Δ the Hartree–Fock gap), signaling the appearance of local moments. At weak coupling, these "local" moments can be defined only at length scales of order $\xi_0 \sim t/\Delta$, which corresponds to the size of bound particle–hole pairs in the Hartree–Fock state.[41]

[38]To obtain (6.118), we use $1/U = \int_{\mathbf{k}} 1/2E_{\mathbf{k}}$ and Eq. (6.A.23).

[39]See footnote 5 on page 402.

[40]This section is slightly more advanced than the previous ones. The quantum NLσM will be discussed in detail in Chapter 12.

[41]ξ_0 is analogous to the BCS coherence length (the size of Cooper pairs) in the theory of superconductivity (Sec. 7.5.2). See footnote 13 on page 408.

Below $T_{\rm N}^{\rm HF}$, it is natural to neglect amplitude fluctuations of the antiferromagnetic order parameter and derive an effective action for the field $\boldsymbol{\Omega}_{\mathbf{r}}$ by integrating out the fermionic degrees of freedom. In this section, we show that at low temperatures, when antiferromagnetic fluctuations have a large correlation length and a small excitation gap, this effective action defines a quantum NLσM.

We start from the action

$$S[\Psi^\dagger,\Psi,\Delta_{\mathbf{r}},\boldsymbol{\Omega}] = \int_0^\beta d\tau\biggl\{\sum_{\mathbf{r}}\left[\frac{1}{U}(\Delta_{c\mathbf{r}}^2+\Delta_{\mathbf{r}}^2)+\Psi_{\mathbf{r}}^\dagger(\partial_\tau-\mu-i\Delta_{c\mathbf{r}}-\Delta_{\mathbf{r}}\boldsymbol{\sigma}\cdot\boldsymbol{\Omega}_{\mathbf{r}})\Psi_{\mathbf{r}}\right] \\ - t\sum_{\langle\mathbf{r},\mathbf{r}'\rangle}(\Psi_{\mathbf{r}}^\dagger\Psi_{\mathbf{r}'}+\text{c.c.})\biggr\} \tag{6.120}$$

introduced in Sec. 6.3.3. In the following, we restrict ourselves to low temperatures $T \ll T_{\rm N}^{\rm HF}$ and approximate $\Delta_{c\mathbf{r}}$ and $\Delta_{\mathbf{r}} \equiv \Delta_{s\mathbf{r}}$ by their $T=0$ saddle-point (Hartree–Fock) values.[42] The charge auxiliary field $\Delta_{c\mathbf{r}} = iU/2$ cancels the chemical potential $\mu = U/2$. In contrast to Sec. 6.4.2, we choose a uniform field $\Delta_{\mathbf{r}} \equiv \Delta$ so that antiferromagnetic fluctuations correspond to fluctuations of $\boldsymbol{\Omega}_{\mathbf{r}}$ with momentum near $\mathbf{Q}=(\pi,\pi)$.

In the zero-temperature limit, when the staggered magnetization is polarized along the $\mathbf{u}_x$ axis, fluctuations of $\boldsymbol{\Omega}_{\mathbf{r}}$ along $\mathbf{u}_z$ with momentum $\mathbf{q}$ couple to fluctuations along $\mathbf{u}_y$ with momentum $\mathbf{q}+\mathbf{Q}$. In the calculation of Sec. 6.4.2, this shows up as a coupling between P_q and Q_{Q+q}. This suggests to parameterize the low-energy fluctuations by[43]

$$\boldsymbol{\Omega}_{\mathbf{r}} = (-1)^{\mathbf{r}}\mathbf{n}_{\mathbf{r}}\sqrt{1-\mathbf{L}_{\mathbf{r}}^2}+\mathbf{L}_{\mathbf{r}} \quad \text{with} \quad \mathbf{n}_{\mathbf{r}}^2=1, \quad \mathbf{n}_{\mathbf{r}}\cdot\mathbf{L}_{\mathbf{r}}=0, \tag{6.121}$$

where $\mathbf{n}_{\mathbf{r}}$ is a slowly varying unit vector (Néel field) taking into account antiferromagnetic fluctuations while the canting field $\mathbf{L}_{\mathbf{r}}$, which is assumed to be small, takes care of ferromagnetic fluctuations. Following Secs. 6.3.3 and 6.4.2, we introduce a fermion field $\Phi_{\mathbf{r}}$ defined by $\Psi_{\mathbf{r}} = R_{\mathbf{r}}\Phi_{\mathbf{r}}$ where the SU(2)/U(1) matrix $R_{\mathbf{r}}$ is defined by $\boldsymbol{\sigma}\cdot\mathbf{n}_{\mathbf{r}} = R_{\mathbf{r}}\sigma^z R_{\mathbf{r}}^\dagger$. The associated matrix $\mathcal{R}_{\mathbf{r}} \in \text{SO(3)}$ maps $\mathbf{u}_z$ onto $\mathbf{n}_{\mathbf{r}}$ and the spin of the fermions $\Phi_{\mathbf{r}}$ is quantized along the axis $\mathbf{n}_{\mathbf{r}}$.

The action can now be written as

$$S[\Phi^\dagger,\Phi,\boldsymbol{\Omega}] = \int_0^\beta d\tau\biggl\{\sum_{\mathbf{r}}\Phi_{\mathbf{r}}^\dagger(\partial_\tau+R_{\mathbf{r}}^\dagger\dot{R}_{\mathbf{r}}-\Delta R_{\mathbf{r}}^\dagger\boldsymbol{\sigma}\cdot\bigl[(-1)^{\mathbf{r}}\mathbf{n}_{\mathbf{r}}\sqrt{1-\mathbf{L}_{\mathbf{r}}^2}+\mathbf{L}_{\mathbf{r}}\bigr]R_{\mathbf{r}})\Phi_{\mathbf{r}} \\ - t\sum_{\langle\mathbf{r},\mathbf{r}'\rangle}(\Phi_{\mathbf{r}}^\dagger R_{\mathbf{r}}^\dagger R_{\mathbf{r}'}\Phi_{\mathbf{r}'}+\text{c.c.})\biggr\}. \tag{6.122}$$

We can simplify this expression by noting that

$$\begin{aligned} R_{\mathbf{r}}^\dagger\boldsymbol{\sigma}\cdot\mathbf{n}_{\mathbf{r}}R_{\mathbf{r}} &= \boldsymbol{\sigma}\cdot\mathcal{R}_{\mathbf{r}}^{-1}(\mathbf{n}_{\mathbf{r}}) = \boldsymbol{\sigma}\cdot\mathbf{u}_z = \sigma^z, \\ R_{\mathbf{r}}^\dagger\boldsymbol{\sigma}\cdot\mathbf{L}_{\mathbf{r}}R_{\mathbf{r}} &= \boldsymbol{\sigma}\cdot\mathcal{R}_{\mathbf{r}}^{-1}(\mathbf{L}_{\mathbf{r}}) = \boldsymbol{\sigma}\cdot\mathbf{l}_{\mathbf{r}}, \end{aligned} \tag{6.123}$$

[42] $\Delta_{c\mathbf{r}} = i(U/2)\langle\Psi_{\mathbf{r}}^\dagger\Psi_{\mathbf{r}}\rangle = iU/2$ is temperature independent at half-filling. This is not true for $\Delta_{\mathbf{r}}$ which vanishes for $T \geq T_{\rm N}^{\rm HF}$ in the Hartree–Fock approximation.

[43] A similar parameterization is made when deriving an NLσM from the quantum Heisenberg model (Sec. 8.4.3).

where $\mathbf{l}_\mathbf{r} = \mathcal{R}_\mathbf{r}^{-1}(\mathbf{L}_\mathbf{r})$ is the rotated canting field. Given that $\mathcal{R}_\mathbf{r}^{-1}(\mathbf{n}_\mathbf{r}) = \mathbf{u}_z$ and $\mathbf{n}_\mathbf{r} \perp \mathbf{L}_\mathbf{r}$, the vector $\mathbf{l}_\mathbf{r}$ lies in the $(\mathbf{u}_x, \mathbf{u}_y)$ plane. This gives

$$S[\Phi^\dagger, \Phi, \mathbf{\Omega}] = \int_0^\beta d\tau \Big\{ \sum_\mathbf{r} \Phi_\mathbf{r}^\dagger \big(\partial_\tau + R_\mathbf{r}^\dagger \dot{R}_\mathbf{r} - \Delta\big[(-1)^\mathbf{r} \sigma^z \sqrt{1 - \mathbf{l}_\mathbf{r}^2} + \mathbf{l}_\mathbf{r} \cdot \boldsymbol{\sigma}\big]\big) \Phi_\mathbf{r} \\ - t \sum_\mathbf{r} \sum_{\mu = x,y} \big(\Phi_\mathbf{r}^\dagger R_\mathbf{r}^\dagger R_{\mathbf{r}+\boldsymbol{\mu}} \Phi_{\mathbf{r}+\boldsymbol{\mu}} + \text{c.c.}\big) \Big\}. \tag{6.124}$$

Assuming $\mathbf{n}_\mathbf{r}$ (i.e. $R_\mathbf{r}$) to be slowly varying, we can take the continuum limit and introduce the gauge fields

$$\begin{aligned} A_{0\mathbf{r}} &= -R_\mathbf{r}^\dagger \dot{R}_\mathbf{r} = \sum_{\nu = x,y,z} A_{0\mathbf{r}}^\nu \sigma^\nu, \\ A_{\mu\mathbf{r}} &= i R_\mathbf{r}^\dagger \partial_\mu R_\mathbf{r} = \sum_{\nu = x,y,z} A_{\mu\mathbf{r}}^\nu \sigma^\nu \qquad (\mu = x, y). \end{aligned} \tag{6.125}$$

$A_{\mu\mathbf{r}}$ $(\mu = 0, x, y)$ is small since it is of order $\partial_\mu \mathbf{n}_\mathbf{r}$, and we can expand the action in these variables. The hopping term gives

$$\Phi_\mathbf{r}^\dagger R_\mathbf{r}^\dagger R_{\mathbf{r}+\boldsymbol{\mu}} \Phi_{\mathbf{r}+\boldsymbol{\mu}} = \Phi_\mathbf{r} \Big(1 - i A_{\mu\mathbf{r}} - \frac{1}{2} A_{\mu\mathbf{r}}^2 - \frac{i}{2} \partial_\mu A_{\mu\mathbf{r}}\Big) \Phi_{\mathbf{r}+\boldsymbol{\mu}} \tag{6.126}$$

to second order in $\partial_\mu \mathbf{n}_\mathbf{r}$. To zeroth order in $\mathbf{l}_\mathbf{r}$ and $A_{\mu\mathbf{r}}$, Eq. (6.124) yields the Hartree–Fock action (6.109) with an antiferromagnetic order polarized along the $\mathbf{u}_x$ axis. The first- and second-order corrections in $A_{\mu\mathbf{r}}$ give paramagnetic and diamagnetic terms, S_p and S_d, S'_d, respectively. The corrections in $\mathbf{l}_\mathbf{r}$ give first- and second-order ferromagnetic fluctuation terms, S_l and S_{l^2}[44]:

$$\begin{aligned} S_p &= -\int_0^\beta d\tau \sum_\mathbf{r} \sum_{\substack{\mu = 0,x,y \\ \nu = x,y,z}} j_{\mu\mathbf{r}}^\nu A_{\mu\mathbf{r}}^\nu, \\ S_d &= \frac{i}{2} \int_0^\beta d\tau \sum_\mathbf{r} \sum_{\substack{\mu = x,y \\ \nu = x,y,z}} A_{\mu\mathbf{r}}^{\nu\,2} j_{\mu\mathbf{r}}^0, \quad S'_d = -\frac{1}{2} \int_0^\beta d\tau \sum_\mathbf{r} \sum_{\substack{\mu = x,y \\ \nu = x,y,z}} j_{\mu\mathbf{r}}^\nu \partial_\mu A_{\mu\mathbf{r}}^\nu, \\ S_l &= -\Delta \int_0^\beta d\tau \sum_\mathbf{r} \sum_{\nu = x,y} l_\mathbf{r}^\nu j_{0\mathbf{r}}^\nu, \quad S_{l^2} = \frac{\Delta}{2} \int_0^\beta d\tau \sum_\mathbf{r} (-1)^\mathbf{r} \mathbf{l}_\mathbf{r}^2 j_{0\mathbf{r}}^z, \end{aligned} \tag{6.127}$$

where the current densities are defined by (6.80).

[44] Note that $A_{\mu\mathbf{r}}^\nu$ $(\mu = x, y)$ is real since $A_{\mu\mathbf{r}}^\dagger = A_{\mu\mathbf{r}}$. We also use $A_{\mu\mathbf{r}}^2 = \sum_{\nu,\nu'} A_{\mu\mathbf{r}}^\nu A_{\mu\mathbf{r}}^{\nu'} \sigma^\nu \sigma^{\nu'} = \sum_\nu A_{\mu\mathbf{r}}^{\nu\,2}$.

We can now derive an effective action for the spin variables $\mathbf{n_r}$ and $\mathbf{L_r}$ by integrating out the fermion field $\Phi_\mathbf{r}$. Keeping terms up to second order in $A^\nu_{\mu\mathbf{r}}$ and $l^\nu_\mathbf{r}$, the effective action is given by first- and second-order cumulants of the five perturbative terms (6.127),

$$S[\mathbf{n},\mathbf{L}] = \langle S_p + S_d + S'_d + S_l + S_{l^2}\rangle - \frac{1}{2}\langle (S_p + S_l)^2\rangle_c, \tag{6.128}$$

where the expectation values $\langle\cdots\rangle$ are to be calculated with the Hartree–Fock action. The evaluation of the first-order cumulants is straightforward,

$$\begin{aligned} &\langle S_p\rangle = -2\frac{\Delta}{U}\int_0^\beta d\tau \sum_\mathbf{r} (-1)^\mathbf{r} A^z_{0\mathbf{r}}, \qquad \langle S_d\rangle = -\frac{K}{4}\int_0^\beta d\tau \sum_\mathbf{r} \sum_{\substack{\mu=x,y\\ \nu=x,y,z}} {A^\nu_{\mu\mathbf{r}}}^2, \\ &\langle S'_d\rangle = 0, \qquad \langle S_l\rangle = 0, \qquad \langle S_{l^2}\rangle = \frac{\Delta^2}{U}\int_0^\beta d\tau \sum_\mathbf{r} \mathbf{l}^2_\mathbf{r}, \end{aligned} \tag{6.129}$$

$K = -2i\langle j^0_{\mu\mathbf{r}}\rangle$ denotes the mean value of the kinetic energy per site. As for the second-order cumulants, we obtain

$$\begin{aligned} \langle S_p^2\rangle_c &= \sum_{q,q'} \sum_{\substack{\mu,\mu'=0,x,y\\ \nu,\nu'=x,y,z}} A^\nu_\mu(-q)\Pi^{\nu\nu'}_{\mu\mu'}(q,q')A^{\nu'}_{\mu'}(q'), \\ \langle S_l^2\rangle_c &= \Delta^2 \sum_{q,q'} \sum_{\nu,\nu'=x,y} l^\nu(-q)\Pi^{\nu\nu'}_{00}(q,q')l^{\nu'}(q'), \\ \langle S_p S_l\rangle_c &= \Delta \sum_{q,q'} \sum_{\substack{\mu=x,y\\ \nu,\nu'=x,y}} A^\nu_\mu(-q)\Pi^{\nu\nu'}_{\mu 0}(q,q')l^{\nu'}(q'), \end{aligned} \tag{6.130}$$

where $\Pi^{\nu\nu'}_{\mu\mu'}$ is the Hartree–Fock correlation function (6.86). At low temperatures, when $T \ll T_\mathrm{N}^\mathrm{HF}$, we can approximate $\Pi^{\nu\nu'}_{\mu\mu'}$ by its $T=0$ value. Since we want to keep terms only up to second order in ∂_μ and $\mathbf{l_r}$, we can set $q = q' = 0$ in the Hartree–Fock correlators in (6.130). Denoting $\Pi^{\nu\nu'}_{\mu\mu'}(q=0,q'=0)$ by $\Pi^{\nu\nu'}_{\mu\mu'}$, we obtain the action

$$\begin{aligned} S[\mathbf{L},\mathbf{n}] = \int_0^\beta d\tau \sum_\mathbf{r} \Bigg\{ &-2\frac{\Delta}{U}(-1)^\mathbf{r} A^z_{0\mathbf{r}} - \frac{K}{4}\sum_{\substack{\mu=x,y\\ \nu=x,y,z}} {A^\nu_{\mu\mathbf{r}}}^2 - \frac{1}{2}\Pi^{xx}_{00}\sum_{\nu=x,y} {A^\nu_{0\mathbf{r}}}^2 \\ &- \frac{1}{2}\Pi^{zz}_{xx}\sum_{\mu=x,y} {A^z_{\mu\mathbf{r}}}^2 + \Delta^2\left[\frac{1}{U} - \frac{1}{2}\Pi^{xx}_{00}\right]\mathbf{l}^2_\mathbf{r} - \Delta\Pi^{xx}_{00}\sum_{\nu=x,y} A^\nu_{0\mathbf{r}} l^\nu_\mathbf{r} \Bigg\}, \end{aligned} \tag{6.131}$$

where we have used the fact that the only non-zero current–current correlation functions are $\Pi^{xx}_{00} = \Pi^{yy}_{00}$ and $\Pi^{zz}_{xx} = \Pi^{zz}_{yy}$ (see Appendix 6.A.2). The term

$$\int_0^\beta d\tau \sum_\mathbf{r} \sum_{\mu=x,y} \left(\frac{K}{2} + \Pi^{zz}_{xx}\right) {A^z_{\mu\mathbf{r}}}^2 \tag{6.132}$$

must vanish as it is not gauge-invariant.[45] Because of the rapidly oscillating factor $(-1)^{\mathbf{r}}$, the so-called Berry phase term[46,47]

$$\langle S_p \rangle = -2\frac{\Delta}{U}\int_0^\beta d\tau \sum_{\mathbf{r}} (-1)^{\mathbf{r}} A^z_{0\mathbf{r}} = -i\frac{\Delta}{U}\int_0^\beta d\tau \sum_{\mathbf{r}} (-1)^{\mathbf{r}} [\dot{\varphi}_{\mathbf{r}} \cos(\theta_{\mathbf{r}}) + \dot{\psi}_{\mathbf{r}}] \tag{6.133}$$

vanishes in two dimensions (see Chapter 8). Using

$$\sum_{\nu=x,y} A^{\nu\,2}_{0\mathbf{r}} = -\frac{1}{4}(\partial_\tau \mathbf{n}_{\mathbf{r}})^2, \qquad \sum_{\nu=x,y} A^{\nu\,2}_{\mu\mathbf{r}} = \frac{1}{4}(\partial_\mu \mathbf{n}_{\mathbf{r}})^2 \qquad (\mu \neq 0),$$
$$\sum_{\nu=x,y} A^{\nu}_{0\mathbf{r}} l^{\nu}_{\mathbf{r}} = \frac{i}{2}(\mathbf{n}_{\mathbf{r}} \times \partial_\tau \mathbf{n}_{\mathbf{r}}) \cdot \mathbf{L}_{\mathbf{r}} \tag{6.134}$$

(see Appendix 6.B), we finally obtain

$$S[\mathbf{L},\mathbf{n}] = \int_0^\beta d\tau \sum_{\mathbf{r}} \biggl\{ -\frac{K}{16}(\boldsymbol{\nabla}\mathbf{n}_{\mathbf{r}})^2 + \frac{1}{8}\Pi^{xx}_{00}(\partial_\tau \mathbf{n}_{\mathbf{r}})^2 + \Delta^2\left[\frac{1}{U} - \frac{1}{2}\Pi^{xx}_{00}\right]\mathbf{L}^2_{\mathbf{r}}$$
$$- \frac{i}{2}\Delta\Pi^{xx}_{00}(\mathbf{n}_{\mathbf{r}} \times \partial_\tau \mathbf{n}_{\mathbf{r}}) \cdot \mathbf{L}_{\mathbf{r}} \biggr\}. \tag{6.135}$$

We are now in a position to derive a low-energy effective action $S[\mathbf{n}]$ by integrating out the field $\mathbf{L}$. The functional integral

$$\int \mathcal{D}[\mathbf{L}] \exp\biggl\{ -\int_0^\beta d\tau \sum_{\mathbf{r}} \left[\Delta^2\left[\frac{1}{U} - \frac{1}{2}\Pi^{xx}_{00}\right]\mathbf{L}^2_{\mathbf{r}} - \frac{i}{2}\Delta\Pi^{xx}_{00}(\mathbf{n}_{\mathbf{r}} \times \partial_\tau \mathbf{n}_{\mathbf{r}}) \cdot \mathbf{L}_{\mathbf{r}} \biggr\} \tag{6.136}$$

must be carried out with the constraint $\mathbf{L}_{\mathbf{r}} \cdot \mathbf{n}_{\mathbf{r}} = 0$ [Eq. (6.121)]. Since $\mathbf{n}_{\mathbf{r}} \times \partial_\tau \mathbf{n}_{\mathbf{r}} \perp \mathbf{n}_{\mathbf{r}}$, the part of $\mathbf{L}_{\mathbf{r}}$ which is parallel to $\mathbf{n}_{\mathbf{r}}$ does not contribute to $(\mathbf{n}_{\mathbf{r}} \times \partial_\tau \mathbf{n}_{\mathbf{r}}) \cdot \mathbf{L}_{\mathbf{r}}$ and the constraint $\mathbf{L}_{\mathbf{r}} \cdot \mathbf{n}_{\mathbf{r}} = 0$ is automatically satisfied. The Gaussian functional integral (6.136) gives

$$\exp\biggl\{ -\frac{1}{8}\frac{\Pi^{xx\,2}_{00}}{\frac{2}{U} - \Pi^{xx}_{00}} \int_0^\beta d\tau \sum_{\mathbf{r}} (\partial_\tau \mathbf{n}_{\mathbf{r}})^2 \biggr\} \tag{6.137}$$

using $(\mathbf{n}_{\mathbf{r}} \times \partial_\tau \mathbf{n}_{\mathbf{r}})^2 = (\partial_\tau \mathbf{n}_{\mathbf{r}})^2$. Putting everything together, we finally obtain

$$S[\mathbf{n}] = \frac{\rho^0_s}{2}\int_0^\beta d\tau \int d^2r \left[(\boldsymbol{\nabla}\mathbf{n}_{\mathbf{r}})^2 + c^{-2}(\partial_\tau \mathbf{n}_{\mathbf{r}})^2 \right], \tag{6.138}$$

where

$$\rho^0_s = -\frac{K}{8}, \qquad c^2 = \frac{4\rho^0_s}{\chi^{\mathrm{RPA}}_\perp}, \qquad \chi^{\mathrm{RPA}}_\perp = \frac{\Pi^{xx}_{00}}{1 - \frac{U}{2}\Pi^{xx}_{00}}. \tag{6.139}$$

[45] A gauge transformation amounts to changing the value of $\psi_{\mathbf{r}}$ in (6.75). (The transverse spin-wave calculation in Secs. 6.3.3 and 6.4.2 was carried out in the gauge $\psi_{\mathbf{r}} = 0$.) Since $A^z_\mu = \frac{1}{2}(\cos\theta\partial_\mu\varphi + \partial_\mu\psi)$ for $\mu = x, y$, the contribution (6.132) to the action (6.131) is not gauge-invariant. In Appendix 6.A.2, we show explicitly that $K/2 + \Pi^{zz}_{xx} = 0$.

[46] See Sec. 8.4.1 for a discussion of the Berry phase term.

[47] In Eq. (6.133), θ and φ refer to the Néel field $\mathbf{n} = (\sin\theta\cos\varphi, \sin\theta\sin\varphi, \cos\theta)$.

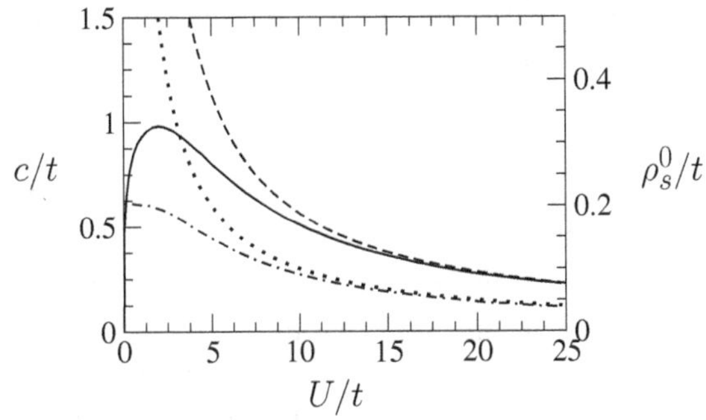

Fig. 6.8 Transverse spin-wave velocity c (solid line) [Eq. (6.118)] and bare stiffness ρ_s^0 (dot-dashed line) obtained from the NLσM [Eq. (6.139)]. The dashed and dotted lines show the results obtained from the two-dimensional antiferromagnetic Heisenberg model with exchange coupling $J = 4t^2/U$. Reprinted from Ref. [29].

Equation (6.138), together with the constraint $\mathbf{n}_\mathbf{r}^2 = 1$, defines a quantum NL$\sigma$M with bare (spin) stiffness ρ_s^0 and velocity c (Fig. 6.8). $\chi_\perp^{\rm RPA}$ is the uniform transverse spin susceptibility $\chi^{xx}(q=0) = \chi^{yy}(q=0)$ in the RPA at zero temperature. c coincides with the velocity obtained in the spin-wave calculation of Sec. 6.4.2.[48] At weak coupling $c \sim t(U/t)^{1/4}$ and $\rho_s^0 \sim t$ whereas $c \simeq \sqrt{2}J$ and $\rho_s^0 \simeq J/4$ when $U \gg t$. At strong coupling $U \gg t$, we recover the quantum NLσM derived from the two-dimensional antiferromagnetic Heisenberg model with exchange coupling $J = 4t^2/U$ (Sec. 8.4.3).

The action $S[\mathbf{n}]$, obtained by taking the continuum limit, must be supplemented with a cutoff Λ in momentum space (and possibly in frequency space). In the strong-coupling limit $U \gg t$, where antiferromagnetic fluctuations are due to local moments, we can take Λ of the order of the inverse lattice spacing. In the weak-coupling limit, the Néel field is ill-defined at length scales smaller than $\xi_0 \sim t/\Delta$ (see the discussion at the beginning of this section). Thus, we choose Λ of the order of $\min(1, \Delta/c)$ (recall that we take the lattice spacing as the unit length).

Let us now discuss the limit of validity of the NLσM at finite temperature. Our derivation is based on the assumption that dominant low-energy fluctuations are transverse spin waves (with respect to the local antiferromagnetic order) with a large correlation length ξ, i.e. a small excitation energy $\sim c/\xi$. The condition $T \ll T_{\rm N}^{\rm HF}$ ensures that amplitude fluctuations of the antiferromagnetic order are frozen at low energies. One should however also verify that the computation of the correlation length ξ within the NLσM is consistent with the assumption of short-range antiferromagnetic order, i.e. $\xi \gg \Lambda^{-1}$. One then finds that the NLσM is valid below a crossover temperature T_X which is of the order of $T_{\rm N}^{\rm HF}$ at weak coupling and of the order of $J = 4t^2/U$ at strong coupling.[49]

The quantum NLσM is studied in Chapter 12. The ground state depends on the value of the coupling constant $g = c/\rho_s^0$ and the cutoff Λ. For g smaller than a critical value

[48]To show the equivalence between (6.118) and (6.139), see footnote 38 on page 426.
[49]See Ref. [29] for a more detailed discussion.

$g_c \equiv g_c(\Lambda)$, the O(3) symmetry of the model is spontaneously broken and there is antiferromagnetic long-range order. At finite temperature, the system is disordered, in agreement with the Mermin–Wagner theorem (Sec. 3.6.4), but the correlation length is exponentially large at low temperatures (renormalized classical regime[50]): $\xi \sim e^{2\pi T/\rho_s}$ where ρ_s is the zero-temperature stiffness. For $g \geq g_c$, quantum fluctuations are sufficiently strong to prevent the occurrence of long-range order in the ground state. For the NLσM derived from the Hubbard model, in the weak-coupling limit $U \ll t$ one finds that $g/g_c \sim e^{-2\pi\sqrt{t/U}}$ is exponentially small so that the ground state is antiferromagnetic. In the strong-coupling limit $U \gg t$, g becomes of the order of g_c. Nevertheless, given the equivalence with the Heisenberg model in this limit, there is no doubt that the ground state of the half-filled Hubbard model on the square lattice is antiferromagnetic.[51]

The NLσM approach discussed in this section allows us to understand the magnetic properties of the system but does not provide us with any information about the fermion properties, in particular in the metallic state at finite temperature where strong antiferromagnetic fluctuations are expected to significantly affect the single-particle excitations. This issue is beyond the scope of the present discussion. Let us nevertheless give a short summary of the phase diagram (Fig. 6.9). Above $T_{\rm N}^{\rm HF}$, spin fluctuations are not important

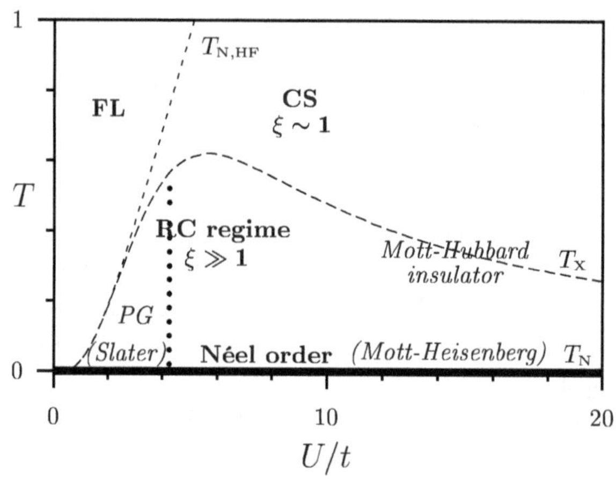

Fig. 6.9 Phase diagram of the two-dimensional half-filled Hubbard model. $T_{\rm N}^{\rm HF} \lesssim T \lesssim 4t$: Fermi liquid (FL) phase; $T_X \lesssim T \lesssim T_{\rm N}^{\rm HF}$: local moments without antiferromagnetic short-range order (Curie spins, $\xi \sim 1$); $0 < T \lesssim T_X$: renormalized classical (RC) regime (local moments with antiferromagnetic short-range order, $\xi \gg 1$); $T = 0$: Slater ($U \ll 4t$) and Mott–Heisenberg ($U \gg 4t$) antiferromagnets. At finite temperature, there is a pseudogap phase ($U \ll 4t$) and a Mott–Hubbard insulator ($U \gg 4t$) separated by a metal–insulator transition (dotted line) defined by the vanishing of the tunneling density of states $N(\omega = 0)$ at zero energy. All lines, except $T_{\rm N} = 0$ (thick solid line), are crossover lines. The NLσM description is valid below T_X in the renormalized classical (RC) regime. Reprinted from Ref. [29].

[50]In the renormalized classical regime, the low-energy physics is dominated by classical fluctuations — associated with the $\omega_\nu = 0$ component of the collective (bosonic) fluctuations — while quantum fluctuations are taken into account *via* a renormalization of the parameters entering the low-energy description (see Chapter 12).

[51]See Chapter 8 for a discussion of the two-dimensional Heisenberg model.

and we expect a Fermi-liquid behavior below a characteristic temperature of the order of the half bandwidth $4t$. Between $T_{\rm N}^{\rm HF}$ and $T_{\rm X}$ (a regime which does not exist in the weak-coupling limit), local moments form but with no antiferromagnetic short-range order (Curie spins: $\xi \sim 1$). Below $T_{\rm X}$, the system enters a renormalized classical regime of spin fluctuations where the antiferromagnetic correlation length becomes exponentially large. Antiferromagnetic long-range order sets in at $T_{\rm N} = 0$. Although there is a smooth evolution of the magnetic properties as a function of U, the physics is quite different for $U \ll 4t$ and $U \gg 4t$. At zero temperature, the system is an antiferromagnet, which evolves from a Slater to a Mott–Heisenberg behavior as U increases. At finite temperature, there is a pseudogap phase for $U \ll 4t$ where the quasi-particle pole of the fermion propagator is suppressed,[52] and a Mott–Hubbard insulator for $U \gg 4t$. These two regimes are separated by a (finite-temperature) metal–insulator transition (dotted line in Fig. 6.9) defined by the vanishing of the tunneling density of states $N(\omega = 0) = -\frac{1}{\pi}\int_{\mathbf{k}} \Im[G^R(\mathbf{k}, \omega = 0)]$ at zero energy.[49]

6.4.3.1 *The attractive Hubbard model at half-filling*

In Sec. 6.1.1, we have seen that the half-filled attractive model ($U < 0$) maps onto the half-filled repulsive model under the particle–hole transformation (6.5). The charge density and pairing operators transform as

$$\begin{aligned} \hat{n}_{\mathbf{r}} &= \sum_\sigma \hat\psi^\dagger_{\mathbf{r}\sigma}\hat\psi_{\mathbf{r}\sigma} \to 2\hat{S}^z_{\mathbf{r}} + 1, \\ \hat\Delta_{\mathbf{r}} &= \hat\psi_{\mathbf{r}\downarrow}\hat\psi_{\mathbf{r}\uparrow} \to (-1)^{\mathbf{r}}\hat{S}^-_{\mathbf{r}}, \qquad \hat\Delta^\dagger_{\mathbf{r}} = \hat\psi^\dagger_{\mathbf{r}\uparrow}\hat\psi^\dagger_{\mathbf{r}\downarrow} \to (-1)^{\mathbf{r}}\hat{S}^+_{\mathbf{r}}, \end{aligned} \tag{6.140}$$

where $\hat{\mathbf{S}}_{\mathbf{r}} = \hat\Psi^\dagger_{\mathbf{r}}(\boldsymbol{\sigma}/2)\hat\Psi_{\mathbf{r}}$ and $\hat{S}^\pm_{\mathbf{r}} = \hat{S}^x_{\mathbf{r}} \pm i\hat{S}^y_{\mathbf{r}}$. Thus, in the attractive half-filled model, $\mathbf{q} = \mathbf{Q}$ charge fluctuations and $\mathbf{q} = 0$ pairing fluctuations combine to form an order parameter with SO(3) symmetry.[53]

The preceding results on the repulsive model can be directly used to deduce the phase diagram of the attractive model (Fig. 6.10). The crossover lines are the same as in Fig. 6.9 but their physical meaning is different. Below the HF transition temperature $T_c^{\rm HF}$, the SO(3) order parameter $(n_{\mathbf{q}=\mathbf{Q}}, \Delta_{\mathbf{q}=0})$ acquires a finite amplitude Δ. This corresponds to the appearance of bound particle–hole and particle–particle pairs with a size $\xi_0 \sim t/\Delta$. Below T_X, directional correlations of the order parameter $(n_{\mathbf{q}=\mathbf{Q}}, \Delta_{\mathbf{q}=0})$ start to grow exponentially (renormalized classical regime) and eventually long-range order sets in at the $T_c = 0$

[52] The pseudogap manifests itself by a strong suppression of spectral weight at low energies in the spectral function $A(\mathbf{k}, \omega) = -\frac{1}{\pi}\Im[G^R(\mathbf{k}, \omega)]$ and the density of states $N(\omega) = \int_{\mathbf{k}} A_\sigma(\mathbf{k}, \omega)$. It becomes a true gap at $T = 0$ when antiferromagnetic long-range order sets in.

[53] Away from half-filling, the degeneracy between charge and pairing fluctuations is lifted (by the magnetic field $\mu - U/2$ in the repulsive model, see Sec. 6.1.1) and the (superconducting) order parameter exhibits SO(2) symmetry. As a result, in the two-dimensional model there is a finite-temperature Berezinskii–Kosterlitz–Thouless (BKT) phase transition to a superconducting state [31].

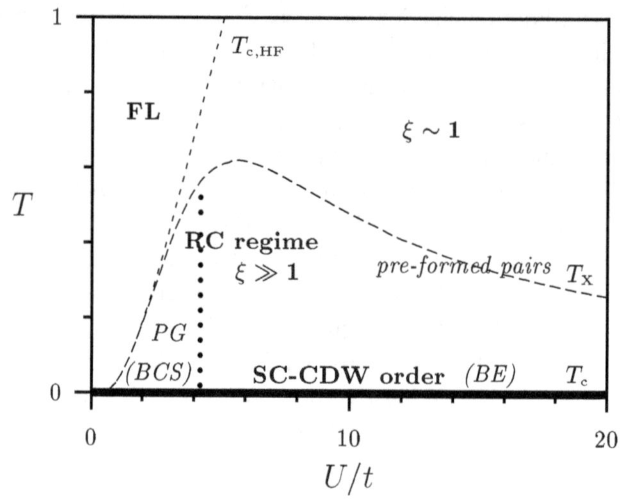

Fig. 6.10 Phase diagram of the two-dimensional half-filled attractive Hubbard model with an attractive interaction $-U \leq 0$. $T \gtrsim T_c^{\rm HF}$: Fermi liquid (FL) phase; $T_X \lesssim T \lesssim T_c^{\rm HF}$: preformed pairs with no superfluid or charge-density-wave order ($\xi \sim 1$); $T \lesssim T_X$: renormalized classical (RC) regime ($\xi \gg 1$); $T = 0$: superconducting (SC) and charge-density-wave (CDW) long-range orders ($U \ll 4t$: BCS limit, $U \gg 4t$: BEC limit). The dotted line is obtained from the vanishing of the tunneling DOS $N(\omega = 0)$ at zero energy. All lines, except $T_c = 0$ (thick solid line), are crossover lines. (Reprinted from Ref. [29].)

phase transition. Because of the SO(3) symmetry, the ground state can have any combination of superconducting and charge-density-wave long-range orders. As $|U|$ increases, the ground state smoothly evolves from the BCS to the BEC limit.[54] In the weak-coupling limit ($|U| \ll 4t$), there is a pseudogap regime at finite temperature due to the directional fluctuations of the SO(3) order parameter. In the strong-coupling limit ($|U| \gg 4t$), between $T_c^{\rm HF}$ and T_X, there is a regime of preformed (local) particle–particle and particle–hole pairs with no superfluid or charge-density-wave short-range order ($\xi \sim 1$). Only below T_X do these bosonic pairs begin to develop short-range order. At $T = 0$, the particle–particle and particle–hole pairs Bose condense and/or localize, thus giving rise to superfluid and/or charge-density-wave long-range orders. The dotted line in Fig. 6.10 is obtained from the vanishing of the tunneling density of states $N(\omega = 0)$ at zero energy.

6.4.4 *Large-U limit of the half-filled Hubbard model*

In Secs. 6.4.2 and 6.4.3, we have seen that the half-filled Hubbard model reproduces the results of the spin-$\frac{1}{2}$ Heisenberg model in the limit $U \gg t$. In the following, we show that the low-energy effective action $S[\boldsymbol{\Omega}]$ in the large-U limit coincides with the action of the Heisenberg model.

[54]See Sec. 7.7 for a discussion of the BCS–BEC crossover in fermionic superfluids.

We start from the action $S = S_{\rm loc} + S_1 + S_2$, where

$$\begin{aligned} S_{\rm loc}[\Phi^\dagger, \Phi] &= \int_0^\beta d\tau \sum_{\mathbf{r}} \Phi_{\mathbf{r}}^\dagger (\partial_\tau - \mu - i\Delta_c - \Delta\sigma^z)\Phi_{\mathbf{r}}, \\ S_1[\Phi^\dagger, \Phi, \mathbf{\Omega}] &= \int_0^\beta d\tau \sum_{\mathbf{r}} \Phi_{\mathbf{r}}^\dagger R_{\mathbf{r}}^\dagger \dot{R}_{\mathbf{r}} \Phi_{\mathbf{r}}, \\ S_2[\Phi^\dagger, \Phi, \mathbf{\Omega}] &= -\int_0^\beta d\tau \sum_{\mathbf{r},\mathbf{r}'} t_{\mathbf{r}\mathbf{r}'} \Phi_{\mathbf{r}}^\dagger R_{\mathbf{r}}^\dagger R_{\mathbf{r}'} \Phi_{\mathbf{r}'} \end{aligned} \tag{6.141}$$

(see Sec. 6.3.3), where we consider the fields Δ_c and Δ at the saddle point level: $\Delta_c = i(U/2)\langle \Phi_{\mathbf{r}}^\dagger \Phi_{\mathbf{r}} \rangle$ and $\Delta = (U/2)\langle \Phi_{\mathbf{r}}^\dagger \sigma^z \Phi_{\mathbf{r}} \rangle$. At half-filling, $\mu + i\Delta_c = 0$ and, in the large-U limit, $\Delta \simeq U/2$. We integrate out the fermion field in a cumulant expansion. To obtain the effective action $S[\mathbf{\Omega}]$ to first order in ∂_τ and second order in t, it is sufficient to retain the first-order cumulant $\langle S_1 + S_2 \rangle$ and the second-order one $\langle S_2^2 \rangle_c$, i.e.

$$S[\mathbf{\Omega}] = \langle S_1 + S_2 \rangle - \frac{1}{2} \langle S_2^2 \rangle_c, \tag{6.142}$$

where the expectation value $\langle \cdots \rangle$ is to be calculated with the local action $S_{\rm loc}$.

The first-order cumulant $\langle S_2 \rangle$ vanishes since $\langle \Phi_{\mathbf{r}\sigma}^* \Phi_{\mathbf{r}'\sigma'} \rangle = 0$ for $\mathbf{r} \neq \mathbf{r}'$. With the gauge choice (6.75),

$$\langle S_1 \rangle = -\frac{i}{2} \int_0^\beta d\tau \sum_{\mathbf{r}} \langle \Phi_{\mathbf{r}}^\dagger \sigma^z \Phi_{\mathbf{r}} \rangle \cos(\theta_{\mathbf{r}}) \dot{\varphi}_{\mathbf{r}} \simeq -\frac{i}{2} \int_0^\beta d\tau \sum_{\mathbf{r}} \cos(\theta_{\mathbf{r}}) \dot{\varphi}_{\mathbf{r}}. \tag{6.143}$$

Introducing the spin-coherent state

$$|\mathbf{\Omega}_{\mathbf{r}}\rangle = R_{\mathbf{r}} |\uparrow\rangle = \cos\left(\frac{\theta_{\mathbf{r}}}{2}\right) e^{-\frac{i}{2}\varphi_{\mathbf{r}}} |\uparrow\rangle + \sin\left(\frac{\theta_{\mathbf{r}}}{2}\right) e^{\frac{i}{2}\varphi_{\mathbf{r}}} |\downarrow\rangle \tag{6.144}$$

(see Sec. 8.4.2), we rewrite (6.143) as

$$\langle S_1 \rangle = \int_0^\beta d\tau \sum_{\mathbf{r}} \langle \mathbf{\Omega}_{\mathbf{r}} | \dot{\mathbf{\Omega}}_{\mathbf{r}} \rangle, \tag{6.145}$$

which we recognize as the Berry-phase term of the spin-coherent-state functional integral (Sec. 8.4.2.2).

The second-order cumulant reads

$$\begin{aligned} \langle S_2^2 \rangle_c = \int_0^\beta d\tau_1 d\tau_2 \sum_{\mathbf{r}_1 \cdots \mathbf{r}_2'} \sum_{\sigma_1 \cdots \sigma_2'} & t_{\mathbf{r}_1\mathbf{r}_1'} t_{\mathbf{r}_2\mathbf{r}_2'} (R_{\mathbf{r}_1}^\dagger(\tau_1) R_{\mathbf{r}_1'}(\tau_1))_{\sigma_1\sigma_1'} (R_{\mathbf{r}_2}^\dagger(\tau_2) R_{\mathbf{r}_2'}(\tau_2))_{\sigma_2\sigma_2'} \\ & \times \langle \Phi_{\mathbf{r}_1\sigma_1}^*(\tau_1) \Phi_{\mathbf{r}_1'\sigma_1'}(\tau_1) \Phi_{\mathbf{r}_2\sigma_2}^*(\tau_2) \Phi_{\mathbf{r}_2'\sigma_2'}(\tau_2) \rangle_c. \end{aligned} \tag{6.146}$$

Using Wick's theorem, we obtain

$$\langle S_2^2\rangle_c = -\int_0^\beta d\tau_1 d\tau_2 \sum_{\mathbf{r}_1,\mathbf{r}_2,\sigma_1,\sigma_2} t_{\mathbf{r}_1\mathbf{r}_2}^2 (R_{\mathbf{r}_1}^\dagger(\tau_1)R_{\mathbf{r}_2}(\tau_1))_{\sigma_1\sigma_2}(R_{\mathbf{r}_2}^\dagger(\tau_2)R_{\mathbf{r}_1}(\tau_2))_{\sigma_2\sigma_1} \times G_{\sigma_1}(\tau_2-\tau_1)G_{\sigma_2}(\tau_1-\tau_2), \tag{6.147}$$

where G_σ is the (local) one-particle propagator corresponding to the action S_{loc}. Since the propagator $G_\sigma(i\omega_n) = (i\omega_n + \sigma\Delta)^{-1}$ is gapped, $G_\sigma(\tau)$ decays exponentially in time over a characteristic time $\sim 1/\Delta \sim 1/U$. If we are interested in the effective action $S[\mathbf{\Omega}]$ for energies much smaller than U, we can approximate $R_{\mathbf{r}_2}^\dagger(\tau_2)R_{\mathbf{r}_1}(\tau_2)$ by $R_{\mathbf{r}_2}^\dagger(\tau_1)R_{\mathbf{r}_1}(\tau_1)$ in (6.147). Using[55]

$$\int_0^\beta d\tau_2\, G_{\sigma_1}(\tau_2-\tau_1)G_{\sigma_2}(\tau_1-\tau_2) = \frac{1}{\beta}\sum_{\omega_n} G_{\sigma_1}(i\omega_n)G_{\sigma_2}(i\omega_n) \simeq -\frac{1}{U}\delta_{\sigma_1,\bar\sigma_2} \tag{6.148}$$

and

$$(R_\mathbf{r}^\dagger R_{\mathbf{r}'})_{\sigma\bar\sigma}(R_{\mathbf{r}'}^\dagger R_\mathbf{r})_{\bar\sigma\sigma} = \frac{1}{2}(1-\mathbf{\Omega}_\mathbf{r}\cdot\mathbf{\Omega}_{\mathbf{r}'}) \tag{6.149}$$

[see Eq. (6.75)], we finally obtain

$$\langle S_2^2\rangle_c = \int_0^\beta d\tau \sum_{\mathbf{r},\mathbf{r}'} \frac{t_{\mathbf{r}\mathbf{r}'}^2}{U}(1-\mathbf{\Omega}_\mathbf{r}\cdot\mathbf{\Omega}_{\mathbf{r}'}). \tag{6.150}$$

From (6.145) and (6.150), we conclude that the low-energy effective action of the half-filled Hubbard model,

$$S[\mathbf{\Omega}] = \int_0^\beta d\tau \Big\{ \sum_\mathbf{r} \langle \mathbf{\Omega}_\mathbf{r}|\dot{\mathbf{\Omega}}_\mathbf{r}\rangle + J\sum_{\langle\mathbf{r},\mathbf{r}'\rangle} \frac{\mathbf{\Omega}_\mathbf{r}\cdot\mathbf{\Omega}_{\mathbf{r}'}-1}{4} \Big\}, \tag{6.151}$$

coincides with the action of the spin-$\frac{1}{2}$ Heisenberg model with an exchange coupling $J = 4t^2/U$. Equation (6.151) is valid for energies and temperatures much smaller than U.

6.4.5 *Two-particle self-consistent theory*

The NLσM approach discussed in Sec. 6.4.3 is valid at low temperatures, below the crossover temperature T_X, in a regime where the antiferromagnetic correlation length is much larger than the lattice spacing. To describe the system near T_X and at higher temperatures, a different method is required. In this section, we discuss the TPSC theory, an approach that allows one to describe, as the temperature decreases, the growth of antiferromagnetic correlations and the formation of the pseudogap in the two-dimensional Hubbard model at weak and intermediate coupling (we consider only the half-filled case).

[55] At low temperatures, i.e. when $T \ll \Delta$, we can evaluate the Matsubara sum in (6.148) in the limit $T = 0$.

6.4.5.1 *The RPA and its limitations*

One of the simplest approaches to the Hubbard model is the RPA (see Sec. 6.2.2), where the charge and spin susceptibilities are given by

$$\begin{aligned}\chi_{\rm ch}(q) &= \frac{\chi_0(q)}{1+\frac{U}{2}\chi_0(q)},\\ \chi_{\rm sp}(q) &= \frac{\chi_0(q)}{1-\frac{U}{2}\chi_0(q)},\end{aligned} \tag{6.152}$$

with $\chi_0(q)$ the non-interacting particle–hole susceptibility [Eq. (6.23)]. As discussed at the beginning of Sec. 6.4, the RPA (or the Hartree–Fock theory) predicts a transition to an antiferromagnetic state at a finite temperature T_c which is forbidden by the Mermin–Wagner theorem in two dimensions.

The RPA suffers from another drawback: It violates the Pauli principle. Let us consider the two diagrams shown in Fig. 6.11(a). For a local interaction, they have opposite sign and cancel (a mere consequence of Pauli principle that prevents two fermions with the same spin σ to be at the same site). This implies that these diagrams do not contribute to the particle–hole response functions $\chi_{\rm ch}$ and $\chi_{\rm sp}$. In the Hubbard model, this is usually automatically taken into account by writing the interaction Hamiltonian as $\hat{H}_{\rm int} = U\sum_{\mathbf{r}} \hat{n}_{\mathbf{r}\uparrow}\hat{n}_{\mathbf{r}\downarrow}$.

More generally, to satisfy the Pauli principle, one must take into account all diagrams that are obtained by interchanging the external legs. This is indeed a necessary condition for the vertices to be antisymmetric under the exchange of two external legs (crossing symmetries). Consider, for instance, the two diagrams of Fig. 6.11(b). The first one is included in the RPA but not the second one. As a result, the RPA violates the crossing symmetries and the Pauli principle.

One can also reach this conclusion by considering

$$\begin{aligned}\frac{1}{\beta N}\sum_q \chi_{\rm ch}(q) &= \langle(\hat{n}_{\mathbf{r}\uparrow}+\hat{n}_{\mathbf{r}\downarrow})^2\rangle - n^2 = n - n^2 + 2\langle\hat{n}_{\mathbf{r}\uparrow}\hat{n}_{\mathbf{r}\downarrow}\rangle,\\ \frac{1}{\beta N}\sum_q \chi_{\rm sp}(q) &= \langle(\hat{n}_{\mathbf{r}\uparrow}-\hat{n}_{\mathbf{r}\downarrow})^2\rangle = n - 2\langle\hat{n}_{\mathbf{r}\uparrow}\hat{n}_{\mathbf{r}\downarrow}\rangle,\end{aligned} \tag{6.153}$$

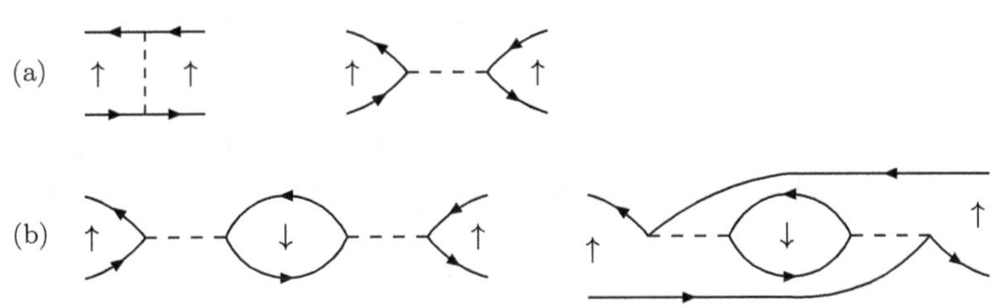

Fig. 6.11 Examples of diagrams obtained by interchanging two incoming legs. In case (a), and for a local interaction, the two diagrams have opposite signs and their sum vanishes.

where we have used $\hat{n}_{\mathbf{r}\sigma}^2 = \hat{n}_{\mathbf{r}\sigma}$, which is a consequence of the Pauli principle, and $\langle \hat{n}_{\mathbf{r}\sigma} \rangle = n/2$ in the paramagnetic phase. Equations (6.153) imply the sum rule

$$\frac{1}{\beta N} \sum_q [\chi_{\rm ch}(q) + \chi_{\rm sp}(q)] = 2n - n^2. \tag{6.154}$$

In the RPA, the lhs of this equation reads

$$\frac{1}{\beta N} \sum_q \left[\frac{\chi_0(q)}{1 + \frac{U}{2}\chi_0(q)} + \frac{\chi_0(q)}{1 - \frac{U}{2}\chi_0(q)} \right]. \tag{6.155}$$

For $U = 0$, this expression must give $2n - n^2$ since the non-interacting susceptibility satisfies the sum rule (6.154). To second order in U, Eq. (6.155) gives

$$\frac{U^2}{2\beta N} \sum_q \chi_0^3(q). \tag{6.156}$$

Using $\omega \chi_0''(\mathbf{q}, \omega) \geq 0$, we find that

$$\chi_0(q) = \int_{-\infty}^{\infty} \frac{d\omega}{\pi} \frac{\chi_0''(\mathbf{q}, \omega)}{\omega - i\omega_\nu} = \int_{-\infty}^{\infty} \frac{d\omega}{\pi} \frac{\omega \chi_0''(\mathbf{q}, \omega)}{\omega^2 + \omega_\nu^2} \tag{6.157}$$

is strictly[56] positive and the $\mathcal{O}(U^2)$ contribution to (6.155) cannot vanish. We conclude that the sum rule (6.154) is violated in the RPA.

6.4.5.2 *Two-particle self-consistent theory*

The TPSC theory can be seen as an improvement of the RPA that preserves the simple mathematical structure of the charge and spin susceptibilities,

$$\begin{aligned} \chi_{\rm ch}(q) &= \frac{\chi^{(1)}(q)}{1 + \frac{U_{\rm ch}}{2}\chi^{(1)}(q)}, \\ \chi_{\rm sp}(q) &= \frac{\chi^{(1)}(q)}{1 - \frac{U_{\rm sp}}{2}\chi^{(1)}(q)}, \end{aligned} \tag{6.158}$$

but, in the spirit of Fermi-liquid theory, with two *a priori* unknown (possibly temperature-dependent) interaction constants $U_{\rm ch}$ and $U_{\rm sp}$. Here, $\chi^{(1)} \sim \int G^{(1)} G^{(1)}$ is a particle–hole pair propagator, and $U_{\rm ch}$ ($U_{\rm sp}$) the 2PI vertex in the charge (spin) particle–hole channel (see Sec. 1.6.3). We assume that it is sufficient at this stage to compute $\chi^{(1)}$ with a one-particle propagator $G^{(1)}$ including only the Hartree–Fock self-energy (we shall come back to this point later). Absorbing the latter in the definition of μ, $\chi^{(1)}(q)$ is the same as the Lindhard function $\chi_0(q)$.[57]

[56] A vanishing $\chi_0(q)$ would imply $\chi_0''(\mathbf{q}, \omega) = 0$ for all frequencies, which is clearly non-physical.
[57] At half-filling, the Hartree–Fock self-energy cancels the chemical potential $\mu = U/2$ (Sec. 6.4.1) so that $\chi^{(1)}$ is equal to χ_0 evaluated with $\mu = 0$.

Substituting Eqs. (6.158) into the two sum rules (6.153), we could determine $U_{\rm ch}$ and $U_{\rm sp}$ if we knew $\langle \hat{n}_{\mathbf{r}\uparrow}\hat{n}_{\mathbf{r}\downarrow}\rangle$. The latter is obtained from the ansatz

$$U\langle \hat{n}_{\mathbf{r}\uparrow}\hat{n}_{\mathbf{r}\downarrow}\rangle = U_{\rm sp}\langle \hat{n}_{\mathbf{r}\uparrow}\rangle\langle \hat{n}_{\mathbf{r}\downarrow}\rangle, \tag{6.159}$$

which is inspired from previous works on the electron gas [37,38] and will be justified in the following in a more formal derivation of the TPSC theory.

One can verify that the TPSC theory does not spoil the good properties of the RPA (due to the fact that it is a conserving approximation, see Sec. 9.1.3.1), namely

$$\chi_{\rm ch,sp}(0, i\omega_\nu) = 0 \quad \text{for} \quad \omega_\nu \neq 0, \tag{6.160}$$

and

$$\begin{aligned}\lim_{|\omega_\nu|\to\infty} \omega_\nu^2 \chi_{\rm ch,sp}(q) &= \int_{-\infty}^{\infty} \frac{d\omega}{\pi}\omega \chi''_{\rm ch,sp}(\mathbf{q},\omega) \\ &= 2\int_{\mathbf{k}} (\epsilon_{\mathbf{k}+\mathbf{q}} + \epsilon_{\mathbf{k}-\mathbf{q}} - 2\epsilon_{\mathbf{k}}) n_F(\xi_{\mathbf{k}}). \end{aligned} \tag{6.161}$$

Equation (6.160) follows trivially from

$$\chi^{(1)}(0, i\omega_\nu) = \chi_0(0, i\omega_\nu) = 0 \quad \text{for} \quad \omega_\nu \neq 0. \tag{6.162}$$

Note that (6.160) would be violated if $\chi^{(1)}$ were to include non-trivial self-energy corrections (see Sec. 9.1.3.1). Equation (6.161) is the analog of the f-sum rule (3.106) in the lattice model and is easily derived,

$$\begin{aligned}\lim_{|\omega_\nu|\to\infty} \omega_\nu^2 \chi_{\rm ch,sp}(q) &= \lim_{|\omega_\nu|\to\infty} \omega_\nu^2 \chi^{(1)}(q) \\ &= -2 \lim_{|\omega_\nu|\to\infty} \omega_\nu^2 \int_{\mathbf{k}} \frac{n_F(\xi_{\mathbf{k}}) - n_F(\xi_{\mathbf{k}+\mathbf{q}})}{i\omega_\nu + \xi_{\mathbf{k}} - \xi_{\mathbf{k}+\mathbf{q}}} \\ &= 2\int_{\mathbf{k}} (\xi_{\mathbf{k}+\mathbf{q}} + \xi_{\mathbf{k}-\mathbf{q}} - 2\xi_{\mathbf{k}}) n_F(\xi_{\mathbf{k}}), \end{aligned} \tag{6.163}$$

using $\lim_{|\omega_\nu|\to\infty} \chi^{(1)}_{\rm ch,sp}(q) = 0$.

Let us now show that the TPSC theory satisfies the Mermin–Wagner theorem. The double occupancy $\langle \hat{n}_{\mathbf{r}\uparrow}\hat{n}_{\mathbf{r}\downarrow}\rangle$ is bounded by its values at $U=0$ and $U=\infty$,

$$\langle \hat{n}_{\mathbf{r}\uparrow}\hat{n}_{\mathbf{r}\downarrow}\rangle_{U=\infty} \leq \langle \hat{n}_{\mathbf{r}\uparrow}\hat{n}_{\mathbf{r}\downarrow}\rangle \leq \langle \hat{n}_{\mathbf{r}\uparrow}\hat{n}_{\mathbf{r}\downarrow}\rangle_{U=0} \quad \text{i.e.} \quad 0 \leq \langle \hat{n}_{\mathbf{r}\uparrow}\hat{n}_{\mathbf{r}\downarrow}\rangle \leq \frac{n^2}{4}, \tag{6.164}$$

where we have used Wick's theorem to obtain $\langle \hat{n}_{\mathbf{r}\uparrow}\hat{n}_{\mathbf{r}\downarrow}\rangle = \langle \hat{n}_{\mathbf{r}\uparrow}\rangle\langle \hat{n}_{\mathbf{r}\downarrow}\rangle = (n/2)^2$ when $U=0$. This implies that $n - 2\langle \hat{n}_{\mathbf{r}\uparrow}\hat{n}_{\mathbf{r}\downarrow}\rangle$ is in the range $[n - n^2/2, n]$. The second of the sum rules (6.153) can be rewritten as

$$\frac{1}{\beta N}\sum_{\mathbf{q}} \chi_{\rm sp}(\mathbf{q}, 0) = n - 2\langle \hat{n}_{\mathbf{r}\uparrow}\hat{n}_{\mathbf{r}\downarrow}\rangle - \frac{1}{\beta N}\sum_{\mathbf{q},\omega_\nu\neq 0} \chi_{\rm sp}(\mathbf{q}, i\omega_\nu), \tag{6.165}$$

where the rhs is obviously finite. Since $\chi_{\rm sp}(\mathbf{q},0)$ is maximum for $\mathbf{q}=\mathbf{Q}=(\pi,\pi)$, we can use

$$\chi^{(1)}(\mathbf{Q}+\mathbf{q},0) \simeq \chi^{(1)}(\mathbf{Q},0) + \frac{1}{2}\mathbf{q}^2 \frac{\partial^2\chi^{(1)}(\mathbf{q},0)}{\partial q_x^2}\bigg|_{\mathbf{q}=\mathbf{Q}} \tag{6.166}$$

for small $\mathbf{q}$ and approximate the static spin susceptibility by

$$\chi_{\rm sp}(\mathbf{Q}+\mathbf{q},0) \simeq \frac{C}{\mathbf{q}^2+\xi^{-2}}, \tag{6.167}$$

where the antiferromagnetic correlation length is defined by

$$\xi^2 = -\frac{U_{\rm sp}}{4}\frac{\partial^2_{q_x}\chi^{(1)}(\mathbf{q},0)|_{\mathbf{q}=\mathbf{Q}}}{1-\frac{U_{\rm sp}}{2}\chi^{(1)}(\mathbf{Q},0)}. \tag{6.168}$$

Thus, the sum rule (6.165) becomes

$$T\int_0^\Lambda \frac{d|\mathbf{q}|}{2\pi}\frac{|\mathbf{q}|C}{\mathbf{q}^2+\xi^{-2}} = n - 2\langle \hat n_{\mathbf{r}\uparrow}\hat n_{\mathbf{r}\downarrow}\rangle - \frac{1}{\beta N}\sum_{\mathbf{q},\omega_\nu\neq 0}\chi_{\rm sp}(\mathbf{q},i\omega_\nu) \equiv C'(T), \tag{6.169}$$

where we have introduced an upper momentum cutoff Λ (below which (6.167) is expected to be valid). Since the rhs is finite, ξ cannot diverge when $T>0$, otherwise the integral in the lhs would be logarithmically divergent.[58] This implies that $U_{\rm sp}$ must decrease at low temperatures to prevent a divergence of $\chi_{\rm sp}(\mathbf{Q},0)$. From (6.169), we deduce[59]

$$\xi \sim \Lambda e^{\frac{2\pi\rho_s(T)}{T}}, \tag{6.170}$$

where $\rho_s(T)=C'(T)/C$ can be seen as an effective spin-wave stiffness. Equation (6.170) is similar to the NLσM result of Sec. 6.4.3 (with however a T-dependent stiffness $\rho_s(T)$), and we can define a crossover temperature T_X below which there is antiferromagnetic short-range order, i.e. $\xi \gg \Lambda^{-1}$ (we shall give a more precise definition of T_X below).[60]

So far, we have assumed that it is sufficient to consider the Hartree–Fock self-energy in the one-particle propagator. This is usually justified for charge and spin excitations, since the collective modes are determined more by conservation laws than by details of the self-energy. The self-energy on the other hand is very sensitive to the collective modes which play an important role at low energies. The second step of the TPSC theory is thus

[58] In three dimensions, the momentum integral remains finite when $\xi^{-1}\to 0$ and nothing prevents ξ to diverge at a non-zero temperature.

[59] A more careful analysis of the non-zero Matsubara-frequency terms would give a temperature-dependent prefactor.

[60] The NLσM approach and the TPSC theory are valid in different, overlapping, temperature ranges. The former requires $T \lesssim T_X$ while the latter is expected to fail, at least quantitatively, when $T \ll T_X$ but remains valid at high temperatures (see the discussion in the following).

to find a better approximation for the self-energy.[61] To do so, we start from the exact expression (1.437). For the Hubbard model, it reads[62]

$$\Sigma_\sigma(k) = -\frac{U}{(\beta N)^2}\sum_{k',q} G_\sigma(k+q)G_{\bar\sigma}(k')G_{\bar\sigma}(k'-q)\Gamma^{(4)}_{\sigma\bar\sigma,\bar\sigma,\sigma}(k+q,k'-q;k',k), \quad (6.171)$$

dropping the Hartree–Fock contribution which is already taken into account by a renormalization of the chemical potential. We can define a particle–hole interaction vertex by $\Gamma^{(4)}_{\rm ph}(1,1';2,2') = \Gamma^{(4)}(1,2;2',1')$, which can be expanded in the charge and spin channels,[63]

$$\begin{aligned}\Gamma^{(4)}_{{\rm ph};\sigma_1\sigma_2,\sigma_3\sigma_4}(k+q,k;k'-q,k') &= \sigma^0_{\sigma_1\sigma_2}\sigma^0_{\sigma_3\sigma_4}\Gamma_{\rm ch}(k+q,k;k'-q,k') \\ &\quad + \boldsymbol{\sigma}_{\sigma_1\sigma_2}\cdot\boldsymbol{\sigma}_{\sigma_3\sigma_4}\Gamma_{\rm sp}(k+q,k;k'-q,k'). \quad (6.172)\end{aligned}$$

In the TPSC theory, one approximates G in (6.171) by the Hartree–Fock propagator $G^{(1)}$ and makes an RPA-like approximation for $\Gamma_{\rm ch,sp}$,[64]

$$\begin{aligned}\Gamma_{\rm ch}(k+q,k;k'-q,k') &= \frac{1}{2}\frac{U_{\rm ch}}{1+\frac{U_{\rm ch}}{2}\chi^{(1)}(q)}, \\ \Gamma_{\rm sp}(k+q,k;k'-q,k') &= -\frac{1}{2}\frac{U_{\rm sp}}{1-\frac{U_{\rm sp}}{2}\chi^{(1)}(q)}.\end{aligned} \quad (6.173)$$

This allows us to rewrite the self-energy (6.171) in the form

$$\Sigma_\sigma(k) = \frac{U}{4\beta N}\sum_q G^{(1)}_\sigma(k+q)[U_{\rm ch}\chi_{\rm ch}(q) + U_{\rm sp}\chi_{\rm sp}(q)]. \quad (6.174)$$

Alternatively, one could define the particle–hole interaction vertex by $\Gamma^{(4)}_{\rm ph}(1,2';2,1') = -\Gamma^{(4)}(1,2;2',1')$, which would lead to

$$\Sigma_\sigma(k) = \frac{U}{2\beta N}\sum_q G^{(1)}_{\bar\sigma}(k+q)U_{\rm sp}\chi_{\rm sp}(q). \quad (6.175)$$

The expression (6.174) corresponds to expanding the vertex $\Gamma^{(4)}$ in (6.171) in the longitudinal (parallel spin) channel as shown in Fig. 6.12 [35]. This takes good care of the singularity of $\Gamma^{(4)}_{\sigma\bar\sigma,\bar\sigma,\sigma}(k+q,k'-q;k',k)$ when $\mathbf{q}$ is near $\mathbf{Q}$. The expression (6.175) amounts to an expansion in the transverse (antiparallel spin) channel and takes care of the singularity of $\Gamma^{(4)}$

61 A similar approach was followed in the study of the electron gas (Chapter 5) where the density-density response function was computed in the random-phase approximation (RPA) with Hartree-Fock propagators before obtaining a better approximation of the self-energy by taking into account the interaction of an electron with collective density excitations (Sec. 5.4.2).

62 For clarity, we write explicitly the spin index even though the system is spin-rotation invariant in the normal phase: $\Sigma_\sigma = \Sigma$ and $G_\sigma = G$.

63 $\boldsymbol{\sigma}_{\sigma_1\sigma_2}\cdot\boldsymbol{\sigma}_{\sigma_3\sigma_4} = 2\delta_{\sigma_1,\sigma_4}\delta_{\sigma_2,\sigma_3} - \delta_{\sigma_1,\sigma_2}\delta_{\sigma_3,\sigma_4}$.

64 To lowest order in U, it is easy to show that $\Gamma_{\rm ch} = U/2$ and $\Gamma_{\rm sp} = -U/2$, which suggests that the TPSC approximation takes the form (6.173).

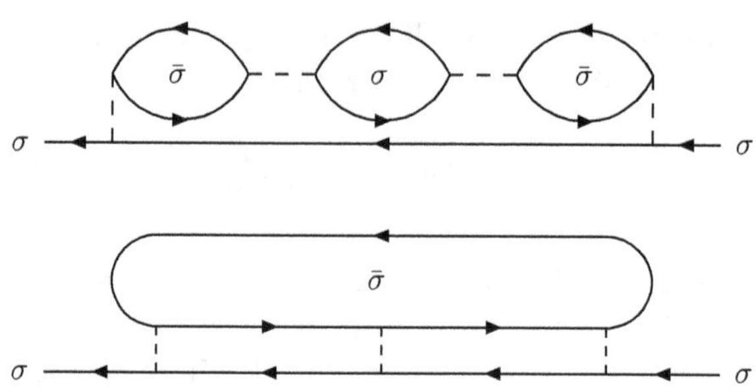

Fig. 6.12 Diagrammatic representation of the self-energy (6.174) (top) and (6.175) (bottom). The dashed line stands for $U_{\rm sp}$ except the one at the far right of each diagram which represents U.

when $\mathbf{k}+\mathbf{q}-\mathbf{k}'$ is near $\mathbf{Q}$. Equation (6.175) being obtained from (6.174) by exchanging the two incoming (or outgoing) legs of the vertex $\Gamma^{(4)}$, one expects that averaging the two expressions leads to a better approximation of the self-energy since this preserves the crossing symmetries. The final result is then

$$\Sigma^{(2)}(k) = U\frac{n}{2} + \frac{U}{8\beta N}\sum_q [U_{\rm ch}\chi_{\rm ch}(q) + 3U_{\rm sp}\chi_{\rm sp}(q)]G^{(1)}(k+q), \tag{6.176}$$

where we have restored in the rhs the Hartree–Fock contribution $\Sigma^{(1)} = Un/2$. This expression is manifestly spin-rotation-invariant, the factor 3 in the rhs coming from the three components of the spin-fluctuation propagator. Note that the TPSC theory, and in particular Eq. (6.176), is expected to fail at low temperatures ($T \ll T_X$) when $\Sigma^{(1)}$ becomes too different from the true self-energy.

At this point, it may appear tempting to use $G^{(2)} = (G_0^{-1} - \Sigma^{(2)})^{-1}$ instead of $G^{(1)}$ in the rhs of (6.176), which would make the approximation conserving (in the sense of Baym and Kadanoff, see Sec. 9.1) when $U_{\rm ch} = U_{\rm sp} = U$. This is, for example, the strategy followed by the fluctuation-exchange approximation (FLEX). But this type of approach does not consider propagators and vertices at the same level of approximation and misses some important physics, e.g. the formation of the pseudogap at low temperatures (see the following). We refer to Ref. [33] and Sec. 9.1.4.2 for a more detailed discussion of this issue.

6.4.5.3 *A more formal derivation of TPSC theory*

We now briefly describe a more formal derivation of the TPSC theory [36]. We start from the relation (1.438) between the self-energy (including the Hartree–Fock term) and the two-particle Green function.[65] For the Hubbard model, it reads

$$\begin{aligned}\sum_{\mathbf{r}'}\int_0^\beta d\tau'\, \Sigma_\sigma(\mathbf{r}_1,\tau_1;\mathbf{r}',\tau')G_\sigma(\mathbf{r}',\tau';\mathbf{r}_2,\tau_2) \\ = -U\langle T_\tau \hat\psi^\dagger_{\mathbf{r}_1\bar\sigma}(\tau_1^+)\hat\psi_{\mathbf{r}_1\bar\sigma}(\tau_1)\hat\psi_{\mathbf{r}_1\sigma}(\tau_1)\hat\psi^\dagger_{\mathbf{r}_2\sigma}(\tau_2)\rangle.\end{aligned} \tag{6.177}$$

[65]Note the sign convention for the propagator in Eq. (1.438) which explains the minus sign in the rhs of (6.177).

When $(\mathbf{r}_2, \tau_2) = (\mathbf{r}_1, \tau_1^+)$, the rhs is simply the double occupancy $U\langle \hat{n}_{\mathbf{r}_1\bar{\sigma}} \hat{n}_{\mathbf{r}_1\sigma} \rangle$ (i.e. the expectation value of the interacting Hamiltonian, see (1.440)). Factorizing as in the Hartree–Fock theory amounts to assuming no correlation: $U\langle \hat{n}_{\mathbf{r}\uparrow} \hat{n}_{\mathbf{r}\downarrow} \rangle \to U\langle \hat{n}_{\mathbf{r}\uparrow} \rangle \langle \hat{n}_{\mathbf{r}\downarrow} \rangle$. In the TPSC theory, one factorizes the rhs of (6.177) as follows:

$$\sum_{\mathbf{r}'} \int_0^\beta d\tau' \, \Sigma_\sigma(\mathbf{r}_1, \tau_1; \mathbf{r}', \tau') G_\sigma(\mathbf{r}', \tau'; \mathbf{r}_2, \tau_2) = U_{\rm sp} G_{\bar{\sigma}}(\mathbf{r}_1, \tau_1; \mathbf{r}_1, \tau_1^+) G_\sigma(\mathbf{r}_1, \tau_1; \mathbf{r}_2, \tau_2), \tag{6.178}$$

with a prefactor $U_{\rm sp}$ chosen such that the exact solution is recovered when $(\mathbf{r}_2, \tau_2) = (\mathbf{r}_1, \tau_1^+)$. This implies

$$U\langle \hat{n}_{\mathbf{r}_1\bar{\sigma}} \hat{n}_{\mathbf{r}_1\sigma} \rangle = U_{\rm sp} \langle \hat{n}_{\mathbf{r}_1\bar{\sigma}} \rangle \langle \hat{n}_{\mathbf{r}_1\sigma} \rangle \quad \text{i.e.} \quad U_{\rm sp} = U \frac{\langle \hat{n}_{\mathbf{r}\uparrow} \hat{n}_{\mathbf{r}\downarrow} \rangle}{\langle \hat{n}_{\mathbf{r}\uparrow} \rangle \langle \hat{n}_{\mathbf{r}\downarrow} \rangle}. \tag{6.179}$$

The solution of (6.178) is

$$\begin{aligned} \Sigma_\sigma^{(1)}(\mathbf{r}_1, \tau_1; \mathbf{r}_2, \tau_2) &= U_{\rm sp} \delta_{\mathbf{r}_1, \mathbf{r}_2} \delta(\tau_1 - \tau_2) G_{\bar{\sigma}}(\mathbf{r}_1, \tau_1; \mathbf{r}_1, \tau_1^+) \\ &= U_{\rm sp} \frac{n}{2} \delta_{\mathbf{r}_1, \mathbf{r}_2} \delta(\tau_1 - \tau_2), \end{aligned} \tag{6.180}$$

which is the self-energy used in the first step of TPSC.

Finally, one must show that $U_{\rm sp}$ coincides with the 2PI vertex $\Phi_{\rm sp}^{(2)}$ in the spin particle–hole channel (which would immediately imply the second equation in (6.158)). This follows from the relation between the (local) vertex $\Phi_{\rm sp}^{(2)}$ and the (local) self-energy [5, 36],

$$\Phi_{\rm sp}^{(2)}(\mathbf{r}, \tau_1, \mathbf{r}, \tau_2; \mathbf{r}, \tau_3, \mathbf{r}, \tau_4) = \frac{1}{2} \left(\frac{\delta \Sigma_\uparrow(\mathbf{r}, \tau_2; \mathbf{r}, \tau_1)}{\delta G_\uparrow(\mathbf{r}, \tau_3; \mathbf{r}, \tau_4)} - \frac{\delta \Sigma_\uparrow(\mathbf{r}, \tau_2; \mathbf{r}, \tau_1)}{\delta G_\downarrow(\mathbf{r}, \tau_3; \mathbf{r}, \tau_4)} \right), \tag{6.181}$$

where $\Sigma_\sigma \equiv \Sigma_\sigma[G]$ is given by (6.180) with $U_{\rm sp} \equiv U_{\rm sp}[G]$ considered as a functional of G *via* Eq. (6.179). The functional derivative of $U_{\rm sp}$ with respect to G_σ cancels in the rhs of (6.181) and we find[66]

$$\Phi_{\rm sp}^{(2)}(\mathbf{r}, \tau_1, \mathbf{r}, \tau_2; \mathbf{r}, \tau_3, \mathbf{r}, \tau_4) = -\frac{U_{\rm sp}}{2} \delta(\tau_1^+ - \tau_2) \delta(\tau_2 - \tau_3) \delta(\tau_3^+ - \tau_4), \tag{6.182}$$

which is the expected result. The 2PI vertex in the charge channel satisfies en equation similar to (6.181) but with a plus sign in the rhs. As a result, the functional derivative $\delta U_{\rm sp}/\delta G_\sigma$ appears in the final expression of $\Phi_{\rm ch}^{(2)}$ which is therefore non-local. In the TPSC theory, $\Phi_{\rm ch}^{(2)}$ is nonetheless assumed to be local and determined from the sum rules (6.153).

[66]Note that for the Hubbard model in the Hartree–Fock approximation, the Luttinger–Ward functional is given by $\Phi[G] = U \int_0^\beta d\tau \sum_{\mathbf{r}} G_\uparrow(\mathbf{r}, \tau; \mathbf{r}, \tau^+) G_\downarrow(\mathbf{r}, \tau; \mathbf{r}, \tau^+)$ and a straightforward calculation gives $\Phi_{\rm sp}^{(2)} = -U/2$.

Let us give a rough derivation of (6.181). The self-energy and the 2PI vertex in the particle–hole channel are defined by

$$\Sigma_{\sigma_1\sigma_2} = \frac{\delta\Phi}{\delta G_{\sigma_2\sigma_1}},$$
$$\Phi^{(2)}_{\sigma_1\sigma_1',\sigma_2\sigma_2'} = \frac{\delta^2\Phi}{\delta G_{\sigma_1\sigma_1'}\delta G_{\sigma_2\sigma_2'}} = \frac{\delta\Sigma_{\sigma_2'\sigma_2}}{\delta G_{\sigma_1\sigma_1'}}, \tag{6.183}$$

where $\Phi \equiv \Phi[G]$ is the Luttinger–Ward functional, and Σ and $\Phi^{(2)}$ are considered as functionals of G (Sec. 1.6.3). We ignore the space and time dependence of the vertices which does not play an important role here. We write G and $\Phi^{(2)}$ in terms of their charge and spin components (Sec. 3.5.4),

$$G_{\sigma_1\sigma_2} = \frac{1}{2}\sigma^0_{\sigma_1\sigma_2}G_{\rm ch} + \frac{1}{2}\sum_{\nu=x,y,z}\sigma^\nu_{\sigma_1\sigma_2}G^\nu_{\rm sp},$$
$$\Phi^{(2)}_{\sigma_1\sigma_1',\sigma_2\sigma_2'} = \sigma^0_{\sigma_1\sigma_1'}\sigma^0_{\sigma_2\sigma_2'}\Phi^{(2)}_{\rm ch} + \sum_{\nu,\nu'=x,y,z}\sigma^\nu_{\sigma_1\sigma_1'}\sigma^{\nu'}_{\sigma_2\sigma_2'}\Phi^{(2)\nu\nu'}_{\rm sp}. \tag{6.184}$$

The 2PI vertex in the spin channel is given by

$$\Phi^{(2)\nu\nu'}_{\rm sp} = \frac{\delta\Sigma^\nu_{\rm sp}}{\delta G^{\nu'}_{\rm sp}} = \frac{\delta^2\Phi}{\delta G^\nu_{\rm sp}\delta G^{\nu'}_{\rm sp}}. \tag{6.185}$$

Assuming that only $G_{\rm ch}$ and $G^z_{\rm sp} \equiv G_{\rm sp}$ are non-zero, so that $G_\sigma = \frac{1}{2}(G_{\rm ch} + \sigma G_{\rm sp})$ and Σ_σ are diagonal in spin indices, one obtains

$$\Sigma_{\rm sp} = \frac{\delta\Phi}{\delta G_{\rm sp}} = \frac{1}{2}\left(\frac{\delta\Phi}{\delta G_\uparrow} - \frac{\delta\Phi}{\delta G_\downarrow}\right) = \frac{1}{2}(\Sigma_\uparrow - \Sigma_\downarrow) \tag{6.186}$$

and

$$\Phi^{(2)}_{\rm sp} = \frac{\delta\Sigma_{\rm sp}}{\delta G_{\rm sp}} = \frac{1}{4}\left(\frac{\delta\Sigma_\uparrow}{\delta G_\uparrow} - \frac{\delta\Sigma_\downarrow}{\delta G_\uparrow} - \frac{\delta\Sigma_\uparrow}{\delta G_\downarrow} + \frac{\delta\Sigma_\downarrow}{\delta G_\downarrow}\right) \tag{6.187}$$

i.e.

$$\Phi^{(2)}_{\rm sp} = \frac{1}{2}\left(\frac{\delta\Sigma_\uparrow}{\delta G_\uparrow} - \frac{\delta\Sigma_\uparrow}{\delta G_\downarrow}\right) \tag{6.188}$$

when $\delta\Sigma_\sigma/\delta G_{\sigma'} = \delta\Sigma_{\bar\sigma}/\delta G_{\bar\sigma'}$.

6.4.5.4 *Pseudogap in the two-dimensional half-filled Hubbard model*

We are now in a position to compute the fermionic spectral function. We start from the expression (6.176) of $\Sigma^{(2)}$ with the Hartree–Fock contribution $Un/2$ included in the chemical potential and ignore the harmless charge fluctuations. To understand the low-frequency structure of $\chi_{\rm sp}(\mathbf{q}, i\omega_\nu)$, it is useful to first consider the retarded response function

$$\chi^{(1)R}(\mathbf{Q},\omega) = -2\int_{\mathbf{k}} \frac{n_F(\epsilon_{\mathbf{k}}) - n_F(\epsilon_{\mathbf{k}+\mathbf{Q}})}{\omega + i0^+ + \epsilon_{\mathbf{k}} - \epsilon_{\mathbf{k}+\mathbf{Q}}}$$
$$= 2\int_{\mathbf{k}} \frac{\tanh(\beta\epsilon_{\mathbf{k}}/2)}{\omega + i0^+ + 2\epsilon_{\mathbf{k}}}. \tag{6.189}$$

The real part gives a (subleading) contribution of order ω^2 in the low-frequency limit whereas the imaginary part reads

$$\begin{aligned}\Im[\chi^{(1)R}(\mathbf{Q},\omega)] &= -2\pi \int d\epsilon\, N(\epsilon)\tanh(\beta\epsilon/2)\delta(\omega+2\epsilon) \\ &= \pi N(\omega/2)\tanh(\beta\omega/4) \\ &\simeq \frac{\pi}{4}N\left(\frac{\omega}{2}\right)\frac{\omega}{T} \quad \text{for} \quad |\omega| \ll T. \end{aligned} \tag{6.190}$$

Combining this expression with (6.166) and (6.168), we obtain

$$\chi^R_{\rm sp}(\mathbf{Q}+\mathbf{q},\omega) \simeq \frac{C}{\mathbf{q}^2+\xi^{-2}(1-i\omega/\omega_{\rm sp})} \tag{6.191}$$

in the low-energy limit. The linear frequency variation in the denominator is valid only for $|\omega| \sim \omega_{\rm sp}$ where the characteristic frequency scale $\omega_{\rm sp}$ is defined by

$$\frac{1}{\omega_{\rm sp}} = \frac{\pi}{8}\frac{U_{\rm sp}}{T}\frac{N(\omega_{\rm sp}/2)}{1-\frac{U_{\rm sp}}{2}\chi^{(1)}(\mathbf{Q},0)}. \tag{6.192}$$

Thus, the Matsubara spin–spin correlation function can be written as

$$\chi_{\rm sp}(\mathbf{Q}+\mathbf{q},i\omega_\nu) \simeq \frac{C}{\mathbf{q}^2+\xi^{-2}(1+|\omega_\nu|/\omega_{\rm sp})}. \tag{6.193}$$

The characteristic frequency scale $\omega_{\rm sp} \sim T\xi^{-2}$ vanishes when ξ diverges. As a result, the spin susceptibility (6.193) is dominated by its classical ($\omega_\nu = 0$) component whenever $2\pi T/\omega_{\rm sp} \sim 2\pi\xi^2 \gg 1$. In the renormalized classical regime, defined by $2\pi T \gg \omega_{\rm sp}$, the TPSC self-energy (6.176) can therefore be approximated by

$$\begin{aligned}\Sigma(k) &= \frac{3}{8}TUU_{\rm sp}\int_{\mathbf{q}} \chi_{\rm sp}(\mathbf{Q}+\mathbf{q},0)G^{(1)}(\mathbf{k}+\mathbf{Q}+\mathbf{q},i\omega_n) \\ &= \frac{3}{8}TUU_{\rm sp}\int_{\mathbf{q}} \frac{C}{\mathbf{q}^2+\xi^{-2}}\frac{1}{i\omega_n+\epsilon_{\mathbf{k}}+\mathbf{v}_{\mathbf{k}}\cdot\mathbf{q}},\end{aligned} \tag{6.194}$$

where we have used $\epsilon_{\mathbf{k}+\mathbf{Q}} = -\epsilon_{\mathbf{k}}$, $\epsilon_{\mathbf{k}+\mathbf{Q}+\mathbf{q}} \simeq -\epsilon_{\mathbf{k}} - \mathbf{v}_k\cdot\mathbf{q}$ and $\mathbf{v}_{\mathbf{k}} = \boldsymbol{\nabla}_{\mathbf{k}}\epsilon_{\mathbf{k}}$. For a particle on the Fermi surface ($\epsilon_{\mathbf{k}_F} = 0$), the imaginary part of the retarded self-energy satisfies[67]

$$\Im[\Sigma^R(\mathbf{k}_F,0)] = -\frac{3}{8}\pi TUU_{\rm sp}\int_{\mathbf{q}} \frac{C}{\mathbf{q}^2+\xi^{-2}}\delta(\mathbf{v}_{\mathbf{k}_F}\cdot\mathbf{q}) \sim -UU_{\rm sp}C\frac{\xi}{\xi^{\rm th}_{\mathbf{k}_F}}, \tag{6.195}$$

where $\xi^{\rm th}_{\mathbf{k}} \sim |\mathbf{v}_{\mathbf{k}}|/T$ is the thermal de Broglie wavelength for a fermion with momentum $\mathbf{k}$. Since ξ diverges exponentially when $T \to 0$, $\Im[\Sigma^R(\mathbf{k}_F,0)]$ grows at low temperatures in contradiction with the Fermi liquid hypothesis $\lim_{T\to 0}\Im[\Sigma^R(\mathbf{k}_F,0)] = 0$. A diverging $\Im[\Sigma^R(\mathbf{k}_F,0)]$ corresponds to a vanishing spectral function $A(\mathbf{k}_F,0)$: Quasi-particles are

[67] Away from half-filling, Eq. (6.195) remains valid at the so-called hot spots where $\xi_{\mathbf{k}_F+\mathbf{Q}} = \xi_{\mathbf{k}_F} = 0$.

suppressed by antiferromagnetic fluctuations at low temperatures. Since $\Im[\Sigma^R(\mathbf{k}_F,0)] \sim \xi/\xi^{\text{th}}_{\mathbf{k}_F}$ it is natural to define the crossover temperature T_X (introduced above) by the Vilk criterion, i.e. as the temperature below which ξ becomes larger than the typical de Broglie wavelength ξ_{th}.[68]

Thus, for $T \gg T_X$ the system behaves as a Fermi liquid whereas for $T \sim T_X$ quasi-particles are suppressed and, as we now discuss, a pseudogap forms. For $|\omega+\epsilon_{\mathbf{k}}| \gg |\mathbf{v}_{\mathbf{k}}|/\xi$,

$$\Re[\Sigma^R(\mathbf{k},\omega)] = \frac{3}{8}TUU_{\text{sp}}\int_{\mathbf{q}}\frac{C}{\mathbf{q}^2+\xi^{-2}}\frac{1}{\omega+\epsilon_{\mathbf{k}}} \equiv \frac{\Delta(T)^2}{\omega+\epsilon_{\mathbf{k}}}, \tag{6.196}$$

where $\Delta(T)^2 = (3/8)UU_{\text{sp}}C\rho_s(T)$ (we have used the sum rule (6.169)). On the other hand,

$$\begin{aligned}\Im[\Sigma^R(\mathbf{k},\omega)] &= -\frac{3}{8}\pi TUU_{\text{sp}}\int_{\mathbf{q}}\frac{C}{\mathbf{q}^2+\xi^{-2}}\delta(\omega+\epsilon_{\mathbf{k}}+\mathbf{v}_{\mathbf{k}}\cdot\mathbf{q}) \\ &\simeq -\frac{T}{4\rho_s(T)}\frac{\Delta(T)^2}{\omega+\epsilon_{\mathbf{k}}}.\end{aligned} \tag{6.197}$$

This yields the spectral function

$$\begin{aligned}A(\mathbf{k},\omega) &= -\frac{1}{\pi}\Im[G^R(\mathbf{k},\omega)] \\ &\simeq \frac{\gamma}{\pi}\frac{|\omega+\epsilon_{\mathbf{k}}|}{(\omega^2-E_{\mathbf{k}}^2)^2+\gamma^2} \quad \text{for} \quad |\omega+\epsilon_{\mathbf{k}}| \gg |\mathbf{v}_{\mathbf{k}}|/\xi,\end{aligned} \tag{6.198}$$

where $E_{\mathbf{k}} = \sqrt{\epsilon_{\mathbf{k}}^2+\Delta(T)^2}$ and $\gamma = T\Delta(T)^2/4\rho_s(T)$. $A(\mathbf{k},\omega)$ exhibits two peaks at $\omega = \pm E_{\mathbf{k}}$ that are precursors of the antiferromagnetic bands that exist at zero temperature in the ordered ground state (Sec. 6.4.1). The width of these peaks is given by $\gamma/\Delta(T)$ for $\epsilon_{\mathbf{k}} \to 0$ (i.e. for $\mathbf{k}$ near the Fermi surface).[69] The formation of the pseudogap at low temperatures is shown qualitatively in Fig. 6.13.

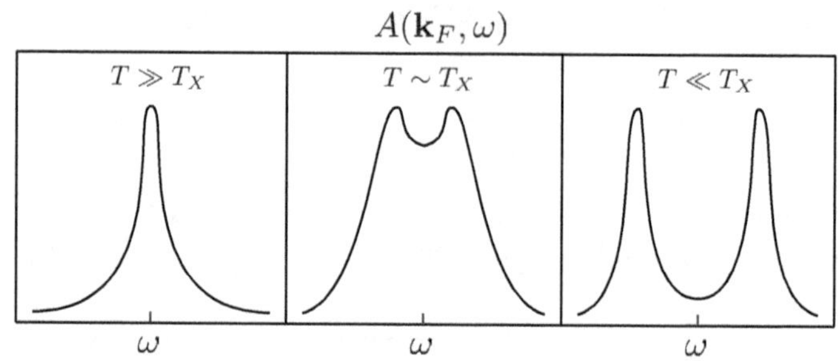

Fig. 6.13 Qualitative sketch of the spectral function $A(\mathbf{k}_F,\omega)$ for a wavevector on the Fermi surface. The pseudogap forms when the temperature becomes of the order of the crossover temperature T_X. For $T \ll T_X$, the two peaks in $A(\mathbf{k}_F,\omega)$ are precursors of the antiferromagnetic bands that exist in the ordered ground state.

[68]Note that the condition $\xi \gg \xi_{\text{th}}$ is more difficult to satisfy than the condition $T \gg \omega_{\text{sp}}$ defining the renormalized classical regime, since $\xi_{\text{th}} \sim 1/T$ while $\omega_{\text{sp}} \sim T\xi^{-2}$.

[69]These results can also be obtained from the NLσM approach discussed in Sec. 6.4.3 [39].

Appendix 6.A Hartree–Fock Correlation Functions

In Fourier space, the Hartree–Fock correlation function $\Pi^{\nu\nu'}_{\mu\mu'}$ is defined by

$$\Pi^{\nu\nu'}_{\mu\mu'}(q,q') = \langle j^{\nu}_{\mu}(q) j^{\nu'}_{\mu'}(-q')\rangle_c, \tag{6.A.1}$$

where

$$j^{\nu}_{\mu}(q) = \frac{1}{\sqrt{\beta N}} \sum_k v^{\mu}_{\mathbf{k},\mathbf{q}} \Phi^{\dagger}_k \sigma^{\nu} \Phi_{k+q},$$
$$v^0_{\mathbf{k},\mathbf{q}} = 1, \qquad v^{\mu}_{\mathbf{k},\mathbf{q}} = \begin{cases} -it(e^{i(k_\mu+q_\mu)} + e^{-ik_\mu}) & \text{if} \quad \mu \neq 0,\ \nu = 0, \\ -it(e^{i(k_\mu+q_\mu)} - e^{-ik_\mu}) & \text{if} \quad \mu \neq 0,\ \nu \neq 0. \end{cases} \tag{6.A.2}$$

The cumulant $\langle\cdots\rangle_c$ in (6.A.1) must be calculated with the Hartree–Fock action S_{HF} [Eq. (6.64) for the ferromagnet and (6.109) for the antiferromagnet]. Using Wick's theorem, one obtains

$$\Pi^{\nu\nu'}_{\mu\mu'}(q,q') = -\frac{1}{\beta N} \sum_{k,k',\sigma,\sigma'} v^{\mu}_{\mathbf{k},\mathbf{q}} v^{\mu'}_{\mathbf{k}',-\mathbf{q}'} \sigma^{\nu}_{\sigma,\sigma'} \sigma^{\nu'}_{\sigma',\sigma} G_\sigma(k'-q',k) G_{\sigma'}(k+q,k'). \tag{6.A.3}$$

In the following, we consider the zero-temperature limit.

6.A.1 *Ferromagnet*

In the ferromagnetic phase, the one-particle Hartree–Fock propagator $G_\sigma(k) = (i\omega_n - E_{\mathbf{k}\sigma})^{-1}$ is diagonal in momentum [see (6.36) or (6.66)] and $\Pi^{\nu\nu'}_{\mu\mu'}(q,q') \propto \delta_{q,q'}$ is diagonal. For $q=0$,

$$\begin{aligned} \Pi^{\nu\nu'}_{\mu\mu'}(0) &= -\frac{1}{\beta N} \sum_{k,\sigma} v^{\mu}_{\mathbf{k},0} v^{\mu'}_{\mathbf{k},0} \sigma^{\nu}_{\sigma,\bar\sigma} \sigma^{\nu'}_{\bar\sigma,\sigma} G_\sigma(k) G_{\bar\sigma}(k) \\ &= -\delta_{\mu,\mu'}\delta_{\nu,\nu'} \frac{2}{\beta N} \sum_k v^{\mu}_{\mathbf{k},0} v^{\mu'}_{\mathbf{k},0} G_\uparrow(k) G_\downarrow(k) \end{aligned} \tag{6.A.4}$$

where we now only consider the case $\nu,\nu' \in \{x,y\}$. Since $v^{\mu}_{\mathbf{k},0} = 2t\sin(k_\mu)$ for $\mu \neq 0$, this gives $\Pi^{xy}_{\mu\mu'}(0) = 0$ and

$$\Pi^{xx}_{\mu\mu}(0) = \Pi^{yy}_{\mu\mu}(0) = \frac{4t^2}{\Delta} \int_{\mathbf{k}} \sin(k_x)^2 [\Theta(-\xi_{\mathbf{k}} + \Delta) - \Theta(-\xi_{\mathbf{k}} - \Delta)] \tag{6.A.5}$$

for $\mu = x, y$ (and $T=0$). Using (6.88) and (6.91), one finds

$$\begin{aligned} \tilde\Pi^{xx}(q) &= -\omega_\nu^2 \Pi^{xx}_{00}(0) + \mathbf{q}^2 \Pi^{xx}_{xx}(0) = \tilde\Pi^{yy}(q), \\ \tilde\Pi^{xy}(q) &= -\omega_\nu^2 \Pi^{xy}_{00}(0) + \mathbf{q}^2 \Pi^{xy}_{xx}(0) = 0 \end{aligned} \tag{6.A.6}$$

to order $\mathcal{O}(q^2)$.

6.A.2 *Antiferromagnet*

In the antiferromagnetic phase, the one-particle propagator is not diagonal since both $G(k) \equiv G_\sigma(k,k)$ and $F(k) \equiv \sigma G_\sigma(k,k+Q)$ are non-zero. As a result, both $\Pi^{\nu\nu'}_{\mu\mu'}(q,q)$ and $\Pi^{\nu\nu'}_{\mu\mu'}(q,q+Q)$ are non-zero.

For the diagonal part $\Pi^{\nu\nu'}_{\mu\mu'}(q) \equiv \Pi^{\nu\nu'}_{\mu\mu'}(q,q)$ at $q=0$, one obtains

$$\begin{aligned}\Pi^{\nu\nu'}_{\mu\mu'}(0) &= -\frac{1}{\beta N}\sum_{k,\sigma}\sigma^{\nu}_{\sigma,\bar\sigma}\sigma^{\nu'}_{\bar\sigma,\sigma}v^{\mu}_{\mathbf{k},0}\left[v^{\mu'}_{\mathbf{k},0}G(k)^2 - v^{\mu'}_{\mathbf{k+Q},0}F(k)^2\right]\\ &= -\delta_{\nu,\nu'}\frac{2}{\beta N}\sum_{k}v^{\mu}_{\mathbf{k},0}\left[v^{\mu'}_{\mathbf{k},0}G(k)^2 - v^{\mu'}_{\mathbf{k+Q},0}F(k)^2\right] \end{aligned} \tag{6.A.7}$$

for $\nu,\nu' \in \{x,y\}$. $\Pi^{\nu\nu}_{\mu\mu'}(0)$ is diagonal in μ,μ'. Using

$$\frac{1}{\beta}\sum_{\omega_n}G^2(k) = -\frac{1}{\beta}\sum_{\omega_n}F^2(k) = -\frac{\Delta^2}{4E^3_{\mathbf{k}}} \qquad (T\to 0), \tag{6.A.8}$$

one finds

$$\Pi^{\nu\nu}_{00}(0) = \int_{\mathbf{k}}\frac{\Delta^2}{E^3_{\mathbf{k}}}, \qquad \Pi^{\nu\nu}_{xx}(0) = \Pi^{\nu\nu}_{yy}(0) = 0 \qquad (\nu = x,y). \tag{6.A.9}$$

A similar calculation gives

$$\Pi^{zz}_{\mu\mu}(0) = -\frac{1}{\beta N}\sum_{k,\sigma}v^{\mu}_{\mathbf{k},0}\left[v^{\mu}_{\mathbf{k},0}G(k)^2 + v^{\mu}_{\mathbf{k+Q},0}F(k)^2\right] = 4t^2\Delta^2\int_{\mathbf{k}}\frac{\sin(k_x)^2}{E^3_{\mathbf{k}}} \tag{6.A.10}$$

for $\mu \neq 0$.

The diagonal part $\Pi^{\nu\nu'}_{\mu\mu'}(Q)$ is given by

$$\Pi^{\nu\nu'}_{\mu\mu'}(Q) = -\delta_{\nu,\nu'}\frac{2}{\beta N}\sum_{k}v^{\mu}_{\mathbf{k},\mathbf{Q}}\left[v^{\mu'}_{\mathbf{k+Q},-\mathbf{Q}}G(k)G(k+Q) - v^{\mu'}_{\mathbf{k},-\mathbf{Q}}F(k)^2\right] \tag{6.A.11}$$

for $\nu,\nu' \in \{x,y\}$. $\Pi^{\nu\nu}_{\mu\mu'}(Q)$ is diagonal in μ,μ'. Using

$$\frac{1}{\beta}\sum_{\omega_n}G(k)G(k+Q) = -\frac{1}{2E_{\mathbf{k}}} + \frac{\Delta^2}{4E^3_{\mathbf{k}}}, \tag{6.A.12}$$

we obtain

$$\Pi^{\nu\nu}_{\mu\mu}(Q) = 4t^2\int_{\mathbf{k}}\cos(k_x)^2\frac{\epsilon^2_{\mathbf{k}}}{E^3_{\mathbf{k}}} \qquad (\nu = x,y,\ \ \mu\neq 0). \tag{6.A.13}$$

Finally, for the non-diagonal part $\Pi^{\nu\nu'}_{\mu\mu'}(Q,0)$, one finds

$$\Pi^{\nu\nu'}_{\mu\mu'}(Q,0) = \frac{1}{\beta N}\sum_{k,\sigma}\sigma\sigma^{\nu}_{\sigma,\bar\sigma}\sigma^{\nu'}_{\bar\sigma,\sigma}v^{\mu}_{\mathbf{k},\mathbf{Q}}\left[v^{\mu'}_{\mathbf{k},0}G(k)F(k+Q) - v^{\mu'}_{\mathbf{k+Q},0}F(k)G(k+Q)\right]. \tag{6.A.14}$$

Using

$$\sum_{\sigma}\sigma\sigma^{\nu}_{\sigma,\bar\sigma}\sigma^{\nu'}_{\bar\sigma,\sigma} = 2i(\delta_{\nu,x}\delta_{\nu',y} - \delta_{\nu,y}\delta_{\nu',x}), \tag{6.A.15}$$

we deduce

$$\Pi_{x0}^{xy}(Q,0) = \frac{8\Delta t}{\beta N}\sum_k \frac{\cos(k_x)\epsilon_{\mathbf{k}}}{(\omega_n^2+E_{\mathbf{k}}^2)^2} = -\frac{2\Delta}{\beta N}\sum_k \frac{\epsilon_{\mathbf{k}}^2}{(\omega_n^2+E_{\mathbf{k}}^2)^2} = -\frac{\Delta}{2}\int_{\mathbf{k}} \frac{\epsilon_{\mathbf{k}}^2}{E_{\mathbf{k}}^3} \tag{6.A.16}$$

and the same expression for $\Pi_{0x}^{yx}(0,Q)$.

The correlation functions

$$\tilde{\Pi}_{\mu,\mu'}(q,q') = \sum_{\mu,\mu'=0,x,y} z_\mu(-q)z_{\mu'}(q')\Pi_{\mu,\mu'}^{\nu\nu'}(q,q') \tag{6.A.17}$$

satisfy

$$\begin{aligned} &\tilde{\Pi}^{\nu\nu}(q) \equiv \tilde{\Pi}^{\nu\nu}(q,q) = -\omega_\nu^2\Pi_{00}^{xx}(0) \\ &\tilde{\Pi}^{\nu\nu}(Q) \equiv \tilde{\Pi}^{\nu\nu}(Q,Q) = 4\int_{\mathbf{k}}\frac{\epsilon_{\mathbf{k}}^4}{E_{\mathbf{k}}^3} = -4K - 4\Delta^2\int_{\mathbf{k}}\frac{\epsilon_{\mathbf{k}}^2}{E_{\mathbf{k}}^3}, \qquad (\nu = x,y) \\ &\tilde{\Pi}^{xy}(q+Q,q) = -4\omega_\nu\Pi_{x0}^{xy}(Q,0) \end{aligned} \tag{6.A.18}$$

to leading order in q, where we have used

$$K = -\int_{\mathbf{k}}\frac{\epsilon_{\mathbf{k}}^2}{E_{\mathbf{k}}}. \tag{6.A.19}$$

6.A.2.1 *Relation between Π_{xx}^{zz} and K*

It is possible to relate the Hartree–Fock correlation function Π_{xx}^{zz} to the mean value of the kinetic energy per site K. Let us consider the integral

$$\int_{\mathbf{k}} \boldsymbol{\nabla}_{\mathbf{k}}\cdot\left(\frac{\epsilon_{\mathbf{k}}}{E_{\mathbf{k}}}\mathbf{v}_{\mathbf{k}}\right) = \oint \frac{\epsilon_{\mathbf{k}}}{E_{\mathbf{k}}}\mathbf{v}_{\mathbf{k}}\cdot(d\mathbf{l}_{\mathbf{k}}\times\mathbf{z}) = 0, \tag{6.A.20}$$

where $v_{\mathbf{k}} = \boldsymbol{\nabla}_{\mathbf{k}}\epsilon_{\mathbf{k}} = 2t(\sin k_x, \sin k_y)$. Here, we have rewritten the momentum integral as a contour integral along the Brillouin zone boundary. The integral vanishes since $\mathbf{v}_{\mathbf{k}}$, being parallel to the Brillouin zone boundary when $\mathbf{k}$ is on the boundary, is perpendicular to $d\mathbf{l}_{\mathbf{k}}\times\mathbf{z}$. Since

$$\boldsymbol{\nabla}_{\mathbf{k}}\cdot\left(\frac{\epsilon_{\mathbf{k}}}{E_{\mathbf{k}}}\mathbf{v}_{\mathbf{k}}\right) = \frac{\epsilon_{\mathbf{k}}}{E_{\mathbf{k}}}\boldsymbol{\nabla}_{\mathbf{k}}\cdot\mathbf{v}_{\mathbf{k}} + \boldsymbol{\nabla}_{\mathbf{k}}\left(\frac{\epsilon_{\mathbf{k}}}{E_{\mathbf{k}}}\right)\cdot\mathbf{v}_{\mathbf{k}} = -\frac{\epsilon_{\mathbf{k}}^2}{E_{\mathbf{k}}} + \frac{\Delta^2}{E_{\mathbf{k}}^3}\mathbf{v}_{\mathbf{k}}^2, \tag{6.A.21}$$

we obtain

$$\int_{\mathbf{k}}\left[-\frac{\epsilon_{\mathbf{k}}^2}{E_{\mathbf{k}}} + 4t^2\Delta^2\frac{\sin(k_x)^2+\sin(k_y)^2}{E_{\mathbf{k}}^3}\right] = 0, \tag{6.A.22}$$

i.e.

$$K + 2\Pi_{xx}^{zz} = 0 \tag{6.A.23}$$

using (6.A.10) and (6.A.19).

Appendix 6.B SU(2) Gauge Fields

In this appendix, we derive Eqs. (6.134). Starting from $\boldsymbol{\sigma}\cdot\mathbf{n}_\mathbf{r} = R_\mathbf{r}\sigma^z R_\mathbf{r}^\dagger$ and $A_\mu = iR^\dagger\partial_\mu R_\mathbf{r} = -i(\partial_\mu R_\mathbf{r}^\dagger)R_\mathbf{r}$, we obtain

$$\begin{aligned}\boldsymbol{\sigma}\cdot\partial_\mu\mathbf{n}_\mathbf{r} &= (\partial_\mu R_\mathbf{r})\sigma^z R_\mathbf{r}^\dagger + R_\mathbf{r}\sigma^z\partial_\mu R_\mathbf{r}^\dagger = -iR_\mathbf{r}A_{\mu\mathbf{r}}\sigma^z R_\mathbf{r}^\dagger + iR_\mathbf{r}\sigma^z A_{\mu\mathbf{r}}R^\dagger \\ &= -i[R_\mathbf{r}A_{\mu\mathbf{r}}R_\mathbf{r}^\dagger, R_\mathbf{r}\sigma^z R_\mathbf{r}^\dagger] = -i[\tilde{A}_{\mu\mathbf{r}}, \boldsymbol{\sigma}\cdot\mathbf{n}_\mathbf{r}],\end{aligned} \tag{6.B.1}$$

where

$$\tilde{A}_{\mu\mathbf{r}} = R_\mathbf{r}A_{\mu\mathbf{r}}R_\mathbf{r}^\dagger = i(\partial_\mu R_\mathbf{r})R_\mathbf{r}^\dagger. \tag{6.B.2}$$

Equation (6.B.1) holds also for $\mu = 0$ in the real-time formalism: $t = -i\tau$, $\partial_0 = \partial_t$ and $A_{0\mathbf{r}} = iR_\mathbf{r}^\dagger\partial_0 R_\mathbf{r}$.

Defining the vectors $\mathbf{A}_{\mu\mathbf{r}} = (A^x_{\mu\mathbf{r}}, A^y_{\mu\mathbf{r}}, A^z_{\mu\mathbf{r}})$ and $\tilde{\mathbf{A}}_{\mu\mathbf{r}} = (\tilde{A}^x_{\mu\mathbf{r}}, \tilde{A}^y_{\mu\mathbf{r}}, \tilde{A}^z_{\mu\mathbf{r}})$, one has

$$A_{\mu\mathbf{r}} = \sum_\nu A^\nu_{\mu\mathbf{r}}\sigma^\nu = \mathbf{A}_{\mu\mathbf{r}}\cdot\boldsymbol{\sigma}, \qquad \tilde{A}_{\mu\mathbf{r}} = \sum_\nu \tilde{A}^\nu_{\mu\mathbf{r}}\sigma^\nu = \tilde{\mathbf{A}}_{\mu\mathbf{r}}\cdot\boldsymbol{\sigma} \tag{6.B.3}$$

and

$$\tilde{\mathbf{A}}_{\mu\mathbf{r}}\cdot\boldsymbol{\sigma} = R_\mathbf{r}\mathbf{A}_{\mu\mathbf{r}}\cdot\boldsymbol{\sigma}R_\mathbf{r}^\dagger. \tag{6.B.4}$$

Since the SO(3) matrix $\mathcal{R}_\mathbf{r}$ associated with the SU(2) matrix $R_\mathbf{r}$ satisfies $R_\mathbf{r}\mathbf{u}\cdot\boldsymbol{\sigma}R_\mathbf{r}^\dagger = \mathcal{R}_\mathbf{r}(\mathbf{u})\cdot\boldsymbol{\sigma}$ for any vector $\mathbf{u}$, Eq. (6.B.4) implies $\tilde{\mathbf{A}}_{\mu\mathbf{r}} = \mathcal{R}_\mathbf{r}(\mathbf{A}_{\mu\mathbf{r}})$.

From (6.B.1), we obtain[70]

$$\boldsymbol{\sigma}\cdot\partial_\mu\mathbf{n}_\mathbf{r} = -i[\tilde{\mathbf{A}}_{\mu\mathbf{r}}\cdot\boldsymbol{\sigma}, \boldsymbol{\sigma}\cdot\mathbf{n}_\mathbf{r}] = 2(\tilde{\mathbf{A}}_{\mu\mathbf{r}}\times\mathbf{n}_\mathbf{r})\cdot\boldsymbol{\sigma} \tag{6.B.5}$$

and in turn

$$\begin{aligned}\tilde{\mathbf{A}}_{\mu\mathbf{r}}\times\mathbf{n}_\mathbf{r} &= \frac{1}{2}\partial_\mu\mathbf{n}_\mathbf{r}, \\ \mathbf{n}_\mathbf{r}\times(\tilde{\mathbf{A}}_{\mu\mathbf{r}}\times\mathbf{n}_\mathbf{r}) = (\mathbf{n}_\mathbf{r}\cdot\mathbf{n}_\mathbf{r})\tilde{\mathbf{A}}_{\mu\mathbf{r}} - (\mathbf{n}_\mathbf{r}\cdot\tilde{\mathbf{A}}_{\mu\mathbf{r}})\mathbf{n}_\mathbf{r} &= \frac{1}{2}\mathbf{n}_\mathbf{r}\times\partial_\mu\mathbf{n}_\mathbf{r}, \\ \tilde{\mathbf{A}}_{\mu\mathbf{r}} = \frac{1}{2}\mathbf{n}_\mathbf{r}\times\partial_\mu\mathbf{n}_\mathbf{r} + (\mathbf{n}_\mathbf{r}\cdot\tilde{\mathbf{A}}_{\mu\mathbf{r}})\mathbf{n}_\mathbf{r} &\equiv \tilde{\mathbf{A}}^\perp_{\mu\mathbf{r}} + \tilde{\mathbf{A}}^\parallel_{\mu\mathbf{r}},\end{aligned} \tag{6.B.6}$$

where $\tilde{\mathbf{A}}^\perp_{\mu\mathbf{r}}$ and $\tilde{\mathbf{A}}^\parallel_{\mu\mathbf{r}}$ denote the components of $\tilde{\mathbf{A}}_{\mu\mathbf{r}}$ perpendicular and parallel to $\mathbf{n}_\mathbf{r}$. Since $\mathcal{R}_\mathbf{r}^{-1}(\mathbf{n}_\mathbf{r}) = \mathbf{u}_z$, $\mathcal{R}_\mathbf{r}^{-1}(\tilde{\mathbf{A}}^\perp_{\mu\mathbf{r}})$ is perpendicular to $\mathbf{u}_z$ while $\mathcal{R}_\mathbf{r}^{-1}(\tilde{\mathbf{A}}^\parallel_{\mu\mathbf{r}})$ is parallel to $\mathbf{u}_z$. Given that $\mathbf{l}_\mathbf{r}$ is perpendicular to $\mathbf{u}_z$, one finds

$$\begin{aligned}\mathbf{A}_{\mu\mathbf{r}}\cdot\mathbf{l}_\mathbf{r} &= \mathcal{R}_\mathbf{r}^{-1}(\tilde{\mathbf{A}}_{\mu\mathbf{r}})\cdot\mathbf{l}_\mathbf{r} = \mathcal{R}_\mathbf{r}^{-1}(\tilde{\mathbf{A}}^\perp_{\mu\mathbf{r}})\cdot\mathbf{l}_\mathbf{r} = \mathcal{R}_\mathbf{r}^{-1}(\tilde{\mathbf{A}}^\perp_{\mu\mathbf{r}})\cdot\mathcal{R}_\mathbf{r}^{-1}(\mathbf{L}_\mathbf{r}) \\ &= \tilde{\mathbf{A}}^\perp_{\mu\mathbf{r}}\cdot\mathbf{L}_\mathbf{r} = \frac{1}{2}(\mathbf{n}_\mathbf{r}\times\partial_\mu\mathbf{n}_\mathbf{r})\cdot\mathbf{L}_\mathbf{r}.\end{aligned} \tag{6.B.7}$$

From (6.B.6), we also deduce

$$\sum_{\nu=x,y} A^{\nu\,2}_{\mu\mathbf{r}} = \mathcal{R}_\mathbf{r}^{-1}(\tilde{\mathbf{A}}^\perp_{\mu\mathbf{r}})^2 = \tilde{\mathbf{A}}^{\perp\,2}_{\mu\mathbf{r}} = \frac{1}{4}(\mathbf{n}_\mathbf{r}\times\partial_\mu\mathbf{n}_\mathbf{r})^2. \tag{6.B.8}$$

Going back to imaginary time $\tau = it$, Eqs. (6.B.7) and (6.B.8) yield (6.134).

[70] We use $[\mathbf{u}\cdot\boldsymbol{\sigma}, \mathbf{v}\cdot\boldsymbol{\sigma}] = 2i(\mathbf{u}\times\mathbf{v})\cdot\boldsymbol{\sigma}$ for any vectors $\mathbf{u}$ and $\mathbf{v}$.

Guide to the bibliography

- For a general introduction to the Hubbard model [1–3], see Ref. [4]. The derivation of the model, starting from the general Hamiltonian of a solid, is discussed in Refs. [4,5].
- The Stoner theory of ferromagnetism and the Slater theory of antiferromagnetism were introduced in Refs. [6, 7].
- Ferromagnetism in (nearly) flat-band systems is discussed in Refs. [8, 9].
- The RPA theory of the antiferromagnetic ground state of the two-dimensional Hubbard model can be found in Refs. [10–13].
- There are numerous papers studying the two-dimensional Hubbard model away from half-filling in the Hartree–Fock/RPA theory, see, e.g., Refs. [14–21]. Some of these papers study the possibility that the ground state of the doped two-dimensional Hubbard model be a spiral spin-density wave (as originally predicted in the t–J model [22, 23]), although the (homogeneous) spiral phase is found to be unstable.
- The NLσM approach to the two-dimensional half-filled Hubbard model is discussed in Refs. [26–29]. Numerical results supporting this approach can be found in Ref. [30].
- The TPSC theory [32–35] is reviewed in Refs. [36] and [5].
- Other approaches to the Hubbard model will be discussed in forthcoming chapters: DMFT (Chapter 9) and RG approach (Chapter 15).

Bibliography

[1] J. Hubbard, Electron correlations in narrow energy bands, *Proc. R. Soc. London*, Ser. A **276**, 238 (1963).
[2] M. C. Gutzwiller, Effect of correlation on the ferromagnetism of transition metals, *Phys. Rev. Lett.* **10**, 159 (1963).
[3] J. Kanamori, Electron correlation and ferromagnetism of transition metals, *Prog. Theor. Phys.* **30**, 275 (1963).
[4] F. Gebhard, *The Mott Metal-Insulator Transition* (Springer-Verlag, Berlin, Heidelberg, 1997).
[5] A.-M. Tremblay, Quantum Material's Theory, from perturbation theory to dynamical-mean field theory. https://www.physique.usherbrooke.ca/pages/tremblay.
[6] E. C. Stoner, Collective electron ferromagnetism in metals and alloys, *Journ. de Phys. et Rad.* **12**, 372 (1951).
[7] J. C. Slater, Magnetic effects and the Hartree-Fock equation, *Phys. Rev.* **82**, 538 (1951).
[8] A. Mielke, Ferromagnetic ground states for the Hubbard model on line graphs, *J. Phys.* A **24**, L73 (1991); Ferromagnetism in the Hubbard model on line graphs and further considerations, *J. Phys.* A **24**, 3311 (1991); Exact ground states for the Hubbard model on the Kagome lattice, *J. Phys.* A **25**, 4335 (1992); Ferromagnetism in the Hubbard model and Hund's rule, *Phys. Lett.* A **174**, 443 (1993); A. Mielke and H. Tasaki, Ferromagnetism in the Hubbard model. Examples from models with degenerate single-electron ground states, *Commun. Math. Phys.* **158**, 341 (1993).
[9] H. Tasaki, Ferromagnetism in the Hubbard models with degenerate single-electron ground states, *Phys. Rev. Lett.* **69**, 1608 (1992); The Hubbard model — an introduction and selected rigorous results, *J. Phys.: Condens. Matter* **10**, 4353 (1998); From Nagaoka's ferromagnetism to flat-band ferromagnetism and beyond, *Prog. Theor. Phys.* **99**, 489 (1998).
[10] T. Moriya, *Spin Fluctuations in Itinerant Electron Magnetism* (Springer-Verlag, Berlin, 1985).
[11] J. R. Schrieffer, X. G. Wen, and S. C. Zhang, Dynamic spin fluctuations and the bag mechanism of high-Tc superconductivity, *Phys. Rev.* B **39**, 11663 (1989).
[12] A. Singh and Z. Tesanović, Collective excitations in a doped antiferromagnet, *Phys. Rev.* B **41**, 614 (1990).
[13] A. V. Chubukov and D. M. Frenkel, Renormalized perturbation theory of magnetic instabilities in the two-dimensional Hubbard model at small doping, *Phys. Rev.* B **46**, 11884 (1992).
[14] H. J. Schulz, Incommensurate antiferromagnetism in the two-dimensional Hubbard model, *Phys. Rev. Lett.* **64**, 1445 (1990).

[15] A. Singh, Z. Tešanović, and J. H. Kim, Doped Hubbard antiferromagnet: Instability and effective interactions, *Phys. Rev.* B **44**, 7757 (1991).
[16] A. V. Chubukov and K. A. Musaelian, Magnetic phases of the two-dimensional Hubbard model at low doping, *Phys. Rev.* B **51**, 12605 (1995).
[17] E. Arrigoni and G. C. Strinati, Doping-induced incommensurate antiferromagnetism in a Mott-Hubbard insulator, *Phys. Rev.* B **44**, 7455 (1991).
[18] M. Dzierzawa, Hartree-Fock theory of spiral magnetic order in the 2-d Hubbard model, *Z. Phys.* B **86**, 49 (1992).
[19] C. Zhou and H. J. Schulz, Quantum fluctuations in the spiral phase of the Hubbard model, *Phys. Rev.* B **52**, R11557(R) (1995).
[20] W. Brenig, Spin-charge fluctuations in itinerant helical magnets, *Ann. Physik* **5**, 123 (1996).
[21] A. P. Kampf, Collective excitations in itinerant spiral magnets, *Phys. Rev.* B **53**, 747 (1996).
[22] B. L. Schraiman and E. D. Siggia, Spiral phase of a doped quantum antiferromagnet, *Phys. Rev. Lett.* **62**, 1564 (1989).
[23] B. L. Schraiman and E. D. Siggia, Excitation spectrum of the spiral state of a doped antiferromagnet, *Phys. Rev.* B **46**, 8305 (1992).
[24] V. Hankevych, B. Kyung, and A.-M. S. Tremblay, Weak ferromagnetism and other instabilities of the two-dimensional $t - t'$ Hubbard model at Van Hove fillings, *Phys. Rev.* B **68**, 214405 (2003).
[25] Y. Nagaoka, Ferromagnetism in a narrow, almost half-filled s band, *Phys. Rev.* **147**, 392 (1966).
[26] Z. Y. Weng, C. S. Ting, and T. K. Lee, Path-integral approach to the Hubbard model, *Phys. Rev.* B **43**, 3790 (1991).
[27] H. J. Schulz, Functional Integrals for Correlated Electrons, in *The Hubbard Model*, edited by D. Baeriswyl (Plenum, New York, 1995).
[28] K. Borejsza and N. Dupuis, Antiferromagnetism and single-particle properties in the two-dimensional half-filled Hubbard model: Slater vs. Mott-Heisenberg, *Europhys. Lett.* **63**, 722 (2003).
[29] K. Borejsza and N. Dupuis, Antiferromagnetism and single-particle properties in the two-dimensional half-filled Hubbard model: A nonlinear sigma model approach, *Phys. Rev.* B **69**, 085119 (2004).
[30] T. Schäfer, F. Geles, D. Rost, G. Rohringer, E. Arrigoni, K. Held, N. Blümer, M. Aichhorn, and A. Toschi, Fate of the false Mott-Hubbard transition in two dimensions, *Phys. Rev.* B **91**, 125109 (2015).
[31] N. Dupuis, Berezinskii-Kosterlitz-Thouless transition and BCS-Bose crossover in the two-dimensional attractive Hubbard model, *Phys. Rev.* B **70**, 134502 (2004).
[32] Y. Vilk, Y. M. Chen, and A.-M. S. Tremblay, Theory of spin and charge fluctuations in the Hubbard model, *Phys. Rev.* B **49**, 13267 (1994).
[33] Y. Vilk and A.-M. S. Tremblay, Non-perturbative many-body approach to the Hubbard model and single-particle pseudogap, *J. Phys. I France* **7**, 1309 (1997).
[34] Y. Vilk and A.-M. S. Tremblay, Destruction of Fermi-liquid quasiparticles in two dimensions by critical fluctuations, *Europhys. Lett.* **33**, 159–164 (1996).
[35] S. Moukouri, S. Allen, F. Lemay, B. Kyung, D. Poulin, Y. M. Vilk, and A.-M. S. Tremblay, Many-body theory versus simulations for the pseudogap in the Hubbard model, *Phys. Rev.* B **61**, 7887 (2000).

[36] A.-M. S. Tremblay, Two-particle-self-consistent approach for the Hubbard model, in *Theoretical Methods for Strongly Correlated Systems*, edited by A. Avella and F. Mancini (Springer Verlag, Berlin, 2011).

[37] K. S. Singwi and M. P. Tosi, Correlations in Electron Liquids, in *Solid State Physics*, edited by H. Ehrenreich, F. Seitz, and D. Turnbull (Academic, New York, 1981).

[38] S. Ichimaru, Strongly coupled plasmas: High-density classical plasmas and degenerate electron liquids, *Rev. Mod. Phys.* **54**, 1017 (1982).

[39] N. Dupuis, Spin fluctuations and pseudogap in the two-dimensional half-filled Hubbard model at weak coupling, *Phys. Rev.* B **65**, 245118 (2002).

Chapter 7

Superfluidity and Superconductivity

Contents

7.1 Superfluidity and Long-Range Current–Current Correlations 458
7.2 Spontaneous U(1) Symmetry Breaking and Superfluidity 462
7.2.1 Off-diagonal long-range order and spontaneous symmetry breaking 462
7.2.2 Quantum hydrodynamics 465
7.2.3 Finite temperature — Two-fluid hydrodynamics 472
7.2.4 Superconductors — Anderson–Higgs mechanism 477
7.3 Gross–Pitaevskii Theory 483
7.3.1 Statics and dynamics of a uniform condensate 484
7.3.2 Vortex 486
7.3.3 Soliton 488
7.4 Bogoliubov Theory 490
7.4.1 Mean-field theory 491
7.4.2 Lee–Huang–Yang correction 492
7.4.3 Transition temperature of the dilute Bose gas 495
7.4.4 Universal thermodynamics of the dilute Bose gas 496
7.4.5 Two-body contact 501
7.5 BCS Mean-Field Theory of Superconductivity 505
7.5.1 Cooper instability and superconducting transition 506
7.5.2 Mean-field theory 508
7.5.3 Ginzburg–Landau theory 516
7.6 Collective Modes in Superconductors 525
7.6.1 BCS correlation functions 528
7.6.2 Anderson–Bogoliubov mode 529
7.6.3 Electrodynamics 533
7.6.4 Higgs amplitude mode 538

7.7 BCS–BEC Crossover 539
7.7.1 Mean-field theory 540
7.7.2 Fluctuations: The Nozières–Schmitt-Rink approach 545
7.7.3 The dilute Fermi gas near a Feshbach resonance 549
Appendix 7.A Elements of Scattering Theory 553
7.A.1 Lippmann–Schwinger equation 553
7.A.2 Central potential 555
7.A.3 Low-energy collisions — Scattering length 557
7.A.4 T matrix and many-body formalism 559
Bibliography 563

To a large extent, superfluidity in boson and fermion systems can be understood on a common basis. In the first part of this chapter, we study these quantum phenomena from general principles with no reference to microscopic theories. Starting from the very definition of superfluidity — the existence of persistent (dissipationless) macroscopic currents — we show that superfluids are characterized by current–current correlations of infinite range. This property implies the Meissner effect in charged superconductors (Sec. 3.4.5).[1] We then introduce the concepts of off-diagonal long-range order and spontaneous U(1) symmetry breaking. The latter enables us to introduce a low-energy effective action for the phase of the superfluid order parameter, from which we can deduce the characteristic equations of a superfluid. The coupling to the electromagnetic field in (charged) superconductors is easily incorporated into this phenomenological approach and leads to the Anderson–Higgs mechanism. In the second part of the chapter, we study microscopic theories of superfluidity and superconductivity: the Gross–Pitaevskii approach and the Bogoliubov theory for boson systems and, for superconductors, the Bardeen–Cooper–Shrieffer (BCS) theory, with particular emphasis on collective modes in both neutral and charged systems. The last section is devoted to the BCS–BEC crossover in fermion superfluids. Unless otherwise specified, we consider only three-dimensional systems.[2]

7.1 Superfluidity and Long-Range Current–Current Correlations

We define a superfluid by its capacity to sustain a persistent macroscopic current. Let us consider a torus-shaped container filled with a fluid and suppose that initially the container and the fluid rotate together with angular velocity ω (Fig. 7.1). At $t = 0$, we stop the container. Because of its viscosity, a normal fluid would rapidly cease its motion and the mean value of the current density $\langle \hat{\mathbf{j}}(\mathbf{r}) \rangle$ in the container would vanish. If the fluid is superfluid, at least some part continues to rotate indefinitely so that $\lim_{t\to\infty} \langle \hat{\mathbf{j}}(\mathbf{r}) \rangle \neq 0$.

---

[1] We refer to fermion superfluids as superconductors regardless of whether or not the fermions carry an electric charge. Neutral and charged superconductors are found in ultracold atomic gases and solids, respectively.

[2] Superfluidity in two- and one-dimensional systems will be discussed in Chapters 10 and 13, respectively (see also Chapter 15 for superconductivity in one-dimensional fermion systems).

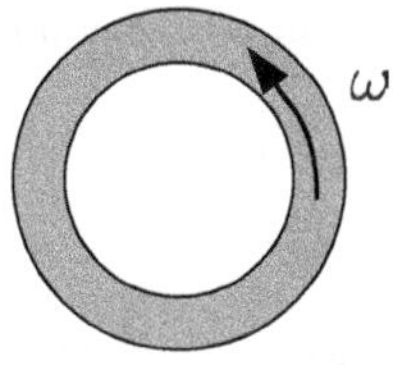

Fig. 7.1 A torus-shaped container and a fluid rotating with angular velocity ω. Due to the absence of viscosity, a superfluid continues to rotate indefinitely when the container is put to rest.

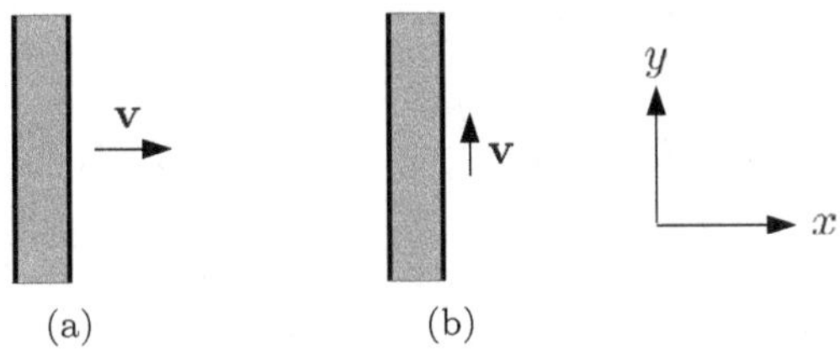

Fig. 7.2 A superfluid confined between two parallel infinite walls. When the walls are accelerated with velocity $\mathbf{v}$, all the fluid moves in case (a) but only the normal component is dragged along in case (b).

It is convenient to consider the analogous experience where the fluid is confined between two infinite parallel planes (Fig. 7.2). The system is initially at rest. Between $t = -\infty$ and $t = 0$, we accelerate the walls of the container to a finite velocity $\mathbf{v}$, i.e. $\mathbf{v}(t) = \mathbf{v}e^{0^+t}$. In the situation depicted in Fig. 7.2(a), all the fluid moves with the walls, so that we must have

$$\langle \hat{\mathbf{j}}(\mathbf{r}) \rangle (t = 0) = n\mathbf{v} \qquad (\mathbf{v} \perp \text{walls}), \tag{7.1}$$

where n is the density of the fluid. When $\mathbf{v}$ is parallel to the walls, only the normal component of the fluid is dragged along, while the superfluid component — due to its absence of viscosity — stays at rest with respect to the laboratory frame,

$$\langle \hat{\mathbf{j}}(\mathbf{r}) \rangle (t = 0) = n_n\mathbf{v} \qquad (\mathbf{v} \parallel \text{walls}), \tag{7.2}$$

where $n_n \leq n$ is the density of the normal component of the fluid. The density $n_s = n - n_n$ of the superfluid component is referred to as the superfluid density. This "two-fluid" picture, where the system is viewed as consisting of a normal fluid and a superfluid, underlies most of the understanding of superfluidity. It will be further discussed in Sec. 7.2.3.

When the velocity $\mathbf{v}$ is small, one can obtain the induced current from linear response theory. In the reference frame moving with the container, the potential due to the container walls is time independent and we can use the usual methods of equilibrium statistical physics. For a single particle, the Hamiltonian in the frame moving with the walls is $\hat{H}' = \hat{\mathbf{p}}'^2/2m$ where $\hat{\mathbf{p}}' = \hat{\mathbf{p}} - m\mathbf{v}$ with $\hat{\mathbf{p}}$ the momentum in the laboratory frame (we momentarily ignore the time dependence of $\mathbf{v}$). Thus, $\hat{H}' = \hat{H} - \hat{\mathbf{p}} \cdot \mathbf{v} + \mathcal{O}(\mathbf{v}^2)$ where $\hat{H} = \hat{\mathbf{p}}^2/2m$. For a many-particle system with a pair-interaction potential $V(\mathbf{r}_i - \mathbf{r}_j)$, $\hat{H}' = \hat{H} - \hat{\mathbf{P}} \cdot \mathbf{v} + \mathcal{O}(\mathbf{v}^2)$ with $\hat{\mathbf{P}}$ the total momentum since the coordinate differences are the same in both frames:

$V(\mathbf{r}_i - \mathbf{r}_j) = V(\mathbf{r}'_i - \mathbf{r}'_j)$. The Hamiltonian to consider is therefore[3]

$$\hat{H} = \hat{H}(\mathbf{v} = 0) - \int d^3r\, \hat{\mathbf{j}}(\mathbf{r}) \cdot m\mathbf{v}e^{0^+t} + \mathcal{O}(\mathbf{v}^2), \tag{7.3}$$

where we now take into account that the velocity is adiabatically switched on. For a Galilean-invariant system, the total momentum $\hat{\mathbf{P}} = m\hat{\mathbf{J}}$ is given by the total current $\hat{\mathbf{J}} = \int d^3r\, \hat{\mathbf{j}}(\mathbf{r})$ with $\hat{\mathbf{j}}(\mathbf{r})$ the current density.[4] At $t = 0$, the current — as measured in the laboratory frame — induced by the moving container is given by

$$\begin{aligned} \langle \hat{\jmath}_\mu(\mathbf{r})\rangle(t=0) &= \int dt' \int d^3r' \sum_\nu \chi^R_{\mu\nu}(\mathbf{r}, t=0; \mathbf{r}', t') m v_\nu e^{0^+t'}, \\ &= \int d^3r' \int \frac{d^3q}{(2\pi)^3} e^{-i\mathbf{q}\cdot\mathbf{r}'} \sum_\nu \chi_{\mu\nu}(\mathbf{q}) m v_\nu, \end{aligned} \tag{7.4}$$

where $\chi^R_{\mu\nu}(\mathbf{r}, t; \mathbf{r}', t') = i\Theta(t-t')\langle[\hat{\jmath}_\mu(\mathbf{r}, t), \hat{\jmath}_\nu(\mathbf{r}', t')]\rangle$ is the retarded current–current correlation function. $\chi^R_{\mu\nu}(\mathbf{q}, \omega)$ is the Fourier transform of $\chi^R_{\mu\nu}(\mathbf{r} - \mathbf{r}', t - t')$.[5] For an infinite system in the three directions of space, one would obtain $\int d^3r\, e^{-i\mathbf{q}\cdot\mathbf{r}} = \mathcal{V}\delta_{\mathbf{q},0} \to (2\pi)^3\delta(\mathbf{q})$. For a finite system, the integral over space gives

$$\int_{-\frac{L_x}{2}}^{\frac{L_x}{2}} dx' e^{-ik_x x'} = \frac{2}{k_x} \sin\left(\frac{k_x L_x}{2}\right) \equiv 2\pi\Delta(k_x), \tag{7.5}$$

(and a similar result for the integral over y' and z') where $\lim_{L_x\to\infty} \Delta(k_x) = \delta(k_x)$. Thus, we obtain

$$\langle \hat{\jmath}_\mu(\mathbf{r})\rangle(t=0) = \int d^3q\, \Delta(q_x)\Delta(q_y)\Delta(q_z) \sum_\nu \chi_{\mu\nu}(\mathbf{q}) m v_\nu. \tag{7.6}$$

The finite size of the system in the x direction is taken into account by first taking the limit $L_y, L_z \to \infty$ and then $L_x \to \infty$ (the container walls are assumed to be perpendicular to the x axis). When $\mathbf{v}$ is perpendicular to the container walls, this gives

$$\langle \hat{\jmath}_x(\mathbf{r})\rangle(t=0) = \lim_{q_x\to 0} \left[\lim_{q_y, q_z \to 0} \chi_{xx}(\mathbf{q})\right] m v_x = \lim_{\mathbf{q}\to 0} \chi_\parallel(\mathbf{q}) m v_x, \tag{7.7}$$

while for $\mathbf{v}$ parallel to the walls

$$\langle \hat{\jmath}_y(\mathbf{r})\rangle(t=0) = \lim_{q_x\to 0} \left[\lim_{q_y, q_z \to 0} \chi_{yy}(\mathbf{q})\right] m v_y = \lim_{\mathbf{q}\to 0} \chi_\perp(\mathbf{q}) m v_y. \tag{7.8}$$

$\chi_\parallel$ and $\chi_\perp$ are the longitudinal and transverse parts of the current–current correlation function[6]

$$\chi_{\mu\nu}(\mathbf{q}) = \frac{q_\mu q_\nu}{\mathbf{q}^2}\chi_\parallel(\mathbf{q}) + \left(\delta_{\mu,\nu} - \frac{q_\mu q_\nu}{\mathbf{q}^2}\right)\chi_\perp(\mathbf{q}). \tag{7.9}$$

[3] Equation (7.3) can also be obtained from (1.C.49).

[4] The result $\hat{\mathbf{P}} = m\hat{\mathbf{J}}$ is a consequence of Noether's theorem for a Galilean-invariant system.

[5] As in Chapter 3, we use the notation $\chi(\mathbf{q}) = \chi^R(\mathbf{q}, \omega = 0)$.

[6] This form follows from the homogeneity and isotropy of the fluid. $\chi_\parallel(\mathbf{q})$ and $\chi_\perp(\mathbf{q})$ are functions of $|\mathbf{q}|$.

Equations (7.1), (7.2) and (7.7), (7.8) imply

$$\lim_{\mathbf{q}\to 0} \chi_\parallel(\mathbf{q}) = \frac{n}{m}, \qquad \lim_{\mathbf{q}\to 0} \chi_\perp(\mathbf{q}) = \frac{n_n}{m}. \tag{7.10}$$

The first of these equations is nothing but the $\mathbf{q} \to 0$ limit of the f-sum rule $\chi_\parallel(\mathbf{q}) = \chi^R_\parallel(\mathbf{q}, 0) = n/m$ ($\mathbf{q} \neq 0$). The latter is a consequence of particle number conservation (or gauge invariance) and was derived in Chapter 3.[7] The system is superfluid if $n_n \neq n$, i.e. if

$$\lim_{\mathbf{q}\to 0} \chi_\parallel(\mathbf{q}) \neq \lim_{\mathbf{q}\to 0} \chi_\perp(\mathbf{q}) \qquad \text{(superfluid)}. \tag{7.11}$$

As noticed in Sec. 3.4.5, this is possible only if $\chi_{\mu\nu}(\mathbf{q})$ is singular for $\mathbf{q} = 0$, i.e. if $\chi_{\mu\nu}(\mathbf{r})$ is of infinite range,

$$\int d^3r \, \chi_{\mu\nu}(\mathbf{r}) \to \pm\infty. \tag{7.12}$$

This leads us to define a superfluid as a system where the current–current correlations are of infinite range.[8] From (7.10), we obtain

$$\chi_{\mu\nu}(\mathbf{q}) = \frac{n_n}{m}\delta_{\mu,\nu} + \frac{n_s}{m}\frac{q_\mu q_\nu}{\mathbf{q}^2} \qquad (\mathbf{q} \to 0) \tag{7.13}$$

and the long-range part of the current–current correlations[9]

$$\lim_{|\mathbf{r}|\to\infty} \chi_{\mu\nu}(\mathbf{r}) = \frac{n_s}{m}\int \frac{d^3q}{(2\pi)^3}\frac{q_\mu q_\nu}{\mathbf{q}^2} e^{i\mathbf{q}\cdot\mathbf{r}} = -\frac{n_s}{4\pi m}\left(\frac{\delta_{\mu,\nu}}{|\mathbf{r}|^3} - 3\frac{r_\mu r_\nu}{|\mathbf{r}|^5}\right). \tag{7.14}$$

Zero-temperature superfluid density

The value of $n_s(T = 0)$ can be deduced from the behavior of $\chi''_\perp(\mathbf{q}, \omega)$. In a Galilean-invariant system, $\hat{\mathbf{j}}(\mathbf{q} = 0, t) = \hat{\mathbf{j}}(\mathbf{q} = 0, t = 0)$ is conserved and $\chi''_\perp(\mathbf{q} = 0, \omega) = 0$. Only low-energy excitations such that $\lim_{\mathbf{q}\to 0} \chi''_\perp(\mathbf{q}, \omega)$ remains finite as $\omega \to 0$ can contribute to

$$\lim_{\mathbf{q}\to 0} \chi_\perp(\mathbf{q}) = \lim_{\mathbf{q}\to 0} \int_{-\infty}^{\infty} \frac{d\omega}{\pi} \frac{\chi''_\perp(\mathbf{q}, \omega)}{\omega} = \frac{n_n}{m}. \tag{7.15}$$

When there is a gap in the transverse excitation spectrum — $\chi''_\perp(\mathbf{q}, \omega)$ vanishes for small ω — $\chi^R_\perp$ decays rapidly in time and the limit $\mathbf{q}, \omega \to 0$ is not singular: $\lim_{\mathbf{q}\to 0} \chi_\perp(\mathbf{q}) = \lim_{\omega\to 0} \chi^R_\perp(0, \omega) = 0$ since $\chi''_\perp(0, \omega) = 0$. We shall see in the following that in a superfluid system, the low-energy excitations are phonons. For reasons of symmetry, a transverse probe

[7] The operator $\hat{\mathbf{j}}$ considered in this chapter is the particle current, while in Chapter 3 it refers to the charge current (hence the factor e^2 in the f-sum rule (3.193)).

[8] In Sec. 3.4.5, we showed that long-range current–current correlations lead to the Meissner effect in a superconductor. Superconductivity is discussed in Secs. 7.2.4 and 7.5–7.7.

[9] Equation (7.14) is derived using $\int_{\mathbf{q}} \frac{q_\mu q_\nu}{\mathbf{q}^2} e^{i\mathbf{q}\cdot\mathbf{r}} = -\partial_{r_\mu}\partial_{r_\nu} \int_{\mathbf{q}} \frac{e^{i\mathbf{q}\cdot\mathbf{r}}}{\mathbf{q}^2} = -\partial_{r_\mu}\partial_{r_\nu} \frac{1}{4\pi|\mathbf{r}|}$.

cannot couple the ground state to a purely longitudinal excitation so that phonons do not contribute to

$$\lim_{T\to 0} \chi_\perp(\mathbf{q}) = 2 \sum_{\substack{n \\ (\epsilon_n \neq \epsilon_0)}} \frac{|\langle 0|\hat{\mathbf{j}}(\mathbf{q}) \cdot \boldsymbol{\eta}_\mathbf{q}|n\rangle|^2}{\epsilon_n - \epsilon_0} \tag{7.16}$$

($\boldsymbol{\eta}_\mathbf{q}$ is a unit vector perpendicular to $\mathbf{q}$). Equation (7.16) follows from the Lehmann representation (3.38); $\{|n\rangle, \epsilon_n\}$ denotes the eigenstates and eigenenergies of the Hamiltonian and $|0\rangle$ the ground state. We conclude that

$$n_n(T=0) = m \lim_{T\to 0} \lim_{\mathbf{q}\to 0} \chi_\perp(\mathbf{q}) = 0 \quad \text{i.e.} \quad n_s(T=0) = n \tag{7.17}$$

in a Galilean-invariant system: The zero-temperature superfluid density is given by the full density of the fluid. Note that this result also holds for an ideal Bose gas, since the current $\mathbf{q}/m$ carried by a particle of momentum $\mathbf{q}$ is longitudinal. The preceding reasoning clearly shows that superfluidity is due to the scarcity of low-energy (transverse) excitations, which (as we shall see) is a consequence of the spontaneous U(1) symmetry breaking in the non-ideal gas. The property $n_s(T=0) = n$ will also be derived in Sec. 7.2.3 and Chapter 13.

7.2 Spontaneous U(1) Symmetry Breaking and Superfluidity

Superfluids are characterized by current–current correlations of infinite range or, equivalently, by a finite superfluid density n_s. Can we also understand superfluidity as the result of spontaneous symmetry breaking and identify the associated order parameter? What are then the fluctuations responsible for the long-range current–current correlations?

7.2.1 *Off-diagonal long-range order and spontaneous symmetry breaking*

7.2.1.1 *Off-diagonal long-range order*

Since an ideal Bose gas has the unique property to undergo a BEC at low temperatures, it is natural to expect a link between BEC and superfluidity. In order to characterize BEC, we consider the one-body density matrix

$$\rho(\mathbf{r}-\mathbf{r}') = \langle \hat\psi^\dagger(\mathbf{r})\hat\psi(\mathbf{r}')\rangle = \int_\mathbf{k} e^{i\mathbf{k}\cdot(\mathbf{r}-\mathbf{r}')} \langle \hat\psi^\dagger(\mathbf{k})\hat\psi(\mathbf{k})\rangle. \tag{7.18}$$

In a normal fluid, $\langle \hat\psi^\dagger(\mathbf{k})\hat\psi(\mathbf{k})\rangle$ is regular when $\mathbf{k} \to 0$ and $\lim_{|\mathbf{r}-\mathbf{r}'|\to\infty} \rho(\mathbf{r}-\mathbf{r}') = 0$. In an ideal Bose gas below the BEC transition temperature, the one-particle state $\hat\psi^\dagger(\mathbf{k}=0)|0\rangle$ is macroscopically occupied, and

$$\langle \hat\psi^\dagger(\mathbf{k})\hat\psi(\mathbf{k})\rangle = N_0 \delta_{\mathbf{k},0} + f(\mathbf{k}), \tag{7.19}$$

where the function $f(\mathbf{k})$ is regular for $\mathbf{k} \to 0$. $N_0 = \langle \hat\psi^\dagger(\mathbf{k}=0)\hat\psi(\mathbf{k}=0)\rangle$ is the number of particles in the condensate. It follows that

$$\rho(\mathbf{r}-\mathbf{r}') = n_0 + f(\mathbf{r}-\mathbf{r}') \to n_0 \quad \text{when} \quad |\mathbf{r}-\mathbf{r}'| \to \infty, \tag{7.20}$$

where $n_0 = N_0/\mathcal{V}$ is the condensate density. n_0 should not be confused with the superfluid density n_s. As we shall see in Sec. 7.4, $n_0(T) < n$ in any interacting system whereas $n_s(T = 0) = n$ in a Galilean-invariant system.[10] The property (7.20), characteristics of BEC, is referred to as off-diagonal long-range order (ODLRO) since it involves the off-diagonal elements ($\mathbf{r} \neq \mathbf{r}'$) of the one-body density matrix $\rho(\mathbf{r}, \mathbf{r}') = \rho(\mathbf{r} - \mathbf{r}')$. We expect ODLRO to be also the fundamental property of superfluidity in interacting Bose fluids.[11]

7.2.1.2 *Spontaneous U(1) symmetry breaking*

At this stage, it is tempting to make the approximation

$$\langle \hat{\psi}^\dagger(\mathbf{r})\hat{\psi}(\mathbf{r}')\rangle \simeq \langle \hat{\psi}^\dagger(\mathbf{r})\rangle\langle \hat{\psi}(\mathbf{r}')\rangle \quad \text{for} \quad |\mathbf{r} - \mathbf{r}'| \to \infty. \tag{7.21}$$

While this is obviously correct in a normal fluid where $\langle \hat{\psi}(\mathbf{r})\rangle = 0$, in a uniform superfluid it implies $\langle \hat{\psi}(\mathbf{r})\rangle = \sqrt{n_0}e^{i\theta}$ (with θ an arbitrary phase) and identifies $\langle \hat{\psi}(\mathbf{r})\rangle$ as the order parameter associated with BEC and superfluidity. $\langle \hat{\psi}(\mathbf{r})\rangle$ is also referred to as the condensate wavefunction.[12] In the superfluid state ($n_0 \neq 0$), the U(1) symmetry, i.e. the invariance under the global U(1) transformation $\hat{\psi}(\mathbf{r}) \to \hat{\psi}(\mathbf{r})e^{i\Lambda}$, is spontaneously broken.

If the Hamiltonian conserves the particle number, $[\hat{H}, \hat{N}] = 0$, we can diagonalize simultaneously $\hat{H}$ and $\hat{N}$. In a state with a given particle number, $\langle \hat{\psi}(\mathbf{r})\rangle = 0$. To understand how we can have $\langle \hat{\psi}(\mathbf{r})\rangle \neq 0$, it is useful to introduce the concept of phase and amplitude of the operator $\hat{\psi}(\mathbf{r})$. Suppose that we write

$$\hat{\psi} = e^{i\hat{\theta}}\sqrt{\hat{n}} \tag{7.22}$$

(for simplicity, we consider a single bosonic degree of freedom) and assume $\hat{\theta}$ to be Hermitian. $\hat{n} = \hat{\psi}^\dagger\hat{\psi}$ is the particle number operator. If (7.22) were legitimate, we would have successfully defined a phase operator $\hat{\theta}$. However, as we shall see, (7.22) is not without problem. The commutation relation $[\hat{\psi}, \hat{\psi}^\dagger] = 1$ implies

$$\left[e^{i\hat{\theta}}, \hat{n}\right] = e^{i\hat{\theta}}. \tag{7.23}$$

This suggests that $\hat{n}$ and $\hat{\theta}$ are conjugate operators,

$$[\hat{n}, \hat{\theta}] = i. \tag{7.24}$$

From (7.24), we indeed deduce $[\hat{n}, \hat{\theta}^p] = ip\hat{\theta}^{p-1}$ and in turn (7.23). Now $[\hat{n}, \hat{\theta}] = i$ gives

$$\langle n|\hat{n}\hat{\theta} - \hat{\theta}\hat{n}|n'\rangle = (n - n')\langle n|\hat{\theta}|n'\rangle = i\delta_{n,n'}, \tag{7.25}$$

which is clearly meaningless. Nevertheless, if we can consider n to be a continuous variable (which is expected to be a good approximation when n is large), we obtain

$$(n - n')\langle n|\hat{\theta}|n'\rangle = i\delta(n - n') \tag{7.26}$$

[10] Note that $n_0(T = 0) = n_s(T = 0) = n$ in an ideal Bose gas.

[11] In Sec. 7.2.3, we shall see that the ideal Bose gas, although it exhibits ODLRO, cannot sustain (dissipationless) superflow.

[12] In the ground state, the phase $\theta = -\mu t$ of the condensate wavefunction is time dependent in the canonical ensemble but time independent in the grand canonical ensemble assumed here (see Sec. 7.3.1).

and therefore $\langle n|\hat{\theta}|n'\rangle = -i\delta'(n-n')$ (using the equality $-x\delta'(x) = \delta(x)$). This should be compared with the analogous result $\langle p|\hat{x}|p'\rangle = i\delta'(p-p')$ for a quantum particle with position x and momentum p. This simple argument suggests that (7.22) with $\hat{\theta}$ Hermitian, although not strictly correct, is meaningful in the limit of large quantum numbers (i.e. a large number of particles).

It is then easy to see that the states $|\theta\rangle$ defined by

$$|\theta\rangle = \frac{1}{\sqrt{2\pi}}\sum_{n=0}^{\infty} e^{in\theta}|n\rangle, \qquad |n\rangle = \int_0^{2\pi}\frac{d\theta}{\sqrt{2\pi}}e^{-in\theta}|\theta\rangle \tag{7.27}$$

are eigenstates of the phase operator $\hat{\theta}$. Since (7.23) implies $e^{i\hat{\theta}}|n\rangle = |n-1\rangle$, we find $e^{i\hat{\theta}}|\theta\rangle = e^{i\theta}|\theta\rangle$ and

$$\hat{\theta}|\theta\rangle = \theta|\theta\rangle. \tag{7.28}$$

A state with a well-defined phase is a coherent superposition of states with different particle numbers.[13]

In a system with many degrees of freedom, we may write $\hat{\psi}(\mathbf{r}) = e^{i\hat{\theta}(\mathbf{r})}\sqrt{\hat{n}(\mathbf{r})}$ with $[\hat{n}(\mathbf{r}),\hat{\theta}(\mathbf{r}')] = i\delta(\mathbf{r}-\mathbf{r}')$ and therefore

$$[\hat{N},\hat{\theta}] = i, \quad \hat{N} = \int d^3r\,\hat{n}(\mathbf{r}) \quad \text{and} \quad \hat{\theta} = \frac{1}{\mathcal{V}}\int d^3r\,\hat{\theta}(\mathbf{r}). \tag{7.29}$$

Equation (7.29) yields the uncertainty relation

$$\Delta N\Delta\theta \gtrsim \frac{1}{2}. \tag{7.30}$$

We are now in a position to understand how spontaneous symmetry breaking can arise despite the conservation of particle number ($[\hat{H},\hat{N}] = 0$) and the uncertainty relation (7.30). Clearly, the phase has to be well defined for $\langle\hat{\psi}(\mathbf{r})\rangle = \langle e^{i\hat{\theta}(\mathbf{r})}\sqrt{\hat{n}(\mathbf{r})}\rangle$ not to vanish. Let us call $|n\rangle$ the ground state of $\hat{H}$ with n particles, and N the number of particles in the system. Broken symmetry is possible when the eigenstates $|n\rangle$ with $n\sim N$ become degenerate in the thermodynamic limit.[14] When N is very large, a small fluctuation $\Delta N/N$, with $1 \ll \Delta N \ll N$, is sufficient to allow the phase to be well-defined without violating the uncertainty relation (7.30). For instance, the state

$$\begin{aligned}|\tilde{\theta}\rangle &= \frac{1}{\sqrt{2\Delta N+1}}\sum_{n=N-\Delta N}^{N+\Delta N} e^{in\theta}|n\rangle \\ &= \frac{e^{iN(\theta-\theta')}}{\sqrt{2\Delta N+1}}\int_0^{2\pi}\frac{d\theta'}{\sqrt{2\pi}}\frac{\sin\left[\left(\Delta N+\frac{1}{2}\right)(\theta-\theta')\right]}{\sin\left(\frac{\theta-\theta'}{2}\right)}|\theta'\rangle\end{aligned} \tag{7.31}$$

is a wave packet centered around θ with a width $\Delta\theta\sim 1/\Delta N$ and corresponds to a mean number of particles $\langle\tilde{\theta}|\hat{N}|\tilde{\theta}\rangle = N$ with fluctuations $\langle\tilde{\theta}|(\hat{N}-N)^2|\tilde{\theta}\rangle \sim \Delta N^2$ in agreement with the uncertainty relation (7.30).

---

[13] For a discussion of amplitude and phase operators in quantum mechanics, see, e.g., Ref. [82].

[14] See the general discussion of spontaneous symmetry breaking in Sec. 3.6.

7.2.1.3 *Functional integral formalism*

In the functional integral formalism, the breakdown of the U(1) symmetry arises very naturally. If we perform a classical (or saddle-point) approximation, we obtain

$$\int \mathcal{D}[\psi^*, \psi]\, e^{iS[\psi^*,\psi]} \to e^{iS[\psi^*,\psi]}, \tag{7.32}$$

where the *classical* field $\psi(\mathbf{r})$ (the order parameter) is determined by the saddle-point equations $\delta S/\delta\psi(\mathbf{r}) = \delta S/\delta\psi^*(\mathbf{r}) = 0$. If $\psi(\mathbf{r}) = \langle\hat{\psi}(\mathbf{r})\rangle \neq 0$, the global U(1) symmetry is broken and the system exhibits ODLRO. Recalling the definition of the functional integral (Chapter 1), the classical approximation (7.32) amounts to assuming that the system is in the coherent state

$$|\psi\rangle = \exp\left\{\int d^3r\, \psi(\mathbf{r})\hat{\psi}^\dagger(\mathbf{r})\right\}|\text{vac}\rangle. \tag{7.33}$$

We therefore come to the conclusion that a superfluid is naturally described by a coherent state. The classical approximation (7.32), (7.33) is the basis of the Gross–Pitaevskii theory; we shall see that it is justified in the low-density limit. It can be improved by including quantum and thermal fluctuations about the classical field $\psi(\mathbf{r})$; this naturally leads to the Bogoliubov theory.

Before exploring the Gross–Pitaevskii and Bogoliubov theories in more detail (Secs. 7.3 and 7.4), we show that most fundamental properties of superfluids and superconductors follow from the broken U(1) symmetry hypothesis, i.e. from the existence of a classical field $\psi(\mathbf{r})$ with a well-defined phase (Secs. 7.2.2–7.2.4).

7.2.2 *Quantum hydrodynamics*

7.2.2.1 *Quadratic Lagrangian*

When the global U(1) symmetry is spontaneously broken, the Goldstone theorem ensures that fluctuations of the phase θ of the field ψ are gapless.[14] Anticipating that the phase mode has a linear dispersion $\omega = c|\mathbf{q}|$ (see Sec. 7.3), the simplest Euclidean effective action — describing the low-energy long-wavelength properties of the system — we can think of reads

$$S[\theta] = \frac{n_s}{2m}\int_0^\beta d\tau \int d^3r \left\{[\boldsymbol{\nabla}\theta(x)]^2 + \frac{\dot{\theta}(x)^2}{c^2}\right\}, \tag{7.34}$$

where m is the mass of the particles. Here, we assume that the phase weakly fluctuates about its equilibrium value and neglect its 2π periodicity. We shall see in the following that n_s, which has the dimension of a density, can be identified with the superfluid density (and therefore the full density n when $T = 0$). The action $S[\theta]$ is invariant under the U(1) transformation $\theta(x) \to \theta(x) + \Lambda$. From Noether's theorem (Chapter 2), we deduce the corresponding particle and current densities,

$$\begin{aligned} n(x) &= -i\frac{\delta S}{\delta\dot\theta(x)} = -i\frac{n_s}{mc^2}\dot\theta(x), \\ \mathbf{j}(x) &= \frac{\delta S}{\delta\boldsymbol{\nabla}\theta(x)} = \frac{n_s}{m}\boldsymbol{\nabla}\theta(x). \end{aligned} \tag{7.35}$$

The continuity equation $i\partial_\tau n(x) + \boldsymbol{\nabla}\cdot\mathbf{j}(x) = 0$ follows from the classical equation of motion $\delta S/\delta\theta(x) = 0$.[15] The expression of $n(x)$ is not satisfying since it vanishes for $\dot\theta(x) = 0$ instead of taking its equilibrium value n. This suggests to add to the action the term $in\int_0^\beta d\tau \int d^3r\,\dot\theta(x)$.[16] Such a term does not modify the classical equation of motion, and should not play an important role for small fluctuations of the phase about its equilibrium value $\theta(x) = \text{const}$. For the time being, we shall therefore proceed with the action (7.34) and interpret the first of Eqs. (7.35) as a definition of the fluctuation $\delta n(x)$ of the density about its equilibrium value rather than the full density $n(x)$. In the following, we shall give a more general discussion of the validity of the action (7.34) and see how we can improve it.

We now calculate the contribution of the phase fluctuations to the current–current correlation function. Using $j_\mu(q) = i\frac{n_s}{m}q_\mu\theta(q)$, we obtain

$$\chi^{\text{phase}}_{\mu\nu}(q) = \langle j_\mu(q) j_\nu(-q)\rangle = \frac{n_s^2}{m^2}q_\mu q_\nu \chi_{\theta\theta}(q) = \frac{n_s}{m}\frac{c^2 q_\mu q_\nu}{\omega_n^2 + c^2\mathbf{q}^2}, \tag{7.36}$$

where

$$\chi_{\theta\theta}(q) = \langle\theta(q)\theta(-q)\rangle = \frac{mc^2}{n_s}\frac{1}{\omega_n^2 + c^2\mathbf{q}^2} \tag{7.37}$$

is the phase–phase correlation function. The current $\mathbf{j}(x)$ being longitudinal [Eq. (7.35)], phase fluctuations give a purely longitudinal contribution to the current–current correlation function $\chi_{\mu\nu}(q)$,

$$\chi^{\text{phase}}_{\parallel}(q) = \frac{n_s}{m}\frac{c^2\mathbf{q}^2}{\omega_n^2 + c^2\mathbf{q}^2}, \qquad \chi^{\text{phase}}_{\perp}(q) = 0. \tag{7.38}$$

Let us now suppose that all other fluctuations are short-range. In the limit $\mathbf{q}, \omega_n \to 0$, they contribute a term $C\delta_{\mu,\nu}$ to $\chi_{\mu\nu}(q)$, so that

$$\chi_{\mu\nu}(\mathbf{q}) = C\delta_{\mu,\nu} + \frac{n_s}{m}\frac{q_\mu q_\nu}{\mathbf{q}^2} = C\left(\delta_{\mu,\nu} - \frac{q_\mu q_\nu}{\mathbf{q}^2}\right) + \left(\frac{n_s}{m} + C\right)\frac{q_\mu q_\nu}{\mathbf{q}^2} \qquad (\mathbf{q}\to 0). \tag{7.39}$$

The constant C is fixed by the f-sum rule $\chi_\parallel(\mathbf{q}) = C + n_s/m = n/m$ (Sec. 3.4.4). The short-range contribution $C\delta_{\mu,\nu} = (n_n/m)\delta_{\mu,\nu}$ comes from the normal component of the fluid. We thus reproduce (7.10) and (7.13) and confirm that the coefficient n_s in the action $S[\theta]$ can be identified with the superfluid density. The latter, defined in Sec. 7.1 from the

---

[15] This is a mere consequence of the fact that the action depends only on the derivatives $\dot\theta$, $\boldsymbol{\nabla}\theta$ of the phase, and not on the phase itself.

[16] $\psi(\beta) = \psi(0)$ implies $\theta(\beta) = \theta(0)$ $[2\pi]$ and $\int_0^\beta d\tau\,\dot\theta$ does not necessarily vanish.

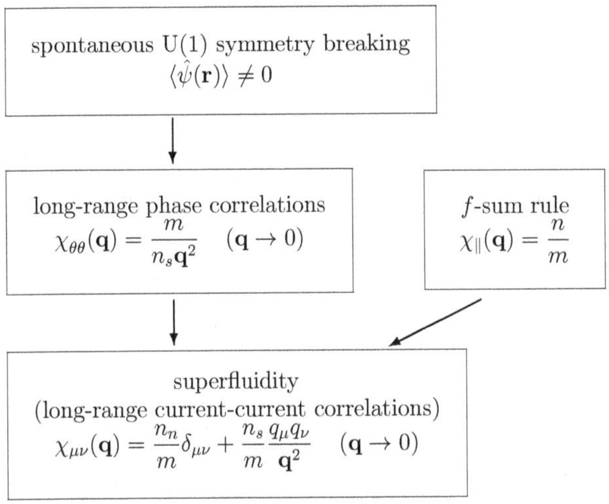

Fig. 7.3 Relation between spontaneous U(1) symmetry breaking and superfluidity. More generally, the figure shows that superfluidity arises from a non-zero phase stiffness n_s and can occur even in the absence of spontaneous U(1) symmetry breaking as in two- and one-dimensional systems.

current–current correlation function, can therefore also be defined form the phase–phase correlation function [Eq. (7.37)],

$$\chi_{\theta\theta}(\mathbf{q}) = \frac{m}{n_s \mathbf{q}^2} \qquad (\mathbf{q} \to 0). \tag{7.40}$$

The relation between spontaneous U(1) symmetry breaking and superfluidity is summarized in Fig. 7.3.

The preceding reasoning actually shows that the necessary condition for superfluidity is a non-zero phase stiffness rather than a spontaneous U(1) symmetry breaking. While the former is a consequence of the latter, the two properties are not equivalent and a non-zero superfluid density is possible without BEC. This is why superfluidity ($n_s > 0$) can exist in two- and one-dimensional systems in spite of a vanishing condensate density ($n_0 = 0$).[17,18]

From (7.40), we can give another (equivalent) definition of the superfluid density. Using the result of Sec. 3.2.7, a slowly varying phase $\langle \theta(\mathbf{r}) \rangle$ (induced by an external source) implies a change

$$\Delta\Omega = \frac{1}{2}\sum_{\mathbf{q}} \chi_{\theta\theta}^{-1}(\mathbf{q})\langle \theta(-\mathbf{q})\rangle\langle \theta(\mathbf{q})\rangle = \frac{n_s}{2m}\int d^3r [\boldsymbol{\nabla}\langle \theta(\mathbf{r})\rangle]^2 \tag{7.41}$$

in the thermodynamic potential [Eq. (3.96)]. Thus, n_s/m appears as the stiffness with respect to to a static twist of the phase of the superfluid order parameter. The various definitions of the superfluid density are summarized in Table 7.1.

[17]See footnote 2 on page 458.

[18]See Sec. 3.6.5.1 for a proof of the absence of BEC in one-dimensional interacting boson systems.

Table 7.1 Various definitions of the superfluid density n_s in a three-dimensional system. The Josephson relation [38], which relates n_s to the one-particle Green function $G(\mathbf{k},0) = -\langle\psi(\mathbf{k},0)\psi^*(\mathbf{k},0)\rangle$ at zero Matsubara frequency, is derived in Ref. [10].

Current–current correlation function	$\frac{n_s}{m} = \lim_{\mathbf{q}\to 0}\left[\chi_\parallel(\mathbf{q}) - \chi_\perp(\mathbf{q})\right]$
Phase–phase correlation function	$\chi_{\theta\theta}(\mathbf{q}) = \frac{m}{n_s\mathbf{q}^2} \qquad (\mathbf{q}\to 0)$
Thermodynamic potential	$\Delta\Omega = \frac{n_s}{2m}\int d^3r[\boldsymbol{\nabla}\langle\theta(\mathbf{r})\rangle]^2$
Josephson relation	$n_s = -\frac{mn_0}{\mathbf{k}^2 G(\mathbf{k},0)} \qquad (\mathbf{k}\to 0)$

The stability of the system requires the phase stiffness to be positive and in turn $n_s \geq 0$. Using the Bogoliubov inequality (Chapter 3), we can also find an upper bound to the superfluid density. Equation (3.280) with $\hat{A} = \hat{\theta}(\mathbf{r},t)$ and $\hat{B} = \partial_t\hat{n}(\mathbf{r},t)$ gives

$$\chi_{\theta\theta}(\mathbf{q})\int_{-\infty}^{\infty}\frac{d\omega}{\pi}\omega\chi''_{nn}(\mathbf{q},\omega) \geq \left|\int_{-\infty}^{\infty}\frac{d\omega}{\pi}\chi''_{\theta n}(\mathbf{q},\omega)\right|^2. \tag{7.42}$$

In the limit $\mathbf{q}\to 0$, $\chi_{\theta\theta}(\mathbf{q})$ is given by (7.40). The integral over ω in the lhs of the inequality (7.42) gives $n\mathbf{q}^2/m$ by the f-sum rule [Eq. (3.106)]. Using $[\hat{n}(\mathbf{r}),\hat{\theta}(\mathbf{r}')] = i\delta(\mathbf{r}-\mathbf{r}')$, we find

$$\int_{-\infty}^{\infty}\frac{d\omega}{\pi}\chi''_{\theta n}(\mathbf{r}-\mathbf{r}',\omega) = 2\chi''_{\theta n}(\mathbf{r}-\mathbf{r}',t=0) = \langle[\hat{n}(\mathbf{r}),\hat{\theta}(\mathbf{r}')]\rangle = i\delta(\mathbf{r}-\mathbf{r}') \tag{7.43}$$

and $\int_{-\infty}^{\infty}\frac{d\omega}{\pi}\chi''_{\theta n}(\mathbf{q},\omega) = i$. Equation (7.42) then gives $n_s \leq n$.[19,20]

Similarly, we can calculate the contribution of the phase fluctuations to the density–density correlation function. Using $\delta n(q) = -\frac{n_s}{mc^2}\omega_n\theta(q)$, we obtain

$$\chi_{nn}(q) = \langle\delta n(q)\delta n(-q)\rangle = -\left(\frac{n_s}{mc^2}\right)^2\omega_n^2\chi_{\theta\theta}(q) = -\frac{n_s}{mc^2}\frac{\omega_n^2}{\omega_n^2 + c^2\mathbf{q}^2} \tag{7.44}$$

and the spectral function

$$\chi''_{nn}(\mathbf{q},\omega) = \Im[\chi^R_{nn}(\mathbf{q},\omega)] = \frac{\pi n_s}{2mc}|\mathbf{q}|[\delta(\omega - c|\mathbf{q}|) - \delta(\omega + c|\mathbf{q}|)]. \tag{7.45}$$

An external scalar potential $i\phi_{\text{ext}}$ that couples to the density $\psi^*\psi$ is taken into account by replacing $\dot{\theta}$ by $\dot{\theta}+\phi_{\text{ext}}$ in the effective action (7.34).[21] Because of the quadratic term in ϕ_{ext}, the linear response to ϕ_{ext} is determined by $-\chi_{nn} - n_s/mc^2$ rather than $-\chi_{nn}$ (Sec. 3.3).

[19] Equation (7.40) and the existence of an upper bound for the phase stiffness are in agreement with the general discussion of Sec. 3.6.3.

[20] The constraint $0 \leq n_s \leq n$ is also obtained for superconductors in Sec. 3.4.5.

[21] This follows from $\psi^*(\partial_\tau + i\phi_{\text{ext}})\psi = in(\phi_{\text{ext}} + \dot{\theta}) + \frac{\dot{n}}{2}$ for a bosonic field $\psi = \sqrt{n}e^{i\theta}$.

Only the former is a genuine response function and satisfies all the properties thereof. Its spectral representation reads

$$\chi_{nn}(\mathbf{q}, z) + \frac{n_s}{mc^2} = -\frac{n_s}{m}\frac{\mathbf{q}^2}{z^2 - c^2\mathbf{q}^2} = \int_{-\infty}^{\infty} \frac{d\omega}{\pi} \frac{\chi''_{nn}(\mathbf{q}, \omega)}{\omega - z}. \tag{7.46}$$

The retarded density–density correlation function $\chi^R_{nn}(\mathbf{q}, \omega)$ exhibits a pole for $\omega = \pm c|\mathbf{q}|$. The Goldstone mode associated with the broken symmetry is thus a sound-wave (or phonon) mode, known as the Bogoliubov mode (Secs. 7.3 and 7.4).[22] From (7.45), we deduce the sum rules

$$\int_{-\infty}^{\infty} \frac{d\omega}{\pi} \omega \chi''_{nn}(\mathbf{q}, \omega) = \frac{n_s \mathbf{q}^2}{m}, \tag{7.47}$$

$$\chi_{nn}(\mathbf{q}) + \frac{n_s}{mc^2} = \int_{-\infty}^{\infty} \frac{d\omega}{\pi} \frac{\chi''_{nn}(\mathbf{q}, \omega)}{\omega} = \frac{n_s}{mc^2}. \tag{7.48}$$

These sum rules hold only in the small-$\mathbf{q}$ limit since they are derived from the low-energy effective action $S[\theta]$ [Eq. (7.34)]. When the superfluid density n_s is equal to the full density n of the fluid, we recover the f-sum rule and, provided that c coincides with the macroscopic sound velocity (which is shown in the following), the compressibility sum rule (see Sec. 3.3). Thus, the phenomenological approach described in this section is most appropriate at zero temperature when all the system is superfluid ($n_s(T = 0) = n$). The effect of finite temperature is discussed in Sec. 7.2.3.

7.2.2.2 *Effective Lagrangian from symmetry constraints*

In the preceding section, we pointed out a drawback of the effective action (7.34). Although it is invariant under a global U(1) transformation, the expression of the particle density $n(x)$ is clearly not correct. We expect $S[\theta]$ to have a term linear in $\dot{\theta}$ in order for $n(x)$ to remain finite in the absence of temporal fluctuations of the phase ($\dot{\theta} = 0$). Difficulties also show up when one considers the Galilean invariance at zero temperature. In a Galilean transformation, $\mathbf{r}' = \mathbf{r} - \mathbf{v}t$ and $t' = t$, the phase θ transforms as

$$\theta'(\mathbf{r}', t') = \theta(\mathbf{r}, t) + \frac{1}{2} m\mathbf{v}^2 t - m\mathbf{v} \cdot \mathbf{r} \tag{7.49}$$

according to (2.61), so that the Lagrangian density

$$\mathcal{L} = -\frac{n_s}{2m} \left\{ [\boldsymbol{\nabla}\theta(x)]^2 - \frac{\dot{\theta}(x)^2}{c^2} \right\} \tag{7.50}$$

(in this section, we use a real-time formalism with $x = (\mathbf{r}, t)$) is not Galilean invariant: $\mathcal{L}[\theta(x)] \neq \mathcal{L}[\theta'(x')]$. Since the combination

$$\partial_t \theta(\mathbf{r}, t) + \frac{1}{2m} [\boldsymbol{\nabla}_{\mathbf{r}} \theta(\mathbf{r}, t)]^2 = \partial_{t'} \theta'(\mathbf{r}', t') + \frac{1}{2m} [\boldsymbol{\nabla}_{\mathbf{r}'} \theta'(\mathbf{r}', t')]^2 \tag{7.51}$$

[22] The fact that the phase mode is also a sound (phonon) mode is due to θ and n being conjugate variables.

is invariant, it is natural to look for a Lagrangian density in the form

$$\mathcal{L} = F\left(\dot{\theta}(x) + \frac{[\boldsymbol{\nabla}\theta(x)]^2}{2m}\right), \tag{7.52}$$

where F is a function to be determined. Expanding F to quadratic order, we obtain

$$\mathcal{L} = -n_s\left\{\dot{\theta}(x) + \frac{[\boldsymbol{\nabla}\theta(x)]^2}{2m}\right\} + \frac{n_s}{2mc^2}\left\{\dot{\theta}(x) + \frac{[\boldsymbol{\nabla}\theta(x)]^2}{2m}\right\}^2, \tag{7.53}$$

where n_s and c are *a priori* unknown coefficients. Retaining only the linear and quadratic terms (which are the leading ones in the low-energy limit), we recover the Lagrangian density (7.50) (which allows us to identify n_s and c as the superfluid density and the phonon velocity, respectively) with the additional (expected) term $-n_s\dot{\theta}$ ensuring that $n(x) = n_s \equiv n$ when $\theta(x) = \text{const}$.

One can also use the fact that the real-time action (1.580) of a boson system is invariant in the (semilocal) transformation $\theta(x) \to \theta(x) + \alpha(t)$ and $\mu \to \mu + \partial_t\alpha(t)$, which implies that the chemical potential must enter the effective Lagrangian in the combination $\mu - \dot{\theta}$. Thus, the function $F(X)$ in (7.53) is of the form $F(X) = P(\mu - X)$. One can then obtain the particle density in the ground state by differentiating the Lagrangian density with respect to μ and setting $\theta(x) = \text{const}$, i.e.

$$n = \left.\frac{\partial\mathcal{L}}{\partial\mu}\right|_{\theta(x)=\text{const}} = P'(\mu) = -F'(0), \tag{7.54}$$

which shows that $P(\mu)$ is the pressure (up to a constant) and $n_s = n$. We also obtain

$$\frac{n_s}{mc^2} = F''(0) = P''(\mu) = n^2\kappa, \tag{7.55}$$

where κ is the compressibility. Comparing with (3.111), we deduce that c coincides with the macroscopic sound velocity $c_s = 1/\sqrt{nm\kappa}$.

The U(1) invariance and Noether's theorem enable us to compute the particle and current densities,

$$\begin{aligned} n(x) &= -\frac{\partial\mathcal{L}}{\partial\dot{\theta}(x)} = n_s - \frac{n_s}{mc^2}\left\{\dot{\theta}(x) + \frac{[\boldsymbol{\nabla}\theta(x)]^2}{2m}\right\}, \\ \mathbf{j}(x) &= -\frac{\partial\mathcal{L}}{\partial\boldsymbol{\nabla}\theta(x)} = n(x)\frac{\boldsymbol{\nabla}\theta(x)}{m}. \end{aligned} \tag{7.56}$$

It is straightforward to verify that the continuity equation $\partial_t n(x) + \boldsymbol{\nabla}\cdot\mathbf{j}(x) = 0$ follows from the classical equation of motion.[23] The expression of the current differs from the previous result (7.35) in that it involves the local density $n(x)$ rather than its equilibrium value n_s. It implies that we can identify

$$\mathbf{v}_s(x) = \frac{\boldsymbol{\nabla}\theta(x)}{m} \tag{7.57}$$

[23]See footnote 15 on page 466.

with the superfluid velocity, i.e. the velocity of the superfluid component of the fluid. This expression is intimately connected to the transformation of the phase in a Galilean transformation. According to (7.49), a superfluid at rest in a frame moving with velocity $\mathbf{v}$ ($\theta'(\mathbf{r}',t') = \text{const}$) has a phase $\theta(\mathbf{r},t) = \text{const} - \frac{1}{2}m\mathbf{v}^2 t + m\mathbf{v}\cdot\mathbf{r}$ in the laboratory frame and a velocity $\mathbf{v} = \boldsymbol{\nabla}\theta(\mathbf{r},t)/m$.

Using $n(x) = -\partial\mathcal{L}/\partial\dot{\theta}(x)$, we can eliminate $\dot{\theta}(x)$ and use a Hamiltonian formulation where the two independent variables are $n(x)$ and $\theta(x)$ instead of $\theta(x)$ and $\dot{\theta}(x)$. Although the Lagrangian and Hamiltonian formulations are equivalent, the latter yields additional insight into the dynamics of the superfluid. The Hamiltonian density is given by

$$\mathcal{H} = -n(x)\dot{\theta}(x) - \mathcal{L} = n(x)\frac{[\boldsymbol{\nabla}\theta(x)]^2}{2m} + \frac{mc^2}{2n_s}[n(x) - n_s]^2. \tag{7.58}$$

The first term in the rhs of (7.58) is the kinetic energy density $\frac{1}{2}mn(x)\mathbf{v}_s^2(x)$. At equilibrium, $\theta(x) = \text{const}$ and $n(x) = n_s = n$. From (7.58), we obtain the compressibility

$$\frac{1}{\kappa} = n^2 \frac{\partial^2\mathcal{H}}{\partial n(x)^2}\bigg|_{n(x)=n} = mnc^2 \tag{7.59}$$

[see Eq. (3.108)] and recover the previous result $c = c_s$. The first of Hamilton's equations of motion

$$\dot{n}(x) = \frac{\partial\mathcal{H}}{\partial\theta(x)} = -\boldsymbol{\nabla}\cdot\left[n(x)\frac{\boldsymbol{\nabla}\theta(x)}{m}\right] = -\boldsymbol{\nabla}\cdot\mathbf{j}(x) \tag{7.60}$$

is nothing but the continuity equation. The second one gives

$$\dot{\theta}(x) = -\frac{\partial\mathcal{H}}{\partial n(x)} = -\frac{1}{2}m\mathbf{v}_s^2(x) - \frac{mc^2}{n}\delta n(x), \tag{7.61}$$

where $\delta n(x) = n(x) - n$. Using (3.110), we can define a local chemical potential

$$\mu(x) = \mu + \frac{1}{n^2\kappa}\delta n(x) = \mu + \frac{mc^2}{n}\delta n(x) \tag{7.62}$$

and rewrite (7.61) as

$$m\partial_t\mathbf{v}_s(x) = -\boldsymbol{\nabla}\left[\frac{1}{2}m\mathbf{v}_s^2(x) + \mu(x)\right]. \tag{7.63}$$

Using the vector identity $\boldsymbol{\nabla}(\frac{1}{2}\mathbf{v}^2) = (\mathbf{v}\cdot\boldsymbol{\nabla})\mathbf{v} + \mathbf{v}\times(\boldsymbol{\nabla}\times\mathbf{v})$ and the irrotational character of superfluid flow, $\boldsymbol{\nabla}\times\mathbf{v}_s = \boldsymbol{\nabla}\times\frac{\boldsymbol{\nabla}\theta}{m} = 0$, we finally obtain

$$m\frac{d\mathbf{v}_s(x)}{dt} = m\partial_t\mathbf{v}_s(x) + m(\mathbf{v}_s(x)\cdot\boldsymbol{\nabla})\mathbf{v}_s(x) = -\boldsymbol{\nabla}\mu(x) \tag{7.64}$$

(d/dt is the usual total derivative), which is analogous to the Euler equation of non-viscous hydrodynamics. Equations (7.60) and (7.63) are often referred to as the two fundamental equations of a superfluid. They can also be derived from the Gross–Pitaevskii equation. Their consequences for the statics and the dynamics of a superfluid system will be considered in Sec. 7.3.

7.2.3 *Finite temperature — Two-fluid hydrodynamics*

The previous general considerations have proven to give a consistent description of a Galilean-invariant system at zero temperature when the superfluid density n_s coincides with the density n of the fluid. In the two-fluid model, one assumes that the sole consequence of a finite temperature is the existence of thermally excited quasi-particles (i.e. phonons at low temperatures) that are assumed to be non-interacting.[24] The latter are identified with the normal fluid and coexist with the superfluid component below the superfluid transition temperature. For a uniform system, the fluid is characterized by the densities ρ_n and ρ_s and the current densities $\mathbf{J}_n$ and $\mathbf{J}_s$,

$$\begin{aligned} \rho &= \rho_n + \rho_s, \\ \mathbf{J} &= \mathbf{J}_n + \mathbf{J}_s = \rho_n \mathbf{v}_n + \rho_s \mathbf{v}_s, \end{aligned} \tag{7.65}$$

where $\mathbf{v}_n$ and $\mathbf{v}_s$ are the velocities of the normal and superfluid components of the fluid, respectively. We follow the standard notation in the literature and use the mass density $\rho = mn$ and mass current density $\mathbf{J} = m\mathbf{j}$.

7.2.3.1 *Superfluid flow: Landau's criterion*

We assume the normal fluid to be at equilibrium in a pipe moving with velocity $\mathbf{v}_n$, while the condensate is moving with velocity $\mathbf{v}_s$. The velocities are measured with respect to the laboratory frame. In the reference frame moving with velocity $\mathbf{v}_s$, the condensate is at rest and the quasi-particles have momentum $\mathbf{k}$ and energy $E_\mathbf{k}$ ($E_\mathbf{k} = c|\mathbf{k}|$ for $\mathbf{k} \to 0$). By Galilean invariance, the quasi-particle excitations, as viewed in the pipe frame, have an energy

$$E'_\mathbf{k} = E_\mathbf{k} + (\mathbf{v}_s - \mathbf{v}_n) \cdot \mathbf{k}, \tag{7.66}$$

where $\mathbf{k}$ is the momentum of the quasi-particles with respect to the frame where the condensate is at rest.[25] The system is stable if the excitation energies in the pipe frame (where the normal fluid is at rest) are positive: $E'_\mathbf{k} \geq 0$ for all $\mathbf{k}$. This requires $|\mathbf{v}_s - \mathbf{v}_n|$ to be smaller than a velocity v_c (Landau's critical velocity) defined by

$$v_c = \min_\mathbf{k} \frac{E_\mathbf{k}}{|\mathbf{k}|}. \tag{7.67}$$

When $|\mathbf{v}_s - \mathbf{v}_n| < v_c$, there is no mechanism by which the condensate can transfer its momentum: The flow is superfluid (i.e. dissipationless). When $|\mathbf{v}_s - \mathbf{v}_n|$ exceeds the critical velocity v_c, the walls create quasi-particle excitations with momenta $\mathbf{k}$ such that $E'_\mathbf{k} < 0$, thus transforming the kinetic energy of the condensate into heat (i.e. incoherent quasi-particle excitations). The superfluid flow is then unstable. In the Gross–Pitaevskii and

[24] In this section, the term "quasi-particles" refers to the elementary excitations whatever their relation to the bare particles. In Chapter 4 on Fermi liquids, we used it in a more restricted sense.

[25] The Hamiltonian in the pipe frame is $\hat{H} - \hat{\mathbf{P}} \cdot (\mathbf{v}_n - \mathbf{v}_s)$, where $\hat{H}$ and $\hat{\mathbf{P}}$ are the Hamiltonian and the total momentum, respectively, in the frame where the condensate is at rest [see Eq. (1.C.49)]; hence (7.66).

Bogoliubov theories (Secs. 7.3 and 7.4), v_c is given by the sound (or phonon) velocity. The instability of the superfluid flow is driven by the creation of phonon excitations. In superfluid ^{4}He, the superfluid flow is limited by roton excitations and v_c is smaller than the sound velocity (see the following).

Although the ideal Bose gas is superfluid according to the criterion $\lim_{\mathbf{q}\to 0}\chi_\perp(\mathbf{q}) = 0$ ($T = 0$), it is unstable against any motion of the fluid since its critical velocity v_c vanishes ($E_\mathbf{k} = \mathbf{k}^2/2m$). This pathological behavior is due to the infinite compressibility of the free Bose gas (hence a vanishing sound velocity). In an interacting Bose gas, the compressibility is finite and the sound velocity non-zero, so that the critical velocity v_c does not vanish.

7.2.3.2 *Calculation of ρ_n*

Since the thermally excited quasi-particles are in equilibrium in the pipe frame, their distribution follows the Bose–Einstein statistics,

$$n_\mathbf{k} = n_B(E'_\mathbf{k}) = n_B[E_\mathbf{k} + (\mathbf{v}_s - \mathbf{v}_n)\cdot\mathbf{k}]. \tag{7.68}$$

In the condensate frame, the current is entirely due to the thermally excited quasi-particles: $\mathbf{J} = \int_\mathbf{k} \mathbf{k} n_B(E'_\mathbf{k})$. In the laboratory frame, we then obtain

$$\mathbf{J} = \rho\mathbf{v}_s + \int_\mathbf{k} \mathbf{k} n_B(E'_\mathbf{k}), \tag{7.69}$$

where $\rho = \rho_n + \rho_s$ is the total density of the gas. Expanding (7.69) to linear order in the velocity and comparing the resulting expression with (7.65), we eventually arrive at

$$\rho_n = -\int_\mathbf{k} \frac{\mathbf{k}^2}{3}\frac{\partial n_B(E_\mathbf{k})}{\partial E_\mathbf{k}} = \int_\mathbf{k} \frac{\mathbf{k}^2}{3}\frac{\beta e^{\beta E_\mathbf{k}}}{(e^{\beta E_\mathbf{k}} - 1)^2} \tag{7.70}$$

in three dimensions. At low temperatures where $E_\mathbf{k} = c|\mathbf{k}|$, (7.70) gives[26]

$$\rho_n = \frac{2\pi^2}{45}\frac{T^4}{c^5}, \tag{7.71}$$

which in particular implies $\rho_s(T = 0) = \rho$. Equation (7.70) is the Landau definition of ρ_n. It is not exact, since it assumes the thermally quasi-particles to form a non-interacting excitation gas. As in Fermi-liquid theory (Chapter 4), the quasi-particle energy $E_\mathbf{k}$ in the condensate frame should be considered as a functional of the distribution function $n_\mathbf{k}$. Since the quasi-particles are in equilibrium in the pipe frame, their energy $E_\mathbf{k}$ in the condensate frame depends on $\mathbf{v}_s - \mathbf{v}_n$. This gives a correction to the current $\mathbf{J}$ [Eq. (7.69)] and in turn to the normal density ρ_n. It can be shown that with this correction, the Landau definition of the normal correction agrees with that based on the transverse current–current response function given by (7.10) [1].

---

[26]We use $\int_0^\infty dx \frac{x^3}{e^x - 1} = \frac{\pi^4}{15}$.

7.2.3.3 *Landau's quasi-particle model of superfluid ^4He*

The two-fluid model was introduced by Landau in order to explain the physical properties of superfluid ^{4}He [18, 19]. Quite remarkably, this model relies only on the quasi-particle excitation spectrum without invoking Bose–Einstein condensation. Landau postulated that the spectrum has the form shown in Fig. 7.4. At small momenta, the energy $E_{\mathbf{k}} = c|\mathbf{k}|$ is linear as expected for a sound mode. At high momenta, the particles should be essentially free and $E_{\mathbf{k}} \sim \mathbf{k}^2/2m$. For $|\mathbf{k}|$ of the order of the inverse interparticle distance, the spectrum shows a minimum,

$$E_{\mathbf{k}} \simeq \Delta + \frac{(|\mathbf{k}| - k_0)^2}{2m_{\text{rot}}}. \tag{7.72}$$

This form of the spectrum was suggested by Landau on the basis of experimental data. The excitations with $|\mathbf{k}| \sim k_0$ are known as rotons. They should be seen as phonon-like excitations accompanied by a backflow. They provide the major contribution to thermodynamics besides that of phonons, and determine the critical velocity $v_c \approx \Delta/k_0$ defined by (7.67).

7.2.3.4 *First and second sound*

At zero temperature, the low-energy excitations are phonons with velocity c. This sound mode is very similar to the zero sound mode of a Fermi liquid (Chapter 4); the restoring force acting on a given particle comes from the average field of the other particles. At finite temperature, it can exist only if its frequency ω is larger than the scattering rate $1/\tau$ of the excitations ($\omega\tau \gg 1$: collisionless regime). The sound velocity is temperature dependent, but the sound mode is otherwise very similar to its zero-temperature counterpart.

Here, we want to investigate the opposite, hydrodynamic, limit $\omega\tau \ll 1$ where collisions are sufficiently frequent to bring about local thermodynamic equilibrium. The driving force behind the sound-wave propagation then comes from the collisions between particles. We shall see that the presence of both normal and superfluid components gives rise to a

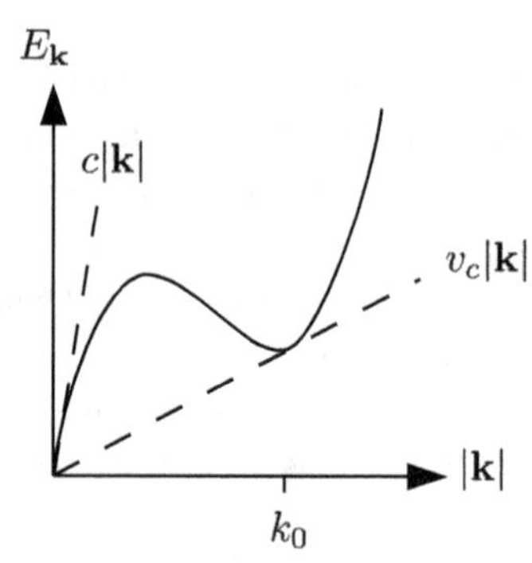

Fig. 7.4 Quasi-particle excitation spectrum in superfluid ^{4}He. c is the sound velocity and v_c Landau's critical velocity.

first-sound mode (similar to its Fermi-liquid counterpart) and a second-sound mode (entropy wave).

In order to derive the hydrodynamic modes of a superfluid, we need to extend the two-fluid model to out-of-equilibrium situations. The two basic equations are

$$\begin{aligned} \rho(x) &= \rho_n(x) + \rho_s(x), \\ \mathbf{J}(x) &= \rho_n(x)\mathbf{v}_n(x) + \rho_s(x)\mathbf{v}_s(x), \end{aligned} \tag{7.73}$$

where all quantities are now space and time dependent.

As in a normal fluid, the density $\rho(x)$ satisfies the continuity equation

$$\partial_t \rho(x) + \boldsymbol{\nabla} \cdot \mathbf{J}(x) = 0, \tag{7.74}$$

and the current $\mathbf{J}(x)$ the dynamic equation

$$\partial_t \mathbf{J}(x) = -\boldsymbol{\nabla} P(x), \tag{7.75}$$

where $P(x)$ is the pressure and $-\boldsymbol{\nabla} P(x)$ the internal force per unit volume. A second conservation law comes from the conservation of the total entropy $S = \int d^3r\, \rho(x)\tilde{s}(x)$ (we neglect any dissipation), where $\tilde{s}(x)$ is the entropy per unit mass. Since the entropy is carried solely by the thermally excited quasi-particles of the normal fluid (the condensate corresponds to a single quantum state), the continuity equation takes the form

$$\partial_t[\rho(x)\tilde{s}(x)] + \boldsymbol{\nabla} \cdot [\rho(x)\tilde{s}(x)\mathbf{v}_n(x)] = 0. \tag{7.76}$$

The last equation that we shall need governs the condensate motion,

$$m\partial_t \mathbf{v}_s(x) = -\boldsymbol{\nabla}\left[\mu(x) + \frac{1}{2}m\mathbf{v}_s^2(x)\right], \tag{7.77}$$

where $\mu(x)$ is the local chemical potential (7.62).

Eliminating $\mathbf{J}$ between (7.74) and (7.75), we obtain a relation between density and pressure,

$$\partial_t^2 \rho(x) = \boldsymbol{\nabla}^2 P(x). \tag{7.78}$$

From (7.74), (7.76) on the one hand and (7.75), (7.77) on the other hand, we deduce

$$\begin{aligned} \partial_t \tilde{s}(x) &= -\frac{\rho_s}{\rho}\tilde{s}\boldsymbol{\nabla} \cdot [\mathbf{v}_n(x) - \mathbf{v}_s(x)], \\ \rho_n \partial_t[\mathbf{v}_s(x) - \mathbf{v}_n(x)] &= -\frac{\rho}{m}\boldsymbol{\nabla}\mu(x) + \boldsymbol{\nabla} P(x), \end{aligned} \tag{7.79}$$

where $\tilde{s}$, ρ_n, ρ_s and ρ denote equilibrium values. Equations (7.79) are valid to linear order in the velocities. The last equation simplifies into

$$\rho_n \partial_t[\mathbf{v}_s(x) - \mathbf{v}_n(x)] = \rho\tilde{s}\boldsymbol{\nabla} T(x) \tag{7.80}$$

by using the Gibbs–Duhem relation $Nd\mu = \mathcal{V}dP - SdT$ and its generalization to local equilibrium situations,

$$\frac{\rho}{m}\boldsymbol{\nabla}\mu(x) = \boldsymbol{\nabla} P(x) - \rho\tilde{s}\boldsymbol{\nabla} T(x). \tag{7.81}$$

By combining (7.79) and (7.80), we obtain

$$\partial_t^2 \tilde{s}(x) = \frac{\rho_s}{\rho_n} \tilde{s}^2 \nabla^2 T(x). \tag{7.82}$$

Considering small deviations of pressure and temperature from equilibrium,

$$\begin{aligned} \delta P(x) &= \left.\frac{\partial P}{\partial \rho}\right|_S \delta\rho(x) + \left.\frac{\partial P}{\partial \tilde{s}}\right|_\rho \delta\tilde{s}(x), \\ \delta T(x) &= \left.\frac{\partial T}{\partial \rho}\right|_S \delta\rho(x) + \left.\frac{\partial T}{\partial \tilde{s}}\right|_\rho \delta\tilde{s}(x), \end{aligned} \tag{7.83}$$

we rewrite (7.78) and (7.82) as two coupled equations for the mass and entropy densities,

$$\begin{aligned} \partial_t^2 \rho(x) - \left.\frac{\partial P}{\partial \rho}\right|_S \nabla^2 \rho(x) - \left.\frac{\partial P}{\partial \tilde{s}}\right|_\rho \nabla^2 \tilde{s}(x) &= 0, \\ \partial_t^2 \tilde{s}(x) - \frac{\rho_s}{\rho_n}\tilde{s}^2 \left[\left.\frac{\partial T}{\partial \rho}\right|_S \nabla^2 \rho(x) + \left.\frac{\partial T}{\partial \tilde{s}}\right|_\rho \nabla^2 \tilde{s}(x) \right] &= 0. \end{aligned} \tag{7.84}$$

The propagating modes satisfy

$$\begin{aligned} 0 &= \omega^4 - \omega^2 \mathbf{q}^2 (c_1^2 + c_2^2) + \mathbf{q}^4 c_2^2 \left(\left.\frac{\partial P}{\partial \rho}\right|_{\tilde{s}} - \left.\frac{\partial \tilde{s}}{\partial T}\right|_\rho \left.\frac{\partial T}{\partial \rho}\right|_{\tilde{s}} \left.\frac{\partial P}{\partial \tilde{s}}\right|_\rho \right) \\ &= \omega^4 - \omega^2 \mathbf{q}^2 (c_1^2 + c_2^2) + \mathbf{q}^4 c_2^2 \left.\frac{\partial P}{\partial \rho}\right|_T \\ &\simeq (\omega^2 - c_1^2 \mathbf{q}^2)(\omega^2 - c_2^2 \mathbf{q}^2), \end{aligned} \tag{7.85}$$

where we have introduced

$$\begin{aligned} c_1^2 &= \left.\frac{\partial P}{\partial \rho}\right|_{\tilde{s}}, \\ c_2^2 &= \frac{\rho_s}{\rho_n} \tilde{s}^2 \left.\frac{\partial T}{\partial \tilde{s}}\right|_\rho = \frac{\rho_s}{\rho_n} \frac{\tilde{s}^2 T}{\tilde{C}_v}, \end{aligned} \tag{7.86}$$

with $\tilde{C}_v = C_v/\mathcal{V}\rho$ the specific heat per unit mass at constant volume. The last line in (7.85) is obtained from the approximation $\partial P/\partial\rho|_T \simeq \partial P/\partial\rho|_S$ (i.e. $\tilde{C}_v \simeq \tilde{C}_p$) valid at low temperatures. In this approximation, one finds two uncoupled modes: a density wave (first sound) with velocity c_1, and an entropy wave (second sound) with velocity c_2. In a first-sound wave, the normal and superfluid move in phase ($\mathbf{v}_s(x) = \mathbf{v}_n(x)$), while the total current vanishes in a second sound wave ($\mathbf{J}(x) = 0$).[27] Since $\rho(x) = \rho$ implies $\delta\tilde{s}(x) \propto \delta T(x)$, a second-sound wave can also be seen as a temperature wave.

[27]These conclusions follow from (7.75) and (7.79).

7.2.4 *Superconductors — Anderson–Higgs mechanism*

With little modification, the analysis of Secs. 7.2.1 and 7.2.2 can be extended to the case of superconductors. In the following, we first consider neutral superconductors before discussing the effect of the coupling to the electromagnetic field in charged systems.

7.2.4.1 *Neutral superconductors*

The considerations on ODLRO and spontaneous symmetry breaking of Sec. 7.2.1 do not directly apply to superconductors as $\langle\hat{\psi}_\sigma(\mathbf{r})\rangle$ necessarily vanishes in a fermion system. Nevertheless, ODLRO can show up in the two-body density matrix

$$\rho_2(\mathbf{r}_1,\mathbf{r}_2;\mathbf{r}'_2,\mathbf{r}'_1) = \langle\hat{\psi}^\dagger_\uparrow(\mathbf{r}_1)\hat{\psi}^\dagger_\downarrow(\mathbf{r}_2)\hat{\psi}_\downarrow(\mathbf{r}'_2)\hat{\psi}_\uparrow(\mathbf{r}'_1)\rangle. \tag{7.87}$$

For simplicity, we consider the case of an s-wave superconductor where Cooper pairs form in a spin-singlet state.[28] ODLRO in superconductors in characterized by

$$\lim_{\substack{|\mathbf{r}_1-\mathbf{r}'_1|\to\infty \\ (\mathbf{r}_1\simeq\mathbf{r}_2,\mathbf{r}'_1\simeq\mathbf{r}'_2)}} \rho_2(\mathbf{r}_1,\mathbf{r}_2;\mathbf{r}'_2,\mathbf{r}'_1) \neq 0. \tag{7.88}$$

We now make the approximation

$$\langle\hat{\psi}^\dagger_\uparrow(\mathbf{r}_1)\hat{\psi}^\dagger_\downarrow(\mathbf{r}_2)\hat{\psi}_\downarrow(\mathbf{r}'_2)\hat{\psi}_\uparrow(\mathbf{r}'_1)\rangle \simeq \langle\hat{\psi}^\dagger_\uparrow(\mathbf{r}_1)\hat{\psi}^\dagger_\downarrow(\mathbf{r}_2)\rangle\langle\hat{\psi}_\downarrow(\mathbf{r}'_2)\hat{\psi}_\uparrow(\mathbf{r}'_1)\rangle \tag{7.89}$$

for $|\mathbf{r}_1-\mathbf{r}'_1|\to\infty$, $\mathbf{r}_1\simeq\mathbf{r}_2$ and $\mathbf{r}'_1\simeq\mathbf{r}'_2$, thus explicitly breaking the U(1) invariance. This suggests that the order parameter is related to $\langle\hat{\psi}_\downarrow(\mathbf{r})\hat{\psi}_\uparrow(\mathbf{r}')\rangle$ with $\mathbf{r}\simeq\mathbf{r}'$. For an s-wave superconductor, a possible choice is

$$\Delta(\mathbf{r}) = \langle\hat{\psi}_\downarrow(\mathbf{r})\hat{\psi}_\uparrow(\mathbf{r})\rangle \tag{7.90}$$

(see Sec. 7.5). In the U(1) transformation $\hat{\psi}_\sigma(\mathbf{r})\to\hat{\psi}_\sigma(\mathbf{r})e^{i\Lambda}$, the order parameter transforms as $\Delta(\mathbf{r})\to\Delta(\mathbf{r})e^{2i\Lambda}$. In the superconducting state, the Goldstone theorem ensures that the phase θ of the fluctuating order parameter $\Delta(x)=|\Delta(x)|e^{i\theta(x)}$ has a gapless excitation spectrum. By analogy with the case of the Bose superfluid, we assume that phase fluctuations are described by the Euclidean effective action

$$S[\theta] = \frac{n_s^*}{2m^*}\int_0^\beta d\tau\int d^3r\left\{[\boldsymbol{\nabla}\theta(x)]^2 + \frac{\dot{\theta}(x)^2}{c^2}\right\}, \tag{7.91}$$

where $m^* = 2m$ is twice the fermion mass and $n_s^* = n_s/2$ half the superfluid fermion density.[29] Invariance of the action under $\theta(x)\to\theta(x)+2\Lambda$ leads us to define the pair

[28] The concept of Cooper pairs, central in the theory of superconductivity, is discussed in more detail in Sec. 7.5.

[29] The reason why m^* and n_s^*, rather than m and n_s, are the natural quantities to consider comes from the fact that superfluidity in a fermion system is due to Cooper pairing. One can also note that θ is the conjugate variable of $n^*(x)=n(x)/2$, i.e. $[\hat{n}^*(\mathbf{r}),\hat{\theta}(\mathbf{r}')] = \frac{1}{2}[\hat{n}(\mathbf{r}),\hat{\theta}(\mathbf{r}')] = i\delta(\mathbf{r}-\mathbf{r}')$ in the operator formalism. In the functional integral formalism, one thus obtain a term $\int_0^\beta d\tau\int d^3r\, i\frac{n}{2}\dot{\theta}$ in the Euclidean action (see, e.g., Ref. [79]).

density $n^*(x)$ and the associated current density $\mathbf{j}^*(x)$,

$$\begin{aligned} \delta n^*(x) &= -i\frac{\delta S}{\delta\dot\theta(x)} = -i\frac{n_s^*}{m^*c^2}\dot\theta(x), \\ \mathbf{j}^*(x) &= \frac{\delta S}{\delta\boldsymbol{\nabla}\theta(x)} = \frac{n_s^*}{m^*}\boldsymbol{\nabla}\theta(x), \end{aligned} \tag{7.92}$$

where $\delta n^*(x)$ denotes the fluctuations of the density about its equilibrium value (see the discussion in Sec. 7.2.2). The fermion density and current are thus defined by $\delta n(x) = 2\delta n^*(x)$ and $\mathbf{j}(x) = 2\mathbf{j}^*(x)$. As in the case of a Bose superfluid, the gapless phase mode $\omega = c|\mathbf{q}|$ derived from (7.91) is a density mode (known as the Anderson–Bogoliubov mode), as can be directly observed on the response function

$$\chi_{nn}(q) + \frac{n_s}{mc^2} = \frac{n_s}{m}\frac{\mathbf{q}^2}{\omega_n^2 + c^2\mathbf{q}^2} \tag{7.93}$$

(see the remark after Eq. (7.45) explaining why we consider $\chi_{nn}(q) + n_s/mc^2$).

Phase fluctuations give a contribution

$$\chi_{\mu\nu}^{\rm phase}(q) = \left(\frac{n_s}{2m}\right)^2 q_\mu q_\nu \chi_{\theta\theta}(q) = \frac{n_s}{m}\frac{c^2 q_\mu q_\nu}{\omega_n^2 + c^2\mathbf{q}^2} \tag{7.94}$$

to the current–current correlation function. This is exactly the result obtained previously for a Bose superfluid [Eq. (7.36)]. By the f-sum rule, this immediately implies

$$\chi_{\mu\nu}(\mathbf{q}) = \frac{n_n}{m}\delta_{\mu,\nu} + \frac{n_s}{m}\frac{q_\mu q_\nu}{\mathbf{q}^2} \qquad (\mathbf{q}\to 0) \tag{7.95}$$

(assuming that only phase fluctuations are long-range) and in turn the superfluidity of the system. The Bogoliubov inequality can be used to show that the superfluid density n_s is bounded by the full fermion density n.

7.2.4.2 *Charged superconductors*

A charged superconductor couples to the electromagnetic field. The action $S[\psi^*, \psi, \phi, \mathbf{A}]$ must be invariant under the electromagnetic gauge transformation (in real time)

$$\begin{aligned} \mathbf{A}(x) &\to \mathbf{A}(x) + \boldsymbol{\nabla}\Lambda(x), \\ \phi(x) &\to \phi(x) - \partial_t\Lambda(x), \\ \psi_\sigma(x) \to \psi_\sigma(x)e^{ie\Lambda(x)}, &\quad \psi_\sigma^*(x) \to \psi_\sigma^*(x)e^{-ie\Lambda(x)} \end{aligned} \tag{7.96}$$

($\Lambda(x)$ is an arbitrary function), where $\phi(x)$ and $\mathbf{A}(x)$ are the scalar and vector electromagnetic potentials and e the fermion charge. The phase of the (fluctuating) superconducting order parameter $\Delta(x)$ transforms as

$$\theta(x) \to \theta(x) + 2e\Lambda(x). \tag{7.97}$$

Electromagnetic gauge invariance therefore implies that the effective action $S[\theta, \phi, \mathbf{A}]$ must be a function of $\dot{\theta}+2e\phi$ and $\boldsymbol{\nabla}\theta - 2e\mathbf{A}$. This leads us to consider the imaginary-time action[30]

$$S[\theta, \phi, \mathbf{A}] = \frac{n_s^*}{2m^*}\int_0^\beta d\tau \int d^3r \left\{[\boldsymbol{\nabla}\theta(x) - e^*\mathbf{A}(x)]^2 + \frac{[\dot{\theta}(x) + e^*\phi(x)]^2}{c^2}\right\} + S_{\text{em}}[\phi, \mathbf{A}], \tag{7.98}$$

where S_{em} is the electromagnetic field action and $e^* = 2e$. In the Coulomb gauge (Chapter 1), we obtain

$$S[\theta, \phi, \mathbf{A}_\perp] = \frac{n_s^*}{2m^*}\int_0^\beta d\tau \int d^3r \left\{[\boldsymbol{\nabla}\theta(x)]^2 + \frac{[\dot{\theta}(x) + e^*\phi(x)]^2}{c^2} + e^{*2}\mathbf{A}_\perp^2(x)\right\} + \frac{\epsilon_0}{2}\int_0^\beta d\tau \int d^3r \left\{[\boldsymbol{\nabla}\phi(x)]^2 + [\partial_\tau \mathbf{A}_\perp(x)]^2 + c_l^2[\boldsymbol{\nabla}\times\mathbf{A}_\perp(x)]^2\right\}, \tag{7.99}$$

where c_l is the velocity of light.

Phase fluctuations couple only to the scalar potential. The action $S[\theta, \phi]$ can be put in the form

$$S[\theta, \phi] = \frac{1}{2}\sum_q (\theta(-q), \phi(-q))\chi^{-1}(q)\begin{pmatrix}\theta(q)\\ \phi(q)\end{pmatrix}, \tag{7.100}$$

where

$$\chi^{-1}(q) = \frac{n_s^*}{m^*c^2}\begin{pmatrix}\omega_n^2 + c^2\mathbf{q}^2 & ie^*\omega_n\\ -ie^*\omega_n & e^{*2}\left(1 + \frac{c^2\mathbf{q}^2}{\omega_p^2}\right)\end{pmatrix} \tag{7.101}$$

and

$$\omega_p = \sqrt{\frac{n_s^* e^{*2}}{\epsilon_0 m^*}} = \sqrt{\frac{n_s e^2}{\epsilon_0 m}}. \tag{7.102}$$

The frequency ω_p is the plasma frequency for a system with density n_s. At zero temperature, when $n_s = n$, it coincides with the plasma frequency introduced in Sec. 3.4. Inverting χ^{-1}, we obtain

$$\chi(q) = \begin{pmatrix}\chi_{\theta\theta}(q) & \chi_{\theta\phi}(q)\\ \chi_{\phi\theta}(q) & \chi_{\phi\phi}(q)\end{pmatrix} = \frac{m^*}{n_s^*\mathbf{q}^2}\frac{1}{\omega_n^2 + \omega_{\mathbf{q}}^2}\begin{pmatrix}\omega_{\mathbf{q}}^2 & -\frac{\omega_p^2}{e^*}i\omega_n\\ \frac{\omega_p^2}{e^*}i\omega_n & \frac{\omega_p^2}{e^{*2}}\left(\omega_n^2 + c^2\mathbf{q}^2\right)\end{pmatrix} \tag{7.103}$$

where

$$\omega_{\mathbf{q}} = \sqrt{c^2\mathbf{q}^2 + \omega_p^2}. \tag{7.104}$$

[30]Recall that $\phi(\mathbf{r}, t) \to i\phi(\mathbf{r}, \tau)$ in the imaginary-time formalism (Chapter 1).

One can also integrate out the scalar potential or, equivalently, eliminate ϕ using the equation $\delta S/\delta\phi(x) = 0$, to obtain the phase-only action

$$S[\theta] = \frac{n_s^*}{2m^*} \sum_q \theta(-q)\theta(q) \frac{\mathbf{q}^2(\omega_n^2 + \omega_p^2)}{\omega_{\mathbf{q}}^2}. \tag{7.105}$$

In the presence of the electromagnetic field, Noether's theorem implies that the particle and current densities are given by[31]

$$\begin{aligned} \delta n(x) &= -2i\frac{\delta S_m}{\delta\dot\theta(x)} = -i\frac{n_s}{2mc^2}[\dot\theta(x) + e^*\phi(x)], \\ \mathbf{j}(x) &= 2\frac{\delta S_m}{\boldsymbol{\nabla}\theta(x)} = \frac{n_s}{2m}[\boldsymbol{\nabla}\theta(x) - e^*\mathbf{A}_\perp(x)] \end{aligned} \tag{7.106}$$

($S_m = S - S_{\rm em}$ denotes the action of the matter), so that

$$\begin{aligned} \chi_{nn}(q) &= -\left(\frac{n_s}{2mc^2}\right)^2 [\omega_n^2\chi_{\theta\theta}(q) - 2ie^*\omega_n\chi_{\theta\phi} + e^{*2}\chi_{\phi\phi}(q)] \\ &= -\frac{n_s}{mc^2}\frac{\omega_n^2 + \omega_p^2}{\omega_n^2 + \omega_{\mathbf{q}}^2}. \end{aligned} \tag{7.107}$$

The spectral functions

$$\begin{aligned} \chi''_{\theta\theta}(\mathbf{q},\omega) &= \Im[\chi^R_{\theta\theta}(\mathbf{q},\omega)] = \frac{2\pi m\omega_{\mathbf{q}}}{n_s\mathbf{q}^2}[\delta(\omega - \omega_{\mathbf{q}}) - \delta(\omega + \omega_{\mathbf{q}})], \\ \chi''_{nn}(\mathbf{q},\omega) &= \Im[\chi^R_{nn}(\mathbf{q},\omega)] = \frac{\pi n_s\mathbf{q}^2}{2m\omega_{\mathbf{q}}}[\delta(\omega - \omega_{\mathbf{q}}) - \delta(\omega + \omega_{\mathbf{q}})] \end{aligned} \tag{7.108}$$

show that the Anderson–Bogoliubov mode of the neutral superconductor has been replaced by a plasmon mode with a gapped spectrum: $\lim_{\mathbf{q}\to 0}\omega_{\mathbf{q}} = \omega_p$.[32] The breakdown of the Goldstone theorem is due to the long-range nature of the Coulomb interaction (see the general discussion in Sec. 3.6.3). Nevertheless, the static ($\omega = 0$) phase–phase correlation function is still long-range since $\chi_{\theta\theta}(\mathbf{q}) = m^*/n_s^*\mathbf{q}^2$ for $\mathbf{q} \to 0$. This ensures that the current–current correlations are of infinite range [Eq. (7.95)] and the system superfluid. In Sec. 3.4.5, we showed that long-range current–current correlations imply the Meissner effect in a superfluid electron system. Thus, the Meissner effect is as fundamental a property as the existence of superflows in a superconductor: Both properties are a direct consequence of the long-range current–current correlations.

As explained in Sec. 7.2.2, the actual response function to an external scalar potential is not χ_{nn} but $\chi_{nn} + n_s/mc^2$. The latter satisfies the spectral representation (7.46). We can

[31] Note that the particle and current densities must be determined from the action (7.98) before choosing a particular gauge.

[32] At finite temperature, we obtain a reduced plasma frequency $\omega_p < ne^2/\epsilon_0 m$, in contradiction with the exact result (Sec. 3.4.1). Again, we find that our phenomenological approach is fully consistent only at zero temperature.

verify that the density–density correlation function satisfies the f-sum rule when $n_s = n$ [see (7.47)] and

$$\left[\chi_{nn}(\mathbf{q}) + \frac{n_s}{mc^2}\right] = \int_{-\infty}^{\infty} \frac{d\omega}{\pi} \frac{\chi''_{nn}(\mathbf{q},\omega)}{\omega} = \frac{\epsilon_0 \mathbf{q}^2}{e^2} \qquad (\mathbf{q} \to 0), \tag{7.109}$$

in agreement with the general result of Sec. 3.4.1 [Eq. (3.137)] provided that we consider $\chi_{nn}(\mathbf{q}) + n_s/mc^2$ instead of $\chi_{nn}(\mathbf{q})$.

Let us now discuss the fate of the electromagnetic fluctuations in the superconductor. Integrating out the phase variable, we obtain the electromagnetic field action

$$\begin{aligned} S[\phi, \mathbf{A}_\perp] = \frac{\epsilon_0}{2} \sum_q \Bigg[&\mathbf{q}^2 \left(1 + \frac{\omega_p^2}{\omega_n^2 + c^2\mathbf{q}^2}\right) \phi(-q)\phi(q) \\ &+ (\omega_n^2 + c_l^2\mathbf{q}^2 + \omega_p^2) \mathbf{A}_\perp(-q) \cdot \mathbf{A}_\perp(q) \Bigg]. \end{aligned} \tag{7.110}$$

From (7.110), we deduce the propagator of the electromagnetic field and in turn the longitudinal and transverse dielectric functions (see Secs. 3.4.1 and 3.4.2),[33]

$$\epsilon_\parallel^R(\mathbf{q},\omega) = 1 - \frac{\omega_p^2}{(\omega + i0^+)^2 - c^2\mathbf{q}^2}, \qquad \epsilon_\perp^R(\mathbf{q},\omega) = 1 - \frac{\omega_p^2}{(\omega + i0^+)^2}. \tag{7.111}$$

We deduce $\epsilon_\parallel^R(\mathbf{q},0) = 1 + \omega_p^2/c^2\mathbf{q}^2$ $(\mathbf{q} \to 0)$, from which follows the compressibility sum rule (3.141) when $n_s = n$. As expected, $\epsilon_\parallel^R(\mathbf{q},\omega)$ vanishes at the plasmon excitation energy $\omega = \omega_\mathbf{q} = (c^2\mathbf{q}^2 + \omega_p^2)^{1/2}$. The propagating modes of the electromagnetic field are determined by $\epsilon_\perp^R(\mathbf{q},\omega)(\omega + i0^+)^2 = c_l^2\mathbf{q}^2$, i.e. $\omega^2 = c_l^2\mathbf{q}^2 + \omega_p^2$ (Fig. 7.5). This result can also be directly read off from the action (7.110). Thus, there is no mode propagating below the plasma frequency. This mechanism, whereby the photon acquires a gap as a consequence of its coupling to the matter, is known as the Anderson–Higgs mechanism. The latter is the analog of the Higgs mechanism in high-energy physics, e.g. in the electroweak theory, where the gauge field (equivalent to the *photon* of the superconductor) and the Goldstone boson (*pairing phase field*) combine to make massive W and Z bosons. The Higgs boson corresponds to the pairing amplitude mode of the superconductor.[34]

The dielectric functions are related to the conductivity by

$$\epsilon_{\parallel,\perp}^R(\mathbf{q},\omega) = 1 + \frac{i}{\epsilon_0(\omega + i0^+)} \sigma_{\parallel,\perp}(\mathbf{q},\omega). \tag{7.112}$$

[33]The longitudinal dielectric constant is related to the density–density correlation function by $\frac{1}{\epsilon_\parallel(q)} = 1 - \frac{e^2}{\epsilon_0 \mathbf{q}^2}\left[\chi_{nn}(q) + \frac{n_s}{mc^2}\right]$.

[34]For this reason, the amplitude mode of the superconductor is also called the Higgs mode (Sec. 7.6.4).

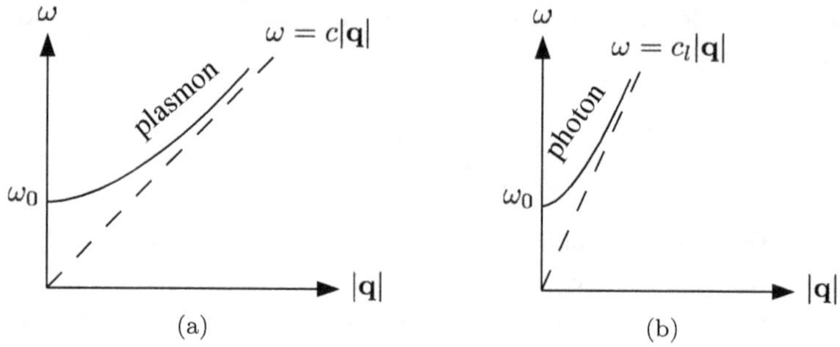

Fig. 7.5 Longitudinal plasmon mode (a) and transverse photon mode (b) in a superconductor. The dashed lines show the spectrum before the coupling between fermions and electromagnetic field (Anderson–Bogoliubov mode $\omega = c|\mathbf{q}|$ (a) and (gapless) photon modes $\omega = c_l|\mathbf{q}|$ (b)).

The conductivity in (7.112) is defined as the response to the local electromagnetic field (Sec. 3.4). We thus obtain[35]

$$\begin{aligned}\sigma_{\parallel}(\mathbf{q},\omega) &= \frac{n_s e^2}{m}\frac{i(\omega+i0^+)}{(\omega+i0^+)^2 - c^2\mathbf{q}^2},\\ \sigma_{\perp}(\mathbf{q},\omega) &= \frac{n_s e^2}{m}\frac{i}{\omega+i0^+}.\end{aligned} \tag{7.113}$$

The conductivity satisfies $\sigma_{\parallel}(\mathbf{q},0)=0$ for $\mathbf{q}\neq 0$ — as required by gauge invariance — and $\sigma_{\parallel}(0,\omega)=\sigma_{\perp}(0,\omega)\equiv\sigma(0,\omega)$. As discussed in Sec. 3.4.4, the Drude (D) and superfluid (D_s) weights are defined as the coefficients of the delta peak in $\Re[\sigma(0,\omega)]$ and $\lim_{\mathbf{q}\to 0}\Re[\sigma_{\perp}(\mathbf{q},\omega)]$,

$$D = D_s = \pi\frac{n_s e^2}{m}. \tag{7.114}$$

The system is therefore both a perfect conductor ($D>0$) and a superconductor ($D_s>0$). As shown in Sec. 3.4.5, the Meissner effect is a direct consequence of the expression (7.113) of the transverse conductivity (see also Sec. 7.6.3).

To conclude this section, we should point out an obvious weakness of the phenomenological approach. The effective action $S[\theta,\phi,\mathbf{A}_{\perp}]$ is expected to be valid in the low-energy long-wavelength limit; there is no guarantee that it satisfactorily describes energies of the order of the plasma frequency. We have also stressed that the phenomenological approach is fully consistent only at zero temperature. At finite temperature, it incorrectly predicts a reduced plasma frequency $\omega_p < (ne^2/\epsilon_0 m)^{1/2}$. Moreover, the fact that the superfluid density n_s enters the longitudinal conductivity $\sigma_{\parallel}(\mathbf{q},\omega)$ and the uniform conductivity $\sigma_{\parallel}(0,\omega) = \sigma_{\perp}(0,\omega)$ is somewhat surprising. In particular, one expects a Drude weight $D=\pi ne^2/m$ in a translation-invariant system even at finite temperature where $n_s < n$.

[35] As in Sec. 3.4.4, the conductivity can also be obtained from the linear response to the electromagnetic potentials.

Response functions in superconductors will be considered again in Sec. 7.6 within a microscopic and more controlled approach.

7.3 Gross–Pitaevskii Theory

Let us start with the real-time action

$$S[\psi^*,\psi] = \int_{-\infty}^{\infty} dt \biggl\{ \int d^3r\, \psi^*(\mathbf{r},t) \left[i\partial_t + \mu + V_{\text{ext}}(\mathbf{r}) + \frac{\boldsymbol{\nabla}^2}{2m} \right] \psi(\mathbf{r},t) \\ - \frac{1}{2} \int d^3r d^3r'\, V(\mathbf{r}-\mathbf{r}')\psi^*(\mathbf{r},t)\psi^*(\mathbf{r}',t)\psi(\mathbf{r}',t)\psi(\mathbf{r},t) \biggr\} \tag{7.115}$$

of a Bose gas in an external potential $V_{\text{ext}}(\mathbf{r})$. The simplest approximation amounts to treating the bosonic field ψ as a classical field. One can also approximate the interaction potential $V(\mathbf{r}-\mathbf{r}')$ between particles by a local potential $g\delta(\mathbf{r}-\mathbf{r}')$ of strength $g = 4\pi a/m$ where a is the s-wave scattering length.[36] In Sec. 7.4, we shall show that this is valid in the dilute limit $na^3 \ll 1$. In such a classical approach, both quantum and thermal fluctuations are neglected, $\psi(x) \equiv \langle \psi(x) \rangle$; all particles are in the condensate and one should normalize the classical field to the total number of particles, i.e. $|\psi(x)|^2 = n(x)$ with $n(x)$ the particle density at point $x = (\mathbf{r},t)$. $\psi(x)$ satisfies the classical equations of motion $\delta S/\delta\psi^*(x) = \delta S/\delta\psi(x) = 0$, which leads to the Gross–Pitaevskii equation[37]

$$\left(-i\partial_t - \mu - \frac{\boldsymbol{\nabla}^2}{2m} + V_{\text{ext}}(\mathbf{r}) + g|\psi(x)|^2 \right)\psi(x) = 0. \tag{7.116}$$

With $\psi(x) = \sqrt{n(x)}e^{i\theta(x)}$, the real and imaginary parts of the Gross–Pitaevskii equation give

$$\dot{n}(x) + \boldsymbol{\nabla}\cdot\left[n(x)\frac{\boldsymbol{\nabla}\theta(x)}{m} \right] = 0, \tag{7.117}$$

$$\dot{\theta}(x) - \mu + V_{\text{ext}}(\mathbf{r}) + gn(x) + \frac{[\boldsymbol{\nabla} n(x)]^2}{8m[n(x)]^2} - \frac{\boldsymbol{\nabla}^2 n(x)}{4mn(x)} + \frac{[\boldsymbol{\nabla}\theta(x)]^2}{2m} = 0. \tag{7.118}$$

The continuity equation (7.117) relates the density $n(x)$ to the current $\mathbf{j}(x) = n(x)\mathbf{v}_s(x) = n(x)\boldsymbol{\nabla}\theta(x)/m$. Equation (7.118) can be rewritten as

$$\dot{\theta}(x) - \mu + V_{\text{ext}}(\mathbf{r}) + gn(x) + \frac{1}{2}m\mathbf{v}_s^2(x) - \frac{1}{2m}\frac{\boldsymbol{\nabla}^2\sqrt{n(x)}}{\sqrt{n(x)}} = 0. \tag{7.119}$$

Taking the gradient of (7.119), we obtain

$$m\partial_t\mathbf{v}_s(x) + \frac{1}{2}m\boldsymbol{\nabla}\mathbf{v}_s^2(x) = -\boldsymbol{\nabla}[gn(x) + V_{\text{ext}}(\mathbf{r})] + \frac{1}{2m}\boldsymbol{\nabla}\left(\frac{\boldsymbol{\nabla}^2\sqrt{n(x)}}{\sqrt{n(x)}} \right). \tag{7.120}$$

[36] See Appendix 7.A for a definition of the s-wave scattering length.

[37] The Gross–Pitaevskii equation is also referred to as a nonlinear Schrödinger equation.

Anticipating that $\mu = gn$ in the ground state [Eq. (7.122)], $g\boldsymbol{\nabla} n(x)$ can be rewritten as the gradient $\boldsymbol{\nabla}\mu(x)$ of the local chemical potential $\mu(x) = gn(x)$. This yields

$$\begin{aligned} m\frac{d\mathbf{v}_s(x)}{dt} &= m\partial_t \mathbf{v}_s(x) + \frac{1}{2}m\boldsymbol{\nabla}\mathbf{v}_s^2(x) \\ &= -\boldsymbol{\nabla}[\mu(x) + V_{\text{ext}}(\mathbf{r})] + \frac{1}{2m}\boldsymbol{\nabla}\left(\frac{\boldsymbol{\nabla}^2\sqrt{n(x)}}{\sqrt{n(x)}}\right), \end{aligned} \tag{7.121}$$

where d/dt is the usual total derivative. We recover the previous result (7.64) except for the external potential $V_{\text{ext}}(\mathbf{r})$ and the so-called quantum pressure term (last term of (7.121)). Like the term $\frac{1}{2}m\mathbf{v}_s^2(x)$, the quantum pressure term comes from the kinetic energy term of the Hamiltonian. It does not give rise to a condensate motion but corresponds to zero-point fluctuations; if we restore the Planck constant in (7.121), we easily find that the quantum pressure term is the only one to depend on $\hbar$. We shall see in the following that it can be ignored in the long-wavelength limit.

7.3.1 *Statics and dynamics of a uniform condensate*

When $V_{\text{ext}} = 0$, the condensate is uniform and we can take $\psi(x) = \sqrt{n}$ real with no loss of generality. Equation (7.119) then gives

$$\mu = gn. \tag{7.122}$$

Note that the condensate at rest is described by a static solution ($\dot\theta = 0$) because the grand canonical Hamiltonian includes the term $-\mu\hat{N}$. Without this chemical potential term in the action, one would obtain a time-dependent phase $\dot\theta(x) = -gn \equiv -\mu$, i.e.

$$\psi(x) = \langle\hat\psi(x)\rangle = \sqrt{n}e^{-i\mu t}. \tag{7.123}$$

The condensate wavefunctions in the canonical and grand canonical ensembles differ by a factor $e^{i\mu t}$. This result is obvious when one considers the matrix element

$$\begin{aligned} &\langle n, N-1|e^{i(\hat H_c - \mu\hat N)t}\hat\psi(\mathbf{r})e^{-i(\hat H_c-\mu\hat N)t}|m, N\rangle \\ &\qquad = e^{i\mu t}\langle n, N-1|e^{i\hat H_c t}\hat\psi(\mathbf{r})e^{-i\hat H_c t}|m, N\rangle \end{aligned} \tag{7.124}$$

between two eigenstates of the system with $N-1$ and N particles. ($\hat H_c$ denotes the Hamiltonian in the canonical ensemble.)

Let us now consider small fluctuations of the condensate about equilibrium. To linear order in $\delta n(x) = n(x) - n$, $\dot\theta(x)$ and $\boldsymbol{\nabla}\theta(x)$, Eqs. (7.117) and (7.118) give

$$\begin{aligned} \delta\dot n(x) + \frac{n}{m}\boldsymbol{\nabla}^2\theta(x) &= 0, \\ \dot\theta(x) + g\delta n(x) - \frac{\boldsymbol{\nabla}^2\delta n(x)}{4mn} &= 0. \end{aligned} \tag{7.125}$$

The linearized equations (7.125) neglect the kinetic energy term $\frac{1}{2}m\mathbf{v}_s^2$ but include part of the quantum pressure term. Eliminating θ, we obtain

$$\delta\ddot{n}(x) - \frac{gn}{m}\boldsymbol{\nabla}^2\delta n(x) + \frac{\boldsymbol{\nabla}^4\delta n(x)}{4m^2} = 0 \tag{7.126}$$

and the spectrum of the density oscillations,

$$\omega_{\mathbf{q}} = \sqrt{\frac{gn\mathbf{q}^2}{m} + \frac{\mathbf{q}^4}{4m^2}} \simeq \begin{cases} c|\mathbf{q}| & \text{for} \quad |\mathbf{q}| \ll \xi^{-1} \qquad \text{(phonon)} \\ \dfrac{\mathbf{q}^2}{2m} + ng & \text{for} \quad |\mathbf{q}| \gg \xi^{-1} \quad \text{(particle excitation)} \end{cases} \tag{7.127}$$

where we have introduced the healing length

$$\xi = \frac{1}{\sqrt{2ngm}} = \frac{1}{\sqrt{8\pi na}}. \tag{7.128}$$

Thus in the long-wavelength limit $|\mathbf{q}| \ll \xi^{-1}$, we recover the Bogoliubov sound mode (phonon mode) — introduced in the phenomenological approach of Sec. 7.2.2 — with a velocity

$$c = \sqrt{\frac{gn}{m}} = \frac{\sqrt{4\pi an}}{m}, \tag{7.129}$$

while short-wavelength excitations ($|\mathbf{q}| \gg \xi^{-1}$) are particle-like and not sensitive to the presence of the condensate. When $|\mathbf{q}| \ll \xi^{-1}$, the low-energy effective action $S[\theta]$ introduced in Sec. 7.2.2.2 gives a reliable description of the long-wavelength properties of the system. In this regime, both the kinetic energy term $\frac{1}{2}m\mathbf{v}_s^2(x)$ and the quantum pressure term can be omitted. The equations of motion reduce to

$$\begin{aligned} \delta\dot{n}(x) + \frac{n}{m}\boldsymbol{\nabla}^2\theta(x) &= 0, \\ \dot{\theta}(x) + g\delta n(x) &= 0 \end{aligned} \tag{7.130}$$

and give a mode with a linear dispersion $\omega = c|\mathbf{q}|$.

The healing length is the distance over which the field ψ tends to its bulk value when subjected to a localized perturbation.[38] Consider, for instance, a system at equilibrium confined in the region $z \geq 0$ and assume that $\psi(\mathbf{r})$ vanishes for $z = 0$: $V_{\text{ext}}(z < 0) = \infty$ and $V_{\text{ext}}(z \geq 0) = 0$. $\psi(\mathbf{r}) = \psi(z)$ satisfies the time-independent Gross–Pitaevskii equation

$$-\frac{\partial_z^2\psi(z)}{2m} + g|\psi(z)|^2\psi(z) = \mu\psi(z) = gn\psi(z), \tag{7.131}$$

i.e.

$$-\frac{\partial_z^2\psi(z)}{2m} + g\left[|\psi(z)|^2 - n\right]\psi(z) = 0, \tag{7.132}$$

[38]The healing length is analogous to the Ginzburg–Landau coherence length (or the BCS coherence $\xi_0 = v_F/\pi\Delta$ at $T = 0$) in superconductors (Sec. 7.5).

with $\psi(z=0)=0$ and $\psi(z\to\infty)=\sqrt{n}$. Equation (7.132) admits the solution

$$\psi(z)=\sqrt{n}\tanh\left(\frac{z}{\sqrt{2}\xi}\right), \tag{7.133}$$

where ξ is the healing length (7.128).

7.3.2 *Vortex*

The Gross–Pitaevskii equation also admits vortex solutions where the density goes to zero at the center of a core region. Vortices are stable in a frame rotating at a sufficiently high angular velocity. (In this section, we consider only time-independent solutions.)

Since $\mathbf{v}_s(\mathbf{r})=\boldsymbol{\nabla}\theta(\mathbf{r})/m$, the flow is irrotational, $\boldsymbol{\nabla}\times\mathbf{v}_s(\mathbf{r})=0$. The integral $\oint_{\mathcal{C}} d\mathbf{l}\cdot\mathbf{v}_s$ must therefore vanish for any closed contour $\mathcal{C}$ in a simply connected system unless the phase of the order parameter $\psi(\mathbf{r})=\sqrt{n(\mathbf{r})}e^{i\theta(\mathbf{r})}$ has one or more singularities. The circulation around a closed contour — possibly enclosing singularities — is given by

$$\Gamma=\oint_{\mathcal{C}} d\mathbf{l}\cdot\mathbf{v}_s=\frac{1}{m}\oint_{\mathcal{C}} d\mathbf{l}\cdot\boldsymbol{\nabla}\theta=\frac{hl}{m} \qquad (l \text{ integer}) \tag{7.134}$$

(we have restored Planck's constant $h=2\pi\hbar$), since $\psi(\mathbf{r})$ has to be single valued. Equation (7.134) expresses the quantization of the circulation in a superfluid.

A vortex solution corresponds to a line-like (point-like in two dimensions) singularity of the phase. For a solution with cylindrical symmetry,

$$\psi(\mathbf{r})=\psi(\rho,\varphi,z)=f(\rho)e^{i\theta(\rho,\varphi)}, \tag{7.135}$$

and we assume the vortex to stand at $\rho=0$ (ρ,φ,z are cylindrical coordinates). The circulation of the phase along a circular contour at a distance ρ for the vortex center is given by

$$\oint_{\mathcal{C}} d\mathbf{l}\cdot\boldsymbol{\nabla}\theta=\rho\int_0^{2\pi} d\varphi\,\boldsymbol{\nabla}\theta\cdot\hat{\mathbf{u}}_\varphi=\int_0^{2\pi} d\varphi\,\partial_\varphi\theta=2\pi l, \tag{7.136}$$

where $\hat{\mathbf{u}}_\varphi$ is a unit vector tangential to the contour. Clearly, the only solution compatible with the cylindrical symmetry is

$$\theta(\varphi)=l\varphi. \tag{7.137}$$

This implies that the velocity distribution is tangential,

$$\mathbf{v}_s(\rho,\varphi)=\frac{\hbar l}{m\rho}\hat{\mathbf{u}}_\varphi. \tag{7.138}$$

The rotational $\boldsymbol{\nabla}\times\mathbf{v}_s$ vanishes for $\rho>0$. Since

$$\int_\Sigma d\boldsymbol{\Sigma}\cdot\boldsymbol{\nabla}\times\mathbf{v}_s=\oint_{\mathcal{C}} d\mathbf{l}\cdot\mathbf{v}_s=\frac{hl}{m} \tag{7.139}$$

for any contour $\mathcal{C}$ enclosing the singularity at $\rho = 0$ (Σ is the surface spanned by the contour), we must have

$$\boldsymbol{\nabla} \times \mathbf{v}_s = \frac{hl}{m}\delta(\boldsymbol{\rho})\hat{\mathbf{z}}, \tag{7.140}$$

where $\boldsymbol{\rho} = \rho\hat{\mathbf{u}}_\rho$ is the projection of $\mathbf{r}$ onto the (x, y) plane.

The amplitude $f(\rho)$ is determined by the Gross–Pitaevskii equation

$$\left(-\frac{\boldsymbol{\nabla}^2}{2m} + gf^2 - \mu\right) f e^{il\varphi} = 0 \tag{7.141}$$

with the boundary conditions $\lim_{\rho\to\infty} f(\rho) = \sqrt{n}$ and $\lim_{\rho\to 0} f(\rho) = 0$. The latter condition is necessary to maintain the kinetic energy finite while $\lim_{\rho\to 0} |\mathbf{v}_s| = \infty$. Since the value of the chemical potential $\mu = gn$ in an infinite system is unchanged by the (localized) vortex solution, (7.141) can be rewritten as

$$\left(\partial_\rho^2 + \frac{\partial_\rho}{\rho}\right) f - \left[\frac{l^2}{\rho^2} - \frac{1}{\xi^2}\left(1 - \frac{f^2}{n}\right)\right] f = 0, \tag{7.142}$$

where ξ is the healing length. Writing $f(\rho) = \sqrt{n}\tilde{f}(x)$ ($x = \rho/\xi$), we finally obtain

$$\partial_x^2 \tilde{f} + \frac{\partial_x \tilde{f}}{x} + \tilde{f}\left(1 - \frac{l^2}{x^2}\right) - \tilde{f}^3 = 0. \tag{7.143}$$

One easily finds the asymptotic behavior

$$\tilde{f}(x) = x^{l^2} \quad (x \to 0), \qquad \tilde{f}(x) = 1 - \frac{l^2}{2x^2} \quad (x \to \infty). \tag{7.144}$$

The dimensionless form of (7.143) shows that the core size is of the order of the healing length ξ (Fig. 7.6).

The main contribution to the energy of the vortex is the kinetic energy. For a vortex of unit length and radius R, one finds

$$E_{\mathrm{v}} \simeq \frac{nm}{2}\int_\xi^R d\rho\, 2\pi\rho \mathbf{v}_s^2(\rho) = \frac{n\pi l^2}{m}\ln\left(\frac{R}{\xi}\right). \tag{7.145}$$

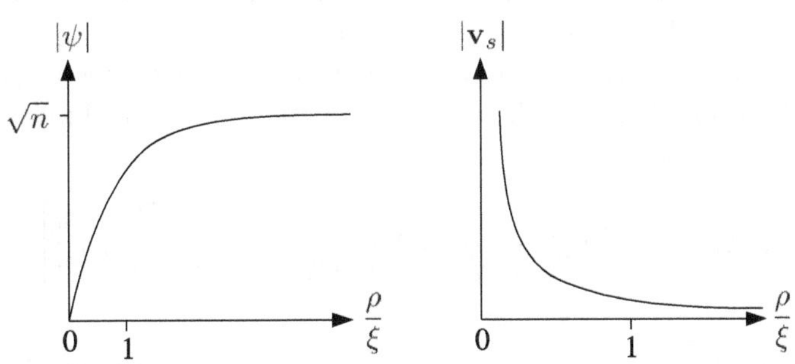

Fig. 7.6 Vortex line with cylindrical symmetry: Amplitudes of the order parameter $\psi(\mathbf{r}) = |\psi(\rho)|e^{i\theta(\varphi)}$ and the velocity $\mathbf{v}_s(\mathbf{r}) = (\hbar l/m\rho)\hat{\mathbf{u}}_\varphi$.

The healing length ξ, which determines the size of the vortex core, provides a natural low-distance cutoff in the integral (7.145). In principle, one should add to (7.145) the change in potential energy due to the vortex. The latter is important only in the vicinity of the vortex core ($|\rho| \lesssim \xi$) and gives a small correction to (7.145).[39]

Let us now suppose that the superfluid is at rest in a cylindrical container rotating about the z axis with angular velocity ω. In the rotating reference frame, the Hamiltonian reads

$$\hat{H}' = \hat{H} - \omega \hat{L}^z, \tag{7.146}$$

where $\hat{L}^z$ is the z component of the angular momentum $\hat{L} = \int d^3r\, \hat{\psi}^\dagger(\mathbf{r})[\mathbf{r} \times (-i\boldsymbol{\nabla})]\hat{\psi}(\mathbf{r})$ as measured in the laboratory frame [Eq. (1.C.57)]. In the coherent state (7.33), the angular momentum takes the mean value

$$\begin{aligned} \langle \psi | \hat{L}^z | \psi \rangle &= \int d^3r\, \psi^*(\mathbf{r})[\mathbf{r} \times (-i\boldsymbol{\nabla})] \cdot \hat{\mathbf{z}} \psi(\mathbf{r}) \\ &= \int dz \int_0^{2\pi} d\varphi \int \rho d\rho\, \psi^*(\rho, \varphi, z)(-i\partial_\varphi)\psi(\rho, \varphi, z), \end{aligned} \tag{7.147}$$

i.e. $l\pi R^2 n$ for a container of unit length and radius R.[40] The vortex energy in the rotating reference frame,

$$E'_{\rm v} = E_{\rm v} - \omega l \pi R^2 n, \tag{7.148}$$

becomes negative for $\omega > \omega_c$, where

$$\omega_c \simeq \frac{l}{mR^2} \ln\left(\frac{R}{\xi}\right). \tag{7.149}$$

When ω exceeds the critical value ω_c, vortices start forming in the superfluid.

7.3.3 *Soliton*

The Gross–Pitaevskii equation has solutions in the nonlinear regime corresponding to solitary waves (also called solitons), i.e. localized disturbances whose shape is constant in time. By extending the time-independent solution (7.133) to the whole space, we obtain a static soliton. Such a soliton is also called a kink since the phase of the order parameter jumps discontinuously by π at the origin $z = 0$ (Fig. 7.7).

The order parameter $\psi(z, t)$ of a soliton moving with velocity v is a function of $z - vt$, since the shape does not change with time. (We consider only one-dimensional solutions.) It is straightforward to verify that the function

$$\psi(z,t) = \sqrt{n}\left[\sqrt{1 - \frac{v^2}{c^2}} \tanh\left(\frac{z - vt}{\sqrt{2}\xi}\sqrt{1 - \frac{v^2}{c^2}}\right) + i\frac{v}{c}\right] \tag{7.150}$$

[39] The energy of the vortex line, obtained from a numerical integration of the Gross–Pitaevskii equation, is $E_{\rm v} \simeq \frac{\pi l^2 n}{m} \ln\left(\frac{1.46R}{\xi}\right)$ [4, 5].

[40] This result can also been obtained classically: $L_z = \int_0^R d\rho\, 2\pi\rho f^2(\rho) m\rho |\mathbf{v}_s(\rho)| = l\pi R^2 n$.

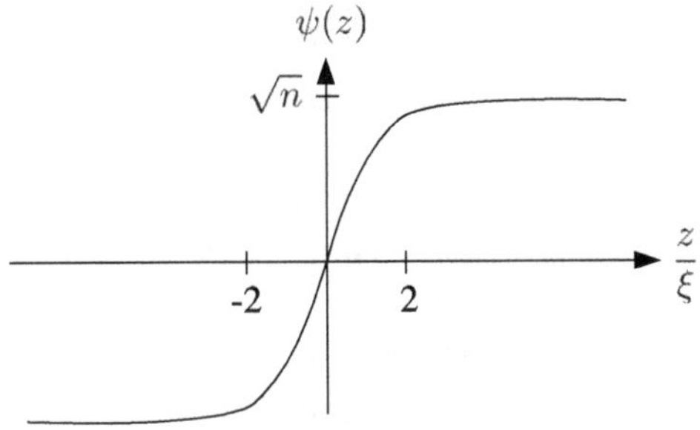

Fig. 7.7 Static soliton (kink) $\psi(z) = \sqrt{n}\tanh(z/\sqrt{2}\xi)$.

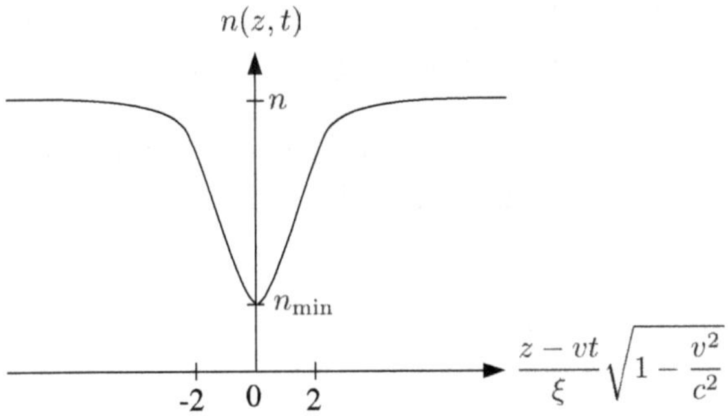

Fig. 7.8 Density profile of a soliton $\psi(z,t)$ described by (7.150) and (7.151).

($v \leq c$) is solution of the Gross–Pitaevskii equation. ξ is the healing length introduced in Sec. 7.3.1. The density profile of the soliton is

$$\begin{aligned} n(z,t) = |\psi(z,t)|^2 &= n\left[\frac{v^2}{c^2} + \left(1 - \frac{v^2}{c^2}\right)\tanh^2\left(\frac{z-vt}{\sqrt{2}\xi}\sqrt{1-\frac{v^2}{c^2}}\right)\right] \\ &= n_{\min} + (n - n_{\min})\tanh^2\left(\frac{z-vt}{\sqrt{2}\xi}\sqrt{1-\frac{v^2}{c^2}}\right), \end{aligned} \tag{7.151}$$

where $n_{\min} = n(z = vt, t) = nv^2/c^2$ is the minimum density (Fig. 7.8). The width of the soliton is given by $\xi/(1 - v^2/c^2)^{1/2}$, and its phase undergoes a change

$$\Delta\theta = \theta(z = \infty) - \theta(z = -\infty) = 2\arcsin\left(\frac{v}{c}\right) - \pi \tag{7.152}$$

as z varies from $-\infty$ to ∞. Because of their signature in optics experiments, solitons corresponding to a density depression are called dark solitons.[41]

[41]See Refs. [4, 5] for a more detailed discussion.

7.4 Bogoliubov Theory

The Bogoliubov theory goes one step beyond the Gross–Pitaevskii approach by including fluctuations of $\psi(x)$ about its mean value $\langle\psi(x)\rangle = \phi_0$ (without loss of generality, we take ϕ_0 real). The condensate density $n_0 = \phi_0^2 \leq \bar{n}$ should not be confused with the superfluid density n_s which, in a Galilean-invariant system at zero temperature, coincides with the full density $\bar{n}$ of the fluid.[42] If the particle–particle interactions are weak, one expects most of the bosons to be in the condensate ($n_0 \sim \bar{n}$) and the fluctuations $\psi'(x) = \psi(x) - \phi_0$ to be small at zero temperature. A Gaussian expansion of the action with respect to ψ', i.e. a one-loop approximation, should then be sufficient for a reliable description of the system beyond the mean-field (Gross–Pitaevskii) approach.

As in the preceding section, we consider a Bose gas with a local two-body interaction $V(\mathbf{r}-\mathbf{r}') = g\delta(\mathbf{r}-\mathbf{r}')$ and introduce an upper momentum cutoff Λ to make the model well defined. However, instead of loosely identifying g with $4\pi a/m$, we carefully determine the s-wave scattering length a as a function of g and Λ by solving the two-body problem in vacuum ($\mu = T = 0$), i.e. by computing the T matrix (see Appendix 7.A). In field theory, the latter is defined by the four-point vertex $\Gamma^{(4)}$ which, in vacuum, reduces to the sum of ladder diagrams shown in Fig. 7.33,

$$T(i\omega_n) = \frac{g}{1+g\Pi(i\omega_n)}, \tag{7.153}$$

where

$$\Pi(i\omega_n) = \int_{\mathbf{k}}\int \frac{d\omega}{2\pi} G_0(\mathbf{k}, i\omega) G_0(-\mathbf{k}, -i\omega + i\omega_n) = \int_{\mathbf{k}} \frac{1}{2\epsilon_{\mathbf{k}} - i\omega_n} \tag{7.154}$$

and $G_0(\mathbf{k}, i\omega) = (i\omega - \epsilon_{\mathbf{k}})^{-1}$ is the non-interacting propagator. The s-wave scattering length a can be obtained from the low-energy limit of the T matrix. For a three-dimensional system

$$T(i\omega_n = 0) = \frac{4\pi a}{m} \quad \text{i.e.} \quad \frac{m}{4\pi a} = \frac{1}{g} + \int_{\mathbf{k}} \frac{1}{2\epsilon_{\mathbf{k}}} = \frac{1}{g} + \frac{m\Lambda}{2\pi^2}, \tag{7.155}$$

where the momentum integral is cut off by $|\mathbf{k}| \leq \Lambda$. The rationale beyond the choice of the local potential $V(\mathbf{r}) = g\delta(\mathbf{r})$ is that at low energies collisions between particles occur only in the s-wave channel and the scattering amplitude is fully determined by $a \equiv a(g, \Lambda)$ (which, in a more realistic model, would be a complicated function of the two-body potential $V(\mathbf{r})$). In practice, one can therefore compute physical quantities in terms of g and Λ and eventually eliminate these two parameters in favor of the s-wave scattering length.[43]

In the following, we compute the effective action $\Gamma[\phi^*, \phi]$ to one-loop order following the general discussion of Sec. 1.7.2. To organize the loop expansion, we introduce a parameter l

[42] In Secs. 7.4.1 and 7.4.2, we denote the mean density of the fluid by $\bar{n}$ whereas n is used for the condensate density (with $n = n_0$ at equilibrium).

[43] The reason why final results can be expressed solely in terms of the s-wave scattering length a (and the boson mass m) in the case of a central potential $V(\mathbf{r}-\mathbf{r}') = V(|\mathbf{r}-\mathbf{r}'|)$ is further discussed in Sec. 7.4.4 (see also Appendix 7.A).

(which will eventually be set to unity) and consider the partition function

$$Z[J^*,J] = \int \mathcal{D}[\psi^*,\psi]\, e^{-l\left(S[\psi^*,\psi]-\int_0^\beta d\tau \int d^3r (J^*\psi + \text{c.c.})\right)} \tag{7.156}$$

in the presence of a complex external source J. The superfluid order parameter is defined by

$$\phi(x) = \frac{1}{l}\frac{\delta \ln Z[J^*,J]}{\delta J^*(x)} = \langle \psi(x)\rangle, \qquad \phi^*(x) = \frac{1}{l}\frac{\delta \ln Z[J^*,J]}{\delta J(x)} = \langle \psi^*(x)\rangle. \tag{7.157}$$

We then introduce the effective action

$$\Gamma[\phi^*,\phi] = -\frac{1}{l}\ln Z[\phi^*,\phi] + \int_0^\beta d\tau \int d^3r (J\phi^* + \text{c.c.}) \tag{7.158}$$

defined as the Legendre transform of the thermodynamic potential $-l^{-1}\ln Z[J^*,J]$. To one-loop order,

$$\Gamma[\phi^*,\phi] = S[\phi^*,\phi] - \frac{1}{2l}\text{Tr}\ln\big(\mathcal{G}[\phi^*,\phi]\big) + \mathcal{O}(l^{-2}), \tag{7.159}$$

where $\mathcal{G}$ is the classical propagator,

$$\mathcal{G}^{-1}[x,x';\phi^*,\phi] = -\begin{pmatrix} \dfrac{\delta^2 S[\phi^*,\phi]}{\delta\phi^*(x)\delta\phi(x')} & \dfrac{\delta^2 S[\phi^*,\phi]}{\delta\phi^*(x)\delta\phi^*(x')} \\ \dfrac{\delta^2 S[\phi^*,\phi]}{\delta\phi(x)\delta\phi(x')} & \dfrac{\delta^2 S[\phi^*,\phi]}{\delta\phi(x)\delta\phi^*(x')} \end{pmatrix}. \tag{7.160}$$

All thermodynamic quantities can be obtained from the effective potential $V(n) = \Gamma[\phi^*,\phi]/\beta\mathcal{V}$ where ϕ is a constant (i.e. uniform and time-independent) field. The U(1) symmetry of the action ensures that $V(n)$ is a function of the condensate density $n = |\phi|^2$. The minimum of $V(n)$ determines the actual condensate density $n_0 = |\langle\psi(x)\rangle|^2$ and the thermodynamic potential $\Omega/\mathcal{V} = V(n_0)$ in the absence of external sources ($J^* = J = 0$). The pressure is then given by

$$P(\mu,T) = -V(n_0). \tag{7.161}$$

7.4.1 *Mean-field theory*

To leading order in $1/l$, $\Gamma[\phi^*,\phi] = S[\phi^*,\phi]$, $V(n) = -\mu n + \frac{g}{2}n^2$, $n_0 = \mu/g$ and $V(n_0) = -\mu^2/2g$. Thus, the condensate density n_0 coincides with the fluid density $\bar{n} = \partial P/\partial\mu$: All particles are in the condensate. From the compressibility $\kappa = \bar{n}^{-2}\partial\bar{n}/\partial\mu = g/\mu^2$, we deduce the macroscopic sound velocity c_s,

$$c_s^2 = \frac{1}{m\bar{n}\kappa} = \frac{\mu}{m} \tag{7.162}$$

(see Sec. 3.3).

For a constant field, the Fourier transformed propagator is defined by

$$\mathcal{G}^{-1}(k,\phi^*,\phi)=\begin{pmatrix} i\omega_n-\xi_{\mathbf{k}}-2g|\phi|^2 & -g\phi^2 \\ -g\phi^{*2} & -i\omega_n-\xi_{\mathbf{k}}-2g|\phi|^2 \end{pmatrix}, \tag{7.163}$$

which gives

$$\det \mathcal{G}^{-1}(k,\phi^*,\phi)=\omega_n^2+E_{\mathbf{k}}(n)^2, \tag{7.164}$$

where

$$E_{\mathbf{k}}(n)=[(\xi_{\mathbf{k}}+2gn)^2-(gn)^2]^{1/2}. \tag{7.165}$$

With $n=n_0=|\phi_0|^2\equiv\mu/g$, we obtain the one-particle excitation energy

$$E_{\mathbf{k}}\equiv E_{\mathbf{k}}(\mu/g)=[\epsilon_{\mathbf{k}}(\epsilon_{\mathbf{k}}+2\mu)]^{1/2}\simeq\begin{cases} c|\mathbf{k}| & \text{if } |\mathbf{k}|\ll k_c, \\ \epsilon_{\mathbf{k}}+\mu & \text{if } |\mathbf{k}|\gg k_c, \end{cases} \tag{7.166}$$

where $c=\sqrt{\mu/m}$ and $k_c^{-1}\sim(2m\mu)^{-1/2}$ is the healing length introduced in Sec. 7.3.1. The one-particle spectrum (7.166) is thus similar to that of the density fluctuations obtained from the Gross–Pitaevskii equation. The velocity c of the phonon mode coincides with the macroscopic sound velocity c_s; this result is actually exact (i.e. valid beyond the mean-field theory) as will be shown in Chapter 13 (see also Sec. 7.2.2).

To compute the superfluid density, we consider a slowly varying field $\phi(x)=\phi_0 e^{i\theta(x)}$ and the corresponding variation of the effective action,

$$\begin{aligned}\Delta\Gamma[\phi^*,\phi]&=\frac{\phi_0^2}{2}\int_0^\beta d\tau d\tau'\int d^3r d^3r'[\mathcal{G}_{12}^{-1}(x,x')+\mathcal{G}_{21}^{-1}(x,x') \\ &\quad -\mathcal{G}_{11}^{-1}(x,x')-\mathcal{G}_{22}^{-1}(x,x')]\theta(x)\theta(x') \\ &=\beta\mathcal{V}\frac{\phi_0^2}{2m}\int_0^\beta d\tau\int d^3r(\boldsymbol{\nabla}\theta)^2, \end{aligned} \tag{7.167}$$

where $\mathcal{G}(x,x')$ is the classical propagator (7.163) evaluated for a constant (real) field $\phi(x)=\phi_0$. Comparing (7.167) with (7.41), we deduce that the superfluid density $n_s=\phi_0^2$ is equal to the full density $\bar{n}\equiv n_0$ of the fluid. Contrary to the result $n_0=\bar{n}$, which is an artifact of the mean-field theory (see the following), the equality between n_s and $\bar{n}$ is exact in a Galilean-invariant system as shown in Sec. 7.1 (see also Chapter 13).

7.4.2 *Lee–Huang–Yang correction*

To determine the effective potential beyond the mean-field level (we consider the zero-temperature limit), one must compute the trace in (7.159),

$$\operatorname{Tr}\ln(\mathcal{G}^{-1})=\sum_k\ln[\mathcal{G}_{11}^{-1}(k)\mathcal{G}_{22}^{-1}(k)-\mathcal{G}_{12}^{-1}(k)\mathcal{G}_{21}^{-1}(k)]. \tag{7.168}$$

Because the 2×2 propagator $\mathcal{G}$ mixes up creation and annihilation fields, we should carefully introduce the convergence factors before computing the Matsubara sum.[44] The correct expression reads[45]

$$\begin{aligned}\operatorname{Tr}\ln(\mathcal{G}^{-1}) &= \sum_k \left[e^{i\omega_n 0^+}\ln[-\mathcal{G}_{11}^{-1}(k)] + e^{-i\omega_n 0^+}\ln[-\mathcal{G}_{22}^{-1}(k)] + \ln\left(1 - \frac{\mathcal{G}_{12}^{-1}(k)\mathcal{G}_{21}^{-1}(k)}{\mathcal{G}_{11}^{-1}(k)\mathcal{G}_{22}^{-1}(k)}\right)\right] \\ &= \sum_k e^{i\omega_n 0^+}\ln\left(\frac{\mathcal{G}_{11}^{-1}(k)}{\mathcal{G}_{22}^{-1}(k)}\det[\mathcal{G}^{-1}(k)]\right), \end{aligned} \tag{7.169}$$

where we have used $\mathcal{G}_{22}^{-1}(k) = \mathcal{G}_{11}^{-1}(-k)$. Using[46]

$$\begin{aligned}\frac{1}{\beta}\sum_{\omega_n} e^{i\omega_n 0^+}\ln[-\mathcal{G}_{11}^{-1}(k)] &= \frac{1}{\beta}\ln\big(1 - e^{-\beta(\xi_{\mathbf{k}}+2gn)}\big), \\ \frac{1}{\beta}\sum_{\omega_n} e^{i\omega_n 0^+}\ln[-\mathcal{G}_{22}^{-1}(k)] &= \frac{1}{\beta}\ln\big(e^{\beta(\xi_{\mathbf{k}}+2gn)} - 1\big), \\ \frac{1}{\beta}\sum_{\omega_n} e^{i\omega_n 0^+}\ln\det[\mathcal{G}^{-1}(k)] &= \frac{2}{\beta}\ln\left[2\sinh\left(\frac{\beta}{2}E_{\mathbf{k}}(n)\right)\right], \end{aligned} \tag{7.170}$$

we obtain the zero-temperature effective potential to one-loop order,

$$V(n) = -\mu n + \frac{g}{2}n^2 + \frac{1}{2l}\int_{\mathbf{k}}(E_{\mathbf{k}}(n) - \xi_{\mathbf{k}} - 2gn) + \mathcal{O}(l^{-2}). \tag{7.171}$$

The condensate density in the equilibrium state is obtained from $V'(n_0) = 0$, i.e.

$$-\mu + gn_0 + \frac{g}{2l}\int_{\mathbf{k}}\left(\frac{2\xi_{\mathbf{k}} + 3gn_0}{E_{\mathbf{k}}(n_0)} - 2\right) = 0, \tag{7.172}$$

so that

$$\begin{aligned} n_0 &= \frac{\mu}{g} - \frac{1}{2l}\int_{\mathbf{k}}\left(\frac{2\epsilon_{\mathbf{k}} + \mu}{E_{\mathbf{k}}} - 2\right) + \mathcal{O}(l^{-2}), \\ V(n_0) &= -\frac{\mu^2}{2g} + \frac{1}{2l}\int_{\mathbf{k}}(E_{\mathbf{k}} - \epsilon_{\mathbf{k}} - \mu) + \mathcal{O}(l^{-2}), \end{aligned} \tag{7.173}$$

where we have set $n_0 = \mu/g$ in the $\mathcal{O}(1/l)$ term ($E_{\mathbf{k}}$ is defined by (7.166)). To eliminate the dependence on the momentum cutoff Λ, we use (7.155) and rewrite the condensate

[44] See the discussion on page 76.

[45] Although the last term in the rhs of (7.169) is convergent, it is convenient to add the factor $e^{i\omega_n 0^+}$.

[46] The first of Eqs. (7.170) is derived in Appendix 1.F (see example 2 and note that $-\mu + 2gn > 0$ for $n \simeq n_0 \simeq \mu/g$). The other two are obtained following the general method given in this appendix.

density as

$$\begin{aligned} n_0 &= \frac{m\mu}{4\pi a} - \int_{\mathbf{k}} \frac{\mu}{2\epsilon_{\mathbf{k}}} - \frac{1}{2}\int_{\mathbf{k}} \left(\frac{2\epsilon_{\mathbf{k}} + \mu}{E_{\mathbf{k}}} - 2 \right) \\ &= \frac{m\mu}{4\pi a} - \frac{\mu}{2}\int_{\mathbf{k}} \left(\frac{1}{\epsilon_{\mathbf{k}}} - \frac{1}{E_{\mathbf{k}}} \right) - \int_{\mathbf{k}} \left(\frac{\epsilon_{\mathbf{k}} + \mu}{E_{\mathbf{k}}} - 1 \right), \end{aligned} \tag{7.174}$$

where we have set $l = 1$. The integral over $\mathbf{k}$ is now convergent and we can take the infinite cutoff limit $\Lambda \to \infty$ with a fixed,

$$\begin{aligned} n_0 &= \frac{m\mu}{4\pi a} - \frac{(m\mu)^{3/2}}{\pi^2} - \frac{2(m\mu)^{3/2}}{3\pi^2} \\ &= \frac{m\mu}{4\pi a}\left(1 - \frac{20}{3\pi}\sqrt{ma^2\mu} \right). \end{aligned} \tag{7.175}$$

The first term in the rhs of the final expression is the mean-field result whereas the second one is known as the Lee–Huang–Yang correction. Similarly, for the pressure, one finds

$$\begin{aligned} P &= \frac{m\mu^2}{8\pi a} - \frac{\mu^2}{4}\int_{\mathbf{k}} \left(\frac{1}{\epsilon_{\mathbf{k}}} - \frac{1}{E_{\mathbf{k}}} \right) - \frac{1}{2}\int_{\mathbf{k}} \left(E_{\mathbf{k}} - \epsilon_{\mathbf{k}} - \mu + \frac{\mu^2}{2E_{\mathbf{k}}} \right) \\ &= \frac{m\mu^2}{8\pi a}\left(1 - \frac{64}{15\pi}\sqrt{ma^2\mu} \right). \end{aligned} \tag{7.176}$$

From (7.176), we obtain the fluid density $\bar{n} = \partial P/\partial\mu$ and the compressibility $\kappa = \bar{n}^{-2}\partial\bar{n}/\partial\mu$,

$$\bar{n} = \frac{m\mu}{4\pi a}\left(1 - \frac{16}{3\pi}\sqrt{ma^2\mu} \right), \tag{7.177}$$

$$\bar{n}^2\kappa = \frac{m}{4\pi a}\left(1 - \frac{8}{\pi}\sqrt{ma^2\mu} \right). \tag{7.178}$$

We see that interactions tend to deplete the condensate: $n_0 < \bar{n}$. The macroscopic sound velocity is now given by

$$c_s^2 = \frac{4\pi a\bar{n}}{m^2}\left(1 + \frac{8}{\pi}\sqrt{ma^2\mu} \right). \tag{7.179}$$

Note that by using the relation (7.177) between μ and $\bar{n}$, one can express the Lee–Huang–Yang correction in terms of $\bar{n}$.

The small expansion parameter of the one-loop calculation is $ma^2\mu \sim \bar{n}a^3$. The Bogoliubov theory is therefore valid in the low-density limit $\bar{n}a^3 \ll 1$ where the condensate depletion

$$\frac{\bar{n} - n_0}{\bar{n}} = \frac{4}{3\pi}\sqrt{ma^2\mu} \simeq \frac{8}{3\sqrt{\pi}}\sqrt{\bar{n}a^3} \tag{7.180}$$

is small.

In principle, one could also compute the two-point vertex $\Gamma^{(2)}$ to one-loop order (see Sec. 1.7.2) to determine the superfluid density n_s and the phonon velocity c. This would allow us to verify that the relations $n_s = \bar{n}$ and $c = c_s$ hold beyond the mean-field theory. This calculation however is made difficult by the appearance of infrared divergences, even though these divergences cancel in physical quantities.[47]

It is straightforward to extend the preceding analysis to finite but low temperatures where, in the dilute limit, both quantum and thermal depletions of the condensate are weak so that the Bogoliubov theory applies. The effective potential reads

$$V(n) = -\mu n + \frac{g}{2}n^2 + \frac{1}{2l}\int_{\mathbf{k}} \left\{ -\xi_{\mathbf{k}} - 2gn + \frac{2}{\beta}\ln\left[2\sinh\left(\frac{\beta}{2}E_{\mathbf{k}}(n)\right)\right]\right\} + \mathcal{O}(l^{-2}), \quad (7.181)$$

which gives the condensate density

$$\begin{aligned} n_0(\mu, T) &= \frac{\mu}{g} + \frac{g}{l}\int_{\mathbf{k}} \left[1 - \frac{2\epsilon_{\mathbf{k}} + \mu}{2E_{\mathbf{k}}} \operatorname{cotanh}\left(\frac{\beta}{2}E_{\mathbf{k}}\right)\right] + \mathcal{O}(l^{-2}) \\ &= n_0(\mu, 0) - \int_{\mathbf{k}} \frac{2\epsilon_{\mathbf{k}} + \mu}{E_{\mathbf{k}}(e^{\beta E_{\mathbf{k}}} - 1)} \end{aligned} \quad (7.182)$$

setting $l = 1$. At low temperatures, when $T \ll \mu$, we can set $E_{\mathbf{k}} = c|\mathbf{k}|$ in the last integral, so that

$$n_0(\mu, T) \simeq n_0(\mu, 0) - \frac{m^{3/2}T^2}{12\sqrt{\mu}}. \quad (7.183)$$

The last term in (7.183) corresponds to the thermal depletion of the condensate.[48] From (7.181) and (7.183), we easily deduce the pressure

$$\begin{aligned} P(\mu, T) &= P(\mu, 0) - \frac{1}{\beta}\int_{\mathbf{k}} \ln(1 - e^{-\beta E_{\mathbf{k}}}) \\ &\simeq P(\mu, 0) + \frac{\pi^2}{90}\left(\frac{m}{\mu}\right)^{3/2} T^4 \end{aligned} \quad (7.184)$$

and the low-temperature entropy density $S/\mathcal{V} = \partial P/\partial T \propto T^3$.

7.4.3 *Transition temperature of the dilute Bose gas*

In this section, we consider an interacting Bose gas near the superfluid transition temperature where interactions between thermally excited quasi-particles must be taken into account. The simplest approach that achieves this goal is the Hartree–Fock–Bogoliubov theory. It improves on the Bogoliubov theory by including the Hartree–Fock contribution

---

[47] We shall come back to this issue in Chapter 13.

[48] The condition $T \ll \mu$ ensures that the thermal depletion is small with respect to $n_0(\mu, 0) \simeq \bar{n}$. When $T \gg \mu$, the thermal depletion varies as $(mT)^{3/2}$ and is small provided that $\bar{n}\lambda_T^3 \gg 1$, i.e. $T \ll T_c$, where $\lambda_T = (2\pi/mT)^{1/2}$ is the thermal de Broglie wavelength and T_c the superfluid transition temperature (see Sec. 7.4.3).

to the self-energy. Above T_c, where the condensate density vanishes, one simply obtains the Hartree–Fock result

$$\Sigma_{\rm HF}(k) = 2gn \tag{7.185}$$

(with n the mean density of particles) so that the boson propagator reduces to $G(\mathbf{k}, i\omega_n) = (i\omega_n - \xi_{\mathbf{k}} - 2gn)^{-1}$. The self-energy correction produces a shift of the chemical potential, $\mu \to \mu' = \mu - 2gn$, and the transition temperature remains unchanged compared to the BEC temperature of the ideal Bose gas

$$T_c^{\rm HF} = T_c^{(0)} = \frac{2\pi}{m} \left[\frac{n}{\zeta(3/2)} \right]^{2/3} \tag{7.186}$$

[see Eq. (1.C.114)]. The Hartree–Fock–Bogoliubov theory is no more useful below T_c, since it predicts a gap in the one-particle spectrum in contradiction with the Goldstone theorem.[49]

Despite the various improvements on the Hartree–Fock–Bogoliubov theory that were developed a long time ago, a satisfactory description of a weakly interacting (i.e. $na^3 \ll 1$) Bose gas near the superfluid transition temperature has remained elusive until very recently. The main issue was to understand how the two-body interaction affects the BEC temperature. It is now agreed that it leads to an increase of T_c proportional to the scattering length,

$$\frac{\Delta T_c}{T_c^{(0)}} = \text{const} \times n^{1/3} a. \tag{7.187}$$

The constant has been estimated from both numerical and analytical methods. It will be further discussed in Chapter 13.

7.4.4 *Universal thermodynamics of the dilute Bose gas*[50]

In this section, we show that the pressure of the d-dimensional dilute Bose gas ($d \geq 2$) can be written in the scaling form

$$\begin{aligned} P(\mu, T) &= \left(\frac{m}{2\pi} \right)^{d/2} \mu^{d/2+1} \mathcal{G}_d \left(\frac{T}{\mu}, \tilde{g}(\mu) \right) \\ &= \frac{T}{\lambda_T^d} \mathcal{F}_d \left(\frac{\mu}{T}, \tilde{g}(T) \right), \end{aligned} \tag{7.188}$$

where $\mathcal{G}_d$ and $\mathcal{F}_d$ are universal scaling functions, $\lambda_T = (2\pi/mT)^{1/2}$ is the thermal de Broglie wavelength, and the energy-dependent dimensionless interaction constant $\tilde{g}(\epsilon)$ is a known function of $ma_d^2\epsilon$ where a_d is the s-wave scattering length. The scaling form (7.188) was explicitly derived in Sec. 7.4.2 for a three-dimensional dilute Bose gas in the weak-interaction limit $ma_3^2\mu \sim na_3^3 \ll 1$ at low temperatures.

---

[49]See the discussion in Sec. 9.1.4.1.

[50]This section is more advanced than the preceding ones and relies on concepts from the theory of critical phenomena (e.g. critical point, universal scaling function, scaling dimension, etc.); see Chapters 10 and 12.

Equation (7.188) implies scaling forms for other thermodynamic quantities, e.g.

$$\begin{aligned} n(\mu,T) &= \frac{\partial}{\partial\mu}P(\mu,T) = \frac{1}{\lambda_T^d}\mathcal{F}_d^{(1,0)}\left(\frac{\mu}{T},\tilde{g}(T)\right), \\ n(\mu,T)^2\kappa(\mu,T) &= \frac{\partial^2}{\partial\mu^2}P(\mu,T) = \frac{1}{T\lambda_T^d}\mathcal{F}_d^{(2,0)}\left(\frac{\mu}{T},\tilde{g}(T)\right), \end{aligned} \tag{7.189}$$

where we use the notation $\mathcal{F}_d^{(i,j)}(x,y) = \partial_x^i\partial_y^j\mathcal{F}_d(x,y)$. The thermodynamics of the dilute Bose gas is thus universal, i.e. fully determined by the scaling function $\mathcal{F}_d$ (or $\mathcal{G}_d$) and only two non-universal quantities, namely the boson mass m and the s-wave scattering length a_d, and is otherwise insensitive to the details of the two-body potential $V(\mathbf{r}-\mathbf{r}') = V(|\mathbf{r}-\mathbf{r}'|)$.

Universality is often associated with the presence of a (quantum) critical point, i.e. a second-order phase transition (see Chapters 10 and 12). The universality of the equation of state of the dilute Bose gas can also be understood from the point of view of the theory of phase transitions. By varying the chemical potential from negative to positive values at $T=0$, one induces a quantum phase transition (QPT) between a state with vanishing pressure and no particles (vacuum) and a superfluid state with a non-zero pressure. This identifies the point $\mu = T = 0$ as a quantum critical point (QCP). In the absence of interactions ($g=0$), the QCP separates the vacuum from a BEC (for $\mu>0$, the system is however not stable since without interactions, nothing prevents an infinite number of bosons to condense into the $\mathbf{k}=0$ one-particle state). A straightforward dimensional analysis of the Gaussian action (corresponding to non-interacting bosons) at $\mu=0$ shows that the field has scaling dimension

$$[\psi] = \frac{1}{2}(d+z-2) = \frac{d}{2}, \tag{7.190}$$

where $z=2$ is the dynamical critical exponent.[51] It follows that

$$[\mu] = 2, \qquad [g] = 2-d. \tag{7.191}$$

The Gaussian fixed point is therefore stable for $d\geq 2$ since g is irrelevant in that case[52] (μ being the only relevant variable when $T=0$); the upper critical dimension is $d_c^+=2$ in agreement with the criterion $d_c^+ + z = 4$ (Chapter 12). For $d\geq 2$, the critical behavior at the QCP is therefore described by non-interacting bosons.

Since the QCP ground state is the vacuum, a particle added to the ground state propagates as a free particle even in the presence of interactions. When $\mu = T = 0$, the one-particle propagator is therefore given exactly by the non-interacting propagator

$$G(k) = (i\omega - \epsilon_{\mathbf{k}})^{-1} \tag{7.192}$$

in any dimension (i.e. not only for $d\geq d_c^+$). We deduce the dynamical critical exponent $z=2$ whereas the anomalous dimension η vanishes. For the same reason, for $\mu<0$

[51] This value follows from the dispersion $\omega = \mathbf{k}^2/2m \sim |\mathbf{k}|^z$ of free bosons.

[52] We shall see that g is marginally irrelevant when $d=2$.

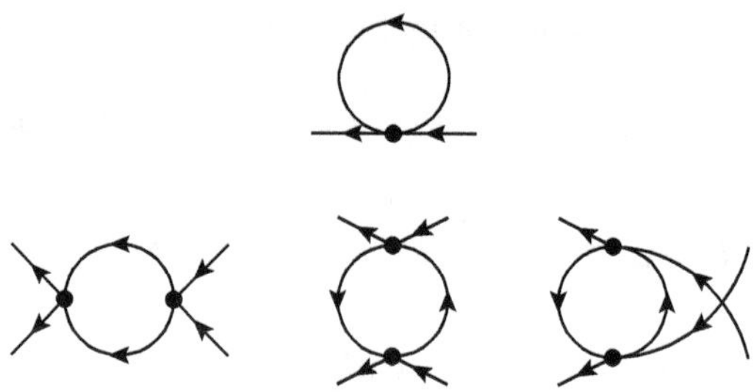

Fig. 7.9 One-loop diagrams contributing to the self-energy (a) and the two-particle vertex (b).

(and $T = 0$), one has the exact result $G(k) = (i\omega - \epsilon_{\mathbf{k}} + \mu)^{-1}$ and the critical exponent associated with the correlation length $\xi = |2m\mu|^{-1/2}$ takes the value $\nu = 1/2$. The renormalized interaction g_R at the QCP is given by the four-point vertex $\Gamma^{(4)}$, i.e. the T matrix in vacuum (using again the fact that the ground state is the vacuum) which, at low energies, is a function of m and a_d (e.g. $g_R = 4\pi a_3/m$ for $d = 3$).

The preceding discussion can be rephrased in the language of the RG. In the Wilson approach, we split the field into slow ($|\mathbf{k}| < \Lambda/s$) and fast ($|\mathbf{k}| > \Lambda/s$) modes and integrate out the latter.[53] The renormalization of μ and g to one-loop order is given by the diagrams of Fig. 7.9 (we restrict the analysis to the QCP $\mu = T = 0$). The self-energy diagram in Fig. 7.9(a) vanishes since

$$\int_{-\infty}^{\infty} \frac{d\omega}{2\pi} G(k) e^{i\omega 0^+} = \int_{-\infty}^{\infty} \frac{d\omega}{2\pi} \frac{e^{i\omega 0^+}}{i\omega - \epsilon_{\mathbf{k}}} = 0. \tag{7.193}$$

This ensures that the one-particle propagator is not renormalized, in agreement with the preceding discussion, so that $z = 2$ and $\eta = 0$. The last two diagrams in Fig. 7.9(b) vanish by virtue of

$$\int_{-\infty}^{\infty} \frac{d\omega}{2\pi} G(k)^2 = \int_{-\infty}^{\infty} \frac{d\omega}{2\pi} \frac{1}{(i\omega - \epsilon_{\mathbf{k}})^2} = 0, \tag{7.194}$$

and only the first diagram gives a contribution to the interaction constant g,

$$g' = g - g^2 \fint_k G(k) G(-k). \tag{7.195}$$

We use the notation $\fint_k$ to indicate that the integration is restricted to fast modes. With the rescaling $\mathbf{r}' = \mathbf{r}/s$, $\tau' = \tau/s^z$ and $\psi'(\mathbf{r}, \tau) = s^{(d+z-2+\eta)/2} \psi(\mathbf{r}, \tau)$, we obtain

$$g(l + dl) = e^{(2-d)dl} \left[g(l) - g(l)^2 m K_d \Lambda^{d-2} dl \right] \tag{7.196}$$

taking $s = e^{dl}$ ($K_d = S_d/(2\pi)^d$ with S_d the surface of the unit hypersphere in d dimensions), i.e.

$$\frac{d\tilde{g}}{dl} = (2 - d)\tilde{g} - \frac{1}{2}\tilde{g}^2, \tag{7.197}$$

[53] See Chapter 10 for a general discussion of the Wilson RG.

where $\tilde{g}(l) = 2mK_d\Lambda^{d-2}g(l)$. Solving Eq. (7.197), one finds

$$\tilde{g}(l) = \begin{cases} \frac{4a_3}{\pi}\Lambda(l) & \text{if } d = 3, \\ -\frac{2}{\ln\left(\frac{\Lambda(l)a_2}{2}\right) + C} & \text{if } d = 2 \end{cases} \tag{7.198}$$

(the result for $d = 3$ holds for $l \gg 1$), where $\Lambda(l) = \Lambda e^{-l}$ and C is the Euler constant. a_3 is defined by (7.155) whereas the two-dimensional scattering length $a_2 = \frac{2}{\Lambda}e^{-2\pi/mg-C}$ [42]. Above the upper critical dimension $d_c^+ = 2$, $\tilde{g}(l)$ vanishes for $l \to \infty$ and the only fixed point is the Gaussian one ($\tilde{g}^* = 0$) which therefore controls the QCP between the vacuum and the superfluid phase. For $d < d_c^+$, the critical behavior is determined by the non-trivial fixed point $\tilde{g}^* = 2(2-d)$; $\tilde{g}$ is an irrelevant variable at the fixed point and μ is the only relevant variable.

It is easy to verify that the one-loop RG equations for the self-energy and the coupling constant are exact at the QCP. Since the propagator

$$G(\mathbf{k}, \tau) = \int_{-\infty}^{\infty} \frac{d\omega}{2\pi} e^{-i\omega 0^+} G(k) = -\Theta(\tau)e^{-\tau\epsilon_{\mathbf{k}}} \tag{7.199}$$

is causal, one immediately sees that all self-energy diagrams vanish. Thus, the result $[\psi] = d/2$, $[\mu] = 2$ and $z = 2$ is exact. Although the fixed point is not Gaussian for $d < d_c^+$, the critical behavior remains trivial. For the same reason, Eq. (7.197) gives the exact value of the two-particle vertex $\Gamma^{(4)}$ in vacuum.[54]

When $d \geq d_c^+ = 2$, there are two relevant perturbations about the Gaussian fixed point $\mu = T = \tilde{g} = 0$: the chemical potential and the temperature. In an RG transformation, they transform as

$$\tilde{\mu}(l) = e^{l/\nu}\tilde{\mu}, \qquad \tilde{T}(l) = e^{zl}\tilde{T} \tag{7.200}$$

near the QCP (i.e. when $|\tilde{\mu}(l)| \lesssim 1$ and $|\tilde{T}(l)| \lesssim 1$). We have introduced the dimensionless variables[55]

$$\tilde{\mu} = \frac{2m\mu}{\Lambda^{1/\nu}}, \qquad \tilde{T} = \frac{2mT}{\Lambda^z}. \tag{7.201}$$

When μ or T are non-zero, the RG equation for $\tilde{g}(l)$ is well approximated by (7.197) as long as $|\tilde{\mu}(l)| \lesssim 1$ and $\tilde{T}(l) \lesssim 1$. This yields two crossover lines defined by $|\tilde{\mu}| \sim \tilde{T}$ (i.e. $|\mu| \sim T$ using $z = 1/\nu = 2$). However, the naive scaling of various physical quantities, valid below the upper critical dimension of the zero-temperature QPT, is modified by the dangerously

[54]If we integrate out fields with momenta in the range $\Lambda(l) \leq |\mathbf{k}| \leq \Lambda$, we obtain the effective interaction $g(l) = g/[1+g\Pi(l)]$ in vacuum (T matrix) with $\Pi(l) = \int_k \Theta(\Lambda - |\mathbf{k}|)\Theta(|\mathbf{k}| - \Lambda(l))G(k)G(-k) = mK_d(\Lambda^{d-2} - \Lambda(l)^{d-2})/(d-2)$. It is easy to verify that $g(l)$ is the solution of Eq. (7.197).

[55]The dimensionless variables $\tilde{\mu}$, $\tilde{T}$ and $\tilde{P}$ are defined according to their scaling dimensions at the fixed point: $[\tilde{\mu}] = 1/\nu$, $[\tilde{T}] = z$ and $[\tilde{P}] = d + z$. With the additional $2m$ factors (and the fact that $z = 1/\nu = 2$), these quantities have no physical units.

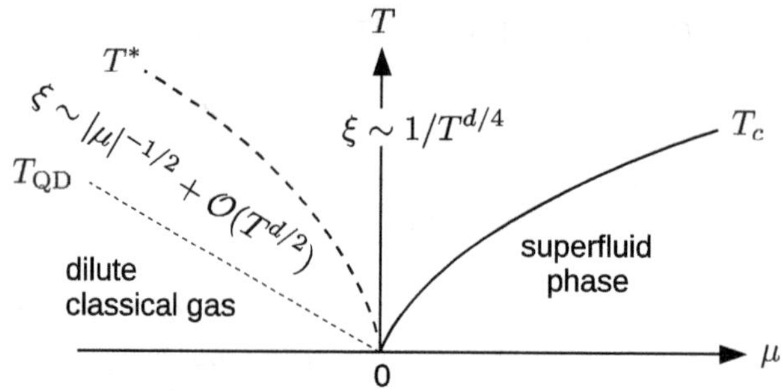

Fig. 7.10 Phase diagram of the dilute Bose gas when $d \geq 2$ ($T_{\rm QD} \sim |\mu|$ and $T^* \sim |\mu|^{2/d}$). The solid line corresponds to the superfluid transition temperature T_c (i.e. the BKT transition temperature for $d = 2$). For $T \to T_c^+$, the correlation length diverges as $(T - T_c)^{-\nu_{\rm cl}}$ with $\nu_{\rm cl}$ the critical exponent of the d-dimensional classical O(2) model ($\nu_{\rm cl} = 1/2$ if $d \geq 4$). The dashed lines show crossovers between regimes characterized by various temperature behaviors of the correlation length ξ. For $\mu < 0$ and $|\mu| \gg T$, the system behaves as a dilute classical gas and a classical Boltzmann description applies ($\xi = |2m\mu|^{-1/2} + \mathcal{O}(e^{-1/T})$).

irrelevant variable $\tilde{g}$.[56] The phase transition and crossovers at finite temperature are shown in Fig. 7.10.[57]

Let us now consider the dimensionless pressure[55]

$$\tilde{P}(\tilde{\mu}, \tilde{T}, \tilde{g}) = \frac{2m}{\Lambda^{d+z}} P(\mu, T), \tag{7.202}$$

expressed in terms of $\tilde{\mu}$, $\tilde{T}$ and $\tilde{g}$. In an RG transformation, $\tilde{P}$ transforms as

$$\tilde{P}(\tilde{\mu}, \tilde{T}, \tilde{g}) = e^{-(d+z)l} \tilde{P}(e^{l/\nu}\tilde{\mu}, e^{zl}\tilde{T}, \tilde{g}(l)), \tag{7.203}$$

provided that $\tilde{T}(l) \ll 1$ and $|\tilde{\mu}(l)| \ll 1$. Equation (7.203) holds for the full pressure since the vanishing of P when $\mu \leq 0$ and $T = 0$ implies that it has no regular part at the transition. Above the upper critical dimension $d_c^+ = 2$, the dangerously irrelevant variable $\tilde{g}$ cannot be omitted and must be included in (7.203). Setting $s = \tilde{T}^{-1/z}$, we obtain

$$\tilde{P}(\tilde{\mu}, \tilde{T}, \tilde{g}) = \frac{\tilde{T}^{d/z+1}}{(4\pi)^{d/2}} \mathcal{F}_d \left(\frac{\tilde{\mu}}{\tilde{T}^{1/\nu z}}, \tilde{g}(T) \right), \tag{7.204}$$

where we use the notation $\tilde{g}(T)$ for $\tilde{g}(l = \ln(\tilde{T}^{-1/z})$. Going back to dimensionful variables and setting $z = 1/\nu = 2$, we finally obtain the expression (7.188) of the pressure to leading order in the limit $\mu, T \to 0$. There are subleading corrections due to higher-order interactions (which are inevitably generated in the RG procedure). In two dimensions, these corrections can be expressed in terms of the scattering length, but this is no longer true in higher dimensions [43]. For example, the contribution of the three-body interactions to the

[56]See Chapter 10 for a discussion of dangerously irrelevant variables. The necessity to consider the boson–boson interactions, even though they are formally irrelevant, can also be deduced from the fact that the ideal Bose gas is unstable for $\mu > 0$.

[57]Figure 7.10 can be obtained from a one-loop RG calculation at finite temperature as done in Chapters 12 and 15. See also Ref. [25].

equation of state of the three-dimensional Bose gas requires the introduction of the so-called three-body parameter.

Three-body losses in a trapped ultracold gas. The spectrum of the two-body interaction potential generally includes many bound states. When three atoms are close to each other, two of them can form a dimer, the third one absorbing part of the energy released by this formation. Since the trap depth is in general much smaller than the released energy, such a three-body recombination expels the three particles out of the trap thus making the Bose gas metastable [44, 45]. Dimensional analysis implies that the loss rate is proportional to a^4 in three dimensions, so that three-body losses become significant when the scattering length a is large, i.e. in the vicinity of a Feshbach resonance (see Appendix 7.A). When $na^3 \ll 1$, the system can reach thermal equilibrium induced by two-body collisions before three-body losses become important.

7.4.5 *Two-body contact*

7.4.5.1 *Thermodynamic definition*

The fact that the equation of state depends only on the scattering length a_d, and not on the details of the interaction potential, suggests do consider a_d as an additional thermodynamic variable besides temperature and chemical potential. The two-body contact C is then essentially defined as the conjugate variable of a_d,

$$d\Omega = -Pd\mathcal{V} - SdT - Nd\mu + \frac{C}{8\pi m}\frac{2^{3-d}}{a_d^{d-1}}da_d, \tag{7.205}$$

where $\mathcal{V} = L^d$ is the volume, i.e.

$$\frac{C}{8\pi m} = \frac{a_d^{d-1}}{2^{3-d}}\frac{\partial\Omega}{\partial a_d}\bigg|_{\mu,T,\mathcal{V}} = \begin{cases} L^3 \frac{\partial P}{\partial(1/a_3)}\Big|_{\mu,T} & (d=3), \\ -\frac{L^2}{2}\frac{\partial P}{\partial \ln a_2}\Big|_{\mu,T} & (d=2). \end{cases} \tag{7.206}$$

C is an extensive quantity with dimension 1/length for $d = 3$ and $1/(\text{length})^2$ for $d = 2$. One can also define the contact from the energy,

$$\frac{C}{8\pi m} = \frac{a_d^{d-1}}{2^{3-d}}\frac{\partial E}{\partial a_d}\bigg|_{N,S,\mathcal{V}}. \tag{7.207}$$

The equivalence between these two definitions follows from the standard thermodynamic relation $\Omega = E - TS - \mu N$.

It is possible to give an alternative definition of the contact using the universal equation of state (7.188) where the dimensionless interaction constant $\tilde{g}(T)$ is a function of ma_d^2T. From (7.206), one then obtains

$$\frac{C}{8\pi m} = -L^d\frac{(2a_d)^{d-2}}{\lambda_T^d}T^2\mathcal{F}_d^{(0,1)}\left(\frac{\mu}{T}, \tilde{g}(T)\right)\tilde{g}'(T). \tag{7.208}$$

On the other hand, the universal scaling function $\mathcal{F}_d$ determines the particle and entropy densities,

$$\begin{aligned} n &= \frac{\partial P}{\partial \mu} = \frac{1}{\lambda_T^d}\mathcal{F}_d^{(1,0)}, \\ s &= \frac{\partial P}{\partial T} = \frac{T}{\lambda_T^d}\left[\left(\frac{d}{2}+1\right)\frac{\mathcal{F}_d}{T} - \frac{\mu}{T^2}\mathcal{F}_d^{(1,0)} + \tilde{g}'(T)\mathcal{F}_d^{(0,1)}\right]. \end{aligned} \tag{7.209}$$

Inverting these relations, we obtain

$$\frac{T}{\lambda_T^d}\mathcal{F}_d^{(1,0)}\tilde{g}'(T) = \frac{1}{T}\left[Ts - \left(\frac{d}{2}+1\right)P + \mu n\right] = \frac{1}{T}\left(\frac{E}{L^d} - \frac{d}{2}P\right), \tag{7.210}$$

so that

$$\frac{C}{8\pi m} = (2a_d)^{d-2}\left(\frac{d}{2}PL^d - E\right). \tag{7.211}$$

In the following, we consider the three-dimensional case ($d = 3$) and denote the scattering length a_3 by a.

7.4.5.2 *Contact and two-body correlations*

A remarkable feature of the contact defined by (7.206), (7.207) is that it is related to the short-range behavior of two-body correlations. Let us consider the two-body density matrix

$$\rho_2(\mathbf{r}_1, \mathbf{r}_2; \mathbf{r}'_1, \mathbf{r}'_2) = \langle \hat{\psi}^\dagger(\mathbf{r}_1)\hat{\psi}^\dagger(\mathbf{r}_2)\hat{\psi}(\mathbf{r}'_2)\hat{\psi}(\mathbf{r}'_1)\rangle. \tag{7.212}$$

The hermiticity property $\rho_2(\mathbf{r}_1, \mathbf{r}_2; \mathbf{r}'_1, \mathbf{r}'_2)^* = \rho_2(\mathbf{r}'_1, \mathbf{r}'_2; \mathbf{r}_1, \mathbf{r}_2)$ implies that the density matrix can be diagonalized and written as

$$\rho_2(\mathbf{r}_1, \mathbf{r}_2; \mathbf{r}'_1, \mathbf{r}'_2) = \sum_i \lambda_i \varphi_i^*(\mathbf{r}_1, \mathbf{r}_2)\varphi_i(\mathbf{r}'_1, \mathbf{r}'_2) \tag{7.213}$$

where the function $\varphi_i(\mathbf{r}_1, \mathbf{r}_2)$ is symmetric in the exchange $\mathbf{r}_1 \leftrightarrow \mathbf{r}_2$ and satisfies $\int d^d r_1 d^d r_2\, \varphi_i^*(\mathbf{r}_1, \mathbf{r}_2)\varphi_j(\mathbf{r}_1, \mathbf{r}_2) = \delta_{i,j}$. The λ_i's are real and satisfy the normalization condition

$$\int d^d r_1 d^d r_2\, \rho_2(\mathbf{r}_1, \mathbf{r}_2; \mathbf{r}_1, \mathbf{r}_2) = \sum_i \lambda_i = \langle \hat{N}^2\rangle - \langle \hat{N}\rangle. \tag{7.214}$$

The hypothesis at the basis of the contact theory is that in a dilute fluid, when $\mathbf{r}_1 \simeq \mathbf{r}_2$ and $\mathbf{r}'_1 \simeq \mathbf{r}'_2$, the functions φ_i that contribute to ρ_2 are determined by two-body physics. The reason is that when two particles are nearby, the probability that a third particle be in the same neighborhood is very low. Thus, we expect

$$\varphi_i(\mathbf{r}_1, \mathbf{r}_2) \sim \psi_0(\mathbf{r}_1 - \mathbf{r}_2), \tag{7.215}$$

where ψ_0 is the two-body wavefunction in the low-energy limit in the s-wave channel.[58] We can therefore write the two-body density matrix as

$$\rho_2(\mathbf{r}_1, \mathbf{r}_2; \mathbf{r}'_1, \mathbf{r}'_2) \simeq \frac{C}{(4\pi a)^2}\psi_0^*(\mathbf{r}_{12})\psi_0(\mathbf{r}'_{12}) \qquad (|\mathbf{r}_{12}|, |\mathbf{r}'_{12}| \ll n^{-1/d}), \tag{7.216}$$

where $n^{-1/d}$ is the average distance between particles and $\mathbf{r}_{12} = \mathbf{r}_1 - \mathbf{r}_2$. Since in three dimensions, $\psi_0(\mathbf{r}) \simeq \mathcal{V}^{-1/2}(a/|\mathbf{r}| - 1)$ for $|\mathbf{r}|$ much larger than the range b of the potential,[59]

$$\rho_2(\mathbf{r}_1, \mathbf{r}_2; \mathbf{r}'_1, \mathbf{r}'_2) \simeq \frac{C}{(4\pi)^2\mathcal{V}}\frac{1}{|\mathbf{r}_{12}|\,|\mathbf{r}'_{12}|} \qquad (b \ll |\mathbf{r}_{12}|, |\mathbf{r}'_{12}| \ll a, n^{-1/3}). \tag{7.217}$$

The two-body contact C, defined by (7.216), measures the probability of two particles to be close together; at short distances, N-body physics intervenes only through this multiplicative constant.

Let us now show that the factor C in (7.216) is the same quantity as that defined in (7.206). We consider the change in the grand potential due to an infinitesimal variation of the interaction potential $V(\mathbf{r})$ between particles,[60,61]

$$\begin{aligned}\delta\Omega &= \frac{1}{2}\int d^3r d^3r'\, \delta V(\mathbf{r}-\mathbf{r}')\langle\hat\psi^\dagger(\mathbf{r})\hat\psi^\dagger(\mathbf{r}')\hat\psi(\mathbf{r}')\hat\psi(\mathbf{r})\rangle \\ &= \frac{1}{2}\int d^3r d^3r'\, \delta V(\mathbf{r}-\mathbf{r}')\rho_2(\mathbf{r}, \mathbf{r}'; \mathbf{r}, \mathbf{r}'). \end{aligned} \tag{7.218}$$

Since $\delta V(\mathbf{r}-\mathbf{r}')$ is non-zero, only for $|\mathbf{r}-\mathbf{r}'| \lesssim b + \delta b$, we can use (7.216), i.e.

$$\rho_2(\mathbf{r}, \mathbf{r}'; \mathbf{r}, \mathbf{r}') \simeq \frac{C}{(4\pi a)^2}\psi_0(\mathbf{r}-\mathbf{r}')^2, \tag{7.219}$$

which gives

$$\delta\Omega = \frac{1}{2}\frac{C}{(4\pi a)^2}\int d^3r d^3r'\, \delta V(\mathbf{r}-\mathbf{r}')\psi_0(\mathbf{r}-\mathbf{r}')^2. \tag{7.220}$$

Using (7.A.34), we finally obtain

$$\delta\Omega = \frac{C}{8\pi m}\frac{\delta a}{a^2} = -\frac{C}{8\pi m}\delta\left(\frac{1}{a}\right). \tag{7.221}$$

Thus, the contact introduced in (7.216) from the short-distance behavior of the two-body density matrix coincides with the thermodynamic quantity defined from the grand potential or the energy in Eqs. (7.206) and (7.207).

---

[58] Recall that at low energies, scattering occurs only in the $\ell = 0$ channel (see Appendix 7.A).

[59] See Appendix 7.A.

[60] We make no assumptions about the potential other than isotropy: $V(\mathbf{r}) = V(|\mathbf{r}|)$.

[61] Equation (7.218) essentially follows from (3.275). It can be also derived from the functional integral defining the partition function $Z = e^{-\beta\Omega}$.

7.4.5.3 *Momentum distribution*

Let us assume that the density matrix $\hat{\rho} = |\Phi\rangle\langle\Phi|$ corresponds to a pure state with N particles. In that case, it is easy to show that the one- and two-body density matrices are given by

$$\rho_1(\mathbf{r}'_1, \mathbf{r}_1) = N \int d^3r_2 \cdots d^3r_N \, \Phi(\mathbf{r}_1, \mathbf{r}_2, \dots, \mathbf{r}_N)\Phi^*(\mathbf{r}'_1, \mathbf{r}_2, \dots, \mathbf{r}_N),$$

$$\rho_2(\mathbf{r}'_1, \mathbf{r}'_2; \mathbf{r}_1, \mathbf{r}_2) = N(N-1) \int d^3r_3 \cdots d^3r_N \, \Phi(\mathbf{r}_1, \mathbf{r}_2, \mathbf{r}_3, \dots, \mathbf{r}_N)\Phi^*(\mathbf{r}'_1, \mathbf{r}'_2, \mathbf{r}_3, \dots, \mathbf{r}_N). \tag{7.222}$$

The momentum distribution $n_{\mathbf{k}} = \langle \hat{\psi}^\dagger_{\mathbf{k}} \hat{\psi}_{\mathbf{k}} \rangle$ can be written as a function of the one-body density matrix,

$$\begin{aligned} n_{\mathbf{k}} &= \frac{1}{\mathcal{V}} \int d^3r_1 d^3r'_1 \, e^{i\mathbf{k}(\mathbf{r}'_1 - \mathbf{r}_1)} \rho_1(\mathbf{r}'_1, \mathbf{r}_1) \\ &= \frac{N}{\mathcal{V}} \int d^3r'_1 d^3r_1 d^3r_2 \cdots d^3r_N \, e^{i\mathbf{k}(\mathbf{r}'_1 - \mathbf{r}_1)} \Phi(\mathbf{r}_1, \mathbf{r}_2, \dots, \mathbf{r}_N)\Phi^*(\mathbf{r}'_1, \mathbf{r}_2, \dots, \mathbf{r}_N). \end{aligned} \tag{7.223}$$

For a non-interacting gas, $n_{\mathbf{k}} = n_B(\epsilon_{\mathbf{k}} - \mu)$ is exponentially suppressed when $\epsilon_{\mathbf{k}} - \mu \gg T$. This indicates that the large-momentum behavior is determined by interactions. For large $\mathbf{k}$, the integral in (7.223) is dominated by the contribution of small distances $\mathbf{r}_1 - \mathbf{r}'_1$. Interactions are expected to be important if one of the other coordinates, e.g. $\mathbf{r}_2$, is close to $\mathbf{r}_1$ and $\mathbf{r}'_1$. In that case, one has

$$\begin{aligned} & N \int d^3r_3 \cdots d^3r_N \, \Phi(\mathbf{r}_1, \mathbf{r}_2, \dots, \mathbf{r}_N)\Phi^*(\mathbf{r}'_1, \mathbf{r}_2, \dots, \mathbf{r}_N) \\ &= \frac{1}{N-1} \rho_2(\mathbf{r}'_1, \mathbf{r}_2; \mathbf{r}_1, \mathbf{r}_2) \\ &\simeq \frac{1}{N-1} \frac{C}{(4\pi)^2 \mathcal{V}} \frac{1}{|\mathbf{r}_1 - \mathbf{r}_2| \, |\mathbf{r}'_1 - \mathbf{r}_2|} \qquad (b \ll |\mathbf{r}_1 - \mathbf{r}_2|, |\mathbf{r}'_1 - \mathbf{r}_2| \ll a, n^{-1/3}) \end{aligned} \tag{7.224}$$

using (7.217). Thus, we find (taking into account a factor $N-1$ due to all possible choices $\mathbf{r}_3, \dots, \mathbf{r}_N$ in addition to $\mathbf{r}_2$)[62]

$$\begin{aligned} n_{\mathbf{k}} &\simeq \frac{1}{\mathcal{V}} \frac{C}{(4\pi)^2 \mathcal{V}} \int d^3r'_1 d^3r_1 d^3r_2 \frac{e^{i\mathbf{k}\cdot(\mathbf{r}'_1 - \mathbf{r}_1)}}{|\mathbf{r}_1 - \mathbf{r}_2| \, |\mathbf{r}'_1 - \mathbf{r}_2|} \\ &\simeq \frac{C}{(4\pi)^2 \mathcal{V}} \left(\int d^3r \frac{e^{i\mathbf{k}\cdot\mathbf{r}}}{|\mathbf{r}|} \right)^2 \\ &\simeq \frac{C}{\mathcal{V}|\mathbf{k}|^4} \qquad (1/a, \, n^{1/3} \ll |\mathbf{k}| \ll 1/b). \end{aligned} \tag{7.225}$$

It is straightforward to extend the derivation of (7.225) to an arbitrary density matrix.

[62]The momentum distribution has the following normalization: $\int_{\mathbf{k}} n_{\mathbf{k}} = N/\mathcal{V}$ (hence the volume appearing in the last result of (7.225)).

7.4.5.4 *Bogoliubov theory*

From the expression (7.176) of the pressure obtained within the Bogoliubov theory (including the Lee–Huang–Yang correction), we deduce the expression of the two-body contact in a three-dimensional Bose gas at zero temperature in the limit $ma^2\mu \sim na^3 \ll 1$,

$$\begin{aligned} C &= \mathcal{V} m^2\mu^2 \\ &= 16\pi^2 a^2 nN \left(1 + \frac{64}{3}\sqrt{\frac{na^3}{\pi}}\right). \end{aligned} \tag{7.226}$$

It is easy to verify that this expression, obtained by taking the derivative of the pressure with respect to $1/a$, can also be obtained from (7.211).

The momentum distribution can be obtained from the propagator (7.163),

$$\begin{aligned} n_{\mathbf{k}} &= -\frac{1}{\beta}\sum_{\omega_n} e^{i\omega_n 0^+} \mathcal{G}_{11}(\mathbf{k}, i\omega_n) \\ &= -\frac{1}{2} + \frac{\epsilon_{\mathbf{k}}}{2[\epsilon_{\mathbf{k}}(\epsilon_{\mathbf{k}} + 2\mu)]^{1/2}} \qquad (T \to 0) \\ &\simeq \frac{m^2\mu^2}{|\mathbf{k}|^4} \qquad (\epsilon_{\mathbf{k}} \gg \mu), \end{aligned} \tag{7.227}$$

in agreement with (7.225) and (7.226).

7.5 BCS Mean-Field Theory of Superconductivity

In this section, we consider a spin-$\frac{1}{2}$ fermion system with the action

$$\begin{aligned} S[\psi^*, \psi] &= \sum_{k,\sigma} \psi^*_\sigma(k)(-i\omega_n - \mu + \epsilon_{\mathbf{k}})\psi_\sigma(k) \\ &\quad - \frac{\lambda}{2\beta\mathcal{V}} \sum_{\substack{k,k',q \\ \sigma,\sigma'}}{}' \psi^*_\sigma(k+q)\psi^*_{\sigma'}(k'-q)\psi_{\sigma'}(k')\psi_\sigma(k), \end{aligned} \tag{7.228}$$

where the ψ's are Grassmann variables. The prime means that the sum is restricted to the low-energy states satisfying $|\xi_{\mathbf{k}}|, |\xi_{\mathbf{k}'}|, |\xi_{\mathbf{k}+\mathbf{q}}|, |\xi_{\mathbf{k}'-\mathbf{q}}| \leq \omega_D$. The range of the fermion–fermion interaction in real space is therefore of order v_F/ω_D (v_F is the Fermi velocity). In the BCS model of conventional superconductors, screening of the Coulomb interaction between electrons as well as between electrons and ions gives rise to an effective attractive interaction at low energies (often seen as arising from the exchange of phonons). Although retardation effects are crucial to understand the origin of the attractive interaction, the latter is well described by the action (7.228) with $\lambda > 0$ and $\omega_D \ll \mu$ the Debye frequency of the phonons. From a more technical point of view, a cutoff (be it ω_D or the fermion bandwidth) is necessary since a strictly local interaction ($\omega_D \to \infty$) would give UV divergences in perturbation theory. In practice, the cutoff ω_D can be sent to infinity whenever it does not significantly modify the final results. In the following, we assume that the interaction is weak, i.e. $\lambda N(0) \ll 1$ ($N(0)$ is the density of state per spin at the Fermi level in the metallic phase).

7.5.1 *Cooper instability and superconducting transition*

The instability of the normal phase is due to the formation of fermion–fermion pairs (Cooper pairs) and is therefore signaled by a divergence of the pairing susceptibility

$$\chi(\mathbf{r}-\mathbf{r}',\tau-\tau') = \langle \psi_\downarrow(\mathbf{r},\tau)\psi_\uparrow(\mathbf{r},\tau)\psi_\uparrow^*(\mathbf{r}',\tau')\psi_\downarrow^*(\mathbf{r}',\tau')\rangle. \tag{7.229}$$

Given the symmetry of the fermion–fermion interaction, we expect superconductivity to occur in the spin-singlet s-wave channel. For $\lambda = 0$, using Wick's theorem, one obtains the bare susceptibility $\chi_0(\mathbf{r}-\mathbf{r}',\tau-\tau') = G_{0,\downarrow}(\mathbf{r}-\mathbf{r}',\tau-\tau')G_{0,\uparrow}(\mathbf{r}-\mathbf{r}',\tau-\tau')$, i.e.

$$\chi_0(q) = \frac{1}{\beta\mathcal{V}}\sum_k G_{0,\downarrow}(k)G_{0,\uparrow}(q-k), \tag{7.230}$$

where $G_{0,\sigma}(k) \equiv G_0(k) = (i\omega_n - \xi_{\mathbf{k}})^{-1}$ is the non-interacting Green function. Performing the sum over Matsubara frequencies in the usual way, we obtain

$$\chi_0 = \int_{\mathbf{k}} \frac{1}{2\xi_{\mathbf{k}}}\tanh\left(\frac{\beta\xi_{\mathbf{k}}}{2}\right) = N(0)\int_0^{\omega_D}\frac{d\xi}{\xi}\tanh\left(\frac{\beta\xi}{2}\right) \tag{7.231}$$

for $\mathbf{q} = 0$ and $\omega_\nu = 0$. In (7.231), we have restricted the integration to the low-energy degrees of freedom $|\xi_{\mathbf{k}}| \leq \omega_D$,[63] and have approximated the density of states $N(\xi) = \int_{\mathbf{k}}\delta(\xi-\xi_{\mathbf{k}})$ by its value at the Fermi level, since in most cases $N(\xi) \simeq N(0)$ when $|\xi| \leq \omega_D \ll \mu$. The remaining integral in (7.231) can be carried out by an integration by part,

$$\begin{aligned}
\chi_0 &= N(0)\tanh\left(\frac{\beta\omega_D}{2}\right)\ln\omega_D - N(0)\frac{\beta}{2}\int_0^{\omega_D} d\xi\frac{\ln\xi}{\cosh^2(\beta\xi/2)} \\
&= N(0)\tanh\left(\frac{\beta\omega_D}{2}\right)\ln\left(\frac{\beta\omega_D}{2}\right) - N(0)\int_0^{\frac{\beta\omega_D}{2}} dx\frac{\ln x}{\cosh^2 x} \\
&\simeq N(0)\ln\left(\frac{2\gamma\omega_D}{\pi T}\right)
\end{aligned} \tag{7.232}$$

for $T \ll \omega_D$. We shall see that the transition temperature is much smaller than ω_D in the weak-coupling limit $\lambda N(0) \ll 1$. We have used $\int_0^{\frac{\beta\omega_D}{2}} dx \ln x \cosh^{-2} x \simeq \int_0^\infty dx \ln x \cosh^{-2} x = -\ln(4\gamma/\pi)$, where $\gamma = e^C \simeq 1.78$ is the exponential of the Euler constant C.

[63] It would be more natural to use the bandwidth as the cutoff in the bare susceptibility. However, when calculating the leading corrections to χ_0 and the transition temperature, only the low-energy degrees of freedom $|\xi_{\mathbf{k}}| \leq \omega_D$ are involved [see, e.g., (7.233) and (7.234)].

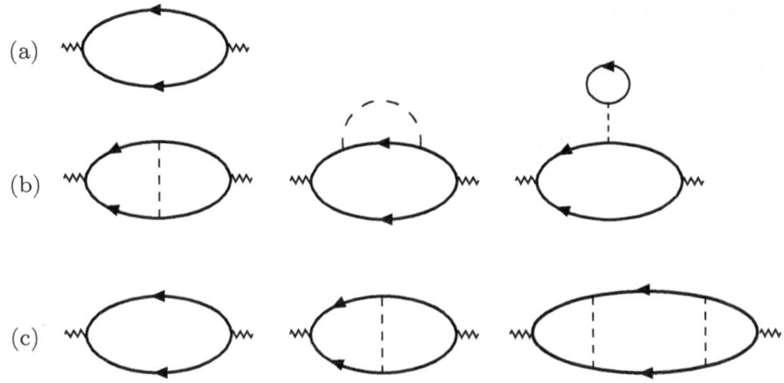

Fig. 7.11 (a) Bare pairing susceptibility χ_0. (b) First-order corrections. (c) Ladder diagrams.

The low-temperature logarithmic divergence of χ_0 also shows up in the perturbation expansion of $\chi \equiv \chi(q = 0)$ in powers of λ. The first-order corrections are shown in Fig. 7.11(b). The contribution of the first diagram reads

$$\frac{\lambda}{(\beta\mathcal{V})^2}\sum_{k,k'} G_{0,\downarrow}(k)G_{0,\uparrow}(-k)G_{0,\downarrow}(k')G_{0,\uparrow}(-k') = \lambda\chi_0^2, \tag{7.233}$$

while the other diagrams can be taken into account by including the Hartree–Fock self-energy in the one-particle propagator $G(k)$. In general, the Hartree–Fock corrections are expected to be the same in both the normal and superconducting phases, and therefore can be omitted (or absorbed into a redefinition of μ) without the comparison between the two phases being affected.

To each order in the perturbation expansion, it is natural to retain the ladder diagram, since they give the highest power of the logarithmic singularity. Summing all these diagrams (Fig. 7.11(c)), which is akin to an RPA, we obtain

$$\chi = \chi_0 + \lambda\chi_0^2 + \lambda^2\chi_0^3 + \cdots = \frac{\chi_0}{1-\lambda\chi_0}. \tag{7.234}$$

Thus, the pairing susceptibility diverges when $1 - \lambda\chi_0 = 0$, which gives the superconducting transition temperature

$$T_c = \frac{2\gamma\omega_D}{\pi}e^{-1/\lambda N(0)} \tag{7.235}$$

in the weak-coupling limit $\lambda N(0) \ll 1$. Above T_c, the susceptibility diverges as

$$\chi = \frac{T_c}{\lambda^2 N(0)(T-T_c)}. \tag{7.236}$$

Note that the divergence occurs only for an attractive interaction ($-\lambda < 0$). For a repulsive interaction, the ladder diagrams reduce the pairing susceptibility with respect to its bare value.

7.5.2 *Mean-field theory*

The divergence of the pairing susceptibility suggests that the order parameter associated with the superconducting transition is $\langle\hat{\psi}_\downarrow(\mathbf{r})\hat{\psi}_\uparrow(\mathbf{r})\rangle$. To set up a mean-field theory, we introduce an auxiliary bosonic field $\Delta(\mathbf{r},\tau)$ by means of the following Hubbard–Stratonovich transformation[64]:

$$e^{-S_{\rm int}[\psi^*,\psi]} = \int \mathcal{D}[\Delta^*,\Delta] \exp\biggl\{-\frac{1}{\lambda}\sum_q |\Delta(q)|^2 + \frac{1}{\sqrt{\beta\mathcal{V}}}\sum_{k,q}[\Delta^*(q)\psi_\downarrow(k)\psi_\uparrow(-k+q) + \text{c.c.}]\biggr\}. \tag{7.237}$$

The integration over Δ in (7.237) reproduces only the spin-singlet part of the interaction ($\sigma' = -\sigma$ in (7.228)). One could consider a more general bosonic field that reproduces both the singlet and triplet parts of the interaction. However, only the singlet part needs to be considered at the mean-field level since a spin-triplet pairing cannot be induced by an interaction with s-wave symmetry.

The mean-field approximation follows from a saddle-point approximation for the functional integral over the auxiliary bosonic field,

$$Z = \int \mathcal{D}[\psi^*,\psi,\Delta^*,\Delta]\, e^{-S[\psi^*,\psi,\Delta^*,\Delta]} \quad\to\quad Z_{\rm MF} = \int \mathcal{D}[\psi^*,\psi]\, e^{-S_{\rm MF}[\psi^*,\psi]}. \tag{7.238}$$

The saddle-point equation $\delta Z_{\rm MF}/\delta\Delta^*(\mathbf{r},\tau) = 0$ then gives

$$\Delta(\mathbf{r},\tau) = \lambda\langle\psi_\downarrow(\mathbf{r},\tau)\psi_\uparrow(\mathbf{r},\tau)\rangle. \tag{7.239}$$

In the following, $\Delta(\mathbf{r},\tau) = \Delta$ is chosen static and uniform (i.e. $\Delta(q) = \sqrt{\beta\mathcal{V}}\delta_{q,0}\Delta$). With no loss of generality, Δ can be taken real and positive. Its value will be determined by solving the gap equation (7.239). The mean-field action reads

$$\begin{aligned} S_{\rm MF}[\psi^*,\psi] &= \beta\mathcal{V}\frac{\Delta^2}{\lambda} + \sum_{k,\sigma}\psi^*_\sigma(k)(-i\omega_n + \xi_{\mathbf{k}})\psi_\sigma(k) - \Delta\sum_k[\psi_\downarrow(k)\psi_\uparrow(-k) + \text{c.c.}] \\ &= \beta\mathcal{V}\frac{\Delta^2}{\lambda} - \sum_k \Psi^\dagger(k)\mathcal{G}^{-1}(k)\Psi(k), \end{aligned} \tag{7.240}$$

where we have introduced the Nambu spinor

$$\Psi(k) = \begin{pmatrix} \psi_\uparrow(k) \\ \psi^*_\downarrow(-k) \end{pmatrix}, \qquad \Psi^\dagger(k) = (\psi^*_\uparrow(k), \psi_\downarrow(-k)) \tag{7.241}$$

[64] See Sec. 6.3.3 for a critical discussion of Hubbard–Stratonovich transformations.

and the inverse propagator

$$\mathcal{G}^{-1}(k) = \begin{pmatrix} i\omega_n - \xi_{\mathbf{k}} & \Delta \\ \Delta & i\omega_n + \xi_{-\mathbf{k}} \end{pmatrix}. \tag{7.242}$$

7.5.2.1 *Ground state and Bogoliubov quasi-particles*

Inverting (7.242), we obtain the propagator

$$\mathcal{G}(k) = -\langle \Psi(k)\Psi^\dagger(k)\rangle = \begin{pmatrix} G(k) & F(k) \\ F^\dagger(k) & -G(-k) \end{pmatrix}, \tag{7.243}$$

where

$$\begin{aligned} G(k) &= -\langle \psi_\sigma(k)\psi^*_\sigma(k)\rangle = \frac{-i\omega_n - \xi_{\mathbf{k}}}{\omega_n^2 + E_{\mathbf{k}}^2}, \\ F(k) &= -\langle \psi_\uparrow(k)\psi_\downarrow(-k)\rangle = \frac{\Delta}{\omega_n^2 + E_{\mathbf{k}}^2}, \\ F^\dagger(k) &= -\langle \psi^*_\downarrow(-k)\psi^*_\uparrow(k)\rangle = F(k) \end{aligned} \tag{7.244}$$

(we have used $\xi_{-\mathbf{k}} = \xi_{\mathbf{k}}$) and

$$E_{\mathbf{k}} = \sqrt{\xi_{\mathbf{k}}^2 + \Delta^2}. \tag{7.245}$$

G and F are referred to as the normal and anomalous Green functions, respectively. F is tied to the broken U(1) symmetry of the superconducting state and vanishes when $\Delta = 0$.

It is convenient to rewrite the Green functions as

$$\begin{aligned} G(k) &= \frac{u_{\mathbf{k}}^2}{i\omega_n - E_{\mathbf{k}}} + \frac{v_{\mathbf{k}}^2}{i\omega_n + E_{\mathbf{k}}}, \\ F(k) &= -u_{\mathbf{k}}v_{\mathbf{k}}\left(\frac{1}{i\omega_n - E_{\mathbf{k}}} - \frac{1}{i\omega_n + E_{\mathbf{k}}}\right), \end{aligned} \tag{7.246}$$

with

$$u_{\mathbf{k}}^2 = \frac{1}{2}\left(1 + \frac{\xi_{\mathbf{k}}}{E_{\mathbf{k}}}\right), \qquad v_{\mathbf{k}}^2 = \frac{1}{2}\left(1 - \frac{\xi_{\mathbf{k}}}{E_{\mathbf{k}}}\right), \qquad u_{\mathbf{k}}v_{\mathbf{k}} = \frac{\Delta}{2E_{\mathbf{k}}}. \tag{7.247}$$

With these notations, the momentum distribution is given by

$$\begin{aligned} n_{\mathbf{k}} = \langle \psi^*_\sigma(\mathbf{k},\tau)\psi_\sigma(\mathbf{k},\tau)\rangle &= \frac{1}{\beta}\sum_{\omega_n} e^{i\omega_n 0^+} G(\mathbf{k}, i\omega_n) \\ &= u_{\mathbf{k}}^2 n_F(E_{\mathbf{k}}) + v_{\mathbf{k}}^2[1 - n_F(E_{\mathbf{k}})] \end{aligned} \tag{7.248}$$

(see Fig. 7.12), and the spectral function reads

$$A_s(\mathbf{k},\omega) = -\frac{1}{\pi}\Im[G(\mathbf{k},\omega + i0^+)] = u_{\mathbf{k}}^2\delta(\omega - E_{\mathbf{k}}) + v_{\mathbf{k}}^2\delta(\omega + E_{\mathbf{k}}). \tag{7.249}$$

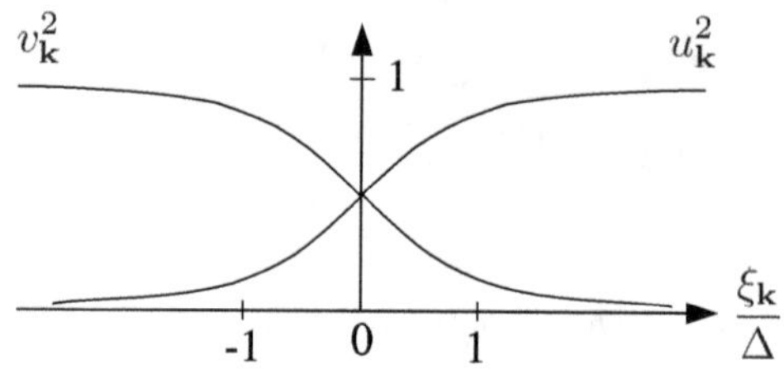

Fig. 7.12 Coefficients $u_{\mathbf{k}}^2$ and $v_{\mathbf{k}}^2$. At $T = 0$, $v_{\mathbf{k}}^2$ determines the momentum distribution $n_{\mathbf{k}}$ [Eq. (7.248)].

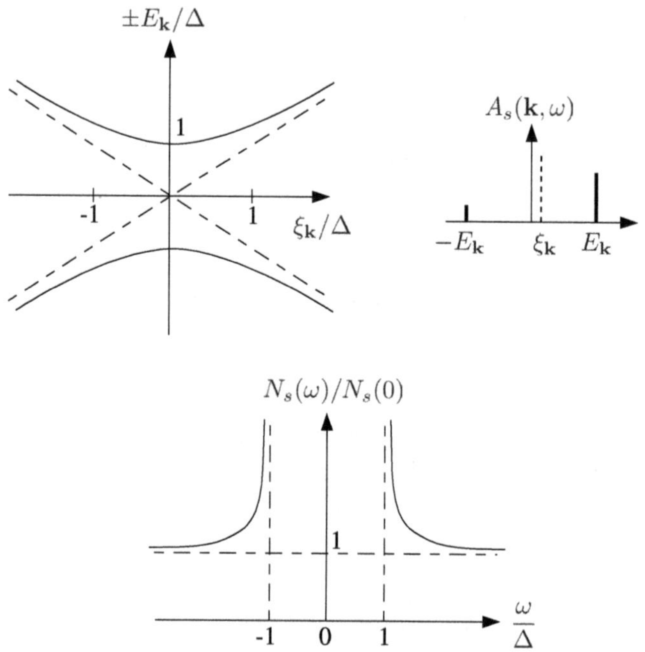

Fig. 7.13 One-particle excitation energy $\pm E_{\mathbf{k}}$, spectral function $A_s(\mathbf{k}, \omega)$ and density of states $N_s(\omega)$ in the superconducting state. The dashed lines show the corresponding results in the normal phase: particle ($\xi_{\mathbf{k}}$) and hole ($-\xi_{\mathbf{k}}$) excitation energies, spectral function $A(\mathbf{k}, \omega) = \delta(\omega - \xi_{\mathbf{k}})$ and density of states $N(\omega) = N(0)$.

For $\Delta = 0$, we recover the spectral function $A(\mathbf{k}, \omega) = \delta(\omega - \xi_{\mathbf{k}})$ of the non-interacting system. In the superconducting state, the excitation at $\omega = \xi_{\mathbf{k}}$ splits into two excitations at $\omega = E_{\mathbf{k}}$ and $\omega = -E_{\mathbf{k}}$ with spectral weight $u_{\mathbf{k}}^2$ and $v_{\mathbf{k}}^2$, respectively (Fig. 7.13). The one-particle excitation spectrum exhibits a gap $\Delta = \min_{\mathbf{k}} E_{\mathbf{k}}$. From (7.249), we obtain the (tunneling) density of states

$$N_s(\omega) = \int_{\mathbf{k}} A_s(\mathbf{k}, \omega) = N(0) \frac{|\omega| \Theta(|\omega| - \Delta)}{\sqrt{\omega^2 - \Delta^2}} \tag{7.250}$$

shown in Fig. 7.13.

The chemical potential is not changed by the superconducting transition, since the particle density is independent of the gap Δ. Indeed, the difference in density between the normal and superconducting phases vanishes when the chemical potential is held fixed:

$$\begin{aligned}
\frac{2}{\beta\mathcal{V}}\sum_k e^{i\omega_n 0^+}\left[G(k)-G_0(k)\right] \\
&= 2\int_{\mathbf{k}}\left[u_{\mathbf{k}}^2 n_F(E_{\mathbf{k}})+v_{\mathbf{k}}^2 n_F(-E_{\mathbf{k}})-n_F(\xi_{\mathbf{k}})\right] \\
&= 2N(0)\int_{-\omega_D}^{\omega_D} d\xi\left[\frac{1}{2}\tanh\left(\frac{\beta\xi}{2}\right)+\frac{\xi}{E}\left(n_F(E)-\frac{1}{2}\right)\right]=0
\end{aligned} \tag{7.251}$$

($E=\sqrt{\xi^2+\Delta^2}$).

To understand the nature of the ground state and the elementary excitations, we diagonalize the mean-field action $S_{\text{MF}}[\psi^*,\psi]$ [Eq. (7.240)]. The Bogoliubov (unitary) transformation

$$\begin{pmatrix}\gamma_\uparrow(k)\\ \gamma_\downarrow^*(-k)\end{pmatrix}=\begin{pmatrix}u_{\mathbf{k}} & -v_{\mathbf{k}}\\ v_{\mathbf{k}} & u_{\mathbf{k}}\end{pmatrix}\begin{pmatrix}\psi_\uparrow(k)\\ \psi_\downarrow^*(-k)\end{pmatrix} \tag{7.252}$$

yields

$$S_{\text{MF}}[\gamma^*,\gamma]=\text{const}-\sum_{k,\sigma}\gamma_\sigma^*(k)(i\omega_n-E_{\mathbf{k}})\gamma_\sigma(k). \tag{7.253}$$

Note that we do not obtain the correct value of the constant term in the rhs of (7.253) because the Nambu formalism mixes creation and annihilation fields (this issue is further discussed below in the section on thermodynamics). Since $E_{\mathbf{k}}>0$, the ground state $|\Psi\rangle$ is defined by $\langle\Psi|\hat\gamma_\sigma^\dagger(\mathbf{k})\hat\gamma_\sigma(\mathbf{k})|\Psi\rangle=\lim_{T\to 0}n_F(E_{\mathbf{k}})=0$ and must therefore satisfy $\hat\gamma_\sigma(\mathbf{k})|\Psi\rangle=0$ for all annihilation operators $\hat\gamma_\sigma(\mathbf{k})$. One readily verifies that the solution is given by the BCS state

$$|\Psi_{\text{BCS}}\rangle=\prod_{\mathbf{k}}\left[u_{\mathbf{k}}+v_{\mathbf{k}}\hat\psi_\uparrow^\dagger(\mathbf{k})\hat\psi_\downarrow^\dagger(-\mathbf{k})\right]|\text{vac}\rangle. \tag{7.254}$$

The elementary excitations with energies $\pm E_{\mathbf{k}}$ are known as Bogoliubov quasi-particles. The corresponding creation operator, $\hat\gamma_\sigma^\dagger(\mathbf{k})$, has both a particle and a hole component [Eq. (7.252)]. In the BCS state, all particles are paired: The states $(\mathbf{k},\uparrow)$ and $(-\mathbf{k},\downarrow)$ are both either occupied (with probability amplitude $v_{\mathbf{k}}$) or empty (with probability amplitude $u_{\mathbf{k}}$). Thus, there are two ways to create an excited state. One can either add a $(\mathbf{k},\sigma)$ fermion or remove a $(-\mathbf{k},-\sigma)$ fermion depending on the occupation of the pair state. In both cases, the final state has one unpaired fermion $(\mathbf{k},\sigma)$ and the energy increases. The operator that creates these one-particle excitations is the operator $\hat\gamma_\sigma^\dagger(\mathbf{k})$: $\hat\gamma_\sigma^\dagger(\mathbf{k})|\Psi_{\text{BCS}}\rangle$ is an excited state with an energy $E_{\mathbf{k}}$ as shown by (7.253). Hence, the peaks at $E_{\mathbf{k}}$ and $-E_{\mathbf{k}}$ appearing in the spectral function $A_s(\mathbf{k},\omega)$ with spectral weights $u_{\mathbf{k}}^2$ and $v_{\mathbf{k}}^2$, respectively.[65]

[65] Recall that the spectral function is non-zero for $\omega=\Delta E$ for a particle-like excitation and $\omega=-\Delta E$ for a hole-like excitation (Chapter 3).

The BCS state as a boson coherent state? Using $[\hat\psi_\uparrow(\mathbf{k})\hat\psi_\downarrow(-\mathbf{k})]^2 = 0$, one can rewrite (7.254) as

$$\begin{aligned} |\Psi_{\rm BCS}\rangle &= C\prod_{\mathbf{k}}\left[1+\Delta\varphi_{\mathbf{k}}\hat\psi^\dagger_\uparrow(\mathbf{k})\hat\psi^\dagger_\downarrow(-\mathbf{k})\right]|{\rm vac}\rangle \\ &= C\exp\Big\{\Delta\sum_{\mathbf{k}}\varphi_{\mathbf{k}}\hat\psi^\dagger_\uparrow(\mathbf{k})\hat\psi^\dagger_\downarrow(-\mathbf{k})\Big\}|{\rm vac}\rangle, \end{aligned} \tag{7.255}$$

where $\varphi_{\mathbf{k}} = v_{\mathbf{k}}/u_{\mathbf{k}}\Delta$. If we introduce the operator

$$\hat\Psi^\dagger(\mathbf{r}) = \int d^3\rho\,\varphi(\boldsymbol{\rho})\hat\psi^\dagger_\uparrow\left(\mathbf{r}+\frac{\boldsymbol{\rho}}{2}\right)\hat\psi^\dagger_\downarrow\left(\mathbf{r}-\frac{\boldsymbol{\rho}}{2}\right) \tag{7.256}$$

and its Fourier transform

$$\hat\Psi^\dagger(\mathbf{q}) = \frac{1}{\sqrt{\mathcal{V}}}\sum_{\mathbf{k}}\varphi_{\mathbf{k}}\hat\psi^\dagger_\uparrow\left(\frac{\mathbf{q}}{2}+\mathbf{k}\right)\hat\psi^\dagger_\downarrow\left(\frac{\mathbf{q}}{2}-\mathbf{k}\right), \tag{7.257}$$

which creates a Cooper pair with wavefunction $\varphi(\boldsymbol{\rho})$ (the Fourier transform of $\varphi_{\mathbf{k}}$), we obtain

$$|\Psi_{\rm BCS}\rangle = Ce^{\sqrt{\mathcal{V}}\Delta\hat\Psi^\dagger(\mathbf{q}=0)}|{\rm vac}\rangle. \tag{7.258}$$

This expression is similar to the one of a Bose condensate and would describe a Bose condensate of Cooper pairs if the operator $\hat\Psi(\mathbf{r})$ were a boson operator.[66] In Sec. 7.7, we shall see that when their size is much smaller than the interparticle distance, the Cooper pairs can be considered as point-like (composite) bosons and the ket $|\Psi_{\rm BCS}\rangle$ can then be seen as describing a Bose condensate.

7.5.2.2 *Superconducting gap* $\Delta(T)$

The value of Δ is determined from (7.239), i.e. by the BCS gap equation

$$\Delta = \lambda\langle\psi_\downarrow(\mathbf{r},\tau)\psi_\uparrow(\mathbf{r},\tau)\rangle = \frac{\lambda}{\beta\mathcal{V}}\sum_k F(k). \tag{7.259}$$

Equation (7.259) is satisfied by either $\Delta = 0$ (metallic phase) or a non-zero solution (superconducting phase) determined by

$$\frac{1}{\lambda} = \frac{1}{\beta\mathcal{V}}\sum_k\frac{1}{\omega_n^2+E_{\mathbf{k}}^2} = \int_{\mathbf{k}}\frac{1}{2E_{\mathbf{k}}}\tanh\left(\frac{\beta E_{\mathbf{k}}}{2}\right). \tag{7.260}$$

For $T\to T_c^-$ ($\Delta\to 0^+$), this equation becomes

$$\frac{1}{\lambda} = \int_{\mathbf{k}}\frac{1}{2\xi_{\mathbf{k}}}\tanh\left(\frac{\beta_c\xi_{\mathbf{k}}}{2}\right) \tag{7.261}$$

($\beta_c = 1/T_c$) so that we recover the transition temperature deduced from the divergence of the pairing susceptibility (Sec. 7.5.1). The temperature dependence of the gap Δ is shown schematically in Fig. 7.14.

---

[66] $\hat\Psi(\mathbf{r})$ and $\hat\Psi^\dagger(\mathbf{r})$ are not canonical bosonic operators since $[\hat\Psi(\mathbf{r}),\hat\Psi^\dagger(\mathbf{r}')]\neq\delta(\mathbf{r}-\mathbf{r}')$.

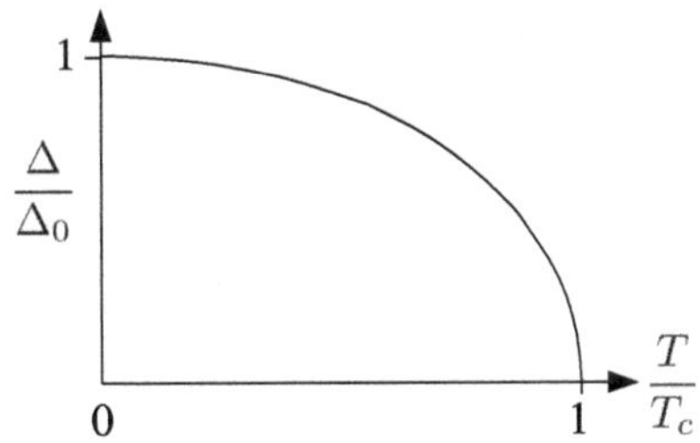

Fig. 7.14 Temperature dependence of the superconducting order parameter Δ $[\Delta_0 = \Delta(T=0)]$.

For $T=0$, the gap equation (7.260) gives

$$\begin{aligned} \frac{1}{\lambda} &= \int_{\mathbf{k}} \frac{1}{2E_{\mathbf{k}}} = N(0) \int_0^{\omega_D} \frac{d\xi}{\sqrt{\xi^2+\Delta^2}} = N(0)\,\mathrm{argsinh}\left(\frac{\omega_D}{\Delta}\right) \\ &\simeq N(0) \ln\left(\frac{2\omega_D}{\Delta}\right) \end{aligned} \tag{7.262}$$

for $\omega_D \gg \Delta$ (i.e. $\lambda N(0) \ll 1$), and

$$\Delta_0 = \Delta(T=0) = 2\omega_D e^{-1/\lambda N(0)}. \tag{7.263}$$

It follows that the ratio

$$\frac{\Delta(T=0)}{T_c} = \frac{\pi}{\gamma} \simeq 1.76 \tag{7.264}$$

takes a universal value. The zero-temperature gap defines the characteristic length scale $\xi_0 = v_F/\pi\Delta_0$ (known as the BCS coherence length), which determines the size of Cooper pairs in real space.

To determine the gap at low temperatures $T \ll T_c$, we rewrite the gap equation (7.260) as

$$\frac{1}{\lambda} = \int_{\mathbf{k}} \frac{1}{2E_{\mathbf{k}}} - \int_{\mathbf{k}} \frac{1}{E_{\mathbf{k}}(e^{\beta E_{\mathbf{k}}}+1)}. \tag{7.265}$$

The term in the lhs is equal to $N(0)\ln(2\omega_D/\Delta_0)$ and the first one in the rhs to $N(0)\ln(2\omega_D/\Delta)$, so that

$$\ln\left(\frac{\Delta_0}{\Delta}\right) = 2\int_0^{\omega_D} \frac{d\xi}{E(e^{\beta E}+1)} \simeq 2\int_0^{\omega_D} \frac{d\xi}{E} e^{-\beta E}, \tag{7.266}$$

where $E = \sqrt{\xi^2+\Delta^2}$. Taking $x = \Delta - E$ as the integration variable, we finally obtain

$$\ln\left(\frac{\Delta_0}{\Delta}\right) \simeq 2\int_\Delta^\infty \frac{dE}{\sqrt{E^2-\Delta^2}} e^{-\beta E} \simeq 2\int_0^\infty \frac{dx}{\sqrt{2\Delta x}} e^{-\beta(\Delta+x)} = \sqrt{\frac{2\pi}{\beta\Delta}} e^{-\beta\Delta} \tag{7.267}$$

(we have used $\int_0^\infty dx\, x^{-1/2} e^{-x} = \sqrt{\pi}$), and

$$\Delta \simeq \Delta_0 - \sqrt{2\pi\Delta_0 T}\, e^{-\Delta_0/T}. \tag{7.268}$$

The value of the gap near the transition temperature T_c is computed in Sec. 7.5.3.

7.5.2.3 *Thermodynamics*

The mean-field action (7.240) being Gaussian, we can integrate out the fermion field and obtain the grand potential

$$\Omega_s = -\frac{1}{\beta}\ln Z_{\mathrm{MF}} = \mathcal{V}\frac{\Delta^2}{\lambda} - \frac{1}{\beta}\sum_k \ln\left[\det \mathcal{G}^{-1}(k)\right]. \tag{7.269}$$

Because the Nambu spinors mix up creation and annihilation fields, we must carefully introduce the convergence factors before computing the Matsubara sum (see the discussion on page 76). The correct expression reads

$$\frac{1}{\beta}\sum_{\omega_n} \ln\left[\det \mathcal{G}^{-1}(k)\right] \equiv \frac{1}{\beta}\sum_{\omega_n}\Biggl\{ e^{i\omega_n 0^+}\ln\left[\mathcal{G}_{11}^{-1}(k)\right] + e^{-i\omega_n 0^+}\ln\left[\mathcal{G}_{22}^{-1}(k)\right] + \ln\left[1 - \frac{\mathcal{G}_{12}^{-1}(k)\mathcal{G}_{21}^{-1}(k)}{\mathcal{G}_{11}^{-1}(k)\mathcal{G}_{22}^{-1}(k)}\right]\Biggr\}. \tag{7.270}$$

Since the last term converges, we can add the factor $e^{i\omega_n 0^+}$ without changing the result. Using $\mathcal{G}_{12}^{-1}(k) = \mathcal{G}_{21}^{-1}(k)$ and $\mathcal{G}_{22}^{-1}(k) = \mathcal{G}_{11}^{-1}(-k)$, we then obtain

$$\frac{1}{\beta}\sum_{\omega_n} \ln\left[\det \mathcal{G}^{-1}(k)\right] \equiv \frac{1}{\beta}\sum_{\omega_n} e^{i\omega_n 0^+}\ln\left[\frac{\mathcal{G}_{11}^{-1}(k)}{\mathcal{G}_{22}^{-1}(k)}\det \mathcal{G}^{-1}(k)\right]. \tag{7.271}$$

From the Matsubara sums[67]

$$\begin{aligned}
\frac{1}{\beta}\sum_{\omega_n} e^{i\omega_n 0^+}\ln\left[-\mathcal{G}_{11}^{-1}(k)\right] &= \frac{1}{\beta}\ln\left(1 + e^{-\beta\xi_{\mathbf{k}}}\right), \\
\frac{1}{\beta}\sum_{\omega_n} e^{i\omega_n 0^+}\ln\left[-\mathcal{G}_{22}^{-1}(k)\right] &= \frac{1}{\beta}\ln\left(1 + e^{\beta\xi_{\mathbf{k}}}\right), \\
\frac{1}{\beta}\sum_{\omega_n} e^{i\omega_n 0^+}\ln\left[\det \mathcal{G}^{-1}(k)\right] &= \frac{2}{\beta}\ln\left[2\cosh\left(\frac{\beta E_{\mathbf{k}}}{2}\right)\right],
\end{aligned} \tag{7.272}$$

we finally obtain

$$\Delta\Omega = \Omega_s - \Omega_n = \mathcal{V}\frac{\Delta^2}{\lambda} - \frac{2}{\beta}\sum_{\mathbf{k}} \ln\left[\frac{\cosh\left(\frac{\beta E_{\mathbf{k}}}{2}\right)}{\cosh\left(\frac{\beta\xi_{\mathbf{k}}}{2}\right)}\right], \tag{7.273}$$

[67]The first two Matsubara sums in (7.272) are derived in Appendix 1.F. The third one can be calculated using the general method described in this appendix.

where

$$\Omega_n = -\frac{2}{\beta}\sum_{\mathbf{k}} \ln\left(1+e^{-\beta\xi_{\mathbf{k}}}\right) \tag{7.274}$$

is the grand potential in the normal phase [Eq. (1.F.14)].

At $T=0$, we find

$$\begin{aligned}\frac{\Delta\Omega(T=0)}{\mathcal{V}} &= \frac{\Delta_0^2}{\lambda} - \int_{\mathbf{k}} [E_{\mathbf{k}} - |\xi_{\mathbf{k}}|] = \frac{\Delta_0^2}{\lambda} - 2N(0)\int_0^{\omega_D} d\xi \left[(\xi^2+\Delta_0^2)^{1/2} - \xi\right] \\ &\simeq \frac{\Delta_0^2}{\lambda} - N(0)\Delta_0^2 \ln\left(\frac{2\omega_D}{\Delta_0}\right) - \frac{N(0)}{2}\Delta_0^2 \\ &= -\frac{N(0)}{2}\Delta_0^2 \end{aligned} \tag{7.275}$$

[we have used (7.263)] in the limit $\Delta_0 \ll \omega_D$. To compute the grand potential at low temperatures, we rewrite (7.273) as

$$\frac{\Delta\Omega}{\mathcal{V}} = \frac{\Delta^2}{\lambda} - \frac{1}{\beta}\int_{\mathbf{k}} \left[2\ln\left(\frac{1+e^{-\beta E_{\mathbf{k}}}}{1+e^{-\beta\xi_{\mathbf{k}}}}\right) + \beta(E_{\mathbf{k}} - \xi_{\mathbf{k}})\right]. \tag{7.276}$$

We then use

$$\begin{aligned}\frac{2}{\beta}\int_{\mathbf{k}} \ln\left(1+e^{-\beta\xi_{\mathbf{k}}}\right) &= \frac{4N(0)}{\beta}\int_0^{\omega_D} d\xi \ln\left(1+e^{-\beta\xi}\right) \simeq \frac{\pi^2}{3}N(0)T^2, \\ \int_{\mathbf{k}} (E_{\mathbf{k}} - \xi_{\mathbf{k}}) &= N(0)\Delta^2\left[\ln\left(\frac{2\omega_D}{\Delta}\right) + \frac{1}{2}\right], \end{aligned} \tag{7.277}$$

and $\Omega_n \simeq \Omega_n(T=0) - \frac{\pi^2}{3}N(0)\mathcal{V}T^2$. This gives

$$\frac{\Omega_s}{\mathcal{V}} = \frac{\Omega_n(T=0)}{\mathcal{V}} - \frac{N(0)}{2}\Delta^2 + N(0)\Delta^2\ln\left(\frac{\Delta}{\Delta_0}\right) - \frac{4N(0)}{\beta}\int_0^{\omega_D} d\xi \ln\left(1+e^{-\beta E}\right). \tag{7.278}$$

In the low-temperature limit $T \ll \Delta$, the remaining integral in (7.278) gives

$$\int_0^{\omega_D} d\xi \ln\left(1+e^{-\beta E}\right) \simeq \int_0^{\infty} d\xi\, e^{-\beta E} = \int_{\Delta}^{\infty} dE \frac{Ee^{-\beta E}}{\sqrt{E^2-\Delta^2}} \simeq \sqrt{\frac{\pi\Delta}{2\beta}}e^{-\beta\Delta}. \tag{7.279}$$

From (7.278) and (7.279) and the low-temperature expression (7.268) of the superconducting gap, we finally obtain

$$\frac{\Omega_s}{\mathcal{V}} \simeq \frac{\Omega_n(T=0)}{\mathcal{V}} - \frac{N(0)}{2}\Delta_0^2 - N(0)\sqrt{8\pi\Delta_0 T^3}\, e^{-\Delta_0/T} \tag{7.280}$$

for $T \ll \Delta$ (i.e. $T \ll T_c$).

We deduce the constant volume specific heat

$$C_s = -T\frac{\partial^2\Omega_s}{\partial T^2} \simeq \mathcal{V}N(0)\sqrt{\frac{8\pi\Delta_0^5}{T^3}}\, e^{-\Delta_0/T} \tag{7.281}$$

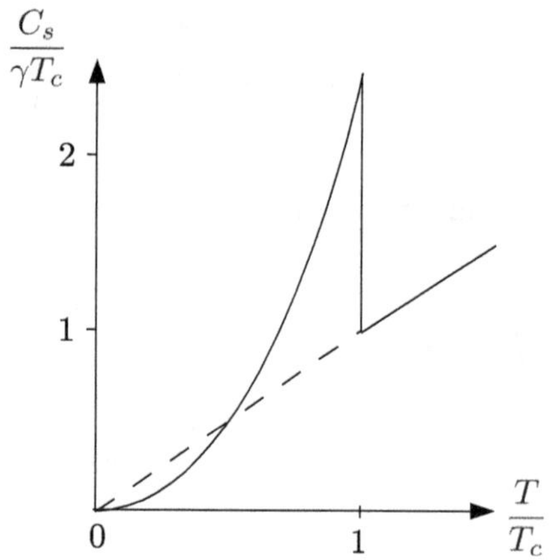

Fig. 7.15 Temperature dependence of the specific heat in a superconductor. The dashed line represents the specific heat $C_n = \gamma T$ in the normal phase.

at low temperatures. Because of the gap in the one-particle excitation spectrum, C_s is exponentially small when $T \ll T_c$ (Fig. 7.15). The specific heat near the transition temperature is discussed in the following section.

7.5.3 *Ginzburg–Landau theory*

In the preceding section, we have derived expressions of the thermodynamic quantities at low temperatures $T \ll T_c$. Near T_c, the order parameter is small, and the grand potential can be expanded in powers of Δ/T. Although we could simply expand the general expression of Ω_s [Eq. (7.273)], it is interesting to compute Ω_s using perturbation theory, since the latter can also be used in more complicated cases, e.g. when the order parameter is not uniform. More generally, the Ginzburg–Landau theory does not need to rely on a microscopic model but merely on a phenomenological expression of the grand potential derived from general principles (symmetries, etc.).

To fourth order in the superconducting order parameter, we obtain the Ginzburg–Landau expansion of the free energy

$$\frac{F_s - F_n}{\mathcal{V}} = A\Delta^2 + \frac{B}{2}\Delta^4. \tag{7.282}$$

To follow the literature, we refer to the thermodynamic potential as the free energy rather than the grand potential.

The coefficients

$$\begin{aligned} A &= \frac{1}{\lambda} - \chi_0 = \frac{1}{\lambda} - \frac{1}{\beta\mathcal{V}} \sum_k G_{0,\uparrow}(k) G_{0,\downarrow}(-k), \\ B &= \frac{1}{\beta\mathcal{V}} \sum_k [G_{0,\uparrow}(k) G_{0,\downarrow}(-k)]^2 \end{aligned} \tag{7.283}$$

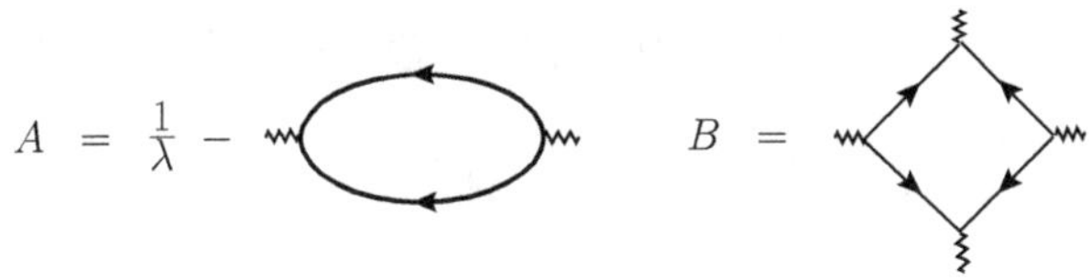

Fig. 7.16 Diagrammatic representation of the Ginzburg–Landau expansion (7.282) of the free energy. The zigzag lines represent the order parameter Δ.

are shown diagrammatically in Fig. 7.16. For $T \simeq T_c$,

$$\begin{aligned} A &= \frac{1}{\lambda} - N(0)\ln\left(\frac{2\gamma\omega_D}{\pi T}\right) \simeq N(0)\frac{T-T_c}{T_c}, \\ B &\simeq \frac{N(0)}{\beta_c}\sum_{\omega_n}\int_{-\infty}^{\infty}\frac{d\xi}{(\omega_n^2+\xi^2)^2} = \frac{N(0)\pi}{2\beta_c}\sum_{\omega_n}\frac{1}{|\omega_n|^3} = N(0)\frac{7\zeta(3)}{8\pi^2T_c^2}, \end{aligned} \tag{7.284}$$

where $\zeta(z)$ is Riemann Zeta function and $\zeta(3) \simeq 1.202$.[68] The value of Δ is obtained from the minimum of the free energy: $\partial F_s/\partial\Delta = 0$. Above T_c ($A > 0$), the gap vanishes, whereas for $T < T_c$ it takes the finite value

$$\Delta = \left(-\frac{A}{B}\right)^{1/2} = \left[\frac{(T_c - T)}{T_c}\frac{8\pi^2T_c^2}{7\zeta(3)}\right]^{1/2}, \tag{7.285}$$

and the free energy is given by

$$\frac{F_s - F_n}{\mathcal{V}} = -\frac{A^2}{2B} = -\frac{4\pi^2}{7\zeta(3)}N(0)(T_c - T)^2. \tag{7.286}$$

From (7.286), we deduce the specific heat below the transition temperature,

$$C_s - C_n = -T\frac{\partial^2(F_s - F_n)}{\partial T^2} = \mathcal{V}\frac{8\pi^2}{7\zeta(3)}N(0)T, \tag{7.287}$$

where $C_n = \mathcal{V}\frac{2\pi^2}{3}N(0)T = \gamma T$ is the specific heat in the normal phase. The specific heat is discontinuous at the transition,

$$\left.\frac{C_s - C_n}{C_n}\right|_{T=T_c^-} = \frac{12}{7\zeta(3)} \simeq 1.43, \tag{7.288}$$

as shown in Fig. 7.15.

When the order parameter is slowly varying in space, we expect a gradient term $\propto [\boldsymbol{\nabla}\Delta(\mathbf{r})]^2$ in the Ginzburg–Landau expansion. This term comes from the $\mathbf{q}$ dependence of

[68] $\sum_{\omega_n}|\omega_n|^{-3}$ is calculated in Appendix 1.F.

the static pairing susceptibility $\chi_0(\mathbf{q})$. An elementary calculation gives

$$\begin{aligned}\chi_0(\mathbf{q}) &= \chi_0 + \frac{1}{\beta\mathcal{V}} \sum_{\mathbf{k},\omega_n} \left[\frac{\mathbf{q}^2/2m}{(i\omega_n - \xi_\mathbf{k})(i\omega_n + \xi_\mathbf{k})^2} - \frac{(\mathbf{k}\cdot\mathbf{q}/m)^2}{(i\omega_n - \xi_\mathbf{k})(i\omega_n + \xi_\mathbf{k})^3} \right] + \mathcal{O}(|\mathbf{q}|^4) \\ &= \chi_0 - \frac{7\zeta(3)}{48\pi^2 T_c^2} N(0) v_F^2 \mathbf{q}^2 + \mathcal{O}(|\mathbf{q}|^4), \end{aligned} \tag{7.289}$$

where $v_F = k_F/m$ is the Fermi velocity. We can therefore generalize (7.282) to non-uniform order parameters by considering the Ginzburg–Landau free energy

$$F_s - F_n = \int d^3r \left[A|\Delta(\mathbf{r})|^2 + D|\boldsymbol{\nabla}\Delta(\mathbf{r})|^2 + \frac{B}{2}|\Delta(\mathbf{r})|^4 \right], \tag{7.290}$$

where

$$D = \frac{7\zeta(3)}{48\pi^2 T_c^2} N(0) v_F^2. \tag{7.291}$$

For the sake of generality, we now allow the order parameter to be complex. Note that Eq. (7.290) is the most general truncation of the free energy to fourth order in Δ and leading order in derivative compatible with the U(1) symmetry.

7.5.3.1 *Free energy in presence of a magnetic field*

In presence of a static magnetic field, the requirement of gauge invariance [Eqs. (7.96)–(7.97)] leads us to replace $-i\boldsymbol{\nabla}$ by $-i\boldsymbol{\nabla} - e^*\mathbf{A}(\mathbf{r})$, where $\mathbf{A}$ is the vector potential and $e^* = 2e$ the charge of the carriers (here, Cooper pairs),

$$F_s = F_{n0} + \int d^3r \left\{ A|\Delta(\mathbf{r})|^2 + D\big|[-i\boldsymbol{\nabla} - e^*\mathbf{A}(\mathbf{r})]\Delta(\mathbf{r})\big|^2 + \frac{B}{2}|\Delta(\mathbf{r})|^4 + \frac{\mathbf{B}^2(\mathbf{r})}{2\mu_0} \right\}. \tag{7.292}$$

F_{n0} is the free energy in the normal phase for $\mathbf{B} = 0$ and $\mathbf{B}^2(\mathbf{r})/2\mu_0$ the magnetic energy density ($\mathbf{B}(\mathbf{r}) = \boldsymbol{\nabla} \times \mathbf{A}(\mathbf{r})$ is the flux density). Here, we neglect the weak paramagnetism or diamagnetism of the normal phase.[69] It is conventional to take a normalization of $\Delta(\mathbf{r})$ so that the coefficient of the gradient term be $1/2m^*$ with $m^* = 2m$ the mass of the Cooper pairs. We therefore introduce

$$\Psi(\mathbf{r}) = \sqrt{2m^* D}\,\Delta(\mathbf{r}) = \left(\frac{7\zeta(3) n}{8\pi^2 T_c^2} \right)^{1/2} \Delta(\mathbf{r}) \tag{7.293}$$

(we have used $N(0) = 3n/2mv_F^2$ in three dimensions) and rewrite the free energy as

$$\begin{aligned} F_s = F_{n0} + \int d^3r \bigg\{ & \tilde{A}|\Psi(\mathbf{r})|^2 + \frac{1}{2m^*}\big|[-i\boldsymbol{\nabla} - e^*\mathbf{A}(\mathbf{r})]\Psi(\mathbf{r})\big|^2 \\ & + \frac{\tilde{B}}{2}|\Psi(\mathbf{r})|^4 + \frac{\mathbf{B}^2(\mathbf{r})}{2\mu_0} \bigg\}, \end{aligned} \tag{7.294}$$

[69] In the microscopic model where the normal phase is described by a non-interacting fermion gas, the normal phase would be (weakly) paramagnetic since Pauli spin paramagnetism dominates over Landau orbital diamagnetism.

where

$$\tilde{A} = \frac{A}{2m^*D} = \frac{6\pi^2}{7\zeta(3)} \frac{T_c(T-T_c)}{\epsilon_F},$$
$$\tilde{B} = \frac{B}{(2m^*D)^2} = \frac{6\pi^2}{7\zeta(3)} \frac{T_c^2}{n\epsilon_F}. \tag{7.295}$$

The free energy F_s can also be derived microscopically from the computation of the diagrams shown in Fig. 7.16 using the expression of the one-particle Green function in a magnetic field. The latter can always be written as [46]

$$G(\mathbf{r},\mathbf{r}';i\omega_n) = e^{ie\int_{\mathbf{r}}^{\mathbf{r}'} d\mathbf{l}\cdot\mathbf{A}} \bar{G}(\mathbf{r}-\mathbf{r}';i\omega_n). \tag{7.296}$$

The microscopic derivation of (7.294) relies on a (semiclassical) eikonal approximation where the translation-invariant function $\bar{G}$ is replaced by the zero-field Green function [12,13]. The function $\bar{G}$ encodes the quantum effects of the field, i.e. the Landau level quantization in an isotropic system. In clean systems, these effects become important at low temperatures when the cyclotron frequency satisfies $\omega_c \gg T$. Some interesting consequences of the quantum effects of the magnetic field have been discussed in the literature [47, 48].

For a superconductor in an applied external field $\mathbf{H}(\mathbf{r})$, the thermodynamic potential to be considered is the Gibbs free energy

$$G(T,\mathbf{H}) = F(T,\mathbf{B}) - \int d^3r\, \mathbf{H}(\mathbf{r})\cdot\mathbf{B}(\mathbf{r}). \tag{7.297}$$

At equilibrium, G must be minimal with respect to variations of $\Psi(\mathbf{r})$ and $\mathbf{A}(\mathbf{r})$: $\delta G/\delta\Psi(\mathbf{r}) = \delta G/\delta\Psi^*(\mathbf{r}) = \delta G/\delta\mathbf{A}(\mathbf{r}) = 0$. An elementary calculation shows that $\Psi(\mathbf{r})$ is solution of the nonlinear Schrödinger equation

$$\frac{1}{2m^*}[-i\boldsymbol{\nabla} - e^*\mathbf{A}(\mathbf{r})]^2\Psi(\mathbf{r}) + \tilde{A}\Psi(\mathbf{r}) + \tilde{B}|\Psi(\mathbf{r})|^2\Psi(\mathbf{r}) = 0 \tag{7.298}$$

while the flux density satisfies

$$\boldsymbol{\nabla}\times[\mathbf{B}(\mathbf{r}) - \mu_0\mathbf{H}(\mathbf{r})] = \boldsymbol{\nabla}\times\mathbf{B}(\mathbf{r}) = \mu_0\mathbf{J}(\mathbf{r}) \tag{7.299}$$

($\boldsymbol{\nabla}\times\mathbf{H}(\mathbf{r}) = \mathbf{J}_{\text{ext}}(\mathbf{r}) = 0$ in the superconductor), where[70]

$$\mathbf{J}(\mathbf{r}) = \frac{-ie^*}{2m^*}[\Psi^*(\mathbf{r})\boldsymbol{\nabla}\Psi(\mathbf{r}) - \boldsymbol{\nabla}\Psi^*(\mathbf{r})\Psi(\mathbf{r})] - \frac{e^{*2}}{m^*}|\Psi(\mathbf{r})|^2\mathbf{A}(\mathbf{r}) \tag{7.300}$$

is the supercurrent density.[71] In deriving (7.298) and (7.299), we have made use of the vanishing of surface integrals due to the usual boundary conditions

$$\mathbf{J}(\mathbf{r})\cdot\mathbf{n} = 0,$$
$$[\mathbf{B}(\mathbf{r}) - \mu_0\mathbf{H}(\mathbf{r})]\times\mathbf{n} = 0 \tag{7.301}$$

[70] As in Chapter 3, we denote by **J** the total current and by **j** the paramagnetic current.

[71] The supercurrent can also be written as $\mathbf{J}(\mathbf{r}) = -\frac{\delta F_m}{\delta\mathbf{A}(\mathbf{r})}$, where $F_m = F - \frac{1}{2\mu_0}\int d^3r\, \mathbf{B}^2(\mathbf{r})$ is the free energy of the matter.

($\mathbf{n}$ is a vector normal to the surface of the superconductor) ensuring that the current normal to the surface vanishes and that the tangential component of the magnetic field is continuous across the surface.

The superfluid density $n_s^*(\mathbf{r}) = n_s(\mathbf{r})/2$ is naturally introduced by identifying the phase stiffness with respect to spatial variations (i.e. the coefficient of $(\boldsymbol{\nabla}\theta)^2$ in (7.294)) with $n_s^*(\mathbf{r})/2m^*$. This yields

$$\Psi(\mathbf{r}) = \sqrt{n_s^*(\mathbf{r})}e^{i\theta(\mathbf{r})}, \tag{7.302}$$

and the current takes the form

$$\mathbf{J}(\mathbf{r}) = \frac{e^* n_s^*(\mathbf{r})}{m^*}[\boldsymbol{\nabla}\theta(\mathbf{r}) - e^*\mathbf{A}(\mathbf{r})], \tag{7.303}$$

which is identical to the expression (7.106) obtained in Sec. 7.2.4. The equation for the transverse part,

$$\mathbf{J}_\perp(\mathbf{r}) = -\frac{e^{*2}}{m^*} n_s^*(\mathbf{r})\mathbf{A}_\perp(\mathbf{r}), \tag{7.304}$$

is the London equation introduced in Sec. 3.4.5. When the superfluid density $n_s^*(\mathbf{r}) = n_s^*$ is uniform, Eqs. (7.299) and (7.304) lead to

$$\boldsymbol{\nabla}^2\mathbf{B}(\mathbf{r}) = \lambda^{-2}\mathbf{B}(\mathbf{r}), \tag{7.305}$$

where

$$\lambda = c\left(\frac{\epsilon_0 m^*}{n_s^* e^{*2}}\right)^{1/2} = c\left(\frac{\epsilon_0 m}{n_s e^2}\right)^{1/2} \tag{7.306}$$

is the London penetration depth. $n_s = 2n_s^*$ denotes the fermion superfluid density. As discussed in Sec. 3.4.5, Eq. (7.305) implies the existence of a supercurrent in a surface layer of width λ that expels the magnetic flux out of the bulk of the superconductor (Meissner effect).

There is an obvious similarity between the Gross–Pitaevskii theory of Bose superfluids and the Ginzburg–Landau theory of superconductors. However, these two approaches differ in some important respects. First, the Ginzburg–Landau expansion is valid (at least quantitatively) only near T_c, whereas the Gross–Pitaevskii equation describes the superfluid in the low-temperature limit where most of the particles are in the condensate. Second, there is a fundamental difference between the order parameters in both theories. In a Bose superfluid, the order parameter $\psi(x) = \sqrt{n(x)}e^{i\theta(x)}$ involves the density $n(x)$ and its conjugate variable $\theta(x)$. By contrast, the order parameter $\Delta(x) = |\Delta(x)|e^{i\theta(x)}$ of a superconductor does not involve the conjugate variable (which is the density) of the phase of the order parameter. In fact, the density was never considered explicitly in our approach to superconductivity. We shall come back to this issue in Sec. 7.7.2.

The Ginzburg–Landau theory, based on the free energy (7.292), is much easier to solve than the full BCS theory in the presence of a magnetic field. As a phenomenological approach, it has also a wider range of validity since it does not rely on a particular mechanism for the pairing between fermions. There are a number of excellent textbooks

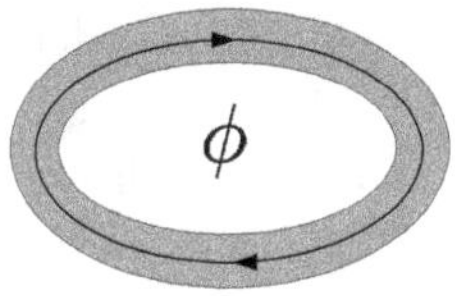

Fig. 7.17 Superconducting ring in a magnetic field (ϕ denotes the magnetic flux through the ring) and integration contour $\mathcal{C}$ used in Eq. (7.307).

discussing the various applications of the Ginzburg–Landau theory, e.g. Refs. [30–32]. In the following, we give a brief introduction to vortices and type I/II superconductors.

7.5.3.2 *Flux quantization*

Let us consider a superconducting ring in a magnetic field as shown in Fig. 7.17. The integration of the supercurrent (7.303) along a closed contour $\mathcal{C}$ lying in the superconductor and enclosing the center of the ring gives

$$\oint_{\mathcal{C}} d\mathbf{l} \cdot \mathbf{J}(\mathbf{r}) = \oint_{\mathcal{C}} d\mathbf{l} \cdot \frac{n_s^*(\mathbf{r})}{m^*} [\boldsymbol{\nabla}\theta(\mathbf{r}) - e^* \mathbf{A}(\mathbf{r})]. \tag{7.307}$$

In the Meissner state, the current vanishes in the bulk of the superconductor and the superfluid density is uniform except in a surface layer whose width is determined by the Ginzburg–Landau coherence length ξ (see the following). If the contour $\mathcal{C}$ lies sufficiently deep inside the superconductor, Eq. (7.307) implies

$$\oint_{\mathcal{C}} d\mathbf{l} \cdot [\boldsymbol{\nabla}\theta(\mathbf{r}) - e^* \mathbf{A}(\mathbf{r})] = 0, \tag{7.308}$$

so that

$$\phi = \int_{\Sigma} d\boldsymbol{\Sigma} \cdot \mathbf{B}(\mathbf{r}) = \oint_{\mathcal{C}} d\mathbf{l} \cdot \mathbf{A}(\mathbf{r}) = \frac{\hbar}{e^*} \oint_{\mathcal{C}} d\mathbf{l} \cdot \boldsymbol{\nabla}\theta(\mathbf{r}) = n\frac{h}{e^*} \qquad (n \text{ integer}) \tag{7.309}$$

(Σ is the surface spanned by the contour), where the last result follows from the single valuedness of the order parameter $\Psi(\mathbf{r})$. We have restored Planck's constant $h = 2\pi\hbar$. The magnetic flux ϕ through the ring is quantized in units of the flux quantum $\phi_0 = h/e^* = h/2e$ associated with the charge $e^* = 2e$.

7.5.3.3 *Type I and type II superconductors*

From the free energy (7.292), we can extract two characteristic length scales. The first one is the London penetration depth λ [Eq. (7.306)]. The second one, the Ginzburg–Landau coherence length

$$\xi = \frac{1}{|2m^* \tilde{A}|^{1/2}} = \left(\frac{7\zeta(3)}{12\pi^2} \frac{\epsilon_F}{m^* T_c |T - T_c|} \right)^{1/2}, \tag{7.310}$$

is obtained by comparing the quadratic term with the gradient term. At $T = 0$, it coincides with the BCS coherence length $\xi_0 = v_F/\pi\Delta_0$ up to a multiplicative constant of order unity.

ξ determines the minimal length over which the order parameter can vary. As in the case of a Bose superfluid (Sec. 7.3.1), this can be illustrated by considering a superconductor confined in the region $z \geq 0$ (with $\mathbf{B} = 0$). From the equation (7.298) satisfied by $\Psi(z)$ with the boundary conditions $\Psi(z=0) = 0$ and $\Psi(z \to \infty) = n_s^* = -\tilde{A}/\tilde{B}$, one easily deduces $\Psi(z) = \sqrt{n_s^*}\tanh(z/2\xi)$. We found the same solution (with ξ the healing length) from the Gross–Pitaevskii equation for a Bose superfluid confined to the half space $z \geq 0$. (We do not detail this calculation which is similar to the one performed in Sec. 7.3.1.)

Let us first compare the Gibbs free energies in the normal and superconducting phases, assuming that the magnetic field is expelled from the bulk in the latter. In a uniform external field $\mathbf{H} = H\hat{\mathbf{z}}$, we have

$$\begin{aligned} G_s(T,H) &= F_s(T,0) \\ G_n(T,H) &= F_n(T,B) - \mathcal{V}BH = F_n(T,0) + \mathcal{V}\frac{B^2}{2\mu_0} - \mathcal{V}BH \\ &\simeq F_n(T,0) - \mathcal{V}\frac{B^2}{2\mu_0} \end{aligned} \tag{7.311}$$

since $\mathbf{B} = 0$ in the superconducting state while $\mathbf{B} = B\hat{\mathbf{z}} \simeq \mu_0 H$ in the metallic state. The Gibbs free energy is the same in both phases, $G_s(T,H_c) = G_n(T,H_c)$, when the magnetic energy is equal to the opposite of the condensation energy,

$$\frac{1}{2}\mu_0 H_c^2 \mathcal{V} = F_n(T,0) - F_s(T,0). \tag{7.312}$$

Equation (7.312) defines the thermodynamic critical field H_c.

Let us now consider the boundary between metallic and superconducting domains for $H = H_c$ (Fig. 7.18). Both phases have the same Gibbs free energy densities. However, because ξ and λ are non-zero, the interface between the two domains gives rise to a surface energy that can be roughly estimated as $\xi\mu_0 H_c^2/2$ (the energy cost due to the suppression of $|\Psi|$ near the interface) plus $-\lambda\mu_0 H_c^2/2$ (the gain in magnetic energy due to the penetration of the flux near the interface). Thus, the surface energy is expected to be negative for $\xi \lesssim \lambda$ and positive for $\xi \gtrsim \lambda$. The exact calculation based on the Ginzburg–Landau free

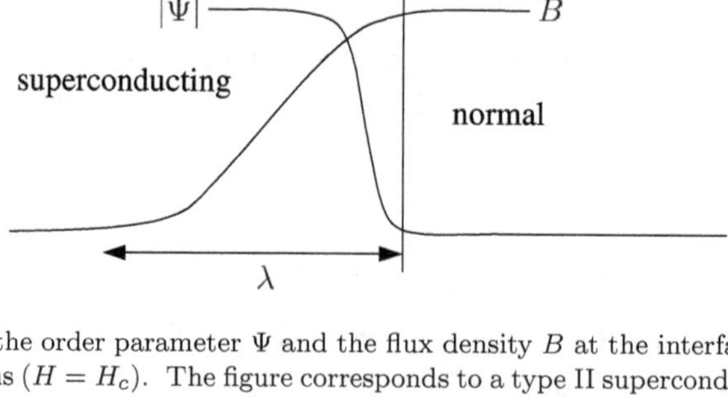

Fig. 7.18 Variation of the order parameter Ψ and the flux density B at the interface between metallic and superconducting domains ($H = H_c$). The figure corresponds to a type II superconductor ($\kappa = \lambda/\xi \geq 1/\sqrt{2}$).

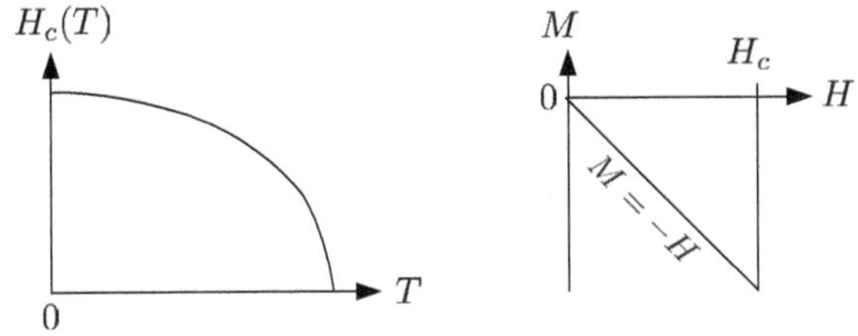

Fig. 7.19 Phase diagram and magnetization $M(H)$ of a type I superconductor.

energy shows that the critical ratio of the Ginzburg–Landau parameter $\kappa = \lambda/\xi$ is $1/\sqrt{2}$. In type I superconductors ($\kappa < 1/\sqrt{2}$), the magnetic flux is always expelled from the bulk. Superconductivity and perfect diamagnetism ($B = 0$ or $M = B/\mu_0 - H = -H$) persist up to $H \leq H_c$. The transition to the metallic state at $H = H_c$ is first-order with a discontinuous jump of the order parameter (Fig. 7.19).

In contrast, in type II superconductors the magnetic flux can penetrate in the form of quantized flux lines (vortices). Vortices in superconductors bear some analogies with those in Bose superfluids (Sec. 7.3.2), with the amplitude of the order parameter vanishing at the center of a core region where the phase is singular. The field H_{c1} at which vortices appear in the system is defined by

$$G_s(T, H_{c1})\Big|_{\text{no flux}} = G_s(T, H_{c1})\Big|_{1\text{ vortex}}, \tag{7.313}$$

or, since $G(T, H) = F(T, B) - H \int d^3r\, B(\mathbf{r})$,

$$\begin{aligned} F_s(T, 0) &= F_s(T, 0) + F_{\text{v}} - H_{c1} \int d^3r\, B(\mathbf{r}) \\ &= F_s(T, 0) + F_{\text{v}} - H_{c1} L\phi, \end{aligned} \tag{7.314}$$

where F_{v} is the change in free energy due to the vortex. L is the length of the vortex line in the sample. The enclosed magnetic flux ϕ is quantized in unit of the flux quantum ϕ_0. From (7.314), we deduce

$$H_{c1} = \frac{F_{\text{v}}}{L\phi}. \tag{7.315}$$

Using cylindrical coordinates $\mathbf{r} = (\rho, \varphi, z)$, the order parameter corresponding to a vortex line parallel to the z axis can be written as

$$\Psi(\mathbf{r}) = f(\rho) e^{i\theta(\rho, \varphi)}, \tag{7.316}$$

where the amplitude $f(\rho)$ vanishes at the vortex center ($\rho = 0$). The size of the vortex core is determined by the Ginzburg–Landau coherence length ξ. We consider the high κ approximation where the vortex core is small and its contribution to the free energy

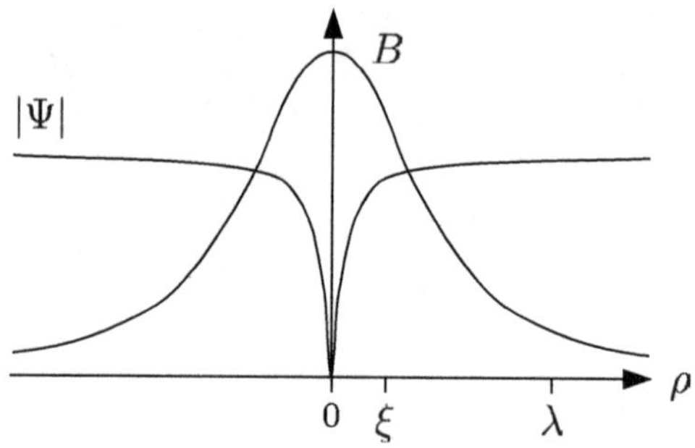

Fig. 7.20 Structure of an isolated vortex in a type II superconductor with $\kappa = \lambda/\xi \gg 1$.

negligible (Fig. 7.20). For $\rho \gtrsim \xi$, the amplitude $f(\rho) \simeq \sqrt{n_s^*}$ is nearly constant. Neglecting the core, the vortex free energy reduces to the kinetic and magnetic contributions

$$\begin{aligned} F_{\mathrm{v}} &= \int d^3r \left\{ \frac{n_s^*}{2m^*}[\boldsymbol{\nabla}\theta(\mathbf{r}) - e^*\mathbf{A}(\mathbf{r})]^2 + \frac{\mathbf{B}^2(\mathbf{r})}{2\mu_0} \right\} \\ &= \frac{1}{2\mu_0}\int d^3r \left[\mathbf{B}^2(\mathbf{r}) + \lambda^2|\boldsymbol{\nabla}\times\mathbf{B}(\mathbf{r})|^2\right] \end{aligned} \tag{7.317}$$

using $\boldsymbol{\nabla}\times\mathbf{B} = \mu_0\mathbf{J}$ and (7.303).

The magnetic flux density is determined by (7.305) with $\mathbf{B}(\mathbf{r}) = B(\rho)\hat{\mathbf{z}}$, i.e.

$$\frac{1}{\rho}\frac{\partial}{\partial\rho}\left(\rho\frac{\partial B(\rho)}{\partial\rho}\right) = \lambda^{-2}B(\rho). \tag{7.318}$$

This equation has the exact solution

$$B(\rho) = \frac{\phi}{2\pi\lambda^2}K_0\left(\frac{\rho}{\lambda}\right), \tag{7.319}$$

where K_0 is the zeroth-order modified Bessel function of the second kind. The arbitrary multiplicative constant in (7.319) is fixed by the flux enclosed by the vortex,[72]

$$\int d^2\rho\, B(\rho) = \frac{\phi}{\lambda^2}\int_0^\infty d\rho\,\rho K_0\left(\frac{\rho}{\lambda}\right) = \phi. \tag{7.320}$$

The two limiting forms of (7.319) are

$$B(\rho) \simeq \begin{cases} \dfrac{\phi}{2\pi\lambda^2}\left(\dfrac{\pi\lambda}{2\rho}\right)^{1/2} e^{-\rho/\lambda} & (\rho \gg \lambda), \\ \dfrac{\phi}{2\pi\lambda^2}\left[\ln\left(\dfrac{\lambda}{\rho}\right) + 0.12\right] & (\xi \lesssim \rho \ll \lambda). \end{cases} \tag{7.321}$$

[72]This follows from $\int_0^\infty dx\, x K_0(ax) = 1/a^2$.

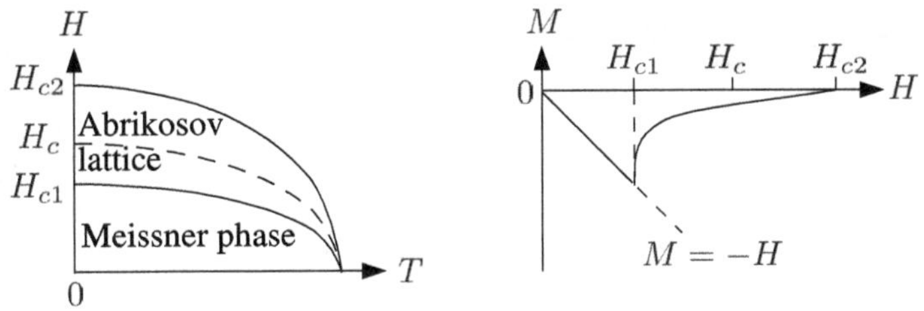

Fig. 7.21 Phase diagram and magnetization $M(H)$ of a type II superconductor.

Rewriting the vortex free energy (7.317) in cylindrical coordinates and integrating by part, we obtain

$$\begin{aligned} F_{\mathrm{v}} &\simeq \frac{L\pi}{\mu_0}\int_{\xi}^{\lambda} d\rho\, \rho\, [B^2 + \lambda^2(\partial_\rho B)^2]\} \\ &= \frac{L\pi}{\mu_0}\int_{\xi}^{\lambda} d\rho\, [\rho B^2 - \lambda^2\left(\partial_\rho B + \rho\partial_\rho^2 B\right) B] + \frac{L\pi\lambda^2}{\mu_0}\rho B\partial_\rho B\Big|_{\xi}^{\lambda}. \end{aligned} \tag{7.322}$$

Using (7.318), the first term in the rhs is seen to vanish, and we are left with

$$F_{\mathrm{v}} \simeq L\frac{\phi^2}{4\pi\mu_0\lambda^2}\ln\left(\frac{\lambda}{\xi}\right). \tag{7.323}$$

Since F_{v} is proportional to the square of the magnetic flux ϕ, the Gibbs free energy is minimized when the flux density penetrates the sample in the form of vortices carrying a single flux quantum ϕ_0. We can substitute (7.323) back into (7.315) to obtain

$$H_{c1} = \frac{\phi_0}{4\pi\mu_0\lambda^2}\ln\left(\frac{\lambda}{\xi}\right). \tag{7.324}$$

For fields higher than H_{c1}, a finite density of vortices appears in the system. Because of their mutual interactions, these vortices form a triangular lattice (Abrikosov lattice). The system undergoes a second-order phase transition to the metallic phase at the upper critical field H_{c2} (Fig. 7.21). Close to H_{c2}, the amplitude of the order parameter $\Psi(\mathbf{r})$ and the magnetization $M = B/\mu_0 - H$ are small; the Ginzburg–Landau equations (7.298) and (7.299) can be linearized thus allowing a complete description of the Abrikosov vortex lattice (see, for instance, Refs. [30–32]).

7.6 Collective Modes in Superconductors

In this section, we study the fluctuations of the superconducting order parameter about its mean-field value at zero temperature (some of the results will be valid at finite temperature).

The action is given by (7.228) and (7.237),

$$S[\psi^*,\psi,\Delta^*,\Delta] = \int_0^\beta d\tau \int d^3r \biggl\{ \frac{\Delta(x)^2}{\lambda} + \sum_\sigma \psi^*_\sigma(x) \left[\partial_\tau - \mu - \frac{\boldsymbol{\nabla}^2}{2m} \right] \psi_\sigma(x) - [\Delta^*(x)\psi_\downarrow(x)\psi_\uparrow(x) + \text{c.c.}] \biggr\}, \tag{7.325}$$

where the cutoff ω_D in momentum space is implicit. Given the spontaneous U(1) symmetry breaking in the BCS state, we expect phase and amplitude fluctuations to exhibit different behaviors. In particular, phase fluctuations should be gapless by virtue of the Goldstone theorem. This suggests to write the fluctuating pairing field as

$$\Delta(x) = [\Delta_0 + \delta|\Delta(x)|]\, e^{i\theta(x)} \tag{7.326}$$

rather than $\Delta(x) = \Delta_0 + \delta\Delta(x)$ (with $\delta\Delta(x)$ complex). The zero-temperature order parameter Δ_0 is assumed real and positive. It is then convenient to perform the gauge transformation

$$\psi_\sigma(x) \to \psi_\sigma(x) e^{\frac{i}{2}\theta(x)}, \tag{7.327}$$

whereby the action becomes

$$\begin{aligned} S[\psi^*,\psi,\delta|\Delta|,\theta] = \int_0^\beta d\tau \int d^3r \biggl\{ &\sum_\sigma \psi^*_\sigma(x) \Bigl[\partial_\tau - \mu - \frac{\boldsymbol{\nabla}^2}{2m} + \frac{i}{2}\dot\theta(x) + \frac{[\boldsymbol{\nabla}\theta(x)]^2}{8m} \\ &- \frac{i}{4m} \boldsymbol{\nabla}\theta(x) \cdot \overleftrightarrow{\boldsymbol{\nabla}} \Bigr] \psi_\sigma(x) + \frac{[\Delta_0 + \delta|\Delta(x)|]^2}{\lambda} \\ &- [\Delta_0 + \delta|\Delta(x)|][\psi_\downarrow(x)\psi_\uparrow(x) + \text{c.c.}] \biggr\}, \end{aligned} \tag{7.328}$$

where the operator $\overleftrightarrow{\boldsymbol{\nabla}} = \overrightarrow{\boldsymbol{\nabla}} - \overleftarrow{\boldsymbol{\nabla}}$ acts only on the fermion field ψ (not on θ). In terms of the Nambu spinors

$$\Psi(x) = \begin{pmatrix} \psi_\uparrow(x) \\ \psi^*_\downarrow(x) \end{pmatrix}, \qquad \Psi^\dagger(x) = (\psi^*_\uparrow(x), \psi_\downarrow(x)), \tag{7.329}$$

we have

$$\begin{aligned} S[\psi^*,\psi,\delta|\Delta|,\theta] = S_{\text{MF}}[\psi^*,\psi] + \int_0^\beta d\tau \int d^3r \biggl\{ &\left[\frac{i}{2}\dot\theta(x) + \frac{[\boldsymbol{\nabla}\theta(x)]^2}{8m} \right] n(x) \\ &+ \frac{\boldsymbol{\nabla}\theta(x)}{2} \cdot \mathbf{j}(x) - \delta|\Delta(x)| j_\Delta(x) + \frac{\delta|\Delta(x)|^2 + 2\Delta_0\delta|\Delta(x)|}{\lambda} \biggr\}, \end{aligned} \tag{7.330}$$

where S_{MF} is the mean-field action (7.240) and

$$\begin{aligned} n(x) &= \Psi^\dagger(x)\tau^z\Psi(x) = \sum_\sigma \psi^*_\sigma(x)\psi_\sigma(x), \\ \mathbf{j}(x) &= -\frac{i}{2m}\Psi^\dagger(x)\overleftrightarrow{\nabla}\Psi(x) = -\frac{i}{2m}\sum_\sigma \psi^*_\sigma(x)\overleftrightarrow{\nabla}\psi_\sigma(x), \\ j_\Delta(x) &= \Psi^\dagger(x)\tau^x\Psi(x) = \psi_\downarrow(x)\psi_\uparrow(x) + \text{c.c.} \end{aligned} \tag{7.331}$$

τ^x, τ^y and τ^z are Pauli matrices acting on Nambu space.

The effective action $S[\delta|\Delta|,\theta]$ governing the fluctuations of the order parameter is obtained by integrating out the fermion field within a cumulant expansion. Assuming the fluctuations to be small, it is sufficient to obtain the action to second order in $\delta|\Delta|$ and θ, which is achieved by retaining only the first- and second-order cumulants. Writing the action (7.330) as

$$\begin{aligned} S[\psi^*,\psi,\delta|\Delta|,\theta] &= S_{\text{MF}}[\psi^*,\psi] + S_\theta[\psi^*,\psi,\theta] + S_{|\Delta|}[\psi^*,\psi,\delta|\Delta|] \\ &\quad + \frac{1}{\lambda}\int_0^\beta d\tau \int d^3r\, \left[2\Delta_0\delta|\Delta(x)| + \delta|\Delta(x)|^2\right] \end{aligned} \tag{7.332}$$

(with obvious notations), we obtain

$$\begin{aligned} Z &= \int \mathcal{D}[\psi^*,\psi,\delta|\Delta|,\theta]\, e^{-S[\psi^*,\psi,\delta|\Delta|,\theta]} \\ &= Z_{\text{MF}} \int \mathcal{D}[\delta|\Delta|,\theta]\, e^{-S[\theta,\delta|\Delta|]}, \end{aligned} \tag{7.333}$$

where

$$\begin{aligned} S[\theta,\delta|\Delta|] &= \frac{1}{\lambda}\int_0^\beta d\tau \int d^3r\, \left[2\Delta_0\delta|\Delta(x)| + \delta|\Delta(x)|^2\right] \\ &\quad + \langle S_\theta + S_{|\Delta|}\rangle - \frac{1}{2}\langle (S_\theta + S_{|\Delta|})^2\rangle_c. \end{aligned} \tag{7.334}$$

The average $\langle\cdots\rangle$ is taken with the BCS action S_{MF}. The first-order cumulants are readily calculated,

$$\begin{aligned} \langle S_\theta\rangle &= \int_0^\beta d\tau \int d^3r \left[\frac{i}{2}\dot\theta(x) + \frac{[\boldsymbol{\nabla}\theta(x)]^2}{8m}\right] n, \\ \langle S_{|\Delta|}\rangle &= -\frac{2\Delta_0}{\lambda}\int_0^\beta d\tau \int d^3r\, \delta|\Delta(x)|, \end{aligned} \tag{7.335}$$

using $\langle n(x)\rangle = n$ (mean density of particles) and $\langle j_\Delta(x)\rangle = 2\Delta_0/\lambda$. As expected, since we expand about the mean-field (saddle-point) solution, the terms linear in $\delta|\Delta|$ cancel in (7.334). The second-order cumulants give

$$\begin{aligned}\langle S_\theta^2\rangle_c = & -\frac{1}{4}\int_0^\beta d\tau d\tau' \int d^3r d^3r'\, \dot\theta(x)\Pi_{nn}(x,x')\dot\theta(x') \\ & + \frac{i}{2}\int_0^\beta d\tau d\tau' \int d^3r d^3r' \sum_\mu \dot\theta(x)\Pi_{nj_\mu}(x,x')\partial_\mu\theta(x') \\ & + \frac{1}{4}\int_0^\beta d\tau d\tau' \int d^3r d^3r' \sum_{\mu,\mu'} \partial_\mu\theta(x)\Pi_{j_\mu j_{\mu'}}(x,x')\partial_{\mu'}\theta(x'), \\ \langle S_{|\Delta|}^2\rangle_c = & \int_0^\beta d\tau d\tau' \int d^3r d^3r'\, \delta|\Delta(x)|\Pi_{j_\Delta j_\Delta}(x,x')\delta|\Delta(x')|, \\ \langle S_\theta S_{|\Delta|}\rangle_c = & -\frac{i}{2}\int_0^\beta d\tau d\tau' \int d^3r d^3r'\, \dot\theta(x)\Pi_{nj_\Delta}(x,x')\delta|\Delta(x')| \\ & - \frac{1}{2}\int_0^\beta d\tau d\tau' \int d^3r d^3r' \sum_\mu \partial_\mu\theta(x)\Pi_{j_\mu j_\Delta}(x,x')\delta|\Delta(x')|\end{aligned} \tag{7.336}$$

(μ runs over the space indices x, y, z). We have introduced the BCS correlation functions

$$\Pi_{jj'}(x,x') = \langle j(x)j'(x')\rangle_c = \langle j(x)j'(x')\rangle - \langle j(x)\rangle\langle j'(x')\rangle, \tag{7.337}$$

where the expectation value $\langle\cdots\rangle$ is computed with the action $S_{\rm MF}[\psi^*,\psi]$.

7.6.1 *BCS correlation functions*

To compute explicitly the action $S[\theta, \delta|\Delta|]$, we must evaluate the correlation functions (7.337). In Fourier space, the current density to be considered reads

$$j_\mu^\nu(q) = \frac{1}{\sqrt{\beta\mathcal{V}}}\sum_k v_\mu(\mathbf{k})\Psi^\dagger(k_-)\tau^\nu\Psi(k_+), \tag{7.338}$$

where $\nu,\mu = 0,x,y,z$ (τ^0 is the identity matrix) and $k = (\mathbf{k}, i\omega_n)$, $q = (\mathbf{q}, i\omega_\nu)$, $k_- = (\mathbf{k}-\mathbf{q}/2, i\omega_n)$ and $k_+ = (\mathbf{k}+\mathbf{q}/2, i\omega_n + i\omega_\nu)$ (ω_ν is a bosonic Matsubara frequency). We have also introduced

$$v_0(\mathbf{k}) = 1, \qquad v_\mu(\mathbf{k}) = \frac{k_\mu}{m} \qquad (\mu \neq 0). \tag{7.339}$$

With these notations, we have $n \equiv j_0^z$, $j_\mu \equiv j_\mu^0$ and $j_\Delta \equiv j_0^x$. Since the mean-field action is Gaussian, one can use Wick's theorem to obtain

$$\begin{aligned}\Pi_{j_\mu^\nu j_{\mu'}^{\nu'}}(q) &= \langle j_\mu^\nu(q) j_{\mu'}^{\nu'}(-q)\rangle_c \\ &= -\frac{1}{\beta\mathcal{V}}\sum_k v_\mu(\mathbf{k})v_{\mu'}(\mathbf{k})\,\mathrm{tr}[\tau^\nu\mathcal{G}(k_+)\tau^{\nu'}\mathcal{G}(k_-)],\end{aligned} \tag{7.340}$$

where tr denotes the trace over the Nambu indices. $\mathcal{G}$ is the mean-field propagator defined in (7.242). Thus, we obtain

$$
\begin{aligned}
\Pi_{nn}(q) &= -\frac{2}{\beta\mathcal{V}}\sum_k [G(k)G(k+q) - F(k)F(k+q)], \\
\Pi_{j_\mu n}(q) &= -\frac{2}{\beta\mathcal{V}}\sum_k v_\mu(\mathbf{k})G(k_-)G(k_+), \\
\Pi_{j_\mu j_{\mu'}}(q) &= -\frac{2}{\beta\mathcal{V}}\sum_k v_\mu(\mathbf{k})v_{\mu'}(\mathbf{k})[G(k_-)G(k_+) + F(k_-)F(k_+)], \\
\Pi_{j_\Delta j_\Delta}(q) &= \frac{1}{\beta\mathcal{V}}\sum_k [G(k)G(-k+q) + G(k)G(-k-q) - 2F(k)F(k+q)], \\
\Pi_{nj_\Delta}(q) &= -\frac{2}{\beta\mathcal{V}}\sum_k F(k)[G(k+q) + G(k-q)], \\
\Pi_{j_\mu j_\Delta}(q) &= -\frac{2}{\beta\mathcal{V}}\sum_k v_\mu(\mathbf{k})F(k_-)[G(k_+) - G(-k_+)].
\end{aligned}
\tag{7.341}
$$

7.6.2 *Anderson–Bogoliubov mode*

7.6.2.1 *Phase-only action in the hydrodynamic limit*

The part of the action that depends on the amplitude fluctuations reads

$$
\begin{aligned}
&\sum_q \delta|\Delta|(-q)\delta|\Delta|(q)\left[\frac{1}{\lambda} - \frac{1}{2}\Pi_{j_\Delta j_\Delta}(q)\right] \\
&\quad + \frac{1}{2}\sum_q \theta(-q)\delta|\Delta|(q)\Big[-\omega_\nu \Pi_{nj_\Delta}(q) - i\sum_\mu q_\mu \Pi_{j_\mu j_\Delta}(q)\Big].
\end{aligned}
\tag{7.342}
$$

By integrating out $\delta|\Delta|$, we obtain a contribution to the phase-only action $S[\theta]$. If one restricts oneself to the second-order contributions in the phase derivatives, which is justified in the hydrodynamic regime $|\mathbf{q}|\xi_0 \ll 1$ and $|\omega_\nu|\xi_0/c \ll 1$ ($\xi_0 = v_F/\pi\Delta_0$ is the BCS coherence length and c the velocity of the phase mode),[73] it is sufficient to evaluate the BCS correlation functions $\Pi_{nj_\Delta}(q)$ and $\Pi_{j_\mu j_\Delta}(q)$ at $q = 0$,

$$
\begin{aligned}
\Pi_{nj_\Delta}(q=0) &= \frac{4N(0)}{\beta}\sum_{\omega_n}\int_{-\omega_D}^{\omega_D} d\xi \frac{\Delta_0(i\omega_n + \xi)}{(\omega_n^2 + \xi^2 + \Delta_0^2)^2} = 0, \\
\Pi_{j_\mu j_\Delta}(q=0) &= \frac{4}{\beta\mathcal{V}}\sum_k \frac{k_\mu}{m}\frac{\Delta_0 i\omega_n}{(\omega_n^2 + \xi_{\mathbf{k}}^2 + \Delta_0^2)^2} = 0.
\end{aligned}
\tag{7.343}
$$

[73]The derivative expansion, i.e. the small-q expansion of the correlation functions $\Pi(q)$, is made possible by the gapped BCS spectrum.

The vanishing of $\Pi_{nj_\Delta}(q=0)$ is due to the particle–hole symmetry which follows from the approximation $N(\xi) \simeq N(0)$ for $|\xi| \leq \omega_D$ (expected to be generally valid when $\omega_D \ll \epsilon_F$). Thus, phase and amplitude fluctuations decouple in the long-wavelength limit, and the phase-only action reads

$$S[\theta] = \frac{1}{8}\sum_q \theta(-q)\theta(q)\left[\frac{n\mathbf{q}^2}{m} + \omega_\nu^2 \Pi_{nn} + 2i\omega_\nu \sum_\mu q_\mu \Pi_{nj_\mu} - \sum_{\mu,\mu'} q_\mu q_{\mu'} \Pi_{j_\mu j_{\mu'}}\right] \tag{7.344}$$

to leading order in derivatives (with $\Pi \equiv \Pi(q=0)$). Here and in the following, we ignore the term proportional to $\dot{\theta}$ that comes from $\langle S_\theta \rangle$ since it does not affect the low-energy long-wavelength fluctuations. Since $\Pi_{nj_\mu}(q=0)$ obviously vanishes, we finally obtain

$$S[\theta] = \frac{n_s}{8m}\int_0^\beta d\tau \int d^3r \left\{ [\boldsymbol{\nabla}\theta(x)]^2 + \frac{\dot{\theta}(x)^2}{c^2} \right\} \tag{7.345}$$

(we have used $\Pi_{j_\mu j_{\mu'}}(q=0) \propto \delta_{\mu,\mu'}$). This result agrees with the phenomenological action proposed in Sec. 7.2.4, with

$$n_s = n - m\Pi_{j_\mu j_\mu}(q=0) \quad \text{and} \quad c = \sqrt{\frac{n_s}{mn^2\kappa_0}} \tag{7.346}$$

the superfluid density and the velocity of the Anderson–Bogoliubov sound (phonon) mode, respectively. $\kappa_0 = \Pi_{nn}(q=0)/n^2 = \lim_{\mathbf{q}\to 0} \Pi^R_{nn}(\mathbf{q}, \omega=0)/n^2$ is the compressibility in the BCS state (we shall see that it coincides with the actual compressibility κ). The definition (7.346) of the superfluid density agrees with (7.10) if we can identify the BCS correlation function $\Pi_{j_\mu j_\mu}(q=0)$ with $\lim_{\mathbf{q}\to 0} \chi^R_\perp(\mathbf{q}, 0)$, where $\chi_\perp$ is the transverse current–current correlation function in the superconductor. We shall see in the following that this is indeed the case.

At zero temperature, we find

$$\frac{1}{\beta}\sum_{\omega_n}[G(k)]^2 = -\frac{1}{\beta}\sum_{\omega_n}[F(k)]^2 = -\frac{\Delta_0^2}{4E_{\mathbf{k}}^3} \qquad (T \to 0) \tag{7.347}$$

so that

$$\begin{aligned} n^2\kappa_0 &= \int_{\mathbf{k}} \frac{\Delta_0^2}{E_{\mathbf{k}}^3} = N(0)\int_{-\omega_D}^{\omega_D} d\xi \frac{\Delta_0^2}{(\xi^2+\Delta_0^2)^{3/2}} \simeq 2N(0), \\ \Pi_{j_\mu j_\mu}(q=0) &= 0. \end{aligned} \tag{7.348}$$

The superfluid density $n_s = n$ is therefore given by the fluid density (as expected for a translation-invariant system at $T=0$), and the velocity of the Anderson–Bogoliubov mode is proportional to the Fermi velocity,[74]

$$c = \left[\frac{n}{2mN(0)}\right]^{1/2} = \frac{v_F}{\sqrt{3}}. \tag{7.349}$$

[74]In d dimensions, one would find $c = v_F/\sqrt{d}$.

Because of the gap in the BCS quasi-particle excitation spectrum, the BCS correlation functions Π are short-range. $\Pi(q)$ is analytic at $q = 0$, thus allowing the derivative expansion leading to (7.345). At finite temperature, due to the presence of thermally excited quasi-particles, $\Pi(q)$ is not analytic at $q = 0$ and the static $(\mathbf{q} \to 0, \omega_\nu = 0)$ and dynamic $(\mathbf{q} = 0, \omega_\nu \to 0)$ limits are different. This non-analyticity, which is associated with the damping of the collective phase mode (known as Landau damping; see Sec. 4.3.3) due to its interaction with thermally excited quasi-particles, prevents an expansion of $\Pi(q)$ about $q = 0$ [52]. Nevertheless, one can still obtain a local action as in (7.345) if one takes the static limit $(\mathbf{q} \to 0, \omega_\nu = 0) \equiv (\mathbf{q} = 0, \omega_\nu = 0)$.[75] This approximation includes the normal fluid contribution to the BCS correlation functions in the static limit [53–55]. Using (7.346), we can then calculate the superfluid density at finite temperature. From

$$
\begin{aligned}
\frac{1}{\beta}\sum_{\omega_n}[G(k)]^2 &= \frac{1}{2E_{\mathbf{k}}^2}\left\{n_F'(E_{\mathbf{k}})(E_{\mathbf{k}}^2+\xi_{\mathbf{k}}^2) + \left[n_F(E_{\mathbf{k}}) - \frac{1}{2}\right]\frac{\Delta_0^2}{E_{\mathbf{k}}}\right\}, \\
\frac{1}{\beta}\sum_{\omega_n}[F(k)]^2 &= \frac{\Delta_0^2}{2E_{\mathbf{k}}^2}\left\{n_F'(E_{\mathbf{k}}) - \left[n_F(E_{\mathbf{k}}) - \frac{1}{2}\right]\frac{1}{E_{\mathbf{k}}}\right\},
\end{aligned}
\tag{7.350}
$$

we deduce

$$
\Pi_{j_\mu j_\mu}(q=0) = -\frac{4}{3}\int_{\mathbf{k}} \frac{\epsilon_{\mathbf{k}}}{m} n_F'(E_{\mathbf{k}}) \tag{7.351}
$$

and

$$
n_s = n - \frac{4}{3}\int_{\mathbf{k}} \epsilon_{\mathbf{k}}[-n_F'(E_{\mathbf{k}})]. \tag{7.352}
$$

At low temperatures $T \ll T_c$, (7.352) predicts an exponentially small reduction of the superfluid density. Nevertheless, we expect the superfluid density to be mainly reduced by the thermally excited phonons in that temperature range. To describe this effect, one must expand the action beyond second order in phase fluctuations. Such a calculation, including the one-loop correction to the phase propagator, gives [55]

$$
n_s = n - \frac{1}{3}\int_{\mathbf{q}} \frac{\mathbf{q}^2}{m}[-n_B'(c|\mathbf{q}|)], \tag{7.353}
$$

which is identical to what we found for a Bose superfluid within the two-fluid model [Eq. (7.70)].

7.6.2.2 *Correlation functions of the neutral superconductor*

The knowledge of the phase-only action $S[\theta]$ allows us to obtain the density–density and current–current response functions of the neutral superconductor. We add to the

[75] The equivalence between $\Pi(q = 0)$ and $\Pi(\mathbf{q} \to 0, i\omega_\nu = 0)$ in the Euclidean formalism is discussed in Sec. 3.2.7.

action (7.325) the term

$$\int_0^\beta d\tau \int d^3r \Big[ie\phi(x)n(x) - e\mathbf{j}(x)\cdot\mathbf{A}(x) + \frac{e^2}{2m}n(x)\mathbf{A}(x)^2\Big], \tag{7.354}$$

where $ie\phi$ and $e\mathbf{A}$ should be seen as fictitious external fields introduced in order to derive the response functions. To make gauge invariance explicit [Eqs. (7.96), (7.97)], we include in the action the diamagnetic term $\propto \mathbf{A}^2$. Fluctuations about the BCS state are studied as before, performing the gauge transformation (7.327). The resulting action is given by (7.332) with $\dot\theta$ and $\boldsymbol{\nabla}\theta$ in the action S_θ replaced by the covariant derivatives $\dot\theta + 2e\phi$ and $\boldsymbol{\nabla}\theta - 2e\mathbf{A}$. The phase-only action, obtained by integrating out the fermions, can be easily deduced from (7.344).[76] Introducing $q_0 = -i\omega_\nu$, $\mathcal{A}_0 = -i\phi$ and $\mathcal{A}_\mu = A_\mu$ ($\mu \neq 0$), we have

$$S[\theta;\mathcal{A}] = \frac{1}{8}\sum_{q,\mu,\mu'} [iq_\mu\theta(-q) + 2e\mathcal{A}_\mu(-q)]\Lambda_{\mu\mu'}(q)[iq_{\mu'}\theta(q) - 2e\mathcal{A}_{\mu'}(q)], \tag{7.355}$$

where μ, μ' now run over $0, x, y, z$ and

$$\Lambda_{\mu\mu'}(q) = \Pi_{j_\mu j'_\mu}(q) - \frac{n}{m}\delta_{\mu,\mu'}(1-\delta_{\mu,0}). \tag{7.356}$$

Integrating out the phase, we obtain

$$\ln Z[\mathcal{A}] = \frac{e^2}{2}\sum_{q,\mu,\mu'} \mathcal{A}_\mu(-q)\mathcal{A}_{\mu'}(q)\left[\Lambda_{\mu\mu'}(q) - \frac{\sum_{\alpha,\beta}\Lambda_{\mu\alpha}(q)q_\alpha q_\beta \Lambda_{\beta\mu'}(q)}{\sum_{\gamma,\delta} q_\gamma \Lambda_{\gamma\delta}(q) q_\delta}\right]. \tag{7.357}$$

From (7.357), we deduce the density–density response function

$$\begin{aligned}\chi_{nn}(q) &= \frac{\delta^2 \ln Z[\mathcal{A}]}{e^2\delta\mathcal{A}_0(-q)\delta\mathcal{A}_0(q)}\bigg|_{\mathcal{A}=0} \\ &= \Pi_{nn}(q) - \frac{\omega_\nu^2\Pi_{nn}^2(q) + i\omega_\nu|\mathbf{q}|\Pi_{nn}(q)[\Pi_{nj_\parallel}(q) + \Pi_{j_\parallel n}(q)] - \mathbf{q}^2\Pi_{nj_\parallel}(q)\Pi_{j_\parallel n}(q)}{\omega_\nu^2\Pi_{nn}(q) + i\omega_\nu|\mathbf{q}|[\Pi_{nj_\parallel}(q) + \Pi_{j_\parallel n}(q)] - \mathbf{q}^2\Pi_{j_\parallel j_\parallel}(q) + \frac{n}{m}\mathbf{q}^2},\end{aligned} \tag{7.358}$$

where we have introduced

$$\begin{aligned}\Pi_{j_\parallel j_\parallel}(q) &= \sum_{\mu,\mu'\neq 0} \frac{q_\mu q_{\mu'}}{\mathbf{q}^2}\Pi_{j_\mu j_{\mu'}}(q), \\ \Pi_{nj_\parallel}(q) &= \sum_{\mu\neq 0}\frac{q_\mu}{|\mathbf{q}|}\Pi_{nj_\mu}(q), \\ \Pi_{j_\parallel n}(q) &= \sum_{\mu\neq 0}\frac{q_\mu}{|\mathbf{q}|}\Pi_{j_\mu n}(q).\end{aligned} \tag{7.359}$$

[76]Since amplitude fluctuations do not couple to phase amplitudes in the hydrodynamic regime (Sec. 7.6.2.1), they do not couple to the covariant derivatives $\dot\theta + 2e\phi$ and $\boldsymbol{\nabla}\theta - 2e\mathbf{A}$ and therefore to the gauge field $(\phi, \mathbf{A})$.

Using $\Pi_{j_\parallel j_\parallel}(q=0) = \Pi_{nj_\parallel}(q=0) = 0$, we obtain

$$n^2\kappa = \lim_{\mathbf{q}\to 0} \chi_{nn}(\mathbf{q},0) = \lim_{\mathbf{q}\to 0} \Pi_{nn}(\mathbf{q},0) \tag{7.360}$$

so that the compressibility κ is given by the BCS state compressibility κ_0.

Taking advantage of the gauge invariance of the action (7.355), we deduce the longitudinal current–current correlation function

$$\chi_\parallel(q) = \frac{n}{m} - \frac{\omega_\nu^2}{\mathbf{q}^2}\chi_{nn}(q) \tag{7.361}$$

[see Eq. (3.185) in Chapter 3]. The transverse correlation function can be obtained from the general expression

$$\begin{aligned}\chi_{j_\mu j_{\mu'}}(q) - \frac{n}{m}\delta_{\mu,\mu'} &= \frac{\delta^2 \ln Z[\mathcal{A}]}{e^2\delta\mathcal{A}_\mu(-q)\delta\mathcal{A}_{\mu'}(q)}\bigg|_{\mathcal{A}=0} \\ &= \Lambda_{\mu\mu'}(q) - \frac{\sum_{\alpha,\beta}\Lambda_{\mu\alpha}(q)q_\alpha q_\beta\Lambda_{\beta\mu'}(q)}{\sum_{\gamma,\delta} q_\gamma\Lambda_{\gamma\delta}(q)q_\delta} \qquad (\mu,\mu'\neq 0).\end{aligned} \tag{7.362}$$

Since $\sum_\alpha \Pi_{j_\mu j_\alpha}(q)q_\alpha = 0$ if $j_\mu(q)$ is transverse, the second term in the rhs of (7.362) does not contribute to the transverse correlation function,

$$\chi_\perp(q) = \Pi_\perp(q). \tag{7.363}$$

As anticipated, $\Pi_\perp$ is the actual (gauge-invariant) transverse current–current correlation function. Thus, the transverse response of the superconductor is entirely determined by the BCS mean-field action and is not affected by phase fluctuations. By contrast, the (gauge-invariant) longitudinal response can be obtained only by including phase fluctuations about the BCS state.

Note that Eqs. (7.358), (7.361) and (7.363) are valid at finite temperature, since they do not rely on a derivative expansion of the action. In the hydrodynamic limit at zero temperature, they simplify into

$$\chi_{nn}(q) = \frac{n}{m}\frac{\mathbf{q}^2}{\omega_\nu^2 + c^2\mathbf{q}^2}, \qquad \chi_\parallel(q) = \frac{n}{m}\frac{c^2\mathbf{q}^2}{\omega_\nu^2 + c^2\mathbf{q}^2}, \qquad \chi_\perp(q) = \mathcal{O}(\mathbf{q}^2) \tag{7.364}$$

using $c^2 = n/m\Pi_{nn}(q=0)$ and again $\Pi_{j_\parallel j_\parallel}(q=0) = \Pi_{nj_\parallel}(q=0) = 0$. The velocity of the Anderson–Bogoliubov mode is equal to the macroscopic sound velocity $c_s = 1/\sqrt{mn\kappa}$. Equation (7.364) satisfy the f-sum rule $\chi_\parallel^R(\mathbf{q},0) = n/m$ and show that the superfluid density $n_s = n$ since $\lim_{\mathbf{q}\to 0}\chi_\perp^R(\mathbf{q},0) = (n-n_s)/m$. They agree with the previous results (7.93) and (7.94) when $n_s = n$.

7.6.3 *Electrodynamics*

In this section, we study the effect of the long-range Coulomb interaction and the coupling to the electromagnetic field in a charged superconductor. To the action (7.355) obtained in the preceding section, we must add the electromagnetic field action (1.635),

$$S_{\rm em}[\phi, \mathbf{A}_\perp] = \frac{\epsilon_0}{2} \sum_{\mathbf{q}} [\mathbf{q}^2 \phi(-q)\phi(q) + (\omega_\nu^2 + c_l^2 \mathbf{q}^2)\mathbf{A}_\perp(-q) \cdot \mathbf{A}_\perp(q)], \tag{7.365}$$

since the gauge field $(\phi, \mathbf{A}_\perp)$ is now dynamical. We work in the Coulomb gauge where the vector potential is transverse. As usual, the $\mathbf{q} = 0$ component of the scalar potential $\phi(q)$ is canceled by the uniform background of positive charges that ensures the global electric neutrality of the system (jellium model). $\mathbf{A}_\perp$ being transverse, it does not couple to the phase and the scalar potential so that the action $S[\theta, \phi, \mathbf{A}_\perp]$ splits into a longitudinal part $S[\theta, \phi]$ and a transverse part $S[\mathbf{A}_\perp]$.

7.6.3.1 *Longitudinal fluctuations*

The action $S[\theta, \phi]$ is given by

$$S[\theta, \phi] = \frac{1}{2} \sum_q (\theta(-q), \phi(-q)) \mathcal{D}^{-1}(q) \begin{pmatrix} \theta(q) \\ \phi(q) \end{pmatrix}, \tag{7.366}$$

where

$$\begin{aligned} \mathcal{D}_{11}^{-1}(q) &= \frac{1}{4}\left(\frac{n\mathbf{q}^2}{m} - \mathbf{q}^2 \Pi_{j_\parallel j_\parallel}(q) + \omega_\nu^2 \Pi_{nn}(q) + 2i\omega_\nu |\mathbf{q}| \Pi_{nj_\parallel}(q)\right), \\ \mathcal{D}_{12}^{-1}(q) &= -\mathcal{D}_{21}^{-1}(q) = \frac{e}{2}[i\omega_\nu \Pi_{nn}(q) - |\mathbf{q}| \Pi_{j_\parallel n}(q)], \\ \mathcal{D}_{22}^{-1}(q) &= e^2 [V_C^{-1}(q) + \Pi_{nn}(q)] \end{aligned} \tag{7.367}$$

and $V_C(q) = e^2/\epsilon_0 \mathbf{q}^2$ is the Fourier transform of the Coulomb interaction. Integrating out the scalar potential yields the following phase-only action:

$$S[\theta] = \frac{1}{8} \sum_q \theta(-q)\theta(q) \left[\frac{n\mathbf{q}^2}{m} - \mathbf{q}^2 \mathcal{L}_{j_\parallel j_\parallel}(q) + \omega_\nu^2 \mathcal{L}_{nn}(q) + 2i\omega_\nu |\mathbf{q}| \mathcal{L}_{nj_\parallel}(q)\right], \tag{7.368}$$

where

$$\begin{aligned} \mathcal{L}_{nn}(q) &= \frac{\Pi_{nn}(q)}{1 + V_C(q)\Pi_{nn}(q)}, \\ \mathcal{L}_{j_\parallel j_\parallel}(q) &= \Pi_{j_\parallel j_\parallel}(q) - \frac{\Pi_{j_\parallel n}(q) V_C(q) \Pi_{nj_\parallel}(q)}{1 + V_C(q)\Pi_{nn}(q)}, \\ \mathcal{L}_{nj_\parallel}(q) &= \frac{\Pi_{nj_\parallel}(q)}{1 + V_C(q)\Pi_{nn}(q)}. \end{aligned} \tag{7.369}$$

The action (7.368) is similar to that of the neutral superconductor provided we replace the BCS correlation function Π by $\mathcal{L}$. The latter exhibits the characteristic form of a screened correlation function as obtained within the RPA (Chapter 5). To obtain the action to order $\mathbf{q}^2$, it is sufficient to set $\mathbf{q} = 0$ in $\mathcal{L}_{j_\parallel j_\parallel}(q)$ and $\mathcal{L}_{nj_\parallel}(q)$.[77] At zero temperature, one then finds

$$S[\theta] = \frac{n_s}{8m} \sum_q \theta(-q)\theta(q)\mathbf{q}^2 \left(1 + \frac{\omega_\nu^2}{\omega_p^2}\right) \tag{7.370}$$

using $\mathcal{L}_{nn}(q) = \epsilon_0 \mathbf{q}^2/e^2$ ($\mathbf{q} \to 0$) and $\mathcal{L}_{j_\parallel j_\parallel}(\mathbf{q} = 0, i\omega_\nu) = \mathcal{L}_{nj_\parallel}(\mathbf{q} = 0, i\omega_\nu) = 0$.[78] $\omega_p^2 = ne^2/\epsilon_0 m$ is the plasma frequency and $n_s = n$ the superfluid density. Note that the quantum term $\propto \omega_\nu^2$ is entirely determined by the long-range Coulomb interaction (i.e. the vanishing of $\mathcal{L}_{nn}(q)$ for $\mathbf{q} \to 0$). The phase–phase correlation function

$$\chi_{\theta\theta}(q) = \frac{4m}{n_s \mathbf{q}^2} \frac{\omega_p^2}{\omega_\nu^2 + \omega_p^2} \tag{7.371}$$

exhibits a pole at the plasma frequency ω_p. The Anderson–Bogoliubov mode of the neutral superconductor has been replaced by a (gapped) plasmon mode but the superfluid density n_s is still finite. We refer the reader to Sec. 7.2.4 for a detailed discussion of the physical meaning of (7.370) and (7.371).

Starting from the action $S[\theta, \phi]$, one can also integrate out the phase to obtain the action of the scalar potential,

$$S[\phi] = \frac{1}{2} \sum_q \phi(-q)\phi(q)e^2 \left[V_C^{-1}(q) + \chi_{nn}^{\text{neutral}}(q)\right], \tag{7.372}$$

where $\chi_{nn}^{\text{neutral}}$ is the density–density response (7.358) function of the neutral superconductor. Note that (7.372) is valid beyond the hydrodynamic limit $|\mathbf{q}|\xi_0 \ll 1$ and at finite temperature, since we have kept the q dependence of the BCS correlation functions.[79] From (7.372), we deduce the longitudinal dielectric function

$$\epsilon_\parallel(q) = 1 + \frac{e^2}{\epsilon_0 \mathbf{q}^2} \chi_{nn}^{\text{neutral}}(q) \simeq 1 + \frac{\omega_p^2}{\omega_\nu^2 + c^2\mathbf{q}^2} \tag{7.373}$$

(see Sec. 3.4.1), where the last result holds in the limit $\mathbf{q} \to 0$ and $T = 0$. Equation (7.373) agrees with the previous result (7.111) and we refer to Sec. 7.2.4 for a discussion of its physical consequences.

7.6.3.2 *Transverse fluctuations*

Let us now derive the action $S[\mathbf{A}_\perp]$ of the transverse field. The calculation is straightforward and gives

$$S[\mathbf{A}_\perp] = \frac{1}{2} \sum_q \mathbf{A}_\perp(-q) \cdot \mathbf{A}_\perp(q) \left[\epsilon_0(\omega_\nu^2 + c_l^2\mathbf{q}^2 + \omega_p^2) - e^2\Pi_\perp(q)\right]. \tag{7.374}$$

---

[77]Note that $\mathcal{L}_{nj_\parallel}(q)$ has no $\mathcal{O}(\mathbf{q})$ term.

[78]This is deduced from $\Pi_{nj_\parallel}(\mathbf{q} = 0, i\omega_\nu) = \Pi_{j_\parallel j_\parallel}(\mathbf{q} = 0, i\omega_\nu) = 0$. The latter result follows from $T\sum_{\omega_n} G(k)G(k+q) = -T\sum_{\omega_n} F(k)F(k+q) = -\Delta_0^2/[E_\mathbf{k}(\omega_\nu^2 + E_\mathbf{k}^2)]$ ($q = (0, i\omega_\nu)$ and $T \to 0$) and $\Pi_{j_\mu j_{\mu'}}(\mathbf{q} = 0, i\omega_\nu) = 0$.

[79]The same holds for the action $S[\mathbf{A}_\perp]$ derived in the following [Eq. (7.374)].

We deduce the transverse dielectric function

$$\epsilon_\perp(q) = 1 + \frac{\omega_p^2}{\omega_\nu^2} - \frac{e^2}{\omega_\nu^2 \epsilon_0}\Pi_\perp(q). \tag{7.375}$$

Comparing this expression with the exact form (3.155), we again see that $\Pi_\perp(q)$ is the actual (gauge-invariant) transverse current–current correlation function. Since $\Pi_\perp(q) = \mathcal{O}(\mathbf{q}^2)$ at zero temperature, the last term in (7.375) can be neglected in the hydrodynamic limit $|\mathbf{q}|\xi_0 \ll 1$ and we recover the expression (7.111) and the gap ω_p of the photon mode (Anderson–Higgs mechanism).

7.6.3.3 *Conductivity and Meissner effect*

The dielectric functions are related to the conductivity (defined as the response to the local electromagnetic field) by (7.112). This yields

$$\begin{aligned}
\sigma_\parallel(\mathbf{q},\omega) &= e^2 \frac{\omega + i0^+}{i\mathbf{q}^2} \chi_{nn}^{\text{neutral } R}(\mathbf{q},\omega) = \frac{e^2}{i(\omega + i0^+)}\left[\chi_\parallel^{\text{neutral } R}(\mathbf{q},\omega) - \frac{n}{m}\right], \\
\sigma_\perp(\mathbf{q},\omega) &= \frac{e^2}{i(\omega + i0^+)}\left[\Pi_\perp^R(\mathbf{q},\omega) - \frac{n}{m}\right].
\end{aligned} \tag{7.376}$$

In the hydrodynamic limit $|\mathbf{q}|\xi_0 \ll 1$ at zero temperature,

$$\begin{aligned}
\sigma_\parallel(\mathbf{q},\omega) &\simeq \frac{ne^2}{m}\frac{i(\omega + i0^+)}{(\omega + i0^+)^2 - c^2\mathbf{q}^2}, \\
\sigma_\perp(\mathbf{q},\omega) &\simeq \frac{ne^2}{m}\frac{i}{\omega + i0^+}.
\end{aligned} \tag{7.377}$$

The zero-temperature uniform conductivity,[80]

$$\sigma_\parallel(0,\omega) = \sigma_\perp(0,\omega) = \frac{ne^2}{m}\frac{i}{\omega + i0^+}, \tag{7.378}$$

shows that the superconductor is a perfect conductor with a Drude weight $D = \pi ne^2/m$. Equation (7.378) actually holds at all temperatures and follows from the translation invariance of the system (see Sec. 3.4.4). Since the zero-frequency peak $D\delta(\omega)$ exhausts all the spectral weight in the conductivity sum rule

$$\int_{-\infty}^{\infty} \frac{d\omega}{2\pi} \Re[\sigma_{\parallel,\perp}(0,\omega)] = \frac{\epsilon_0}{2}\omega_p^2 \tag{7.379}$$

(see Chapter 3), there is no electromagnetic absorption at frequencies higher than 2Δ due to creation of Bogoliubov quasi-particle/quasi-hole excitations (as one could have naively expected). In a disordered system where translation invariance is broken, we expect part of the spectral weight to be transferred from the Drude peak to frequencies $|\omega| \geq 2\Delta$ (Fig. 7.22). This is indeed the case in the so-called dirty superconductors where the mean

[80] We use $\Pi_{j_\mu j_\mu}(\mathbf{q} = 0, i\omega_\nu) = 0$, i.e. $\Pi_\parallel^R(0,\omega) = \Pi_\perp^R(0,\omega) = 0$.

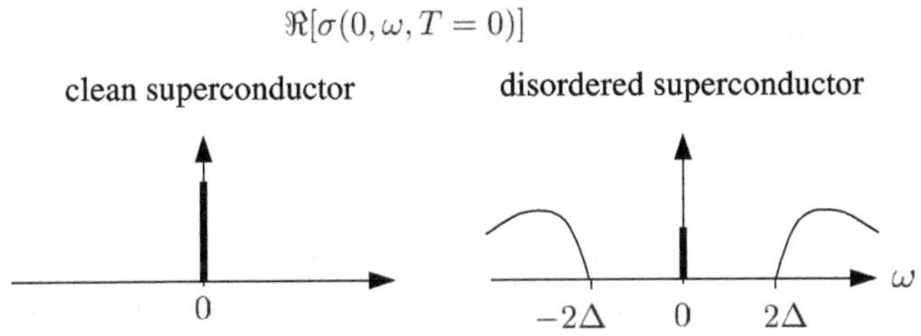

Fig. 7.22 Real part of the uniform conductivity in a clean (translation-invariant) and disordered superconductor in the dirty limit. The spectral weight of the zero-frequency peak is given by the Drude weight D.

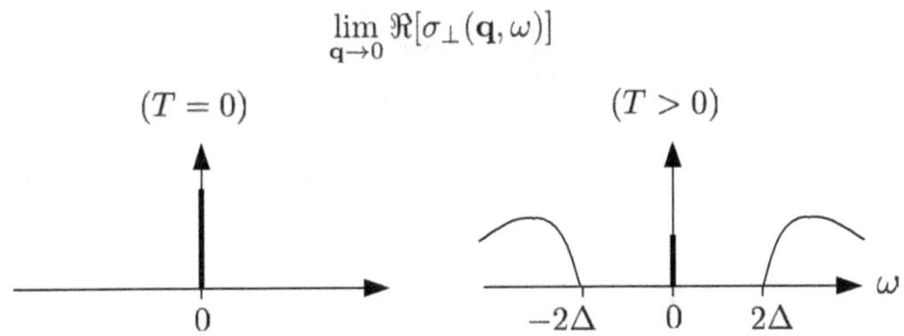

Fig. 7.23 Real part of $\lim_{\mathbf{q}\to 0}\sigma_\perp(\mathbf{q},\omega)$ in a clean (translation-invariant) superconductor at $T = 0$ and $T > 0$. The spectral weight of the zero-frequency peak is given by the superfluid weight $D_s = \pi n_s e^2/m$.

free path l is smaller than the BCS coherence length ξ_0; the width $1/\tau = v_F/l$ of the Drude peak in the conductivity $\sigma(\omega) = \sigma_0/(1+\omega^2\tau^2)$ of the metallic phase is then larger than 2Δ.

The superfluid weight $D_s = \pi n_s e^2/m$ is defined as the weight of the zero-frequency peak in $\lim_{\mathbf{q}\to 0}\Re[\sigma_\perp(\mathbf{q},\omega)]$ (Chapter 3). From (7.376), we thus recover the expression $n_s = n - m\lim_{\mathbf{q}\to 0}\Pi^R_\perp(\mathbf{q},0)$ of the superfluid density obtained earlier [Eq. (7.346)].[81] At zero temperature ($n_s = n$), the zero-frequency peak exhausts all the spectral weight in the sum rule (7.379). At finite temperature, we expect part of the spectral weight to be transferred to frequencies $|\omega| \geq 2\Delta$ in agreement with the reduction of the superfluid density (Fig. 7.23).

Last, the transverse current–current correlation function also determines the response to a static magnetic field,

$$e\langle \mathbf{J}(\mathbf{q},0)\rangle = e^2\left[\Pi^R_\perp(\mathbf{q},0) - \frac{n}{m}\right]\langle \mathbf{A}_\perp(\mathbf{q},0)\rangle \tag{7.380}$$

(see Chapter 3). For $|\mathbf{q}|\xi_0 \ll 1$, we can set $\mathbf{q} = 0$ in the rhs of (7.380), i.e.

$$e\langle \mathbf{J}(\mathbf{q},0)\rangle = -\frac{n_s e^2}{m}\langle \mathbf{A}_\perp(\mathbf{q},0)\rangle = -\frac{1}{\mu_0\lambda^2}\langle \mathbf{A}_\perp(\mathbf{q},0)\rangle, \tag{7.381}$$

and recover the London equation and in turn the Meissner effect (Sec. 3.4.5). Equation (7.381) can be used when the London penetration depth λ, which determines the typical length scale for the variation of the magnetic field, is much larger than ξ_0. In contrast, when $\lambda \lesssim \xi_0$, the response to the vector potential becomes non-local and the full $\mathbf{q}$

[81] Since the BCS excitation spectrum is gapped, $\Pi_{j_\mu j_\mu}(q)$ is regular for $q \to 0$ and $\lim_{\mathbf{q}\to 0}\Pi^R_\perp(\mathbf{q},0) = \Pi_{j_\mu j_\mu}(q=0)$.

dependence of the transverse current–current correlation function in (7.380) must be kept. This limit is known as the Pippard (non-local) limit [12, 13].

7.6.4 *Higgs amplitude mode*

The part of the action involving amplitude fluctuations is given by (7.342). In Sec. 7.6.2.1, we showed that there is no coupling between amplitude and phase fluctuations in the low-energy long-wavelength limit $|\mathbf{q}|\xi_0, |\omega_\nu|\xi_0/c \ll 1$. Since

$$\begin{aligned} \Pi_{nj_\Delta}(\mathbf{q}=0, i\omega_\nu) &= \frac{2N(0)}{\beta}\sum_{\omega_n}\int_{-\omega_D}^{\omega_D} d\xi \frac{\Delta_0(2i\omega_n + i\omega_\nu + 2\xi)}{(\omega_n^2+E^2)[(\omega_n+\omega_\nu)^2+E^2]} = 0, \\ \Pi_{j_\mu j_\Delta}(\mathbf{q}=0, i\omega_\nu) &= \frac{2}{\beta\mathcal{V}}\sum_k \frac{k_\mu}{m}\frac{\Delta_0(2i\omega_n + i\omega_\nu)}{(\omega_n^2+E_{\mathbf{k}}^2)[(\omega_n+\omega_\nu)^2+E_{\mathbf{k}}^2]} = 0 \end{aligned} \tag{7.382}$$

($E^2 = \xi^2 + \Delta_0^2$), the condition $\mathbf{q}\to 0$ is actually sufficient to make the coupling vanish. The action governing amplitude fluctuations is thus given by

$$S[\delta|\Delta|] = \sum_q \delta|\Delta|(-q)\delta|\Delta|(q)\left[\frac{1}{\lambda} - \frac{1}{2}\Pi_{j_\Delta j_\Delta}(q)\right], \tag{7.383}$$

to order $\mathcal{O}(\mathbf{q}^2)$ and the corresponding spectrum is obtained from the equation

$$\frac{1}{\lambda} - \frac{1}{2}\Pi^R_{j_\Delta j_\Delta}(\mathbf{q},\omega) = 0. \tag{7.384}$$

From (7.341) and

$$\left.\begin{aligned} \frac{1}{\beta}\sum_{\omega_n} G(k)G(-k+q) &= \frac{E_{\mathbf{k}}^2 + \xi_{\mathbf{k}}^2 + i\omega_\nu\xi_{\mathbf{k}}}{E_{\mathbf{k}}(\omega_\nu^2 + 4E_{\mathbf{k}}^2)} \\ \frac{1}{\beta}\sum_{\omega_n} F(k)F(k+q) &= \frac{\Delta_0^2}{E_{\mathbf{k}}(\omega_\nu^2 + 4E_{\mathbf{k}}^2)} \end{aligned}\right\} \quad (\mathbf{q}=0,\ T\to 0), \tag{7.385}$$

we deduce

$$\Pi_{j_\Delta j_\Delta}(\mathbf{q}=0, i\omega_\nu) = 4\int_{\mathbf{k}} \frac{\xi_{\mathbf{k}}^2}{E_{\mathbf{k}}(\omega_\nu^2 + 4E_{\mathbf{k}}^2)}. \tag{7.386}$$

We then immediately see that

$$\frac{1}{\lambda} - \frac{1}{2}\Pi^R_{j_\Delta j_\Delta}(\mathbf{q}=0, \omega=2\Delta_0) = \frac{1}{\lambda} - \int_{\mathbf{k}} \frac{1}{2E_{\mathbf{k}}} = 0 \tag{7.387}$$

using the gap equation (7.262). We deduce the existence of a "Higgs" amplitude mode[82] with a gap $2\Delta_0$ in the limit $\mathbf{q}\to 0$. A more detailed analysis leads to the following dispersion law [50, 51]:

$$\omega \simeq 2\Delta_0 + (0.2369 - 0.2956i)\frac{v_F^2\mathbf{q}^2}{4\Delta_0} \qquad (\mathbf{q}\to 0), \tag{7.388}$$

[82]The reason for this terminology is explained in Sec. 7.2.4.

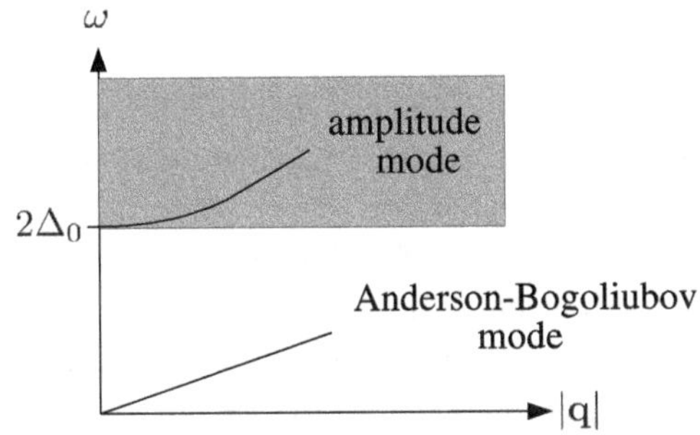

Fig. 7.24 Phase and amplitude modes in a superconductor. The shaded area shows the continuum of Bogoliubov quasi-particle/quasi-hole excitations. When long-range Coulomb interaction is included, the Anderson–Bogoliubov mode becomes gapped ($\omega = c|\mathbf{q}| \rightarrow (c^2\mathbf{q}^2 + \omega_p^2)^{1/2}$).

corresponding to an overdamped mode. The imaginary part is due to the coupling of the amplitude mode to the BCS quasi-particle/quasi-hole excitations with energy $E_{\mathbf{k}+\mathbf{q}} + E_{\mathbf{k}} \geq 2\Delta_0$. The spectrum of a superconductor, with both amplitude and phase modes, is shown in Fig. 7.24.

7.7 BCS–BEC Crossover

In the preceding sections, we have discussed two different theories of superfluidity:

- The BEC of bosons. At a fundamental level, these bosons are always composite objects consisting of an even number of fermions. They usually form at a very high temperature T_0, and condense at a much lower temperature $T_c \ll T_0$. In the superfluid phase, the low-energy physics is dominated by the center of mass motion of the bosons (Bogoliubov sound mode).
- The BCS theory, which describes superfluidity in a degenerate Fermi liquid. Cooper pairs form and condense at the same temperature $T_c \ll \epsilon_F$, and are strongly overlapping even at $T = 0$. Their size, given by the BCS coherence length ξ_0, is much larger than the average distance between particles. In a conventional (low-T_c) superconductor, $k_F\xi_0 \sim 10^3 - 10^4$. The transition temperature and the thermodynamics are controlled by pair breaking, i.e. excitations of Bogoliubov quasi-particles. The pair center of mass motion (Anderson–Bogoliubov sound mode in a neutral system, plasmon in a charged system) plays a minor role in the physical properties of a BCS superconductor.

In this section, we consider the crossover from the weak-coupling (BCS) limit to the strong-coupling (BEC) limit as observed in superfluid ultracold fermion gases. As the attractive interaction between fermions increases, we expect the size of the Cooper pairs to shrink and eventually become much smaller than the average distance between fermions. In that limit, the underlying fermionic nature of the pairs, which is important when two pairs are nearby, should not matter and we therefore expect superfluidity to originate from the BEC of preformed composite (point-like) bosons.

7.7.1 *Mean-field theory*

7.7.1.1 *s-wave scattering length — Two-particle bound state in vacuum*

The scattering between two particles (atoms) in vacuum is conveniently characterized by the s-wave scattering length a. We shall slightly deviate from the model used until now and regularize the action (7.228) by a UV cutoff acting on the fermion dispersion, $\epsilon_{\mathbf{k}} \leq \Lambda$, rather than $\xi_{\mathbf{k}}$.[83] The s-wave scattering length $a \equiv a(\lambda, \Lambda)$ is a function of Λ and the strength $-\lambda < 0$ of the attractive interaction between fermions. Since the low-energy properties discussed in the following depend only on the (physical) parameter a, we shall work at fixed a and take the infinite cutoff limit $\Lambda \to \infty$ whenever possible.

The scattering length a can be obtained from the low-energy limit of the two-particle vertex. In vacuum, Γ coincides with the T matrix and is given by the sum of the ladder diagrams. These diagrams have already been considered in Sec. 7.4 when discussing the Bose gas in the low-density limit.[84] Γ is given by

$$\Gamma(q) = -\lambda + \lambda\Pi(q)\Gamma(q) = \frac{-\lambda}{1 - \lambda\Pi(q)}, \tag{7.389}$$

where

$$\begin{aligned} \Pi(q) &= \int_{\mathbf{k}} \int_{-\infty}^{\infty} \frac{d\omega}{2\pi} \frac{1}{(i\omega - \epsilon_{\mathbf{k}})(i\omega_\nu - i\omega - \epsilon_{\mathbf{q}-\mathbf{k}})} \\ &= \int_{\mathbf{k}} \frac{1}{-i\omega_\nu + \epsilon_{\mathbf{k}} + \epsilon_{\mathbf{q}-\mathbf{k}}} \end{aligned} \tag{7.390}$$

is the particle–particle pair propagator in vacuum.[85] In three dimensions, the s-wave scattering length is defined by

$$\frac{m}{4\pi a} = \frac{1}{\Gamma(0,0)} = -\frac{1}{\lambda} + \int_{\mathbf{k}} \frac{1}{2\epsilon_{\mathbf{k}}} = -\frac{1}{\lambda} + \frac{(m^3\Lambda)^{1/2}}{\sqrt{2}\pi^2} \tag{7.391}$$

and is shown *vs* λ in Fig. 7.25. It is negative at small coupling, diverges when $\lambda = \lambda_c = \sqrt{2}\pi^2/(m^3\Lambda)^{1/2}$, and is positive for $\lambda > \lambda_c$. The resonance at $\lambda = \lambda_c$ corresponds to the unitary limit. As is well known from scattering theory, the divergence of the scattering length signals the appearance of a two-body bound state.[84] The bound state energy can be obtained from the poles of the retarded two-particle vertex $\Gamma(0, \omega + i0^+)$. Using

$$\Pi(0, \omega + i0^+) = \int_{\mathbf{k}} \frac{1}{-\omega - i0^+ + 2\epsilon_{\mathbf{k}}} \simeq \frac{(m^3\Lambda)^{1/2}}{\sqrt{2}\pi^2} - \frac{(m^3|\omega|)^{1/2}}{4\pi} \tag{7.392}$$

for $\omega < 0$ and $\Lambda \gg |\omega|$, we obtain

$$\frac{1}{\Gamma(0, \omega + i0^+)} = \frac{m}{4\pi a}\left[1 - a(m|\omega|)^{1/2}\right]. \tag{7.393}$$

---

[83] All cutoffs lead to the same low-energy physics when expressed in terms of a.

[84] See also Appendix 7.A.

[85] $\Pi(q)$ is equal to the pairing susceptibility $\chi_0(q)$ evaluated for $\mu = T = 0$.

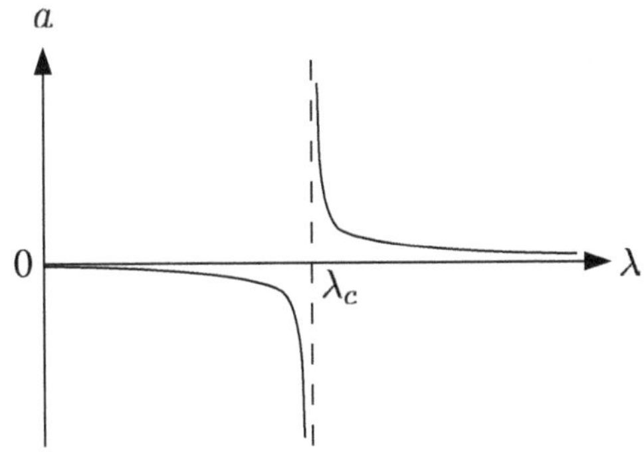

Fig. 7.25 s-wave scattering length a in vacuum *vs* λ for a fixed value of the cutoff Λ ($\lambda_c = \sqrt{2}\pi^2/(m^3\Lambda)^{1/2}$).

For $a > 0$ (i.e. $\lambda > \lambda_c$), the T matrix admits a pole at the negative energy $\omega = -E_b = -1/ma^2$. a determines not only the energy E_b but also the extension $\sim 1/(mE_b)^{1/2}$ of the bound state.[86]

7.7.1.2 *Mean-field equations*

After these preliminaries, we are in a position to determine the evolution of the order parameter Δ and the chemical potential μ as a function of the scattering length a within the BCS theory. As the momentum distribution can be significantly modified in the strong-coupling limit where the pairs become local, we cannot assume that the chemical potential is given by the Fermi energy ϵ_F of the non-interacting system but must solve the equation $N = n\mathcal{V} = -\partial\Omega_{\rm MF}/\partial\mu$ together with the gap equation $\partial\Omega_{\rm MF}/\partial\Delta = 0$ [Eq. (7.260)] where $\Omega_{\rm MF} = -\frac{1}{\beta}\ln Z_{\rm MF}$. Using (7.391) to eliminate the explicit dependence on λ and Λ, we obtain

$$\frac{m}{4\pi a} = \int_{\mathbf{k}} \left[\frac{1}{2\epsilon_{\mathbf{k}}} - \frac{1}{2E_{\mathbf{k}}}\tanh\left(\frac{\beta E_{\mathbf{k}}}{2}\right)\right], \tag{7.394}$$

$$n = \int_{\mathbf{k}} \left[1 - \frac{\xi_{\mathbf{k}}}{E_{\mathbf{k}}}\tanh\left(\frac{\beta E_{\mathbf{k}}}{2}\right)\right]. \tag{7.395}$$

Since the sums over $\mathbf{k}$ are convergent, we can take the limit $\Lambda \to \infty$ (with a fixed).

7.7.1.3 *Ground state*

The chemical potential and the gap derived from a numerical integration of (7.394) and (7.395) at zero temperature are shown in Fig. 7.26. The BCS–BEC crossover at $T = 0$ is very sharp and occurs for $-1 \lesssim 1/k_F a \lesssim 1$.

[86]Experimentally, the crossover from $a < 0$ to $a > 0$ in an ultracold gas can be induced by a Feshbach resonance, see Appendix 7.A.

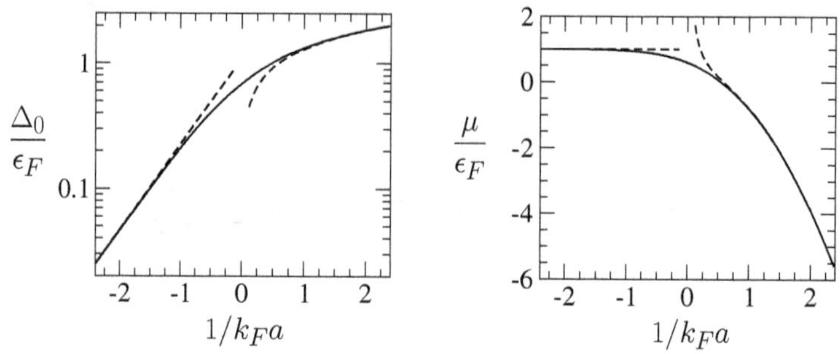

Fig. 7.26 Order parameter Δ_0 and chemical potential μ *vs* $1/k_F a$ at $T = 0$. Analytical results for the weak- and strong-coupling limit are shown as dashed lines.

Analytical results can be obtained in the weak- and strong-coupling limit. In the BCS limit $1/k_F a \to -\infty$, $\Delta_0/\mu \to 0$, we find

$$\begin{aligned} n = 2\int_{\mathbf{k}} v_{\mathbf{k}}^2 &= \int_{-\mu}^{\infty} d\xi\, N(\xi) \left[1 - \frac{\xi}{|\xi|} + \frac{\xi}{|\xi|} - \frac{\xi}{(\xi^2 + \Delta_0^2)^{1/2}} \right] \\ &= \int_{-\mu}^{\infty} d\xi\, N(\xi) \left[2\Theta(-\xi) + \frac{\xi}{|\xi|} - \frac{\xi}{(\xi^2 + \Delta_0^2)^{1/2}} \right] \\ &\simeq 2\int_{-\mu}^{\infty} d\xi\, N(\xi)\Theta(-\xi) \end{aligned} \tag{7.396}$$

($v_{\mathbf{k}}^2$ is defined by (7.247)), where $N(\xi)$ is the (non-interacting) density of states. This shows that the chemical potential μ equals its value $\epsilon_F = k_F^2/2m$ in the normal state. The gap is given by

$$\Delta_0 = 8e^{-2}\epsilon_F e^{-\pi/2k_F|a|}. \tag{7.397}$$

This result is obtained by writing the mean-field equation (7.394) in terms of the hypergeometric function and then using $\mu/\Delta_0 \to \infty$ [69]. Apart from the prefactor, it can be directly obtained from the previous result (7.263) by replacing $-1/\lambda N(0)$ by $\pi/2k_F a + \text{const}$, using $N(0) = mk_F/2\pi^2$ and (7.391).[87]

In the strong-coupling limit $1/k_F a \gg 1$, we anticipate that $\mu < 0$ and $\Delta_0^2/\mu^2 \sim na^3$. To leading order in $na^3 \sim (k_F a)^3$, we then find

$$\begin{aligned} \frac{m}{4\pi a} &= \frac{1}{2}\int_{\mathbf{k}} \left(\frac{1}{\epsilon_{\mathbf{k}}} - \frac{1}{\xi_{\mathbf{k}}} + \frac{\Delta_0^2}{2\xi_{\mathbf{k}}^3} \right), \\ n &= \int_{\mathbf{k}} \frac{\Delta_0^2}{2\xi_{\mathbf{k}}^2}. \end{aligned} \tag{7.398}$$

[87] Equations (7.263) and (7.391) give $\Delta_0 = 2\Lambda e^{\pi/2k_F a - \sqrt{2m\Lambda}/k_F}$. Since the Fermi energy acts as an effective high-energy cutoff for the pairing, by taking the limit $\Lambda \to \epsilon_F$ (with a fixed), we obtain $\Delta_0 \sim \epsilon_F e^{\pi/2k_F a}$ in agreement with (7.397). A similar argument gives $T_0 \sim \epsilon_F e^{\pi/2k_F a}$ for the pair-formation temperature [Eq. (7.403)].

Using

$$\int_{\mathbf{k}} \left(\frac{1}{\epsilon_{\mathbf{k}}} - \frac{1}{\xi_{\mathbf{k}}} \right) = \frac{m^{3/2}|\mu|^{1/2}}{\sqrt{2}\pi}, \qquad \int_{\mathbf{k}} \frac{1}{\xi_{\mathbf{k}}^p} = m^{3/2} c_p |\mu|^{3/2-p},$$
$$c_p = \frac{(p-\frac{5}{2})(p-\frac{7}{2})\cdots\frac{1}{2}}{(p-1)!} c_2 \quad (p \geq 3), \qquad c_2 = \frac{1}{2\sqrt{2}\pi} \tag{7.399}$$

(these relations are deduced from one another by applying $\partial_{|\mu|}$), we obtain

$$\frac{m}{4\pi a} = \frac{m^{3/2}|\mu|^{1/2}}{2\sqrt{2}\pi} + \frac{\Delta_0^2 m^{3/2}|\mu|^{-3/2}}{32\sqrt{2}\pi}, \qquad n = \frac{m^{3/2}\Delta_0^2}{4\sqrt{2}\pi|\mu|^{1/2}}. \tag{7.400}$$

From (7.400), we finally deduce

$$\Delta_0 = \left(\frac{4\pi n}{m^2 a} \right)^{1/2} = \left(\frac{4\delta\mu}{ma^2} \right)^{1/2},$$
$$\mu = -\frac{E_b}{2} + \delta\mu \quad \text{with} \quad \delta\mu = \frac{\pi n a}{m} \ll E_b \tag{7.401}$$

to leading order in na^3. From these results, we deduce the zero-temperature phase diagram shown in Fig. 7.27.

The momentum distribution $n_{\mathbf{k}} = v_{\mathbf{k}}^2$ clearly shows the change in behavior as the system evolves from the BCS to the BEC limit. In the weak-coupling limit, $n_{\mathbf{k}} \simeq \Theta(k_F - |\mathbf{k}|)$ is little affected by the gap Δ_0. With increasing coupling, the pairs become more tightly bound and $n_{\mathbf{k}} - \Theta(k_F - |\mathbf{k}|)$ broader (Fig. 7.28).

The gap in the one-particle excitation spectrum is given by the minimum of the Bogoliubov quasi-particle energy,

$$E_{\text{gap}} = \min_{\mathbf{k}} \left(\xi_{\mathbf{k}}^2 + \Delta_0^2 \right)^{1/2} = \begin{cases} \Delta_0 & \text{for } \mu > 0, \\ (\mu^2 + \Delta_0^2)^{1/2} & \text{for } \mu < 0. \end{cases} \tag{7.402}$$

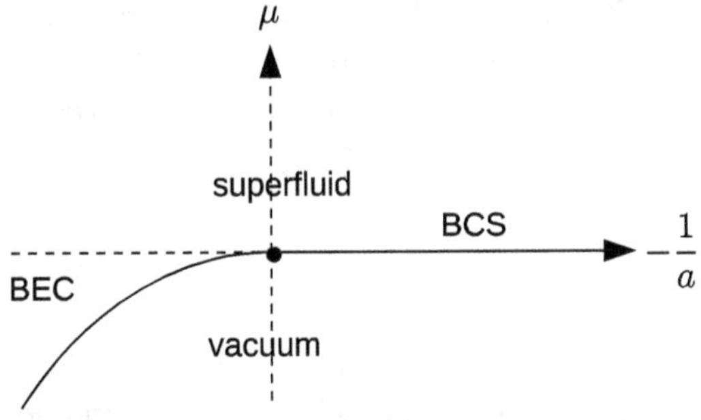

Fig. 7.27 Zero-temperature phase diagram of a three-dimensional Fermi gas near a Feshbach resonance. $1/a$ measures the detuning from the resonance. The vacuum state has no particles and is determined by $\mu \leq 0$ if $a < 0$ (BCS side) and $\mu < -1/2ma^2$ if $a > 0$ (BEC side). The density of particles vanishes continuously at the transition between the superfluid phase and the vacuum. As discussed in Sec. 7.7.3, the quantum multicritical point at $\mu = 1/a = 0$ (shown as a black dot) controls the universal physics of the Fermi gas near the Feshbach resonance.

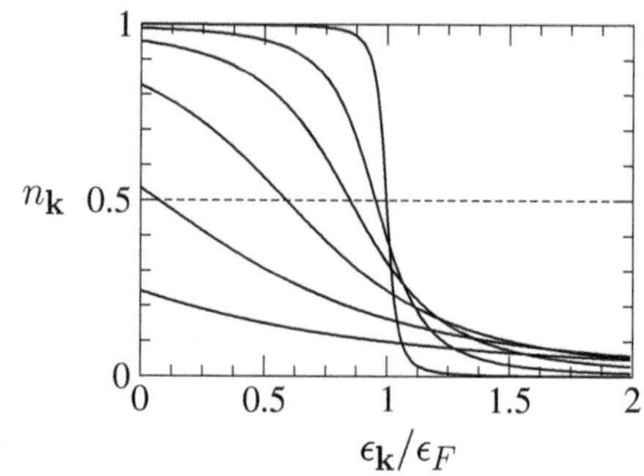

Fig. 7.28 Momentum distribution $n_{\mathbf{k}}$ *vs* $\epsilon_{\mathbf{k}}$ for $1/k_Fa = -2, -1, -0.5, 0, 0.5$ and 1 (from top to bottom). $n_{\mathbf{k}}$ intersects the dotted line $n_{\mathbf{k}} = 1/2$ when $\epsilon_{\mathbf{k}} = \mu$.

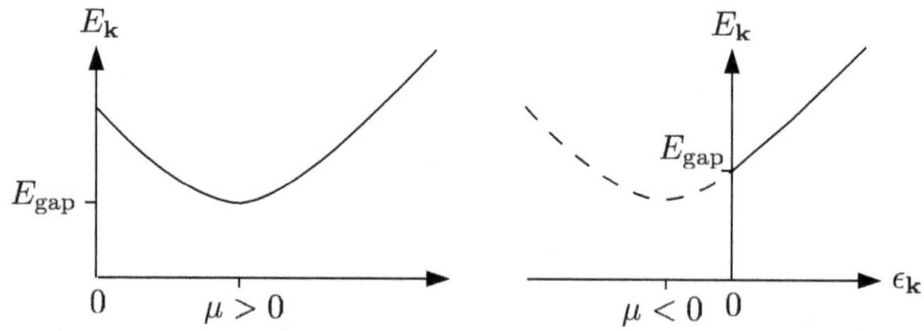

Fig. 7.29 Bogoliubov quasi-particle energy $E_{\mathbf{k}} = (\xi_{\mathbf{k}}^2 + \Delta_0^2)^{1/2}$ *vs* $\epsilon_{\mathbf{k}}$ for $\mu > 0$ and $\mu < 0$.

When μ is positive and lies in the fermion band, E_{gap} is given by Δ_0. The BCS–BEC crossover roughly occurs when μ becomes negative and E_{gap} starts depending both on Δ_0 and μ (Fig. 7.29). In the strong-coupling limit, where the chemical potential lies below the bottom of the band with $|\mu| \gg \Delta_0$, the one-particle excitation gap is mainly determined by $\mu \simeq -E_b/2$. In this limit, the bound state energy E_b is much larger than the Fermi energy ϵ_F. The minimum energy to break a Cooper pair, $2E_{\text{gap}} \simeq 2|\mu|$, is then equal to the bound state energy E_b of the composite boson. Contrary to the BCS limit, the gap does not originate from the broken symmetry but merely from the existence of a two-body bound state.

7.7.1.4 *The pair-formation temperature T_0*

The temperature T_0 below which the gap Δ is non-zero, obtained from Eqs. (7.394) and (7.395), is shown in Fig. 7.31. At the mean-field level, the symmetric phase ($\Delta = 0$) can only describe a normal state consisting of essentially non-interacting fermions. Thus, T_0 is the temperature at which the fermions form two-body bound states. At weak coupling, the superconducting transition is controlled by pair breaking so that T_c coincides with T_0 (up to small corrections due to thermal fluctuations) as we shall see in the following section. By contrast, in the strong-coupling limit T_0 is much larger than T_c.

Analytical expressions for T_0 can be obtained in the two limiting cases $1/k_Fa \to -\infty$ and $1/k_Fa \to \infty$. In the BCS limit, one has

$$\mu(T_0) \simeq \epsilon_F, \qquad T_0 \simeq \frac{8e^{-2}\gamma}{\pi}\epsilon_F e^{-\pi/2k_F|a|}, \tag{7.403}$$

where γ is the exponential of the Euler constant. Except for the value of the prefactor, this result can be deduced from the previous result (7.235).[87]

In the strong-coupling limit, the mean-field equations for $T = T_0$ can be written as

$$\begin{aligned} \frac{m}{4\pi a} &= \frac{1}{2}\int_{\mathbf{k}}\left[\frac{1}{\epsilon_{\mathbf{k}}} - \frac{1}{\xi_{\mathbf{k}}}\tanh\left(\frac{\beta\xi_{\mathbf{k}}}{2}\right)\right] \simeq \frac{1}{2}\int_{\mathbf{k}}\left(\frac{1}{\epsilon_{\mathbf{k}}} - \frac{1}{\xi_{\mathbf{k}}}\right) = \frac{m^{3/2}|\mu|^{1/2}}{2\sqrt{2}\pi}, \\ n &= \int_{\mathbf{k}}\left[1 - \tanh\left(\frac{\beta\xi_{\mathbf{k}}}{2}\right)\right] \simeq 2\int_{\mathbf{k}} e^{-\beta\xi_{\mathbf{k}}} = \frac{1}{\sqrt{2}}\left(\frac{m}{\beta\pi}\right)^{3/2} e^{\beta\mu}, \end{aligned} \tag{7.404}$$

using $\xi_{\mathbf{k}} = \epsilon_{\mathbf{k}} + |\mu| \gg T$ and $\int_0^\infty dx\sqrt{x}e^{-x} = \sqrt{\pi}/2$. In the limit $1/k_Fa \to \infty$, Eq. (7.404) give

$$\mu(T_0) = -\frac{E_b}{2}, \qquad T_0 \sim \frac{E_b}{\ln(E_b/\epsilon_F)}. \tag{7.405}$$

7.7.2 *Fluctuations: The Nozières–Schmitt-Rink approach*

To study the occurrence of superconductivity, one must take into account the center of mass motion of the pairs. The latter corresponds to fluctuations of Δ about its average value $\langle\Delta\rangle$.

7.7.2.1 *Gaussian fluctuations in the normal phase*

Let us first consider the normal phase where $\langle\Delta\rangle = 0$. Starting from the action (7.325) (with a cutoff Λ acting on the dispersion $\epsilon_{\mathbf{k}}$) and integrating out the fermions, we obtain

$$\begin{aligned} Z &= Z_{\rm MF}\det(\lambda)\int\mathcal{D}[\Delta^*,\Delta]\,e^{-S[\Delta^*,\Delta]}, \\ S[\Delta^*,\Delta] &= \sum_q|\Delta(q)|^2\left[\frac{1}{\lambda} - \chi_0(q)\right] \equiv \sum_q|\Delta(q)|^2\Gamma(q)^{-1} \end{aligned} \tag{7.406}$$

if we retain only the Gaussian fluctuations of the pairing field Δ. $\chi_0(q)$ is the pairing susceptibility introduced in Sec. 7.5.1. Carrying out the functional integral in (7.406), we obtain the thermodynamic potential

$$\begin{aligned} \Omega &= -\frac{1}{\beta}\ln[Z_{\rm MF}\det(\lambda\Gamma)] = \Omega_{\rm MF} - \frac{1}{\beta}\sum_q\ln[\lambda\Gamma(q)]e^{i\omega_\nu 0^+} \\ &= \Omega_{\rm MF} - \frac{1}{\beta}\sum_q\sum_{n=1}^\infty\frac{[\lambda\chi_0(q)]^n}{n}e^{i\omega_\nu 0^+}, \end{aligned} \tag{7.407}$$

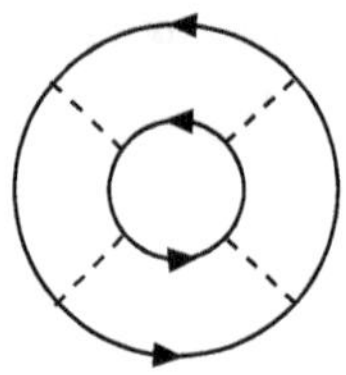

Fig. 7.30 A typical diagram describing the contribution of pair fluctuations to the thermodynamic potential [Eq. (7.407)].

where we have added the usual convergence factor $e^{i\omega_\nu 0^+}$. The last expression shows that the pair fluctuation contribution to the thermodynamic potential is given by the sum of ladder diagrams (Fig. 7.30).[88]

The Gaussian action $S[\Delta^*, \Delta]$ can only describe the normal phase ($\langle\Delta\rangle = 0$). The transition to the superconducting phase is signaled by a divergence of the bosonic propagator $\Gamma(q)$ at zero momentum and frequency ($q = 0$). Using the definition of the scattering length in (7.391), the inverse propagator reads[89]

$$\begin{aligned}\Gamma(q)^{-1} &= -\frac{m}{4\pi a} + \int_{\mathbf{k}} \frac{1}{2\epsilon_{\mathbf{k}}} - \chi_0(q) \\ &= -\frac{m}{4\pi a} + \int_{\mathbf{k}} \left[\frac{1}{2\epsilon_{\mathbf{k}}} + \frac{1 - n_F(\xi_{\mathbf{k}}) - n_F(\xi_{\mathbf{q}-\mathbf{k}})}{i\omega_\nu - \xi_{\mathbf{k}} - \xi_{\mathbf{q}-\mathbf{k}}}\right],\end{aligned} \tag{7.408}$$

so that the criterion $\Gamma(q = 0)^{-1} = 0$ reproduces the mean-field gap equation [Eq. (7.394) with $\Delta \to 0$]. Non-trivial effects beyond mean-field theory come from the equation $N = n\mathcal{V} = -\partial\Omega/\partial\mu$,

$$n = n_{\mathrm{MF}} + \frac{\partial}{\partial\mu}\frac{1}{\beta\mathcal{V}}\sum_q \ln\left[\lambda\Gamma(q)\right] e^{i\omega_\nu 0^+}, \tag{7.409}$$

where $n_{\mathrm{MF}} = \int_{\mathbf{k}}[1 - \tanh(\beta\xi_{\mathbf{k}}/2)]$. The superconducting transition obtained from (7.408) and (7.409) is shown in Fig. 7.31. As already emphasized, $T_c \simeq T_0$ in the BCS limit, whereas $T_c \ll T_0$ is the BEC limit.

In the strong-coupling limit, when $T \sim T_c \ll T_0$, $\Gamma(q)$ is well approximated by its zero-temperature value since T_c is much smaller than the mean-field gap set by $|\mu|$. We then obtain

$$\begin{aligned}\Gamma(q)^{-1} &\simeq -\frac{m}{4\pi a} + \int_{\mathbf{k}} \left[\frac{1}{2\epsilon_{\mathbf{k}}} - \frac{1}{2\xi_{\mathbf{k}}} + \frac{\mathbf{q}^2}{8m\xi_{\mathbf{k}}^2} - \frac{(\mathbf{k}\cdot\mathbf{q})^2}{8m^2\xi_{\mathbf{k}}^3} - \frac{i\omega_\nu}{4\xi_{\mathbf{k}}^2}\right] \\ &= \mathbf{q}^2\frac{ma}{32\pi} - i\omega_\nu\frac{m^2 a}{8\pi} - \frac{m^2 a}{4\pi}\delta\mu\end{aligned} \tag{7.410}$$

[88]The connection between the Gaussian approximation (for the action of an auxiliary bosonic field) and the RPA (summation of ladder and/or bubble graphs) has also been noticed in Chapter 5.

[89]Note that $\Gamma(0, 0)$ in (7.391), besides being defined for $\mu = T = 0$, has the sign opposite to $\Gamma(q)$ in (7.406).

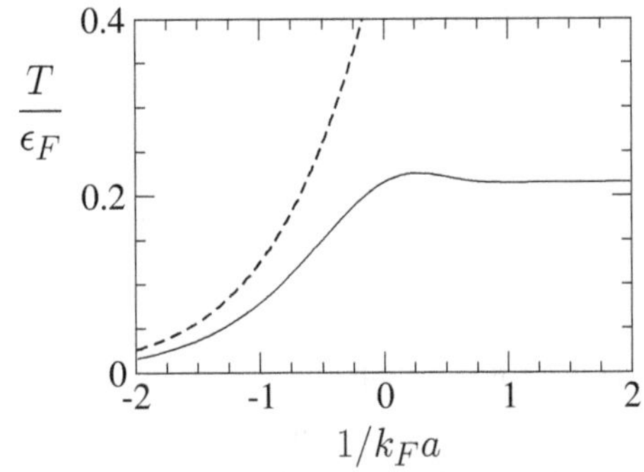

Fig. 7.31 Qualitative phase diagram of the BCS–BEC crossover: pair-formation temperature T_0 (dashed line) and superconducting transition temperature T_c (solid line) *vs* $1/k_Fa$. (Figure reproduced from Ref. [71].)

to lowest order in $\mathbf{q}^2$, ω_ν and $\delta\mu = \mu + E_b/2$. We have used (7.399) and $\int_{\mathbf{k}} (\mathbf{k}\cdot\mathbf{q})^2/8m^2\xi_{\mathbf{k}}^3 = \mathbf{q}^2 ma/32\pi$. The isolated pole $i\omega_\nu = \mathbf{q}^2/4m - 2\mu - E_b$ in (7.410) corresponds to the two-body bound state (i.e. the pair center of mass motion). At higher energies, we expect $\Gamma(q)$ to exhibit a branch cut reflecting the continuum of two-particle excitations due to pair breaking. The criterion $\Gamma(q=0)^{-1} = 0$ together with (7.409) gives

$$\begin{aligned} \mu &= -\frac{E_b}{2}, \\ n &= -\frac{2}{\beta\mathcal{V}} \sum_q \frac{e^{i\omega_\nu 0^+}}{i\omega_\nu + 2\mu + E_b - \frac{\mathbf{q}^2}{4m}} = 2\int_{\mathbf{q}} n_B\left(\frac{\mathbf{q}^2}{4m} - 2\mu - E_b\right) \end{aligned} \tag{7.411}$$

(noting that $n_{\rm MF}$ is negligible when $T_c \ll |\mu|$) if we use the low-energy limit (7.410). These equations show that in the strong-coupling limit, the superconducting transition corresponds to the BEC of bosons of mass $m_B = 2m$ and density $n_B = n/2$,[90] i.e.

$$T_c = \frac{2\pi}{m_B}\left(\frac{n_B}{\zeta(3/2)}\right)^{2/3} = \frac{\pi}{m}\left(\frac{n}{2\zeta(3/2)}\right)^{2/3} \simeq 0.218\epsilon_F. \tag{7.412}$$

Interactions between bosons (not considered here) do not substantially affect the condensation temperature, the correction being of order $n^{1/3}a$ (Sec. 7.4.3).

In the strong-coupling limit, one can also derive an explicit expression of the action $S[\Delta^*, \Delta]$. By introducing the rescaled field

$$\Psi(q) = \left(\frac{m^2 a}{8\pi}\right)^{1/2} \Delta(q), \tag{7.413}$$

[90]Recall that BEC of free bosons occurs when the chemical potential coincides with the minimum value of the dispersion (i.e. $2\mu = -E_b$ here).

we obtain from (7.406) the action of a free boson gas,

$$S[\Psi^*, \Psi] = \sum_q |\Psi(q)|^2 \Big(-i\omega_\nu + \frac{\mathbf{q}^2}{2m_B} - E_b - \mu_B \Big), \tag{7.414}$$

with mass m_B, dispersion $\omega_{\mathbf{q}} = -E_b + \mathbf{q}^2/2m_B$ and chemical potential $\mu_B = 2\mu$.

7.7.2.2 *Beyond Gaussian fluctuations*

In order to describe the superfluid phase below T_c, one should (at least) include the quartic term $S^{(4)}$ in the bosonic action $S[\Delta^*, \Delta]$. The calculation of $S^{(4)}$ is similar to that of the quartic term in the Ginzburg–Landau expansion of the mean-field free energy in the superconducting (see the diagram for the coefficient B in Fig. 7.16),

$$\begin{aligned} S^{(4)}[\Delta^*, \Delta] &= \frac{1}{2(\beta\mathcal{V})^2} \sum_{k,q_1,q_2,q_3} \Delta^*(q_1)\Delta(q_2)\Delta^*(q_3)\Delta(q_1 - q_2 + q_3) \\ &\quad \times G_{0,\uparrow}(k)G_{0,\downarrow}(q_1 - k)G_{0,\downarrow}(q_2 - k)G_{0,\uparrow}(q_3 - q_2 + k) \\ &\simeq \frac{1}{2\beta\mathcal{V}} \sum_k [G_{0,\uparrow}(k)G_{0,\downarrow}(-k)]^2 \int_0^\beta d\tau \int d^3r |\Delta(x)|^4. \end{aligned} \tag{7.415}$$

Using

$$\frac{1}{\beta\mathcal{V}} \sum_k [G_{0,\uparrow}(k)G_{0,\downarrow}(-k)]^2 = \int_{\mathbf{k}} \frac{1}{4\xi_{\mathbf{k}}^3} \tag{7.416}$$

and (7.399) (with $\mu = -E_b/2$ to leading order), we finally obtain

$$S^{(4)}[\Psi^*, \Psi] = \frac{1}{2}\frac{4\pi a_B}{m_B} \int_0^\beta d\tau \int d^3r |\Psi(x)|^4. \tag{7.417}$$

In the low-energy limit, the composite bosons interact *via* a repulsive interaction characterized by a (boson) scattering length $a_B = 2a$. This result should be compared with the exact value $a_B \simeq 0.6a$ obtained from the solution of the four-body problem [80]. The higher-order terms in the Ψ field are not expected to be important even at $T = 0$. Indeed, the expansion of the action in powers of Δ is controlled by the parameter $|\Delta|^2/|\mu|^2$ whose mean-field estimate (7.401) is of order na^3.[91] From the bosonic action defined by (7.414) and (7.417), we immediately conclude that the low-energy excitations are phonons (Bogoliubov sound mode) with a velocity

$$c = \frac{\sqrt{4\pi a_B n_B}}{m_B} = \frac{\sqrt{\pi a n}}{m} \tag{7.418}$$

to leading order in na^3 (see Sec. 7.4). A more complete analysis of the BCS–BEC crossover shows that the Anderson–Bogoliubov sound mode of the BCS superconductor continuously evolves into the Bogoliubov sound mode of the composite bosons in the BEC limit [71].

[91] The coefficient of the $2p$th-order term is proportional to $\frac{1}{\beta\mathcal{V}}\sum_k [G(k)G(-k)]^p \sim \int_{\mathbf{k}} \xi_{\mathbf{k}}^{1-2p} \sim m^{3/2}|\mu|^{-2p+5/2}$. The $2p$th-order term in the action is then of order $m^{3/2}|\mu|^{-2p+5/2}|\Delta|^{2p} \sim (na^3)^p/ma^5$ using the mean-field values of μ and Δ [Eq. (7.401)].

It is quite remarkable that a bosonic effective action in the BCS–BEC crossover can be derived from the underlying fermion degrees of freedom. Nevertheless, our approach is not fully satisfying for the following reason. In a boson system, the amplitude of the field $\Psi_B(x) = \sqrt{n_B(x)}e^{i\theta(x)}$ is related to the particle density $n_B(x)$. Since the field $\Psi(x)$ was obtained by rescaling the pairing field $\Delta(x) = |\Delta(x)|e^{i\theta(x)}$, there is no guarantee that it coincides with $\sqrt{n_B(x)}e^{i\theta(x)} = \sqrt{n(x)/2}e^{i\theta(x)}$ although this turns out to be true in the BEC limit at the mean-field level: $|\Psi(x)|^2 = n/2$ when $\Delta(x) = \Delta_0$ [Eqs. (7.401) and (7.413)]. In a fermion system, there is *a priori* no simple relation between the amplitude of the pairing field and the density. This suggests that a minimal description of the BCS–BEC crossover should include both the pairing ($\Delta(x)$) and the density ($n(x)$) fields from the outset. Such an approach does indicate that the pairing and amplitude fluctuations are not independent in the BEC limit but satisfy

$$|\Delta(x)|^2 \approx \frac{4\pi}{m^2 a} n(x), \tag{7.419}$$

thus making possible a complete description in terms of the bosonic field $\Psi(x) = \sqrt{n(x)/2}e^{i\theta(x)}$ [78, 79]. Away from the strong-coupling limit, density fluctuations affect pairing fluctuations and should be (at least in principle) taken into account. The very fact that density and phase (of the pairing field) are conjugate variables should have led us to consider density and pairing fluctuations on an equal footing [78]. In the BCS limit, the effect of density fluctuations turns out to be weak and can be neglected.[92]

As a final remark, we note that we have expanded the action $S[\Delta^*, \Delta]$ about $\Delta = 0$. Instead, we could (and should rather) expand about the mean-field value of Δ. Below the pair-formation temperature T_0, although the order parameter $\langle\Delta\rangle$ vanishes for $T \geq T_c$ due to the lack of phase coherence, the mean amplitude $\langle|\Delta|\rangle$ becomes finite, reflecting the formation of local pairs.[93] A thorough analysis along these lines gives much better results from a quantitative point of view [81]. In particular, the boson scattering length $a_B \simeq 0.57a$ is found to be much closer to the exact four-body prediction $a_B \simeq 0.6a$.

7.7.3 *The dilute Fermi gas near a Feshbach resonance*

The absence of an intrinsic length scale at the unitary limit, where $1/a = 0$, implies that the physics of the dilute Fermi gas near a Feshbach resonance is universal and controlled by only three parameters: μ, T and the detuning ν from the resonance. This universal physics arises from the existence of a multicritical point at $\mu = T = 1/a = 0$.

7.7.3.1 *Equation of state*

Let us consider the two-body problem in vacuum from the RG point of view. This problem has already been discussed in the bosonic case in Sec. 7.4.4. The interaction constant

[92] For instance, in the BCS limit density, fluctuations change the velocity of the Anderson–Bogoliubov mode $c = \frac{v_F}{\sqrt{3}}$ into $c = \frac{v_F}{\sqrt{3}}[1 - \lambda N(0)]^{1/2}$.

[93] In the equivalent magnetic problem, $\langle|\Delta|\rangle$ would determine the amplitude of the local moments that form below T_0 and order at the much lower temperature T_c.

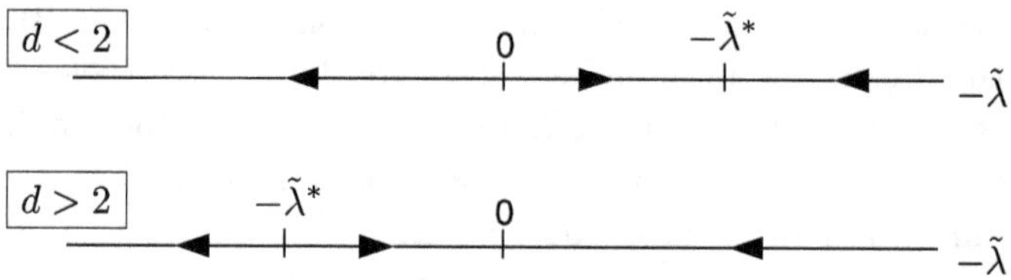

Fig. 7.32 RG flow obtained from (7.420). For $d < 2$, the stable fixed point $-\tilde\lambda^* > 0$ describes bosons or fermions with a repulsive interaction. For $d > 2$, the unstable fixed point $-\tilde\lambda^* < 0$ describes fermions at a Feshbach resonance.

$\tilde\lambda = K_d(2m)^{d/2}\Lambda^{(d-2)/2}\lambda$ satisfies the RG equation[94]

$$\frac{d\tilde\lambda}{dl} = (2-d)\tilde\lambda + \frac{1}{2}\tilde\lambda^2 \tag{7.420}$$

in d dimensions. Apart from the Gaussian fixed point $\tilde\lambda^* = 0$, this equation admits a stable fixed point for repulsive interactions ($-\lambda > 0$) when $d < 2$ and an unstable fixed point for attractive interactions ($-\lambda < 0$) when $d > 2$ (Fig. 7.32). The latter is not physical for bosons as the system would not be stable in the presence of attractive interactions. For fermions, it describes particles at a Feshbach resonance as one can see by noting that the fixed point is reached by fine-tuning λ to λ_c; there is then no intrinsic scale since $a \to \infty$. $\tilde\lambda \to 0$ for $l \to \infty$ when $\lambda < \lambda_c$, whereas $-\tilde\lambda \to -\infty$ when $\lambda > \lambda_c$, which corresponds to a strong interaction between fermions giving rise to a bound state with energy $-E_b = -1/ma^2$.

Thus for fermions at zero temperature with attractive interactions, in addition to the chemical potential μ, there is a second relevant perturbation $\nu \propto \tilde\lambda^* - \tilde\lambda$ (not to be confused with the critical exponent ν) when $d > 2$ corresponding to the detuning from the Feshbach resonance. By linearizing the RG equation (7.420), one finds that this perturbation is associated with the eigenvalue $d-2$,[95] i.e.

$$[\mu] = 2, \qquad [\nu] = d-2. \tag{7.421}$$

The corresponding zero-temperature phase diagram is shown in Fig. 7.27 for $d = 3$.[96] Since the temperature has scaling dimension $[T] = z = 2$ at the multicritical point $\mu = \nu = 0$,[95] we deduce that the pressure of the dilute Fermi gas near a Feshbach resonance can be expressed in the scaling form

$$P(\mu, T, \nu) = \left(\frac{m}{2\pi}\right)^{d/2} T^{d/2+1}\mathcal{F}_d\left(\frac{\mu}{T}, \frac{\nu}{(mT)^{(d-2)/2}}\right). \tag{7.422}$$

[94]The different signs in (7.197) and (7.420) come from the fact that we denote by $-\lambda$ the interaction in the fermion case. Note also that Λ now denotes the energy cutoff (and not the momentum cutoff).

[95]The scaling dimension $[\mu] = 2$ and the dynamical critical exponent $z = 2$ are deduced from the fact that the ground state at the multicritical point is the vacuum; see the discussion in Sec. 7.4.4.

[96]For arbitrary $d > 2$, the phase diagram is similar with the horizontal axis labeled by the detuning ν from resonance.

In three dimensions, it is customary to define $\nu = -1/a$ so that

$$P(\mu, T, \nu) = \left(\frac{m}{2\pi}\right)^{3/2} T^{5/2} \mathcal{F}\left(\frac{\mu}{T}, \frac{1}{\sqrt{mTa}}\right) \tag{7.423}$$

with a slight redefinition of the scaling function $\mathcal{F} \equiv \mathcal{F}_3$ to eliminate a minus sign. This expression is similar to that of the pressure of the three-dimensional Bose gas derived in Sec. 7.4.4. The reason for the dependence of the universal scaling function on $\sqrt{mTa}$ is however different in the two cases. In the Bose gas case, it is due to the interaction constant λ being dangerously irrelevant at the Gaussian fixed point while in the fermion case it is due to λ being a relevant perturbation at the multicritical point.

Equation (7.422) implies that the pressure of the unitary Fermi gas, i.e. at a Feshbach resonance, expressed in units of $m^{d/2}T^{d/2+1}$ is a function of μ/T only,

$$P(\mu, T) = \left(\frac{m}{2\pi}\right)^{d/2} T^{d/2+1} \mathcal{F}_d\left(\frac{\mu}{T}, 0\right) \qquad \text{(unitary limit)}. \tag{7.424}$$

From the particle and entropy densities, $n = \partial P/\partial\mu$ and $S/\mathcal{V} = \partial P/\partial T$, one then obtains

$$P\mathcal{V} = \frac{2}{d} E \tag{7.425}$$

using $-P\mathcal{V} = E - TS - \mu N$. The phase transition from the normal to the superfluid phase is associated with a singularity of the scaling function $\mathcal{F}_d(x, 0)$ at x_c so that the ratio $T_c/\mu = 1/x_c$ is universal. Since $n = (mT/2\pi)^{d/2} \partial_x \mathcal{F}_d(x, 0)|_{x=x_c}$, it follows that the ratio T_c/ϵ_F between the transition temperature and the Fermi energy ϵ_F of the non-interacting gas[97] is also universal. At zero temperature, it is convenient to write the pressure in the scaling form,

$$P(\mu) = \left(\frac{m}{2\pi}\right)^{3/2} \mu^{5/2} \mathcal{G}\left(\frac{1}{\sqrt{m\mu a}}\right), \tag{7.426}$$

where we now consider a three-dimensional fluid and assume $\mu > 0$ (which is satisfied near the unitary limit). Computing the density $n = \partial P/\partial\mu$ and the energy $E = -P\mathcal{V} + \mu n$, one finds that

$$\frac{E}{N} = \frac{3}{5} \epsilon_F \xi \tag{7.427}$$

at the unitary limit, where the Bertsch parameter

$$\xi = \frac{4}{\pi^{1/3}} \left(\frac{2}{15\mathcal{G}(0)}\right)^{2/3} \tag{7.428}$$

is a universal number ($\xi = 1$ for free fermions).

In the BCS theory, $T_c/\epsilon_F \simeq 0.5$ and $\xi \simeq 0.59$. In the Nozières–Schmitt-Rink approach to the three-dimensional BCS–BEC crossover discussed in Sec. 7.7.2, one finds $T_c/\epsilon_F \simeq 0.23$ [85]. These results can be improved in various ways [73,83], but diagrammatic theories

[97] $\epsilon_F = (3\pi^2 n)^{2/3}/2m$ for $d = 3$.

do not yield a precise estimate of universal ratios at the unitary limit and more involved theories are required.[98]

7.7.3.2 *Two-body contact*

As for the case of bosons (Sec. 7.4.5), the two-body contact C can be defined from the equation of state by taking a derivative with respect to the inverse of the scattering length,

$$\frac{C}{4\pi m} = -\left.\frac{\partial \Omega}{\partial(1/a)}\right|_{\mu,T,\mathcal{V}} = -\left.\frac{\partial E}{\partial(1/a)}\right|_{N,S,\mathcal{V}}. \tag{7.429}$$

Note that this definition differs from (7.206) and (7.207) by a factor 1/2. From the equation of state (7.423), one then obtains

$$\frac{C}{4\pi m} = 2a\left(\frac{3}{2}P\mathcal{V} - E\right) \tag{7.430}$$

for a three-dimensional system. Since $P\mathcal{V} = (3/2)E$ at the unitary limit, it is tempting to see the contact as a measure of deviations from the unitary limit (or scale invariance). C however converges to a non-zero value when a diverges [see Eq. (7.437)].

The contact determines the short-distance behavior of the two-body density matrix

$$\begin{aligned}\rho_2(\mathbf{r}_1, \mathbf{r}_2; \mathbf{r}'_2, \mathbf{r}'_1) &= \langle \hat\psi^\dagger_\uparrow(\mathbf{r}_1)\hat\psi^\dagger_\downarrow(\mathbf{r}_2)\hat\psi_\downarrow(\mathbf{r}'_2)\hat\psi_\uparrow(\mathbf{r}'_1)\rangle \\ &\simeq \frac{C}{(4\pi)^2\mathcal{V}}\frac{1}{|\mathbf{r}_{12}|\,|\mathbf{r}'_{12}|} \qquad (b \ll |\mathbf{r}_{12}|, |\mathbf{r}'_{12}| \ll a, n^{-1/3})\end{aligned} \tag{7.431}$$

and the large-momentum behavior of the distribution function[99]

$$n_{\mathbf{k}\sigma} = \langle \hat\psi^\dagger_\sigma(\mathbf{k})\hat\psi_\sigma(\mathbf{k})\rangle \simeq \frac{C}{\mathcal{V}|\mathbf{k}|^4} \qquad (1/a,\, n^{1/3} \ll |\mathbf{k}| \ll 1/b). \tag{7.432}$$

On the BCS side of the BCS–BEC crossover, the exponentially small gap can be ignored when computing the grand potential. The leading correction to the non-interacting result comes from the Hartree–Fock term,

$$\Omega = \Omega_0 - \mathcal{V}\lambda\frac{n^2}{4} = \Omega_0 + \mathcal{V}\frac{4\pi a}{m}\frac{n^2}{4}, \tag{7.433}$$

where the last result is obtained by expressing $-\lambda$ as a function of the scattering length [Eq. (7.391)] and retaining only the leading term in the limit $a \to 0^-$. We thus deduce

$$C = (2\pi a)^2 nN = \frac{4}{3}(k_F a)^2 N k_F, \tag{7.434}$$

which is reminiscent of the Bogoliubov result (7.226) for bosons.

---

[98]Diagrammatic determinant Monte Carlo methods give $T_c/\epsilon_F = 0.171(5)$ [86] and $T_c/\epsilon_F = 0.152(7)$ [87] whereas lattice Monte Carlo calculations yield $\xi \simeq 0.366(16)$ [88].

[99]The momentum distribution has the following normalization: $\int_{\mathbf{k}} n_{\mathbf{k}\sigma} = N/2\mathcal{V}$.

On the BEC side, the system consists of composite bosons (dimers) made of opposite-spin fermions. The low-energy behavior is governed by the effective action defined by (7.414) and (7.417). At the mean-field level,

$$\begin{aligned} \Omega &= -\frac{\mathcal{V}}{2g_B}(2\mu + E_b)^2, \\ E &= -\frac{N}{2}E_b + \frac{\mathcal{V}}{2}g_B n_B^2, \end{aligned} \tag{7.435}$$

where $g_B = 4\pi a_B/m_B$ with $m_B = 2m$ and a_B the dimer–dimer scattering length. The energy is given by the energy of $N/2$ dimers, $-(N/2)E_b = -N/2ma^2$, and a correction due to their condensation into a superfluid state. This leads to

$$C = \frac{4\pi N}{a} + \pi^2 a^2 n N \frac{\partial a_B}{\partial a}, \tag{7.436}$$

where $\partial a_B/\partial a = 2$ in the simple approach of Sec. 7.7.2 or $\partial a_B/\partial a \simeq 0.6$ if we take for a_B the exact scattering length associated with dimer–dimer scattering.

At the unitary limit, one can obtain an expression of the contact in term of the scaling function $\mathcal{G}$ appearing in the zero-temperature equation of state (7.426). A straightforward calculation, assuming that $\mathcal{G}(x) - \mathcal{G}(0)$, varies linearly near $x = 0$, gives

$$\begin{aligned} C &= \mathcal{V}\sqrt{\frac{2}{\pi}}(m\mu)^2\mathcal{G}'(0) \\ &= \frac{8(2\pi)^{5/6}}{5(15)^{1/3}}\frac{\mathcal{G}'(0)}{\mathcal{G}(0)^{4/3}}Nk_F, \end{aligned} \tag{7.437}$$

thus showing that the ratio C/Nk_F is universal. There are no analytic calculations of C/Nk_F, but quantum Monte Carlo simulations give $C/Nk_F = 3.396(9)$ [89] and 2.95(10) [90].

Appendix 7.A Elements of Scattering Theory

In this appendix, we briefly review the basic notions of scattering theory.[100]

7.A.1 *Lippmann–Schwinger equation*

Let us consider two particles of mass m interacting *via* a potential $V(\mathbf{r}_1 - \mathbf{r}_2)$ in three-dimensional space. It is convenient to introduce the center-of-mass variables,

$$\mathbf{R} = \frac{\mathbf{r}_1 + \mathbf{r}_2}{2}, \qquad \mathbf{P} = \mathbf{p}_1 + \mathbf{p}_2 \tag{7.A.1}$$

and those of the relative motion

$$\mathbf{r} = \mathbf{r}_1 - \mathbf{r}_2, \qquad \mathbf{p} = \frac{\mathbf{p}_1 - \mathbf{p}_2}{2}. \tag{7.A.2}$$

[100] For a more detailed discussion, see the references in the bibliography section.

The center-of-mass Hamiltonian $\hat{\mathbf{P}}^2/4m$ describes the motion of a free particle of mass $2m$ and is independent of the relative-motion Hamiltonian

$$\hat{H} = \frac{\hat{\mathbf{p}}^2}{2m_r} + V(\hat{\mathbf{r}}), \tag{7.A.3}$$

which describes a particle of reduced mass $m_r = m/2$ subjected to the potential $V(\mathbf{r})$.

We look for a solution $|\psi_{\mathbf{k}}\rangle$ of the Schrödinger equation

$$(\epsilon - \hat{H})|\psi\rangle = (\epsilon - \hat{H}_0 - \hat{V})|\psi\rangle = 0, \tag{7.A.4}$$

which coincides with the plane-wave state $|\mathbf{k}\rangle$ when $V \to 0$. Here, $\hat{H}_0 = \hat{\mathbf{p}}^2/2m_r$ denotes the Hamiltonian of the relative motion when $\hat{V} = 0$. It is easy to verify that

$$|\psi_{\mathbf{k}}\rangle = |\mathbf{k}\rangle + (\epsilon_{\mathbf{k}} - \hat{H}_0)^{-1}\hat{V}|\psi_{\mathbf{k}}\rangle \tag{7.A.5}$$

is solution of (7.A.4) with the eigenenergy $\epsilon = \epsilon_{\mathbf{k}} \equiv \mathbf{k}^2/2m_r$. To make Eq. (7.A.5) well defined, we add to $\epsilon_{\mathbf{k}}$ an infinitesimal imaginary part, which leads to the Lippmann–Schwinger equation

$$|\psi_{\mathbf{k}}\rangle = |\mathbf{k}\rangle + \hat{G}_0(\epsilon_{\mathbf{k}})\hat{V}|\psi_{\mathbf{k}}\rangle, \tag{7.A.6}$$

where

$$\hat{G}_0(\epsilon) = (\epsilon + i0^+ - \hat{H}_0)^{-1} \tag{7.A.7}$$

is the retarded Green operator for the relative motion in the absence of potential. By choosing the retarded (causal) operator, we ensure that $|\psi_{\mathbf{k}}\rangle$ describes the scattering of an incoming plane wave by the potential V. The stationary scattering states $|\psi_{\mathbf{k}}\rangle$ form an orthonormal set of states. A basis of the Hilbert state can be obtained by adding to this set the possible bound states of $\hat{H}$.

In the coordinate representation, the Lippmann–Schwinger equation reads[101]

$$\begin{aligned} \psi_{\mathbf{k}}(\mathbf{r}) &= \frac{e^{i\mathbf{k}\cdot\mathbf{r}}}{\sqrt{\mathcal{V}}} - \frac{m_r}{2\pi}\int d^3r' \frac{e^{ik|\mathbf{r}-\mathbf{r}'|}}{|\mathbf{r}-\mathbf{r}'|} V(\mathbf{r}')\psi_{\mathbf{k}}(\mathbf{r}') \\ &\simeq \frac{1}{\sqrt{\mathcal{V}}}\left\{ e^{i\mathbf{k}\cdot\mathbf{r}} + f(\mathbf{k}',\mathbf{k})\frac{e^{ikr}}{r} \right\} \qquad (r \to \infty \text{ and } \mathbf{k}' = k\hat{\mathbf{r}}), \end{aligned} \tag{7.A.8}$$

where[102]

$$f(\mathbf{k}',\mathbf{k}) = -\sqrt{\mathcal{V}}\frac{m_r}{2\pi}\int d^3r'\, e^{-i\mathbf{k}'\cdot\mathbf{r}'} V(\mathbf{r}')\psi_{\mathbf{k}}(\mathbf{r}') = -\mathcal{V}\frac{m_r}{2\pi}\langle\mathbf{k}'|\hat{V}|\psi_{\mathbf{k}}\rangle \tag{7.A.9}$$

is the scattering amplitude. The term $e^{i\mathbf{k}\cdot\mathbf{r}}$ in (7.A.8) describes the incoming wave while the scattered part corresponds to a spherical wave e^{ikr}/r with an amplitude $f(\mathbf{k}',\mathbf{k})$ which

[101] We use $\langle\mathbf{r}|\hat{G}_0(\epsilon_{\mathbf{k}})|\mathbf{r}'\rangle = -\frac{m_r}{2\pi}\frac{e^{ik|\mathbf{r}-\mathbf{r}'|}}{|\mathbf{r}-\mathbf{r}'|}$ and the notations $k = |\mathbf{k}|$, $r = |\mathbf{r}|$.

[102] The presence of the volume $\mathcal{V}$ in Eqs. (7.A.9) and (7.A.36) comes from the normalization of the state $|\mathbf{k}\rangle$: $\langle\mathbf{k}|\mathbf{k}'\rangle = \delta_{\mathbf{k},\mathbf{k}'}$ and $\langle\mathbf{r}|\mathbf{k}\rangle = \frac{1}{\sqrt{\mathcal{V}}}e^{i\mathbf{k}\cdot\mathbf{r}}$.

depends on the scattering direction $\mathbf{k}' = k\hat{\mathbf{r}}$. Choosing $\mathbf{k}$ as the polar axis, $f(\mathbf{k}',\mathbf{k}) \equiv f_k(\theta,\varphi)$ becomes a function of k and the azimuthal and polar angles defining the direction $\mathbf{k}'$ of the scattered wave. The scattering amplitude determines the differential and total cross sections[103]

$$\sigma_k(\theta,\varphi) = |f_k(\theta,\varphi)|^2, \qquad \sigma_k^{\rm tot} = \int d\Omega\, \sigma_k(\theta,\varphi). \tag{7.A.10}$$

Not all functions are eligible as a scattering amplitude. In particular, $f_k(\theta,\varphi)$ must satisfy the optical theorem

$$\Im[f(\mathbf{k},\mathbf{k})] = \frac{k}{4\pi}\sigma_k^{\rm tot}, \tag{7.A.11}$$

which follows from the unitary evolution in quantum mechanics, i.e. the conservation of the particle flux.

7.A.2 *Central potential*

In the case of a central potential $V(\mathbf{r}) = V(r)$, the Hamiltonian commutes with the angular momentum operator $\hat{\mathbf{L}} = \hat{\mathbf{r}} \times \hat{\mathbf{p}}$ and one can look for solutions

$$\psi_{k\ell m}(\mathbf{r}) = \psi_{k\ell m}(r,\theta,\varphi) = \frac{1}{r} u_{k\ell}(r) Y_\ell^m(\theta,\varphi) \tag{7.A.12}$$

of the Schrödinger equation that are also eigenstates of $\hat{\mathbf{L}}^2$ and $\hat{L}^z$,

$$\hat{\mathbf{L}}^2 Y_\ell^m = \ell(\ell+1) Y_\ell^m, \quad \hat{L}^z Y_\ell^m = m Y_\ell^m \quad \text{with} \begin{cases} \ell \geq 0, \\ -\ell \leq m \leq \ell \end{cases} \tag{7.A.13}$$

(Y_ℓ^m is a spherical harmonic). The Schrödinger equation then reads

$$\left[-\frac{1}{2m_r}\frac{d^2}{dr^2} + \frac{\ell(\ell+1)}{2m_r r^2} + V(r) \right] u_{k\ell}(r) = \frac{k^2}{2m_r} u_{k\ell}(r) \tag{7.A.14}$$

writing the energy as $\epsilon = k^2/2m_r$. Since the Laplacian is not defined for $r=0$, one must ensure that $u_{k\ell}(r)$ is sufficiently regular at the origin for $\psi_{k\ell m}$ to be an eigenfunction. This is achieved by imposing $u_{k\ell}(0) = 0$.[104] For given k and ℓ, Eq. (7.A.14) corresponds to a one-dimensional problem for a particle subjected to an effective potential $V_{\rm eff}(r) = V(r) + \ell(\ell+1)/2m_r r^2$ equal to the sum of $V(r)$ and a centrifugal potential associated with the angular momentum (one can impose the condition $V_{\rm eff}(r) \to \infty$ for $r<0$ to ensure that $u_{k\ell}(r=0) = 0$).

---

[103] For an incident flux F_i of particles with energy $\epsilon_{\mathbf{k}}$, the number of particles detected by a detector per unit time in the solid angle $d\Omega$ about the direction (θ,φ) is $dn = \sigma_k(\theta,\varphi) F_i d\Omega$.

[104] It can be shown that $u_{k\ell}(r) \sim r^{\ell+1}$ $(r \to 0)$ for physical solutions of the Schrödinger equation.

In the absence of the potential, the solutions for $\epsilon > 0$ (k real) are given by

$$\psi^{(0)}_{k\ell m}(\mathbf{r}) = Ckj_\ell(kr)Y_\ell^m(\theta,\varphi)$$
$$\simeq \frac{C}{2i}\left[\frac{1}{r}e^{i(kr-\ell\pi/2)} - \frac{1}{r}e^{-i(kr-\ell\pi/2)}\right]Y_\ell^m(\theta,\varphi) \qquad (r\to\infty), \tag{7.A.15}$$

where $j_\ell(x) = \sqrt{\pi/2x}J_{\ell+\frac{1}{2}}(x)$ is a spherical Bessel function and C a normalization constant.[105] At large distance, $\psi^{(0)}_{k\ell m}(\mathbf{r})$ behaves as the sum of an incoming and outgoing spherical wave with a relative phase difference $\ell\pi$. The asymptotic form of the solution of the radial equation (7.A.14), obtained by neglecting $V_{\text{eff}}(r)$, can be written in the form

$$u_{k\ell}(r) = Ae^{ikr} + Be^{-ikr} = C\sin\left(kr - \ell\frac{\pi}{2} + \delta_\ell(k)\right) \tag{7.A.16}$$

(we have used $|A| = |B|$, which follows from particle number conservation), i.e.

$$\psi_{k\ell m}(\mathbf{r}) \simeq \frac{C}{2i}\left[\frac{1}{r}e^{i(kr-\ell\pi/2-\delta_\ell(k))} - \frac{1}{r}e^{-i(kr-\ell\pi/2-\delta_\ell(k))}\right]Y_\ell^m(\theta,\varphi) \quad (r\to\infty). \tag{7.A.17}$$

At large distance, the only effect of the potential is to introduce an additional phase shift $2\delta_\ell(k)$ between the incoming and outgoing spherical waves. It can be shown that the phase shift behaves as $\delta_\ell(k) \sim k^{2\ell+1}$ for $k \to 0$ (except at the unitary limit, see the following).

By expressing the stationary scattering state $\psi_\mathbf{k}$ in terms of the $\psi_{k\ell m}$'s, it is possible to relate the scattering amplitude to the phase shifts,

$$f_k(\theta) = \frac{1}{k}\sum_{\ell=0}^{\infty}(2\ell+1)e^{i\delta_l(k)}\sin\delta_\ell(k)P_\ell(\cos\theta), \tag{7.A.18}$$

where P_ℓ is a Legendre polynomial. Note that for a central potential, $f(\mathbf{k}',\mathbf{k}) = f_k(\theta)$ depends only on the angle θ between $\mathbf{k}$ and $\mathbf{k}'$. This yields the differential cross section

$$\sigma_k(\theta) = \frac{1}{k^2}\left|\sum_{\ell=0}^{\infty}(2\ell+1)e^{i\delta_l(k)}\sin\delta_\ell(k)P_\ell(\cos\theta)\right|^2 \tag{7.A.19}$$

and the total cross section

$$\sigma_k^{\text{tot}} = \frac{4\pi}{k^2}\sum_{\ell=0}^{\infty}(2\ell+1)\sin^2\delta_\ell(k). \tag{7.A.20}$$

The contribution of the partial wave ℓ is positive and bounded by $(2\ell+1)4\pi/k^2$, which is the maximum value allowed by unitarity. In that case, $\delta_\ell(k) \to \pm\pi/2\,[\pi]$ instead of being linear in k.

When the two particles involved in the collision are indistinguishable, the state vector must be symmetric (bosons) or antisymmetric (fermions). For spin-zero bosons, the wavefunction must therefore satisfy $\psi_{k\ell m}(\mathbf{r}) = \psi_{k\ell m}(-\mathbf{r})$ (the wavefunction of the center of mass is clearly invariant under the exchange of the two particles since $\mathbf{R}\to\mathbf{R}$), which implies

[105] $j_\ell(x) \sim x^\ell$ for $x \to 0$ and $j_\ell(x) \simeq \frac{1}{x}\sin(x - \ell\pi/2)$ for $x \to \infty$.

that only even values of ℓ are allowed. For spin-$\frac{1}{2}$ fermions, the wavefunction must be symmetric (antisymmetric) if the two particles are in a spin-singlet (triplet) state. Only even (odd) values of ℓ are allowed in the former (latter) case. In all cases, the total cross section is multiplied by a factor two. We do not further discuss this issue since the (anti)symmetry properties of the wavefunction are automatically taken care of in the many-body approach described in Sec. 7.A.4.

7.A.3 *Low-energy collisions — Scattering length*

If the potential vanishes beyond a certain scale, its range b is simply defined as the width of the potential: $V(r) = 0$ if $r > b$. For a potential $V(r) \sim -C/r^n$ behaving as a power law at large distances, the range of the potential can be defined by considering a wave packet of size σ localized at a distance $\sim \sigma$ of the center of the potential. The potential energy is significant with respect to the kinetic energy related to the localization if

$$\frac{C}{\sigma^n} \gtrsim \frac{1}{m\sigma^2} \quad \text{i.e.} \quad \sigma \lesssim b \sim (mC)^{1/(n-2)}, \tag{7.A.21}$$

which defines the range b of the potential. The latter can be defined only if $n > 2$.[106]

In the absence of the potential, a particle coming from infinity bounces on the centrifugal barrier at the point $r_{k\ell}$ defined by $k^2/2m_r \sim \ell(\ell+1)/2m_r r_{k\ell}^2$, i.e. $kr_{k\ell} \sim [\ell(\ell+1)]^{1/2}$. This means that for a finite-range potential, only partial waves with $\ell \leq \ell_M$, $kb \sim [\ell_M(\ell_M+1)]^{1/2}$, are affected by the potential. At low energies ($kb \ll 1$), scattering therefore occurs only in the s-wave channel ($\ell = 0$).[107] Since the typical energy of a particle is determined by the temperature, $k^2/2m \sim T$ i.e. $1/k \sim \lambda_T = \sqrt{2\pi/mT}$ (λ_T is the thermal de Broglie wavelength), all phase shifts, except $\delta_0(k)$, vanish when $b \ll \lambda_T$.

When scattering occurs only in the s-wave channel, $f_k(\theta) \equiv f_k$ is isotropic and

$$\frac{1}{f_k} = \frac{k}{\tan\delta_0(k)} - ik, \tag{7.A.22}$$

where the $-ik$ term comes from the optical theorem and ensures the unitarity of the scattering process (i.e. the conservation of the particle flux). For $k \to 0$, $\tan\delta_0(k) \sim k$ and the proportionality coefficient defines the s-wave scattering length,

$$a = -\lim_{k\to 0} \frac{\tan\delta_0(k)}{k}, \tag{7.A.23}$$

so that[108]

$$f_k = -\frac{a}{1+ika} \qquad (k \to 0). \tag{7.A.24}$$

[106] The van der Waals interaction, which decreases as $-C/r^6$ and plays an important role in ultracold gases, can therefore be considered as short-range for all practical purposes.

[107] For two spin-$\frac{1}{2}$ fermions in a spin-triplet state, the s-wave scattering channel is not allowed and scattering primarily occurs in the triplet channel $\ell = 1$.

[108] To next order, the expansion of the scattering amplitude reads $1/f_k = -1/a - ik + \frac{1}{2}r_e k^2$, where r_e is the effective range. We shall always assume a vanishing effective range and describe the low-energy collisions only in terms of a.

The scattering length can be positive or negative and can even diverge. The case $1/a = 0$, i.e. $f_k \to i/k$, corresponds to the unitary limit where $\delta_0(k) \to \pm\pi/2$ $[\pi]$ and the cross section $\sigma_k^{\rm tot} = 4\pi|f_k|^2 \to 4\pi/k^2$ takes the maximum allowed value. In ultracold gases, the scattering length can be controlled by a Feshbach resonance due to the coupling of two atoms in an open scattering channel to the bound state of a closed channel. The resonance can be tuned by a magnetic field when the magnetic moments of the separated atoms (in the open channel) and the magnetic moment of the bound state in the closed channel are different [20].

When $r \gg b$, $u_k(r)$ satisfies the equation

$$u_k(r) = C\sin\big(kr + \delta_0(k)\big) \simeq Ck(r-a) \qquad (k \to 0), \tag{7.A.25}$$

i.e.[109]

$$\psi_k(r) \simeq \frac{1}{\sqrt{\mathcal{V}}}\left(1 - \frac{r}{a}\right) \qquad (r \gg b \text{ and } k \to 0). \tag{7.A.26}$$

Thus, the scattering length fully determines the behavior of $\psi_k(\mathbf{r})$ for $r \gg b$ in the low-energy limit $k \to 0$.

So far, we have considered only scattering states. For a bound state with negative energy $\epsilon = -\kappa^2/2m_r$, $u''_\kappa(r) - \kappa^2 u_\kappa(r) = 0$ when $r \gg b$, which gives $u_\kappa(r) \sim e^{-\kappa r}$. When the bound-state energy tends to zero, i.e. $\kappa \to 0$, by continuity we expect to recover the behavior $u_k(r) \sim 1 - r/a$ of a scattering state with vanishing energy ($k \to 0$). Thus for a bound state with small energy, $\kappa = 1/a$. The scattering length is then necessarily positive and determines the extension of the state. It diverges (and changes sign) when the bound state disappears ($\kappa = 1/a \to 0$).

Let us derive a relation between the change δa in the scattering length and the change $\delta V(r)$ in the potential.[110] We start from the Schrödinger equation satisfied by the zero-energy wavefunction $\psi_0(r) = u_0(r)/r$ in the s-wave channel $\ell = 0$,

$$-\frac{1}{2m_r}u''_0(r) + V(r)u_0(r) = 0, \tag{7.A.27}$$

with

$$u_0(0) = 0, \qquad u_0(r) \simeq \frac{1}{\sqrt{\mathcal{V}}}(r-a) \quad (r \gg b). \tag{7.A.28}$$

When $V(r)$ changes by $\delta V(r)$, $u_0(r)$ changes by $\delta u_0(r)$ and to first order

$$-\frac{1}{2m_r}\delta u''_0(r) + V(r)\delta u_0(r) + \delta V(r)u_0(r) = 0, \tag{7.A.29}$$

with $u_0(0) + \delta u_0(0) = 0$ and $u_0(r) + \delta u_0(r) \simeq (r - a - \delta a)/\sqrt{\mathcal{V}}$ for $r \gg b + \delta b$ (δb is the variation of the potential range), i.e.

$$\delta u_0(0) = 0, \qquad \delta u_0(r) = -\frac{\delta a}{\sqrt{\mathcal{V}}} \quad (r \gg b). \tag{7.A.30}$$

[109]The prefactor $1/\sqrt{\mathcal{V}}$ in (7.A.26) can be deduced from (7.A.8).

[110]This relation will be useful when discussing the two-body contact in Sec. 7.4.5.

We now multiply (7.A.29) by $u_0(r)$ and integrate over r up to some value $r_{\rm max} \gg b$,

$$\int_0^{r_{\rm max}} dr \left[-\frac{1}{2m_r} u_0(r)\delta u_0''(r) + V(r)u_0(r)\delta u_0(r) + \delta V(r) u_0(r)^2 \right] = 0. \qquad (7.A.31)$$

With a double integration by part, we obtain

$$\begin{aligned} \int_0^{r_{\rm max}} dr\, u_0(r)\delta u_0''(r) &= u_0 \delta u_0' \Big|_0^{r_{\rm max}} - \delta u_0 u_0' \Big|_0^{r_{\rm max}} + \int_0^{r_{\rm max}} dr\, u_0''(r)\delta u_0(r) \\ &= \frac{\delta a}{\mathcal{V}} + 2m_r \int_0^{r_{\rm max}} dr\, V(r) u_0(r) \delta u_0(r) \end{aligned} \qquad (7.A.32)$$

using (7.A.28) and (7.A.30), so that

$$\int_0^{r_{\rm max}} dr\, \delta V(r) u_0(r)^2 = \frac{\delta a}{2m_r \mathcal{V}}. \qquad (7.A.33)$$

Taking the limit $r_{\rm max} \to \infty$, we finally obtain

$$\mathcal{V} \int d^3r\, \delta V(r) \psi_0(r)^2 = \frac{4\pi \delta a}{m}. \qquad (7.A.34)$$

7.A.4 *T matrix and many-body formalism*

The transition matrix (or T matrix) is defined by

$$\begin{aligned} \hat{T}(\epsilon) &= \hat{V} + \hat{V}\hat{G}_0(\epsilon)\hat{T}(\epsilon) \\ &= \hat{V} + \hat{V}\hat{G}_0(\epsilon)\hat{V} + \hat{V}\hat{G}_0(\epsilon)\hat{V}G_0(\epsilon)\hat{V} + \cdots \end{aligned} \qquad (7.A.35)$$

and allows one to express the scattering amplitude as[102]

$$f(\mathbf{k}', \mathbf{k}) = -\mathcal{V}\frac{m_r}{2\pi} \langle \mathbf{k}' | \hat{T}(\epsilon_{\mathbf{k}}) | \mathbf{k} \rangle. \qquad (7.A.36)$$

In the many-body formalism, the T matrix is defined by the four-point vertex $\Gamma^{(4)}$ in vacuum (vanishing density and zero temperature: $\mu = T = 0$) corresponding to a pair of particles with zero total momentum. It is obtained by summing the ladder diagrams shown in Fig. 7.33 since these are the only diagrams that contribute when $\mu = T = 0$.[111] For two spin-zero bosons, the T matrix satisfies the Bethe–Salpeter equation

$$\begin{aligned} T_{\mathbf{k}',\mathbf{k}}(i\omega) &= V(\mathbf{k}' - \mathbf{k}) \\ &\quad - \int_{\mathbf{k}''} \int \frac{d\omega''}{2\pi} V(\mathbf{k}' - \mathbf{k}'') G(\mathbf{k}'', i\omega'') G(-\mathbf{k}'', i\omega - i\omega'') T_{\mathbf{k}'',\mathbf{k}}(i\omega), \end{aligned} \qquad (7.A.37)$$

[111]This is due to the one-particle propagator $G(\mathbf{k}, \tau) = -\Theta(\tau)e^{-\epsilon_{\mathbf{k}}}$ being causal when $\mu = T = 0$ [Eq. (1.200)].

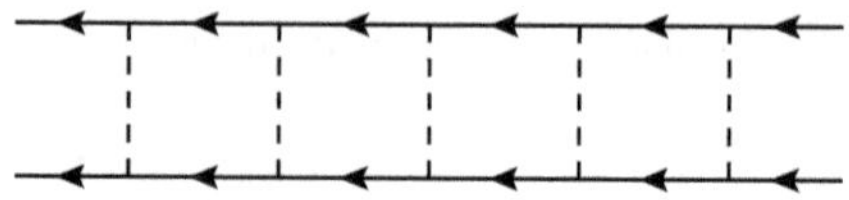

Fig. 7.33 Diagrammatic representation of the T matrix.

where ω is a Matsubara frequency,[112] $V(\mathbf{k}'-\mathbf{k}) = \mathcal{V}V_{\mathbf{k}',\mathbf{k}}$ the Fourier transform of the potential, $G(\mathbf{k}'', i\omega'') = (i\omega'' - \epsilon_{\mathbf{k}''})^{-1}$ and

$$\int \frac{d\omega''}{2\pi} G(\mathbf{k}'', i\omega'')G(-\mathbf{k}'', i\omega - i\omega'') = \frac{1}{2\epsilon_{\mathbf{k}''} - i\omega}. \tag{7.A.38}$$

Note that $\epsilon_{\mathbf{k}} = \mathbf{k}^2/2m$ now denotes the kinetic energy of a particle with mass m. The retarded T matrix, obtained by analytic continuation $i\omega \to \omega + i0^+$, satisfies

$$T^R_{\mathbf{k}',\mathbf{k}}(\omega) = V(\mathbf{k}'-\mathbf{k}) + \int_{\mathbf{k}''} V(\mathbf{k}'-\mathbf{k}'') \frac{1}{\omega + i0^+ - 2\epsilon_{\mathbf{k}''}} T^R_{\mathbf{k}'',\mathbf{k}}(\omega), \tag{7.A.39}$$

which is nothing but Eq. (7.A.35). At low energies, i.e. when $\epsilon_{\mathbf{k}}, \epsilon_{\mathbf{k}'}, \omega \to 0$, scattering occurs in the s-wave channel and the T matrix gives the value of the scattering length,

$$\lim_{\mathbf{k},\mathbf{k}'\to 0} T^R_{\mathbf{k}',\mathbf{k}}(\omega = k^2/m) = -\lim_{\mathbf{k},\mathbf{k}'\to 0} \frac{4\pi}{m} f(\mathbf{k}, \mathbf{k}') = \frac{4\pi a}{m}. \tag{7.A.40}$$

Since the T matrix coincides with the two-particle vertex in vacuum, it also provides us with information about the two-particle spectrum. In particular, its poles give the energies of the bound states of the two-body problem.

The T matrix can be computed in two dimensions. In that case, the scattering amplitude vanishes logarithmically at low energies,[113]

$$f_k = \frac{1}{\ln\frac{ka}{2} + C - i\frac{\pi}{2}} \qquad (k \to 0), \tag{7.A.41}$$

where C is the Euler constant. This equation can be considered as the definition of the scattering length a in two dimensions. The scattering amplitude can be obtained from the T matrix using $f_k = -(m/2\pi)T^R(\omega = k^2/m)$.

Spin-$\frac{1}{2}$ fermions can be treated similarly. For two particles in a spin-singlet state, the two lines in the diagram of Fig. 7.33 correspond to spin-up and spin-down particles, but the Bethe–Salpeter equation (7.A.39) is unchanged. The s-wave scattering length is defined by (7.A.40) for a three-dimensional system.

It is of course difficult to compute the T matrix for a realistic potential $V(\mathbf{r}) = V(|\mathbf{r}|)$. However, since its low-energy limit is fully determined by the s-wave scattering length (and possibly the effective range), one can consider a simple model where the T matrix

[112] At zero temperature, the Matsubara frequency $\omega_n \equiv \omega$ becomes a continuous variable.

[113] In two dimensions, the scattering states satisfy $\psi_{\mathbf{k}}(\mathbf{r}) \sim e^{i\mathbf{k}\cdot\mathbf{r}} + i\frac{\pi}{2} f_k H_0^{(1)}(kr)$ with $H_0^{(1)}$ a Hankel function of the first kind: $H_0^{(1)}(x) \simeq \sqrt{\frac{2}{\pi x}} e^{i(x-\pi/4)}$ for $x \gg 1$ and $H_0^{(1)}(x) \simeq \frac{2i}{\pi} \ln x$ for $0 < x \ll 1$.

can be easily computed. Once the scattering length has been expressed in terms of the (few) parameters of the model, one can compute other physical quantities within the same model and express them in terms of the scattering length. This is the path followed in Sec. 7.4 to determine the equation of state of a dilute Bose gas and in Sec. 7.7 to study the BCS–BEC crossover. In these sections, the potential $V(\mathbf{r}) = g\delta(\mathbf{r})$ is assumed to be local and the model is regularized by a UV momentum cutoff. The equation for the T matrix reads

$$T(i\omega) = g + g\int_{\mathbf{k}} \frac{1}{i\omega - 2\epsilon_{\mathbf{k}}} T(i\omega), \tag{7.A.42}$$

where

$$\begin{aligned}\int_{\mathbf{k}} \frac{1}{i\omega - 2\epsilon_{\mathbf{k}}} &= \frac{1}{2\pi^2}\int_0^\Lambda dk \frac{k^2}{i\omega - k^2/m} \\ &= -\frac{m\Lambda}{2\pi^2} + \frac{1}{4\pi^2}\int_{-\Lambda}^{\Lambda} dk \frac{im\omega}{i\omega - k^2/m} \\ &= -\frac{m\Lambda}{2\pi^2} + \frac{m^{3/2}}{4\pi}\sqrt{-i\omega}\end{aligned} \tag{7.A.43}$$

(we have taken the limit $\Lambda \to \infty$ in the last integral). We thus obtain

$$T(i\omega) = \frac{g}{1 - g\left(-\frac{m\Lambda}{2\pi^2} + \frac{m^{3/2}}{4\pi}\sqrt{-i\omega}\right)} = \frac{4\pi/m}{\frac{1}{a} - \sqrt{-im\omega}}, \tag{7.A.44}$$

where the scattering length is defined by

$$\frac{m}{4\pi a} = \frac{1}{g} + \frac{m\Lambda}{2\pi^2}. \tag{7.A.45}$$

Equation (7.A.44) gives the universal low-energy form of the T matrix in a three-dimensional system. The expression (7.A.24) of the scattering amplitude is obtained from $f_k = -(m/4\pi)T^R(\omega = k^2/m)$.

In two dimensions,

$$\int_{\mathbf{k}} \frac{1}{i\omega - 2\epsilon_{\mathbf{k}}} = -\frac{m}{4\pi}\ln\left(-\frac{\Lambda^2}{im\omega}\right) \tag{7.A.46}$$

for $|\omega| \ll \Lambda$, which gives the T matrix

$$T(i\omega) = \frac{g}{1 + \frac{mg}{4\pi}\ln\left(-\frac{\Lambda^2}{im\omega}\right)} = -\frac{2\pi/m}{\ln\frac{a}{2}\sqrt{-i\omega m} + C}, \tag{7.A.47}$$

where the scattering length is defined by

$$a = \frac{2}{\Lambda}e^{-2\pi/mg - C}. \tag{7.A.48}$$

The scattering amplitude (7.A.41) is recovered from (7.A.47) using $f_k = -(m/2\pi)T^R(\omega = k^2/m)$.

Guide to the bibliography

- There are a number of excellent textbooks and reviews on superfluidity [1–14]. Many of them discuss the Gross-Pitaevskii [15, 16] and Bogoliubov [17] theories as well as Landau's two-fluid model of superfluid Helium [18, 19]. An experimental overview of superfluid Helium is provided by Tilley and Tilley [2]. BEC and superfluidity in ultracold atomic gases is discussed in Refs. [4–6, 20–22]. Reference [23] contains a detailed discussion of various approximations (Hartree–Fock–Bogoliubov, Beliaev, Popov, etc.) beyond the Bogoliubov theory. An historical overview of BEC is given by Griffin [24].
- There are also a number of excellent textbooks on superconductivity [2, 3, 7, 9, 12, 13, 29–31]. Most provide a detailed presentation of the BCS mean-field theory [33]. The Ginzburg–Landau approach is discussed in depth in Refs. [30, 31] and an excellent review can be found in the two volumes edited by Parks (which contain many remarkable articles on the BCS theory) [32].
- Off-diagonal long-range order and the concept of spontaneous U(1) symmetry breaking in superfluids and superconductors are discussed in Anderson's book [8] and in Refs. [9, 34–37]. For a discussion of superfluidity that does not rely on spontaneous symmetry breaking, see Leggett's book [6].
- Effective actions of superfluid systems are discussed in Refs. [74, 75]. Effective actions for superconductors including both the pairing and density fields can be found in Refs. [76–79].
- For an early review on collective modes in superconductors, see the article by Martin in Ref. [32]. The Anderson–Bogoliubov mode was first discussed in Ref. [56]. The amplitude mode is discussed in Refs. [49–51].
- The BCS–BEC crossover was first discussed by Eagles and Leggett [66, 67]. The Nozières-Schmitt–Rink approach [68] is reviewed in Ref. [71] and diagrammatic theories in Ref. [83]. For a thorough discussion of the BCS–BEC crossover, see Refs. [72, 73].
- Scattering theory is discussed in many textbooks, see, e.g., Refs. [57–59].
- The two-body contact was introduced in Ref. [60] and is also reviewed in Refs. [61–65].
- Universality in the unitary Fermi gas is discussed in Refs. [25, 84] from the point of view of phase transitions.

Bibliography

[1] P. Nozières and D. Pines, *The Theory of Quantum Liquids, Vol. II: Superfluid Bose Liquids* (Addison-Wesley, Reading, Massachusetts, 1990).

[2] D. R. Tilley and J. Tilley, *Superfluidity and Superconductivity,* 3rd edition (Adam Hilger, Bristol, 1990).

[3] J. F. Annett, *Superconductivity, Superfluids and Condensates* (Oxford University Press, Oxford, 2004).

[4] L. Pitaevskii and S. Stringari, *Bose–Einstein Condensation* (Clarendon Press, Oxford, 2003).

[5] C. J. Pethick and H. Smith, *Bose–Einstein Condensation in Dilute Gases* (Cambridge University Press, Cambridge, 2002).

[6] A. J. Leggett, *Quantum Liquids: Bose Condensation and Cooper Pairing in Condensed-Matter Systems* (Oxford Graduate Texts, Oxford, 2006).

[7] B. Svistunov, E. Babaev, and N. Prokof'ev, *Superfluid States of Matter* (CRC Press, New York, 2015).

[8] P. W. Anderson, *Basic Notions of Condensed Matter Physics* (Advanced Book Classics, Menlo Park, California, 1984).

[9] D. Forster, *Hydrodynamic Fluctuations, Broken Symmetry, and Correlation Functions,* Chapters 10 and 12 (Addison-Wesley, Reading, Massachusetts, 1975).

[10] G. Baym, The microscopic description of superfluidity, *Mathematical Methods in Solid State and Superfluid Theory: Scottish Universities' Summer School,* edited by R. C. Clark and G. H. Derrick (Springer, Berlin, 1968), pp. 121–156.

[11] V. N. Popov, *Functional Integrals in Quantum Field Theory and Statistical Physics* (Reidel, Dordrecht, 1983).

[12] A. L. Fetter and J. D. Walecka, *Quantum Theory of Many-Particle Systems,* Chapters 6, 13 and 14 (Dover, Mineola, New York, 2003).

[13] A. A. Abrikosov, L. P. Gor'kov, and I. E. Dzyaloshinski, *Methods of Quantum Field Theory in Statistical Physics,* Chapters 5 and 7 (Dover, Mineola, New York, 1975).

[14] J. O. Andersen, Theory of the weakly interacting Bose gas, *Rev. Mod. Phys.* **76**, 599 (2004).

[15] E. P. Gross, Structure of a quantized vortex in boson systems, *Nuovo Cimento* **20**, 454 (1961).

[16] L. P. Pitaevskii, Vortex lines in an imperfect Bose gas, *Sov. Phys. JETP* **13**, 451 (1961).

[17] N. N. Bogoliubov, On the theory of superfluidity, *J. Phys. USSR* **11**, 23 (1947); reprinted in D. Pines, *The Many-Body Problem* (W. A. Benjamin, New York, 1961), p. 292.

[18] L. D. Landau, The theory of superfluidity of helium II, *J. Phys. USSR* **5**, 71 (1941).

[19] L. D. Landau, On the theory of superfluidity of helium II, *J. Phys. USSR* **11**, 91 (1947).

[20] H. T. C. Stoof, K. B. Gubbels, and D. B. M. Dickerscheid, *Ultracold Quantum Fields* (Springer, Berlin, 2009).

[21] A. J. Leggett, Bose–Einstein condensation in the alkali gases: Some fundamental concepts, *Rev. Mod. Phys.* **73**, 307 (2001).

[22] F. Dalfovo, S. Giorgni, L. P. Pitaevskii, and S. Stringari, Theory of Bose–Einstein condensation in trapped gases, *Rev. Mod. Phys.* **71**, 463 (1999).

[23] H. Shi and A. Griffin, Finite-temperature excitations in a dilute Bose-condensed gas, *Phys. Rep.* **304**, 1 (1998).

[24] A. Griffin, A brief history of our understanding of BEC: from Bose to Beliaev, in *Bose–Einstein Condensation in Atomic Gases*, edited by M. Inguscio, S. Stringari, and C. Wieman (Italian Physical Society, 1999), arXiv:cond-mat/9901123.

[25] S. Sachdev, *Quantum Phase Transitions* (Cambridge University Press, 2011).

[26] T. D. Lee and C. N. Yang, Many-body problem in quantum mechanics and quantum statistical mechanics, *Phys. Rev.* **105**, 1119 (1957).

[27] T. D. Lee, K. Huang, and C. N. Yang, Eigenvalues and eigenfunctions of a Bose system of hard spheres and its low-temperature properties, *Phys. Rev.* **106**, 1135 (1957).

[28] T. T. Wu, Ground state of a Bose system of hard spheres, *Phys. Rev.* **115**, 1390 (1959).

[29] J. R. Schrieffer, *Theory of Superconductivity* (Benjamin/Cummings, Reading, MA).

[30] P. G. de Gennes, *Superconductivity of Metals and Alloys* (Addison Wesley, Reading, Massachusetts, 1966).

[31] M. Thinkham, *Introduction to Superconductivity*, 2nd edition (McGraw-Hill, Malabar, Florida, 1980).

[32] *Superconductivity*, edited by R. D. Parks (Marcel Dekker, New York, 1969).

[33] J. Bardeen, L. N. Cooper, and J. R. Schrieffer, Theory of superconductivity, *Phys. Rev.* **108**, 1175 (1957).

[34] O. Penrose and L. Onsager, Bose–Einstein condensation and liquid helium, *Phys. Rev.* B **104**, 576 (1956).

[35] C. N. Yang, Concept of off-diagonal long-range order and the quantum phases of liquid He and of superconductors, *Rev. Mod. Phys.* **34**, 694 (1962).

[36] M. Greiter, Is electromagnetic gauge invariance spontaneously violated in superconductors? *Ann. Phys.* **319**, 217 (2005).

[37] N. R. Poniatowski, Superconductivity, broken gauge symmetry, and the Higgs mechanism, *American Journal of Physics* **87**, 436 (2019).

[38] B. D. Josephson, Relation between the superfluid density and order parameter for superfluid He near T_c, *Phys. Lett.* **21**, 608 (1966).

[39] N. M. Hugenholtz and D. Pines, Ground-state energy and excitation spectrum of a system of interacting bosons, *Phys. Rev.* **116**, 489 (1959).

[40] J. Gavoret and P. Nozières, Structure of the perturbation expansion for the Bose liquid at zero temperature, *Ann. Phys.* **28**, 349 (1964).

[41] G. Baym, J. P. Blaizot, M. Holzmann, F. Laloë, and D. Vautherin, Bose–Einstein transition in a dilute interacting gas, *Eur. Phys. J.* B **24**, 107 (2001).

[42] A. Rançon and N. Dupuis, Universal thermodynamics of a two-dimensional Bose gas, *Phys. Rev.* A **85**, 063607 (2012).

[43] S. K. Adhikari, T. Frederico, and I. D. Goldman, Perturbative renormalization in quantum few-body problems, *Phys. Rev. Lett.* **74**, 487 (1995).

[44] F. Chevy and C. Salomon, Strongly correlated Bose gases, *J. Phys.* B **49**, 192001 (2016).

[45] E. Braaten and H.-W. Hammer, Efimov physics in cold atoms, *Ann. Phys.* **322**, 120 (2007).
[46] L. W. Gruenberg and L. Gunther, Effect of orbital quantization on the critical field of type-II superconductors, *Phys. Rev.* **176**, 606 (1968).
[47] M. Rasolt and Z. Tešanović, Theoretical aspects of superconductivity in very high magnetic fields, *Rev. Mod. Phys.* **64**, 709 (1992).
[48] N. Dupuis, Mean-field theory of a quasi-one-dimensional superconductor in a high magnetic field, *J. Phys. I (France)* **5**, 1577 (1995).
[49] P. B. Littlewood and C. M. Varma, Amplitude collective modes in superconductors and their coupling to charge-density waves, *Phys. Rev.* B **26**, 4883 (1982).
[50] H. Kurkjian, S. N. Klimin, J. Tempere, and Y. Castin, Pair-breaking collective branch in BCS superconductors and superfluid Fermi gases, *Phys. Rev. Lett.* **122**, 093403 (2019).
[51] Y. Castin and H. Kurkjian, Collective excitation branch in the continuum of pair-condensed Fermi gases: Analytical study and scaling laws, *Comptes Rendus, Physique, Tome* **21**, 253 (2020).
[52] E. Abrahams and T. Tsuneto, Time variation of the Ginzburg-Landau order parameter, *Phys. Rev.* **152**, 416 (1966).
[53] I. J. R. Aitchison, G. Metikas, and D. J. Lee, Finite-temperature time-dependent effective theory for the Goldstone field in a BCS-type superfluid, *Phys. Rev.* B **62**, 6638 (2000).
[54] S. G. Sharapov, H. Beck, and V. M. Loktev, Finite-temperature time-dependent effective theory for the phase field in two-dimensional d-wave neutral superconductors, *Phys. Rev.* B **64**, 134519 (2001).
[55] L. Benfatto, A. Toschi, and S. Caprara, Low-energy phase-only action in a superconductor: A comparison with the XY model, *Phys. Rev.* B **69**, 184510 (2004).
[56] P. W. Anderson, Random-phase approximation in the theory of superconductivity, *Phys. Rev.* **112**, 1900 (1958).
[57] A. Messiah, *Quantum Mechanics*, Vol. I (North-Holland, Amsterdam, 1962).
[58] J. J. Sakurai, *Modern Quantum Mechanics* (Addison Wesley, Reading, 1994).
[59] J. Dalibard, *Interactions between Atoms in Quantum Gases (I)* (Collège de France, lecture series 2020–2021), https://pro.college-de-france.fr/jean.dalibard/CdF/2021/total_en.pdf.
[60] S. Tan, Energetics of a strongly correlated Fermi gas, *Ann. Phys.* **323**, 2952 (2008); Generalized virial theorem and pressure relation for a strongly correlated Fermi gas, *Ann. Phys.* **323**, 2987 (2008); Large momentum part of a strongly correlated Fermi gas, *Ann. Phys.* **323**, 2971 (2008).
[61] J. Dalibard, *Interactions between Atoms in Quantum Gases (II)* (Collège de France, lecture series 2021–2022), https://pro.college-de-france.fr/jean.dalibard/CdF/2022/total_en.pdf.
[62] E. Braaten and L. Platter, Exact relations for a strongly interacting Fermi gas from the operator product expansion, *Phys. Rev. Lett.* **100**, 205301 (2008).
[63] S. Zhang and A. J. Leggett, Universal properties of the ultracold Fermi gas, *Phys. Rev.* A **79**, 023601 (2009).
[64] R. Combescot, F. Alzetto, and X. Leyronas, Particle distribution tail and related energy formula, *Phys. Rev.* A **79**, 053640 (2009).
[65] F. Werner and Y. Castin, General relations for quantum gases in two and three dimensions: Two-component fermions, *Phys. Rev.* A **86**, 013626 (2012); General relations for quantum gases in two and three dimensions. II. Bosons and mixtures, *Phys. Rev.* A **86**, 053633 (2012).
[66] D. M. Eagles, Possible pairing without superconductivity at low carrier concentrations in bulk and thin-film superconducting semiconductors, *Phys. Rev.* **186**, 456 (1969).

[67] A. J. Leggett, Diatomic molecules and Cooper pairs, in *Modern Trends in the Theory of Condensed Matter*, edited by A. Peralski and J. Przystawa (Springer-Verlag, Berlin, 1980).
[68] P. Nozières and S. Schmitt-Rink, Bose condensation in an attractive fermion gas: From weak to strong coupling superconductivity, *J. Low Temp. Phys.* **59**, 195 (1985).
[69] R. Haussmann, Crossover from BCS superconductivity to Bose–Einstein condensation: A self-consistent theory, *Z. Phys.* B **91**, 291 (1993).
[70] J. R. Engelbrecht, M. Randeria, and C. A. R. Sá de Melo, BCS to Bose crossover: Broken-symmetry state, *Phys. Rev.* B **55**, 15153 (1997).
[71] M. Randeria, Crossover from BCS theory to Bose–Einstein condensation, in *Bose–Einstein Condensation*, edited by A. Griffin, D. W. Snoke, and S. Strinagari (Cambridge University Press, Cambridge, 1995).
[72] The BCS-BEC crossover and the unitary Fermi gas, in *Lecture Notes in Physics*, edited by W. Zwerger (Springer, Berlin, 2012).
[73] G. C. Strinati, P. Pieri, G. Röpke, P. Schuck, and M. Urban, The BCS–BEC crossover: From ultra-cold Fermi gases to nuclear systems, *Phys. Rep.* **738**, 1 (2018).
[74] M. Greiter, F. Wilczek, and E. Witten, Hydrodynamic relations in superconductivity, *Mod. Phys. Lett.* B **3**, 903 (1989).
[75] D. Son and M. Wingate, General coordinate invariance and conformal invariance in nonrelativistic physics: Unitary Fermi gas, *Ann. Phys.* **321**, 197 (2006).
[76] M. Stone, Dynamics of $T = 0$ condensates, *Int. J. Mod. Phys.* B **9**, 1359 (1995).
[77] I. J. R. Aitchison, P. Ao, D. J. Thouless, and X. M. Zhu, Effective Lagrangians for BCS superconductors at $T = 0$, *Phys. Rev.* B **51**, 6531 (1995).
[78] S. De Palo, C. Castellani, C. Di Castro, and B. K. Chakraverty, Effective action for superconductors and BCS-Bose crossover, *Phys. Rev.* B **60**, 564 (1999).
[79] N. Dupuis, Effective action for superfluid Fermi systems in the strong-coupling limit, *Phys. Rev.* A **72**, 013606 (2005).
[80] D. S. Petrov, C. Salomon, and G. V. Shlyapnikov, Weakly bound dimers of fermionic atoms, *Phys. Rev. Lett.* **93**, 090404 (2004).
[81] H. Hu, X.-J. Liu, and P. D. Drummond, Equation of state of a superfluid Fermi gas in the BCS-BEC crossover, *Europhys. Lett.* **74**, 574 (2006).
[82] P. Carruthers and M. M. Nieto, Phase and angle variables in quantum mechanics, *Rev. Mod. Phys.* **40**, 411 (1968).
[83] M. Pini, P. Pieri, and G. C. Strinati, Fermi gas throughout the BCS-BEC crossover: Comparative study of t-matrix approaches with various degrees of self-consistency, *Phys. Rev.* B **99**, 094502 (2019).
[84] P. Nikolic and S. Sachdev, Renormalization-group fixed points, universal phase diagram, and $1/N$ expansion for quantum liquids with interactions near the unitarity limit, *Phys. Rev.* A **75**, 033608 (2007).
[85] P.-A. Pantel, D. Davesne, and M. Urban, Polarized Fermi gases at finite temperature in the BCS-BEC crossover, *Phys. Rev.* A **90**, 053629 (2014); Erratum *Phys. Rev.* A **94**, 019901 (2016).
[86] O. Goulko and M. Wingate, Thermodynamics of balanced and slightly spin-imbalanced Fermi gases at unitarity, *Phys. Rev.* A **82**, 053621 (2010).
[87] E. Burovski, N. Prokof'ev, B. Svistunov, and M. Troyer, Critical temperature and thermodynamics of attractive fermions at unitarity, *Phys. Rev. Lett.* **96**, 160402 (2006).

[88] M. G. Endres, D. B. Kaplan, J.-W. Lee, and A. N. Nicholson, Lattice Monte Carlo calculations for unitary fermions in a finite box, *Phys. Rev.* A **87**, 023615 (2013).

[89] S. Gandolfi, K. E. Schmidt, and J. Carlson, BEC-BCS crossover and universal relations in unitary Fermi gases, *Phys. Rev.* A **83**, 041601(R) (2011); Erratum *Phys. Rev.* A **83**, 049901 (2011).

[90] J. E. Drut, T. A. Lähde, and T. Ten, Momentum distribution and contact of the unitary Fermi gas, *Phys. Rev. Lett.* **106**, 205302 (2011).

Chapter 8

Quantum Magnetism

Contents

8.1 The Heisenberg Model 570
 8.1.1 The ferromagnet 571
 8.1.2 The antiferromagnet 573
8.2 Random-Phase Approximation 573
 8.2.1 Classical dynamics 573
 8.2.2 Spin susceptibilities in the RPA 575
8.3 Spin Waves: Holstein–Primakoff Bosons 577
 8.3.1 The ferromagnet 577
 8.3.2 The antiferromagnet 578
8.4 Spin-Coherent-State Functional Integral 581
 8.4.1 Berry phase 581
 8.4.2 Functional integral 586
 8.4.3 Quantum nonlinear σ model 591
8.5 Schwinger-Boson Mean-Field Theory 597
 8.5.1 Schwinger bosons 597
 8.5.2 Mean-field theory of the antiferromagnetic Heisenberg model 598
 8.5.3 SU($\mathcal{N}$) Heisenberg model 603
Appendix 8.A A Single Spin in an External Field 604
 8.A.1 Free energy 605
 8.A.2 Spin–spin correlation function 605
Appendix 8.B Spin-Coherent States 606
Bibliography 609

This chapter is devoted to quantum magnetism. Our aim is not to give an exhaustive introduction to the subject but to discuss a few key concepts and theoretical methods. After an introduction to the Heisenberg model, its classical dynamics and its mean-field solution, we discuss spin waves in ferromagnets and antiferromagnets using both an RPA and Holstein–Primakoff bosons (Secs. 8.1–8.3). Then, we introduce the spin-coherent-state functional integral and show that the low-energy behavior of an antiferromagnet is governed by a quantum NLσM with a topological Berry phase term (Sec. 8.4). Finally, we discuss a mean-field theory of the Heisenberg model based on a Schwinger-boson representation of the spin operator (Sec. 8.5).

8.1 The Heisenberg Model

The Heisenberg model is defined by the Hamiltonian

$$\hat{H} = -\frac{1}{2} \sum_{\mathbf{r},\mathbf{r}'} J_{\mathbf{r},\mathbf{r}'} \hat{\mathbf{S}}_{\mathbf{r}} \cdot \hat{\mathbf{S}}_{\mathbf{r}'}, \tag{8.1}$$

where $\{\mathbf{r}\}$ denotes the N sites of a d-dimensional hypercubic lattice and the lattice spacing is taken as the unit length. $\hat{\mathbf{S}}_{\mathbf{r}} = (\hat{S}^x_{\mathbf{r}}, \hat{S}^y_{\mathbf{r}}, \hat{S}^z_{\mathbf{r}})$ is a spin-S operator satisfying $\hat{\mathbf{S}}^2_{\mathbf{r}} = S(S+1)$ and the commutation relations

$$[\hat{S}^\nu_{\mathbf{r}}, \hat{S}^{\nu'}_{\mathbf{r}'}] = i \sum_{\nu''=x,y,z} \epsilon_{\nu\nu'\nu''} \delta_{\mathbf{r},\mathbf{r}'} \hat{S}^{\nu''}_{\mathbf{r}}. \tag{8.2}$$

In the following, we shall consider mainly the case where the spin–spin interaction $J_{\mathbf{r},\mathbf{r}'}$ equals J if $\mathbf{r}$ and $\mathbf{r}'$ are nearest neighbors and vanishes otherwise. The Fourier transform of $J_{\mathbf{r},\mathbf{r}'}$ is then given by

$$J(\mathbf{q}) = \frac{1}{N} \sum_{\mathbf{r},\mathbf{r}'} e^{-i\mathbf{q}\cdot(\mathbf{r}-\mathbf{r}')} J_{\mathbf{r},\mathbf{r}'} = 2J \sum_{\mu=1}^{d} \cos q_\mu. \tag{8.3}$$

Solving the Heisenberg model is in general a very difficult task. The problem simplifies in the large-S limit where the spin operators $\hat{\mathbf{S}}_{\mathbf{r}}$ can be treated as classical spin variables of length $|\mathbf{S}_{\mathbf{r}}| = S$, since the commutator $[\hat{S}^\nu_{\mathbf{r}}, \hat{S}^{\nu'}_{\mathbf{r}'}] = \mathcal{O}(S)$ [Eq. (8.2)] can be neglected with respect to the product $\hat{S}^\nu_{\mathbf{r}} \hat{S}^{\nu'}_{\mathbf{r}'}$, the latter being $\mathcal{O}(S^2)$. Finding the ground state of the system then amounts to minimizing the classical energy

$$E_{\mathrm{cl}} = -\frac{1}{2} \sum_{\mathbf{r},\mathbf{r}'} J_{\mathbf{r},\mathbf{r}'} \mathbf{S}_{\mathbf{r}} \cdot \mathbf{S}_{\mathbf{r}'} = -\frac{1}{2} \sum_{\mathbf{q}} J(\mathbf{q}) \mathbf{S}_{-\mathbf{q}} \cdot \mathbf{S}_{\mathbf{q}} \tag{8.4}$$

with the constraint $|\mathbf{S}_{\mathbf{r}}| = S$. In the simple case we are considering [Eq. (8.3)], the solution is straightforward. For $J > 0$, the ground state is ferromagnetic with all spins pointing in the same direction, e.g. $\mathbf{S}_{\mathbf{r}} = S\mathbf{u}_z$. For $J < 0$, the ground state is antiferromagnetic with nearest-neighbor spins antiparallel, e.g. $\mathbf{S}_{\mathbf{r}} = S(-1)^{\mathbf{r}} \mathbf{u}_z$ (Néel state).[1]

More generally, for arbitrary lattice and interactions, the ground state is determined by the maximum of $J(\mathbf{q})$. For example, if $J(\mathbf{q})$ is maximum at $\pm\mathbf{q}_0$, one has

$$\mathbf{S}_{\mathbf{r}} = \frac{1}{\sqrt{N}} [\mathbf{S}_{\mathbf{q}_0} e^{i\mathbf{q}_0\cdot\mathbf{r}} + \mathbf{S}_{-\mathbf{q}_0} e^{-i\mathbf{q}_0\cdot\mathbf{r}}] \tag{8.5}$$

[1]We use the notation $(-1)^{\mathbf{r}} \equiv e^{i\mathbf{Q}\cdot\mathbf{r}}$ with $\mathbf{Q} = (\pi, \ldots, \pi)$.

with $\mathbf{S}_{-\mathbf{q}_0} = \mathbf{S}^*_{\mathbf{q}_0}$ since $\mathbf{S}_\mathbf{r}$ is real, i.e.

$$\mathbf{S}_\mathbf{r} = S[\mathbf{a}\cos(\mathbf{q}_0 \cdot \mathbf{r} + \theta) + \mathbf{b}\sin(\mathbf{q}_0 \cdot \mathbf{r} + \theta)], \tag{8.6}$$

where $\mathbf{a}$ and $\mathbf{b}$ are perpendicular unit vectors (thus ensuring that the condition $|\mathbf{S}_\mathbf{r}| = S$ is fulfilled) and θ an arbitrary phase. Ferromagnetism and antiferromagnetism correspond to $\mathbf{q}_0 = 0$ and $\mathbf{q}_0 = \mathbf{Q} = (\pi, \dots, \pi)$, respectively. In other cases, Eq. (8.6) corresponds to spiral (or helicoidal) order.

The large-S limit provides us with a convenient starting point to study the effect of quantum fluctuations. Most of the discussion in this chapter, except the Schwinger-boson theory in Sec. 8.5, is based on a $1/S$ expansion.

8.1.1 *The ferromagnet*

In the case of the ferromagnet, it turns out that the classical spin configurations with lowest energy also determine the various possible quantum ground states. Assuming the polarization to be along the z axis, the ground state reads

$$|0\rangle = \bigotimes_\mathbf{r} |S\rangle_\mathbf{r}, \tag{8.7}$$

where $|S\rangle_\mathbf{r}$ denotes the eigenstate of $\hat{S}^z_\mathbf{r}$ with the highest possible eigenvalue $S^z = S$. By writing the Hamiltonian as

$$\hat{H} = -J \sum_{\langle \mathbf{r},\mathbf{r}' \rangle} \left(\hat{S}^z_\mathbf{r} \hat{S}^z_{\mathbf{r}'} + \frac{1}{2} \hat{S}^+_\mathbf{r} \hat{S}^-_{\mathbf{r}'} + \frac{1}{2} \hat{S}^-_\mathbf{r} \hat{S}^+_{\mathbf{r}'} \right) \tag{8.8}$$

(the sum is over all pairs of nearest neighbors) where $\hat{S}^\pm_\mathbf{r} = \hat{S}^x_\mathbf{r} \pm i \hat{S}^y_\mathbf{r}$, one immediately sees that $|0\rangle$ is an eigenstate of $\hat{H}$ with an eigenvalue given by the classical energy $E_{\rm cl} = -NJS^2d$ since $\hat{S}^+_\mathbf{r} \hat{S}^-_{\mathbf{r}'} |0\rangle = \hat{S}^-_\mathbf{r} \hat{S}^+_{\mathbf{r}'} |0\rangle = 0$.[2] To show that $E_{\rm cl}$ is the lowest eigenvalue of $\hat{H}$, let us consider an eigenstate $|0'\rangle$ with eigenvalue $E'_0 = \langle 0'|\hat{H}|0'\rangle = -J\sum_{\langle \mathbf{r},\mathbf{r}' \rangle} \langle 0'|\hat{\mathbf{S}}_\mathbf{r} \cdot \hat{\mathbf{S}}_{\mathbf{r}'}|0'\rangle$. The largest value of $\langle 0'|\hat{\mathbf{S}}_\mathbf{r} \cdot \hat{\mathbf{S}}_{\mathbf{r}'}|0'\rangle$ is at most equal to the largest eigenvalue of $\hat{\mathbf{S}}_\mathbf{r} \cdot \hat{\mathbf{S}}_{\mathbf{r}'}$, that is S^2, so that $E_{0'} \geq -J\sum_{\langle \mathbf{r},\mathbf{r}' \rangle} S^2 = E_{\rm cl}$.[3]

It is also possible to find some of the low-lying excited states of the ferromagnetic Heisenberg model. Let us consider the normalized excited state

$$|\mathbf{r}\rangle = \frac{1}{\sqrt{2S}} \hat{S}^-_\mathbf{r} |0\rangle \tag{8.9}$$

obtained by decreasing S^z from S to $S-1$ for the spin at site $\mathbf{r}$. This state is not an eigenstate of the Hamiltonian since

$$(\hat{S}^-_{\mathbf{r}_1} \hat{S}^+_{\mathbf{r}_2} + \hat{S}^+_{\mathbf{r}_1} \hat{S}^-_{\mathbf{r}_2})|\mathbf{r}\rangle = 2S(\delta_{\mathbf{r},\mathbf{r}_2}|\mathbf{r}_1\rangle + \delta_{\mathbf{r},\mathbf{r}_1})|\mathbf{r}_2\rangle \qquad (\mathbf{r}_1 \neq \mathbf{r}_2). \tag{8.10}$$

---

[2] $\hat{S}^\pm|S^z\rangle = [(S \mp S^z)(S + 1 \pm S^z)]^{1/2}|S^z \pm 1\rangle$ implies that $\hat{S}^+|S\rangle = 0$.

[3] Write $\hat{\mathbf{S}}_\mathbf{r} \cdot \hat{\mathbf{S}}_{\mathbf{r}'} = \frac{1}{2}[(\hat{\mathbf{S}}_\mathbf{r} + \hat{\mathbf{S}}_{\mathbf{r}'})^2 - \hat{\mathbf{S}}^2_\mathbf{r} - \hat{\mathbf{S}}^2_{\mathbf{r}'}]$ and notice that $\hat{\mathbf{S}}^2_\mathbf{r} = S(S+1)$ whereas $(\hat{\mathbf{S}}_\mathbf{r} + \hat{\mathbf{S}}_{\mathbf{r}'})^2$ has eigenvalues $F(F+1)$ with $0 \leq F \leq 2S$. It follows that the largest and smallest eigenvalues of $\hat{\mathbf{S}}_\mathbf{r} \cdot \hat{\mathbf{S}}_{\mathbf{r}'}$ when $\mathbf{r} \neq \mathbf{r}'$ are S^2 and $-S(S+1)$, respectively.

But it is easy to show that the linear combination

$$|\mathbf{q}\rangle = \frac{1}{\sqrt{N}} \sum_{\mathbf{r}} e^{i\mathbf{q}\cdot\mathbf{r}} |\mathbf{r}\rangle, \qquad \hat{H}|\mathbf{q}\rangle = (E_{\rm cl} + \omega_{\mathbf{q}})|\mathbf{q}\rangle \tag{8.11}$$

is an eigenstate with an excitation energy

$$\omega_{\mathbf{q}} = S[J(0) - J(\mathbf{q})] \tag{8.12}$$

which behaves as $JS\mathbf{q}^2$ in the long wavelength limit $|\mathbf{q}| \ll 1$. To give a physical interpretation of the state $|\mathbf{q}\rangle$, let us consider the transverse spin correlation function[4]

$$\langle \mathbf{q}|\hat{\mathbf{S}}^{\perp}_{\mathbf{r}} \cdot \hat{\mathbf{S}}^{\perp}_{\mathbf{r}'}|\mathbf{q}\rangle = \frac{1}{2}\langle \mathbf{q}|\hat{S}^{-}_{\mathbf{r}} \hat{S}^{+}_{\mathbf{r}'} + \hat{S}^{+}_{\mathbf{r}} \hat{S}^{-}_{\mathbf{r}'}|\mathbf{q}\rangle = \frac{2S}{N} \cos[\mathbf{q} \cdot (\mathbf{r} - \mathbf{r}')] \qquad (\mathbf{r} \neq \mathbf{r}'), \tag{8.13}$$

where $\hat{\mathbf{S}}^{\perp}_{\mathbf{r}} = (\hat{S}^{x}_{\mathbf{r}}, \hat{S}^{y}_{\mathbf{r}})$ and we have used (8.10). Thus, on average, the orientations of the transverse (i.e. perpendicular to the magnetization) component of two spins separated by a distance $\mathbf{r} - \mathbf{r}'$ differ by an angle $\mathbf{q} \cdot (\mathbf{r} - \mathbf{r}')$. One can therefore see the state $|\mathbf{q}\rangle$ as a state containing a spin wave, or “magnon”, of wavevector $\mathbf{q}$ and energy $\omega_{\mathbf{q}}$.

One can define creation and annihilation operators for magnons *via* $|\mathbf{q}\rangle = \hat{a}^{\dagger}_{\mathbf{q}}|0\rangle$,[5] i.e.

$$\hat{a}^{\dagger}_{\mathbf{q}} = \frac{1}{\sqrt{2S}} \hat{S}^{-}_{-\mathbf{q}}, \qquad \hat{a}_{\mathbf{q}} = \frac{1}{\sqrt{2S}} \hat{S}^{+}_{\mathbf{q}}. \tag{8.14}$$

These operators are not bosonic operators since the commutator

$$[\hat{a}_{\mathbf{q}}, \hat{a}^{\dagger}_{\mathbf{q}'}] = \frac{1}{NS} \sum_{\mathbf{r}} e^{-i(\mathbf{q}-\mathbf{q}')\cdot\mathbf{r}} \hat{S}^{z}_{\mathbf{r}} \tag{8.15}$$

does not reduce to $\delta_{\mathbf{q},\mathbf{q}'}$. However, given that $[\hat{a}_{\mathbf{q}}, \hat{a}^{\dagger}_{\mathbf{q}'}]|0\rangle = \delta_{\mathbf{q},\mathbf{q}'}|0\rangle$, one can assume the magnons to obey bosonic statistics if their density is low enough, which is expected to be the case at low temperatures. Furthermore, when their density is small, one can neglect the interactions between magnons. The mean number of magnons with momentum $\mathbf{q}$ is then given by the Bose–Einstein statistics, $\langle \hat{a}^{\dagger}_{\mathbf{q}} \hat{a}_{\mathbf{q}} \rangle = n_B(\omega_{\mathbf{q}})$, and it is straightforward to compute thermodynamic quantities. For example, at finite temperature, the magnetization per site is given by

$$\langle \hat{S}^{z}_{\mathbf{r}} \rangle_T = \langle \hat{S}^{z}_{\mathbf{r}} \rangle_{T=0} - \frac{1}{N} \sum_{\mathbf{q}} n_B(\omega_{\mathbf{q}}) = S - \int_{\mathbf{q}} \frac{1}{e^{\beta\omega_{\mathbf{q}}} - 1}. \tag{8.16}$$

At low temperatures, $T \ll JS$, the momentum integral in (8.16) is dominated by small values of $|\mathbf{q}|$ and one can approximate $\omega_{\mathbf{q}}$ by $JS\mathbf{q}^2$. In three dimensions, this leads to the Bloch $T^{3/2}$ law: $S - \langle \hat{S}^{z}_{\mathbf{r}} \rangle_T \propto T^{3/2}$. In two dimensions, the integral in (8.16) is infrared divergent, which suggests the absence of long-range order at finite temperature, in agreement with the Mermin–Wagner theorem (Sec. 3.6.4). A similar calculation shows that in three dimensions, the energy varies as $T^{5/2}$ and the specific heat as $T^{3/2}$.

[4]Note that $\langle \mathbf{q}|\hat{\mathbf{S}}^{\perp}_{\mathbf{r}}|\mathbf{q}\rangle = 0$, which is the reason why we consider the correlation function (8.13).

[5]Fourier transformed operators are defined by $\hat{a}_{\mathbf{q}} = \frac{1}{\sqrt{N}} \sum_{\mathbf{r}} e^{-i\mathbf{q}\cdot\mathbf{r}} \hat{a}_{\mathbf{r}}$, $\hat{a}^{\dagger}_{\mathbf{q}} = \frac{1}{\sqrt{N}} \sum_{\mathbf{r}} e^{i\mathbf{q}\cdot\mathbf{r}} \hat{a}^{\dagger}_{\mathbf{r}}$ and $\hat{\mathbf{S}}_{\mathbf{q}} = \frac{1}{\sqrt{N}} \sum_{\mathbf{r}} e^{-i\mathbf{q}\cdot\mathbf{r}} \hat{\mathbf{S}}_{\mathbf{r}}$.

8.1.2 *The antiferromagnet*

In the case of the antiferromagnet, the classical spin configuration with minimum energy yields the Néel state (assuming the polarization along the z axis)

$$|0\rangle = \bigotimes_{\mathbf{r}} |(-1)^{\mathbf{r}} S\rangle_{\mathbf{r}}, \tag{8.17}$$

which is not a ground state of the quantum Hamiltonian. Contrary to the ferromagnet, the operator $\hat{S}^+_{\mathbf{r}} \hat{S}^-_{\mathbf{r}'}$ acting on nearest-neighbor spins does not always vanish since $\hat{S}^+_{\mathbf{r}} \hat{S}^-_{\mathbf{r}'} |-S\rangle_{\mathbf{r}} \otimes |S\rangle_{\mathbf{r}'} \propto |-S+1\rangle_{\mathbf{r}} \otimes |S-1\rangle_{\mathbf{r}'}$. This process can be seen as a manifestation of quantum fluctuations. Is there spontaneous symmetry breaking and long-range order in the antiferromagnetic Heisenberg model?[6] What are the low-lying excitations? We shall try to answer these questions in the following sections.

The only case where the ground state of the Heisenberg model is known exactly is the one-dimensional spin-$\frac{1}{2}$ chain: Spin-rotation is not broken in the ground state but the $q = \pi$ spin–spin correlation function decays algebraically (not exponentially). The elementary excitations form a particle–hole pair continuum $\omega_q = \epsilon_{q+k} - \epsilon_k$, obtained from fundamental excitations with dispersion law $\epsilon_k = \frac{\pi}{2}|J \sin k|$ which are usually called spinons.[7] More generally, for half-integer spin chains, there exist excited states whose energy vanishes when N is even and $N \to \infty$ (Lieb, Schultz and Mattis theorem [23]). We shall see that integer spin chains are fundamentally different and exhibit a gapped spectrum. In higher dimensions, whether S is integer or half-integer does not play an important role (Sec. 8.4.3).

8.2 Random-Phase Approximation

A first insight into the dynamics of the (anti)ferromagnet can be obtained from a purely classical analysis or the RPA. For the antiferromagnet, these two approaches make sense only if quantum fluctuations do not suppress long-range order; we shall see that this is indeed the case for a hypercubic lattice in dimension $d \geq 2$.

8.2.1 *Classical dynamics*

We write the classical energy (8.4) as

$$E_{\text{cl}} = -\frac{1}{2} \sum_{\mathbf{r}} \mathbf{h}_{\mathbf{r}} \cdot \boldsymbol{\mu}_{\mathbf{r}}, \tag{8.18}$$

where $\mathbf{h}_{\mathbf{r}} = -\frac{1}{g\mu_B} \sum_{\mathbf{r}'} J_{\mathbf{r},\mathbf{r}'} \mathbf{S}_{\mathbf{r}'}$ is the effective magnetic field acting on the magnetic moment $\boldsymbol{\mu}_{\mathbf{r}} = -g\mu_B \mathbf{S}_{\mathbf{r}}$.[8] From classical mechanics, the rate of change of the angular momentum $\mathbf{S}_{\mathbf{r}}$

[6] For a discussion about the meaning of spontaneous symmetry breaking in antiferromagnets, see Sec. 3.6.2.2.

[7] The spinon is not a spin wave, although it has a similar dispersion (see Sec. 8.3.2), since it does not correspond to small fluctuations about a ground state with spontaneously broken symmetry.

[8] μ_B is the Bohr magneton and g the Landé factor.

is given by $\boldsymbol{\mu}_{\mathbf{r}} \times \mathbf{h}_{\mathbf{r}}$, i.e.

$$\frac{d}{dt}\mathbf{S}_{\mathbf{r}} = \boldsymbol{\mu}_{\mathbf{r}} \times \mathbf{h}_{\mathbf{r}} = \sum_{\mathbf{r}'} J_{\mathbf{r},\mathbf{r}'} \mathbf{S}_{\mathbf{r}} \times \mathbf{S}_{\mathbf{r}'}. \tag{8.19}$$

Note that a factor 1/2 is introduced in (8.18) to avoid double counting of the spin–spin interactions. Assuming small deviations from the ground state, we linearize this equation about $\mathbf{S}_{\mathbf{r}} = S\mathbf{u}_z$ (ferromagnet) or $\mathbf{S}_{\mathbf{r}} = (-1)^{\mathbf{r}} S\mathbf{u}_z$ (antiferromagnet). For the ferromagnet, one obtains

$$\begin{aligned} \frac{d}{dt} S^x_{\mathbf{q}} &= S[J(0) - J(\mathbf{q})] S^y_{\mathbf{q}}, \\ \frac{d}{dt} S^y_{\mathbf{q}} &= S[J(\mathbf{q}) - J(0)] S^x_{\mathbf{q}}. \end{aligned} \tag{8.20}$$

Looking for plane wave solutions $S^{x,y}_{\mathbf{q}} \sim e^{-i\omega t}$, one recovers the magnon excitation energy (8.12) vanishing quadratically for $\mathbf{q} \to 0$ (Fig. 8.1).

In the case of the antiferromagnet,

$$\begin{aligned} \frac{d}{dt} S^x_{\mathbf{q}} &= S[J(\mathbf{Q}) - J(\mathbf{q}+\mathbf{Q})] S^y_{\mathbf{q}+\mathbf{Q}}, \\ \frac{d}{dt} S^y_{\mathbf{q}+\mathbf{Q}} &= -S[J(\mathbf{Q}) - J(\mathbf{q})] S^x_{\mathbf{q}}, \end{aligned} \tag{8.21}$$

where $\mathbf{Q} = (\pi, \ldots, \pi)$ and $J(\mathbf{q}+\mathbf{Q}) = -J(\mathbf{q})$, which gives

$$\omega_{\mathbf{q}} = S\sqrt{J(0)^2 - J(\mathbf{q})^2}. \tag{8.22}$$

Unlike ferromagnetic spin waves, there are two modes, $\mathbf{q} = 0$ and $\mathbf{q} = \mathbf{Q}$, where the energy vanishes. Furthermore, the energy vanishes linearly with momentum, i.e. $\omega_{\mathbf{q}} = c|\mathbf{q}|$ for $\mathbf{q} \to 0$ or $\omega_{\mathbf{q}} = c|\mathbf{q} - \mathbf{Q}|$ for $\mathbf{q} \to \mathbf{Q}$, with a velocity $c = 2\sqrt{d}|J|S$ (Fig. 8.1).

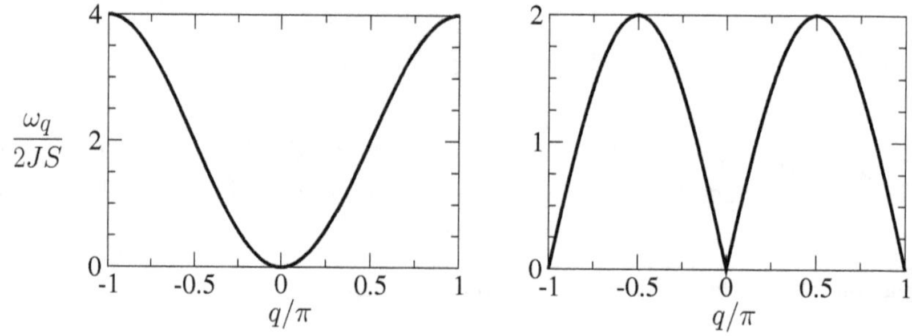

Fig. 8.1 Spin-wave dispersion ω_q in ferromagnets (left) and antiferromagnets (right). The dispersion is shown for $d = 1$.

8.2.2 Spin susceptibilities in the RPA

The spin-wave modes can also be obtained from an RPA. Let us start with a static mean-field approximation obtained by decoupling the spin–spin interaction,[9]

$$\begin{aligned}\hat{H}_{\text{MF}} &= -\frac{1}{2}\sum_{\mathbf{r},\mathbf{r}'} J_{\mathbf{r},\mathbf{r}'}\left(\langle\hat{\mathbf{S}}_{\mathbf{r}}\rangle\cdot\hat{\mathbf{S}}_{\mathbf{r}'} + \hat{\mathbf{S}}_{\mathbf{r}}\cdot\langle\hat{\mathbf{S}}_{\mathbf{r}'}\rangle - \langle\hat{\mathbf{S}}_{\mathbf{r}}\rangle\cdot\langle\hat{\mathbf{S}}_{\mathbf{r}'}\rangle\right)\\ &= -\sum_{\mathbf{r}} \mathbf{h}_{\mathbf{r}}\cdot\hat{\mathbf{S}}_{\mathbf{r}} + \text{const},\end{aligned} \tag{8.23}$$

where

$$\mathbf{h}_{\mathbf{r}} = \sum_{\mathbf{r}'} J_{\mathbf{r},\mathbf{r}'}\langle\hat{\mathbf{S}}_{\mathbf{r}'}\rangle = mJ(\mathbf{q})e^{i\mathbf{q}\cdot\mathbf{r}}\mathbf{u}_z \tag{8.24}$$

is the effective field acting on the spin located at site $\mathbf{r}$. The last expression in (8.24) is obtained by assuming $\langle\hat{\mathbf{S}}_{\mathbf{r}}\rangle = me^{i\mathbf{q}\cdot\mathbf{r}}\mathbf{u}_z$ with $\mathbf{q}=0$ (ferromagnet) or $\mathbf{q}=\mathbf{Q}$ (antiferromagnet). The mean-field Hamiltonian (8.23) describes a set of decoupled spins and can be easily solved. The magnetization m is obtained from the self-consistent equation

$$\langle\hat{S}^z_{\mathbf{r}}\rangle = \frac{\text{Tr}\left(e^{-\beta\hat{H}_{\text{MF}}}\hat{S}^z_{\mathbf{r}}\right)}{\text{Tr}\left(e^{-\beta\hat{H}_{\text{MF}}}\right)}, \tag{8.25}$$

which gives (see Appendix 8.A)

$$m = SB_S(\beta hS), \tag{8.26}$$

where $h = mJ(\mathbf{q})$ and

$$B_S(x) = \frac{2S+1}{2S}\,\text{cotanh}\left(\frac{2S+1}{2S}x\right) - \frac{1}{2S}\,\text{cotanh}\left(\frac{x}{2S}\right) \tag{8.27}$$

is the Brillouin function. At zero temperature, $m = SB_S(\infty) = S$, which is the expected result in a mean-field theory neglecting quantum fluctuations. For an infinitesimal magnetization, using $B_S(x) = x(S+1)/3S + \mathcal{O}(x^3)$, one can linearize Eq. (8.26) to obtain the transition temperature

$$T_c = \frac{S(S+1)}{3}J(\mathbf{q}) = \frac{2}{3}|J|dS(S+1). \tag{8.28}$$

Note that T_c takes the same expression for the ferromagnet ($J>0$) and the antiferromagnet ($J<0$).

[9] Equation (8.23) is obtained by linearizing the Hamiltonian in the fluctuation operator $\delta\hat{\mathbf{S}}_{\mathbf{r}} = \hat{\mathbf{S}}_{\mathbf{r}} - \langle\hat{\mathbf{S}}_{\mathbf{r}}\rangle$.

Let us now consider the system in the presence of a time-dependent external field $\mathbf{h}_{\mathbf{r}}^{\rm ext}(t) = \mathbf{h}^{\rm ext} e^{i(\mathbf{q}\cdot\mathbf{r}-\omega t)} + {\rm c.c.}$ (with $\mathbf{q}$ now arbitrary),

$$\hat{H} = -\frac{1}{2}\sum_{\mathbf{r},\mathbf{r}'} J_{\mathbf{r},\mathbf{r}'}\hat{\mathbf{S}}_{\mathbf{r}}\cdot\hat{\mathbf{S}}_{\mathbf{r}'} - \sum_{\mathbf{r}} \mathbf{h}_{\mathbf{r}}^{\rm ext}(t)\cdot\hat{\mathbf{S}}_{\mathbf{r}}. \tag{8.29}$$

The RPA amounts to linearizing the Hamiltonian with respect to the fluctuations $\delta\hat{\mathbf{S}}_{\mathbf{r}}(t) = \hat{\mathbf{S}}_{\mathbf{r}} - \langle\hat{\mathbf{S}}_{\mathbf{r}}\rangle(t)$ of the spin operator,

$$\hat{H}_{\rm RPA} = -\sum_{\mathbf{r},\mathbf{r}'} J_{\mathbf{r},\mathbf{r}'}\langle\hat{\mathbf{S}}_{\mathbf{r}}\rangle(t)\cdot\hat{\mathbf{S}}_{\mathbf{r}'} - \sum_{\mathbf{r}} \mathbf{h}_{\mathbf{r}}^{\rm ext}(t)\cdot\hat{\mathbf{S}}_{\mathbf{r}} + {\rm const.} \tag{8.30}$$

Since $\langle\hat{\mathbf{S}}_{\mathbf{r}}\rangle(t) = \langle\hat{\mathbf{S}}_{\mathbf{r}}\rangle + \delta\langle\hat{\mathbf{S}}_{\mathbf{r}}\rangle(t)$, with $\langle\hat{\mathbf{S}}_{\mathbf{r}}\rangle$ the mean value of the spin operator when $\mathbf{h}^{\rm ext} = 0$, we obtain the RPA Hamiltonian

$$\hat{H}_{\rm RPA} = \hat{H}_{\rm MF} - \sum_{\mathbf{r}} \mathbf{h}_{\mathbf{r}}(t)\cdot\hat{\mathbf{S}}_{\mathbf{r}}, \tag{8.31}$$

where

$$\mathbf{h}_{\mathbf{r}}(t) = \mathbf{h}_{\mathbf{r}}^{\rm ext}(t) + \sum_{\mathbf{r}'} J_{\mathbf{r},\mathbf{r}'}\delta\langle\hat{\mathbf{S}}_{\mathbf{r}'}\rangle(t) \tag{8.32}$$

is the time-dependent effective field acting on the spin located at site $\mathbf{r}$. Thus, the RPA Hamiltonian boils down to a system of decoupled spins subject to a self-consistent time-dependent field. The linear response to the field is given by the retarded spin–spin correlation function χ_0^R obtained from the mean-field Hamiltonian $\hat{H}_{\rm MF}$,[10]

$$\delta\langle\hat{S}_{\mathbf{r}}^{\nu}\rangle(t) = \int_{-\infty}^{t} dt' \sum_{\mathbf{r}',\nu'} \chi_0^{R\,\nu\nu'}(\mathbf{r},\mathbf{r}';t-t') h_{\mathbf{r}'}^{\nu'}(t'). \tag{8.33}$$

In matrix form, we can rewrite this equation as

$$\delta\langle\hat{S}\rangle = \chi_0^R h = \chi_0^R (J\delta\langle\hat{S}\rangle + h^{\rm ext}) \tag{8.34}$$

and obtain the retarded RPA susceptibility χ^R,

$$\delta\langle\hat{S}\rangle = (1-\chi_0^R J)^{-1}\chi_0^R h^{\rm ext} \equiv \chi^R h^{\rm ext}, \qquad \chi^R = (1-\chi_0^R J)^{-1}\chi_0^R. \tag{8.35}$$

The spin-wave spectrum is obtained from the divergence of the spin susceptibility, i.e. $\det \chi^{R-1} = \det(\chi_0^{R-1} - J) = 0$. The $T = 0$ susceptibility χ_0^R is obtained from the Matsubara correlation function χ_0 calculated in Appendix 8.A [Eqs. (8.A.6), (8.A.9)],

$$\begin{aligned} \chi_0^{R\,xx}(\mathbf{q},\mathbf{q},\omega) &= \chi_0^{R\,yy}(\mathbf{q},\mathbf{q},\omega) = \frac{S h_{\rm MF}}{h_{\rm MF}^2-\omega^2}, \\ \chi_0^{R\,xy}(\mathbf{q},\mathbf{q}+\mathbf{Q},\omega) &= -\chi_0^{R\,yx}(\mathbf{q},\mathbf{q}+\mathbf{Q},\omega) = \frac{i\omega S}{h_{\rm MF}^2-\omega^2}, \end{aligned} \tag{8.36}$$

[10] $\chi_0(\mathbf{r},\mathbf{r}';t-t') \propto \delta_{\mathbf{r},\mathbf{r}'}$ is obviously local, since the spins are decoupled at the mean-field level, but depends on $\mathbf{r} = \mathbf{r}'$ if $\langle\hat{\mathbf{S}}_{\mathbf{r}}\rangle$ is not uniform.

where $\mathbf{Q} = 0$ (ferromagnet) or $\mathbf{Q} = (\pi, \dots, \pi)$ (antiferromagnet), and $h_{\rm MF} = SJ(\mathbf{Q})$ [Eq. (8.24)]. The zz component $\chi_0^{R\,zz}$ of the spin susceptibility does not couple to the other components and is not considered here since we are interested only in the transverse (spin-wave) excitations. Inverting the transverse part of χ_0^R, one finds that the spin-wave spectrum is determined by

$$\begin{vmatrix} S[J(\mathbf{Q}) - J(\mathbf{q})] & -i\omega \\ i\omega & S[J(\mathbf{Q}) - J(\mathbf{q}+\mathbf{Q})] \end{vmatrix} = 0, \tag{8.37}$$

which reproduces the previous result for the ferromagnet [Eq. (8.12)] and agrees with the classical analysis for the antiferromagnet [Eq. (8.22)].

8.3 Spin Waves: Holstein–Primakoff Bosons

In the large-S limit, spin operators become commuting classical variables and the ground state coincides with the classical ground state. It is possible to take into account quantum fluctuations in a $1/S$ expansion. A convenient way to do so is to write the spin operator in terms of Holstein–Primakoff bosons. The large-S expansion justifies the spin-wave results obtained previously from the classical dynamics or the RPA.

8.3.1 *The ferromagnet*

Let us consider the ferromagnetic ground state $|0\rangle$ with all spins pointing in the $\mathbf{u}_z$ direction. We parameterize deviations from the ground state by means of boson operators $\hat a_{\mathbf{r}}$ and $\hat a^\dagger_{\mathbf{r}}$,

$$\begin{aligned} \hat S^z_{\mathbf{r}} &= S - \hat n_{\mathbf{r}}, \\ \hat S^+_{\mathbf{r}} &= \sqrt{2S} f(\hat n_{\mathbf{r}}) \hat a_{\mathbf{r}}, \qquad \hat n_{\mathbf{r}} = \hat a^\dagger_{\mathbf{r}} \hat a_{\mathbf{r}}, \qquad f(\hat n_{\mathbf{r}}) = \left(1 - \frac{\hat n_{\mathbf{r}}}{2S}\right)^{1/2}. \\ \hat S^-_{\mathbf{r}} &= \sqrt{2S} \hat a^\dagger_{\mathbf{r}} f(\hat n_{\mathbf{r}}), \end{aligned} \tag{8.38}$$

It is straightforward to show that the boson commutation relations $[\hat a_{\mathbf{r}}, \hat a^\dagger_{\mathbf{r}'}] = \delta_{\mathbf{r},\mathbf{r}'}$, $[\hat a_{\mathbf{r}}, \hat a_{\mathbf{r}'}] = [\hat a^\dagger_{\mathbf{r}}, \hat a^\dagger_{\mathbf{r}'}] = 0$ imply $\hat S^2_{\mathbf{r}} = S(S+1)$ and the usual spin commutation relations,

$$\begin{aligned} [\hat S^+_{\mathbf{r}}, \hat S^-_{\mathbf{r}}] &= 2S[f(\hat n_{\mathbf{r}}) \hat a_{\mathbf{r}}, \hat a^\dagger_{\mathbf{r}} f(\hat n_{\mathbf{r}})] = 2(S - \hat n_{\mathbf{r}}) = 2\hat S^z_{\mathbf{r}}, \\ [\hat S^+_{\mathbf{r}}, \hat S^z_{\mathbf{r}}] &= \sqrt{2S}[f(\hat n_{\mathbf{r}}) \hat a_{\mathbf{r}}, S - \hat n_{\mathbf{r}}] = -\sqrt{2S} f(\hat n_{\mathbf{r}}) \hat a_{\mathbf{r}} = -\hat S^+_{\mathbf{r}}, \\ [\hat S^-_{\mathbf{r}}, \hat S^z_{\mathbf{r}}] &= \sqrt{2S}[\hat a^\dagger_{\mathbf{r}} f(\hat n_{\mathbf{r}}), S - \hat n_{\mathbf{r}}] = \sqrt{2S} \hat a^\dagger_{\mathbf{r}} f(\hat n_{\mathbf{r}}) = \hat S^-_{\mathbf{r}}, \end{aligned} \tag{8.39}$$

i.e. $\hat{\mathbf{S}}_{\mathbf{r}} \times \hat{\mathbf{S}}_{\mathbf{r}} = i \hat{\mathbf{S}}_{\mathbf{r}}$. Since $S^z \in [-S, S]$, the physical Hilbert space for the bosons corresponds to $n_{\mathbf{r}} = \langle \hat a^\dagger_{\mathbf{r}} \hat a_{\mathbf{r}} \rangle \le 2S$. Given that $f(n = 2S) = 0$, the spin operators, expressed in terms of the boson operators [Eqs. (8.38)], do not couple the physical space $n_{\mathbf{r}} \le 2S$ to the nonphysical one $n_{\mathbf{r}} > 2S$.

As a function of the boson operators, the Heisenberg Hamiltonian (8.8) reads

$$\hat{H} = -\frac{1}{2}\sum_{\mathbf{r},\mathbf{r}'} J_{\mathbf{r},\mathbf{r}'}\left[(S-\hat{n}_\mathbf{r})(S-\hat{n}_{\mathbf{r}'}) + Sf(\hat{n}_\mathbf{r})\hat{a}_\mathbf{r}\hat{a}^\dagger_{\mathbf{r}'}f(\hat{n}_{\mathbf{r}'}) + S\hat{a}^\dagger_\mathbf{r}f(\hat{n}_\mathbf{r})f(\hat{n}_{\mathbf{r}'})\hat{a}_{\mathbf{r}'}\right]. \quad (8.40)$$

We can now consider the large-S limit and expand $\hat{H}$ in $1/S$. However, if we truncate the expansion (i.e. the function $f(n)$) to a finite order, then the spin operators $\hat{S}^\pm_\mathbf{r}$ will couple the physical space $n_\mathbf{r} \leq 2S$ to the nonphysical one $n_\mathbf{r} > 2S$. For the expansion to make sense, we must therefore require $n_\mathbf{r} \ll 2S$, which is indeed the case since the ground state is close to the classical state when $S \gg 1$.

To leading (non-trivial) order, using $f(\hat{n}_\mathbf{r}) = 1 + \mathcal{O}(1/S)$, one finds

$$\begin{aligned}\hat{H} &= -\frac{1}{2}\sum_{\mathbf{r},\mathbf{r}'} J_{\mathbf{r},\mathbf{r}'}\left[S^2 - S(\hat{n}_\mathbf{r}+\hat{n}_{\mathbf{r}'}) + S(\hat{a}_\mathbf{r}\hat{a}^\dagger_{\mathbf{r}'} + \hat{a}^\dagger_\mathbf{r}\hat{a}_{\mathbf{r}'})\right] \\ &= E_\mathrm{cl} + \sum_\mathbf{q} \omega_\mathbf{q}\hat{a}^\dagger_\mathbf{q}\hat{a}_\mathbf{q},\end{aligned} \quad (8.41)$$

where

$$\omega_\mathbf{q} = S[J(0) - J(\mathbf{q})] \equiv 2dJS(1-\gamma_\mathbf{q}), \qquad \gamma_\mathbf{q} = \frac{J(\mathbf{q})}{J(0)} = \frac{1}{d}\sum_{\mu=1}^d \cos q_\mu \quad (8.42)$$

and $E_\mathrm{cl} = -\frac{1}{2}J(0)NS^2$ is the ground state energy. The Hamiltonian (8.41) describes non-interacting bosons with a dispersion which becomes quadratic in the long wavelength limit: $\omega_\mathbf{q} \simeq SJ\mathbf{q}^2$ for $|\mathbf{q}| \ll 1$. Since Eq. (8.41) was obtained using $f(\hat{n}_\mathbf{r}) = 1$, we have $\hat{S}^-_\mathbf{q} = \sqrt{2S}\hat{a}^\dagger_{-\mathbf{q}}$ to leading order of the $1/S$ expansion, and we recover the fact that the magnon creation operator is $\hat{S}^-_{-\mathbf{q}}/\sqrt{2S}$ [Eq. (8.14)]. The ground state is the vacuum of bosons, $\langle\hat{a}^\dagger_\mathbf{q}\hat{a}_\mathbf{q}\rangle = 0$, which corresponds to the classical spin configuration $\langle\hat{S}^z_\mathbf{r}\rangle = S - \langle\hat{a}^\dagger_\mathbf{r}\hat{a}_\mathbf{r}\rangle = S$.

The $\mathcal{O}(1/S^0)$ term, omitted in Eq. (8.41),

$$\hat{H}_\mathrm{int} = -\frac{1}{2}\sum_{\mathbf{r},\mathbf{r}'} J_{\mathbf{r},\mathbf{r}'}\left(\hat{n}_\mathbf{r}\hat{n}_{\mathbf{r}'} - \frac{1}{2}\hat{n}_\mathbf{r}\hat{a}_{\mathbf{r}'}\hat{a}^\dagger_{\mathbf{r}'} - \frac{1}{2}\hat{a}_\mathbf{r}\hat{a}^\dagger_{\mathbf{r}'}\hat{n}_{\mathbf{r}'}\right), \quad (8.43)$$

is quartic in the boson operators and describes interactions between magnons.

8.3.2 *The antiferromagnet*

The Néel state corresponds to $\langle\hat{S}^z_\mathbf{r}\rangle = S$ on sublattice A and $\langle S^z_\mathbf{r}\rangle = -S$ on sublattice B.[11] It is convenient to define the Holstein–Primakoff bosons as follows:

$$\begin{cases}\hat{S}^z_\mathbf{r} = S - \hat{n}_\mathbf{r} \\ \hat{S}^+_\mathbf{r} = \sqrt{2S}f(\hat{n}_\mathbf{r})\hat{a}_\mathbf{r} \\ \hat{S}^-_\mathbf{r} = \sqrt{2S}\hat{a}^\dagger_\mathbf{r}f(\hat{n}_\mathbf{r})\end{cases} (\mathbf{r}\in \mathrm{A}), \qquad \begin{cases}\hat{S}^z_\mathbf{r} = -S + \hat{n}_\mathbf{r} \\ \hat{S}^+_\mathbf{r} = \sqrt{2S}\hat{a}^\dagger_\mathbf{r}f(\hat{n}_\mathbf{r}) \\ \hat{S}^-_\mathbf{r} = \sqrt{2S}f(\hat{n}_\mathbf{r})\hat{a}_\mathbf{r}\end{cases} (\mathbf{r}\in \mathrm{B}), \quad (8.44)$$

[11] Let us recall that we consider a d-dimensional hypercubic lattice, which can be divided into two sublattices A and B, with all sites in A having their nearest neighbors in B and *vice versa*.

so that the (classical) Néel state corresponds to the vacuum of bosons. The function $f(\hat{n})$ is defined as before [Eq. (8.38)] and one easily verifies that the spin commutation relations are satisfied on both sublattices.

To leading (non-trivial) order in the $1/S$ expansion,

$$\begin{aligned}\hat{H} &= E_{\rm cl} + \frac{1}{2}\sum_{\mathbf{r},\mathbf{r}'} SJ_{\mathbf{r},\mathbf{r}'}\left[(-1)^{\mathbf{r}+\mathbf{r}'}(\hat{n}_{\mathbf{r}} + \hat{n}_{\mathbf{r}'}) - (\hat{a}_{\mathbf{r}}\hat{a}_{\mathbf{r}'} + \text{h.c.})\right] \\ &= E_{\rm cl} - \frac{1}{2}\sum_{\mathbf{q}} S\left[2J(0)\hat{a}^\dagger_{\mathbf{q}}\hat{a}_{\mathbf{q}} + J(\mathbf{q})(\hat{a}_{\mathbf{q}}\hat{a}_{-\mathbf{q}} + \text{h.c.})\right], \end{aligned} \tag{8.45}$$

where $E_{\rm cl} = -\frac{1}{2}NS^2J(\mathbf{Q})$ is the energy of the classical Néel state and $\mathbf{Q} = (\pi,\dots,\pi)$. The Hamiltonian can be diagonalized by the following Bogoliubov transformation:

$$\begin{aligned}\hat{\alpha}^\dagger_{\mathbf{q}} &= u_{\mathbf{q}}\hat{a}^\dagger_{\mathbf{q}} - v_{\mathbf{q}}\hat{a}_{-\mathbf{q}}, \\ \hat{\alpha}_{-\mathbf{q}} &= -v_{\mathbf{q}}\hat{a}^\dagger_{\mathbf{q}} + u_{\mathbf{q}}\hat{a}_{-\mathbf{q}},\end{aligned} \tag{8.46}$$

where $\hat{\alpha}_{\mathbf{q}}, \hat{\alpha}^\dagger_{\mathbf{q}}$ are bosonic operators. We assume that $u_{\mathbf{q}}$ and $v_{\mathbf{q}}$ are real and satisfy $u_{\mathbf{q}} = u_{-\mathbf{q}}$, $v_{\mathbf{q}} = v_{-\mathbf{q}}$. The condition $u^2_{\mathbf{q}} - v^2_{\mathbf{q}} = 1$ ensures that the transformation (8.46) is unitary: $[\hat{\alpha}_{\mathbf{q}}, \hat{\alpha}^\dagger_{\mathbf{q}'}] = \delta_{\mathbf{q},\mathbf{q}'}$ and $[\hat{\alpha}_{\mathbf{q}}, \hat{\alpha}_{\mathbf{q}'}] = [\hat{\alpha}^\dagger_{\mathbf{q}}, \hat{\alpha}^\dagger_{\mathbf{q}'}] = 0$. The Hamiltonian

$$\hat{H} = E_{\rm cl} - \frac{1}{2}J(0)S\sum_{\mathbf{q}}(\hat{\alpha}^\dagger_{\mathbf{q}}, \hat{\alpha}_{-\mathbf{q}})\begin{pmatrix} A_{\mathbf{q}} & B_{\mathbf{q}} \\ B_{\mathbf{q}} & A_{\mathbf{q}}\end{pmatrix}\begin{pmatrix}\hat{\alpha}_{\mathbf{q}} \\ \hat{\alpha}^\dagger_{-\mathbf{q}}\end{pmatrix}, \tag{8.47}$$

where

$$A_{\mathbf{q}} = u^2_{\mathbf{q}} + v^2_{\mathbf{q}} + 2\gamma_{\mathbf{q}}u_{\mathbf{q}}v_{\mathbf{q}}, \qquad B_{\mathbf{q}} = 2u_{\mathbf{q}}v_{\mathbf{q}} + \gamma_{\mathbf{q}}(u^2_{\mathbf{q}} + v^2_{\mathbf{q}}), \tag{8.48}$$

becomes diagonal if we choose $u_{\mathbf{q}} = \cosh\theta_{\mathbf{q}}$ and $v_{\mathbf{q}} = \sinh\theta_{\mathbf{q}}$ with $\tanh(2\theta_{\mathbf{q}}) = -\gamma_{\mathbf{q}}$:

$$\hat{H} = E_{\rm cl} - \frac{1}{2}J(0)S\sum_{\mathbf{q}}(1-\gamma^2_{\mathbf{q}})^{1/2} + \sum_{\mathbf{q}}\omega_{\mathbf{q}}\hat{\alpha}^\dagger_{\mathbf{q}}\hat{\alpha}_{\mathbf{q}}. \tag{8.49}$$

We recover the dispersion $\omega_{\mathbf{q}} = |J(0)|S(1-\gamma^2_{\mathbf{q}})^{1/2}$ of the antiferromagnetic spin wave obtained from the classical dynamics and the RPA (Sec. 8.2). The ground state is defined by the vacuum of bosons $\hat{\alpha}, \hat{\alpha}^\dagger$ but differs from the vacuum of Holstein–Primakoff bosons: $\langle\hat{\alpha}^\dagger_{\mathbf{q}}\hat{\alpha}_{\mathbf{q}}\rangle = 0$ but $\langle\hat{a}^\dagger_{\mathbf{q}}\hat{a}_{\mathbf{q}}\rangle \neq 0$. Because of the quantum fluctuations of the Holstein–Primakoff bosons, the ground state energy is given by

$$E_0 = E_{\rm cl} - \frac{1}{2}J(0)S\sum_{\mathbf{q}}(1-\gamma^2_{\mathbf{q}})^{1/2} \tag{8.50}$$

to order $\mathcal{O}(S)$. The staggered magnetization $(-1)^{\mathbf{r}}\langle\hat{S}^z_{\mathbf{r}}\rangle$, which is equal to S in the classical Néel state, is reduced by quantum fluctuations,

$$\begin{aligned}(-1)^{\mathbf{r}}\langle\hat{S}^z_{\mathbf{r}}\rangle &= S - \langle\hat{a}^\dagger_{\mathbf{r}}\hat{a}_{\mathbf{r}}\rangle \\ &= S - \int_{\mathbf{q}}\langle u^2_{\mathbf{q}}\hat{\alpha}^\dagger_{\mathbf{q}}\hat{\alpha}_{\mathbf{q}} + v^2_{\mathbf{q}}\hat{\alpha}_{\mathbf{q}}\hat{\alpha}^\dagger_{\mathbf{q}} + u_{\mathbf{q}}v_{\mathbf{q}}(\hat{\alpha}^\dagger_{\mathbf{q}}\hat{\alpha}^\dagger_{-\mathbf{q}} + \text{h.c.})\rangle. \end{aligned} \tag{8.51}$$

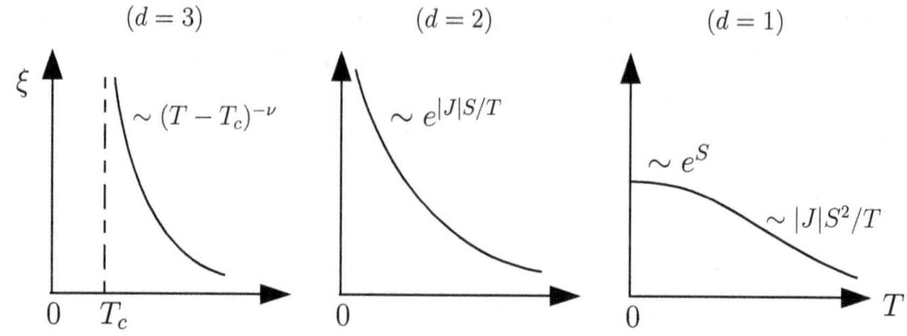

Fig. 8.2 Correlation length $\xi(T)$ in the Heisenberg model on a hypercubic lattice as predicted by the spin-wave expansion. These predictions are qualitatively correct except for half-integer spins in one dimension.

Using $\langle \hat\alpha^\dagger_{\mathbf q}\hat\alpha^\dagger_{-\mathbf q}\rangle = \langle \hat\alpha_{\mathbf q}\hat\alpha_{-\mathbf q}\rangle = 0$ and $\langle \hat\alpha^\dagger_{\mathbf q}\hat\alpha_{\mathbf q}\rangle = n_B(\omega_{\mathbf q})$, one finally obtains

$$(-1)^{\mathbf r}\langle \hat S^z_{\mathbf r}\rangle = S - \frac{1}{2}\int_{\mathbf q}\left[-1 + \frac{2n_B(\omega_{\mathbf q})+1}{(1-\gamma^2_{\mathbf q})^{1/2}}\right]. \tag{8.52}$$

We discuss below the consequences of Eq. (8.52) in one, two and three dimensions (Fig. 8.2).

8.3.2.1 *One dimension*

In one dimension, the infrared divergence of the momentum integral in (8.52) suggests that there is no long-range order. An estimate of the zero-temperature correlation length can be obtained from the (crude) criterion

$$(-1)^{\mathbf r}\langle \hat S^z_{\mathbf r}\rangle_{T=0} \sim S - \text{const}\int_{\xi^{-1}}^{\Lambda}\frac{dq}{q} = 0 \tag{8.53}$$

expressing the fact that fluctuations above the momentum scale ξ^{-1} are responsible for the suppression of long-range order.[12] Λ denotes an upper momentum cutoff, of the order of the inverse lattice spacing, such that $1-\gamma_q \simeq q^2/2$ for $|q| \lesssim \Lambda$. Equation (8.53) yields $\xi \sim \Lambda^{-1}e^{\text{const}\times S}$, which turns out to be correct in the large-S limit for integer spins but not for half-integer spins (see Sec. 8.4.3).

At finite temperature, we expect the correlation length to be reduced by thermal fluctuations. The latter correspond to fluctuations with momentum smaller than T/c. If the correlation length satisfies $\xi \gg c/T$ (the system is then primarily disordered by thermal fluctuations), it can be estimated from the criterion[12]

$$(-1)^{\mathbf r}\langle \hat S^z_{\mathbf r}\rangle_{T} \sim S - \frac{T}{\pi c}\int_{\xi^{-1}}^{T/c}\frac{dq}{q^2} = 0, \tag{8.54}$$

using $n_B(\omega_q) \sim T/\omega_q \sim T/c|q| \gg 1$ when $|q| \ll T/c$ ($c = 2|J|S$). When $S \gg 1$, Eq. (8.54) gives $\xi \sim Sc/T \sim |J|S^2/T \gg c/T$.

---

[12]The momentum integrals in (8.53), (8.54) stand for contributions coming from both $q \simeq 0$ and $q \simeq \pi$.

8.3.2.2 *Two dimensions*

In two dimensions, Eq. (8.52) gives

$$(-1)^{\mathbf{r}}\langle \hat{S}^z_{\mathbf{r}}\rangle_{T=0} \simeq S - 0.19660. \tag{8.55}$$

For $S = 1/2$, one finds $(-1)^{\mathbf{r}}\langle \hat{S}^z_{\mathbf{r}}\rangle_{T=0} \simeq 0.303$, which is in very good agreement with numerical calculations [2], somewhat surprisingly since the spin-wave expansion is justified only for $S \gg 1$. To next orders, taking into account interactions between spin waves [2],

$$(-1)^{\mathbf{r}}\langle \hat{S}^z_{\mathbf{r}}\rangle_{T=0} \simeq S - 0.19660 - \frac{0.00068}{S^2} + \mathcal{O}\left(\frac{1}{S^3}\right). \tag{8.56}$$

The vanishing of the $\mathcal{O}(1/S)$ term as well as the small coefficient of the $\mathcal{O}(1/S^2)$ may explain the success of the spin-wave expansion for the Heisenberg model on a square lattice.

The infrared divergence of the momentum integral in (8.52) when $T > 0$ suggests the absence of long-range order at finite temperature. From Eq. (8.54), with dq/q^2 replaced by $d|\mathbf{q}|/|\mathbf{q}|$, we obtain the correlation length $\xi \sim e^{\mathrm{const}\times S/T}$. This result will be confirmed in Sec. 8.4.3.

8.3.2.3 *Three dimensions*

In three dimensions, $(-1)^{\mathbf{r}}\langle \hat{S}^z_{\mathbf{r}}\rangle_{T=0} \simeq S - 0.07836$. The momentum integral in (8.52) remains infrared convergent at finite temperature. At low temperatures, one finds $(-1)^{\mathbf{r}}(\langle S^z_{\mathbf{r}}\rangle - \langle S^z_{\mathbf{r}}\rangle_{T=0}) \propto T^2$. Thermodynamic quantities can easily be computed. For instance, the energy $E(T) - E(0) \propto T^4$ yields a T^3-dependent specific heat. At high temperatures, the system should be disordered by thermal fluctuations. The phase transition to the antiferromagnetic state at the Néel temperature T_N is expected to be in the universality class of the O(3) model, the correlation length $\xi \sim (T - T_N)^{-\nu}$ diverging as a power law with a critical exponent ν corresponding to three-dimensional Wilson–Fisher fixed point (see Chapter 10).

8.4 Spin-Coherent-State Functional Integral

In Chapter 1, we showed how to write the partition function of interacting quantum particles, be they bosons or fermions, as a functional integral. In this section, we show that this is also possible for quantum spins with the help of suitably defined spin-coherent states. The spin-coherent-state functional integral allows us to recover the spin-wave dynamics when the ground state exhibits a spontaneous breaking of spin-rotation invariance. It also enables to go beyond the spin-wave analysis and show that the low-energy behavior of antiferromagnetic quantum spin systems is governed by a quantum NLσM with a topological Berry phase term, from which important results can be deduced (Sec. 8.4.3).

8.4.1 *Berry phase*

Let us first discuss the concept of Berry phase, which turns out to play a crucial role in the spin-coherent-state functional integral. We consider a system with a Hamiltonian

$\hat{H}(t) \equiv \hat{H}(\mathbf{R})$ depending on a parameter $\mathbf{R} \equiv \mathbf{R}(t) = (R_1(t), R_2(t), \ldots)$ which varies with time. The evolution of the system as a function of time is given by the Schrödinger equation

$$i\frac{d}{dt}|\psi(t)\rangle = \hat{H}(\mathbf{R})|\psi(t)\rangle. \tag{8.57}$$

Let us assume that the instantaneous eigenstates of $\hat{H}(\mathbf{R})$ are known,

$$\hat{H}(\mathbf{R})|\psi_l(\mathbf{R})\rangle = \epsilon_l(\mathbf{R})|\psi_l(\mathbf{R})\rangle, \tag{8.58}$$

and expand $|\psi(t)\rangle$ on the basis $\{|\psi_l(\mathbf{R})\rangle\}$,

$$|\psi(t)\rangle = \sum_l c_l(t)|\psi_l(\mathbf{R})\rangle. \tag{8.59}$$

Inserting (8.59) into (8.57), we find[13]

$$i\sum_l [\dot{c}_l(t)|\psi_l(\mathbf{R})\rangle + c_l(t)\dot{\mathbf{R}} \cdot |\boldsymbol{\nabla}\psi_l(\mathbf{R})\rangle] = \sum_l c_l(t)\epsilon_l(\mathbf{R})|\psi_l(\mathbf{R})\rangle \tag{8.60}$$

and in turn, with a left multiplication by $\langle\psi_n(\mathbf{R})|$,

$$i\dot{c}_n(t) = c_n(t)\epsilon_n(\mathbf{R}) - \sum_l \alpha_{nl}(t)c_l(t), \tag{8.61}$$

where $\alpha_{nl}(t) = i\dot{\mathbf{R}} \cdot \langle\psi_n(\mathbf{R})|\boldsymbol{\nabla}\psi_l(\mathbf{R})\rangle$.

8.4.1.1 *The adiabatic approximation*

Introducing

$$\tilde{c}_n(t) = c_n(t)e^{i\int_0^t dt'\epsilon_n(\mathbf{R}')} \qquad (\mathbf{R}' = \mathbf{R}(t')), \tag{8.62}$$

Eq. (8.61) becomes

$$i\dot{\tilde{c}}_n(t) = -\sum_l e^{i\int_0^t dt'\omega_{nl}(t')}\alpha_{nl}(t)\tilde{c}_l(t) \quad \text{with} \quad \omega_{nl}(t) = \epsilon_n(\mathbf{R}) - \epsilon_l(\mathbf{R}). \tag{8.63}$$

We assume that at time $t = 0$, the system is in the state $|\psi_n(\mathbf{R}(0))\rangle$: $c_l(t=0) = \delta_{n,l}$. If $\mathbf{R}(t) = \mathbf{R}(0)$ is constant, then $\alpha_{nl} = 0$ and Eq. (8.63) gives

$$c_l(t) = \delta_{n,l}e^{-i\int_0^t dt'\epsilon_n(\mathbf{R})} = \delta_{n,l}e^{-i\epsilon_n(\mathbf{R})t} \qquad (\mathbf{R}(t) = \mathbf{R}(0)). \tag{8.64}$$

When $\mathbf{R}$ varies slowly in time, we can look for a solution in powers of $\dot{\mathbf{R}}$. To lowest order,

$$i\dot{\tilde{c}}_l(t) = -e^{i\int_0^t dt'\omega_{ln}(t')}\alpha_{ln}(t) \quad \text{for} \quad l \neq n, \tag{8.65}$$

with $\tilde{c}_l(0) = 0$. If α_{ln} and ω_{ln} were time independent, Eq. (8.65) would give

$$\tilde{c}_l(t) = \frac{\alpha_{ln}}{\omega_{ln}}(e^{i\omega_{ln}t} - 1) \qquad (l \neq n). \tag{8.66}$$

---

[13] We use the notation $\dot{c}_l(t) = dc_l(t)/dt$ and $|\boldsymbol{\nabla}\psi_l(\mathbf{R})\rangle = \boldsymbol{\nabla}_{\mathbf{R}}|\psi_l(\mathbf{R})\rangle$.

One would then conclude that the probability amplitude $\tilde{c}_l$ remains small at any time provided that $|\alpha_{ln}/\omega_{ln}| \ll 1$. In the general case, where ω_{ln} and α_{ln} vary with time, we expect that when the condition

$$\frac{\max_t|\alpha_{ln}(t)|}{\min_t|\omega_{ln}(t)|} \ll 1 \tag{8.67}$$

is fulfilled, the system will follow adiabatically the state $|\psi_n(\mathbf{R})\rangle$, i.e. $c_n(t) \simeq 1$.[14] In that case, the time evolution of the system is entirely determined by $c_n(t)$. Note that the condition (8.67) is violated whenever the state $|\psi_n(\mathbf{R})\rangle$ becomes momentarily degenerate with another state $|\psi_l(\mathbf{R})\rangle$.

8.4.1.2 *Dynamic and geometric phases*

We now assume the adiabatic approximation to be valid: $|\psi(t)\rangle \simeq c_n(t)|\psi_n(\mathbf{R}\rangle$ with

$$i\dot{c}_n(t) = c_n[\epsilon_n(\mathbf{R}) - \dot{\mathbf{R}}\cdot\mathbf{A}_n(\mathbf{R})] \quad \text{and} \quad \mathbf{A}_n(\mathbf{R}) = i\langle\psi_n(\mathbf{R})|\boldsymbol{\nabla}\psi_n(\mathbf{R})\rangle, \tag{8.68}$$

where the real vector $\mathbf{A}_n(\mathbf{R})$, called the Berry connection, is analogous to a "vector potential".[15] Thus,

$$c_n(t) = e^{-i\int_0^t dt'\epsilon_n(\mathbf{R}')+i\gamma(t)}, \tag{8.69}$$

where the first term in the exponential is the dynamic phase and the second one,

$$\gamma(t) = \int_0^t dt'\,\dot{\mathbf{R}}'\cdot\mathbf{A}_n(\mathbf{R}') = \int_{\mathbf{R}(0)}^{\mathbf{R}(t)} d\mathbf{R}\cdot\mathbf{A}_n(\mathbf{R}), \tag{8.70}$$

the geometric (or Berry) phase. The last integral in (8.70) is taken along the line $\mathbf{R}(t')$ for $0 \le t' \le t$. The Berry phase depends only on the "trajectory" followed by the parameter $\mathbf{R}$ and is independent of the time t to go from $\mathbf{R}(0)$ to $\mathbf{R}(t)$ (provided that the adiabatic approximation remains valid): $\gamma(t)$ is purely topological.

Since two states differing by a phase describe the same physical state, one can wonder whether the Berry phase $\gamma(t)$ carries a physical meaning or, on the contrary, can be removed by a gauge transformation. In the gauge transformation

$$|\psi_n(\mathbf{R})\rangle \to |\psi_n'(\mathbf{R})\rangle = e^{i\Lambda(\mathbf{R})}|\psi_n(\mathbf{R})\rangle, \tag{8.71}$$

the Berry connection changes as

$$\mathbf{A}_n(\mathbf{R}) \to \mathbf{A}_n'(\mathbf{R}) = i\langle\psi_n'(\mathbf{R})|\boldsymbol{\nabla}\psi_n'(\mathbf{R})\rangle = \mathbf{A}_n(\mathbf{R}) - \boldsymbol{\nabla}\Lambda(\mathbf{R}). \tag{8.72}$$

Consider now a closed path $\mathcal{C}$: $\mathbf{R}(0) \to \mathbf{R}(t) \to \mathbf{R}(T) = \mathbf{R}(0)$. The Berry phase is clearly invariant since

$$\gamma(\mathcal{C}) \to \gamma'(\mathcal{C}) = \gamma(\mathcal{C}) - \oint_{\mathcal{C}} d\mathbf{R}\cdot\boldsymbol{\nabla}\Lambda(\mathbf{R}) = \gamma(\mathcal{C}). \tag{8.73}$$

---

[14]In a system with a large number of degrees of freedom, it is possible that $|c_l(t)|$ $(l \neq n)$ remains small even when $|c_n(t)|$ differs significantly from unity.

[15]The fact that $\mathbf{A}_n(\mathbf{R})$ is real follows from $\langle\psi_n|\psi_n\rangle = 1$ and $\langle\boldsymbol{\nabla}\psi_n|\psi_n\rangle + \langle\psi_n|\boldsymbol{\nabla}\psi_n\rangle = 0$, so that $\langle\boldsymbol{\nabla}\psi_n|\psi_n\rangle$ is purely imaginary.

We use the notation $\gamma(\mathcal{C})$ to emphasize that the Berry phase depends only on $\mathcal{C}$ and not on the dynamics of the motion. Thus, the Berry phase cannot be removed by a mere gauge transformation and has a true physical meaning as we shall see in the following.

When the parameter $\mathbf{R} = (R_1, R_2, R_3)$ is three-dimensional, the Berry phase

$$\gamma(\mathcal{C}) = \oint_{\mathcal{C}} d\mathbf{R} \cdot \mathbf{A}_n(\mathbf{R}) = \int_{\Sigma} \boldsymbol{\mathcal{B}}_n(\mathbf{R}) \cdot d\boldsymbol{\Sigma} \tag{8.74}$$

can be expressed as an integral over a surface Σ defined by the contour $\mathcal{C}$. The Berry curvature

$$\boldsymbol{\mathcal{B}}_n(\mathbf{R}) = \boldsymbol{\nabla} \times \mathbf{A}_n(\mathbf{R}) = i\langle \boldsymbol{\nabla}\psi_n(\mathbf{R})| \times |\boldsymbol{\nabla}\psi_n(\mathbf{R})\rangle, \tag{8.75}$$

analogous to a "magnetic field", is gauge invariant (this follows from (8.72)) like the Berry phase $\gamma(\mathcal{C})$ for a closed path, but is a local quantity.

8.4.1.3 *Example: spin-$\frac{1}{2}$ in a time-varying external field*

We consider a spin-$\frac{1}{2}$ in an external field $\mathbf{B}(t) = B\boldsymbol{\Omega}(t)$ whose direction $\boldsymbol{\Omega}(t)$ varies slowly in time ($\boldsymbol{\Omega}(t)^2 = 1$ and $B > 0$),

$$\hat{H}(t) = -B\boldsymbol{\Omega}(t) \cdot \hat{\mathbf{S}}, \tag{8.76}$$

with $\hat{\mathbf{S}} = \boldsymbol{\sigma}/2$ the spin operator and $\boldsymbol{\sigma} = (\sigma^x, \sigma^y, \sigma^z)$ the Pauli matrices. We assume that the adiabatic approximation is valid and that the system remains in the (instantaneous) ground state of $\hat{H}(t)$. The latter can be written as[16]

$$\begin{aligned} |\boldsymbol{\Omega}\rangle &= \hat{U}(\boldsymbol{\Omega})|\uparrow\rangle = e^{-\frac{i}{2}\varphi\sigma^z} e^{-\frac{i}{2}\theta\sigma^y} e^{-\frac{i}{2}\psi\sigma^z} |\uparrow\rangle \\ &= \cos\left(\frac{\theta}{2}\right) e^{-\frac{i}{2}(\varphi+\psi)}|\uparrow\rangle + \sin\left(\frac{\theta}{2}\right) e^{\frac{i}{2}(\varphi-\psi)}|\downarrow\rangle, \end{aligned} \tag{8.77}$$

where $|\uparrow\rangle$ and $|\downarrow\rangle$ denote the eigenstates of σ^z with eigenvalue 1 and -1, respectively. θ and φ are the polar and azimuthal angles associated with $\boldsymbol{\Omega} = (\sin\theta\cos\varphi, \sin\theta\sin\varphi, \cos\theta)$, while ψ is arbitrary and depends on the gauge choice. $\hat{U}(\boldsymbol{\Omega})$ is an SU(2) rotation operator such that the state $|\boldsymbol{\Omega}\rangle$ is an eigenstate of $\hat{\mathbf{S}} \cdot \boldsymbol{\Omega}$ with eigenvalue 1/2.

For a closed path $\boldsymbol{\Omega}(T) = \boldsymbol{\Omega}(0)$, the Berry phase is given by (8.74) with $\mathbf{R} \equiv \boldsymbol{\Omega}$. An elementary calculation gives[17]

$$\boldsymbol{\mathcal{B}}(\boldsymbol{\Omega}) = -\frac{1}{2}\boldsymbol{\Omega} \quad \text{and} \quad \gamma(\mathcal{C}) = -\frac{1}{2}\int_{\Sigma} \boldsymbol{\Omega} \cdot d\boldsymbol{\Sigma} = -\frac{1}{2}\omega(\mathcal{C}), \tag{8.78}$$

where $\omega(\mathcal{C})$ is the oriented solid angle (seen from the origin) defined by the closed path $\mathcal{C}$ on the unit sphere. This expression of the "magnetic field" $\boldsymbol{\mathcal{B}}$ shows that the latter is not created by a current distribution but by a magnetic monopole located at the origin.

---

[16] $|\boldsymbol{\Omega}\rangle$ is an example of a spin-coherent state; see Sec. 8.4.2.

[17] We use spherical coordinates to compute $\boldsymbol{\mathcal{B}}(\boldsymbol{\Omega})$ and $\mathbf{A}(\boldsymbol{\Omega})$.

This monopole has a charge q_m defined by $\nabla \cdot \mathcal{B} = 4\pi q_m \delta(\mathbf{B})$, i.e. $q_m = -\frac{1}{2}$.[18,19] Note that there is an ambiguity in the expression (8.78) of the Berry phase $\gamma(C)$ since there are two possible choices for the surface Σ corresponding to the interior of the closed path C on the unit sphere. However, since the two solid angles differ by 4π, the resulting Berry phases differ by 2π and are therefore equivalent.

There is actually a subtle point here. The state $|\mathbf{\Omega}\rangle$ is not defined everywhere on the sphere. For instance, in the gauge $\psi = -\varphi$, $|\mathbf{\Omega}\rangle$ is not defined at the south pole since $\sin(\frac{\theta}{2})e^{i\varphi}$ does not approach a unique value when $\theta \to \pi$; the Berry connection $\mathbf{A}(\mathbf{\Omega})$ and the Berry curvature $\mathcal{B}(\mathbf{\Omega})$ are then defined everywhere on the unit sphere except at the south pole. An explicit calculation gives[17]

$$\mathbf{A}(\mathbf{\Omega}) = i\langle \mathbf{\Omega}|\nabla\mathbf{\Omega}\rangle = -\frac{1-\cos\theta}{2\sin\theta}\mathbf{u}_\varphi, \tag{8.79}$$

where $\mathbf{u}_\varphi = (-\sin\varphi, \cos\varphi, 0)$. From (8.79), one easily sees that the line integral of the vector potential along a tiny loop around the south pole,

$$\lim_{\theta\to\pi}\int_0^{2\pi} d\varphi\, \mathbf{A}(\mathbf{\Omega})\cdot\mathbf{u}_\varphi = \lim_{\theta\to\pi} A_\varphi 2\pi\sin\theta = -2\pi, \tag{8.80}$$

gives the full monopole flux $4\pi q_m$. Thus, the "vector potential" $\mathbf{A}$ describes not a monopole at the origin but a tiny flux line (the Dirac string) along the negative-z axis, which brings the entire flux to the origin from where it emanates radially.[20] Using (8.79), one recovers the result (8.78),

$$\gamma(C) = \oint_C d\mathbf{\Omega}\cdot\mathbf{A}(\mathbf{\Omega}) = \int_0^T dt\frac{\dot\varphi}{2}(\cos\theta - 1) \equiv -\frac{1}{2}\omega(C), \tag{8.81}$$

where now $\omega(C)$ is defined such as excluding the south pole.

The location of the Dirac string depends on the definition of $|\mathbf{\Omega}\rangle$. In the gauge $\psi = \varphi$, the Berry connection[21,22]

$$\mathbf{A}'(\mathbf{\Omega}) = \frac{1+\cos\theta}{2\sin\theta}\mathbf{u}_\varphi \tag{8.82}$$

is not defined at the north pole. The Berry phase reads

$$\gamma'(C) = \oint_C d\mathbf{\Omega}\cdot\mathbf{A}'(\mathbf{\Omega}) = \int_0^T dt\frac{\dot\varphi}{2}(\cos\theta + 1) = -\frac{1}{2}\omega'(C) = \gamma(C) + 2\pi, \tag{8.83}$$

where $\omega'(C) = \omega(C) - 4\pi$ is defined by the surface Σ' excluding the north pole.

[18] If, more generally, we had also allowed the amplitude of the external field $\mathbf{B}(t) = B(t)\mathbf{\Omega}(t)$ to vary in time, we would have found a Berry curvature $\mathcal{B}(\mathbf{B}) = -\mathbf{B}/2|\mathbf{B}|^3$.
[19] Here, we use CGS units. In SI units, $\nabla\cdot\mathcal{B} = q_m\delta(\mathbf{B})$ or $\nabla\cdot\mathcal{B} = \mu_0 q_m \delta(\mathbf{B})$.
[20] Recall that any vector potential describing a monopole has a string.
[21] $\mathbf{A}$ and $\mathbf{A}'$ [Eqs. (8.79), (8.82)] are related by the gauge transformation $|\mathbf{\Omega}'\rangle = e^{\frac{i}{2}\varphi}|\mathbf{\Omega}\rangle$.
[22] Another common gauge choice is $\psi = 0$, which gives $\mathbf{A}(\mathbf{\Omega}) = \frac{1}{2\tan\theta}\mathbf{u}_\varphi$.

The preceding analysis can be generalized to an arbitrary spin S (see Sec. 8.4.2). Again, one can use different gauges to calculate the Berry phase. For instance, Eqs. (8.81) and (8.83) become $\gamma(\mathcal{C}) = -S\omega(\mathcal{C})$ and $\gamma'(\mathcal{C}) = \gamma(\mathcal{C}) + 4S\pi$. Demanding that $\gamma(\mathcal{C})$ be defined modulo 2π leads to the familiar requirement that $2S$ be an integer.

8.4.2 *Functional integral*

8.4.2.1 *Spin-coherent states*

We consider a spin S and denote by $|S^z = -S, \dots, S\rangle$ the $2S+1$ eigenstates of $\hat{S}^z$. Spin-coherent states are defined by

$$|\mathbf{\Omega}\rangle = e^{-i\varphi\hat{S}^z} e^{-i\theta\hat{S}^y} e^{-i\psi\hat{S}^z} |S\rangle, \tag{8.84}$$

where $\mathbf{\Omega}(\theta, \varphi)$ is the unit vector introduced in the preceding section. $|\mathbf{\Omega}\rangle$ is obtained from $|S\rangle$ by a rotation of angle θ about the y axis followed by a rotation of angle φ about the z axis. The angle ψ is arbitrary and corresponds to a gauge freedom (Sec. 8.4.1). Since $\hat{\mathbf{S}}$ transforms like a vector in a spin rotation (Sec. 2.2.3),

$$\langle\mathbf{\Omega}|\hat{\mathbf{S}}|\mathbf{\Omega}\rangle = S\mathbf{\Omega}. \tag{8.85}$$

Note however that $|\mathbf{\Omega}\rangle$ is not an eigenstate of $\hat{\mathbf{S}}$. In Appendix 8.B, we show that the scalar product between two coherent states is given by

$$\langle\mathbf{\Omega}'|\mathbf{\Omega}\rangle = \left(\cos\frac{\theta'}{2}\cos\frac{\theta}{2} + e^{i(\varphi-\varphi')}\sin\frac{\theta'}{2}\sin\frac{\theta}{2}\right)^{2S}, \tag{8.86}$$

and obtain the following resolution of the identity:

$$(2S+1)\int\frac{d\Omega}{4\pi}|\mathbf{\Omega}\rangle\langle\mathbf{\Omega}| = 1, \tag{8.87}$$

where the "1" in the rhs denotes the identity operator.

8.4.2.2 *Spin-coherent-state functional integral*

Let us consider a single spin S in an external field with Hamiltonian

$$\hat{H} = -\mathbf{h}\cdot\hat{\mathbf{S}}. \tag{8.88}$$

We can write the partition function

$$Z = \operatorname{Tr} e^{-\beta\hat{H}} = (2S+1)\int\frac{d\Omega}{4\pi}\langle\mathbf{\Omega}|e^{-\beta\hat{H}}|\mathbf{\Omega}\rangle \tag{8.89}$$

as a functional integral over spin-coherent states following the general method described in Chapter 1. First, we split the imaginary "time" β into M infinitesimal steps $\epsilon = \beta/M$ and introduce $M-1$ times the resolution of the identity (8.87),

$$Z = \left(\frac{2S+1}{4\pi}\right)^M \int \prod_{k=1}^M d\Omega_k \prod_{k=1}^M \langle \boldsymbol{\Omega}_k | e^{-\epsilon \hat{H}} | \boldsymbol{\Omega}_{k-1} \rangle, \tag{8.90}$$

with $\boldsymbol{\Omega}_0 = \boldsymbol{\Omega}_M$. We then use

$$\begin{aligned} \langle \boldsymbol{\Omega}_k | e^{-\epsilon \hat{H}} | \boldsymbol{\Omega}_{k-1} \rangle &= \langle \boldsymbol{\Omega}_k | 1 - \epsilon \hat{H} | \boldsymbol{\Omega}_{k-1} \rangle + \mathcal{O}(\epsilon^2) \\ &= \langle \boldsymbol{\Omega}_k | \boldsymbol{\Omega}_{k-1} \rangle - \epsilon \langle \boldsymbol{\Omega}_k | \hat{H} | \boldsymbol{\Omega}_{k-1} \rangle + \mathcal{O}(\epsilon^2). \end{aligned} \tag{8.91}$$

Setting $\tau = k\epsilon$ and considering $\tau \in [0, \beta]$ as a continuous variable in the limit $\epsilon = \beta/M \to 0$, we finally obtain[23,24]

$$\begin{aligned} \langle \boldsymbol{\Omega}_k | e^{-\epsilon \hat{H}} | \boldsymbol{\Omega}_{k-1} \rangle &= \langle \boldsymbol{\Omega} | \boldsymbol{\Omega} \rangle - \epsilon \langle \boldsymbol{\Omega} | \partial_\tau \boldsymbol{\Omega} \rangle - \epsilon \langle \boldsymbol{\Omega} | \hat{H} | \boldsymbol{\Omega} \rangle + \mathcal{O}(\epsilon^2) \\ &= e^{-\epsilon \langle \boldsymbol{\Omega} | \partial_\tau \boldsymbol{\Omega} \rangle - \epsilon H[\boldsymbol{\Omega}]} + \mathcal{O}(\epsilon^2), \end{aligned} \tag{8.92}$$

where $H[\boldsymbol{\Omega}] = \langle \boldsymbol{\Omega} | \hat{H} | \boldsymbol{\Omega} \rangle$ and we have used $\langle \boldsymbol{\Omega} | \boldsymbol{\Omega} \rangle = 1$. Inserting (8.92) into (8.90) gives

$$\begin{aligned} Z &= \left(\frac{2S+1}{4\pi}\right)^M \int \prod_{k=1}^M d\Omega_k \exp\left\{ -\epsilon \sum_{k=1}^M (\langle \boldsymbol{\Omega} | \partial_\tau \boldsymbol{\Omega} \rangle + H[\boldsymbol{\Omega}]) \right\} \\ &\equiv \int_{\boldsymbol{\Omega}(\beta) = \boldsymbol{\Omega}(0)} \mathcal{D}[\boldsymbol{\Omega}] \exp\left\{ -\mathcal{S}_B[\boldsymbol{\Omega}] - \int_0^\beta d\tau\, H[\boldsymbol{\Omega}] \right\}, \end{aligned} \tag{8.93}$$

where the action

$$\mathcal{S}_B[\boldsymbol{\Omega}] = \int_0^\beta d\tau \langle \boldsymbol{\Omega} | \partial_\tau \boldsymbol{\Omega} \rangle = -i \int_0^\beta d\tau\, \dot{\boldsymbol{\Omega}} \cdot \mathbf{A}(\boldsymbol{\Omega}) \tag{8.94}$$

is the Berry phase term introduced in Sec. 8.4.1.3 in the particular case $S = \frac{1}{2}$. To compute the "vector potential" $\mathbf{A}(\boldsymbol{\Omega}) = i \langle \boldsymbol{\Omega} | \boldsymbol{\nabla} \boldsymbol{\Omega} \rangle$ for an arbitrary value of S, we use

$$|\boldsymbol{\nabla} \boldsymbol{\Omega}\rangle = \mathbf{u}_\theta \frac{\partial}{\partial \theta} |\boldsymbol{\Omega}\rangle + \mathbf{u}_\varphi \frac{1}{\sin\theta} \frac{\partial}{\partial \varphi} |\boldsymbol{\Omega}\rangle \tag{8.95}$$

and

$$\langle \boldsymbol{\Omega} | \partial_\theta \boldsymbol{\Omega} \rangle = 0, \qquad \langle \boldsymbol{\Omega} | \partial_\varphi \boldsymbol{\Omega} \rangle = iS(1 - \cos\theta), \tag{8.96}$$

where $\mathbf{u}_\varphi = (-\sin\varphi, \cos\varphi, 0)$ and $\mathbf{u}_\theta = (\cos\theta\cos\varphi, \cos\theta\sin\varphi, -\sin\theta)$ (see Appendix 8.B). We thus obtain

$$\begin{aligned} \mathbf{A}(\boldsymbol{\Omega}) &= -S \frac{1 - \cos\theta}{\sin\theta} \mathbf{u}_\varphi, \\ \mathcal{S}_B[\boldsymbol{\Omega}] &= iS \int_0^\beta d\tau\, \dot{\varphi}(1 - \cos\theta), \end{aligned} \tag{8.97}$$

[23] We use the notation $|\partial_\tau \boldsymbol{\Omega}\rangle = \partial_\tau |\boldsymbol{\Omega}\rangle$.

[24] As pointed out in Chapter 1, the continuous-time limit in a functional integral may be ill-defined and in case of doubt, one should always return to the original discrete-time formulation.

which agrees with (8.79) when $S = 1/2$. Using the results of Sec. 8.4.1, we conclude that the Berry phase term

$$\mathcal{S}_B[\mathbf{\Omega}] = -i\int_0^\beta d\tau\, \dot{\mathbf{\Omega}} \cdot \mathbf{A}(\mathbf{\Omega}) = iS\omega(\mathcal{C}) \tag{8.98}$$

is simply determined by the solid angle $\omega(\mathcal{C})$ corresponding to the closed path $\mathcal{C}$: $\mathbf{\Omega}(0) \to \mathbf{\Omega}(\beta)$ on the unit sphere.

As in the case of the coherent-state functional integral for bosons and fermions (Chapter 1), we have replaced finite differences by derivatives as if the paths $\{\mathbf{\Omega}_k\}$ were smooth. There is however in general no reason why the difference $\mathbf{\Omega}_k - \mathbf{\Omega}_{k-1}$ should be small in the limit $\epsilon \to 0$. The only exception is the large-S limit. When $S \to \infty$, the scalar product $\langle \mathbf{\Omega}_k | \mathbf{\Omega}_{k-1} \rangle$ is non-zero only if $\mathbf{\Omega}_k \to \mathbf{\Omega}_{k-1}$.[25] Thus, for $S \gg 1$, which is usually the limit where this formalism is used, the Berry phase term ensures that the typical paths are continuous.

We are now in a position to determine the classical spin dynamics of a single spin in an external field $\mathbf{h}$:

$$\frac{\delta}{\delta \mathbf{\Omega}(\tau)} \left\{ \mathcal{S}_B[\mathbf{\Omega}] - S\int_0^\beta d\tau\, \mathbf{h} \cdot \mathbf{\Omega}(\tau) \right\} = 0. \tag{8.99}$$

To compute the first functional derivative, we consider the variation of the Berry phase term (8.94) due to a change $\delta\mathbf{\Omega}(\tau)$,

$$\delta\mathcal{S}_B = -i\int_0^\beta d\tau \left(\delta\dot{\Omega}_\alpha A_\alpha + \dot{\Omega}_\alpha \frac{\partial A_\alpha}{\partial \Omega_\gamma} \delta\Omega_\gamma \right) \tag{8.100}$$

(with an implicit sum over repeated indices). Adding and subtracting $(\partial A_\alpha / \partial \Omega_\gamma) \dot{\Omega}_\gamma \delta\Omega_\alpha$, we obtain

$$\delta\mathcal{S}_B = -i\int_0^\beta d\tau \left(\frac{\partial A_\alpha}{\partial \Omega_\gamma} (\dot{\Omega}_\alpha \delta\Omega_\gamma - \dot{\Omega}_\gamma \delta\Omega_\alpha) + \delta\dot{\Omega}_\alpha A_\alpha + \frac{\partial A_\alpha}{\partial \Omega_\gamma} \dot{\Omega}_\gamma \delta\Omega_\alpha \right). \tag{8.101}$$

The last two terms give $\partial_\tau (A_\alpha \delta\Omega_\alpha)$ and their contribution vanishes because of the periodic boundary conditions in imaginary time. Thus,

$$\delta\mathcal{S}_B = -i\int_0^\beta d\tau\, \epsilon_{\alpha\gamma\gamma'} \frac{\partial A_\alpha}{\partial \Omega_\gamma} (\dot{\mathbf{\Omega}} \times \delta\mathbf{\Omega})_{\gamma'} = i\int_0^\beta d\tau (\mathbf{\nabla} \times \mathbf{A}) \cdot (\dot{\mathbf{\Omega}} \times \delta\mathbf{\Omega}). \tag{8.102}$$

Using the fact that the Berry connection $\boldsymbol{\mathcal{B}} = \mathbf{\nabla} \times \mathbf{A} = -S\mathbf{\Omega}$ (Sec. 8.4.1), we finally obtain

$$\delta\mathcal{S}_B = -iS\int_0^\beta d\tau\, \delta\mathbf{\Omega} \cdot (\mathbf{\Omega} \times \dot{\mathbf{\Omega}}). \tag{8.103}$$

The classical equation of motion (8.99) then reads

$$i\dot{\mathbf{\Omega}} = \mathbf{\Omega} \times \mathbf{h}. \tag{8.104}$$

[25] Equation (8.86) implies $|\langle \mathbf{\Omega}' | \mathbf{\Omega} \rangle| = \left(\frac{1 + \mathbf{\Omega}' \cdot \mathbf{\Omega}}{2} \right)^S$.

Performing a Wick rotation to real time, $t = -i\tau$, we recover the classical equation of motion of a spin $\mathbf{S} = S\mathbf{\Omega}$ in an external field: $d\mathbf{S}/dt = \mathbf{S} \times \mathbf{h}$ [Eq. (8.19)].

The generalization of the spin-coherent-state functional integral to a many-spin Hamiltonian defined on a lattice is straightforward. Spin-coherent states $|\mathbf{\Omega}\rangle$ are now defined by a unit vector field $\mathbf{\Omega}_\mathbf{r}$ defined at each lattice site $\mathbf{r}$, and the resolution of the identity becomes

$$\left(\frac{2S+1}{4\pi}\right)^N \int \prod_\mathbf{r} d\mathbf{\Omega}_\mathbf{r} |\mathbf{\Omega}\rangle\langle\mathbf{\Omega}| = 1. \tag{8.105}$$

The partition function can be written as a functional integral

$$Z = \int_{\mathbf{\Omega}(\beta)=\mathbf{\Omega}(0)} \mathcal{D}[\mathbf{\Omega}]\, e^{-\mathcal{S}_B[\mathbf{\Omega}] - \int_0^\beta d\tau\, H[\mathbf{\Omega}]}, \tag{8.106}$$

where the Berry phase term

$$\mathcal{S}_B[\mathbf{\Omega}] = \int_0^\beta d\tau \langle\mathbf{\Omega}|\partial_\tau \mathbf{\Omega}\rangle = \int_0^\beta d\tau \sum_\mathbf{r} \langle\mathbf{\Omega}_\mathbf{r}|\partial_\tau \mathbf{\Omega}_\mathbf{r}\rangle \tag{8.107}$$

is simply the sum of the single-spin contributions. The Hamiltonian expectation value $H[\mathbf{\Omega}] = \langle\mathbf{\Omega}|\hat{H}|\mathbf{\Omega}\rangle$, e.g.

$$H[\mathbf{\Omega}] = -\frac{S^2}{2} \sum_{\mathbf{r},\mathbf{r}'} J_{\mathbf{r},\mathbf{r}'} \mathbf{\Omega}_\mathbf{r} \cdot \mathbf{\Omega}_{\mathbf{r}'} \tag{8.108}$$

for the Heisenberg model (8.1), is obtained using standard properties of the spin-coherent states. The equation of motion $\delta\mathcal{S}[\mathbf{\Omega}]/\delta\mathbf{\Omega}_\mathbf{r}(\tau) = 0$, with the action $\mathcal{S}[\mathbf{\Omega}] = \mathcal{S}_B[\mathbf{\Omega}] + \int_0^\beta d\tau\, H[\mathbf{\Omega}]$, reproduces the classical dynamics, e.g. using (8.108),

$$\dot{\mathbf{\Omega}}_\mathbf{r} = S \sum_{\mathbf{r}'} J_{\mathbf{r},\mathbf{r}'} \mathbf{\Omega}_\mathbf{r} \times \mathbf{\Omega}_{\mathbf{r}'} \tag{8.109}$$

in real time. In Sec. 8.2.1, we have seen that linearizing this equation about the classical solution $\mathbf{\Omega}_\mathbf{r}^{\rm cl} = \mathbf{u}_z$ (ferromagnet) or $\mathbf{\Omega}_\mathbf{r}^{\rm cl} = (-1)^\mathbf{r}\mathbf{u}_z$ (antiferromagnet) yields the spin-wave excitations. Spin waves are discussed in more detail in the following section.

8.4.2.3 *Spin waves*

In the limit $S \to \infty$, the action is of order S^2 and a saddle-point approximation becomes exact.[26] The variable $\mathbf{\Omega}_\mathbf{r}(\tau)$ does not fluctuate and coincides with the classical solution $\mathbf{\Omega}_\mathbf{r}^{\rm cl}$ obtained by minimizing the action,

$$\left.\frac{\delta\mathcal{S}[\mathbf{\Omega}]}{\delta\mathbf{\Omega}_\mathbf{r}(\tau)}\right|_{\mathbf{\Omega}_\mathbf{r}(\tau)=\mathbf{\Omega}_\mathbf{r}^{\rm cl}} = 0. \tag{8.110}$$

[26] Formally, we can justify the saddle-point approximation and the loop expansion by rescaling the time variable: $\tau \to \bar\tau = S\tau$ and $\beta \to \bar\beta = S\beta$. The action $\mathcal{S}[\mathbf{\Omega}]$ then becomes proportional to S and a loop expansion, controlled by the small parameter $1/S$, becomes possible (Sec. 1.7).

In the large-S limit, we expect the functional integral to be dominated by small fluctuations $\delta\mathbf{\Omega}_\mathbf{r} = \mathbf{\Omega}_\mathbf{r} - \mathbf{\Omega}_\mathbf{r}^{\rm cl}$ about the classical state. We use the parameterization

$$P_\mathbf{r} = S\delta\mathbf{\Omega}_\mathbf{r} \cdot \mathbf{u}_\theta, \qquad Q_\mathbf{r} = \delta\mathbf{\Omega}_\mathbf{r} \cdot \mathbf{u}_\varphi. \tag{8.111}$$

For small fluctuations, the spin-coherent-state functional integral becomes an integral over $P_\mathbf{r}$ and $Q_\mathbf{r}$, i.e. $\mathcal{D}[\mathbf{\Omega}] \to \mathcal{D}[P,Q]$, and the domain of integration over $P_\mathbf{r}$ and $Q_\mathbf{r}$ can be extended to $]-\infty,\infty[$. We shall see that the spin-wave dynamics is encoded in the action $\mathcal{S}[P,Q]$ to quadratic order.

Let us first consider the ferromagnetic case with $\mathbf{\Omega}_\mathbf{r}^{\rm cl} = \mathbf{u}_x$. One then has

$$P_\mathbf{r} = -S\cos\theta_\mathbf{r} \quad \text{and} \quad Q_\mathbf{r} = \sin\theta_\mathbf{r}\sin\varphi_\mathbf{r} \simeq \varphi_\mathbf{r}. \tag{8.112}$$

To determine the Berry phase term to quadratic order in P, Q, we use (8.94), which gives[27,28]

$$\mathcal{S}_B = -i\frac{S}{2}\int_0^\beta d\tau \sum_\mathbf{r} \mathbf{\Omega}_\mathbf{r}^{\rm cl} \cdot (\delta\dot{\mathbf{\Omega}}_\mathbf{r} \times \delta\mathbf{\Omega}_\mathbf{r}) = \frac{i}{2}\int_0^\beta d\tau \sum_\mathbf{r} (P_\mathbf{r}\dot{Q}_\mathbf{r} - \dot{P}_\mathbf{r}Q_\mathbf{r}) \tag{8.113}$$

to quadratic order. As for the Hamiltonian, we obtain

$$\begin{aligned} H[\mathbf{\Omega}] &= -JS^2 \sum_{\langle \mathbf{r},\mathbf{r}'\rangle} \left[\cos\theta_\mathbf{r}\cos\theta_{\mathbf{r}'} + \sin\theta_\mathbf{r}\sin\theta_{\mathbf{r}'}\cos(\varphi_\mathbf{r} - \varphi_{\mathbf{r}'})\right] \\ &= E_{\rm cl} + \frac{JS^2}{2}\sum_{\langle \mathbf{r},\mathbf{r}'\rangle}\left[\frac{(P_\mathbf{r} - P_{\mathbf{r}'})^2}{S^2} + (Q_\mathbf{r} - Q_{\mathbf{r}'})^2\right] \end{aligned} \tag{8.114}$$

to quadratic order in P and Q. In Fourier space, we can therefore write the action as (dropping the ferromagnetic ground state energy $E_{\rm cl}$)

$$\mathcal{S}[P,Q] = \frac{1}{2}\sum_q (P_{-q}, Q_{-q}) \begin{pmatrix} 2Jd(1-\gamma_\mathbf{q}) & \omega_\nu \\ -\omega_\nu & 2JdS^2(1-\gamma_\mathbf{q}) \end{pmatrix} \begin{pmatrix} P_q \\ Q_q \end{pmatrix}, \tag{8.115}$$

with $q = (\mathbf{q}, i\omega_\nu)$ and $\gamma_\mathbf{q}$ defined in (8.42). The dispersion of the spin-wave modes is obtained from the vanishing of the determinant of the 2×2 matrix in (8.115) after analytic continuation $i\omega_\nu \to \omega + i0^+$, i.e. $\omega_\mathbf{q} = 2dJS(1-\gamma_\mathbf{q})$, in agreement with the results of Sec. 8.3.1.

In the antiferromagnetic case ($J<0$), we choose $\mathbf{\Omega}_\mathbf{r}^{\rm cl} = (-1)^\mathbf{r}\mathbf{u}_x$ so that

$$P_\mathbf{r} = -S\cos\theta_\mathbf{r} \quad \text{and} \quad Q_\mathbf{r} = \begin{cases} \sin\theta_\mathbf{r}\sin\varphi_\mathbf{r} \simeq \varphi_\mathbf{r} & \text{if } \mathbf{r}\in A, \\ -\sin\theta_\mathbf{r}\sin\varphi_\mathbf{r} \simeq \varphi_\mathbf{r} + \pi & \text{if } \mathbf{r}\in B. \end{cases} \tag{8.116}$$

---

[27]Equation (8.113) follows from (for a single site and with an implicit sum over repeated indices): $\mathcal{S}_B \simeq -i\int_0^\beta d\tau\, \delta\dot{\mathbf{\Omega}}_\alpha \frac{\partial A_\alpha}{\partial \mathbf{\Omega}_\gamma}\big|_{\rm cl}\delta\mathbf{\Omega}_\gamma = -\frac{i}{2}\int_0^\beta d\tau \frac{\partial A_\alpha}{\partial \mathbf{\Omega}_\gamma}\big|_{\rm cl}(\delta\dot{\mathbf{\Omega}}_\alpha\delta\mathbf{\Omega}_\gamma - \delta\mathbf{\Omega}_\alpha\delta\dot{\mathbf{\Omega}}_\gamma) = \frac{i}{2}\int_0^\beta d\tau(\boldsymbol{\nabla}\times\mathbf{A})\cdot(\delta\dot{\mathbf{\Omega}}\times\delta\mathbf{\Omega})$ with $\boldsymbol{\nabla}\times\mathbf{A} = -S\mathbf{\Omega}^{\rm cl}$.

[28]Alternatively, Eq. (8.113) can be obtained from an explicit expression of $\mathbf{A}(\mathbf{\Omega})$ and $\mathcal{S}_B[\mathbf{\Omega}]$ in a given gauge, e.g. Eq. (8.97).

The Berry phase term is unchanged [Eq. (8.113)] and the Hamiltonian is given by

$$H[\mathbf{\Omega}] = E_{\rm cl} + \frac{|J|S^2}{2} \sum_{\langle \mathbf{r},\mathbf{r}'\rangle} \left[\frac{(P_{\mathbf{r}} + P_{\mathbf{r}'})^2}{S^2} + (Q_{\mathbf{r}} - Q_{\mathbf{r}'})^2 \right] \tag{8.117}$$

to quadratic order in P and Q. This yields the action (dropping the energy $E_{\rm cl}$ of the classical antiferromagnetic state)

$$\mathcal{S}[P,Q] = \frac{1}{2} \sum_q (P_{-q}, Q_{-q}) \begin{pmatrix} 2|J|d(1+\gamma_{\mathbf{q}}) & \omega_\nu \\ -\omega_\nu & 2|J|dS^2(1-\gamma_{\mathbf{q}}) \end{pmatrix} \begin{pmatrix} P_q \\ Q_q \end{pmatrix}, \tag{8.118}$$

and in turn the spin-wave dispersion $\omega_{\mathbf{q}} = 2d|J|S(1-\gamma_{\mathbf{q}}^2)^{1/2}$, in agreement with the results of Sec. 8.3.2.

8.4.3 *Quantum nonlinear σ model*

The spin-wave expansion is valid provided that there is long-range order. In this section, we consider a more general situation where the system exhibits short-range antiferromagnetic order. The antiferromagnetic correlation length ξ is then much larger than the lattice spacing but not necessary infinite as it would be in the presence of long-range order. We expect spin-wave fluctuations about the local antiferromagnetic order to be the dominant fluctuations at length scales smaller than ξ. We shall see that the low-energy behavior is described by a quantum NLσM with a Berry phase term.[29]

To capture short-range antiferromagnetic order, it is tempting to write $\mathbf{\Omega}_{\mathbf{r}} \simeq (-1)^{\mathbf{r}} \mathbf{n}_{\mathbf{r}}$ where the Néel field $\mathbf{n}_{\mathbf{r}}$ is expected to be slowly varying. But the spin-wave expansion about the classical Néel state shows that antiferromagnetic fluctuations with momentum near $\mathbf{Q} = (\pi, \dots, \pi)$ couple to long-wavelength ferromagnetic fluctuations.[30] Thus, short-range antiferromagnetic order is captured by the following parameterization[31]:

$$\mathbf{\Omega}_{\mathbf{r}} = (-1)^{\mathbf{r}} \mathbf{n}_{\mathbf{r}} \left(1 - \frac{\mathbf{L}_{\mathbf{r}}^2}{S^2}\right)^{1/2} + \frac{\mathbf{L}_{\mathbf{r}}}{S}, \tag{8.119}$$

where

$$\mathbf{n}_{\mathbf{r}}^2 = 1 \quad \text{and} \quad \mathbf{n}_{\mathbf{r}} \perp \mathbf{L}_{\mathbf{r}} = 0, \tag{8.120}$$

so that $\mathbf{\Omega}_{\mathbf{r}}^2 = 1$. Both the Néel field $\mathbf{n}_{\mathbf{r}}$ and the canting field $\mathbf{L}_{\mathbf{r}}$, which describes ferromagnetic fluctuations of the spins, are assumed to be slowly varying (an assumption that should hold at least in the large-S limit). The prefactor $1/S$ has been associated with $\mathbf{L}_{\mathbf{r}}$ so that the spatial integral of $\mathbf{L}_{\mathbf{r}}$ over any region is precisely the total magnetization in

[29] For a general discussion of the quantum NLσM, see Chapter 12.

[30] In Eq. (8.118), fluctuations $P_{\mathbf{q}\simeq 0}$ and $Q_{\mathbf{q}\simeq 0}$ correspond to ferromagnetic fluctuations along $\mathbf{u}_z$ and antiferromagnetic fluctuations along $\mathbf{u}_y$, respectively (the reverse is true for $\mathbf{q} \simeq \mathbf{Q}$).

[31] We shall see that $\mathbf{L}_{\mathbf{r}}$ and $\mathbf{n}_{\mathbf{r}} \times \dot{\mathbf{n}}_{\mathbf{r}}$ are conjugate fields [Eq. (8.128)], which shows *a posteriori* the necessity to introduce $\mathbf{L}_{\mathbf{r}}$.

that region. Moreover, $\mathbf{L_r}$ is assumed to be small, $|\mathbf{L_r}|/S \ll 1$, so that spins on nearby sites are predominantly antiparallel.

We can therefore look for an effective action $\mathcal{S}[\mathbf{n},\mathbf{L}]$ in the continuum limit and to leading order in a derivative expansion. This is done by writing the partition function as a functional integral over the fields $\mathbf{n_r}$ and $\mathbf{L_r}$ using[32]

$$\mathcal{D}[\boldsymbol{\Omega}] = \mathcal{D}[\mathbf{n},\mathbf{L}]\mathcal{J}[\mathbf{n},\mathbf{L}]\delta(\mathbf{n}^2-1)\delta(\mathbf{n}\cdot\mathbf{L}), \tag{8.121}$$

where $\mathcal{J}[\mathbf{n},\mathbf{L}]$, the Jacobian of the transformation (8.119), is a constant to leading order in $1/S$.[33] In the transformation (8.119), it seems that we have replaced 2 degrees of freedom per site (e.g. the polar and azimuthal angles, $\theta_\mathbf{r}$ and $\varphi_\mathbf{r}$, of the spin variable $\boldsymbol{\Omega}_\mathbf{r}$) by 4 (6 degrees of freedom for $\mathbf{n_r}$ and $\mathbf{L_r}$ minus the 2 constraints (8.120)). This increase in the number of degrees of freedom is due to the decomposition (8.119) being not unique; nevertheless, it has no effect on the low-energy effective action of the Néel field we wish to derive.[34] To rewrite the Heisenberg Hamiltonian in terms of $\mathbf{n_r}$ and $\mathbf{L_r}$, we use

$$\sum_{\langle\mathbf{r},\mathbf{r}'\rangle} \boldsymbol{\Omega}_\mathbf{r}\cdot\boldsymbol{\Omega}_{\mathbf{r}'} = \frac{1}{2}\sum_{\langle\mathbf{r},\mathbf{r}'\rangle}(\boldsymbol{\Omega}_\mathbf{r}+\boldsymbol{\Omega}_{\mathbf{r}'})^2 + \text{const} \tag{8.122}$$

and

$$\sum_{\langle\mathbf{r},\mathbf{r}'\rangle}(\boldsymbol{\Omega}_\mathbf{r}+\boldsymbol{\Omega}_{\mathbf{r}'})^2 \simeq \sum_{\langle\mathbf{r},\mathbf{r}'\rangle}\left[(\mathbf{n_r}-\mathbf{n_{r'}})^2 + \frac{1}{S^2}(\mathbf{L_r}+\mathbf{L_{r'}})^2\right] \tag{8.123}$$

if we ignore terms of order $|\mathbf{L_r}|^3$, $\mathbf{L}_\mathbf{r}^2(\mathbf{n_r}-\mathbf{n_{r'}})$ and $(-1)^\mathbf{r}(\mathbf{L_r}+\mathbf{L_{r'}})\cdot(\mathbf{n_r}-\mathbf{n_{r'}})$.[35] The last term in (8.123) gives a contribution to the Hamiltonian

$$-\frac{J}{2}\sum_{\langle\mathbf{r},\mathbf{r}'\rangle}(\mathbf{L_r}+\mathbf{L_{r'}})^2 = -\frac{1}{2}\sum_{\mathbf{r},\mathbf{r}'}\mathbf{L_r}\cdot\mathbf{L_{r'}}(2dJ\delta_{\mathbf{r},\mathbf{r}'}+J_{\mathbf{r},\mathbf{r}'}) \equiv \frac{1}{2}\sum_{\mathbf{r},\mathbf{r}'}\chi^{-1}_{\perp\mathbf{r},\mathbf{r}'}\mathbf{L_r}\cdot\mathbf{L_{r'}}, \tag{8.124}$$

where

$$\chi_\perp(\mathbf{q}) = \frac{1}{J(\mathbf{Q})-J(\mathbf{q})} \tag{8.125}$$

is the static ($\omega=0$) transverse susceptibility in the Néel state obtained in the RPA.[36] To zeroth order in $\mathbf{L}$, the Berry phase term is equal to[37]

$$\mathcal{S}'_B[\mathbf{n}] = \int_0^\beta d\tau\sum_\mathbf{r}\langle(-1)^\mathbf{r}\mathbf{n_r}|\partial_\tau(-1)^\mathbf{r}\mathbf{n_r}\rangle = \int_0^\beta d\tau\sum_\mathbf{r}(-1)^\mathbf{r}\langle\mathbf{n_r}|\partial_\tau\mathbf{n_r}\rangle. \tag{8.126}$$

[32]The constraint $\delta(\mathbf{n}^2-1)$ in (8.121) should be understood as $\prod_{\mathbf{r},\tau}\delta(\mathbf{n_r}(\tau)^2-1)$ (and similarly for $\delta(\mathbf{n}\cdot\mathbf{L})$).

[33]In the large-S limit, we can parameterize the small fluctuations of $\mathbf{n_r}$ and $\mathbf{L_r}$ by $P_\mathbf{r}$ and $Q_\mathbf{r}$ [Eqs. (8.111)] and let these variables vary between $-\infty$ and ∞, which shows that the Jacobian is a mere S-dependent constant.

[34]We could impose the correct number of degrees of freedom by reducing the Brillouin zone by a factor two.

[35]In the continuum limit, these terms are $\mathcal{O}(|\mathbf{L}|^3, \mathbf{L}^2\boldsymbol{\nabla}\mathbf{n}, \mathbf{L}\cdot\boldsymbol{\nabla}^2\mathbf{n})$.

[36]The interpretation of $\chi_\perp(\mathbf{q})$ as the transverse susceptibility in the Néel state follows from $\chi^{-1}=\chi_0^{-1}-J$ in the RPA (in a matrix sense) with χ_0 defined by (8.36).

[37]We use $\int_0^\beta d\tau\langle-\mathbf{n_r}|-\partial_\tau\mathbf{n_r}\rangle = -\int_0^\beta d\tau\langle\mathbf{n_r}|\partial_\tau\mathbf{n_r}\rangle + 4i\pi S$ and ignore the term $4i\pi S$ (which does not modify the functional integral) to obtain the last expression in (8.126).

To obtain the $\mathcal{O}(\mathbf{L})$ correction, we use Eq. (8.103) with $\boldsymbol{\Omega}_\mathbf{r} = (-1)^\mathbf{r}\mathbf{n}_\mathbf{r}$ and $\delta\boldsymbol{\Omega}_\mathbf{r} = \mathbf{L}_\mathbf{r}/S$, which gives

$$\mathcal{S}_B[\mathbf{n},\mathbf{L}] = \mathcal{S}'_B[\mathbf{n}] - i\int_0^\beta d\tau \sum_\mathbf{r} \mathbf{L}_\mathbf{r}\cdot(\mathbf{n}_\mathbf{r}\times\partial_\tau\mathbf{n}_\mathbf{r}). \tag{8.127}$$

We thus obtain the action

$$\mathcal{S}[\mathbf{n},\mathbf{L}] = \mathcal{S}_B[\mathbf{n},\mathbf{L}] + \frac{1}{2}\int_0^\beta d\tau \sum_{\langle\mathbf{r},\mathbf{r}'\rangle}\Big[|J|S^2(\mathbf{n}_\mathbf{r}-\mathbf{n}_{\mathbf{r}'})^2 + \chi^{-1}_{\perp\mathbf{r},\mathbf{r}'}\mathbf{L}_\mathbf{r}\cdot\mathbf{L}_{\mathbf{r}'}\Big]. \tag{8.128}$$

The functional integral over $\mathbf{L}$ gives

$$\begin{aligned}\int\mathcal{D}[\mathbf{L}]\exp\bigg\{&-\frac{1}{2}\int_0^\beta d\tau\sum_{\mathbf{r},\mathbf{r}'}\chi^{-1}_{\perp\mathbf{r},\mathbf{r}'}\mathbf{L}_\mathbf{r}\cdot\mathbf{L}_{\mathbf{r}'} + i\int_0^\beta d\tau\sum_\mathbf{r}\mathbf{L}_\mathbf{r}\cdot(\mathbf{n}_\mathbf{r}\times\partial_\tau\mathbf{n}_\mathbf{r})\bigg\}\\ &= \exp\bigg\{-\frac{1}{2}\int_0^\beta d\tau\sum_{\mathbf{r},\mathbf{r}'}\chi_{\perp\mathbf{r},\mathbf{r}'}(\mathbf{n}_\mathbf{r}\times\partial_\tau\mathbf{n}_\mathbf{r})\cdot(\mathbf{n}_{\mathbf{r}'}\times\partial_\tau\mathbf{n}_{\mathbf{r}'})\bigg\}. \end{aligned}\tag{8.129}$$

Since the scalar product $\mathbf{L}_\mathbf{r}\cdot(\mathbf{n}_\mathbf{r}\times\partial_\tau\mathbf{n}_\mathbf{r})$ involves only the component of $\mathbf{L}_\mathbf{r}$ which is perpendicular to $\mathbf{n}_\mathbf{r}$, we can ignore the constraint $\mathbf{L}_\mathbf{r}\perp\mathbf{n}_\mathbf{r}$ when performing the integration over $\mathbf{L}_\mathbf{r}$. To second order in $\partial_\tau\mathbf{n}_\mathbf{r}$, we obtain the contribution to the action[38]

$$\frac{1}{2}\int_0^\beta d\tau\sum_\mathbf{r}\chi_\perp(\mathbf{n}_\mathbf{r}\times\partial_\tau\mathbf{n}_\mathbf{r})^2 = \frac{1}{2}\int_0^\beta d\tau\sum_\mathbf{r}\chi_\perp(\partial_\tau\mathbf{n}_\mathbf{r})^2 \tag{8.130}$$

with $\chi_\perp \equiv \chi_\perp(\mathbf{q}=0) = 1/4|J|d$ [Eq. (8.125)]. Taking the continuum limit, we finally obtain[32]

$$Z = \int\mathcal{D}[\mathbf{n}]\,\delta(\mathbf{n}^2-1)e^{-\mathcal{S}[\mathbf{n}]}, \tag{8.131}$$

where

$$\mathcal{S}[\mathbf{n}] = \mathcal{S}'_B[\mathbf{n}] + \frac{\rho_s^0}{2}\int_0^\beta d\tau\int d^dr\Big[(\boldsymbol{\nabla}\mathbf{n})^2 + \frac{1}{c^2}(\partial_\tau\mathbf{n})^2\Big]. \tag{8.132}$$

This low-energy description must be supplemented with a UV momentum cutoff $\Lambda \ll 1$ below which the continuum approximation is valid (in practice, one often takes a cutoff of the order of the inverse lattice spacing). Equations (8.131) and (8.132) define a quantum NLσM with an additional Berry phase term $\mathcal{S}'_B$.[29] $\rho_s^0 = |J|S^2$ is the (bare) stiffness and the (bare) velocity $c = (\rho_s^0/\chi_\perp)^{1/2} = 2\sqrt{d}|J|S$ coincides with the velocity of the spin-wave modes derived in the preceding sections.[39]

---

[38] $\int_0^\beta d\tau\sum_{\mathbf{r},\mathbf{r}'}\chi_{\perp\mathbf{r},\mathbf{r}'}(\mathbf{n}_\mathbf{r}\times\partial_\tau\mathbf{n}_\mathbf{r})\cdot(\mathbf{n}_{\mathbf{r}'}\times\partial_\tau\mathbf{n}_{\mathbf{r}'}) = \sum_q\chi_\perp(\mathbf{q})(\mathbf{n}\times\partial_\tau\mathbf{n})_{-q}\cdot(\mathbf{n}\times\partial_\tau\mathbf{n})_q \simeq \sum_q\chi_\perp(\mathbf{q}=0)(\mathbf{n}\times\partial_\tau\mathbf{n})_{-q}\cdot(\mathbf{n}\times\partial_\tau\mathbf{n})_q$ to second order in derivatives, hence Eq. (8.130).

[39] It should be noted that the mapping of the Heisenberg model onto the NLσM model (8.132), which relies on the spin-coherent-state functional integral, is fully justified only in the large-S limit; see the discussion of the validity of the functional integral, in particular the assumption that the paths $\{\boldsymbol{\Omega}_k\}$ are smooth, in Sec. 8.4.2.2.

8.4.3.1 *One-dimensional antiferromagnet*

Let us rewrite the Berry phase term $\mathcal{S}'_B$ in the following form:

$$\begin{aligned}\mathcal{S}'_B[\mathbf{n}] &= -iS\sum_i(\omega[\mathbf{n}_{2i+1}] - \omega[\mathbf{n}_{2i}]) \\ &\simeq -iS\sum_i\int_0^\beta d\tau(\mathbf{n}_{2i+1} - \mathbf{n}_{2i})\cdot\frac{\delta\omega[\mathbf{n}_{2i}]}{\delta\mathbf{n}_{2i}(\tau)},\end{aligned} \tag{8.133}$$

assuming that $\mathbf{n}_i$ varies smoothly in space. Here, $\mathbf{n}_i(\tau)$ stands for $\mathbf{n}_{x_i}(\tau)$ and $\omega[\mathbf{n}_i]$ denotes the solid angle defined by the closed trajectory of $\mathbf{n}_i(\tau)$ between times 0 and β [see Eq. (8.98)]. Using $\delta\omega[\mathbf{n}]/\delta\mathbf{n}(\tau) = -\mathbf{n}(\tau)\times\partial_\tau\mathbf{n}(\tau)$ [Eq. (8.103)], we obtain

$$\begin{aligned}\mathcal{S}'_B &= iS\int_0^\beta d\tau\sum_i(\mathbf{n}_{2i+1} - \mathbf{n}_{2i})\cdot(\mathbf{n}_{2i}\times\partial_\tau\mathbf{n}_{2i}) \\ &\simeq \frac{i}{2}S\int_0^L dx\int_0^\beta d\tau\,\partial_x\mathbf{n}\cdot(\mathbf{n}\times\partial_\tau\mathbf{n})\end{aligned} \tag{8.134}$$

in the continuum limit. This result can be rewritten as $\mathcal{S}'_B = i2\pi S\Theta$, where

$$\Theta[\mathbf{n}] = \frac{1}{4\pi}\int_0^L dx\int_0^\beta d\tau\,\mathbf{n}\cdot(\partial_\tau\mathbf{n}\times\partial_x\mathbf{n}) \tag{8.135}$$

is the topological winding number (or Pontryagin index) of the mapping $\mathbf{n}:\ [0,\beta]\times[0,L]\to S^2_{\rm int}$ with $S^2_{\rm int}$ the surface of the unit sphere, i.e. the mapping from two-dimensional spacetime to the unit sphere $S^2_{\rm int}$. If we assume that the field approaches a constant value on the boundary of its domain of definition, then the spacetime is topologically equivalent to the unit sphere and $\mathbf{n}$ realizes a mapping from $S^2_{\rm phys}$ to $S^2_{\rm int}$.[40] It turns out that $\Theta[\mathbf{n}]$ counts how many times spacetime has been wrapped around the sphere and must therefore be an integer.

To see this, let us introduce the two-dimensional coordinate $(x_1, x_2)\equiv(\tau, x)$ (simply denoted by x in the following):

$$\begin{aligned}\Theta[\mathbf{n}] &= \frac{1}{8\pi}\int d^2x\,\epsilon_{\mu\nu}\mathbf{n}\cdot(\partial_\mu\mathbf{n}\times\partial_\nu\mathbf{n}) \\ &= \frac{1}{8\pi}\int d^2x\,\epsilon_{\mu\nu}\epsilon_{abc}n^a\frac{\partial n^b}{\partial x_\mu}\frac{\partial n^c}{\partial x_\nu} \\ &= \frac{1}{8\pi}\int d^2x\,\epsilon_{\mu\nu}\epsilon_{abc}n^a\frac{\partial n^b}{\partial\xi_r}\frac{\partial\xi_r}{\partial x_\mu}\frac{\partial n^c}{\partial\xi_s}\frac{\partial\xi_s}{\partial x_\nu}\end{aligned} \tag{8.136}$$

---

[40]Here, we distinguish between the unit sphere in physical space ($S^2_{\rm phys}$) and the one in internal space ($S^2_{\rm int}$).

(a sum over repeated indices is implied) where $\xi = (\xi_1, \xi_2) \equiv (\theta, \varphi)$ with θ and φ the polar and azimuthal angles corresponding to the unit vector $\mathbf{n}$. Now

$$d^2\xi = d^2x\, \epsilon_{\mu\nu} \frac{\partial \xi_1}{\partial x_\mu} \frac{\partial \xi_2}{\partial x_\nu}, \qquad \epsilon_{rs} d^2\xi = d^2x\, \epsilon_{\mu\nu} \frac{\partial \xi_r}{\partial x_\mu} \frac{\partial \xi_s}{\partial x_\nu} \tag{8.137}$$

so that

$$\Theta[\mathbf{n}] = \frac{1}{8\pi} \int d^2\xi\, \epsilon_{abc} \epsilon_{rs} n^a \frac{\partial n^b}{\partial \xi_r} \frac{\partial n^c}{\partial \xi_s}. \tag{8.138}$$

Consider now an element $d\mathbf{S}_{\text{int}} = (dS^1_{\text{int}}, dS^2_{\text{int}}, dS^3_{\text{int}})$ of the surface of the unit sphere S^2_{int}. Noting that

$$dn^b dn^c = \epsilon_{rs} \frac{\partial n^b}{\partial \xi_r} \frac{\partial n^c}{\partial \xi_s} d^2\xi, \tag{8.139}$$

one has

$$dS^a_{\text{int}} = \frac{1}{2} \epsilon_{abc} \epsilon_{rs} \frac{\partial n^b}{\partial \xi_r} \frac{\partial n^c}{\partial \xi_s} d^2\xi. \tag{8.140}$$

Comparing (8.138) and (8.140), we obtain

$$\Theta[\mathbf{n}] = \frac{1}{4\pi} \int dS^a_{\text{int}} n^a = \frac{1}{4\pi} \int dS_{\text{int}}. \tag{8.141}$$

Since the unit sphere S^2_{int} has area 4π, we clearly see that $\Theta[\mathbf{n}]$ gives the number of times the internal sphere is traversed as we span the coordinate space as compactified into S^2_{phys}.

We are therefore led to distinguish integer and half-integer spins. In the former case, $e^{-\mathcal{S}'_B} = 1$ and the effective action $\mathcal{S}[\mathbf{n}]$ reduces to that of the one-dimensional quantum NLσM. At zero temperature, the latter is equivalent to the classical two-dimensional NLσM. The correlation length is finite,[41]

$$\xi \sim \Lambda^{-1} e^{\frac{2\pi\rho_s^0}{c}} \simeq \Lambda^{-1} e^{\pi S}, \tag{8.142}$$

in agreement with the conclusions of Sec. 8.3.2.1. The finite correlation length implies a non-zero gap in the spectrum (the so-called Haldane gap): $\Delta = c/\xi \sim c\Lambda e^{-\pi S}$.

For half-integer spins, the Berry phase term $\mathcal{S}'_B = i2\pi S\Theta$ cannot be ignored. The model defined by (8.131) and (8.132) is difficult to study but shows that the low-energy behavior of all half-integer spin chains is the same. From the Lieb, Schultz and Mattis theorem [23], we know that the spectrum is gapless in the thermodynamic limit. In the case $S = \frac{1}{2}$, the spectrum is given exactly by the Bethe ansatz solution. The low-energy behavior can be obtained from "bosonization" (a low-energy effective approach). The spin-$\frac{1}{2}$ Heisenberg chain belongs to a class of one-dimensional quantum fluids known as "Luttinger liquids", characterized by a gapless spectrum and power-law decaying correlation functions.[42]

---

[41] Equation (8.142) will be derived in Chapter 10.

[42] Besides spin chains, Luttinger liquids include one-dimensional interacting bosons and fermions (see Chapters 13 and 15).

This has led to the conjecture that half-integer-spin chains, regardless of the value of S, are gapless while integer-spin chains are gapped (Haldane's conjecture) [14]. No counter-examples have invalidated this conjecture so far.

8.4.3.2 *Two-dimensional antiferromagnet*

Let us first ignore the Berry phase term. At zero temperature, the two-dimensional quantum NLσM is equivalent to the classical three-dimensional NLσM. There is a QPT between an ordered state with spontaneously broken spin-rotation invariance and a disordered paramagnetic state. The RG approach to one-loop order predicts the transition to occur at the critical value $\tilde{g}_c = 4\pi^2$ of the dimensionless coupling constant $\tilde{g} = gc\Lambda$ of the NLσM (Chapter 10). In the case of interest here, $g = 1/\rho_s^0$ and $\tilde{g} = 2\sqrt{2}\Lambda/S$. For large S, $\tilde{g} \ll \tilde{g}_c$ and the ground state exhibits long-range antiferromagnetic order. For S of order one, it is difficult to conclude since the UV momentum cutoff Λ is not precisely known. Nevertheless, it is known from numerical calculations that even for $S = \frac{1}{2}$, the ground state of the Heisenberg on the square lattice has antiferromagnetic long-range order [2]. At finite temperature and $g < g_c$, the quantum NLσM is disordered by thermal fluctuations. This low-energy behavior is well described by a classical NLσM but with a stiffness renormalized by quantum fluctuations (see Chapter 12). The antiferromagnetic correlation length diverges at low temperatures as[43]

$$\xi \sim \frac{c}{T} e^{2\pi\rho_s/T}, \tag{8.143}$$

where ρ_s is the zero-temperature stiffness ($\rho_s < \rho_s^0$), in agreement with the conclusions of Sec. 8.3.2.2.

Consider now the Berry phase term $\mathcal{S}'_B$ for a smooth spacetime configuration of $\mathbf{n}(\mathbf{r},\tau)$ and evaluate (8.126) row by row on the square lattice. Apart from an oscillating sign (which we momentarily discard), for each row, we recover the one-dimensional result, namely a result quantized in integer multiples of $2\pi S$. Since $\mathbf{n}$ varies slowly in space, the contribution to the Berry phase term must change smoothly as we move from one row to the neighboring one. This is compatible with the quantization only if each row yields precisely the same integer. Taking into account the oscillating sign, we conclude that $\mathcal{S}'_B$ must vanish. Thus, a non-zero Berry phase can only be due to singular configurations. For a three-component vector field, the only topologically stable possibility is the so-called hedgehog singularity. The hedgehog events do not play any role in the Néel phase ($g < g_c$) so that all previous conclusions regarding the two-dimensional Heisenberg model on the square lattice are valid even when $\mathcal{S}'_B$ is considered. But for $g > g_c$, they proliferate and lead to a spontaneously broken lattice symmetry unless S is an even integer (in which case the Berry phase term $\mathcal{S}'_B$ has no effect). The ground state, referred to as a valence bond solid (or a spin-Peierls order), can be characterized by the quantity $P_{\mathbf{r},\mathbf{r}'} = \langle \hat{\mathbf{S}}_\mathbf{r} \cdot \hat{\mathbf{S}}_{\mathbf{r}'} \rangle$, which is a scalar under spin

[43] Equation (8.143) is the one-loop RG result. A more elaborate treatment gives a temperature-independent pre-exponential factor [17].

rotations so that a non-zero value does not break spin-rotation invariance. In the valence bond solid, $P_{\mathbf{r},\mathbf{r}'}$ is such that it breaks the lattice translation invariance.

Since the Néel state and the valence bond solid break different symmetries (spin-rotation invariance and lattice translation symmetry, respectively), Landau's theory of phase transitions predicts the transition to be first order. The Berry phase term allows for a different scenario where the theory contains an emergent gauge field and deconfined degrees of freedom (spinons) at the QCP, whereas both phases are characterized by conventional "confining" order parameters.[44]

8.5 Schwinger-Boson Mean-Field Theory

The symmetric phases of the Heisenberg model are easier to describe using a representation in which the rotational invariance remains manifest. In this section, we discuss a mean-field solution of the antiferromagnetic Heisenberg model where the spin operators are represented by Schwinger bosons.

8.5.1 *Schwinger bosons*

A spin-S operator can be represented by two boson operators $\hat{b}_\uparrow$ and $\hat{b}_\downarrow$,

$$\hat{\mathbf{S}} = \frac{1}{2}\hat{b}^\dagger \boldsymbol{\sigma} \hat{b}, \qquad \hat{b} = \begin{pmatrix} \hat{b}_\uparrow \\ \hat{b}_\downarrow \end{pmatrix} \tag{8.144}$$

($\boldsymbol{\sigma} = (\sigma^x, \sigma^y, \sigma^z)$ stands for the Pauli matrices), satisfying the constraint

$$\hat{b}^\dagger \hat{b} = \sum_\sigma \hat{b}^\dagger_\sigma \hat{b}_\sigma = 2S. \tag{8.145}$$

More explicitly, the spin operator is expressed as

$$\begin{aligned} \hat{S}^+ &= \hat{S}^x + i\hat{S}^y = \hat{b}^\dagger_\uparrow \hat{b}_\downarrow, \\ \hat{S}^- &= \hat{S}^x - i\hat{S}^y = \hat{b}^\dagger_\downarrow \hat{b}_\uparrow, \\ \hat{S}^z &= \frac{1}{2}(\hat{b}^\dagger_\uparrow \hat{b}_\uparrow - \hat{b}^\dagger_\downarrow \hat{b}_\downarrow). \end{aligned} \tag{8.146}$$

It is easy to verify that the relation $\hat{\mathbf{S}}^2 = S(S+1)$ and the spin commutation relations $\hat{\mathbf{S}} \times \hat{\mathbf{S}} = i\hat{\mathbf{S}}$ follow from the bosonic commutation relations $[\hat{b}_\sigma, \hat{b}^\dagger_{\sigma'}] = \delta_{\sigma,\sigma'}$ and $[\hat{b}_\sigma, \hat{b}_{\sigma'}] = [\hat{b}^\dagger_\sigma, \hat{b}^\dagger_{\sigma'}] = 0$. For a spin-$\frac{1}{2}$, the two physical states (satisfying the constraint (8.145)) are $|\uparrow\rangle = \hat{b}^\dagger_\uparrow|\text{vac}\rangle$ and $|\downarrow\rangle = \hat{b}^\dagger_\downarrow|\text{vac}\rangle$, where $|\text{vac}\rangle$ is the vacuum of Schwinger bosons.

---

[44] We refer to Chapter 19 of Ref. [9] for a further discussion of the role of the Berry phase term in two-dimensional antiferromagnets and a discussion of deconfined criticality.

8.5.2 *Mean-field theory of the antiferromagnetic Heisenberg model*

Defining Schwinger bosons $\hat{b}_{\mathbf{r}}, \hat{b}^\dagger_{\mathbf{r}}$ at each lattice site (we assume a d-dimensional hypercubic lattice),[45] we can rewrite the Heisenberg model as

$$\hat{H} = -\frac{J}{2} \sum_{\langle \mathbf{r},\mathbf{r}' \rangle} (2S^2 - \hat{\mathcal{A}}^\dagger_{\mathbf{rr}'} \hat{\mathcal{A}}_{\mathbf{rr}'}), \tag{8.147}$$

with $\hat{\mathcal{A}}_{\mathbf{rr}'} = \hat{b}_{\mathbf{r}\uparrow}\hat{b}_{\mathbf{r}'\downarrow} - \hat{b}_{\mathbf{r}\downarrow}\hat{b}_{\mathbf{r}'\uparrow}$. For spins-$\frac{1}{2}$, $\hat{\mathcal{A}}^\dagger_{\mathbf{rr}'}$ creates a singlet on the bond $(\mathbf{r}, \mathbf{r}')$,

$$\hat{\mathcal{A}}^\dagger_{\mathbf{rr}'}|\text{vac}\rangle = (\hat{b}^\dagger_{\mathbf{r}\uparrow}\hat{b}^\dagger_{\mathbf{r}'\downarrow} - \hat{b}^\dagger_{\mathbf{r}\downarrow}\hat{b}^\dagger_{\mathbf{r}'\uparrow})|\text{vac}\rangle = |\uparrow\rangle_{\mathbf{r}}|\downarrow\rangle_{\mathbf{r}'} - |\downarrow\rangle_{\mathbf{r}}|\uparrow\rangle_{\mathbf{r}'}. \tag{8.148}$$

Thus, we expect that short-range antiferromagnetic order will be characterized by a non-zero value of $\langle \hat{\mathcal{A}}_{\mathbf{rr}'} \rangle$, which suggests to decouple the product $\hat{\mathcal{A}}^\dagger\hat{\mathcal{A}}$ in (8.147) following the usual mean-field procedure.

At this point, it is convenient to make a spin rotation of angle π about the y axis on one of the two sublattices (e.g. sublattice B). This is equivalent to the change of variables $\hat{b}_\downarrow \to \hat{b}_\uparrow$ and $\hat{b}_\uparrow \to -\hat{b}_\downarrow$,[46] which is a canonical transformation preserving the bosonic commutation relations and the constraint (8.145). This leaves the form (8.147) of the Hamiltonian unchanged, but the bond operator is now defined by

$$\hat{\mathcal{A}}_{\mathbf{rr}'} = \hat{b}_{\mathbf{r}\uparrow}\hat{b}_{\mathbf{r}'\uparrow} + \hat{b}_{\mathbf{r}\downarrow}\hat{b}_{\mathbf{r}'\downarrow}. \tag{8.149}$$

We solve the Hamiltonian in a mean-field approximation which will be justified in Sec. 8.5.3. On the one hand, we impose the constraint (8.145) only on average, $\langle \hat{b}^\dagger_{\mathbf{r}}\hat{b}_{\mathbf{r}} \rangle = 2S$. This is done by introducing a Lagrange multiplier $\lambda \equiv \lambda_{\mathbf{r}}$. On the other hand, we decouple the quartic bosonic term of the Hamiltonian: $\hat{\mathcal{A}}^\dagger\hat{\mathcal{A}} \to \langle \hat{\mathcal{A}}^\dagger \rangle \hat{\mathcal{A}} + \hat{\mathcal{A}}^\dagger \langle \hat{\mathcal{A}} \rangle - \langle \hat{\mathcal{A}}^\dagger \rangle \langle \hat{\mathcal{A}} \rangle$. This yields the Hamiltonian

$$\hat{H}_{\text{MF}} = E_0 + \lambda \sum_{\mathbf{r}} \hat{b}^\dagger_{\mathbf{r}}\hat{b}_{\mathbf{r}} + \sum_{\langle \mathbf{r},\mathbf{r}' \rangle} (Q^* \hat{\mathcal{A}}_{\mathbf{rr}'} + \text{h.c.}), \tag{8.150}$$

where

$$Q = \frac{J}{2}\langle \hat{\mathcal{A}}_{\mathbf{rr}'} \rangle, \qquad Q^* = \frac{J}{2}\langle \hat{\mathcal{A}}^\dagger_{\mathbf{rr}'} \rangle, \qquad E_0 = -\lambda 2SN - \frac{z}{J}|Q|^2 N \tag{8.151}$$

($z = 2d$ is the number of nearest neighbors). In Fourier space,

$$\hat{H}_{\text{MF}} = E_0 + \lambda \sum_{\mathbf{q}} \hat{b}^\dagger_{\mathbf{q}}\hat{b}_{\mathbf{q}} + \frac{z}{2} \sum_{\mathbf{q},\sigma} \gamma_{\mathbf{q}} Q (\hat{b}_{\mathbf{q}\sigma}\hat{b}_{-\mathbf{q}\sigma} + \text{h.c.}), \tag{8.152}$$

where, with no loss of generality, we now take Q real. $\hat{H}_{\text{MF}}$ can be diagonalized by the (canonical) Bogoliubov transformation

$$\hat{b}_{\mathbf{q}\sigma} = \cosh(\theta_{\mathbf{q}})\hat{\alpha}_{\mathbf{q}\sigma} + \sinh(\theta_{\mathbf{q}})\hat{\alpha}^\dagger_{-\mathbf{q}\sigma}, \tag{8.153}$$

---

[45] Schwinger bosons on different sites commute.

[46] From Eqs. (8.146), we verify that the transformation of the boson operators implies $\hat{S}^x \to -\hat{S}^x$, $\hat{S}^y \to \hat{S}^y$ and $\hat{S}^z \to -\hat{S}^z$.

where we assume $\theta_{\mathbf{q}} = \theta_{-\mathbf{q}}$. The Hamiltonian

$$\hat{H}_{\mathrm{MF}} = E_0 + \sum_{\mathbf{q},\sigma} \Big\{ \frac{1}{2}[\lambda \sinh(2\theta_q) + zQ\gamma_{\mathbf{q}} \cosh(2\theta_{\mathbf{q}})](\hat{\alpha}_{\mathbf{q}\sigma}\hat{\alpha}_{-\mathbf{q}\sigma} + \text{h.c.}) + [\lambda \cosh(2\theta_{\mathbf{q}}) + zQ\gamma_{\mathbf{q}} \sinh(2\theta_q)]\hat{\alpha}^\dagger_{\mathbf{q}\sigma}\hat{\alpha}_{\mathbf{q}\sigma} + \lambda \sinh^2(\theta_{\mathbf{q}}) + \frac{z}{2}Q\gamma_{\mathbf{q}} \sinh(2\theta_q) \Big\} \tag{8.154}$$

takes a simple form if we choose $\tanh(2\theta_{\mathbf{q}}) = -zQ\gamma_{\mathbf{q}}/\lambda$,

$$\hat{H}_{\mathrm{MF}} = -N\Big[\lambda(2S+1) + \frac{z}{J}Q^2\Big] + \sum_{\mathbf{q},\sigma} \omega_{\mathbf{q}} \Big(\hat{\alpha}^\dagger_{\mathbf{q}\sigma}\hat{\alpha}_{\mathbf{q}\sigma} + \frac{1}{2} \Big), \tag{8.155}$$

where

$$\omega_{\mathbf{q}} = (\lambda^2 - z^2 Q^2 \gamma_{\mathbf{q}}^2)^{1/2} \tag{8.156}$$

is the dispersion of the bosons $\hat{\alpha}$. The spectrum has a gap $\Delta = (\lambda^2 - z^2Q^2)^{1/2}$ and the minimum excitation energy is reached for $\mathbf{q} = 0$ and $\mathbf{q} = \mathbf{Q}$. $\omega_{\mathbf{q}} \simeq (\Delta^2 + c^2\mathbf{q}^2)^{1/2}$ for $\mathbf{q} \to 0$ and $\omega_{\mathbf{q}} \simeq (\Delta^2 + c^2(\mathbf{q} - \mathbf{Q})^2)^{1/2}$ for $\mathbf{q} \to \mathbf{Q}$ where $c = \sqrt{2}z|Q|$. If $\lambda/z|Q| \to 1$, the gap vanishes and we expect BEC, i.e. antiferromagnetic long-range order (see the discussion of the two-dimensional antiferromagnet in the following). A non-zero gap Δ implies a finite antiferromagnetic correlation length ξ. The boson propagator decays exponentially in space with a characteristic length c/Δ.[47] The spin–spin correlation function is a two-particle correlation function and should therefore decay as the square of the boson propagator, i.e. $\xi = c/2\Delta$.[48]

Since the Hamiltonian reduces to a sum of decoupled harmonic oscillators, the free energy is given by

$$F = -N\Big[\lambda(2S+1) + \frac{z}{J}Q^2\Big] + \frac{2}{\beta}\sum_{\mathbf{q}} \ln\Big[2\sinh\Big(\frac{\beta}{2}\omega_{\mathbf{q}}\Big)\Big]. \tag{8.157}$$

The values of λ and Q are obtained from $\langle \hat{b}^\dagger_{\mathbf{r}}\hat{b}_{\mathbf{r}} \rangle = 2S$ and $Q = \frac{J}{2}\langle \hat{A}_{\mathbf{r}\mathbf{r}'} \rangle$ or, equivalently, by requiring $\partial F/\partial\lambda = \partial F/\partial Q = 0$:

$$\begin{aligned} S + \frac{1}{2} &= \int_{\mathbf{q}} \Big(n_{\mathbf{q}} + \frac{1}{2} \Big) \frac{\lambda}{\omega_{\mathbf{q}}}, \\ \frac{Q}{|J|} &= \int_{\mathbf{q}} \Big(n_{\mathbf{q}} + \frac{1}{2} \Big) zQ\frac{\gamma_{\mathbf{q}}^2}{\omega_{\mathbf{q}}}, \end{aligned} \tag{8.158}$$

where $n_{\mathbf{q}} = n_B(\omega_{\mathbf{q}})$.

[47]This follows from $\omega_{\mathbf{q}} \simeq (\Delta^2 + c^2\mathbf{q}^2)^{1/2}$ for $\mathbf{q} \to 0$.

[48]For an explicit calculation of the spin–spin correlation function, see [1, 25].

8.5.2.1 *One-dimensional antiferromagnet*

In one dimension and for $T = 0$, the first mean-field equation (8.158) gives

$$S + \frac{1}{2} = \frac{1}{2} \int_{-\pi}^{\pi} \frac{dq}{2\pi} \frac{1}{\sqrt{1 - \eta^2 \cos(q)^2}} = \frac{1}{\pi} K(\eta), \tag{8.159}$$

where $\eta = 2|Q|/\lambda$ and K is the complete elliptic integral of the first kind.[49] Equation (8.159) determines the correlation length $\xi = \eta/2\sqrt{1-\eta^2}$ as a function of S.

In the large-S limit, using $K(\eta) \simeq \ln(4/\sqrt{1-\eta^2})$ for $\eta \to 1$, one finds

$$\xi \simeq \frac{e^{\pi S}}{8} \qquad (S \gg 1), \tag{8.160}$$

which agrees with the result obtained from the NLσM for integer S [Eq. (8.142)]. The Schwinger-boson mean-field theory is not adequate for half-integer spin chains since it fails to obtain a gapless spectrum. Subtracting the first mean-field equation from the second one in (8.158), and using $\lambda \simeq z|Q|$ when $S \gg 1$, one obtains

$$c - 2|J| \left(S + \frac{1}{2} \right) = -|J| \int_q \frac{\lambda}{\omega_q} \sin(q)^2 \simeq -\frac{2}{\pi} |J|, \tag{8.161}$$

and in turn

$$c \simeq 2|J| \left(S + \frac{1}{2} - \frac{1}{\pi} \right) \simeq 2|J| S, \tag{8.162}$$

in agreement with the spin-wave result.[50]

At finite temperature, when $T \gg \Delta$ (i.e. $\xi \gg c/T$), we can use $n_q \simeq T/\omega_q$ so that the mean-field equations become

$$\begin{aligned} S + \frac{1}{2} &= \int_q \frac{\lambda T}{\omega_q^2} = \frac{T}{\Delta}, \\ \frac{1}{|J|} &= 2T \int_q \frac{\cos^2 q}{\omega_q^2} = \frac{2T}{c^2} \left(\frac{\sqrt{\Delta^2 + c^2}}{\Delta} - 1 \right). \end{aligned} \tag{8.163}$$

In the large-S limit, one finds $\Delta \simeq T/S$ and $c \simeq 2|J|S$, i.e. a correlation length $\xi \sim |J|S^2/T \gg c/T$ in agreement with the conclusions drawn from the spin-wave expansion in Sec. 8.3.2.1.

8.5.2.2 *Two-dimensional antiferromagnet*

For $d = 2$ and $T = 0$, the first mean-field equation gives

$$S + \frac{1}{2} = \frac{1}{2} \int_{-1}^{1} d\gamma \frac{\rho(\gamma)}{\sqrt{1 - \eta^2 \gamma^2}}, \tag{8.164}$$

---

[49] $K(k) = \int_0^{\pi/2} \frac{d\theta}{\sqrt{1-k^2 \sin^2\theta}}$.

[50] Let us recall that the spin-wave analysis is not justified in one dimension since it assumes long-range order.

where $\eta = 4|Q|/\lambda$ and

$$\rho(\gamma) = \int_{\mathbf{q}} \delta(\gamma - \gamma_{\mathbf{q}}) = \frac{2}{\pi^2} K(\sqrt{1-\gamma^2}) \tag{8.165}$$

is the density of states associated with the dispersion $\gamma_{\mathbf{q}}$. The rhs in (8.164) is maximum for $\eta = 1$ and then equal to ~ 0.6966. For $S \geq 0.1966$, Eq. (8.164) has therefore no solution and we must allow for BEC in the states $\mathbf{q} = 0$ and $\mathbf{q} = (\pi, \pi)$ (the two momenta where $\omega_{\mathbf{q}} = 0$ when $\eta = 1$). We consider a condensate defined by

$$\langle \hat{b}_{\mathbf{r}\sigma} \rangle = \sqrt{n_0} e^{i\theta} \delta_{\sigma,\uparrow}, \tag{8.166}$$

with $n_0 = |\langle b_{\mathbf{r}\uparrow} \rangle|^2$ the condensate density and θ an arbitrary phase. This corresponds to a staggered magnetization along the z axis,[51]

$$\langle \hat{S}^z_{\mathbf{r}} \rangle = \frac{n_0}{2} (-1)^{\mathbf{r}}. \tag{8.167}$$

In the absence of quantum fluctuations, all bosons would be in the condensate, $|\langle \hat{b}_{\mathbf{r}\uparrow} \rangle|^2 = 2S$, which would give a staggered magnetization $m_0 = n_0/2 = S$ as in the classical Néel state.

The condensate yields the additional contribution

$$N n_0 \left[\lambda + \frac{z}{2} (Q^* e^{2i\theta} + \text{c.c.}) \right] \tag{8.168}$$

to the mean-field Hamiltonian (8.150), where we now consider a complex order parameter Q. If we take $\theta = 0$, the mean-field free energy is minimized for a real and negative order parameter Q. The mean-field equations become

$$\begin{aligned} S + \frac{1}{2} &= \frac{n_0}{2} + \int_{\mathbf{q}} \left(n_{\mathbf{q}} + \frac{1}{2} \right) \frac{\lambda}{\omega_{\mathbf{q}}}, \\ \frac{Q}{|J|} &= -\frac{n_0}{2} + \int_{\mathbf{q}} \left(n_{\mathbf{q}} + \frac{1}{2} \right) 4Q \frac{\gamma_{\mathbf{q}}^2}{\omega_{\mathbf{q}}}, \end{aligned} \tag{8.169}$$

where the boson dispersion $\omega_{\mathbf{q}} = \lambda(1 - \gamma_{\mathbf{q}}^2)^{1/2}$ vanishes for $\mathbf{q} = 0$ and $\mathbf{q} = (\pi, \pi)$. The first equation gives the staggered magnetization in the ground state,

$$m_0 = \frac{n_0}{2} = S + \frac{1}{2} - \frac{1}{2} \int_{\mathbf{q}} \frac{1}{1 - \gamma_{\mathbf{q}}^2} \simeq S - 0.1966, \tag{8.170}$$

and reproduces the result obtained from the spin-wave expansion (Sec. 8.3.2). From (8.169), one also obtains

$$\frac{|Q|}{|J|} = S + \frac{1}{2} - \frac{1}{2} \int_{\mathbf{q}} \sqrt{1 - \gamma_{\mathbf{q}}^2}, \tag{8.171}$$

[51] Equation (8.167) is obtained by noting that $\hat{S}^z_{\mathbf{r}} = \frac{1}{2}(\hat{b}^\dagger_{\mathbf{r}\uparrow}\hat{b}_{\mathbf{r}\uparrow} - \hat{b}^\dagger_{\mathbf{r}\downarrow}\hat{b}_{\mathbf{r}\downarrow})$ for $\mathbf{r} \in$ A and $\hat{S}^z_{\mathbf{r}} = -\frac{1}{2}(\hat{b}^\dagger_{\mathbf{r}\uparrow}\hat{b}_{\mathbf{r}\uparrow} - \hat{b}^\dagger_{\mathbf{r}\downarrow}\hat{b}_{\mathbf{r}\downarrow})$ for $\mathbf{r} \in$ B in terms of the rotated boson operators defined before (8.149).

i.e.

$$\frac{c}{2\sqrt{2}|J|S} = 1 + \frac{1}{2S}\Big[1 - \frac{2}{\pi^2}\int_{-1}^{1} d\gamma \sqrt{1-\gamma^2} K(\sqrt{1-\gamma^2})\Big]$$
$$\simeq 1 + \frac{0.07897}{S}. \tag{8.172}$$

The last term in (8.172) corresponds to a correction to the leading-order spin-wave velocity $c = 2\sqrt{2}|J|S$ (Sec. 8.3.2).

At finite temperature, the integral

$$\int_{\mathbf{q}} \frac{n_{\mathbf{q}}}{\omega_{\mathbf{q}}} \sim T \int_{\mathbf{q}} \frac{1}{\omega_{\mathbf{q}}^2} \tag{8.173}$$

is logarithmically divergent if the boson dispersion is gapless: $\omega_{\mathbf{q}} \simeq c|\mathbf{q}|$ for $\mathbf{q} \to 0$ and $\omega_{\mathbf{q}} \simeq c|\mathbf{q} - \mathbf{Q}|$ for $\mathbf{q} \to \mathbf{Q} = (\pi, \pi)$. In agreement with the Mermin–Wagner theorem, BEC is not possible at finite temperature in two dimensions. Thus, we must have a vanishing condensate density, $n_0 = 0$, and a gapped spectrum, $\eta < 1$. Let us rewrite the mean-field equations as

$$\begin{aligned} S + \frac{1}{2} &= \frac{1}{2}\int_{-1}^{1} d\gamma\, \rho(\gamma)(1-\eta^2\gamma^2)^{-1/2} \operatorname{cotanh}\left(\frac{\lambda}{2T}\sqrt{1-\eta^2\gamma^2}\right), \\ S + \frac{1}{2} - \frac{\lambda\eta^2}{4|J|} &= \frac{1}{2}\int_{-1}^{1} d\gamma\, \rho(\gamma)(1-\eta^2\gamma^2)^{1/2} \operatorname{cotanh}\left(\frac{\lambda}{2T}\sqrt{1-\eta^2\gamma^2}\right) \end{aligned} \tag{8.174}$$

(the second one is obtained by combining the two equations (8.169)). Following Ref. [25], we rewrite the first equation as

$$\begin{aligned} S + \frac{1}{2} &= \frac{2T}{\pi\lambda}\left[\frac{1}{2\eta}\ln\left(\frac{1+\eta}{1-\eta}\right) - \ln\left(\frac{2\lambda}{T}\right)\right] \\ &\quad + \frac{1}{2}\int_{-1}^{1} d\gamma\, \rho(\gamma)(1-\eta^2\gamma^2)^{-1/2} + \mathcal{O}(T^3), \end{aligned} \tag{8.175}$$

which gives (noting that $\eta \simeq 1$ at low temperatures)

$$\sqrt{1-\eta} \simeq \frac{T}{\lambda\sqrt{2}} e^{-\pi m_0 \lambda/2T}, \tag{8.176}$$

where m_0 is the ground state magnetization [Eq. (8.170)]. The second equation (8.174) gives [25]

$$S + \frac{1}{2} - \frac{\lambda\eta^2}{4|J|} = \frac{1}{2}\int_{-1}^{1} d\gamma\, \rho(\gamma)(1-\eta^2\gamma^2)^{1/2} + \mathcal{O}(T^3) \tag{8.177}$$

i.e.

$$\frac{\lambda\eta^2}{4|J|} \simeq S + \frac{1}{2}\left[1 - \int_{-1}^{1} d\gamma\, \rho(\gamma)\sqrt{1-\eta^2\gamma^2}\right] \equiv Z_c S. \tag{8.178}$$

We therefore obtain $\lambda \simeq 4|J|Z_cS$ (since $\eta \simeq 1$) and, from (8.176),

$$\sqrt{1-\eta} = \frac{T}{4\sqrt{2}|J|Z_cS} e^{-2\pi m_0|J|Z_cS/T} \tag{8.179}$$

as well as

$$\xi = \frac{\eta}{2\sqrt{2}\sqrt{1-\eta^2}} \simeq \frac{c}{2T} e^{2\pi m_0|J|Z_cS/T}, \tag{8.180}$$

which reproduces the temperature dependence of the correlation length obtained from the quantum NLσM (Sec. 8.4.3). By analogy with (8.143), we infer that the stiffness takes the form

$$\rho_s = m_0 Z_c S|J| \tag{8.181}$$

in the zero-temperature ordered ground state. We recover $\rho_s \simeq \rho_s^{(0)} = |J|S^2$ in the large-S limit.

8.5.3 SU($\mathcal{N}$) Heisenberg model

In this section, we show that the Schwinger-boson mean-field theory can be put on a firm basis by extending the spin symmetry of the Heisenberg model from SU(2) to an SU($\mathcal{N}$). This is done by considering an $\mathcal{N}$-flavor operator $\hat{b}_{\mathbf{r}m}$ ($m = 1, \dots, \mathcal{N}$) and redefining the bond operator (8.149) as

$$\hat{\mathcal{A}}_{\mathbf{rr}'} = \sum_{m=1}^{\mathcal{N}} \hat{b}_{\mathbf{r}m}\hat{b}_{\mathbf{r}'m} \quad \text{with} \quad \sum_{m=1}^{\mathcal{N}} \hat{b}^\dagger_{\mathbf{r}m}\hat{b}_{\mathbf{r}m} = \mathcal{N}S. \tag{8.182}$$

The Hamiltonian (8.147) becomes

$$\hat{H} = -\frac{J}{\mathcal{N}} \sum_{\langle \mathbf{r},\mathbf{r}'\rangle} (\mathcal{N}S^2 - \hat{\mathcal{A}}^\dagger_{\mathbf{rr}'}\hat{\mathcal{A}}_{\mathbf{rr}'}) \tag{8.183}$$

after a proper rescaling[52] of the exchange interaction J, and is invariant in an SU($\mathcal{N}$) rotation of the boson field $\hat{b}_{\mathbf{r}} = (\hat{b}_{\mathbf{r}1}, \dots, \hat{b}_{\mathbf{r}\mathcal{N}})$.[53]

We can write the partition function as a functional integral

$$Z = \int \mathcal{D}[b^*, b, \lambda]\, e^{-\int_0^\beta d\tau \left\{\sum_{\mathbf{r},m}\left[b^*_{\mathbf{r}m}\partial_\tau b_{\mathbf{r}m} + i\lambda_{\mathbf{r}}(b^*_{\mathbf{r}m}b_{\mathbf{r}m} - S)\right] + H[b^*,b]\right\}} \tag{8.184}$$

over a complex bosonic field $b_{\mathbf{r}}$ and a Lagrange multiplier $\lambda_{\mathbf{r}}$ which enforces the constraint $\hat{b}^\dagger_{\mathbf{r}}\hat{b}_{\mathbf{r}} = \mathcal{N}S$. We now decouple the Hamiltonian by means of a Hubbard–Stratonovich transformation,

$$e^{-\int_0^\beta d\tau\, H[b^\dagger,b]} = \int \mathcal{D}[Q^*, Q]\, e^{-\int_0^\beta d\tau \sum_{\langle \mathbf{r},\mathbf{r}'\rangle}\left[\frac{\mathcal{N}}{|J|}|Q_{\mathbf{rr}'}|^2 + Q^*_{\mathbf{rr}'}\mathcal{A}_{\mathbf{rr}'} + Q_{\mathbf{rr}'}\mathcal{A}^*_{\mathbf{rr}'}\right]}, \tag{8.185}$$

---

[52]This rescaling is necessary to have a well-defined large-$\mathcal{N}$ limit.

[53]The Hamiltonian (8.183) is invariant under suitably defined SU($\mathcal{N}$) "rotations". See Chapter 16 of Ref. [1] for a discussion.

where the auxiliary field $Q_{\mathbf{rr}'}$ is defined for each pair $(\mathbf{r}, \mathbf{r}')$ of nearest neighbors. The action is now a function of b^*, b, λ and Q^*, Q. Integrating out the boson field, we obtain

$$\int \mathcal{D}[b^*, b]\, e^{-\int_0^\beta d\tau \sum_m \left\{\sum_{\mathbf{r}} b^*_{\mathbf{r}m}(\partial_\tau + i\lambda_{\mathbf{r}})b_{\mathbf{r}m} + \sum_{\langle \mathbf{r},\mathbf{r}'\rangle}(Q^*_{\mathbf{rr}'} b_{\mathbf{r}m} b_{\mathbf{r}'m} + \text{c.c.})\right\}} = (\det \mathcal{G})^{\mathcal{N}/2}, \quad (8.186)$$

where the boson propagator $\mathcal{G} \equiv \mathcal{G}[Q^*, Q, \lambda]$ is defined by

$$\mathcal{G}^{-1}_{\mathbf{rr}'}(\tau - \tau') = \begin{pmatrix} (\partial_\tau + i\lambda_{\mathbf{r}})\delta_{\mathbf{r},\mathbf{r}'} & Q_{\mathbf{rr}'} \\ Q^*_{\mathbf{rr}'} & (-\partial_\tau + i\lambda_{\mathbf{r}})\delta_{\mathbf{r},\mathbf{r}'} \end{pmatrix} \delta(\tau - \tau'), \quad (8.187)$$

and $Q^{(*)}_{\mathbf{rr}'}$ vanishes if $\mathbf{r}$ and $\mathbf{r}'$ are not nearest neighbors. Putting everything together, we obtain the partition function

$$Z = \int \mathcal{D}[Q^*, Q, \lambda]\, e^{-\mathcal{N}\mathcal{S}[Q^*, Q, \lambda]}, \quad (8.188)$$

where

$$\mathcal{S}[Q^*, Q, \lambda] = \int_0^\beta d\tau \left[\frac{1}{|J|}\sum_{\langle \mathbf{r},\mathbf{r}'\rangle} |Q_{\mathbf{rr}'}|^2 - iS\sum_{\mathbf{r}} \lambda_{\mathbf{r}}\right] - \frac{1}{2}\operatorname{Tr}\ln \mathcal{G}[Q^*, Q, \lambda] \quad (8.189)$$

is independent of $\mathcal{N}$. In the limit $\mathcal{N} \to \infty$, the saddle-point approximation becomes exact (see Sec. 1.7). The saddle-point equations are

$$\begin{aligned} 0 &= \frac{\delta \ln Z}{\delta \lambda_{\mathbf{r}}(\tau)} = -i\sum_m \langle b^*_{\mathbf{r}m} b_{\mathbf{r}m}\rangle + i\mathcal{N}S, \\ 0 &= \frac{\delta \ln Z}{\delta Q^*_{\mathbf{rr}'}(\tau)} = -\frac{\mathcal{N}}{|J|} Q_{\mathbf{rr}'} - \langle \mathcal{A}_{\mathbf{rr}'}\rangle, \\ 0 &= \frac{\delta \ln Z}{\delta Q_{\mathbf{rr}'}(\tau)} = -\frac{\mathcal{N}}{|J|} Q^*_{\mathbf{rr}'} - \langle \mathcal{A}^*_{\mathbf{rr}'}\rangle. \end{aligned} \quad (8.190)$$

Assuming uniform and static fields: $\lambda_{\mathbf{r}}(\tau) = i\lambda$, $Q^{(*)}_{\mathbf{rr}'}(\tau) = Q^{(*)}$, we recover the mean-field equations studied in Sec. 8.5.2 for $\mathcal{N} = 2$.

The mean-field theory is exact only in the limit $\mathcal{N} \to \infty$. For $\mathcal{N} < \infty$, some non-physical states which violate the local constraint $\hat{b}^\dagger_{\mathbf{r}}\hat{b}_{\mathbf{r}} = \mathcal{N}S$ are included. In particular, the mean field "quasi-particles" (i.e. the bosons $\hat{\alpha}$ in Sec. 8.5.2) involve non-physical density fluctuations of Schwinger bosons. Actual excitations are composites of an even number of quasi-particles. The local constraint is enforced through the dynamical fluctuations of the field λ and can be, at least in principle, reintroduced into the theory *via* a systematic $1/\mathcal{N}$ expansion [1].

Appendix 8.A A Single Spin in an External Field

We consider a single spin in an external field parallel to the z axis, with Hamiltonian $\hat{H} = -h\hat{S}^z$.

8.A.1 *Free energy*

The partition function is easily calculated using the basis $\{|m\rangle\}$ of eigenstates of $\hat{S}^z$, $\hat{S}^z|m\rangle = m|m\rangle$ $(-S \le m \le S)$,

$$Z = \mathrm{Tr}\, e^{\beta h \hat{S}^z} = \sum_{m=-S}^{S} e^{\beta h m} = \frac{\sinh\left[\beta h(S+\frac{1}{2})\right]}{\sinh\left(\frac{\beta h}{2}\right)}. \tag{8.A.1}$$

The free energy is given by $F = -\frac{1}{\beta}\ln Z$ and the magnetization

$$m = -\frac{\partial F}{\partial h} = SB_S(\beta h S) \tag{8.A.2}$$

can be expressed in terms of the Brillouin function $B_S(x)$ [Eq. (8.27)].

8.A.2 *Spin–spin correlation function*

In imaginary time, the connected propagator is defined by

$$G^{\nu\nu'}(\tau) = \langle T_\tau \hat{S}^\nu(\tau)\hat{S}^{\nu'}(0)\rangle - \langle \hat{S}^\nu\rangle\langle \hat{S}^{\nu'}\rangle, \tag{8.A.3}$$

where $\hat{S}^\nu(\tau) = e^{\tau\hat{H}}\hat{S}^\nu e^{-\tau\hat{H}}$ is the spin operator in the Heisenberg picture. $G^{\nu\nu'}$ is calculated by using the resolution of the identity $\sum_m |m\rangle\langle m| = 1$. For $\tau > 0$,

$$\begin{aligned} G^{zz}(\tau) &= \frac{1}{Z}\sum_m \langle m|e^{-(\beta-\tau)\hat{H}}\hat{S}^z|m\rangle\langle m|e^{-\tau\hat{H}}\hat{S}^z|m\rangle - \langle \hat{S}^z\rangle^2 \\ &= \frac{1}{Z}\sum_m m^2 e^{\beta m h} - \left(\frac{1}{Z}\sum_m m\, e^{\beta m h}\right)^2, \end{aligned} \tag{8.A.4}$$

i.e.

$$G^{zz}(\tau) = -\frac{1}{\beta}\frac{\partial^2 F}{\partial h^2} = S^2 B_S'(\beta h S) \qquad (\tau > 0). \tag{8.A.5}$$

We deduce

$$G^{zz}(i\omega_n) = \int_0^\beta d\tau\, e^{i\omega_n \tau} G^{zz}(\tau) = S^2 B_S'(\beta h S)\beta\delta_{\omega_n,0}. \tag{8.A.6}$$

A similar calculation gives[54]

$$\begin{aligned} G^{-+}(\tau) &= \frac{1}{Z}\sum_m e^{(\beta-\tau)mh + \tau(m+1)h}|\langle m+1|\hat{S}^+|m\rangle|^2 \\ &= \frac{e^{\tau h}}{e^{\beta h}-1} 2SB_S(\beta h S) \end{aligned} \tag{8.A.7}$$

[54] $\hat{S}^\pm = \hat{S}^x \pm i\hat{S}^y$ and $\hat{S}^\pm|m\rangle = [S(S+1) - m(m\pm 1)]^{1/2}|m\pm 1\rangle$.

for $\tau > 0$, i.e.

$$G^{-+}(i\omega_n) = G^{+-}(-i\omega_n) = \frac{2S}{i\omega_n + h} B_S(\beta h S). \tag{8.A.8}$$

From (8.A.8), we deduce

$$\begin{aligned} G^{xx}(i\omega_n) = G^{yy}(i\omega_n) &= \frac{h}{\omega_n^2 + h^2} S B_S(\beta h S), \\ G^{xy}(i\omega_n) = -G^{yx}(i\omega_n) &= -\frac{\omega_n}{\omega_n^2 + h^2} S B_S(\beta h S). \end{aligned} \tag{8.A.9}$$

Appendix 8.B Spin-Coherent States

It is useful to give a slightly different definition of the spin-coherent state $|\mathbf{\Omega}\rangle$ [Eq. (8.84)]. First we write

$$|\mathbf{\Omega}\rangle = e^{i\theta \mathbf{u}\cdot\hat{\mathbf{S}}}|S\rangle = e^{\frac{\theta}{2}e^{i\varphi}\hat{S}^- - \frac{\theta}{2}e^{-i\varphi}\hat{S}^+}|S\rangle, \tag{8.B.1}$$

where $\mathbf{u} = (\sin\varphi, -\cos\varphi, 0)$ and we have chosen the gauge $\psi = 0$. The SU(2) spin rotation operator in (8.B.1) corresponds to a rotation of angle $-\theta$ about the $\mathbf{u}$ axis. We then use[55]

$$e^{\frac{\theta}{2}e^{i\varphi}\hat{S}^- - \frac{\theta}{2}e^{-i\varphi}\hat{S}^+} = e^{z\hat{S}^-} e^{-\ln(1+|z|^2)\hat{S}^z} e^{-z^*\hat{S}^+} \quad \text{with} \quad z = \tan\left(\frac{\theta}{2}\right) e^{i\varphi} \tag{8.B.2}$$

and $e^{-z^*\hat{S}^+}|S\rangle = |S\rangle$ (which follows from $\hat{S}^+|S\rangle = 0$) to obtain

$$|\mathbf{\Omega}\rangle = e^{z\hat{S}^-} e^{-\ln(1+|z|^2)\hat{S}^z}|S\rangle = (1+|z|^2)^{-S} e^{z\hat{S}^-}|S\rangle. \tag{8.B.3}$$

Using

$$\hat{S}^-|S^z\rangle = [(S - S^z + 1)(S + S^z)]^{1/2}|S - 1\rangle, \tag{8.B.4}$$

we finally obtain

$$|\mathbf{\Omega}\rangle = (1+|z|^2)^{-S} \sum_{p=0}^{2S} \sqrt{C_{2S}^p} z^p |S - p\rangle, \tag{8.B.5}$$

where $C_{2S}^p = (2S)!/[p!(2S-p)!]$.

It is then easy to show that the scalar product between two coherent states is given by

$$\begin{aligned} \langle \mathbf{\Omega}'|\mathbf{\Omega}\rangle &= \frac{(1 + z'z)^{2S}}{(1+|z'|^2)^S(1+|z|^2)^S} \\ &= \left(\cos\frac{\theta'}{2}\cos\frac{\theta}{2} + e^{i(\varphi-\varphi')}\sin\frac{\theta'}{2}\sin\frac{\theta}{2}\right)^{2S} \end{aligned} \tag{8.B.6}$$

[55] See Appendix A in Ref. [24] for a derivation of (8.B.2).

and the resolution of the identity reads

$$(2S+1)\int \frac{dz^*dz}{2i\pi}\frac{|z\rangle\langle z|}{(1+|z^2|)^2} = (2S+1)\int \frac{d\Omega}{4\pi}|\mathbf{\Omega}\rangle\langle\mathbf{\Omega}| = 1, \tag{8.B.7}$$

where $dz^*dz = 2id\Re(z)d\Im(z)$. From (8.B.5), we also deduce

$$\begin{aligned}
\frac{\partial}{\partial\theta}|\mathbf{\Omega}\rangle &= (1+|z|^2)^{-S}\sum_{p=0}^{2S}\sqrt{C_{2S}^p}\left[-S|z| + \frac{p}{2|z|\left(\cos\frac{\theta}{2}\right)^2}\right]|S-p\rangle, \\
\frac{\partial}{\partial\varphi}|\mathbf{\Omega}\rangle &= (1+|z|^2)^{-S}\sum_{p=0}^{2S}\sqrt{C_{2S}^p}ipz^p|S-p\rangle,
\end{aligned} \tag{8.B.8}$$

which leads to (8.96).

Guide to the bibliography

- For a general reference on quantum magnetism and the Heisenberg model, see Ref. [1]. The spin-$\frac{1}{2}$ Heisenberg model is reviewed in Ref. [2].
- Spin waves in ferromagnets and antiferromagnets are discussed in a number of textbooks [1–3]. See also Ref. [25].
- Holstein–Primakoff bosons [4] are reviewed in Ref. [1].
- For a discussion of the Berry phase [5], see Refs. [6–8].
- Spin-coherent states and the spin-coherent-state functional integral are discussed in Refs. [1, 9–12].
- The derivation of the NLσM from the Heisenberg model [13–15] can be found in Refs. [1,9]. The role of the Berry phase term in one dimension is addressed in Ref. [16]. The two-dimensional quantum NLσM is discussed in Chapter 12 and Ref. [17]. For a discussion of the role of the Berry phase term and the transition between the Néel state and the valence bond solid in two-dimensional Heisenberg antiferromagnets, see Refs. [9, 18].
- Quantum spin chains and the concept of Luttinger liquid are discussed in Refs. [9, 13, 14, 19, 20].
- The Schwinger-boson mean-field theory of the Heisenberg model is described in Refs. [1, 21, 22].

Bibliography

[1] A. Auerbach, *Interacting Electrons and Quantum Magnetism* (Springer-Verlag, New York, 1994).

[2] E. Manousakis, The spin-1/2 Heisenberg antiferromagnet on a square lattice and its application to the cuprous oxides, *Rev. Mod. Phys.* **63**, 1 (1991).

[3] C. Kittel, *Introduction to Solid State Physics* (John Wiley & Sons, Inc., New York, London, 1996); *Quantum Theory of Solids* (John Wiley & Sons, Inc., New York, London, 1963).

[4] T. Holstein and H. Primakoff, Field dependence of the intrinsic domain magnetization of a ferromagnet, *Phys. Rev.* **58**, 1098 (1940).

[5] M. V. Berry, Quantal phase factors accompanying adiabatic changes, *Proc. Roy. Soc. London* A **392**, 45–57 (1984).

[6] R. Shankar, *Principles of Quantum Mechanics* (Plenum Press, New York, 1994).

[7] A. Messiah, *Quantum Mechanics*, Vol. II (North-Holland, Amsterdam, 1962).

[8] A. Shapere and F. Wilczek, *Geometric Phases in Physics* (World Scientific, Singapore, 1989).

[9] S. Sachdev, *Quantum Phase Transitions* (Cambridge University Press, Cambridge, 2011).

[10] R. Shankar, *Quantum Field Theory and Condensed Matter: An Introduction* (Cambridge University Press, Cambridge, 2017).

[11] J. R. Kauder and B. Skagerstam, eds., *Coherent States* (World Scientific, Singapore, 1985).

[12] J. R. Klauder, Path integrals and stationary-phase approximations, *Phys. Rev.* D **19**, 2349 (1979).

[13] F. D. M. Haldane, Continuum dynamics of the 1-D Heisenberg antiferromagnet: Identification with the O(3) nonlinear sigma model, *Phys. Lett.* A **93**, 464 (1983).

[14] F. D. M. Haldane, Nonlinear field theory of large-spin Heisenberg antiferromagnets: Semiclassically quantized solitons of the one-dimensional easy-axis Néel state, *Phys. Rev. Lett.* **50**, 1153 (1983).

[15] F. D. M. Haldane, Nonlinear σ model and the topological distinction between integer- and half-integer-spin antiferromagnets in two dimensions, *Phys. Rev. Lett.* **61**, 1029 (1988).

[16] R. Shankar and N. Read, The $\theta = \pi$ nonlinear sigma model is massless, *Nucl. Phys.* B **336**, 457 (1990).

[17] S. Chakravarty, B. I. Halperin, and D. R. Nelson, Two-dimensional quantum Heisenberg antiferromagnet at low temperatures, *Phys. Rev.* B **39**, 2344 (1989).

[18] T. Senthil, A. Vishwanath, L. Balents, S. Sachdev, and M. P. A. Fisher, Deconfined quantum critical points, *Science* **303**, 1490 (2004).

[19] I. Affleck, Quantum spin chains and the Haldane gap, *J. Phys. Condensed Matter* **1**, 3047 (1989).

[20] T. Giamarchi, *Quantum Physics in One Dimension* (Oxford University Press, 2004).

[21] D. P. Arovas and A. Auerbach, Functional integral theories of low-dimensional quantum Heisenberg models, *Phys. Rev.* B **38**, 316 (1988).

[22] A. Auerbach and D. P. Arovas, Spin dynamics in the square-lattice antiferromagnet, *Phys. Rev. Lett.* **61**, 617 (1988).

[23] E. Lieb, T. D. Schultz, and D. C. Mattis, Two soluble models of an antiferromagnetic chain, *Ann. Phys.* **16**, 407 (1961).

[24] F. T. Arecchi, E. Courtens, R. Gilmore, and H. Thomas, Atomic coherent states in quantum optics, *Phys. Rev.* A **6**, 2211 (1972).

[25] M. Takahashi, Modified spin-wave theory of a square-lattice antiferromagnet, *Phys. Rev.* B **40**, 2494 (1989).

Chapter 9

Baym–Kadanoff Formalism

Contents

9.1 Conserving Approximations . . . 612
  9.1.1 Symmetries, conservation laws and Ward identities . . . 612
  9.1.2 Conserving approximations from generating functionals . . . 615
  9.1.3 Perturbative conserving approximations . . . 619
  9.1.4 Limitations of the Baym–Kadanoff approach . . . 623
  9.1.5 Thermodynamic self-consistency . . . 627
9.2 Mott Transition and Dynamical Mean-Field Theory . . . 628
  9.2.1 Brinkman–Rice theory of the Mott transition . . . 629
  9.2.2 Dynamical mean-field theory . . . 630
9.3 Self-Energy Functional Theory . . . 637
  9.3.1 Variational cluster approximation . . . 640
  9.3.2 Dynamical mean-field theory recovered . . . 641
  9.3.3 Phase diagram of the half-filled two-dimensional Hubbard model . . . 643
Appendix 9.A Brinkman–Rice Theory: Gutzwiller Approximation . . . 645
  9.A.1 Kotliar–Ruckenstein slave bosons . . . 649
Bibliography . . . 653

In Chapter 1, we have shown how to calculate the n-particle Green functions $G^{(n)}$ perturbatively using Feynman diagrams. On the other hand, we know that conservation laws, which follow from continuous symmetries, imply relations between Green functions (or, equivalently, between 1PI/2PI vertices) known as Ward identities (Sec. 2.3). How can we ensure that a given diagrammatic approximation respects these conservation laws? In this chapter, we show that the Baym–Kadanoff formalism, based on the Luttinger–Ward functional introduced in Sec. 1.6.3, provides us with a natural framework to set up conserving approximations, i.e. approximations that satisfy conservation laws (Sec. 9.1).

Perturbative conserving approximations are obtained by truncating the Luttinger–Ward functional $\Phi[G]$, i.e. by typically retaining only a few low-order diagrams contributing to $\Phi[G]$. In Sec. 9.1.3, we reconsider the RPA theory of the electron gas and the BCS theory of superconductivity, previously discussed in Chapters 5 and 7, and show that they are conserving (perturbative) approximations.

In Sec. 9.1.4, we discuss some shortcomings of the Baym–Kadanoff formalism, in particular the multivaluedness of the Luttinger–Ward functional in Hubbard-like models and the breakdown at strong coupling of the perturbation expansion in terms of the interaction strength and the fully dressed propagator.

The Baym–Kadanoff formalism is not limited to perturbative approximations. In Secs. 9.2 and 9.3, we consider two examples of non-perturbative approaches in the framework of the Hubbard model: the dynamical mean-field theory (DMFT) and the self-energy functional theory. DMFT gives a consistent picture of the Mott metal-to-insulator transition in fermion systems but neglects magnetic correlations. The self-energy functional theory is a versatile approximation scheme which recovers DMFT and allows for extensions which take into account short-range magnetic fluctuations.

9.1 Conserving Approximations

9.1.1 *Symmetries, conservation laws and Ward identities*

Let us briefly recall an important result derived in Chapter 2: Continuous symmetries imply conservation laws and relations between Green functions known as Ward identities. To be more specific, we consider a system whose action

$$S[\psi] = \int_0^\beta d\tau \int d^d r\, \mathcal{L}(\psi_j, \partial_\mu \psi_j) \tag{9.1}$$

is invariant in the infinitesimal continuous transformation

$$\psi_j(x) \to \psi_j(x) + i\epsilon F_j(x). \tag{9.2}$$

Here, $x = (\mathbf{r}, \tau)$ and j labels the different fields which can be bosonic or fermionic. j includes a spin index σ as well as a charge index $c = \pm$ such that $\psi_j = \psi_\sigma$ (ψ^*_σ) if $c = -$ $(+)$. Classically, the invariance of the action implies the existence of a conserved charge $Q = \int d^d r\, j_0(x)$ and a corresponding continuity equation, $i\partial_\tau j_0 + \boldsymbol{\nabla} \cdot \mathbf{j} = 0$, where the Noether current densities j_0 and $\mathbf{j}$ in the imaginary-time (Euclidean) formalism are defined by (2.134) (see Appendix 1.A.3). In the quantum theory, the invariance of the action in the infinitesimal continuous transformation implies

$$i\partial_\tau \langle j_0(x)\rangle + \boldsymbol{\nabla} \cdot \langle \mathbf{j}(x)\rangle + i \sum_j J_j(x) \langle F_j(x)\rangle = 0, \tag{9.3}$$

where the average $\langle \cdots \rangle$ is taken in the presence of an external source J_j that couples linearly to the field ψ_j [Eq. (2.136)]. Ward identities, i.e. relations between Green functions that follow from the symmetry of the action, are obtained by taking functional derivatives of Eq. (9.3) with respect to J_j and then setting the external source to zero [Eq. (2.137)].

Setting $J_j = 0$ in (9.3), one simply obtains

$$i\partial_\tau \langle j_0(x)\rangle + \boldsymbol{\nabla}\cdot\langle \mathbf{j}(x)\rangle = 0, \tag{9.4}$$

where the expectation value $\langle\cdots\rangle$ is now taken with a vanishing external source. For a system in equilibrium, Eq. (9.4) is in general trivially satisfied since $\langle j_0(x)\rangle$ = const and $\langle \mathbf{j}(x)\rangle = 0$. Taking the functional derivative of (9.3) with respect to $J_{j_3}(x_3)$ and $J_{j_4}(x_4)$ and setting $J_j = 0$, one obtains[1]

$$\begin{aligned} &i\partial_\tau \langle \psi_{j_4}(x_4)\psi_{j_3}(x_3)j_0(x)\rangle + \boldsymbol{\nabla}\cdot\langle \psi_{j_4}(x_4)\psi_{j_3}(x_3)\mathbf{j}(x)\rangle \\ &\quad + i\delta(x-x_3)\langle \psi_{j_4}(x_4)F_{j_3}(x_3)\rangle + i\delta(x-x_4)\langle F_{j_4}(x_4)\psi_{j_3}(x_3)\rangle = 0 \end{aligned} \tag{9.5}$$

[see Eqs. (2.137)]. Since F is generally a linear function of the field ψ, this equation relates the two-point, $G \equiv G^{(2)}$, and four-point, $G^{(4)}$, Green functions. Thus, G and $G^{(4)}$ cannot be approximated independently if the Ward identity (9.5) is to be satisfied: Equation (9.5) is a necessary requirement for the approximation to be conserving.

As an example, let us consider a boson or fermion system invariant in the U(1) transformation

$$\begin{aligned} \psi_\sigma(x) &\to e^{i\epsilon}\psi_\sigma(x) = \psi_\sigma(x) + i\epsilon\psi_\sigma(x) + \mathcal{O}(\epsilon^2), \\ \psi^*_\sigma(x) &\to e^{-i\epsilon}\psi^*_\sigma(x) = \psi^*_\sigma(x) - i\epsilon\psi^*_\sigma(x) + \mathcal{O}(\epsilon^2). \end{aligned} \tag{9.6}$$

Since $F_\sigma(x) = \psi_\sigma(x)$ and $F^*_\sigma(x) = -\psi^*_\sigma(x)$, the current densities (2.134) take the familiar form for a quantum fluid[2]

$$j_0(x) = \sum_\sigma \psi^*_\sigma(x)\psi_\sigma(x), \qquad \mathbf{j}(x) = -\frac{i}{2m}\sum_\sigma [\psi^*_\sigma(x)\boldsymbol{\nabla}\psi_\sigma(x) - \text{c.c.}], \tag{9.7}$$

and (9.5) can be written in the form[3]

$$\left[\partial_{\tau_1} + \partial_{\tau_2} - \frac{1}{2m}(\boldsymbol{\nabla}_1^2 - \boldsymbol{\nabla}_2^2)\right] G^{(4)}(1,3,;4,2)\Big|_{2=1^+} = [\delta(1,3) - \delta(1,4)]G(3,4), \tag{9.8}$$

where we use the notation $1 = (x_1, \tau_1, j_1)$, $1^+ = (x_1, \tau_1 + 0^+, j_1)$, etc. It is straightforward to derive the Ward identities corresponding to the conservation of other quantities, such as the total (angular) momentum.

We have assumed so far that the action $S[\psi]$ is defined by a Lagrangian density $\mathcal{L}(\psi_j, \partial_\mu\psi_j)$ [Eq. (9.1)] and used the form (2.137) of the Ward identities. This assumption is by no means necessary. If we write the action as in (2.138), with the non-interacting part defined by the propagator G_0, then the Ward identities (9.4) and

[1]We do not discuss the equation obtained from a single functional derivative, which is relevant only in Bose systems with spontaneous U(1) symmetry breaking.

[2]Equation (9.7) holds provided that the interacting part of the action is invariant in the local U(1) transformation $\psi_\sigma(x) \to e^{i\epsilon(x)}\psi_\sigma(x)$ and $\psi^*_\sigma(x) \to e^{-i\epsilon(x)}\psi^*_\sigma(x)$.

[3]For non-interacting particles, Eq. (9.8) follows from the equation of motion of the one-particle Green function: $(-\partial_\tau + \mu + \frac{\boldsymbol{\nabla}^2}{2m})G_0(x,x') = \delta(x-x')$ and $(\partial_{\tau'} + \mu + \frac{\boldsymbol{\nabla}'^2}{2m})G_0(x,x') = \delta(x-x')$.

(9.5) become

$$\int d^d r' \int_0^\beta d\tau' \sum_{j,j'} G^{-1}_{0,jj'}(x,x')\langle F_j(x)\psi_{j'}(x')\rangle = 0, \tag{9.9}$$

and

$$\int d^d r' \int_0^\beta d\tau' \sum_{j,j'} G^{-1}_{0,jj'}(x,x')\langle \psi_{j_4}(x_4)\psi_{j_3}(x_3)F_j(x)\psi_{j'}(x')\rangle$$
$$+ \delta(x-x_3)\langle \psi_{j_4}(x_4)F_{j_3}(x_3)\rangle + \delta(x-x_4)\langle F_{j_4}(x_4)\psi_{j_3}(x_3)\rangle = 0 \tag{9.10}$$

[see Eqs. (2.141)]. Equation (9.8) follows from (9.10) when the full action is U(1) invariant (and its interacting part invariant in a local U(1) transformation).

Ward identities are crucial when calculating the linear response χ to an external source which couples to an operator bilinear in the ψ field. In that case, χ if fully determined by the four-point Green function $G^{(4)}$. For example, the linear response to a scalar potential is given by the density–density correlation function (Sec. 3.3). If the system sustains a propagating sound mode with dispersion $\omega_{\mathbf{q}} = c|\mathbf{q}|$ (with c the sound velocity), the retarded density–density correlation function $\chi^R(\mathbf{q},\omega)$ will exhibit a pole at the frequency $\omega = \omega_{\mathbf{q}}$; if the sound mode is damped, the pole will appear at an imaginary frequency $\omega_{\mathbf{q}} - i\Gamma_{\mathbf{q}}$ but the spectral function $\chi''(\mathbf{q},\omega) = \Im[\chi^R(\mathbf{q},\omega)]$ will still exhibit a strong resonance for $\omega \simeq \omega_{\mathbf{q}}$. Satisfying the local charge conservation (i.e. the continuity equation) is crucial to obtain the sound mode.[4] If the Ward identity (9.8) following from the U(1) invariance is violated, it is likely that the sound-mode pole will not appear in the density–density correlation function and the computation of the latter will not be reliable. Similarly, computing the conductivity (essentially given by the current–current correlation function; see Sec. 3.4.4.1) without satisfying the local charge conservation and the corresponding Ward identity (9.8) is meaningless.

We have seen in Sec. 3.4.4.3 that the electromagnetic response function $K_{\mu\nu}(x,x')$ is transverse as a consequence of the U(1) symmetry or, equivalently, the conservation of particles. Let us show that this property follows from the Ward identity (9.5) (with $F_\sigma(x) = \psi_\sigma(x)$ and $F^*(x) = -\psi^*_\sigma(x)$). We set $\sigma_3 = \sigma_4$, $\tau_4 = \tau_3^+$, and successively $\mathbf{r}_3 = \mathbf{r}_4$, $\mathbf{r}_3 = \mathbf{r}_4 + \boldsymbol{\delta}$ and $\mathbf{r}_4 = \mathbf{r}_3 + \boldsymbol{\delta}$. This yields

$$\langle (i\partial_\tau j_0(x) + \boldsymbol{\nabla}\cdot\mathbf{j}(x))\psi^*_{\sigma_3}(x_3)\psi_{\sigma_3}(x_3)\rangle = 0 \tag{9.11}$$

and

$$\begin{aligned} &\langle [i\partial_\tau j_0(x) + \boldsymbol{\nabla}\cdot\mathbf{j}(x)]\psi^*_{\sigma_4}(x_4)\psi_{\sigma_4}(x_4+\delta)\rangle \\ &\quad + i[\delta(x-x_4-\delta) - \delta(x-x_4)]\langle \psi^*_{\sigma_4}(x_4)\psi_{\sigma_4}(x_4+\delta)\rangle = 0, \\ &\langle [i\partial_\tau j_0(x) + \boldsymbol{\nabla}\cdot\mathbf{j}(x)]\psi^*_{\sigma_3}(x_3+\delta)\psi_{\sigma_3}(x_3)\rangle \\ &\quad + i[\delta(x-x_3) - \delta(x-x_3-\delta)]\langle \psi^*_{\sigma_3}(x_3+\delta)\psi_{\sigma_3}(x_3)\rangle = 0, \end{aligned} \tag{9.12}$$

[4]See the elementary discussion of the sound mode at the end of Sec. 3.3.

where $\delta = (\boldsymbol{\delta}, 0)$ and a sum over σ_3 or σ_4 is implied. Taking the difference between the last two equations and the limit $\boldsymbol{\delta} \to 0$, we obtain

$$\begin{aligned} 0 &= \langle [i\partial_\tau j_0(x) + \boldsymbol{\nabla} \cdot \mathbf{j}(x)][\psi^*_{\sigma'}(x')\partial_{x'_\mu}\psi_{\sigma'}(x') - \text{c.c.}]\rangle \\ &\quad - 2i\partial_{x_\mu}\delta(x - x')\langle \psi^*_{\sigma'}(x')\psi_{\sigma'}(x')\rangle \\ &= \langle [i\partial_\tau j_0(x) + \boldsymbol{\nabla} \cdot \mathbf{j}(x)] j_\mu(x')\rangle - \frac{n}{m}\partial_\mu \delta(x - x') \end{aligned} \tag{9.13}$$

for $\mu \neq 0$, where $n = \langle j_0(x)\rangle$ is the mean density of particles. This equation, together with (9.11), can be rewritten as

$$i\partial_\tau K_{0\nu}(x - x') + \sum_{\mu \neq 0} \partial_\mu K_{\mu\nu}(x - x') = 0 \qquad (\nu = 0, x, \ldots), \tag{9.14}$$

where

$$K_{\mu\nu}(x - x') = \langle j_\mu(x) j_\nu(x')\rangle - \frac{n}{m}\delta_{\mu,\nu}(1 - \delta_{\mu,0})\delta(x - x') \tag{9.15}$$

is the electromagnetic response function (Sec. 3.4.4.3).

9.1.2 *Conserving approximations from generating functionals*

The relations (2.137) and (2.141) are one form of the Ward identities associated with the invariance of the action in a continuous transformation. Another form is given by relations between the 1PI or 2PI vertices. The latter are defined as functional derivatives of a generating functional, the effective action $\Gamma[\phi]$ for the 1PI vertices and the Luttinger–Ward functional $\Phi[G]$ for the 2PI ones (Chapter 1). In Chapter 2, we have seen that some of the Ward identities follow from the symmetry properties of the generating functional inherited from the symmetries of the microscopic action. This means that these Ward identities will be automatically satisfied in any calculation based on an approximate generating functional that fulfills the required symmetries.

9.1.2.1 *1PI formalism*

The effective action

$$\Gamma[\phi] = -\ln Z[J] + \int_0^\beta d\tau \int d^d r \sum_j J_j(x)\phi_j(x) \tag{9.16}$$

is defined as the Legendre transform of the generating functional $\ln Z[J]$ of connected Green functions, with $\phi_j(x) = \langle \psi_j(x)\rangle$ (Sec. 1.6.2). For any infinitesimal continuous transformation (9.2), with F_j a linear function of ψ_j, which leaves both the action and the measure of the functional integral invariant, the effective action (9.16) is invariant in the infinitesimal transformation

$$\phi_j(x) \to \phi_j(x) + i\epsilon\langle F_j(x)\rangle. \tag{9.17}$$

In Sec. 2.3.3, we showed that the global Ward identities for the 1PI vertices (and in particular the Goldstone theorem when the symmetry is spontaneously broken) are a mere consequence of the invariance of $\Gamma[\phi]$ in the transformation (9.17). Global Ward identities will therefore be automatically satisfied in any approximate calculation of the effective action which preserves this invariance. For example, the loop expansion discussed in Sec. 1.7.2 satisfies the global Ward identities.

On the other hand, there is in general no systematic way to guarantee that the local Ward identities are satisfied in the 1PI formalism. Local Ward identities were indeed obtained by considering the Legendre transform of (2.140) but were not related to a symmetry property of the effective action $\Gamma[\phi]$. To illustrate this issue, consider the one-loop approximation,

$$\Gamma[\phi] = S[\phi] - \frac{1}{2}\mathrm{Tr}\ln S^{(2)}[\phi], \tag{9.18}$$

where

$$S^{(2)}_{jj'}[x, x'; \phi] = \frac{\delta^2 S[\phi]}{\delta\phi_j(x)\delta\phi_{j'}(x')} \tag{9.19}$$

denotes the second functional derivative of the action $S[\phi]$. In the absence of spontaneous symmetry breaking, the state of the system corresponds to $\phi_j(x) = 0$. The self-energy deduced from (9.18) is then given by the one-loop (tadpole) diagram and the four-point vertex $\Gamma^{(4)}$ reduces to the bare two-body interaction (see Sec. 1.7.2). The four-point Green function is then given by the diagrams shown in Fig. 9.1(b) [see Eq. (1.416)]. This approximation is not conserving since an infinite number of diagrams is missing for (9.5) to be satisfied. The 2PI formalism shows that when the self-energy is given by the one-loop diagram, the bare vertex corresponds to the 2PI vertex and not the full four-point vertex.[5]

This does not necessarily imply that it is never possible to set up a conserving approximation in the 1PI formalism. Suppose, for instance, that we are interested in the electromagnetic response function (density–density and current–current correlation functions). As pointed out above, it is crucial to satisfy the Ward identity associated with the global U(1) invariance of the microscopic action. We can make the global U(1) invariance a local one by introducing a scalar and a vector potential $(A_0, \mathbf{A})$ and replacing the derivatives in the action $S[\psi]$ by covariant derivatives: $\partial_\mu \to \partial_\mu - iA_\mu$ $(\mu = 0, x, \ldots)$. Here, $(A_0, \mathbf{A})$ is not a dynamical gauge field but a mere external source which allows us to define the density

$$J_\mu(x) = -\frac{\delta S[\psi]}{\delta A_\mu(x)} \tag{9.20}$$

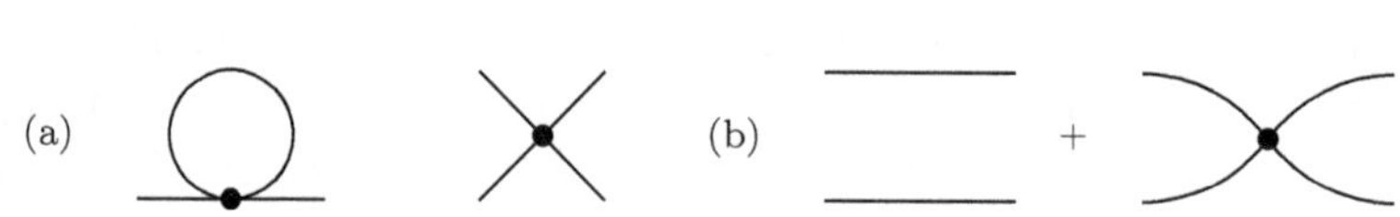

Fig. 9.1 (a) Self-energy and four-point vertex obtained from the one-loop effective action $\Gamma[\phi]$ in the absence of broken symmetry [Eq. (9.18)]. (b) Four-point Green function $G^{(4)}$ to one-loop order.

[5]See, e.g., the discussion of the RPA theory of the electron gas in Sec. 9.1.3.1.

($-iJ_0$ and $J_{\mu\neq 0}$ correspond to the particle and current densities, respectively). The microscopic action $S[\psi^*,\psi;A]$ is invariant in the gauge transformation

$$\begin{aligned} A'_\mu(x) &= A_\mu(x) + \partial_\mu \Lambda(x), \\ \psi'_\sigma(x) = e^{i\Lambda(x)}\psi_\sigma(x), &\qquad \psi'^*_\sigma(x) = e^{-i\Lambda(x)}\psi^*_\sigma(x), \end{aligned} \tag{9.21}$$

where $\Lambda(x)$ is an arbitrary function and we now consider the fields $\psi^*_\sigma, \psi_\sigma$ rather than ψ_j. It is easy to show that the effective action is also invariant in the local gauge transformation (9.21),[6] i.e.

$$\Gamma[\phi^*,\phi;A] = \Gamma[\phi'^*,\phi';A'], \tag{9.22}$$

where $\phi'_\sigma(x) = e^{i\Lambda(x)}\phi_\sigma(x)$ and $\phi'^*_\sigma(x) = e^{-i\Lambda(x)}\phi^*_\sigma(x)$. This property ensures that the electromagnetic response function

$$\tilde{K}_{\mu\nu}(x,x') = \left.\frac{\delta^2 \ln Z[J^*,J;A]}{\delta A_\mu(x)\delta A_\nu(x')}\right|_{J^*=J=A=0} \tag{9.23}$$

computed from $\Gamma[\phi^*,\phi;A]$ satisfies all the (global and local) Ward identities associated with the U(1) invariance of the microscopic action.

To illustrate this point, let us consider the microscopic action

$$S = \int_0^\beta d\tau \int d^d r \sum_\sigma \psi^*_\sigma \left[\partial_\tau - \mu - iA_0 + \frac{1}{2m}(-i\boldsymbol{\nabla} - \mathbf{A})^2\right]\psi_\sigma + S_{\rm int} \tag{9.24}$$

for quantum particles in the presence of the external potential A_μ. S is invariant in the gauge transformation (9.21) and the effective action satisfies (9.22). It is convenient to use the two-component real field $(\phi_1,\phi_2)^T$ defined by $\phi = \frac{1}{\sqrt{2}}(\phi_1 + i\phi_2)$ so that a gauge transformation now corresponds to the rotation

$$\begin{pmatrix}\phi'_1 \\ \phi'_2\end{pmatrix} = \begin{pmatrix}\cos\Lambda & -\sin\Lambda \\ \sin\Lambda & \cos\Lambda\end{pmatrix}\begin{pmatrix}\phi_1 \\ \phi_2\end{pmatrix}. \tag{9.25}$$

Differentiating (9.22) with respect to $\Lambda(x)$ and setting $\Lambda(x)=0$ gives

$$\sum_{i,j} \frac{\delta\Gamma[\phi^*,\phi;A]}{\delta\phi_i(x)}\epsilon_{ij}\phi_j(x) - \sum_\mu \partial_\mu \frac{\delta\Gamma[\phi^*,\phi;A]}{\delta A_\mu(x)} = 0. \tag{9.26}$$

If we now differentiate with respect to $A_\nu(x')$, we obtain

$$\sum_\mu \partial_\mu \frac{\delta^2\Gamma[\phi^*,\phi;A]}{\delta A_\mu(x)\delta A_\nu(x')} = \sum_{i,j} \frac{\delta^2\Gamma[\phi^*,\phi;A]}{\delta\phi_i(x)\delta A_\nu(x')}\epsilon_{ij}\phi_j(x). \tag{9.27}$$

In a fermion or a normal boson system, the physical state corresponds to $\phi_i = 0$ and we deduce from (9.27) that the electromagnetic response

$$\tilde{K}_{\mu\nu}(x-x') = -\left.\frac{\delta^2\Gamma[\phi^*,\phi;A]}{\delta A_\mu(x)\delta A_\nu(x')}\right|_{\phi^*=\phi=A_\mu=0} \tag{9.28}$$

[6]In Eq. (9.22), it should be understood that Γ has the same functional form in the lhs and rhs.

is transverse, i.e.

$$\sum_{\mu=0,x,\ldots} \partial_\mu \tilde{K}_{\mu\nu}(x-x') = 0, \tag{9.29}$$

in agreement with (9.14).[7]

In a superfluid boson system where the physical state is defined by $|\phi| = \sqrt{n_0}$, with n_0 the condensate density, the electromagnetic response is not given by (9.28) (there are additional terms due to the non-zero condensate density). A careful analysis shows again that $K_{\mu\nu}$ is transverse.[8]

9.1.2.2 *2PI formalism*

The Luttinger–Ward functional $\Phi[G]$ is defined from the 2PI effective action

$$\begin{aligned}\Gamma[G] &= -\ln Z[J] - \frac{1}{2}\int d\tau_1 d\tau_2 \int d^d r_1 d^d r_2 \sum_{j_1,j_2} J_{j_1 j_2}(x_1,x_2) G_{j_1 j_2}(x_1,x_2) \\ &= -\frac{\zeta}{2}\mathrm{tr}\ln(-G) + \frac{\zeta}{2}\mathrm{tr}(G_0^{-1}G - 1) + \Phi[G]\end{aligned} \tag{9.30}$$

of the free energy $-\ln Z[J]$. Here, $J_{j_1 j_2}(x_1,x_2) = \zeta J_{j_2 j_1}(x_2,x_1)$ is a bilinear external source and

$$G_{j_1 j_2}[x_1,x_2;J] = -\langle \psi_{j_1}(x_1)\psi_{j_2}(x_2)\rangle = -\frac{\delta \ln Z[J]}{\delta J_{j_1 j_2}(x_1,x_2)} \tag{9.31}$$

stands for the one-particle Green function. For $J=0$, the latter corresponds to the stationary point of $\Gamma[G]$.[9]

Consider now the local transformation

$$\psi_j(x) \to \psi_j(x) + i\epsilon(x)F_j(x) \tag{9.32}$$

where

$$F_j(x) = \int_0^\beta d\tau' \int d^d r' \sum_{j'} t_{jj'}(x,x')\psi_{j'}(x') \tag{9.33}$$

is a linear function $\psi_j(x)$. The one-particle Green function then changes as

$$G(1,2) \to G(1,2) + i\sum_3 [\epsilon(x_1)t(1,3)G(3,2) + \epsilon(x_2)t(2,3)G(1,3)] \tag{9.34}$$

and the Luttinger–Ward functional is invariant if the transformation (9.32) leaves the interacting part $S_{\rm int}$ of the action invariant (Sec. 2.3.4). The remarkable feature of the 2PI formalism is that both global and local Ward identities satisfied by the 2PI vertices are a mere

[7] $\tilde{K}_{\mu\nu}$ slightly differs from the response function $K_{\mu\nu}$ introduced in (9.15) (this is due to iA_0 appearing in (9.24), which implies $J_0 = i\psi^*\psi$), i.e. $\tilde{K}_{00} = -K_{00}$, $\tilde{K}_{\mu 0} = iK_{\mu 0}$ and $\tilde{K}_{\mu\nu} = K_{\mu\nu}$ ($\mu,\nu \neq 0$), so that Eqs. (9.14) and (9.29) agree.

[8] See Ref. [51].

[9] In Chapters 1 and 2, we denoted $G[J=0]$ by $\bar{G}$. In the present discussion, we use the same notation for G (as the dummy variable of $\Gamma[G]$ or $\Phi[G]$) and $\bar{G}$.

consequence of the invariance of $\Phi[G]$. This implies that any calculation based on an approximate Luttinger–Ward functional ("Φ-derivable" approximation) which remains invariant in the transformation (9.34) is conserving. In particular, any diagrammatic approximation where the Luttinger–Ward functional is truncated by retaining only a subset of diagrams is conserving since all diagrams contributing to $\Phi[G]$ are invariant in the transformation (9.34) (Sec. 2.3.4). Two simple examples, where the Luttinger–Ward functional is truncated to the lowest-order diagram, are discussed in the following section.

9.1.3 *Perturbative conserving approximations*

9.1.3.1 *RPA theory of the electron gas*

Let us show that the RPA theory of the electron gas discussed in Chapter 5 is conserving and can be reformulated in the 2PI formalism. We thus consider a three-dimensional system of electrons interacting *via* the Coulomb interaction $v(\mathbf{r}) = e^2/4\pi\epsilon_0|\mathbf{r}|$. The global neutrality of the system is maintained by a uniform background of positive charge (jellium model) and is simply taken into account by setting the Fourier transform $v(\mathbf{q}) = e^2/\epsilon_0\mathbf{q}^2$ of $v(\mathbf{r})$ equal to zero when $\mathbf{q} = 0$ (Sec. 5.1). The lowest-order contribution to the Luttinger–Ward functional is shown in Fig. 9.2. Retaining only the Hartree term, we obtain

$$\Phi[G] = \frac{1}{2}\int_0^\beta d\tau \int d^3r d^3r'\, v(\mathbf{r}-\mathbf{r}') \sum_{\sigma,\sigma'} G_\sigma(\mathbf{r},\tau;\mathbf{r};\tau^+) G_{\sigma'}(\mathbf{r}',\tau;\mathbf{r}';\tau^+) \tag{9.35}$$

(with $G_\sigma(\mathbf{r},\tau';\mathbf{r}',\tau') = -\langle\psi_\sigma(\mathbf{r},\tau)\psi^*_\sigma(\mathbf{r}',\tau')\rangle\rangle$ and the Hartree self-energy

$$\begin{aligned}\Sigma_\sigma(\mathbf{r},\tau;\mathbf{r}',\tau') &= \frac{\delta\Phi[G]}{\delta G_\sigma(\mathbf{r}',\tau';\mathbf{r},\tau)} \\ &= \delta(\mathbf{r}-\mathbf{r}')\delta(\tau-\tau'^+) \int d^3r''\, v(\mathbf{r}-\mathbf{r}'') \sum_{\sigma''} G_{\sigma''}(\mathbf{r}'',\tau;\mathbf{r}'',\tau^+),\end{aligned} \tag{9.36}$$

where $G^{-1} = G_0^{-1} - \Sigma$ with G_0 the non-interacting one-particle Green function. Equation (9.36) is a self-consistent equation for Σ. If we look for a solution which preserves space and time translation, i.e. $\Sigma(\mathbf{r},\tau;\mathbf{r}',\tau') = \Sigma(\mathbf{r}-\mathbf{r}',\tau-\tau')$, we find

$$\Sigma(k) = v(\mathbf{q}=0)\frac{1}{\beta\mathcal{V}}\sum_{k',\sigma'} G_{\sigma'}(k')e^{i\omega_{n'}0^+} = v(\mathbf{q}=0)n, \tag{9.37}$$

where n is the density of particles. The Hartree self-energy vanishes since $v(\mathbf{q}=0) = 0$ in the jellium model.

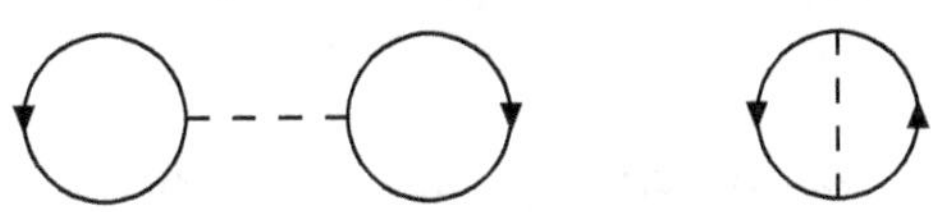

Fig. 9.2 Lowest-order contributions to the Luttinger–Ward functional of the electron gas. These two contributions give rise to the Hartree and Fock self-energies, respectively.

The two-particle 2PI vertex

$$\begin{aligned}\Phi^{(2)}_{\sigma_1,\sigma_2}(\mathbf{r}_1,\tau_1,\mathbf{r}'_1,\tau'_1;\mathbf{r}_2,\tau_2,\mathbf{r}'_2,\tau'_2)\\ &= \frac{\delta^2\Phi[G]}{\delta G_{\sigma_1}(\mathbf{r}_1,\tau_1;\mathbf{r}'_1,\tau'_1)\delta G_{\sigma_2}(\mathbf{r}_2,\tau_2;\mathbf{r}'_2,\tau'_2)}\\ &= \delta(\tau_1-\tau_2)\delta(\tau'_1-\tau_1^+)\delta(\tau'_2-\tau_2^+)\delta(\mathbf{r}'_1-\mathbf{r}_1)\delta(\mathbf{r}'_2-\mathbf{r}_2)v(\mathbf{r}_1-\mathbf{r}_2)\end{aligned} \tag{9.38}$$

is nothing but the bare vertex in the particle–hole channel corresponding to the first diagram of Fig. 1.10. The particle–hole pair propagator

$$\begin{aligned}W^{(2)}_{\sigma_1,\sigma_2}(\mathbf{r}_1,\tau_1,\mathbf{r}'_1,\tau'_1;\mathbf{r}_2,\tau_2,\mathbf{r}'_2,\tau'_2) &= \langle\psi_{\sigma_1}(\mathbf{r}_1,\tau_1)\psi_{\sigma_2}(\mathbf{r}_2,\tau_2)\psi^*_{\sigma_2}(\mathbf{r}'_2,\tau'_2)\psi^*_{\sigma_1}(\mathbf{r}'_1,\tau'_1)\rangle\\ &\quad - G_{\sigma_1}(\mathbf{r}_1,\tau_1;\mathbf{r}'_1;\tau'_1)G_{\sigma_2}(\mathbf{r}_2,\tau_2;\mathbf{r}'_2;\tau'_2)\end{aligned} \tag{9.39}$$

is obtained from the Bethe–Saltpeter equation $W^{(2)} = \Pi - \Pi\Phi^{(2)}W^{(2)}$ where[10]

$$\Pi_{\sigma_1,\sigma_2}(\mathbf{r}_1,\tau_1,\mathbf{r}'_1,\tau'_1;\mathbf{r}_2,\tau_2,\mathbf{r}'_2,\tau'_2) = -\delta_{\sigma_1,\sigma_2}G_{\sigma_1}(\mathbf{r}_1,\tau_1;\mathbf{r}'_2;\tau'_2)G_{\sigma_2}(\mathbf{r}_2,\tau_2;\mathbf{r}'_1;\tau'_1). \tag{9.40}$$

From this, we can obtain the density–density response function

$$\chi_{nn}(\mathbf{r},\tau;\mathbf{r}',\tau') = \sum_{\sigma,\sigma'} W^{(2)}_{\sigma,\sigma'}(\mathbf{r},\tau,\mathbf{r},\tau^+;\mathbf{r}',\tau',\mathbf{r}',\tau'^+). \tag{9.41}$$

In Fourier space, this gives

$$\chi_{nn}(q) = \frac{\chi^0_{nn}(q)}{1+v(\mathbf{q})\chi^0_{nn}(q)}, \tag{9.42}$$

where

$$\chi^0_{nn}(q) = -\frac{2}{\beta\mathcal{V}}\sum_{k,\sigma} G_\sigma(k)G_\sigma(k+q), \tag{9.43}$$

which reproduces the RPA result (5.11).

Since the RPA is conserving, it must satisfy two properties that follow from particle number conservation,

$$\begin{aligned}\lim_{|\omega_n|\to\infty}\omega_n^2\chi_{nn}(\mathbf{q},i\omega_n) = \int_{-\infty}^{\infty}\frac{d\omega}{\pi}\omega\,\chi''_{nn}(\mathbf{q},\omega) = \frac{n\mathbf{q}^2}{m} \qquad (f\text{-sum rule}),\\ \chi_{nn}(\mathbf{q}=0,i\omega_n) = 0 \quad \text{for} \quad \omega_n \neq 0\end{aligned} \tag{9.44}$$

[see Eqs. (3.106) and (3.195)]. The second equation is a direct consequence of the same property in the non-interacting system: $\chi^0_{nn}(\mathbf{q}=0,i\omega_n) = 0$ for $\omega_n \neq 0$, whereas the first one comes from

$$\lim_{|\omega_n|\to\infty}\omega_n^2\chi_{nn}(\mathbf{q},i\omega_n) = \lim_{|\omega_n|\to\infty}\omega_n^2\chi^0_{nn}(\mathbf{q},i\omega_n) = \frac{n\mathbf{q}^2}{m}. \tag{9.45}$$

[10] We follow the notations of Sec. 1.6.3.

Let us show how a naive "improvement" of the RPA can make it non-conserving. Using (1.F.24), we can write

$$\chi^0_{nn}(\mathbf{q}, i\omega_n) = 2\int_{\mathbf{k}}\int_{-\infty}^{\infty} d\epsilon d\epsilon' A(\mathbf{k},\epsilon)A(\mathbf{k}+\mathbf{q},\epsilon')\frac{(\epsilon'-\epsilon)[n_F(\epsilon)-n_F(\epsilon')]}{\omega_n^2+(\epsilon-\epsilon')^2} \tag{9.46}$$

as a function of the electron spectral function $A(\mathbf{k},\epsilon) = -\frac{1}{\pi}\Im[G^R(\mathbf{k},\epsilon)]$. With the non-interacting (or Hartree) Green function, $A(\mathbf{k},\epsilon) = \delta(\epsilon-\xi_{\mathbf{k}})$, so that $\chi^0_{nn}(\mathbf{q} = 0, i\omega_n) = 0$ for $\omega_n \neq 0$. If we try to improve the RPA by using a propagator that includes a non-trivial self-energy (but still using $v(\mathbf{q})$ as the particle–hole irreducible vertex), $A(\mathbf{k},\epsilon)$ is then not a delta function anymore and $\chi^0_{nn}(\mathbf{q} = 0, i\omega_n)$ cannot vanish since the integrand in (9.46) is positive. We obtain a non-conserving approximation since the second of Eqs. (9.44) is violated. The only way to restore particle conservation is to include "vertex corrections" in the calculation of the irreducible part Π_{nn} (approximated by χ^0_{nn} in the RPA) of the density–density response function χ_{nn}. Diagrammatically, these corrections are represented by Coulomb lines joining the particle and hole lines in Π_{nn} and can be seen as corrections to the interaction vertex $v(\mathbf{q})$.[11]

9.1.3.2 *BCS superconductors*

The BCS theory of superconductivity can also be reformulated in the 2PI formalism. In a superfluid system with spontaneous U(1) symmetry breaking, the lowest-order Luttinger–Ward functional includes, in addition to the two contributions shown in Fig. 9.2, the "anomalous" term shown diagrammatically in Fig. 9.3. BCS theory is reproduced by retaining only the latter,

$$\Phi[F, F^\dagger] = -\lambda\int_0^\beta d\tau\int d^3r\, F^\dagger(\mathbf{r},\tau;\mathbf{r},\tau)F(\mathbf{r},\tau;\mathbf{r},\tau), \tag{9.47}$$

where $F(\mathbf{r},\tau;\mathbf{r}',\tau') = -\langle\psi_\uparrow(\mathbf{r},\tau)\psi_\downarrow(\mathbf{r}',\tau')\rangle$ and $F^\dagger(\mathbf{r},\tau;\mathbf{r}',\tau') = -\langle\psi^*_\downarrow(\mathbf{r},\tau)\psi^*_\uparrow(\mathbf{r}',\tau')\rangle$ are the anomalous propagators and $-\lambda < 0$ is the strength of the attractive interaction (Sec. 7.5). The self-consistent equation obtained from the self-energy deduced from (9.47) reproduces the gap equation of BCS theory,

$$\Delta = \frac{\lambda}{\beta\mathcal{V}}\sum_k F(k), \tag{9.48}$$

where $\Delta = \lambda\langle\psi_\downarrow(\mathbf{r},\tau)\psi_\uparrow(\mathbf{r},\tau)\rangle$.

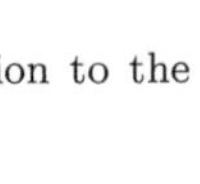

Fig. 9.3 Lowest-order "anomalous" contribution to the Luttinger–Ward functional in a superconductor [Eq. (9.47)].

[11]The second diagram in Fig. 5.1 corresponds to a vertex correction, the following ones to self-energy corrections.

To derive (9.48), it is convenient to use the notation of Sec. 1.6.3. We introduce a charge index $c = \pm$ such that $\psi_+ = \psi^*$ and $\psi_- = \psi$ and use the opposite sign convention for the one-particle Green function: $F \equiv -G_{-\uparrow,-\downarrow}$, $F^\dagger \equiv -G_{+\downarrow,+\uparrow}$. One then has

$$\Phi[G] = -\lambda \int_0^\beta d\tau \int d^3r\, G_{+\downarrow,+\uparrow}(\mathbf{r},\tau;\mathbf{r},\tau)G_{-\uparrow,-\downarrow}(\mathbf{r},\tau;\mathbf{r},\tau) \tag{9.49}$$

and the self-energies

$$\begin{aligned}\Sigma_{-\uparrow,-\downarrow}(\mathbf{r},\tau;\mathbf{r}',\tau') &= -\lambda\delta(\tau-\tau')\delta(\mathbf{r}-\mathbf{r}')G_{+\downarrow,+\uparrow}(\mathbf{r},\tau;\mathbf{r},\tau),\\ \Sigma_{+\downarrow,+\uparrow}(\mathbf{r},\tau;\mathbf{r}',\tau') &= -\lambda\delta(\tau-\tau')\delta(\mathbf{r}-\mathbf{r}')G_{-\uparrow,-\downarrow}(\mathbf{r},\tau;\mathbf{r},\tau),\end{aligned} \tag{9.50}$$

with $\Sigma_{-\uparrow,-\downarrow} = -\Sigma_{-\downarrow,-\uparrow}$ and $\Sigma_{+\downarrow,+\uparrow} = -\Sigma_{+\uparrow,+\downarrow}$. With our notation, Dyson's equation reads $G^{-1} = G_0^{-1} + \Sigma$. The propagator is obtained by diagonalizing the quadratic form

$$(\psi_{+\downarrow}(-k),\psi_{-\uparrow}(-k))\mathcal{G}^{-1}(k)\begin{pmatrix}\psi_{-\downarrow}(k)\\ \psi_{+\uparrow}(k)\end{pmatrix}, \tag{9.51}$$

where[12]

$$\mathcal{G}^{-1}(k) = \begin{pmatrix} G^{-1}_{0;+\downarrow,-\downarrow}(k) & \Sigma_{+\downarrow,+\uparrow}(k)\\ \Sigma_{-\uparrow,-\downarrow}(k) & G^{-1}_{0;-\uparrow,+\uparrow}(k)\end{pmatrix} = \begin{pmatrix} -i\omega_n+\xi_{\mathbf{k}} & \Delta\\ \Delta^* & -i\omega_n-\xi_{\mathbf{k}}\end{pmatrix} \tag{9.52}$$

with the gap Δ defined by $\Delta = \lambda G_{-\downarrow,-\uparrow}(\mathbf{r},\tau;\mathbf{r},\tau)$ and $\Delta^* = \lambda G_{+\uparrow,+\downarrow}(\mathbf{r},\tau;\mathbf{r},\tau)$. Inverting (9.52) gives

$$\mathcal{G}(k) = \frac{1}{\omega_n^2+E_{\mathbf{k}}^2}\begin{pmatrix} i\omega_n+\xi_{\mathbf{k}} & \Delta\\ \Delta^* & i\omega_n-\xi_{\mathbf{k}}\end{pmatrix}, \tag{9.53}$$

where $E_{\mathbf{k}} = \sqrt{\xi_{\mathbf{k}}^2+|\Delta|^2}$. The gap equation (9.48) then follows from (9.50).

The two-particle 2PI vertex

$$\begin{aligned}&\Phi^{(2)}_{+\downarrow,+\uparrow;-\uparrow,-\downarrow}(\mathbf{r}_1,\tau_1,\mathbf{r}'_1,\tau'_1;\mathbf{r}_2,\tau_2,\mathbf{r}'_2,\tau'_2)\\ &\quad= \frac{\delta^2\Phi[G]}{\delta F^\dagger(\mathbf{r}_1,\tau_1;,\mathbf{r}'_1,\tau'_1)\delta F(\mathbf{r}_2,\tau_2;\mathbf{r}'_2,\tau'_2)}\\ &\quad= -\lambda\delta(\tau_1-\tau_2)\delta(\tau_1-\tau'_1)\delta(\tau_2-\tau'_2)\delta(\mathbf{r}_1-\mathbf{r}_2)\delta(\mathbf{r}_1-\mathbf{r}'_1)\delta(\mathbf{r}_2-\mathbf{r}'_2)\end{aligned} \tag{9.54}$$

corresponds to the bare vertex in the particle–particle channel. The particle–particle pair propagator

$$\begin{aligned}W^{(2)}(\mathbf{r}_1,\tau_1,\mathbf{r}'_1,\tau'_1;\mathbf{r}_2,\tau_2,\mathbf{r}'_2,\tau'_2) &= \langle\psi_\downarrow(\mathbf{r}_1,\tau_1)\psi_\uparrow(\mathbf{r}'_1,\tau'_1)\psi^*_\uparrow(\mathbf{r}_2,\tau_2)\psi^*_\downarrow(\mathbf{r}'_2,\tau'_2)\rangle\\ &\quad - F(\mathbf{r}'_1,\tau'_1;\mathbf{r}_1,\tau_1)F^\dagger(\mathbf{r}'_2,\tau'_2;\mathbf{r}_2,\tau_2)\end{aligned} \tag{9.55}$$

---

[12]We use $G^{-1}_{0;+\sigma,-\sigma}(k) \equiv -G_0^{-1}(k) = -(i\omega_n-\xi_{\mathbf{k}})$ and $G^{-1}_{0;-\sigma,+\sigma}(k) = -G^{-1}_{0;+\sigma,-\sigma}(-k)$.

can be obtained from the Bethe–Saltpeter equation $W^{(2)} = \Pi - \Pi\Phi^{(2)}W^{(2)}$ where

$$\Pi(\mathbf{r}_1, \tau_1, \mathbf{r}_1', \tau_1'; \mathbf{r}_2, \tau_2, \mathbf{r}_2', \tau_2') = -G_\downarrow(\mathbf{r}_1, \tau_1; \mathbf{r}_2, \tau_2)G_\uparrow(\mathbf{r}_1', \tau_1'; \mathbf{r}_2', \tau_2'). \tag{9.56}$$

Such a calculation [52], which is beyond the scope of our discussion, is in essence an RPA. The collective mode spectrum obtained from the poles of $W^{(2)}$ (corresponding to amplitude and phase fluctuations of the superconducting order parameter) agrees with the results obtained in Sec. 7.6. In particular, the fact that the Goldstone theorem is satisfied in a neutral superconductor (the phase mode is gapless, see Sec. 7.6.2) is a consequence of the RPA being conserving.

9.1.4 *Limitations of the Baym–Kadanoff approach*

In this section, we discuss two examples where the Baym–Kadanoff formalism fails to provide a correct physical description. We then discuss a more general difficulty with Φ-derivable approximations: the multivaluedness of the Luttinger–Ward functional $\Phi[G]$.

9.1.4.1 *Superfluid Bose gas*

A key property of a superfluid Bose gas is the existence of a mode with linear dispersion showing up as a pole of the one-particle propagator: the Bogoliubov sound mode (Sec. 7.4). The existence of this gapless mode is a consequence of the Hugenholtz–Pines theorem relating the chemical potential to the $k = 0$ value of the self-energy (Chapter 13). In the 1PI formalism, whenever the effective action $\Gamma[\phi^*, \phi]$ is invariant in a global U(1) transformation, the Goldstone and in turn the Hugenholtz–Pines theorem is satisfied. As discussed in Sec. 9.1.2.1, there is however no guarantee that conservation laws are fulfilled, i.e. that the four-point correlation function satisfies the Ward identities associated with the U(1) invariance of the action.

By contrast, Φ-derivable theories automatically fulfill Ward identities. This ensures, for instance, that the density–density correlation function of the superfluid Bose gas is gapless (since this property is a consequence of conservation laws). But there is no guarantee that the one-particle spectrum is gapless as required by the Hugenholtz–Pines theorem. For instance, the Luttinger–Ward functional $\Phi[G, \phi]$ shown in Fig. 9.4 leads to a gapped one-particle spectrum even though the density fluctuation spectrum obtained from χ_{nn} is gapless. We refer to Ref. [7] for a general discussion of gapless *vs* conserving approximations for superfluid Bose systems.

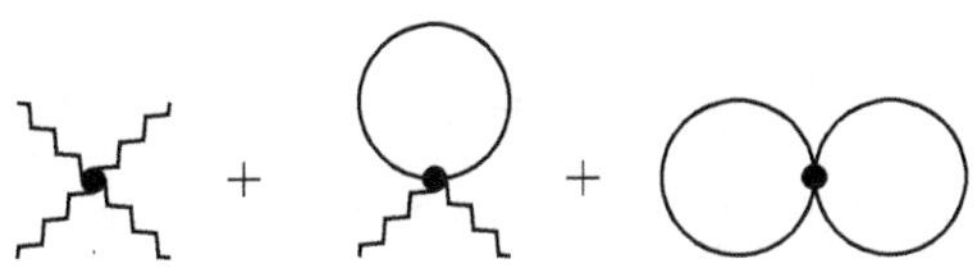

Fig. 9.4 Lowest-order Luttinger–Ward functional $\Phi[G, \phi]$ in a superfluid Bose system. The zigzag line stands for the order parameter $\phi = \langle\psi\rangle$ and the solid one for the connected one-particle propagator (see Sec. 1.6.3).

9.1.4.2 *Fluctuation-exchange approximation*

The ground state of the two-dimensional half-filled Hubbard model is antiferromagnetic (Sec. 6.4). Long-range order is suppressed by fluctuations at finite temperature in agreement with the Mermin–Wagner theorem. An important issue is to understand the behavior of the fermions at low temperatures. Is the system a Fermi liquid at any finite temperatures, in which case the spectral function $A(\mathbf{k}, \omega) = -\frac{1}{\pi}\Im[G^R(\mathbf{k}, \omega)]$ exhibits a well-defined peak at the energy $\xi_\mathbf{k}$ of the quasi-particle excitations (Sec. 4.4.1)? An alternative is that the spectral function shows no sign of quasi-particle excitations but two broadened peaks separated by a pseudogap. A simple argument shows that the latter scenario is the correct one. At zero temperature, we expect $A(\mathbf{k}, \omega)$ to exhibit two peaks corresponding to gapped quasi-particle excitations as in the Hartree–Fock theory (Sec. 6.4.1). These two peaks are separated by a gap due to the presence of antiferromagnetic long-range order. At any finite temperature, the long-range order disappears in two dimensions. However, by continuity, the two-peak structure in $A(\mathbf{k}, \omega)$ cannot disappear as soon as we raise the temperature. The only possible scenario is then that at finite but low temperatures, the fermion spectral function exhibits two broadened peaks, which are precursors of the zero-temperature quasi-particle excitations, separated by a pseudogap: $A(\mathbf{k}_F, \omega)$ then shows a minimum at $\omega = 0$ for any wavevector $\mathbf{k}_F$ on the Fermi surface in marked contrast with Fermi-liquid theory.[13]

The spectral function of the two-dimensional half-filled Hubbard model has been computed in the fluctuation-exchange approximation (FLEX), which is based on the Luttinger–Ward functional shown diagrammatically in Fig. 9.5. The ladder and bubble diagrams describe longitudinal and transverse spin fluctuations.[14] The corresponding self-energy diagrams, obtained by removing a propagator line, describe the interaction of the fermions with collective spin fluctuations. It turns out that FLEX fails to see the pseudogap. The reason is that a whole class of diagrams, including the so-called vertex corrections, is absent.

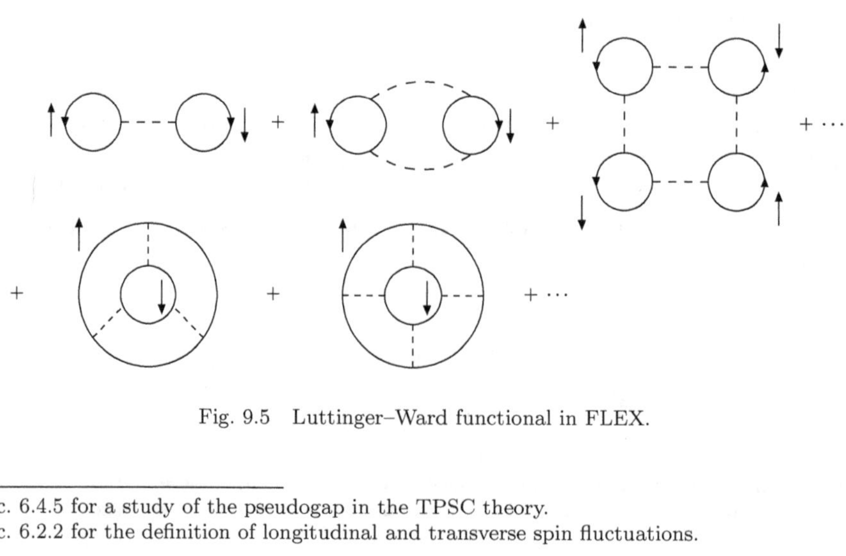

Fig. 9.5 Luttinger–Ward functional in FLEX.

[13] See Sec. 6.4.5 for a study of the pseudogap in the TPSC theory.

[14] See Sec. 6.2.2 for the definition of longitudinal and transverse spin fluctuations.

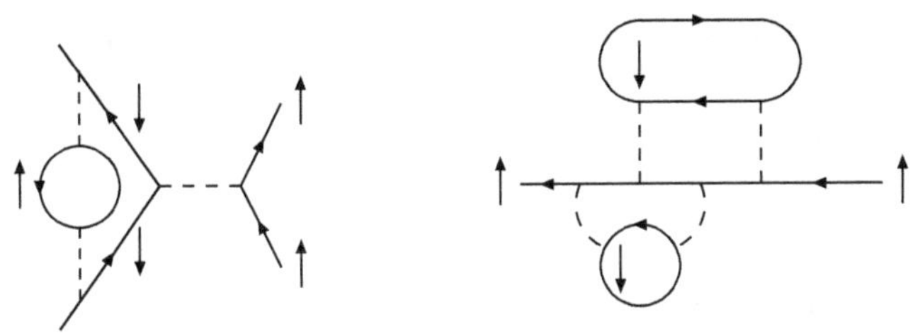

Fig. 9.6 (Left) An example of a vertex correction in the Hubbard model (i.e. a diagram that cannot be taken into account by a self-energy correction in a lower-order diagram). (Right) An example of a self-energy diagram with vertex correction that is not included in FLEX.

An example of such a diagram is shown in Fig. 9.6. A well-known case where vertex corrections are neglected is the Eliashberg (strong-coupling) theory of superconductivity: For the electron–phonon interaction, vertex corrections are controlled by the ratio m/M between the electron mass and the ion mass and can be safely neglected (Migdal's theorem). For fermion–fermion interactions, there is no Migdal's theorem and vertex corrections play a crucial role. In other words, self-consistency in the propagators (coming from the self-consistent equation $\Sigma = \delta\Phi/\delta G$) without self-consistency in vertices inhibits the existence of the pseudogap. We refer to Ref. [14] for a detailed discussion of the shortcomings of FLEX.

9.1.4.3 *Multivaluedness of the Luttinger–Ward functional*

In some cases, the Luttinger–Ward functional $\Phi[G]$ turns out to be a multivalued functional of G. To illustrate this point, we follow Ref. [15] and consider a zero spacetime dimensional toy model defined by the partition function

$$Z = \int \prod_{\sigma=\uparrow,\downarrow} d\psi_\sigma^* d\psi_\sigma \, e^{\mu_\uparrow \psi_\uparrow^* \psi_\uparrow + \mu_\downarrow \psi_\downarrow^* \psi_\downarrow - U \psi_\uparrow^* \psi_\uparrow \psi_\downarrow^* \psi_\downarrow}, \tag{9.57}$$

where ψ_σ and ψ_σ^* are Grassmann variables. Z is easily calculated by expanding the exponential,

$$Z = \mu_\uparrow \mu_\downarrow - U, \tag{9.58}$$

and the propagator reads

$$G_\sigma = -\langle \psi_\sigma \psi_\sigma^* \rangle = \frac{\partial \ln Z}{\partial \mu_\sigma} = \frac{\mu_{\bar\sigma}}{\mu_\uparrow \mu_\downarrow - U}, \tag{9.59}$$

where we use the notation $\bar\sigma = \downarrow (\uparrow)$ if $\sigma = \uparrow (\downarrow)$. For $U = 0$,

$$G_{0,\sigma} = \frac{1}{\mu_\sigma}. \tag{9.60}$$

The self-energy is then obtained from Dyson's equation,

$$\Sigma_\sigma = G_{0,\sigma}^{-1} - G_\sigma^{-1} = \frac{U}{\mu_{\bar\sigma}}. \tag{9.61}$$

From now on, we assume $\mu_\sigma > 0$ and $U < 0$ to avoid any singularity in the propagator as U varies.

Let us now express the self-energy as a function of $G = \{G_\uparrow, G_\downarrow\}$ instead of $\mu = \{\mu_\uparrow, \mu_\downarrow\}$ (with U fixed). To do so, we must write G_0 as a function of G. Eliminating μ_σ between (9.59) and (9.60) gives

$$G_\sigma = \frac{G_{0,\sigma}}{1 - UG_{0,\uparrow}G_{0,\downarrow}}. \tag{9.62}$$

For a given G, one finds two solutions

$$G_{0,\sigma}^{(\pm)}(G) = \frac{-1 \pm (1 + 4UG_\uparrow G_\downarrow)^{1/2}}{2UG_{\bar{\sigma}}}, \tag{9.63}$$

which are real provided we restrict ourselves to $1 + 4UG_\uparrow G_\downarrow \geq 0$. From (9.60), (9.61) and (9.63), we deduce

$$\Sigma_\sigma^{(\pm)}(G) = \frac{-1 \pm (1 + 4UG_\uparrow G_\downarrow)^{1/2}}{2G_\sigma}. \tag{9.64}$$

Thus, $\Sigma_\sigma(G)$ is a double-valued function of G. The exact self-energy (9.61) is obtained from $\Sigma_\sigma^{(s)}(G)$ if one takes $s = \operatorname{sgn}(\mu_\uparrow\mu_\downarrow + U) = \operatorname{sgn}(\mu_\uparrow\mu_\downarrow - |U|)$.

Since the self-energy $\Sigma_\sigma(G) = \partial\Phi(G)/\partial G_\sigma$ can be obtained as the derivative of the Luttinger–Ward functional $\Phi(G)$ (actually a function in our toy model), the latter must necessarily also be double-valued. $\Phi(G)$ is defined by the 2PI functional

$$\begin{aligned}\Gamma(G) &= -\ln Z(\mu) + \sum_\sigma \mu_\sigma G_\sigma \\ &= \ln(G_\uparrow G_\downarrow) + 2 + \Phi(G),\end{aligned} \tag{9.65}$$

where $\ln(G_\uparrow G_\downarrow)+2$ corresponds to the limit $U = 0$ and $Z(\mu)$ is given by (9.58). Here, μ must be considered as a function of G obtained by inverting $G_\sigma = \partial \ln Z(\mu)/\partial\mu_\sigma$. From (9.63) and $G_{0,\sigma} = 1/\mu_\sigma$, one obtains

$$\mu_\sigma^{(\pm)} = \frac{1 \pm (1 + 4UG_\uparrow G_\downarrow)^{1/2}}{2G_\sigma}, \tag{9.66}$$

which gives

$$\Phi^{(\pm)}(G) = \ln\left(\frac{-1 \pm (1 + 4UG_\uparrow G_\downarrow)^{1/2}}{2U}\right) \pm (1 + 4UG_\uparrow G_\downarrow)^{1/2} - \ln(G_\uparrow G_\downarrow) - 1. \tag{9.67}$$

The multivaluedness of $\Gamma(G)$ and $\Phi(G)$ comes from the fact that the map $\mu \to G$ or, equivalently, the map $G_0 \to G$, is two-to-one and therefore not invertible — this follows from the mapping $(\mu_\uparrow, \mu_\downarrow) \to \ln Z(\mu)$ being neither convex nor concave. One easily checks that

$$\frac{\partial\Phi^{(\pm)}(G)}{\partial G_\sigma} = \Sigma_\sigma^{(\pm)}(G). \tag{9.68}$$

The physical self-energy is obtained from $\Phi^{(+)}(G)$ when $\mu_\uparrow\mu_\downarrow + U > 0$ and $\Phi^{(-)}(G)$ when $\mu_\uparrow\mu_\downarrow + U < 0$.[15]

[15] We verify that $\Phi^{(+)}(G)$, which gives the physical self-energy for small $|U|$, vanishes for $U \to 0$.

The multivaluedness of $\Phi(G)$ has drastic consequences for the convergence of the diagrammatic series $\Sigma_{\text{bold}}(G)$ in terms of the fully dressed propagator G and the bare interaction U [15].[16] $\Sigma_{\text{bold}}(G(\mu))$ always converges to $\Sigma_\sigma^{(+)}(G(\mu))$ (with $G(\mu)$ the exact propagator) which is the physical self-energy for $|U| < \mu_\uparrow\mu_\downarrow$ but not for $|U| > \mu_\uparrow\mu_\downarrow$.[17]

It turns out the multivaluedness of the Luttinger–Ward functional is not only a curiosity of the zero spacetime dimensional toy model (9.57) but also a property of fermionic models with Hubbard on-site interactions. For the Hubbard model close to half-filling, it was found numerically that the map $G_0 \to G$ is two-to-one and therefore not invertible [16]. As a result, $\Sigma[G]$ is a double-valued functional of G as in the toy model considered above. Moreover, for U below a critical value U_c, the diagrammatic series $\Sigma_{\text{bold}}(G)$ converges to the physical solution but for $U > U_c$, it converges to the nonphysical solution. Given the "universality" of the Luttinger–Ward functional,[18] this conclusion applies to all models with Hubbard-like interaction. Thus, skeleton expansions generically break down at strong coupling without however being ill-behaved at weak coupling. On the other hand, the bare series in terms of G_0 (obtained from standard perturbation theory in U, see Chapter 1) always seems to converge to the physical solution [16].

9.1.5 *Thermodynamic self-consistency*

There is another property of Φ-derivable approximations which is worth mentioning. Some thermodynamic quantities can be obtained from both the thermodynamic potential $\Omega = -\frac{1}{\beta}\ln Z$ and the one-particle Green function. For a Φ-derivable approximation, the answer is unique. Let us consider, for instance, the total number of particles $N = -\partial\Omega/\partial\mu$ in a system with chemical potential μ in the grand canonical ensemble. From (1.447) and (1.449), we can write the thermodynamic potential as

$$\beta\Omega = -\zeta\operatorname{Tr}\ln(-G) + \zeta\operatorname{Tr}(\Sigma G) + \Phi[G], \tag{9.69}$$

where $G^{-1} = G_0^{-1} - \Sigma$ and

$$\Sigma(1,2) = -\zeta\frac{\delta\Phi[G]}{\delta G(2,1)}, \tag{9.70}$$

assuming for simplicity that the U(1) symmetry is not broken. We then obtain

$$\begin{aligned}\frac{\partial}{\partial\mu}\Phi[G] &= \int d1d2\,\frac{\delta\Phi[G]}{\delta G(1,2)}\frac{\partial G(1,2)}{\partial\mu}\\ &= -\zeta\int d1d2\,\Sigma(2,1)\frac{\partial G(1,2)}{\partial\mu} = -\zeta\operatorname{Tr}\left(\Sigma\frac{\partial G}{\partial\mu}\right)\end{aligned} \tag{9.71}$$

[16] $\Sigma_{\text{bold}}(G)$ includes only skeleton or bold-line diagrams, i.e. diagrams that do not contain self-energy insertions: See footnote 62 on page 96.

[17] $\Sigma_{\text{bold}}(G)$ is the $U \to 0$ Taylor series of $\Sigma^{(+)}(G)$. It cannot be the Taylor series of $\Sigma^{(-)}(G)$ since the latter does not vanish at $U = 0$.

[18] i.e. the fact that the functional $\Phi[G]$ is the same for all models sharing the same interaction Hamiltonian.

and

$$\beta\frac{\partial\Omega}{\partial\mu} = -\zeta\text{Tr}\left(G^{-1}\frac{\partial G}{\partial\mu}\right) + \zeta\text{Tr}\left(\frac{\partial\Sigma}{\partial\mu}G\right)$$
$$= \zeta\text{Tr}\left(\frac{\partial G^{-1}}{\partial\mu}G\right) + \zeta\text{Tr}\left(\frac{\partial\Sigma}{\partial_\mu}G\right)$$
$$= \zeta\text{Tr}\left(\frac{\partial G_0^{-1}}{\partial\mu}G\right)$$
$$= \zeta\text{Tr}(G) \tag{9.72}$$

since $\partial G_0^{-1}/\partial\mu$ is the unit matrix, so that finally

$$N = -\frac{\partial\Omega}{\partial\mu} = -\frac{\zeta}{\beta}\text{Tr}(G) \tag{9.73}$$

as it should be.

9.2 Mott Transition and Dynamical Mean-Field Theory

The Hubbard model has been discussed in Chapter 6 with an emphasis on its magnetic instabilities. In this section, we would like to discuss the Mott metal-to-insulator transition, a transition which is fully induced by the charge degrees of freedom and *a priori* has nothing to do with magnetic ordering (although in many cases, e.g. the Hubbard model on a hypercubic lattice, Mott localization and antiferromagnetic correlations are closely intertwined; see the following and in particular Sec. 9.3.3).

Early descriptions of the Mott transition started from the atomic limit at half-filling, which can be seen as an extreme case of a Mott insulator (Sec. 6.1.3).[19] In the presence of intersite single-electron hopping, the two atomic levels broaden into two bands (the upper and lower Hubbard bands) of half-width W and separated by a gap of order $U - 2W$. As U is reduced, there is a critical value U_c for which the two bands merge and the system becomes metallic. The Hubbard picture [46] of the Mott transition is thus associated with the closure of a gap. This description is essentially valid in the insulator but fails to provide us with a view of the metal consistent with Fermi-liquid theory.

In Sec. 9.2.1, we discuss an approach due to Brinkman and Rice, based on earlier work of Gutzwiller, which starts from the metallic phase. In this theory, the metal-to-insulator transition is driven by the disappearance of the quasi-particles (their spectral weight z_{k_F} vanishes) and the vanishing of the density of doubly occupied sites. In Sec. 9.2.2, we present the DMFT of the Mott transition, an approach which reconciles the Hubbard and Brinkman–Rice pictures.

[19] In this section and the following ones, we sometimes adopt the language of solid-state physics where the Hubbard model describes electrons in a solid although our discussion may also apply to fermionic atoms. The atomic limit then refers to the local ($t = 0$) limit of the model.

9.2.1 Brinkman–Rice theory of the Mott transition

The (non-normalized) Gutzwiller state is defined by

$$|\Psi\rangle = g^{\hat{D}}|\Psi_0\rangle, \tag{9.74}$$

where $|\Psi_0\rangle$ is a Slater determinant (Hartree–Fock state), $\hat{D} = \sum_{\mathbf{r}} \hat{D}_{\mathbf{r}} = \sum_{\mathbf{r}} \hat{n}_{\mathbf{r}\uparrow}\hat{n}_{\mathbf{r}\downarrow}$ counts the number of doubly occupied sites ($\hat{n}_{\mathbf{r}\sigma} = \hat{\psi}^\dagger_{\mathbf{r}\sigma}\hat{\psi}_{\mathbf{r}\sigma}$), and g is a variational parameter which must be determined by minimizing the energy

$$E = \frac{\langle\Psi|\hat{H}|\Psi\rangle}{\langle\Psi|\Psi\rangle}, \tag{9.75}$$

where $\hat{H}$ is the Hamiltonian of the Hubbard model [Eq. (6.1)]. Since $\hat{P}_{\mathbf{r}} = 1 - \hat{D}_{\mathbf{r}}$ is a projector, $\hat{P}^2_{\mathbf{r}} = \hat{P}_{\mathbf{r}}$, one has

$$g^{\hat{D}_{\mathbf{r}}} = g^{1-\hat{P}_{\mathbf{r}}} = g + (1-g)\hat{P}_{\mathbf{r}} = 1 - (1-g)\hat{D}_{\mathbf{r}} \tag{9.76}$$

and

$$|\Psi\rangle = \prod_{\mathbf{r}} [1 - (1-g)\hat{D}_{\mathbf{r}}]|\Psi_0\rangle. \tag{9.77}$$

For $g = 1$, one recovers the Hartree–Fock state $|\Psi_0\rangle$ whereas for $g = 0$, the ground state $|\Psi\rangle = \prod_{\mathbf{r}} \hat{P}_{\mathbf{r}}|\Psi_0\rangle$ is obtained from $|\Psi_0\rangle$ by projection onto the subspace without double occupation.

The computation of the variational energy E relies on the Gutzwiller approximation and is detailed in Appendix 9.A.[20] At half-filling (one electron per site on average) and for a non-magnetic solution

$$\epsilon = \frac{E}{L} = q_\sigma \epsilon_0 + Ud, \tag{9.78}$$

where

$$q_\sigma = 16d\left(\frac{1}{2} - d\right) \tag{9.79}$$

and $\epsilon_0 < 0$ is the average kinetic energy per electron in the non-correlated state $|\Psi_0\rangle$. Here, L is the number of sites in the lattice and the density of doubly occupied sites $d = \langle\hat{D}_{\mathbf{r}}\rangle/L$ is now used as the variational parameter.[21]

Minimizing ϵ with respect to d (with $d \geq 0$), one finds

$$d = \frac{8\epsilon_0 + U}{32\epsilon_0} \quad \text{and} \quad \epsilon_{\text{min}} = \epsilon_0\left(1 - \frac{U}{U_c}\right)^2 \tag{9.80}$$

---

[20] In Appendix 9.A, we also show that the Brinkman–Rice theory can be obtained from a mean-field theory using the Kotliar–Ruckenstein slave bosons.

[21] This is achieved by first expressing g as a function of d.

for $U \leq U_c = 8|\epsilon_0|$ while $d = 0$ and $\epsilon_{\rm min} = 0$ for $U \geq U_c$. Thus for $U < U_c$, the state $|\Psi\rangle$ describes a metallic phase with a non-zero kinetic energy (see Appendix 9.A),

$$\frac{\langle\Psi|\hat{\psi}^\dagger_\sigma(\mathbf{k})\hat{\psi}_\sigma(\mathbf{k})|\Psi\rangle}{\langle\Psi|\Psi\rangle} = q_\sigma\langle\hat{\psi}^\dagger_\sigma(\mathbf{k})\hat{\psi}_\sigma(\mathbf{k})\rangle_0, \tag{9.81}$$

where $\langle\cdots\rangle_0$ denotes the expectation value with respect to $|\Psi_0\rangle$. This equation implies that the momentum distribution calculated from $|\Psi\rangle$ has a discontinuity q_σ on the Fermi surface, which we identify with the quasi-particle weight z_{k_F} (Sec. 4.4.1). The density of doubly occupied sites is also non-zero and is equal to 1/4 for $U = 0$ as expected for non-interacting electrons. Thus, the Brinkman–Rice theory provides us with a picture of the low-energy properties of the metallic state which is consistent with Fermi-liquid theory (Chapter 4). However, it does not account for the high-energy excitations forming the Hubbard bands, which should be present already in the metallic state (see Sec. 9.2.2).

When $U \to U_c^-$, the system undergoes a "Brinkman–Rice" transition to a Mott insulator with a vanishing density of doubly occupied sites and a vanishing kinetic energy. This description is of course very crude since charge fluctuations should not be fully suppressed in the Mott insulator except when $U \to \infty$. Furthermore, the Brinkman–Rice insulator lacks all antiferromagnetic correlations between localized electrons which, for $U \gg U_c$, are described by the Heisenberg model (Sec. 6.4.4).

9.2.2 *Dynamical mean-field theory*

9.2.2.1 *Electrons in infinite dimensions*

Let us first discuss the simplifications which arise in the Hubbard model in the limit of infinite dimension. The non-interacting density of state (per site) reads

$$N_0(\epsilon) = \int_{\mathbf{k}} \delta(\epsilon - \epsilon_{\mathbf{k}}) = \int_{-\infty}^{\infty} \frac{ds}{2\pi} \int_{\mathbf{k}} e^{-is(\epsilon-\epsilon_{\mathbf{k}})}, \tag{9.82}$$

where $\epsilon_{\mathbf{k}} = -2t\sum_{i=1}^d \cos k_i$ is the dispersion of the free electrons for a d-dimensional hypercubic lattice. Summing over $\mathbf{k} \in [-\pi, \pi]^d$ gives

$$N_0(\epsilon) = \int_{-\infty}^{\infty} \frac{ds}{2\pi} e^{-is\epsilon} \prod_{i=1}^d \int_{-\pi}^{\pi} \frac{dk_i}{2\pi} e^{-i2st\cos k_i} = \int_{-\infty}^{\infty} \frac{ds}{2\pi} e^{-is\epsilon}[J_0(2st)]^d, \tag{9.83}$$

where J_0 is the zeroth-order Bessel function. To obtain a finite result when $d \to \infty$, we must write the hopping amplitude as[22]

$$t = \frac{t^*}{\sqrt{d}} \tag{9.84}$$

and consider t^* fixed. Using $J_0(x) = 1 - x^2/4 + \mathcal{O}(x^4)$, we find

$$\lim_{d\to\infty} N_0(\epsilon) = \int_{-\infty}^{\infty} \frac{ds}{2\pi} e^{-is\epsilon - s^2 t^{*2}} = \frac{1}{\sqrt{4\pi}t^*} e^{-\epsilon^2/4t^{*2}}. \tag{9.85}$$

[22]This ensures that the variance $\overline{\epsilon_{\mathbf{k}}^2} = \int_{-\infty}^{\infty} d\epsilon\, \epsilon^2 N_0(\epsilon)$ of the energy is finite in the limit $d \to \infty$.

Alternatively, Eq. (9.85) can be obtained by considering k_i as a random variable uniformly distributed in the interval $]-\pi,\pi]$. $X_i = \sqrt{2}\cos k_i$ is then a random variable with zero mean and unit variance. By the central-limit theorem, the random variable $X_d = d^{-1/2}\sum_{i=1}^{d} X_i$ converges for $d\to\infty$ to a normal distributed random variable with zero mean and unit variance, i.e. with a probability distribution function $f(x) = e^{-x^2/2}/\sqrt{2\pi}$. $N_0(\epsilon)$ can then be regarded as the distribution function of the random variable $-t\sqrt{2d}X_d$, i.e.

$$N_0(\epsilon) = \int_{\mathbf{k}} \delta(\epsilon + t\sqrt{2d}X_d) = \frac{1}{t\sqrt{2d}} f\left(-\frac{\epsilon}{\sqrt{2d}t}\right) = \frac{1}{t\sqrt{4\pi d}} e^{-\epsilon^2/4dt^2}, \qquad (9.86)$$

in agreement with (9.85).

Consider now the propagator $G(\mathbf{r},\mathbf{r}';i\omega_n)$ in the limit $d \gg 1$. Since $t = t^*/\sqrt{d} \to 0$ for $d\to\infty$, we can compute G by an expansion about the local ($t=0$) limit. It follows that

$$G(\mathbf{r},\mathbf{r}';i\omega_n) = \mathcal{O}(t^{\|\mathbf{r}-\mathbf{r}'\|}) = \mathcal{O}(d^{-\frac{1}{2}\|\mathbf{r}-\mathbf{r}'\|}), \qquad (9.87)$$

where $\|\mathbf{r}-\mathbf{r}'\|$, the Manhattan distance (New York metric) between sites $\mathbf{r}$ and $\mathbf{r}'$, is the minimum number of hoppings necessary to go from $\mathbf{r}$ to $\mathbf{r}'$.

Let us then consider a diagram contributing to the skeleton expansion of the self-energy.[23] For any two interaction vertices, located at sites $\mathbf{r}$ and $\mathbf{r}'$, there are at least three lines joining $\mathbf{r}$ and $\mathbf{r}'$. If there were only one, the diagram would not be 1PI (and therefore not an admissible self-energy diagram). One also easily convinces oneself that the diagram would not be 2PI if only two paths were joining the two sites. The Green functions associated with these three paths contribute at most to order $\mathcal{O}(d^{-\frac{3}{2}\|\mathbf{r}-\mathbf{r}'\|})$. To evaluate the diagram, one must first sum over all sites $\mathbf{r}'$ at the same Manhattan distance R of $\mathbf{r}$ (the distance R can be summed over afterwards): This yields a factor corresponding to the number of sites $\mathbf{r}'$ at the same distance R of $\mathbf{r}$, which is clearly at most of order $\mathcal{O}(d^R)$. Thus, the contribution to the diagram for a given distance R is suppressed by a factor $\mathcal{O}(d^{-\frac{3}{2}R}d^R) = \mathcal{O}(d^{-\frac{R}{2}})$, so that only diagrams with $R=0$ survive in the limit $d\to\infty$. The reasoning holds also when the vertex at site $\mathbf{r}$ or $\mathbf{r}'$ is one of the external vertices, i.e. a vertex attached to an external leg (there is then no sum over $\mathbf{r}$ or $\mathbf{r}'$ in that case). We therefore conclude that in the limit $d\to\infty$, a diagram in the skeleton expansion of the self-energy has all its vertices located at the same lattice site. This implies in particular that the self-energy

$$\Sigma_\sigma(\mathbf{r},\mathbf{r}';i\omega_n) = \delta_{\mathbf{r},\mathbf{r}'}\Sigma_\sigma(i\omega_n) \qquad (9.88)$$

is local.

9.2.2.2 *Dynamical mean-field theory*

The position-space collapse of diagrams [Eq. (9.88)] has an important consequence for the Luttinger–Ward functional $\Phi[G]$. All diagrams contributing to Φ being 2PI (or skeleton),

[23] Skeleton diagrams (aka bold-line or 2PI diagrams) do not contain self-energy insertions (Sec. 1.6.3).

they are purely local in the limit $d \to \infty$ and one can write

$$\Phi[G] = \sum_{\mathbf{r}} \Phi_{\text{imp}}[G(\mathbf{r},\mathbf{r})], \tag{9.89}$$

where Φ_{imp} is the Luttinger–Ward functional of a single-site model (or impurity model) with interaction $\hat{H}_{\text{int}} = U\hat{n}_\uparrow\hat{n}_\downarrow$.[24] The impurity model is defined by its action,

$$S_{\text{imp}}[\psi^*,\psi] = -\int_0^\beta d\tau d\tau' \sum_\sigma \psi^*_\sigma(\tau)\mathcal{G}^{-1}_{0,\sigma}(\tau-\tau')\psi_\sigma(\tau') + U\int_0^\beta d\tau\, n_\uparrow(\tau)n_\downarrow(\tau'). \tag{9.90}$$

To determine the "bare" propagator $\mathcal{G}_0$, let us consider the (local) self-energy of the infinite-dimensional Hubbard model (which is also the self-energy of the impurity model),

$$\Sigma_\sigma(i\omega_n) = \beta\frac{\delta\Phi[G]}{\delta G_\sigma(\mathbf{r},\mathbf{r};i\omega_n)} \equiv \beta\frac{\delta\Phi_{\text{imp}}[G]}{\delta G_\sigma(i\omega_n)}. \tag{9.91}$$

The Green function $G_{\text{loc},\sigma}(i\omega_n) \equiv G_\sigma(\mathbf{r},\mathbf{r};i\omega_n)$ which enters the self-energy diagrams in the Hubbard model is then

$$G_{\text{loc},\sigma}(i\omega_n) = \int_{\mathbf{k}} G_\sigma(\mathbf{k},i\omega_n) = \int_{\mathbf{k}} \frac{1}{i\omega_n+\mu-\epsilon_{\mathbf{k}}-\Sigma_\sigma(i\omega_n)}. \tag{9.92}$$

To ensure that the impurity model is equivalent to the Hubbard model, one must impose that G_{loc} coincides with the propagator $\mathcal{G}$ obtained from S_{imp}, i.e.

$$\mathcal{G}_\sigma(i\omega_n) = \frac{1}{\mathcal{G}^{-1}_{0,\sigma}(i\omega_n)-\Sigma_\sigma(i\omega_n)} \equiv G_{\text{loc},\sigma}(i\omega_n). \tag{9.93}$$

Equations (9.92) and (9.93) are the two self-consistent equations of DMFT.[25]

Single-impurity Anderson model. The action (9.90) can be seen as the action of the single-impurity Anderson model. The latter is defined by the Hamiltonian

$$\hat{H} = \epsilon_d\sum_\sigma \hat{d}^\dagger_\sigma\hat{d}_\sigma + U\hat{n}_{d\uparrow}\hat{n}_{d\downarrow} + \sum_{\mathbf{k},\sigma}[(\epsilon_{\mathbf{k}}-\mu)\hat{c}^\dagger_{\mathbf{k}\sigma}\hat{c}_{\mathbf{k}\sigma} + (V_{\mathbf{k}}\hat{d}^\dagger_\sigma\hat{c}_{\mathbf{k}\sigma} + \text{h.c.})] \tag{9.94}$$

and describes an impurity embedded in a metal. Here, ϵ_d is the energy of a localized electronic state (impurity) and U is the Coulomb energy when the impurity is occupied by two electrons with opposite spins. $V_{\mathbf{k}}$ is the hybridization matrix element with the conduction electron states of the metal. Writing the partition function as a functional integral and integrating out $c_{\mathbf{k}\sigma}(\tau)$ yields an effective action for the field $d_\sigma(\tau)$ that has the form (9.90) with

$$\mathcal{G}^{-1}_{0,\sigma}(i\omega_n) = i\omega_n - \epsilon_d - \sum_{\mathbf{k}}\frac{|V_{\mathbf{k}}|^2}{i\omega_n-\epsilon_{\mathbf{k}}+\mu}. \tag{9.95}$$

[24] This is a direct consequence of (9.88) and the fact that $\Phi[G]$ depends only on the interaction Hamiltonian (and not on the kinetic part of the Hamiltonian); see Sec. 1.6.3.

[25] There is an alternative derivation of the DMFT equations, known as the cavity method, which consists in considering a given site of the lattice and integrating out the degrees of freedom located at the other sites [26].

Thus, DMFT essentially replaces a (complicated) lattice model by a (simpler) effective single-site model where the electrons interact with a bath (defined by $\mathcal{G}_0$) which must be determined self-consistently. In this sense, DMFT is a mean-field theory which bears similarities with the mean-field theory of lattice spin models where a single spin interacts with an effective magnetic field which must be determined self-consistently (Chapter 10). In contrast to these mean-field theories however, DMFT is a "dynamical" theory to the extent where the "Weiss" field $\mathcal{G}_0$ is frequency dependent: Spatial fluctuations are frozen but local quantum fluctuations (i.e. temporal fluctuations between the different quantum states at a given site) are fully taken into account.

The main difficulty is now to find the self-energy Σ for a given propagator $\mathcal{G}_0$. In practice, Eqs. (9.92) and (9.93) are solved iteratively. Starting from an initial guess for $\mathcal{G}_0$, the impurity Green function $\mathcal{G}$ is calculated by using an appropriate "solver" for the impurity model. The self-energy is then obtained from $\Sigma = \mathcal{G}_0^{-1} - \mathcal{G}^{-1}$ and used to obtain the Green function $G_{\rm loc}$ of the lattice model [Eq. (9.92)]. An update function $\mathcal{G}_0$ is finally obtained from (9.93) and injected into the impurity solver. The procedure is iterated until convergence is reached.

It is readily seen that DMFT is exact in the non-interacting and single-site limits. For $U = 0$, one finds $\Sigma_\sigma(i\omega_n) = 0$. Equations (9.92) and (9.93) then give $\mathcal{G}_{0,\sigma}(i\omega_n) = G_{{\rm loc},\sigma}(i\omega_n) = \int_{\mathbf{k}}(i\omega_n + \mu - \epsilon_{\mathbf{k}})^{-1}$. For $t = 0$, $\epsilon_{\mathbf{k}} = 0$. Equation (9.92) then implies that $G_{\rm loc}$ is the Green function of the single-site Hubbard model, and Eq. (9.93) gives $\mathcal{G}_{0,\sigma}^{-1}(i\omega_n) = \mathcal{G}_\sigma^{-1}(i\omega_n) + \Sigma_\sigma(i\omega_n) = i\omega_n + \mu$. Thus, the impurity model reduces to the single-site Hubbard model as it should be.

In the limit of infinite dimensions, the density of states on the hypercubic lattice has an infinite bandwidth[26] for nearest-neighbor hopping and this may appear problematic. It is possible to construct a density of states with a finite bandwidth but this typically requires long-range hopping. In general, however, one simply uses DMFT as an approximate solution of a finite-dimensional (e.g. three-dimensional) model obtained by assuming that the problem can be mapped onto an effective impurity model [Eq. (9.89)]. In other words, we use the DMFT equations (9.92) and (9.93) directly for the finite-dimensional model.[27]

9.2.2.3 *Paramagnetic solution*

Let us now discuss the paramagnetic DMFT solution in the half-filled Hubbard model [26]. For small U, i.e. U much smaller than the effective (non-interacting) half-bandwidth W defined by $W^2 \sim \int d\epsilon\, \epsilon^2 N_0(\epsilon)$, one finds a self-energy typical of a Fermi liquid (Sec. 4.4.1),

$$\begin{aligned} \Re[\Sigma^R(\omega)] &= \Sigma^R(0) + \left(1 - \frac{1}{z}\right)\omega + \mathcal{O}(\omega^3), \\ \Im[\Sigma^R(\omega)] &= -B\omega^2 + \mathcal{O}(\omega^4), \end{aligned} \tag{9.96}$$

[26] i.e. $N(\omega) \neq 0$ for $-\infty < \omega < \infty$.

[27] Let us stress that within DMFT, the space dimension plays a role only *via* the non-interacting density of states $N_0(\epsilon)$; two systems with the same $N_0(\epsilon)$ will share the same physical properties.

where $\Sigma^R(0) = U/2$ is fixed by particle–hole symmetry at half-filling (where $\mu = U/2$). The interacting density of states[28] $N(\omega = 0)$ at zero energy is pinned at its non-interacting value as required by the Luttinger theorem for a momentum-independent self-energy [24]. When $U \ll W$, the quasi-particle weight (which is $\mathbf{k}$ independent for a local self-energy) $z \simeq 1$ and $N(\omega)$ is only weakly modified by interactions. For larger values of U, a narrow quasi-particle peak of width $\sim zW$ and spectral weight z is formed at the Fermi level $\omega = 0$. In addition to this low-energy quasi-particle peak, the one-particle spectrum exhibits high-energy excitations around $\omega = \pm U/2$ (lower and upper Hubbard bands) (Fig. 9.7). These excitations are associated with atomic-like transitions, corresponding to the addition and the removal of an electron on a site, which broaden into bands in the solid. The metallic solution disappears at U_{c2} when the quasi-particle weight z vanishes.

For large U, the solution of the DMFT equations describes an insulator with a vanishing density of states at low energies: $N(\omega)$ exhibits two bands separated by an energy gap Δ (Fig. 9.7). The self-energy has the low-energy behavior

$$\Sigma(i\omega_n) = \frac{U}{2} + \frac{\rho_2}{i\omega_n} + \mathcal{O}(\omega_n) \quad \text{with} \quad \frac{1}{\rho_2} = \int_{-\infty}^{\infty} d\epsilon \frac{N(\epsilon)}{\epsilon^2}, \tag{9.97}$$

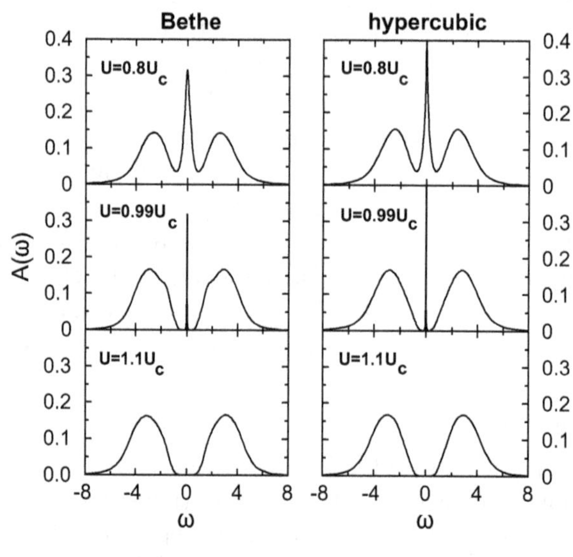

Fig. 9.7 Zero-temperature density of states $N(\omega)$ obtained from DMFT for Bethe and hypercubic lattices and various values of U ($t^* = 1/\sqrt{2}$). For the Bethe lattice with infinite coordination number, the non-interacting density of states is given by $N_0(\omega) = \Theta(2 - |\omega|)\sqrt{4 - \omega^2}/2\pi$ whereas for the infinite-dimensional hypercubic lattice, $N_0(\omega)$ is given by (9.85). In both cases, the narrow quasi-particle peak which develops at the Fermi level in the metallic phase (and coexists with the lower and upper Hubbard bands) vanishes when $U = U_c$. Reprinted from Ref. [47].

[28] In the presence of interactions, the single-particle density of states is defined by $N(\omega) = -\frac{1}{\pi}\int_{\mathbf{k}} \Im[G^R(\mathbf{k}, \omega)] = -\frac{1}{\pi}\Im[G^R_{\text{loc}}(\omega)]$.

which is similar to the result in the single-site limit ($t = 0$) (Sec. 6.1.3). The insulating solution exists for $U > U_{c1}$. Since $U_{c1} < U_{c2}$, both the metallic and insulating solutions exist for $U_{c1} \leq U \leq U_{c2}$. One finds that the metallic solution has always a smaller free energy. Thus, the metal–insulator transition is second order and takes place at $U = U_{c2}$ when the quasi-particle weight z vanishes. The spectral gap is already formed ($\Delta > 0$) at the transition.

At finite temperature, it is not possible to distinguish *stricto sensu* a metal from an insulator. Nevertheless at low (but finite) temperatures, there is a region of the phase diagram defined by $U_{c1}(T) \leq U \leq U_{c2}(T)$ where one finds two paramagnetic solutions, one of which being continuously connected to the $T = 0$ metallic solution while the other one is connected to the insulating solution. The metal–insulator transition is first order at finite temperature and takes place at $U_c(T)$ when the free energies of the two solutions cross. The first-order transition line $U_c(T)$ ends at a second-order critical point where the two lines $U_{c1}(T)$ and $U_{c2}(T)$ merge. At higher temperatures, there is a rapid crossover from a metallic-like solution to an insulating-like one (Fig. 9.8, left panel).

It is interesting to note that the slope of the transition line $U_c(T)$ is negative. This means that, at constant U, the metallic phase can be reached from the insulator by cooling the system. This anomalous behavior is due to the paramagnetic insulating solution having a macroscopic entropy at $T = 0$, $S_{\text{ins}} = k_B \ln 2$ (per electron), while the entropy S_{met} is of order T in a Fermi liquid. This is clearly an artifact due to the neglect of the antiferromagnetic exchange coupling $J \sim t^2/U$ between the electron spins in the insulator. This antiferromagnetic coupling between localized spins must lead to a vanishing entropy at $T = 0$ and one expects a positive difference $\Delta S = S_{\text{met}} - S_{\text{ins}}$ at low temperatures eventually leading to a positive slope dU_c/dT.

DMFT makes it clear that the Mott transition is due to the freezing of charge degrees of freedom and has nothing to do with antiferromagnetism. The cost in Coulomb energy (described by the repulsive on-site interaction in the Hubbard model) becomes too large in comparison with the gain in kinetic energy for the electrons to delocalize as in the metallic

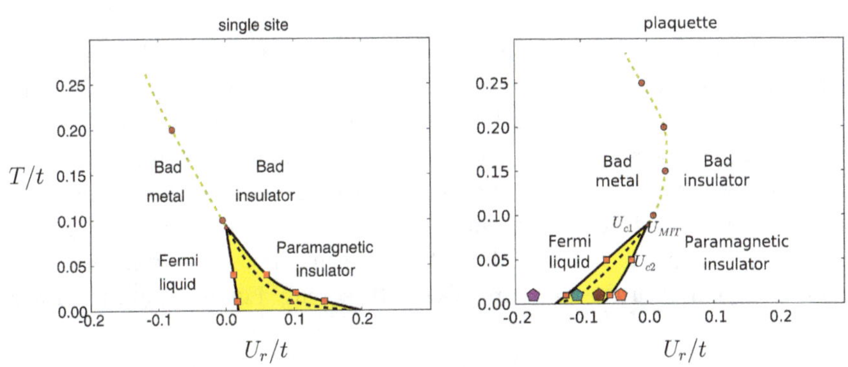

Fig. 9.8 Phase diagram of the two-dimensional Hubbard model at half-filling on the square lattice. (left) Single-site DMFT, (right) C-DMFT with $L_c = 4$. $U_r = (U - U_{\text{MIT}})/U_{\text{MIT}}$ with $U_{\text{MIT}} = 9.35t$ (DMFT) and $U_{\text{MIT}} = 6.05t$ (C-DMFT). Reprinted from Ref. [48].

phase. On the other hand, what is clearly lacking in the paramagnetic DMFT solution we have discussed so far is the magnetism associated with the electron spins. In the insulator, the electrons behave as local moments and nearest neighbors interact antiferromagnetically with an exchange coupling constant $J \sim t^2/U$ (in the limit $U \gg W$, the Hubbard model maps onto the Heisenberg model with $J = 4t^2/U$, see Sec. 6.4.4). Thus, the paramagnetic solution with no antiferromagnetic correlations is realized only at finite temperature when $T \gg J$: The system is then in a paramagnetic state with fluctuating local moments characterized by a Curie–Weiss susceptibility $\chi \sim 1/T$ and a large spin entropy. When the system is cooled below $T \sim J$, the entropy is suppressed and the local moments order antiferromagnetically in three dimensions (Mott–Heisenberg insulator);[29] this temperature regime is not properly described by DMFT.

In the small-U limit ($U \ll W$), magnetism also plays a crucial role since the nesting of the Fermi surface at half-filling leads to a spin-density-wave instability (Slater antiferromagnetism, see Sec. 6.4). In this description, magnetism is the cause of the gap opening; charge fluctuations are usually taken into account at the Hartree–Fock (mean-field) level and are therefore essentially ignored. In Sec. 6.4, we have shown that Slater's theory actually describes both the small- and large-U limit and reproduces the results obtained from the Heisenberg model when $U \gg W$. In the absence of magnetism, the system would remain metallic which is clearly wrong in the large-U limit at half-filling.

The single-site DMFT can easily be amended, by distinguishing between two sublattices A and B, to allow for a mean-field-like antiferromagnetic solution. Not surprisingly, one finds an antiferromagnetic phase at low temperatures, with a Néel temperature T_N which increases with U at small U and decreases as $J \sim 4t^2/U$ at large U. The finite-temperature metal–insulator transition occurring at $T_c(U)$ discussed above is completely hidden by the antiferromagnetic phase as shown in Fig. 9.9; the onset of antiferromagnetism at finite temperature makes the metal–insulator transition within the paramagnetic phase discussed earlier completely irrelevant. The situation is of course different in the presence of frustration.[29] In this case, the Néel temperature may be significantly suppressed so that the finite-temperature metal–insulator transition in the paramagnetic phase becomes relevant.

Although the two-sublattice version of DMFT allows one to describe phases with long-range antiferromagnetic order, short-range fluctuations are not taken into account. More generally, because of its local nature, DMFT remains a mean-field theory, as far as spatial fluctuations are concerned, and is unable to include short-range fluctuations as well as the feedback of these fluctuations on the one-electron Green function. This severely limits the applicability of DMFT to low-dimensional models. For example, DMFT predicts long-range antiferromagnetic order at finite temperature in the two-dimensional Hubbard model in contradiction with the Mermin–Wagner theorem. Furthermore, the restriction to

[29] Antiferromagnetic long-range order can be suppressed by frustration due, e.g., to next-nearest-neighbor hopping — antiferromagnetism is then favored by nearest-neighbor hopping but not by that between next-nearest neighbors — or a non-bipartite lattice.

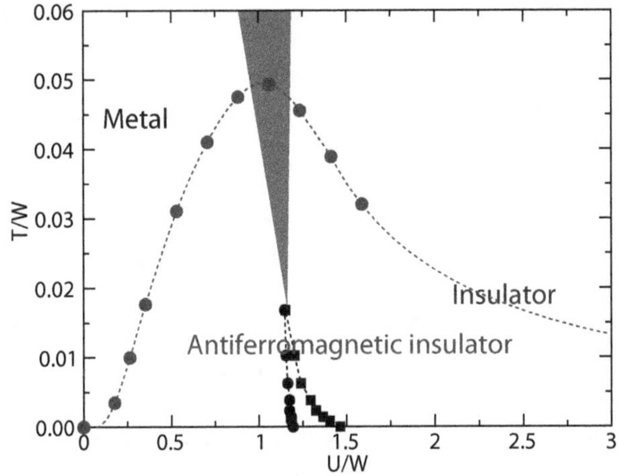

Fig. 9.9 DMFT phase diagram of the half-filled Hubbard model on a cubic lattice (with non-interacting bandwidth $W = 12t$). Also shown are the two transition lines $U_{c1}(T)$ and $U_{c2}(T)$ obtained when one artificially ignores the antiferromagnetic solution. Reprinted from Ref. [49].

a single-site impurity problem implies that DMFT cannot describe symmetry-broken phases with a non-local order parameter, such as a d-wave superconducting order parameter.[30]

A natural improvement of DMFT is to extend the self-consistent embedding of a single site to an embedding of a cluster consisting of several sites. Cluster extensions of DMFT include the dynamical cluster approximation (where the cluster extension is formulated in reciprocal space), the cellular dynamical mean-field theory (C-DMFT), the variational cluster approximation (VCA), etc. Many of these approaches can be formulated in the framework of the self-energy functional theory, which is the subject of Sec. 9.3.

9.3 Self-Energy Functional Theory

So far, we have considered approximation schemes where the self-energy equation $\Sigma = -\zeta\delta\Phi[G]/\delta G$ is solved exactly but with a non-exact Luttinger–Ward functional. In the perturbative approach, one simply truncates $\Phi[G]$ by retaining only a few low-order diagrams. In the DMFT, one takes into account only local diagrams, i.e. diagrams where all interaction vertices refer to the same lattice site.

In this section, we discuss an alternative approach based on a slightly different strategy. Rather than solving the equation $\Sigma = -\zeta\delta\Phi/\delta G = G_0^{-1} - G^{-1}$ starting from an approximated Luttinger–Ward functional, we make no assumption on $\Phi[G]$ but find the self-energy within a restricted space of self-energies taken from a reference system (typically a system consisting of disconnected clusters).

[30]The (possible) existence of d-wave superconductivity is a key issue in the two-dimensional Hubbard model (Chapter 15).

Let us start from the 2PI effective action (we restrict ourselves to fermion systems and assume that the U(1) symmetry is not broken)

$$\Gamma[G] = \mathrm{Tr}\ln(-G) - \mathrm{Tr}(G_0^{-1}G - 1) + \Phi[G] \tag{9.98}$$

and the ensuing definition of the self-energy $\Sigma[G]$,

$$\Sigma_{\sigma\sigma'}[\mathbf{r},\tau;\mathbf{r}',\tau';G] = \frac{\delta\Phi[G]}{\delta G_{\sigma'\sigma}(\mathbf{r}',\tau';\mathbf{r},\tau)}. \tag{9.99}$$

Assuming the functional $\Sigma[G]$ to be invertible, we can construct the Legendre transform

$$F[\Sigma] = \Phi[G] - \mathrm{Tr}(\Sigma G) \tag{9.100}$$

of the Luttinger–Ward functional. With (9.99), we immediately find

$$\frac{\delta F[\Sigma]}{\delta \Sigma_{\sigma\sigma'}(\mathbf{r},\tau;\mathbf{r}',\tau')} = -G_{\sigma'\sigma}[\mathbf{r}',\tau';\mathbf{r},\tau;\Sigma], \tag{9.101}$$

where $G[\Sigma]$ is here a functional of the self-energy. The self-energy functional[31] is defined by

$$\Gamma[\Sigma] = \mathrm{Tr}\ln\big(-(G_0^{-1} - \Sigma)^{-1}\big) + F[\Sigma] \tag{9.102}$$

and satisfies

$$\frac{\delta\Gamma[\Sigma]}{\delta \Sigma_{\sigma\sigma'}(\mathbf{r},\tau;\mathbf{r}',\tau')} = (G_0^{-1} - \Sigma)^{-1}_{\mathbf{r}',\tau',\sigma';\mathbf{r},\tau,\sigma} - G_{\sigma'\sigma}[\mathbf{r}',\tau';\mathbf{r},\tau;\Sigma] \tag{9.103}$$

[using (9.101)]. Thus, the physical self-energy, defined by $\Sigma = G_0^{-1} - G^{-1}$ (Dyson's equation), is given by the stationarity points of the self-energy functional $\Gamma[\Sigma]$.

We can write the Hamiltonian of the Hubbard model as

$$\hat{H}(\mathbf{t},U) = \hat{H}_0(\mathbf{t}) + \hat{H}_U, \tag{9.104}$$

where $\mathbf{t} \equiv (t,\mu)$ stands for the hopping amplitude and the chemical potential, and $\hat{H}_U$ is the interaction Hamiltonian. We know that $\Phi[G]$ is independent of $\mathbf{t}$; the same is true for $F[\Sigma]$.[32] Two systems with respective Hamiltonians $\hat{H}(\mathbf{t},U) = \hat{H}_0(\mathbf{t}) + \hat{H}_U$ and $\hat{H}(\mathbf{t}',U) = \hat{H}_0(\mathbf{t}') + \hat{H}_U$ will therefore share the same functional $F[\Sigma]$ and one has

$$\begin{aligned} \Gamma_{\mathbf{t}}[\Sigma] &= \mathrm{Tr}\ln\big(-(G_0^{-1} - \Sigma)^{-1}\big) + F[\Sigma], \\ \Gamma_{\mathbf{t}'}[\Sigma] &= \mathrm{Tr}\ln\big(-(G_0'^{-1} - \Sigma)^{-1}\big) + F[\Sigma], \end{aligned} \tag{9.105}$$

where $G_0 \equiv G_0(\mathbf{t})$ and $G_0' \equiv G_0(\mathbf{t}')$ are the non-interacting Green functions obtained from $\hat{H}_0(\mathbf{t})$ and $\hat{H}_0(\mathbf{t}')$, respectively. The system with Hamiltonian $\hat{H}(\mathbf{t}',U)$ will be referred to

[31] Equation (9.102) means that we can now consider the grand potential $\Omega[\Sigma] = \frac{1}{\beta}\Gamma[\Sigma]$ as a functional of the self-energy.

[32] This property is obvious in the perturbation expansion of $\Phi[G]$; more generally (and non-perturbatively), it follows from the definition of the 2PI functional $\Gamma[G]$ as a Legendre transform (Sec. 1.6.3).

as the reference system. The unknown functional $F[\Sigma]$ can be eliminated by taking the difference of the two preceding equations,

$$\Gamma_{\mathbf{t}}[\Sigma] = \Gamma_{\mathbf{t}'}[\Sigma] + \mathrm{Tr}\ln\big(-(G_0^{-1} - \Sigma)^{-1}\big) - \mathrm{Tr}\ln\big(-(G_0'^{-1} - \Sigma)^{-1}\big). \tag{9.106}$$

This equation is exact. To make the determination of the stationarity points of $\Gamma_{\mathbf{t}}[\Sigma]$ possible, one must restrict the space of possible self-energies. A natural approximation is to evaluate $\Gamma_{\mathbf{t}}[\Sigma]$ for the (physical) self-energies $\Sigma(\mathbf{t}')$ obtained from the reference system (assuming that the reference Hamiltonian $\hat{H}(\mathbf{t}', U)$ can be solved exactly, at least numerically). The functional $\Gamma_{\mathbf{t}}[\Sigma(\mathbf{t}')] \equiv \Gamma_{\mathbf{t}}(\mathbf{t}')$ then becomes a function of $\mathbf{t}'$,

$$\Gamma_{\mathbf{t}}(\mathbf{t}') = \beta\Omega(\mathbf{t}') + \mathrm{Tr}\ln\big(-(G_0^{-1} - \Sigma(\mathbf{t}'))^{-1}\big) - \mathrm{Tr}\ln\big(-(G_0'^{-1} - \Sigma(\mathbf{t}'))^{-1}\big), \tag{9.107}$$

where $\Omega(\mathbf{t}') = \frac{1}{\beta}\Gamma_{\mathbf{t}'}[\Sigma(\mathbf{t}')]$ is the exact grand potential of the reference system. The stationarity condition $\delta\Gamma_{\mathbf{t}}[\Sigma]/\delta\Sigma = 0$ is now simply

$$\frac{\partial}{\partial \mathbf{t}'}\Gamma_{\mathbf{t}}(\mathbf{t}') = 0, \tag{9.108}$$

i.e.

$$\begin{aligned} &\beta\frac{\partial}{\partial \mathbf{t}'}\Omega(\mathbf{t}') + \mathrm{Tr}\big(G_0^{-1} - \Sigma(\mathbf{t}')\big)^{-1}\frac{\partial\Sigma(\mathbf{t}')}{\partial \mathbf{t}'} \\ &- \mathrm{Tr}\big(G_0'^{-1} - \Sigma(\mathbf{t}')\big)^{-1}\left(\frac{\partial\Sigma(\mathbf{t}')}{\partial \mathbf{t}'} - \frac{\partial G_0'^{-1}}{\partial \mathbf{t}'}\right) = 0. \end{aligned} \tag{9.109}$$

Using the stationarity of $\Gamma_{\mathbf{t}'}[\Sigma]$ for $\Sigma = \Sigma(\mathbf{t}')$,[33]

$$\begin{aligned} \beta\partial_{\mathbf{t}'}\Omega(\mathbf{t}') &= \frac{\delta\Gamma_{\mathbf{t}'}[\Sigma]}{\delta\Sigma}\bigg|_{\Sigma=\Sigma(\mathbf{t}')} + \frac{\partial}{\partial \mathbf{t}'}\mathrm{Tr}\ln\big(-(G_0'^{-1} - \Sigma)^{-1}\big)\Big|_{\Sigma=\Sigma(\mathbf{t}')} \\ &= -\mathrm{Tr}\big(G_0'^{-1} - \Sigma(\mathbf{t}')\big)^{-1}\frac{\partial G_0'^{-1}}{\partial \mathbf{t}'}, \end{aligned} \tag{9.110}$$

we finally obtain

$$\mathrm{Tr}\left\{\Big[\big(G_0^{-1} - \Sigma(\mathbf{t}')\big)^{-1} - \big(G_0'^{-1} - \Sigma(\mathbf{t}')\big)^{-1}\Big]\frac{\partial\Sigma(\mathbf{t}')}{\partial \mathbf{t}'}\right\} = 0. \tag{9.111}$$

Equation (9.111) is the fundamental equation of the self-energy functional theory. When both the system and the reference systems are translation invariant, it can be rewritten as[34]

$$\frac{1}{\beta L}\sum_{\mathbf{k},\omega_n,\sigma}\left\{\big[G_{0,\sigma}^{-1}(\mathbf{k}, i\omega_n) - \Sigma_\sigma(\mathbf{k}, i\omega_n)\big]^{-1} - G'_\sigma(\mathbf{k}, i\omega_n)\right\}\frac{\partial\Sigma_\sigma(\mathbf{k}, i\omega_n)}{\partial \mathbf{t}'} = 0 \tag{9.112}$$

(assuming the self-energy to be diagonal in spin indices), where $\Sigma \equiv \Sigma(\mathbf{t}')$ and L is the number of sites.

---

[33] Here, we use the second of Eqs. (9.105) and the fact that $F[\Sigma]$ is independent of $\mathbf{t}'$.

[34] Equation (9.112) also holds when spin rotation is broken provided that the self-energy is diagonal in the spin indices.

9.3.1 *Variational cluster approximation*

A natural choice for the reference system is to take a set of independent clusters obtained from the original model by switching off the hopping tunneling amplitudes between some sites. Examples are given in Fig. 9.10. The hopping amplitude $\mathbf{t}'$ within each cluster should be considered as a variational parameter (and may not be uniform within the cluster). One can also consider the site energy $\epsilon'_{\mathbf{r}}$ (which vanishes in the Hubbard model) as variational parameters; equivalently, this amounts to taking the cluster chemical potential μ' as a variational parameter (provided that the site energy $\epsilon'_{\mathbf{r}} = \epsilon'_0$ is uniform) so that $\mathbf{t}' \equiv (t', \mu')$.[35] The self-consistency equation (9.111) becomes

$$\frac{1}{\beta} \sum_{\omega_n,\sigma,\mathbf{r},\mathbf{r}'}^{\text{(cluster)}} \left\{ \left[G_{0,\sigma}^{-1}(i\omega_n) - \Sigma_\sigma(i\omega_n) \right]^{-1} - G'_\sigma(i\omega_n) \right\}_{\mathbf{r},\mathbf{r}'} \frac{\partial \Sigma_\sigma(\mathbf{r}', \mathbf{r}; i\omega_n)}{\partial \mathbf{t}'} = 0 \tag{9.113}$$

(the sum is restricted to the sites $\mathbf{r}$ and $\mathbf{r}'$ of a single cluster), since the self-energy $\Sigma(\mathbf{t}')$ of the reference system has no inter-cluster components.

Let us consider the (approximate) Green function $G^{-1} = G_0^{-1} - \Sigma(\mathbf{t}')$ of the Hubbard model. If, instead of deducing the optimal value of $\mathbf{t}'$ from (9.113), one simply chooses $\mathbf{t}' = \mathbf{t}$, one obtains the cluster-perturbation theory (CPT) Green function. In particular for

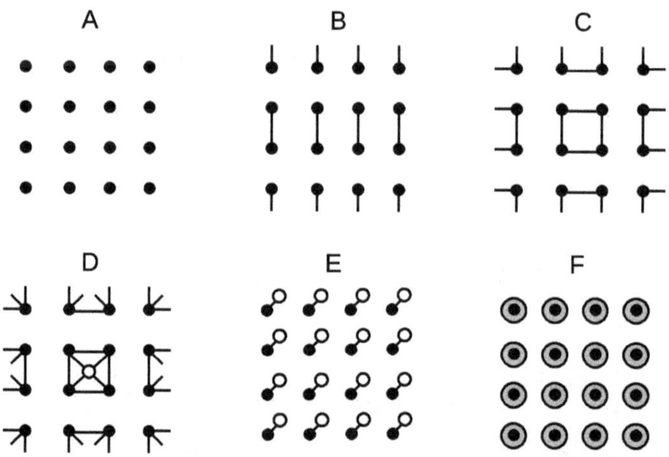

Fig. 9.10 Examples of reference systems used in the VCA. Filled circles correspond to correlated sites (with onsite interaction U) and a solid line joining two correlated sites stands for the hopping amplitude t'. $\hat{H}'_A$, $\hat{H}'_B$ and $\hat{H}'_C$ generate VCA with $L_c = 1, 2, 4$, respectively. In the lower figures, the open circles show uncorrelated "bath" sites with $U = 0$ (there are L_b bath sites attached to each correlated site) while the gray area indicates an infinite number of bath sites ($L_b = \infty$). $\hat{H}'_E$ yields the DIA with $L_b = 1$ and $\hat{H}'_F$ the DMFT. $\hat{H}'_D$ corresponds to an intermediate approximation with one additional bath site per cluster. Reprinted from Ref. [37].

[35]Note however that the correct physical interpretation of μ' is a non-zero site energy $\epsilon'_0 = \mu - \mu'$ since the reference system should be in the same macroscopic state as the original system, i.e. at the same temperature T and chemical potential μ.

a single-site cluster, one finds

$$G_\sigma^{-1}(\mathbf{k}, i\omega_n) = i\omega_n + \mu - \epsilon_{\mathbf{k}} - \Sigma_\sigma(i\omega_n) = G_{\mathrm{loc},\sigma}^{-1}(i\omega_n) - \epsilon_{\mathbf{k}}, \tag{9.114}$$

i.e.

$$G_\sigma(\mathbf{k}, i\omega_n) = \frac{G_{\mathrm{loc},\sigma}(i\omega_n)}{1 - \epsilon_{\mathbf{k}} G_{\mathrm{loc},\sigma}(i\omega_n)}, \tag{9.115}$$

where $\Sigma_\sigma(i\omega_n)$ and $G_{\mathrm{loc},\sigma}(i\omega_n)$ are the self-energy and Green function in the single-site Hubbard model (Sec. 6.1.3). Equation (9.115) is known as the Hubbard I approximation.[36]

When the cluster has more than one site, the CPT (as well as the VCA discussed in the following) runs into the following difficulty: The cluster self-energy Σ does not preserve the translation symmetry of the original lattice. One solution to this problem is to periodize the self-energy by replacing $\Sigma(\mathbf{k}, \mathbf{k}')$ by $\delta_{\mathbf{k},\mathbf{k}'}\Sigma(\mathbf{k}, \mathbf{k})$. We refer to Ref. [41] for a justification of this procedure in the framework of the self-energy functional theory and to the bibliography at the end of the chapter for other ideas.

The variational cluster approximation (VCA) improves the CPT by considering $\mathbf{t}' = (t', \mu')$ as a variational parameter. An interesting aspect of VCA is that one can add variational parameters in the cluster Hamiltonian that are not present in the original Hamiltonian, in particular parameters which lead to a lower symmetry. For example, the cluster Hamiltonian may include an additional magnetic-field term,

$$\Delta \hat{H}_0' = B' \sum_{\mathbf{r}} z_{\mathbf{r}} (\hat{n}_{\mathbf{r}\uparrow} - \hat{n}_{\mathbf{r}\downarrow}), \tag{9.116}$$

with $z_{\mathbf{r}} = 1$ for a ferromagnetic field — $z_{\mathbf{r}} = 1$ on one of the two sublattices and -1 on the other one for an antiferromagnetic (staggered) field. $\Delta\hat{H}_0'$ is a valid contribution to the Hamiltonian of the reference system since it does not change the interacting part of the Hamiltonian. Let us stress that there is no magnetic field in the Hamiltonian of the original system. The latter would correspond to a contribution $\Delta\hat{H}_0 = B\sum_{\mathbf{r}} z_{\mathbf{r}}(\hat{n}_{\mathbf{r}\uparrow} - \hat{n}_{\mathbf{r}\downarrow})$ to the Hamiltonian $\hat{H}_0$. The value of B' follows from the stationarity condition (9.111). Spontaneous symmetry breaking of the spin-rotation invariance is obtained if $B' \neq 0$ (while $B = 0$).

There are many applications of VCA, e.g. to antiferromagnetic spiral phases, ferromagnetism, and d-wave superconductivity. We refer to the bibliography at the end of the chapter (see also Sec. 9.3.3).

9.3.2 *Dynamical mean-field theory recovered*

When constructing the reference system, nothing prevents us to add degrees of freedom. Since the interacting part of $\hat{H}'$ must be the same as that of $\hat{H}$, the only possibility is to add new sites (or "orbitals") where the Hubbard U vanishes. These are called "bath sites". Figure 9.10 shows a few examples. In Fig. 9.10(e), the reference system consists of

[36]The Hubbard I approximation is discussed in detail in Chapter 14 in the framework of the Bose–Hubbard model where it describes the superfluid–Mott-insulator transition of lattice bosons.

disconnected single-site clusters with an additional bath site in each cluster. More generally, to each correlated site of the many-site clusters considered in the VCA, we can add L_b bath sites. Figure 9.10(e) corresponds to $L_c = L_b = 1$ and the reference system Hamiltonian reads

$$H' = \sum_{\mathbf{r}} \Big\{ -\mu' \sum_{\sigma} \hat{\psi}^\dagger_{\mathbf{r}\sigma} \hat{\psi}_{\mathbf{r}\sigma} + U \hat{n}_{\mathbf{r}\uparrow} \hat{n}_{\mathbf{r}\downarrow} + \sum_{\sigma} [\epsilon_b \hat{a}^\dagger_{\mathbf{r}\sigma} \hat{a}_{\mathbf{r}\sigma} + (V \hat{\psi}^\dagger_{\mathbf{r}\sigma} \hat{a}_{\mathbf{r}\sigma} + \text{h.c.})] \Big\}, \tag{9.117}$$

where the fermionic operator $\hat{a}_{\mathbf{r}\sigma}$ corresponds to the bath degrees of freedom. The variational parameters of $\hat{H}'_0$ are now $\mathbf{t}' = (\mu', \epsilon_b, V)$. The Hamiltonian (9.117) generates a local self-energy $\Sigma_\sigma(\mathbf{r}, \mathbf{r}'; i\omega_n) = \delta_{\mathbf{r},\mathbf{r}'} \Sigma_\sigma(i\omega_n)$ but compared to the single-site VCA, the variational parameter space is larger. The resulting approximation is called the dynamical impurity approximation (DIA) with $L_b = 1$.

Interestingly, the DIA allows one to recover DMFT. Starting from a single-site cluster with $L_b = 1$ and adding more and more bath sites, one ends up with a continuum of bath degrees of freedom ($L_b = \infty$) in each cluster. The corresponding Hamiltonian reads

$$H' = \sum_{\mathbf{r}} \Big\{ -\mu' \sum_{\sigma} \hat{\psi}^\dagger_{\mathbf{r}\sigma} \hat{\psi}_{\mathbf{r}\sigma} + U \hat{n}_{\mathbf{r}\uparrow} \hat{n}_{\mathbf{r}\downarrow} + \sum_{k=1}^{L_b} \sum_{\sigma} [\epsilon_k \hat{a}^\dagger_{\mathbf{r}k\sigma} \hat{a}_{\mathbf{r}k\sigma} + (V_k \hat{\psi}^\dagger_{\mathbf{r}\sigma} \hat{a}_{\mathbf{r}k\sigma} + \text{h.c.})] \Big\} \tag{9.118}$$

and corresponds to a set of disconnected single-impurity Anderson models (see page 632). Since the trial self-energy is local, Eq. (9.111) gives

$$\frac{1}{\beta} \sum_{\omega_n, \sigma, \mathbf{r}} \Big\{ \big[G^{-1}_{0,\sigma}(i\omega_n) - \Sigma_\sigma(i\omega_n) \big]^{-1} - G'_\sigma(i\omega_n) \Big\}_{\mathbf{r},\mathbf{r}} \frac{\partial \Sigma_\sigma(\mathbf{r}, \mathbf{r}; i\omega_n)}{\partial \mathbf{t}'} = 0. \tag{9.119}$$

This equation is solved if one can find variational parameters (ϵ_k, V_k, μ'), i.e. an impurity self-energy $\Sigma_\sigma(i\omega_n)$, such that the curly bracket in (9.119) vanishes for any ω_n. Explicitly, the condition reads

$$\big[G^{-1}_{0,\sigma}(i\omega_n) - \Sigma_\sigma(i\omega_n) \big]^{-1}_{\mathbf{r},\mathbf{r}} - G'_\sigma(i\omega_n) = 0, \tag{9.120}$$

i.e.

$$\int_{\mathbf{k}} \frac{1}{i\omega_n + \mu - \epsilon_{\mathbf{k}} - \Sigma_\sigma(i\omega_n)} - G'_\sigma(i\omega_n) = 0, \tag{9.121}$$

which is just the self-consistency equation of DMFT (with $G' \equiv \mathcal{G}$ the Green function of the impurity model) [Eq. (9.93)]. Note that for any finite number of degrees of freedom, it is impossible to satisfy the DMFT self-consistency condition since the impurity Green function G' has a finite number of poles when $L_b < \infty$ while the local lattice Green function $\big[G^{-1}_{0,\sigma}(i\omega_n) - \Sigma_\sigma(i\omega_n) \big]^{-1}_{\mathbf{r},\mathbf{r}}$ has obviously branch cuts. A solution to Eq. (9.119) can nevertheless always be found when $L_b < \infty$. In that case, one can see (9.119) as the projection of the equation $(G_0^{-1} - \Sigma)^{-1} - G' = 0$ onto the hypersurface $\mathcal{S} = \{\Sigma \,/\, \exists \mathbf{t}' : \Sigma = \Sigma(\mathbf{t}')\}$ obtained by taking the scalar product with vectors $\partial_{\mathbf{t}'} \Sigma(\mathbf{t}')$ tangential to the hypersurface.

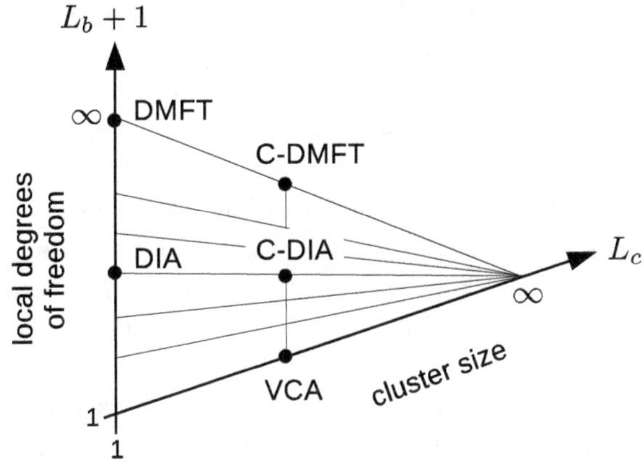

Fig. 9.11 Summary of the various approximations derived from the self-energy functional theory. L_c is the number of correlated sites in a cluster of the reference system and L_b is the number of bath sites attached to each correlated site. VCA is obtained for finite clusters with $L_c > 1$ but without bath sites ($L_b = 0$). DIA is obtained for $L_b \geq 1$ with a single-site cluster ($L_c = 1$); C-DIA corresponds to $L_b \geq 1$ and $L_c > 1$. An infinite number of bath sites, ($L_b = \infty$) generates DMFT ($L_c = 1$) and C-DMFT ($L_c > 1$). Any approximation with $L_c = \infty$, irrespective of the number of local degrees of freedom, gives the exact solution. Reprinted from Ref. [37].

A similar reasoning can be done with L_c-site clusters and L_b bath sites attached to each site in any cluster. When $L_b \to \infty$, Eq. (9.113) can be solved by imposing the condition

$$\left\{ \left[G_{0,\sigma}^{-1}(i\omega_n) - \Sigma_\sigma(i\omega_n) \right]^{-1} - G'_\sigma(i\omega_n) \right\}_{\mathbf{r},\mathbf{r}'} = 0, \tag{9.122}$$

where $\mathbf{r}$ and $\mathbf{r}'$ refer to the L_c sites of the cluster. Equation (9.122) is the self-consistency condition of C-DMFT [30, 31]. For a finite number of bath sites ($L_b < \infty$), one obtains the cellular dynamical impurity approximation (C-DIA).

The various approximations differ by the numbers of cluster sites (L_c) and bath sites (L_b) attached to each correlated site of the clusters as summarized in Fig. 9.11.

9.3.3 *Phase diagram of the half-filled two-dimensional Hubbard model*

To conclude this chapter, we further discuss the respective role of antiferromagnetic and charge fluctuations in the half-filled two-dimensional Hubbard model.

In Fig. 9.8, we compare the phase diagram obtained from single-site DMFT ($L_c = 1$ and $L_b = \infty$) and C-DMFT with four-site clusters ($L_c = 4$ and $L_b = \infty$). Recall that DMFT predicts the finite-temperature metal–insulator transition line $U_c(T)$ to have a negative slope dU_c/dT as a result of the macroscopic entropy of the $T = 0$ insulator due to the neglect of antiferromagnetic correlations between localized electrons (Fig. 9.8, left panel). C-DMFT cures this defect by taking into account antiferromagnetic fluctuations with range smaller than the cluster size. Thus C-DMFT still predicts a finite-temperature metal–insulator transition but now with a positive slope dU_c/dT due to the strong suppression of the entropy by antiferromagnetic fluctuations in the low-temperature insulating phase (Fig. 9.8, right panel). There is another important difference with single-site DMFT:

The finite-temperature transition line $U_c(T)$ remains first order down to zero temperature, i.e. the $T = 0$ transition is discontinuous rather than continuous as in DMFT.

But is C-DMFT sufficient to understand the physics of the half-filled Hubbard model? In the three-dimensional model, the finite-temperature transition between the metal and the paramagnetic insulator is an artifact of C-DMFT and is hidden by the antiferromagnetic transition once the possibility of antiferromagnetic long-range order is properly taken into account (see Fig. 9.9 and Sec. 9.2.2). What about the two-dimensional case where antiferromagnetic long-range order is forbidden by the Mermin–Wagner theorem? Is the concept of a paramagnetic Mott transition, driven by the blocking of charge degrees of freedom, relevant?

The two-dimensional Hubbard model has been studied by a variety of methods including the VCA, the dynamical vertex approximation (DΓA) — a diagrammatic extension of DMFT which includes short- and long-range correlations beyond DMFT on the same footing — and lattice quantum Monte Carlo simulations. While an extensive review is beyond the scope of our discussion, let us summarize some of the main results following Ref. [50].

VCA shows that short-range antiferromagnetic fluctuations reduce U_c and the width of the coexistence region significantly with respect to DMFT (in agreement with the C-DMFT results shown in Fig. 9.8). The smaller U is, the greater the effect of long-range antiferromagnetic fluctuations become. VCA is however limited by the finite size (L_c) of the clusters and to fully address the role of antiferromagnetic fluctuations, it is necessary to rely on other approaches. The results obtained from DΓA and lattice quantum Monte Carlo simulations are shown in Fig. 9.12 where the red dashed line marks the interaction $U_c(T)$

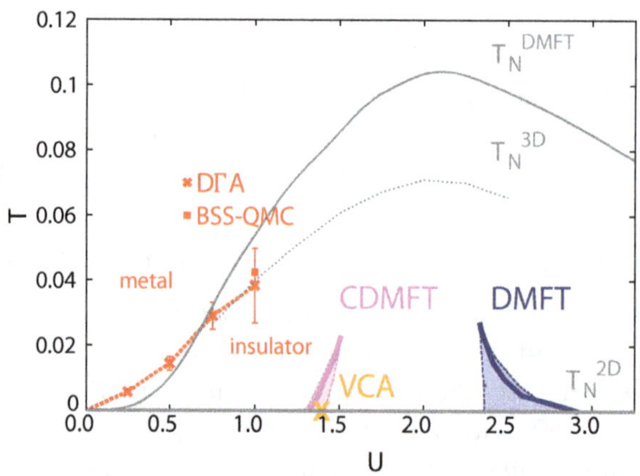

Fig. 9.12 Phase diagram of the half-filled Hubbard model on a square lattice obtained from various approaches. When short-range antiferromagnetic correlations are taken into account (*via* C-DMFT or VCA), the DMFT transition line $U_c(T)$ is shifted towards lower interaction values and the coexistence regions (hatched regions) shrink. The inclusion of long-range fluctuations leads to a vanishing $U_c(T)$ in the low-temperature regime (crosses and red dashed line: DΓA; solid red box: quantum Monte Carlo simulations). Also shown are the DMFT and DΓA three-dimensional Néel temperatures (light gray dotted lines) as well as the DΓA two-dimensional one (gray line at $T = 0$) which fulfills the Mermin–Wagner theorem. $4t = 1$ sets the energy scale. Reprinted from Ref. [50].

above which a spectral gap opens due to a strong enhancement of the electron scattering rate at low frequencies. These results suggest that at sufficiently low temperatures, antiferromagnetic fluctuations always open a spectral gap even at arbitrarily small U: There is no metal-to-insulator at finite temperature (while the ground state exhibits antiferromagnetic long-range order for any non-zero U). This means that the "idealized" DMFT picture of the Mott transition, driven by the blocking of charge degrees of freedom, is completely overturned in the two-dimensional half-filled Hubbard model by strong, spatially extended antiferromagnetic fluctuations.

Furthermore, DΓA shows that the spin–spin correlation function indicates the presence of strong antiferromagnetic fluctuations at low temperatures and the concomitant enhancement of the electron scattering rate, which leads to a destruction of the Fermi surface (spectral gap). At weak coupling, the destruction of the metallic state upon decreasing T is accompanied by a slight reduction of the potential energy $U\langle \hat{n}_{\mathbf{r}\uparrow}\hat{n}_{\mathbf{r}\downarrow}\rangle$ (of order 1% for $U = 4t$ and $T/t \simeq 0.1 - 0.4$ while the antiferromagnetic correlation length $\xi \gg 100$ is much larger than the lattice spacing). This rules out any particular role of prelocalization by the blocking of charge degrees of freedom. In fact, the slight decrease of the potential energy and the large correlation length is a hallmark of the Slater-like nature of the antiferromagnetic fluctuations. The overall picture that emerges is thus consistent with the low-T NLσM approach[37] (Sec. 6.4.3) and the weak-coupling TPSC theory (Sec. 6.4.5).

Appendix 9.A Brinkman–Rice Theory: Gutzwiller Approximation

In this appendix, we detail the calculation of the variational energy (9.75). We consider a L-site lattice and denote by N_σ the number of spin-σ electrons. A complete basis is given by the states

$$|\mathcal{N}_\uparrow, \mathcal{N}_\downarrow\rangle = \prod_{\mathbf{r}\in\mathcal{N}_\uparrow, \mathbf{r}'\in\mathcal{N}_\downarrow} \hat{\psi}^\dagger_{\mathbf{r}\uparrow}\hat{\psi}^\dagger_{\mathbf{r}'\downarrow}|\text{vac}\rangle, \tag{9.A.1}$$

where $\mathcal{N}_\sigma$ is a subset of lattice sites and $|\text{vac}\rangle$ denotes the vacuum state. We can write

$$|\Psi_0\rangle = \sum_{\mathcal{N}_\uparrow, \mathcal{N}_\downarrow} C_{\mathcal{N}_\uparrow, \mathcal{N}_\downarrow} |\mathcal{N}_\uparrow, \mathcal{N}_\downarrow\rangle, \tag{9.A.2}$$

where the sum is over all sets $\mathcal{N}_\sigma$ containing $|\mathcal{N}_\sigma| = N_\sigma$ sites since $|\Psi_0\rangle$ corresponds to N_σ electrons of spin σ.

We are now in a position to determine the ground state energy (9.75). One easily finds

$$\langle\Psi|\Psi\rangle = \langle\Psi_0|g^{2\hat{D}}|\Psi_0\rangle = \sum_{\mathcal{N}} g^{2D(\mathcal{N})}|C_{\mathcal{N}}|^2, \tag{9.A.3}$$

[37] The NLσM approach predicts a finite-temperature metal–insulator transition between a pseudogap phase and a Mott–Hubbard insulator as the strength of the Coulomb interaction increases: At a critical value $U_c(T)$ (with $U_c(T \to 0) \simeq 4.25t$), the density of states at zero energy $N(\omega = 0)$ vanishes and the pseudogap becomes a true gap [53, 54]. Note however that for $T \to 0$ and $U \ll U_c(T)$, the density of states $N(\omega = 0)$ is exponentially suppressed so that the physical picture is not much different from that advocated in [50].

where we use the notation $\mathcal{N} = \{\mathcal{N}_\uparrow, \mathcal{N}_\downarrow\}$ and the sum is restricted to sets such $|\mathcal{N}_\sigma| = N_\sigma$. $D(\mathcal{N})$ denotes the number of doubly occupied sites in the state $|\mathcal{N}\rangle = |\mathcal{N}_\uparrow, \mathcal{N}_\downarrow\rangle$. Similarly, we have

$$\begin{aligned}\langle\Psi|\hat{D}_{\mathbf{r}}|\Psi\rangle &= \sum_{\mathcal{N},\mathcal{N}'} C^*_{\mathcal{N}'} C_{\mathcal{N}} \langle\mathcal{N}'|g^{\hat{D}} \hat{D}_{\mathbf{r}} g^{\hat{D}}|\mathcal{N}\rangle \\ &= \sum_{\mathcal{N}} |C_{\mathcal{N}}|^2 g^{2D(\mathcal{N})} \langle\mathcal{N}|\hat{D}_{\mathbf{r}}|\mathcal{N}\rangle \\ &= \sum_{\substack{\mathcal{N} \\ (\mathbf{r}\in\mathcal{N}_\uparrow\cap\mathcal{N}_\downarrow)}} |C_{\mathcal{N}}|^2 g^{2D(\mathcal{N})}. \end{aligned} \tag{9.A.4}$$

Finally, for the expectation value of the hopping term, we find

$$\begin{aligned}\langle\Psi|\hat{\psi}^\dagger_{\mathbf{r}\uparrow}\hat{\psi}_{\mathbf{r}'\uparrow}|\Psi\rangle &= \sum_{\mathcal{N},\mathcal{N}'} C^*_{\mathcal{N}'} C_{\mathcal{N}} \langle\mathcal{N}'|g^{\hat{D}} \psi^\dagger_{\mathbf{r}\uparrow}\hat{\psi}_{\mathbf{r}'\uparrow} g^{\hat{D}}|\mathcal{N}\rangle \\ &= \sum_{\mathcal{N},\mathcal{N}',\mathcal{N}''} C^*_{\mathcal{N}'} C_{\mathcal{N}} g^{D(\mathcal{N})+D(\mathcal{N}')} \langle\mathcal{N}'|\hat{\psi}^\dagger_{\mathbf{r}\uparrow}|\mathcal{N}''\rangle\langle\mathcal{N}''|\hat{\psi}_{\mathbf{r}'\uparrow}|\mathcal{N}\rangle. \end{aligned} \tag{9.A.5}$$

The set $\mathcal{N}''$ must satisfy $\mathcal{N}'_\uparrow = \mathcal{N}''_\uparrow \cup \mathbf{r}$, $\mathcal{N}_\uparrow = \mathcal{N}''_\uparrow \cup \mathbf{r}'$ and $\mathcal{N}''_\downarrow = \mathcal{N}'_\downarrow = \mathcal{N}_\downarrow$, so that (9.A.5) can be rewritten as

$$\langle\Psi|\hat{\psi}^\dagger_{\mathbf{r}\uparrow}\hat{\psi}_{\mathbf{r}'\uparrow}|\Psi\rangle = \sum_{\substack{\mathcal{N}_\downarrow,\mathcal{N}''_\uparrow \\ (\mathbf{r},\mathbf{r}'\notin\mathcal{N}''_\uparrow)}} C^*_{\mathcal{N}''_\uparrow\cup\mathbf{r},\mathcal{N}_\downarrow} C_{\mathcal{N}''_\uparrow\cup\mathbf{r}',\mathcal{N}_\downarrow} g^{D(\mathcal{N}''_\uparrow\cup\mathbf{r},\mathcal{N}_\downarrow)+D(\mathcal{N}''_\uparrow\cup\mathbf{r}',\mathcal{N}_\downarrow)}, \tag{9.A.6}$$

where $|\mathcal{N}''_\uparrow| = N_\uparrow - 1$.

To compute the expectation values (9.A.3), (9.A.4) and (9.A.6), we must make further approximations. In the Gutzwiller approximation, one assumes that the coefficients $|C_{\mathcal{N}}|^2$ in (9.A.3) are independent of the special choice of $\mathcal{N}$. This means that all local distributions of electrons have the same probability in the uncorrelated state $|\Psi_0\rangle$. The value of $|C_{\mathcal{N}}|^2$ is simply deduced from the normalization condition[38]

$$\langle\Psi_0|\Psi_0\rangle = \sum_{\mathcal{N}} |C_{\mathcal{N}}|^2 = 1 \quad \text{i.e.} \quad |C_{\mathcal{N}}|^2 = \frac{1}{C_L^{N_\uparrow} C_L^{N_\downarrow}} \equiv P(L, N_\uparrow)P(L, N_\downarrow), \tag{9.A.7}$$

which gives

$$\begin{aligned}\langle\Psi|\Psi\rangle &= P(L, N_\uparrow)P(L, N_\downarrow) \sum_{\mathcal{N}} g^{2D(\mathcal{N})} \\ &= P(L, N_\uparrow)P(L, N_\downarrow) \sum_{D} g^{2D} G_D(L, N_\uparrow, N_\downarrow), \end{aligned} \tag{9.A.8}$$

where $G_D(L, N_\uparrow, N_\downarrow)$ is the number of configurations with D doubly occupied sites with the constraint that there are N_σ spin-σ electrons. A given configuration is defined by D,

[38] We use the notation $C_n^p = n!/p!(n-p)!$.

L_σ (number of singly occupied sites with spin-σ electrons) and E (number of empty sites). One obviously has $L = D + L_\uparrow + L_\downarrow + E$, $N_\sigma = L_\sigma + D$ and

$$G_D(L, N_\uparrow, N_\downarrow) = \frac{L!}{D!L_\uparrow!L_\downarrow!E!} = \frac{L!}{D!(N_\uparrow - D)!(N_\downarrow - D)!(L - N_\uparrow - N_\downarrow + D)!}. \tag{9.A.9}$$

Similarly, from (9.A.4), one obtains

$$\begin{aligned}\langle\Psi|\hat{D}_{\mathbf{r}}|\Psi\rangle &= P(L, N_\uparrow)P(L, N_\downarrow) \sum_{\substack{\mathcal{N} \\ (\mathbf{r}\in\mathcal{N}_\uparrow\cap\mathcal{N}_\downarrow)}} g^{2D(\mathcal{N})} \\ &= P(L, N_\uparrow)P(L, N_\downarrow) \sum_{\substack{\mathcal{N}' \\ (\mathbf{r}\notin\mathcal{N}'_\uparrow\cup\mathcal{N}'_\downarrow)}} g^{2D(\mathcal{N}'_\uparrow\cup\mathbf{r},\mathcal{N}'_\downarrow\cup\mathbf{r})} \\ &= P(L, N_\uparrow)P(L, N_\downarrow) \sum_D g^{2D+2} G_D(L-1, N_\uparrow - 1, N_\downarrow - 1) \\ &= P(L, N_\uparrow)P(L, N_\downarrow) \sum_D g^{2D+2} \frac{(N_\uparrow - D)(N_\downarrow - D)}{L(L - N_\uparrow - N_\downarrow + D + 1)} G_D(L, N_\uparrow, N_\downarrow) \\ &\simeq P(L, N_\uparrow)P(L, N_\downarrow) \sum_D g^{2D+2} \frac{(N_\uparrow - D)(N_\downarrow - D)}{L(L - N_\uparrow - N_\downarrow + D)} G_D(L, N_\uparrow, N_\downarrow).\end{aligned} \tag{9.A.10}$$

To calculate the expectation value of the hopping term, one assumes that the coefficients C and C^* in (9.A.6) are independent of $\mathcal{N}$ and $\mathcal{N}'$. Their values are determined by considering the non-interacting case,

$$\begin{aligned}\langle\hat{\psi}^\dagger_{\mathbf{r}\uparrow}\hat{\psi}_{\mathbf{r}'\uparrow}\rangle_0 \equiv \langle\Psi_0|\hat{\psi}^\dagger_{\mathbf{r}\uparrow}\hat{\psi}_{\mathbf{r}'\uparrow}|\Psi_0\rangle &= \sum_{\substack{\mathcal{N}_\downarrow,\mathcal{N}''_\uparrow \\ (\mathbf{r},\mathbf{r}'\notin\mathcal{N}''_\uparrow)}} C^*_{\mathcal{N}''_\uparrow\cup\mathbf{r},\mathcal{N}_\downarrow} C_{\mathcal{N}''_\uparrow\cup\mathbf{r}',\mathcal{N}_\downarrow} \\ &= [P(L-2, N_\uparrow - 1)P(L, N_\downarrow)]^{-1} C^*_{\mathcal{N}''_\uparrow\cup\mathbf{r},\mathcal{N}_\downarrow} C_{\mathcal{N}''_\uparrow\cup\mathbf{r}',\mathcal{N}_\downarrow},\end{aligned} \tag{9.A.11}$$

so that

$$\begin{aligned}\langle\Psi|\hat{\psi}^\dagger_{\mathbf{r}\uparrow}\hat{\psi}_{\mathbf{r}'\uparrow}|\Psi\rangle &= \langle\hat{\psi}^\dagger_{\mathbf{r}\uparrow}\hat{\psi}_{\mathbf{r}'\uparrow}\rangle_0 P(L-2, N_\uparrow - 1)P(L, N_\downarrow) \\ &\quad\times \sum_{\substack{\mathcal{N}_\downarrow,\mathcal{N}''_\uparrow \\ (\mathbf{r},\mathbf{r}'\notin\mathcal{N}''_\uparrow)}} g^{D(\mathcal{N}''_\uparrow\cup\mathbf{r},\mathcal{N}_\downarrow)+D(\mathcal{N}''_\uparrow\cup\mathbf{r}',\mathcal{N}_\downarrow)}.\end{aligned} \tag{9.A.12}$$

Since the sites $\mathbf{r}$ and $\mathbf{r}'$ can be either occupied or not by a spin-down electron, we must distinguish four cases ($\mathbf{r},\mathbf{r}' \notin \mathcal{N}_\downarrow$, $\mathbf{r} \in \mathcal{N}_\downarrow$ and $\mathbf{r}' \notin \mathcal{N}_\downarrow$, $\mathbf{r} \notin \mathcal{N}_\downarrow$ and $\mathbf{r}' \in \mathcal{N}_\downarrow$, $\mathbf{r},\mathbf{r}' \in \mathcal{N}_\downarrow$), which gives

$$\langle\Psi|\hat{\psi}^\dagger_{\mathbf{r}\uparrow}\hat{\psi}_{\mathbf{r}'\uparrow}|\Psi\rangle = \langle\hat{\psi}^\dagger_{\mathbf{r}\uparrow}\hat{\psi}_{\mathbf{r}'\uparrow}\rangle_0 P(L-2,N_\uparrow-1)P(L,N_\downarrow)$$
$$\times\sum_D\Big[g^{2D}G_D(L-2,N_\uparrow-1,N_\downarrow)+2g^{2D+1}G_D(L-2,N_\uparrow-1,N_\downarrow-1)$$
$$+\,g^{2D+2}G_D(L-2,N_\uparrow-1,N_\downarrow-2)\Big]. \tag{9.A.13}$$

This equation can be rewritten as

$$\langle\Psi|\hat{\psi}^\dagger_{\mathbf{r}\uparrow}\hat{\psi}_{\mathbf{r}'\uparrow}|\Psi\rangle = \langle\hat{\psi}^\dagger_{\mathbf{r}\uparrow}\hat{\psi}_{\mathbf{r}'\uparrow}\rangle_0 P(L,N_\uparrow)P(L,N_\downarrow)\frac{L^2}{N_\uparrow(L-N_\uparrow)}$$
$$\times\sum_D g^{2D}G_D(L,N_\uparrow,N_\downarrow)\left[\frac{(N_\uparrow-D)(L-N_\uparrow-N_\downarrow+D)}{L^2}+2g\frac{(N_\uparrow-D)(N_\downarrow-D)}{L^2}\right.$$
$$\left.+\,g^2\frac{(N_\uparrow-D)(N_\downarrow-D)^2}{L^2(L-N_\uparrow-N_\downarrow+D)}\right], \tag{9.A.14}$$

where, again, we have neglected numbers of order unity in the quotients.

Consider now the norm $\langle\Psi|\Psi\rangle$ given by (9.A.8) and (9.A.9),

$$\langle\Psi|\Psi\rangle = P(L,N_\uparrow)P(L,N_\downarrow)\sum_D g^{2D}\frac{L!}{D!(N_\uparrow-D)!(N_\downarrow-D)!(L-N_\uparrow-N_\downarrow+D)!}. \tag{9.A.15}$$

In the thermodynamic limit, the summand in this equation is a sharply peaked Gaussian function of the intensive variable $d = D/L$. We can therefore replace the sum by its dominant term multiplied by a factor $W_D(L,N_\uparrow,N_\downarrow)$, which comes from the Gaussian weight in D,

$$\langle\Psi|\Psi\rangle \simeq W_D(L,N_\uparrow,N_\downarrow)P(L,N_\uparrow)P(L,N_\downarrow)$$
$$\times\frac{g^{2D}L!}{D!(N_\uparrow-D)!(N_\downarrow-D)!(L-N_\uparrow-N_\downarrow+D)!}, \tag{9.A.16}$$

where the rhs is evaluated at the stationary value D determined by

$$\frac{\partial}{\partial_D}\frac{g^{2D}}{D!(N_\uparrow-D)!(N_\downarrow-D)!(L-N_\uparrow-N_\downarrow+D)!}=0. \tag{9.A.17}$$

Using Stirling's formula, $n! \simeq n^n$, one finds

$$g^2 = d\frac{1-n_\uparrow-n_\downarrow+d}{(n_\uparrow-d)(n_\downarrow-d)}, \tag{9.A.18}$$

where $d = D/L$, $n_\sigma = N_\sigma/L$ are the densities of doubly occupied sites and spin-σ electrons, respectively. The same reasoning holds for $\langle\Psi|\hat{D}_{\mathbf{r}}|\Psi\rangle$ [Eq. (9.A.10)]. In the thermodynamic limit, the sum over D is dominated by a single term where the value of D is obtained by minimizing $g^{2D}G_D(L-1, N_\uparrow - 1, N_\downarrow - 1)$, which again gives (9.A.18). Thus, we find

$$\langle\Psi|\hat{D}_{\mathbf{r}}|\Psi\rangle = \langle\Psi|\Psi\rangle g^2 \frac{(n_\uparrow - d)(n_\downarrow - d)}{1 - n_\uparrow - n_\downarrow + d} = \langle\Psi|\Psi\rangle d \quad \text{i.e.} \quad \frac{\langle\Psi|\hat{D}_{\mathbf{r}}|\Psi\rangle}{\langle\Psi|\Psi\rangle} = d. \tag{9.A.19}$$

The last matrix element to consider is given by (9.A.14). It is readily seen that in the thermodynamic limit, each of the free summands is maximum for the same value of d, given by (9.A.18). Thus, one finds

$$\begin{aligned}\langle\Psi|\hat{\psi}^\dagger_{\mathbf{r}\uparrow}\hat{\psi}_{\mathbf{r}'\uparrow}|\Psi\rangle &= \langle\hat{\psi}^\dagger_{\mathbf{r}\uparrow}\hat{\psi}_{\mathbf{r}'\uparrow}\rangle_0 \frac{\langle\Psi|\Psi\rangle}{n_\uparrow(1-n_\uparrow)}\Big[(n_\uparrow - d)(1 - n_\uparrow - n_\downarrow + d) \\ &\quad + 2g(n_\uparrow - d)(n_\downarrow - d) + g^2\frac{(n_\uparrow - d)(n_\downarrow - d)^2}{1 - n_\uparrow - n_\downarrow + d}\Big] \\ &= \langle\hat{\psi}^\dagger_{\mathbf{r}\uparrow}\hat{\psi}_{\mathbf{r}'\uparrow}\rangle_0\langle\Psi|\Psi\rangle q_\uparrow,\end{aligned} \tag{9.A.20}$$

where

$$q_\sigma = \frac{1}{n_\sigma(1-n_\sigma)}\left\{[(n_\sigma - d)(1 - n_\uparrow - n_\downarrow + d)]^{1/2} + [d(n_{\bar{\sigma}} - d)]^{1/2}\right\}^2 \tag{9.A.21}$$

(with $n_{\bar{\sigma}} = n_{-\sigma}$).

Gathering the preceding results, we finally obtain the variational energy (9.75),

$$E = -t\sum_{\langle\mathbf{r},\mathbf{r}'\rangle,\sigma}[q_\sigma\langle\hat{\psi}^\dagger_{\mathbf{r}\sigma}\hat{\psi}_{\mathbf{r}'\sigma}\rangle_0 + \text{c.c.}] + U\sum_{\mathbf{r}} d. \tag{9.A.22}$$

At half-filling, where $n_\uparrow = n_\downarrow = 1/2$ for a non-magnetic solution, one obtains (9.78) and (9.79).

9.A.1 *Kotliar–Ruckenstein slave bosons*

We now show how the Brinkman–Rice theory can be reformulated as a standard mean-field approximation. To do so, we enlarge the Hilbert space by introducing four bosonic operators ($\hat{e}_{\mathbf{r}}$, $\hat{p}_{\mathbf{r}\uparrow}$, $\hat{p}_{\mathbf{r}\downarrow}$ and $\hat{d}_{\mathbf{r}}$) in addition to the fermionic operator (now denoted by $\hat{f}_{\mathbf{r}\sigma}$) in such a way that the four states corresponding to site $\mathbf{r}$ can be written as

$$\begin{aligned}|\text{vac}\rangle &= \hat{e}^\dagger_{\mathbf{r}}|\widetilde{\text{vac}}\rangle, \\ \hat{\psi}^\dagger_{\mathbf{r}\sigma}|\text{vac}\rangle &= \hat{f}^\dagger_{\mathbf{r}\sigma}\hat{p}^\dagger_{\mathbf{r}\sigma}|\widetilde{\text{vac}}\rangle, \\ \hat{\psi}^\dagger_{\mathbf{r}\uparrow}\hat{\psi}^\dagger_{\mathbf{r}\downarrow}|\text{vac}\rangle &= \hat{f}^\dagger_{\mathbf{r}\uparrow}\hat{f}^\dagger_{\mathbf{r}\downarrow}\hat{d}^\dagger_{\mathbf{r}}|\widetilde{\text{vac}}\rangle,\end{aligned} \tag{9.A.23}$$

where $|\widetilde{\text{vac}}\rangle$ denotes the vacuum of the enlarged Hilbert space. To make this new representation meaningful, we must impose constraints to ensure that only the physically relevant part of the enlarged Hilbert space is considered:

$$\hat{e}^\dagger_{\mathbf{r}}\hat{e}_{\mathbf{r}} + \hat{p}^\dagger_{\mathbf{r}\uparrow}\hat{p}_{\mathbf{r}\uparrow} + \hat{p}^\dagger_{\mathbf{r}\downarrow}\hat{p}_{\mathbf{r}\downarrow} + \hat{d}^\dagger_{\mathbf{r}}\hat{d}_{\mathbf{r}} = 1, \tag{9.A.24}$$

$$\hat{f}^\dagger_{\mathbf{r}\sigma}\hat{f}_{\mathbf{r}\sigma} = \hat{p}^\dagger_{\mathbf{r}\sigma}\hat{p}_{\mathbf{r}\sigma} + \hat{d}^\dagger_{\mathbf{r}}\hat{d}_{\mathbf{r}}. \tag{9.A.25}$$

The first constraint ensures that a site is either empty, singly occupied or doubly occupied. The second one means that the presence of a spin-σ electron at site $\mathbf{r}$ is due to either a simply occupied site or a doubly occupied site. It is easy to verify that the physical electron operator in the enlarged Hilbert space is now defined as

$$\hat{\psi}_{\mathbf{r}\sigma} \equiv \hat{z}_{\mathbf{r}\sigma} \hat{f}_{\mathbf{r}\sigma} \quad \text{with} \quad \hat{z}_{\mathbf{r}\sigma} = \hat{e}^\dagger_{\mathbf{r}} \hat{p}_{\mathbf{r}\sigma} + \hat{p}^\dagger_{\mathbf{r}\bar{\sigma}} \hat{d}_{\mathbf{r}} \tag{9.A.26}$$

and the Hamiltonian reads

$$\hat{H} = -t \sum_{\langle \mathbf{r},\mathbf{r}' \rangle,\sigma} \hat{z}^\dagger_{\mathbf{r}\sigma} \hat{f}^\dagger_{\mathbf{r}\sigma} \hat{f}_{\mathbf{r}'\sigma} \hat{z}_{\mathbf{r}'\sigma} + U \sum_{\mathbf{r}} \hat{d}^\dagger_{\mathbf{r}} \hat{d}_{\mathbf{r}}. \tag{9.A.27}$$

In the mean-field approximation, one simply replaces the boson operators $\hat{e}_{\mathbf{r}}, \hat{p}_{\mathbf{r}\sigma}, \hat{d}_{\mathbf{r}}$ by c-numbers $e_{\mathbf{r}}, p_{\mathbf{r}\sigma}, d_{\mathbf{r}}$ and from now on, we consider a translation-invariant non-magnetic solution: $e_{\mathbf{r}} = e$, $p_{\mathbf{r}\sigma} = p$ and $d_{\mathbf{r}} = d$ (assumed real with no loss of generality). The constraints (9.A.24) and (9.A.25) reduce to

$$e^2 + 2p^2 + d^2 - 1 = 0, \tag{9.A.28}$$

$$n_\sigma = \langle \hat{f}^\dagger_{\mathbf{r}\sigma} \hat{f}_{\mathbf{r}\sigma} \rangle = p^2 + d^2, \tag{9.A.29}$$

where $n_\sigma = \langle \hat{\psi}^\dagger_{\mathbf{r}\sigma} \hat{\psi}_{\mathbf{r}\sigma} \rangle$ is the density of spin-σ electrons. The mean-field Hamiltonian takes the simple form

$$\hat{H}_{\mathrm{MF}} = -tz^2 \sum_{\langle \mathbf{r},\mathbf{r}' \rangle,\sigma} (\hat{f}^\dagger_{\mathbf{r}\sigma} \hat{f}_{\mathbf{r}'\sigma} + \mathrm{h.c.}) + ULd^2, \tag{9.A.30}$$

where

$$z^2 = p^2(e+d)^2. \tag{9.A.31}$$

$\hat{H}_{\mathrm{MF}}$ describes free electrons moving on the lattice with a renormalized hopping amplitude tz^2. The density of doubly occupied sites is $\langle \hat{d}^\dagger_{\mathbf{r}} \hat{d}_{\mathbf{r}} \rangle = \langle \hat{f}^\dagger_{\mathbf{r}\uparrow} \hat{f}_{\mathbf{r}\uparrow} \hat{f}^\dagger_{\mathbf{r}\downarrow} \hat{f}_{\mathbf{r}\downarrow} \rangle$. In the mean-field approximation, this gives $d^2 = \langle \hat{f}^\dagger_{\mathbf{r}\uparrow} \hat{f}_{\mathbf{r}\uparrow} \rangle \langle \hat{f}^\dagger_{\mathbf{r}\downarrow} \hat{f}_{\mathbf{r}\downarrow} \rangle = n_\sigma^2 = n^2/4$ (using Wick's theorem) and, using (9.A.28) and (9.A.29), $p^2 = n/2 - n^2/4$ and $e^2 = (1-n/2)^2$. From (9.A.31), we conclude that the hopping amplitude t is renormalized by a U-independent factor $z^2 = n/2 - n^2/4$ so that we do not recover the expected result $z = 1$ in the non-interacting limit.

The failure of the mean-field theory when $U = 0$ can be circumvented by noting that the operators $\hat{z}_{\mathbf{r}\sigma}$ can be replaced by

$$\hat{\tilde{z}}_{\mathbf{r}\sigma} = \hat{e}^\dagger_{\mathbf{r}} \hat{L}_{\mathbf{r}\sigma} \hat{R}_{\mathbf{r}\sigma} \hat{p}_{\mathbf{r}\sigma} + \hat{p}^\dagger_{\mathbf{r}\bar{\sigma}} \hat{L}_{\mathbf{r}\sigma} \hat{R}_{\mathbf{r}\sigma} \hat{d}_{\mathbf{r}}, \tag{9.A.32}$$

where

$$\hat{L}_{\mathbf{r}\sigma} = (1 - \hat{p}^\dagger_{\mathbf{r}\sigma} \hat{p}_{\mathbf{r}\sigma} - \hat{d}^\dagger_{\mathbf{r}} \hat{d}_{\mathbf{r}})^{-1/2}, \qquad \hat{R}_{\mathbf{r}\sigma} = (1 - \hat{p}^\dagger_{\mathbf{r}\bar{\sigma}} \hat{p}_{\mathbf{r}\bar{\sigma}} - \hat{e}^\dagger_{\mathbf{r}} \hat{e}_{\mathbf{r}})^{-1/2}, \tag{9.A.33}$$

since this does not change the Hamiltonian in the physical Hilbert space. The reason is that $\hat{L}_{\mathbf{r}\sigma}$ is the unit operator in the space of empty sites while $\hat{R}_{\mathbf{r}\sigma}$ is the unit operator in the space of singly occupied sites. It is obvious that there are infinitely many Hamiltonians in the enlarged Hilbert space which are equivalent when restricted to the physical subspace. This arbitrariness presents no difficulty as long as the constraints are handled exactly. But any approximation which relaxes the constraint is sensitive to the precise choice of $\hat{H}$. The advantage of using $\hat{\tilde{z}}_{\mathbf{r}\sigma}$ rather than $\hat{z}_{\mathbf{r}\sigma}$ is that the mean-field approximation now reproduces correctly the $U = 0$ limit: $1/L^2 = 1 - n/2$ and $1/R^2 = n/2$ implies $\tilde{z}^2 = (zLR)^2 = 1$; the hopping amplitude $t\tilde{z}^2 = t$ is not renormalized.

The mean-field Hamiltonian is now

$$\hat{H}_{\text{MF}} = -t\tilde{z}^2 \sum_{\langle \mathbf{r},\mathbf{r}'\rangle,\sigma} (\hat{f}^\dagger_{\mathbf{r}\sigma}\hat{f}_{\mathbf{r}'\sigma} + \text{h.c.}) + ULd^2, \tag{9.A.34}$$

where

$$\tilde{z}^2 = \frac{p^2(e+d)^2}{(1-p^2-d^2)(1-p^2-e^2)} = \frac{n/2 - d^2}{n/2(1-n/2)}\left(\sqrt{1-n+d^2}+d\right)^2 \equiv q_\sigma, \tag{9.A.35}$$

making use of the constraints (9.A.28) and (9.A.29). q_σ was introduced in Eq. (9.A.21).[39] $\hat{H}_{\text{MF}}$ is now expressed only in terms of the fermionic operator $\hat{f}_{\mathbf{r}\sigma}$ and the mean-field parameter d. The latter is determined by minimizing the energy

$$\langle \hat{H}_{\text{MF}}\rangle = -t \sum_{\langle \mathbf{r},\mathbf{r}'\rangle,\sigma} q_\sigma(\langle \hat{f}^\dagger_{\mathbf{r}\sigma}\hat{f}_{\mathbf{r}'\sigma}\rangle + \text{c.c.}) + ULd^2. \tag{9.A.36}$$

Since $\hat{H}_{\text{MF}}$ is a non-interacting Hamiltonian for the $\hat{f}$ fermions, $\langle \hat{f}^\dagger_{\mathbf{r}\sigma}\hat{f}_{\mathbf{r}'\sigma}\rangle$ is equal to its value $\langle \hat{f}^\dagger_{\mathbf{r}\sigma}\hat{f}_{\mathbf{r}'\sigma}\rangle_0$ in the non-interacting state at $U = 0$.[40] We thus recover the variational energy of the Brinkman–Rice theory.[39]

[39] Note that d^2 corresponds to what was denoted by d in the Brinkman–Rice theory.

[40] The occupation numbers $\langle \hat{f}^\dagger_\sigma(\mathbf{k})\hat{f}_\sigma(\mathbf{k})\rangle$ and $\langle \hat{f}^\dagger_\sigma(\mathbf{k})\hat{f}_\sigma(\mathbf{k})\rangle_0$ are equal for a fixed density.

Guide to the bibliography

- The Baym–Kadanoff formalism and the Luttinger–Ward functional (the generating functional of 2PI vertices) were introduced in Refs. [1–4] in condensed-matter physics and generalized to relativistic field theories in Ref. [5]. For a general introduction, see Ref. [6].
- Gapless *vs* conserving approximations in superfluid Bose gases are discussed in Ref. [7]. For approximations that are both gapless and conserving, see, e.g., Refs. [8–10].
- The FLEX for the two-dimensional Hubbard model has been introduced in Refs. [11–13]. For a critical discussion of its shortcomings, see Ref. [14].
- The multivaluedness of the Luttinger–Ward functional is discussed in Refs. [15–17].
- The Brinkman–Rice theory of the Mott transition [18], based on the Gutzwiller approximation [19], is thoroughly discussed in Refs. [20, 21].
- Fermions in infinite dimensions are studied in Refs. [22–24]. DMFT [25] is reviewed in Refs. [26–28]. For the derivation of DMFT equations in the Baym–Kadanoff formalism, see Ref. [29]. C-DMFT is discussed in Refs. [30, 31]. Extension of DMFT to the Bose–Hubbard model can be found in Refs. [32–35].
- The self-energy functional theory [36] is reviewed in Refs. [37–40]. Originally proposed for the (Fermi–)Hubbard model, it was extended to the Bose–Hubbard model in Refs. [41–44].
- Many introductory texts to the physics of strongly correlated fermions and the Hubbard model can be found in the proceedings of the Autumn School on Correlated Electrons (Forschungszentrum Jülich) [45].

Bibliography

[1] J. M. Luttinger, Ground-state energy of a many-fermion system. II, *Phys. Rev.* **118**, 1417 (1960).

[2] G. Baym and L. P. Kadanoff, Conservation laws and correlation functions, *Phys. Rev.* **124**, 287 (1961).

[3] G. Baym, Self-consistent approximations in many-body systems, *Phys. Rev.* **127**, 1391 (1962).

[4] C. De Dominicis and P. C. Martin, Stationary entropy principle and renormalization in normal and superfluid systems. I. Algebraic formulation, *J. Math. Phys.* **5**, 14 (1964); Stationary entropy principle and renormalization in normal and superfluid systems. II. Diagrammatic formulation, *J. Math. Phys.* **5**, 31 (1964).

[5] J. M. Cornwall, R. Jackiw, and E. Tomboulis, Effective action for composite operators, *Phys. Rev.* D **10**, 2428 (1974).

[6] G. Stefanucci and R. van Leeuwen, *Nonequilibrium Many-Body Theory of Quantum Systems: A Modern Introduction* (Cambridge University Press, Cambridge, 2013).

[7] P. C. Hohenberg and P. C. Martin, Microscopic theory of superfluid helium, *Ann. Phys.* **34**, 291 (1964).

[8] T. Kita, Conserving gapless mean-field theory for weakly interacting Bose gases, *J. Phys. Soc. Jpn* **75**, 044603 (2006).

[9] V. I. Yukalov and H. Kleinert, Gapless Hartree-Fock-Bogoliubov approximation for Bose gases, *Phys. Rev.* A **73**, 063612 (2006).

[10] V. I. Yukalov, Representative statistical ensembles for Bose systems with broken gauge symmetry, *Ann. Phys.* **323**, 461 (2008).

[11] N. E. Bickers, D. J. Scalapino, and S. R. White, Conserving approximations for strongly correlated electron systems: Bethe-Salpeter equation and dynamics for the two-dimensional Hubbard model, *Phys. Rev. Lett.* **62**, 961 (1989).

[12] N. E. Bickers and D. J. Scalapino, Conserving approximations for strongly fluctuating electron systems. I. Formalism and calculational approach, *Ann. Phys.* **193**, 206 (1989).

[13] N. E. Bickers and S. R. White, Conserving approximations for strongly fluctuating electron systems. II. Numerical results and parquet extension, *Phys. Rev.* B **43**, 8044 (1991).

[14] Y. Vilk and A.-M. S. Tremblay, Non-perturbative many-body approach to the Hubbard model and single-particle pseudogap, *J. Phys. I France* **7**, 1309 (1997).

[15] R. Rossi and F. Werner, Skeleton series and multivaluedness of the self-energy functional in zero space-time dimensions, *J. Phys.* A **48**, 485202 (2015).

[16] E. Kozik, M. Ferrero, and A. Georges, Nonexistence of the Luttinger-Ward functional and misleading convergence of skeleton diagrammatic series for Hubbard-Like models, *Phys. Rev. Lett.* **114**, (2015).

[17] A. Stan, P. Romaniello, S. Rigamonti, L. Reining, and J. A. Berger, Unphysical and physical solutions in many-body theories: From weak to strong correlation, *New J. Phys.* **17**, 093045 (2015).

[18] W. F. Brinkman and T. M. Rice, Application of Gutzwiller's variational method to the metal-insulator transition, *Phys. Rev.* B **2**, 4302 (1970).

[19] M. C. Gutzwiller, Correlation of electrons in a narrow s band, *Phys. Rev.* **137**, A1726 (1965).

[20] V. Z. Vulović and E. Abrahams, Gutzwiller approach to the Anderson lattice model with no orbital degeneracy, *Phys. Rev.* B **36**, 2614 (1987).

[21] J. Bünemann, The Gutzwiller approximation for degenerate bands: A formal derivation, *Eur. Phys. J.* B **4**, 29 (1998).

[22] W. Metzner and D. Vollhardt, Correlated lattice fermions in $d = \infty$ dimensions, *Phys. Rev. Lett.* **62**, 324 (1989).

[23] E. Müller-Hartmann, Correlated fermions on a lattice in high dimensions, *Z. Phys.* B **74**, 507 (1989).

[24] E. Müller-Hartmann, The Hubbard model at high dimensions: Some exact results and weak coupling theory, *Z. Phys.* B **76**, 211 (1989).

[25] A. Georges and G. Kotliar, Hubbard model in infinite dimensions, *Phys. Rev.* B **45**, 6479 (1992).

[26] A. Georges, G. Kotliar, W. Krauth, and M. J. Rozenberg, Dynamical mean-field theory of strongly correlated fermion systems and the limit of infinite dimensions, *Rev. Mod. Phys.* **68**, 13 (1996).

[27] G. Kotliar, S. Y. Savrasov, K. Haule, V. S. Oudovenko, O. Parcollet, and C. A. Marianetti, Electronic structure calculations with dynamical mean-field theory, *Rev. Mod. Phys.* **78**, 865 (2006).

[28] D. Vollhardt, Dynamical mean-field theory of electronic correlations in models and materials, in *Lectures on the Physics of Strongly Correlated Systems XIV*, edited by A. Avella and F. Mancini (American Institute of Physics, Melville, New York, 2010); https://arxiv.org/abs/1004.5069v3.

[29] A. Georges, Strongly correlated electron materials: Dynamical mean-field theory and electronic structure, in *Lectures on the Physics of Highly Correlated Electron Systems VIII*, edited by A. Avella and F. Mancini (American Institute of Physics, Melville, New York, 2004); https://arxiv.org/abs/cond-mat/0403123.

[30] G. Kotliar, S. Y. Savrasov, G. Pálsson, and G. Biroli, Cellular dynamical mean field approach to strongly correlated systems, *Phys. Rev. Lett.* **87**, 186401 (2001).

[31] T. Maier, M. Jarrell, T. Pruschke, and M. H. Hettler, Quantum cluster theories, *Rev. Mod. Phys.* **77**, 1027 (2005).

[32] P. Anders, E. Gull, L. Pollet, M. Troyer, and P. Werner, Dynamical mean field solution of the Bose-Hubbard model, *Phys. Rev. Lett.* **105**, 096402 (2010).

[33] P. Anders, E. Gull, L. Pollet, M. Troyer, and P. Werner, Dynamical mean-field theory for bosons, *New J. Phys.* **13**, 075013 (2011).

[34] K. Byczuk and D. Vollhardt, Correlated bosons on a lattice: Dynamical mean-field theory for Bose–Einstein condensed and normal phases, *Phys. Rev.* B **77**, 235106 (2008).

[35] J. Panas, A. Kauch, J. Kuneš, D. Vollhardt, and K. Byczuk, Numerical calculation of spectral functions of the Bose-Hubbard model using bosonic dynamical mean-field theory, *Phys. Rev.* B **92**, 045102 (2015).

[36] M. Potthoff, Self-energy-functional approach to systems of correlated electrons, *Eur. Phys. J.* B **32**, 429 (2003).

[37] M. Potthoff, Self-energy-functional theory, in *Theoretical Methods for Strongly Correlated Systems*, edited by A. Avella and F. Mancini (Springer, Heidelberg, 2011); https://arxiv.org/abs/1108.2183.

[38] M. Potthoff, Static and dynamic variational principles for strongly correlated electron systems, *AIP Conf. Proc.* **1419**, 199 (2011); https://arxiv.org/abs/1202.4907.

[39] M. Potthoff, Making use of self-energy functionals: The variational cluster approximation, in *Lecture Notes of the Autumn School on Correlated Electrons 2014*, edited by E. Pavarini, E. Koch, D. Vollhardt, and A. Lichtenstein (Forschungszentrum Jülich, Jülich, 2014); https://arxiv.org/abs/1407.4065.

[40] M. Potthoff, Cluster extensions of dynamical mean-field theory, in *Lecture Notes of the Autumn School on Correlated Electrons 2018*, edited by E. Pavarini, E. Koch, A. Lichtenstein, and D. Vollhardt (Forschungszentrum Jülich, Jülich, 2018).

[41] W. Koller and N. Dupuis, Variational cluster perturbation theory for Bose-Hubbard models, *J. Phys.: Condens. Matter* **18** 9525 (2006).

[42] M. Knap, E. Arrigoni, and W. von der Linden, Spectral properties of strongly correlated bosons in two-dimensional optical lattices, *Phys. Rev.* B **81**, 024301 (2010).

[43] M. Knap, E. Arrigoni, and W. von der Linden, Variational cluster approach for strongly-correlated lattice bosons in the superfluid phase, *Phys. Rev.* B **83**, 134507 (2011).

[44] E. Arrigoni, M. Knap, and W. von der Linden, Extended self-energy functional approach for strongly correlated lattice bosons in the superfluid phase, *Phys. Rev.* B **84**, 014535 (2011).

[45] https://www.cond-mat.de/events/correl.html.

[46] J. Hubbard, Electron correlations in narrow energy bands III. An improved solution, *Proc. Roy. Soc. (London)* A **281**, 401 (1964).

[47] R. Bulla, Zero temperature metal-insulator transition in the infinite-dimensional Hubbard model, *Phys. Rev. Lett.* **83**, 136 (1999).

[48] H. Park, K. Haule, and G. Kotliar, Cluster dynamical mean field theory of the Mott transition, *Phys. Rev. Lett.* **101**, 186403 (2008).

[49] T. Pruschke, Antiferromagnetism and metal insulator transition in the frustrated Hubbard model, *Prog. Theor. Phys. Suppl.* **160**, 274 (2005).

[50] T. Schäfer, F. Geles, D. Rost, G. Rohringer, E. Arrigoni, K. Held, N. Blümer, M. Aichhorn, and A. Toschi, Fate of the false Mott-Hubbard transition in two dimensions, *Phys. Rev.* B **91**, 125109 (2015).

[51] F. Rose and N. Dupuis, Nonperturbative functional renormalization-group approach to transport in the vicinity of a $(2+1)$-dimensional $O(N)$-symmetric quantum critical point, *Phys. Rev.* B **95**, 014513 (2017).

[52] P. W. Anderson, Random-phase approximation in the theory of superconductivity, *Phys. Rev.* **112**, 1900 (1958).

[53] K. Borejsza and N. Dupuis, Antiferromagnetism and single-particle properties in the two-dimensional half-filled Hubbard model: Slater vs. Mott-Heisenberg, *Europhys. Lett.* **63**, 722 (2003).

[54] K. Borejsza and N. Dupuis, Antiferromagnetism and single-particle properties in the two-dimensional half-filled Hubbard model: A nonlinear sigma model approach, *Phys. Rev.* B **69**, 085119 (2004).

Index

Φ-derivable approximation, 619
n-particle irreducible (nPI), 86
1PI diagram, 86
1PI vertex, 85
2PI diagram, 96
2PI vertex, 96

Abrikosov vortex lattice, 525
action, 134, 140
adiabatic approximation, 582
adiabatic continuity, vii, 316, 362
algebraic long-range order, 292
amplitude operator, 463
analytic continuation, 27, 41, 115, 254
Anderson–Bogoliubov mode, 283, 478, 480, 529, 530, 548
Anderson-Higgs mechanism, 481, 536
annihilation operator, 31
anomalous dimension, 497
anomaly, 218
antiferromagnetism, 294, 301, 305, 362, 405, 407, 421, 573
antilinear operator, 206
antiunitary operator, 194, 206

backflow, 474
Baker-Hausdorff formula, 7
Baym-Kadanoff formalism, 611–628
BCS channel, 351
BCS coherence length, 513
BCS gap equation, 512, 621
BCS loop, 351
BCS state, 511
BCS theory, 505–516, 539, 621
BCS-BEC crossover, 434, 539–553
BEC, 168, 305, 462, 539
BEC transition temperature, 168, 496
Berry connection, 583
Berry curvature, 584
Berry phase, 435, 581–586, 593
Bertsch parameter, 551
Bethe ansatz, 595
Bethe-Salpeter equation, 98, 350, 559, 620, 623
bilinear source, 94, 101, 226, 614, 618
bipartite lattice, 294, 421, 636
BKT phase transition, 433, 500
Bloch law, 572
Bogoliubov quasi-particle, 511, 539
Bogoliubov sound mode, 225, 469, 485, 539, 548
Bogoliubov theory, 465, 490–495
Bogoliubov transformation, 511, 579, 598
Bogoliubov's inequality, 291, 468
Boltzmann equation, 325
Boltzmann's constant, xix
Boltzmann-Gibbs distribution, 155, 165
Bose-Einstein distribution function, xx, 167
Bose-Hubbard model, 641, 652

bosonization (spin chains), 595
bound particle-hole pair, 408, 426, 433
bound particle-particle pair, 433
Brückner, 411
Brillouin function, 575, 605
Brinkman-Rice theory, 629, 645–651
broken symmetry, vii

C-DIA, 643
C-DMFT, 637, 643
canonical ensemble, 156
Cauchy-Riemann conditions, 104
causality, 248
charge susceptibility, 253, 392
charge-density wave, 434
classical electrodynamics, 123, 144
classical field theory, 139–146
classical ideal gas, 165
cluster decomposition property, 295
coherent state
  bosons, 43
  electromagnetic field, 129
  fermions, 48
  harmonic oscillator, 45
  spins, 586, 606
coherent-state functional integral, 21
collective excitation, 245
collective mode, 245
collision integral, 325
collisional regime, 260, 266
collisionless regime, 331, 333, 336, 474
combinatorial factor, 63
composite boson, 512, 539–548
composite field, 227
compressibility, 259
  charged system, 264
  electron gas, 382, 387
  Fermi liquid, 322
  isentropic, 260
  isothermal, 259
compressibility sum rule, 260, 469, 481
  electron gas, 382
  Fermi liquid, 336
condensate, 105, 120, 462, 512
condensate density, 122, 168, 304, 463, 467
condensate depletion, 494, 495
condensate wavefunction, 463, 484
condensed matter, vii–x
conductivity, 273
conservation laws, 94, 144, 199, 209, 221, 325, 385, 394, 440, 612–625
conserved charge, 143, 215, 612
conserved current, 143
conserving approximation, 94, 385, 394, 439, 612–628
continuity equation, 143, 215–217, 221, 297, 612
Cooper instability, 506
Cooper pair, 477, 506, 539
Cooper pairing, 477
correlation energy, 386
correlation function, 240–258
correlation hole, 394
correlation length, 292
Coulomb energy, 124
Coulomb gauge, 124, 261, 273, 479, 534
Coulomb interaction, 131
  regularization of, 371
covariant derivative, 217
CPT, 640
creation operator, 31
critical exponent, 497, 500, 550, 581
critical phenomena, 496
critical point, 497, 635
cross section, 555
crossing symmetry, 437
cumulants of a probability distribution, 80
Curie-Weiss susceptibility, 636

damping, 266, 316, 329, 334, 338, 381, 409, 414, 531, 539, 614
dangerously irrelevant variable, 500

de Broglie wavelength, 13, 165, 445, 495, 557
Debye frequency, 505
deconfined criticality, 597
density of states, 156
  Fermi liquid, 317
  Hubbard model, 401
  quantum ideal gas, 167
  superconductor, 510
density operator, 151–160
density-density response function, 247, 258–260
  charged system, 262
  electron gas, 253, 371–377, 620
  Fermi liquid, 333, 360
  Hubbard model, 407, 414
  superconductor, 532
  superfluid, 468, 480, 623
DΓA, 644
DIA, 642
diamagnetic term, 133, 217, 267, 269, 333, 428, 532
dilute Bose gas, 495–501
dilute Fermi gas, 549
Dirac string, 585
dirty limit (superconductor), 536
dissipation, 243, 249, 283
DMFT, 631–637, 641
drag current, 321
Drude weight, 279, 283, 482
  superconductor, 536
dynamic phase, 583
dynamical cluster approximation, 637
Dyson's equation, 86, 90, 95, 102
Dyson-Beliaev equation, 90, 106

effective action, 83
  1PI, 84, 89, 102, 108–115, 222–225, 490, 615–618
  2PI, 94, 99, 618, 638
effective mass, 317, 322, 344, 389, 402
effective range, 557

Ehrenfest's equation, 148
eikonal approximation, 519
electrodynamics of superconductors, 482, 533–538
electromagnetic response function, 261–283, 615, 617
  Fermi liquid, 336, 357
  superconductor, 482, 533–538
electroweak theory, 481
Eliashberg's theory of superconductivity, 625
Elitzur's theorem, 295
energy-momentum tensor, 144, 326, 330
entropy wave, 475
Euclidean action, 13, 52
Euclidean Lagrangian, 13
Euclidean time, 13
Euler's equation, 260, 471
Euler-Lagrange equations, 6, 134, 140
  electromagnetism, 123, 145
exchange energy, 386
exchange hole, 393
external source, 79, 251

f-sum rule, 259, 265, 336, 382, 439, 461, 469, 481, 620
Fermi liquid, 313–362, 388
Fermi momentum, 315, 362
Fermi surface, 315, 362, 402
  nesting, 362, 406, 636
Fermi velocity, 315, 317
Fermi-Dirac distribution function, xx, 167
Fermi-liquid theory, 313–362
  one dimension, 362
  two dimensions, 362
ferromagnetism, 226, 291, 301, 324, 395, 403, 407–409, 571
  flat-band systems, 412, 451
Feshbach resonance, ix, 501, 541, 549, 558
Feynman diagrams, viii, 60, 120, 187

amputated, 86
connected, 66
field operator, 34
Fierz ambiguity, 416
first sound, 260, 329, 475
velocity, 260, 323
FLEX, 442, 624, 652
fluctuation-dissipation theorem, 249, 302, 304
flux quantization, 521
Fock space, 23, 28, 150, 158
Friedel oscillations, 385
functional derivatives, 168–170
functional integral
bosons and fermions, 51
electromagnetic field, 130
quantum harmonic string, 25
quantum spins, 435, 586, 589
functionals, 168–170

Galilean invariance, 205, 301
Galilean transformation, 139, 204, 231
gauge field, 213, 428, 450, 481, 534, 597
gauge principle, 213
gauge transformation, 213
Gaussian functional integrals, 176–178
Gaussian integrals, 170–178
Gell-Mann and Low theorem, 184
generating functional, 78
1PI vertices, 84
2PI vertices, 96
connected Green functions, 79
generator, 138, 144, 196
Gibbs free energy, 84, 258, 301, 519
Gibbs-Duhem relation, 475
Ginzburg-Landau coherence length, 485, 521, 523
Ginzburg-Landau expansion, 182, 516
Ginzburg-Landau free energy, 518
Ginzburg-Landau parameter, 523
Ginzburg-Landau theory, 516–525
Glauber, 46
Goldstone modes, 225, 298
counting of, 299
type A and type B, 300
Goldstone's theorem, 222, 227, 296
grand canonical ensemble, 158
grand potential, 159
graphene, ix
Grassmann algebra, 46
Grassmann variable, 43, 46, 173
Green function, 27, 38, 53, 67–72
connected, 78–81
imaginary time, 40
many particles, 42
normal and anomalous, 78, 212, 284, 509, 621
one particle, 39, 284
real time, 42
retarded or advanced, 28
Gross-Pitaevskii equation, 483
Gross-Pitaevskii theory, 465, 483–489, 520
Gutzwiller approximation, 629, 646

Haldane's conjecture, 596
Haldane's gap, 595
Hamilton's equations of motion, 137
Hamiltonian, 6, 137
Hamiltonian density, 141
harmonic string, 21–28
Hartree self-energy, 376
Hartree-Fock theory, 286, 388, 409, 415, 422, 496
Hartree-Fock-Bogoliubov theory, 495
healing length, 485, 492
Heaviside step function, xx
heavy-fermion systems, ix
hedgehog, 596
Heisenberg model, 291, 294, 570
with SU(N) symmetry, 603
large-U limit of the Hubbard model, 405, 426, 434
Heisenberg picture, 149, 183

Heisenberg's correspondence principle, 148
helicoidal order, 571
helium ^{3}He, 314
helium ^{4}He, 473, 474, 562
Helmholtz free energy, 84, 158
Higgs boson, ix, 481
Higgs mechanism, 481
Higgs mode in superconductors, ix, 481, 538
high-T_c superconductors, ix
Holstein-Primakoff bosons, 577–581
hot spot, 445
Hubbard bands, 628, 630, 634
Hubbard I approximation, 641
Hubbard model, 400–405, 624, 628, 638
  attractive, 401, 433
Hubbard-Stratonovich field, 375, 416
Hubbard-Stratonovich transformation, 131, 375, 415, 508, 603
Hugenholtz-Pines theorem, 623
hydrodynamic regime, 260, 266, 330, 335, 474, 529

ideal Fermi gas, 315
imaginary time, 12
impurity solver, 633
interaction picture, 183, 187
interaction vertex, 62, 88
interpolating field/operator, 297

jellium model, 261, 370, 534, 619
Josephson relation, 468

Källén-Lehmann representation, *see* Lehmann representation
Kanamori, 411
kinetic equation, 325–355
kink, 488
Klein-Gordon Lagrangian, 21
Kotliar-Ruckenstein slave bosons, 629, 649
Kramers' degeneracy, 210
Kramers-Kronig relations, 248, 285
Kubo's formula, 239
  conductivity, 273

Lagrangian, 6, 134, 140
  electromagnetic field, 123, 144
  superfluid, 469
Lagrangian density, 140, 214
Landau damping, 329, 334, 381, 531
Landau diamagnetism, 271, 280, 337, 518
Landau function f, 318, 349
Landau parameters, 319
Landau's criterion, 472
Landau's critical velocity, 472
large-N expansion, 604
large-S expansion, 570, 577, 590
Lee-Huang-Yang correction, 494
Lehmann representation, *see* spectral representation
Lie algebra, 196
Lie group, 195
Lieb, Schultz and Mattis theorem, 595
lifetime, 245, 294, 316, 328, 381
Lindhard function, 378
linked cluster theorem, 66, 188
Liouville's equation, 164
Liouville's theorem, 164
Liouville-von Neumann equation, 152
Lippmann-Schwinger equation, 554
local field factor, 394
London equation, 281, 520, 537
London limit, 281
London penetration depth, 282, 520, 521, 537
long-range order, 292, 295, 303
longitudinal conductivity, 272, 275
longitudinal dielectric function, 262, 371
longitudinal-conductivity sum rule, 265, 278, 382, 536

loop expansion, 105–115
Lorentz invariance, ix, 299
Luttinger liquid
  one-dimensional fermions, 362, 395
  spin chains, 595
Luttinger theorem, 342, 362, 389
Luttinger-Ward functional, 94–103, 227, 360, 618
  infinite dimension, 631
  multivaluedness, 95, 625

macrostate, 151
magnetic interactions, 133, 261, 273
magnetic monopole, 584
magnon, 572, 578
many-particle Green function, 42
Marshall's theorem, 294
massless modes, 298
Matsubara frequencies, 25, 41
  summation, 178–183
maximum statistical entropy principle, 154, 165
mean-field theory, viii
Meissner effect, 271, 282, 480, 520, 537
Mermin-Wagner theorem, 302
microcanonical ensemble, 156
microstate, 151
Migdal's theorem, 625
minimal coupling, 217
moments of a probability distribution, 80
Mott insulator, 405, 628, 630
Mott transition, 628–637, 643
Mott-Heisenberg antiferromagnet, 405, 421
Mott-Heisenberg insulator, 405, 636
multi-pair excitation, 329, 338, 346, 381
multicritical point, 543, 549

Néel state, 294, 570, 573, 578, 591, 601
Néel temperature, 405, 421
Nagaoka's theorem, 412
Nambu spinor, 508, 526
NLσM, 426–433, 591–593
Noether current, 215, 217, 612
Noether density, 215, 217, 612
Noether's theorem, 135, 142, 215
non-perturbative functional RG, 5, 83
nonlinear Schrödinger equation, 483, 519
normal-ordered operator, 45
Nozières-Schmitt–Rink approach, 545–549, 551

ODLRO, 463, 477
optical lattice, ix
optical theorem, 555
order parameter, 292
organic conductors, ix

pairing fluctuations, 433, 549
pairing operator/field, 433, 526, 545, 549
pairing susceptibility, 181, 506, 518
paramagnetic term, 127, 130, 217, 261
paramagnon, 409
parity, 199, 242, 287
particle number conservation and gauge invariance, 274
particle-hole channel, 96, 101, 229, 351, 408, 444, 620
particle-hole interaction vertex, 230, 340–356, 441, 621
particle-hole pair, 96, 100, 316, 351, 408, 426, 433
particle-hole pair continuum, 380, 414, 573
particle-hole pair excitation, 266, 269, 329, 337, 372–382, 409–415
particle-hole pair propagator, 66, 94, 98, 352–358, 438, 620
particle-hole response function, 346
particle-hole susceptibility, 180, 437
particle-hole symmetry, 530, 634
particle-hole transformation, 400, 433
particle-particle channel, 97, 101, 416, 622

particle-particle pair, 100, 351, 433
particle-particle pair propagator, 101, 540, 622
partition function, 154–167
path integral, 7–21
Pauli matrices, xx
Pauli paramagnetism, 324, 518
Pauli principle (violation of), 437
Pauli susceptibility, 281
perfect-screening sum rule, 265, 382
persistent current, 458
perturbation theory, viii, 59–78
  self-energy and 1PI vertices, 87–91
  self-energy and 2PI vertices, 95–103
  zero temperature, 115–123
Pfaffian, 175
phase operator, 463
phonon, 23, 469, 472, 530
photon, 128, 269, 481, 536
Pippard limit, 281, 538
Planck's constant, xix
plasma frequency, 133
plasmon, 263, 266, 335, 372, 381, 382
  superconductor, 480, 535
Poisson bracket, 137
Pomeranchuk instability, 320, 322, 409
preformed particle-hole pair, 434
preformed particle-particle pair, 434
principle of least action, 134
projective representation, 195
propagator, 7, 39
  imaginary time, 12
  retarded or advanced, 39
pseudogap, 433, 444–446, 624, 645
pure state, 151

QCP, 497, 597
QPT, ix, 497, 596
quantum Boltzmann equation, 354
quantum chromodynamics, ix
quantum electrodynamics, 232
quantum field, 150
quantum field theory, ix
quantum fluctuations and temporal field fluctuations, 27
quantum Hall systems, ix
quantum ideal gas, 167
quantum pressure, 484
quasi-classical state, 46
quasi-hole, 316
quasi-long-range order, 292
quasi-particle, 314, 342
quasi-particle distribution function, 316, 317
quasi-particle lifetime, 317, 342–347, 364, 393
quasi-particle weight, 340, 342, 359, 393

reduced density operator, 152
renormalized classical regime, 432, 445
residual symmetry group, 296
response function, 238
  imaginary time, 251
  retarded or advanced, 238
RG, viii
  dilute Bose gas, 498
  dilute Fermi gas, 549
  Fermi liquid, 345, 359, 365
roton, 473, 474
RPA
  bubble and ladder diagrams, 66
  electron gas, 371–393, 619
  Fermi-liquid theory, 318
  Heisenberg model, 575
  Hubbard model, 407
  Slater's theory, 421
  Stoner's theory, 412
  superconductor, 507, 534, 546, 623
RPA self-energy, 390

s-wave scattering length, 483, 490, 540, 548, 557
saddle-point approximation, 105
scaling dimension, 496, 550

scaling function, 551
scattering amplitude, 554
scattering rate, 340, 346, 393, 474, 645
scattering theory, 553–561
Schrödinger equation, 149
Schrödinger field, 150
Schrödinger picture, 149, 183
Schwinger bosons, 597
Schwinger-Dyson equation, 82
screening, 262, 268, 270, 383
second quantization, 28–38, 150
second sound, 475
self-energy, 86, 95, 285
  normal or anomalous, 287
self-energy functional theory, 637–643
single-impurity Anderson model, 632, 642
skeleton diagram, 96, 228, 627
Slater antiferromagnet, 421, 636, 645
Slater determinant, 31, 116
Slater's theory, 421
SO(2), 433
SO(3), 144, 202, 418, 427, 433, 434, 450
soft modes, 298
soliton, 488
spacetime symmetry, 212, 301
specific heat
  Fermi liquid, 322
  Heisenberg model, 572, 581
  superconductor, 517
spectral function, 245, 251
  Fermi liquid, 341
  one-particle Green function, 284
spectral representation, 244, 252
  one-particle Green function, 284
speed of light, xix
spin chains, 573, 595, 600
spin susceptibility, 306, 324, 405, 431, 437, 576
spin wave, 226, 298, 332, 409, 413, 417, 424
  antiferromagnet, 574, 577, 578, 590
  ferromagnet, 572, 574, 577, 590
spin-coherent-state functional integral, 581
spin-density wave, 403, 408, 421
  incommensurate, 406, 408
  spiral, 451
spin-triplet pairing, 508
spinon, 573, 597
spiral order, 571, 641
spontaneous symmetry breaking, 224, 291, 294
staggered magnetization, 294, 422, 579, 601
stationary scattering state, 554
stationary-phase approximation, 104
statistical entropy, 153, 163, 165
statistical Fermi-liquid theory, 365
statistical field theory, ix
steepest-descent approximation, 103
stiffness, 258, 294, 301
Stoner criterion, 408
Stoner's theory, 409
stress-energy tensor, 144
structure constants (Lie group), 196
structure factor, 249, 334
SU(2), 203, 291, 295, 405, 417, 427, 450, 584, 606
SU(N), 603
sum rules, 250, 285
superconducting coherence length, 281
superconductivity, 281, 434, 477
  d-wave, 637, 641
  type I and type II, 523
superfluid density, 281, 462, 467, 530
superfluid weight, 282, 482
  superconductor, 537
superfluidity, 304, 458
susceptibility, 238, 246
symmetry factor, 63
symmetry transformation, 135, 138, 194

active *vs* passive point of view, 135, 194
continuous, 195
internal and external, 212
symmetry-restoring variable, 297

T matrix, 490, 540, 559–561
t-J model, 451
thermodynamic entropy, 163
thermodynamic self-consistency, 627
thermodynamic sum rule, 260
Thomas-Fermi screening, 383
three-body losses, 501
three-body parameter, 501
three-body recombination, 501
time-dependent Hartree approximation, 394
time-dependent Hartree-Fock approximation, 394
time-evolution operator, 183, 208
time-ordered product of operators, 11, 54
time-ordering operator, 11, 40
time-reversal symmetry, 207, 218, 242, 287
topological materials, ix
tower of states, 294
TPSC theory, 394, 438–446, 645
transport scattering time, 279
transverse conductivity, 272, 337
transverse delta function, 128
transverse dielectric function, 268
transverse-conductivity sum rule, 283, 536
two-body contact, 501–505, 552
two-fluid hydrodynamics, 459, 472
two-fluid model, 472
two-particle Green function, 60, 70, 98, 289, 350
two-particle propagator, 227

U(1) symmetry
and gauge transformation, 212
and particle-number conservation, 213
and superfluidity, 462
ultracold atomic gases, vii–x, 314, 458, 501, 539, 557, 558
umklapp processes, 279
unitary Fermi gas, 551
unitary limit, 540, 549, 558
unitary operator, 194
universality, viii
dilute Bose gas, 496–501
dilute Fermi gas, 549
upper critical dimension, 497

vacuum fluctuation diagram, 62, 120, 187
vacuum state, xx, 28
valence bond solid, 596
van der Waals interaction, 557
Van Hove singularity, 407
VCA, 637
vertex correction, 359, 395, 621, 624
Vilk's criterion, 446
viscosity, 458
von Neumann entropy, 153
vortex, 486, 523

W boson, 481
Ward identities, 219, 355, 612
conductivity, 614
density-density correlation function, 614
linear response, 614
Watanabe-Brauner matrix, 299
Weiss field, 633
Wick's rotation, 12, 117
Wick's theorem, 59, 120, 170, 171, 177, 186
Wigner crystal, 304, 373, 395
Wigner distribution function, 353
Wigner operator, 194
Wigner's theorem, 194
Wilson-Fisher fixed point, 581
winding number, 594

Yukawa potential, 384

Z boson, 481
zero sound, 266, 329, 332, 334, 338, 409, 415
ZS channel, 350
ZS loop, 351
ZS′ channel, 351
ZS′ loop, 351

www.ingramcontent.com/pod-product-compliance
Lightning Source LLC
LaVergne TN
LVHW081932200926
848484LV00009B/2407
* 9 7 8 1 8 0 0 6 1 3 9 0 4 *